Support for the
International Symposium and Educational Workshop
on Fish-Marking Techniques
and for publication of this proceedings
was provided by

Sport Fish Restoration Act Funds

administered by the

**U.S. Fish and Wildlife Service,
Division of Federal Aid**

Conference Steering Committee

Nick C. Parker, *Chair*

Program Committee and Session Chairs

Albert Giorgi	Douglas Jester	Fred Utter
Roy Heidinger	Nick Parker	Gary Winans
	Eric Prince	

Session Cochairs

Leslie Holland-Bartels
Gordon McFarlane
Robin Waples
Richard Wydoski

Arrangements Committee

Anthony Novotny
Diana Nielsen
Rae Jean Sielen

Advisors

Robert Sousa   Christopher Dlugokenski

Fish-Marking Techniques

Fish-Marking Techniques

Edited by

Nick C. Parker
Albert E. Giorgi
Roy C. Heidinger
Douglas B. Jester, Jr.
Eric D. Prince
Gary A. Winans

Proceedings of the International Symposium and
Educational Workshop on Fish-Marking Techniques
Held at the University of Washington, Seattle, Washington, USA
June 27–July 1, 1988

American Fisheries Society Symposium 7

Bethesda, Maryland
1990

The American Fisheries Society Symposium series is a registered serial. Suggested citation formats follow.

Entire book

Parker, N. C., A. E. Giorgi, R. C. Heidinger, D. B. Jester, Jr., E. D. Prince, and G. A. Winans, editors. 1990. Fish-marking techniques. American Fisheries Society Symposium 7.

Article within the book

Haegele, C. W. 1990. Anchor tag return rates for Pacific herring in British Columbia. American Fisheries Society Symposium 7:127–133.

Library of Congress Catalog Card Number: 90-84827
ISBN 0-913235-59-8   ISSN 0892-2284

Address orders to

American Fisheries Society
5410 Grosvenor Lane, Suite 110
Bethesda, Maryland 20814, USA

Contents

ELECTRONIC TAGS

GENETIC MARKS

CHEMICAL MARKS

DESIGN, PROGRAM CONTRIBUTION, AND ANALYSIS

Preface

The Sport Fish Restoration Program was enacted into U.S. law in 1950. In an effort to infuse additional support into sport fisheries, Congress imposed a user-pay excise tax on fishing rods, reels, creels, artificial lures, baits, and flies. Assessed at the manufacturers' level and passed on to consumers, this 10% tax totaled about $500,000,000 from 1951 to 1984. In 1984, the federal Wallop–Breaux amendments added excise taxes on most other sportfishing equipment, taxes attributed to motorboat fuels, and import duties on fishing equipment and yachts and pleasure craft to the program. Revenues from the program are collected by the U.S. Internal Revenue Service and deposited in an interest-bearing account in the U.S. Treasury called the Sport Fish Restoration account of the Wallop–Breaux Trust Fund. In the first 4 years since passage of the Wallop–Breaux amendments, collections have exceeded the cumulative revenues of 1951–1984.

Annually, these funds are apportioned to each of the 50 states under a formula grant based 60% on the number of licenses sold by a state relative to licenses sold by all other states and 40% on the land and water areas of that state relative to other states. No state receives less than 1% nor more than 5% of the total funds. States are given 2 years to obligate their share of the apportionment and are reimbursed 75% of all preapproved project expenditures.

The U.S. Fish and Wildlife Service is authorized to deduct up to 6% of the total annual tax collections for administration of the Sport Fish Restoration Program within the Division of Federal Aid. A portion of these administrative funds was used to fund the International Symposium and Educational Workshop on Fish-Marking Techniques.

With the large increase of funds available for sport-fish restoration resulting from the Wallop–Breaux amendments, it soon became apparent that the states and the Fish and Wildlife Service would have to provide enhanced services and accountability to the angling public that pays the tax. Using the most efficient techniques and resources available, we have to eliminate false starts, wasted time, and unnecessary duplication of effort in order to achieve the best fisheries management possible. During a review of critical subjects within the Division of Federal Aid, one of the first to emerge involved fish tagging and marking. We asked the American Fisheries Society to mobilize their expertise to assist us in developing a world-class workshop and conference. We felt that an international symposium would be the best vehicle to bring together researchers, technical specialists, and fish theorists to transfer the most current fish-marking information and techniques to fishery managers and field biologists.

In addition to advancing fishery science, our goal was to provide an effective means for technology transfer. Consequently, there are three elements to this project. The first was the conference itself, which was attended by more than 400 fishery specialists who heard 79 papers and saw 30 posters presentations. It created the environment for the critical exchange of technical information as well as an opportunity for synergistic discussions among the participants. Benefits yet unknown will result from those contacts and exchanges.

The second component is this proceedings volume. This document will serve as a record of the presentations and an important reference for researchers who wish to learn specific details relating to a particular fish-marking or tagging technique.

Third, and probably the most important from a fishery manager's viewpoint, will be the production of a field techniques handbook. Soon to be published, it will set a standard for fish-marking techniques for the rest of this century.

We believe that the U.S. Fish and Wildlife Service, through the Division of Federal Aid, has taken a positive step toward assisting states in their effort to manage more efficiently the nation's fishery resources. Toward that end, we are proud to have worked cooperatively with the American Fisheries Society to jointly sponsor this conference.

ROBERT J. SOUSA

Division of Federal Aid
U.S. Fish and Wildlife Service
Washington, D.C. 20240, USA

Acknowledgments

A conference as large and successful as the International Symposium and Educational Workshop on Fish-Marking Techniques does not happen through its own inertia. Many people and institutions contributed their time and resources to make the symposium a benchmark in the transfer of fishery management techniques.

Excellent facilities at the University of Washington and a fine social program were organized by Tony Novotny (National Marine Fisheries Service) and his Arrangements Committee in Seattle. Diana Nielsen (University of Washington) and her assistants on the University's conference management staff administered the conference with skill, resourcefulness, and grace. Rae Jean Sielen (Biosonics, Inc.) generated the symposium's announcements, program, and other publicity. Stan Smith (U.S. Fish and Wildlife Service) handled the myriad details associated with the trade show. All these people did much more than this short list suggests to create flawless arrangements that were enjoyed by all who attended the conference.

The editors of this volume formed the nucleus of the Program Committee, but we would have floundered without the sustained help of our co-chairs: Leslie Holland-Bartels (U.S. Fish and Wildlife Service, La Crosse, Wisconsin), Sandy McFarlane (Department of Fisheries and Oceans, Nanaimo, British Columbia), Robin Waples (National Marine Fisheries Service, Seattle), and Dick Wydoski (U.S. Fish and Wildlife Service, Denver, Colorado). With more than 200 preliminary abstracts to sort, over 200 peer reviews to coordinate, and 150 final manuscripts and poster presentations to compile, we could not have coped without them. Fred Utter (National Marine Fisheries Service, Seattle) contributed greatly to the early planning of conference and program. The many, many people who volunteered manuscript reviews assured the high technical quality of the symposium and of this proceedings.

Biosonics, Inc., and Floy Tag and Manufacturing, Inc., both in Seattle, contributed funds, materials, and personnel in support of the conference. We particularly appreciate the personal hospitality given to the Steering Committee by Russell Amick of Floy Tag. Robert Stickney extended to the conference the services of the University of Washington's School of Fisheries, of which he is Director. The National Marine Fisheries Service hosted our initial planning session in Seattle.

All of us shared our conference workloads with secretaries and assistants. Sara Clements, who helped Chairman Parker coordinate the symposium from the U.S. Fish and Wildlife Service laboratory in Marion, Alabama, accepted an especially heavy burden and handled it admirably, but our staff support was broad and much appreciated. In the editorial office of the American Fisheries Society, Sally Kendall, Bera Knipling, Meredith Donovan, Rod Gabel, Dan Guthrie, Eleanor Brown, Mel Crystal, and Robert Kendall brought the proceedings to fruition.

The idea for this conference originated in the U.S. Fish and Wildlife Service's Division of Federal Aid. Robert Sousa was especially instrumental in galvanizing the support of the American Fisheries Society and the broader fisheries profession for the project, and in arranging the Federal Aid Division's generous support through Sport Fish Restoration Act funds. The Division's Chris Dlugokenski has facilitated our efforts from first to last.

We have been gratified throughout by the cooperation and goodwill shown to us by the many authors and speakers who contributed to the conference and to its published record. It is their work, singly and collectively, that has made this project worthwhile.

NICK C. PARKER
Conference Chair
U.S. Fish and Wildlife Service
Lubbock, Texas

ALBERT E. GIORGI
National Marine Fisheries Service
Seattle, Washington

ROY C. HEIDINGER
Southern Illinois University
Carbondale, Illinois

DOUGLAS B. JESTER, JR.
Michigan Department of Natural Resources
Lansing, Michigan

ERIC D. PRINCE
National Marine Fisheries Service
Miami, Florida

GARY A. WINANS
National Marine Fisheries Service
Seattle, Washington

American Fisheries Society Symposium 7:1–4, 1990
© Copyright by the American Fisheries Society 1990

OVERVIEWS

Fish Marking and the Magnuson Act

WILLIAM G. GORDON[1]

Rural Delivery 1, Box 174
Howard, Pennsylvania 16841, USA

Abstract.—The Magnuson Act brought fisheries within 320 km of the United States under U.S. jurisdiction in 1976. Most of the benefits of the Magnuson Act have derived from reduction of foreign fishing on U.S. fish stocks. Most of the improved stock assessment that supported fishery management under the Magnuson Act was based on data collected by U.S. observers who had to be carried by foreign vessels when they fished under U.S. jurisdiction. Observers are not required on domestic vessels, which have almost completely replaced foreign vessels in recent years (often under joint-venture arrangements with foreign companies). Stocks subject largely to domestic fishing have generally declined or remained low under the Magnuson Act. Fishery assessment and management advice can be expected to become less reliable with the loss of fisheries observer data. Large-scale marking programs in which marked fish are recovered from landings may provide a way to manage the newly domestic fisheries. Even when catch sampling can be done, fish marking may be necessary to obtain sufficient precision in data and in management.

When fisheries jurisdiction in the USA was extended to 320 km (200 miles) off territorial coasts under terms of the Fisheries Conservation and Management Act of 1976, known as the Magnuson Act, offshore fisheries that had been largely unmanaged came under management by new regional fishery management councils and the U.S. government. In the early implementation of the Magnuson Act, we instituted preliminary fishery management plans for many of the fisheries within our jurisdiction. These plans sought to remedy over two decades of overfishing, or what was believed to be rampant overfishing. Foreign fishing was permitted under the Magnuson Act at first, but was sharply curtailed for the overexploited stocks. Some stocks were not depleted or seemed to be improving; catch levels established under bilateral agreements were reduced to allow continued rebuilding of these stocks. Permissible catches in some healthy fisheries also were reduced to protect stocks of other species that had been harvested as incidental catches.

These actions were based either on informed opinion or best guesses. We had little hard evidence to act upon, and often suspected that foreign fishers operating here withheld information from our scientists. But by and large, foreign fishing for most stocks was only slightly curtailed during the first years of extended jurisdiction. The overall catch level was gradually reduced from pre-Magnuson levels to protect clearly depleted or overfished stocks and some of the species formerly harvested by the foreign fleets were totally reserved for U.S. fishermen. The U.S. fishing industry, however, did not have the infrastructure to exploit and process many of the resources that were suddenly available to them within our fisheries zone. As a consequence, foreign fishers were able to continue to fish under terms of the Magnuson Act.

In 1980, the U.S. Congress, somewhat concerned by the slow increase in domestic participation in our fisheries, amended the Magnuson Act with a new allocation policy that became known as the "Fish and Chips" policy—you do something for us and we'll do something for you. Under this amendment, foreign harvest allocations were based on the extent to which a foreign nation contributed to development of the U.S. fishing industry. Contributions could be in the form of tariff reductions on U.S. fishery products in the foreign country, purchase of U.S. products, or participation in joint ventures. Joint venture agreements blossomed particularly in the Pacific

---

[1] William Gordon is a recent Assistant Administrator for Fisheries, responsible for the National Marine Fisheries Service, within the National Oceanic and Atmospheric Administration. He is currently Executive Vice President, New Jersey Marine Science Consortium, Sandy Hook, New Jersey.

Northwest, although several were established along the Atlantic coast. The practice under most joint ventures was that American fishermen caught the fish and sold them over the side to foreign processing vessels.

The "Fish and Chips" amendment to the Magnuson Act greatly accelerated the removal of all foreign fisheries from the U.S. economic zone and the development of our domestic fisheries. In the Gulf of Alaska, for example, catch was taken almost exclusively by foreign trawlers prior to 1980. In 1980, following 2 years of limited operation, joint-venture fisheries began to displace the foreign freezer-trawlers. By 1983, joint-venture operations were taking approximately 76% of the total trawl catch of walleye pollock *Theragra chalcogramma*, the biggest fishery in the region. Although some foreign processors continue to receive and process the landings from joint-venture fisheries, U.S. trawlers have displaced foreign harvest vessels in virtually all jurisdictional waters of the USA.

The rapid change from foreign to domestic harvesters is profoundly affecting our management regime. From 1977, when the Magnuson Act went into effect, until the early 1980s, management efforts were directed at reducing perceived overfishing and reducing the catch by foreign fisheries. Catch levels were reduced or held constant and efforts were made to transfer catches to joint ventures or domestic fishermen. Early joint ventures were encouraged and minimally regulated. The regulations were often exemplified by conservation-oriented restrictions, such as time–area closures and limits on incidental catches, that were applied only to foreign operations, not to joint ventures.

The Magnuson Act gave us an ability to greatly improve the quality of stock assessments and fishery management advice. Most of the improvements came about as a result of the foreign fisheries observer program. Under the Magnuson Act, all foreign fishing vessels must carry a U.S. observer on board, if requested to do so, while they operate under U.S. jurisdiction. These observers estimate catch levels and obtain biological samples. These data have permitted analyses that were previously impossible.

The "domestication" of U.S. fisheries may negate these recent advances in knowledge of the fishery resources. Observers are not mandatory on domestic vessels, and only a few are being carried under a voluntary program. Unless the observer program is expanded and includes sampling of the major portion of the domestic catch, or unless other means are found to get good data on catch and stock composition, future fishery assessments and management advice will become increasingly unreliable because of lack of data.

Early Effects of the Magnuson Act

With all the new knowledge arising from Magnuson Act programs, one would think that fish and shellfish stocks off the USA would now be in good shape. This is true in some, but not in all areas, and I offer a quick tour of U.S. fisheries to highlight current situations.

Our groundfish resource off Alaska has passed through a decline affecting the major exploited species groups. The subsequent increase in fish abundance has been partly due to catch reductions or harvest stabilization and partly to high levels of recruitment in the mid 1970s. Stocks off Alaska seem to be healthy; those that had been depleted continue to rebuild. Except for Pacific salmon *Oncorhynchus* spp., the same could be said for Washington and Oregon stocks. Anadromous salmon are exceptions because they are affected by problems with freshwater habitats and less by the fisheries than most of the other species.

In the northwest Atlantic off New England, and along the mid-Atlantic coasts, I cannot offer a bright picture. Virtually every stock of groundfish is near or below historic low levels. Since 1976, fishing effort has more than doubled and the major traditional stocks are severely depressed. For example, haddock *Melanogrammus aeglefinus*, which prior to the arrival of foreign fishing produced annual yields of 45,000–50,000 tonnes for 30 years, now produces less than 2,000 tonnes. We cannot blame the foreigners, only ourselves, for this. Furthermore, there has been a radical shift in biomass composition. No longer do the gadoid species—the cods *Gadus* spp., haddock, and hakes *Urophycis* spp.—make up a major portion of the biomass. Sharks and skates—species with little appeal in U.S. markets—now constitute about 70% of the biomass.

There were no foreign fisheries of consequence in the south Atlantic and Gulf of Mexico, so people there felt no need to rush into management regimes under the Magnuson Act. Since 1976, however, there has been a major growth in recreational fishing and in some commercial ventures in the region. As a result, such species as swordfish *Xiphias gladius*, Spanish mackerel *Scomberomorus maculatus*, king mackerel *Scombero-*

morus cavalla, and golden redfish *Sebastes marinus* are believed to be seriously overfished. There are now major controversies over management and allocation of these species. No one disputes that overfishing has reduced these stocks to substantially lower levels than in the prior decade.

It is worthwhile to mention Alaska again after this brief tour. The North Pacific Management Council was fortunate in having to deal principally with foreign fisheries for most of the past decade; it would have been difficult to curtail effort in a domestic fishery. Elsewhere, the domestic industry now consists of several user groups who all vie for the same limited resource, further complicating allocation and conservation issues. This is happening, or will happen, off Alaska. The Alaskan groundfish fishery has become domestic. The number of vessels has grown so large that seasons as short as a few days or weeks have to be set on stocks that were fished all year in the past. There is also increasing competition among user groups who petition for special allocations, areas, seasons, and so on. With this strong competition for a common resource, domestic fishermen are reluctant to provide fisheries information or allow observers onboard for fear that the information collected will be used to reduce their allocations or fishing time.

Needs for Fish Marking under the Magnuson Act

The emerging problem of gathering reliable information on marine fisheries leads me to the subject of this conference: fish marking. I believe we urgently need to increase our fish-marking programs. I wish that I had stood here 15 years ago, talking about the need for reliable marking systems and models to convert and interpret mark returns. Can we replace the data obtained by surveys and observers with recovery of tagged fish from landings? Can we obtain credible information on stock composition, migration, age at capture, stock size, and so on? This is indeed a major challenge.

During my career I was often faced with dilemmas over how to deal with various fish stocks. Here is a sampling.

• Alaska plaice *Pleuronectes quadrituberculatus*. Now the controversy rages: what is going on with Alaskan plaice in the so-called doughnut hole—that area beyond our extended jurisdiction in the north Pacific? Are the fish there of U.S. stocks? Russian? Whose are they? How are we to allocate the harvest?

• Bluefin tuna *Thunnus thynnus*. I still bear the scars of trying to manage Atlantic bluefin tuna when the controversy raged over whether the fish were oceanic or resident along the U.S. east coast. Or were they from the south Atlantic? From off Europe? Where did they come from? Where did they go to? And who was overfishing them? We still do not know very much about them.

• Atlantic menhaden *Brevoortia tyrannus*. One of the earliest and largest tagging efforts I know of involved Atlantic menhaden. Over a million fish were tagged with coded wire tags. The returns established that there was a single stock of Atlantic menhaden ranging from Florida all the way to the Canadian border. It was pretty clear who was harvesting them by where they were processed. The value of a robust tagging system showed very quickly.

• Atlantic cod *Gadus morhua*. Georges Bank had a "king" cod. At one time it was not controversial. With the decision of the World Court to divide Georges Bank between the USA and Canada, that stock became split by a political boundary line. Fishermen of both countries claim most of the fish on Georges Bank. We still do not know very much about stock composition, or about who is taking what of Atlantic cod (or of haddock, or of yellowtail flounder *Limanda ferruginea*) off Georges Bank.

• Pollock *Pollachius virens*. Pollock in the Atlantic Ocean is another transboundary species. We do not know where it comes from or goes to, or who is harvesting what.

• Atlantic mackerel *Scomber scombrus*. Perhaps there are two stocks of Atlantic mackerel, one that spawns off Canada and another off the USA. Nobody knows for sure and stock identification is sorely needed.

• King mackerel. I have already alluded to king mackerel in the south Atlantic and the Gulf of Mexico. Some people argue that the stock is shared by Mexico, yet there is no major tagging effort under way that might settle the matter.

• Golden redfish. The golden redfish on the Gulf of Mexico—where it is harvested, where does it come from?

• Atlantic salmon *Salmo salar*. Like Pacific salmon, the Atlantic salmon is subject to a fairly intensive tagging effort. Returns are beginning to reflect the species' stock composition, streams of origin, and the like.

• Atlantic herrings *Alosa* spp. For each of the Atlantic coast herrings, perhaps there are many different stocks that use various streams and estuaries and intermix in the ocean. Nobody knows for sure.

• Striped bass *Morone saxatilis*. Problems of striped bass have attracted much attention. Striped bass are now being tagged very aggressively to determine which stocks contribute to harvest in the ocean—those from Chesapeake Bay, the Hudson River, or elsewhere—and in what proportions.

• Sea turtles. Some species of sea turtles are endangered and all are threatened. Very little is known about the early life history of turtles, and it is a formidable challenge to learn more—females do not return to the beach for 7–10 years or perhaps longer, and the males never come back ashore. But we face real need to understand their movements and their stock structure.

Resource managers would have a much better understanding of the biology of many of these species if we had good ways of marking animals, recovering the marks or the marked animals, and making sense of the data. As most fisheries mature, we will need more precise data and more precise management. Marking may be the most cost-effective and reliable way to obtain the necessary precision.

Acknowledgments

Douglas B. Jester, Jr. invited this address and edited the transcript into readable form.

American Fisheries Society Symposium 7:5–7, 1990

Value of Fish Marking in Fisheries Management

RAY HILBORN

Fisheries Research Institute WH-10, University of Washington
Seattle, Washington 98195, USA

CARL J. WALTERS

Resource Ecology, University of British Columbia
Vancouver, British Columbia V6T 1W5, Canada

DOUGLAS B. JESTER, JR.

Michigan Department of Natural Resources, Box 30028, Lansing, Michigan 48909, USA

Abstract.—World fisheries production has leveled off. Aquaculture plays an increasingly important role in fish production. In many parts of the world, recreational fishing is increasingly important. These trends present five challenges to fisheries managers: rebuilding of depleted stocks, economic rationalization of fisheries, control of management costs, allocation of stocks fished by multiple types of exploiters or in multiple jurisdictions, and evaluation and refinement of aquacultural practices. Fish marking will be an important practice as fisheries managers address these challenges. We estimate that the aggregate increased value of North American fisheries that can be obtained through fish marking exceeds US$1 × 10^9 per year.

This symposium broadly addresses the techniques and applications of fish marking in fisheries management. We have been asked to provide some perspective by estimating the economic value of fish marking.

It is impossible to estimate in any detail the value of fish marking that already has occurred. The data required are simply unavailable. We have chosen, therefore, to provide an informal estimate of the aggregate value that might result from fish marking programs in the foreseeable future. We arrive at our estimate via a brief review of current trends in fisheries and fisheries management, a summary of the challenges for fisheries managers that we see in those trends, and some commentaries on the role of fish marking in meeting these challenges.

Current Trends in Fisheries and Fisheries Management

Royce (1987) reviewed the status of the world's fisheries, and discerned three current transitions. First, fisheries are moving from a search-and-find phase to better management of existing stocks. Second, aquaculture is becoming increasingly important. Third, recreational fishing is growing in importance.

Development of the world's fisheries was very rapid from roughly 1950 to 1970; annual growth was roughly 6% as total annual world landings rose from 20 to 70 million tonnes. Since the early 1970s, growth of annual landings has declined essentially to zero. Current landings now are 90–100 million tonnes per year. Estimates of total world maximum sustainable yield range from 100 to 130 million tonnes per year. More than half of the world's fish stocks are estimated to be overexploited. Any further gain in total fisheries production of traditional fish species must come primarily from better use of already exploited fish stocks rather than from discovery of new stocks. With the potential exception of totally new products like krill, we must now start doing a better job with what we have instead of looking around for something new.

Aquacultural production, now in the range of 7–9 million tonnes per year, is not a large volume in the grand scheme of things but is of exceptionally high value. Further, aquaculture is now growing rapidly in some parts of the world. Increasingly, fisheries managers are stocking cultured fish to sustain harvested stocks even in the oceans.

Recreational fishing has become the major means of fish harvest in some cases. This is particularly true in North America, where recreational fishing is now as big as or bigger than commercial fishing by some economic measures. In 1985, direct expenditures on sport fishing were US$27 × 10^9 in the USA (U.S. Fish and Wildlife Service 1988) and Can$4.5 × 10^9 in Canada (Canada Department of Fisheries and Oceans 1988). This is several times the landed value of commercial fishes in both countries. This is not an entirely appropriate comparison but it does demonstrate the relative size of the recreational fishery. Not only are recreational fisheries big, they compete with commercial fishing. As we discuss later, the demands for fisheries management information are much more intense when there is conflict between user groups than when there are single user groups. The development of recreational fisheries is going to pressure management agencies to know much, much more about what fish are being caught and who is catching them.

Major Challenges to Fisheries Managers

Fisheries managers face five major challenges, which follow from the preceding trends.

The first challenge is to rebuild the roughly half (Royce 1987) of the world's stocks that are currently overexploited. On the Pacific coast of North America, chinook salmon *Oncorhynchus tshawytscha* provide a classic example of this challenge. People who are not involved cannot imagine the energy that the state, provincial, and federal agencies in the USA and Canada have put into trying to rebuild chinook salmon stocks and what this means in terms of data requirements— nor can they imagine the rehabilitation challenges facing fisheries managers elsewhere, with different species.

The second major challenge, which probably is the hardest, is rational economic management of fisheries. The politics of fisheries management is such that rationalization founders on the uncertainty associated with management of aquatic stocks. Our institutions cannot cope well with short-term costs and conflicts that will be certain if the potential benefits are uncertain.

The third challenge is the economic rationalization of fisheries monitoring and regulation. A fundamental problem in many fisheries, certainly on the Pacific North American coast, is that management agencies spend a significant portion of the landed value in management costs. Sometimes they even spend more than the fishery is worth. Fisheries managers must learn how to manage stocks cost-effectively.

The fourth challenge is management and allocation of transboundary or migratory stocks. Fisheries managers encounter incredible conflict when two jurisdictions fish the same stocks, whether the boundaries are national or tribal. Most of the great European fisheries disasters of

the last 20 years occurred because multiple jurisdictions fighting over the same stock were unable to agree on some of the data and hence on the allocation of the stocks. We estimate that 30–50% of total world production is subject to interjurisdictional conflicts.

The fifth challenge is use of aquacultural production or enhancement and ocean ranching of fish. In many cases, we are attempting to produce many more fish than can or will be produced naturally. The costs of production and the technical problems of stock enhancement require us to closely assess the results of our efforts.

Role of Marking in Fisheries Management

We now consider the role of fish marking in meeting these challenges. We were reminded of how large this role can be when two of us recently worked in Australia on stock assessment in various fisheries; time after time it seemed that 50–80% of what was known about a fishery came from marking data.

Marks have proven their value in at least five important fisheries measurement problems: stock contribution or use, fish growth, fish movement, fish survival, and population estimation. Most fisheries harvest from multiple fish stocks and most fish stocks are harvested by multiple groups of fishermen; recovery of marked fish of known origin is almost the only way to sort out such interactions. Growth analyses depend on accurately knowing the passage of time as well as fish size; marks are often the best way to measure time intervals between observations of fish sizes or to validate other aging methods. Patterns of movement usually cannot be discerned without substantial ambiguity unless mark recovery data, especially multiple recoveries of the same mark, are available. Marking methods for population and survival estimates are well-known; most of the statistical literature on fish marking treats these estimation problems. Each of these uses of marks contributes in important ways to fisheries management. We observe, however, that much greater benefit is likely to come from measurement of stock use, growth, and movement than from population estimation as we face the challenges of the future.

Few if any studies have estimated the effects that marking data have had on fisheries management decisions or the value of those effects. However, we can offer numerous examples in which marking data helped resource agencies meet the challenges of fisheries management.

First, we turn to the challenge of depleted stocks. Using the example of chinook salmon, almost all our knowledge about this species on the west coast of North America comes from the analysis of marking data. We basically have marking data, catch records, and escapement counts (fish that escape the fisheries to reach spawning streams). We cannot understand the catch data without the marking data in order to determine where the fish come from. The entire management system is driven by the analysis of marking data. Without these data, we would not know what was going on and we certainly would not have the political and institutional will to do anything about it. Only when we began to have marking data for chinook salmon were we able to identify stocks in the ocean catches, to determine how the competing jurisdictions were catching each other's fish, and to institute appropriate political changes.

The second challenge is economic rationalization of the fisheries. Compared to most other data that fisheries managers can obtain, marking data are highly certain. This certainty adds considerably to the political will of governments faced with the need to tighten fishery regulation. Furthermore, well-designed marking programs often lead to clearer understanding of when and how fish are being harvested. Efforts to recover salmon tags from anglers in the Pacific Northwest rapidly led to recognition that angler harvest was much larger and more economically important than previously believed.

The third challenge is cheaper fisheries management. By providing better information about fish, marking programs can help managers identify and eliminate misdirected or irrelevant (but costly) management activities. The costs of marking programs themselves must be lowered as well, however; presently, most such programs are expensive. The South Pacific Commission, for example, recently funded a tagging study of yellowfin tuna *Thunnus albacares* that will cost about US$5,000 for each tag recovered. A lot of innovation is needed to reduce the price of marking programs, especially the cost of mark recovery. Improvements must come not only from new technologies but also from more effective sampling designs and data analyses.

The fourth challenge is management and allocation of transboundary and migratory stocks. In this challenge more than any other, marking has played and will continue to play one of its most important roles. Marking is well-suited for the estimation of fish movement, which usually is what transboundary fisheries disputes are all about. In dozens of cases, marking has provided the critical evidence that convinces political masters that we actually know something about what the fish are doing and that one country is taking another country's fish. In the Pacific, allocations of Pacific salmon, Pacific halibut *Hippoglossus stenolepis*, and yellowfin tuna depend on marking data. Until managers have good data on who is catching the fish, it is nearly impossible to get any kind of political movement toward interjurisdictional regulations.

The fifth challenge to which marking can contribute is fisheries enhancement and fish ranching. All aquaculture leading to release of fish in the wild is evaluated only by marking programs, which tells us whether or not the fish actually contribute to the fisheries. All of our hatchery programs depend upon marking to determine the best hatchery practices, sizes, and times of release, water temperatures, rearing densities, and release numbers.

Value of Fish Marking

We hope we have made a strong case that fish marking has important and often major roles to play in the challenges faced by fisheries managers. We cannot assign a firm economic value to fish marking. We estimate, however, that U.S. and Canadian fisheries combined are worth about 10×10^9 per year. If we could improve their value by 10% through well-crafted marking programs, fish marking in North America would be worth 1×10^9 per year. We believe this is a very conservative (hence realistic) estimate in view of the important roles marking will play in meeting the current challenges of fisheries management.

Acknowledgments

Ulrike Hilborn made life pleasant while we talked through this paper. We also appreciate the continuing professional stimulation provided by Ralf Yorque.

References

Canada Department of Fisheries and Oceans (DFO). 1988. Sport fishing in Canada 1985. DFO, Ottawa.

Royce, W. F. 1987. Fishery development. Academic Press, Orlando, Florida.

U.S. Fish and Wildlife Service (FWS). 1988. 1985 National survey of fishing, hunting, and wildlife associated recreation. FWS, Washington, D.C.

American Fisheries Society Symposium 7:9–29, 1990

EXTERNAL TAGS AND MARKS

Historical Review of the Development of External Tags and Marks

G. A. McFarlane

Department of Fisheries and Oceans, Biological Sciences Branch
Pacific Biological Station, Nanaimo, British Columbia V9R 5K6, Canada

Richard S. Wydoski

U.S. Fish and Wildlife Service, Division of Federal Aid
Denver Federal Center, Denver, Colorado 80225, USA

Eric D. Prince

National Marine Fisheries Service, Southeast Fisheries Center
75 Virginia Beach Drive, Miami, Florida 33149, USA

Abstract.—External tags and marks have been used extensively on aquatic organisms in studies of movement, abundance, age and growth, mortality, and behavior. We examine the history of the development and use of external tags, with emphasis on attachment tags, mutilation, dyes and pigments, branding, and morphological and meristic techniques. In general, tags and marks have been used to increase our understanding of the biology of the tagged organism and to help develop rational management strategies. Major advances in tagging methodology occurred when investigators applied their understanding of the target animal's biology to the design or modification of a tag.

Marking aquatic animals is an important means to increase our biological understanding of these organisms and to obtain information for rational resource management. Fish and other animals are marked to provide information on stock identity, movements and migration (including rates and routes), abundance, age and growth (including validation of age-determination methods), mortality, behavior, and stocking success.

In this review, we examine the development and use of external tags and marks in the aquatic environment. We could not consider every study published over the last 100 years or evaluate the success of all tags that have been used. However, we have reviewed approximately 900 papers published since 1884 to document the evolution of external tags and marks and to determine when and why major advances in tagging methodology occurred. We believe these advances occurred when investigators brought a thorough understanding of the biology of an animal to the design and development of a tag.

Historical Overview

Recognition marking was attempted on birds and fish before the 19th century (Delany 1978).

The earliest marking of an animal probably occurred between 218 and 201 B.C. when a besieged garrison sent a swallow, taken from her nestlings, to a Roman officer. The officer tied a thread, knotted to indicate the date of a planned relief attack, to the swallow's leg and released the bird, which then returned to its nest (Delany 1978). Between the 13th and 18th centuries, other recognition marks were employed on birds, primarily to prove ownership. As early as the 13th century, falconers marked their birds with name plates or bands. In the seventeenth century, landowners used metal collars to distinguish their domesticated birds from those of other landowners.

Although it is uncertain when fish were first marked, Izaak Walton reported in *The Compleat Angler*, first published in 1653, that private individuals marked juvenile Atlantic salmon *Salmo salar* by tying ribbons to their tails (Walton and Cotton 1898). These marking "programs" demonstrated that Atlantic salmon returned from the sea to the same part of their natal river. Calderwood (1902) reported instances in which Atlantic salmon were marked with fin clips or wire tied around the tail. The Experimental Committee of Tweed Commissioners conducted two marking programs with Atlantic salmon during 1851–1865 and 1870–1873 (Delany 1978). The commissioners achieved some success by clipping fins, attaching pieces of wire to tails and jaws, and attaching numbered labels to tails and opercles.

The first fish tagging in North America occurred in 1873 when Charles G. Atkins marked Atlantic salmon in the Penobscot River, Maine, with a dangler-type tag attached through the muscles in front of the dorsal fin (Rounsefell and Kask 1945).

Since the late 1800s, numerous tags and marks have been used on fish (Emery and Wydoski 1987). Initial tagging emphasis was on salmonids but attention expanded quickly to marine species. As early as 1894, the Petersen disc was used successfully to tag flatfish and cod. Tagging of other demersal marine species was not attempted until after the First World War. It was not until the 1930s that small pelagic species such as Pacific herring *Clupea harengus pallasi* were successfully tagged with metal body cavity tags (Rounsefell and Dahlgren 1933; Jakobsson 1970). Larger pelagic fishes were not successfully tagged until the early 1950s, when spaghetti loop tags (Wilson 1953) and dart tags (Yamashita and Waldron 1958; Mather 1963) were introduced.

Early marking or tagging studies (1800s to the 1930s) were undertaken primarily to determine movements and identify stocks. Since the 1930s, tagging and marking studies have expanded to provide information on age and growth and to estimate population size or mortality and survival rates (Ricker 1956). Most large-scale marking programs have developed since 1945 as various tags have been more critically evaluated with respect to retention or recovery, ease of application, and effects on fish.

Development and Use of Tags and Marks

Detailed descriptions of many external tags and marks were presented by Rounsefell and Kask (1945), Jakobsson (1970), Laird and Scott (1978), Stonehouse (1978), Jones (1979), and Wydoski and Emery (1983). The 900 or so studies considered in our review were published between 1884 and 1986. The majority (65%) of them were conducted on marine species.

Physical tags were used in 65% of the studies reviewed (Figure 1), followed in frequency by dyes and pigments (14%), mutilation (13%), branding (6%), and meristic and morphometric characters including ageing structures (2%). Of the physical tags used, the Petersen disc, spa-

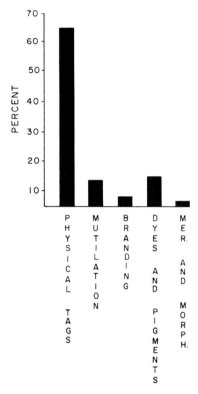

FIGURE 1.—Relative usage of tags and marks, by general category, in approximately 900 studies published between 1884 and 1986. "Mer. and morph." refers to meristic and morphometric characters.

ghetti loop, strap, anchor, Atkins, and dart tags have accounted for 82% (Figure 2).

Many tags and marks developed initially for freshwater, marine, or anadromous fishes have been applied subsequently to other species regardless of habitat. Refinements or modifications to improve tag or mark retention, to reduce adverse marking effects on the fish, and to meet particular research or management needs have continued. Furthermore, tags that were developed for other groups of animals (e.g., birds, mammals) have been applied to or adapted for fish. Tags and marks have evolved over a long time period (Table 1). This evolution has accelerated in recent years, corresponding with greater information needs for various species and life stages.

The basic considerations for selecting an external mark or tag are the objectives of the study; effect on survival, behavior and growth; permanency of the mark; number and size of the organisms to be marked; stress of capture, handling,

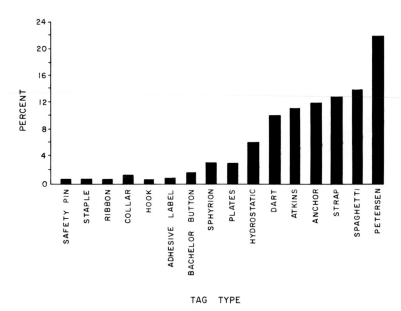

FIGURE 2.—Relative uses of external physical tags, 1884–1986.

and marking the organisms; skills required by project personnel; cost of conducting the experiment; recovery of tagged fish; and coordination required among agencies, states, or countries (Wydoski and Emery 1983). We have grouped tags and marks into the general categories of "natural" tags, which include meristic and morphometric characters and scale patterns, and "artificial" tags, which include mutilation, branding, dyes, paints, pigments, and physical tags.

Natural Tags

Meristic, morphometric, and scale characteristics have been used intermittently as natural tags since the early 1900s. Natural marks include meristic counts, proportional sizes of body parts, shape and coloration, and circulus patterns of scales. The use of meristic and morphometric marks is limited because these traits are subject to environmental and genetic influences, which frequently cause counts and measures to overlap among populations. The shape, size, and circulus patterns of scales are the most frequently used natural marks.

Godsil (1948) compared body proportions of yellowfin tuna *Thunnus albacares* and albacore *Thunnus alalunga* to differentiate Japanese, "American," and Hawaiian stocks. Recent attempts to differentiate stocks of fish by truss network analysis have had some success (Schwei-

gert 1990, this volume). A truss network divides a fish's body into contiguous geometric cells; the cells uniquely define each fish's shape and are amenable to multivariate analysis.

Bulkley (1963) used variations in spotting, other coloration, and hyoid tooth counts in an effort to identify stocks of cutthroat trout in Yellowstone Lake, Wyoming. Due to natural variation in these characters, he was not able to distinguish any of the cutthroat trout subspecies that had been reported by other workers. He suggested that the use of these characters to delineate stocks be reevaluated. Abnormal pigmentation was used by de Veen (1969) to distinguish populations of plaice in the North Sea. He suggested this method was useful for assessing mortality and recruitment rates from different nursery areas. Probably the greatest success with natural marks has been the use of scale pattern analysis to differentiate stocks of Pacific salmon *Oncorhynchus* spp. on the high seas (Major et al. 1972; Bilton and Messinger 1975) and in mixed-stock fisheries (Cook and Guthrie 1987; Marshall et al. 1987). Perhaps the two most famous early studies using scales were those of Hjort (1914) and Gilbert (1924). Hjort first reported on the famous 1904 year class of Atlantic herring *Clupea harengus harengus*. Scales of these fish possessed a unique mark formed in the third year of life that permitted scientists to follow the year class throughout its existence. Gilbert's

TABLE 1.—Historical development of tags and marks to identify fish.

Tag or mark	Year	Species[a]	Remarks	Reference
Attachment tags: ribbons, threads, wires				
Ribbons; colored wool	1653	Atlantic salmon	Attached by private landowners	Walton and Cotton (1898)
Wire in tails or jaws	1851	Atlantic salmon	Attached by private landowners	Delany (1978)
Archer tag	1888	Atlantic salmon	Flat strip of silver with ends pointed and at right angles attached to dorsal fin	Rounsefell and Kask (1945)
Silver wire loops	1905	Tay salmon		Delany (1978)
Collar tag	1872	Atlantic salmon	Rubber band encircling caudal peduncle	Marukawa and Kamiya (1930)
Collar tag	1911	Bluefin tuna	Copper chain encircling caudal peduncle	Sella (1924)
Collar tag	1925	Mackerel	Celluloid poultry tag band around caudal peduncle	Rounsefell and Kask (1945)
Hook tag	1927	Tuna	Numbered hooks to indicate locality	Rounsefell and Kask (1945)
Attachment tags: plates, discs				
Numbered labels	1851	Atlantic salmon	Attached to tail or opercle by private landowners	Delany (1978)
Atkins tag	1873	Atlantic salmon	Flat metal plates held by wire or threads	Rounsefell and Kask (1945)
Petersen disc	1894	Plaice	Circular discs held by wire	Petersen (1896)
Archer tag	1900	Atlantic salmon	Two flat plates held together with two nickel pins	Rounsefell and Kask (1945)
Bachelor button	1908	Pacific salmon	Flat discs held together by shaft	Greene (1911)
Bachelor button	1932	Haddock	Cupped plates rather than flat plates	Rounsefell and Everhart (1953)
Lofting tag	1901	Atlantic salmon	Metal discs with attached flat arms soldered to one disc	Rounsefell and Kask (1945)
Heinicke stud tag	1903	Plaice	Disc with pointed shaft; rubber disc on opposite side held by enlargement of shaft	Rounsefell and Everhart (1953)
Sturgeon tag	1932	Sturgeon	Two discs attached by wire through two holes in discs	Russell (1932)
Opercle clip tag	1932	Tuna	Single plate with attached wires that pass through an opercle hole and are spread apart	Rounsefell and Everhart (1953)
Petersen disc	1930s	Various marine species	Nickel pins substituted for wire	Rounsefell and Everhart (1953); Jensen (1958)
Petersen disc	1950s	Various marine species	Monel metal pins substituted for nickel	Jensen (1958)
Safety pin tag	1905	Plaice	Aluminum plate with bent pin that fits into groove in plate	Rounsefell and Kask (1945); Thomson (1962)
Attachment tags: barbs				
Barb tag	1932	Tuna, lobster, brown trout	Held by barbs in flesh; barbed shaft on disc or metal/plastic strip with barbs	Rounsefell and Kask (1945)

TABLE 1.—Continued.

Tag or mark	Year	Species[a]	Remarks	Reference
Attachment tags: danglers				
Atkins tag	1873	Atlantic salmon	Flat metal plate held by wire or thread in a loop	Rounsefell and Everhart (1953)
Carlin tag	1955	Atlantic salmon	Flat metal plate held by 2 wires that pass through muscles and are twisted on opposite side	Carlin (1955)
Fingerling tag	1960s	Salmonids	Plastic disc attached by monofilament nylon fastened by a knot	Howett Plastics, Molalla, Oregon
Internal anchor tag	1930 1936	Pacific salmon Flounder	Rectangular metal plate with rounded ends placed in body cavity; flexible thread or chain allowed attachment of a small external disc or plate	Rounsefell and Kask (1945)
Hydrostatic tag	1930s		Capsule that contains legend attached through body with wire yolk	Rounsefell and Everhart (1953)
Spring anchor tag	1963	Freshwater species	Stainless steel anchor attached to external dangling legend	Lawler (1963)
Sphyrion tag	1963	Salmon and lobsters	Anchor tag designed after parasitic copepod *Sphyrion lumpi*	Scarratt and Elson (1965)
Attachment tags: carapace tag				
Carapace tag		Lobsters	Metal or plastic plate attached around carapace with a wire	Rounsefell and Everhart (1953)
Attachment tags: rings, bands, straps				
Ring tag	1884	Atlantic salmon	Silver ring with number attached to adipose fin	Rounsefell and Everhart (1953)
Ring tag	1894	Atlantic salmon	Horseshoe-shaped ring	Rounsefell and Everhart (1953)
Heinicke ring tag	1902	Plaice	Attached through muscle posterior to dorsal fin	Rounsefell and Everhart (1953)
Strap tag	1922	Pacific salmon	Cattle ear tags attached to upper lobe of caudal fin; used on marine fishes from 1923	Rounsefell and Kask (1945)
Jaw tag	1940s		Metal rings (bird leg bands) attached to jaws	Rounsefell and Everhart (1953)
Jaw tag	1950s	Lingcod	Red celluloid strip that forms a ring and overlaps the jaw	Rounsefell and Everhart (1953)
Opercle tag	1925	Halibut	Cattle ear tags attached to opercle	Rounsefell and Everhart (1953)
Opercle tag	1930	Trout	Small strap tags applied to opercle, upper or lower jaw	Rounsefell and Everhart (1953)
Opercle	1934	Tuna	Strap tag attached to preopercle with red celluloid disc; none recovered	Rounsefell and Everhart (1953)
Opercle tag	1971	Bream	Applied to preopercle	Goldspink and Banks (1971)

TABLE 1.—Continued.

Tag or mark	Year	Species[a]	Remarks	Reference
Attachment tags: internal (detected externally)				
Visible implant (VI tag)	1987	Salmonids, herring, others	Injected under living transparent tissue usually around eye socket; can be read externally; tags are sequential in a roll form with alphanumeric legends	Northwest Marine Technology, Inc., Shaw Island, Washington
Attachment tags: vinyl tubing				
Spaghetti tag	1953	Tuna	Vinyl tubing with legend attached through muscles of back and secured with a knot	Wilson (1953)
Spaghetti tag	1957	Largemouth bass	First used on freshwater fish	Tebo (1957)
Dart tag	1950s	Fresh- and saltwater species	Vinyl tubing attached with a nylon barb	Rounsefell (1975)
Herring tag	1963	Herring	Vinyl tubing attached through musculature with ends fastened with a nylon "V"	Watson (1963)
Anchor tag	1968	Salmonids	Vinyl tubing attached to nylon T-bars in a "clip" and applied with a clothing tag applicator	Dell (1968)
Lock tag	1970s	Salmonids	Vinyl tubing with a nylon insert that has a locking mechanism	Floy Tag Co., Seattle, Washington
Cinch-up tag	1970s	Salmonids	Similar to lock tag but nylon insert can be "cinched up" at various locations on the nylon	Floy Tag Co., Seattle, Washington
Meristic characters				
Scales	1914	Herring	Unique mark during 3rd year of life to track year-classes	Hjort (1914)
Scales		Salmonids	Spacing of circuli used to differentiate stocks	Koo (1962); Bilton and Messinger (1975)
Natural spotting	1950s	Cutthroat trout	Use of spotting as markers for separating spawning groups in a lake	Cope (1957)
Hyoid teeth	1960s	Cutthroat trout	Used to separate various spawning groups in a lake	Bulkley (1963)
Gill rakers	1960s	Char		Frost (1963)
Mutilation				
Punched holes in fins	1896	Plaice		Petersen (1896)
Fin clipping	1829	Atlantic salmon		Calderwood (1902)
Fin clipping		Various	Review paper	Stuart (1958)
Spine clips	1976	Tilapia	Clipped tips of dorsal and anal fin spines using a code	Rinne (1976)
Scarring soft fin rays	1977	Lake trout, Arctic char, white sucker	Clipped soft rays in a pattern; forms a round scar	Welch and Mills (1981)

TABLE 1.—Continued.

Tag or mark	Year	Species[a]	Remarks	Reference
Pigments, dyes, stains				
Paint	1920s, 1930s		Used on invertebrates	Gowanlock (1927); Federighi (1931)
Dyes and chemicals	1950s	Various	Review paper	Arnold (1966)
Dye on fin clip	1965	Various	Combination method	USDA (1971)
Dyes in food	1966	Brown trout	Developing eggs and fry dyed for identification	Bagenal (1967)
Radioactive elements	1950s	Various		Griffin (1952)
Fluorescent pigment	1972	Coho salmon	Mass marking of cohorts	Phinney and Mathews (1973)
Coloured liquid latex	1955	Warmwater species	Applied with injector	Davis (1955)
Chemical mark	1975	Channel catfish	Applied mark with a silver nitrate pencil	Thomas (1975)
Brands				
Cold-branding	1950s	Salmonids	Use of dry ice and acetone; later use of liquid nitrogen	Buss (1953); Everest and Edmundson (1967)
Cold-branding	1970s	Channel catfish	Applied to warmwater species	Coutant (1972)
Cold-branding	1970s	European flounder, plaice	Applied to marine species	Dando and Ling (1980)
Hot-branding	1908	Salmon	Applied with cauterization iron	Marsh and Cobb (1908)
Hot-branding	1970s	Channel catfish	Applied to warmwater species	Joyce and El-Ibiary (1977)
Laser branding	1970s		Use of lasers to apply a brand	USDA (1971)

[a]Species not previously cited in the text: bluefin tuna *Thunnus thynnus*, plaice *Pleuronectes platessa*, Pacific salmon *Oncorhynchus* spp., haddock *Melanogrammus aeglefinus*, brown trout *Salmo trutta*, lingcod *Ophiodon elongatus*, halibut *Hippoglossus* spp., bream *Abramis brama*, largemouth bass *Micropterus salmoides*, cutthroat trout *Oncorhynchus clarki* (formerly *Salmo clarki*), char *Salvelinus willughbii*, lake trout *S. namaycush*, Arctic char *S. alpinus*, white sucker *Catostomus commersoni*, coho salmon *Oncorhynchus kisutch*, channel catfish *Ictalurus punctatus*, European flounder *Platichthys flesus*.

pioneering studies of sockeye salmon *Oncorhynchus nerka* from the Fraser River led to the regular identification of major stocks of this species as they entered commercial fisheries in Canada and the USA.

Artificial Tags

Mutilation.—Mutilation of fins or other bony parts to mark fishes for studies of migration or growth has been practiced for over 100 years. Mutilation is generally achieved by clipping or hole-punching fins or other body parts. This simple and quick method has been extremely popular for batch-marking cohorts for long-term studies or short-term abundance studies. However, the identification of individual animals is limited. For example, Everhart et al. (1975) reported only 10 possible combinations of marks when two of the

paired fins are used. Calderwood (1902) referred to studies conducted in Scotland as early as 1829 in which adipose fins were cut off young Atlantic salmon. In 1892, Fulton (1893) marked Atlantic herring by punching triangular holes in their tails. However, the hole had to be small, hence inconspicuous, to avoid excessive tissue damage and the method was unsuccessful because it was difficult to identify the marks. Petersen (1896) reported the first successful marking of a marine species by cutting holes in the caudal fins of plaice. During the early 1900s, mutilation was used intermittently with limited success, due to the investigator's inability to recognize large numbers of individual fish. Since the late 1930s, mutilation has been used extensively on many different types of organisms—particularly as a mark to identify fish with internal tags. It has also been used extensively to batch-mark small fish of a specific cohort or size so that stocking success could be determined.

Several authors have marked individual organisms by mutilation. Cagle (1939) clipped shells and toes of turtles in specific combinations for individual identification. Wilder (1953) punched holes in the telsons of American lobsters *Homarus americanus* to examine growth and mortality rates, and to calculate minimum size limits for the fishery. Ropes and Merrill (1969) notched the shells of surf clams to identify individuals for growth determinations, and cited a study by Mend and Barnes who used this technique for soft-shell clams in 1904. Anderson (1973) clipped flesh in a specific pattern to examine the behavior and ecology of the nudibranch *Rostanga pulchra*. More recently, clipping or scarring of individual rays of fish in specific patterns has allowed individual identification (Rinne 1976; Welch and Mills 1981).

Brands.—Both hot- and cold-branding techniques have been applied to organisms with mixed success. Branding is the process of marking an organism by placing either a hot or cold instrument with numbers or letters against the body for a few seconds. The brand displaces or concentrates pigment at the site. Both hot- and cold-branding have been used on fish, crustaceans, amphibians, reptiles, and cetaceans (Stonehouse 1978). In general, cold-branding has been more successful than hot-branding. The primary disadvantage of these techniques is that the brands become illegible with time, so they are of limited value for long-term studies. In addition, "closed" characters or numbers (e.g., "p" or "b") cannot be easily identified because the pigment fills the "closed" area.

Although livestock were branded with hot irons in the western USA during the 1800s, fish were not hot-branded until the 20th century. Marsh and Cobb (1908) used a cauterization instrument to apply hot brands to Pacific salmon fry, but they found the S-shaped marks disappeared quickly. Heat-branding was used again with limited success in the late 1950s and 1960s (Watson 1961; Groves and Novotny 1965; Brock and Farrell 1977). The introduction and use of freeze-branding in the 1950s and 1960s was more successful (K. Buss, Pennsylvania Fish Commission, unpublished 1953 report; Everest and Edmundson 1967; Fujihara and Nakatani 1967). Raleigh et al. (1973) successfully branded salmonids and centrarchids with liquid nitrogen at −196°C. Coutant

(1972) and Brock and Farrell (1977) reported similar results for non-salmonids. More recently, Dando and Ling (1980) reported that freeze brands on plaice and European flounder could be identified 4 years after application. Freeze-branding has also been successfully conducted on amphibians (Clark 1971; Daugherty 1976), reptiles (Lewke and Stroud 1974), and slugs (Richter 1973).

Dyes and pigments.—Dyes and pigments have been used extensively since the 1930s. This category includes any dyes, stains, inks, paints, liquid latex or plastics, metallic compounds, tetracycline antibiotics, and radioactive isotopes that are administered by immersion, injection, tattooing, or feeding (Klima 1965; Arnold 1966). The advantage of using dyes and pigments is the ease with which many animals can be marked quickly. The major disadvantages are that individuals cannot be identified and mark retention is generally brief. Of the various chemical marks, paint was first used (almost exclusively) on shellfish. Gowanlock (1927) filed the shell in a characteristic pattern, filled in the grooves using a glass-marking pencil, and painted a portion of the shell with waterproof ink; his success was limited, however, due to severe abrasion of shells under natural conditions. Federighi (1931) marked oyster drills with oil paints, wax pencil, and sealing wax to determine short-term migrations.

Tattooing of fish was first attempted in 1929 by Landolfi and Heryold (according to Hickling 1945), who marked six European eels *Anguilla anguilla* in the "oriental fashion" using bundles of needles and Indian ink. Later, Hickling (1945) used this method on plaice, sole, and rays, but the marks were unrecognizable after a few weeks. Tattooing was successfully used by Woodbury (1948), who marked reptiles with an electric tattooing device, and by Dunstan and Bostick (1956), who tattooed juvenile Pacific salmon with dyes and inorganic metal oxides.

The most frequently used and successful of the chemical marking procedures has been the staining of organisms by immersion in or injection of dyes. Initially, immersion was used on invertebrates such as starfish (Loosanoff 1937; Feder 1955). Dawson (1957) achieved some success by immersing shrimp in various biological stains, which led to their use for other species. One unique use of staining was reported by Karlsson and Sisson (1973), who immersed the swimmerets and setae of American lobsters in hematoxylin to stain residual egg cement. This technique allowed

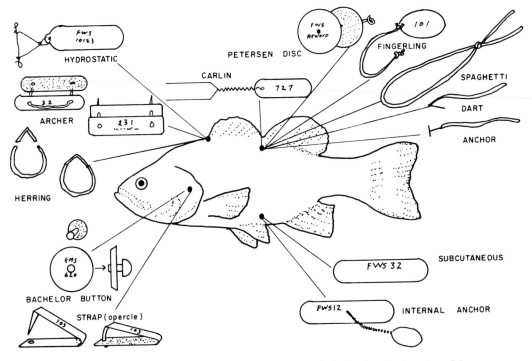

FIGURE 3.—Principal types of external tags and anatomical sites for attachment on fish.

conservation officers to identify lobsters whose eggs had been forcibly removed, and it was used to prosecute lobster poachers. Immersion has been used successfully for marking large numbers of small fish that could not be marked by other methods (Deacon 1961; Ward and Verhoeven 1963; Lawler and Fitz-Earle 1968). Koval (1969) used neutral red dye to mark the young of 11 species of fish whose movements were being studied in the Dnepr River. Injected India ink was used as a mark in the 1930s and 1940s (Kask 1936; Dunn and Coker 1951), and various biological stains have been used to mark fish since that time. Injection of fluorescent pigments with compressed air has been used successfully to mark large numbers of fish (Phinney et al. 1967). Strange and Kennedy (1982) reported 5-year retention of fluorescent particles on rainbow trout *Oncorhynchus mykiss* (formerly *Salmo gairdneri*), and Nielson (1990, this volume) found that fluorescent granules could be identified on cutthroat trout after 12 years.

Physical tags.—Although mutilation, branding, and application of pigments or stains were applied with some success, the development of external tags offered biologists the versatility necessary to address many biological questions. Many of the

tags currently in use have the same design as those used 35 or 40 years ago, although the materials used have changed. Other tag designs are relatively new.

Tags are made conspicuous by their color, shape, size, or attachment location (Figure 3). Physical tags are made from a variety of materials and can be loosely grouped into three categories (Jakobsson 1970).

(1) Button and disk tags, such as the Petersen disc, Archer tag, Bachelor button, and Heinke stud, are fixed firmly and close to the body by wire pins, rivets, or elastic studs.

(2) Dangling tags include a large variety of tags that dangle freely at the end of some form of attachment; attachments may be wire bridles, nylon loops, or some form of internal anchor. This group includes the Atkins tag, Carlin tag, hydrostatic tag, internal anchor tag, plastic flags, Alcathem tags, Scottish polythene tags, and plastic roll tags. We have also included the dart tags and "Floy" anchor tags in this group even though these do not "dangle freely."

(3) All other tags form a heterogeneous group; they have little in common except they lack any specific attachment mechanism. This

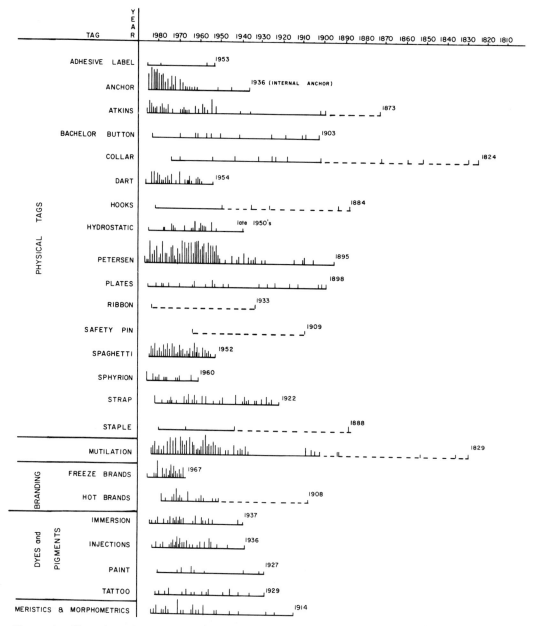

FIGURE 4.—Chronology of various tags and marks used in approximately 900 studies published between 1884 and 1986. Dotted lines indicate studies referenced in more recent papers. Date indicates year of first use of the tag. Histograms (vertical lines) indicate relative number of studies in which the tag or mark was used.

group includes arrow and strap tags, collar and ring tags, hook tags, and ''spaghetti'' loop tags.

We do not treat the evolution of each tag, but we examine the general development of physical tags (Table 1; Figure 4) and highlight some of the major advances in tagging technology.

The first physical tags were developed and used when the need for individual marks was recognized. The first large-scale tagging study was conducted by Atkins on Atlantic salmon in 1873. The tag consisted of a bead or plate attached by a wire that pierced the tissue to form a loop. The Atkins tag was used almost exclusively on salmonids. It was followed by the development of the Petersen disc (Petersen 1896), which is discussed in detail later in this review. Collar tags (a ring of any material that encircles a constriction of the body, usually the caudal peduncle for fishes) were used in the early 1900s. Early collar tags included rubber bands (Marukawa and Kamiya 1930), copper chains (Sella 1924), and wire (Calderwood 1902). In some cases, the collar tag had a label for individual identification. The use of this tag has been, for the most part, unsuccessful. Plate tags are similar to Petersen discs in design and method of attachment. The difference is that only one plate is attached to the organism instead of two. The earliest use of this tag was reported by Bumpus (1901), who tagged American lobsters with copper plates. Gilbert (1924) used brass plates and brass wire for tagging salmon in 1922. During the 1920s and 1930s, other metals were used, such as aluminum on salmon and silver on invertebrates. By the late 1940s, metal plates were replaced by celluloid and plastic.

The next group of tags to be used were strap tags (Gilbert 1924) and stud or button tags (Greene 1911; Rounsefell and Kask 1945). Both button and stud tags had low recoveries compared to Petersen discs and their use was not widespread. Strap tags have been used with reasonable success on demersal fish since development of those tags in 1922.

Dangler tags, attached by a wire or nylon bridle have been popular over the last century. The most frequently used tags of this type are the Carlin and hydrostatic tags, which have a low loss rate when they are attached by wire bridle or internal anchor. These tags require a great deal of time to apply, however, and traditional dangler tags have been replaced by modern anchor tags in recent studies.

Metal subcutaneous tags have been used in livestock for several years. The first known use of this tag on fish was by Le Cren (1954), who placed metal tags under the skin of char *Salvelinus* sp. near the dorsal fin. Butler (1957) designed a plastic tag that was visible through the skin for use on trout. A more recent modification of this tag was a small piece of diazo film (Heugel et al. 1977) that was placed just below the skin anterior to the dorsal fin of small poeciliid fishes. The major drawback of subcutaneous tags is that they must be surgically implanted, which limits the number of fish that can be tagged within a given time.

During 1987, personnel at the Northwest Marine Technology Laboratory, Shaw Island, Washington, developed a small (1.0 by 2.5 mm) subcutaneous tag that is injected under transparent tissue such as that found near the orbit of the eye in some fish species. This tag can be injected easily from a roll with consecutive numbers and read under magnification without killing the fish. The small wound heals quickly and eliminates chronic wound problems associated with tags that protrude through the skin.

Several other tags have been introduced over the years with limited or no success. These have included safety pins (Rounsefell and Kask 1945; Thompson 1962), ribbons (Rounsefell and Dahlgren 1933), labelled hooks (Malimgren 1884; Jarvi 1949), and staples (Jordan and Smith 1968; Yap and Furtado 1980). The staple tag is almost identical to the Archer tag (Rounsefell and Kask 1945), and has been used mainly on salmon.

The most recent addition (early 1950s) to the external tag inventory has been tags made from vinyl tubing. They are relatively inexpensive and are available in a variety of sizes and colors. The most well known of these are the spaghetti, dart, and anchor tags. The spaghetti loop tag is simply a loop of vinyl tubing, with a legend printed directly upon it, that passes through the organism and is secured by a knot or metal crimp. Wilson (1953) first used this tag on tuna; within a few years, it was applied to freshwater fish (Tebo 1957), and since has been used on a wide variety of species (Jensen 1963; Hartt 1963). Several agencies use it as a standard tag. Modifications to this tag include lock-on and cinch-up tags that have either a nylon locking device instead of a knot or a nylon fastener that can be "cinched" up to fit various sizes of fish.

Dart tags were introduced in the early 1950s (Mather 1963). These tags consist of straight, flexible shafts, one end of which is barbed or shaped into an "arrow" or "harpoon" head. Metal darts were used first, but nylon barbs were developed (Yamashita and Waldron 1958) when it was reported that the nylon dart heads were recovered more frequently than metal ones. Since then, return rates have been variable for both types of dart head, and some investigators of large, marine pelagic fish species have returned to the metal dart head (Squire and Nielsen 1983). Smaller dart tags have been used widely on other fish and invertebrates including soles, salmonids, lobsters, and crabs.

The modern anchor tag is similar to the dart tag except the barbed head has been replaced by a nylon T-bar. Vinyl tubing is attached sequentially to a "clip" of nylon T-bar heads of the type used to attach price tags to clothing (Dell 1968). These tags are applied to fish quickly with a hand-operated applicator. The retention of the early anchor tags by fish was poor because the cement used to attach the nylon T-bar to the vinyl tubing deteriorated. New cements have corrected this problem. When properly applied, the retention rate is high for most species. These tags are extremely popular because of the ease with which large numbers of fish can be tagged with individual serial numbers, the low cost, and the good retention rates.

After the early 1950s, there was a tremendous increase in the number of studies involving external tags and marks (Figure 4). This was correlated with a world-wide requirement to develop effective management strategies for fish stocks. In addition, postwar industries were making rapid technological advances in the development and manufacture of new materials, particularly plastics, that allowed many existing tags to be modified or new tags to be developed, and the versatility and usefulness of tags increased accordingly. However, we believe that the major advances in tag design and application methods were made when biologists developed tags to fit the biology of an animal. It became increasingly evident that tags were needed that could remain with the organisms for several years, would not stress the animal to the point of altering behavior or growth, could be applied easily to a large number of individuals, would not become lost or entangled when the animal was captured during normal fishing operations, and could be easily identified.

Development of Selected External Tags

Petersen Disc

We have selected examples of tag development that illustrate some major advances in tag design. The first of these major advances was the development of the Petersen disc (Petersen 1896). Shortly after beginning his experiments using mutilation techniques, Petersen realized the need for individual identification of fish to validate ages and estimate growth rates. His first attempts were unsuccessful because the tags lasted only a few months. From these initial experiments Petersen determined that a fixed disc tag would provide the retention rate necessary. He originally used two bone discs, one each side of the fish, connected by a silver wire running through the dorsal musculature and twisted at each end. Although reasonably

successful (some tags were recovered after 1 year), the numbers and letters burned into the bone discs became illegible and the discs gradually deteriorated in sea water. Brass discs with numbers stamped on them were then used in conjunction with or instead of the bone discs.

Garstang (1905) and Hjort (1914) used silver discs in place of bone. Ebonite was substituted for bone or brass as early as 1905 (Schmidt 1931) and the use of silver wire was common. By 1934, celluloid discs replaced ebonite; later, various plastic discs provided a large variety of colors and sizes. Concurrent with disc development, experiments were conducted with other wire materials for fasteners, most notably with nickel, monel, and stainless steel (Jensen 1958). The use of stainless steel pins has been prevalent in studies since the early 1950s.

The most recent modification to the Petersen disc was reported by McFarlane and Beamish (1986). Because information indicated that spiny dogfish *Squalus acanthias* live longer than 70 years, tags were required that would remain on these sharks at least 20 years so that long-term movements could be monitored. Previous attempts to tag spiny dogfish with plastic anchor tags and Petersen discs attached with stainless steel and other wire pins had proved unsatisfactory because few recoveries were made after 2–3 years. The modified tags consist of two elongate plastic discs attached by two titanium pins inserted at the base of the first dorsal fin. Recoveries of these tags are still being made after 9 years (McFarlane, unpublished data).

A major drawback of the Petersen disc tag is the restriction of substantial growth in body thickness during long-term experiments. The Petersen disc tag was quite adequate for most flatfish species except halibut (for which the strap tag was adopted), and for short-term experiments with salmonids and crustaceans. However, for most roundfish studies and any long-term experiments with crustaceans, hydrostatic tags, other dangler tags, and anchor tags have been substituted for Petersen discs.

Internal Anchor Tag

The development of the internal anchor tag (Figure 5a) was the next major advance in external marking. It was developed by Rounsefell in 1936 (Rounsefell and Kask 1945) to ". . . meet the need for an externally visible tag in which the wound could heal completely and the fish could undergo a very large increase in size without

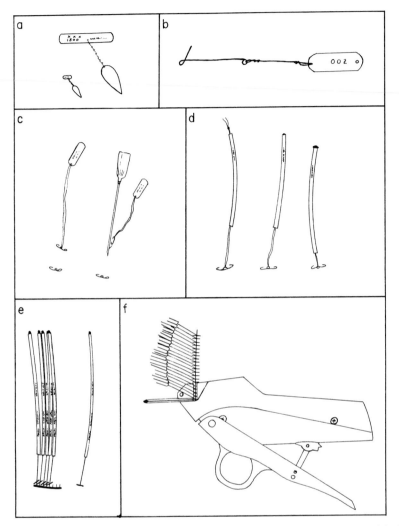

FIGURE 5.—Evolution of the anchor tag: (a) internal anchor, (b) Lawler's spring anchor, (c) original sphyrion, (d) modified sphyrion, (e) present "Floy" anchor, (f) tag applicator.

losing the tag". This tag was the "prototype" for all anchor tags and its development really emphasized the method of attachment rather than tag type. Anchor attachments have been used in conjunction with the hydrostatic tag, the plastic roll tag and others. The anchor originally was inserted through a small incision in the abdominal wall. In later studies, it was pushed through an incision in the dorsal musculature and lodged in the flesh, presumably against the far side of the interneural bones. Both these attachments were reported to provide return rates that were similar to or better than those of stud tags or Petersen discs. Anchor tags were used successfully on eels, cod, haddock, halibut, sardine, salmon, and other

fish species; they also worked well on crabs and lobsters when they were inserted into the chitonous dermis. The next development was introduced by Lawler (1963), who needed a tag that could be used to estimate population parameters of freshwater fish. The tag had to "... provide good anchorage, be easy to apply to fish of all sizes, and not have any protuberances which might tend to catch on nets or other objects, thus increasing the chances of the fish being caught." Lawler's tag consisted of an identifying disc attached by an anchor of stainless steel, tantalum wire, or nylon (Figure 5b). A hypodermic needle was inserted through the flesh below the dorsal fin, between the interneurals and out the other side of the fish. The spring end of the anchor was compressed and inserted in the hollow needle; then the needle was withdrawn, allowing the spring to open in position between the interneurals. Lawler reported recapture rates of 18% after 2 years with this tag.

At the same time Lawler developed his internal anchor, Scarratt and Elson (1965) were looking for a tag that could be applied to young fish and be carried by the fish through to maturity. They focussed their attention on the method by which the parasitic copepod *Sphyrion lumpi* is attached to its host and designed the *"sphyrion"* tag (Figures 5c,d) to approximate the hooklike attachment mechanism of the parasite. The tag was inserted with a hypodermic needle into the dorsal musculature of a fish between the interneural bones and was used successfully on salmon and shad *Dorosoma* sp. It was particularly valuable for lobster studies because it remained on the animals through moulting.

Scarratt (1970) modified the sphyrion tag in 1966 by replacing the flat oval discs with extruded vinyl "spaghetti" fused to the anchor (Figure 5d). This tag was the prototype for the modern anchor tag (Figure 5e; Dell 1968). Anchor tags are inserted with an applicator (Figure 5f) that is loaded with a clip of sequentially marked tags, allowing large numbers of fish to be marked individually, quickly, and precisely even by inexperienced taggers. The anchor tag have been used successfully on a wide variety of fish and invertebrates since its introduction in 1967.

Recently, a new version of the internal anchor tag has been introduced that has an elongate plastic disc as the anchor and a vinyl tube streamer for recognition. The anchor, with a legend printed on it, is inserted into the body cavity. This tag comes in various sizes and can be used on a wide variety of fish. Preliminary results of this tagging procedure have been encouraging for some species (Fable 1990, this volume).

Dart Tags for Pelagic Fish

An excellent example of tag development to meet specific needs is found in the tagging history of marine pelagic fish, particularly tunas (Wilson 1953). The first attempt to tag a pelagic species was made by Sella in 1911 (Rounsefell and Kask 1945); the tag was a piece of copper chain fastened around the caudal peduncle. Several other tags were tried thereafter, none of them successfully: hooks (Sella 1924), leather and bronze straps (Frade and Dentinho 1935), celluloid discs (West-

man and Neville 1942; Scagel 1949; Partlo 1951), and Petersen discs and plastic strips (Powell et al. 1952; Schaefers 1952). It became apparent to biologists that a new tag should be developed for pelagic species. Using an experimental water tunnel similar to that of Alverson and Chenowith (1951), Wilson (1953) developed the "spaghetti" loop tag (Figure 6a). The discovery that this tag had low resistance in flowing water led to the first successful tagging experiments on yellowfin tuna and albacore. The spaghetti tag has subsequently been used on many fish species.

Although the spaghetti tag was successful on smaller tunas, there was no satisfactory tag for larger fish that could not be taken aboard the boat. In 1954, Mather (1963) designed the first dart tag (Figure 6b), which resembled a "miniature harpoon" and could be applied to the fish while it was still in the water. This tag was quickly adopted for individually marking large fish of many marine pelagic species. The harpoon-like dart of this tag was made of stainless steel and was driven into the musculature of the fish at an angle that would allow the spaghetti streamer to lie alongside the fish as it swam. Yamashita and Waldron (1958) modified this tag by using a nylon barb (Figure 6c) and reported significantly higher returns from tagged skipjack tuna *Euthynnus pelamis*. Subsequent studies in which both stainless steel and nylon darts (Figure 6d) were used have shown either similar return rates (Mather 1963; Baglin et al. 1980) or significantly higher returns of the stainless steel dart (Squire and Nielson 1983). In the latter case, the very tough skin of the test animals (billfish, Istiophoridae) could have contributed to the higher return rates of stainless steel darts. Some authors attributed the higher return rates to the "in water" application technique used in these experiments, but other studies that involved only soft-skinned scombrids indicated no significant difference in shedding rates of the two dart types (Baglin et al. 1980). Therefore, ease of dart penetration is an important consideration when a tagging program is planned In the Cooperative Shark Tagging Program along the U.S. Atlantic coast, for example, project staff sharpen the stainless steel dart heads before issuing them to cooperators so that very tough shark skins can be more readily penetrated (J. Casey, U.S. National Marine Fisheries Service, personal communication).

Baglin et al. (1980) concluded from their double-tagging experiment with bluefin tuna that shedding rates did not differ significantly between

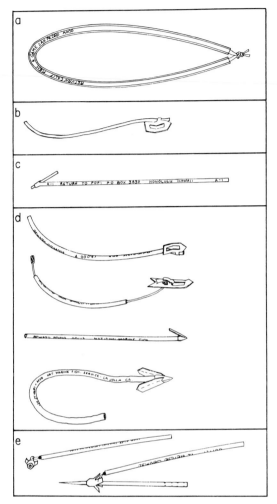

FIGURE 6. Evolution of vinyl tubing dart tags: (**a**) "spaghetti" tag developed by Wilson (1953), (**b**) dart tag with stainless steel head developed in 1954 by Mather (1963), (**c**) dart tag with nylon head developed by Yamashita and Waldron (1958), (**d**) common dart tags currently in use (Squire and Nielsen 1983), (**e**) most recent dart tag (Hallprint, personal communication).

lar tags with two large barbs, and were injection-moulded from hydroscopic nylon for use on school-size bluefin tuna (less than 45 kg). The intent was to develop a tag that would encourage the adhesion of tissue to the nylon anchor and thus reduce the long-term component of tag shedding. To attach the information portion of the tag, the monofilament section was run through a hole in the side of the head and then burned to form a small bubble. This form of attachment was not totally satisfactory because burnt monofilament crystallized with age and shedding of tags remained high. An experiment to test these new tags was cancelled when the seine fishery for bluefin tuna was greatly reduced because of landing quotas in 1982 and large numbers of fish could not be obtained.

At about the same time, reports from the Co-operative Gamefish Tagging Program (Scott et al. 1990) indicated that numerous recaptured bluefin tuna and billfish retained only the monofilament portion of the standard tag, having lost the yellow vinyl section containing identification numbers and the return address. This led to the discovery that the brass crimp used to attach the vinyl sleeve corroded over time. This known source of shedding was corrected in 1981 by doubling over the monofilament and using shrink tubing to secure the end.

The next step in development of a better dart tag occurred when it became popular among recreational anglers to tag king mackerel *Scomberomorus cavalla* in the Atlantic Ocean (Fable 1990). The nylon head developed for school-size bluefin tuna was again miniaturized (to 6.5 mm in diameter) for tagging smaller scombrids. The use of nylon heads was preferred because the skin of scombrids is easily penetrated. The anchor portion also was modified by elimination of the side hole for the monofilament attachment. Instead, a hole was drilled in the base portion of the anchor and liquid phenol was used to bond the nylon head to the monofilament section. This approach was distinctly better than burning the monofilament or gluing the nylon heads to the vinyl portion of dart tags; Bayliff and Holland (1986), for example, reported cases in which such glue joints did not hold.

Experiments to evaluate the 6.5-mm dart tags on king mackerel were delayed, but these tags were tried on red drum *Sciaenops ocellatus* in the Gulf of Mexico (Gutherz et al. 1990, this volume). After a procedure was developed that allowed insertion of the nylon darts under the heavy

stainless steel and plastic darts, but that shedding rates did vary over time. In addition, they felt that the tag-shedding rates encountered with both tags were unacceptably high and recommended further development. In response to this need, a more reliable plastic head was made in 1982 at the Southeast Fisheries Center, U.S. National Marine Fisheries Service, by miniaturizing the large plastic head used to tag giant bluefin tuna (more than 132 kg) in the Atlantic Ocean (Scott et al. 1990, this volume). The new heads (reduced from 17 to 10 mm in diameter) were designed as intermuscu-

TABLE 2.—Tagging programs of selected international fisheries commissions or agencies.

Commission or agency	Area	Target species	Tag type	Date started
International Pacific Halibut Commission (IPHC)	Northwest Pacific Ocean	Pleuronectids	Strap	1925
International Pacific Salmon Fisheries Commission (IPSFC)	Pacific Ocean	Salmonids	Petersen disc, fin clip	1938
Inter-American Tropical Tuna Commission (IATTC)	Pacific Ocean	Scombrids	Dart (nylon anchor)	1955
Great Lakes Fisheries Commission (GLFC)	Great Lakes, North America	Salmonids, percids	Fin clip, coded wire, jaw tag	1960
International Commission for the Conservation of Atlantic Tunas (ICCAT)	Atlantic Ocean	Scombrids, istiophorids	Dart (nylon and steel anchors)	1966
South Pacific Commission (SPC)	South Pacific Ocean	Scombrids	Dart	1977
European Development Fund (EDF)	Indian Ocean	Scombrids	Dart	1988
Indo-Pacific Tuna Development and Management Programme (IPTP)	Indian and Pacific oceans	Scombrids	Dart	1988

cycloid scales of red drum, several thousand fish were tagged and released, and a portion of these fish were retained in holding ponds for observation. Preliminary analysis indicated that muscle tissue encapsulated the entire insertion area and adhered to the nylon head. Tags encapsulated in this manner were almost impossible to take out by hand and had to be cut out of the fish with a knife (Gutherz et al. 1990). However, further evaluation is necessary before definitive conclusions can be made concerning long- and short-term shedding rates.

The most recent development of dart tags is currently being conducted in Australia (M. Hall, Hallprint Ltd., personal communication), where very sharp stainless steel applicators are being used to insert nylon heads with four strong barbs (Figure 6e). This approach takes advantage of the best characteristics of both materials to reduce shedding rates: the ease and consistency with which stainless steel penetrates and the susceptibility of nylon to tissue adherence. Results of these studies are not yet available.

Most large tagging studies now incorporate either nylon or stainless steel darts, depending on the intended target species. Although some studies have shown lower returns with dart tags than with spaghetti tags, there is no doubt that the development and use of dart tags on larger marine pelagic species has provided considerable information on the movement of these animals that would have been unobtainable with other conventional tags.

International Tagging Programs

Most large-scale international tagging programs have been undertaken since the Second World War (Table 2). The International Pacific Halibut Commission, however, started its program in 1925. Recognition of the over-exploitation of stocks of Pacific halibut *Hippoglossus stenolepis* in the early 1920s was primarily responsible for the formation of this commission and its subsequent management programs, including tagging efforts. The strap tag used on Pacific halibut is attached to the jaw and has an acceptable level of retention. However, most international agencies target either salmonids or scombrids and the disc tag and dart tag have been the most popular choices for these species groups (Table 2). The International Commission for the Conservation of Atlantic Tunas is another agency that chose a tag for a specific purpose. This commission recently added istiophorids to its tagging program (Miyake 1990, this volume) and chose the stainless steel dart tag, rather than the nylon dart it uses on scombrids, because of the ease with which the metal head penetrates the tough skin of billfishes (Scott et al. 1990).

Overview

No single tag is suitable for all organisms. The basic considerations for selection of an external mark or tag are the objectives of the study; the effect of the tag on survival, behavior, or growth; the permanency and recognition of the mark; the number and size of the organisms to be marked; the stress of capture, handling, and marking on the organisms; the ease of skin penetration or application of the tag; the availability of project personnel; and the costs and benefits of the study. Of particular importance is the length of time that an organism must retain the tag or mark; this time can range from days to decades depending on the species involved and the purpose of the study.

Each tag or mark has a different set of capabilities and limitations, and the performance of each varies from species to species. Life history information about the species to be marked should be considered carefully when experiments are planned. Major advances in tagging have occurred when biologists used a thorough knowledge of an animal's biology to design or select a tag.

Experimental design and analysis are covered elsewhere in this volume, but it is appropriate to mention here that credibility of results from a tagging study rests on demonstrations that major assumptions and biases have been met or accounted for. When using a natural tag such as a unique mark on bony structures, the biologist must prove that the mark remains with the fish throughout its life. Tag loss is probably the most serious problem that must be dealt with when results from tagging experiments are quantified. It may be necessary to use special techniques, such as double-tagging, to estimate tag or mark loss (Myhre 1966; Bayliff and Mobrand 1972; Kirkwood 1981; Beamish and McFarlane 1988). The tag should not change the survival rate, growth rate, or behavior of the marked fish or affect the catchability of the fish by various gear types; the validity of these basic premises should be evaluated for each tagging program. The recovery of tags and tagged fish, whether by project staff or by cooperating fishermen and anglers, is subject to numerous sampling and other biases; these too must be addressed during program design and analysis. Bayliff and Holland (1986) discussed in detail the recovery mechanisms for marked tunas.

Historically, tags have been developed to examine specific questions about the dynamics of a single species or stock. An increasing number of studies, however, are addressing multispecies and multidisciplinary issues, such as the interrelationships of physical and biological oceanography and fisheries biology. These studies require even more detailed information, much of which will be obtained from a better integration of several techniques. Within the field of tagging, recent developments in internal, electronic, genetic, and

chemical markers look promising. In particular, the recent use of nuclear DNA to discern lineages of fish illustrates the level of study that will be available to us.

For the most part, the overall effects of tagging have not been dealt with effectively. If we continue to increase our use of tags, we also must increase our knowledge of how tags affect the growth, mortality, behavior, and other basic components of a tagged animal's life history. Further, we must give more attention to the factors that affect our interpretation of tagging data, particularly tag loss and nonreporting of recovered tags, and ways of standardizing the data in terms of biomass.

Tags may be replaced by other, still undeveloped, techniques for certain types of studies. Given the continued development of new tags and tagging methods, however, there surely will be major changes in the way we conduct and interpret our tagging studies during the next 10–20 years.

Acknowledgments

We thank Kirstin Urdahl who carried out the literature review and spent long hours tracking down many of the early papers. We also thank W. Bayliff, R. Beamish, B. Leaman, M. Saunders, and S. Westrheim for helpful suggestions on earlier drafts of this paper.

References

Alverson, D. L., and H. H. Chenowith. 1951. Experimental testing of fish tags on albacore in a water tunnel. U.S. Fish and Wildlife Service, Commission of Fisheries Review 13(8):1–7.

Anderson, E. 1973. A method for marking nudibranchs. Veliger 16(1):121–122.

Arnold, D. E. 1966. Marking fish with dyes and other chemicals. U.S. Fish and Wildlife Service Technical Paper 10.

Bagenal, T. B. 1967. A method of marking fish eggs and larvae. Nature (London) 214:113.

Baglin, E. R., Jr., M. I. Farber, W. H. Lenarz, and J. M. Mason, Jr. 1980. Shedding rates of plastic and metal dart tags from Atlantic bluefin tuna, *Thunnus thynnus*. U.S. National Marine Fisheries Service Fishery Bulletin 78:279–285.

Bayliff, W. H., and K. N. Holland. 1986. Material and methods for tagging tuna and billfishes; recovering the tags and handling the recapture data. FAO (Food and Agriculture Organization of the United Nations) Fisheries Technical Paper 279.

Bayliff, W. H., and L. M. Mobrand. 1972. Estimates of the rates of shedding of dart tags from yellowfin tuna. Inter-American Tropical Tuna Commission Bulletin 15:441–462.

Beamish, R. J., and G. A. McFarlane. 1988. Resident and dispersal behaviour of sablefish (*Anoplopoma fimbria*) in the slope waters off Canada's west coast. Canadian Journal of Fisheries and Aquatic Sciences 45:152–164.

Bilton, H. T., and H. B. Messinger. 1975. Identification of major British Columbia and Alaska runs of age 1.2 and 1.3 sockeye from their scale characters. International North Pacific Fisheries Commission Bulletin 32:109–129.

Brock, J. A., and R. K. Farrell. 1977. Freeze and lazer marking of channel catfish. Progressive Fish-Culturist 39:138.

Bulkley, R. V. 1963. Natural variation in spotting, hyoid teeth counts, and coloration of Yellowstone cutthroat trout (*Salmo clarki lewisi*) (Girard). U.S. Fish and Wildlife Service Special Scientific Report—Fisheries 460.

Bumpus, H. C. 1901. On the movement of certain lobsters liberated at Woods Hole during the summer of 1898. Bulletin of the U.S. Fisheries Commission 19(1899):225–230.

Butler, R. L. 1957. The development of a vinyl plastic subcutaneous tag for trout. California Fish and Game 43:201–212.

Cagle, F. R. 1939. A system of marking turtles for future identification. Copeia 1939:170–173.

Calderwood, W. L. 1902. A contribution to the life history of the salmon, as observed by means of marking adult fish. Annual Report of the Fisheries Board for Scotland 20(2):55–100.

Carlin, B. 1955. Tagging of salmon smolts in the River Lagan. Pages 57–74 *in* Annual report for 1954. Institute of Freshwater Research, Drottningholm, Sweden.

Clark, D. R., Jr. 1971. Branding as a marking technique for amphibians and reptiles. Copeia 1971:148–151.

Cook, R. C., and I. Guthrie. 1987. In-season stock identification of sockeye salmon (*Oncorhynchus nerka*) using scale pattern recognition. Canadian Special Publication of Fisheries and Aquatic Sciences 96:327–334.

Cope, O. B. 1957. Races of cutthroat trout in Yellowstone Lake. U.S. Fish and Wildlife Service Special Scientific Report—Fisheries 208:74–84.

Coutant, C. C. 1972. Successful cold branding of nonsalmonids. Progressive Fish-Culturist 34:131–132.

Dando, P. R., and R. Ling. 1980. Freeze-branding of flatfish: flounder, *Platichthys flesus*, and plaice, *Pleuronectes platessa*. Journal of the Marine Biological Association of the United Kingdom 60:741–748.

Daugherty, C. H. 1976. Freeze-branding as a technique for marking anurans. Copeia 1976:836–838.

Davis, C. S. 1955. The injection of latex solution as a fish marking technique. Investigations of Indiana Lakes and Streams 4:111–116.

Dawson, C. E. 1957. Studies on the marking of commercial shrimp with biological stains. U.S. Fish and Wildlife Service Special Scientific Report—Fisheries 231.

Deacon, J. E. 1961. A staining method for making large numbers of small fish. Progressive Fish-Culturist 23:41–42.

Delany, M. J. 1978. Introduction: marking animals for research. Pages 3–10 *in* B. Stonehouse, editor. Animal marking: recognition marking of animals in research. University Park Press, Baltimore, Maryland.

Dell, M. B. 1968. A new fish tag and rapid, cartridge-fed applicator. Transactions of the American Fisheries Society 97:57–59.

de Veen, J. F. 1969. Abnormal pigmentation as a possible tool in the study of populations of the plaice (*Pleuronectes platessa* L.). Journal du Conseil, Conseil International pour l'Exploration de la Mer 32:344–383.

Dunn, A., and C. M. Coker. 1951. Notes on marking live fish with biological stains. Copeia 1951:28–31.

Dunstan, W. A., and W. E. Bostick. 1956. New tattooing devices for marking juvenile salmon. Washington State Department of Fisheries, Fisheries Research Paper 1(4):70–79.

Emery, L., and R. Wydoski. 1987. Marking and tagging of aquatic animals: an indexed bibliography. U.S. Fish and Wildlife Service Resource Publication 165.

Everest, F. H., and E. H. Edmundson. 1967. Cold branding for field use in marking juvenile salmonids. Progressive Fish-Culturist 29:175–176.

Everhart, W., A. Eipper, and W. Youngs. 1975. Principles of fishery science. Cornell University Press, Ithaca, New York.

Fable, W. A. 1990. Summary of king mackerel tagging in the southeastern USA: mark–recapture techniques and factors influencing tag returns. American Fisheries Society Symposium 7:161–167.

Feder, H. M. 1955. The use of vital stains in marking Pacific coast starfish. California Fish and Game 41:245–246.

Federighi, H. 1931. Studies on the oyster drill (*Urosalpine cinerea*). U.S. Bureau of Fisheries Bulletin 47(4):85–111.

Frade, F., and S. Dentinho. 1935. Marcao de Atuns no Carta do Algarve. Boletim da Sociedade Portuguesa de Ciencias Naturais 12(10).

Frost, W. W. 1963. The homing of charr (*Salvelinus willughbii*). (Gunther) in Windermere. Animal Behaviour 11:74–82.

Fujihara, M. P., and R. F. Nakatani. 1967. Cold and mild heat marking of fish. Progressive Fish-Culturist 29:172–174.

Fulton, T. W. 1893. An experimental investigation on the migrations and rate of growth of the food fishes. Annual Report of the Fishery Board of Scotland 11(3):176–196.

Garstang, W. 1905. Report on experiments with marked fish during 1902–1903. Pages 13–44 *in* First report on fishery and hydrographical investigations in the North Sea and adjacent waters (southern area). Conducted for His Majesty's Government by the Marine Biological Association of the United Kingdom, 1902–1903, volume 1. Darling and Sons, Edinburgh, Scotland.

Gilbert, C. H. 1924. Experiment in tagging adult red salmon, Alaska peninsula fisheries conservation, summer of 1922. U.S. Bureau of Fisheries Bulletin 39:39–50.

Godsil, H. C. 1948. A preliminary population study of the yellowfin tuna and the albacore. California Division of Fish and Game, Fisheries Bulletin 70.

Goldspink, C. R., and J. W. Banks. 1971. A readily recognizable tag for marking bream *Abramis brama* (L.). Journal of Fish Biology 3:407–411.

Gowanlock, J. N. 1927. Contributions to the study of marine gastropods II. The intertidal life of *Buccinum undatum*, a study in non-adaptation. Contributions to Canadian Biology 3(5).

Greene, C. W. 1911. The migration of salmon in the Columbia River. U.S. Bureau of Fisheries Bulletin 29:129–148.

Griffin, D. R. 1952. Radioactive tagging of animals under natural conditions. Ecology 33:329–335.

Groves, A. B., and A. J. Novotny. 1965. A thermal-marking technique for juvenile salmonids. Transactions of the American Fisheries Society 94:386–389.

Gutherz, E. J., B. A. Rohr, and R. V. Minton. 1990. Use of hydroscopic molded nylon dart and internal anchor tags on red drum. American Fisheries Society Symposium 7:152–160.

Hartt, A. C. 1963. Problems in tagging salmon at sea. International Commission for the Northwest Atlantic Fisheries Special Publication 4:144–145.

Heugel, B. R., G. R. Jaswick, and W. S. Moore. 1977. Subcutaneous Diazo film tag for small fishes. Progressive Fish-Culturist 39:98–99.

Hickling, C. F. 1945. Marking fish with the electric tattooing needle. Journal of the Marine Biological Association of the United Kingdom 26:166–169.

Hjort, J. 1914. Fluctuations in the great fisheries of northern Europe. Rapports et Proces-Verbaux, Conseil Permanent International pour l'Exploration de la Mer 20:3–228.

Jakobsson, J. 1970. On fish tags and tagging. Oceanography and Marine Biology: An Annual Review 8:457–499.

Jarvi, T. H. 1949. A review of the results of salmon marking in the Baltic region. Journal du Conseil, Conseil International pour l'Exploration de la Mer 16:100–112.

Jensen, A. C. 1958. Corrosion resistance of fish tagging pins. U.S. Fish and Wildlife Service Special Scientific Report—Fisheries 262.

Jensen, A. C. 1963. Further field experiments with tags with haddock. International Commission of the Northwest Atlantic Fisheries Special Publication 4:194–203.

Jones, R. 1979. Materials and methods used in marking experiments in fishery research. FAO (Food and Agricultural Organization of the United Nations) Fisheries Technical Paper 190.

Jordan, F. P., and H. D. Smith. 1968. An aluminum staple tag for population estimates of salmon smolts. Progressive Fish-Culturist 30:230–234.

Joyce, J. A., and H. M. El-Ibiary. 1977. Persistency of hot brands and their effects on growth and survival of fingerling channel catfish. Progressive Fish-Culturist 39:112–114.

Karlsson, J., and R. Sisson. 1973. A technique for detection of brushed lobsters by staining of cement on swimmerets. Transactions of the American Fisheries Society 102:847–848.

Kask, J. L. 1936. The experimental marking of halibut. Science (Washington, D.C.) 83:435–436.

Kirkwood, G. P. 1981. Generalized models for the estimation of rates of tag shedding by southern bluefin tuna (*Thunnus thunnus*). Journal du Conseil, Conseil International pour l'Exploration de la Mer 39:256–260.

Klima, E. F. 1965. Evaluation of biological stains, inks, and fluorescent pigments as marks for shrimp. U.S. Fish and Wildlife Service Special Scientific Report—Fisheries 511.

Koo, T. S. Y. 1962. Differential scale characters among species of Pacific salmon. Pages 123-135 *in* T. S. Y. Koo, editor. Studies of Alaska red salmon. University of Washington Press, Seattle.

Koval, N. V. 1969. Experimental use of neutral red dye to tag young fish. Journal of Hydrobiology 5(6):95–99.

Laird, L. M., and B. Scott. 1978. Marking and tagging. IBP (International Biological Programme) Handbook 3:84–100.

Lawler, G. H. 1963. Spring stainless steel anchor tag. Journal of the Fisheries Research Board of Canada 20:1553.

Lawler, G. H., and M. Fitz-Earle. 1968. Marking small fish with stains for estimating populations in Heming Lake, Manitoba. Journal of the Fisheries Research Board of Canada 25:255–266.

Le Cren, E. D. 1954. A subcutaneous tag for fish. Journal du Conseil, Conseil International pour l'Exploration de la Mer 20:72–82.

Lewke, R. E., and R. K. Stroud. 1974. Freeze-branding as a method of marking snakes. Copeia 1974:997–1000.

Loosanoff, V. L. 1937. Use of Nile blue sulfate in marking starfish. Science (Washington, D.C.) 85:412.

McFarlane, G. A., and R. J. Beamish. 1986. A tag suitable for assessing long-term movements of spiny dogfish and preliminary results of a study utilizing this tag. North American Journal of Fisheries Management 6:69–76.

Major, R. L., K. H. Mosher, and J. E. Mason. 1972. Identification of stocks of Pacific salmon by means of scale features. Pages 209–231 *in* R. C. Simon and P. A. Larkin, editors. The stock concept in Pacific salmon. H. R. MacMillan Lectures in Fisheries, University of British Columbia, Vancouver, Canada.

Malimgren, A. J. 1884. The migrations of the salmon (*Salmo salar* L.) in the Baltic. U.S. Fish Commission Bulletin 4:322–328.

Marsh, M. C., and J. N. Cobb. 1908. The fisheries of Alaska in 1908. U.S. Bureau of Fisheries Bulletin Document 645.

Marshall, S., and nine coauthors. 1987. Application of scale pattern analysis to the management of Alaska's sockeye salmon (Oncorhynchus nerka) fisheries. Canadian Special Publication of Fisheries and Aquatic Sciences 96:307–326.

Mather, F. J., III. 1963. Tags and tagging techniques for large pelagic fishes. International Commission for the Northwest Atlantic Fisheries Special Publication 4:288–293.

Marukawa, H., and T. Kamiya. 1930. Some results of the migration of important food fish. Journal of the Imperial Fisheries Experimental Station of Japan 1(1):5–38.

Miyake, P. M. 1990. History of the ICCAT tagging program, 1971-1986. American Fisheries Society Symposium 7:746–764.

Myhre, R. J. 1966. Loss of tags from Pacific halibut, as determined by double-tag experiments. International Pacific Halibut Commission Report 41.

Nielson, B. R. 1990. Twelve-year overview of fluorescent grit marking of cutthroat trout in Bear Lake, Utah–Idaho. American Fisheries Society Symposium 7:42–46.

Partlo, J. M. 1951. A report on the 1950 albacore fishing of British Columbia (Thunnus alalunga). Fisheries Research Board of Canada, Pacific Biological Station, Circular 23, Nanaimo.

Petersen, C. G. J. 1896. The yearly immigration of young plaice into the Limfjord from the German Sea. Report of the Danish Biological Station to the Board of Agriculture (Copenhagen) 6:5–30.

Phinney, D. E., and S. B. Mathews. 1973. Retention of fluorescent pigment by coho salmon after two years. Progressive Fish-Culturist 35:161–163.

Phinney, D. E., D. M. Miller, and M. L. Kahlberg. 1967. Mass-marking young salmonids with fluorescent pigment. Transactions of the American Fisheries Society 96:157–162.

Powell, D. F., D. L. Alverson, and R. Livingston. 1952. North Pacific albacore tuna exploration—1950. U.S. Fish and Wildlife Service Fisheries Leaflet 402.

Raleigh, R. F., J. B. McLaren, and D. R. Graff. 1973. Effects of topical location, branding techniques, and changes in hue on recognition of cold brands in centrachid and salmonid fish. Transactions of the American Fisheries Society 102:637–641.

Richter, K. O. 1973. Freeze-branding for individually marking the banana slug (Ariolimax columbianus G.). Northwest Science 47:109–113.

Ricker, W. E. 1956. Uses of marking animals in ecological studies: the marking of fish. Ecology 37:665–670.

Rinne, J. N. 1976. Coded spine clipping to identify individuals of the spiny-rayed fish Tilapia. Journal of the Fisheries Research Board of Canada 33:2626–2629.

Ropes, J. W., and A. S. Merrill. 1969. Marking surf clams. Proceedings of the National Shellfish Association 60:99–106.

Rounsefell, G. A. 1975. Marking as a tool for research and management. Pages 223–234 in G. A. Rounsefell, editor. Ecology, utilization, and management of marine fishes. Mosby, St. Louis, Missouri.

Rounsefell, G. A., and E. H. Dahlgren. 1933. Tagging experiments on the Pacific herring, Clupea pallasi. Journal du Conseil, Conseil International pour l'Exploration de la Mer 8:371–384.

Rounsefell, G. A., and W. H. Everhart. 1953. Fishery science, its methods and applications. Wiley, New York.

Rounsefell, G. A., and J. L. Kask. 1945. How to mark fish. Transactions of the American Fisheries Society 73:320–363.

Russell, E. S. 1932. Guide to the fish marks used by members of the International Council for the Exploration of the Sea and by some non-participant countries. Journal du Conseil, Conseil International pour l'Exploration de la Mer 7:133–165.

Scagel, R. F. 1949. Report on the investigations of albacore (Thunnus alalunga). Pages 21–22 in Fisheries Research Board of Canada, Pacific Biological Station, Circular 17, Nanaimo.

Scarratt, D. J. 1970. Laboratory and field tests of modified sphyrion tags on lobsters (Homarus americanus). Journal of the Fisheries Research Board of Canada 17:257–264.

Scarratt, D. J., and P. F. Elson. Preliminary trials of a tag for salmon and lobsters. Journal of the Fisheries Research Board of Canada 22:421-423.

Schaefers, E. A. 1952. North Pacific albacore tuna exploration—1951. Commercial Fisheries Review 14(5):1–12.

Schmidt, J. 1931. Summary of the Danish marking experiments on cod, 1904–1929, at the Faroes, Iceland and Greenland. Rapports et Process-verbaux des Reunions, Conseil Permanent International pour l'Exploration de la Mer 72:2–13.

Schweigert, J. B. 1990. Comparison of morphometric and meristic data against truss networks for describing Pacific herring stocks. American Fisheries Society Symposium 7:47–62.

Scott, E. L., E. D. Prince, and C. D. Goodyear. 1990. History of the cooperative game fish tagging program in the Atlantic Ocean, Gulf of Mexico, and Caribbean Sea, 1954–1987. American Fisheries Society 7:841–853.

Sella, M. 1924. Migrations and habitat of the tuna (Thunnus thynnus L.), studied by the method of the hooks, with observations on growth, on the operation of the fisheries. Memoira R. Comitato Talassografico Italiano 156:1–24. English translation: U.S. Fish and Wildlife Service, Special Scientific Report—Fisheries 76:1–20.

Squire, J. C., and D. V. Nielsen. 1983. Results of a tagging program to determine migration rates and patterns for black marlin, Makaira indica, in the southwest Pacific Ocean. NOAA (National Oceanic and Atmospheric Administration) Technical Report NMFS (National Marine Fisheries Service) SSRF 772.

Stonehouse, B., editor. 1978. Animal marking: Recognition marking of animals in research. University Park Press, Baltimore, Maryland.

Strange, C. D., and J. A. Kennedy. 1982. Evaluation of fluorescent pigment marking of brown trout (*Salmo trutta* L.) and Atlantic salmon (*Salmo salar* L.). Fisheries Management 13:89–95.

Stuart, T. A. 1958. Marking and regeneration of fins. Freshwater Salmon Fisheries Research, Report 22, Edinburgh, Scotland.

Tebo, L. B., Jr. 1957. Preliminary experiments on the use of spaghetti tags. Proceedings of the Annual Conference Southeastern Association of Game and Fish Commissions 10:77–80.

Thomas, A. E. 1975. Marking channel catfish with silver nitrate. Progressive Fish-Culturist 37:250–252.

Thomson, G. M. 1962. The tagging and marking of marine animals in Australia. International Commission of Northwest Atlantic Fisheries Special Publication 4:50–58.

USDA (U.S. Department of Agriculture). 1971. Laser marks in fish. U.S. Department of Agriculture Research Service (Report) 19.

Walton, I., and C. Cotton. 1898. The compleat angler, or the contemplative man's recreation. Little, Brown, Boston.

Ward, F. J., and L. A. Verhoeven. 1963. Two biological stains as markers for sockeye salmon fry. Transactions of the American Fisheries Society 92:379–383.

Watson, J. E. 1961. The branding of sea herring as a short-term mark. Progressive Fish-Culturist 23:105.

Watson, J. E. 1963. A method for tagging immature herring. U.S. Fish and Wildlife Service Special Scientific Report—Fisheries 451.

Welch, H. E., and K. H. Mills. 1981. Marking fish by scarring soft fin rays. Canadian Journal of Fisheries and Aquatic Sciences 38:1168–1170.

Westman, J. B., and W. C. Neville. 1942. The tuna fishery of Long Island, New York. Board of Supervisors, Nassau County, New York.

Wilder, D. G. 1953. The growth rate of the American lobster (*Homarus americanus*). Journal of the Fisheries Research Board of Canada 10:371–412.

Wilson, R. C. 1953. Tuna marking, a progress report. California Fish and Game 39:429–442.

Woodbury, A. M. 1948. Marking reptiles with an electric tattooing outfit. Copeia 1948:127–128.

Wydoski, R. S., and L. Emery. 1983. Tagging and marking. Pages 215–237 in L. Nielsen and D. Johnson, editors. Fisheries techniques. American Fisheries Society, Bethesda, Maryland.

Yamashita, D., and K. Waldron. 1958. An all-plastic dart-type fish tag. California Fish and Game 44:311–317.

Yap, S. Y., and J. I. Furtado. 1980. Evaluation of two tagging/marking techniques and their practical application in *Osteochilus hasseltie* C + V. (Cyprinidae) population estimates and movement at Swbang Reservoir, Malaysia. Hydrobiologia 38:35–47.

American Fisheries Society Symposium 7:30–35, 1990

Freeze Branding with CO_2: An Effective and Easy-to-Use Field Method to Mark Fish

M. D. Bryant, C. A. Dolloff[1], P. E. Porter, and B. E. Wright

Forestry Sciences Laboratory, U.S. Forest Service
Post Office Box 20909, Juneau, Alaska 99802, USA

Abstract.—In 1980, Bryant and Walkotten described an easy-to-use apparatus to freeze-brand fish with compressed CO_2, and the apparatus has been used extensively in southeast Alaska. Here we describe improvements in marking techniques that have developed since 1980 and present additional data on the effects and durability of freeze brands on fish. Among the improvements are a decrease in branding time (over 300 fish/h can be branded) and a reduction in brand size. We also discuss application of the method in a field study of juvenile salmonids in several habitat types in Maybeso Creek, Prince of Wales Island, Alaska.

The release of marked fish into natural populations is used for studies of population densities, movement, and growth. Mark–recapture techniques are used by fisheries biologists in both management and research. A variety of marking methods is available, but relatively few methods have been developed that permit an observer to identify external marks on large numbers of small or juvenile fish, particularly under field conditions.

Although both hot and cold brands have been used to mark fish, freeze branding generally has been more effective (Fujihara and Nakatani 1967; Smith 1973; Joyce and El Ibiary 1977). Freeze branding may be achieved by application of liquid nitrogen (Raleigh et al. 1973; Smith 1973; Turner et al. 1974), dry ice (solid CO_2) (Everest and Edmundson 1967; Fujihara and Nakatani 1967), or Freon and laser beams (Brock and Farrell 1977). The coolants used for freeze branding are difficult to transport to or to store in remote field locations without refrigeration. Separate containers generally are used to hold the coolant (either dry ice in alcohol or acetone, or liquid nitrogen) and the brand. Brands must be both legible and durable (i.e., retained for at least 1 year).

To hold the brand in liquid nitrogen, Busack (1985) used a Dewar thermos flask. Sorensen et al. (1983) used an open container filled with liquid nitrogen to hold multiple brands that included a combination of numbers and letters for marking American eels *Anguilla rostrata*. These multiple brands were clear, legible, and durable.

---

[1]Present address: Southeastern Forest Experiment Station, Department of Fisheries and Wildlife Science, Virginia Polytechnic Institute and State University, Blacksburg, Virginia 24061, USA.

In southeast Alaska, we have used freeze branding, as developed by Bryant and Walkotten (1980), to mark juvenile coho salmon *Oncorhynchus kisutch*, steelhead *O. mykiss*, cutthroat trout *O. clarki*, and Dolly Varden *Salvelinus malma* ranging in size from about 35 mm up to 200 mm fork length (FL). Some of our studies last only a few days or weeks to assess short-term changes in survival and movements among contiguous habitats; however, other studies are conducted to monitor seasonal and annual production and migration patterns across distinct habitat types. Because most of our field studies are conducted in remote areas, our equipment must be lightweight, easy to use, and reliable under harsh conditions.

In this paper, we describe the design and operation of freeze-branding equipment that uses compressed CO_2 as a coolant. In addition, we present the results of freeze-branding tests conducted to determine mark retention and survival for juvenile coho salmon held in captivity and for juvenile salmonids in the field.

Methods

Equipment.—The equipment described by Bryant and Walkotten (1980) is shown in Figure 1 and the specifications are given in Table 1. We use a 9.1-kg fire extinguisher that carries 2.27 kg of CO_2. This container is easily transportable and otherwise suitable for field use. A larger fire extinguisher may be used but could be difficult to carry. The fittings between the CO_2 cylinders and the hose should be standardized so that tanks are interchangeable. Construction and assembly of the hose and nozzle, as well as of the brand tip,

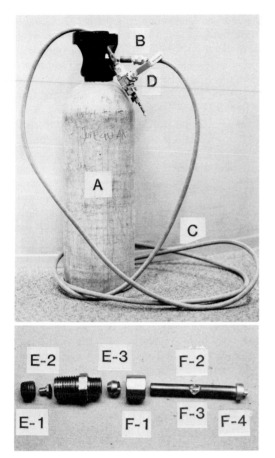

FIGURE 1.—Freeze-branding equipment for use with compressed CO_2 as a coolant (see Table 1 for equipment specifications and accessories needed for operation).

TABLE 1.—Parts, accessories, and specifications for freeze-branding equipment for use with compressed CO_2 (see Figure 1).

Items[a]	Specifications; number needed
Freeze-branding equipment	
Fire extinguisher (A)	9.1 kg; 1
Filter (B)	10 μm; 1
Hose (C)	Medium pressure; 0.5-cm inside diameter; 1
Ball valve (D)	Standard; 1
Nozzle assembly (E)	
(1) Hex-head plug	0.125 cm (national pipe thread); 1
(2) M0298 nozzle	0.1448-mm orifice; 1
(3) Male compression tube fitting	For 0.635-cm tube; 1
Brand assembly (F)	
(1) Sleeve	0.635 cm; 1
(2) Copper tube with relief slot	0.635 cm; 1
(3) Silver rod (inserted into copper tube)	5.12 mm × 38 mm; 1
(4) Silver brand	1 mm × 5 mm or 2 mm × 10 mm; 1
Accessories	
Adjustable wrench	30.5 cm; 1
Adjustable wrench	20.3 cm; 1
Adjustable wrench	10.2 cm; 1
Small copper wire brush	Toothbrush-size; 1
Teflon pipe-thread sealer	Roll; 1
Allen wrench	To fit 0.125-cm hex-head plug; 1
Drill bits	0.1448 mm, to clean nozzle orifice; 3–4
M0298 nozzles	0.1448 mm; 4–6
Drilled and tapped hex-head plugs	0.125 cm; 2–3
Acetone	500 mL

[a] Letters and numbers in parentheses relate to those in Figure 1.

are described by Bryant and Walkotten (1980). The specialized pieces can be constructed by most machine shops.

The nozzle orifice is a critical element. A diameter larger than 0.14 mm releases too much CO_2 and will not sufficiently cool the brand tip. A smaller opening tends to clog with impurities in the gas. An in-line filter eliminates some but not all impurities, and the nozzle still may clog. During field use, we have found that it is quicker and easier to change nozzles and continue to brand than to clean the nozzle during the branding process. Nozzles are relatively easy to clean with acetone and a small-diameter drill bit. A small tool kit (Table 1) should be included with the field equipment.

Originally, we used a brand that was 2 mm wide and 10 mm high; we later used a brand that was 1 mm × 5 mm for young coho salmon less than 50 mm total length. We used distinctly different letters for each size of brand so that individual brands would not be confused due to incomplete marking. All were straight-line letters: T, V, X, U, or I. We did not use the U and V brands in proximate locations because, over a period of a few months, growth of the fish or a poor brand could cause a V to be confused with a U. By changing the orientation of these letters (e.g., normal, inverted, or sideways toward the head or tail), 15 distinct marks could be made. More combinations may be obtained by using different sides of the fish or, for larger fish (>100 mm), by using different locations on the same side of the fish, such as behind or forward of the dorsal fin. Although other letters, such as B, A, Z, F, M, etc., could be used, an A could be confused with an inverted V, and an incomplete F with a T or I. With larger brands, there is less chance of incomplete marking, and perhaps more complex symbols could be used effectively.

Marking procedure.—The cooling effect is caused by the rapid drop of pressurized CO_2 from

54.4 atmospheres to atmospheric pressure. As this occurs, dry ice begins to form behind the silver rod in the brand assembly (Figure 1), and the temperature of the brand is lowered rapidly. The CO_2 is released for 10–15 s to cool the brand and turned off, and then the brand is applied to the fish. A leather glove should be used when the brand assembly is held, to prevent frostbite.

Before marking and handling them, we anesthetize fish with tricaine; other appropriate anesthetics may be used. The fish then are arranged on a board in rows and columns; the cooled brand is applied to each fish for about 2 s. Original tests (Bryant and Walkotten 1980) were done with 4-s applications, but we have found that a shorter application time works as well. With the shorter application time, we have been able to effectively brand more than 30 fish before the brand must be recooled. Under normal circumstances, up to 2,000 fish can be branded within 2 h, and one 9.1-kg tank of CO_2 can be used to brand 3,500 fish. In warmer weather (i.e., above 22°C), the brand tip will have to be cooled more frequently, and more CO_2 will be required than at lower temperatures. The brand is effective if a white frost mark appears where the brand was applied. When this does not occur, the brand must be recooled and the fish remarked. The frost mark will disappear, and the brand will not be visible until scar tissue forms about 48 h later.

If the brand sticks to the fish and cannot be removed easily, the brand and fish should be submerged in the recovery water (i.e., in the tank where fish recover from anesthesia), and the brand will release. Dipping the brand in acetone before and after each cooling and marking sequence will clean and dry the brand and thus reduce sticking. The brand should be cleaned with acetone and a dry cloth after about 30 fish are branded or if it begins to stick. Excess water should be removed from the fish by blotting or gently shaking the fish before they are branded to prevent ice build-up on the brand. Ice on the brand tip may result in an incomplete mark. Although the equipment can be used in inclement weather, rain hitting the brand tip will result in rapid ice build-up and warming of the brand.

Test marking.—Initial tests in 1979 were conducted in Juneau, Alaska, and at the School of Fisheries, University of Washington, Seattle (Bryant and Walkotten 1980). In 1985, 152 juvenile coho salmon (70–120 mm FL) were marked and held in aquaria for 6 weeks at the National Marine Fisheries Service's Auke Bay Laboratory near Juneau. Application times of 1, 2, 3, 4, and 5 s were used on seven groups of fish; there were two replicates of the 1- and 3-s branding times. All groups were held in the same tank. Treatments were differentiated by brand codes (e.g., $V_L = 4$ s, $V_R = 3$ s; R = the right side of the fish and L = the left side). The fish were checked periodically. At the end of the test period, brands were read and classified into three subjective groups based on legibility. In the highest rated class, group 1, the brand was distinct and clear. Group 2 had brands that showed some blurring or were incomplete but legible. In group 3, the brand was blurred and indistinct, but readable. The marking procedure was the same one that we use in the field.

Results and Discussion

Consistency and Duration

In the tests conducted at Juneau and at the University of Washington, fish with brands appeared to be unaffected by the mark, and all marks were visible a few days after application. These results were discussed further by Bryant and Walkotten (1980). Tests of branding times indicated that application for at least 2 s gave the best results (Table 2). Furthermore, when the brand was applied for 2 s, only two brands had a quality rating lower than 1. Two fish were found that had no discernible mark, and we were unable to tell if they actually had been branded.

No mortality occurred during or immediately after the marking procedures. Although 25 deaths eventually were recorded, these occurred several weeks after the treatment, and only two fish showed fungus around the brand scar. Mortality did not appear to be related to the length of time the brand was applied to the fish and was more likely due to aquarium conditions than to marking procedures. These observations, combined with previous experiments, our field studies, and studies of other freeze-branding methods, provided good evidence that the branding procedure itself does not impose unusual stress on the fish if proper care is exercised as the fish are handled.

The 1-mm × 5-mm brand produced legible and durable marks on the smaller fish with no adverse effects. The larger brand size (2 mm × 10 mm) was more legible on larger fish and did not blend with natural markings. When the smaller brand was used on larger fish, it was readable but not as easy to see without close inspection.

TABLE 2.—Effect of branding time on freeze-brand legibility 6 weeks after application to juvenile coho salmon. N = 25 for each row. Branding times of 1 and 3 s each had two replicates.

Branding time (s)	Number of highly visible brands	Number of visible but indistinct brands	Number of obscure brands	Deaths
1	9	7	3	6
	10	8	3	4
2	19	1	1	6
3	21	0	2	2
	22	2	0	1
4	24	0	0	1
5	20	0	0	5

The hatchery and aquarium studies previously described demonstrated that the brands were clearly visible after 6 months. In these studies, it was not feasible to hold the fish any longer due to demand for other uses of the facilities. Brand retention for longer periods has been verified by field studies on juvenile salmonids in southeast Alaska and on other species. Sorenson et al. (1983) reported a legible brand on an American eel recaptured 1 year after marking. In our studies, juvenile salmonids with marks applied up to a year earlier have been recovered. None have been reported on returning adult salmon; however, we have not specifically looked for them.

Because a brand results from formation of scar tissue (Laird et al. 1975), it will dissipate as the fish grows. Most of the older brands that we have observed are easily distinguished from more recent brands. The older brands are often distorted. A V may look like a U, and an older brand is usually more difficult to see. The process may be compared to marking a deflated balloon, inflating it, and then reading the mark. This limits the effectiveness of the mark when it is used on fast-growing fish. For example, the small brand size (1 mm × 5 mm) on a 40-mm coho salmon fry may not be visible when the fish is 65 mm long a year later; however, the larger brand (2 mm × 10 mm) applied to a 50-mm parr may be visible a year later when the fish is 90 mm long.

Application of the Method

In a typical field application of the method, we differentiated fish from several habitat types over several seasons, frequently capturing and marking 300–900 fish/d. The study was conducted in the Maybeso Creek basin on Prince of Wales Island. We use results from three sample periods—July and October 1985, and March 1986—to demon-strate that the method can be used to mark a large number of fish quickly and that marks are easily recognized after several months.

We branded 4,313 juvenile coho salmon, Dolly Varden, and steelhead in two study sections of Maybeso Creek—MBS-1 and MBS-3—in July 1985 (Table 3). Fish ranged from 40 to 140 mm in total length and were branded on the right side. In October, 3,722 juvenile salmonids were branded on the left side in the two study sections (Table 3). Fifteen habitat types were sampled, including a beaver pond and stream complex (Puyallup) associated with section MBS-1. Separate codes were applied to fish from each habitat type within each study section. The following March, all locations were resampled, and marked fish were identified by code and time of mark (i.e., a brand on the left or right side of the fish). A mark–recapture experiment was conducted during each sampling pe-

TABLE 3.—Number of salmonids freeze-branded in Maybeso Creek, Prince of Wales Island, Alaska, during July and October 1985. The number of habitat sites sampled within each location was six, seven, and two in July and six, five, and two in October for MBS-1, MBS-2, and Puyallup pond–tributary, respectively.

Location and species	Number of fish branded	
	July	October
MBS-1		
Coho salmon	228	307
Dolly Varden	538	132
Steelhead	123	463
MBS-3		
Coho salmon	941	762
Dolly Varden	971	251
Steelhead	413	805
Puyallup pond–tributary		
Coho salmon	644	596
Dolly Varden	414	208
Steelhead	41	198
All (total fish branded)	4,313	3,722

TABLE 4.—Number of freeze-branded salmonids recovered in Maybeso Creek, Prince of Wales Island, Alaska, during October 1985 and March 1986.

Location and species	Branding period		
	Sep 1984	Jul 1985	Oct 1985
Recovered in October 1985			
MBS-1			
Coho salmon	9	53	
Dolly Varden	0	31	
Steelhead	0	2	
MBS-3			
Coho salmon	0	102	
Dolly Varden	0	12	
Steelhead	0	8	
Puyallup pond–tributary			
Coho salmon	9	34	
Dolly Varden	0	6	
Steelhead	0	1	
All (total fish recovered)	18	249	
Recovered in March 1986			
MBS-1			
Coho salmon	0	17	100
Dolly Varden	0	13	34
Steelhead	0	0	20
MBS-3			
Coho salmon	0	19	59
Dolly Varden	0	8	25
Steelhead	0	2	30
Puyallup pond–tributary			
Coho salmon	0	4	77
Dolly Varden	1	0	72
Steelhead	0	0	17
All (total fish recovered)	1	63	434

riod. The sampling sequence in the two Maybeso Creek study sections was accomplished in 8 d during each sampling period.

Brands applied during July 1985 were easily identified during the October 1985 sampling period, when 249 July-branded fish (5.8%) were recovered from all locations; 18 coho salmon that had been marked in September 1984 were recovered as well (Table 4). As expected, more of the branded fish recovered in March 1986 had been branded in October 1985 (434; 11.7%) than in July 1985 (63; 1.5%) (Table 4).

Because separate brands had been applied to fish captured from each habitat type within each study section, we were able to detect movement among habitat types. Of the 53 coho salmon recaptured in October 1985 in section MBS-1, 40 (75%) were recaptured in the same site where they were branded (Table 5). Similar results were found in the MBS-3 section, where 72% of the marked coho salmon were recovered in their original habitat, and most were found in specific habitat types, especially the backwater areas. The recovery rate of coho salmon branded in October 1985 was high in the off-channel habitat during the March 1986 sampling period.

We were able to use different brand codes for the six habitat types in the MBS-1 study section and for two sections in a nearby tributary. Different brands were used in all but two habitat types in the MBS-3 section. Duplicate brands were used in the two least productive (in terms of number of fish captured) habitat sites. Because sections MBS-1 and MBS-3 were separated by more than 3 km, we did not expect much fish movement between them. In fact, no fish branded in the upstream MBS-3 section were recaptured in the downstream MBS-1 section or vice versa during the course of the study. Less than five fish that were branded in the tributary system have been recaptured in the MBS-1 section during other sampling periods.

The low percentage of total brands recovered was not unexpected because the study was conducted over a relatively long period of time in a stream system in which fish could move into and out of study sections. Movement and natural mortality of both marked and unmarked fish, and recruitment of unmarked fish into the study sections was expected to dilute the number of

TABLE 5.—Number (percent) of freeze-branded coho salmon recovered by habitat type in section MBS-1 of Maybeso Creek, October 1985.

Original branding habitat	Habitat where fish were recovered			
	Backwater	Backwater-pool	Left channel debris	All
Backwater	29 (54.7)[a]	0 (0)	2 (3.8)	31 (58.5)
Backwater-pool	4 (7.5)	4 (7.5)[a]	1 (1.9)	9 (16.9)
Main channel	0 (0)	0 (0)	0 (0)	0 (0)
Midchannel debris	2 (3.8)	0 (0)	0 (0)	2 (3.8)
Right channel	0 (0)	1 (1.9)	1 (1.9)	2 (3.8)
Left channel debris	3 (5.7)	0 (0)	6 (11.3)[a]	9 (17.0)
All	38 (71.7)	5 (9.4)	10 (18.9)	53 (100.0)

[a]Recovered in same habitat type in which branded.

marked fish. In the more stable and isolated habitat types, such as backwaters and backwater pools, recovery rates were high (Table 5). Over 50% of the coho salmon parr branded in the MBS-1 backwater in 1985 were recaptured in the same habitat in March 1986. In the previous year, there was nearly a 90% recapture rate in the same habitat (Bryant, unpublished data). These results indicated that branding did not adversely affect survival of the parr.

The good brand visibility and the number of different brand codes made it possible to detect movement among several different habitat types over several seasons. In addition, because we were able to differentially mark fish within a study site during discrete sampling periods, we will be able to use a multiple mark–recapture technique in future studies to evaluate fish survival and migration. With our equipment, multiple marks that consist of a combination of numerals and letters could be used to identify small numbers of fish individually.

References

Brock, J. A., and R. K. Farrell. 1977. Freeze and laser marking of channel catfish. Progressive Fish-Culturist 39:138.

Bryant, M. D., and W. J. Walkotten. 1980. Carbon dioxide freeze branding device for use on juvenile salmonids. Progressive Fish-Culturist 42:55–56.

Busack, C. 1985. A simplified cold-branding apparatus. Progressive Fish-Culturist 47:127–128.

Everest, F. H., and E. H. Edmundson. 1967. Cold branding for field use in marking juvenile salmonids. Progressive Fish-Culturist 29:175–176.

Fujihara, M. P., and R. E. Nakatani. 1967. Cold and mild heat marking of fish. Progressive Fish-Culturist 29:172–174.

Joyce, J. A., and H. M. El Ibiary. 1977. Persistency of hot brands and their effect on growth and survival of fingerling channel catfish. Progressive Fish-Culturist 39:112–114.

Laird, L. M., R. J. Roberts, W. M. Shearer, and J. F. McArdle. 1975. Freeze branding of juvenile salmon. Journal of Fish Biology 7:167–171.

Raleigh, R. F., J. B. McLaren, and D. G. Graff. 1973. Effects of topical location, branding techniques, and changes in hue on recognition of cold brands in centrarchid and salmonid fish. Transactions of the American Fisheries Society 102:637–641.

Smith, J. R. 1973. Branding chinook, coho, and sockeye salmon fry with hot and cold metal tools. Progressive Fish-Culturist 35:94–96.

Sorensen, P. W., M. Bianchini, and H. E. Winn. 1983. Individually marking American eels by freeze branding. Progressive Fish-Culturist 45:62–63.

Turner, S. E., G. W. Proctor, and R. L. Parker. 1974. Rapid marking of rainbow trout. Progressive Fish-Culturist 36:172–174.

American Fisheries Society Symposium 7:36–37, 1990

Cold-Branding Techniques for Estimating Atlantic Salmon Parr Densities

Alexis E. Knight

U.S. Fish and Wildlife Service, Federal Building, Room 124
Laconia, New Hampshire 03246, USA

Abstract.—Cold-brands, produced with liquid nitrogen as the coolant, have been used for identifying Atlantic salmon parr *Salmo salar* during electrofishing mark–recapture surveys. Reliable, discrete letter or numeral brands applied to fish greater than 7 cm total length are distinguishable within 2 to 14 d. A small percentage of the brands are retained until the following year. Development time for a recognizable brand appears to be temperature dependent. Because the brands are dark colored, they should be applied to lightly pigmented areas on the fish.

State and federal natural resource agencies are engaged in a program to restore depleted or extirpated anadromous Atlantic salmon *Salmo salar* populations to the rivers of New England. Studies of the instream performance of young Atlantic salmon stocked in vacant habitats within the Merrimack River basin require intensive monitoring of growth and survival of these fish. Estimates of population densities and survival through marking and subsequent recapture by electrofishing required a relatively efficient short-term marking method. Cold-branding, with liquid nitrogen as a coolant, was chosen as a rapid marking method that has minimal impact on juvenile Atlantic salmon.

Methods

A 5-L commercial Dewar flask was fitted with a 10-mm-diameter copper rod inserted to within 2 cm of the flask bottom. The protruding 7 cm of the bar was bent at a 90° angle and threaded to accommodate a brass or copper nut that was silver-soldered to the brand (Figure 1). Brass brands, 20 mm square and 12 mm thick, were engraved with block letters 6 mm high. The letters were undercut 3 mm, which reduced the thickness of the brand, other than the letter, to 9 mm. Liquid nitrogen in the flask chilled the brand through the copper rod. Contact time for application of the brand to the middorsal region of the Atlantic salmon parr was 1–2 s (Raleigh et al. 1973); the minimum size of fish that could be branded was about 7 cm.

Seven habitat zones for juvenile Atlantic salmon were measured at normal summer flows within the streams, and the upstream and downstream ends of these zones were marked permanently to provide a constant area for obtaining estimates of population density. The actual areas of these sites varied from 14.2 m^2 to 43.8 m^2, depending on stream configuration. Fish were collected by electrofishing with pulsed DC at 300–500 V and 60 Hz. Fish were placed in a black container, anesthetized with tricaine at a concentration of 50–60 mg/L, and then gently pressed against the brand for 1–2 s. Population estimates were derived from the modified Chapman formula, and 95% confidence limits were estimated from the variance (Ricker 1975).

Results and Discussion

Valid population estimates derived from mark–recapture operations require that sufficient numbers of marked individuals be recaptured to provide reasonable results. Cold-brands have been used to obtain estimates of juvenile Atlantic salmon densities in streams since 1976. In 1986 and 1987, ventral fin clips were used in conjunction with cold-brands to assess the effectiveness of cold-brands applied to 2,781 Atlantic salmon parr. Brands applied in September 1986 were distinct within 2–3 d following application at a rather stable stream temperature of 8°C. In the fall of 1987, 1,796 Atlantic salmon parr were processed in the seven study sites, and all brands were recognizable within 3–14 d after application. Development time for a good mark was extended when the stream temperature was above 10°C and underwent diurnal fluctuations. Repetitive sampling for estimates of population density usually was conducted over a short period of time, and a second or third sample of fish was taken within 2–3 weeks. Year-to-year mark retention, although not an objective, resulted in recognition of 26 to 244 brands in 1985 and only 2 of 243 brands in 1986. In addition, larger Atlantic salmon parr

FIGURE 1.—Atlantic salmon parr marked with a stationary brand affixed to a Dewer flask containing liquid nitrogen.

migrated to the sea in the spring following brand application.

Liquid nitrogen, at a temperature of −196°C, will produce persistent marks that may be recognizable up to 1 year from the date of application. The procedures outlined for cold-branding by Raleigh et al. (1973) will result in reliable, discrete identification of Atlantic salmon parr (Figure 2). Fish must exceed 7 cm total length to accommodate the 6-mm-high brand. Cold-brands on juvenile Atlantic salmon develop as dark marks and therefore should be applied to lightly pigmented areas. Fish recaptured for population estimates

FIGURE 2.—Atlantic salmon parr with a fully developed cold-brand.

within the stream study sites were anesthetized in dark-colored containers and then examined in light-colored containers where brands could be seen more readily.

Liquid nitrogen has produced satisfactory marks on other fish species as well (Fay and Pardue 1985; Myers and Iwamoto 1986; Emery and Wydoski 1987). The development of a satisfactory cold- or freeze-brand most likely is related to the heat-sinking capability of the brand as well as to its mass. Dermal tissue of the fish must be affected in the short 1–2 s duration of application.

The hazards of working with liquid nitrogen require strict safety precautions. Liquid nitrogen must be transported outside of the passenger-carrying area of a vehicle, and it requires the use of standard containers and transfer devices. Bryant and Walkotten (1979) reported the success of freeze-branding juvenile salmonids with carbon dioxide contained in a fire extinguisher, which I believe is a less hazardous method. These authors reported that the release of pressurized carbon dioxide at approximately −105°C provided adequate cooling for the brand.

Acknowledgments

I thank all those persons who assisted in the preparation of this manuscript, especially the field biologists of the New Hampshire Fish and Game Department.

References

Bryant, M. D., and W. J. Walkotten. 1979. Carbon dioxide freeze-branding device for use on juvenile salmonids. Progressive Fish-Culturist 42:55–56.

Emery, L., and R. Wydoski. 1987. Marking and tagging of aquatic animals: an indexed bibliography. U.S. Fish and Wildlife Service Resource Publication 165.

Fay, C. W., and G. B. Pardue. 1985. Freeze brands and submandibular latex injections as identifying marks on rainbow trout. North American Journal of Fisheries Management 5:248–251.

Myers, J. M., and R. N. Iwamoto. 1986. Evaluation of thermal and chemical marking for tilapia. Progressive Fish-Culturist 48:288–289.

Raleigh, R. F., J. B. McLaren, and D. R. Graff. 1973. Effects of topical location, branding techniques and changes in hue on recognition of cold brands in centrarchid and salmonid fish. Transactions of the American Fisheries Society 102:637–641.

Ricker, W. E. 1975. Computation and interpretation of biological statistics of fish populations. Fisheries Research Board of Canada Bulletin 191.

American Fisheries Society Symposium 7:38-41, 1990

Tattoo-Ink Marking Method for Batch-Identification of Fish

Jeffrey C. Laufle

U.S. Army Engineer District, Seattle, Environmental Resources Section
Post Office Box C-3755, Seattle, Washington 98124, USA

Llew Johnson

915 118th Avenue, SE, Suite 200
Bellevue, Washington 98005, USA

Cindy L. Monk

Seattle City Light, Power Supply and Planning Division
1015 3rd Avenue, Seattle, Washington 98104, USA

Abstract.—A batch-marking technique was developed for a study of short-term nearshore movement of anadromous fishes in an arctic environment. The method included the use of tattoo artists' pigments and a jet-spray inoculator to apply colored dots to selected areas on individual fish. This method met project needs for ease of application as well as for durability and variety of possible marks. Marking equipment was easily portable, and no special equipment was necessary for reading the marks. In this paper, we describe the marking method and discuss its utility in comparison to previously described marking methods. We also describe the tests needed to fully prove this method's usefulness.

It is often desirable to mark fish in batches to distinguish different groups for fisheries research and management. Kelly (1967) stated that an ideal marking dye should "be long-lasting, . . . be permanent on every fish, be nontoxic and nonirritating, have no effect on normal growth, require little or no extra formulation, allow rapid marking, be inexpensive, be nonencumbering (not interfere with normal feeding or swimming), be easily visible to an untrained observer, provide a number of different mark combinations, and require a minimum of specialized equipment." Although various methods that use pigments or dyes have been employed, development of this ideal marking method has been an elusive goal.

Fluorescent pigment often is used to mark trout (Salmonidae) before they are released for put-and-take sport fisheries. The pigment (Pierson and Bayne 1983; Strange and Kennedy 1984; Evenson and Ewing 1985) is sprayed onto many fish at a time and is probably the quickest method available for application; however, it requires a source of compressed air. Furthermore, the marks are not visible to the unaided eye, but require an ultraviolet light source, which in turn requires a power source.

Freeze branding and submandibular latex injections were used by Fay and Pardue (1985) to differentiate between strains of rainbow trout *On-corhynchus mykiss* (formerly *Salmo gairdneri*) taken in a sport fishery. Freeze-branding equipment is fairly portable and provides a great variety of possible marks, but it may require some consistency in pressure and length of application to individual fish for legible retention of brands. When applied properly, freeze branding does have proven field utility. Submandibular latex injections are not applied quickly.

Duncan and Donaldson (1968) described an injection method using fluorescent pigments. Marks remained readable longer with ultraviolet light than with daylight. Kelly (1967) also tested a large variety of dyes applied with a syringe. He narrowed these to two that he found satisfactory in terms of durability, nontoxicity, and speed of application. Injection has a disadvantage because it is nonversatile. Volz and Wheeler (1966) used an injection device driven by an electric motor that used a 12-V battery and a set of paint pigments. Although the pigments had an apparently uniform base, the necessity for even this portable power source could be a disadvantage in the field. Hart and Pitcher (1969) used a jet inoculator, but with a nonuniform pigment base. Galloway and Britch (1983) used a jet inoculator and a single pigment, Alcian Blue, to serve as a recognition marker for freeze-branded fish. They achieved speedy application, and they were not hindered

FIGURE 1.—Dental inoculator and empty glass cartridge used in fish marking experiment. The inoculator has a cartridge containing ink fitted near the left end. The jet-spray nozzle is located at the upper left end of the inoculator. The trigger is halfway down the handle, and the spray volume adjustment is at the right end.

by having only one color because the freeze-branding provided a good variety of marks.

In this paper, we describe a method that meets many of Kelly's (1967) criteria and point out those areas necessary to test further to fully prove the method. Our method has promise for short-term and possibly long-term batch-marking of fish. It was tested and used in a field study of short-term nearshore movement of anadromous fish at Prudhoe Bay, Alaska, in the summer of 1983. The primary species studied and marked were Arctic char *Salvelinus alpinus*, least cisco *Coregonus sardinella*, and Arctic cisco *C. autumnalis*.

Methods

Marking procedure.—Fish were marked with tattoo artist's ink[1] intracutaneously injected near the insertions of paired and anal fins and on the side of the caudal and peduncle. The marking instrument was a Syrijet Mark II dental inoculator[2,3] (Figure 1). This is a hand-held instrument that employs a hand-pumped jet spray with

[1]Spaulding and Rogers Manufacturing, Route 85, New Scotland Road, Voorheesville, New York 12186, USA.

[2]Made by Mizzy, Inc., and available from National Keystone, 616 Hollywood Avenue, Cherry Hill, New Jersey 08002, USA.

[3]Use of trade names does not imply endorsement by the U.S. government.

a gradually variable volume (about 0.04–0.20 mL). The ink, supplied in powdered form, was mixed with water; doing so required a trace of 70% isopropyl alcohol as a wetting agent. The ink was mixed at a concentration to yield an opaque suspension that was not too thick to flow through the spray unit. The mixture was placed in glass cartridges that were furnished by the maker of the inoculator. These cartridges then were fitted into the inoculator. Ink colors used were red, blue, green, and yellow.

For marking, the fish were held with the ventral side up and injected with ink at the selected body site (Figure 2). Distance of the inoculator from the

FIGURE 2.—A cisco after dye marking. The mark appears as a dark dot near the pelvic fin insertion.

fish varied from directly touching to 2–3 cm away, depending on the size of the fish. Larger fish were given slightly larger marks by holding the inoculator off the body of the fish and by slightly increasing the volume of the injection. Optimum volume and distance were determined by practice. Marks were generally about 3–5 mm in diameter, though sometimes the ink spread subcutaneously to two or three times that diameter.

Routine cleaning of the inoculator was accomplished by repeated discharge of a mixture of water and a trace of alcohol aimed at a blank sheet of paper until no further color was observed. This was necessary any time a switch of colors was made. The instrument was also immersible in fresh water, a maintenance requirement in a marine environment.

Laboratory tests.—Before the field study, an initial test of the method for mark durability and its effect on fish mortality was carried out at the University of Washington's fish hatchery in Seattle. Mark retention for 1 week was deemed adequate for study purposes. Ten hatchery-reared rainbow trout, about 30 cm in length, were allowed to acclimate in a tank for 48 h at ambient Lake Washington surface temperatures (about 15–18°C). Each fish was anesthetized with MS-222 (tricaine) then marked with a dot of red tattoo ink adjacent to a pectoral or ventral fin or the anal fin. The fish were held for a week after marking and fed maintenance rations. Fish were observed for mortality or outward signs of stress or other abnormalities; however, no histological or pathological examination was made.

Field study.—The intent of the field study was to determine short-term (same day) nearshore movement of Arctic char, Arctic cisco, and least cisco at Prudhoe Bay, Alaska, on the Beaufort Sea. Fish larger than about 100 mm fork length (FL) were marked either at the capture site, where fyke nets were used, or at a work shed. Fish smaller than 100 mm FL exhibited a high injury rate and high mortality from the marking process and therefore were not used. At the fyke nets, fish were transferred to floating net-pens, attached to the side of the work boat, by raising the trap and opening it into the pen. This procedure required taking the fish out of the water briefly. If fish were to be transferred to the work shed, they were moved in 114-L plastic barrels full of seawater. In the shed, fish were held in the barrels, which were aerated with one or more standard aquarium pumps. For measuring and marking, fish were anesthetized with MS-222 in

dishpans, 19-L buckets, or the large barrels, depending on the size of the fish. Anesthetizing containers were aerated with aquarium pumps or, if marking was done away from electrical power, water was exchanged regularly. After the fish were measured, they were marked with a colored dot and then were placed into another aerated barrel of seawater to recover from anesthesia. Immediately following recovery, the fish were moved to floating net-pens to await release, which occurred when sufficient numbers (100 or more) were available for the experiment or within 3 d, whichever came first. Several pens about 1.7 m³ were used to hold an average of about 25 fish each, depending on the size of the fish (average length, about 30 cm). The fish were not fed. An attempt was made to recapture these fish with lampara seines on the same day the fish were released to determine movement patterns.

Results and Discussion

Laboratory fish suffered no mortality and had no outward evidence of ill effects from marking. The tattoo ink remained bright until the fish were released at the end of 1 week. In the field, the different colors showed up well, except for yellow, which was not distinct enough against the white skin of the anadromous fish and was discontinued.

All field-released fish exhibited good marks after the holding period of up to 3 d. The release experiments were disappointing in that very few fish were recaptured, even on the same day that they were released; however, the marks on recaptured individuals were readily identifiable. Individuals were recaptured by other researchers as far away as 10 km (Arctic char and Arctic cisco) and as long after release as 30 d (Arctic char), and again the marks were quite recognizable (K. Critchlow, Woodward-Clyde Consultants, personal communication). The same was true for two recaptures, a broad whitefish *Coregonus nasus* and an Arctic cisco, made in July 1984, about 1 year after the fish were released.

Our study was limited by the available time, funding, and number of test subjects, and our initial testing was aimed at determining that the technique was adequate for our purposes. Our intent here was simply to show that the tattoo-ink marking method fulfilled several of the criteria listed by Kelly (1967). We stress that additional tests should be conducted to establish whether the technique meets other criteria, so it is proven more fully.

The ink requires little mixing because the water–powder–alcohol mixture is formulated very easily. The marking technique is rapid; however, our field study involved measuring fish as well as marking them, so we did not mark fish as quickly as possible. We conservatively estimate that one person with little experience could mark 200 or more anesthetized fish per hour. If a second person rapidly provided fish and full ink cartridges to the marker, the rate could be considerably higher.

The technique does not encumber the fish as a tag might. However, tests should be performed to determine if irritation occurs from the ink that might inhibit movement or diminish the general vigor of the fish. In fact, there is a definite need to perform detailed toxicologic tests to determine whether the inks cause mortality, lesions, increased disease potential, stress, or inhibition of feeding, growth, or reproduction.

The inoculator costs several hundred dollars and therefore should be affordable for most studies. The tattoo ink is inexpensive. Both items are easily obtainable by mail order.

The equipment is very portable. The actual marking equipment and supplies, including anesthetic, will fit in a hand-carried hinged box with trays, such as a fishing tackle box or artist's carrying case. The limiting factor will be the size of containers needed by the researcher for anesthetizing and holding fish.

Mark retention was judged adequate for the needs of this study (i.e., for a few days); however, because fish were not otherwise marked, we cannot rule out the possibility that some fish were recaptured without being recognized. Because recognizable recaptures were made as long as a year later, this seemed unlikely, but long-term testing to determine the true longevity of the marks should be done under controlled conditions with fin clips or some other means of duplicate marking. Marks were easily recognizable by all observers on those fish that had them.

Tattooing fish with more than one color or in more than one location, or both, would provide a large variety of marks to distinguish batches of fish, or, if numbers were small enough, individuals. More colors could be explored for distinguishability, including light colors for testing against dark backgrounds.

We believe that tattoo-ink marking has good potential for field use if evaluation of untested criteria shows positive results.

References

Duncan, R. N., and I. J. Donaldson. 1968. Tattoo-marking of fingerling salmonids with fluorescent pigments. Journal of the Fisheries Research Board of Canada 25:2233–2236.

Evenson, M. D., and R. D. Ewing. 1985. Long-term retention of fluorescent pigments marks by spring chinook salmon and summer steelhead. North American Journal of Fisheries Management 5:26–32.

Fay, C. W., and G. B. Pardue. 1985. Freeze brands and submandibular latex injections as identifying marks on rainbow trout. North American Journal of Fisheries Management 5:248–251.

Galloway, B. J., and R. P. Britch, editors. 1983. Environmental studies (1982) for the Endicott development, volume 3. Report to Sohio Alaska Petroleum Company, Anchorage, Alaska.

Hart, P. J. B., and T. J. Pitcher. 1969. Field trials of fish marking using a jet inoculator. Journal of Fish Biology 1:383–385.

Kelly, W. H. 1967. Marking freshwater and a marine fish by injected dyes. Transactions of the American Fisheries Society 96:163–175.

Pierson, J. M., and D. R. Bayne. 1983. Long-term retention of fluorescent pigment by four fishes used in warmwater culture. Progressive Fish-Culturist 45:186–188.

Strange, C. D., and G. J. A. Kennedy. 1984. Fluorescent pigment marking of cyprinids. Fisheries Management 15:67–69.

Volz, C. D., and C. O. Wheeler. 1966. A portable fish-tattooing device. Progressive Fish-Culturist 28:54–56.

American Fisheries Society Symposium 7:42–46, 1990

Twelve-Year Overview of Fluorescent Grit Marking of Cutthroat Trout in Bear Lake, Utah–Idaho

BRYCE R. NIELSON

Utah Division of Wildlife Resources
Box 231, Garden City, Utah 84028, USA

Abstract.—Bear Lake cutthroat trout *Oncorhynchus clarki utah* (formerly *Salmo clarki utah*) used for stocking were marked annually with fluorescent polystyrene grit from 1975 to 1987 at Bear Lake, Utah–Idaho. Fish were 125, 175, and 225 mm total length (TL) when marked. Pigment was applied with a portable, gasoline-powered air compressor and sandblaster after the fish were placed in a screened marking box. The quantity of pigment used varied inversely with the size of fish marked. Red, yellow, and light green grits were evaluated. In subsequent years, marked fish accounted for 90% of the total gill-net catch. Detection rates of red and yellow grits were similar (91%), and that of light green grit was slightly less (86%). An average of 96% of fish marked at 225 mm TL that were recaptured in gill nets had retained pigment marks after 6 years; 86% of fish marked at 125 mm and 175 mm TL that were recaptured in gill nets had retained their pigment marks after a maximum of 12 years. Among fish double-marked with grit and fin clips, fin-clip loss (5%) was less than pigment loss (10%). An average of 81% of spawning fish with fin clips had grit marks. Maximum mark-retention time was 12 years, the length of the study period. Fluorescent pigment was an acceptable, permanent, long-term mark that was cost-effective with no apparent negative effects on the fish.

The Bear Lake Cutthroat Trout Enhancement Project, initiated in 1973 by the Utah Division of Wildlife Resources and funded by Federal Aid in Sportfish Restoration, required that millions of young cutthroat trout *Oncorhynchus clarki utah* (formerly *Salmo clarki utah*) be marked with an external mark that would be recognizable over a long period. These cutthroat trout are long-lived (>10 years) and inhabit the pelagic zone of a 28,200-hectare lake. Marking of fish was necessary to assess survival of annually stocked cohorts and to provide known-age information for age and growth studies. The marking technique had to be inexpensive, result in readily detectable marks, and not influence growth or survival. The marking procedures considered included the use of fin clips, tags, and fluorescent grit.

Fin removal (fin clipping) is a common fish-marking technique. Marks are readily detected in the field without specialized equipment and typically are permanent. However, fin clipping is labor-intensive, not useful for smaller salmonids, and requires anesthesia. In addition, increased mortality among fin-clipped fish has been reported (Nicola and Cordone 1973; Mears and Hatch 1976). Fin regeneration by salmonids also has been reported (Mears 1976) and may introduce bias in long-term studies. Few combinations of fins can be clipped without severely affecting the mobility of the fish, so the number of groups of

fish within a population that can be marked is limited.

Tagging involves a variety of marker types, such as dart, spaghetti, cinch, disk, jaw, and coded wire tags. The array of colors and numbers that can be used with tags provides definitive separation among year-classes or individual fish on a long-term basis. Tag application and detection is time-consuming, however and requires skilled personnel and specialized equipment. Tag loss may be substantial and fish health may be impaired (Rawstron 1973). Tagging also becomes more difficult as fish size decreases.

Mass-marking of fish with fluorescent grit was described initially by Jackson (1959) and refined further by Phinney et al. (1967). This technique involves embedding fluorescent pigment granules into the epidermal layer of fish with a sandblaster and compressed air. Although color combinations are limited, the technique apparently has a minor effect on the fish and can be used for inexpensive marking of many fish. Early studies with fluorescent pigment typically were of limited duration, and retention time for these marks was unknown when the Bear Lake project began. Evenson and Ewing (1985) have since reported pigment retention up to 54 months for summer steelhead *Oncorhynchus mykiss* (formerly *Salmo gairdneri*).

In this paper, I report on the longevity of fluorescent grit marks on cutthroat trout. Percentages of marked fish captured in gill nets and spawning traps were evaluated, and comparisons of pigment (grit marks) and fin-clip identification and retention were made.

Methods

Between 1975 and 1987, approximately 7,750,000 Bear Lake cutthroat trout were marked with fluorescent grit pigment and stocked in Bear Lake. Marking methods were similar for all lots of fish. Four single colors and one color combination were used: blue, red, yellow, light green, and red–blue. All batches of grit were supplied by Scientific Marking Materials of Seattle, Washington. The technique employed was a modification of that described by Phinney et al. (1967). Marking equipment consisted of a Speedaire gasoline-powered compressor with a 4.5-kW engine fitted to a 114-L tank that generated air flows of 0.65 m^3/min at a pressure of 12.3 kg/cm^2. A Port-O-Blast sandblaster was connected to the compressor with 15.2 m of low-pressure hose. The jet orifice of the sandblaster was enlarged to 2.8 mm. Pressure during spraying ranged from 8.8 to 10.5 kg/cm^2 and was gauged at the tank. McAfee (1982) indicated that gauge pressures do not represent gun pressure; therefore, delivery force of air from the spray nozzle also was measured on occasion. These pressures ranged between 300 and 400 g/cm^2 at a distance of 40 cm. Grit was sifted with a standard flour sifter and funneled into 0.95-L canisters to break up compacted grit and prevent clogging. Canisters were filled to 75% of capacity, attached to the gun, and replaced when 10% of the grit remained.

Groups of fish marked during the study averaged between 8 and 55 fish/kg. Fish were crowded upstream from a screen placed in the middle of the raceway. A marking box, 61 cm × 91 cm × 20 cm, with a slotted aluminum bottom was placed on boards across the raceway at that point. The fish then were crowded downstream toward this screen with a second screen. When enough fish were crowded together, they were netted directly into the marking box, sprayed with pigment, and tipped over the screen into the downstream section of the raceway. Fish were sprayed at a distance of 40 cm with one circular pass and one side-to-side pass before being returned to the water. During spraying, the sandblaster was shaken vigorously to facilitate pigment flow. If clogging occurred, fish were remarked. Grit obstructions usually were cleared with back-pressure. The marking operation was most efficient

with a four-person crew: one sprayer, one netter, one crowder, and one person who emptied the marking box after spraying.

To determine the effectiveness of grit-marking, pelvic fin clips were combined with grit marks on 1,379,914 fish from 1981 to 1983. Fish were grit-marked initially, held for at least 1 week, anesthetized, and then fin-clipped. The fish subsequently were treated with a 1-hr drip of 50% Hyamine 3500 (quaternary ammonia) at 2.0 mg/L for 3 d in the raceway to prevent infection. Fish usually were stocked within 10 d of marking.

Marks were detected by two methods. Gill-netted fish were returned to the lab and inspected in a darkened room under ultraviolet light with a UVL-56, 365-nm-wavelength, hand-held lamp (UVP, Inc., San Gabriel, California). Spawners captured in the spawning traps were examined in a suspended tent inside the trap building that had a model XX-15L, 365-nm, display lamp located inside. Fish were considered marked if at least one granule was readily recognizable and unmovable when touched. Fish were examined for marks by fisheries biologists experienced in grit-mark detection.

Gill-netting occurred annually at various locations throughout the lake during the study period. All cutthroat trout collected in gill nets from 1976 to 1987 were examined for marks. Data were collected on pigment color and associated fin clips, if appropriate. Fish exhibiting no grit marks or fin clips were aged immediately by examination of their otoliths, which were extracted and observed three-dimensionally under a binocular dissection scope without sectioning or staining. Data were summarized separately for marked and unmarked fish, and data for marked fish were separated further according to pigment color and presence or absence of fin clips.

Determinations of mark retention in the spawning population began in 1981 when the first stocked cohorts matured. Observations were made on live, anesthetized fish before they spawned. Grit colors and fin clips were noted; however, the percentage of the population derived from stocked cohorts was not determined due to the inability to age live, unmarked fish before they were released from the traps.

Results and Discussion

Grit-marking of cutthroat trout in Bear Lake began in 1975 with the use of red pigment on 125-mm fish (all lengths given in this paper are total lengths). Concurrent trials with blue grit and

TABLE 1.—Color scheme for marking different cohorts of cutthroat trout in Bear Lake, 1975–1987. From 1981 to 1983, fish were double-marked by clipping either the right or left pelvic fin in addition to grit-marking.

Years (cohorts)	Grit color		
	Red	Light green	Yellow
1975, 1978, 1981, 1984, 1987	X		
1976, 1979, 1982, 1985		X	
1977, 1980, 1983, 1986			X

a red–blue grit combination also were initiated but were abandoned in 1977 due to difficulty in detecting blue grit. Dark green pigment was not available at the beginning of the project and could not be tested. Grit quality varied throughout the study period: hues of the same color varied between years, and although no actual measurements were taken, grit coarseness also appeared to differ among lots of material.

All Bear Lake cutthroat trout stocked between 1975 and 1987 were marked with fluorescent grit. Smaller fish required more grit; 1 kg of grit marked 545 kg, 847 kg, and 750 kg of 125-mm, 175-mm and 225-mm cutthroat trout, respectively. The same color was used to mark similar stocks each year; a red–light green–yellow sequence was followed throughout the project period. Fin clips were combined with grit marks on larger fish (175 and 225 mm) from 1981 to 1983 (Table 1).

There is natural recruitment into the population of cutthroat trout in Bear Lake; however, the magnitude is unknown and believed to be low. For analysis, all cutthroat trout sampled were assumed to have been stocked and marked prior to introduction. The percentage of fish marked for stocking usually averaged 95%.

Altogether, 1,668 cutthroat trout were captured in gill nets during the study period. Grit marks were retained in an average of 90% of these fish (range, 22% in 1976 to 94% in 1984; Figure 1). For all sizes of fish, performance of the three colors was similar: red and yellow pigments were detected in 91% of fish marked with those colors; light green was retained in 86%. Mark retention was related to size of fish at the time of marking (Figure 2). Age-2 (225-mm) cutthroat trout marked with grit and fin clips and stocked from 1981 to 1983 exhibited a 96% mark retention after a maximum of 6 years. Age-1 cohorts that averaged 125 mm and 175 mm at the time of marking had lower retention rates of 87% and 86%, respectively, after a maximum of 12 years.

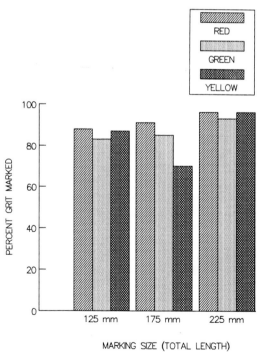

FIGURE 2.—Percentage of cutthroat trout gill-netted in Bear Lake during 1976–1987 that had retained grit marks applied to the fish before stocking from 1975 to 1987. Marking size is the size of the fish at the time of marking.

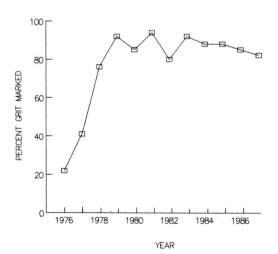

FIGURE 1.—Percentage of grit-marked cutthroat trout in gill-net catches from Bear Lake during 1976–1987. All cutthroat trout stocked in Bear Lake from 1975 to 1987 were marked with fluorescent grit.

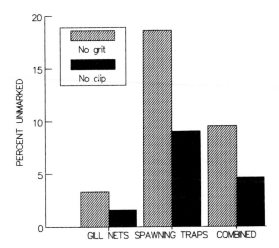

FIGURE 3.—Incidence of mark loss in Bear Lake cutthroat trout marked with both fin clips and fluorescent grit. Fish were sampled with gill nets and spawning traps to determine mark retention for up to 6 years after marking in 1981–1983.

Fish double-marked with a fin clip and grit were evaluated to assess long-term retention of both types of marks. Fish were caught in gill nets and spawning traps and separated into three groups: fish with grit marks only, fish with fin clips only, and fish with both types of marks. Gill-netted fish within these categories were aged by otoliths to identify the cohort and estimate retention time. Figure 3 illustrates the percentages of fish that lost either or both marks. Overall, fin clips were lost from a lower percentage of fish (5%) than grit (10%). Retention of both marks on fish caught in gill nets was excellent, whereas grit marks on sexually mature fish caught in spawning traps were not retained as well as fin clips. Evenson and Ewing (1985) made similar observations on other salmonids *Oncorhynchus* spp. They reported decreased grit detention on spawning fish as a result of skin changes and a darker body coloration in mature fish. Males typically did not retain marks as well as females.

Maximum retention time of grit has not been fully assessed. For spawning Bear Lake cutthroat trout, distinguishable marks have been observed on age-12 fish marked with red or light green pigment, and on age-11 fish marked with yellow pigment. Additional data will be collected on long-lived adults until the cohort leaves the population.

Conclusion

Retention of red and yellow grits was similar and slightly better than light green; retention of all three colors was good. Pigment retention improved with increasing size of fish marked. Salmonids as small as 25 mm have been experimentally grit-marked (Bandow 1987), but an increase in surface area with increasing fish size obviously results in more fluorescent granules being embedded in the epidermal layer. In addition, stress to the fish during marking due to pigment abrasion and epidermal drying was reduced with larger fish. Pigment was most often retained on the back anterior to the dorsal fin, on the caudal peduncle, and near the eyes. Marks with less than five pigment granules rarely were detected. Imbedded pigment was immovable to the touch. Care should be exercised to assure that pigment granules that might fall from equipment or clothing are not interpreted as marks if observed on fish.

Grit marks compared favorably with fin clips as a long-term marking method. Fin clips were easier to observe in the field and may have caused biologists to spend more time searching for a pigment mark on double-marked fish. Comparisons of growth and survival of fish marked with grit or fin clips were not within the scope of this study.

Minor variations exist in application techniques among workers, and improvements are being made to reduce this variation so that marks are more consistent. Although the effects of grit size were not addressed in this paper, a report by Bandow (1987) indicated that minimizing variation in grit size resulted in more effective marks.

The limited number of colors available for grit-marking can result in confusion when grit is used to identify cohorts of such long-lived species as Bear Lake cutthroat trout or lake trout *Salvelinus namaycush*. Such species may have 3–4 cohorts with the same color mark in the population simultaneously. Reliable aging techniques must be available to distinguish cohorts marked with the same color if grit marking is to be used with confidence.

Color quality control should be stressed. Yellow grit fluoresces orange when it is embedded within the epidermal layer and can be confused with red by inexperienced observers if the original pigment lots are not true red and yellow.

This study indicated that fluorescent grit is an acceptable permanent marking technique for long-term studies with salmonids. Mark retention has been demonstrated to age 12 for cutthroat trout in Bear Lake, and longer retention times may be possible. There was no indication that grit pigments fade or that their fluorescent qualities diminish with time. Long-term marks can be obtained by proper application before stocking, but detection of marks on older fish requires experienced biologists working under optimum conditions with appropriate equipment.

References

Bandow, F. 1987. Fluorescent pigment marking of seven Minnesota fish species. Minnesota Department of Natural Resources, Federal Aid in Fish Restoration, F-26-R, Study 306, Completion Report, St. Paul.

Evenson, M. D., and R. D. Ewing. 1985. Long-term retention of fluorescent pigment marks by spring chinook salmon and summer steelhead. North American Journal of Fisheries Management 5:26–32.

Jackson, C. F. 1959. A technique of mass-marking fish by means of compressed air. New Hampshire Fish and Game Department, Technical Circular 17, Concord.

McAfee, M. E. 1982. Fluorescent pigment spray marking: do we know where we stand? Proceedings of the Annual Meeting Colorado–Wyoming Chapter, American Fisheries Society 17:79–85. (Colorado State University, Fort Collins.)

Mears, H. C. 1976. Overwinter regeneration of clipped fins in fingerling brook trout. Progressive Fish-Culturist 38:73.

Mears, H. C., and R. W. Hatch. 1976. Overwinter survival of fingerling brook trout with single and multiple fin clips. Transactions of the American Fisheries Society 105:669–674.

Nicola, S. J., and A. J. Cordone. 1973. Effects of fin removal on survival and growth of rainbow trout in a natural environment. Transactions of the American Fisheries Society 102:753–758.

Phinney, D. E., D. M. Miller, and M. L. Dahlberg. 1967. Mass-marking young salmonids with fluorescent pigment. Transactions of the American Fisheries Society 96:157–162.

Rawstron, R. R. 1973. Comparisons of disk dangler, trailer, and internal anchor tags on three species of salmonids. California Fish and Game 59:266–280.

American Fisheries Society Symposium 7:7:47–62, 1990

Comparison of Morphometric and Meristic Data against Truss Networks for Describing Pacific Herring Stocks

J. F. Schweigert

Department of Fisheries and Oceans, Biological Sciences Branch
Pacific Biological Station, Nanaimo, British Columbia, V9R 5K6, Canada

Abstract.—Truss measurements and traditional morphometric and meristic data were collected on Pacific herring *Clupea harengus pallasi* from four geographically separate areas along the British Columbia coast. Univariate and multivariate statistical analyses were used to compare these methods for routine stock identification. Both methods produced similar results based on principal components analysis, discriminant function analysis, and canonical variates analysis. Small sample sizes prevented the detection of significant differences among fish except for those from the west coast of Vancouver Island. Stepwise discriminant function analysis yielded a reduced variable set that identified significant differences among all four areas with the two data types. However, no meaningful interpretation of the variables responsible for separation of the groups was possible. Truss data are collected easily and rapidly, and provide a more comprehensive description of fish shape than traditional measurements, except for body width. Results from truss analysis also are interpreted more readily in ecological terms, which eventually may provide a functional stock definition leading to testable hypotheses of underlying genetic differentiation. Truss analysis is recommended strongly for future stock identification studies, provided adequate sample sizes can be obtained for statistical validity.

The development of a stock definition scheme that facilitates the logistical management of a fishery and the maintenance of biological and genetic integrity of its population units is crucial to the long-term survival of the fish stocks and to the fishery itself. Ricker (1973) was the first to hypothesize that a fishery operating on a mixture of stocks of unequal productivity would lead to the eventual demise of the least productive subunits. Sinclair et al. (1985) subsequently demonstrated the occurrence of this phenomenon in some of the Atlantic herring stocks *Clupea harengus harengus* in the Gulf of Maine. Various techniques have been developed to describe functionally, if not genetically, discrete stocks of fish (Ihssen et al. 1981a). Traditional univariate and multivariate analyses of morphometric and meristic data have been the favored approaches for describing the structure of fish stocks for many species (Sharp et al. 1978; McGlade and MacCrimmon 1979; Casselman et al. 1981; Ihssen et al. 1981b; Meng and Stocker 1984; King 1985; MacCrimmon and Claytor 1985). Although the results of some of these studies have been useful for stock management, there has been little guidance for determining which characters are useful stock discriminators and should be used routinely in new studies. Instead, the approach generally has been to measure as many characters as time and energy permit and hope that there is enough information to describe statistically and, presumably, genetically distinct stocks. The result has been a suboptimal selection of variables that often are aligned with the longitudinal axis of the body and so do not adequately measure body depth or characters that frequently are localized in one part of the fish, such as the head (Strauss and Bookstein 1982). Additionally, many characters extend over considerable portions of the body, so traditional measurements neglect local variation in shape. Meristic characters are environmentally labile and are affected during development by factors such as water temperature (Tester 1938).

A new technique, known as truss network analysis, attempts to overcome many of these deficiencies by developing a framework that can be applied routinely to describe morphometric variation of fish of any shape (Strauss and Bookstein 1982; Bookstein et al. 1985). In this paper, I present the results of applying conventional morphometric and meristic analysis and truss network analysis to samples of Pacific herring *Clupea harengus pallasi* from various areas along the British Columbia coast. I also discuss the utility and deficiencies of each method for routine stock identification.

Methods

Samples and morphometric and meristic data collection.—During February and March of 1986,

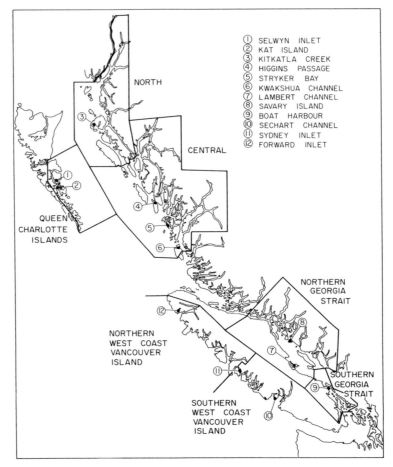

FIGURE 1.—Distribution of sampling locations for collections of Pacific herring for morphometric, meristic, and truss analysis in British Columbia.

12 samples of about 100 fish each were collected from the seven stock groups identified for separate management in the Pacific herring sac-roe fishery in British Columbia (Figure 1; Table 1).Twelve morphometric measurements and four meristic characters identified by Meng and Stocker (1984) as useful for stock identification (Figure 2; Table 2) were obtained from 50 fish in

TABLE 1.—Sources and numbers of Pacific herring collected along the British Columbia coast for morphometric, meristic, and truss analyses.

Sample	Statistical and management area		Locality	N	Date (1986)
1	2E	Queen Charlotte Islands	Selwyn Inlet	10	Mar 29
2	2E	Queen Charlotte Islands	Kat Island	10	Mar 26
3	5	North coast	Kitkatla Creek	10	Mar 18
4	7	Central coast	Higgins Passage	5	Mar 29
5	7	Central coast	Stryker Bay	10	Mar 20
6	8	Central coast	Kwakshua Channel	10	Mar 17
7	14	Georgia Strait	Lambert Channel	10	Mar 15
8	15	Georgia Strait	Savary Island	10	Mar 16
9	17	Georgia Strait	Boat Harbour	10	Mar 15
10	23	West coast Vancouver Island	Sechart Channel	10	Mar 3
11	25	West coast Vancouver Island	Sydney Inlet	10	Feb 28
12	27	West coast Vancouver Island	Forward Inlet	10	Mar 6

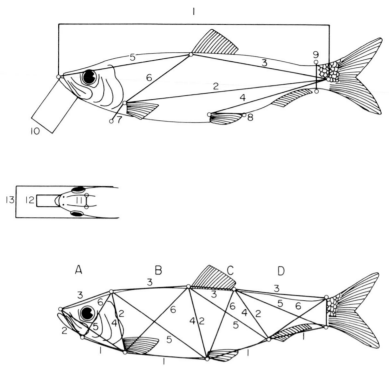

FIGURE 2.—Morphometric characters measured on individual Pacific herring with the measuring points associated with each variable indicated (above; numbers are defined in Table 2), and truss landmarks and associated characters for each of the four cells in a stylized herring (below).

each sample. Truss measurements (Figure 2) also were taken on every fifth fish, and only these specimens were used in subsequent analyses. Traditional morphometric distances were measured with calipers to the nearest 0.1 mm. Counts of fin rays and gill rakers were made with a binocular dissecting microscope, and counts of vertebrae (including the urostyle) were made from X-ray plates (DuPont Cronex Hi-Plus UI film). The fish also were weighed and sexed, and scale samples were taken for age determination.

Truss measurements.—Ten landmark points were selected to define measurements for the truss data (Figure 2). The location of landmarks on herring was difficult because of the scarcity of definitive body segments. The landmarks were, in clockwise order, (1) the snout, (2) the dorsal surface of the head on a line at the posterior edge of the skull connecting the rear margins of the opercular flaps, the (3) the anterior and posterior insertions of the dorsal fin, the (5) dorsal and (6) ventral insertions of the caudal fin, (7) the anterior insertion of the anal fin, the anterior insertions of the (8) pelvic and (9) pectoral fins, and (10) the posterior end of the lower maxilla.

As fish were processed for morphometric and meristic measurements, every fifth individual was placed right side down on a sheet of water-resistant paper, and pins were placed in the paper at each landmark along the periphery of the fish; a block of styrofoam under the paper facilitated the pinning procedure. Fish and sample number were recorded on each sheet as the fish was pinned. The landmark data were transferred to computer from the paper with an X–Y coordinate digitizing pad (GTCO Corporation, Rockville, Maryland); measurements were to the nearest 0.25 mm. The Euclidean distances between landmarks were calculated by computer according to the Pythagorean theorem. The truss network measurements form a series of four cells or contiguous quadrilaterals along the body form (A–D in Figure 2). The cells and truss characters are defined similarly to the scheme of Strauss and Bookstein (1982), whereby the distance between landmarks 1 and 2 is truss character A1 in cell A (Figure 2).

Statistical analysis.—The first step in the analysis of the data was a screening and verification procedure. All statistical analyses were performed with the Statistical Analysis System (SAS

TABLE 2.—Acronyms for morphometric and meristic characters used for analysis of Pacific herring stocks. Variable numbers for morphometric characters are those shown in Figure 2.

Variable number	Acronym	Physical description[a]
	Morphometric characters	
1	Length	Standard length
2	Peccaud	Insertion of pectoral fin to CMP
3	Dorcaud	Insertion of dorsal fin to CMP
4	Pelcaud	Insertion of pelvic fin to CMP
5	Sndors	Tip of snout to insertion of dorsal fin
6	Pectdors	Insertion of pectoral fin to insertion of dorsal fin
7	Pectw	Distance between pectoral fin insertions
8	Pelvhei	Pelvic fin height
9	Peduncle	Peduncle height
10	Snoutmax	Maxillary length
11	Inorbw	Interorbital width
12	Maxw	Maxillary width
13	Headw	Head width
	Meristic characters	
	Pectfin	Pectoral fin rays
	Analfin	Anal fin rays
	Gillow	Lower gill rakers
	Vert	Vertebrae

[a]CMP = caudal measuring point (Hourston and Miller 1980).

1985a, 1985b). The SAS procedure UNIVARI-ATE was used to generate histograms and normal probability plots for all the morphometric, meristic, and truss data. Whenever extreme values appeared in the histogram or in box and whisker plots, the values were checked for typographical accuracy and for consistency.

Next, the morphometric and truss measurements were adjusted for differences in fish size. The most popular procedure is to use the slopes of the regressions between the total fish length and each morphometric character (Thorpe 1976). The morphometric and truss measurements were adjusted to the overall mean length of all fish in the samples according to the following relationship:

$$ADJCHR_i = CHR_i - [SLOPE_i(SL - MSL)];$$

$ADJCHR_i$ is the value of the size-adjusted character i; CHR_i is the value of the unadjusted character i; $SLOPE_i$ is the regression slope for total length versus character i regression line; SL is the standard length of each fish; and MSL is the overall mean standard length. Ordinary least-squares regression lines were calculated in all instances because the relationships over the available fish-size range were not noticeably curvilinear. Because of the small number of individuals for which truss data were available, data from the same geographical areas (Table 1) were combined for subsequent analyses. In particular, the north

(mainland) coast and Queen Charlotte Islands samples were combined, as were the central (mainland) coast, Georgia Strait, and west coast Vancouver Island samples. Meristic characters generally are not significantly related to total length (Meng and Stocker 1984) and so were not adjusted for this study.

Means and standard deviations were calculated for meristic and adjusted morphometric and truss variables for each sampling area. Each character was subjected to a one-way analysis of variance among areas to test for significant differences. Principal components analysis (PCA) then was performed on the combined morphometric and meristic data and on the truss data. Plots of the principal components were generated to assess whether any stock structure was evident from these data. For conciseness, only plots of the first two components are presented here. Although there was a tendency for fish from the same area to group together, there was significant overlap among fish from different areas.

No obvious structuring of the data was evident from the PCA analysis. I used discriminant function analysis and canonical correlation analysis on the data from the four geographical groupings to test for statistical differences. The SAS discriminant analysis routine (DISCRIM) calculates linear discriminant functions that maximally separate the groups based on a measure of generalized distance (Mahalanobis D^2) among them. It also provides a summary of the classification accuracy of the discriminant functions for assigning individual fish to the correct groups of origin. The SAS canonical correlation analysis routine (CANDISC) calculates a series of $k - 1$ canonical vectors (k is the number of groups) that maximize the correlation among the morphometric data and a series of dummy variables representing the groups. The vectors are uncorrelated with each other but have maximum multiple correlation with the groups. Tests of the equality of group means are provided by both routines and can be calculated from the D^2 values directly (Lachenbruch 1975). The canonical variates were plotted to discern the distribution of individuals in a reduced dimensional space, and their separation was approximated by identifying group means and an associated estimate of their standard deviation (Seal 1964; Pimentel 1979). The relationship between the canonical variates and the original variables was determined by cross-correlating them; this is the only method for reliably comparing the two sets of variables (Gittins 1979).

The minimum number of variables necessary to distinguish groups was determined. I used stepwise and forward selection followed by backward elimination discriminant function analysis to identify the morphometric, meristic, and truss characters that were most important in determining differences between areas (Claytor and MacCrimmon 1987). I then ran the discriminant analysis and canonical variates analysis together on the characters identified by the stepwise discriminant analysis as significant variables for stock separation to evaluate any changes in stock resolution. Finally, I ran the discriminant function analysis on the morphometric and meristic characters separately for males and females to assess whether there was any improvement in the ability of the discriminant functions to classify fish by area of origin for each sex.

Results

Morphometric and Meristic Measurements

The slopes of the regression between the morphometric characters and standard length are shown in Table 3. The correlations approach or exceed 0.90 for all characters except interorbital width and pectoral width. These two characters are difficult to measure, and the larger scatter of points about their regressions may be due to measurement errors. The means, standard deviations, and F-values from a one-way analysis of variance for the adjusted morphometric and meristic characters for the four areas are presented in Table 4. The standard deviations for each character are similar between areas with a couple of exceptions. The standard deviation for Sndors (see Table 2 for definitions of variables) in the central coast samples and Snoutmax in the Queen Charlotte Islands samples are unusually high, and the estimate of the standard deviation about the mean vertebral count is higher for Georgia Strait than for the other areas. Histograms of the adjusted morphometric characters show aberrant values that may have been outliers from the character–standard length regression lines; however, these could not be deleted from the data set. These values would account for the high standard deviations in morphometric measurements from the central coast and Queen Charlotte Islands. The high standard deviation for the vertebral count remains unexplained because this variable was not adjusted for fish size.

Two of the 12 morphometric variables—pectoral and interorbital widths—and one of the four meristic variables—gill raker count—differed significantly among areas. The differences among the other morphological characters and the general pattern of variation among these variables between areas did not show any clear trends.

The principal components analysis of the morphometric and meristic data reduced the variation in the 16 variables to 16 principal components, of which the first few accounted for most of the variation. Generally, the first principal component is regarded as a size component and the second and third as shape components. In this study, the first component accounted for 31% of the variance, and the second accounted for an additional 24%, for a total of only 55% (Table 5). Generally, principal components analysis is able to explain more of the variability in the data. The lack of resolution resulting from this procedure for our data was evident in the plot of the first two components (Figure 3). There was little patterning evident, and the points from individual areas overlapped to such a degree that no stock separation was apparent.

The discriminant analysis of the morphometric data indicated that, based on the F-statistic, there was a significant difference in classification success between the samples from the west coast of Vancouver Island and each of the other three areas, but not among the latter (Table 6). However, the classification of fish to their area of origin did not reflect this difference because fish

TABLE 3.—Regression coefficients used to adjust morphometric and truss measures to average fish size (mean standard length = 203.41 mm). Acronyms for morphometric and meristic characters are described in Table 2, and notation for the truss characters is shown in Figure 2.

Character	Slope	Intercept	Correlation
Morphometric characters			
Pectcaud	0.8096	−3.1385	0.9816
Dorcaud	0.4629	9.7945	0.9492
Pelcaud	0.4407	7.9825	0.9251
Sndors	0.4610	10.5418	0.8976
Pectdors	0.3964	−10.5572	0.9483
Pectw	0.1025	−6.4524	0.7440
Pelvhei	0.0972	1.0280	0.8867
Peduncle	0.0687	2.2131	0.9004
Snoutmax	0.0991	5.9839	0.8494
Inorbw	0.0299	1.2497	0.7738
Maxw	0.0290	0.9022	0.8767
Headw	0.1028	−0.7469	0.9101
Truss characters			
A1	0.1026	5.0728	0.6208
A2	0.0958	7.2542	0.8191
A3	0.1191	9.1639	0.7741
A4	0.1631	6.9772	0.8539
A5	0.1958	11.1371	0.8341
A6	0.1211	8.6959	0.8494
B1	0.3914	−9.3234	0.8650
B3	0.3371	2.3957	0.9136
B4	0.2976	−9.2621	0.8818
B5	0.4686	−2.5127	0.9341
B6	0.3875	−4.9489	0.9206
C1	0.2228	−2.2385	0.8174
C3	0.1158	4.7841	0.7656
C4	0.2149	−0.5982	0.8823
C5	0.3166	1.0529	0.9150
C6	0.2733	−6.0271	0.8538
D1	0.1418	9.9798	0.7533
D3	0.2815	5.7827	0.8675
D4	0.0701	3.2764	0.8407
D5	0.2917	7.7921	0.8746
D6	0.1877	8.6219	0.8344

for all three areas were classified correctly almost as well as those from the west coast of Vancouver Island (Table 6). The canonical variates analysis provided some insight into this contradiction (Figure 4). Although the centroid for the west coast of Vancouver Island was distinct from the other areas, there was considerable scatter and overlap among individuals from all areas. The canonical correlations for the first two canonical variates were significantly different from zero and accounted for 56% and 26% of the variation in the data. The third canonical correlation was not significant ($P = 0.09$). An examination of the correlations between the first two canonical variates and the adjusted morphological and meristic variables did not reveal any clear pattern (Table 7). The variables Dorcaud, Pectw, Peduncle, and Headw were significantly ($P < 0.01$) and positively correlated with the first canonical variate, whereas the Inorbw and Gillow were negatively

correlated with the first canonical variate (Table 7). Pectw, Maxw, Headw, and Vert were positively correlated with the second canonical variate, whereas Pectcaud, Pectfin, and Peduncle were negatively related to this variate. In addition, the magnitude of the correlations were all similar and made it impossible to determine which of the variables were most important for sample separation.

Discriminant analysis with stepwise and forward selection and backward elimination identified 9 of 16 variables as significant for separating the samples of fish from the four areas. They were, in the order entered by the forward selection option: Inorbw, Pectw, Peduncle, Gillow, Pectfin, Pectcaud, Headw, Pelcaud, and Vert. There was consensus by the three variable selection options for all characters but Vert, which was chosen by only two options; the third chose Pelvhei in this case. The discriminant analysis

TABLE 4.—Means, standard deviations (in parentheses), and F-values for a one-way analysis of variance for morphometric, meristic, and truss characters of Pacific herring from four regions of British Columbia. Acronyms for morphometric and meristic characters are described in Table 2, and notation for truss characters is shown in Figure 2. Asterisks denote $P < 0.05*$ or $P < 0.01**$.

Character	North coast ($N = 30$)	Central coast ($N = 25$)	Georgia Strait ($N = 30$)	West coast of Vancouver Island ($N = 30$)	F-value
Morphometric characters					
Pectcaud	161.0(2.75)	162.4(2.25)	160.6(3.29)	162.2(3.17)	2.62
Dorcaud	103.4(2.96)	102.9(2.07)	104.0(2.43)	105.3(3.47)	3.81*
Pelcaud	98.0(3.75)	98.0(3.20)	97.6(2.81)	97.1(3.81)	0.51
Sndors	104.7(3.92)	104.1(6.17)	104.0(2.58)	104.3(4.25)	0.15
Pectdors	69.9(2.50)	70.3(1.92)	69.4(2.42)	70.6(2.89)	1.26
Pectw	14.5(1.81)	13.7(1.43)	13.8(1.22)	15.3(1.79)	6.16**
Pelvhei	20.8(0.76)	20.5(1.08)	21.2(1.05)	20.8(0.87)	2.55
Peduncle	15.9(0.61)	16.3(0.54)	16.3(0.48)	16.4(0.74)	3.57*
Snoutmax	26.2(1.91)	26.1(0.83)	26.1(0.76)	26.2(0.78)	0.11
Inorbw	7.5(0.34)	7.4(0.42)	7.4(0.42)	7.0(0.53)	6.19**
Maxw	6.8(0.36)	6.7(0.28)	6.9(0.25)	6.8(0.31)	1.40
Headw	20.3(0.94)	19.8(0.58)	20.2(0.69)	20.4(1.05)	2.97*
Meristic characters					
Pectfin	16.9(0.58)	17.2(0.47)	17.3(0.79)	16.9(0.64)	2.37
Analfin	16.5(0.77)	16.2(0.78)	16.5(1.01)	16.3(0.92)	0.43
Gillow	44.9(1.53)	45.2(1.50)	44.1(1.60)	43.7(1.69)	5.79**
Vert	53.1(0.66)	52.6(0.76)	52.6(1.52)	52.7(1.05)	1.74
Truss characters					
A1	26.4(1.54)	25.6(2.31)	26.8(1.52)	24.8(3.49)	4.27**
A2	26.6(1.04)	26.7(1.20)	26.5(1.52)	27.0(1.27)	0.88
A3	33.7(2.18)	32.8(1.61)	33.4(1.55)	33.4(1.91)	1.21
A4	40.3(1.94)	39.3(1.60)	40.2(1.38)	40.4(2.34)	1.97
A5	51.4(1.92)	50.6(2.50)	51.5(1.72)	50.1(3.25)	2.20
A6	33.4(1.70)	32.6(1.14)	33.3(1.16)	33.8(1.40)	3.58*
B1	69.2(4.37)	72.7(3.32)	68.8(4.28)	70.7(4.18)	5.14**
B3	71.3(2.91)	71.0(3.36)	70.9(1.90)	70.5(3.17)	0.39
B4	50.8(3.11)	52.1(2.29)	50.0(3.28)	52.2(2.77)	3.78*
B5	92.3(3.49)	94.0(2.92)	92.2(3.53)	92.6(3.43)	1.63
B6	73.5(3.25)	74.8(2.79)	72.5(2.57)	74.7(3.32)	3.79*
C1	43.0(3.32)	42.5(2.00)	43.3(3.18)	43.2(3.16)	0.42
C3	28.1(1.97)	28.0(1.77)	28.3(1.58)	28.7(2.02)	0.96
C4	42.9(2.57)	42.6(1.94)	43.0(1.97)	43.7(2.07)	1.37
C5	65.0(3.04)	64.6(2.57)	65.6(2.14)	66.3(2.57)	2.24
C6	49.2(2.94)	50.3(2.47)	48.3(3.79)	50.3(2.92)	2.76*
D1	38.2(1.90)	38.9(2.59)	38.6(2.66)	39.5(2.11)	1.69
D3	62.0(3.41)	62.6(2.57)	62.9(2.44)	64.4(3.25)	3.44*
D4	17.2(0.95)	17.5(0.71)	17.6(0.60)	17.7(1.02)	1.74
D5	66.0(3.18)	66.3(2.40)	67.0(2.83)	68.8(3.05)	5.43**
D6	46.3(2.15)	47.3(2.50)	46.5(2.32)	47.1(2.42)	1.18

TABLE 5.—Principal components (PC), canonical variates (CV), and significance of the canonical correlations for morphological and truss variates of Pacific herring from British Columbia. Asterisks denote $P < 0.01**$.

Statistic	PCI	PCII	CVI	CVII
Morphometric and meristic characters				
Eigenvalue	20.08	15.77	0.6766	0.3119
Proportion variance	0.31	0.24	0.5552	0.2559
Canonical correlation			0.6353	0.4876
Likelihood ratio			0.3696**	0.6197**
Truss characters				
Eigenvalue	48.95	28.46	0.7876	0.4084
Proportion variance	0.30	0.18	0.5950	0.3085
Canonical correlation			0.6638	0.5385
Likelihood ratio			0.3522**	0.6297

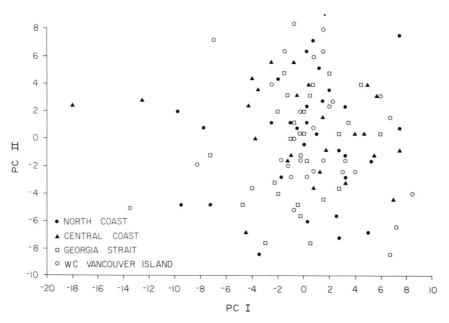

FIGURE 3.—Plot of the first and second principal components (PC) from the analysis of morphometric and meristic characters for Pacific herring from four geographical areas along the British Columbia coast.

was repeated with only these nine variables to obtain the results in Table 8. In most cases, the Mahalanobis distances between areas were larger than with all variables and, with the greater number of degrees of freedom in the reduced model, the differences between areas were significant in all instances except between Georgia Strait and the central coast. However, the classification accuracy was not markedly different from that obtained by using all the variables. It was higher for

fish from the north coast and lower for those from the west coast of Vancouver Island.

Because of the small sample sizes, it was not possible to attempt statistically valid analyses between areas by sex. However, as a rough estimate of the importance of sex differences in discrimination success, I ran the discriminant analysis on males and females separately for the four areas (Table 9). The results indicate that the discriminant functions were able to separate the

TABLE 6.—Percent classification success (PCS), generalized Mahalanobis distances (D^2), and F-statistics for the linear discriminant functions for all morphometric and meristic (MM) and truss characters (TR) for Pacific herring from four areas of British Columbia. Asterisks denote $P < 0.05^*$ or $P < 0.01^{**}$.

		Classified as							
		North coast		Central coast		Georgia Strait		West coast of Vancouver Island	
Area	Statistic	MM	TR	MM	TR	MM	TR	MM	TR
North coast	PCS	50	50	20	13	17	20	13	17
	D^2			2.53	2.81	2.04	1.14	4.17	5.22
	F			1.55	1.14	1.42	0.53	2.90**	2.44**
Central coast	PCS	16	12	72	64	8	12	4	12
	D^2					2.32	3.00	4.76	4.55
	F					1.42	1.21	2.91**	1.84
Georgia Strait	PCS	17	23	10	17	55	50	17	10
	D^2							2.99	3.78
	F							2.08*	1.77
West coast Vancouver Island	PCS	10	3	7	17	13	13	70	67

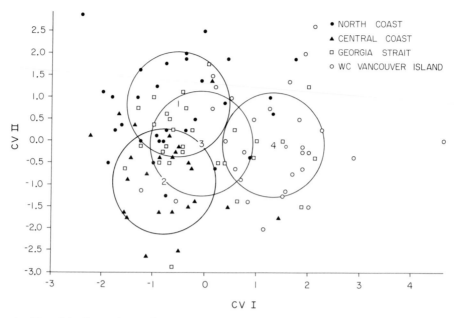

FIGURE 4.—Plot of the first and second canonical variates (CV) from the analysis of morphometric and meristic characters for Pacific herring from four geographical areas along the British Columbia coast. Numbers identify the centroids for each area (in the order of the legend) and circles denote one standard deviation about the centroid.

fish in all areas as well as or better than they could when the sexes were combined (compare Table 6).

Truss Characters

The truss measures were allied with individual morphometric characters in several instances, but no direct comparisons of individual variables were possible because of the multivariate nature of most of the analyses. Instead, I examined the whole suite of truss characters with the intent of making some general comparisons between the two data sets. Note that sides 2 and 4 on each adjoining cell are redundant so that the original 24 variables actually were reduced to 21 for the four cells. The correlations between the truss measurements and standard length approximate 0.85 and are similar to the results for morphometric data (Table 3). In particular, correlations for A1, A3, C3, and D1 all are below 0.80, and an examination of Figure 2 suggests that there may have been some difficulty in identifying landmarks, which possibly resulted in erroneous values for some fish.

The means and standard deviations of the adjusted truss characters are presented in Table 4. Standard deviations between areas differ more in these data than in the morphometric data. Anomalously high standard deviations were

found for A1 and A5 in the data for the west coast of Vancouver Island, for B3 in the central coast data, and for C6 in the Georgia Strait data. As for the morphometric data, the only explanation that seems likely is that some individuals were outliers from the character–standard length regression.

The one-way analysis of variance tests detected significant between-area differences for A1, A6, B1, B4, B6, C6, D3, and D5 (Table 4). In a very general way, these variables described the body depth and length of the tail section of the fish. There was no consistency in the differences in these variables for the different areas.

In principal components analysis of truss data, the first principal axis accounted for 30% of the variance, and the second for 18%, for a total of only 48%, slightly less than for the morphometric and meristic analysis (Table 5). As with the morphometric and meristic data, the plot of the principal components revealed no stock structure and almost complete overlap in the groups of fish from the four areas (Figure 5).

The discriminant function analysis of the truss data revealed a significant difference only between fish from the north coast and the west coast of Vancouver Island (Table 6). The accuracy with which fish were classified to the four areas was very similar for morphometric and

TABLE 7.—Pearson product-moment correlation coefficients (ρ) and significance levels (*P*) for the adjusted morphometric, meristic, and truss characters and the first (CVI) and second (CVII) canonical variates for Pacific herring samples from British Columbia. Acronyms for morphometric and meristic characters are described in Table 2, and notation for truss characters is shown in Figure 2.

Variables	CVI		CVII	
	ρ	*P*	ρ	*P*
Morphometric characters				
Pectcaud	0.122	0.1960	−0.323	0.0005
Dorcaud	0.473	0.0001	0.094	0.3217
Pelcaud	−0.181	0.0545	0.023	0.8074
Sndors	−0.003	0.9710	0.106	0.2595
Pectdors	0.145	0.1248	−0.113	0.2326
Pectw	0.482	0.0001	0.297	0.0013
Pelvhei	0.084	0.3752	0.209	0.0258
Peduncle	0.301	0.0011	−0.445	0.0001
Snoutmax	0.046	0.6265	0.054	0.5659
Inorbw	−0.585	0.0001	0.161	0.0865
Maxw	0.110	0.2433	0.283	0.0023
Headw	0.299	0.0012	0.396	0.0001
Meristic characters				
Pectfin	−0.137	0.1453	−0.254	0.0063
Analfin	−0.016	0.8679	0.186	0.0480
Gillow	−0.528	0.0001	−0.049	0.6076
Vert	−0.064	0.4966	0.391	0.0001
Truss characters				
A1	−0.420	0.0001	0.231	0.0128
A2	0.192	0.0401	−0.066	0.4855
A3	−0.024	0.8010	0.301	0.0011
A4	0.084	0.3716	0.391	0.0001
A5	−0.312	0.0007	0.200	0.0325
A6	0.252	0.0066	0.436	0.0001
B1	0.164	0.0796	−0.616	0.0001
B3	−0.148	0.1145	−0.023	0.8097
B4	0.300	0.0011	−0.381	0.0001
B5	0.020	0.8288	−0.380	0.0001
B6	0.261	0.0049	−0.392	0.0001
C1	0.054	0.5631	0.176	0.0596
C3	0.212	0.0230	0.137	0.1452
C4	0.254	0.0062	0.157	0.0932
C5	0.286	0.0019	0.264	0.0043
C6	0.242	0.0092	−0.327	0.0004
D1	0.301	0.0011	−0.102	0.2799
D3	0.431	0.0001	0.052	0.5784
D4	0.259	0.0052	−0.005	0.9610
D5	0.515	0.0001	0.171	0.0676
D6	0.167	0.0739	−0.246	0.0080

meristic data and truss networks, although slightly more fish were misclassified with the truss data. The canonical correlation analysis produced two canonical variates that accounted for 60% and 31% of the variance in the truss data, which was slightly more than for the morphometric and meristic data (56% and 26%, respectively); only the first canonical correlation was significantly different from zero (Table 5). The plot of the first two canonical variates was almost identical to that from the morphometric and meristic data: fish from the west coast of Vancouver Island were segregated from the other three areas (Figure 6). Fish from the

central coast again were slightly but not significantly segregated from those of the north coast and Georgia Strait.

The correlations between the truss characters and the canonical variates suggested some interesting differences. The first canonical variate was significantly correlated (*P* < 0.01) with A6, B4, B6, C4, C5, C6, D1, D3, D4, and D5, and negatively related to A1 and A5. Variables in the first set of variables are indicators of body depth and length of the caudal region. The two variates in the second set were inversely related to the length of the head. The highest correlations in the group were with D3 and D5, which measure the length of

TABLE 8.—Percent classification success (PCS), generalized Mahalanobis distances (D^2), and F-statistics for the linear discriminant functions for the morphometric and meristic (MM) and truss characters (TR) identified by stepwise discriminant analysis as providing the most parsimonious separation of Pacific herring from four areas in British Columbia. Asterisks denote $P < 0.05^*$ or $P < 0.01^{**}$.

| Area | Statistic | Classified as | | | | | | | |
| | | North coast | | Central coast | | Georgia Strait | | West coast of Vancouver Island | |
		MM	TR	MM	TR	MM	TR	MM	TR
North coast	PCS	80	33	7	20	3	20	10	27
	D^2		3.59		1.13	4.10	0.16	4.46	1.85
	F			4.62**	3.64**	5.90**	0.58	6.41**	6.54**
Central coast	PCS	12	8	68	68	20	16	0	8
	D^2					1.08	1.39	4.29	2.60
	F					1.39	4.46**	5.51**	8.35**
Georgia Strait	PCS	14	30	24	23	45	30	17	17
	D^2							2.56	1.65
	F							3.67**	5.86**
West coast Vancouver Island	PCS	13	3	10	13	20	17	57	67

fish from the rear of the dorsal fin to the tail. The correlations with the second canonical variate were significant and positive for A3, A4, A6, and C5, and negative for B1, B4, B5, B6, C6, and D6. Variables in the first group are measures of the head and anterior body depth; those in the second set were inversely related to the length of the body between the head and dorsal fin, and to the body depth in general. The largest correlation was negative for the distance between the pectoral and pelvic fins, and positive for the head depth. Again, although a couple of the correlations were quite large, most of them were of similar magnitude making it difficult to identify the specific variables that represent the canonical variates.

The stepwise discriminant analysis for the truss measures was straightforward and selected only four characters by each option that were based on the significance of the F-statistic. The variables for the most parsimonious model for separating fish from the four areas were A1, A6, B1, and D5 (Table 7). Interestingly, three of these variables were most different between areas in the one-way analysis of variance. They appeared to be general reflections of head depth and overall body size. Discriminant analysis performed on only these four variables yielded a significant difference among all of the means for the four areas except between fish from the north coast and Georgia Strait (Table 8). The classification accuracy of the reduced data set declined for the north coast and

TABLE 9.—A posteriori classification accuracy based on linear discriminant function analysis of morphometric and meristic characters for male (M, $N = 52$) and female (F, $N = 56$) Pacific herring from four areas in British Columbia. Values represent the percentage of fish classified into each geographic area.

| Area | Sex | Classified as | | | |
		North coast	Central coast	Georgia Strait	West coast of Vancouver Island
North coast	M	76	6	18	0
	F	75	17	0	8
Central coast	M	8	85	0	8
	F	0	67	17	17
Georgia Strait	M	0	9	55	36
	F	8	8	69	15
West coast of Vancouver Island	M	0	9	27	64
	F	5	5	16	74

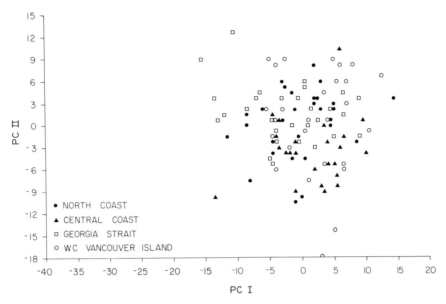

FIGURE 5.—Plot of the first and second principal components (PC) from the analysis of truss measures for Pacific herring from four geographical areas along the British Columbia coast.

Georgia Strait, but remained similar for the other two areas.

Discussion

Early attempts to describe discrete stocks of Pacific herring in British Columbia relied primarily on vertebral counts (Tester 1937, 1949), although a few morphometric and other meristic characters were examined as well. The statistical analyses and computing facilities available at that time limited the degree to which inferences of stock structure could be made with these data. Additional studies have focused on the types of variables that are useful for Pacific herring stock identification (Schweigert 1981; Meng and Stocker 1984; King 1985). All of these studies have indicated some promise in the use of morphometric and meristic characters for the definition and delineation of discrete stocks of Pacific herring. However, there continue to be persistent overlaps in most variables between populations and an inability to identify characters that can segregate groups of fish consistently for management purposes.

A few attempts at applying truss network analysis to pollock *Pollachius virens* and haddock *Melanogrammus aeglefinus* (McGlade and Boulding 1986), Pacific salmon *Oncorhynchus* spp. (Winans 1984), and sculpins *Cottus* spp. (Strauss and Bookstein 1982) have suggested some potential for

obtaining more useful results. Winans (1984) and Strauss and Bookstein (1982) have done limited comparisons of morphometric and truss data sets that suggested only slightly superior performance by the truss network approach. Results from my study were similar to their findings.

In this study, I found both morphometric and meristic characters and truss measures to be highly correlated with standard length (Table 3). It is evident, therefore, that adjustment for size variation among individuals is required before further ordination analyses are performed. There has been considerable debate in the literature about adjustment for size differences. Often, it is argued that, in the results of PCA, the first principal component contains most of the size information and, by eliminating it, one retains the shape differentiation in the second and third components (Thorpe 1976). A technique for "shearing" the principal components to adjust for some of the size differences also has been developed (Humphries et al. 1981). Additional debate has centered on whether the adjustment for size differences should be done separately for each purported population in the study or with the total size variation among all groups. Claytor and Mac-Crimmon (1987) have shown that there usually is little to be gained by partitioning size according to within-group variation and that shearing of principal components to extract some of the size

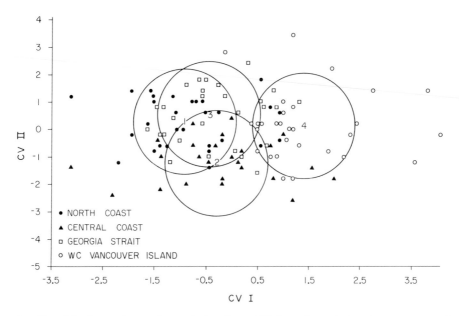

FIGURE 6.—Plot of the first and second canonical variates (CV) from the analysis of truss measures for Pacific herring from four geographical areas along the British Columbia coast. Numbers identify the centroids for each area (in the order of the legend) and circles denote one standard deviation about the centroid.

information most often does not increase resolution of between-group differences. Thus, I used the pooled slopes of the character–standard length regression lines for all population samples to adjust individual characters and the pooled covariance matrix in all discriminant and canonical variates analyses.

The results of the one-way analysis of variance of the size-adjusted variables were disconcerting because Meng and Stocker (1984) had shown significant between-sample differences in all of the morphometric characters I collected. The probable explanation is that my results are based on much smaller sample sizes. Another possible explanation is that, in my study, there may be some between-observer variation that was not present in the study of Meng and Stocker (1984).

Small sample size also may be responsible for the apparent similarity in the truss data between areas that showed only eight significant between-group differences, yet had a standard error comparable to that of the morphometric data. An explanation for the high variability in measurements may be difficulty in correctly identifying landmarks and the inability to properly straighten out individual fish, particularly the smaller ones, for pinning. In part, these errors can be adjusted for by correcting the landmark positions until they are all coplanar in a procedure known as flattening

the truss. Strauss and Bookstein (1982) described the method for implementing this procedure mathematically, and McGlade and Boulding (1986) provided software to perform a slightly modified version of the analysis.

The principal components analysis indicated no significant differences in the results from the two types of data, which is consistent with the results presented by both Winans (1984) and Strauss and Bookstein (1982). The major difference in my study is in the small amount of variation accounted for by the first two principal components. For the combined morphometric and meristic data set and the truss data set seven and nine components, respectively, were required to explain 90% of the variation. In most other studies cited, the first two components frequently explained at least this much of the variability. The covariances of my data were quite large, which may stem partly from small sample size or, quite possibly, from high intercharacter correlations. In any event, the principal components did not reveal any a priori stock separation.

As with any ordination technique, if no structure is evident from the data, it becomes necessary to impose a structure and test the associated hypotheses. For example, the principal components analysis showed no stock separation, but there may have been statistically significant differences between the fish collected from the four areas. Discriminant and canonical correlation analysis are methods designed to test these hypotheses. The estimated distance between groups was not significant for any but those from the west coast of Vancouver Island, and the classification success (50–70%) was not much better than could be achieved by chance. These results are comparable to those of Meng and Stocker (1984), who found a maximum classification success of 85–90% for fish from northern areas (generally similar to samples 1–6, Table 1) and southern areas (samples 7–12); their results were based on 14 morphometric and 3 meristic characters. Both my study and theirs yielded similar Mahalanobis distances (3.0–5.0) between areas, and both indicated a minor reduction in classification success from discriminant function analysis based on a minimum number of classification variables identified through stepwise discriminant function analysis. The optimal variables identified were not identical in the two studies, but 8 of the 9 identified here were within the first 11 identified by Meng and Stocker (1984) as the best stock separators. The classification results for the truss data were a bit poorer than for the morphometric and meristic data. The reason for this appears to be that a number of variables important for classification accuracy are associated with width measurements (Inorbw, Pectw, Headw), which cannot be incorporated readily into the truss network analysis.

The correlations between the canonical variates and adjusted variables suggest that the morphometric characters interorbital width, head width, peduncle depth, dorsal caudal length, and gill raker numbers on the lower limb account for most of the separation that is detectable along the first canonical variate. The same correlations on the truss measures suggest that measures of body depth and size of the caudal region are the important features differing among areas, as reflected in the first canonical variate. There is a marked difference in the characters that the two methods suggest as important for sample separation. However, a strength of the truss network approach is that it identifies aspects of shape that are important for distinguishing stocks and that may lead to testable hypotheses about the ecology of the different stocks. This has more relevance to a biologist than the distance from the snout to the insertion of the dorsal fin.

The greatest advantage of truss analysis is that a large number of variables can be generated relatively quickly. In this study, I had four cells. The addition of new cells would add another five variables each. It is important to include as many cells as possible to capture the localized variations in shape that may offer the best separation of fish among areas (McGlade and Boulding 1986). Although it is useful to have as many variables as possible, this requires one to collect more fish from each area to get reasonable degrees of freedom to test for significant between-area differences.

Because truss data are easily collected, it is logical to select the most useful variables with stepwise discriminant analysis for further analysis of a reduced character set. In the present study, only 4 of the 21 truss characters were identified as important for stock separation. These four characters are diagonal elements that incorporate both length and depth characteristics of the fish. In contrast, nine conventional morphometric and meristic characters were significant stock separators in a stepwise discriminant analysis; Meng and Stocker (1984) found all 16 variables to be important. However, if some of those characters, such as head width, could be incorporated into a three-dimensional truss network, perhaps with the use of electronic calipers to automatically digitize the measurements on head and body width at points along the body, truss network analysis might perform better for identifying stocks than conventional morphometric and meristic variables.

In summary, the present analysis concurs with earlier studies by Winans (1984) and Strauss and Bookstein (1982) that indicated no substantial differences in the ability of truss networks and traditional morphometric and meristic data to differentiate among groups of fish assumed to constitute genetically distinct stocks or populations. However, the truss approach provides a more efficient means for comprehensively characterizing and archiving the shapes of individual fish, and these data may be evaluated as evidence for stock or population separation for fisheries management. The results of the present study, although not encouraging from a fisheries management perspective, are consistent with other studies of pelagic species in which morphological and meristic characteristics are not well differentiated at the localized spatial scale required for fisheries management (Sharp et al. 1978; Meng and Stocker 1984; King 1985). Similarly, genetic differentiation of stocks of pelagic species by electrophoretic studies have not been particularly successful (Smith and Jamieson 1986). Only a few comprehensive studies of either electrophoretic or morphometric variation have been done at the localized spatial scale necessary for describing discrete stocks. Truss analysis may provide a cost-effective means to produce a better understanding of any stock structuring if it exists for these species.

Acknowledgments

The morphometric and meristic data and truss measures were obtained under contract from Envirocon Ltd., Vancouver, British Columbia. C. Haegele provided constructive comments on an earlier draft of this paper.

References

Bookstein, F. L., B. Chernoff, R. L. Elder, J. M. Humphries, G. R. Smith, and R. E. Strauss. 1985. Morphometrics in evolutionary biology. Special Publication Academy of Natural Sciences of Philadelphia 15.

Casselman, J. M., J. J. Collins, E. J. Crossman, P. E. Ihssen, and G. R. Spangler. 1981. Lake whitefish (*Coregonus clupeaformis*) stocks of the Ontario waters of Lake Huron. Canadian Journal of Fisheries and Aquatic Sciences 38:1772–1789.

Claytor, R. R., and H. R. MacCrimmon. 1987. Partitioning size from morphometric data: a comparison of five statistical procedures used in fisheries stock identification research. Canadian Technical Report of Fisheries and Aquatic Sciences 1531.

Gittins, R. 1979. Ecological applications of canonical analysis. Pages 309–535 *in* L. Orloci, C. R. Rao, and W. M. Stiteler, editors. Multivariate methods in ecological work. International Cooperative Publishing House, Fairland, Maryland.

Hourston, A. R., and D. C. Miller. 1980. Procedures for sampling herring at the Pacific biological station. Canadian Manuscript Report of Fisheries and Aquatic Sciences 1554.

Humphries, J. M., F. L. Bookstein, B. Chernoff, G. R. Smith, R. L. Elder, and S. G. Poss. 1981. Multivariate discrimination by shape in relation to size. Systematic Zoology 30:291–308.

Ihssen, P. E., H. E. Booke, J. M. Casselman, J. M. McGlade, N. R. Payne, and F. M. Utter. 1981a. Stock identification: materials and methods. Canadian Journal of Fisheries and Aquatic Sciences 38:1838–1855.

Ihssen, P. E., D. O. Evans, W. J. Christie, J. A. Reckahn, and R. J. Desjardine. 1981b. Life history, morphology, and electrophoretic characteristics of five allopatric stocks of lake whitefish (*Coregonus clupeaformis*) in the Great Lakes region. Canadian Journal of Fisheries and Aquatic Sciences 38:1790–1807.

King, D. P. F. 1985. Morphological and meristic differences among spawning aggregations of north-east Atlantic herring, *Clupea harengus* L. Journal of Fish Biology 26:591–607.

Lachenbruch, P. A. 1975. Discriminant analysis. Hafner Press, New York.

MacCrimmon, H. R., and R. R. Claytor. 1985. Meristic and morphometric identity of Baltic stocks of Atlantic salmon. Canadian Journal of Zoology 63:2032–2037.

McGlade, J. M., and E. G. Boulding. 1986. The truss: a geometric and statistical approach to the analysis of form in fishes. Canadian Technical Report of Fisheries and Aquatic Sciences 1457.

McGlade, J. M., and H. R. MacCrimmon. 1979. Taxonomic congruence in three populations of Quebec brook trout, *Salvelinus fontinalis* (Mitchill). Canadian Journal of Zoology 57:1998–2009.

Meng, H. J., and M. Stocker. 1984. An evaluation of morphometrics and meristics for stock separation of Pacific herring (*Clupea harengus pallasi*). Canadian Journal of Fisheries and Aquatic Sciences 41:414–422.

Pimentel, R. A. 1979. Morphometrics, the multivariate analysis of biological data. Kendall/Hunt, Dubuque, Iowa.

Ricker, W. E. 1973. Two mechanisms that make it impossible to maintain peak-period yields from stocks of Pacific salmon and other fishes. Journal of the Fisheries Research Board of Canada 30:1275–1286.

SAS. 1985a. SAS user's guide: basics, version 5 edition. SAS Institute, Cary, North Carolina.

SAS. 1985b. SAS user's guide: statistics, version 5 edition. SAS Institute, Cary, North Carolina.

Schweigert, J. F. 1981. Pattern recognition of morphometric and meristic characters as a basis for herring stock identification. Canadian Technical Report of Fisheries and Aquatic Sciences 1021.

Seal, H. L. 1964. Multivariate statistical analysis for biologists. Methuen, London.

Sharp, J. C., K. W. Able, and W. C. Leggett. 1978. Utility of meristic and morphometric characters for identification of capelin (*Mallotus villosus*) stocks in Canadian Atlantic waters. Journal of the Fisheries Research Board of Canada 35:124–130.

Sinclair, M., V. C. Anthony, T. D. Iles, and R. N. O'Boyle. 1985. Stock assessment problems in Atlantic herring (*Clupea harengus*) in the northwest Atlantic. Canadian Journal of Fisheries and Aquatic Sciences 42:888–898.

Smith, P. J., and A. Jamieson. 1986. Stock discreteness in herrings: a conceptual revolution. Fisheries Research (Amsterdam) 4:223–234.

Strauss, R. E., and F. L. Bookstein. 1982. The truss: body form reconstructions in morphometrics. Systematic Zoology 31:113–135.

Tester, A. L. 1937. Populations of herring (*Clupea pallasi*) in the coastal waters of British Columbia. Journal of the Biological Board of Canada 3:108–144.

Tester, A. L. 1938. Variation in the mean vertebral count of herring (*Clupea pallasii*) with water temperature. Journal du Conseil, Conseil International pour l'Exploration de la Mer 13:71–75.

Tester, A. L. 1949. Populations of herring along the

west coast of Vancouver Island on the basis of mean vertebral number, with a critique of the method. Journal of the Fisheries Research Board of Canada 7:403–420.

Thorpe, R. S. 1976. Biometric analysis of geographic variation and racial affinities. Biological Reviews of the Cambridge Philosophical Society 51:407–452.

Winans, G. A. 1984. Multivariate morphometric variability in Pacific salmon: technical demonstration. Canadian Journal of Fisheries and Aquatic Sciences 41:1150–1159.

American Fisheries Society Symposium 7:63–70, 1990

Bias and Variation in Stock Composition Estimates Due to Scale Regeneration

Curtis M. Knudsen

Washington Department of Fisheries, 115 General Administration Building
Olympia, Washington 98504, USA

Abstract.—Scale samples collected from adult coho salmon *Oncorhynchus kisutch* showed between-population differences in rates of scale regeneration that were as great as 41%. Increasing the number of scales collected per fish from one to six decreased the average difference in between-population regeneration rates from 18 to 10% and doubled the average proportion of usable fish for scale analysis. In 6 of 10 populations, scale samples showed significant dependence among adjacent scales in the probability of scale regeneration. Therefore, scale sampling should be spread out over as large an area of a fish's body as possible within a region of consistent scale patterns. Results from two-stock stimulations showed that bias in estimates of stock contribution declined from 23 to 11% when the number of scales collected per fish was increased from one to six. The average number of fish included in simulated mixture samples nearly doubled, whereas variation in stock composition estimates due to scale regeneration declined, when the number of scales collected per fish was increased from one to six. Increasing the number of scales collected per fish reduces classification bias when regeneration rates differ between stocks; increases precision in stock composition estimates when regeneration rates are greater than zero; and makes more efficient use of sampling effort by increasing the proportion of usable fish within a sample. A method for correcting classification estimates to account for unequal regeneration rates between populations is demonstrated.

Fish regenerate scales that are dislodged from the epidermis and lost due to injury, disease, and stress (Neave 1940). When a scale is lost, the life history information contained in its circulus patterns is lost and not reproduced on the regenerated scale.

The proportion of fish eliminated from scale pattern analyses due to scale regeneration can be quite large and may constitute 50% or more of an entire sample (Bohn and Jensen 1971; Harris et al. 1981; Rogers et al. 1983; INPFC 1987), although this information generally is not reported in the literature. Therefore, the collection of fish measured is a subset of the original sample. If the scale sample represents several stocks that contribute to a fishery, and if the proportion of fish deleted from the sample differs among stocks due to stock-specific regeneration rates, those stocks with the lowest regeneration rates will be overrepresented, and classification results will be biased.

In this paper, I examine the effect of increasing the number of scales collected per fish on the estimated proportions of usable fish in 10 populations of coho salmon *Oncorhynchus kisutch*, the difference in estimated regeneration rates between these groups, the effective sample sizes (measurable fish) for each group, and the magnitude of bias in simulated classification results. In

addition, a technique for correcting classification results for bias due to unequal regeneration rates between stocks is presented and demonstrated on simulated stock composition estimates.

Scale Collections

Carcasses of spawned-out coho salmon were sampled for scales at seven western Washington locations between 1985 and 1987 (Table 1), and acetate impressions of the scales were made (Koo 1962). Six scales were taken from the left side of each fish in the area of the fish's body shown in Figure 1. This area includes the International North Pacific Fisheries Commission's (INPFC) preferred area A and half of the dorsal-anterior INPFC area B (Major et al. 1972). When adjacent scales are collected, scale samples should exhibit a high degree of dependence in the probability of regeneration because phenomena causing regeneration will affect patches of adjacent scales similarly. Thus, if a scale taken from fish has been regenerated, the probability that adjacent scales also are regenerated probably is increased. Therefore, it seems reasonable that scales should be collected from over as large an area of the body as possible to minimize dependence in regeneration rates, as long as variation in scale patterns between body areas is controlled. Knudsen (1985)

TABLE 1.—Percentages of coho salmon with at least one usable scale ($p = 1 - q$) for scale pattern analysis when one and six scales were collected per fish. The values of $1 - q^6$ represent the expected proportion of fish with at least one usable scale when six scales per fish are sampled, given that the binomial distribution and independence of events are true.

| | | Percentage of fish possessing at least one usable scale | | |
| | | One scale collected per fish | Six scales collected per fish | |
Origin of fish	N			$1 - q^6$
Skykomish Hatchery 1985	85	57.8	97.7	99.4
Skykomish Hatchery 1986	29	63.8	96.6	99.8
Skykomish Hatchery 1987	109	59.8	94.5	99.6
Skagit Hatchery 1985	131	57.4	93.9	99.4
Skagit Hatchery 1987	172	53.1	95.4	98.9
Simpson Hatchery 1987	124	44.1	88.7	96.9
Bingham Creek wild 1987	168	29.1	78.6	87.3
Grizzly Creek wild 1986	29	25.3	75.9	82.6
Lewis Creek wild 1986	24	34.7	87.5	92.2
Deschutes River wild 1985	62	23.4	71.0	79.8
Average for all		44.9	88.0	93.6

showed that measurements of scales collected from within the body area described in Figure 1 did not differ significantly in chinook salmon *Oncorhynchus tshawytscha* and are suitable for stock separation analyses, and Clutter and Whitesel (1956) found that circulus counts of scales from within this body area differed by less than 0.3 circuli for sockeye salmon *O. nerka*. Although comparable analyses have not been done with coho salmon, it seems reasonable to assume the same principle holds for this species.

Scale impressions were viewed with a microfiche reader at 24 and 48× magnification, and scales were characterized as either regenerated or usable according to the size and shape of the scale focus. In general, small circular or slightly oblong foci are characteristic of usable scales. Slightly regenerated scales (scales regenerated early in life) lack small portions of life history information and have irregularly shaped, larger foci, whereas scales regenerated much later in life are deformed

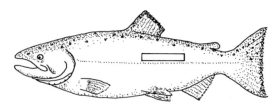

FIGURE 1.—The area of each coho salmon's body sampled for scales is indicated by the box just dorsal to the lateral line. Sampling was spread out over the approximately 100 scales covering this area to reduce dependence in the probability of scale regeneration.

and lack large amounts of life history information (Neave 1940; Mosher 1968).

Regeneration Rate

Estimating Usable Fish

In the following text, "usable fish" is defined as any fish yielding at least one nonregenerated or usable scale within the sample of n scales ($1 \leq n \leq 6$), "proportion of usable fish" is the number of usable fish divided by the total number of fish sampled, and "regeneration rate" is the proportion of fish yielding only regenerated scales within the sample. Estimates of the proportion of usable fish when one scale per fish is taken were made for each of the 10 populations by summing the number of usable scales and dividing by the total number of scales collected. Estimates of the proportion of usable fish when six scales per individual were collected were made by determining the total number of usable fish and dividing by the total number of fish sampled. The mean overall between-group difference in the usable proportion of fish was calculated in the following manner. First, the absolute values of all between-group pairwise differences in the usable proportion of fish were determined; this was done for each of the two sampling plans: one or six scales collected per fish. The mean was then determined for each plan.

To test whether there was dependence in the probability of regeneration of scales collected from one side of a fish, the observed number of fish possessing n regenerated scales ($0 \leq n \leq 6$)

TABLE 2.—Observed frequency distributions of the percentage of coho salmon with n usable scales ($0 \leq n \leq 6$) for scale pattern analysis when six scales per fish were collected.

Origin of fish	n						
	0	1	2	3	4	5	6
Skykomish Hatchery 1985	2	5	18	27	22	19	7
Skykomish Hatchery 1986	3	10	14	3	24	31	14
Skykomish Hatchery 1987	6	6	14	15	29	18	12
Skagit Hatchery 1985	6	8	17	17	19	21	12
Skagit Hatchery 1987	5	11	21	20	20	16	8
Simpson Hatchery 1987	11	18	26	22	10	10	3
Bingham Creek wild 1987	21	27	23	18	7	4	0
Grizzly Creek wild 1986	24	31	25	14	3	3	0
Lewis Creek wild 1986	13	25	25	25	4	8	0
Deschutes River wild 1985	29	34	15	14	6	2	0

out of six was compared to the expected number, under the assumption of independence of events, with a chi-square test. The expected frequencies were based on a binomial distribution, q being the probability that a scale is regenerated and p the probability that a scale is usable ($1-q$) when the number of trials (scales collected per fish) is six. If there is dependence between events, expected values of p^x and q^x (x = number of scales collected per fish; $x > 1$) are actually less than observed values. Thus, the tails of the expected frequency distribution of fish, which represent fish with large or small numbers of regenerated scales, will underestimate the observed values, and the expected middle frequencies will overestimate the observed values. Expected cell values were pooled so that no expected value was less than five (Zar 1984).

Usable Fish from Wild and Hatchery Stocks

Estimates of the proportion of usable fish from each stock when one and six scales per fish were collected are given in Table 1. A comparison of the overall means for each sampling routine shows that sample sizes can, on average, be

TABLE 3.—Results of chi-square tests comparing the expected frequency of regenerated scales per coho salmon to observed values for each of 10 samples.

Origin of fish	Chi-square	df	P
Skykomish Hatchery 1985	3.65	2	0.16
Skykomish Hatchery 1986	3.22	1	0.07
Skykomish Hatchery 1987	12.69	2	<0.01
Skagit Hatchery 1985	43.58	3	<0.01
Skagit Hatchery 1987	32.51	3	<0.01
Simpson Hatchery 1987	41.58	4	<0.01
Bingham Creek wild 1987	20.94	3	<0.01
Grizzly Creek wild 1986	1.41	2	0.49
Lewis Creek wild 1986	0.69	1	0.41
Deschutes River wild 1985	9.51	2	<0.01

doubled (range, 1.5–3.0 times) by increasing the number of scales taken per fish from one to six. The between-group differences in the proportion of usable scales were 18% when one scale per fish was taken but only 10% when six scales per fish were collected. In theory, the proportion of usable fish will increase as the number of scales collected per fish increases to some upper limit determined by the proportion of fish in the population that possesses only regenerated scales within the body region sampled.

The observed percentages of fish with n usable scales ($0 \leq n \leq 6$) for each of the 10 populations are given in Table 2, and the results of the chi-square tests are given in Table 3. Table 2 shows a clear difference in the frequency distributions of hatchery and wild coho salmon: the distributions are skewed to the right (lower regeneration rates) for hatchery fish and skewed to the left (higher regeneration rates) for wild fish. Six of the 10 groups in Table 3 had frequency distributions that were significantly different from expected values based on the assumption of independence of trials. Collecting scales over the body area described in Figure 1 did not eliminate dependence in scale regeneration in all groups.

In all 10 populations, the expected proportion of fish in the tails of the frequency distribution, which represented fish with either all or nearly all usable or regenerated scales, underestimated the observed proportions. Thus, dependence is present at significant levels in six of the samples and at low levels in the other samples. The proportion of fish in the 10 populations that would be usable when six scales per fish are collected could be estimated from the binomial distribution and data from one scale per fish. However, the proportion of fish that have all regenerated scales would always be underestimated, and consequently, the

number of usable fish would always be overestimated. This is demonstrated in Table 1, which
shows that the difference between expected and
observed proportions was as high as 9%.

Simulated Two-Stock Classification Analyses

Classification of Usable Fish

A common method of estimating the contributions of stocks of fish to a mixed-stock fishery
involves classifying each individual in a fishery
sample to its stock of origin. When the contributing stocks are segregated, either before or after
the fishery, baseline scale samples are collected
and the scale patterns are measured. A classification rule is derived from these measurements and
then used to classify fish in mixed-stock samples.
An estimate of the confusion matrix C (McLachlan and Basford 1988), in which each C_{ij} represents the probability of classifying an individual
from stock i to stock j, can be made from the
baseline data with a leaving-one-out technique
(Lackenbruch 1975). The vector of classification
results R, in which each R_i is the proportion of the
fishery sample classified as stock i, can then be
corrected for errors due to bias in the classification rule to get U, a vector made up of the nearly
unbiased estimates of the proportion of each stock
i in the fishery (Cook 1983), defined as

$$U = C^{-1}R. \qquad (1)$$

A two-stock classification model, in which each
fish possessed a unique stock-specific identifier
resulting in 100% classification accuracy, was
constructed for the stimulation analyses. Therefore, C in equation (1) equals the identity matrix,
and there is no bias or variation in classification
results due to the classification rule. (Appendix 1
describes the simulation process in detail.) In the
following text, the construction and classification
of a mixed-stock sample constitutes one simulation. Six separate runs were made, each composed of 500 simulations, and each run was based
on a different number of scales collected per fish.

Each simulation constructed a fishery sample
($N = 100$) consisting of equal numbers of fish from
the two stocks (1:1). Thus, R was held constant
over all simulations. In each simulation, the proportion of usable fish from each stock was estimated from random samples of fish and scales
selected with replacement from the 1986 Skykomish Hatchery (stock 1) and 1985 Deschutes
River (stock 2) samples. This simulated the sampling of fish and scales from each stock within the

fishery. Within each simulation, fish with at least
one usable scale were identified from each stock
in the fishery sample. Those fish that were not
usable were eliminated, and the remaining usable
fish were classified. This simulated the processes
of screening fish for regenerated scales after collection from the fishery, measuring usable scales,
and classifying the usable fish. The mean and
standard deviation of the number of usable fish in
each fishery sample, the total number of usable
fish from each stock in the fishery sample, and the
classification point estimates for each stock over
the 500 simulations were then calculated. Variation in the point estimates for each stock is due to
variation in the number of usable fish from each
stock in the mixed-stock sample; the number of
usable fish is determined by the simulated scale
regeneration data for the two stocks. Because C is
the identity matrix in these analyses, equation (1)
reduces to

$$U = R \qquad (2)$$

However, R is not necessarily the true proportion
of fish from each stock present in the mixed-stock
sample but rather represents the proportion of
usable fish from each stock present in the usable
fishery sample.

Simulation Results

In the following, bias is defined as the absolute
value of the difference between the true proportion of each stock in the mixed-stock sample (0.5)
and the estimated proportion derived from the
usable fish in the simulated fishery sample. As the
number of scales collected per fish increased from
one to six, the total fishery sample size increased
by 72% (Table 4), and bias in stock composition
estimates declined from 23 to 11% (Table 5).
Stock-1 and stock-2 sample sizes in the fishery
sample increased an average of 50 and 150%,
respectively, whereas the difference between the
two stocks in the number of usable fish declined
from 20.3 initially to 16.5 (Table 4).

Bias Correction for Scale Regeneration

Correction Procedure

Scale pattern analysts tacitly make the assumption that scale regeneration rates of stocks are
equal to zero. However, it is clear from the data
presented above that this may not always be true
and that when stocks have unequal regeneration
rates classification results are biased. Therefore,
scale regeneration is an additional source of vari-

TABLE 4.—Mean and SD of the number of usable fish for scale pattern analyses from each of two coho salmon stocks in 500 simulated mixed-stock samples ($N = 100$ fish per mixture). Classification was based on scale patterns for n scales per fish ($1 \leq n \leq 6$). Initially, each stock was equally represented (1:1) in the mixed-stock sample; fish were then eliminated if they did not have at least one usable scale (see Appendix 1).

| Number of scales collected per fish | Number of usable fish | | | | | |
| | Mixed-stock sample | | Stock 1 | | Stock 2 | |
	Mean	SD	Mean	SD	Mean	SD
1	43.9	4.6	32.1	3.5	11.8	3.1
2	57.8	4.3	39.4	2.8	18.4	3.3
3	64.9	4.4	42.4	2.5	22.5	3.6
4	69.6	4.1	44.3	2.2	25.3	3.5
5	72.7	4.1	45.2	2.0	27.5	3.5
6	75.5	3.9	46.0	2.0	29.5	3.5

ation in stock composition estimates derived from scale pattern analysis.

Step 9 of Appendix 1 describes in detail the simulated correction process used in this analysis. In general, we can define a square i by i matrix (i equals the number of stocks), G, in which the off-diagonals are equal to zero and each G_{ii} equals the usable proportion of fish in the entire fishery sample divided by stock i's proportion of usable fish. Thus, G can be used to account for variation and bias due to regeneration of scales (see Appendix 2). Multiplying G by U gives vector U', the proportion of usable fish from each stock i in the usable fishery population, or

$$U' = U G. \tag{3}$$

Beginning with equation (1), in which scale analysts make the assumption of no scale regeneration, and substituting U' for U results in

$$U' = C^{-1}R. \tag{4}$$

Substituting from equation (3) into equation (4) and solving for U gives

$$U = C^{-1}RG^{-1}. \tag{5}$$

Equation (5) will account for the effects of scale regeneration on classification results. Because C is the identity matrix in the present simulation analyses, equation (5) reduces to

$$U = RG^{-1}. \tag{6}$$

Equation (6) can be used to account for the effects of scale regeneration when a maximum likelihood estimation (MLE) technique (Millar 1988), which does not include a classification rule bias correction factor, is used to estimate R.

Due to random errors in sampling the fishery and baseline populations, and thus in estimating G, the resulting estimates of U in equations (5) or (6) will not necessarily conform to the restriction that they sum to unity. (Point estimates for a fishery sample summing to more or less than 1.0 are not realistic.) A simple method to make the values of U conform to this restriction involves calculating values of U corrected for scale regeneration bias with either equation (5) or (6) and then dividing each resulting U_i by the summation of U over i.

TABLE 5.—Estimates of the mean percent and SD for stock composition estimates ($N = 500$) based on scale patterns of two coho salmon stocks present in simulated mixed-stock samples that were (1) uncorrected for bias due to unequal regeneration rates; (2) corrected for regeneration rate bias but not constrained; and (3) bias corrected and constrained to sum to unity. n = the number of scales collected per fish in the simulated mixed-stock samples. Each stock actually represented 50% of the mixed-stock sample.

| | Biased point estimates | | | | Bias corrected | | | | Corrected and constrained | | | |
| | Stock 1 | | Stock 2 | | Stock 1 | | Stock 2 | | Stock 1 | | Stock 2 | |
n	Mean	SD	Mean	SD	Mean	SD	Mean	SD	Mean	SD	Mean	SD
1	73	5.7	27	5.7	51	8.5	54	20.9	50	9.9	50	9.9
2	68	4.1	32	4.1	50	5.2	52	14.0	50	7.1	50	7.1
3	66	3.9	34	3.9	50	4.2	50	11.3	50	5.9	50	5.9
4	64	3.5	36	3.5	50	3.7	51	9.9	50	5.3	50	5.3
5	62	3.2	38	3.2	50	3.2	51	9.7	50	4.9	50	4.9
6	61	3.0	39	3.0	50	3.0	50	8.9	50	4.6	50	4.6

Corrected Stock Composition Estimates

Table 5 gives the mean percent composition estimate and standard deviation for each of the two stocks over the 500 simulations for each value of n ($1 \leq n \leq 6$) scales sampled per fish. Results are presented for the biased estimates, estimates corrected for regeneration rate bias, and the bias-corrected estimates constrained to sum to unity. The bias-corrected and constrained means did not differ from the true value of 50%. The biased estimate means differed by as much as 23%, whereas the bias-corrected means differed by 4% or less from the true value. The bias-corrected and constrained means were more accurate, though less precise (larger SDs), than the uncorrected means. All standard deviations declined as the number of scales per fish increased; therefore, precision of the stock composition estimates increased as the variation in the proportion of usable fish decreased.

Discussion

Coho, chinook, and sockeye salmon and steelhead *Oncorhynchus mykiss* (formerly *Salmo gairdneri*) have scale regeneration rates of 27% or greater (Bilton 1984; INPFC 1987), whereas regeneration rates for chum salmon *O. keta* appear to be about 10% or less (J. Sneva, Washington Department of Fisheries, unpublished data; INPFC 1987). Increasing the number of scales collected per fish for coho, chinook, and sockeye salmon and for steelhead to six or more should significantly increase the proportion of usable fish within a sample and also will reduce bias due to between-stock differences in regeneration rates, if it exists. Increasing baseline and fishery sample sizes will, in turn, increase the precision of classification. If brood-year or age-class differences in scale patterns warrant the construction of different classification models to separate different stocks or groups, then regeneration rates for each baseline group need to be estimated.

The constrained bias-correction technique reduced bias in stock composition estimates relative to those made from just the usable proportion of fish in the usable mixture sample, and yielded more accurate results with some increased variation. This technique appears to be useful for correcting composition estimates when regeneration rates differ between stocks.

An alternative method of accounting for fish with all regenerated scales would be to use an MLE technique for estimating stock composition and creating a new scale variable, defined as the proportion of regenerated scales from each fish. The MLE can deal with individual fish that have incomplete data sets, such as unmeasurable fish with all regenerated scales, and will determine the stock composition that is most likely to be represented by a mixed-stock sample (Millar 1988). Data such as those presented in Table 3 suggest that this new scale variable would increase discrimination between hatchery and wild coho salmon stocks. Rearing conditions at the three hatcheries are typically stable, and there are no serious problems with crowding, food, water quality, predation, or disease. Scale regeneration in these 10 groups of coho salmon typically occurred just after or sometime before migration of smolts to salt water. This probably accounts for the lower scale regeneration rates in the hatchery groups; naturally reared groups experience conditions during this period of life that are generally more variable and hazardous. The stable rearing conditions at the Skykomish and Skagit hatcheries also are reflected in the relatively consistent regeneration data over time (Table 2). It is also possible that a hatchery with poor rearing conditions would produce fish with high scale regeneration rates.

Unequal between-group scale regeneration rates is a potential problem in aging analyses as well, and can occur either between stocks or between age-classes and cohorts of the same stock (Bilton 1984). This will result in biased age composition estimates. Stocks and different age-classes within a stock can experience different environmental influences that affect regeneration rates. It may also be true that older fish, exposed to factors causing scale regeneration over a longer period of time, have higher regeneration rates than younger cohorts. These age-related problems are areas deserving further study.

When scale regeneration rates are greater than zero, there is an additional source of variation and possible bias in scale pattern analyses that is not currently accounted for in stock composition estimates. From the results reported in this paper, it is clear that (1) when regeneration rates differ between stocks, classification results will be biased; (2) as the number of scales collected per fish increases, the proportion of usable fish increases, and bias and variation in classification results due to scale regeneration decrease; (3) scale regeneration is an additional source of variation in stock composition estimates; and (4) bias in stock composition estimates can be corrected with estimates of the proportion of usable fish in the fishery and baseline populations.

Acknowledgments

I thank John Sneva of the Washington Department of Fisheries for his help in sample collection and for sharing his expertise in scale pattern analysis. I appreciate constructive criticisms from two anonymous reviewers.

References

Bilton, H. T. 1984. The incidence of regenerated scales among Fraser River and Skeena River sockeye salmon (Oncorhynchus nerka). Canadian Manuscript Report of Fisheries and Aquatic Sciences 1783.

Bohn, B. R., and H. E. Jensen. 1971. Investigation of scale patterns as a means of identifying races of spring chinook salmon in the Columbia River. Research Reports of the Fish Commission of Oregon 3:28–36.

Clutter, R. I., and L. E. Whitesel. 1956. Collection and interpretation of sockeye salmon scales. International Pacific Salmon Fisheries Commission Bulletin 9.

Cook, R. C. 1983. Simulation and application of stock composition estimators. Canada Journal of Fisheries and Aquatic Sciences 40:2113–2118.

Harris, C. K., R. H. Conrad, K. W. Myers, R. W. Tyler, and R. L. Burgner. 1981. Monitoring migration and abundance of salmon at sea—1980. International North Pacific Fisheries Commission Annual Report 1980:71–90.

INPFC (International North Pacific Fisheries Commission). 1987. Report of the scale pattern working group. INPFC Vancouver, British Columbia.

Knudsen, C. M. 1985. Chinook salmon scale character variability due to body area sampled and possible effects on stock separation studies. Master's thesis. University of Washington, Seattle.

Koo, T. S. Y. 1962. Age and growth of red salmon scales by graphical means. Pages 49–122 in T. S. Y. Koo, editor. Studies of Alaska red salmon. University of Washington Press, Seattle.

Lackenbruch, P. A. 1975. Discriminant analysis. Hafner Press, New York.

Major, R. L., K. H. Mosher, and J. E. Mason. 1972. Identification of stocks of Pacific salmon by means of their scale features. Pages 209–231 in R. C. Simon and P. A. Larkin, editors. The stock concept in Pacific salmon. H. R. MacMillan Lectures in Fisheries, University of British Columbia, Vancouver.

McLachlan, G. J., and K. E. Basford. 1988. Mixture models: inference and applications to clustering. Marcel Dekker, New York.

Millar, R. B. 1988. Statistical methodology for composition estimation of high seas salmonid mixtures using scale analysis. University of Washington, Fisheries Research Institute, FRI-UW-8801, Seattle.

Mosher, K. H. 1968. Photographic atlas of sockeye salmon scales. U.S. Fish and Wildlife Service Fishery Bulletin 67:243–280.

Neave, F. 1940. On the histology and regeneration of the teleost scale. Quarterly Journal of Microscopical Science 84:541–568.

Rogers, D. E., K. J. Bruya, K. W. Myers, C. M. Knudsen, N. D. Davis, and T. Nishida. 1983. Origins of chinook salmon in the area of the Japanese mothership fishery. University of Washington, Fisheries Research Institute, FRI-UW-8311, Seattle.

Zar, J. H. 1984. Biostatistical analysis, 2nd edition. Prentice-Hall, Englewood Cliffs, New Jersey.

Appendix 1: Outline of Simulated Classification Analyses

(1) Randomly select, with replacement, a fish from stock 1. For simulations, we have assumed that the 1986 Skykomish Hatchery (stock 1) and 1985 Deschutes River (stock 2) groups coho salmon are representative of two stocks that might appear in a mixed fishery. Associated with each fish is a unique stock-specific identifier and six scale values, each equal to zero or one; zeros represent regenerated scales and ones represent usable scales.

(2) Randomly select, with replacement, n scale values ($1 \leq n \leq 6$) from the fish selected in step (1). Values of n remain constant for all fish selected in a given mixed-stock sample. It is assumed that the six scales are representative of the approximately 100 scales from the fish's body area described in Figure 1.

(3) Repeat steps (1) and (2) through 50 iterations

to obtain a complete sample from stock 1. (Each stock is represented by 50 of 100 fish in the mixed-stock sample for each simulation.)

(4) Repeat steps (1)–(3) for stock 2.

(5) Determine which fish in the mixed-stock sample have at least one usable scale. This simulates the process of viewing each fish's scales after they were collected in a mixed-stock sample and determining which are usable.

(6) Read the unique stock-specific code for each usable fish in the usable mixed-stock sample. Calculate the number of usable fish in the sample by stock. This completes one simulation.

(7) Repeat steps (1) through (6) 500 times keeping n (the number of scales selected per fish) constant. This completes one run.

(8) For each run, calculate the mean and standard deviation of the number and proportion of usable fish and the total number of usable fish in the mixed-stock sample by stock over the 500 simulations.

(9) For the bias-correction procedure, repeat steps (1)–(4) above for each run and independently estimate the proportion of usable fish from each stock. This simulates sampling fish on the spawning grounds when they are segregated by stock, and it is assumed that regeneration rates on the spawning grounds are representative of rates found in the fishery for each stock. The spawning grounds are sampled 4–5 months after commercial fishing for coho salmon begins. Scale regeneration

by coho salmon occurs primarily during the first 2 years of life and rarely during the final year's growth (unpublished data). Thus, fish sampled for scales in the fishery should have regeneration rates representative of fish from the same population sampled a few months later on the spawning grounds.

Within each simulation, the estimate of the proportion of usable fish in the mixed-stock sample and the proportion of usable fish from each baseline stock sample is used to construct **G** (equations 5 and 6; Appendix 2) which then is used to correct classification results for the effects of scale regeneration.

Appendix 2: Bias Correction for Scale Regeneration

The number of stock-i fish in the mixed-stock fishery sample (N_i) will be reduced, due to scale regeneration, by the proportion of usable fish from stock i (P_i), and their product ($N_i P_i$) equals the proportion of usable stock-i fish in the fishery (U_i') multiplied by the number of usable fish in the fishery sample (N'):

$$N_i P_i = U_i' N'. \qquad (A2.1)$$

We wish to solve for U_i from equation (A2.1). We can estimate P_i by sampling baseline populations on the spawning grounds to get $\hat{P}_i$ if we assume that each population's regeneration rate in the fishery is representative of its regeneration rate on the spawning grounds (see Appendix 1, step 9). Equation (A2.1) can be used to solve for

U_i by dividing both sides by $\Sigma_i N_i$, the total number of fish in the fishery, and P_i. Thus,

$$U_i = \frac{N_i}{\sum_i N_i} = \frac{U_i' N'}{P_i \sum_i N_i}$$

N' divided by $\Sigma_i N_i$ is the proportion of usable fish in the fishery, F, and we can estimate this quantity, $\hat{F}$, from the fishery sample. U_i' is estimated with equation (4) to give $\hat{U}_i'$. Therefore, the estimate from the mixed-stock sample of the proportion of stock-i in the fishery when the effects of scale regeneration are accounted for is

$$\hat{U}_i = \frac{\hat{U}_i' \hat{F}}{Pi};$$

$\hat{F}$ and $\hat{P}_i$ are used to estimate G_{ii} in **G** from equation (5).

American Fisheries Society Symposium 7:71–77, 1990
© Copyright by the American Fisheries Society 1990

Use of Scale Patterns and Shape as Discriminators between Wild and Hatchery Striped Bass Stocks in California

WILLIAM R. ROSS

BioSonics, Inc., 4520 Union Bay Place, N.E., Seattle, Washington 98105, USA

ALAN PICKARD

California Department of Fish and Game
1701 Nimbus Road, Suite B, Rancho Cordova, California 95670, USA

Abstract.—Determining the contribution of hatchery production to a fishery requires methods for discriminating between individuals of wild and hatchery origin. Differences in scale shape and circulus spacing between wild and hatchery fish can serve as innate tags. These scale features were used to successfully discriminate (87–91% accuracy) between yearling hatchery and wild striped bass *Morone saxatilis* from the San Francisco Bay area. Hatchery fish were also successfully assigned to their hatchery of origin (66–88% accuracy) by scale shape and circulus spacing.

The California Department of Fish and Game (CDFG) began a hatchery evaluation program for striped bass *Morone saxatilis* in 1983. The study objectives have been to determine the survival of hatchery-reared striped bass and to determine their contribution to the sport fishery. The ability to discriminate between fish of hatchery and wild origin has been essential to the hatchery evaluation. The program has expanded considerably since 1983, and in 1987 the CDFG used coded wire tags to mark about 216,000 (44%) of the approximately 490,000 hatchery-raised yearling striped bass released that year. Although coded wire tags are a satisfactory method of identifying hatchery fish, the process of tag implantation, detection, and recovery is costly. Analysis of innate or induced scale features may provide a cost-effective alternative. Circulus spacing patterns (Cook 1982) and scale shape (Jarvis et al. 1978) have been used to identify stocks of other species, and Riley and Margraf (1983) used scale shape to determine the area of origin for several striped bass stocks from the Atlantic coast of the USA. The objectives of this study were to evaluate scale analysis as an alternative method for determining the origin of California striped bass and to determine whether the combined use of circulus spacing and scale shape provided greater classification accuracy than either used alone.

Methods

Scale collection and preparation.—Between May 1 and May 18, 1987, 400 yearling striped bass from four California hatcheries were collected by dip net from hatchery trucks on the day of stock-

ing. One hundred striped bass were collected from each hatchery. The hatcheries sampled were Arrowhead Fisheries (Arrowhead) near Red Bluff, Namakan West Fisheries (Namakan) near Los Banos, The Fishery near Sacramento, and CDFG's Central Valley Hatchery (CVH), also near Sacramento (Figure 1).

On July 1, 1987, 60 wild striped bass were selected from a sample of 100 yearling-size fish collected from the fish screens at the intake of the State Water Project, California Aqueduct (Figure 1). These fish were determined to be yearlings by visual examination of scales. Fish were considered of wild origin if they did not contain a coded wire tag. Because only 44% of the 490,000 hatchery-reared yearling striped bass released in May 1987 contained coded wire tags, it was possible that some of the 60 wild striped bass were actually of hatchery origin. We therefore performed a sensitivity analysis to estimate the number of untagged hatchery striped bass in our sample of 60 wild fish.

An estimate of abundance for wild yearlings was not available for 1987, but abundance estimates for 1982 through 1986 ranged from 1.6 to 5.0 million. If the abundance of wild yearling striped bass in 1987 was in that range, then the 274,000 untagged hatchery yearlings would make up between 5 and 15% of the total wild and untagged yearling striped bass population. We would then expect that, of the 60 wild yearlings collected for this study, between 3 and 9 of them would actually be of hatchery origin. The effect of erroneously including 3–9 hatchery striped bass in the wild sample would be to introduce some "noise"

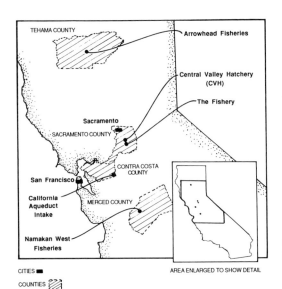

FIGURE 1.—Location of hatcheries in central California. Inset shows enlarged area in relation to entire state.

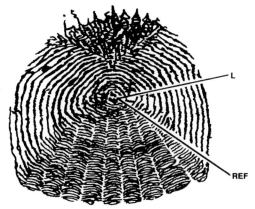

FIGURE 2.—Stylized California striped bass scale showing the luminance extraction line (L) and the reference line (REF).

into the discriminant function and to decrease the resulting classification accuracy. We concluded that the potential inclusion of 3–9 untagged hatchery striped bass in the wild sample would not seriously affect the objectives of this study.

Scales were taken from the left side of each fish midway between the spinous dorsal fin and the lateral line. Fork lengths were recorded for each fish. In the laboratory, five nonregenerated scales were selected from each sample and cleaned by an initial soaking in fresh water, then a gentle scrubbing with a toothbrush, if necessary, and a final rinse with fresh water. Wet scales were positioned between two glass slides that were then taped together as a permanent mount. Each slide was labeled with the origin and fork length of the fish. Eleven CVH slides were damaged in transit, so the sample size for CVH was 89 rather than 100.

Data Acquisition

BioSonics' Optical Pattern Recognition System (OPRS), consisting of a Compaq Deskpro microcomputer, video digitizer (frame-grabber) circuit board, closed-circuit video camera and monitor, and proprietary feature-extraction software, was used to acquire data. The video camera was connected to an Olympus BH-2 microscope equipped with a 2.5×-magnification photocompensating eyepiece and an MTV-3 parfocalizing adapter. Circulus data were collected with a 4× objective lens, and scale shape data with a 2× objective lens.

The video camera converted the optical image to an analog electrical signal and transmitted it to the video digitizer. After enhancing the contrast of the analog signal by adjusting its gain and offset (amplification), the video digitizer converted the analog signal to a set of discrete values ranging between 0 (black) and 255 (white). The entire video image was digitized every 30th of a second, and the resulting values were stored in a 512 by 512 array of pixels or picture elements. To compensate for the nonsymmetric aspect ratio of the original video image, the pixel array was mapped to a symmetric coordinate system. This symmetry-corrected array of pixels was used for all subsequent image measurements. The digitized image was converted back to an analog signal and displayed on the closed-circuit video monitor to provide feedback to the operator as the image was focused, manipulated, or measured.

The OPRS command sequences used to acquire circulus and shape data were preprogrammed as macros by the operator. Most steps for data acquisition were then performed automatically; the program paused only for those operations requiring manual input from the operator. Both circulus and shape data could be acquired from a scale in less than 2 min.

Circulus spacing measurement.—Circulus spacing was measured in the dorsolateral field of each scale. All measurements were in meters. One scale from each slide was selected and positioned in the field of view with the ctenii pointing upwards. To ensure that data were collected

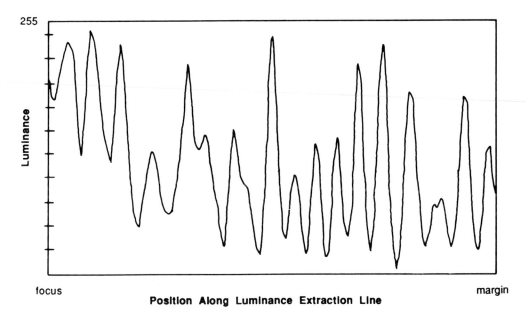

focus   **Position Along Luminance Extraction Line**   margin

FIGURE 3.—Typical luminance profile extracted from a California striped bass scale. Luminance was measured on an arbitrary scale of 0 (black) to 255 (white).

consistently from all scales, a reference line was established along the boundary between the widely spaced circuli in the dorsolateral field and the closely spaced circuli of the anterior field (Figure 2). This location was chosen because the boundary between the closely and widely spaced circuli is a convenient scale landmark that can be consistently and reliably identified on all scales.

Once the reference line was established, the operator drew a line from the scale's focus to the scale's margin at an angle of 45° counterclockwise from the reference line. The luminance profile, a

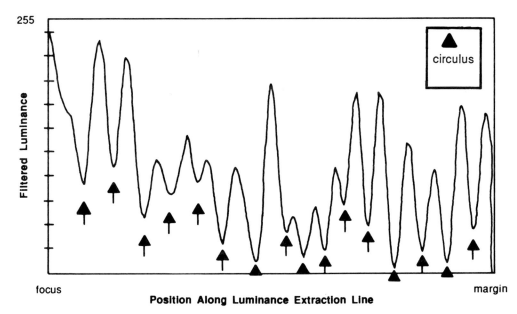

focus   **Position Along Luminance Extraction Line**   margin

FIGURE 4.—Local minima of filtered luminance values that were identified as circuli for a typical California striped bass scale. Luminance was measured on an arbitrary scale of 0 (black) to 255 (white).

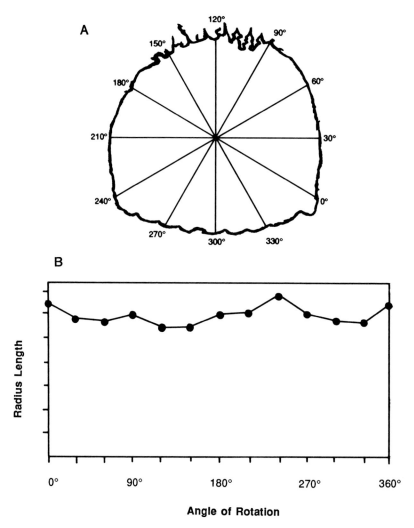

FIGURE 5.—(A) Typical outline of a California striped bass scale showing radius measurements from the centroid to the boundary at 30° increments from the reference line. (B) Plot of radius length from the centroid to the boundary as a function of angle of rotation about the centroid.

vector of the luminance (brightness) values along the line (Figure 3), was extracted and filtered with a weighted moving average to identify local luminance minima corresponding to circuli (Figure 4). The weighted moving average filter used the following expression to compute a new luminance value, L_i, for each point along the line: L_i (new) = $(L_{i-3} + 3L_{i-2} + 6L_{i-1} + 9L_i + 6L_{i+1} + 3L_{i+2} + L_{i+3})/29$. A luminance minimum on the filtered luminance profile was marked as a circulus if the difference in luminance between the minimum and the preceding maximum was greater than 15. These circulus positions were displayed on the video monitor, and the operator could interac-

tively add or delete circulus marks as necessary. The operator also placed a circulus mark at the scale margin to provide a measurement from the focus to the margin. When the operator was satisfied that all circuli were correctly marked, the intercirculus distances were computed and recorded in a data file.

Data from the wild group and each of the hatcheries were stored in separate data files. The intercirculus distances were used to compute other scale features for analysis. These features were the total length of the line from focus to edge, average intercirculus spacing, total number of circuli, and average spacing of the first five circuli. Variables containing the distances from the focus to the first three checks (constrictions or local minima in circulus spacing) were also created by applying a second weighted moving average filter to the intercirculus distances. This is similar to the method used by Koo (1962) to identify annuli.

Scale shape measurement.—Fourier techniques (Jarvis et al. 1978; Riley and Margraf 1983; Bird et al. 1986) were used to provide quantitative descriptors of scale shape. After extracting circulus spacing data from a scale as described above, the operator applied a binary threshold function to the video image. This function set all pixel values in the image to either black or white according to whether the pixel's original value was below or above a threshold value. The operator adjusted the amplification applied to the image until the perimeter of the scale was well-defined. An automatic boundary-tracking "bug" based on Roberts' (1965) edge-detection algorithm was then used to trace counterclockwise along the perimeter of the scale. The starting and ending point of the trace was the point where the reference line met the edge of the scale.

The centroid or center of gravity of the area defined by the boundary trace was located at the mean X- and mean Y-coordinates of all points on the boundary. The distance from the centroid to the scale boundary was measured at 128 points spaced at equal angular increments around the boundary. These measurements produced a vector of radius lengths as a function of angle about the centroid (Figure 5). The angular increment was 2.81° (i.e., 360°/128 samples). The fast Fourier transform (Brigham 1974) was used to generate a Fourier series from this vector. Each harmonic or term of the series consisted of a cosine coefficient and a sine coefficient. Each harmonic describes a fundamental shape component, and the amplitude of a harmonic represents its relative contribution to the overall shape (Rayner 1971). The amplitude of a harmonic was computed at the square root of the sum of squares of its cosine and sine coefficients. Harmonic amplitudes were the shape descriptors used in this study.

Data Analysis

Discriminant analysis was used to classify fish into hatchery or wild groups. In particular, the direct density estimation procedure of Cook (1982), modified for the assumptions of the linear discriminant function (Cook 1987), was employed. The leaving-one-out procedure of Lachenbruch (1967) allowed the entire set of known-origin scales to be used as both the training set and as the test set for evaluating the performance of the discriminant analyses without introducing bias. Classification arrays (Cook and Lord 1978) were used to tabulate and display the results of the discriminant analysis. The set of scale features providing the best discrimination was determined iteratively.

Results

Seven features were derived from circulus measurements: length from focus to edge, average intercirculus spacing, average spacing of the first five circuli, distance to the first check, distance to the second check, distance to the third check, and total number of circuli. Sixty-four Fourier harmonic amplitudes were derived and designated as harmonic 1, harmonic 2, and so on. Only the first 10 harmonics were used for analysis.

Hatchery versus Wild

Classification of fish as either hatchery or wild origin when all seven circulus features and harmonics 2, 3, 4, 5, 6, 7, and 8 were used was 91% correct for hatchery fish and 87% correct for wild fish (Table 1). Classification based only on length from focus to edge, average intercirculus spacing, and average spacing of the first five circuli was 84% correct for hatchery fish and 77% correct for wild fish. Classification based only on harmonics 2, 4, 7, and 8 was 72% correct for hatchery fish and 77% correct for wild fish.

Hatchery versus Hatchery

When classification of hatchery fish to hatchery of origin was based on average intercirculus spacing, average spacing of the first five circuli, dis-

TABLE 1.—Discriminant classification of hatchery or wild striped bass according to scale characters.

True origin	Percent classified as	
	Hatchery	Wild
Based on circulus features and shape harmonics		
Hatchery	91	13
Wild	9	87
Based on circulus features only		
Hatchery	84	23
Wild	16	77
Based on shape harmonics 2, 4, 7, and 8		
Hatchery	72	23
Wild	28	77

TABLE 2.—Discriminant classifications of striped bass to hatchery of origin based on scale characters.

True hatchery of origin	Percent classified as			
	Arrowhead	Central Valley	Namakan	The Fishery
Based on circulus features and shape harmonics				
Arrowhead	66	8	22	13
Central Valley	6	88	10	4
Namakan	20	4	66	3
The Fishery	8	0	2	80
Based on circulus features alone				
Arrowhead	53	2	34	25
Central Valley	3	74	14	25
Namakan	27	6	33	17
The Fishery	17	16	19	33
Based on shape harmonics only				
Arrowhead	45	18	21	9
Central Valley	18	74	11	5
Namakan	31	4	66	2
The Fishery	6	4	2	84

tance to the first check, distance to the third check, total number of circuli, and harmonics 2, 3, 4, 5, 6, 7, and 8, it was 66% correct for Arrowhead, 88% correct for CVH, 66% correct for Namakan, and 80% correct for The Fishery (Table 2). Classification based only on length from focus to edge, average intercirculus spacing, average spacing of the first five circuli, and distance to the first check was 53% correct for Arrowhead, 74% correct for CVH, 33% correct for Namakan, and 33% correct for The Fishery. Classification based only on harmonics 2, 3, 4, 5, 6, 7, and 8 was 45% correct for Arrowhead, 74% correct for CVH, 66% correct for Namakan, and 84% correct for The Fishery.

Discussion

The classification accuracies achieved were quite encouraging. The highest accuracies for both hatchery–wild and hatchery–hatchery discriminations were produced by using both circulus features and scale shape data. It is worth noting that the classification error rate for the wild group was 13%, which was within the range expected if 5–15% of the wild group were actually of hatchery origin. Accuracies were degraded when either circulus features or scale shape were used alone. When used alone, circulus features provided the best discrimination for hatchery versus wild, and shape features provided the best overall discrimination among hatcheries.

Although this study did not investigate genetic or environmental factors responsible for variability in scale features, the differences among hatcheries in the ambient water temperatures of rearing pens deserves discussion. All the hatcheries use well water for their supply, but Arrowhead and The Fishery incorporate retention reservoirs to provide the head for gravity-flow systems, whereas Namakan and CVH pump directly from the well to the rearing facilities. Water temperature in the rearing facilities is influenced by the use of retention reservoirs in which the water is subject to cooling (Table 3). These temperature differences may have induced changes in circulus spacing and scale shape.

This first investigation examined only yearling fish and innate (i.e., noninduced) scale features. The results have encouraged us to plan additional studies to validate this technique with 2-, 3-, and 4-year-old striped bass scales collected in the sport fishery. We hope this technique will prove effective for identifying striped bass of hatchery origin that are collected in the sport fishery and ultimately eliminate the necessity of tagging hatchery-produced striped bass. Studies are also being planned to determine if unique signatures can be induced in the scales of hatchery-reared striped bass by modulation of water temperatures, feeding rates, and crowding densities. Such signa-

TABLE 3.—Water temperatures (°C) at striped bass hatcheries.

Hatchery	Temperature at	
	Source well	Rearing facility
Arrowhead	19.4	12.8–19.4
Central Valley	20.0	20.0–20.5
Namakan	20.0	16.7–21.1
The Fishery	22.2	18.9–20.5

tures would increase classification accuracy, particularly among hatchery fish.

Acknowledgments

Several people contributed to the successful completion of this study. Jeff Condiotty provided overall project management and coordination, and Dave Parks acquired and analysed the OPRS data. This study was funded in part by the California Department of Fish and Game. Participation in this study does not constitute an endorsement of the Biosonics Optical Pattern Recognition System by the California Department of Fish and Game.

References

Bird, J. L., D. T. Eppler, and D. M. Checkly, Jr. 1986. Comparisons of herring otoliths using Fourier series shape analysis. Canadian Journal of Fisheries and Aquatic Sciences 43:1228–1234.

Brigham, E. O. 1974. The fast Fourier transform. Prentice Hall, Englewood Cliffs, New Jersey.

Cook, R. C. 1982. Stock identification of sockeye salmon with scale pattern recognition. Canadian Journal of Fisheries and Aquatic Sciences 39:611–617.

Cook, R. C. 1987. In-season stock identification using scale pattern recognition. Canadian Special Publication of Fisheries and Aquatic Sciences 96:327–334.

Cook, R. C., and G. E. Lord. 1978. Identification of stocks of Bristol Bay sockeye salmon, *Oncorhynchus nerka,* by evaluating scale patterns with a polynomial discriminant method. U.S. National Marine Fisheries Service Fishery Bulletin 76:415–423.

Jarvis, R. S., H. F. Klodowski, and S. S. Sheldon. 1978. New method of quantifying scale shape and an application to stock identification in walleye. Transactions of the American Fisheries Society 107:528–534.

Koo, T. S. Y. 1962. Age and growth studies of red salmon scales by graphical means. Pages 23–49 *in* T. S. Y. Koo, editor. Studies of Alaska red salmon. University of Washington Press, Seattle.

Lachenbruch, P. A. 1967. An almost unbiased method of obtaining confidence intervals for the probability of misclassification in discriminant analysis. Biometrics 23:639–645.

Riley, L. M., and F. J. Margraf. 1983. Scale shape as an innate tag for the identification of striped bass stocks. Final Report (Contract 14-16-0009-81-036) to U.S. Fish and Wildlife Service, Washington, D.C.

Rayner, J. N. 1971. An introduction to spectral analysis. Pion, London.

Roberts, L. G. 1965. Machine perception of three-dimensional solids. Pages 159–197 *in* J. T. Tippet, editor. Optical and electro-optical information processing. MIT Press, Cambridge, Massachusetts.

American Fisheries Society Symposium 7:78–83, 1990
© Copyright by the American Fisheries Society 1990

Stocking Checks on Scales as Marks for Identifying Hatchery Striped Bass in the Hudson River

MICHAEL HUMPHREYS,[1] RICHARD E. PARK, JAMES J. REICHLE,
AND MARK T. MATTSON

Normandeau Associates, Inc., 25 Nashua Road, Bedford, New Hampshire 03102, USA

DENNIS J. DUNNING AND QUENTIN E. ROSS

New York Power Authority, 123 Main Street, White Plains, New York 10601, USA

Abstract.—Acetate scale impressions from 171 hatchery-reared striped bass *Morone saxatilis*, which had been marked with magnetic binary-coded wire tags, stocked into the Hudson River, and recaptured 1–31 months later, were compared with scale impressions from wild Hudson River striped bass. This comparison was made to identify characteristic growth patterns that would provide an alternative method for separation of wild and stocked fish. Most scales from hatchery-reared fish had thick, widely spaced circuli near the focus that corresponded to rapid hatchery growth, followed by an abrupt growth check resulting from handling, tagging, and adaptation to natural food sources after release into the river. This "stocking check" pattern was present in varying degrees on scales of all hatchery fish that had been in the wild for at least 60 d ($N = 107$). Scales from wild fish showed more uniform circulus spacing corresponding to a more stable growth rate. In controlled tests, three experienced scale readers were able to correctly identify the origin (hatchery or wild) of young-of-year and yearling fish an average of 89% and 95% of the time, respectively. The hatchery striped bass detection efficiency achieved in these tests is close to the short-term (10-week) tag retention rates presently obtained with magnetic binary-coded wire tags.

Introduction of hatchery-reared fingerling striped bass *Morone saxatilis* into many historically important striped bass nursery areas has been undertaken on a large scale to supplement wild stocks. To evaluate the contribution of stocking efforts, hatchery-raised fish must be distinguishable from fish spawned naturally. An ideal mark for hatchery-produced striped bass fingerlings would be permanent, unmistakably recognizable, inexpensive, and easy to apply, and would have no effect on the fish's growth, mortality, behavior, or liability to capture by predators or fishing gear (Laird and Stott 1978). A method satisfying all of these criteria has not been available. Striped bass fingerlings stocked into the Hudson River estuary under contract with utility companies are currently marked with magnetic binary-coded wire tags. The primary drawbacks of this tag are that it requires sophisticated tagging and detection equipment, and each fish must be individually handled during tagging, which increases labor costs and may increase mortality.

During age and growth studies of Hudson River striped bass, we observed a consistent difference in the pattern of first-year scale growth between wild and recaptured hatchery fish before formation of the first annulus. Hatchery fish were characterized by a "stocking check" that appeared at approximately the time when fish were stocked and was believed to be the result of a brief cessation in growth caused by handling and tagging stresses and feeding changes associated with stocking. It became apparent that a consistently identifiable stocking check on scales of hatchery fish might provide an alternative to magnetic binary-coded wire tags. The objective of the current investigation was to evaluate the use of this stocking check as a reliable and consistent means of identifying recaptured hatchery striped bass up to 2 years after they were stocked in the river.

Methods

Scale samples from several year-classes of recaptured, magnetic-tagged hatchery striped bass were obtained from research programs conducted in the Hudson River estuary during 1986 and 1987. All scale samples were removed from the area midway between the lateral line and the notch between the spinous and soft dorsal fins. Date,

---

[1]Present address: Connecticut Department of Environmental Protection, Bureau of Fish and Wildlife, Inland Fisheries Division, State Office Building, 165 Capitol Avenue, Room 255, Hartford, Connecticut 06106, USA.

TABLE 1.—Frequency of occurrence (and percent) of diagnostic features used to identify scales of recaptured hatchery-produced striped bass from the Hudson River, New York.

| | | | Number (percent) of fish with | | | | |
Cohort	Recapture month	N	Change in circulus spacing	Deformed circuli	Dead-end circuli	All features weak	No growth after stocking
Young of year	Aug	14	2 (14)	1 (7)	2 (14)	2 (14)	9 (64)
	Sep	31	13 (42)	13 (42)	14 (45)	0 (0)	16 (52)
	Oct	14	13 (93)	13 (93)	12 (86)	0 (0)	0 (0)
	Nov	5	2 (40)	2 (40)	4 (80)	0 (0)	1 (20)
	Dec	8	4 (50)	2 (25)	6 (75)	1 (13)	0 (0)
	Jan	12	11 (92)	11 (92)	12 (100)	0 (0)	0 (0)
	Feb	6	6 (100)	6 (100)	6 (100)	0 (0)	0 (0)
	Mar	12	12 (100)	12 (100)	12 (100)	0 (0)	0 (0)
	Apr	4	4 (100)	4 (100)	4 (100)	0 (0)	0 (0)
Total	Aug–Apr	106	67 (63)	64 (60)	72 (68)	3 (3)	26 (25)
Age-1 fish	Dec–Apr	60	51 (85)	31 (52)	34 (57)	6 (10)	0 (0)
Age-2 fish	Dec–Apr	5	2 (40)	1 (12)	2 (40)	3 (60)	0 (0)

location where fish were collected, and total length of each fish were recorded. Acetate impressions of four to six scales from each fish were examined with the aid of a microfiche projector equipped with 47× magnification. First-year growth patterns on each set of scale impressions were examined to determine the features of the stocking check that were common to hatchery fish scales but absent from scales of wild fish. After preliminary determination of the best set of characteristics for identification of the stocking check, 171 hatchery fish scale samples were evaluated ($N = 106$, 60, and 5 fish of ages 0, 1, and 2, respectively). Samples were rated on the strength of each of these characteristics, and frequency of occurrence of each type of identifying feature was then calculated.

Scale radius and stocking check measurements (recorded to the nearest mm at 47× magnification) were made along a line from the focus to the scale's anterior edge. Measurements were recorded to permit back-calculation of fish length at time of check formation. This was done to verify that the check actually formed at the length these fish were stocked. The regression equation for the body length–scale radius relationship was calculated. The ordinate intercept from this equation, which is an estimate of the length at which scales begin to form, was used in the modified direct-proportion formula (Fraser 1916) to back-calculate the length at check formation:

$$L_n - a = \frac{S_n}{S}(L - a);$$

a = ordinate intercept from the regression equation;
L_n = length of fish at check formation;
L = length of fish when scale sample was obtained;
S_n = radius of check;
S = total scale radius.

Three individuals experienced at determining ages of Hudson River striped bass from scale impressions were trained to recognize scales of hatchery fish from the identified stocking check characteristics. After training, these individuals were each presented with two sets of test scales to evaluate their abilities to identify hatchery fish from scales alone. The first test set consisted of scales from 95 wild and 22 age-0 hatchery striped bass, and the second test set contained scales from 76 wild and 19 hatchery yearlings. The three test subjects had no prior knowledge of fish age or origin.

Results

Scales from 80 of 106 (75%) young-of-year hatchery striped bass exhibited some degree of check formation (Table 1). The 26 fish that showed no check were recaptured within 60 d after stocking and had apparently not grown enough since stocking to form the check. A check was observed on scales from 60 of 61 (98%) young of year recaptured after September, and on scales of all 42 (100%) young of year collected after

November. All 60 yearlings and 5 age-2 fish exhibited some degree of check formation before the first annulus.

The regression equation for the body length–scale radius relationship was

$$L = 1.95S + 10.97;$$

L = total fish length (mm);
S = total scale radius (mm) $\times$ 47 (microfiche magnification).

Back-calculation of length at check formation yielded mean lengths of 83, 77, and 79 mm for 1984, 1985, and 1986 year-classes, respectively. These back-calculated lengths were similar to the median stocking lengths (77–81 mm) for the three years (EA 1985, 1986, 1987). From these results, it appears likely that the growth check on scales of hatchery fish occurred at approximately the time of stocking.

Three scale features were associated with the stocking check on the majority of hatchery fish scales that exhibited a check. The strongest diagnostic characteristic was a change in circulus spacing before and after the check (Figure 1). Circuli formed during hatchery growth before stocking were thick and widely spaced. This pattern corresponded to rapid hatchery growth (2–3 mm/d) sustained by automatic 24-h feeding. After the check, circuli were usually thinner and more closely spaced. This pattern corresponded to reduced feeding and slower growth after stocking. This abrupt change in growth was strong in 85% ($N = 79$) of young-of-year hatchery fish that exhibited growth after stocking. This feature was also present in 85% ($N = 60$) of the yearling hatchery fish and 40% ($N = 5$) of age-2 hatchery fish examined (Table 1). The change in growth appeared unique to stocked hatchery striped bass in the Hudson River and did not occur in the vast majority of the more than 3,000 wild Hudson River striped bass scales examined during concurrent age and growth studies.

A second feature that aided in the identification of the stocking check was a distinct line of distorted and fragmented circuli through the scale's anterior field (Figure 1). This line separated the zones of hatchery and river growth and corresponded to the actual check in growth associated with reduced feeding before tagging, handling and tagging stress, and adaptation to new food sources and feeding behavior after stocking. This feature was strong on 67% of the hatchery striped bass

scales with growth after stocking. Fragmented and distorted circuli are often used to identify annuli and are not unique to stocking check formation in hatchery striped bass.

Cutoff or "dead-end" circuli were associated with the stocking check in the lateral fields of many of the scales from hatchery fish. This feature, typically seen as several short circuli abutting the next complete outer circulus (Figure 1), was identified on scales from 74% of the hatchery fish with growth past the check. Dead-end circuli were slightly different from circulus cross-over, which occurs in the same area of the scale during annulus formation. Cross-over is the fusion of long lateral circuli with an annulus that angles across their posterior ends. In contrast, dead-end circuli are short and terminate where the anterior ends abut an outer circulus without fusing with it.

The three test subjects were able to correctly identify the origin (hatchery or wild) of test fish an average of 89% and 95% of the time for young-of-year and yearling striped bass, respectively (Table 2). Of 54 errors, 38 (70%) were the result of assigning hatchery origin to wild fish. This type of error generally occurred when weak natural growth checks were present on scales of wild fish. A disproportionately large number of these errors (50%) were attributable to one overzealous individual who was less skilled at hatchery-versus-wild discrimination during the first test (young-of-year fish). This individual improved markedly on the second test and correctly identified the origin of all but four yearling fish (96% correct).

Sixteen errors resulted from failure to observe the stocking check on scales of hatchery fish. Missed stocking checks generally had weak diagnostic characteristics or little growth beyond the check for comparison of circulus spacing. No errors resulted from mistaking stocking checks for annuli or vice versa.

Discussion

The stocking check on scales from the majority of hatchery-produced Hudson River striped bass offers potential as a means of recognition of hatchery fish. This method has several advantages over magnetic binary-coded wire tags. Fish scales are mass-marked by normal hatchery procedures. No additional time, equipment, manpower, or fish handling are required. The stocking check is permanent and may persist throughout the life of a fish. Although the oldest hatchery fish examined

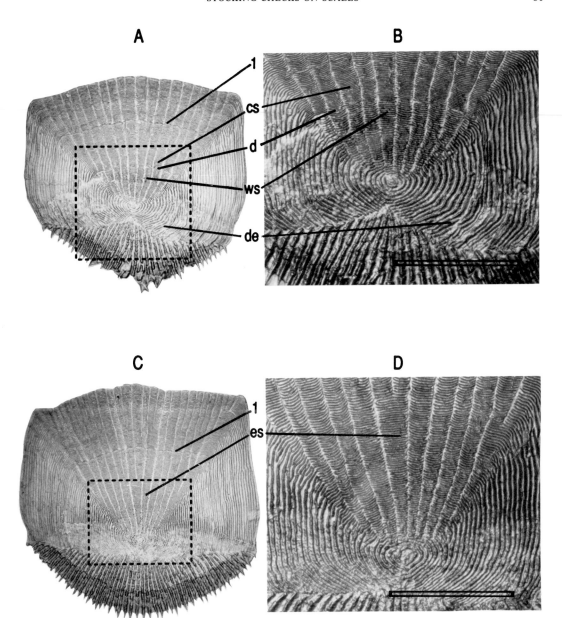

FIGURE 1.—Scales from hatchery and wild striped bass. (A) Scale from a hatchery-produced striped bass that was stocked into the Hudson River as a young of year and recaptured as a yearling. Features used to identify scales of hatchery striped bass are indicated: de = dead-end circuli associated with the stocking check; ws = widely spaced circuli that correspond to rapid growth in the hatchery; d = deformed circuli that correspond to the stocking check; cs = closely spaced circuli that indicate reduced growth rate after stocking; 1 = first annulus. (B) Enlargement of the center of scale A (indicated by square). Bar = 1 mm. (C) Scale from a wild yearling Hudson River striped bass with evenly spaced circuli (es). The first annulus (1) is also indicated. (D) Enlargement of the center of scale C (indicated by square). Bar = 1 mm.

were only age 2, it appears unlikely that the clarity of stocking checks will diminish with time. In contrast, magnetic tag retention studies indicate that 6.5–14.7% of hatchery fish lose their tags within the first 10 weeks after stocking (EA 1985, 1986, 1987). The long-term tag shedding rate is unknown. The number of tags that may lose their magnetism over time and thus go undetected also is unknown.

TABLE 2.—Assignment of hatchery or wild origin to age-0 and age-1 Hudson River striped bass by examination of scale patterns.

Cohort	N	Assigned by[a]	Number of fish assigned to		Number of fish assigned incorrectly as		Error rate (%)
			Hatchery	Wild	Hatchery	Wild	
Young of year	117	Reader A	26	91	6	2	7
		Reader B	23	94	6	5	10
		Reader C	39	78	18	1	16
		Actual origin	22	95			
Age-1 fish	95	Reader A	16	79	1	4	5
		Reader B	24	71	6	1	7
		Reader C	17	78	1	3	4
		Actual origin	19	76			

[a]Readers are persons trained in aging fish from scale patterns.

Detection of magnetic tags in the field requires electronic tag-detection devices. Suspected hatchery recaptures must be sacrificed and dissected to verify the presence of a magnetic tag because fishhooks and fish with magnetic tags from other release sources may be present (Mattson et al. 1990, this volume). Recognition of stocking checks requires only conventional scale-aging equipment, and fish may be released alive after a scale sample is removed.

Binary-coded magnetic tags have some advantages over the stocking check for recognizing hatchery striped bass. The binary code permits identification of different batches of stocked fish within the same year-class. Thus, evaluation of growth and survival of fingerlings stocked at different locations and on different dates is possible. The stocking check, in combination with conventional aging techniques, only permits identification of hatchery year-classes. Recently stocked young-of-year hatchery fish are difficult to identify by examination of scales. Significant growth after stocking is necessary for recognition of the stocking-check pattern. Some fish do not show significant scale growth for up to 60 d after release. Thus, with the stocking-check method, the contribution of hatchery fish could not be accounted for in young-of-year abundance indices. Young-of-year magnetic-tagged fish, however, are no more difficult to detect than older magnetic-tagged hatchery fish.

Individuals responsible for identification of stocking checks must be experienced at determining ages of striped bass from scales. They must also be carefully trained to recognize the combination of features that characterize the stocking check. Large numbers of wild as well as hatchery fish scales must be examined in the laboratory in order to identify the hatchery fish. In contrast, field detection of magnetic tags is fast and requires little technical skill; only suspected recaptures must be further examined in the laboratory.

The strength and consistency of the stocking check probably depends on hatchery conditions, which may vary from year to year. Factors such as disease or overcrowding may reduce the growth rate and make the stocking-check pattern less recognizable. Elimination of the magnetic-tag application procedure itself may weaken the stocking-check characteristics. Conversely, purposeful manipulation of hatchery conditions (e.g., cold shock, a brief period of starvation, or both) may enhance the resolution of stocking-check characteristics and improve recognition accuracy.

Acknowledgments

Funding for this project was provided by Central Hudson Gas and Electric Corporation, Consolidated Edison Company of New York, Inc., the New York Power Authority, Niagara Mohawk Power Corporation, and Orange and Rockland Utilities under terms of the Hudson River Cooling Tower Settlement Agreement. This project was developed in cooperation with the New York State Department of Environmental Conservation. Field and laboratory work was conducted by Normandeau Associates, Inc., and verification of hatchery fish was conducted by EA Science and Technology.

References

EA (Ecological Analysts Inc.). 1985. Hudson River striped bass hatchery 1984 overview. Report to

Central Hudson Gas and Electric Corporation (and others), Poughkeepsie, New York.

EA (Ecological Analysts Inc.). 1986. Hudson River striped bass hatchery 1985 overview. Report to Central Hudson Gas and Electric Corporation (and others), Poughkeepsie, New York.

EA (Ecological Analysts Inc.). 1987. Hudson River striped bass hatchery 1986 overview. Report to Central Hudson Gas and Electric Corporation (and others), Poughkeepsie, New York.

Fraser, C. M. 1916. Growth of the spring salmon. Pages 29–39 in Transactions of the Pacific Fisheries Society Seattle, for 1915. Seattle, Washington.

Laird, L. M., and B. Stott. 1978. Marking and tagging. IBP (International Biological Programme) Handbook 3:84–100.

Mattson, M. T., B. R. Friedman, D. J. Dunning, and Q. E. Ross. 1990. Magnetic tag detection efficiency in Hudson River striped bass. American Fisheries Society Symposium 7:267–271.

American Fisheries Society Symposium 7:84–93, 1990

Evaluation of Snow Crab Tags Retainable through Molting

G. V. Hurley

Hurley Fisheries Consulting Ltd.
45 Alderney Drive, Suite 815, Dartmouth, Nova Scotia B2Y 2N6, Canada

R. W. Elner

Department of Fisheries and Oceans, Biological Sciences Branch
Post Office Box 550, Halifax, Nova Scotia B3J 2S7, Canada

D. M. Taylor

Department of Fisheries and Oceans, Biological Sciences Branch
Post Office Box 5667, St. John's, Newfoundland A1C 5X1, Canada

R. F. J. Bailey

Department of Fisheries and Oceans, Biological Sciences Branch
Maurice Lamontagne Institute, Mont-Joli, Quebec G5H 3Z4, Canada

Abstract.—Laboratory tests were performed to evaluate T-bar and magnetic tags for morphometrically immature male snow crabs *Chionoectes opilio*. Molting mortality for T-bar-tagged crabs ranged from 100% for crabs with deeply implanted tags to 16% for shallow insertions. Many crabs that completed ecdysis had severe deformities, such as missing or malformed legs and protruding gills. In comparison, over 90% of the untagged controls molted successfully. Magnetic tags, injected into the dactylus of a walking leg, resulted in negligible molting mortality. However, after ecdysis, the magnetically tagged leg was usually deformed, and in many cases the tag was either left behind in the cast shell or lost through rotting of the dactylus. There was no significant difference in the mean percentage growth per molt between crabs tagged with either of the two tag types or between tagged crabs and the untagged control group. We conclude that unless a correction is made for the poor effective retention rate for both tag types, biases could result in estimates of biomass and exploitation rate for snow crabs.

The snow crab *Chionoecetes opilio* is a large, deepwater spider crab found in the northwest Atlantic and north Pacific oceans and the Sea of Japan. Snow crabs are a valuable resource in eastern Canada and the subject of much research (Elner 1982; Elner and Bailey 1986; Bailey and Elner 1989).

Tags have been extensively used in studies of snow crab growth and movement over the past 20 years (Tanino and Ito 1968a; 1968b; Watson 1970; Taylor 1982). In addition, estimates of biomass and exploitation rates obtained from mark–recapture studies have been used as a basis for management of snow crab fisheries (Bailey 1978; Elner and Robichaud 1983; Bailey and Dufour 1987). However, crabs are difficult to tag because they periodically molt. All exoskeletal parts are lost at ecdysis; thus, conventional marks are of limited value, and a tag is required that can be retained through successive molts. Also, the tag should be suitable for rapid application in the

field, not injure the crab, and be readily identifiable at recapture (see Farmer 1981 for review).

Two types of tags with the potential to be retained through ecdysis have been used on snow crabs. The T-bar tag, although anchored within the crab's body, has a visible external leader. The tag is applied through the posterior ecdysial suture between the carapace and the first abdominal segment (Anonymous 1970). In comparison, the magnetic tag is minute and is injected into the dactylus muscle of a walking leg (Bailey and Dufour 1987).

Information on snow crab growth provided by T-bar and magnetic tags has been scanty (McBride 1982; Bailey and Dufour 1987), and there is concern that this may be due to problems with the actual tagging methods. Problems that can affect the overall effective retention rate of tags include mortality, tag loss, migration, injuries and deformities, lack of cooperation by fishermen in retrieving tags, and inadequate monitoring of landings for tagged animals. Discrimination and quantification of the relative importance of each of these problems is impractical in the field. Biologists, when estimating biomass from tag returns, have either discounted the returns entirely (Elner and Robichaud 1983) or compensated for them with a single correction factor (Bailey and Dufour 1987).

Laboratory studies are useful for testing tags because individual crabs can be followed through molting, and controlled experiments can be conducted to compare tagged and untagged crabs. Despite the widespread use of tags on crabs, there have been few validation studies conducted in the laboratory (for *Callinectes sapidus*, Fannaly 1978 and van Montfrans et al. 1986; for *Cancer pagurus*, Edwards 1965). In a preliminary study (Fujita and Takeshita 1979), T-bar-tagged snow crabs were observed molting in aquaria. One of three crabs tagged through the posterior ecdysial suture died, and five of seven crabs tagged through the carapace lost tags, died, or were deformed after ecdysis. On the basis of their results, Fujita and Takeshita (1979) recommended additional research on the tag and insertion method. Apart from limited work by Bailey and Dufour (1987), there has been no laboratory testing of magnetic tags for snow crabs.

The purpose of our study was to evaluate T-bar and magnetic tags for snow crabs under laboratory conditions. Tags were evaluated in terms of their retention properties and possible effects on growth and mortality.

Methods

Male snow crabs were collected by trawling in the Bay of Chaleur, Gulf of St. Lawrence, on November 28, 1984. Once in the laboratory, crabs were held in filtered, aerated, flowing seawater of approximately 30‰ salinity. They were kept communally until mid-January 1985, when they were transferred to individual compartments (0.4 × 0.4 × 0.15 m deep).

Water temperature decreased from 8° to below 3°C between November 1984 and January 1985, and remained at 4–6°C between February and June 1985. Temperatures fluctuated up to as high as 9–10°C in late June and mid-September 1985. Crabs were fed ad libitum every second day with pelletized commercial lobster food supplemented by occasional feedings of mussels and clams.

Tagging took place between January 31 and April 24, 1985. Because snow crabs may undergo a terminal molt at morphometric maturity, only morphometrically immature males (according to

the chela allometry criteria described in Conan and Comeau 1986) were used so that retention of the tags through the molt could be assessed. The size of each crab was recorded by measuring carapace width (to the nearest 0.1 mm with vernier calipers) at tagging and after molting when the shell had hardened.

The major experiment was to evaluate the standard T-bar and magnetic tags and application methods used in Canada. Fifteen snow crabs were tagged with each type of tag, and 15 untagged crabs of similar size were maintained as controls. Subsequently, additional snow crabs were tagged according to modification of the standard methods (described in Table 1). The modified methods were designed to either verify observations from the major experiment or improve the effective retention rate of the tag.

The T-bar tag (Floy Tag and Manufacturing, Seattle, Washington), consists of a vinyl toggle (10 mm long, 1.2 mm in diameter) and shaft (16 mm long, 0.5 mm in diameter) that connect to a 4-mm-long tapered cone and a 55-mm length of number 20 vinyl tubing printed with an individual serial number. A clip of 25 tags loads into a tagging gun equipped with a hollow injector needle that is 20 mm long and 1.8 mm in diameter. The injector needle was pushed through the posterior ecdysial suture (epimeral line), 2–6 mm from either the right or left coxopodite of the last walking leg, and the tag was inserted. For all T-bar-tagged crabs except the standard T-bar group, the anchor and shaft portions of the tag were painted silver, and each crab was X-rayed immediately after tagging and at death (Figure 1A). Crabs in the other T-bar-tagged groups were tagged with either deeply and shallowly implanted tags and tested for differences in survivorship. In the deep T-bar treatment, toggles were positioned in the basal leg musculature on the left side of the crab; in the shallow T-bar treatment, tags were placed in the dorsal muscle above the interabdominal skeleton.

The magnetic tags consisted of 2-mm segments cut from a stainless steel wire 0.356 mm in diameter. The tags were magnetized then implanted with an automatic injector (Northwest Marine Technology, Shaw Island, Washington). The presence of the tag in the crab was checked after tagging and periodically throughout the study with a detector that emitted an acoustical signal. All crabs except the standard magnetic group were X-rayed to determine the exact position of the tag after tagging, molting, and death. The standard magnetic crabs were X-rayed only after death.

TABLE 1.—Summary of tagging techniques, number and size of crabs, tagging dates, and tagging-to-molt interval for morphometrically immature male snow crabs tagged with magnetic and T-bar tags.

Treatment group	Location of tag	N	Carapace width (range in mm)	Date tagged	Mean interval (d) between tagging and molting (range)
Control, untagged (premolt)		15	60.7–72.5		
Standard magnetic (premolt)	Right side, dactylus of third periopod	15	58.3–71.7	Feb 10	14.9 (3–29)
Dactylus-L magnetic (premolt)	Left side, dactylus of third periopod	8	62.7–75.2	Mar 7	9.4 (4–19)
Dactylus-L magnetic (postmolt)	Left side, dactylus of third periopod	8	69.2–81.2	Apr 9	
Body magnetic (postmolt)	Left side, basal leg muscles	8	59.9–90.7	Apr 9	
Standard T-bar (premolt)	Right branchial region	15	60.4–72.3	Jan 31	22.5 (5–46)
Deep T-bar (premolt)	Left side, basal leg muscles	8	67.4–81.0	Mar 7[a] Mar 11[a]	7.9 (3–18)
Shallow T-bar (premolt)	Right side dorsal muscle	6	61.2–83.4	Mar 19[a] Mar 29[b]	8.0 (3–18)
Shallow T-bar (postmolt)	Right side, dorsal muscle	8	67.6–80.4	Apr 24	

[a]Four crabs tagged on this date.
[b]Two crabs tagged on this date.

Survival was analyzed statistically with contingency tables; percent growth was analyzed with regression and Bartlett's test for homogeneity of group variances. Because the regressions of percent growth against size were not significantly different but the variances among groups were, the growth data was analyzed with pairwise t-tests for independent samples with unequal variances (Steel and Torrie 1980).

Results

All snow crabs molted in the laboratory with the exception of three crabs (one in each of the three major groups) that died before ecdysis.

Mortality

Seven of 14 standard T-bar snow crabs died while molting. Three of the deaths were due to errors in tagging technique (Figure 1B, C); these crabs could not pull their tags free to complete ecdysis. Three additional crabs managed only to extricate their legs on the untagged side of the body and died in a twisted position (Figure 1D). Dissection of these three crabs revealed tags that were deeply embedded in the basal leg muscles with the toggle penetrating a leg septum. Another crab backed most of

its carapace out of the old shell, but failed to molt further and died (Figure 1E).

In the deep T-bar group, all eight crabs died during molting. Most of these progressed no further than to swell and lift their old carapace before dying, whereas others died in a twisted position, and only their legs on the right side were removed from the old shell. These results contrasted markedly with the lower molt mortalities of the shallow T-bar (premolt) group (1 death in 6) and the untagged control crabs (1 death in 14) (Figure 2). In the shallow T-bar group, one of the five surviving crabs lost its two distal walking legs during molting, and the single crab that died failed to completely rupture its pericardial sac (Figure 1H). One control crab died at ecdysis when it was caught under a tank wall. A comparison of the control, standard T-bar, and deep T-bar groups revealed that survival through the molt was significantly different among the groups ($P < 0.05$). There was no significant differences ($P > 0.05$) in survival rate between magnetic-tagged crabs and control crabs.

The standard magnetic and dactylus-L (premolt) groups experienced molt mortalities of 2 deaths in 14 and 0 deaths in 8, respectively, versus 1 death in 14 for the control group (Figures 3 and 4).

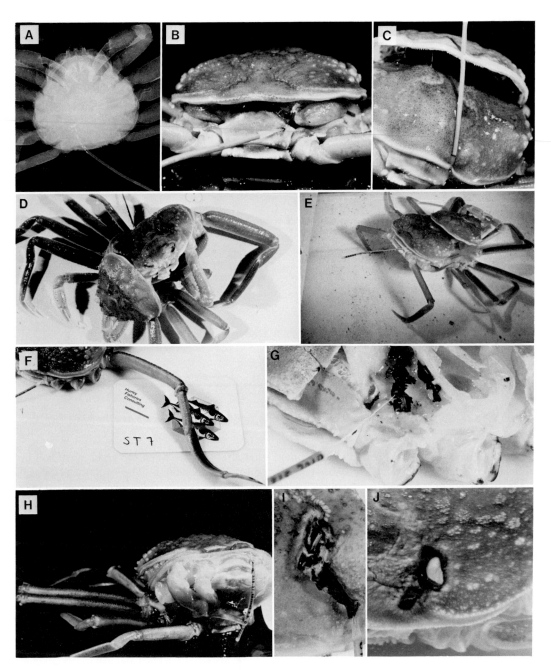

FIGURE 1.—Location of T-bar tags and problems associated with them in morphometrically immature male snow crabs. Length of tag cone is 4 mm; length of vinyl leader is 55 mm. (A) X-ray photograph showing implanted T-bar tag. Note position of toggle and shank of tag. (B) Tag inserted through the first abdominal segment, rather than through epimeral line, preventing retraction of new carapace at ecdysis. (C) Tag wedged in the margin of the old carapace due to damage at tagging, preventing retraction of new carapace. (D) Crab unable to withdraw legs from old exoskeleton on tagged side during molting. (E) Crab failing to shed old shell completely during ecdysis, possibly because the tag delayed the rate of withdrawal. (F) Deformities (exposed gill above leg and grossly malformed leg) plus three missing legs of tagged crab that completed ecdysis. (G) Dissection of crab that died during ecdysis because the toggle was trapped in the midlateral septum between the hemocoel and basal muscle of the third leg. Note the black deposit around the toggle. (H) Death caused by failure of the crab to rupture its pericardial sack and withdraw from the old shell. (I) Dissection 3 months after successful molting, showing cementation of shallowly implanted T-bar tag to the inside of the carapace. Note black deposits around tag shank and part of the toggle. Encapsulation by such deposits prevents tag retention in a subsequent molt. (J) External view of a carapace 3 months after successful molting with a shallowly implanted T-bar tag. Note the black deposit outlining the internal portion of the tag and the muscle protruding through a hole in the carapace.

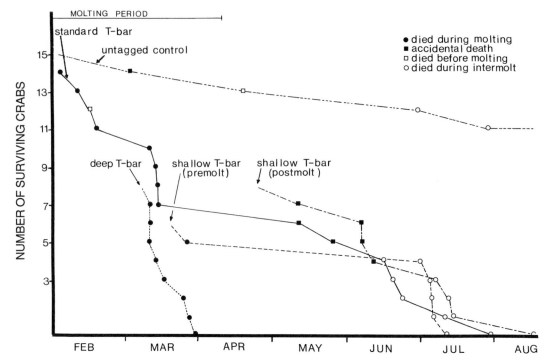

FIGURE 2.—Survival of T-bar-tagged and control snow crabs. Tag treatments are described in Table 1.

After molting, there was a period (April–June) of relatively stable survival. Five T-bar-tagged crabs (two standard T-bar and three shallow T-bar [postmolt]) were killed when their tanks accidentally drained (Figure 2). The deaths of most of the surviving tagged crabs and two control crabs were coincident with a rise in water temperature to 9–10°C (Figures 2 and 3). For T-bar-tagged and control crabs, mortality during the period of elevated water temperature was significantly different ($P < 0.05$); however, for magnetically tagged and control crabs, there was no significant difference ($P > 0.05$).

Tag Loss

In the T-bar groups, there was no loss of tags during the postmolt period.

Although only 2 of 22 crabs marked with magnetic tags (14 standard and 8 dactylus-L crabs) died while molting, 15 of the remaining 20 crabs lost tags during this period (Figure 3). Tag loss occurred when the tagged leg was lost at ecdysis or when the tag was left behind in the cast shell (Figure 5).

The magnetically tagged crabs continued to lose tags through rotting of the dactylus after molting (Figure 5). Losses occurred within a 4-week pe-

riod after ecdysis when the crabs were still in a soft-shelled condition. The bacteria responsible for the rot was a coldwater marine strain of *Pseudomonas* sp.

Abnormalities

Percentages of abnormalities associated with T-bar tags are shown in Figure 6. Seven crabs in the standard T-bar group molted successfully but were left with abnormalities such as missing or deformed legs and gills outside the carapace (Figure 1F). At dissection, it was observed that these crabs were not tagged as deeply as those that had died during molting. Most of the tag anchors were encapsulated with a black deposit in the hemocoel region above the third and fourth walking legs (Figure 1G).

Many of the crabs tagged with magnetic tags had abnormalities related to the tagging process (Figure 4). For example, legs lost at ecdysis were always on the tagged side of the body and always included the tagged leg. Individual crabs lost up to three legs during molting. Five of the 14 standard magnetic and one of the eight dactylus-L (premolt) crabs lost legs during the molt. Seven crabs in the standard magnetic group retained all their legs through the molt; however,

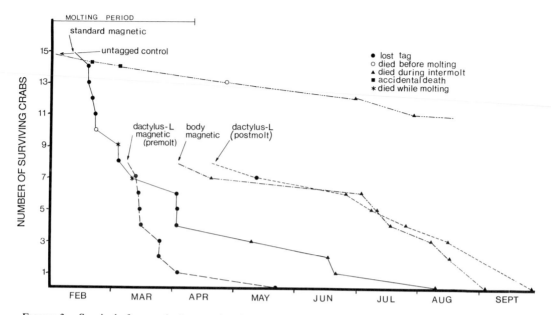

FIGURE 3.—Survival of magnetically tagged and control snow crabs. Tag treatments are described in Table 1.

five had a deformed tagged dactylus (Figure 7A). The tag pierced the surface of the dactylus in three crabs (Figure 7B). Three crabs were double-tagged; one of these died at molting, and the other two lost their tagged leg at molting. Another crab lost its tagged leg and was left with two deformed legs that were flattened and twisted at the merus.

Tag Encapsulation

Most T-bar-tagged crabs that died during postmolt had the toggle and shaft portions of their tags encrusted with a black deposit. The deposit often accumulated on the crab itself at the point of tag insertion, on the internal septa, and on the carapace. The most extensive accumulations were observed in some of the shallow T-bar (premolt) group, especially where the tag toggle adhered to the underside of the carapace. In such cases, the tag shaft and toggle became cemented to the shell (Figure 1I) and holes could be produced in the carapace (Figure 1J).

As with the T-bar tags, magnetic tags could become cemented to the dactylus shell. Blackening around the tag was often visible externally and sometimes extended to other parts of the leg (Figure 7C); however, in a few cases the deposit was white. Tags that were entirely surrounded by dactylus or basal leg muscles (body magnetic group) were free of deposit.

Growth per Molt

The slopes of the regression lines of percentage growth per molt versus size were not significantly different from zero ($P > 0.05$) for crabs marked with either of the two tag types or the control crabs (Figure 8; Table 2). Although there were significant but unexplainable differences in variance ($P < 0.01$), the mean increase in size was not significantly different among the three groups ($P > 0.05$).

Discussion

Our tests indicated that tagged crabs suffer higher mortalities than untagged crabs, that tags are frequently lost, and that abnormalities induced by tagging may decrease the vulnerability of crabs to recapture. These factors, unless corrected for, would effectively decrease the ratio of tagged to untagged crabs in samples and result in overestimates of population abundance and underestimates of exploitation rate (Ricker 1975).

The shallow-implanted T-bar and magnetic tags may be useful for studies of growth per molt. We observed no significant difference in percentage increase in carapace width between tagged crabs and the control group. Despite high variability in the data, which could be tag-induced, our growth values are similar to those from field studies with magnetic (Bailey, unpublished data) and T-bar

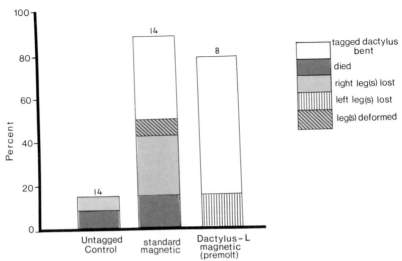

FIGURE 4.—Mortalities and abnormalities of snow crabs caused by magnetic tags. Sample size is above each bar; tag treatments are described in Table 1.

tags (Taylor 1982, unpublished data; Elner, unpublished data), as well as those from laboratory observations (Miller and Watson 1976; O'Halloran 1985). However, because of the low effective retention rates, investigators using T-bar or magnetic tags in the field should make ample allowance for tag losses so that meaningful quantities of growth data are retrieved. A similar allowance, plus consideration of the effects of tag-induced abnormalities on mobility, would apply for field studies into long-term movements.

The identified problems influencing the effective retention rate are difficult to correct for, and a reliable mark that can be retained through the molt is still needed for snow crabs. However, modifications in the design or method of application of the existing tags could lead to improved retention in the future. For example, alternative

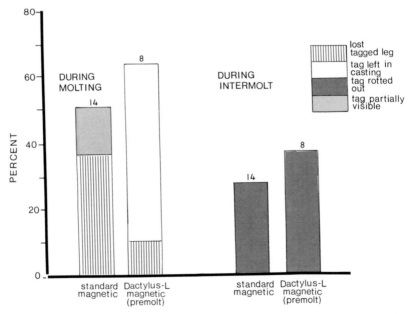

FIGURE 5.—Percent loss of magnetic tags from snow crabs during molting and the subsequent intermolt period. Sample size is above each bar; tag treatments are described in Table 1.

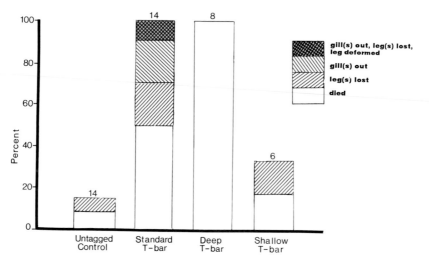

FIGURE 6.—Mortalities and abnormalities in snow crabs caused by T-bar tags. Sample size is above each bar; tag treatments are described in Table 1.

toggle designs may reduce the incidence of cementation of shallowly implanted T-bar tags to the carapace wall. While improved molt-retainable marks are being developed, we suggest that the simple spaghetti tag, consisting of vinyl tubing tied around the cephalothorax, could be reintroduced. The benefits of spaghetti tags are that they can be used without internal injury, and they are conspicuous, economical, and rapidly applied. A major disadvantage, their patent lack of retention through molting, may not be as severe a problem as previously perceived. O'Halloran (1985) and

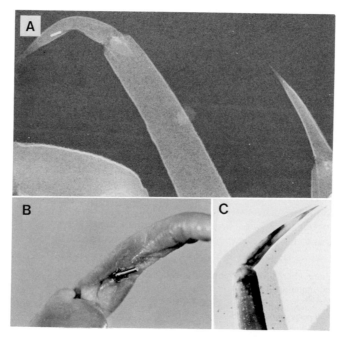

FIGURE 7.—Abnormalities in snow crabs caused by magnetic tags. (A) X-ray photograph showing a normal untagged dactylus and tag inside a deformed dactylus after molting. (B) Tag protruding through a deformed dactylus 3 months after molting; the crab was tagged 8 d before ecdysis. (C) Black discoloration on a tagged dactylus 3 months after tagging; the crab had been tagged shortly after ecdysis. Tags are 2 mm long.

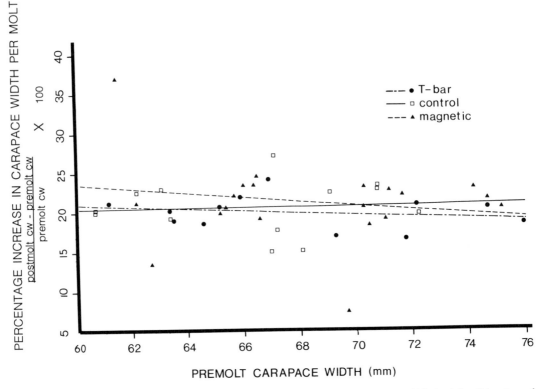

FIGURE 8.—Regression of percentage growth per molt versus premolt carapace width (cw) for T-bar-tagged, magnetically tagged, and control snow crabs.

Conan and Comeau (1986) showed that male snow crabs undergo a terminal molt at morphometric maturity. If morphometrically mature males can be identified from their chelal morphometry, it is feasible to use spaghetti tags as long-term marks on such crabs to assess movement patterns, biomass, and exploitation rates. Male snow crabs with spaghetti tags have been recovered up to 7 years after release off eastern Cape Breton Island (Elner and D. A. Robichaud, unpublished data) and after 16 years in Trinity Bay, Newfoundland (Taylor, unpublished data).

The lessons from our study are broad and applicable to all mark–recapture experiments.

There should be rigorous laboratory testing of tags before they are used in the field, and potential biases in the effective retention rate should be understood and quantified so results from tagging studies can be interpreted accurately.

Acknowledgments

R. K. O'Dor, T. P. (Aero) Foyle, and S. Koshio of Dalhousie University gave assistance and helpful suggestions. N. Balch and the staff of the Aquatron Laboratory provided facilities. A. Dorion, Department of Fisheries and Oceans, and Captain A. Gionet were instrumental in capturing the snow crabs. M. Primrose and L. V. Colpitts of

TABLE 2.—Least-squares regressions for percent growth per molt versus premolt carapace width in millimeters for tagged and untagged immature male snow crabs.

Tag type	Mean percent growth per molt (SD)	N	Regression parameters		r^2
			Intercept	Slope (P^a)	
T-bar	20.0 (2.1)	12	29.04	−0.134 (>0.05)	0.09
Magnetic	21.3 (5.4)	20	40.51	−0.282 (>0.05)	0.05
None	20.8 (3.4)	13	19.26	0.023 (>0.05)	0.001

[a]Null hypothesis, slope = 0.

Dalhousie University were responsible for photographic processing and graphics, respectively. The X-rays were made at the Center for Marine Geology, Dalhousie University, and the Victoria General Hospital, Halifax. The research was performed under an unsolicited proposal award from the Department of Fisheries and Oceans and Department of Supply and Services, Canada, to Hurley Fisheries Consulting Ltd.

References

Anonymous. 1970. Tanner crab tagged successfully for first time. U.S. National Marine Fisheries Service Commercial Fisheries Review 33(3):19.

Bailey, R. 1978. Analysis of the snow crab population in northwestern Cape Breton, 1978. Canadian Atlantic Fisheries Scientific Advisory Committee, Research Document 78/41, Dartmouth, Canada.

Bailey, R. F. J., and R. Dufour. 1987. Field use of an injected ferromagnetic tag on the snow crab (*Chionoecetes opilio* O. Fab.). Journal du Conseil, Conseil International pour L'Exploration de la Mer 43:237–244.

Bailey, R. F. J., and R. W. Elner. 1989. Northwest Atlantic snow crab fisheries: lessons in research and management. Pages 261–280 *in* J. F. Caddy, editor. Marine invertebrate fisheries: their assessment and management. Wiley, New York.

Conan, G. Y., and M. Comeau. 1986. Functional maturity and terminal molt of snow crab, *Chionoecetes opilio*. Canadian Journal of Fisheries and Aquatic Sciences 43:1710–1719.

Edwards, E. 1965. Observations on growth of the edible crab (*Cancer pagurus*). Rapports et Procès-Verbaux des Réunions, Conseil International pour L'Exploration de la Mer 156:62–70.

Elner, R. W. 1982. Overview of the snow crab *Chionoecetes opilio* fishery in Atlantic Canada. Pages 3–19 *in* Proceedings of the international symposium on the genus *Chionoecetes*. Lowell Wakefield Fisheries Symposia Series. University of Alaska, Alaska Sea Grant Report 82-10, Fairbanks.

Elner, R. W., and R. F. J. Bailey. 1986. Differential susceptibility of Atlantic snow crab, *Chionoecetes opilio*, stocks to management. Canadian Special Publication of Fisheries and Aquatic Sciences 92:335–346.

Elner, R. W., and D. A. Robichaud. 1983. Status of the snow crab resource off the Atlantic coast of Cape Breton Island, 1982. Canadian Atlantic Fisheries Scientific Advisory Committee, Research Document 83/5, Dartmouth, Canada.

Fannaly, M. T. 1978. A method for tagging immature blue crabs (*Callinectes sapidus* Rathbun). North-east Gulf Science 2:124–126.

Farmer, A. S. D. 1981. A review of crustacean marking methods with particular reference to penaeid shrimp. Kuwait Bulletin of Marine Science 2:167–183.

Fujita, H., and K. Takeshita. 1979. Tagging technique for tanner crab long-term tag. Bulletin of the Far Seas Fisheries Research Laboratory 17:223–226.

McBride, J. 1982. Tanner crab tag development and tagging experiments 1978–1982. Pages 383–403 *in* Proceedings of the international symposium on the genus *Chionoecetes*. Lowell Wakefield Fisheries Symposia Series. University of Alaska, Alaska Sea Grant Report, 82-10, Fairbanks.

Miller, R. J., and J. Watson. 1976. Growth per molt and limb regeneration in the spider crab, *Chionoecetes opilio*. Journal of the Fisheries Research Board of Canada 33:1644–1649.

O'Halloran, M. J. 1985. Molt cycle changes and the control of molt in male snow crab, *Chionoecetes opilio*. Master's thesis, Dalhousie University, Halifax, Canada.

Ricker, W. E. 1975. Computation and interpretation of biological statistics of fish populations. Fisheries Research Board of Canada Bulletin 191.

Steel, R. G. D., and J. H. Torrie. 1980. Principles and procedures of statistics. McGraw-Hill, New York.

Tanino, Y., and K. Ito. 1968a. Studies on the tagging experiments of the zuwai-crab, *Chionoecetes opilio* O. Fabricius, in the Japan Sea. I. Comparisons of the tagging methods. Bulletin of the Japan Sea Regional Fisheries Research Laboratory 20:35–41.

Tanino, Y., and K. Ito. 1968b. Studies on the tagging experiments of the zuwai-crab, *Chionoecetes opilio* O. Fabricius, in the Japan Sea. II. Considerations on the validity of a currently used tagging method in relation to the limb loss and regeneration. Bulletin of the Japan Sea Regional Fisheries Research Laboratory 20:43–48.

Taylor, D. M. 1982. A recent development in tagging studies on snow crab, *Chionoecetes opilio*, in Newfoundland—retention of tags through ecdysis. Pages 405–417 *in* Proceedings of the international symposium on the genus *Chionoecetes*. Lowell Wakefield Fisheries Symposia Series. University of Alaska, Alaska Sea Grant Report 82-10, Fairbanks.

van Montfrans, J., J. Capelli, R. J. Orth, and C. K. Ryer. 1986. Use of microwire tags for tagging juvenile blue crabs (*Callinectes sapidus* Rathbun). Journal of Crustacean Biology 6:370–376.

Watson, J. 1970. Tag recaptures and movements of adult male snow crabs *Chionoecetes opilio* (O. Fabricius) in the Gaspé region of the Gulf of St. Lawrence. Fisheries Research Board of Canada Technical Report 204.

American Fisheries Society Symposium 7:94–100, 1990

Effectiveness of the Australian Western Rock Lobster Tag for Marking Juvenile American Lobsters along the Maine Coast

Jay S. Krouse and Glenn E. Nutting

Maine Department of Marine Resources, West Boothbay Harbor, Maine 04575, USA

Abstract.—The suitability of the western rock lobster tag for long-term (>1 year) identification of juvenile American lobsters *Homarus americanus* was evaluated under laboratory and field conditions. This spaghetti tag's oblong plastic anchor was inserted beneath the dorsal musculature lying between the cephalothorax and abdomen. In the laboratory, small juveniles (25–39 mm in carapace length) marked with miniature tags retained 97% of the tags after undergoing an average of 2.3 molts (maximum, 5). Larger juveniles (39–80 mm) marked with standard-size tags had an overall retention of 89% (2.7 molt average). The tag caused no adverse effect on growth and survival. However, if the pericardial sac or hepatopancreas was ruptured during tag implantation, death was likely. From 1979 through 1987, we tagged and released 10,191 lobsters (25–80 mm) of which 1,858 (18.2%) were recaptured by research personnel and commercial fishermen. Of these recoveries, 1,193 lobsters had molted at least once, 28 showed carapace length increases of at least 100% (an increase indicating at least five molts), and 58 had been at liberty for more than 3 years. These observations indicate that the western rock lobster tag is an effective means of marking juvenile American lobsters for extended periods (up to 56 months) necessary for the study of growth and movement patterns.

Since the early 1970s, when researchers developed (Scarratt and Elson 1965) and successively field-tested (Cooper 1970; Scarratt 1970) the sphyrion tag, an external spaghetti tag with a small hook-shaped wire anchor, the American lobster *Homarus americanus* has been the subject of numerous tagging studies along the northeast coasts of the USA and Canada (see reviews by Krouse 1980 and Stasko 1980). The majority of lobsters marked in these studies were larger than 81 mm in carapace length, the most prevalent minimum legal size, and only a small number were less than 65 mm (all lobster lengths in this paper are carapace lengths). This apparent selection for adult-sized lobsters may be attributed to (1) early research interests oriented more toward older life stages; (2) the difficulty of collecting lobsters less than 50 mm, in contrast to the readily available legal-sized animals; and (3) the perceived unsuitability of the standard sphyrion tag for use on juvenile lobsters (Scarratt 1970).

During the early 1980s, we discovered that small American lobsters (<40 mm) could be collected by hand at a beach near our Boothbay Harbor (Maine) Laboratory in sufficient numbers to permit mark–recapture studies of migration, growth, and survival. Therefore, we set out to find a satisfactory tagging technique for juvenile lobsters. On the basis of favorable results we had experienced with the western rock lobster tag in preliminary trials with adult American lobsters, it

was felt that the design of this tag, with minor modification, would be suitable for marking small animals. The western rock lobster tag, which is identical to the sphyrion tag except for the plastic anchor, was developed in Australia by Chittleborough (1974) for the western rock lobster *Panulirus longipes cygnus* and has been widely used on other species of spiny lobsters.

In the study reported here, we tagged American lobsters ranging from 25 to 80 mm with miniature and full-sized western rock lobster tags. The suitability of this tag design for the long-term marking of juvenile American lobsters was evaluated on the basis of laboratory experiments and field releases.

Methods

Study area and collection methods.—All lobsters tagged in this study were either captured by hand, with wire lobster traps (2.54-cm-square mesh), or by scuba diving in the vicinity of Pratt's Island, southwest of Boothbay Harbor near the mouth of the Sheepscot River estuary (Figure 1). Hand collections were usually made monthly (April–November) during low spring tides at Pratt's Island beach, a coarse sand beach with overlying rocks and bedrock. After capture, each lobster was carefully placed in an 18.9-L bucket and separated from other lobsters by seaweed to minimize the chance of injury. Before tagging, which usually took place within hours of capture,

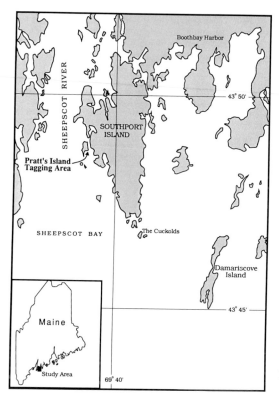

FIGURE 1.—Location of the Pratt's Island study area near Boothbay Harbor, Maine.

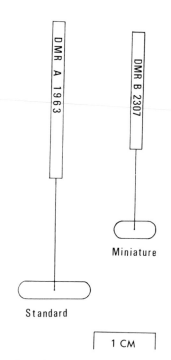

FIGURE 2.—Miniature and full-sized western rock lobster tags used to tag juvenile American lobsters.

the lobsters were refrigerated in moist seaweed at 4–8°C. Carapace length (to the nearest millimeter), sex, and claw condition of the lobsters were recorded and the lobsters were then tagged and assigned to one of the laboratory tests or released in the area of original capture at Pratt's Island.

Tagging technique.—The western rock lobster tag used in this study consisted of a serially numbered yellow spaghetti tube (2 mm in diameter) attached by 0.3-mm-diameter plastic thread to an oblong plastic toggle. One of two tag sizes was used, depending on the lobster's size. For individuals larger than 50 mm, we applied tags with anchors measuring 3 × 12 mm and tubes 2.54–3.18 cm long. For smaller animals, the anchors and spaghetti tube were reduced to 3 × 7 mm and 1.91–2.54 cm, respectively (Figure 2). The distance between the anchor and tube was about 13 and 18 mm for the miniature and standard-sized tags, respectively. If necessary, this distance could be increased for larger animals by gently pulling the thread and anchor away from the tube. This adjustment was possible because the tags, which we assembled, had no adhesives.

Unlike those available from a manufacturer, our tags depended on a friction-fit union for securing the thread to the tube. This was accomplished by heating the thread to create a bead and then pulling the beaded thread about 6 mm into the tube.

Tagging was conducted by a two-person team. One individual held the lobster with one hand on the carapace while the other hand grasped the abdomen and exerted an opposing force to fully expose the dorsal junction of the carapace and tail. The other person used a scalpel to cut a small lateral incision (3–4 mm) well off the midline on the dorsal surface of the integument between carapace and abdomen. Next, the tag's anchor was carefully inserted with a forceps through the incision at about a 45° angle, directed anteriorly, and in most cases was implanted beneath the dorsal extensor muscle. Proper positioning of the anchor was further ensured by simultaneously pulling and twisting the tag.

Laboratory experiments.—Tagged and untagged (control) lobsters were individually held in the open-seawater system of our wet laboratory or in partitioned 38.8-L aquaria, where temperatures ranged throughout the year from 1 to 16°C and 17 to 23°C, respectively. In the wet-labora-

TABLE 1.—Criteria used to classify tag attachment condition of American lobsters tagged with western rock lobster tags.

Classification	Description
Excellent	Tag well secured; no evidence of infection or necrotic tissue; integument unbroken around tag
Good	Tag well secured; slight trace of infection or necrotic tissue; integument broken around tag
Fair	Tag fairly well attached (portion of tag anchor may be visible); either scarred tissue or infection at tag insertion site
Poor	Anchor almost completely exposed and barely attached; extensive infection or necrosis at tag insertion site

TABLE 2.—Retention of western rock lobster tags by American lobsters after one to five molts in aquaria and open-system seawater tanks.

	Aquaria		Tanks	
Molt	Number molting	% of tags retained	Number molting	% of tags retained
I	16	100	58	96.6
II	8	100	47	87.2
III	7	85.7	32	87.5
IV	4	100	14	78.6
V	2	100	8	75.0

The aquarium and wet-laboratory tests were concluded after 18 and 30 months, respectively; however, those wet-laboratory lobsters with tags still intact were kept under observation for up to another 4 years.

Field releases.—From 1979 through 1987, 10,191 juvenile lobsters (25–80 mm) were tagged and released at the Pratt's Island study area. Intertidal hand collections, which occurred at about monthly intervals (exclusive of winter), accounted for 5,351 lobsters. Another 3,975 juveniles were caught with traps, and 865 were captured by Scuba divers. Recaptured tagged lobsters were returned to the laboratory where sex, carapace length, claw status (missing or regenerated), tag condition, and tag number were recorded. Recaptures were then re-released.

Local fishermen were requested to immediately return to the ocean any tagged sublegal lobsters (81 mm, minimum legal size in Maine, 1958–1987) that they caught and to report and furnish for our inspection any tagged legal recaptures. To encourage cooperation of the fishermen, all individuals providing tag recapture information became entrants in an annual trap lottery.

tory experiment, 120 lobsters (60 per treatment), ranging from 39 to 80 mm (averaging 64.4 and 67.3 for the tagged and control groups, respectively), were maintained in 10 454-L tanks (12 animals per tank). Each tank measured 1.2 m long by 1.2 m wide by 0.3 m high and was divided into 12 equal-sized compartments by 2.54-cm-square vinyl-coated wire mesh. For the aquarium trials, 20 lobsters (10 per treatment) were initially maintained in four aquaria (5 per tank). However, as deaths occurred, test animals were replaced with 10 newly tagged and 6 untagged lobsters for a total of 36 animals (20 tagged and 16 controls). Tagged and control lobsters had mean lengths of 32.4 mm (range, 25–39 mm) and 32.6 mm (range, 27–39 mm), respectively.

Aquarium lobsters were fed a diet of minced blue mussels *Mytilus edulis*, softshells *Mya arenaria*, and common periwinkles *Littorina littorea*, usually twice a week. The generally larger wet-laboratory animals were fed crushed softshells and pieces of Atlantic herring *Clupea harengus harengus* and alewives *Alosa pseudoharengus*. Newly molted lobsters were allowed to consume their cast shells and were not measured for at least a week after ecdysis to reduce the chance of autotomy and to ensure a more accurate measurement. Observations were made daily for molting, mortality, and any abnormal behavior.

To supplement our evaluation of tag retention, we carefully examined the sites of tag attachment after ecdysis. Based on the condition of the integument surrounding the tag's thread and the degree of anchor exposure (index of tag securement), the lobster was assigned to one of four tag attachment classifications (Table 1).

Results and Discussion

Laboratory Study

Tag Retention.—Aquarium-held lobsters, which ranged in size from 25 to 39 mm, showed tag retentions of 100% (total $N = 30$) after all of the first five molts except molt III, when one of seven tags was lost (Table 2). In the wet-laboratory trials, tag loss was greater for the considerably larger lobsters (mean, 64.4 mm). Retention for the first three molts varied from 87.2 to 96.6% and decreased to 75.0–78.6% for molts IV and V.

The lower retention rates of the wet-laboratory lobsters were believed to be the result of several factors. Although lobster size per se was not clearly related to tag retention, tag loss was higher

TABLE 3.—Percentages and numbers (in parentheses) of American lobsters assigned to one of four tag conditions after one to five molts in aquarium and wet-laboratory experiments. Each asterisk denotes a lost tag.

Molt	Aquaria				Wet laboratory			
	Excellent	Good	Fair	Poor	Excellent	Good	Fair	Poor
I	88 (7)	12 (1)			28 (15)	35 (19)	33 (18)	4 (2) *
II	75 (6)		13 (1)	13 (1) *	51 (19) *	35 (13)	11 (4) *****	3 (1)
III	67 (4)	17 (1)	17 (1)		56 (15)	33 (9)	7 (2) **	4 (1) *
IV	100 (4)				25 (3) **	25 (3) *	33 (4) *	17 (2)
V	100 (2)				83 (5)		17 (1) **	
Total	82 (23)	7 (2)	7 (2)	4 (1)	42 (57)	32 (44)	21 (29)	5 (6)

for larger lobsters (mean length of animals that lost tags was 86.8 mm). Two factors should be considered here: the tag may have been removed by the lobster, or the tag may have fallen off during molting. Under laboratory conditions, we have observed tagged lobsters tugging on their own tag with the chelated second and third walking legs. Even though we have never witnessed tag removal by this means, it seems reasonable to believe that any undue strain might unseat the anchor and result in tag loss. This process might have greater significance for larger animals, which would be capable of exerting more force. The other factor that might have contributed to tag loss was filament length (distance between the tag and anchor). If the filament was too short, the proximal end of the spaghetti tube could have become ensnarled with the old carapace and subsequently lost. The probability of this type of loss increased as tagged lobsters passed through succeeding molts and grew to a size for which their tag length was inadequate. Ennis (1986) also suggested that tag loss might be related to tag length. Another important cause of tag loss in the wet-laboratory trials was the tag damage inflicted by other lobsters. Lobsters were sometimes able to tamper with the tag of the neighboring lobster by using their maxillipeds or clawed pereiopod to reach through the wire partition. Several tags in this study had been chewed; in many cases, only the thread remained intact. Similar tag destruction and loss were reported by Cooper (1970) and Ennis (1986) for American lobsters marked with sphyrion tags.

Aquarium-held lobsters had less necrosis or infection at the tag-insertion site and apparently firmer tag attachment than the wet-laboratory animals (Table 3); this may have accounted for the better tag retention in aquaria. For aquarium lobsters, tag condition was excellent after 82% of the molts. This was true after only 42% of the molts for the wet-laboratory animals. In comparison, of 506 legal-sized lobsters recaptured by fishermen, 28% had tags in excellent condition, 23% were good, 37% fair, and 12% poor.

Although it was not possible to positively identify the factors influencing tag condition, the generally poorer status of the tags on wet-laboratory and field-recaptured lobsters indicated that any external pressure on the tag might adversely affect its condition and hasten its loss. For instance, Ennis (1986) found that some lobsters in the wild lost tags from entanglement with the substrate. As mentioned earlier, lobsters tampering with either their own tags or those of other lobsters were a problem in the wet-laboratory study. This behavior may have contributed to tag loss from four lobsters with good or excellent tag condition. Otherwise, tags with these ratings would be expected to be retained through the molt. Those tags rated fair or poor showed the highest loss. It was apparent that unless anchors remain firmly embedded beneath the thoracoabdominal musculature, the tag will be lost.

TABLE 4.—Number (and percent) of American lobsters that died from various causes in aquarium and wet-laboratory tagging experiments.

	Aquaria		Wet laboratory	
Cause of death	Tagged	Control	Tagged	Control
Unexplained	5 (25)	9 (56)	5 (8)	3 (5)
Cannibalism	4 (20)	2 (13)	2 (3)	2 (3)
Support system failure	4 (20)	3 (18)	6 (10)	10 (17)
Total	13	14	13	15

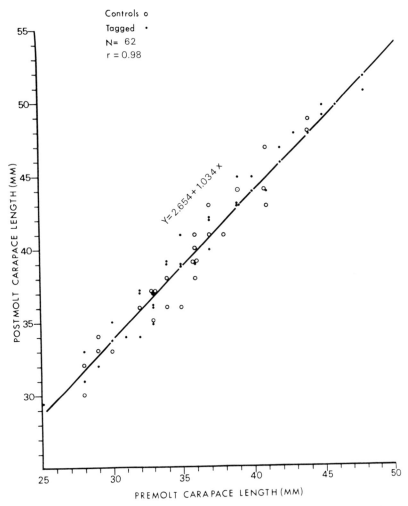

FIGURE 3.—Relation of postmolt (*Y*) to premolt (*X*) carapace length for American lobsters that were tagged and untagged (controls) in aquarium tagging trials with western rock lobster tags.

Mortality.—Throughout the study, tagged and untagged lobsters died at similar rates ($P > 0.05$; *t*-test) either from cannibalism, failure of some component of the seawater system, or unexplained causes (Table 4). Because of the long mean times to death from unexplained causes (90 and 340 d for aquaria and wet-laboratory tests), it was not likely that tagging contributed to the mortality seen in this study. In our experience with both sphyrion and western rock lobster tags, tag-induced mortality occurred within 24 hr of tagging (J. S. Krouse, unpublished data). Moreover, the behavior of tagged animals was not noticeably different from that of control animals.

Growth.—The growth of lobsters is a function of molt increment and frequency. The growth increments of tagged and untagged American lobsters were analyzed by regressing carapace length against premolt length for each experiment (Figures 3, 4). Analysis of covariance indicated no statistical differences ($P > 0.05$) between the coefficients of the linear regressions, so the data for the two treatments were combined. Because the total longevity of lobsters in each of the tests was not the same, comparisons of group molt frequencies were not considered valid. Instead, the elapsed time from tagging to the first molt and subsequent intermolt intervals up to the third molt were compared by analysis of covariance (Table

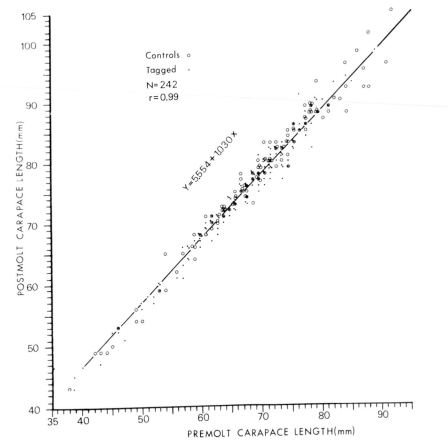

FIGURE 4.—Relation of postmolt (*Y*) to premolt (*X*) carapace lengths for American lobsters that were tagged and untagged (controls) in wet-laboratory tagging trials with western rock lobster tags.

5). Because the molt history of newly tagged lobsters was unknown, the time from tagging to first molt was not analyzed statistically. After the first molt, the intermolt periods were similar ($P > 0.05$) for tagged and control lobsters in aquaria, but significant differences ($P < 0.05$) were appar-ent in the wet-laboratory tests. The intermolt period between the first and second molts was longer for the tagged lobsters, whereas the con-verse was true in the last intermolt period. The consistent aquarium results and the offsetting inconsistencies of the wet-laboratory findings in-

TABLE 5.—Molt-period data for tagged and control (untagged) American lobsters used in laboratory experiments. Data are mean ± SE (*N*). Asterisks indicate significant differences between tagged and control animals ($P < 0.05$).

Experiment	Group	Days from tagging to first molt[a]	Days from first to second molt	Days from second to third molt
Aquaria	Tagged	50.8±9.9 (13)	127.0±24.1 (11)	162.1±17.6 (7)
	Control	52.0±10.4 (9)	86.0±18.0 (10)	106.8±21.3 (4)
Wet laboratory	Tagged	82.5±8.9 (58)	420.5±25.3* (45)	363.6±9.8* (21)
	Control	69.2±5.3 (57)	389.4±30.8 (41)	435.5±32.4 (14)

[a]Not analyzed statistically because molt history before tagging was unknown.

dicate that western rock lobster tags should be useful for obtaining estimates of growth of tagged lobsters recaptured in the wild.

Field Releases

Of 10,191 sublegal-sized lobsters (25–80 mm) tagged and released in the Sheepscot River estuary from 1979 through 1987, 1,858 (18.2%) were recaptured by research personnel (hand, trap, and scuba catches) and commercial fishermen. Commercial returns, which only included those tagged lobsters recruited into the legal-sized phase (≥81 mm) of the fishery, accounted for 619 of the recaptures. Molting, which was determined on the basis of size increase and tag condition, had occurred at least once for 1,193 of the recaptured lobsters. Twenty-eight animals showed carapace length increases of 100% or more; based on molt increments of 15–20% (Krouse, unpublished data), these lobsters had molted at least five times. One recaptured lobster, at liberty for 52 months, had grown from 29 to 95 mm (228%) as a result of an estimated eight molts. The longest time between tagging and recapture was 56 months for an individual that increased in length by 183%; 58 recaptured lobsters had been free more than 3 years and had an average size increase of 92.1%. These data indicate the tag's persistence through numerous molts and suitability for marking lobsters over several years.

Recommendations

In view of the favorable results of our laboratory and field tagging trials, the western rock lobster tag was determined to be an effective means of marking juvenile American lobsters. Lobster behavior, growth, and survival were not adversely affected by the tag. However, if an internal organ such as the heart or hepatopancreas were accidentally ruptured during tag application, death would probably ensue. The likelihood of such accidents diminishes with greater tagger experience.

The persistence of the tag, which was amply demonstrated by the considerable growth and time at liberty of tagged lobsters, qualifies it for use in growth and movement studies that require observation over prolonged periods.

Acknowledgements

We are grateful to the many biologist aides who assisted with the field collections and tag and release operations. We also express thanks to those lobster fishermen and dealers who provided us with tag recapture information.

References

Chittleborough, R. G. 1974. Development of a tag for the western rock lobster. Australia Commonwealth Scientific and Industrial Research Organization Division of Fisheries and Oceanography Report 56.

Cooper, R. A. 1970. Retention of marks and their effects on growth, behavior and migrations of the American lobster, *Homarus americanus*. Transactions of the American Fisheries Society 99:409–417.

Ennis, G. P. 1986. Sphyrion tag loss from the American lobster *Homarus americanus*. Transactions of the American Fisheries Society 115:914–917.

Krouse, J. S. 1980. Summary of lobster, *Homarus americanus*, tagging studies in American waters (1898–1978). Canadian Technical Report of Fisheries and Aquatic Sciences 932:135–140.

Scarratt, D. J. 1970. Laboratory and field tests of modified sphyrion tags on lobsters (*Homarus americanus*). Journal of the Fisheries Research Board of Canada 27:257–264.

Scarratt, D. J., and P. F. Elson. 1965. Preliminary trials of a tag for salmon and lobsters. Journal of the Fisheries Research Board of Canada 22:421–423.

Stasko, A. B. 1980. Tagging and lobster movements in Canada. Canadian Technical Report of Fisheries and Aquatic Sciences 932:141–150.

American Fisheries Society Symposium 7:101–108, 1990

Development and Field Evaluation of a Mini-Spaghetti Tag for Individual Identification of Small Fishes

LAUREN J. CHAPMAN AND DAVID J. BEVAN[1]

Department of Biology, McGill University
1205 Avenue Docteur Penfield, Montreal, Quebec H3A 1B1, Canada

Abstract.—Miniature spaghetti tags were used in an 11-month field study of *Poecilia gillii*, a molly inhabiting highly seasonal dry-forest streams of northwest Costa Rica. Small handmade tags were individually coded with beads of nail polish and put through the dorsal musculature with a fine hypodermic needle. Tagged fish were larger than 3 cm in total length and were generally released into small, disconnected pools. Minimum estimates of tag retention were obtained from field recapture data. At least 44% of the tags were retained after 25–31 d, and at least 18% were retained after 82–90 d. Some tagged fish were still found after 197–205 d (4%). In three of the four sample periods, tagged fish weighed significantly less than untagged fish of the same length. The differences in adjusted mean weights between tagged and untagged fish for the four sample periods ranged from 0 to 9%. In two aquarium experiments with relatively large fish, specific growth did not differ significantly between tagged and untagged fishes.

The recognition of individuals is important in many studies of behavior, growth, and demography. A requisite of any marking or tagging method is that it should have a minimal effect on the individual's behavior, growth, and survival. Various methods have been reported that are suitable for fishes and involve either applied marks or tags (Laird 1978; Wydoski and Emery 1983). However, the individual identification of small fishes is particularly difficult. Although several methods for marking small fishes have been described (e.g., Hart and Pitcher 1969; Heugel et al. 1977; Thresher and Gronell 1978; McNicol and Noakes 1979; Patzner 1984; Power 1984; Busack 1985; Crumpton 1985; Thompson et al. 1986), they have several potential problems, such as low retention times, limited numbers of unique marks, high mortality, and high expense.

This paper describes a method for producing a miniature spaghetti tag and presents an initial evaluation of its use in a study of *Poecilia gillii* (Poeciliidae), a molly inhabiting a tropical dry-forest stream in Costa Rica. Tags were used to assess the seasonal movements of individual fish larger than 3 cm (all measurements as total length) and to estimate the density and total population size of 16 isolated pools.

Study Site

The study was carried out in a small temporary stream (Quebrada Jicote) in Santa Rosa National Park, Costa Rica, between April 1987 and March 1988. In Santa Rosa, the rivers and streams run only in the wet season, in association with major rain events. During much of the dry season, the water in the study sector of the stream consisted primarily of a large permanent pool (Source Pool, 18 m in diameter) at the base of a 20-m waterfall and approximately 0.5 km of dry streambed with 19 smaller pools (size range, 2–10 m in diameter). Between April 1987 and March 1988, the entire stream or large sections flowed only during two short periods (approximately August 2–12, 1987 and October 2–14, 1987). During these floods, many fish moved into previously dry areas. As water levels fell, many fish returned to a permanent pool, whereas others were trapped in desiccating pools and died.

Methods

Tag production.—Tags were made by first winding transparent nylon monofilament (invisible sewing thread, 0.20 mm in diameter) several times around a board. Subsequently, small drops of nail polish were applied to each thread with a fine needle to form individual beads (0.4 mm in diameter, 1.0 mm in length). The monofilament was raised slightly from the board to allow circular beads to form. Each tag consisted of 1–4 adjacent beads, and eight colors of nail polish were used (red, orange, yellow, blue, pink, green,

[1]Present address: Department of Animal and Poultry Science, University of Guelph, Guelph, Ontario N1G 2W1, Canada.

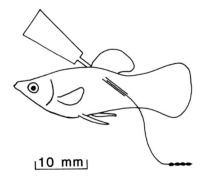

|10 mm|

FIGURE 1.—Method of applying mini-spaghetti tags to small fishes.

purple, and silver). The total number of individual combinations available with eight colors is 4,680. By attaching the tag to either the left or right side, the potential number of individual tags was 9,360. The time to make 100 tags was approximately 2 h, and the material cost of the tags was low.

Tags were attached according to a method modified from Patzner (1984). A fine hypodermic needle (27-gauge, 1.27 cm long) was passed through the dorsal musculature near the dorsal fin on a diagonal. The tag was placed on the diagonal to reduce drag (Figure 1). A tag of approximately 3 cm was threaded through the needle, and the needle was then removed to leave the bead combination on one side (Figure 1). The excess thread on the unbeaded side was melted to form a small knob. In most cases, a very small dab of nail polish was applied to this knob to increase its size. Approximately 30 fish could be weighed, measured, and tagged in an hour. Tags eventually worked their way through the fish in variable periods of time. Generally, fish observed in the field healed quickly, so there was little or no external sign of the tag after its passage through the fish.

Approximately 3,500 fish were tagged according to this procedure over the term of study. Generally, only fish longer than 3 cm were tagged.

Field evaluation.—To assess the effect of the tag on fish under field conditions, we compared the weight–length relationship of tagged and untagged fish recaptured during the same sampling period. Results are presented in this paper for fish initially sampled and marked from the Source Pool during four periods: September 2–5, 1987; November 10, 1987; December 26–29, 1987; and February 13–14, 1988. Recaptured tagged fish included those fish tagged at any point during the field season up to 3 weeks before the sampling

period. Fish were captured in minnow traps and held in plastic buckets before being returned to the site of capture. Fish were weighed (to the nearest 0.05 g) with a portable electronic balance and measured for total length (to the nearest 0.5 mm). Analysis of covariance was used to test for a difference in weight–length relationships between tagged and untagged fish; the Y-axis was weight and the X-axis was length. The data were $\log_{10}$-transformed to stabilize the variance and linearize the data. Only female mollies were used in the analyses because males constituted an average of only 24% of the samples, and there was some evidence to suggest differences in weight–length relationships between males and females. For each sample period, data were tested for homogeneity of slopes before potential differences in the Y-intercepts were considered (Sokal and Rohlf 1981).

We did not evaluate tag retention directly in the field; however, some insight was gained from repetitive sampling, which provided minimum estimates of tags retained. We examined recapture data from 287 fish tagged in the source pool between June 7 and June 12, 1987 (mean total length ± SD, 38 ± 7 mm; range, 30–63 mm). Most of the permanent pools were sampled several times between late June and December 1987 to obtain data for mark–recapture estimates of density and total population size. The numbers of fish recaptured during a particular sample period or any subsequent sampling period were recorded for all sampling periods from late June to December. This was expressed as a percentage of the original 287 fish and represents the minimum percentage of tags remaining in the system at each sample interval.

Laboratory evaluation.—Two aquarium experiments were conducted at the field station in Costa Rica as a preliminary examination of the effect of mini-spaghetti tags on the growth of *P. gillii*.

In experiment 1, 20 females (44.5 ± 3.5 mm) were paired by size and placed in 10 compartments (12 × 20 × 30 cm) in two aquaria. Compartments within aquaria were separated by fiberglass mesh. Fish were measured and weighed, and one of each pair was tagged. Fish were fed ad libitum once daily and remeasured after 31 d. A paired t-test was used to test for a significant difference in the specific growth (percent increase per day) of the two groups.

In experiment 2, the potential confounding effects of dominance-related behavior in *P. gillii* were reduced by placing 18 females (40.2 ± 7.8

TABLE 1.—Analysis of covariance for the effect of tagging on the weight–length relationship of *Poecilia gillii* for four sampling periods. Data were $\log_{10}$-transformed. $\log_{10}$(body weight, g) was considered the dependent variable and $\log_{10}$(total length, cm) the independent variable. The adjusted mean represents the antilog of $\log_{10}$ mean body weight adjusted for $\log_{10}$ total length.

Statistic	Tagged fish	Untagged control fish	F	P
Sample period 1, Sep 2–5				
Slope	3.07	2.94	2.68	0.103
Y-intercept (g)	−1.97	−1.85	20.66	0.000
Adjusted mean (g)	0.81	0.88		
N	95	163		
Sample period 2, Nov 10				
Slope	2.97	2.88	0.56	0.457
Y-intercept (g)	−1.93	−1.84	27.28	0.000
Adjusted mean (g)	0.74	0.81		
N	83	89		
Sample period 3, Dec 26–29				
Slope	2.92	2.70	3.22	0.075
Y-intercept (g)	−1.86	−1.70	10.39	0.002
Adjusted mean (g)	0.95	1.00		
N	86	77		
Sample period 4, Feb 13–14				
Slope	3.25	3.07	2.39	0.124
Y-intercept (g)	−2.04	−1.93	0.01	0.923
Adjusted mean (g)	1.02	1.02		
N	84	82		

mm) in individual compartments (12 × 15 × 20 cm). Two aquaria had six compartments, each separated by glass barriers, and a third aquarium contained six mesh bags. For paired comparisons, two fish of approximately the same size were placed in adjacent compartments. Fish were measured and weighed, and one of each pair was tagged. Fish were fed ad libitum and were remeasured after 48 d (two pairs after 39 d) and again 24 d later.

Results

Field Evaluation

Analyses of covariance of weight–length regressions for tagged versus untagged fishes in the field indicated that for all fish tagged during the four initial sample periods, there were no significant differences in the slopes of weight on length between tagged and untagged fishes (Table 1, Figure 2). For three of the four sampling periods, the Y-intercepts of the tagged fish were significantly less than those of the untagged fish. This difference suggested that tagging resulted in a loss of condition. For sample period 4, there was no difference in the Y-intercepts, so during this pe-

riod, tagged fish showed no significant loss of condition. The differences in the weight (adjusted for length) between tagged and untagged fish were 7%, 9%, 6%, and 0% for sample periods 1, 2, 3, and 4, respectively.

Figure 3 indicates the percentage of tagged fish from the original 287 fish tagged in the Source Pool in June that were recaptured during a given sample interval or during any sample thereafter until December. Thus, the figure represents the minimum percentage of tagged fish remaining in the system a specified number of days after tagging. This estimate of tag retention is very conservative. For each period, the percentage of recaptured fish from the original sample represents those fish that did not die, lose their tags, or disperse to areas not sampled, and were recaptured. A minimum of 44% of the tags were retained after 25–31 d, a minimum of 18% were retained after 82–90 d, and tagged fish were still found after 197–205 d (4%). Tags were retained during and subsequent to major rain events when fish were exposed to strong and turbulent currents, and were retained by fish that dispersed from the dry-season pools.

There was no obvious effect of tags on the behavior of the fish. Tagged fish were observed courting, chasing, and feeding without any sign of irregular behavior. No tagged fish were recaptured in the field with signs of fungal infection at the tag site. However, there was an accumulation of algae on the beads. The degree of algal accumulation varied between pools and times of the year and was most severe in pools that had high light intensity. In general, the algae could be removed from the tag without removing the tag from the fish and did not prevent identification of the fish.

Mortality of tagged fish in the field is unknown; however, death of fish during the tagging process was very low. In general, fish were placed in buckets subsequent to tagging and held until all of the fish in one or more of the buckets were tagged. Of the first 1,912 fish that were tagged, less than 1% died during or immediately after handling procedures.

Laboratory Evaluation

In experiment 1, there was no significant difference in mean specific growth of tagged (1.40 ± 0.38 %/d) and untagged fish (1.55 ± 0.26 %/d; $t =$ 1.38 $P > 0.20$). There was no tag loss over the 31 d of the experiment.

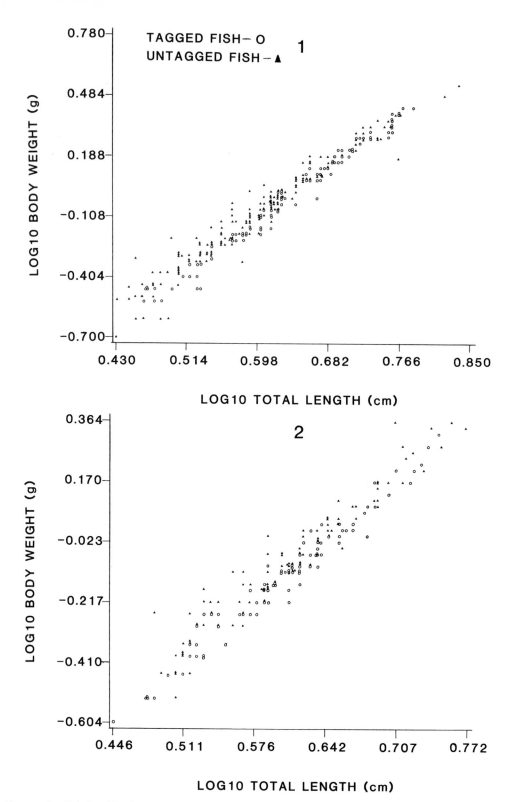

FIGURE 2.—Relationships between $\log_{10}$(body weight, g) and $\log_{10}$(body length, cm) for mollies tagged with mini-spaghetti tags and untagged fish during four sampling periods: (1) September 2–5, 1987; (2) November 10, 1987; (3) December 26–29, 1987; and (4) February 13–14, 1988.

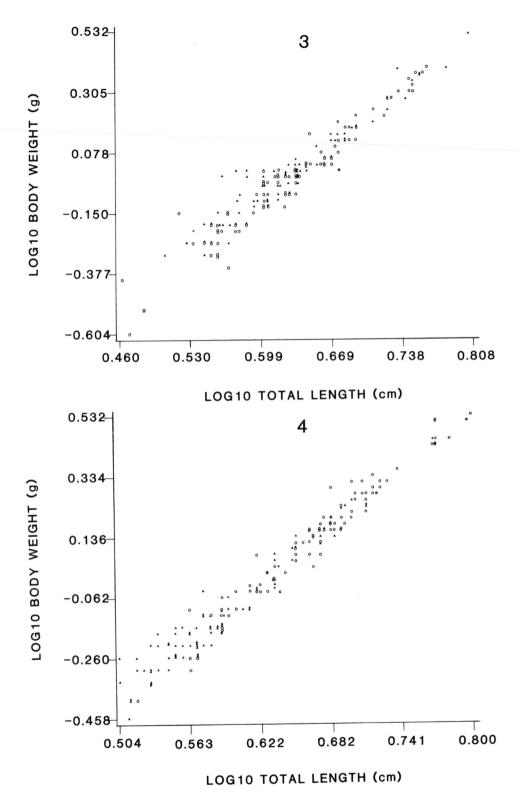

FIGURE 2.—Continued.

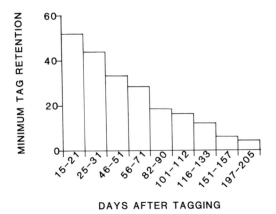

FIGURE 3.—Retention rates of mini-spaghetti tags for mollies. Minimum tag retention is the percent of tagged fish recaptured during each sampling interval (days after tagging) or any sample period thereafter. Returns were from 287 fish tagged initially.

In experiment 2, there was no significant difference in mean specific growth of tagged (1.31 ± 0.69 %/d) and untagged fish (1.57 ± 0.51 %/d; $t = 1.18$, $P > 0.20$) after 48 d. Similarly, there was no significant difference in mean specific growth of tagged (1.29 ± 0.76 %/d) and untagged fish (1.42 ± 0.56 %/d; $t = 0.44$, $P > 0.65$) after an additional 24 d. In this experiment, there was one tag loss 45 d after tagging. One tagged fish died 49 d after tagging.

Tag retention in the aquarium experiments was much longer than the minimum estimates derived from the field data (100% in experiment 1 after 31 d, 88% in experiment 2 after 45 d). In two additional aquaria, 12 tagged fish and 8 tagged fish were held with comparable numbers of untagged fish. Tag retention for these fish was 90% over 31 d.

Discussion

The field and laboratory data support the continued use of our mini-spaghetti tags for studying certain areas of fish biology. The tags described are simple and inexpensive to produce in large numbers with different individual color combinations. Although actual tag retention in the field is unknown, a minimum of 44% of the tags remained after 25–31 d, and a minimum of 18% remained after 82–90 d. Tagged fish were still found after 197–205 d (4%). The estimates of tag retention obtained in the aquarium experiments were less biased and higher than those obtained from the field data, and compare favorably with short-term

tag retention in other studies in which individually coded tags were used for small fishes. Leary and Murphy (1975) reported a tag loss of 14.7% after 30 d for anchovies *Stolephorus purpureus* (mean standard length, 45 mm) tagged with implanted small stainless steel wires. Crumpton (1985) internally tagged fingerling largemouth bass *Micropterus salmoides* with magnetic microwire tags and reported tag retention rates of 25% and 10% for vomerine-tagged fish (mean total length, 119 mm) and forebrain-tagged fish (mean total length, 106 mm) after 69 d. With respect to long-term tag retention, our minimum estimates obtained from field data (a minimum of 4% of the tags remained in the system after 197–205 d) are relatively low compared to estimates for external anchor tags or Carlin tags (e.g., Carline and Brynildson 1972; O'Grady 1984). Relatively low long-term tag retention in this study may be related in part to the small size of the fish tagged. Yap and Furtado (1980), who evaluated external staple tags on *Osteochilus hasselti*, suggested a higher loss of tags in smaller fish.

Our estimate of tag retention in the field is conservative and represents the percentage of fish from the original sample that did not die or lose their tags, and were recaptured. A thorough evaluation of tag retention would require a comparison of mortality obtained from tag recaptures with estimates from an independent source (e.g., cohort analyses or double marking). These analyses were beyond the scope of this preliminary evaluation. Considering the ease of capture of many small fishes, the ease and expense of producing the tags, and the time required to put on the tags, the degree of tag loss is unlikely to be a serious constraint in the application of these tags to other field situations.

One must be cautious when using miniature spaghetti tags for estimates of condition and growth in small fishes. The field data suggest that tagging can result in loss of condition. In three of the four sample periods, the tagged fish of a given length weighed significantly less than untagged fish of the same length. In the aquarium experiments, there were no significant differences in specific growth. However, because the laboratory experiments were limited primarily to relatively large mollies, further evaluation is required for smaller size-classes.

Algal accumulation on tags was a problem in the field, although heavy accumulation was restricted to certain months and pools. Algal accumulation decreased the efficiency of tag reading, and increased drag from the algae may have contributed to tag loss and a reduction in fish condition.

Poecilia gillii is a relatively hearty fish with respect to handling, and thus one might anticipate higher mortality when applying this tagging procedure to more sensitive species. However, in support of this method for use on other small fishes, the miniature spaghetti tags have been used successfully with reasonably high retention and low mortality in behavioral studies of zebra danios *Danio (Brachydanio) rerio* under laboratory conditions (J. Grant, personal communication).

There is a variety of other methods available for marking small fishes. Heugel et al. (1977) described small subcutaneous diazo film tags surgically implanted into the dorsal musculature of small poeciliid fishes. Tag retention was 62.5% after 18 months in the laboratory, and there was no mortality attributable to tagging. However, it seems necessary to dissect the tags out of the fish for identification, so the possibility of multiple recaptures or external identification for behavioral studies is precluded. Caudal fin branding of small fish for individual recognition has been used by McNicol and Noakes (1979), and brands were retained for at least 8 weeks in salmonids. Laird et al. (1975) obtained mark retention times of at least 16 weeks for juvenile salmon (13 cm) freeze-branded with liquid nitrogen. There has been a variety of studies in which small fishes were marked by applying pigments, dyes, fluorescent plastic chips, latex, or fluorescent pigments through a tattooing procedure or subcutaneous injection. Hill and Grossman (1987) reported 50% retention over a 25-week laboratory study for marks of acrylic paints injected subcutaneously into a variety of small stream fishes. Thompson et al. (1986) reported 96% retention after 6 months for fluorescent plastic chips pressure-injected into *Etheostoma* sp. (42 mm, mean standard length). Wigley (1952) tattooed larval lampreys (as small as 30 mm) with aqueous solutions of dyes, and marks were still visible after 1.5 years. In many cases, applied marks have been used to examine population parameters and movement patterns, or for the differentiation of stocks; however, in some studies, combinations of colors, marking sites on the fish, or both have been used for the identification of individuals (e.g., Thresher and Gronell 1978; Thompson et al. 1986; Hill and Grossman 1987). There are, however, potential problems with applied marks for individual identification of small fishes, such as the blending of colors with the fishes' coloration or differential fading of different colors on the same fish.

The best method for marking small fishes depends to some degree on the objectives of the study. Our miniature spaghetti tags are easily applied, are inexpensive to produce, and provide individual identification over multiple recaptures. These tags may also be of benefit to other behavioral studies. The color codes on tags are clearly visible in an aquarium, and preliminary trials with fluorescent beads suggest that individuals are easily followed under field conditions. Therefore, these tags provide an opportunity to look at individual movement patterns and home range parameters in small fishes. In addition, the tag is suitable for the estimation of population parameters, such as population size and density.

Possibilities for future tag refinement include producing a smoother, smaller tag with less available surface area for algal accumulation and drag. This could be achieved by minimizing the gap between successive beads and coating the beads with a transparent layer. Other refinements include the use of enamel paint (e.g., modeling paint) to provide a better selection of color combinations.

Acknowledgments

We thank the National Parks Service of Costa Rica for permission to work in Santa Rosa National Park. Colin Chapman and Debbie Kohn gave invaluable assistance during the study. Financial support was provided by a Natural Sciences and Engineering Research Council (NSERC) of Canada operating grant to D. L. Kramer and an NSERC postgraduate scholarship to L.J.C.

References

Busack, C. 1985. A simplified cold-branding apparatus. Progressive Fish-culturist 47:127–128.

Carline, R. F., and O. M. Brynildson. 1972. Effects of the Floy anchor tags on the growth and survival of brook trout (*Salvelinus fontinalis*). Journal of the Fisheries Research Board of Canada 29:458–460.

Crumpton, J. E. 1985. Effects of micromagnetic wire tags on the growth and survival of fingerling largemouth bass. Proceedings of the Annual Conference Southeastern Association of Fish and Wildlife Agencies 37:391–394.

Hart, P. J. B., and T. J. Pitcher. 1969. Field trials of fish marking using a jet inoculator. Journal of Fish Biology 1:383–385.

Heugel, B. R., G. R. Joswiak, and W. S. Moore. 1977. Subcutaneous diazo film tag for small fishes. Progressive Fish-Culturist 39:98–99.

Hill, J., and G. D. Grossman. 1987. Effects of subcutaneous marking on stream fishes. Copeia 1987:492–495.

Laird, L. M. 1978. Marking fish. Pages 95–101 in B. Stonehouse, editor. Animal marking. MacMillan, London.

Laird, L. M., R. J. Roberts, W. M. Shearer, and J. F. McArdle. 1975. Freeze branding of juvenile salmon. Journal of Fish Biology 7:167–171.

Leary, D. F., and G. I. Murphy. 1975. A successful method for tagging the small, fragile engraulid, *Stolephorus purpureus*. Transactions of the American Fisheries Society 104:53–55.

McNicol, R. E., and D. L. Noakes. 1979. Caudal fin branding fish for individual recognition in behavior studies. Behavior Research Methods and Instrumentation 11:95–97.

O'Grady, M. F. 1984. The effects of fin-clipping, Floy-tagging and fin-damage on the survival and growth of brown trout (*Salmo trutta* L.) stocked in Irish lakes. Fisheries Management 15:49–58.

Patzner, R. A. 1984. Individual tagging of small fish. Aquaculture 40:251–253.

Power, M. E. 1984. Depth distributions of armored catfish: predator-induced resource avoidance? Ecology 65:523–528.

Sokal, R. R., and F. J. Rohlf. 1981. Biometry, 2nd edition. Freeman, San Francisco.

Thompson, K. W., L. A. Knight, and N. C. Parker. 1986. Color-coded fluorescent plastic chips for marking small fishes. Copeia 1986:544–546.

Thresher, R. E., and A. M. Gronell. 1978. Subcutaneous tagging of small reef fishes. Copeia 1978:352–353.

Wigley, R. I. 1952. A method of marking larval lampreys. Copeia 1952:203–204.

Wydoski, R., and L. Emery. 1983. Tagging and marking. Pages 215–237 in L. A. Nielsen and D. L. Johnson, editors. Fisheries techniques. American Fisheries Society, Bethesda, Maryland.

Yap, S., and J. I. Furtado. 1980. Evaluation of two tagging/marking techniques, and their practical application in *Osteochilus hasselti* C. and V. (Cyprinidae) population estimates and movement at Subang Reservoir, Malaysia. Hydrobiologia 18:35–47.

American Fisheries Society Symposium 7:109–116, 1990

Marking and Tagging Intertidal Fishes: Review of Techniques

JOHN R. MORING

U.S. Fish and Wildlife Service, Maine Cooperative Fish and Wildlife Research Unit
Department of Zoology, University of Maine, Orono, Maine 04469 USA

Abstract.—The diversity of form among intertidal fishes precludes the use of any one type of tag or mark for all members of the fish community. The small size of most such fishes makes the use of numbered tags difficult. In addition, the specific rock and algal habitat requirements of many species, particularly those with anguilliform body shape (e.g., Pholidae, Stichaeidae, Blenniidae) and those with ventral attachment structures (e.g., Cyclopteridae and Gobiesocidae), restrict effective use of external tags because such fishes move within narrow confines of rocks or algae. Studies of movement patterns and homing behavior of intertidal fishes have required effective techniques for individual or group identification. Marks and tags are reviewed from 46 studies in the USA, Canada, South Africa, United Kingdom, Japan, and France; they generally fall within one of four categories: (1) clips of fins and fin rays; (2) dyes, paints, and vital stains; (3) external beads or threads; and (4) natural markings or scars. The most common techniques for group marking have involved clipping of fins and fin rays (53% of all studies examined), but various combinations of small, colored embroidery beads have been regularly used to identify individual cottids. A method of individually fin-clipping rock gunnels *Pholis gunnellus* and other anguilliform species is discussed, as are several other marking techniques for possible use in future studies.

Tidepool fishes are found throughout the world and include year-round resident species, seasonal species, and transient species that can be uncommon visitors to tidepools. Intertidal fishes include species in all of the above categories as well as those species foraging in the intertidal zone at flood tides. From a practical standpoint, these flood-tide species are difficult to study except with certain types of gill nets and traps that have inherent problems if used for anything but gross interpretations of movements. As a consequence, most studies of intertidal fishes have dealt with the more easily studied fishes found in tidepools at ebb tide. It is for these species that marks and tags have been used to study territoriality, movements, and population size.

No one technique is applicable to all intertidal fish species because these species have such diverse forms. A review of 46 studies around the world indicates that intertidal fish communities in temperate areas may contain up to 30 species and those in tropical areas up to 124 species. The diversity of fish shape and habitat is broad. Shapes may be anguilliform (such as in the Anguillidae, Pholidae, and Stichaeidae), fusiform (Clupeidae, Gadidae, Scorpaenidae), restricted fusiform (Cyclopteridae), elongate fusiform (Blenniidae, Clinidae, Cryptacanthodidae), saggitiform (*Liparis* spp. of the Cyclopteridae), extremely compressed (Embiotocidae), extremely depressed (Gobiesocidae), or extremely compressed with the body rotated (Bothidae and Pleuronectidae); other important groups, such as the Cottidae, do not fall within a traditional category of form (Norman and Greenwood 1963). In addition, habitats vary among tidepool and other intertidal fishes. Clingfishes (Gobiesocidae) adhere to the undersides of rocks, whereas anguilliform fishes hide amongst algal clumps or in spaces below rocks. Cottids are often exposed on the bottom, but are partially camouflaged by their markings, and seasonal visitors, such as juvenile Atlantic herring *Clupea harengus harengus* and rockfishes (*Sebastes* spp.), are often free-swimming in pools.

Because of this diversity, researchers have used a variety of marking techniques in an attempt to obtain needed data on intertidal movements and behavior without employing marks that make fish overly susceptible to avian and aquatic predators. Although studies of biological aspects of intertidal fishes and fish communities have been relatively common in many parts of the world (an excellent review has been presented by Gibson 1982), comparatively few studies have used marks or tags to identify individual fish or groups of fish. A review of these latter studies worldwide indicates that the marking techniques used generally fall within one of four categories: (1) fin clips and fin ray clips; (2) dyes, paints, and vital stains; (3) external beads or threads; and (4) natural marks or scars. The objective of this paper is to review these techniques and discuss other marks little used or not previously used for intertidal fishes that may be suitable for future species or community studies. These techniques will be discussed in conjunction with the "Guidelines for Use of Fishes in Field Research" (ASIH et al. 1988).

Fin and Fin Ray Clips

The most common and easiest technique for marking intertidal fishes is clipping of fins and fin rays. These marking procedures were used in 53% of the studies examined. Although dozens of species have been marked in experiments worldwide, there is no consistent fin that is preferred by all researchers.

The paired fins are particularly important to the Cottidae because these species lack swim bladders and rest on the bottom triplane-like on the pectoral and pelvic fins, and the ventral part of the caudal fin. Although Gersbacher and Denison (1930) concluded (in only limited tests) that fin-clipping the caudal fin was impractical for the tidepool sculpin *Oligocottus maculosus* because of its small size, Moring (1976) successfully marked 14 species of fishes, including seven species of cottids as small as 20 mm in total length, with partial caudal and left and right pectoral fin clips. Preliminary laboratory tests in that study indicated careful, partial fin clips, even of pectoral fins, did not significantly impair movements nor increase mortality. Richkus (1978) also marked pelvic and pectoral fins of the woolly sculpin *Clinocottus analis* less than 35 mm long, and Green (1971) successfully clipped pelvic fins and posterior portions of dorsal and anal fins of tidepool sculpins.

Rather than remove a portion of an entire fin, some researchers have removed only fin rays. Gibson (1967) excised one to two fin rays from dorsal and anal fins of several European species, and Marsh et al. (1978) identified individuals of six species of clinid fishes by clipping fin rays of pectoral fins. For several anguilliform fishes of North America, the fin clip of choice has generally been the tip of the caudal fin (Barton 1974; Moring 1976; Burgess 1978).

Biologists should exercise care when clipping fins of intertidal fishes, many of which are small in size. Fine scissors or clippers are necessary, and only a portion of fins should be excised. The ASIH et al. (1988) recommended not only the use of an anesthetic or immobilization by chilling for such marking, but the use of a topical antiseptic after clipping. However, the technique has the advantage of speed and limited mortality due to the mark. Regeneration of fins is a distinct disadvantage if long-term studies (in excess of several months, depending on the species) are necessary. Marsh et al. (1978) found that pectoral fin clips remained identifiable on clinid fishes after 3 months, but Gibson (1967) noted regeneration of fins in 6–7 months. Similarly, there was complete regeneration of a fin of a rock gunnel *Pholis gunnellus* recaptured after almost 23 months (J. R. Moring, unpublished data).

Dyes, Paints, and Vital Stains

Several researchers have experimented with various types of paints or stains as a means of marking groups or species and for identifying individual fish. Injected dyes probably have been most useful for intertidal fishes, whereas other techniques have had only limited success, at best.

Gersbacher and Denison (1930) attempted to mark tidepool sculpins with red and copper-colored paint, but the paint did not dry quickly enough. Of greater potential use, however, are injected dyes. Smith (1970) successfully injected fluorescent pigment beneath the skin of the brook stickleback *Culaea inconstans* and the ninespine stickleback *Pungitius pungitius* (a tidepool inhabitant along the Atlantic coast); marks were still apparent 8–11 months later. Gibson (1967) used vital stains to mark intertidal and subtidal fishes; only Bismarck brown and neutral red were useful. Other stains were either lethal or became indistinguishable too rapidly. Davis (1971) also found Bismarck brown to be useful for marking fry of Atlantic herring, a visitor to New England tidepools. Although the purpose of marking was to examine growth features of scales rather than movements, Ichikawa and Hiyama (1954) injected lead acetate into the yellowfin goby *Acanthogobius flavimanus* and subsequently recovered 7% of the fish.

In a similar manner, Kelly (1967) and Misitano (1976; personal communication) marked winter flounder *Pseudopleuronectes americanus* and juvenile English sole *Parophrys vetulus,* respectively, with National fast blue injected under the skin of the blind side. Hydrated chromium oxide was also successfully injected into winter flounder by Kelly. Winter flounder are found in sand-bottomed tidepools along the north Atlantic coast of North America, and English sole are found in mud-bottomed tidepools of the Pacific coast from central California northward (Moring 1972; C. L. Hubbs, Scripps Institution of Oceanography, personal communication). Dyes were clearly visible

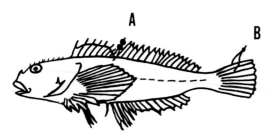

FIGURE 1.—Embroidery bead (A) and thread (B) attachment sites for marking individuals or groups of tidepool sculpin.

on winter flounder and English sole up to 4 months after marking, at the termination of experiments (Kelly 1967; D. A. Misitano, National Marine Fisheries Service, personal communication).

Dye injection can be effective, although virtually no information is available about how such marks might affect the susceptibility of fish to predation or additional delayed mortality due to stress or disease from handling and injection. Kelly (1967) was unable to document that National fast blue or hydrated chromium oxide had any effect on survival of winter flounder. These marks are apparently quite effective for marking pleuronectiform fishes because the marked blind side is generally hidden from predators, yet the dye mark is quite obvious to biologists. However, before the use of any such dyes, paints, or vital stains with intertidal fishes, a small pilot study should be conducted to assure that neither the stain nor the injection has any adverse effect on survival, behavior, or attraction to predators. This concern may not be unfounded. Studies with small minnows have shown that injected vital stains can cause changes in swimming behavior and potentially make such fish overly susceptible to predation (S. T. Ross, University of Southern Mississippi, unpublished data).

External Beads or Threads

Because of the small size of most tidepool and many intertidal fishes, external tags must be quite small yet visible if behavioral observations are to be made. Gersbacher and Denison (1930) inserted a white thread through the caudal fin of tidepool sculpins as a means of identifying fish for homing experiments (Figure 1). Although the fish were not individually identified with such thread, the technique was effective as a short-term (2–3 week) marker of fish subsequently recaptured in home pools.

A more popular method of marking has been the use of colored beads attached to the backs of fishes (Figure 1). This method was used as early as 1949, but its first use with intertidal fishes was probably by Williams (1957) for identifying woolly sculpins and juvenile opaleyes *Girella nigricans*. Subsequently, several researchers have used variations of the method to tag woolly sculpins (Richkus 1978), mosshead sculpins *Clinocottus globiceps* (Green 1973), and tidepool sculpins (Green 1971; Khoo 1974; Craik 1981).

In most cases, embroidery or "Indian" beads are attached to nylon monofilament line passed through the dorsal musculature midway between the first and second dorsal fins (of sculpins) or simply along the dorsal surface (other species). Fish can be individually marked by using different color combinations of one or two beads that are generally 1–2 mm in diameter. A variation of this method was used by Stephens et al. (1970) to mark several blenny species *Hypsoblennius* spp. Modified Petersen discs were constructed with glass beads heat-melted on nylon monofilament line. Again, beads were color-coded, and laboratory experiments showed that the tags lasted for several months without a loss. Field experiments indicated even longer retention, up to 14 months for some fishes.

Natural Markings or Scars

Some researchers have suggested the use of natural markings or color patterns, or natural scars as a means of identifying individual tidepool fishes, much in the manner that flukes were used by Katona and Whitehead (1981) to identify individual humpback whales *Megaptera novaeangliae*. Carlisle (1961) observed intertidal behavior of two British fishes, the sea bass *Morone labrax* and the mullet *Mugil labrosus*, by using distinctive scars, color markings, and damaged fins.

Natural marks have only limited application for behavioral studies and even less value for movement or homing research. Natural marks have an advantage over artificially applied forms of identification in that they do not contribute any additional mortality or changes in behavior. However, such identifying marks are only practical when a limited number of fish are present (Carlisle's observations involved just two sea bass and seven grey mullet) and when fish are being observed in a confined area.

Potential Techniques

Marking and tagging of intertidal fishes are important for the study of certain biological and

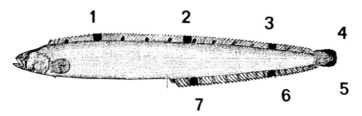

FIGURE 2.—Fin-clip locations used for marking rock gunnels, modified from Jordan and Evermann (1900). For practical application, no more than three marks should be used, so there would be 63 possible unique combinations. With an additional indicator mark (e.g., clip on right or left pectoral fin) the number of unique marks is increased.

population characters but have received limited evaluation. Some new procedures are being explored for use with these fishes.

Fin-Coding for Anguilliform Fishes

Anguilliform (eel-like) fishes, particularly those in the families Pholidae and Stichaeidae, are difficult to mark because of their unique habitat amongst heavy patches of marine algae, beneath rocks, or burrowed into sand. External tags are impractical because of these habitat requirements, and partial clipping of the caudal fin does not allow for identification of individual fish.

A fin-clip coding procedure has been used experimentally by Moring (unpublished data) to identify individual specimens of rock gunnels for movement studies. For most pholid or stichaeid fishes, the pectoral fins are quite small, but the dorsal and anal fins are long. Together with the dorsal and ventral portions of the caudal fin, seven distinct locations are available for clipping (Figure 2). There are 63 potential combinations of one, two, or three marks. A small indicator clip on the left or right pectoral or a brand could double or triple the potential number of mark combinations.

Fin clips can be applied by creating a half-circle in the fin with a hand-held paper punch or V-notches can be made with small scissors; U-notches can be made with fine, curved scissors. Preliminary experiments in 1980 with 36 marked rock gunnels suggested that this is a practical method for following movements of individual fish in a hostile environment for up to 2 months. Fish were recaptured up to 6 months after marking, and one individual (with regenerated fins) was recovered almost 2 years later. Although fins regenerate after 6 months, a regenerated mark can be discerned with a field lens or microscope.

External Tags

Habitat restrictions aside, most numbered external tags are too large for intertidal fishes be- cause of the small size of the animals. Some tags, however, are exceptions. Numbered subcutaneous tags of vinyl plastic have been used successfully for salmonid species (Butler 1957) and could have application for intertidal fishes in short-term experiments (several weeks to several months). A variation of the subcutaneous tag is the V.I. (visible implant) tag recently introduced by Northwest Marine Technology, Inc. A 1.0- × 2.5- × 0.1-mm numbered tag is inserted under transparent tissue with a hand-held tag injector; however, this tag may not be economical for intertidal fish experiments involving relatively small numbers of fish because of the cost of application equipment.

The cunner *Tautogolabrus adspersus* is a common inshore fish of the northwestern Atlantic Ocean. Juvenile cunners are found in tidepools, particularly those in salt marshes, but larger individuals (95–320 mm in total length) are active intertidal forage species during flood tides (Edwards et al. 1982; Ojeda 1987). Pottle and Green (1979) used Floy anchor tags (model FD67) for marking subtidal cunners, and Olla et al. (1974, 1975) used anchor tags (model 67C) and ultrasonic transmitters to track cunners and tautogs *Tautoga onitis* in subtidal waters. Obviously, fish equipped with radio or ultrasonic tags must be large in size. Even the use of anchor tags generally requires intertidal fish to be at least 150 mm in length and nonconcealed in behavior, but these tags can be used for such intertidal visitors as adult cabezon *Scorpaenichthys marmoratus* of the Pacific coast and adult shorthorn sculpins *Myoxocephalus scorpius* of the Atlantic coast.

Experimental Techniques

With the need to mark large numbers of intertidal fishes, new techniques will be developed. Branding (hot or cold) is a viable technique that has worked well with non-intertidal species and could be used with some species of Cottidae and

Pleuronectiformes. Although branding has some logistical constraints, its use as a mark should be explored. The guidelines of ASIH et al. (1988) recommend freeze-branding over heat- or electrical branding for experiments exceeding a few months, although electrically applied marks have been successful with Salmonidae (Moring and Fay 1984). The guidelines conclude that branding can provide a mark that heals rapidly and seldom becomes infected. Another technique, radioactive marking, was used by Hoss (1967) to mark young of three species of flounders *Paralichthys* spp. with radioactive cerium and cobalt; mark retention was good for up to several weeks for these fish averaging 26–49 mg in size. Obviously, this technique has several environmental, health, and logistical constraints.

Some researchers have applied fluorescent pigment to coastal marine fishes. Knudsen and Herke (1978), for example, successfully sprayed fluorescent pigments on juvenile Atlantic croaker *Micropogonias undulatus,* a species that enters coastal marshes; marks were retained for 2 months. This spraying technique can work particularly well for scaled fishes in short-term experiments (up to 3 months), but many species of intertidal fishes lacking scales, including cottids, would be inappropriate subjects for such marking. Moodie and Salfert (1982), however, have had some success with sprayed pigment to mark a freshwater scaleless species, the brook stickleback. Retention of the mark was 93% after 8 d; initial mortality was low but increased over time. There has also been some concern expressed that small juvenile fishes can be harmed by fluorescent spraying. Some of the smallest Atlantic croakers sprayed by Knudsen and Herke (1978) suffered significant damage to fin rays and integument. This damage could have affected their subsequent survival (E. E. Knudsen, U.S. Fish and Wildlife Service, unpublished data).

Perhaps an even more useful method for marking intertidal fishes was that of Thresher and Gronell (1978), who used syringes to inject dye into small reef fishes while diving. In general, India ink and procion yellow (a vital stain) were too watery for underwater use, and these marks were not permanent. Injected tempera paints proved to be effective short-term marks (up to 3–5 weeks), but injected acrylic paints were effective for longer-term studies (longer than 3–5 weeks).

A final method to consider is the use of parasitic marks. Though less distinctive than an artificial mark or tag, these biological markers can have applications under special situations; for example, continent-specific parasites have been used for identifying stocks of Atlantic salmon *Salmo salar* (Moring and Fay 1984). Along the north Atlantic coasts of Europe and North America, many species of inshore fishes, including intertidal species such as cunner (Sekhar and Threlfell 1970) and rock gunnel (Gorman and Moring 1982), harbor metacercarial infections of the digenetic fluke *Cryptocotyle lingua.* Because species of the gastropod *Littorina* are the first intermediate host of the fluke, it has been suggested by Gorman and Moring (1982) that the severity of infection of a fish by this parasite may be a function of time spent in the intertidal zone and potential exposure to *Littorina* spp. An index of metacercariae can be an indicator of subtidal–intertidal movement patterns or evidence of horizontal movement between sites.

Conclusions

The diversity of form and habitat among intertidal fishes precludes the effective use of one type of mark for all members of a community. Although each marking technique has inherent advantages and disadvantages (Table 1), several conclusions can be drawn. The mark of choice still appears to be fin clipping. In terms of limited mortality and ease and speed of marking, this technique is the most appropriate for the diverse families and body forms of intertidal fish communities.

Although a fin-coding procedure has been used for individually marking anguilliform fishes, the clipping of fins does not normally allow for the recognition of individual fish. Because of regeneration of fins, marks are not always evident in studies exceeding several months. But, these disadvantages aside, the procedure is still the most widely adaptable to intertidal studies.

Although external tags can be used for many intertidal species of sufficient length, most strictly tidepool fishes adapt some sort of coloration or behavior to avoid detection by terrestrial and avian predators. Many of these fishes hide beneath or between rocks or amongst clumps of algae, and external tags can hinder movements. To take advantage of such habitat, most tidepool fishes are small in size, and thus numbered external tags cannot be used. A promising approach for studying such fishes may be the use of the new V.I. subcutaneous tag or a modification of some small subcutaneous tags currently available.

TABLE 1.—Advantages and disadvantages of various marking and tagging techniques for intertidal fishes.

Method	Advantages	Disadvantages
Natural markings or scars	No handling of fish; no added mortality or changes in behavior due to marking	Useful for only limited numbers of fish of certain species in confined areas
External beads or threads	Useful for small fish; moderate ease in application; retention up to 14 months	Beads have limited numbers of combinations for individually tagged fishes; not suitable for species using rock crevices; threads have even fewer combinations but may have applications with hiding-type fishes
Paints (topical)	Ease of application	Not retained by fishes because of drying time
Injected dyes and stains	Ease of application; marks visible up to 11 months; some vital stains have little adverse effect; effective for blind-side marking of pleuronectiform fishes; injected tempera paints effective for short-term experiments; injected acrylic paints effective for longer-term experiments	Some stains and dyes are lethal or can affect swimming behavior; some stains become indistinguishable within days; some dyes too watery to use
Fin and fin ray clips	Ease of application; fishes can be rapidly marked; fishes as small as 20 mm have been successfully marked with partial clips	Can have an adverse effect on swimming behavior if too much of a fin is excised; anesthetic should be used, but could cause mortality; regeneration of fins after few months
Fin codes for anguilliform fishes	Can identifiably mark fishes with fin clip; ease of application	Fins regenerate; long-term mortality and effects on swimming behavior unknown; anesthetic should be used
Subcutaneous tags	Can identifiably mark fishes	Cannot use with small fishes (<80 mm); anesthetic should be used; possible mortality due to insertion; no information available for intertidal fishes
Visible implant tag	Can identifiably mark fishes	No information available for intertidal fishes
Floy anchor tags	Can identifiably mark fishes; tags can be observed from a distance	Can only use with large intertidal fishes; some mortality due to tagging; anesthetic should be used; potential susceptibility to predators
Ultrasonic transmitters	Can follow continuous movements of individual fish	Fishes must be quite large; no information available for intertidal fishes
Brands	Ease of application; marks recognizable for several months; marks heal rapidly and seldom become infected	Logistical problems for freeze, thermal, or electrical branding; may not be suitable for many intertidal species; limited information available for intertidal fishes
Radioactive marks	Retention good for several weeks	Environmental, health, and logistical constraints; limited information available for intertidal fishes
Sprayed fluorescent pigment	Ease of application; rapid marking of large numbers of fish; mark retention up to 2 months, perhaps longer	May kill small or juvenile fishes; may cause fin ray or skin damage; generally will not work with scaleless species; limited information available for intertidal fishes
Parasitic markers	Natural mark; no mortality associated with marking	Generally must kill fishes at recapture; only for broad geographical identification; parasite ranges may overlap; no information available for intertidal fishes

In the future, techniques involving injected dyes, paints, and stains will probably increase in use, but such increased use will first require a series of experiments to evaluate the suitability of such material for individual species. Not all dyes and stains are appropriate. Some may cause mortality or a change in swimming behavior, whereas others are too watery, do not leave an identifiable mark, or do not last for an extended period of time.

There is a great need to evaluate marking techniques for intertidal fishes. Although Moring (1976) fin-clipped 1,124 intertidal fishes in a northern California fish community and Green (1973) marked 86 mosshead sculpins with embroidery beads along the coast of British Columbia, most studies have used relatively small numbers of fish. As a consequence, the death of only a few individuals due to the use of an inappropriate mark or tag can have a significant effect on the results of field studies. Virtually nothing is known about the efficacy of marks and tags on intertidal fishes.

Once appropriate procedures have been tested in pilot studies, anesthetics should be regularly used to avoid unnecessary stress and injury to fish. This may be particularly critical in warmer waters, where Overstreet (1988) has shown pollutants and other environmental factors can accelerate the effect of stress or injury on fish.

Acknowledgments

Cooperators in the Maine Cooperative Fish and Wildlife Research Unit include the University of Maine, U.S. Fish and Wildlife Service, Maine Department of Inland Fisheries and Wildlife, and the Wildlife Management Institute. Funding was primarily by the National Geographic Society (grant 3145-85), The Nature Conservancy (Maine Chapter), and a Faculty Research Award from the University of Maine. The late Carl Hubbs provided copies of his field notebooks containing data on pleuronectiform fishes in tidepools. Richard Sayers and Cynthia Perry, University of Maine, kindly reviewed the manuscript, and three anonymous referees provided useful unpublished information.

References

ASIH (American Society of Ichthyologists and Herpetologists), American Fisheries Society, and American Institute of Fishery Research Biologists. 1988. Guidelines for use of fishes in field research. Fisheries (Bethesda) 13(2):16–23.

Barton, M. G. 1974. Studies on the intertidal vertical distribution, food habits, and movements of five species of eel blennies (Pisces: Stichaeidae and Pholidae) at San Simeon, California. Master's thesis. California State University, Fullerton.

Burgess, T. J. 1978. The comparative ecology of two sympatric polychromatic populations of *Xererpes fucorum* Jordan & Gilbert (Pisces: Pholididae) from the rocky intertidal zone of central California. Journal of Experimental Marine Biology and Ecology 35:43–58.

Butler, R. L. 1957. The development of a vinyl plastic subcutaneous tag for trout. California Fish and Game 43:201–212.

Carlisle, D. B. 1961. Intertidal territory in fish. Animal Behaviour 9:106–107.

Craik, G. J. S. 1981. The effects of age and length on homing performance in the intertidal cottid, *Oligocottus maculosus* Girard. Canadian Journal of Zoology 59:598–604.

Davis, C. W. 1971. Marking small Atlantic herring with biological stains. Progressive Fish-Culturist 33:160–162.

Edwards, D. C., D. O. Conover, and F. Sutter III. 1982. Mobile predators and the structure of marine intertidal communities. Ecology 63:1175–1180.

Gersbacher, W. M., and M. Denison. 1930. Experiments with animals in tide pools. Publications of the Puget Sound Biological Station, University of Washington 7:209–215.

Gibson, R. N. 1967. Studies on the movements of littoral fish. Journal of Animal Ecology 36:215–301.

Gibson, R. N. 1982. Recent studies on the biology of intertidal fishes. Oceanography and Marine Biology: An Annual Review 20:363–414.

Gorman, A. M., and J. R. Moring. 1982. Occurrence of *Cryptocotyle lingua* (Creplin, 1825) (Digenea: Heterophyidae) in the rock gunnel, *Pholis gunnellus* (L.). Canadian Journal of Zoology 60:2526–2528.

Green, J. M. 1971. High tide movements and homing behaviour of the tidepool sculpin *Oligocottus maculosus*. Journal of the Fisheries Research Board of Canada 28:383–389.

Green, J. M. 1973. Evidence for homing in the mosshead sculpin (*Clinocottus globiceps*). Journal of the Fisheries Research Board of Canada 30:129–130.

Hoss, D. E. 1967. Marking post-larval paralichthid flounders with radioactive elements. Transactions of the American Fisheries Society 96:151–156.

Ichikawa, R., and Y. Hiyama. 1954. Scale growth rate of the common goby assured by the lead-acetate injection method. Japanese Journal of Ichthyology 3:49–52.

Jordan, D. S., and B. W. Evermann. 1900. The fishes of North and Middle America. U.S. National Museum Bulletin 47 (Part IV).

Katona, S. K., and H. P. Whitehead. 1981. Identifying humpback whales using their natural markings. Polar Record 20:439–444.

Kelly, W. H. 1967. Marking freshwater and a marine fish by injected dyes. Transactions of the American Fisheries Society 96:163–175.

Khoo, H. W. 1974. Sensory basis of homing in the intertidal fish *Oligocottus maculosus* Girard. Canadian Journal of Zoology 52:1023–1029.

Knudsen, E. E., and W. H. Herke. 1978. Growth rate of marked juvenile Atlantic croakers, *Micropogon undulatus*, and length of stay in a coastal marsh nursery in southwest Louisiana. Transactions of the American Fisheries Society 107:12–20.

Marsh, B., T. M. Crowe, and W. R. Siegfried. 1978. Species richness and abundance of clinid fish (Teleostei; Clinidae) in intertidal rock pools. Zoologica Africana 13:283–291.

Misitano, D. A. 1976. Size and stage of development of larval English sole, *Parophrys vetulus*, at time of entry into Humboldt Bay. California Fish and Game 62:93–98.

Moodie, G. E. E., and I. G. Salfert. 1982. Evaluation of fluorescent pigment for marking a scaleless fish, the brook stickleback. Progressive Fish-Culturist 44:192–195.

Moring, J. R. 1972. Check list of intertidal fishes of Trinidad Bay, California, and adjacent areas. California Fish and Game 58:315–320.

Moring, J. R. 1976. Estimates of population size for tidepool sculpins, *Oligocottus maculosus*, and other intertidal fishes, Trinidad Bay, Humboldt County, California. California Fish and Game 62:65–72.

Moring, J. R., and C. W. Fay. 1984. Identification of U.S. stocks of Atlantic salmon in non-U.S: waters: non-tag marking techniques. Final Report to the National Marine Fisheries Service, Woods Hole, Massachusetts.

Norman, J. R., and P. H. Greenwood. 1963. A history of fishes, 2nd edition. Ernest Benn, London.

Ojeda, F. P. 1987. Rocky subtidal community structure in the Gulf of Maine: the role of mobile predators. Doctoral dissertation. University of Maine, Orono.

Olla, B. L., A. J. Bejda, and A. D. Martin. 1974. Daily activity, movements, feeding, and seasonal occurrence in the tautog, *Tautoga onitis*. U.S. National Marine Fisheries Service Fishery Bulletin 72:27–35.

Olla, B. L., A. J. Bejda, and A. D. Martin. 1975. Activity, movements, and feeding behavior of the cunner, *Tautogolabrus adspersus,* and comparison of food habits with young tautog, *Tautoga onitis,* off Long Island, New York. U.S. National Marine Fisheries Service Fishery Bulletin 73:895–900.

Overstreet, R. M. 1988. Aquatic pollution problems, southeastern U.S. coasts: histopathological indicators. Aquatic Toxicology (Amsterdam) 11:213–239.

Pottle, R. A., and J. M. Green. 1979. Territorial behaviour of the north temperate labrid, *Tautogolabrus adspersus.* Canadian Journal of Zoology 57: 2337–2347.

Richkus, W. A. 1978. A quantitative study of intertide-pool movement of the wooly sculpin *Clinocottus analis.* Marine Biology (Berlin) 49:277–284.

Sekhar, S. C., and W. Threlfall. 1970. Infection of the cunner, *Tautogolabrus adsperus* (Walbaum), with metacercariae of *Cryptocotyle lingua* (Creplin, 1825). Journal of Helminthology 44:189–198.

Smith, R. J. F. 1970. A technique for marking small fish with injected fluorescent dyes. Journal of the Fisheries Research Board of Canada 27:1889–1891.

Stephens, J. S., Jr., R. K. Johnson, G. S. Key, and J. E. McCosker. 1970. The comparative ecology of three sympatric species of California blennies of the genus *Hypsoblennius* Gill (Teleostomi, Blennidae). Ecological Monographs 40:213–233.

Thresher, R. E., and A. M. Gronell. 1978. Subcutaneous tagging of small reef fishes. Copeia 1978: 352–353.

Williams, G. C. 1957. Homing behavior of California rocky shore fishes. University of California Publications in Zoology 59:249–284.

American Fisheries Society Symposium 7:117–120, 1990

Effects of Abdominally Implanted Internal Anchor Tags on Largemouth Bass[1]

Kenneth C. Weathers, Sylvia L. Morse,
Mark B. Bain, and William D. Davies

U.S. Fish and Wildlife Service, Alabama Cooperative Fish and Wildlife Research Unit[2]
Department of Fisheries and Allied Aquacultures, Auburn University, Alabama 36849, USA

Abstract.—Within the last two decades, fishery biologists have increased the use of surgically implanted abdominal anchor tags for fish marking, primarily because these tags are thought to have a high retention rate. Most studies have been based on the assumption of no tag-induced mortality, even though the surgical procedure may increase the incidence of infection and damage internal organs in some species of fish. In this study, the incidence of tag-induced mortality was examined, as was the rate of tag loss. Altogether, 383 largemouth bass *Micropterus salmoides* were used in 30-d and 90-d pond experiments. There was no statistically significant tag-induced mortality in the 30-d experiment consisting of nine replicates. The unreplicated 90-d experiment yielded survival rates of 63% and 80% for tagged and untagged fish, respectively. We did not find any evidence of tag loss in either experiment.

Since the invention of the body cavity tag in the late 1920s by Robert Nesbit (Rounsefell and Everhart 1953; Rounsefell 1975), fishery biologists have attempted to identify individual fish with such devices. Nesbit's tags consisted of a colored strip of celluloid inserted into a vertical incision through the abdominal wall. In the 1930s, George Rounsefell and Edwin Dahlgren modified this design by using magnetic tags so that electromagnets could be used to recover tags as the fish were processed (Rounsefell and Everhart 1953).

Internal anchor tags have been greatly modified since their introduction, and their importance as a fisheries research tool is well documented in many fishery techniques manuals (Rounsefell and Everhart 1953; Bagenal 1978; Wydoski and Emery 1983). Despite improvement in tag technology, tag retention can still be a problem with the most commonly used anchor tags (injected tags anchored below the dorsal spine). Using largemouth bass *Micropterus salmoides,* Gilbert and Hightower (1981) examined the retention rates of the popular FD-68 T-bar anchor tag (Floy Tag and Manufacturing, Seattle), which is inserted at the

base of the dorsal spine, and they estimated loss rates of 48–69% within 6 months, depending on the position of the tag. Wilbur and Duchrow (1973) also tested several variations of the dorsal anchor tag on largemouth bass in hatchery ponds. Using Floy tag models FD-67, FD-67C, and FD-68B, they found loss rates of 53%, 25%, and 12%, respectively, over 14 weeks. Elam (1971) found that abdominally inserted internal anchor tags had a 9-month loss rate of 13% for red drum *Sciaenops ocellata* and black drum *Pogonias cromis.*

In addition to tag retention, tag-induced mortality has always been a major concern in fish-tagging experiments. In an aquarium study of rock bass *Ambloplites rupestris* tagged with internal capsules and external streamers, Topp (1967) observed five tagged fish and four untagged fish for 76 d and reported no adverse effects from the tags. Winters (1977a, 1977b) found that healing was rapid in Atlantic herring *Clupea harengus harengus* tagged with abdominally inserted magnetic disks.

Recently, some fishery biologists have been using an internal anchor tag (Floy Tag and Manufacturing), which consists of an oval disk and a hollow vinyl streamer. The tags used in our experiment consisted of a 16 mm × 5 mm × 1 mm disk with a hole in the center through which the hollow vinyl streamer (50–60 mm long) was inserted. One end of the streamer was enlarged to hold it in place against the disk. The disk portion of the tag was surgically implanted into the abdominal cavity, and the streamer remained out-

[1]Contribution number 8-881683P of the Alabama Agricultural Experiment Station.

[2]Cooperators include the U.S. Fish and Wildlife Service, Game and Fish Division of the Alabama Department of Conservation and Natural Resources, Wildlife Management Institute, Auburn University (Alabama Agricultural Experiment Station, Department of Fisheries and Allied Aquacultures, Department of Zoology and Wildlife Science).

TABLE 1.—Size categories, length ranges, and numbers of largemouth bass used in a 30-d pond experiment to evaluate internal anchor tags.

Size category	Length range (mm)	Number tagged	Number untagged
Small	131–228	38	35
Medium	229–406	38	36
Large	407–568	14	19

side the body. This type of internal anchor tag has been assumed to involve negligible tag loss and surgery-related mortality (Matlock and Weaver 1979). Our objectives in this study were to determine statistically if the insertion of this tag into the abdominal cavity caused significant mortality over 30 d, to document any incidence of tag loss, and to observe any indications of tag-induced mortality over 90 d.

Methods

The study consisted of two experiments with tagged and untagged largemouth bass. The first experiment was conducted in three identical 0.1-hectare ponds at the Auburn University Fisheries Research Station. Each pond was stocked with 10 tagged and 10 untagged fish (controls). To account for size-related effects, experimental fish were divided into small, medium, and large size categories (Table 1) and allocated to treatments and ponds by randomly selecting three fish from each length-group and a tenth fish of any length. The ponds were drained 1 month after stocking, and a complete count of survivors was made. Tag loss and presence of gross external lesions were determined visually. The 30-d pond experiment included three repetitions (roughly September, October, November 1987) of the three-pond trials for a total of nine replicates. The experimental ponds

contained forage consisting primarily of small bluegills *Lepomis macrochirus*.

The second experiment consisted of a single stocking of tagged and untagged fish in a 0.5-hectare pond. The pond previously contained only populations of *Lepomis* sp. On 15 October 1987, 99 tagged fish and 104 untagged control fish were stocked. On 15 January 1988, the pond was drained and all surviving experimental fish were recorded and inspected for tag loss.

The fish were captured from farm ponds by electrofishing. Before being stocked in the experimental ponds, all fish were anesthetized with tricaine (MS-222), measured, and randomly assigned to treatment groups. Fish in the control group received no incision. Fish in the tagged group were held on their back as an incision was made through the abdominal wall into the ventral side of the abdominal cavity. The incision was made parallel to the vertebral column, slightly anterior to the anus and off the midline. A number 11 scalpel blade was used to make an incision slightly longer than the width of the tag disk. The streamer was folded down against the disk as the disk was inserted and twisted to ensure attachment.

For the replicated 30-d pond experiment, parametric tests were used for statistical analyses after arcsine transformation of percent survival data (as described in Sokal and Rohlf 1981). All tests were repeated with equivalent nonparametric tests applied to untransformed data. Only the results of the parametric tests are reported because both types of analysis resulted in the same conclusion.

Results

In our replicated 30-d experiment, the overall survival of tagged (86%) and untagged fish (82%)

TABLE 2.—Parametric statistical test results[a] obtained from arcsine-transformed percent survival data for tagged and untagged largemouth bass in a 30-d pond experiment to evaluate internal anchor tags.

Test	Factors evaluated	df	P
One-tailed paired *t*-test	Treatments (tagging, pooled data)	17	0.9575
Two-way analysis of variance	Treatments	1	0.1010
	Size category[b]	2	0.0202
	Treatment × size interaction	2	0.1890
Two-way analysis of variance	Treatments	1	0.7639
	Month[c]	2	0.0780
	Treatment × month interaction	2	0.5893

[a]Similar P-values were obtained with nonparametric tests.
[b]Size categories are described in Table 1.
[c]September, October, and November 1987.

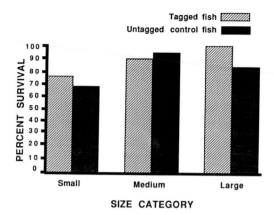

FIGURE 1.—Percent survival within size-classes (Table 1) of tagged and untagged largemouth bass in a 30-d pond experiment to evaluate internal anchor tags. Data represent three monthly replicates combined.

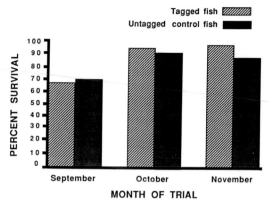

FIGURE 2.—Percent survival of tagged and untagged largemouth bass in three monthly replicates of a 30-d pond experiment to evaluate internal anchor tags.

did not differ significantly (Table 2). Survival varied significantly among size categories but had no relationship to treatment. Survival exceeded 80% for large- and medium-size fish and was slightly less for small fish (Figure 1). The variability in survival by month approached significance but had no relation to treatment (Table 2). Lower survival rates occurred in September (Figure 2), when water temperatures were warm (approximately 22.5°C at harvest). The 90-d unreplicated experiment yielded survival rates for tagged and untagged fish of 63% and 80%, respectively. In these two experiments, 189 fish were tagged and 194 fish were used as controls; 139 tagged fish and 157 control fish survived, and very few dead fish remains were observed, presumably due to scavengers. All of the untagged fish recovered were examined for any indication of tag loss (open wounds or scars), but no such evidence were found. In some fish, the surgery area was inflamed and swollen, although in no case was there gross tissue deterioration. Also, we did not observe any fish with tags that were migrating into the body cavity as previously noted by Topp (1962) with older versions of internal tags.

Discussion

Our results suggested that tagging surgery does not cause significant mortality over 30 d. Monthly mortality as high as 32% (Figure 2) did occur for both treatment groups combined, and handling stress and water temperatures were suspected as the major contributing factors. Because the fish used in our study varied considerably in size, we

suspected that some mortality of small fish was due to predation by the larger fish, but the tagging procedure had no effect on survival of fish for any size-class (Figure 1). Visual observation of each tag wound showed no evidence of serious infection in either the 30-d or the 90-d experiment.

Vogelbein and Overstreet (1987) provided a very thorough description of the pathological tissue changes observed in spot *Leiostomus xanthurus* and spotted seatrout *Cynoscion nebulosus* after the fish were tagged with internal anchor tags in the abdominal cavity. They found that both species exhibited acute inflammation and irritation around the tag wound 4 h to 4 d posttagging, and the tag disk eventually became encapsulated by fibrogranular tissue. Numerous complications were observed with the tagged spot and included secondary mycotic and bacterial infections, gross peritonitis, penetration of the ovaries, and displacement of the tag disk into the swim bladder or the intestines. Complications with the spotted seatrout were less severe. Melanomacrophage centers and small capsular granulomas were the only signs of secondary infection; there was no evidence of tag dislocation or expulsion. In Winters' (1977a) experiment with Atlantic herring marked with abdominally inserted magnetic tags, there was no evidence of infection after 12 months. Small cuts of the testes were reported in 2 of the 75 observed herring.

In addition to negligible tag-induced mortality for largemouth bass, there are several other advantages offered by the surgically implanted anchor tags we evaluated. One of the most important is high tag retention over extended time periods. We did not find any cases of tag loss during our experiments. In an experiment involving double-tagged largemouth bass and voluntary tag returns by anglers, tag loss from internal anchor tags was reported as less than 2% (S. Quinn, Georgia Department of Natural Resources, personal communication). Another advantage is the ability to print information on both the internal and external portion of the tag. One disadvantage of surgically implanted tags is the time required to anesthetize fish; this step is sometimes omitted in field situations.

Acknowledgments

This study is part of the Tennessee River fishery research program funded by the Alabama Game and Fish Division. Our initial use of surgically implanted anchor tags resulted from a recommendation by William Reeves of the Alabama Game and Fish Division. We thank the following Auburn University students who volunteered many hours of field help: William Bridges, Alan Kinsolving, Kedric Nutt, Steve Meyers, and Dennis Pridgen. We also thank Vern Minton of the Alabama Marine Resources Division for his instruction and training in the tagging method.

References

Bagenal, T. 1978. Methods for assessment of fish production in freshwaters. IBP (International Biological Programme) Handbook 3.

Elam, L. 1971. Evaluation of fish tagging methods. Texas Parks and Wildlife Department, Coastal Fisheries Project, Report 39-55, Austin.

Gilbert, R. J., and J. E. Hightower. 1981. Assessment of tag loss and mortality of largemouth bass in an unfished reservoir. University of Georgia, Georgia Cooperative Fish and Wildlife Research Unit, Athens.

Matlock, G. C., and J. E. Weaver. 1979. Fish tagging in Texas bays during November 1975–September 1976. Texas Parks and Wildlife Department, Management Data Series 1, Austin.

Rounsefell, G. A. 1975. Ecology, utilization, and management of marine fisheries. Mosby, St. Louis, Missouri.

Rounsefell, G. A., and W. H. Everhart. 1953. Fisheries science, its methods and applications. Wiley, New York.

Sokal, R. R., and F. J. Rohlf. 1981. Biometry. Freeman, San Francisco.

Topp, R. W. 1962. The tagging of fishes in Florida, 1962 Program. Florida State Board of Conservation, Marine Laboratory Professional Papers Series 5, St. Petersburg.

Topp, R. W. 1967. An internal capsule fish tag. California Fish and Game 53:288–289.

Vogelbein, W. K., and R. M. Overstreet. 1987. Histopathology of the internal anchor tag in spot and spotted seatrout. Transactions of the American Fisheries Society 116:745–756.

Wilbur, R. L., and R. M. Duchrow. 1973. Differential retention of five Floy tags on largemouth bass (*Micropterus salmoides*) in hatchery ponds. Proceedings of the Annual Conference Southeastern Association of Game and Fish Commissioners, 26:407–413.

Winters, G. H. 1977a. Estimates of tag extrusion and initial mortality in Atlantic herring (*Clupea harengus harengus*) released with abdominally inserted magnetic tags. Journal of the Fisheries Research Board of Canada 34:354–359.

Winters, G. H. 1977b. Healing of wounds and location of tags in Atlantic herring (*Clupea harengus harengus*) released with abdominally inserted magnetic tags. Journal of the Fisheries Research Board of Canada 34:2402–2404.

Wydoski, R., and L. Emery. 1983. Tagging and marking. Pages 215–237 in L. A. Nielsen and D. L. Johnson, editors. Fisheries techniques. American Fisheries Society, Bethesda, Maryland.

American Fisheries Society Symposium 7:121–126, 1990
© Copyright by the American Fisheries Society 1990

Abrasion and Protrusion of Internal Anchor Tags in Hudson River Striped Bass

MARK T. MATTSON

Normandeau Associates, Incorporated
25 Nashua Road, Bedford, New Hampshire 03102, USA

JOHN R. WALDMAN

Hudson River Foundation
122 East 42nd Street, Suite 1901, New York, New York 10168, USA

DENNIS J. DUNNING AND QUENTIN E. ROSS

New York Power Authority
123 Main Street, White Plains, New York 10601, USA

Abstract.—During March–May 1984, November 1985–May 1986, and December 1986–May 1987, 737, 18,488, and 9,473 Hudson River striped bass *Morone saxatilis* were captured, marked with internal-anchor–external-streamer tags, and released. None of the 400 fish recaptured within 7 months after release had tags that were abraded or anchors that were protruding from the insertion site. During the 1986–1987 tagging period, 130 striped bass were recaptured with tags in place 7 months or longer after marking. The tag legend was abraded but legible on 28% of these fish; on 13%, the tag number was abraded so that one or more digits were illegible. In addition, 5% of the recaptured striped bass had an infected tag-insertion site, and 11% had the anchor protruding. More than 7% of 958 striped bass tags returned by fishermen had an illegible tag number, and this percentage generally increased with time. At least some dermal irritation around the tag-insertion was observed by 26% of the 489 fishermen who caught tagged striped bass. Illegible return addresses probably reduced tag returns from fishermen; however, this reduction was partially offset by advertising and promotion of the tag-return program. We believe the observed tag abrasion and anchor protrusion was the result of reduced flexibility of the vinyl streamer as the tubing aged. Because of the tag design and insertion site, striped bass swimming close to the river bottom or other structures would subject the anterior, distal portion of the streamer to abrasion. Additionally, leverage applied to the stiffened streamer as the fish swims may cause anchor protrusion.

A fundamental assumption of any mark–recapture program is that the marks are recognized and reported (Cormack 1968; Ricker 1975; Seber 1982). Factors that differentially affect mark recognition and reporting should be identified and systematically examined to quantify the direction and degree of bias. Biases that are quantified can generally be reconciled with the mark–recapture program objectives either by statistically adjusting recapture rates, by restricting hypotheses, or by using a better mark.

Internal anchor tags with external vinyl tubing bearing the tag information are now widely used because they have high visibility and relatively long retention rates (Wydoski and Emery 1983; Dunning et al. 1987). Additionally, chemical treatment can reduce fouling problems that inhibit legibility of external streamer information after long periods (years) between release and recapture (Wydoski and Emery 1983). The high visibility and

long retention of internal-anchor–external-streamer tags has undoubtedly minimized the problems of mark recognition and nonreporting experienced with other physical marks, such as clipped fins. However, in this study, we observed two mechanisms by which internal-anchor–external-streamer-tag information can be lost: abrasion of tag legends and tag shedding due to anchor protrusion.

Loss of the tag, the return address, or the reward information may substantially reduce the quantity and quality of information recovered from fishermen. Loss of the unique tag number due to abrasion would prevent linking of release and recapture data for returned tags. Anchor protrusion was recently reported for spot *Leiostomus xanthurus* but not for spotted seatrout *Cynoscion nebulosus* (Vogelbein and Overstreet 1987), so this phenomenon may be specific to individual fish species. The objectives of our study were to (1) determine the tag abrasion rate for Hudson River striped bass *Morone saxatilis,* (2) observe the extent of tag shedding caused by anchor protrusion, and (3) evaluate the effects of tag abrasion and anchor protrusion on tag returns from fishermen.

Methods

A tagging program was conducted from March 1984 to May 1987 to provide information about the movements and population dynamics of Hudson River striped bass. Using otter trawls and Scottish seines, we tagged and released 737 stripped bass between 300 and 1,064 mm long (all lengths as total length) during a 5-d/week effort from March through May 1984. Sampling was conducted in the Hudson River between 37 and 100 km upriver from the southern tip of Manhattan Island, New York. In addition, from November 1985 through May 1986 and December 1986 through April 1987, we tagged and released predominantly immature striped bass (200–450 mm) in the lower Hudson River (adjacent to Manhattan Island).

All striped bass at least 200 mm long and in good condition at the time of capture were tagged and released with Floy internal-anchor–external-streamer tags (Floy Tag and Manufacturing, Seattle, Washington). Each tag is assembled from two components: a flexible external vinyl tube about 2 mm in diameter and 65 mm long, and a rigid, oval plastic anchor with a central 2-mm hole through which the tube is fitted. Two anchor sizes were used: a 16-mm by 5-mm anchor for tagging all fish 200–299 mm long and a 26-mm by 6-mm anchor for tagging fish at least 300 mm long. This tag was previously reported to be retained for relatively long times in Hudson River striped bass (2 years) and to have high absolute retention (98%) for fish held up to 180 d in outdoor pools (Dunning et al. 1987). The anchor of the tag was inserted internally by removing a scale located midway between the anus and the distal tip of the depressed pelvic fins and five or six scale rows dorsolaterally from the ventral midline. The selection of this anchor-insertion site was based on gross anatomical examination of organ placement in the peritoneal cavity. A horizontal incision about 5 mm long was made with a hooking movement of a curved-blade scalpel. The incision was made though the musculature but not through the peritoneum. The anchor was pushed through the incision until it punctured the peritoneum and entered the abdominal cavity and was set by a gentle pull of the external streamer. When set, the long axis of the anchor aligned with the long axis of the fish because the peritoneal cavity is narrow

compared to the length of the anchor. The incision was treated with a merbromin-based topical antiseptic, and the fish was released.

All striped bass caught during the tag-and-release phases of the study were first examined for tag wounds and tags. If a wound was observed at the tag insertion site, the condition of the wound was recorded (infected or healed). If a tag from this or other studies was present, the unique number and condition of the tag legend, as well as the condition of the insertion site, were observed and recorded. Recaptured fish were measured and released again if they were healthy and the tag legend was legible. The tag streamer was cut off close to the insertion site and retained for examination if the legend was judged difficult to read, particularly if the number was partly or completely missing. A new tag was placed in healthy recaptured fish from which abraded tags were removed, and these fish were released again. Fish in poor condition were killed and examined to determine the condition of the anchor-insertion area inside the peritoneal cavity. Tags that were difficult to read were examined to classify the condition of three parts of the legend: number, reward information, and return address. If one or more printed letters or numbers on the legend was missing or illegible, the legend was classified as partly or completely missing.

The yellow external streamer of each tag contained two lines of printed legend: a unique five-digit number and reward value (US $) on the first line (e.g., 12345 REWARD $5–$1000) and a return address on the second (MAIL TO HRF PO BOX 1731 GCS NY, NY 10163). The legend was left-justified towards the proximal end of the tag so that line 1 ended about 10 mm from the distal end, and line 2 ended 2–3 mm from the distal end. The tag-return program was advertised locally (New York–New Jersey) by posters, presentations, and direct mailing of brochures to fishing clubs, marinas, and bait and tackle shops. Monthly press releases, advertisements, and popular articles in fishing magazines and newspapers also promoted the tagging program along the U.S. east coast. Each fisherman who returned a tag to the Hudson River Foundation (HRF) received the minimum reward value ($5 or $10) and a questionnaire. When the questionnaire was completed and returned, the individual was entered into an annual drawing from all tags returned that year for nine prizes (five $100, two $500, and two $1,000). Tags returned by fishermen were also examined to classify the condition of the legend, and the fish-

TABLE 1.—Incidence of tag abrasion and internal anchor protrusion for Hudson River striped bass tagged during three tagging periods from 1984 through 1987 and recaptured either within the same tagging period or during later tagging periods.

Variable	Tagging year		
	1984	1985–1986	1986–1987
Release information			
Tagging period within year	9 Apr–7 Jun	11 Nov–12 May	21 Dec–8 May
Total length of tagged fish (mm)	≥300	≥200	≥200
Number of fish tagged	737	18,488	9,473
Number of fish examined for tags	1,253	19,682	10,069
Fish recaptured within the same tagging period			
Number of fish recaptured	0	249	151
Days at large		0–146	0–95
Number of fish with abraded tags		0	0
Number of fish with anchor protrusion		0	0
Fish recaptured from previous tagging periods			
Number of fish recaptured		2	113[a]
Days at large		717	228–495
Number of fish with abraded tags		0	20
Number of fish with anchor protrusion		0	14
Fish recaptured with missing tags or tag numbers			
Number of fish with abraded numbers (one or more digits missing)	0	0	17
Number of fish with tag wounds	0	0	4

[a] Three fish from previous tagging periods were recaptured twice in 1986–1987.

ermen were questioned about the condition of the tag-insertion site.

Proportions were compared statistically with the chi-square test (Snedecor and Cochran 1980; Sokal and Rohlf 1981).

Results

Tag Abrasion and Anchor Protrusion

Although the external streamer is relatively flexible when new, tags recovered from fish recaptured at least 7 months after release ($N = 72$) had stiffened and exhibited a set curvature such that the distal tip of the streamer oriented towards the posterior of the fish. Abraded tags exhibited a consistent pattern of wear on the anterior edge of the curvature of the external streamer. Little or no abrasion was seen at or near the proximal end of the streamer (<10 mm from the insertion site). Between about 11 and 40 mm from the insertion site, there was a region of deep, longitudinal scratches that lessened and graded into a finely abraded surface at the distal end.

Anchor protrusion was typically observed ($N = 12$) with the anterior edge of the anchor protruding 1–4 mm through the body wall about 5–10 mm anterior to the tag insertion site. For two fish, the entire anchor had worked through the body wall, and the tag was retained by a small loop of flesh.

The site of anchor protrusion was either well healed or exhibited some degree of inflammation or infection. Examination of the gross anatomy of the anchor-protrusion site of four fish revealed an adhesion of scar tissue encasing the anchor. The gross anatomy of the adhesion was similar for fish with or without anchor protrusion. Fish with well-healed tag-insertion or -protrusion sites typically exhibited scar tissue in direct contact and firmly binding most or all of the anchor. Fish with inflamed or infected sites had a fluid-filled area within the adhesion in the vicinity of the anchor.

Tagging Program

Tag abrasion or anchor protrusion was not observed among 400 stripped bass tagged, released, and recaptured within the same tagging period (Table 1). In the 1985–1986 tagging period, tags from 249 fish released and recaptured after 0–146 d (mean, 36 d; SD, 37 d) exhibited no legend abrasion or anchor protrusion. Likewise, during the 1986–1987 tagging period, tags from 151 fish released and recaptured after 0–95 d (mean, 24 d; SD, 24 d) exhibited no abrasion or anchor protrusion.

Both tag abrasion and anchor protrusion were observed among 115 recaptured striped bass that were verified as tagged and released in prior tagging periods (Table 1). In 1985–1986, two re-

TABLE 2.—Anchor position and condition of the anchor-insertion site for 130 Hudson River striped bass tagged with internal anchor tags in 1984 and 1985–1986 and recaptured during the 1986–1987 tagging period.[a]

| | Condition anchor-insertion site | | | |
| | Healed | | Infected | |
Anchor position	N	%	N	%
Internal	109	84	7	5
Protruding	2	2	12	9
Total	111	86	19	14

[a]Three fish were recaptured twice during 1986–1987; 17 fish were recaptured with missing tag numbers and were assumed to have been tagged and released before 1986–1987.

captured fish had tags from 1984. Both of these fish had been released nearly 2 years before recapture (717 d for each fish), and the tags did not exhibit abrasion or anchor protrusion. In 1986–1987, 113 fish were recaptured from 1985–1986 after 228–495 d (mean, 358 d; SD, 57 d). Tags from 20 of the 113 fish exhibited abraded legends but legible tag numbers (all digits readable), and 14 of the 113 fish exhibited anchor protrusion. Recaptured fish with abraded tags in 1986–1987 were at large longer (range, 361–495 d; mean, 419 d; SD, 40 d) than recaptured fish without abraded tags (range, 228–483 d; mean, 346 d; SD, 53 d). An additional 17 fish were recaptured in 1986–1987 that had tag numbers partly or completely missing, and four fish without tags had wounds at the tag-insertion site. If the fish with missing tag numbers were tagged before 1986–87, 28.5% (37/130) of the Hudson River striped bass recaptured at least 7 months after release lost part or all of the tag legend, and 13.1% (17/130) of the fish lost part or all of the unique tag number due to abrasion.

An uninfected or well-healed tag-insertion site without anchor protrusion was observed on all of the 400 striped bass tagged, released, and recaptured within the same tagging period and on most (84%) of the 130 fish recaptured from prior periods (Table 2). Significantly more fish had infected

anchor-insertion sites if the anchor also protruded ($P < 0.001$).

Typically, only line 1 (tag number and reward information) or line 2 (return address) of the streamer legend was abraded so one or more numbers or letters were missing. If line 1 was illegible, line 2 was legible, and vice versa. Among 29 abraded tags from the 1986–1987 tagging period that were classified for legibility of the two lines of the external streamer legend, only 17% had partly or completely missing legends on both lines (Table 3). Significantly ($P < 0.01$) more of these abraded tags had line 1 of the legend partly or completely missing (69%), whereas line 2 was frequently unabraded and legible (55%).

Returns from Fishermen

Abrasion of both the tag number and other parts of the tag legend generally increased with time. Abraded numbers were observed on line 1 of 7.4% of the 958 tags returned to HRF by fishermen from fish tagged and released during 1985–1986 (Table 4). Although anchor protrusion was not specifically identified, 26.4% of the 489 fishermen who responded to a question about the condition of the tag-insertion site indicated that at least some infection or irritation was observed. Abrasion of one or more digits on part or all of the tag legend was observed on 25.7% (179/697) of the tag returns received by HRF from fishermen between March 1986 and February 1987 from fish tagged and released in 1985–1986 (Table 5).

Discussion

Tag abrasion was time-dependent, occurred cumulatively, and caused complete or partial loss of the tag legend from 28.5% of the Hudson River striped bass recaptured more than 7 months after tagging. Days at large for fish recaptured with partially or completely illegible tag numbers could not be determined. Three observations from this study suggest that tag abrasion becomes significant after 7 months: the absence of abraded tags on 400

TABLE 3.—Number (N) and percentage of internal anchor tags exhibiting legend abrasion for 29 Hudson River striped bass classified by the condition of tag number and reward information (line 1 on the vinyl tubing) and return address (line 2) for the 1986–1987 annual tagging period.

| | Portion of tag legend examined | | | | | |
| | Line 1 | | Line 2 | | Both lines | |
Condition	N	%	N	%	N	%
Not abraded, legible	2	7	16	55	0	0
Abraded, legible	7	24	3	10	0	0
Partly or completely missing	20	69	10	35	5	17

TABLE 4.—Number and percentage of abraded, illegible tag numbers by recapture month for internal anchor tags returned by fishermen from 958 Hudson River striped bass tagged and released during 1985–1986 and recaptured during March 1986–September 1987.

Recapture year and month	Proportion (abraded/ recovered)	Percent abraded	Cumulative percent abraded
1986			
Mar	0/5	0.0	0.0
Apr	0/78	0.0	0.0
May	0/121	0.0	0.0
Jun	0/83	0.0	0.0
Jul	1/40	2.5	0.3
Aug	1/48	2.1	0.6
Sep	1/52	1.9	0.7
Oct	7/82	8.5	2.0
Nov	12/83	14.5	3.7
Dec	1/43	2.3	3.6
1987			
Jan	0/4	0.0	3.6
Feb	2/7	28.6	3.9
Mar	1/11	9.1	4.0
Apr	9/72	12.5	4.8
May	11/66	16.7	5.8
Jun	14/57	24.6	7.0
Jul	2/42	4.8	6.9
Aug	6/28	21.4	7.4
Sep	3/36	8.3	7.4
Mar 86–Sep 87	71/958	7.4	

fish tagged less than 146 d before recapture, the presence of some abraded but legible tags on 115 fish tagged at least 361 d before recapture, and the increase in the percentage of abraded tags returned by fishermen after 7 months.

We believe tag abrasion occurs in striped bass because they are found in close proximity to a variety of hard substrates, including rock, gravel, and sand. The horizontal pressure that is applied in an anterior-to-posterior direction as the fish swims and reduced flexibility of the external streamer

TABLE 5.—Number and percentage of internal anchor tags with illegible digits or letters from Hudson River striped bass tagged in 1985–1986. Tag returns are arranged by century groupings according to the order they were received from fishermen.

Tag grouping	Number of illegible tags/number of tags returned as		
	Proportion (illegible/recovered)	Percent illegible	Cumulative percent illegible
1–99	0/99	0.0	0.0
100–199	0/97	0.0	0.0
200–299	1/97	1.0	0.3
300–399	16/98	16.3	4.3
400–499	38/100	38.0	11.2
500–599	50/99	50.5	17.8
600–699	61/90	67.8	24.4
700–799	13/17	76.5	25.7
Total	179/697	25.7	

over time appears to create a permanent curvature; therefore, the anterior edge of the curved external streamer is subjected to abrasion. Differences in abrasion observed along the length of the streamer are consistent with a decline in resistance from the anchor-insertion site to the distal end of the streamer tubing. Firm resistance close to the anchor-insertion site would produce deep scratches, and lessened resistance and a more horizontal position due to curvature of the streamer near the distal end would result in fine abrasion. The unique number, located on the streamer near the distal end, would be abraded and lost if oriented on the anterior edge of the curvature.

Anchor protrusion may be caused by several factors. Leverage on the external streamer caused by reduced flexibility, water resistance, and contact with substrates could press the anterior edge of the tag anchor against the peritoneal wall. The rigid plastic anchor has a sharp, thin edge that could then cut through the body wall musculature anterior to the tag insertion site. The irritation caused by constant pressure also may reduce the ability of some fish to heal the wound and thus facilitate anchor protrusion. Anchor protrusion and active histological expulsion of internal anchor tags was observed in spot by Vogelbein and Overstreet (1987), and those authors suggested that fish such as spot (or striped bass) with laterally compressed bodies have relatively small, narrow peritoneal cavities, which may facilitate anchor-induced irritation.

Some tags are undoubtedly lost as a result of complete protrusion of the anchor. Partial anchor protrusion appears to be an early stage of tag shedding. Wounds observed at the tag site of four fish in this study may have been caused either by tag shedding or removal of the streamer by fishermen participating in the tag-return program; however, all fish with tag wounds observed during the tagging programs were not killed and examined for the presence of the internal anchor in the abdominal cavity. Therefore, we could not discriminate between tag wounds resulting from shedding (no anchor present) and tag wounds from streamers cut by fisherman and mailed to HRF (anchor present). In a subsequent striped bass tagging program (1988–1989), all fish caught with wounds at the tag insertion site were killed and examined; 38% (10 of 26) had anchors present, indicating the external streamer had been cut by fishermen.

It remains uncertain if all tags associated with anchor protrusion are ultimately shed. Tags with well-healed insertion sites and anchor protrusion wounds may not be shed because they are held in place by scar tissue. Therefore, it was not possible to calculate an accurate long-term tag shedding rate based on the presence of tag wounds and anchor protrusion.

Tag abrasion may result in the loss of the tag number and other portions of the external streamer legend. Without the unique tag number, recapture data for fish caught by fishermen, by state or federal biologists, or in this tagging study cannot be linked with release data. Therefore, at least time and distance traveled and growth data are lost. In this program, 13% of the fish recaptured after 7 months lost part or all of the tag number due to abrasion. The apparent time dependence of abrasion and the presence of at least some abrasion on 28% of the fish recaptured after 7 months suggested that an increase in the proportion of abraded tags would occur over time.

It was probably not coincidental that 13% of the tags recovered after 7 months had missing numbers, whereas about twice that many tags (28%) exhibited some abrasion on the legend. The incidence of illegible tag numbers and reward values on line 1 was inversely related to illegible return addresses on line 2. Equal spacing in the placement of line 1 and line 2 around the circumference of the tag streamer probably precluded simultaneous abrasion of both lines. If line 1 was oriented to the anterior of the fish and subjected to abrasion, line 2 was posterior and not abraded, and vice versa. If tags returned by fishermen had illegible numbers on line 1, they would generally not have illegible return addresses. Tags with illegible return addresses would not be returned to the HRF unless the fishermen were aware of the tagging program through promotional efforts and advertising. However, all HRF tags recovered during the annual tagging periods would be recognized and reported by the tagging crews. Therefore, the percentage of tags recovered by the tagging crews with missing numbers on line 1 was not affected by abrasion and could be used to estimate the expected reduction in tag returns from fishermen due to legend abrasion. If both fishermen and the tagging crews sampled the same population of tagged and untagged striped bass, about 13% of the fishermen tag returns should exhibit abrasion on line 1. However, only 7.4% of the tags returned by fishermen had illegible numbers on line 1. Therefore, promotional efforts and advertising appeared to offset the effects of missing return addresses on tag returns from fishermen by 5.7%.

Conclusions

Tag abrasion and anchor protrusion results in the loss of movement and age information about recaptured fish. Mark–recapture estimators that are not adjusted for tag loss due to anchor protrusion and tag number loss due to abrasion will be biased because the number of recaptured fish attributed to each release cohort will be underestimated. A redesigned tag would be the best solution to the tag abrasion and anchor protrusion problems.

Acknowledgments

Funding for this project was provided by Central Hudson Gas and Electric Corporation, Consolidated Edison Company of New York, Inc., the New York Power Authority, Niagara Mohawk Power Corporation, and Orange and Rockland Utilities under terms of the Hudson River Cooling Tower Settlement Agreement. This project was developed in cooperation with the New York State Department of Environmental Conservation (NYSDEC); biologists from NYSDEC and the Striped Bass Emergency Committee suggested the use of the internal anchor tag. Field work was conducted by Normandeau Associates, Inc., under the direction of Michael J. Ricci. We are also grateful for the constructive comments provided by the editor, two anonymous reviewers, and Eric D. Prince.

References

Cormack, R. M. 1968. The statistics of capture–recapture methods. Oceanography and Marine Biology: An Annual Review 6:455–506.

Dunning, D. J., Q. E. Ross, J. R. Waldman, and M. T. Mattson. 1987. Tag retention and tagging mortality in Hudson River striped bass. North American Journal of Fisheries Management 7:535–538.

Ricker, W. E. 1975. Computation and interpretation of biological statistics of fish populations. Fisheries Research Board of Canada Bulletin 191.

Seber, G. A. F. 1982. The estimation of animal abundance, 2nd edition. Charles Griffin, London.

Snedecor, G. W., and W. G. Cochran. 1980. Statistical methods, 7th edition. Iowa State University Press, Ames.

Sokal, R. R., and F. J. Rohlf. 1981. Biometry. Freeman, San Francisco.

Vogelbein, W. K., and R. M. Overstreet. 1987. Histopathology of the internal anchor tag in spot and spotted sea trout. Transactions of the American Fisheries Society 116:745–756.

Wydoski, R., and L. Emery. 1983. Tagging and marking. Pages 215–238 in L. A. Nielsen and D. L. Johnson, editors. Fisheries techniques. American Fisheries Society, Bethesda, Maryland.

American Fisheries Society Symposium 7:127–133, 1990

Anchor Tag Return Rates for Pacific Herring in British Columbia

C. W. Haegele

Department of Fisheries and Oceans, Biological Sciences Branch, Pacific Biological Station
Nanaimo, British Columbia V9R 5K6, Canada

Abstract.—Pacific herring *Clupea harengus pallasi* were tagged with external anchor tags over a 4-year period (1979–1983), and tag recoveries were obtained until 1986. There were 1,824 tags returned from 310,068 tagged fish. It was estimated that initial tagging mortality was 60% and that 50% of tags were shed within 2 weeks of release. Annual mortality of tagged fish was considerably higher than natural mortality. A further 25% of tags were shed within 1 year of release. Approximately 35% of the tags in the catch were reported. Fisheries were restricted temporally and spatially, and some of the tagged stocks were not fished. The tag and techniques used in this study were not suitable for the fish or the fisheries. Any further tagging studies require better fish-handling techniques to reduce tagging mortality; a more suitable tag with better retention, lower tag-induced mortality, and higher detection; and, ideally, more substantial fisheries over a wider spectrum of time and space.

Tagging studies to determine stock distribution and movements of Pacific herring *Clupea harengus pallasi* were initiated in British Columbia in winter 1979. An external tag seemed advisable for these studies because the British Columbia Pacific herring catch from spring fisheries is processed manually for sac-roe extraction, and other catches, mostly in winter, are also handled as they are sorted for size, filleted, etc. Experiments with Atlantic herring *C. h. harengus* from 1974 to 1978 showed that anchor tags were the most suitable external tag; the recovery rate was 4.6%, and anchor tags had the highest tag application rate of several tags evaluated (Nakashima and Winters 1984). Therefore, the anchor tag was chosen for the British Columbia tagging studies.

During a tag-recovery trial in November 1978, in which fish in a floating live-pond were tagged and processed by sorting for size 3 d later, 27% of tags were recovered (Hay et al. 1979). It was estimated that 50% of the tags were lost in the pond or during transport and that about 50% of tagged fish entering the processing plant were reported. In April 1979, tag reporting was estimated for fish processed for roe by tagging fish during unloading from a packing barge (Hay and Mitchell 1979). Of 298 tagged fish, 84% were recovered. Follow-up tag-recovery trials were also conducted during this tagging study. In November 1981, tagged Pacific herring were introduced into the seine catches from 33 vessels fishing for food herring. The catch was transported and delivered to 13 processing plants by 48 packer vessels (Haegele 1982). The average tag-return rate for these trials was 37% (95% confi-

dence interval, 29–45%). It was estimated that 25% of tags were lost during pumping of the catch from the seine to the packer vessel. In March 1983, tagged Pacific herring were introduced into five seine catches and in the holds of nine packer vessels transporting fish caught in gill nets (Haegele 1983). These fish were processed for roe in nine processing plants, and the average tag-return rate for reddish-colored (international orange, orange, and pink) tags was 34% (95% confidence interval, 20–47%). Anchor-tag retention by Pacific herring held in captivity appeared to be acceptable (Hay 1981). Fish were tagged in July 1979 and kept in 3-m³ fiberglass tanks supplied with constantly running seawater. Ten percent of tags were shed during a 6-month test, whereas 22% and 26% of tags were shed during two 18-month tests. There were no estimates of tagging mortality made before this tagging study, although it was recognized from experience in handling fish for transport to the laboratory that handling mortality may exceed 50%.

As the British Columbia Pacific herring tagging studies progressed, it became apparent that tag returns were substantially lower than expected from the information available on tagging mortality, tag loss, and tag reporting. Returns, especially for periods in excess of 3 months, were too few to satisfy the experimental design or permit the reliable interpretation of results. In 1982, the tagging studies were restricted geographically to address more limited questions of stock identity and movement. Tagging ended in 1983, whereas tag returns were actively solicited until 1986. Release and return information from the

1979–1983 tag releases are analyzed in this paper to determine expected tag-return rates for anchor tags on Pacific herring. Probable causes for low returns are identified.

Methods

Tagging.—Pacific herring were tagged with Floy anchor tags (type FD-68BC). In 1979–1981, only long tags colored international orange were used. In 1982–1983, the same long tags and short tags in a variety of colors were used. Long tags had a 42-mm-long plastic portion bearing the legend and tag number. This plastic portion was connected to a 10-mm-long toggle by a 15-mm-long piece of monofilament. The plastic portion and monofilament of short tags were 32 and 8 mm long respectively. Tags were inserted with an applicator gun in the dorsal musculature near the posterior margin of the dorsal fin in 1979–1981 and near the anterior margin of the dorsal fin in 1982–1983.

Fish were captured by seine except during fall taggings, when trawls were also used. When captured by seine, fish were dipnetted from the seine into a plastic tub partially filled with seawater. Fish were taken individually from the tub, tagged, and returned to the seine, except in 1979 and 1980, when fish were placed in a holding pond secured to the side of the vessel. After tagging was completed, the fish were released from the seine or, when holding ponds were used, the tagged fish in the pond were released in synchrony with the fish from the seine. When trawls were used, the codend was unlaced along part of its length, the floating codend was secured to the side of the vessel, and fish were dipnetted from it as from seines. Fish tagged from trawls were placed in tanks supplied with running seawater, held until the completion of tagging, and then released. Between 500 and 2,000 Pacific herring were tagged per set or tow. Fish were held in the seine between 1 and 2 h while tagging was done. All Pacific herring for spring releases were captured with commercial seines, which were 411 m long and 73 m deep. Most Pacific herring for winter releases were captured with a smaller seine, 274 m long and 58 m deep, which effectively fished approximately one-third the volume of the commercial seines. In spring releases, several commercial seine vessels were deployed at the same time, and much of the tagging was done by the vessel crew supervised by one technical representative. Winter tagging was done mostly from a

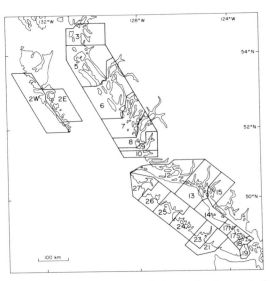

FIGURE 1.—Fisheries statistical areas for the coast of British Columbia.

research vessel, and technical personnel assisted in the fishing and did most of the tagging.

Recovery.—Posters describing the tagging program were displayed in fish processing plants and fisheries offices, where postage-paid tag-return envelopes were also made available. A Can$2 reward was paid to the finder from 1979 to 1981; thereafter, returns were entered in a semiannual prize drawing.

Analysis.—The British Columbia coast is divided into statistical areas for the collection of catch and biological data. An adaptation of this system was used in the treatment of the tagging data (Figure 1). Each area generally encompasses bodies of water and shorelines that support what may be considered stocks of Pacific herring on the basis of consistent records of spawn over time and spawn timing (Hay 1986).

Migratory movements and fisheries of Pacific herring are seasonal and conform to the following seasons, which were used for compiling all tagging data: spring (February 1–April 30), summer (May 1–July 31), fall (August 1–October 31), and winter (November 1–January 31). Pacific herring generally reside inshore near the spawning grounds in the spring, and most spawning occurs in March (Hourston and Haegele 1980). The roe fishery occurs just before or at spawning. Spring tagging and release of Pacific herring usually also preceded spawning, and when there was a fishery in the area, releases of tagged fish usually (not always) were made before the fishery. Immedi-

TABLE 1.—Release and return information for anchor-tagged Pacific herring released in the spring near spawning grounds. Sp = spring; Su = summer; Fa = fall; Wi = winter.

| Release | | | Number of returns by season after release | | | | | | | | | | | | | |
Area	Year	Number of fish tagged	0 Sp	1 Su	2 Fa	3 Wi	4 Sp	5 Su	6 Fa	7 Wi	8 Sp	9 Su	10 Fa	11 Wi	12 Sp	Total returns
2W	1980	2,858	0	0	0	1	0	0	0	0	2	0	0	0	0	3
2W	1981	4,459	0	0	0	0	0	0	0	0	1	0	0	0	1	2
2E	1980	4,861	2	0	0	0	4	1	0	1	1	0	0	0	1	10
2E	1981	7,887	7	2	0	0	14	0	0	0	2	0	0	0	2	27
3	1980	2,432	1	2	0	0	1	0	0	0	0	0	0	0	0	4
3	1981	2,972	0	1	0	7	1	0	0	0	0	0	0	0	0	9
3	1982	8,281	26	1	0	0	1	0	0	0	1	0	0	0	0	29
5	1980	4,846	0	1	0	2	0	0	0	0	0	0	0	0	0	3
5	1981	9,932	12	0	0	3	1	0	0	0	1	0	0	0	0	17
5	1982	9,162	33	0	0	0	3	0	0	0	1	0	0	0	0	37
6	1980	2,470	0	0	0	0	0	0	0	0	0	0	0	0	1	1
6	1981	4,953	1	1	0	2	1	0	0	0	0	0	0	0	0	5
7	1980	3,617	0	2	0	0	1	0	0	0	0	0	0	0	0	3
7	1981	4,939	0	1	0	0	8	0	0	0	1	0	0	0	1	11
8	1980	1,949	0	0	0	0	0	0	0	0	1	0	0	0	0	1
8	1981	2,448	0	0	0	0	0	0	0	0	0	0	0	0	0	0
9	1980	987	0	0	0	0	0	0	0	0	0	0	0	0	0	0
9	1981	990	0	0	0	0	1	0	0	0	0	0	0	0	0	1
10	1980	982	0	0	0	0	0	0	0	0	0	0	0	0	0	0
10	1981	993	0	0	0	0	0	0	0	0	0	0	0	0	0	0
12	1980	1,879	0	2	0	0	0	0	0	0	0	0	0	0	0	2
12	1981	3,455	1	2	0	0	0	0	0	0	0	0	0	0	0	3
13	1980	1,037	6	0	1	0	0	0	0	0	0	0	0	0	0	7
13	1981	2,941	57	2	0	0	0	0	0	0	1	0	0	0	0	60
14	1980	3,474	0	2	0	0	0	0	0	0	0	0	0	0	0	2
14	1981	8,364	1	0	0	1	6	0	0	0	0	0	0	0	1	9
15	1980	1,676	3	4	0	0	0	0	0	0	0	0	0	0	0	7
15	1981	1,987	0	1	1	1	2	0	0	0	0	0	0	0	0	5
17N	1980	497	0	0	0	0	0	0	0	0	1	0	0	0	0	1
17N	1981	3,492	1	0	0	1	0	0	0	0	0	0	0	0	0	2
17S	1980	1,913	0	0	0	0	1	0	0	1	3	0	0	0	0	5
17S	1981	2,945	1	0	0	1	0	0	0	0	1	0	0	0	0	3
18	1980	782	0	1	0	0	0	0	0	0	0	0	0	0	0	1
18	1981	2,971	0	0	0	2	0	0	0	0	0	0	0	0	2	4
23	1980	2,913	0	1	0	1	1	0	0	0	3	0	0	0	0	6
23	1981	2,964	13	0	0	0	3	0	1	0	0	0	0	0	0	17
23	1982	17,860	216	1	0	0	26	0	0	0	5	0	0	0	0	248
23	1983	6,438	68	0	0	0	4	0	0	0	0	0	0	0	0	72
24	1980	3,333	0	0	0	0	4	1	0	0	2	0	0	0	0	7
24	1981	4,437	9	0	0	3	2	0	0	0	2	0	0	0	0	16
24	1983	6,336	0	0	0	0	2	0	0	0	0	0	0	0	0	2
25	1980	1,537	0	0	0	0	1	0	0	0	0	0	0	0	0	1
25	1981	1,992	0	0	0	0	0	0	0	0	0	0	0	0	0	0
25	1983	3,669	21	0	0	0	2	0	0	0	0	0	0	0	0	23
26	1980	948	0	1	0	0	0	0	0	0	0	0	0	0	0	1
27	1980	1,956	0	0	0	0	0	0	0	0	0	0	0	0	0	0
27	1983	1,757	0	0	0	0	1	0	0	0	0	0	0	0	0	1
Total		175,571	479	28	2	25	91	2	1	2	29	0	0	0	9	668

ately after spawning, Pacific herring migrate to offshore feeding areas, where they remain until early fall. There were no major Pacific herring fisheries in the summer or fall. Fall releases were made in September, before migration to inshore holding areas, where Pacific herring reside for the winter. There are food and bait fisheries in the winter on these aggregations. Pacific herring were tagged both before and after this fishery. Minor catches are made in late winter, mostly for bait

and aquarium food. In late winter and early spring, Pacific herring migrate to the vicinity of the spawning grounds.

Only the returns from roe fisheries were used in analyzing return rates because these were the only coastwide fisheries. Returns in the season of release were excluded for spring releases because some of these releases were made after the fishery. The analysis was also done for selected releases in each of the three seasons. These

TABLE 2.—Release and return information for anchor-tagged Pacific herring released in the fall on offshore feeding grounds. Sp = spring; Su = summer; Fa = fall; Wi = winter.

Release			Number of returns by season after release													
Area	Year	Number of fish tagged	0 Fa	1 Wi	2 Sp	3 Su	4 Fa	5 Wi	6 Sp	7 Su	8 Fa	9 Wi	10 Sp	11 Su	12 Fa	Total returns
21	1980	13,726	1	4	7	0	0	3	4	0	0	0	2	0	0	21
21	1981	10,191	1	3	24	0	0	1	11	0	0	0	1	0	0	41
Total		23,917	2	7	31	0	0	4	15	0	0	0	3	0	0	62

represent the highest recovery rates obtained and are for 1981 area 21 fall releases, 1981 area 17S winter releases, and 1982 area 23 spring releases. For the 1982 area 23 spring releases, returns in the season of release are included because tagging preceded the fishery.

Results

The catch in tagging sets was larger for spring releases than for winter releases. Spring catches ($N = 218$) ranged from 0.2 to 400 tonnes and averaged 31 tonnes with a median catch of 10 tonnes. Winter catches ($N = 98$) ranged from 0.1 to 150 tonnes, averaged 17 tonnes, and had a median of 5 tonnes.

From winter 1979 to winter 1983, 310,068 tagged Pacific herring were released. There were 1,824 returns within 3 years of release (Haegele 1986). Returns were mostly from roe fisheries in the spring (64.1%) and from winter food and bait fisheries (33.7%). The remaining 2.2% were returned from summer and fall fisheries targeting other species. Spring releases were made coastwide (Table 1), fall releases were made only in area 21 (Table 2), and winter releases were made principally in areas 13 and 17S (Table 3). Pacific herring fisheries did not occur in all areas during the study period. The food and bait fishery was confined to areas 5 and 17S, whereas roe fisheries were restricted to different areas in different years. There were no roe fisheries in seven areas; in eight areas, there were roe fisheries in only one or two years; in four areas, there were roe fisheries in most years; and there were roe fisheries in all years only in areas 2E and 14.

TABLE 3.—Release and return information for anchor-tagged Pacific herring released in the winter in inshore holding areas before and after the winter food fishery. Sp = spring; Su = summer; Fa = fall; Wi = winter.

Release			Number of returns by season after release													
Area	Year	Number of fish tagged	0 Wi	1 Sp	2 Su	3 Fa	4 Wi	5 Sp	6 Su	7 Fa	8 Wi	9 Sp	10 Su	11 Fa	12 Wi	Total returns
Before food fishery																
17S	1979	1,264	25	3	0	0	0	0	0	0	0	0	0	0	0	28
17S	1980	3,176	18	7	0	0	4	1	0	1	0	0	0	0	0	31
17S	1981	9,171	317	85	1	0	8	20	0	0	0	2	0	0	0	433
17S	1982	11,039	114	105	0	0	1	3	0	0	0	0	0	0	0	223
17S	1983	1,985	6	2	0	0	0	0	0	0	0	0	0	0	0	8
18	1980	2,466	5	4	0	0	0	5	0	0	0	1	0	0	0	15
Total		29,101	485	206	1	0	13	29	0	1	0	3	0	0	0	738
After food fishery																
5	1980	2,982	1	3	0	0	1	0	0	0	0	1	0	1	0	7
13	1980	3,888	7	0	0	0	0	0	0	0	0	1	0	0	0	8
13	1982	11,881	6	19	1	0	0	3	0	0	0	0	0	0	0	29
13	1983	17,619	1	72	0	0	0	0	0	0	0	0	0	0	0	73
17S	1979	2,601	2	2	1	0	0	0	0	0	0	0	0	0	0	5
17S	1980	2,958	0	4	0	0	1	0	0	0	0	0	0	0	0	5
17S	1981	10,485	13	81	0	0	4	12	0	0	0	1	0	0	0	111
17S	1982	7,460	6	33	0	0	0	1	0	0	0	0	0	0	0	40
17S	1983	17,680	36	26	0	0	0	0	0	0	0	0	0	0	0	62
19	1982	3,925	0	16	0	0	0	0	0	0	0	0	0	0	0	16
Total		81,479	72	256	2	0	6	16	0	0	0	3	0	1	0	356

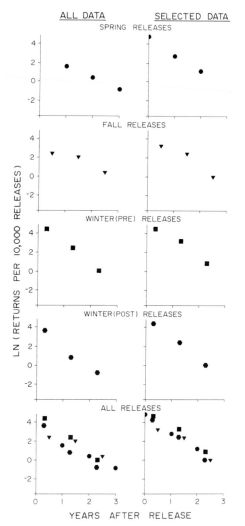

FIGURE 2.—Relationship between log$_e$ of tag returns (Y-axis) and time (X-axis) for different seasons of release of anchor-tagged Pacific herring. "Pre" and "Post" refer to releases during periods before and after the winter food fishery off British Columbia.

Rates of return were low and mortality was higher (Figure 2) than expected from the estimated instantaneous natural mortality (0.36) used in an age-structured stock assessment model and the average fishing mortality (0.20) (Haist et al. 1986). For example, for all spring releases, tag returns after 4, 8, and 12 seasons (1, 2, and 3 years at large) were 0.052, 0.016, and 0.005%, respectively. The highest rate of return was obtained for 1982 area 23 spring releases, for which 1.129, 0.171, and 0.026% of tags were returned after 0, 4, and 8 seasons. There was no fishery after 12

seasons. These return rates are lower than the occurrence of tags in the fishery. It was estimated with tag-recovery trials, during which a known number of tagged fish were introduced into the catches, that 34% of tags in the roe catch are reported (Haegele 1983). Nevertheless, even when adjusting for this rate of reporting, I estimated that only 3.39% (95% confidence interval, 2.07–4.70%) of tags from spring 1982 releases in area 23, 4–27 days before the roe fisheries in area 23, were recovered in the catch. The estimated catch for this fishery was 16% of available spawner biomass (22,635 tonnes); this estimate was based on escapement estimates obtained from scuba-diving surveys of spawn (Haegele and Schweigert 1984). Hence, immediate tag loss, either due to mortality or tag shedding, was estimated to be about 80% in 1982 in area 23, for which the expected-to-actual recovery ratio averaged 0.21 (Table 4).

Discussion

Certain criteria must be met for a tagging study to be successful. Handling mortality should be low and tag retention high. There should be substantial fisheries over a wide spectrum of time and space and high reporting of tags in the catch. The limited success of the Pacific herring anchor-tag study can be attributed to a failure to fully satisfy any of these requirements.

In a previous tagging study, Pacific herring were tagged in British Columbia from 1937 to 1956 with internal body cavity tags (Stevenson 1954). In these studies, small seines 100 m long and 4 m deep and even smaller beach seines were used to obtain fish for tagging, and tag placement required an incision in the abdominal wall. Tagging mortality was estimated at 60%. Recovery rates for these body cavity tags, which were trapped by magnets or removed by tag detectors, were higher than recovery rates for anchor tags in this study. By comparing Figures 2 and 3, it can be seen that recovery rates of body cavity tags were from 5 times (selected anchor-tag data) to 14 times (all anchor-tag data) higher. However, catches were higher during the study of body cavity tags, when fisheries were for meal reduction, and up to 60% of stocks were harvested annually in coastwide fisheries in the later years of that study (Taylor 1964). By contrast, the Pacific herring fisheries during the anchor-tag study were restricted to a few areas in which the catch was generally less than 20% of estimated biomass.

TABLE 4.—Spring tag releases of Pacific herring, by tagging set, and returns from the ensuing roe fishery (4–27 d later) in statistical area 23 in 1982. Actual recoveries are based on 0.34 of tags in catch being returned, and expected recoveries are based on a catch of 3,613 tonnes from a stock of 22,635 tonnes.

Set	Number of fish released	Number of tags returned	Recoveries in catch		
			Actual	Expected	Actual/expected
1	849	2	5.9	135.5	0.043
2	1,770	14	41.2	282.5	0.146
3	1,783	13	38.2	284.6	0.134
4	1,787	47	138.2	285.2	0.485
5	884	22	64.7	141.1	0.459
6	859	6	17.6	137.1	0.129
7	432	4	11.8	69.0	0.171
8	1,586	17	50.0	253.2	0.198
9	1,790	16	47.1	285.7	0.165
10	1,779	21	61.8	284.0	0.218
11	880	4	11.8	140.5	0.084
12	1,685	16	47.1	269.0	0.175
13	1,776	34	100.0	283.5	0.353
Total	17,860	216	635.3	2,850.8	0.212

Handling mortality and initial tag shedding appear to have been 80% in this study. High mortality from handling is not uncommon with herring. Hay (1981) estimated 20% handling mortality even 4 months after capture for fish held in captivity. Nakashima and Winters (1984) attributed lower tag returns to holding Atlantic herring in captivity for 1 and 2 d after capture, although they made no estimate of tagging mortality. Experimental impoundment studies have shown up to 52% mortality, which was mostly attributed to handling (Kreiberg et al. 1984). Documentation of commercial Pacific herring ponding operations showed up to 82% mortality after 10 d (Shields et

al. 1985), and the average mortality after 8–13 d was 37% (95% confidence interval, 23–59%). If anchor-tagging mortality is assumed to have been 60%, as was estimated for body cavity tags, then up to one-half of tags would have been shed within a short time after release to account for the 80% tag loss. In laboratory studies, initial tag loss was low, although approximately 25% of tags were shed within 1 year (Hay 1981). However, in the field, where a large number of fish are tagged in a short time, often during inclement weather, some tags may have been improperly applied.

Anchor-tag return rates for winter releases were higher than for spring releases only in the first year for all data; return rates were similar for winter and spring for selected data. Hence, the size of the net and catch and the experience of the taggers did not appear to have influenced tagging mortality and tag loss. Pacific herring are accessible only to large nets at most times because they generally school at depths of 30–60 m. The exception is at spawning, when Pacific herring may be captured with small beach seines and thus experience a minimum of trauma. Other tags would probably result in the same initial tagging mortality from handling. However, if tag retention and reporting of tags is increased and long-term tagging mortality decreased, it may be feasible to tag Pacific herring for long-term homing and migration studies. Due to the small number of tag returns in this study, it could not be determined whether tag color and size affected mortality, retention, or tag reporting, or whether the offering of a $2 reward versus a prize drawing affected tag reporting, although the reporting of reddish-col-

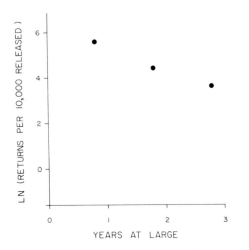

FIGURE 3.—Relationship between $\log_e$ of tag returns (Y-axis) and years at large (X-axis) for the internally tagged Pacific herring returns from the winter reduction fishery off British Columbia.

ored tags was about 50% higher than for tags of other colors (Haegele 1983). A coded wire tag was found suitable for implantation in Pacific herring; 400 fish could be tagged per hour per tagging machine (Haegele 1984), but there have been no follow-up studies on tag retention and recovery.

References

Haegele, C. W. 1982. Recovery of external anchor tags during processing of food herring in British Columbia. Canadian Manuscript Report of Fisheries and Aquatic Sciences 1669.

Haegele, C. W. 1983. Recovery rates for external anchor tags in British Columbia herring fisheries. Canadian Manuscript Report of Fisheries and Aquatic Sciences 1745.

Haegele, C. W. 1984. Tagging of herring in British Columbia during the 1983–84 herring season. Canadian Industry Report of Fisheries and Aquatic Sciences 1534.

Haegele, C. W. 1986. Returns from anchor taggings of herring in British Columbia, 1979 to 1985. Canadian Data Report Fisheries and Aquatic Sciences 582.

Haegele, C. W., and J. F. Schweigert. 1984. Herring stock estimates from a diving survey of spawn in Barkley Sound in 1982, with comparisons to other estimates. Canadian Manuscript Report of Fisheries and Aquatic Sciences 1798.

Haist, V., J. F. Schweigert, and M. Stocker. 1986. Stock assessment for British Columbia herring in 1985 and forecasts of the potential catch in 1986. Canadian Manuscript Report of Fisheries and Aquatic Sciences 1889.

Hay, D. E. 1981. Retention of tags and survival of tagged Pacific herring held in captivity. Canadian Technical Report of Fisheries and Aquatic Sciences 1050.

Hay, D. E. 1986. A stock hypothesis based on spawn and winter distribution. Canadian Manuscript Report of Fisheries and Aquatic Sciences 1871:145–148.

Hay, D. E., C. W. Haegele, and D. C. Miller, 1979. The feasibility of tagging and recovering herring with external body tags—a pilot study. Canada Fisheries and Marine Service Manuscript Report 1505.

Hay, D. E., and J. L. Mitchell. 1979. Recovery of tagged fish from a roe processing plant—a pilot study. Canada Fisheries and Marine Service Manuscript Report 1544.

Hourston, A. S., and C. W. Haegele. 1980. Herring on Canada's Pacific coast. Canadian Special Publication of Fisheries and Aquatic Sciences 48.

Kreiberg, H., J. R. Brett, and A. Solmie. 1984. Roe herring impoundment research—report on the 1982/1983 studies. Canadian Technical Report of Fisheries and Aquatic Sciences 1261.

Nakashima, B. S., and G. H. Winters. 1984. Selection of external tags for marking Atlantic herring (*Clupea harengus harengus*). Canadian Journal of Fisheries and Aquatic Sciences 41:1341–1348.

Shields, T. L., G. S. Jamieson, and P. E. Sproat. 1985. Spawn-on-kelp fisheries in the Queen Charlotte Islands and northern British Columbia coast—1982 and 1983. Canadian Technical Report of Fisheries and Aquatic Sciences 1372.

Stevenson, J. C. 1954. The movement of herring in British Columbia waters as determined by tagging, with a description of tagging and tag recovery methods. Document 55 *in* Herring tagging and results. Conseil International pour l'Exploration de la Mer, Special Scientific Meeting, Copenhagen.

Taylor, F. H. C. 1964. Life history and present status of British Columbia herring stocks. Fisheries Research Board of Canada Bulletin 143.

American Fisheries Society Symposium 7:134–141, 1990

Evaluation of Tagging Techniques for Shortnose Sturgeon and Atlantic Sturgeon

THEODORE I. J. SMITH

South Carolina Wildlife and Marine Resources Department
Post Office Box 12559, Charleston, South Carolina 29412, USA

SCOTT D. LAMPRECHT

South Carolina Wildlife and Marine Resources Department, Wildlife and Fresh Water Fisheries
Division, Post Office Drawer 190, Bonneau, South Carolina 29431, USA

J. WAYNE HALL

South Carolina Wildlife and Marine Resources Department
Post Office Box 12559, Charleston, South Carolina 29412, USA

Abstract.—During 1985–1988, numerous tagging studies were conducted in an effort to identify suitable long-term tags for use on Atlantic coast sturgeons, namely the shortnose sturgeon *Acipenser brevirostrum* and the Atlantic sturgeon *A. oxyrhynchus*. The purpose of the research was to identify reliable tags for use in population estimation and assessment of stock enhancement programs. Tests were conducted in ponds with cultured juvenile and wild-caught adult shortnose sturgeons, and wild-caught juvenile Atlantic sturgeons. Tagging studies with cultured juvenile shortnose sturgeons were also conducted in tanks. Marks and tags tested were India ink tattoo, subcutaneous injection of acrylic paint, dart tag with nylon T-anchor, Carlin tag, Archer tag, an internal anchor tag, monel strap tag, and passive integrated transponder (PIT) tag. Different attachment sites were tested for some of these tags. Three types of tags appear suitable for use with sturgeons. Of the external tags, the Carlin tag that attached through the base of the dorsal fin and the abdominal internal anchor tag both were retained well. The internal anchor tag was suitable for shortnose sturgeons if the fish were held in fresh water after tagging. If they were returned to brackish water before healing was completed, serious lesions and eventual breakdown of the body wall occurred. For a permanent internal tag, the PIT tag shows promise, but additional testing is required. However, specialized electronic equipment is required to read the tag, and tag cost is relatively high.

North American sturgeons were important items of commerce during colonial days; large-scale fisheries for the Atlantic coast species developed during the latter quarter of the 19th century from Canada to Florida (Smith 1985, 1990). Fishing intensity and landings increased rapidly for these fish and peaked about 1890. Shortly thereafter, all major fisheries for North American sturgeons began to collapse due to a combination of factors including overfishing, construction of dams on spawning rivers, and water pollution (Harkness and Dymond 1961; Leland 1968; Priegel and Wirth 1977; Galbreath 1985; Pasch and Alexander 1986). Early fishery managers attempted to maintain the fisheries through development of hatcheries to produce cultured fish for restocking native waters (Ryder 1890; Cobb 1900; Stone 1900). Spawning efforts were substantial but results were largely unsuccessful (Dean 1894; Leach 1920). By 1920, most hatchery

efforts were abandoned and the stocks were severely depleted (Smith and Dingley 1984). Today, commercial harvesting of the larger Atlantic sturgeon *Acipenser oxyrhynchus* is prohibited in most states, and the shortnose sturgeon *A. brevirostrum* is listed as an endangered species in the USA (Smith 1990).

Over the past decade, the U.S. Fish and Wildlife Service and the State of South Carolina have been involved in a cooperative program aimed at collection of life history and ecological data for Atlantic and shortnose sturgeons in the southeastern USA (Smith et al. 1982; Smith and Dingley 1984; Smith 1985). In addition, efforts have focused on developing a hatchery and nursery technology to evaluate possible stock enhancement programs for these sturgeons.

Substantial progress has been achieved in development of hatchery and nursery systems for shortnose sturgeon (Smith et al. 1985; 1988) and small-scale stock enhancement activities have been initiated (Smith, in press). However, evaluation of success is difficult due to the lack of information on suitable tags and tagging techniques for long-term identification of stocked sturgeons, to the general unavailability of juvenile shortnose sturgeons (whether tagged or untagged) for recapture, and to the long time period required for juveniles to grow to adults that can be recaptured during spawning migrations.

Research was begun to identify long-term tags to mark wild-caught adults and juveniles for mark–recapture studies of life history and population abundance and to tag cultured juveniles permanently for assessment of stock enhancement programs. This paper summarizes our experiences with tagging Atlantic and shortnose sturgeons.

Methods

During 1985–1988, tagging studies were conducted with juvenile and adult shortnose sturgeons and juvenile Atlantic sturgeons. The shortnose sturgeons were obtained from two sources. The juveniles, 300–600 mm long (all lengths cited herein are total lengths), were produced from our hatchery and nursery systems (Smith et al. 1988); the adults, 600–800 mm, were wild fish captured incidentally by commercial gill-net fishermen fishing for shad in the Savannah River. The Atlantic sturgeons were also wild fish caught incidentally by shad fishermen. However, these fish were all juveniles and had a size range similar to that of the wild adult shortnose sturgeons, 600–800 mm.

The tagging studies were conducted in both tank and pond culture systems. The cultured shortnose sturgeons were tested in indoor cylindrical fiberglass tanks (3.7 m × 0.8 m deep), in outdoor cylindrical tanks (6.2 m × 1.4 m deep), and in ponds at the South Carolina Wildlife and Marine Resources Department's Waddell Mariculture Center (WMC). The tank culture systems all used recirculated water, with occasional partial water exchanges. In these studies, fresh water was typically employed, although some studies were conducted in brackish water (3–10‰). Tagging studies of the wild adult shortnose sturgeons and the wild juvenile Atlantic sturgeons were conducted only in rectangular ponds ranging in size from 0.1 to 0.5 hectare and located at WMC. These ponds had a mean water depth of 1.4 m (range, 1.2–1.8 m). Salinity ranged from 3 to 6‰ during most pond studies. Water quality was maintained at satisfactory levels in all studies and no known mortality occurred due to poor water

quality. However, the sturgeon ponds were often used by alligators, herons, and ospreys, which attempted to feed on the sturgeons. Some shortnose sturgeons were taken but overall survival was greater than 80%. Survival of Atlantic sturgeons was 100%.

Fish were fed during all studies; however, only the cultured fish were known to feed actively on the trout ration (38% protein) provided. Trout pellets of different sizes were provided to the pond-held fish in addition to an occasional supplement of chopped squid. Upon drainage, the pond bottoms contained numerous small feeding depressions, indicating that the sturgeons fed on the bottom fauna and flora, and perhaps on some of the trout pellets and supplemental rations. Growth of fish in all groups was obvious during the various studies.

Thirty-two different tag-type, tag-placement, and fish-group studies were conducted during 1985–1988. Descriptions of the tags and tagging techniques are presented in Table 1 and Figure 1. In initial studies we tested India ink tattoos and subcutaneous injection of acrylic paint on the white abdomens of the sturgeons. Next, Archer-style tags and Carlin tags were attached through the scutes with monel wire passed through drilled holes. Carlin tags were also attached to the base of the dorsal fin. Some Carlin tags were modified by attachment of the disc to a separate loop of monel wire (Figure 1). Monel strap tags were fastened to the anterior base of the dorsal fin. Dart tags were also attached to the scutes and near the base of the dorsal fin, and internal anchor tags were inserted in the gut cavity through a surgical incision. In all cases, external tag materials consisted of plastic and monel wire. Passive integrated transponder (PIT) tags were embedded dorsally in the musculature between two posterior scutes; these studies began in early 1985 with cultured and wild shortnose sturgeons and juvenile Atlantic sturgeons. At that time, a porous plastic material was used to encapsulate the tag. Later it was learned that premature tag failure occurred in salmonids due to infiltration of body fluids into the tag. On the sturgeons, 50% of the tags were still readable after 34 months, but in some cases the signal was extremely weak, which suggests that tag failure may have been a problem. Data from these early PIT-tag studies are not included in the present report. Only data from a recent study of shortnose sturgeons with the newer glass-encapsulated PIT tags are reported.

TABLE 1.—Description of tags and tagging techniques used on Atlantic and shortnose sturgeons.

Tag type or method	Description	Attachment
Tattoo with India ink	Veterinary tattoo gun	Numbers written on white abdomen
Acrylic paint	Commercial acrylic paint	Different colors injected subcutaneously on white abdomen
Archer tag to scute	Plastic plates, 6 mm × 25 mm, attached with 0.96-mm-diameter monel wire	Identification plates attached to both sides of a dorsal scute with monel wire inserted through two predrilled holes
Carlin tag[a] to scute	Plastic disc, 6 mm × 19 mm, attached with 0.38–0.52-mm-diameter monel wire	Disc attached with monel wire passed through one predrilled hole in scute
Carlin tag[a] to dorsal fin	Plastic disc, 6 mm × 19 mm, attached with 0.38–0.52-mm-diameter monel wire	Disc attached to base of dorsal fin with monel wire passed through two predrilled holes
Monel strap	V-shaped monel strap	Crimped onto anterior base of dorsal fin
Dart tag to scute	T-anchor, 9-mm nylon, with 85-mm streamer	T-bar inserted in hole drilled in dorsal scute
Dart tag to dorsal fin	T-anchor, 9-mm nylon, with 85-mm streamer	T-bar injected near base of dorsal fin musculature
Internal anchor tag	Plastic anchor disc, 5 mm × 16 mm, with 55-mm plastic streamer (2.0 mm in diameter)	Disc inserted through surgical incision in abdominal cavity
Passive integrated transponder (PIT)	Glass capsule, 10 mm × 2.1 mm in diameter	Embedded in musculature between dorsal scutes

[a]All Carlin tags were modified (unless stated otherwise) such that the disc was attached by a separate loop (1 cm in diameter) of monel wire. This allowed freer movement of the disc and less stress on the attachment site.

The number of fish involved in the studies ranged from 1 to 61; most studies involved 10–20 animals. Variation in the number of test animals was directly related to availability of animals and facilities, and logistical considerations (personnel, sampling schedules, etc.) rather than to experimental design. During some studies, fish were simultaneously tagged with several types of tags,

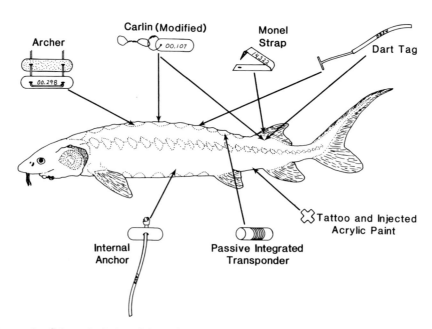

FIGURE 1.—Schematic design of the various types of tags tested on Atlantic and shortnose sturgeons.

TABLE 2.—Summary of tagging studies with cultured juvenile shortnose sturgeons (300–600 mm total length) reared in tanks.

Type of tag	Number of fish	Duration (d)	Retention (%)	Comments
Tattoo with India ink	12	120	100	Barely visible by end of study
Acrylic paint injection	5	120	100	Barely visible by end of study
Dart tag to dorsal fin	12	127	25	Produced excess trauma
Carlin tag to scute	61	170	71	Tags easily tangled in nets
	13	150	58	Tags easily tagged in nets
	32	175	18	Tags dropped off in tank
	16	142	69	Tags dropped off in tank
Carlin tag to dorsal fin	7	150	100	Good retention in spite of repeated netting of fish
	16	175	91	Good retention in spite of repeated netting of fish
	16	175	82	Good retention in spite of repeated netting of fish
	16	142	94	Good retention in spite of repeated netting of fish
Internal anchor tag	12	130	100	Anchor discs not incorporated in tissue
	16	142	100	Anchor discs not incorporated in tissue

or with the same type of tag at different sites. In most cases, tagging studies were not directly replicated, although often several groups of similarly tagged fish were reared in different systems. In other cases, the same type of tag was concurrently tested on both Atlantic and shortnose sturgeons.

Duration of the tagging studies depended on tag performance, injury to the fish, and availability of culture facilities. At times, tagged and untagged sturgeons were held together in ponds. Sometimes, during sampling of fish used in concurrent tagging studies or from ponds containing untagged fish, some fish could not be accounted for, and some had no tags. In such cases, it was often impossible to determine whether a fish had shed its tag or a tagged fish had died. For purposes of data analysis we assumed that tag losses were due to poor tag retention rather than to mortality. Thus, the lack of adjustment for mortality may result in slight underestimation of some retention rates. However, due to the generally high survival rates (81–100%), such underestimation of rates

should have been low. Fish were netted or seined from tanks and ponds, and some entanglement of external tags did occur (especially of the modified Carlin tags). Care was exercised to minimize tag loss and fish injury due to sampling, which occurred at 4-week intervals during most tank studies and at about 6-month intervals during the pond studies.

Results

Cultured Juvenile Shortnose Sturgeons

Many tagging studies were focused on the cultured shortnose sturgeons (Tables 2 and 3) because these fish were readily available and would be used in the initial stock enhancement programs. Tattoos and injected acrylic paints were easy to apply but were only observable for about 4 months. Dart anchor tags attached to the dorsal fin had a low retention rate and caused tissue damage. Carlin tags attached to a dorsal scute also were not retained long in the tank and pond tests (Tables 2 and 3). In the tanks, the highest reten-

TABLE 3.—Summary of tagging studies with cultured juvenile shortnose sturgeons (300–600 mm total length) reared in ponds. Each fish was tagged with the three types of tags.

Type of tag	Number of fish	Duration (months)	Retention (%)[a]	Comments
Carlin tag to scute	26	19	0	19% retention after 8 months
Carlin tag to dorsal fin	26	19	81	
Internal anchor tag	26	19	77	Anchor discs not incorporated in tissue

[a]Nineteen percent of the fish were missing at end of the study.

TABLE 4.—Results of tagging studies with wild adult shortnose sturgeons held in ponds.

Type of tag	Number of fish	Duration (d)	Retention (%)	Comments
Archer to scute	34	90		Caused serious tissue damage especially at 18‰ salinity
Carlin tag[a] to scute	10	454	30[b]	
Carlin tag[a] to dorsal fin	23	454	83[b]	
		1,008	79[c]	
	5	342	100	
	14	510	100	
	15	468	100	
		603	87	Some tag loss may have been due to mortality of fish
	24	188	100	
		323	88	Some tag loss may have been due to mortality of fish
Monel strap	1	150	0	
Passive integrated transponder	4	60	100	

[a]Carlin tags were not modified. The disc attached directly to the attachment wire.
[b]Four sturgeons died; they had unknown tag types. No retention-rate adjustment was made for these lost fish.
[c]Retention rate was based on the number of fish alive on day 454 (19 fish) divided by the number of live fish retaining tags on day 1,008.

tion rate was 71% after 170 d for Carlin tags (attached to a dorsal scute); in the pond there was only a 19% retention rate after 8 months.

Fish with Carlin tags attached to the base of the dorsal fin and internal anchor tags exhibited high tag-retention rates. In tank tests, these Carlin tags were 82–100% retained, over test periods of 142 to 175 d. In the pond studies, 81% of the fish retained these tags after 19 months (Table 3). The internal anchor tags were 100% retained during the 130–142-d tank tests in fresh water. In the pond test, 77% retention was recorded after 19 months (Table 3). The lower retention rate of the internal anchor tags in the pond study may have been due to mortality of tagged fish (Table 3), because severe tissue damage resulted when the sturgeons were placed in brackish water after tagging. Lesions occurred around the site of insertion and eventually developed into 2–3-cm holes extending through the abdominal wall. In tank tests, no tissue damage occurred if fish tagged with the internal anchor tags were placed in fresh water for 2–4 weeks before exposure to brackish water.

Wild Adult Shortnose Sturgeons

Archer-style tags attached through a dorsal scute were the first tags tested with wild shortnose sturgeons. After 90 d, however, many tags were lost or in the process of being shed, especially in water of higher salinity (Table 4). These tags caused severe tissue damage and resulted in scute

deformation, bleeding, and, in one case, scute loss. Unmodified Carlin tags attached to a scute did not cause severe damage and were retained better than the Archer tags, but only 30% of these tags were retained after 454 d (Table 4). Of the Carlin tags attached to base of the dorsal fin, 100 to 79% were retained over periods of 342 to 1,008 d, respectively. In the single test of a monel strap tag, this tag was lost by day 150. Limited testing of PIT tags showed these tags were 100% retained, at least up to 60 d (Table 4).

Wild Juvenile Atlantic Sturgeons

Some of the tagging studies conducted on juvenile Atlantic sturgeons were similar to those conducted on shortnose sturgeons (Table 5). The Archer-style tags caused some tissue damage but it was much less severe than that observed in the shortnose sturgeons over the same time period. However, some tissue damage did occur and it seemed likely that tags would eventually be lost. Of monel strap tags, 75% were retained over 620 d, and some tissue growth developed around the tags. However, the retention rate decreased to 50% by day 754. Dart tags attached to dorsal scutes and to the base of the dorsal fin resulted in low initial rates of retention (40% after 180 d) but once established, the dart tags remained attached for 454 d (Table 5). Carlin tags attached to the scutes were 82% retained up to day 678, although erosion around the attachment hole was apparent. In con-

TABLE 5.—Results of tagging studies of wild juvenile Atlantic sturgeons (650–1,000 mm total length) reared in ponds.

Type of tag	Number of fish	Duration (d)	Retention (%)	Comments
Archer tag to scute	24	90		Caused tissue damage
Monel strap	4	620	75	Some tissue overgrowth
		754	50	
Dart tag to scute	25	180	40	Newly tagged
	4	454	75	Previously tagged, well attached
Dart tag to dorsal fin	25	180	40	Newly tagged
	6	454	100	Previously tagged, well attached
Carlin tag to scute	22	454	100	Unmodified tags used, erosion
		678	82	of attachment site obvious
		1,021	64	
Carlin tag to dorsal fin	5	678	100	Unmodified tags used
		1,021	80	

trast, the Carlin tags attached to the base of the dorsal fin were 80% retained even after 1,021 d.

Discussion

There is limited information on tagging of North American sturgeons. For the most part, large slow-growing fish or adults have been previously tagged with a variety of dangling disc tags attached through or near the base of the dorsal fin, or through or under scutes. With such tags, studies have been conducted on movements, life history, population dynamics, etc. (Miller 1972; Dadswell 1979; Kolhorst 1979; Wooley and Crateau 1982; Threader and Brousseau 1986). However, there have been no long-term studies to evaluate retention rates or damage to the sturgeons with different tags. The work discussed here is the most detailed evaluation of tagging techniques for any North American sturgeon, yet additional studies are needed.

Atlantic sturgeons have larger, thicker scutes and tougher skins than do shortnose sturgeons, and they grow at a much slower rate, at least in captivity (Smith et al. 1981, 1988). In part, these considerations may explain why the various tags were generally retained longer on juvenile Atlantic sturgeons than on cultured juvenile or adult shortnose sturgeons.

The overall results show that the Carlin tag mounted on the base of the dorsal fin is a dependable type of tag for use on Atlantic and shortnose sturgeons. When properly attached, the Carlin tag has a long retention time and causes little damage to the fish. Major disadvantages of this tag include fouling and a high susceptibility to entanglement in commercial and recreational gill nets. In South Carolina, there have been recapture rates of 50–80% of Carlin-tagged sturgeons released in areas intensively fished by commercial gill-net fishermen.

The internal anchor tag was also retained well by shortnose sturgeons, but did not result in increased susceptibility to gill-net capture. However, several correctable problems were encountered. The identification number on the streamer portion often became illegible, but improved manufacturing techniques should extend the readable life of the tag. Typical foreign-body tissue encapsulation (Vogelbein and Overstreet 1987) of the anchor was not observed. Therefore, the tag rotated, which resulted in continued irritation at the incision site. Occasionally, the streamer portion of the tag slipped into the abdominal cavity, so that only the end protruded. However, the streamer could easily be pulled back out. Stabilization of the streamer attachment to the anchor should reduce both irritation and slippage. The worst problem occurred when tagged shortnose sturgeons were returned immediately to brackish water. Extensive tissue damage resulted, similar to that reported for spots *Leiostomus xanthunus* (Vogelbein and Overstreet 1987). However, if the shortnose sturgeons were returned to fresh water for 2–4 weeks immediately after tagging, no wounds occurred. These fish could then be placed safely in brackish water without further damage. Presumably, internal anchor tags will also be satisfactory for use on Atlantic sturgeons, but juvenile Atlantic sturgeons were not available for testing.

Presently, pond studies are underway in South Carolina to examine the use of PIT tags on Atlantic and shortnose sturgeons. Preliminary indications are that these tags cause little injury and have high retention rates initially. Tag-evaluation studies of salmonids, striped bass *Morone saxatilis*, and red drums *Sciaenops ocellatus* suggest that PIT tags will become important in fisheries research (Prentice and Park 1984; Monan 1985; Jenkins and Smith 1990, this volume). The major disadvantages of the PIT tags include their high cost, the requirement of specialized detection equipment, and the lack of a readily visible external mark on tagged fish. Nevertheless, these tags are capable of providing detailed information on individual fish that was not previously available to fishery managers.

Summary

A broad range of external marks and tags was examined for use on Atlantic and shortnose sturgeons. Of these, two types of tags, Carlin and internal anchor tags, had high rates of retention for long periods of time. However, it should be noted that shortnose sturgeons tagged with internal anchor tags need time to heal in fresh water before they are released if there is a possibility that tagged fish can move into brackish water. The internal anchor tag was not tested on Atlantic sturgeons, but the retention rate should be similarly high. Only limited retention-rate data are available on the suitability of the new glass-encapsulated PIT tags for use on sturgeons. However, our preliminary information is promising and this tag is now being extensively evaluated on shortnose sturgeons in South Carolina.

Acknowledgments

We thank our staff, especially Al Stokes, Robert Smiley, Bill Oldland, Richard Hamilton, and Wally Jenkins for their assistance throughout these studies. Linda K. Greene typed the manuscript and Karen Swanson prepared the figure. This is contribution 246 from the South Carolina Marine Resources Center. Research support was provided by the U.S Fish and Wildlife Service under contract S.C. AFS-11 and cooperative agreement 14-16-0004-87-949, and by the State of South Carolina. Trade names mentioned herein do not imply endorsement.

References

Cobb, J. N. 1900. The sturgeon fishery of Delaware River and Bay. Pages 369–380 *in* Report of the Commissioner for 1899. U.S. Commission of Fish and Fisheries, Washington, D.C.

Dadswell, M. J. 1979. Biology and population characteristics of the shortnose sturgeon, *Acipenser brevirostrum* LeSueur 1818 (Osteichthyes Acipenseridae), in the Saint John estuary, New Brunswick, Canada. Canadian Journal of Zoology 57: 2186–2210.

Dean, B. 1894. Recent experiments in sturgeon hatching on the Delaware River. U.S. Fish Commission Bulletin 13:335–339.

Galbreath, J. L. 1985. Status, life history, and management of Columbia River white sturgeon, *Acipenser transmontanus*. Pages 119–126 *in* F. P. Binkowski and S. I. Doroshov, editors. North American sturgeons: biology and aquaculture potential. Dr. W. Junk, Dordrecht, The Netherlands.

Harkness, W. J. K., and J. R. Dymond. 1961. The lake sturgeon, the history of its fishery and problems of conservation. Ontario Department of Lands and Forestry, Toronto.

Jenkins, W. E., and T. I. J. Smith. 1990. Use of PIT tags to individually identify striped bass and red drum brood stocks. American Fisheries Society Symposium 7:341–345.

Kolhorst, D. W. 1979. Effect of first pectoral fin ray removal on survival and estimated harvest rate of white sturgeon in the Sacramento–San Joaquin estuary. California Fish and Game 65:173–177.

Leach, G. C. 1920. Artificial propagation in sturgeon. Part 1. Review of sturgeon culture in the United States. U.S. Fish Commission Report 1919:3–5.

Leland, J. G., III. 1968. A survey of the sturgeon fishery of South Carolina (Wadmalaw Island, South Carolina). Contributions from Bears Bluff Laboratories 47:1–27.

Miller, L. W. 1972. White sturgeon population characteristics in the Sacramento–San Joaquin estuary as measured by tagging. California Fish and Game 58:94–101.

Monan, G. E. 1985. Advances in tagging and tracking hatchery salmonids: coded wire tags, multiple-coded and miniature radio tags, and the passive integrated transponder tag. NOAA (National Oceanic and Atmospheric Administration) Technical Report NMFS (National Marine Fisheries Service) 27:33–37.

Pasch, R. W., and C. M. Alexander. 1986. Effects of commercial fishing on paddlefish populations. Pages 46–53 *in* J. Dillard, L. Graham, and T. Russell, editors. The paddlefish: status, management, and propagation. American Fishery Society, North Central Division, Special Publication 7, Bethesda, Maryland.

Prentice, E. F., and D. L. Park. 1984. A study to determine the biological feasibility of a new fish tagging system. Bonneville Power Administration, Annual Report, Portland, Oregon.

Priegal, G. R., and T. L. Wirth. 1977. The lake sturgeon: its life history, ecology and management. Wisconsin Department of Natural Resources, Publication 4-3600 (77), Madison.

Ryder, J. A. 1890. The sturgeons and sturgeon industries of the eastern coast of the United States, with an account of experiments bearing upon sturgeon culture. U.S. Fish Commission Bulletin 8:231–328.

Smith, T. I. J. 1985. The fishery, biology and management of Atlantic sturgeon, *Acipenser oxyrhynchus*, in North America. Environmental Biology of Fishes 14:61–72.

Smith, T. I. J. 1990. Culture of North American sturgeons for fishery enhancement. NOAA (National Oceanic and Atmospheric Administration) Technical Report NMFS (National Marine Fisheries Service) 85:19–27.

Smith, T. I. J., and E. K. Dingley. 1984. Review of biology and culture of Atlantic *(Acipenser oxyrhynchus)* and shortnose sturgeon *(A. brevirostrum)*. Journal of the World Mariculture Society 15:210–218.

Smith, T. I. J., E. K. Dingley, R. D. Lindsey, S. B. Van Sant, R. A. Smiley, and A. D. Stokes. 1985. Spawning and culture of shortnose sturgeon, *Acipenser brevirostrum*. Journal of the World Mariculture Society 16:104–113.

Smith, T. I. J., E. K. Dingley, and D. E. Marchette. 1981. Culture trials with Atlantic sturgeon, *Acipenser oxyrhynchus*, in the United States of America. Journal of the World Mariculture Society 12(2):78–87.

Smith, T. I. J., D. E. Marchette, and R. A. Smiley. 1982. Life history, ecology, culture and management of Atlantic sturgeon, *Acipenser oxyrhynchus oxyrhynchus* Mitchill, in South Carolina. South Carolina Wildlife and Marine Resources Research Department. Final Report, Project AFS-9, U.S. Fish and Wildlife Service, Atlanta.

Smith, T. I. J., W. Oldland, R. Hamilton, and W. E. Jenkins. 1988. Development of nursery systems for shortnose sturgeon, *Acipenser brevirostrum*. Proceedings of the Annual Conference Southeastern Association of Fish and Wildlife Agencies 40:143–151.

Stone, L. 1900. The spawning habits of the lake sturgeon *(Acipenser rubicundus)*. Transactions of the American Fishery Society 29:118–128.

Threader, R. W., and C. S. Brousseau. 1986. Biology and management of the lake sturgeon in the Moose River, Ontario. North American Journal of Fisheries Management 6:383–390.

Wooley, C. M., and E. J. Crateau. 1982. Observations of Gulf of Mexico sturgeon *(Acipenser oxyrhynchus desotoi)* in the Apalachicola River, Florida. Florida Scientist 45:244–248.

Vogelbein, W. K., and R. M. Overstreet. 1987. Histopathology of the internal anchor tag in spot and spotted seatrout. Transactions of the American Fisheries Society 116:745–756.

American Fisheries Society Symposium 7:142–146, 1990
© Copyright by the American Fisheries Society 1990

Evaluation of Various External Marking Techniques for Atlantic Salmon

K. A. COOMBS,[1] J. K. BAILEY, C. M. HERBINGER,[2] AND G. W. FRIARS[3]

Salmon Genetics Research Program, Atlantic Salmon Federation
Box 429, St. Andrews, New Brunswick E0G 2X0, Canada

Abstract.—The identification of groups of fish, and of individuals reared together, is important in many research studies, particularly genetic programs. In the Salmon Genetics Research Program, at the Atlantic Salmon Federation, a variety of external marking techniques has been used to identify families and individuals of Atlantic salmon *Salmo salar* with variable success. Six groups of parr reared together were identified with adipose and pelvic fin clips, singly or in combinations. Regeneration of the adipose fin was negligible, whereas regeneration of the pelvic fins may have resulted in the erroneous classification of fish in these groups. Families were marked with thermal brands, in 42 specific combinations of brand position and orientation. Errors in reading brands amounted to 0.63%. The loss of brands, usually those located on the posterior and ventral locations (below the lateral line), was 27% 8 months after brands were applied. The loss of numbered operculum tags has decreased over the years from 30 to 12% by improvement of application techniques. However, 3% of operculum tags were lost during handling procedures. Jet injection of Alcian Blue dye into the fin rays of parr produced 25 individual marks in 28 families. These marks were of good quality and satisfactory for small-scale projects, but required too much labor for large-scale genetic programs.

Many studies involving fish require techniques for identification of different groups that are reared in the same tank. The ability to distinguish individuals over long time intervals is a particular requirement of genetic studies.

Fish-marking techniques have to satisfy a number of requirements to maximize effectiveness; such requirements include efficacy on small fish, minimal effect on fish behavior or physiology, relatively low cost, low input of labor and training, suitable retention time, and accuracy.

The Salmon Genetics Research Program (SGRP) has used a variety of marking techniques with variable success. Coded wire nose tags were used in the past. These tags were accurate, but it was necessary to either kill or X-ray the fish to decode the tags. Because data were collected on individual fish that were identified by family at various stages of development, X-ray-readable tags proved unsatisfactory. Cold branding with liquid nitrogen was discontinued because the procedure was too laborious and the marks did not

persist as long as desired. Cold branding in combination with fin clips, however, has been used quite successfully in other genetic programs (Gunnes and Reftsie 1980). Carlin and spaghetti tags were also time-consuming to apply, became detached too quickly, and could only be used effectively on larger fish.

Marking techniques presently preferred at the SGRP include fin clips, thermal brands, opercular tags, and jet injection of dye. Fin clips are used to mark groups of fish reared together. Fin clips have been used to identify fish, and have little effect on growth and mortality (Armstrong 1947; Shetter 1950; Coble 1967). Brands that require the application of heat have also been used to mark families of fish quite successfully (Groves and Jones 1969; Niggol 1969; Smith 1973). Opercular tags are used to mark larger individual fish, already identified by family with brands and fin clips. The jet injection of dye was evaluated by various workers (Hart and Pitcher 1969; Starkie 1975; Pitcher and Kennedy 1977) and was proven effective as a long-lasting batch mark that caused minimal damage to fish. The jet injection of dye is presently being used in smaller-scale experiments.

In this paper we describe the techniques currently used at the SGRP, together with the results and relative success of marking Atlantic salmon *Salmo salar* at different stages of development.

---

[1]Present address: Department of Fisheries and Aquaculture, Box 5001, St. Stephen, New Brunswick E3L 2X4, Canada.

[2]Present address: Department of Biology, Dalhousie University, Halifax, Nova Scotia B3H 3J1, Canada.

[3]Holds a cross-appointment at the University of New-Brunswick, Saint John, New Brunswick, Canada.

Techniques were evaluated by examination of the degree of error in mark identification, loss of identification, and possible erroneous classification of fish.

Methods

Fin clipping.—Eight-four full-sibling Atlantic salmon families, from the 1985 year-class, were reared indoors in individual 1.1 m diameter hatchery tanks. Three months after hatching, when they averaged 5.5 cm long (all lengths given in this paper are fork lengths), the fish in each tank were counted, fin-clipped, and transferred outdoors to 7.5-m-diameter tanks. Fin clipping involved removal of the entire fin, as close as possible to the body of the fish, with surgical scissors. Twenty-eight tanks were used, of which half were fiberglass and half were concrete. The fish in each family were evenly divided, and placed in replicate groups in each type of tank. Each tank contained a sample of parr from six families.

To distinguish among families reared in the same tank, each family was given one of five different fin-clip combinations, or no clip. The marks used were adipose fin only (A), right pelvic fin only (RP), left pelvic fin only (LP), adipose fin plus right pelvic fin (A–RP), adipose fin plus left pelvic fin (A–LP), and no clip (NC). The numbers of fish per tank were approximately equal (mean, 2,935; range, ±298); individual family size ranged from 85 to 1,500 fish within a tank. Between 7 and 9 months after they were clipped and transferred outdoors, the fish in each tank were sorted into their respective families on the basis of fin clips.

Similar procedures were followed for the 1986 year-class. Fewer families were involved and only A, LP, RP, and NC marks were used. Three months after fin clipping (October), each fish in two of the tanks was examined for fin regeneration and assigned to one of four subjective categories that ranged from no regeneration (score = 0) to almost complete regeneration, but still recognizable (score = 3).

Thermal branding.—Parr and presumptive smolts (based on a minimum length of 13 cm) of the 1985 year-class were counted within families and tanks. Presumptive smolts were all batch-marked with hot wire brands to distinguish among the 84 families that were combined and destined for sea cages. The branding tip was made of high-resistance, stainless steel wire 0.5 mm in diameter. A rheostatically controlled transformer was used to convert alternating current to direct current, and electrical energy was converted to

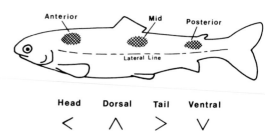

FIGURE 1.—Branding positions above the lateral line of Atlantic salmon. Positions below the lateral line and on both sides of the fish also were used. V-symbols pointing in one of four directions were applied.

heat at the branding tip, which glowed red hot. Brands were applied to each fish for less than 1 s.

After sorting, each fish in each of the 84 families (1985 year-class) was given a unique mark. A V-shaped brand was used, and families were distinguished by brand position and orientation. Six branding sites were established on each side of each fish (three above and three below the lateral line), and the V was pointed in one of four directions (Figure 1). Thus, 48 unique marks were possible with a single brand (2 sides × 6 positions × 4 directions). The additional use of adipose fin clips (present or absent) doubled the number of unique marks. In this case, only 42 of the brands were used, with and without an adipose clip to give 84 marks.

Fish were branded between January and March (mean length, 16.5 cm) and placed in sea cages. They were examined 5 and 8 months later, in August and November.

Opercular tagging.—Sequentially numbered opercular tags (wing tags for day-old chickens; Ketchum Manufacturing Sales, Ottawa, Ontario) were applied to Atlantic salmon that had been reared in sea cages for approximately 13 months. The aluminum tags (2 cm long) were clamped to the operculum of each of 3,397 fish (mean length, 54.8 cm) in 45 branded families of the 1984 year-class. The fish were examined 6 months later and a new tag was given to any fish that had lost a tag or had a loose tag.

Dye injection.—Marking by dye injection involved the jet injection of a dot of dye into the fin rays of Atlantic salmon parr. A Madajet® (Mada Medical Products) dental inoculator loaded with an aqueous solution (65 mg/L) of Alcian Blue dye (Sigma Chemical) was used. This technique was similar to that used by Herbinger (1987). Fins were marked either singly or in combination to

TABLE 1.—Numbers of Atlantic salmon used to evaluate marking techniques, by year-class and age at marking.

	Number of fish by year class and age				
	1984, 25 months	1985		1986	
Mark		3 months	10 months	3 months	16 months
Fin clip		82,000		60,000	
Thermal brand			5,040		
Operculum tag	3,397				
Jet injection					700

identify individual fish, which had also been identified by family with thermal brands. By means of one dot on the left and right pelvic and pectoral fins, and two dots on the caudal fin, 25 individual marks were produced in each of 28 families ($N = 700$). These fish averaged 8.1 cm long. The dots were evaluated 3 months later.

Results and Discussion

Table 1 summarizes the stages of development, year-classes, and numbers of fish used to evaluate the marking techniques. In the foregoing procedures, only one identifying mark was applied to each fish. Thus, the proportion of marks that were incorrectly classified could not be determined precisely. One must assume that unrecorded mortality (that is, the number originally marked minus the number at evaluation) and unidentifiable marks occurred at random. Therefore, the relative levels of erroneous classification can only be inferred from deviations from expected values.

Fin Clipping

A consistent pattern of apparent erroneous classification of the fin clips in the 1985 year-class emerged. The observed number of fish (total number of fish in each fin-clip group at sorting) was compared with the number expected for each fin-clip group. The expected number of fish for each group in Table 2 was derived by subtracting

the number of dead and unidentifiable fish from the original number that were clipped 7 months previously. A proportion was used to express the number of dead and unidentifiable fish, because the number of clipped fish per group varied. Groups with an adipose or right pelvic clip or no clip (A, RP, or NC) had more fish than expected (Table 2). Those groups with A-LP, A-RP, and LP clips had fewer fish than expected. It is likely that some of the originally clipped pelvic fins had regenerated. Therefore, those fish were incorrectly identified and assigned to the groups with no pelvic fin clip (A or NC groups, Table 2). The RP clips were not classified incorrectly, because they yielded 2.2% more than expected.

Fin clipping does not appear to significantly affect the growth of fish (Armstrong 1947; Shetter 1950; Coble 1967). This is important in genetic studies because results otherwise would be biased. It is particularly important for our work because we do not clip the largest group of fish in each tank.

The regeneration of fins clipped from fish in the 1986 year-class, other than the adipose fin, proved to be a serious problem. Three months after fin clipping, only 0.2% of the adipose fins examined showed signs of regeneration, whereas 46.4% of left pelvic and 53.4% of right pelvic fins had regenerated almost completely (Table 3). In most cases, identification was still possible, but careful examination was required.

TABLE 2.—Numbers of Atlantic salmon (1985 year-class) observed at sorting and the numbers expected in each category of fin clip (A = adipose fin; LP, RP = left and right pelvic fins; NC = no clip).

Fin clip	Observed number	Expected number[a]	Deviation (%)
A	8,801	7,696	+14.4
A + LP	7,423	7,696	−3.5
A + RP	6,529	7,361	−11.3
NC	34,381	33,623	+2.3
LP	14,803	15,668	−5.5
RP	4,969	4,864	+2.2

[a]Expected number is derived by subtraction of the number of dead and unidentifiable fish from the original number of fish that were clipped per group, 7 months previously. Dead and unidentifiable fish were assumed to be randomly distributed among fin-clip categories.

TABLE 3.—Regeneration of fins by Atlantic salmon (1986 year-class) 3 months after fins had been clipped. A score of 0 denotes no regeneration; a score of 3 denotes almost complete regeneration.

Fin clipped	Number	Regeneration score			
		0	1	2	3
Left pelvic	887	2.4%	10.7%	40.5%	46.4%
Right pelvic	219	12.8%	16.9%	16.9%	53.4%
Adipose	1,846	99.8%	0.2%		
Unidentifiable	81				

Thermal Branding

Errors in reading brands were negligible for fish that retained their brands. Error refers to the proportion of brands that were read but did not originally exist, and families (brand categories) in which there were more fish than originally marked. Brands in certain areas were more difficult to read or did not persist as well as others. The number of fish with unidentifiable brands was 17% in August and 27% in November (Table 4). At this time, opercular tags were applied to individual fish to avoid further loss of family identification and to identify individuals within families. Brands that were difficult or impossible to read may have resulted from faulty application. Brand distortion may occur if there are differences in scar formation or tissue growth in certain areas on the fish. Brands located below the lateral line were observed less frequently than those above. This difference may be due to the lack of contrast in the light-colored ventral area of Atlantic salmon, particularly when brands were examined in bright sunlight. Above the lateral line there was more contrast, and brands were easier to see. Brands on the caudal peduncle (posterior brands)

TABLE 4.—Mean fork lengths, number of fish observed, and errors associated with brands on Atlantic salmon branded in March, placed in sea cages, and examined in August and November.

	Sampling period		
Variable	March (branding)	August	November
Fork length (cm)	16.3	29.1	44.8
Total number	5,189	4,943	4,843
Error (%)[a]		0.42	0.21
Brand categories underrepresented[b]		23	21
Unidentifiable (%)		17.2	27.0

[a] Percentage of brands that were read but did not exist, and of fish counted that exceeded the number put in the sea cages with each brand.

[b] The number of brand categories (out of 42) in which there were fewer fish than the expected on the basis of the average number of fish per family, with the assumption of random mortality.

were also noticed less frequently than those positioned at the middle and anterior sites. In the caudal peduncle region, the space available for branding was more limited than in other areas, and brands were more difficult to apply.

Other workers have advocated the use of a Z-shaped brand after testing shapes such as V, O, I, S, F, and T (Nahhas and Jones 1980). However, a Z-shaped brand can only be oriented unambiguously in two directions. A V-shaped brand has proven to be the best for marking large numbers of fish, when brand orientation is critical. Generally, brand persistence is very good and brands can be read 2–3 years after application as long as care is taken to apply even pressure during the initial branding operation and as long as fish are of an adequate size.

Opercular Tagging

Opercular tags were easily applied and provided a rapid method for the individual identification of larger fish. In a 6-month interval, approximately 12% of the fish lost their tags. However, 25% of the lost tags (3% overall) were recovered in the anesthetic and recovery tubs. This indicates that handling procedures were a major factor in the loss of tags. Careful netting and handling may minimize tag loss, and fish with loose tags should be retagged to reduce loss in the future. To lessen the possibility of a tag working itself loose, it should be placed securely against the curve of the operculum in an area where movement is minimal. Aluminum tags tend to foul with algae when water temperatures are high. It is unknown whether fouled tags affect growth, feeding, or respiration. However, lesions were observed on the flank posterior to the tag on some fish, and tended to be more common on those fish with long algal streamers. Algal problems can be minimized by application of tags in the fall, after water temperatures begin to drop. In other years, loss of tags has been as high as 30% over a 6-month period.

Dye Injection

Dye injection proved to be a satisfactory procedure, producing easily read, unambiguous marks. Marks at all fin positions were equally identifiable. Over a 3-month period 0.3% of the jet-injected dots disappeared; however, the remaining dots appeared to persist for at least 6 months. These results are comparable to those of a previous study (Herbinger 1987).

Dye injection in a large-scale operation proved to be impractical. The procedure was time-consuming, and injectors could not keep pace with branders. The development of an automatically repeating injector might increase marking speed. Nevertheless, when used in combination with branding, this technique proved successful and accurate for smaller-scale studies.

Conclusion

These marking techniques all proved to be satisfactory. Regeneration of fins other than the adipose fin presented a problem, possibly causing the erroneous classification of some fish. Thermal brands were reliable; however, the loss of marks was substantial, particularly of posterior and ventral brands. Opercular tags and jet injection of dye were dependable for individual identification of fish, but previous brands and fin clips were used to identify families. A reliable method of marking small fish is necessary. Fin clipping is the first mark used to identify families of fish at the SGRP; all subsequent marking is based on correct identification of fin clips. Large numbers of fish are involved; therefore, overall erroneous classification is small. With better marking techniques, fewer fish may need to be marked, and incorrect classification of fish could be avoided.

References

Armstrong, G. C. 1947. Mortality, rate of growth and fin regeneration of marked and unmarked lake trout fingerlings at the provincial fish hatchery, Port Arthur, Ontario. Transactions of the American Fisheries Society 77:129–131.

Coble, D. W. 1967. Effects of fin-clipping on mortality and growth of yellow perch with a review of similar investigations. Journal of Wildlife Management 31:173–180.

Groves, A. B., and I. W. Jones. 1969. Permanent thermal branding of coho salmon, Oncorhynchus kisutch. Transactions of the American Fisheries Society 98:334–335.

Gunnes, K., and T. Reftsie. 1980. Cold-branding and fin-clipping for marking of salmonids. Aquaculture 19:292–299.

Hart, P. J. B., and T. J. Pitcher. 1969. Field trials of fish marking using a jet inoculator. Journal of Fish Biology 1:383–385.

Herbinger, C. M. 1987. A study of Atlantic salmon (Salmo salar) maturation using individually identified fish. Doctoral dissertation. Dalhousie University, Halifax, Canada.

Nahhas, R., and N. V. Jones. 1980. The application of the freeze-branding technique to trout fry. Fisheries Management 11:23–28.

Niggol, K, 1969. Thermal marking of adult chinook salmon, Oncorhynchus tshawytscha. Transactions of the American Fisheries Society 98:331–332.

Pitcher, T. J., and G. J. A. Kennedy. 1977. The longevity and quality of fin marks made with a jet inoculator. Fisheries Management 8:16–18.

Shetter, D. S. 1950. The effect of fin removal on fingerling lake trout (Cristivomer namaycush). Transactions of the American Fisheries Society 80:260–277.

Smith, J. M. 1973. Branding chinook, coho, and sockeye salmon fry with hot and cold metal tools. Progressive Fish-Culturist 35:94–96.

Starkie, A. 1975. A note on the use of a jet inoculator for marking fish. Fisheries Management 6:48–49.

American Fisheries Society Symposium 7:147–151, 1990

Tagging Demersal Marine Fish in Subzero Temperatures along the Canadian Atlantic Coast

DOUGLAS CLAY[1]

Department of Fisheries and Oceans, Marine Fish Division, Bedford Institute of Oceanography
Dartmouth, Nova Scotia B2Y 4A2, Canada

Abstract.—Tagging in air temperatures of −15 to −25°C was successfully carried out by reducing the exposure time of the fish. The fish were maintained fully submerged in ambient-temperature sea water in large tanks during the tagging operation. Indirect evaluation of this technique was conducted by comparison of the tag-recovery results with similar studies in other seasons. The data confirm that the tag-recovery rate and observed movement of Atlantic cod *Gadus morhua* tagged in winter are within the expected ranges. Inshore Atlantic cod tagged in winter appeared to move northeast and southwest along the shore from Halifax harbor, Nova Scotia. This observation was also made during earlier studies conducted in more favorable seasons.

Tagging is frequently one of the first research tools used when management of a fishery is initially considered. By tagging fish, the manager hopes to delineate the exploited stock, identify movement and interactions with other stocks, and obtain information on growth. For many studies, it is desirable to tag in all seasons to identify individual population components. However, each season brings tagging problems. Summer tagging often results in increased mortality associated with induced thermal shock as captured fish are raised through the thermocline; in the sea, surface-to-bottom temperature differentials of 15–20°C can occur. In winter, thermoclines are generally not a problem, nor are vertical temperature differences as extreme; however, subzero air temperatures during tagging can damage the exposed fish and result in high mortality.

Tagging is often an expensive and time-consuming process. Tagging at sea is particularly expensive because of the high cost of the tagging platform. To reduce this cost component, government research vessels were used between January and March, a time when the vessels were often idle due to storms and severe cold, which made "distant waters" work hazardous and unattractive.

This study was designed to test modified tagging procedures on these research vessels in subzero air temperatures. The aim was to minimize mortality of the fish due to exposure to ambient air temperature.

Methods

The research vessel used for this project was the 39-m stern trawler RV *E.E. Prince*. A yankee-36 trawl with a 9-mm-mesh liner in the cod end was towed on smooth bottom for 10-min intervals. The tow period was kept short in order to keep catches small and minimize the time the fish were in the net. Tows were made during daylight hours in the approaches to Halifax harbor between 5 and 30 km offshore. The net was retrieved and the fish were quickly spilled into a fiberglass holding tank (1.23 m × 1.23 m × 1.23 m) on deck. This tank was lined with a 76-mm-mesh (stretched) net that could be pulled to the surface to retrieve fish for transfer to the tagging tank. Ambient-temperature surface seawater was pumped into the tanks and drained through an external standpipe to keep the water clean and well oxygenated. Bathythermograph casts indicated no thermocline and a maximum temperature differential between surface and bottom of 4°C. The narrowness of the differential was due to the relatively shallow depths (<60 m) and the strong winter winds, which facilitate vertical mixing of the water column.

Fish remained in the holding tank from 5 min to 1 h; the time depended on the size of the catch and the backlog of fish awaiting processing. Dead, injured, or weak fish were discarded. A scoop net was used to move the fish from this holding tank to a fiberglass tagging rank (0.92 m × 1.23 m ×

[1]Present address: Marine and Anadromous Fish Division, Department of Fisheries and Oceans, Gulf Fisheries Center, Post Office Box 5030, Moncton, New Brunswick E1C 9B6, Canada.

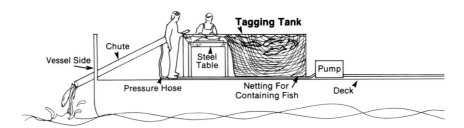

FIGURE 1.—Cross-sectional view of tagging tank, with fish-return chute, steel table, and net liner.

2.46 m).[2] Air temperatures often ranged from −10 to −25°C; these temperatures, combined with winds approaching 40 km/h produced wind-chill equivalents of −30 to −50°C. To prevent surface freezing of the fish, it was imperative that they not be exposed to the air for more than 5 s.

The fiberglass tagging tank contained a metal table with the top submerged 15 cm beneath the water surface (Figure 1). This tank was lined with a 76-mm-mesh net that could be pulled to the surface to spill the fish over the surface of the table. The netting was sewn to the perimeter of

_____

[2]One recommendation that would simplify this work and reduce stress for the fish would be to include an open chute (15 cm deep × 30 cm wide) from the holding tank to the tagging tank. This chute would hang from a 15-cm spout cut into the top edge of the holding tank. The stand-pipe drain would no longer be necessary because the seawater would overflow down the chute to the tagging tank. When the fish were required they could be "pursed" with the net liner and spilled down the chute.

the table surface to prevent the fish from moving beneath the table.

As the fish were spilled onto the table they were guided over one of the two stainless-steel meter sticks fastened to the table top. Length was measured to the nearest centimeter. A dart tag with a nylon T-bar (Floy Tag and Manufacturing[3]) was inserted at the base of the dorsal fin so as to lodge the T-bar between adjacent hyponeural processes. Each fish was injected with 100 mg of oxytetracycline per kilogram of fish as part of an ongoing age-validation study. Immediately after tagging, the fish were returned to the sea through a chute (polyvinylchloride pipe, 3 m long and 25 cm in diameter). This chute extended down from the edge of the tagging tank through the scuppers, and ended about 1 m beyond the side of the vessel and 1 m above the surface (Figure 2). To maintain

_____

[3]The mention of any specific product does not imply an endorsement by the Department of Fisheries and Oceans.

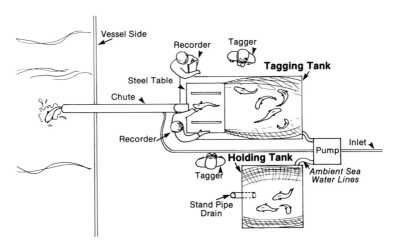

FIGURE 2.—Top view of stern deck of RV *E.E. Prince,* showing tanks for holding and tagging fish, return chute, pump for seawater aeration, and relative positions of the taggers and recorders.

TABLE 1.—Numbers of four species tagged, released and later recaptured during three test winter tagging cruises on the RV *E.E. Prince*. Numbers in parentheses are the percentages recaptured. Recovery values are from similar areas during warmer-season tagging studies are given for comparison (W. Stobo, Department of Fisheries and Oceans, unpublished).

Tagging date	Activity	Atlantic cod	American plaice	Winter flounder	Yellowtail flounder	Total
			Species			
Jan 1979	Tagging	75		29	5	127
	Recapture	10(13)		1(3)	1(20)	12(9)
Feb 1979	Tagging	506	43	221	3	787
	Recapture	79(16)	4(9)	6(3)	0	89(11)
Feb 1980	Tagging	890	1,065	109	12	2,820
	Recapture	115(13)	50(5)	3(3)	0	169(6)
Total	Tagging	1,471	1,108	359	20	3,734
	Recapture	194(14)	54(5)	10(3)	1(3)	270(7)
Nonwinter studies		(10–15)	(6.6)	(2.2)	(0.7)	

a wet, relatively warm surface in this chute, a 2.5-cm pressure hose sprayed ambient-temperature seawater through a hole near the upper end of the pipe. The round fish developed enough speed during their exit to shoot them 1–2 m beneath the surface. Flatfish or skates tended to bellyflop on the surface and then swim away.

Atlantic cod *Gadus morhua* were the most common species caught and tagged (about 40%), and American plaice *Hippoglossoides platessoides* were the next most common (about 30%). Winter flounder *Pseudopleuronectes americana* and yellowtail flounder *Limanda ferruginea* accounted for most of the remainder. All species caught were tagged, to test the technique as completely as possible.

Three cruises were conducted during which this equipment was used. In January and February 1979 and February 1980, 127, 787, and 2,820 fish were tagged, respectively. During these periods the surface water temperature ranged from 1.5 to −0.5°C; loose pan ice and slush ice often were present. To combat this cold water and permit maximum dexterity for gentler handling of the fish, the taggers wore three layers of thin gloves, the inner and outer layers being cotton and the middle layer being surgical latex. Some taggers filled the inner gloves with hot water to extend their working time.

Results and Discussion

The actual tagging experiments were conducted on the open aft deck of the research vessel. Due to subzero temperatures, most persons could only tag for 10–15 min. During this time 30–60 fish could be processed; the data were recorded by a second person. After the taggers reached the limit of their endurance, they retired to the onboard laboratory to warm up for 5–10 min, and then exchanged duties with the recorders. With this routine, each tagging team of two persons could tag 90–180 fish per hour, usually considerably faster than the vessel was landing them.

Quantitative evaluation of this technique requires estimates of natural mortality and fishing mortality during the tag-recovery period, for comparison with similar data sets from other seasons. Neither of those variables is known with any degree of accuracy. Qualitative assessment of this technique is possible by comparison of the results of the tag recoveries to data from past studies in other seasons in nearby areas.

Less than 10% of the fish coming on board were discarded because of weakness or damage that we thought would have seriously affected survival after tagging. The remaining 3,734 fish, comprising four commercially important species, were tagged, injected, and released without incident (Table 1). Of these fish, 270 (7%) were recovered by various sectors of the commercial fishing industry in the 5 years after tagging. The tag-return rate varied among species: 14% for Atlantic cod and 3–5% for the three flatfish species. Tag-recovery rates depend upon many factors, the three major ones being tag-induced mortality (including tag loss), fishing mortality, and intrinsic natural mortality. In this study, only 5% of the total number of tags returned were recovered in the quarter of the year (months 1–3) in which the tagging was carried out (Table 2). This return was

TABLE 2.—Numbers of tagged fish of four species recovered during and after the year of tagging, by post-tagging quarter-year.

Quarter after tagging	Atlantic cod	American plaice	Winter flounder	Yellowtail flounder
First year (0–12 months) after tagging				
1	7	5	0	0
2	19	7	1	0
3	37	10	1	0
4	21	17	0	0
Second year (13–24 months) after tagging				
1	39	8	0	0
2	6	1	1	1
3	9	0	1	0
4	6	0	1	0
Third year (25–36 months) after tagging				
1	2	0	0	0
2	1	0	0	0
3	5	1	0	0
4	1	0	0	0
Fourth and fifth years after tagging				
1	3	0	0	0
2	1	0	0	0
3	1	0	0	0
4	0	0	0	0

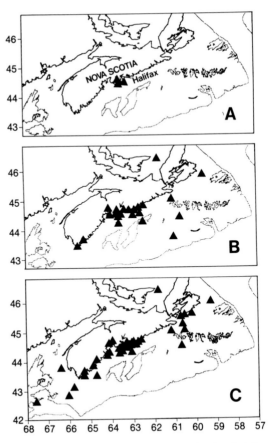

FIGURE 3.—Recovery locations (black triangles) for Atlantic cod tagged in (A) January 1979, (B) February 1979, and (C) February 1980. The triangles can represent more than one tag. The dotted line represents the 183-m depth contour.

probably due to the severe weather that limits the inshore commercial fishery in winter. This pattern implies that a higher rate of return in the first quarter (and therefore overall) would have occurred if tagging had taken place during a more active period of the fishery, as is often the case with tagging work.

The tag-recovery levels achieved in this study fall within the range of values observed for this dart tag in similar tagging studies carried out during other seasons on the Nova Scotia shelf over the previous 10 years (Table 1) (W. Stobo, Department of Fisheries and Oceans, unpublished data). In tagging studies of Atlantic cod, in which there were much higher rates of return—from 20 to over 30% (McKenzie 1956; Templeman 1979)—tagging sites and times often coincided with major commercial fisheries. In addition, the majority of returns in those studies occurred in the same year of tagging, on the same and neighboring fishing grounds. In many other studies there have been lower rates of return of tagged Atlantic cod—8% (McKenzie 1934) to 10% (McKenzie 1956).

Distribution and movement of our Atlantic cod were similar to those observed in a series of summer and fall tagging operations in the approaches to Halifax harbor, during 1934, 1935, and 1936 (McKenzie, 1956). At that time, 942 fish were tagged in May and June. Of those, 25–30% were recaptured, 75% of them in the first year. About 25% of the recaptured fish had moved more than 19 km from the tagging site. Also in that study, 158 fish were tagged in October and November. For those fish, there was only a 5% tag-recovery rate. This rate may have been a result of the method used to capture the fish for tagging (gill nets) rather than of the season. These fish tagged in October and November showed about the same degree of movement as those tagged in the summer.

About 70% of the Atlantic cod tagged in our work were caught within 20 km of the nearshore tagging area (Figure 3). It is notable that, as observed by McKenzie (1934), 75% of the remain-

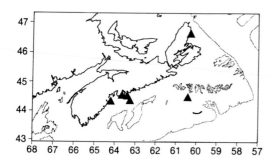

FIGURE 4.—Recovery locations for American plaice tagged in February 1979 and February 1980. The triangles can represent more than one tag. The dotted line represents the 183-m depth contour.

ing recaptured fish were caught within 20 km of shore; some of these were recovered hundreds of kilometers from the tagging site. Fish that did move offshore included two recovered in the southern Gulf of St. Lawrence (>600 km away) and one in Georges Basin (>400 km).

Although no relevant data on historic distribution patterns are available for the flatfish species tagged, the recovery rates were comparable to those in recent studies. All 10 winter flounders *Pseudopleuronectes americanus* and the single yellowtail *Limanda ferruginea* recovered were within a few kilometers of the tagging site. Four (10%) of the recaptured American plaice had moved more than 20 km, two to the offshore, one

southwest along the shore (60 km), and one 400 km northeast along the shore (Figure 4).

These data on tag-recovery rate, recovery time span, and dispersal fall within the range of data from earlier studies conducted during other seasons. Thus, it seems that by keeping the fish completely submerged, fish can be tagged in subzero air temperatures as successfully as in any other season.

Acknowledgments

I thank the captain and crew of the RV *E.E. Prince* for their help in this work. I especially appreciate those persons who dipped their hands in ice water for several hours each day: Diane Beanlands, Bev Fowler, Geraldine Young, Fred Rahey, and Heather Clay. W. Stobo, Marine Fish Division, Dartmouth, and R. Alexander, Marine and Anadromous Fish Division, Moncton, provided valuable editorial reviews of earlier versions of this report.

References

McKenzie, R. A. 1934. Cod movements on the Canadian Atlantic coast. Contributions to Canadian Biology and Fisheries 8(31):434–458.
McKenzie, R. A. 1956. Atlantic cod tagging off the southern Canadian mainland. Fisheries Research Board of Canada Bulletin 105.
Templeman, W. 1979. Migration and intermingling of stocks of Atlantic cod, *Gadus morhua*, of the Newfoundland and adjacent areas from tagging in 1962–66. International Commission of Northwest Atlantic Fisheries Research Bulletin 14:15–48.

American Fisheries Society Symposium 7:152–160, 1990

Use of Hydroscopic Molded Nylon Dart and Internal Anchor Tags on Red Drum

ELMER J. GUTHERZ AND BENNIE A. ROHR

National Marine Fisheries Service, Southeast Fisheries Center, Mississippi Laboratories
Pascagoula Facility, Post Office Drawer 1207, Pascagoula, Mississippi 39567, USA

R. VERNON MINTON

Alabama Department of Conservation and Natural Resources, Marine Resources Division
Post Office Drawer 458, Gulf Shores, Alabama 36542, USA

Abstract.—We conducted an experiment to determine tag shedding by, and mortality of, red drums *Sciaenops ocellatus* marked with dart tag heads (hydroscopic molded nucleated nylon) and internal anchor tags. Large red drums (3.8–12.1 kg) were collected in 20-m depths off Mississippi and transferred to five 0.1-hectare brackish-water holding ponds in Gulf Shores, Alabama. After a 30-d acclimatization period, the fish were tagged. They were removed and inspected 117 d after tagging. Condition of tagging sites, loss of tags, and deaths were recorded. No significant differences were found in shedding rates or mortality of fish with different tag types. The overall mean tag-loss rate of fish in ponds was 18.2%, with a 95% confidence interval of 12.1–30.8%. Concurrently, adult red drums were tagged offshore for life history information and population estimates. Data gained from double-tagged fish caught by commercial and recreational fishermen indicated a 25% tag-loss rate. From the offshore tagging project, two tagged wild red drums were recaptured, after 166 and 349 d at large, and frozen for analysis of tissue adherence to the hydroscopic molded nylon dart tag heads. Preliminary examination suggested that tissue adhered both to the hydroscopic molded nylon tag head and the overlying connective tissue forming the encapsulation. Tag location and placement were also evaluated in terms of injury to the fish; we make recommendations for tag placement to minimize injury.

Many terrestrial and aquatic animals have been studied with mark–recapture methods, but the suitability of tagging techniques for aquatic animals often depends on the species studied (Rounsefell and Everhart 1953; Royce 1972; Ricker 1975). Adverse effects of tagging may include mortality and injury due to tagging, slow growth, and stress from excessive handling of fish, which can bias estimates of growth, movement, and population size.

Red drums *Sciaenops ocellatus* are important recreational and commercial fish in the south Atlantic region of the USA (Simmons and Brewer 1962; Matlock 1980; Mercer 1984; USNMFS 1986). Management of red drum stocks in the Gulf of Mexico reached near-crisis status when a small fleet of highly efficient purse seiners began taking large catches of offshore adult red drums. With the apparent decline in stock size, better information on all aspects of red drum population biology was sought by the Gulf of Mexico Fishery Management Council. In response, the Southeast Area Monitoring and Assessment Program developed a 3-year cooperative red drum research program to obtain information on size, age structure, and movement patterns of adult red drums offshore. Little informa-

tion is presently available on these stocks, which may represent the spawning brood stock (Matlock 1980, Mercer 1984). It is this offshore spawning stock that contributes most heavily to the numbers of inshore juveniles and prerecruits that are caught by recreational fishermen (Gulf of Mexico Fishery Management Council 1987).

The recent commercial exploitation of this offshore spawning stock prompted a tagging project to provide information on offshore red drum population biology. The desirability of long-term tag retention and the large size of offshore adults necessitated a tag that could be easily and quickly applied, with minimal damage to the fish and few lost tags. Concurrent with the tagging project, an experiment was conducted in holding ponds to evaluate tag shedding, mortality, adherence of tissue to the tag heads, and optimal tag placement. After completion of tagging, a tag-recovery program was undertaken by the National Marine Fisheries Service (NMFS) laboratory in Pascagoula, Mississippi, to obtain data on movements of red drums offshore and to compute a population estimate based on proportions of tagged fish recaptured.

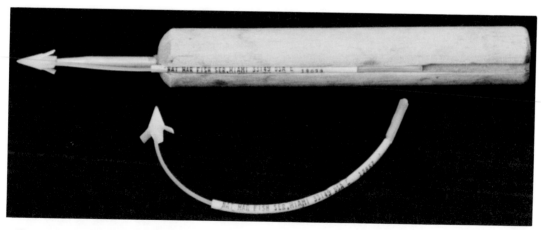

FIGURE 1.—Molded nylon dart tag and its applicator; tag head was constructed of hydroscopic molded nucleated nylon welded to the monofilament streamer with liquid phenol.

Methods

Pond study.—To determine rates of tag shedding and fish mortality, a cooperative experiment was conducted by the NMFS and the Alabama Department of Conservation and Natural Resources, Marine Resources Division, at the Claude Peteet Mariculture Center in Gulf Shores, Alabama. A commercial purse-seine vessel was used to collect 157 adult red drums on September 22, 1986. The fish were collected off Mississippi in 20-m depths. The fish were transported from the collection site in two 41-m³ holding tanks. Water quality was maintained by constant exchanges of seawater. Red drums stressed by a distended air bladder were punctured with a hypodermic needle to deflate the air bladder. Ten fish died during transport. When the vessel docked, bottom sediments and muddy water entered the holding tanks. Shortly thereafter, the fish experienced stress and 22 fish died. The water was oxygenated until signs of stress were alleviated and mortality was curtailed.

The process of unloading and stocking fish in the five holding ponds caused stress, as evidenced by rapid movement and slapping actions, which resulted in bruising and loss of scales. Fish were then held for a 30-d acclimatization period during which 31 additional fish died. All pretagging mortality of pond-held fish occurred within the first week. Each 0.1-hectare holding pond held approximately 1.06 million liters of brackish water (12–17‰ salinity). After stocking the fish, ponds were monitored daily for temperature, salinity, dissolved oxygen, and fish mortality. Fish were selected randomly for placement into each holding pond. Water quality differed only slightly among ponds.

A beach seine was used to collect fish from the ponds for tagging. Each pond was seined only once, and the escaped fish represented an untagged control group. Red drums were removed individually, placed on top of a by-catch separator, and allowed to slide to the tagging area. The tagging team included a tagger, fish holder, and recorder. Tagging procedures were the same as those in the offshore tagging operations. Nylon dart tags (Figure 1) were inserted intramuscularly between the second and third scale rows, below the anterior dorsal fin rays; the internal anchor tags (Figure 2) were inserted in the abdominal musculature or cavity. Nylon dart tag heads were constructed from hydroscopic molded nucleated nylon, which was welded with liquid phenol to a monofilament streamer. These tags were 171.5 mm long, and consisted of a 155-mm monofilament streamer and a dart head 16.5 mm × 6.5 mm. The dart head was constructed with two 5.5-mm-long anchor tabs (Figure 1). This tag was designed as an intramuscular attachment device and was intended to encourage the adhesion of tissue to the tag head to improve long-term retention (E. D. Prince, NMFS, personal communication). Streamers for nylon dart tags and the internal anchor tags were constructed by Floy Tag and Manufacturing (Seattle, Washington)[1] and the nylon tag heads by Master Tool Company (Miami

[1]The National Marine Fisheries Service does not approve, recommend, or endorse any commercial product mentioned in the publication.

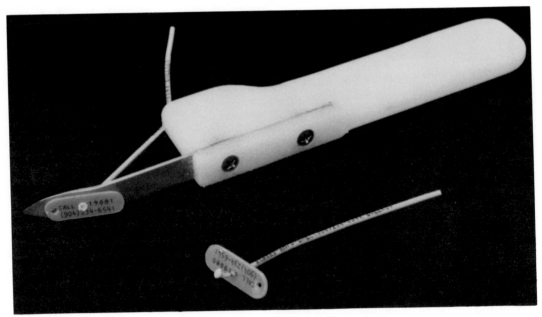

FIGURE 2.—Internal anchor tag and its applicator.

Lakes, Florida). Nylon dart tags were assembled and applicators were provided by NMFS personnel. The wooden nylon dart tag applicator consisted of a 150-mm wooden handle and a 45-mm stainless steel rod to which the tag was attached (Figure 1). Internal anchor tags were applied to the abdominal musculature or cavity near the distal end of the pectoral fin, or above and anterior to the anal vent. Internal anchor tags near the pectoral fin were applied with an applicator (Figure 2), whereas internal anchor tags near the vent were inserted after an incision was made in the abdominal wall with a scalpel. Internal anchor tags consisted of a streamer, 90 mm × 1.5 mm, and an anchor tab, 25 mm × 6 mm. The tab had a 1.5-mm hole to attach the tag to the applicator. The applicator consisted of a nylon or polypro-polene handle (150 mm × 32 mm, 35 mm at the highest end). The applicator blade was 125 mm long, of which 65 mm was exposed and 60 mm was recessed and anchored into the handle (Figure 2). Internal anchor tags and their application were described by Fable et al. (1988).

After tagging, red drums were returned to their respective ponds. Sixty-one tagged fish ranged from 661 to 971 mm fork length (all fish lengths in this paper are fork lengths). Tag type and number of tagged fish in each pond varied (Table 1). Seined fish were tagged in one of three ways: (1) single dart (20 fish); (2) double darts, both on the same side of the fish (18 fish); or (3) single internal anchor (22 fish). Tagged fish were then held undisturbed for 117 d, after which they were recaptured. Recapture data included tag type and

TABLE 1.—Summary data on tagged red drums and their recapture, Claude Peteet Mariculture Center, Gulf Shores, Alabama. Numbers in parenthesis are numbers of nylon dart tags. Control fish were not tagged.

Pond	Number of fish tagged on 23 Oct 1986			Number of tagged fish recaptured on 17 Feb 1987			Number of control fish	Number of tags lost		Deaths (tagged plus control fish)
	Single dart	Double dart	Internal anchor	Single dart	Double dart	Internal anchor		Dart	Internal anchor	
D/3	5	5 (10)	5	5	5 (9)	3	8	1	1	2
D/4	5	5 (10)	5	5	5 (8)	5	9	2	0	0
D/5	6[a]	2 (4)	3	3	1 (2)	3	6	5	0	0
D/6	2	3 (6)	5	2	2 (2)	5	4	4	0	1
D/8	3	3 (6)	4	3	3 (6)	2	5	0	1	1
Total	20	18 (36)	22	18	16 (27)	18	32	12	2	4

[a]One tagged fish was previously tagged on the *Captain Grumpy* on September 18, 1986.

number, condition of tag site (healed, red, or lesioned), and condition of fish.

Mortality rates were compared between fish with different tag types (dart versus internal anchor) and tagged versus untagged fish. Shedding rates of the two tag types were also examined. Pearson's chi-square test for equality of proportions (Z-test for equality of two proportions) was used to test the null hypothesis that the proportions were not different. The 95% confidence interval on the unknown true proportions (Clopper–Pearson confidence interval) was computed. A $(1 - \alpha)$ 100% confidence interval (α is the probability of a type I error) on the actual proportion (P_L = lower limit and P_U = upper limit) is

$$P_L(n,x) \leq P \leq P_U(n,x);$$

x denotes the number of dead fish (or tags lost) out of n number of fish. Here

$$P_L(n,x) = \frac{x}{x + (n - x + 1) F_{0.5\alpha}},$$

$F_{0.5\alpha}$ being the $(1 - 0.5\alpha)$100th percentile point in the F-distribution with $2(n - x + 1)$ and $2x$ degrees of freedom,

$$P_U(n,x) = 1 - \frac{n - x}{(n - x) + (x + 1) F_{0.5\alpha}},$$

$F_{0.5\alpha}$ being the $(1 - 0.5\alpha)$100th percentile point in the F-distribution with $2(x + 1)$ and $2(n - x)$ degrees of freedom (Brownlee 1965; Odeh and Owen 1983).

Offshore study.—To obtain population estimates of red drums offshore as well as data on their life history and movement patterns, an offshore tagging project was conducted. Tags and their application were similar to those described for the pond experiment. Tagging operations began in September 1986 and ended in June 1987; tag-recovery operations began in July 1987 and continued until October 1987. All tagging activities in the field were conducted aboard a chartered commercial purse-seine vessel. Tagging and tag-recovery operations were similar in that fish were pursed, brailed, and placed onto a separator grate. Red drums were then either tagged or examined for tags and released over the side. About every 20th fish was removed to obtain biological information, including data on age and growth, stock identity, and maturation. Tagging sites were examined, classified into one of three categories (healed, red, and lesioned), and tabulated. In addition, the tags of recaptured fish were tugged to determine if they were firmly attached or loose.

Other studies.—A preliminary evaluation of tissue adherence to the nylon tag head and tag placement was conducted. Adherence of tissue to the dart tag head was evaluated for two recaptured red drums from the offshore tagging project. Tags were in place for 166 d (857-mm fish tagged April 23 and recaptured October 9, 1987) and 349 d (821-mm fish tagged September 17, 1986 and recaptured September 9, 1987). Tags and surrounding tissue were removed from the fish and the tissue was then carefully cut away from the tag. Degree of adherence was defined by microscopic examination of the tissue and the nylon tag head. Photographs were taken of the tag sites and placement in the musculature, and a diagrammatic illustration was drawn of the encapsulated tag head.

Tag placement was evaluated by placing dart and internal anchor tags in several positions on a large red drum. Dart tags were inserted in the dorsal musculature above the lateral line at the juncture of the second, third, or fourth scale rows. Internal anchor tags were placed in the abdominal musculature or cavity above the ventral midline, between the distal end of the pectoral fin and the vent. The fish was then dissected and the tag pathway and surrounding tissue were examined for damage.

Results

Pond Study

Data on dart and internal anchor tags were combined because there was insufficient evidence to conclude that mortality or shedding rates were different. Statistical tests showed no significant differences in mortality of tagged and untagged fish (Table 2). Overlap of 95% confidence intervals for mortality rates of fish in the five tag categories substantiated the absence of significant differences. Among tag-loss rates, no significant differences were noted (Table 2). Chi-square values, Z values, and associated probabilities for tag type and tag placement indicated that there were no significant differences between tag types in proportions of tags lost.

Significant differences between tag types among the five ponds were not observed; therefore, data from the ponds were pooled and a single overall tag-loss rate was computed. However, some evidence pointed to different tag-loss rates for the five ponds because tag loss from two ponds dif-

TABLE 2.—Tag loss and mortality for red drums in the pond experiment; and tag loss for 39 double-tagged fish returned by commercial and sport fishermen and the tag recovery project from tagged offshore red drums. SD = single dart, DD = double dart, PA = pectoral internal anchor, VA = vent internal anchor, C = control untagged, and I = internal anchor.

Variable	Tag type						
	Ponds					Offshore	
	SD	DD	PA	VA	C	DD	I[a]
Tag loss							
Lost	3	9	0	2		12	4
Recovered	18	27	10	8		52	10
Percent lost	14	25	0	20		19	29
Mortality							
Alive	21	21	10	10	30		
Dead	0	0	1	1	2		
Percent dead	0	0	9	9	6		

[a] Double-tagged fish with one internal anchor tag and one nylon dart tag.

fered (Table 1). We judged that the difference was not great enough and the sample size was too small to warrant separate calculations of tag-loss rate for each pond. Respective tag-loss rates for dart and internal anchor tags were not considered important (Table 2); therefore, an overall mean tag loss rate for all ponds and for both tag types was calculated as 18.2%, with an associated 95% confidence interval of 12.1–30.8%.

Tag-induced lesions were evaluated for 64 sites (45 darts, 19 internal anchors) when fish were removed from the holding ponds (Table 3). Most tags appeared to be firmly anchored, but 62% of the sites showed redness or lesions. We did not test these data statistically due to their subjective nature. Although the sample size was small, the differing incidence of lesions between dart and internal anchor tags may be meaningful.

Offshore Study

Observations on tag-induced lesions were obtained from fish returned from recreational and commercial fishermen and from our recapture study on the commercial purse seiner. Seventy-seven dart and 14 internal anchor tag sites were examined (Table 3). Most tags were firmly anchored in place, and 73% of sites had healed. In contrast, only 38% of tagging sites on pond fish had healed when examined. Water conditions may have been better offshore than in the ponds (Table 3).

Tag-loss rates based only on recovery of double-tagged red drums offshore indicated no great differences between loss of nylon dart tags (24%) and internal anchor tags (30%).

Other Studies

Dart tags were completely encapsulated in both of the dissected red drums that had been recaptured offshore. Tag entrance sites were healed and tags were firmly attached to the fish; removal required dissection. Tissue adherence was noted in the encapsulation of the dart head (Figure 3). The fish that had been at large for 349 d showed a greater degree of encapsulation and development of tissue surrounding the rigid monofilament streamer than the fish recaptured after 166 d. Encapsulation consisted of a whitish gray mass of cystic fibrous connective tissue. The underlying pinkish tissue may have been muscle or vascularized tissue. When a portion of the whitish gray tissue overlying the tag head was removed, a

TABLE 3.—Condition of tissue in proximity of tagging sites for nylon dart and internal anchor tags from large red drums in ponds and offshore. Condition is defined as (1) healed, no apparent deleterious effects; (2) red, some redness around entry site but no open lesion; and (3) lesion, open ulcerated tissue that may be associated with some necrotic tissue.

Tag type	Total number of tag sites	Tag site condition					
		Healed		Red		Lesion	
		Number	Percent	Number	Percent	Number	Percent
Ponds							
Dart	45	16	36	8	18	21	47
Internal anchor	19	8	42	5	26	6	26
Total	64	24	38	13	20	27	42
Offshore							
Dart	77	58	75	14	18	5	6
Internal anchor	14	8	57	4	29	2	14
Total	91	66	73	18	20	7	8

FIGURE 3.—Nylon dart tag embedded in the dorsal musculature of a red drum. Note the tissue sheath encapsulating the molded nylon tag head and a portion of the monofilament streamer.

series of small tufts of pinkish tissue were observed (Figure 4). This tissue appeared to attach to the tag head and the overlying whitish gray connective tissue, and gave the tag head a granulated appearance. These observations are preliminary and the recapture of additional specimens will be necessary before more definitive statements can be made.

Tags placed in large red drums were evaluated for possible internal injury to the fish. Tags were dissected to determine the extent of damage at the tagging site and along the tag's pathway (Table 4). Proper tag placement should decrease tissue damage and increase the probability of tissue adherence to the nylon dart tag head.

Discussion

To reduce mortality after tagging, fish should be held if possible to identify and remove injured or weakened fish before tagging (Hein and Shepard 1980; Garrison 1971). Mortality can result from stress due to the method of capture, handling, transport, or injury while tagging (Parker et al. 1963). Survival of tagged red drums, indicated by tag recoveries averaging 10.3%, was nearly twice as great as that of other sciaenid species tagged off Texas (Garrison 1971; Simmons and Brewer 1976). Garrison (1971) reported that tagging red drums with internal anchor tags did not increase mortality over that of untagged control fish. This

result was also seen in the pattern of mortality during our pond experiment (Table 2).

Offshore red drums had a higher tag-retention rate than pond fish (Table 2). Retention of nylon dart and internal anchor tags was not significantly different in the pond study, although a higher percentage of the dart tags was lost (Table 2). Recovery of offshore fish indicated a higher retention rate for nylon dart tags than internal anchor tags. During the offshore tagging study, 15,349 fish were tagged with 19,570 tags (15,771 nylon dart and 3,799 internal anchor tags). Slightly less than 1% (136) of the tagged fish have been recovered with 179 tags (161 nylon dart and 18 internal anchor tags). Return data consisted of 89.9% nylon dart and 10.1% internal anchor tags. Nylon dart tags constituted only 80.6% of the tags placed in offshore red drums; the retention rate of nylon dart versus internal anchor tags was thus higher. The low tag-recovery rate (0.9%) may mask the true loss rate, if tagging mortality was high or if the tag implantation site healed completely when tags were lost and tagging scars went unnoticed. For double-tagged offshore red drums (39 fish) only, there was a tag-shedding rate of 30% for internal anchor tags, 24% for nylon dart tags, and 25% for the combined tags. These rates were within the computed 95% confidence interval of 12–31% from the pond study. Mean tag loss from the pond experiment was 18.2%, which is reason-

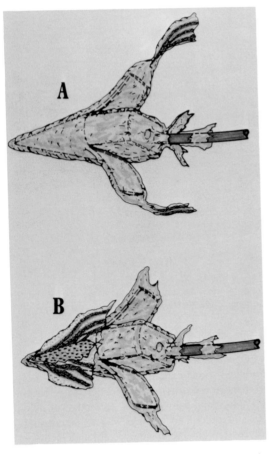

FIGURE 4.—Diagrammatic representation of the tissue-encapsulated head of a dart tag removed from a 821-mm red drum. The tag was embedded in the red drum for 349 d; note granulated appearance of muscle tufts on the surgical nylon tag head (B).

ably similar to the combined loss rate for offshore fish. Tag retention of internal anchor tags in red drums off Texas was high (Garrison 1971; Weaver 1976; Hein and Shepard 1980). Floy internal anchor tags used in blueback herring *Alosa aestivalis* by Bulak (1983) and in yellow perch *Perca flavescens* by Dunning et al. (1987) also had high retention rates. The shedding rate of dart tags by yellow perch was not determined by Stobo (1972), but he stated that retention was generally good. However, Dunning et al. (1987) noted poor retention of anchor and dart tags. Nylon dart tags were recovered from red drums double-tagged offshore and free for 1–395 d. Tag retention did not decrease over time. Time units were defined as 100 d with tag loss as follows: 0–100 d, 20 tags returned, 3 lost; 101–200 d, 8 returned, 1 lost; 201–300 d, 10 returned, 6 lost; and 301–395 d, 14 returned, 2 lost. These loss rates suggest that tags were well locked in position, that tag encapsulation and tissue adhesion enhanced retention, or that red drums are very tolerant of intramuscular tags and do not have a high rejection rate.

Injuries to tagged fish may have resulted from poor tagging techniques that allowed tags to move in the musculature, or from poor tag design (Yamashita and Waldron 1958; Saunders 1968; Stobo 1972; Weaver 1976; Hein and Shepard 1980). Sites from which tags were lost often were associated with necrotic tissue around the implantation site (Stobo 1972; Hein and Shepard 1980). Tags with attached marine growth may chafe the fish's side (Saunders 1968). Weaver (1976), who used Floy anchor tags in red drums, stated that little tissue damage was evident at the implantation site and

TABLE 4.—Observations on the penetration and effects of six nylon dart tags placed in a large red drum.

| Tag | Tag location (number of oblique scale rows) | | Penetration depth of dart head | Effect on body tissue |
	Rows to tag	Rows from tag to lateral line		
1	1	4	Shallow penetration, 8 mm plus dart head	Dart head near base of second dorsal fin in red muscle mass; wound subject to bleeding
2	3	3	Penetration to 20 mm past plane of neural spines	Dart head lodged in white muscle mass between vertebrae and pterygiophores
3	5	1	Penetration to top of a vertebra	Dart head in red muscle mass immediately above vertebra; wound subject to bleeding
4	2	4	Diversion of dart head toward tail	Dart deflected by a neural spine, disrupting both red and white muscle; wound subject to bleeding
5	3	3	Penetration into fillet on opposite side	Dart head lodged in white muscle mass
6	4	2	Penetration to between neural spines and a vertebral centrum	Dart head lodged in central red muscle mass at lateral line; wound subject to bleeding

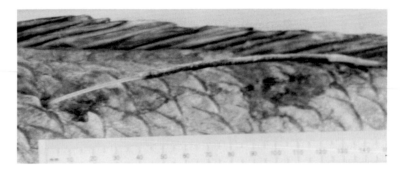

FIGURE 5.—Dart-tag entrance site on a red drum and discolored area along the distal portion of the monofilament streamer. Note algae-like growth on the sheath over the monofilament streamer.

that healing was usually complete after 62 d. Wounds observed on offshore red drums may have resulted from movement of the rigid monofilament streamer or from an accumulation of marine growth on the streamer. It is difficult to state with certainty that the tag caused the loss of scales and tissue. However, skin discoloration matched a portion of the streamer that was also discolored (Figure 5).

Information on tissue attachment to the nylon dart tag head is not available; however, the idea of a tag biologically acceptable to a fish is intriguing. These tags were initially designed for use with bluefin tuna *Thunnus thynnus* and king mackerel *Scomberomorus cavalla* (E. D. Prince, NMFS, personal communication). They were then placed in other large pelagic fish and red drums. Large pelagic species have provided few tag returns thus far, and no opportunity to examine tissue adherence, but the recapture of large red drums has provided an opportunity to examine tissue reactions. The nylon dart head was designed for intramuscular placement to encourage adherence of tissue for reduction of long-term shedding. The time required for tissue to begin attachment is unknown, but the entire tag head and a portion of the monofilament streamer were encapsulated after 166 d in the specimens we examined. Tag encapsulation has been reported for fish with internal anchor tags in the abdominal cavity (Rounsefell and Everhart 1953; Vogelbein and Overstreet 1987) and in skipjack tuna *Euthynnus pelamis* with dart tags in the musculature (Yamashita and Waldron 1958). Encapsulation appears to develop quickly, as in spots *Leiostomus xanthurus;* within 48 h, internal anchor tags were encapsulated by and sequestered within a thin capsule, which then thickened by the fourth day (Vogelbein and Overstreet 1987). In skipjack

tuna, the intramuscular dart tags were encapsulated within 30 d but were not fused to the tag (Yamashita and Waldron 1958). In red drums tagged with nylon dart tags, tissue appeared to become attached to both the tag head and the overlying connective tissue. This connective tissue formed the cyst encasing the tag head and part of the rigid monofilament streamer. Histological examination of this tissue is required to clarify the degree, if any, of adherence, and the process of encapsulation. Tissue adherence to or encapsulation of the nylon dart tag head would presumably enhance retention of this tag, and long-term tag shedding might be reduced.

Tag placement is critical for long retention and reduction of mortality and injury. Musculature in fish consists of both red and white muscle. The white muscle fibers are larger and provide most of the muscle mass in fish. Red muscle is designed for aerobic metabolism; therefore, it has a rich blood supply provided by many small capillaries (Bone and Marshall 1982). Intramuscular tags placed in white rather than red muscle would do less damage. Placement of tags in the abdominal cavity must be done with care to avoid injury to internal organs or blood vessels. During the height of the spawning period, red drum gonads were occasionally cut during offshore tagging efforts, as evidenced by eggs or milt on the applicator.

Acknowledgments

We extend special thanks to the crew of the *Captain Grumpy* for their efforts in capturing and maintaining adult red drums in good condition during transit; and to personnel of the Claude Peteet Mariculture Center for their efforts in maintaining ponds and red drums, and for assistance in tagging and recovery operations. We also thank Gary M. Russell for collection and tagging,

Arvind Shah for statistical analysis, Chris Gledhill, Ren Lohoefener, Scott Nichols, and Andy Kemmerer for their critical reviews, and Dale Burgin for typing the manuscript.

References

Bone, Q., and N. B. Marshall. 1982. Biology of fishes. Blackie and Sons, Glascow, Scotland.

Brownlee, K. A. 1965. Statistical theory and methodology in science and engineering. Wiley, New York.

Bulak, J. S. 1983. Evaluation of Floy anchor tags for short term mark–recapture studies with blueback herring. North American Journal of Fisheries Management 3:91–94.

Dunning, D. J., Q. E. Ross, J. R. Waldman, and M. T. Mattson. 1987. Tag retention by, and tagging mortality of, Hudson River striped bass. North American Journal of Fisheries Management 7:535–538.

Fable, W. A., Jr., L. Trent, G. W. Bane, and S. W. Ellsworth. 1988. Movement of king mackerel (*Scomberomorus cavalla*) tagged in southeast Louisiana, 1983–1985. U.S. National Marine Fisheries Service Marine Fisheries Review 49(2):98–101.

Garrison, C. T. 1971. Evaluation of fish tagging methods. Pages 39–56 *in* Coastal fisheries project report 1971. Texas Parks and Wildlife Department, Austin.

Gulf of Mexico Fishery Management Council. 1987. Report of the red drum stock assessment group meeting. Red Drum Scientific Assessment Group, Tampa, Florida.

Hein, S., and J. Shepard. 1980. A preliminary tagging study on red drum, *Sciaenops ocellata*, in quarter acre ponds. Louisiana Department of Wildlife and Fisheries. Pages 33–39 *in* Contributions of the marine research laboratory, 1978. Louisiana Department of Wildlife and Fisheries Technical Bulletin 31, Grand Terre Island.

Matlock, G. C. 1980. History and management of the red drum fishery. Pages 37–53 *in* Proceeding of the red drum and seatrout colloquium. Gulf States Marine Fisheries Commission, Ocean Springs, Mississippi.

Mercer, L. P. 1984. A biological and fisheries profile of red drum, *Sciaenops ocellatus*. North Carolina Department of Natural Resources and Community Development, Division of Marine Fisheries, Special Scientific Report 41, Raleigh.

Odeh, R. E., and D. B. Owen. 1983. Attribute sampling plans, tables of tests and confidence limits for proportions. Marcel Dekker, New York.

Parker, R. R., E. C. Black, and P. A. Larken. 1963. Some aspects of fish marking mortality. International Commission for the Northwest Atlantic Fisheries Special Publication 4:117–122.

Ricker, W. E. 1975. Computation and interpretation of biological statistics of fish populations. Fisheries Research Board of Canada Bulletin 191.

Rounsefell, G. A., and W. H. Everhart. 1953. Fishery science, its methods and applications. Wiley, New York.

Royce, W. F. 1972. Introduction to the fishery sciences. Academic Press, New York.

Saunders, R. L. 1968. An evaluation of two methods of attaching tags to Atlantic salmon smolts. Progressive Fish-Culturist 30:104–108.

Simmons, E. G., and J. P. Brewer. 1962. A study of redfish, *Sciaenops ocellata* Linnaeus, and black drum, *Pogonias cromis* Linnaeus. Publications of the Institute of Marine Science, University of Texas, 8:184–211.

Simmons, E. G., and J. P. Brewer. 1976. Fish tagging on the Texas coast. Pages 66–82 *in* Coastal fisheries project report 1976. Texas Parks and Wildlife Department, Austin.

Stobo, W. T. 1972. The effect of dart tags on yellow perch. Transactions of the American Fisheries Society 101:365–366.

USNMFS (U.S. National Marine Fisheries Service). 1986. Proposed secretarial fishery management plan, regulatory impact review, initial regulatory flexibility analysis, and draft environmental impact statement for the red drum fishery of the Gulf of Mexico. USNMFS, Washington, D.C.

Vogelbein, W. K., and R. M. Overstreet. 1987. Histopathology of the internal anchor tag in spot and spotted seatrout. Transactions of the American Fisheries Society 116:745–756.

Weaver, J. C. 1976. Retention of the Floy tag in red drum, black drum, and sheephead. Pages 59–65 *in* Coastal fisheries project report 1976. Texas Parks and Wildlife Department, Austin.

Yamashita, D. T., and K. D. Waldron. 1958. An all-plastic dart type fish tag. California Fish and Game 44:311–317.

American Fisheries Society Symposium 7:161–167, 1990

Summary of King Mackerel Tagging in the Southeastern USA:
Mark–Recapture Techniques and
Factors Influencing Tag Returns

WILLIAM A. FABLE, JR.

National Marine Fisheries Service, Southeast Fisheries Center, Panama City Laboratory
3500 Delwood Beach Road, Panama City, Florida 32408, USA

Abstract.—Between 1975 and 1988, biologists with the Florida Department of Natural Resources and the National Marine Fisheries Service tagged more than 26,000 king mackerel *Scomberomorus cavalla* in the Atlantic Ocean and Gulf of Mexico, from North Carolina to Yucatan, Mexico. Various types of tags were used, but internal anchor tags were most effective and provided over 1,400 recoveries. Most fish were caught for tagging by trolling with hook and line; tagging the fish in V-shaped tagging cradles aboard vessels proved most efficient. Overall operational costs have increased markedly during these studies mainly due to the increased cost of purchasing fish for tagging. Tagging studies in the 1970s provided the basis for establishing separate management units of king mackerel in the Atlantic Ocean and Gulf of Mexico, and subsequent tagging studies contributed to further definition of fish in the Gulf of Mexico into distinct eastern and western groups. Increased fishery regulation has resulted in decreased tag reporting because fishermen express their resentment of the controls by not reporting tags. The most effective incentive for fishermen to return recovered tags is a high monetary reward.

King mackerel *Scomberomorus cavalla* is a coastal, pelagic scombrid that ranges from the Gulf of Maine to Rio de Janeiro, Brazil (Briggs 1958). The fish reach a maximum size of 173 cm fork length and 45 kg in weight (Collette and Nauen 1983), but most king mackerel taken commercially range between 60 and 90 cm (Trent et al. 1983). In 1985, commercial fishermen landed 2.4 million kg and recreational anglers landed 5.3 million kg of this fish (USNMFS 1986a, 1986b).

In the early 1960s, king mackerel was the species most desired by private boat anglers, and it was the staple of Florida's charter fleet (Moe 1963). At that time, a small troll fishery in southeast Florida was the only commercial fishery of any size for this species in the USA. With the advent of the power block in 1963 (Beaumariage 1973), gillnetting and aerial fish spotting increased yearly commercial landings to a high of over 4.7 million kg in 1974 (Gulf of Mexico and South Atlantic Fishery Management Council 1985). Since that time, stock assessments by National Marine Fisheries Service (NMFS) personnel have indicated that the Gulf of Mexico migratory group (those fish off southern Florida in the winter months) has declined due to overexploitation. Catch quotas were set in 1982 and were lowered in the ensuing years. In December 1987, catch quotas had been met for both commercial and recreational fisheries on the eastern zone of the Gulf

migratory group, and all but catch-and-release, recreational fishing for king mackerel stopped.

The first tagging of king mackerel took place in 1963 when 640 fish were tagged with spaghetti tags and 47 were tagged with internal anchor tags (Beaumariage 1964, 1969; Beaumariage and Wittich 1966; Moe 1966). Only six tagged fish were recovered and little migration information resulted. However, it generally was accepted by fishermen that, in Florida waters, these fish migrated northward in spring and southward in fall.

When landings of king mackerel peaked in 1975, a cooperative mark–recapture study was undertaken by the Florida Department of Natural Resources (FDNR) and the NMFS. Between 1975 and 1979, 17,042 king mackerel were tagged by biologists with these agencies and 1,171 tags were returned (Sutherland and Fable 1980; R. O. Williams and M. F. Godcharles, FDNR, unpublished 1984 report). Besides the accepted north–south migrations, Williams and Godcharles identified two stocks or migratory groups, a Gulf group and an Atlantic group. The ranges of the two groups roughly coincided with the boundaries of the Gulf of Mexico and the Atlantic coast of the southeastern USA, but the Gulf group extended into waters off southeast Florida in the winter months. All stock assessments and catch allocations for king mackerel since 1985 have been made separately for these two migratory groups.

Since 1980, tagging by the NMFS and various cooperating agencies has been aimed at more detailed discrimination of king mackerel stocks. Two hundred thirty-two tags have been returned from 9,122 king mackerel tagged since the end of the FDNR–NMFS cooperative study. Information from this tagging has helped divide the previously described Gulf group into eastern and western segments (Fable et al. 1987) and is providing more information on the Atlantic coast fish and the king mackerel moving into Mexican waters.

This report describes and discusses the mark-recapture methods used for king mackerel studies in the southeastern USA and Mexico and discusses factors influencing the tag returns.

Methods

The earliest king mackerel tagging studies in the 1960s (Moe 1966), used spaghetti tags (Floy Tag and Manufacturing) because of successes with these tags reported by Wilson (1953) and Clemens (1961) who used them for tagging other large scombrids. All tags had a legend on them that advertised a reward and gave a return address. Fish were caught from small boats with commercial handline techniques and placed in a padded tagging trough. Moe (1966) stressed that speed was the most important factor for successful tagging of king mackerel and that a maximum limit of 40 s out of the water would ensure survival of this species.

When the FDNR–NMFS cooperative study began in 1975, FDNR biologists obtained fish for tagging in southern Florida by purchasing live king mackerel from commercial handline fishermen. Fish were brought aboard the boat by the fishermen, and the hook was pulled over a de-hooking bar and held upside down until the fish fell off into a box. With wet, gloved hands, the tagger lifted the fish into a padded tagging cradle and recorded its fork length. Fishermen were paid market price or slightly more for all tagged fish.

The FDNR biologists chose an internal anchor tag (Figures 1, 2) for use on king mackerel. Although 47 of these tags were used on king mackerel in 1964 and 1965 (Beaumariage and Wittich 1966; and Beaumariage 1969) and none were returned, Topp (1963) stressed this tag's permanence and nonirritating qualities. In 1968, 60 king mackerel were tagged with internal anchor tags, and four of the tags were returned (Williams and J. A. Huff, FDNR, unpublished 1976 report). When tagging began in the 1970s, the tags were

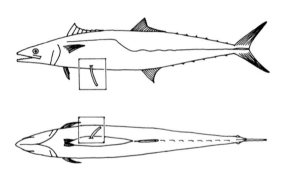

FIGURE 1.—Lateral and ventral views of king mackerel with placement of internal anchor tags indicated.

inserted by making a 6-10-mm incision in the anterior abdomen (Figure 1) with a scalpel, inserting the disk of the internal anchor tag by hand, and returning the fish to the water (Williams, FDNR, unpublished 1976 report).

National Marine Fisheries Service biologists tagged king mackerel where commercial handline fisheries for the species did not exist: in North Carolina, northwest Florida, and Texas (Sutherland and Fable 1980). Fish were caught by the taggers with a hook and line (either handline or rod and reel) and tagged with a single-barb dart tag (Figure 2). Sutherland and Fable (1980) reported that Fry and Roedel (1949) found a minimum of 23% mortality in chub (Pacific) mackerel *Scomber japonicus* tagged with a body cavity tag,

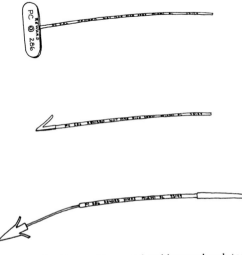

FIGURE 2.—Types of tags used on king mackerel: top, internal anchor tag; middle, single-barb dart tag; bottom, double-barb dart tag.

and that Yamashita and Waldron (1958) reported favorable results with single-barb dart tags.

About 1,100 fish were tagged initially in 1975 in the water alongside the boat by inserting the single-barb dart tag in the dorsal musculature with a tagging needle on a pole. The leader was then cut as close to the hooks as possible. Later, fish were brought aboard the boat and held in a V-shaped tagging cradle or on a wet deck so hooks could be removed, fork length could be measured, and the tag could be inserted more carefully (Sutherland and Fable 1980). Like the internal anchor tags, the single-barb dart tags had a legend indicating that a reward was offered and the address to which the tag should be returned. Throughout the FDNR–NMFS cooperative study, a reward was offered that varied from a minimum of $1 (raised to $5 in 1976) to a maximum of $25, depending on the serial number of the tag. Fishermen were familiarized with the mark–recapture work by various methods, including posters at marinas and fish dealers, newspaper articles and press releases, and personal contacts.

In 1983, when king mackerel tagging resumed after the conclusion of the FDNR–NMFS study from 1975 to 1979, it was in response to a new hook-and-line fishery developing in Louisiana. Internal anchor tags were used exclusively, and fish were caught on handlines from government boats or were purchased live from commercial handline fishermen. As reported by Fable et al. (1987), fish initially were held down on deck to immobilize them for tagging. Later, on commercial vessels, fish were unhooked over a dehooking bar and tagged in a padded tagging cradle. Rewards were set uniformly at $10 for each returned tag and remained at that amount through 1985. In 1986, a yearly $1,000 drawing from returned tags was initiated as an added incentive to anglers.

From 1983 through 1987, king mackerel mark–recapture work took place from North Carolina to Mexico. The initiation of tagging in a given area depended on the need for data from that region (usually emphasized by a fishery management council request), the accessibility of adequate numbers of king mackerel, and adequate funding for the work.

In Louisiana, on the east coast of Florida, and to some extent in North Carolina, most tagging was done from commercial vessels. The limiting factor for the number of fish tagged was the cost of the fish. In the 10-year period from 1975 to 1985, the ex-vessel price for king mackerel increased from around US$0.40 per pound to $1.50

or more. This often meant that each tagged fish cost $20.

Encouraging recreational anglers to tag and release their unwanted king mackerel catch has been the most economical approach to mackerel tagging. This method has been used with some success in Texas and North Carolina, but recreational anglers find it difficult to handle king mackerel for insertion of internal anchor tags. In response to this problem, a double-barb dart tag, originally developed for small bluefin tuna *Thunnus thunnus* by E. Prince (NMFS, Miami), was adapted for king mackerel (Figure 2). This tag was used for tagging king mackerel in Panama City, Florida, in 1986, and has been used by anglers participating in the cooperative game-fish tagging program at the NMFS Southeast Fisheries Center in Miami, Florida.

The most challenging mark–recapture work has been done in Mexico by biologists from the Mote Marine Laboratory under contract to the National Marine Fisheries Service and the Instituto Nacional de Pesca, the Mexican fisheries agency. Hook-and-line trolling methods have been used successfully, but vessels are usually small, open outboard launches and frequently are idled by inclement weather. The most effective method for obtaining king mackerel in Mexican waters has been the use of trap nets (termed almadrabas) found in the Veracruz area. These nets were fished twice a day by the local fishermen. Live king mackerel were dipnetted or simply picked up out of the net when it was raised. Often, 30 fish or more were tagged in as many minutes. The fish were brought aboard launches, placed in a tagging cradle, measured, tagged, and released.

Results and Discussion

The FDNR–NMFS tagging study conducted between 1975 and 1979 resulted in 17,042 tagged king mackerel and 1,171 returned tags. Internal anchor tags used by FDNR personnel yielded a much higher return rate (8.1%) than the single-barb dart tags (1.1%) used by NMFS biologists (Table 1). In the initial year of the NMFS tagging study (1975), over 1,100 king mackerel were tagged in the waters off northwestern Florida, and no tags were returned (Sutherland and Fable 1980). After that, fish were brought aboard the boat and tagged, and returns resulted, but the return rate never approached that achieved by FDNR biologists using internal anchor tags.

In a tag-retention experiment (Williams, FDNR, unpublished 1978 report), 1,354 king

TABLE 1.—Numbers of king mackerel tagged, the number returned, and return rates from mark–recapture studies during the 1970s and 1980s.

Tagging years	Organization[a] or area	Number of fish		Return rate (%)
		Tagged	Returned	
1975–1979	FDNR	14,137	1,139	8.1
1975–1979	NMFS	2,905	32	1.1
1983–1987	All areas combined	9,122	232	2.5
1983–1987	North Carolina	2,196	32	1.5
1985–1986	NE Florida	891	31	3.5
1987	SE Florida	1,005	40	4.0
1983–1987	NW Florida	1,174	32	2.7
1985–1987	Louisiana	2,362	59	2.5
1985–1987	Texas	341	8	2.3
1984–1987	Mexico	1,153	30	2.6

[a]FDNR = Florida Department of Natural Resources; NMFS = National Marine Fisheries Service.

mackerel were marked in equal numbers with either an internal anchor tag or a single-barb dart tag. There was little difference in return rates for the first 180 d; however, no dart tags were recovered after that time. After 480 d, 39 internal anchor tags and only 14 dart tags had been returned. Both Williams (unpublished) and Sutherland and Fable (1980) reported that failure (i.e., loss) of the single-barb dart tag may have been the cause of this difference. Evidence indicated that the plastic streamer detached from the nylon dart after 3–4 months due to breakdown of the adhesive connecting them (Bruger 1981).

Adding to this problem, at least during the initial tagging of 1,100 king mackerel in northwest Florida, fish were tagged in the water and hooks were left in the fishes's mouths (Sutherland and Fable 1980). In Texas that same year (1975), 282 king mackerel were tagged with the same type of single-barb dart tag, but they were brought aboard the boat and the hooks were removed. Three returns resulted from these 282 tagged fish. Sutherland and Fable (1980) believed that failure to implant the tag properly in moving fish, as well as dangling hooks contributed to the poor results.

In the FDNR–NMFS mark–recapture study in the 1970s, the two agencies used different methods and different tags, and the FDNR achieved an excellent return rate. When the FDNR returns were combined with returns from tagging by NMFS in the Gulf of Mexico, a great deal of valuable migration information resulted. No extensive comparative tag testing took place (except that each group used a different type of tag), and tag retention and tagging mortality were not measured.

When tagging resumed in the 1980s, internal anchor tags were used almost exclusively because of the 8.1% return rate and the successes reported

by Williams and Godcharles (unpublished). From 1980 to 1988, NMFS and cooperating agencies tagged 9,122 more king mackerel and received 232 tag returns (a 2.5% return rate). However, new influences affected tag return rates after 1982. In the 1970s, king mackerel tagging was new to the fishermen, and most welcomed the opportunity to participate, both by tagging fish and returning tags, because they had observed the species' movements for years and wanted more information about these movements. At that time, fishery allocations and quotas did not exist and landings were higher than ever before, although high catches by large gill-net vessels were starting to worry some fishermen. In the following years, catches varied greatly, and quarrels between fishermen with different gear types broke out. By 1982, the South Atlantic and Gulf of Mexico Fishery Management Councils were exerting influence on fishermen, and in May of 1983, the commercial hook-and-line fishery was closed for about 2 months when the quota was reached. By 1987, combined recreational and commercial catch quotas for both the Atlantic Ocean and Gulf of Mexico migratory groups for the entire southeastern USA were reduced to 5.4 million kg. In the winters of 1985–1987, the commercial fishery in south Florida met its quota each year. The recreational fishery also was closed in 1987. Most fishermen resent these closures, and many react by not reporting tag recoveries.

The nonreporting of tag recoveries is evident in Table 1, where the overall return rate in the 1980s was 2.5% versus 8.1% in the 1970s for internal anchor tags. Some of this difference is probably due to tagging in areas where an intense commercial fishery did not exist, such as northwestern Florida and Texas, but even when tagging took place in the midst of heavy fishing pressure, such

TABLE 2.—Results of tests to compare the effectiveness of internal anchor tags (IAT) and double-barb dart tags (DT) for king mackerel in northwest Florida.

Number and type of tags	N	Number of tags per fish	Number of fish returned	Number of tags returned		Number of tags lost	
				IAT	DT	IAT	DT
1 IAT	139	1	6	6		0	
2 IATs	139	2	5	8		2	
1 DT	139	1			1		0
1 IAT + 1 DT	139	2	3	3	1	0	2
Total	556		15	17	2	2	2

as in northeastern and southeastern Florida, return rates were less than half of the return rate for the 1970s.

Nonreporting of internal anchor tags has been discussed by Matlock (1981) and Green et al. (1983). They found that only 28–29% of tags implanted in sciaenids landed in Texas waters were reported. The most frequent reason for not reporting a tag, according to the anglers, was their failure to find it. These authors suggested that reporting might be increased if the tag were more visible and larger monetary rewards were offered. (Their rewards varied between $1 and $25.)

Little more can be done to make tags more visible (NMFS biologists have used international orange tags since 1983), but rewards could make a substantial difference. In the 1970s, rewards varied from a minimum of $5 to a maximum of $25. In the 1980s, the reward was set at $10, and a $1,000 drawing from returned tags was held each year in 1986 and 1987. In the 1970s, king mackerel were purchased from commercial fishermen for about $5 each. In the mid-1980s, the tagging of about 1,900 king mackerel on the east coast of Florida cost over $36,000 for the fish purchases alone (about $19 each). Seventy-one of these fish were recovered, and $710 was spent on rewards besides the $1,000 spent each year for the drawing. Thus, less than $3,000 was spent in monetary rewards to get returns on over $36,000 worth of fish. Most fishermen can afford to give up a $10 reward, but if the reward were raised to $50 or more, few fishermen would ignore or discard tags. If the reward had been $50, the total spent on rewards (at the same return rate) would still have been under $6,000.

Other mark–recapture studies at the Southeast Fisheries Center (billfish tagging, redfish tagging) have had $5 rewards, and raising the reward on king mackerel tags past $10 has not been authorized. In one instance, however, a substantial reward ($100) was offered for the return of tetra-cycline-tagged king mackerel in 1983 and 1984. Posters advertising this were distributed, and 216 fish were marked and released. Seventeen tags were returned. This represented a 7.9% return rate—a rate close to that realized by Williams and Godcharles (unpublished) in the 1970s. This indicated that a high reward will induce more fishermen to return tags.

Most of the mark–recapture work done in the 1980s was not designed for experimentation with various tagging techniques or testing of different tag types. When fish were purchased for $15 to $20 each aboard commercial vessels, the tagging technique and tag type that had proven successful in the past was employed. In waters adjacent to the NMFS's laboratory in Panama City, Florida, however, alternative marking techniques could be tested. Because king mackerel can be caught in fairly large numbers in the summer months and are found relatively close to shore, and because sea conditions allow outboard boats to be employed by NMFS personnel to catch the fish, my colleagues and I did some tag testing and double tagging in 1986 and 1987.

A double-barb dart tag (Figure 2) was developed for king mackerel from the design by Prince (unpublished) for bluefin tuna. This tag was tested in 1986 by comparing it to the internal anchor tag. Five hundred fifty-six king mackerel were tagged with either one internal anchor, two internal anchors, one double-barb dart, or one double-barb dart and one internal anchor. In all, 278 double-barb dart tags and 556 internal anchor tags were used. Seventeen internal anchor tags were returned and two had fallen out, whereas two double-barb dart tags were returned and two had fallen out (Table 2). Returns were greatest when fish were tagged with one internal anchor tag and were least with one dart tag. The double-barb dart tags have been in use for king mackerel since 1986 in the cooperative game-fish tagging program centered at the NMFS's Miami laboratory. Alto-

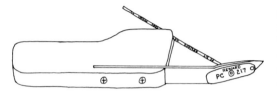

FIGURE 3.—Internal anchor tag inserter. (Designed by L. Trent.)

gether, 208 king mackerel had been tagged in this program by 1988, and two tags have been recovered.

The double-barb dart tag was somewhat less effective than the internal anchor for tagging king mackerel during the limited 1986 testing, but the difficulties encountered during insertion of anchor tags have limited their use by recreational anglers. A stainless steel knife-like inserter designed by L. Trent (NMFS, Panama City, Florida) holds an appropriately modified internal anchor tag while it slices under the abdominal skin and flesh of a king mackerel (Figure 3). This device may allow an untrained tagger to insert these tags easily. These inserters are being used on sciaenids in North Carolina and were being studied by Floy Tag and Manufacturing for possible commercial production.

In 1987, 402 king mackerel taken in Panama City all were double-tagged with internal anchor tags. As of May 1, 1988, only one fish had been recovered. The return rate had declined to almost zero, largely due to a closed commercial fishery and to zero bag limits in the recreational fishery of south Florida, where Gulf king mackerel winter. When these fisheries re-open in mid-1988, some tag returns are anticipated despite the fishermen's resentment of fishery management practices.

Conclusions

There are several lessons to be learned from king mackerel mark–recapture studies that have been done in the southeastern USA. Most important is the need for comparative tag testing (i.e., tag retention and tagging mortality determinations) before long-term, expensive mark–recapture programs begin. Such testing is not easily or inexpensively done on a scombrid such as king mackerel, but the time and money would be well spent.

Once tagged fish are freed, maximum effort should be used to recover tags. Although anglers accepted tagging in the 1970s and monetary rewards seemed adequate to gain their cooperation, fisheries regulations imposed in the 1980s resulted

in resentment of governmental interference by anglers, and they became reluctant to return tags. To counteract this resentment, two measures seem appropriate. The first is to increase public awareness of the purpose of the regulations. Although the numerous fishermen interviewed seemed to understand the purpose of the regulations, a second measure—a high monetary reward for returned tags—probably would be more effective. When the tagged fish cost up to $20 each, the minimum reward should be $50 when return rates are well under 10%.

References

Beaumariage, D. S. 1964. Returns from the 1963 Schlitz tagging program. Florida Board of Conservation Marine Research Laboratory Technical Series 43.

Beaumariage, D. S. 1969. Returns from the 1965 Schlitz tagging program including a cumulative analysis of previous results. Florida Department of Natural Resources Marine Research Laboratory Technical Series 59.

Beaumariage, D. S. 1973. Age, growth, and reproduction of king mackerel, *Scomberomorus cavalla*, in Florida. Florida Marine Research Publications 1.

Beaumariage, D. S., and A. C. Wittich. 1966. Returns from the 1964 Schlitz tagging program. Florida Board of Conservation Marine Research Laboratory Technical Series 47.

Briggs, J. C. 1958. A list of Florida fishes and their distribution. Bulletin of the Florida State Museum Biological Series 2(8).

Bruger, G. E. 1981. Comparison of internal anchor tags and Floy FT-6B dart tags for tagging snook, *Centropomus undecimalis*. Northeast Gulf Science 4:119–122.

Clemens, H. B. 1961. The migration, age, and growth of Pacific albacore (*Thunnus germo*), 1951–1958. California Department of Fish and Game, Fish Bulletin 115.

Collette, B. B., and C. E. Nauen. 1983. FAO species catalog, volume 2. Scombrids of the world. FAO (Food and Agriculture Organization of the United Nations) Fisheries Synopsis 125.

Fable, W. A., Jr., L. Trent, G. W. Bane, and S. W. Ellsworth. 1987. Movements of king mackerel (*Scomberomorus cavalla*) tagged in southeast Louisiana, 1983–1985. U.S. National Marine Fisheries Service Marine Fisheries Review 49(2):98–101.

Fry, D. H., and P. M. Roedel. 1949. Tagging experiments on the Pacific mackerel (*Pneumatophorus diego*). California Department of Fish and Game, Fish Bulletin 73.

Green, A. W., G. C. Matlock, and J. E. Weaver. 1983. A method for directly estimating the tag-reporting rate of anglers. Transactions of the American Fisheries Society 112:412–415.

Gulf of Mexico and South Atlantic Fishery Management Councils. 1985. Final amendment 1 fishery management plan environmental impact statement for the coastal migratory pelagic resources (mackerels) in the Gulf of Mexico and South Atlantic region. Gulf of Mexico Fishery Management Council, Tampa, Florida.

Matlock, G. C. 1981. Nonreporting of recaptured tagged fish by saltwater recreational boat anglers in Texas. Transactions of the American Fisheries Society 110:90–92.

Moe, M. A., Jr. 1963. A survey of offshore fishing in Florida. Florida Board of Conservation Marine Laboratory Professional Paper Series 4.

Moe, M. A., Jr. 1966. Tagging fishes in Florida offshore waters. Florida Board of Conservation Marine Research Laboratory Technical Series 49.

Sutherland, D. F., and W. A. Fable, Jr. 1980. Results of a king mackerel (*Scomberomorus cavalla*) and Atlantic Spanish mackerel (*Scomberomorus maculatus*) migration study, 1975–79. NOAA (National Oceanic and Atmospheric Administration) Technical Memorandum NMFS (National Marine Fisheries Service) SEFC-12.

Topp, R. 1963. The tagging of fishes in Florida 1962 program. Florida Board of Conservation Marine Research Laboratory Professional Papers Series 5.

Trent, L., R. O. Williams, R. G. Taylor, C. H. Saloman, and C. H. Manooch III. 1983. Size, sex ratio, and recruitment in various fisheries of king mackerel, *Scomberomorus cavalla*, in the southeastern United States. U.S. National Marine Fisheries Service Fishery Bulletin 81:709–721.

USNMFS (U.S. National Marine Fisheries Service). 1986a. Fisheries of the United States, 1985. U.S. National Marine Fisheries Service Current Fishery Statistics 8380.

USNMFS (U.S. National Marine Fisheries Service). 1986b. Marine recreational fishery statistics survey, Atlantic and Gulf coasts, 1985. U.S. National Marine Fisheries Service Current Fishery Statistics 8327.

Wilson, R. C. 1953. Tuna marking, a progress report. California Fish and Game 39:429–442.

Yamashita, D. T., and K. D. Waldron. 1958. An all plastic dart type fish tag. California Fish and Game 44:311–317.

American Fisheries Society Symposium 7:168–172, 1990

Underwater Tagging and Visual Recapture as a Technique for Studying Movement Patterns of Rockfish

KATHLEEN R. MATTHEWS[1] AND ROBERT H. REAVIS

Fisheries Research Institute, School of Fisheries
University of Washington, Seattle, Washington 98195, USA

Abstract.—Researchers using scuba studied the movement patterns of rockfishes *Sebastes* spp. in water 5–20 m deep by tagging and visually resighting ("recapturing") fish under water. Divers anesthetized rockfish with methomidate hydrochloride and then captured them with a net. Each fish was tagged with a nylon T-bar anchor tag modified by the addition of a numbered Petersen disc. This tag enabled divers to identify individual rockfishes under water. This underwater tag–recapture technique has several advantages over conventional techniques (hook and line, nets, electroshock, etc.). The fish were resighted during scuba surveys; thus, we began to obtain recapture information immediately without waiting months or years for recovered tags to be sent in from fisheries. This method provided more data than conventional methods, which are often hampered by low numbers of tag returns. Furthermore, on-site capture allowed release of fish at a precise location and depth, and in a particular condition. With this capture method, we knew for sure whether fish released over the side of a boat survived the pressure change and returned safely to the bottom. As with all external tags, some shedding occurred. Finally, resightings provided ongoing multiple "recapture" information on seasonal movements and site tenacity of individual fish over 2 years.

Tag–recapture studies in fisheries research are widespread: a computerized literature search of "fish/tagging" in the 1982–1986 Aquatic Sciences and Fisheries Abstracts data base (Cambridge Scientific Abstracts 1987) identified 434 entries. Tagging studies are instrumental to our knowledge of growth, movement, and habitat use (Wydoski and Emery 1983). In general, tagging studies have relied upon hook and line, various types of nets (either active or passive), or electroshocking for the capture of fish (Hayes 1983; Hubert 1983; Reynolds 1983). All of these methods require that the fish be brought to the surface for tagging. After tagging, they are typically released at the surface. Recapture is usually contingent upon catching and reporting of tagged fish by anglers, sport divers, and commercial fishermen (Wydoski and Emery 1983).

Several problems result from these procedures. Fish are stressed by capture and handling. Of greater consequence is the trauma suffered when captured fish are raised to the surface from depths where pressure is greater than one atmosphere (Wydoski and Emery 1983). Release of stressed fish at the surface may subject them to increased predation before they can reach the bottom, and

leaves the researcher without an accurate estimate of survivorship (Jones 1979). Furthermore, because most fish are recaptured by sport and commercial fishermen, information (location, depth, time) may be inaccurate and may require months or years to collect. Finally, because recaptured fish are usually killed, opportunities for multiple recaptures are limited.

In a recent study of movements of demersal rockfish *Sebastes* spp. in Puget Sound, Washington, we used an underwater tag–recapture technique to overcome some limitations of conventional tag–recapture studies. In this paper we describe this technique, modifications we incorporated, and the method's advantages relative to conventional tagging techniques.

Methods

Equipment.—The use of scuba requires that researchers work in pairs. In our team, one diver carried a hand-held fishing net. The other diver carried a container of a new fish anesthetic, methomidate hydrochloride (Wildlife Laboratories, Fort Collins, Colorado).[2] The powdered anesthetic was mixed with water (2 g/L). For use in salt water, we dissolved the powder in fresh water; the solution then appeared cloudy and was

---

[1]Present address: Department of Fisheries and Oceans, Pacific Biological Station, Nanaimo, British Columbia, V9R 5K6, Canada.

[2]Reference to trade names in this paper does not imply endorsement by the authors.

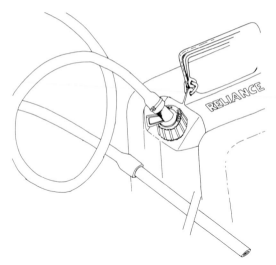

FIGURE 1.—Water bottle used to transport anesthetic; note on–off spigot and extender made of surgical tubing with acrylic tubing.

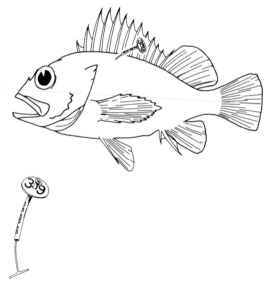

FIGURE 2.—Tagged rockfish and close-up of anchor tag showing numbered Petersen disc modification.

therefore more visible under water. To transport the liquid anesthetic, we used a Reliance 10-L collapsible plastic water jug with an on–off spigot (Figure 1). The anesthetic was dispensed when the diver was approximately 1 m away, to avoid eliciting escape responses from fish. To facilitate application, an extender was added to the jug, and consisted of 0.5 m of surgical tubing secured to the spigot by an inelastic pull tie and 1 m of clear acrylic tubing. The tubing served as a conduit to deliver anesthetic to individual fish. The rest of the equipment, carried by either diver, included a plexiglass slate, tags, and a tagging gun. The slate was used to measure fish and record data, which comprised tag number and color, species, total length, and exact location of capture and release. All this equipment was attached with lanyards to the divers.

The tag was a modified nylon T-bar anchor tag (FD-67, Floy Tag and Manufacturing, Seattle, Washington). For individual recognition underwater, a numbered and laminated disc, 1.4 cm in diameter, was added to the end of the anchor tag (Figure 2). The anchor portion of the tag was a nylon T-bar. A hole drilled in the disc center was slipped over the tag, and the end of the tag was burned and spread to ensure that the disc did not slip off. The tag was burned outdoors with a warm teflon-coated iron to minimize the inhalation of toxic gas. The laminated disc was an improvement over McElderry's (1980) handmade plastic tags marked by indelible ink. Initially we used

McElderry's tag design, but the "indelible" ink faded over time and the tags became unreadable. The manufactured tags and discs we used were available in a variety of colors, and each disc had two characters stamped on it. Various combinations of colored tags and discs could be coded for different study sites or experiments. These tags were embedded in the dorsal musculature of the fish with a standard tagging gun.

Tagging technique.—A diver approached an individual fish, the conduit was placed as close (<15 cm) to the fish's mouth as possible, and the anesthetic was dispensed by squeezing the plastic bottle. These rockfishes were generally found in holes or crevices, or on the bottom. Fishes in holes or crevices were completely sedated and gently prodded out of their holes into the net. Fishes resting on the substrate were only mildly sedated; the net was then positioned overhead and brought down over the fish. The quantity of anesthetic used was determined by the fish's response; usually an individual fish required two or three squirts from the anesthetic jug. After 1–2 min a fish was sufficiently narcotized for capture. Occasionally a fish was caught without the anesthetic, but attempts to catch moving fish were usually unsuccessful.

Once in the net, fish were kept away from the rock substrate to prevent injury, because sometimes they were still active. Fish were placed on the slate, measured, and tagged. One diver re-

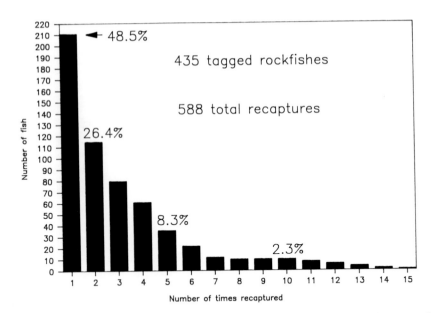

FIGURE 3.—Number of fish recaptured from 1 to 15 times from July 1986 through January 1988.

corded the colors of tag and disc, and the disc characters. The other diver released the fish at the point of capture and the fish was observed for 1 min to assess its condition. The entire tagging procedure took approximately 3–5 min.

Fish were moved 50 m to 6.4 km from their sites of capture in a series of displacement experiments. Fish in one group were captured and tagged as described, placed in a canvas bag to eliminate visual cues, carried along the bottom for 50–100 m, and released. These displacements were carried out at a constant pressure (depth). Other fish were moved greater distances (400 m to 6.4 km). These fish were captured as described, slowly brought to the surface (approximately 6 m/min) in a bag, tagged on the surface, and transported by boat. At the release site, fish were taken to the bottom by divers, their condition was observed, and the fish were released. If the displaced fish were stressed or dead, they were not included in the study.

Results

From July 1986 through January 1988, 435 copper *Sebastes caurinus*, quillback *S. maliger*, and brown *S. auriculatus* rockfishes were tagged on six separate reefs in central Puget Sound. We worked at depths of 5–20 m, with horizontal visibility ranging from 3 to 10 m. On average, 6.5 fish per dive were tagged during 67 dives; each dive lasted about 40 min. A maximum of 13 fish

were tagged during one dive. Through January 1988, 211 (48.5%) rockfishes have been resighted at least once (Figure 3). There have been 588 recaptured fish, including up to 15 multiple recaptures of individual fish.

From recaptured fish, we were able to collect information on site tenacity at different reefs. For example, on one reef, multiple recaptures often took place at the same rock were the fish were originally tagged, indicating that the fish were sedentary. By surveying adjacent reefs, we were able to detect movement between reefs. Thirteen rockfishes tagged on an inshore natural reef during the summer of 1986 were resighted during the winter of 1987 on an artificial reef 400 m away. During the following summer of 1987, 6 of these 13 rockfishes were again seen on the inshore natural reef. Similarly, 18 rockfishes tagged at the artificial reef were seen during the summer on the inshore natural reef.

In our displacement experiments, we moved 137 tagged rockfishes from 50 m to 6.4 km away from their sites of original capture. In subsequent scuba surveys we identified 87 individuals that had returned. One copper rockfish displaced 6.4 km was resighted during the next dive at the original location, 22 d later.

Discussion

In this tag–recapture study we had a significantly higher percentage of fish recaptured

(48.5%) than in other studies of rocky reef fishes by traditional methods. Only 484 (3%) of 14,795 black rockfish *S. melanops* tagged off Washington and Oregon were reported by Culver (1987). Coombs (1979) tagged 1,098 rockfish on offshore reefs in Oregon and recaptured 2.5%. Mathews and Barker (1983) tagged 342 quillback rockfish in northern Puget Sound and recovered 11 tags (3.2%). In contrast, McElderry's (1980) underwater tag–recapture study in Barkley Sound produced resightings of 67% of 90 tagged black rockfish and 93% of 14 tagged china rockfish *S. nebulosus*. He resighted a higher proportion of fish than we did because he conducted an intensive survey over a short time period (2 months) and limited area. By comparison, we have tag–recapture data for 19 consecutive months over a larger area, and tag–recapture efforts are still continuing.

By resighting our tagged fish, we eliminated a common problem plaguing tag–recapture studies: failure to report recaptured fish with tags. Although lack of reporting is a recognized problem, it is difficult to quantify. Barker (1979) attempted an estimate of nonreporting in conjunction with his well-publicized tagging study. He personally distributed 150 tags to anglers, requested anglers to return the tags to him, and used a reward as an incentive. Only 20 tags (13.3%) were returned. By circumventing this problem in our study, we produced a higher percentage of recaptured fish from which we can draw more meaningful conclusions.

We returned tagged fish directly to the bottom; hence, we were able to determine not only the exact location and depth, but the condition of the fish as well. Fish tagged on site (under water) did not experience pressure trauma. Fish brought to the surface during displacement experiments experienced swim bladder problems but recovered when divers returned them to the bottom; some (<5%) were so stressed that we did not include them in our study. Pressure trauma sustained during conventional tagging may contribute to low tag returns. Gowan (1983) noted that 21 quillback rockfish he tagged and released had difficulty remaining upright; none of these fish were ever recovered.

As with all external tagging, shedding occurred with our modified anchor tag. Culver (1987) provided evidence that tag shedding by rockfish can be substantial (20–37.5%). Because rockfishes frequently reside under rock ledges and rocky outcrops, there is a greater potential for shedding than would be found in a pen or aquarium. There-

fore, experiments in which fish are held in pens or aquaria will not provide an accurate estimate of the amount of shedding. We saw tags caught in rock crevices, evidence of tag shedding by our fish. The added drag of the numbered disc probably causes more snagging and shedding than a standard anchor tag. Because shedding occurs, it is important to draw conclusions only from sightings of tagged fishes, and not to draw conclusions when tagged fishes are not seen. For example, if a tagged fish is not sighted, it should not be assumed that the fish has moved or died.

In traditional tag–recapture studies, conclusions about fish movement patterns are generally drawn from recapture of 3–10% of the fish. Although these results may indicate one sort of habit (e.g., sedentary, mobile), the other 90–97% of fish unreported may be doing something very different. Moreover, in traditional studies, a fish is usually recaptured only once. Therefore, if a fish is tagged in one location and recaptured at the same location months or years later, the assumption is often made that no interim movement occurred. Our multiple-resighting information has shown that this assumption is not always true, and that seasonal movements do occur.

One of our overall objectives in our study of fish movements was to determine whether rockfishes return to the capture site when displaced. After the rockfishes had been displaced, we concentrated our efforts at the point of removal to determine if the fish had returned. We were able to estimate how long it took displaced rockfishes to return, and whether they returned to the exact location. Had we been required to wait for sportfishermen to return our tagged fish for this information, fish might have returned long before they were caught. Furthermore, sportfishermen are often unable to give accurate information on where they caught tagged fish. In contrast, we were able to document within meters the locations to which fish returned.

Despite the many advantages of this underwater tagging and recapturing method, its uses are limited to relatively small areas and sedentary species. The method would not be effective with highly mobile fish that have large home ranges or for studying fish over large geographical areas. The visual observations of tagged fish are only conducted during dives, which may be separated in time by days, cannot be performed during periods of strong current, and are unreliable at night. Furthermore, because the return trips took 10–127 days, we were unable to determine the return routes. By combining this study with one involving ultrasonic transmitters (Matthews et al. 1990, this volume), we overcame some of these limitations and provided a more complete picture of rockfish movement.

Methomidate hydrochloride was an effective anesthetic for rockfish. A small dose was enough to slow the fish for capture, tagging, and release. The chemical acted rapidly and fish apparently recovered within a few minutes. Although there are no published reports comparing this new anesthetic to common fish anesthetics (tricaine, quinaldine), this drug seems less caustic and is easier to handle.

In conclusion, this technique and study design overcame many limitations of traditional tagging studies when applied to shallow-water, sedentary fishes within safe diving ranges. It provided detailed data and a high percentage of recaptured fish from which meaningful information could be derived, and did not rely on fishermen for tag returns.

Acknowledgments

We thank Bruce Miller and Thomas Quinn for providing funding, advice, and constructive criticism. The Washington Department of Fisheries provided boats, vehicles, and support. Debra Murie, University of Victoria, Victoria, British Columbia, kindly shared the underwater tagging technique. Vince Macurdy spent countless hours assisting with our diving, tagging, and recapturing. Tom Butler drew the illustrations. This is contribution 764 from the University of Washington Fisheries Research Institute.

References

Barker, M. W. 1979. Population and fishery dynamics of recreationally exploited marine bottomfish of northern Puget Sound. Doctoral dissertation, University of Washington, Seattle.

Cambridge Scientific Abstracts. 1987. Aquatic sciences and fisheries abstracts, compact disc, volume 2.05, 1982–1986. Cambridge Scientific Abstracts, Bethesda, Maryland.

Coombs, C. I. 1979. Reef fishes near Depoe Bay, Oregon: movement and the recreational fishery. Master's thesis. Oregon State University, Corvallis.

Culver, B. N. 1987. Results from tagging black rockfish (*Sebastes melanops*) off the Washington and northern Oregon coast. Pages 231–240 *in* Proceedings of the International Rockfish Symposium. University of Alaska Sea Grant Report 87-2, Fairbanks.

Gowan, R. E. 1983. Population dynamics and exploitation rates of rockfish (*Sebastes* spp.) in central Puget Sound, Washington. Doctoral dissertation. University of Washington, Seattle.

Hayes, M. L. 1983. Active fish capture methods. Pages 123–145 *in* L. A. Nielsen and D. L. Johnson, editors. Fisheries techniques. American Fisheries Society, Bethesda, Maryland.

Hubert, W. A. 1983. Passive capture techniques. Pages 95–122 *in* L. A. Nielsen and D. L. Johnson, editors. Fisheries techniques. American Fisheries Society, Bethesda, Maryland.

Jones, R. 1979. Materials and methods used in marking experiments in fishery research. FAO (Food and Agriculture Organization of the United Nations) Fisheries Technical Paper 190.

Mathews, S. B. and M. W. Barker. 1983. Movements of rockfish (*Sebastes*) tagged in northern Puget Sound. U.S. National Marine Fisheries Service Fishery Bulletin 82:916–922.

Matthews, K. R., T. P. Quinn, and B. S. Miller. 1990. Use of ultrasonic transmitters to track demersal rockfish movements on shallow rocky reefs. American Fisheries Society Symposium 7:375–379.

McElderry, H. I. 1980. A comparative study of the movement habits and their relationship to buoyancy compensation in two species of shallow reef rockfish (Pisces, Scorpaenidae). Master's thesis. University of Victoria, Victoria, Canada.

Reynolds, J. B. 1983. Electrofishing. Pages 147–163 *in* L. A. Nielsen and D. L. Johnson, editors. Fisheries techniques. American Fisheries Society, Bethesda, Maryland.

Wydoski, R. and L. Emery. 1983. Tagging and marking. Pages 215–237 *in* L. A. Nielsen and D. L. Johnson, editors. Fisheries techniques. American Fisheries Society, Bethesda, Maryland.

American Fisheries Society Symposium 7:173–182, 1990

INTERNAL TAGS AND MARKS

Internal Extrinsic Identification Systems: Overview of Implanted Wire Tags, Otolith Marks, and Parasites

RAYMOND M. BUCKLEY AND H. LEE BLANKENSHIP

Washington State Department of Fisheries, 115 General Administration Building
Olympia, Washington 98504, USA

Abstract.—Implanted wire tags, otolith marks (induced by variations in temperature, feeding, or photoperiod), and natural parasites are internal extrinsic systems used to identify fish for recovery of specific information after considerable time and growth. The validity of these systems depends upon the assumptions that identified fish are representative of the species with regard to behavior, biological functions, and mortality factors, and thus provide unbiased data. The histopathology of fish tagging, and the long-term failure of percutaneous devices, indicate that internal identification systems are superior to external tags for validating the assumptions and for long-term recovery of information. These three internal identification systems can be applied to fish of almost any size, although otolith marks require retaining fish in controlled environments for limited periods. Coded wire tags have a large data capacity and can identify groups or individuals in all fishery situations. Tags implanted in transparent tissues are externally detectable without the use of an external indicator. Otolith marks and parasites enable the identification of groups of fish, most practically in confined fishery situations. All three systems can be used for within-season management of fisheries. The lack of widespread use of internal extrinsic identification systems for many species is based on a preconceived idea that small, obscure identifiers are not recoverable in fisheries. This bias appears to be based on the historic use of external identifiers and not on any in-depth analysis of their reliability and accuracy.

The identification of fish over time, either as individuals or as members of a group, is one of the basic requirements for studies related to ecology, biology, population dynamic, or fisheries management. Both intrinsic (related to the real nature of the fish) and extrinsic (dependent on external circumstances) identification systems are used by researchers in these fields. Each of these systems has various applications. The validity of either system relies on the basic assumption that an identified fish is representative of its species, and thus provides unbiased data, within the context of a particular study. In many cases, it seems that the choice or acceptability of these identification systems is related more to historic use than to proven reliability and accuracy based on in-depth analysis.

Extrinsic identification systems are either totally external or include an external component. The extrinsic methods most commonly used for fisheries research involve either tagging or marking of individuals, the latter through alteration, mutilation, or coloring of specific body parts. Internal extrinsic identification systems utilize tags, marks, or natural parasites that are com-

pletely enclosed within the tissues of the fish. External extrinsic identification systems, such as external tags, fin clips, or external parasites, are not completely enclosed and thus are continually affected by physical, chemical, and biological features of the environment.

The basic assumptions underlying the use of all extrinsic identification systems are that the identified fish will show normal behavior and biological functions, be subjected to normal predation and other mortality factors, and provide accurate specific information, ideally after considerable time and growth. Although these assumptions are a common part of fisheries research, few studies, among the many that involve fish tags and marks (more than 1,400 were listed in a review by Emery and Wydoski 1987), adequately validated the use of these identification systems in relation to the basic assumptions.

Comparisons of specific internal and external extrinsic identification systems have been made by a number of researchers in relation to the objectives of particular studies, but these usually involved discussions rather than in-depth analyses (e.g., Isaksson and Bergman 1978: internal and external tags; Bergman et al. 1968, Thrower and Smoker 1984, and Elrod and Schneider 1986: internal tags and fin marks; Volk et al. 1987: otolith marks and other extrinsic identifiers; MacKenzie 1983: internal and external parasites). There have been no specific analyses to determine the merits of internal and external extrinsic identification systems in general, or to determine for which fishery research applications each system is best suited. However, for biological factors, intuition suggests that benign, internal extrinsic identification systems are superior to external identification systems, because it is less likely that the assumption of biological normality will be violated. This likelihood is true for all internal systems, provided the relative sizes of the identification mechanism and the fish are physically compatible, and the internal location of the identification mechanism is not disruptive to the normal biological functions of the fish. Internal mechanisms that are relatively large in comparison to the size of the fish (i.e., up to 2.0% of body weight) can be expelled from some species of fish during the course of normal biological functions (Chisholm and Hubert 1985; Marty and Summerfelt 1986).

Opinions on the accuracy of the information resulting from internal and external extrinsic identification systems are highly varied, and identification methods other than fish tags have been little discussed. Information obtained from fish tagging (primarily with external tags) has been specifically criticized by Gulland (1983), who stated that "only a minority of tagging experiments can be used with any confidence to estimate fishing mortality" because of difficulties in ensuring that tagged fish respond normally. A review of philosophical considerations and experimental results also led P. K. Bergman (Northwest Marine Technology, Inc., unpublished) to conclude that the benefits of using external tags— retrieval of information from live fish, and identification of individual fish—may be offset by costs associated with added tag effects that can compromise measurement of critical variables.

In this paper, our objectives are (1) to discuss the biological merits of using internal extrinsic identification systems in fisheries research, and (2) to present an overview of current applications and advances for three of these systems: implanted coded wire tags; otolith marks induced by variations in temperature, feeding, and photoperiod; and natural parasites. These three systems were selected because they represent applications

of distinctly different extrinsic factors—the addition of foreign bodies, the manipulation of environmental variables, and the occurrence of foreign organisms.

Biological Considerations

Internal versus External Systems

If we assume that acceptable internal extrinsic identification systems are benign, then the greatest advantage of these systems is that they do not interact directly with the external environment. The potential for adverse effects on the identified fish is present in all external extrinsic identification systems, especially those that create significant and extended disruptions of the integument. Mutilation of a specific body part (e.g., fin clips) is usually a single event that has reduced potential for continual irritation of the integument, and therefore the wound has a greater potential to heal. However, extended mechanical irritation of the integument can often be caused by the design of external tags. External fish tags have an internal anchor, that passes through the tissues to the exterior of the fish and is attached to an external indicator segment.

The potential biological effects on fish identified with external tags recently came under consideration during the development of a new identification system based on internal tags (Bergman, unpublished). The incomplete healing of the integument (tag wound) during the life of the fish causes chronic wounds, which affect the biological normality of the identified fish as well as long-term tag retention. These considerations are worthy of further discussion here, because the negative aspects of external fish tags define the merits of all internal extrinsic identification systems, especially internal fish tags.

Research findings on the histopathology of fish tagging (Roberts et al. 1973a, 1973b, 1973c; Morgan and Roberts 1976), and reports in the medical literature on the long-term failure of percutaneous devices—objects that penetrate permanently through a defect in the skin—in mammalian and other tissues (Hall et al. 1984; von Recum 1984), seem to show that biological normality and long-term data recovery are affected by almost all external fish tagging techniques. The sinus created by the internal anchor of the tag allows microbial agents free access from the surrounding environment to the internal tissues (Roberts et al. 1973c). Problems continue unless an increase in the girth of the fish completely embeds the external portion of the tag (Roberts et al. 1973b). There has been no success in repeatedly and reproducibly maintaining percutaneous implants in humans or other mammals for periods longer than 3 months; failures are caused primarily by extrusion due to marsupialization (formation of tissue pouches around the implant), permigration, infection and abcess formation, and avulsion (von Recum 1984). These results suggested that failures would occur regardless of the materials used for the percutaneous device, or the animal species involved. Other research has shown that extrinsic and intrinsic forces that cause shearing and tearing at the skin—implant interface also result in failure of percutaneous implants; extrinsic forces are those applied to the skin or the implant by the external environment, and intrinsic forces are those that have to do directly or indirectly with the body's growth and cell maturation, such as formation of scar tissue (Hall et al. 1984). All of the foregoing factors suggest that certain basic assumptions—that loss of external fish tags and mortality related to tagging decrease after the initial tagging period, and that the external tag wound heals—are incorrect. A further consideration is that external fish tags are almost always applied under septic conditions.

Advantages of Internal Systems

The use of internal fish tags does not result in the problems associated with percutaneous external fish tags, and, therefore, tagged fish are more likely to be normal, especially if tag material does not react with tissue and if tags are very small. Small tags minimize both the disruption of integument and tissues during implantation and the time required for implantation wounds to heal. Small internal tags can also be implanted in a wide variety of locations in fish of all sizes.

Internal marks that are created by controlled variations in external environmental factors do not result in percutaneous problems. Because these marks are intended for later recovery, levels of variation that create the marks must be within acceptable ranges for a given species in order not to have any lasting biological effects. Further, if no physical objects are added to fish identified by this method, it is reasonable to conclude that these internal marks do not affect the biological normality of identified fish.

The assumption of biological normality is somewhat less certain when internal parasites are used as fish tags. There is usually some evolutionary compatibility between the fish species and the

parasite, so the presence of the parasite is the "normal" condition for the fish; however, parasites with pathological effects can affect behavior and mortality of fish (MacKenzie 1983). If only parasites that are symbiotic with the host, or at least not significantly injurious, are used as fish tags, there is little possibility for adverse biological effects on the identified fish.

Implanted Coded Wire Tags

Capabilities of the System

The basic coded-wire-tag identification system was described by Bergman et al. (1968). Wire tags can be applied to small as well as to large fish; fish as small as 0.25 g have been successfully tagged, and tags have been recovered from fish exceeding 2.0 kg (Thrower and Smoker 1984). The binary coding system is etched on the surface of the wire tag in a repeating code that has a nearly unlimited capacity for data, giving a capability to internally code replicate experimental groups. In common production usage, the coding on the wire tags identifies an experimental group, although identification of individual fish is now possible with wire that has continuously changing, sequential codes (R. D. Fralick, Northwest Marine Technology, Inc., personal communication).

The coded-wire-tag identification system has been tested for management and research applications on 27 genera of fishes (Table 1). Histological examination of the tag material (type 302 stainless steel wire) has shown that interactions with tissue are minimal (Bergman et al. 1968; Fletcher et al. 1987), and direct effects from the tag and its application are relatively minor (Isaksson and Bergman 1978; Thrower and Smoker 1984; Elrod and Schneider 1986). However, coded wire tags implanted in improper locations can cause significant problems (e.g., in olfactory nerves of small chum salmon *Ocorhynchus keta*: Morrison and Zajac 1987). The greatest use of coded wire tags is in the management of salmon fisheries in the northeastern Pacific Ocean, where fishery agencies in the USA and Canada collectively tagged 42 million juvenile fish in 1986 (K. L. Johnson, Pacific Marine Fisheries Commission, unpublished). This extensive experience in tagging, with species of similar body form and internal anatomy, has resulted in production-size releases that average less than 5% tag loss 30 d after tagging, and do not show significant increases in tag loss 200–300 d after tagging (Blankenship 1990, this volume). With other genera (Table 1),

TABLE 1.—Genera of fishes successfully tagged with implanted coded wire tags.

Fishes	Tag location
Anchovies (Stolephorus)	Nasal cartilage
Catfishes (Ictalurus)	Nasal cartilage, opercular muscle, nape muscle
Char (Salvelinus)	Nasal cartilage
Cichlids (Tilapia)	Nasal cartilage, body muscle
Codfishes (Gadus, Theragra)	Opercular muscle
Drums (Sciaenops)	Opercular muscle
Goatfishes (Mulloidichthys)	Opercular muscle, adipose eyelid
Grayling (Thymallus)	Nasal cartilage
Greenlings (Ophiodon)	Nasal cartilage
Herrings (Clupea)	Nape muscle
Jacks (Caranx)	Opercular muscle, adipose eyelid
Minnows (Notemigonus, Ptychocheilus)	Nasal cartilage, opercular muscle, nape muscle
Paddlefishes (Polyodon)	Rostrum
Perches (Stizostedion)	Nasal cartilage, opercular muscle
Pikes (Esox)	Opercular muscle
Salmon (Oncorhynchus, Salmo)	Nasal cartilage
Scorpionfishes (Sebastes)	Opercular muscle
Snappers (Lutjanus)	Opercular muscle, adipose eyelid, cornea
Snooks (Centropomus)	Opercular muscle
Sturgeons (Acipenser)	Nasal cartilage, first dorsal scute
Sunfishes (Micropterus, Lepomis)	Opercular muscle, interorbital dermis
Temperate basses (Morone)	Nasal cartilage, body muscle
Whitefishes (Coregonus)	Nasal cartilage

tag loss rates in excess of 5% are unusual in experimental and management programs; when they do occur, they can be related to inexperience in applying the coded wire tag to new species. Excessive tag loss has been corrected by change of the tagging technique and implantation at more appropriate locations on the fish. For example, Elrod and Schneider 1986 reported tag loss from lake trout Salvelinus namaycush was reduced from 11 to 3% when the tagging technique and tag location were changed. C. E. Bordner, University of California, Davis (unpublished), found that tag loss from white sturgeons Acipenser transmontanus was reduced from 40 to 0% when the tag location was changed. Klar and Parker (1986) reported tag loss from striped bass Morone saxatilis dropped from 77 to 0% after a change in tag location.

The coded wire tag is normally implanted internally in cartilage, connective tissue, or muscle in fish. It is injected with a hypodermic needle by automated or manual techniques. Coded wire tags can be implanted in specific anatomical parts of fish of almost any size, and later detected electronically by their magnetic field. Visual detection of coded-wire-tagged fish is usually accomplished

through the use of secondary external indicators. For salmonids, it is common for the tag to be implanted in the nasal cartilage and for the adipose fin to be removed; the external indicator is lack of the fin (Bergman et al. 1968). Experimental tagging of white sturgeons has been successful, and involves implantation of the tag in the nasal cartilage or under the first dorsal scute; amputation of a barbel has shown promise as an external indicator (C. E. Bordner, unpublished). In recent studies on a variety of pelagic and demersal fishes, in temperate and tropical environments, researchers examined the use of healed nodules on cut soft fin rays (e.g., Welch and Mills 1981) and implanted dye marks (e.g., Gollmann et al. 1986) as external indicators for coded-wire-tag injection locations in the opercular musculature, interorbital dermis, and adipose eyelid tissue (Blankenship, unpublished).

An increase in the utility of coded wire tags resulted from the use of transparent tissues as tag-implantation locations, as recommended by Bergman (unpublished) in the development of the visible implanted tag (see Haw et al. 1990 this volume). In such a location, the coded wire tag is clearly visible, eliminating the need for an external indicator. Coded wire tags implanted in the transparent interorbital dermis of small pumpkinseeds (Lepomis gibbosus) were clearly visible after 19 months (Haw et al. 1990). Experiments with rainbow trout Oncorhynchus mykiss (formerly Salmo gairdneri) and brown bullheads Ictalurus nebulosus has indicated that the adipose fin is a promising injection site for internal coded wire tags that remain visible (F. Haw, Northwest Marine Technology, Inc., personal communication).

Fishery Recovery of Tags

The recovery of coded wire tags is an important aspect of their use in fishery research. Recovery is accomplished in the commercial and recreational salmon fisheries in the northeast Pacific through the collection of the heads or snouts of fish with missing adipose fins, in both voluntary and structured tag-recovery programs (Buckley and Haw 1978). Innovative education techniques for recognition of the secondary external indicator and monthly tag-recovery reward programs overcome problems normally associated with the recovery of internal fish tags.

A similar voluntary tag-recovery program was used in the recreational fishery for lingcod Ophiodon elongatus in Puget Sound. Juvenile lingcod were released with coded wire tags implanted in the nasal cartilage, but the tagged fish were not given a secondary external indicator; instead, heads were collected from all lingcod that were smaller than the maximum size predicted for the age of the tagged fish (Buckley, unpublished). In contrast, automated recovery of coded wire tags from high-volume catches was used in the North Sea fishery for Atlantic herring *Clupea harengus harengus*, where an automatic tag detector removed tagged fish from the conveyor line transporting the catch to the processing plant (Corten 1985).

The coded wire tag seems to be compatible with a variety of fishes, and with a range of fishery research and management objectives. Historically, the coded wire tag was mainly used as a management tool in salmonid fisheries, and was recently applied in the management of striped bass (several million tagged per year). A worldwide inventory of tag and release programs for marine fishes, 1980–1985 (IGFA 1987), shows that there were 55 tagging programs that involved species other than salmonids and 24 programs that involved salmon. This list excludes the larger oceanic species that may require capture, handling, and release conditions not readily adaptable to coded-wire-tag methods. Coded wire tags were used only in the Washington program that tagged lingcod, and in 16 salmon-tagging programs; with one exception, all of the other tagging was done with external tags.

Recovery of Tags

In normal applications, coded wire tags are dissected from dead fish. Recovery of a small tag (1.0 mm long and 0.25 mm in diameter) is enhanced by the electronic detection of the tags' magnetic field, and by tag implantation in a specific anatomical part of the fish. Decoding the information on the tag requires only a standard binocular microscope. The average time for a coded wire tag to be removed from a fish and decoded, and for the information to be entered into a computer information system, is 4 min (Washington State Department of Fisheries, unpublished). The coded-wire-tag system allows within season management of the fishery with a 7- to 10-day lag from recovery of tagged fish to final data analysis.

In some fishery applications, researchers wish to avoid sacrificing identified fish. Shallow implantation of coded wire tags in transparent tissues makes the tags visible, and facilitates benign surgical recovery of the tags (Haw et al. 1990).

This type of tag removal appears to have little effect on the fish, and the fish can be immediately retagged in adjacent, similar tissue. Initial multiple tagging of each fish and sequential recovery of the tags during the life of the fish is also possible.

Otolith Marks

Capabilities of the System

Methods for inducing specific marks in fish otoliths, through manipulations of environmental temperature, feeding rates, and photoperiod, were described by a number of researchers (e.g., Brothers 1985; Volk et al. 1987). The strengths of this system include easy application to otoliths of embryos en masse, during the earliest development stages of the otolith, as well as to otoliths at any time during the growth period of the fish. Otolith microstructural features are permanent and can be viewed and analyzed in fish of any age (Brothers 1985). Variations in the specific marks on the otoliths can produce a number of unique identification patterns. The most practical use of this system is to identify large groups of fish from controlled environments (e.g., artificial production).

In transmitted light, growth increments in otoliths appear as opaque and translucent bands progressing outward from the center. Distinct variations in the width of these bands are known to be caused by a number of endogenous factors, such as metamorphosis and spawning, as well as by atypical extrinsic factors, such as increases or decreases in normal temperature regimes and feeding rates, (for reviews see Taubert and Coble 1977; Brothers 1981; Campana and Neilson 1982; Victor 1982). The identification system consists of specific incremental banding patterns in the otolith, which correspond to planned changes in extrinsic factors during periods of artificial manipulation.

Experimental manipulations of environmental temperature, feeding rate, and photoperiod have been used primarily to induce marks in otoliths in order to validate the periodicity of incremental growth (Victor 1982; Volk et al. 1987). Experiments testing these manipulations to mark fish en masse for release and subsequent fishery recovery have been limited to hatchery production of salmonids. From these experiments, it has been found that temperature changes, or alternating cycles of feeding and starvation, produce the best results (Brothers 1985; Volk et al. 1990, this volume); variations in photoperiod had no effect on otolith microstructure, or the effects were masked by the stronger effects of temperature and feeding (Washington State Department of Fisheries, unpublished). Current management applications are limited to planned releases of lake trout that have been subjected to cyclic temperature changes during embryonic development (Brothers 1985), and to production-size releases of chinook salmon *Oncorhynchus tschawytscha* (E. C. Volk, Washington State Department of Fisheries, unpublished), coho salmon *O. kisutch*, and chum salmon *O. keta* that have been subjected to repeated, sudden changes in water temperatures during embryonic and pre–yolk absorption stages of development (Volk et al. 1990).

The recommended difference between cyclic high and low temperatures is at least 5°C, and preferably 10°C, to optimize the differences between the opaque and translucent bands in lake trout otoliths (Brothers 1985). Experiments with chum salmon embryos involved temperature treatments in which temperature differences ranged from 1.8 to 5.8°C. All treatments produced good banding patterns in the otoliths, and the bands were as distinct at the lower temperature range as at the higher range (Volk et al. 1987). Production releases of chinook salmon were subjected to incubation temperatures alternating between ambient and 2.0°C below ambient temperature at intervals of 1 to 5 d (E. C. Volk, unpublished). Production coho salmon, later released, were subjected to similar treatment at 3.5°C above ambient temperature at 48-h intervals (Volk et al. 1990). In experiments with chinook salmon fingerlings, researchers used alternating 5-d periods of feeding and starvation (Washington State Department of Fisheries, unpublished).

Fishery Applications

The release and fishery recovery of fish with marked otoliths will be of greatest value in the management of fishes that undergo artificial production. These fishes are readily available for manipulations of incubation or rearing temperatures, or feeding rates, under the controlled circumstances required to produce specific banding patterns in the otoliths. This mass identification system will be especially useful in the management of terminal-area salmonid fisheries that harvest mixed stocks, and require stock definitions to control exploitation rates (Volk et al. 1987). In fisheries that harvest a large proportion of unmarked fish, a secondary external indicator may be necessary to identify otolith-marked fish in order to reduce the number of otoliths that need to be recovered and analyzed.

Mark Recovery

Otoliths are specific anatomical parts and, therefore, readily located and recovered from dead fish by standard dissection techniques. The previously time-consuming processes of otolith preparation and evaluation by means of manual sectioning and polishing, and microscope-aided interpretation, have been improved considerably through the use of automated sectioning devices and computerized image-analysis techniques (Volk et al. 1990). These improvements allow the simultaneous preparation of large numbers of otoliths by casting them in resin blocks, which are ground to the appropriate thickness for analysis (Washington State Department of Fisheries, unpublished). The average time for an otolith to be removed, prepared, and evaluated, is 7 min (Volk, personal communication). When it is perfected, this identification system should have the capability to provide information for within-season management.

Parasites

Parasites as Tags

The basic methods for using natural parasites as biological "tags" in an internal identification system have been most recently reviewed and described by MacKenzie (1983). The strengths of this system include the symbiotic nature of the parasites that are acceptable for use as tags, and the ability the system provides to gain a variety of biological, ecological, and evolutionary information about the host fish. Parasites are present in almost all developmental stages of fish, and those acceptable as tags provide a number of unique identifications usable in population-level studies in fisheries management and research (Margolis 1982; MacKenzie 1983).

The composition of the parasite fauna in fish depends upon ecological factors that determine the opportunities for contact between parasite and host, and evolutionary factors that determine the physiological and morphological adaptations of parasite and host (Margolis 1965). Fish parasites can be so highly specialized that they are able to develop successful parasite–host relationships with only a single fish species or closely related species, or parasites can be relatively unspecialized and unrestricted and occur in a wide variety of unrelated fishes. "The greatest barrier to the

TABLE 2.—Taxonomic groups of fish parasites used as biological tags.

Group	Usual parasitic mode			Fish role in life cycle		
	Endoparasitic	Ectoparasitic	Systemic	Only host	Intermediate host	Definitive host
Fungi		×	×			×
Protoza	×			×		
Monogenea		×		×		
Digenea	×				×	×
Cestoda	×				×	×
Nematoda	×				×	×
Acanthocephala	×				×	×
Crustacea		×		×		

efficient use of parasites as biological tags for fish is lack of knowledge of the biology of fish parasites, particularly in relation to their life cycles and their life spans" (MacKenzie 1983).

MacKenzie (1983) proposed the following general criteria to assess the probable value of a parasite as a biological tag in fish population studies; these criteria can be modified and expanded to accommodate particular species of fish (e.g., MacKenzie 1985 for the Atlantic herring). (1) The parasite should cause significantly different levels of infection in the subject host species in different parts of the study area. (2) No ectoparasite that is easily detached, leaving no evidence of its past presence on the host, should be considered, because these may be lost during capture and handling. (3) The method of examination should involve the minimum of dissection. A high degree of site-specificity on the part of the parasite is an advantage. (4) The parasite should be easily detectable and identifiable. (5) The parasite must have a life span, or remain in an identifiable form, in the subject host long enough to cover the time scale of the investigation. (6) The parasite should have no marked pathological effects on the subject host.

Although a wide variety of parasites is found in fish (e.g., 175, and possibly 180, species are known from Pacific salmon within the natural range of this fish group: Margolis 1982), relatively few have current value in internal identification systems; the pathological effects, biology, and distributions of most parasites are unknown. Parasites from eight major taxonomic groups that occur naturally in fish have been investigated for use as biological tags (Table 2). The selection of the appropriate parasite or combination of parasites for a particular study is based on a number of factors that relate the study objective to aspects of various host–parasite and interparasite relationships; basic information is lacking on many of these relationships, as well as on parasite taxonomy (Margolis 1982). Earlier researchers proposed that the value of a parasite as a biological tag should include the stability of the infection from year to year (Sindermann 1961; Kabata 1963); however, this condition is almost impossible to fulfill because most parasite populations seem to be basically unstable (Kennedy 1977). MacKenzie (1983) noted that the problems posed by annual variations in infection rates can be overcome by examination of single year classes of the host fish over several consecutive years, an amount of time that would be required in any case to verify that significant variations do actually occur. Independent verification of the age of parasite-tagged fish is often required in order to use data from parasites accurately.

Fishery Applications

The use of parasites as biological tags in studies of fish populations is best suited to the separation of relatively self-contained stocks of fish. MacKenzie (1983) presented four areas of study and gave general types of data that could be provided by parasitized fish. MacKenzie's list can be summarized as follows: (1) stock separation—definition of geographical boundaries of populations of adult fish; (2) seasonal migrations—delineation of movements between regions, for spawning or feeding; (3) recruitment—delineation of migrations of juvenile fish from nursery grounds to adult feeding and spawning grounds; (4) age-dependent migrations of adults—delineation of movements of sexually mature fish to different habitats as they grow older.

Beverley-Burton and Pippy (1977) suggested that the use of morphometric variations of a single parasite can be useful in stock identification and in determining a number of host-dependent factors. Parasites have been used successfully to separate sea-run and nonmigrating stocks of Arctic char *Salvelinus alpinus* (Dick and Belosevic 1981), to determine the continent of origin of steelhead (anadromous *Oncorhynchus mykiss*) harvested at sea (Margolis 1965), to trace recruitment migrations of Atlantic herring (MacKenzie 1985), and to determine the stock structure of adult sockeye salmon *O. nerka* harvested in a commercial fishery (Margolis 1982). In the sockeye salmon study, stock composition was determined weekly, 12–24 h prior to each weekly opening of the commercial fishery. Subsequent use of this fishery management technique, which was based on patterns of parasite prevalence in catches and escapements, enabled within-season determination of run-timing differences for adults, effectiveness of harvest regulations, and variations in weekly stock-specific exploitation rates (K. D. Hyatt, Canadian Department of Fisheries and Oceans, unpublished).

Parasite Recovery

Internal parasites used as biological tags are usually recovered by standard dissection techniques to obtain tissue or organ samples from dead fish. The ability to recover the parasites is enhanced if parasites are associated with a specific anatomical feature of the fish. The identification and enumeration procedures range from direct counting with a microscope (e.g., three species encysted on the outer surfaces of pyloric caeca in Atlantic herring: MacKenzie 1985) to several steps of chemical and physical treatment of the samples prior to recovery of the parasites for counting (e.g., 16 species and juveniles of 4 taxa located systemically in sockeye salmon: Bailey and Margolis 1987). The average time for the recovery, preparation, and evaluation of the parasites used for the within-season management of a sockeye salmon fishery is about 12 min per fish with a procedure that involves an average of three steps to determine the presence or absence of the parasite (Hyatt, personal communication).

Discussion

The lack of widespread use of internal extrinsic identification systems, which are far more compatible with basic assumptions of biological normality and accurate information recovery than the more popular external extrinsic identification systems, leads to the conclusion that researchers have a preconceived idea that small, obscure identification systems cannot be recovered in fisheries. It can be further concluded that this bias is related more to the historic use, and thereby the implied acceptability, of external identification systems, than to any proven reliability and accuracy based on in-depth analysis. This situation is

most clearly illustrated in consideration of the use of external and internal fish tags, as tags are in widespread use in both experimental and management programs.

The idea that fish tags need to be very visible to the human eye for effective recovery was historically pervasive in fisheries research. For species other than salmonids, this idea remains evident in current tagging programs to such a degree that any resulting negative effects seem to be ignored. For example, Jones (1977) said, "For many of the commercially important marine fish species, it is necessary to rely on commercial fishermen for returning tags. The tag must be attached externally to the fish and it is important that it should be as large and conspicuous as possible without interfering with swimming." Although this type of tag may be appropriate for a few short-term studies of large fish, the concepts presented by Jones (1977) relating to tag size, tag location, and tag recovery, are not valid. It seems unlikely that fish carrying large tags, with adherent growths of hydroids, barnacles, etc., could be normal in behavior or biological functions. Tags that are large and conspicuous can be prime targets for predators, and can cause entanglement in nets. Such entanglement results in selective biases in the capture of tagged fish. Also, it has been shown that reliance on fishermen for tag recoveries involves problems in deliberate (Paulik 1961; Butler 1962) and inadvertent (Konstantinov 1978) failure to report tags.

There are obviously some research situations in which external fish tags are either appropriate or overridingly convenient, and measurements of tag effect can be determined to develop correction factors for related data analysis. Nor are all external fish tags large and conspicuous. However, even the smaller ones have the inherent problems of percutaneous application, as well as the problems caused by the interactions with the external environment. These serious problems are not given full consideration, and could compromise the measurement of critical variables.

It can be reasoned that information about the potential for percutaneous problems has been "hidden" in the medical literature and therefore has not been readily available to fishery researchers. However, it can also be reasoned that careful consideration of the basic assumptions of fish tagging cannot support the continued widespread use of external fish tags in situations where they clearly have the potential to invalidate the basic assumptions upon which research is based. Available internal fish tags can match the identification attributes of external fish tags (i.e., visual detection, individual identification, and information recovery from live fish), and they are applicable to most situations and most fishes. It seems that what is needed to realize the advantages of internal extrinsic identification systems more fully is more innovation and less bias on the part of fishery researchers.

Acknowledgments

We thank E. C. Volk and S. L. Schroder for information on otolith marking techniques, and K. D. Hyatt, G. J. Steer, T. Dalton, and P. M. Rounds for providing information on the use of fish parasites as biological tags.

References

Bailey, R. E., and L. Margolis. 1987. Comparisons of parasite fauna of juvenile sockeye salmon (*Oncorhynchus nerka*) from southern British Columbia and Washington State lakes. Canadian Journal of Zoology 65:420–431.

Bergman, P. K., K. B. Jefferts, H. F. Fiscus, and R. L. Hager. 1968. A preliminary evaluation of an implanted coded wire fish tag. Washington Department of Fisheries, Fisheries Research Paper 3:63–84.

Beverly-Berton, M., and J. H. C. Pippy. 1977. Morphometric variations among larval *Anisakis simplex* (Nemotoda: Ascaridoidea) from fishes of the North Atlantic and their uses as biological indicators of host stocks. Environmental Biology of Fishes 2:309–314.

Blankenship, H. L. 1990. Effects of time and fish size on coded wire tag loss from chinook and coho salmon. American Fisheries Society Symposium 7:237–243.

Brothers, E. B. 1981. What can otolith microstructure tell us about daily and subdaily events in the early life history of fish? Rapports et Procès-Verbaux des Réunions, Conseil International pour L'Exploration de la Mer 178:393–394.

Brothers, E. B. 1985. Otolith marking techniques for the early life history stages of lake trout. Great Lakes Fishery Commission, Research Completion Report, Ann Arbor, Michigan.

Buckley, R. M., and F. Haw. 1978. Enhancement of Puget Sound populations of resident coho salmon (*Oncorhynchus kisutch*). Pages 93–103 *in* Proceedings of the 1977 northeast Pacific chinook and coho salmon workshop. Canada Department of Fisheries and Environment, Vancouver.

Butler, R. L. 1962. Recognition and return of trout tags by California anglers. California Fish and Game 43:5–18.

Campana, S. E., and J. D. Neilson. 1982. Daily growth increments in otoliths of starry flounder (*Platichthys stellatus*) and the influence of some environmental variables in their production. Canadian Journal of Fisheries and Aquatic Sciences 39:937–942.

Chisholm, I. M., and W. A. Hubert. 1985. Expulsion of dummy transmitters by rainbow trout. Transactions of the American Fisheries Society 114:766–767.

Corten, A. 1985. Stock assessment of herring in the northwestern North Sea by micro-wire tagging. Netherlands Institute for Fishery Investigations, Final Report, Ijmuiden, The Netherlands.

Dick, T. A., and M. Belosevic. 1981. Parasites of Arctic charr *Salvelinus alpinus* (Linnaeus) and their use in separating sea-run and non-migrating charr. Journal of Fish Biology 18:339–347.

Elrod, J. H., and C. P. Schneider. 1986. Evaluation of coded wire tags for marking lake trout. North American Journal of Fisheries Management 6:264–271.

Emery, L., and R. Wydoski. 1987. Marking and tagging of aquatic animals: an indexed bibliography. U.S. Fish and Wildlife Service Resource Publication 165.

Fletcher, D. H., F. Haw, and P. K. Bergman. 1987. Retention of coded wire tags implanted into cheek musculature of largemouth bass. North American Journal of Fisheries Management 7:436–439.

Gollmann, H. P., E. Kainz, and O. Fuchs. 1986. Marking and tagging of fish with particular regard to the application of dyes, especially of Alcian Blue 8 GS. Oesterreichs Fisherei 39(11–12):340–345.

Gulland, J. A. 1983. Fish stock assessment. Wiley, New York.

Hall, C. W., P. A. Wilcox, S. R. McFarland, and J. J. Ghidoni. 1984. Some factors that influence prolonged interfacial continuity. Journal of Biomedical Materials Research 18:383–393.

Haw, F., P. K. Bergman, R. D. Fralick, R. M. Buckley, and H. L. Blankenship. 1990. Visible implanted fish tag. American Fisheries Society Symposium 7:311–315.

IGFA (International Game Fish Association). 1987. World record game fishes: freshwater, saltwater, and fly fishing. IGFA, Fort Lauderdale, Florida.

Isaksson, A., and P. K. Bergman. 1978. An evaluation of two tagging methods and survival rates of different age and treatment groups of hatchery-reared Atlantic salmon smolts. Journal of Agricultural Research in Iceland 10(1):74–99.

Jones, R. 1977. Tagging: theoretical methods and practical difficulties. Pages 46–66 *in* J. A. Gulland, editor. Fish population dynamics. Wiley, New York.

Kabata, Z. 1963. Parasites as biological tags. International Commission for the Northwest Atlantic Fisheries Special Publication 4:31–37.

Kennedy, C. R. 1977. The regulation of fish parasite populations. Pages 63–110 *in* G. W. Esch, editor. Regulation of parasite populations. Academic Press, New York.

Klar, G. T., and N. C. Parker. 1986. Marking fingerling striped bass and blue tilapia with coded wire tags and microtaggants. North American Journal of Fisheries Management 6:439–444.

Konstantinov, K. G. 1978. Modern methods of fish tagging. Journal of Ichthyology 17:924–938.

MacKenzie, K. 1983. Parasites as biological tags in fish population studies. Advances in Applied Biology 7:251–331.

MacKenzie, K. 1985. The use of parasites as biological tags in population studies of herring (*Clupea harengus* L.) in the North Sea and to the north and west of Scotland. Journal du Conseil, Conseil International pour l'Exploration de la Mer 42:33–64.

Margolis, L. 1965. Parasites as an auxiliary source of information about the biology of Pacific salmons (genus: *Oncorhynchus*). Journal of the Fisheries Research Board of Canada 22:1387–1395.

Margolis, L. 1982. Parasitology of Pacific salmon—an overview. Pages 135–226 *in* E. Meerovitch, editor. Aspects of parasitology. McGill University, Montreal.

Marty, G. D., and R. C. Summerfelt. 1986. Pathways and mechanisms for expulsion of surgically implanted dummy transmitters from channel catfish. Transactions of the American Fisheries Society 115:577–589.

Morgan, R. I. G., and R. J. Roberts. 1976. The histopathology of salmon tagging. IV. The effect of severe exercise on the induced tagging lesion in salmon parr at two temperatures. Journal of Fish Biology 8:289–292.

Morrison, J., and D. Zajac. 1987. Histological effect of coded wire tagging in chum salmon. North American Journal of Fisheries Management 7:439–441.

Paulik, G. J. 1961. Detection of incomplete reporting of tags. Journal of the Fisheries Research Board of Canada 18:817–832.

Roberts, R. J., A. MacQueen, W. M. Shearer, and H. Young. 1973a. The histopathology of salmon tagging. I. The tagging lesion in newly tagged parr. Journal of Fish Biology 5:497–503.

Roberts, R. J., A. MacQueen, W. M. Shearer, and H. Young. 1973b. The histopathology of salmon tagging. II. The chronic tagging lesion in returning adult fish. Journal of Fish Biology 5:615–619.

Roberts, R. J., A. MacQueen, W. M. Shearer, and H. Young. 1973c. The histopathology of salmon tagging. III. Secondary infections associated with tagging. Journal of Fish Biology 5:621–623.

Sindermann, C. J. 1961. Parasite tags for marine fish. Journal of Wildlife Management 25:41–47.

Taubert, B. D., and D. W. Coble. 1977. Daily rings in otoliths of three species of *Lepomis* and *Tilapia mossambica*. Journal of the Fisheries Research Board of Canada 34:332–340.

Thrower, F. P., and W. W. Smoker. 1984. First adult returns of pink salmon tagged as emergents with binary-coded wires. Transactions of the American Fisheries Society 113:803–804.

Victor, B. C. 1982. Daily otolith increments and recruitment in two coral reef wrasses, *Thalassoma bifasciatum* and *Halichoeres bivittatus*. Marine Biology 71:203–208.

Volk, E. C., S. L. Schroder, and K. L. Fresh. 1987. Inducement of banding patterns on the otoliths of juvenile chum salmon (*Oncorhynchus keta*). Pages 206–212 *in* P. Rigby, editor. Proceedings of the 1987 northeast Pacific pink and chum salmon workshop. Alaska Department of Fish and Game, Juneau.

Volk, E. C., S. L. Schroder, and K. L. Fresh. 1990. Inducement of unique otolith banding patterns as a practical means to mass-mark juvenile Pacific salmon. American Fisheries Society Symposium 7:203–215.

von Recum, A. F. 1984. Applications and failure modes of percutaneous devices: a review. Journal of Biomedical Materials Research 18:323–336.

Welch, H. E., and K. H. Mills. 1981. Marking fish by scarring soft fin rays. Canadian Journal of Fisheries and Aquatic Sciences 38:1168–1170.

American Fisheries Society Symposium 7:183–202, 1990

Otolith Marking

EDWARD B. BROTHERS

EFS Consultants, 3 Sunset West, Ithaca, New York 14850, USA

Abstract.—The microstructural and chemical properties of teleost otoliths provide multiple opportunities for the interpretation of natural marks and the induction of artificial ones. Natural microstructural features (e.g., increment spacing patterns) may result in signature growth patterns that are specific to habitats, water bodies, or even climatic events. Natural chemical fingerprints can result from the deposition of distinctive isotopic and trace-element compositions. The practical extension of these phenomena is that the early life stages of fish—embryos, larvae, and juveniles—can be marked en masse. A pilot project on lake trout *Salvelinus namaycush* demonstrated the feasibility of mass-marking all of these stages with otolith "codes" that could be designed to place distinctive marks on multiple lots. The experimental design included comparison of the effectiveness of exposure to tetracycline, acetazolamide, strontium, rare earth elements, and variable environmental temperatures. Temperature manipulations can create a large variety of unique and easily visible marks in all stages. The method is easily applied to mass-marking in either open or closed rearing systems, and procedures for the preparation and viewing of the marks are relatively simple compared to those for other techniques examined. Similar results have been obtained for several other salmonids and a variety of freshwater and marine species.

The otoliths (in particular the sagittae and lapilli) are the first calcified structures to appear in embryonic or larval stages of teleosts. Once formed, the early otolith growth structures remain intact and unchanged for the entire life of the fish (Brothers 1984). Chemical and microstructural features formed during early ontogeny persist throughout the life of the fish and may be analyzed and viewed in fish of any age. Because the otoliths are the only permanent and persistent structures present in the very earliest life stages, they offer the only possibility for producing a unique and endogenous mark at those stages—i.e., a mark that is distinctive but not dependent upon the insertion, attachment, or application of some internal or external tag, brand, marker, dye, etc.

Marking, tagging, and identification of fish populations, stocks, and even individuals is useful in a variety of fisheries applications in the laboratory and field. Techniques are available for tagging of emergent fry of most salmonids; however, the costs are relatively high for massive programs and induced mortality and tag loss are still problems. Embryos, alevins, and the smaller fry of some species such as lake trout *Salvelinus namaycush* cannot be tagged with coded wires or microtags (e.g., see Elrod and Schneider 1986; Klar and Parker 1986). These techniques are also not technically feasible for marking the young stages (especially larvae) of the vast majority of teleosts because the comparable stages are considerably smaller and more delicate than they are for salmonids.

Many current fish-stocking efforts are directed at the early stages of a wide variety of freshwater and marine fishes; assessment of the success of stocking of early life stages may require discrimination of lots reared under different conditions, or released at different sizes, ages, times of day, or localities. Species with small, pelagic larvae produce enormous numbers of young and have high mortality in the hatchery and wild. Experiments and stocking programs often deal with extremely large numbers of animals. Marking of different lots (e.g., of known parentage or genetic background) allows lots to be combined to control for environmental variables in laboratory or culture experiments. Mass-marking techniques are required under such circumstances, in order to mark a sufficient number of individuals and to minimize the effects of handling these sensitive stages. Otolith marking offers several clear advantages that make it the best possible way to mark fish when such limitations arise. (1) It is applicable to the very youngest and smallest stages of all species, including embryos. (2) It produces a permanent mark. (3) It is accomplished in batches with minimal or no manipulation or handling of the fish. (4) Groups or lots can be uniquely marked. Obvious disadvantages or limitations of otolith marking include the following. (1) Fish must be killed to remove and examine otoliths in order to detect the presence of a mark, if there are no external marks or other indications to identify stocked fish. (2) Otolith marking does

183

not allow recognition or coding of individuals. (3) The production of marks and the preparation of otoliths for viewing those marks requires development of techniques specially suited to the species and life stages involved. This requirement is general for all marking studies and not unique to otolith-based systems; however, the skills involved and some of the equipment may be novel to some fisheries programs. (4) Otolith marking is not easily applied to the marking of wild fish in the field.

Classification and Review of Otolith Marks

The structural (gross and microstructural) and chemical properties of teleost otoliths provide potentially rich opportunities for the development of marking procedures or the recognition of existing natural marks. For the purposes of this paper, a "mark" is broadly defined as any feature useful for identifying groups, populations, or individuals within a fish species. Otolith marks may be considered to be internal tags that are visually or chemically distinctive identifying features. Information derived from otolith analysis (e.g., growth rate) is also potentially useful for identifying population structure; however, discussion of such applications is beyond the scope of this paper.

Otolith marks can be classified by two basic types of causation: natural marks and induced marks. Each of these classes can be divided further into categories of chemically or structurally recognizable marks. Natural marks are produced in wild fish in their natural habitat. In contrast, artificially induced marks are caused by human manipulation of the fish or its environment, either in the laboratory (hatchery) or field. This manipulation may be intentional and designed to produce a specific mark, or it may be the result of normal handling and transfer in captivity or at the time of release. Otolith marks may be compositional, and observed by chemical analysis; or structural, and observed visually. The compositional and structural categories are not really mutually exclusive because structural features such as microstructural growth increments are the result of small localized differences in relative composition. Some of the literature citations included in the following review are not specific to otoliths, but deal with chemical marks in bones, scales, or both. These calcified tissues have many chemical properties similar to those of otoliths and therefore are useful models for otolith responses, or may point to areas where otolith experiments are warranted.

Natural Chemical Marks

The examination of natural marks in otoliths and other calcified structures most commonly occurs as part of efforts to identify stocks (Ihssen et al. 1981). These studies may or may not be completely successful in their own right, but nevertheless demonstrate the variation in otolith composition and structure within a species, and highlight the environmental influences that produce them. Understanding of these phenomena provides a basis for artificial manipulation to induce marks. Therefore, the study of natural marks has great value in several contexts.

The analysis of the chemical composition of otoliths and other calcified tissues is a promising and potentially powerful technique for identification of populations. The basic premises are that many materials are incorporated in small amounts in tissues and crystalline matrices and that the relative and absolute deposition rates and resultant concentrations of these materials are functions of their availability in the local environment. Furthermore, it is assumed that, once deposited, the mix of trace materials is not altered substantially, and chemical composition may be analyzed to yield a "chemoprint" or locality-specific compositional profile. In practice, the materials to be analyzed are usually at the isotopic or elemental level (although, theoretically, various compounds could also be examined) and may be referred to as trace elements, because the ones of greatest interest are likely to be in low concentrations. These elements may or may not be active biologically; the ones of lesser activity are more valuable to the analysis because they are less likely to be metabolically concentrated and transferred in or out of tissues.

Several studies have included trace-element analysis for the purpose of stock or population identification; some researchers have examined all detectable elements and statistically compared multiple peaks in X-ray spectra (Calaprice 1971; Calaprice et al. 1971; Moreau and Barbeau 1979; Mulligan et al. 1983, 1987; E. Brothers, unpublished); others have concentrated on one or a few elements or isotopes such as zinc, iron, strontium, oxygen, or certain rare earth elements (Miller 1963; Van Coillie et al. 1974; Bagenal et al. 1976; Gauldie and Nathan 1977; Papadopoulou et al. 1978, 1980; Lapi and Mulligan 1981; Moreau et al. 1983; Gauldie et al. 1986; Belanger et al. 1987). In all of these examples, the investigators discovered compositional differences in otoliths, scales, bones, or whole organisms that were correlated with the locality of fish collection. The relative concentrations of many of these elements in tissues are functions of availability in the environment plus temperature-mediated physical and biological processes (i.e., the rate of incorporation or substitution in tissues is a temperature-dependent process). A complete understanding of these phenomena may allow for back-calculations of temperature from elemental composition, or deductions of spatial distribution and origin (Degans et al. 1969; Gauldie and Nathan 1977; Gauldie et al. 1980; Schneider and Smith 1982; Radtke and Targett 1984). However, temperature effects also can confuse analysis of stock identification by obscuring differences related to availability and seasonal fluctuations.

To date, most chemical analyses of otoliths have been done on marine fishes (see also Dannevig 1956), organisms that are exposed to a relatively rich and diverse chemical bath—seawater. Concentrations of elements such as iron, magnesium, manganese, strontium, sodium, and zinc in otoliths range from a few to several micrograms per gram. Exposure and incorporation levels in freshwater fish are likely to be considerably lower. Analyses may require very sophisticated and sensitive analytic techniques such as X-ray microanalysis (energy-dispersive and wavelength-dispersive), tube-excited X-ray fluorescence spectroscopy, atomic absorption spectroscopy, neutron activation analysis, inductively coupled argon-plasma atomic emission spectrometry, and related mass spectrometry methods. The major lesson to be learned from what we know about the chemical composition of otoliths is that they contain detectable and naturally variable levels of many elements. The variation is interpretable in terms of specificity to localities or bodies of water. The natural composition also suggests elements that are commonly and easily incorporated and thus may be employed for artificial marking.

Natural Structural Marks

Natural structural marks include internal microstructural features as well as some characters based on relative otolith size or shape. The latter are more straightforward because morphometric characteristics of the whole otolith or the central element termed the "nucleus" are used in the same way as other morphometric or meristic characters in stock identification (Messieh 1972; McKern et al. 1974; Casselman et al. 1981; Ihssen et al. 1981; Neilson et al. 1985b). Internal micro-

structural features of otoliths are used in ways analogous to the traditional methods of scale analysis for stock identification (see Ihssen et al. 1981 for review). For scales, the primary units are the circuli and how they are grouped and separated in a higher-level pattern consisting of the annuli. Spacing and patterns of circuli and annuli can be used to discriminate between stocks that differ in geographic origin, spawning time, or hatchery versus wild origin. Otolith microstructure (reviewed by Gjosaeter et al. 1984; Campana and Neilson 1985; Garland 1987) has the daily growth increment as its basic unit. In addition, these units show considerable variation in width and optical density (chemical composition), which are related to the influence of varying environmental or physiological factors. Daily growth increments may be organized into distinctive higher-order temporal patterns, which are referable to endogenous and exogenous events in the life of a fish. Annual zones also occur, but most characteristic marks are formed over a much shorter time period. Because of the fine temporal record in the otolith, and the sensitivity of otolith growth patterns to environmental conditions (e.g., Campana and Neilson 1982; Neilson and Geen 1982, 1985; Volk et al. 1984; Neilson et al. 1985a; Rosa and Re 1985; Gutierrez and Morales-Nin 1986; Mosegaard 1986), the potential for the production of characteristic or unique marks is much greater for otoliths than scales. Also, the literature presents no unambiguous evidence for otolith resorption or modification of internal structure over time. It is clear from many studies that the microstructure represents a permanent record of both natural and induced marks. A large portion of the research on otolith microstructure deals with counts and measurements of daily growth increments, and interpretive analysis of patterns formed. The number of published works is large, and growing at a rapid rate. Here, I wish only to highlight the ways in which microstructure may serve to group wild fish into smaller units based on differences in their environment, and in the chronology of life history events such as migration, spawning, and settlement. A few examples can illustrate what these marks may look like. Categories of potentially useful marks include the following.

Migration marks.—Migration marks are discrete transitions in growth patterns that reflect a change of environment during the growth of the fish (and the otolith). Examples are movement from a headwater stream to a large river, from a

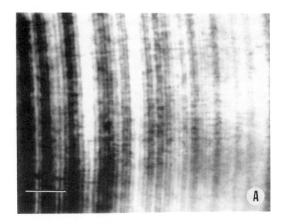

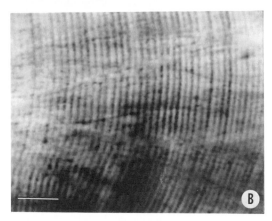

FIGURE 1.—Comparison of otolith growth patterns for a cutthroat trout that resided in a stream and then in Hungry Horse Reservoir, Montana. Photomicrographs are from a video monitor; bars = 20 μm. (A) Pattern formed during stream residence in summer of the fourth year of life. Note the fine daily growth increments of varying width and optical density due to stream temperatures that changed in response to daily weather patterns. (B) Pattern formed during reservoir residence in the summer of the fifth year of life. The daily growth increments are wider and more regular.

river to an estuary, from an estuary to the open ocean, and from a stream to a lake. The primary requirements for occurrence of a good mark under such circumstances are that the fish must be relatively young or growing (to ensure a relatively complete otolith record) and that there must be fairly large differences in the conditions of life for the fish in the two environments (especially temperature and feeding conditions). For example, there is a distinct difference between the daily increment patterns formed during stream versus reservoir residency. Figure 1 illustrates this difference for a cutthroat trout *Oncorhynchus clarki*

(formerly *Salmo clarki*) in its fourth and fifth years of life.

Temperature marks and other habitat "signatures".—Although temperature may be a prime proximate cause of a characteristic microstructural pattern in migration marks, sometimes local habitats may have distinctive characteristics related to temperature, pH, food abundance, etc. that also induce characteristic otolith patterns or marks. For example, some streams in a watershed might receive water from warm springs whereas others do not. A particular species may spawn in either a lake or in inlet or outlet rivers in the same system, that differ in thermal characteristics. In comparison with offshore environments, coastal lagoons may exhibit large temporal (e.g., diel, seasonal) variations in thermal and other characteristics depending on the level of terrestrial input or tidal exchange with the sea. Localized differences in weather conditions (e.g., time of snowmelt and runoff) may also impose characteristic signatures on the daily increment record. Microstructural patterns caused by daily temperature variations may be useful as marks, as illustrated by two cutthroat trout that originated from a tributary of Hungry Horse Reservoir, Montana (Figure 2). The otolith regions illustrated were formed immediately preceding downstream migration in June 1987. Note the precise correspondence of the patterns from the two fish. Dark and light increments correlate with cool and warm air and stream temperatures on the respective days. Figure 3 shows a dramatic example of a natural mark caused by a 4-d period of unusually warm temperatures (14–17 June 1983) that coincided with a temporary cessation of hypolimnial releases from the Cannonsville Reservoir into the Delaware River, New York. An age-1 brown trout *Salmo trutta* collected in August 1983 (Figure 3A) and an age-2 individual collected in June of the following year (Figure 3B) both showed this mark. The presence of this mark indicated which fish were present in the river at that time. Fish that were in tributaries during the period did not have this mark in their otoliths.

A common related application of this concept is the discrimination of wild- from hatchery-produced individuals. Even if no special effort is made to mark the otoliths of stocked fish, the hatchery environment can be sufficiently different from the conditions experienced by wild fish that the otoliths are easily distinguishable, in my experience with a variety of salmonids. In addition, it is also possible to determine the date of stocking

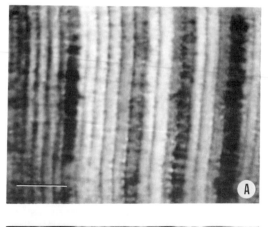

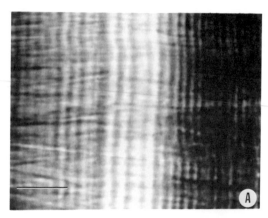

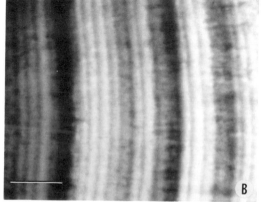

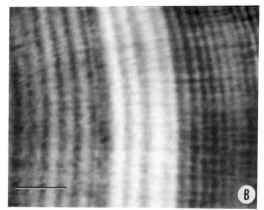

FIGURE 2.—Similar temperature-correlated daily growth patterns in otoliths of two cutthroat trout (A and B) just before the fish migrated from the same stream to Hungry Horse Reservoir during their fourth year of life. Photomicrographs are from a video monitor; bars = 10 μm.

FIGURE 3.—Temperature-induced marks in otoliths of brown trout from the Delaware River, New York. The four "light" increments were formed during a 4-d interruption of cold hypolimnial releases from a reservoir. Photomicrographs are from a video monitor; bars = 10 μm. (A) June 1983 growth in an age-1 fish collected August 1983. (B) June 1983 growth in an age-2 fish collected June 1984.

from the change in microstructural pattern that occurs when the fish enter the stream environment. Date of stocking can be determined by counting back from the otolith margin when a complete record exists, or by identifying a calendar date by correlation of daily water- or air-temperature records with patterns of day-to-day variation in increment width and optical density in the period after the "stocking mark." Although this level of accuracy may be extremely useful in salmonid fisheries management, it may not be achievable for many other kinds of fish in other environments. Figure 4 illustrates otolith features related to the release of hatchery fish into streams; similar examples could be shown for fish stocked into lakes. This example really combines the use of natural marks with artificially induced

(but not intentional) ones and a kind of human-induced "migration."

Spawning date and duration of the larval stage.— Microstructural analysis can be used to reconstruct life histories of fish larvae, such as the duration of the larval period. To the extent that there are geographic or seasonal differences in this character, fish can be classified on this basis. For young fish, it is usually possible to determine the total age in days, so one can determine a spawning date with reasonable certainty. This may allow partitioning of fish into temporal or geographic cohorts if there is sufficient information on the reproductive biology of the species and populations.

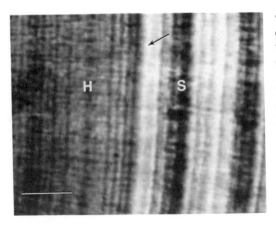

FIGURE 4.—Contrast between growth in hatchery (H) and stream (S) in a brown trout otolith from Owasco Inlet, New York. The arrow points to the increment formed on the stocking day. Note the opaque, narrow, and regular increments to the left. Photomicrograph is from a video monitor; bar = 10 μm.

Artificial or Induced Marks

Otolith marking of fish has been accomplished successfully by several techniques in a variety of fishes (for larvae and juveniles, some examples are Taubert and Coble 1977; Brothers 1981 and unpublished; Tanaka et al. 1981; Campana and Neilson 1982; Mugiya and Muramatsu 1982; Victor 1982; Hettler 1984; Schmitt 1984; Tsukamoto 1985; Wilson et al. 1987). These experiments were often performed during studies of otolith growth or as part of laboratory work to validate the occurrence of daily growth increments, and not to produce marked fish for release into the wild. Some otolith marking in adults has been used to mark fish for release and subsequent recapture (Lanzing and Hynd 1966; Blackler 1974; Wild and Foreman 1980; Brothers, unpublished). However, these fish also carried externally visible tags; the otolith work was still geared to validation of age. Otolith-marking experiments have been carried out as a kind of tagging procedure only recently (Mosegaard 1986; Volk et al. 1990, this volume; Bergstedt et al. 1990, this volume; Brothers, Great Lakes Fishery Commission, unpublished).

Induced Chemical Markers

Chemical markers involve alterations of "normal" otolith growth patterns with chemicals and drugs, or incorporation of chemicals into the otolith structure. Individual injections or mass-marking by feeding or immersion are possible for larvae or older stages. Special problems arise

when embryos are to be marked because passage of chemicals through the egg membranes is likely to be difficult for many or most species. Some examples of chemical markers follow.

Tetracycline and other fluorescent compounds.—Tetracycline is a well-known marker for calcified structures (including otoliths) in fish (Weber and Ridgway 1962, 1967; Choate 1964; Kobayashi et al. 1964; Odense and Logan 1974; Bilton 1986; Dabrowski and Tsukomoto 1986; Koenings et al. 1986). Tetracycline is one member of a family of antibiotic drugs (oxytetracycline, chlorotetracycline, and demethylchlortetracycline), which all become incorporated in growing calcified tissues and produce characteristic fluorescent emissions when illuminated with ultraviolet (UV) light. Other compounds such as fluorescein, calcein, and related materials (Rahn and Perrin 1970) also produce characteristic fluorescent marks in bone and otoliths (Wilson et al. 1987; Beckman et al. 1990, this volume). Multiple exposures or combinations of different markers (Suzuki and Mathews 1966; Kariyama et al. 1969) could be used to produce a series of marks unique to separate batches of fish. For larger fish, the drugs are usually injected. Addition of the compounds to food, or immersion of fish in various tetracycline solutions, has also been successful, especially for young marine fishes (Lanzing and Hynd 1966; Hettler 1984; Schmitt 1984; Gleason and Recksiek 1990, this volume). Immersion is the only practical procedure for batches of fish in life stages too young to feed or too small to inject; however, there is very little published work on immersion marking of young freshwater fish (Tsukamoto 1985). Viewing of a marked otolith, particularly if the specimen is from an adult fish, requires grinding or sectioning of the otolith and microscopic examination with incident UV illumination. Tetracycline marks are known to be somewhat labile in light (especially for scales); however, good marks are known to persist for at least several years in internal structures such as vertebrae and otoliths (Blackler 1974).

Acetazolamide.—Successful marking of the otoliths of young fishes has been obtained with the carbonic anhydrase inhibitor acetazolamide (Mugiya 1977; Mugiya and Muramatsu 1982; Brothers, unpublished; R. Radtke and W. Haake, University of South Carolina, personal communication). The mark produced has been described as a growth disruption or interruption caused by an abrupt cessation of aragonite crystallite deposition and delineated by a thin protein layer. The mark can be seen with the light microscope, but confirmation and optimal viewing requires the use of the scanning electron microscope (SEM). Normally occurring growth interruptions (e.g., at hatching in some salmonids) may make recognition of the mark difficult. To my knowledge, acetazolamide has been administered only by injection; however, it is water-soluble and immersion is a possible method of exposure.

Elemental marking.—In the same way that naturally occurring trace elements can produce marks in the otoliths of wild fish, intentional exposure to certain elements can produce distinctive chemical marks in the otoliths of captive fish. The analytical procedures are also the same, but the search is limited to specific elements with concentrations much higher than those of naturally deposited materials. Exposure may be by immersion or injection; however, toxicity is of definite concern for a number of compounds at higher concentrations. Elements of interest include lead (Jensen and Cumming 1967), strontium (Ophel and Judd 1968; Behrens Yamada et al. 1979; Behrens Yamada and Mulligan 1982, 1990, this volume; in squid statoliths: Hurley et al. 1985), and various others, singly or in combination (Miller 1963; Calaprice and Calaprice 1970, Muncy and D'Silva 1981; Muncy et al. 1990, this volume). Fluoride, copper, zinc, and various rare earth elements are also of particular interest. Incorporation of radioactive isotopes such as calcium-45 can be detected with autoradiography (Yoklavich and Boehlert 1987) or mass spectrometry. For nonradioactive elements, analysis of single or perhaps multiple exposures and resulting deposition layers could involve use of the electron microprobe and X-ray spectrometry in conjunction with the SEM. Marks should appear as discrete bands superimposed over the SEM image when the probe is used in the elemental "map" mode. If the marking element is of sufficient concentration and different average atomic mass, the bands can be visualized with a back-scattered electron detector (e.g., Hurley et al. 1985). Figure 5 illustrates a strontium-rich band deposited in the otoliths of postlarvae of several marine species as the result of a 1-d immersion in strontium-rich seawater. Similar results for lake trout are illustrated in a later example. An alternative way to detect marked otoliths would be to use instrumentation that analyzes whole, pulverized, or digested otoliths. In this case, multiple marks would have to involve unique mixtures of added elements.

Induced Structural Marks

As discussed previously, there is a wealth of experimental evidence that demonstrates the individual and combined influences of both endogenous and exogenous factors on the formation of growth increments in teleost otoliths over the course of a day, or even hours (see Campana and Neilson 1985; and other reviews cited in the section on natural marks). The exogenous variables that have been investigated most thoroughly are temperature, light, and food availability. Other factors such as pH, current, and physical disturbance also can affect otolith growth; however, the effects are less pronounced or less controllable. Manipulations of light and feeding are likely to produce distinct otolith marks, but observations on salmonids and other fishes demonstrate that manipulations of temperature produce the most reliable and striking results (Wilson and Larkin 1980; Campana 1983; Volk et al. 1984; Mosegaard 1986; Brothers, unpublished) (Figure 6). Manipulations of this type can affect animals en masse in either closed or open systems, with minimal handling or damage to even the smallest of fishes or early life stages. Temperature or thermal marking ("otolithography") is also easily accomplished before hatching for those fishes (e.g., salmonids) that have large embryos with well-developed otoliths. The types of marks caused by temperature manipulations are clearly distinguishable with light microscopy, and can be modified to create a large variety of distinctive batch marks. The analyses required are the least critical, quickest, and most economical, particularly in comparison to analyses of chemically based marks. The only preparation needed is grinding or sectioning of the otolith, in order to view the microstructure formed during the earlier marking stages (Bergstedt et al. 1990; Volk et al. 1990). The procedures could be semiautomated with the incorporation of image-analysis instrumentation and preprogrammed pattern-recognition software. Thermal marking appears to have the greatest flexibility and ease of application of all the otolith-marking procedures investigated to date.

Case Study: Assessment of Otolith-Marking Techniques for Early Life Stages of Lake Trout

During the fall of 1984 and the spring of 1985 I performed a series of experiments on the eggs, alevins, and fry of hatchery stocks of lake trout in order to determine the feasibility of producing

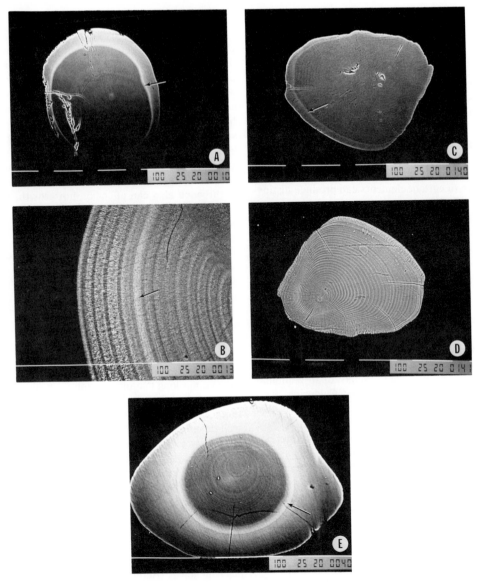

FIGURE 5.—Strontium marks in otoliths of postlarval tropical marine fishes. Arrows point to the marked daily increment. Residual strontium deposition may persist for several days. Scanning electron micrographs were taken in the back-scattered electron mode; bright areas indicate the presence of strontium enrichment compared to the typical calcium carbonate otolith material. White bars = 100 μm. (A) Unidentified goby: ground, polished, and unetched otolith. The fish was sacrificed 4.5 d after the marking day. (B) Same specimen as in A, but after the otolith was etched with acid. (C) Foureye butterflyfish *Chaetodon capistratus:* ground, polished, and unetched lapillus. (D) Same (but rotated) specimen as in C, but after etching. (E) Speckled worm eel *Myrophis punctatus:* saggitta from a late leptocephalus. The sagitta showed a remarkable rate of growth that corresponded to settlement and transformation of the leptocephalus that occurred in the 7-d period after capture and marking.

distinctive marks in their otoliths. The research was supported by the Great Lake Fishery Commission in anticipation of major stocking programs that would be designed to evaluate the success rate of planting early-life-stage lake trout on historic spawning reefs. These studies (see Bergstedt et al. 1990) would require that fish be marked as eggs (embryos), or sac fry (eleuthero-embryos), that marks be "encodable" in order to identify multiple groups of fish stocked in different

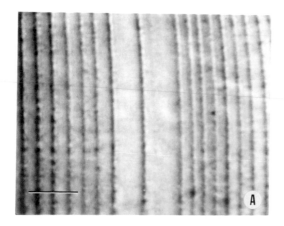

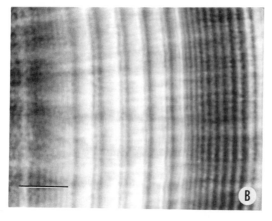

FIGURE 6.—Marks produced in the otoliths of captive rainbow trout *Oncorhynchus mykiss* (formerly *Salmo gairdneri*) by manipulations of the temperature regime in a stream-simulation tank. Photomicrographs are from a video monitor; bars = 10 μm. (A) Mark produced over a 4-d period by omission of two "thermal nights." A thermal night is the 12-h period of declining temperatures that makes up half of a typical diel cycle. The mean daily temperature was higher for the days represented left of the mark than for those on the right. (B) Marks produced over a period of about 2 weeks of radically altered temperature cycles. In this case long "thermal days" (52 h of heating) were alternated with 12-h heating times. Each heating time was bounded by 12 h of cooling.

years and locations, that the marking technique be easily and quickly applied to large numbers of fish (i.e., hundreds of thousands to millions), and that the marks be persistent and distinguishable over time periods comparable to the life span of lake trout (in excess of 20 years). Some method of otolith marking seemed to be the approach that would be most likely to meet all of these requirements. A complete report on the materials, meth-

ods, and results is available from the Great Lakes Fishery Commission (1451 Green Road, Ann Arbor, Michigan 48105); for this paper I summarize the study, illustrate some of the marks produced, and discuss the results in terms of recommendations for otolith-marking programs.

Lake trout eggs were obtained at the eyed stage and partitioned into 85 lots (approximately 100 eggs per lot). Several hundred eggs of Atlantic salmon *Salmo salar* were given similar treatment. The lots were separately maintained, monitored, and subsampled during the course of the 5-month study period. Some were held in standard hatching jars for the entire time, and served both as controls and sources for new experiments. Most were subjected to single or multiple exposures to temperature, chemical markers, or both. Except during experimental treatments, the fish were maintained in a constant temperature (9 ± 0.1°C), and controlled light environment (14 h light:10 h darkness). These conditions were similar to those in a typical hatchery environment and were exactly the same as those experienced by the fish maintained at the Tunison Laboratory of Fish Nutrition (U.S. Fish and Wildlife Service), where the research was carried out. Mortality in all groups was monitored at 2-d intervals or less. Control fish received sham treatments and were maintained for comparison with experimental fish. The progress of marking experiments was continuously monitored during the course of the work by periodic (weekly or more frequent) subsampling of experimental lots and also by examination of otoliths several days after a marking procedure was performed.

Marks were produced by the following treatments. (1) For temperature treatment, fish were held in a closed system for periods of 1 d to several weeks. Temperature was manipulated on precise schedules by combinations of heating and refrigerating units controlled by timers and thermostats. Temperature conditions were continuously recorded. A typical diel temperature cycle consisted of 12 h of warming and 12 h of cooling, with a mean of 10°C and range of 12°C. Other temperature regimes involved patterns of longer and shorter times of warming, cooling, and stasis. (2) Chemical treatments (Table 1) involved exposures by injection or immersion of eggs, alevins, and fry; sometimes they were combined with temperature manipulations. Feeding experiments were carried out only on fry in the open system.

Otoliths were prepared and examined with light microscopy, incident fluorescent microscopy,

TABLE 1.—Summary of treatments for induction of chemical marks in lake trout otoliths.

Compound or solution	Method of exposure	Range of concentrations	Other substances in combination[a]	Purpose
Oxytetracycline hydrochloride (both liquid and powdered, terramycin)	Injection Immersion Diet	50–100 mg/kg 250–1,000 mg/L 0.2–1% of feed	NaCl solution DMSO, sea salt Food	Otolith marker
Dimethyl sulfoxide (DMSO)	Immersion	0–10%	Tetracycline, TbCl₂	Acceleration of uptake of markers
Acetazolamide sodium	Injection	100 mg/kg	NaCl solution	Otolith marker
Sodium chloride (in distilled water)	Injection	1.3%	All injections	Saline medium for injections
Strontium chloride	Injection Immersion	100 mg/kg 150–300 mg/L	NaCl solution DMSO	Otolith marker
Terbium chloride hydrate	Injection Immersion	100 mg/kg 150 mg/L	NaCl solution DMSO	Otolith marker
Europium chloride	Immersion	150 mg/L		Otolith marker
Sea salt (artificial)	Immersion	0.6%	Tetracycline	Acceleration of uptake of markers

[a] DMSO = dimethyl sulfoxide.

SEM, and electron microprobe. Methods generally followed procedures described and referenced in Brothers (1985).

Results and Discussion

Mortality of both control and experimental fish was similar to that experienced by fish of the same brood and maintained by other researchers at the Tunison Laboratory under the standard "hatchery" conditions. The only exceptions were due to closed-system treatments with extended high temperatures (18°C for 2 d) and the resultant elevated nitrite levels, or to exposure to dimethyl sulfoxide (DMSO) concentrations exceeding 5%. The only other noticeable effect of the experimental manipulations was the retardation or acceleration of growth and development, as a result of extended exposure to mean temperatures lower or higher than those to which control fish were exposed.

The otoliths (sagittae and lapilli) appeared in the otic vesicles of embryos just before the eyes became pigmented and visible without the aid of a microscope (the "eyed" stage). This stage corresponds to the end of Balon's (1980) E^36 stage, and is attained about 20–30 d after fertilization if lake trout are reared at about 9°C. The first calcified otolith elements are a series of separate, optically dense primordia and their surrounding translucent "cores" (each 5–20 μm in diameter, Figure 7A). Calcification is present if the structures show birefringence when viewed with two (crossed) polarizing lenses and transmitted illumination. The separate elements quickly consolidate into singular sagittae and lapilli once incremental growth begins (Figure 7B, C). It was at this point that the earliest marking experiments were started. In control fish, growth increments formed prior to hatching were expressed comparatively weakly and were somewhat irregular or uneven in width, contrast, or both. At or close to the time of hatching, many fish developed one to three prominent growth increments, characterized by optically dense and discrete protein-rich subunits (Figure 8).

As defined in this study, a microstructural growth increment is a bipartite structure composed of two subunits or zones. One subunit appears more optically dense or opaque in transmitted light than the other (Figure 9). With preparation for the SEM by an etching technique, the opaque microstructural zones are more deeply etched and appear as "valleys"; the intervening "ridges" correspond to the translucent microstructural zones. It is generally agreed that the opaque zones are relatively rich in proteinaceous matrix, whereas calcium carbonate (aragonite) dominates the chemical composition of the translucent zones. These zones may be characterized as matrix-dominant versus calcium-dominant. Alex Wild (Inter-American Tropical Tuna Commission, personal communication) has suggested using the term M-zone and C-zone, respectively,

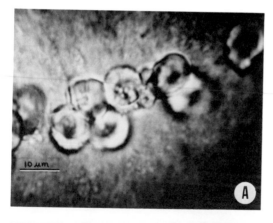

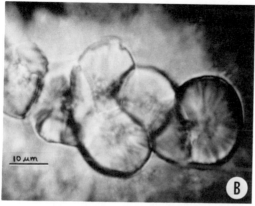

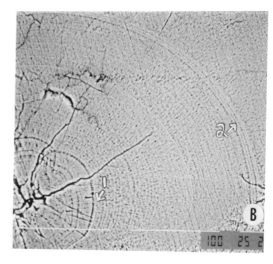

FIGURE 8.—Hatching marks and growth increments in otoliths of lake trout subjected to a constant-temperature treatment. (A) Otolith from a control alevin; the arrow points to the hatching mark. Photomicrograph is from a video monitor; bar = 20 μm. (B) Otolith from an experimental fish that was in a constant-temperature environment except for 2 d of diel cycles combined with exposure to terbium. The arrows point to (1) hatching and (2) the day of a terbium–diel temperature treatment. The scanning electron micrograph was taken in the back-scattered electron mode; white bars = 100 μm.

FIGURE 7.—Early ontogeny of sagittae in lake trout embryos prior to hatching. Photomicrographs are from a video monitor; bars = 10 μm. (A) In situ view of multiple primordia in the noncoalesced core in the otic capsule of an embryo at the "eyed stage." (B) Coalesced core with connected primordia in a later embryo. (C) Young saggita after a 3-d exposure to a diel temperature cycle.

and I will adopt this nomenclature here. It should be noted that these terms are synonymous in usage (but not structural implication) to "discontinuous zone" and "incremental zone" as defined and first introduced by Mugiya et al. (1981). Finally, observations made during this research confirmed earlier reports (Brothers 1981; Campana and Neilson 1985) that the M-zone is formed at night or during times of depressed (declining) water temperatures, and the C-zone is deposited

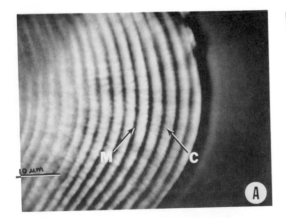

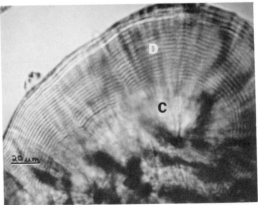

FIGURE 10.—Comparison of daily otolith growth increments formed under a diel temperature cycle (D) and under constant temperature (C) in lake trout alevin; compare this figure with Figure 8A. Photomicrograph is from a video monitor; bar = 20 μm.

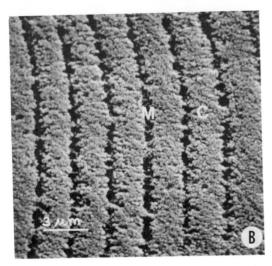

FIGURE 9.—Daily growth increments formed in lake trout otolith under a diel temperature cycle; C = calcium-dominant zone; M = matrix-dominant zone. (A) Photomicrograph taken from a video monitor; bar = 10 μm. (B) Scanning electron micrograph taken in the secondary electron mode; bar = 3 μm.

during daylight hours or when water temperatures are elevated (rising). Strong temperature effects generally mask the effect of photoperiod.

Temperature marking.—Otolith microstructure (increment width and relative protein content) varied in predictable ways with respect to temperature variations for all fish life stages. For fish held at constant temperatures, otolith increments were more difficult to distinguish and there was less contrast between the C- and M-zones, compared to fish exposed to diel fluctuations (Figure 10). A second class of temperature effects resulted from exposure to temperature cycles with a longer period (more than 4 d) of gradually rising then

falling temperatures. Daily growth increments were either very faint or absent under these conditions and areas of deposition of more translucent and opaque materials corresponded to rising and falling temperatures, respectively. Such treatments generally produced characteristic wide zones, or "marks," but were not nearly as effective in creating obvious and unmistakable marks as the shorter-term temperature cycles (6 h to 2 d). The third class of temperature effects can be seen as alterations in the width of growth increments.

Daily growth increments formed under diel temperature cycles exhibited very high contrast between the C- and M-zones, and there was an exact correspondence between the number of M-zones and the number of daily temperature drops ("thermal nights") experienced by a fish. This condition obtained whether the fish were on a diel cycle synchronized with the light cycle, or on a diel cycle that was out of phase with the light cycle by 6, 12, or 18 h. Similarly, if fish experienced thermal nights every 36, 48, 60, 72, or 96 h, they produced prominent M-zones only at those times, even if a diel light cycle was still in effect. These results appear to be applicable to other salmonids, on the basis of my results with Atlantic salmon, rainbow trout, brook trout *Salvelinus fontinalis,* and brown trout. Mosegaard (1986) confirmed and extended these findings to Arctic char *Salvelinus alpinus,* and Volk et al. (1990) and others report basically the same responses in Pacific salmon. Otolith microstructure of nonsalmonid fishes is also strongly influenced by

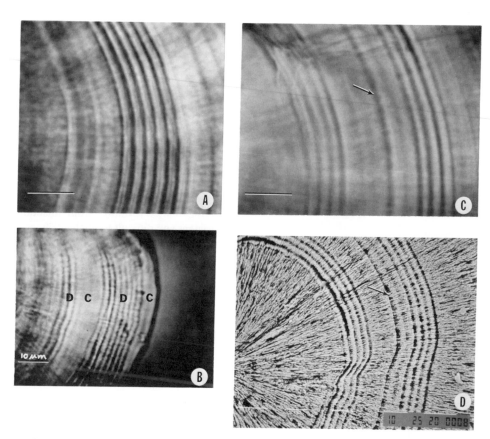

FIGURE 11.—Temperature marks produced in otoliths of lake trout alevins by short-term exposure to diel temperature cycles alternating with constant temperature. (A) Marks from a 6-d exposure to a diel temperature cycle, bounded by deposition at constant temperature. Photomicrograph is from a video monitor; bar = 10 μm. (B) Deposition during alternating periods of diel (D) and constant (C) temperatures. Photomicrograph is from a video monitor; bar = 10 μm. (C) Growth patterns from two 4-d exposures to diel temperature cycles. The arrow points to a hatching mark formed during the intervening constant-temperature period. Photomicrograph is from a video monitor; bar = 10 μm. (D) Same experiment as in C. Scanning electron micrograph was taken in the back-scattered electron mode; white bars = 10 μm.

temperature (Brothers, unpublished); however, there has been nothing published on mass-marking as a direct application.

The foregoing observations suggest the following several methods to distinctively mark the otoliths of young salmonids and other fishes.

(1) The thermal regime can be shifted from constant thermal environment (e.g., that typical of many spring-fed hatcheries) to a diel temperature cycle for a given number of days, and then back to constant conditions. The result is a known number of high-contrast daily growth increments (Figure 11A).

(2) The thermal regime can be shifted back and forth between the constant thermal environment and the cycling diel regime. Each shift should last for at least 3 d. The result is groupings of high-

contrast increments (the number in each group depends on the length of exposure to the cycled regime) alternating with areas of very low-contrast daily increments (again, the width of these areas depends upon the length of exposure to the constant temperature). Such alternate treatments can be varied to produce a wide selection of distinctively marked otoliths (Figure 11B–D). Of course, these shifts in thermal regime could be accomplished by changing or blending water sources, use of heating elements, etc.; the fish themselves do not have to be physically moved.

(3) The most complicated and variable marks can be produced by controlling the relative lengths of the heating and cooling portions of the temperature cycle. These heating times of variable length may be interspersed with standard

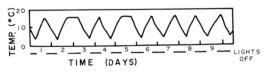

FIGURE 12.—Temperature marks produced in a lake trout otolith by a temperature cycle of varying period and a light cycle of constant period. (A) Temperature and light cycles. (B) Otolith response to the temperature and light cycles in A; wide C-zones were produced by the two extended warm periods. Photomicrograph is from a video monitor; bar = 10 μm.

12-h cooling times. A simple experiment was performed in which the diel cycle was modified to produce the temperature pattern illustrated in Figure 12A. The resultant otolith response (Figure 12B) produced doubly wide C-zones during the two 24-h heating periods. This type of response, in which the width of the C-zone varies with the duration of the warming period was developed to produce a digitally coded marking system. "Dots," "dashes," and "spaces" were produced as follows: a heating and cooling cycle of 6 h:12 h yielded a narrow C-zone, or "dot;" a 12:12-h cycle produced an intermediate-width C-zone or "space;" a 24:12-h cycle produced a wide C-zone or "dash." Letter combinations in the International Morse Code (such as "LT" for lake trout and "GLFC" for Great Lakes Fishery Commission: Figure 13) or any other binary code could be produced with these elements.

Temperature marking proved to be a simple yet highly reliable and effective method to produce unique marks in the otoliths of lake trout at all stages of development from embryos to fry. Viewing the marks required only midsagittal sectioning and examination with the light microscope. Responses in the lapilli were similar to those seen in the sagittae, so there are four copies of the mark in each fish, two in the sagittae and two in the lapilli. Temperature-related alteration of the development rate of the fish may be largely avoided by controlling the diel range so that the mean temperature approximates any desired value.

Chemical marking.—Chemical marking of the otoliths was accomplished by the following variety of procedures.

(1) Acetazolamide injections produced otolith marks in alevins and fry, but seemed to cause elevated mortality at hatching when administered to embryos. The mark in alevins and fry was characterized by two or three markedly narrower and lower-contrast daily increments, followed by about a week of gradual recovery (Figure 14). Immersion experiments were not attempted because the marks resulting from injection were inconsistent and difficult to distinguish, especially in fish held at constant temperatures. The mark was also not unique, in that very similar microstructural patterns could be produced by temperature manipulations or were observed in the otoliths of wild fish (i.e., similar marks could be produced by natural environmental perturbations).

(2) Tetracycline injections consistently produced fluorescent marks in the otoliths and bony tissues. The otolith marks corresponded to 1–3 d of growth and were relatively faint and sometimes difficult to distinguish. Bones, especially fin rays and head bones, were marked more strongly. Immersion experiments with relatively high concentrations (500–1,000 mg/L for 2–3 d, with or without DMSO) did produce marked otoliths in alevins and fry; however marks were invariably fainter than those observed in injected fish. Embryos did not pick up the tetracycline, although the egg membrane fluoresced. Fish exposed to tetracycline added to the diet did not show distinct fluorescence in the otoliths or bones.

In general, tetracycline exposure appears to be an inferior otolith-marking technique compared to temperature manipulation. Although preparation of otoliths for viewing is similar to that needed for thermal marks, incident-light fluorescence microscopy is also required. The marks themselves can be faint and difficult to distinguish, and the number of unique marking patterns is much more limited. Immersion, the only practical marking procedure, is limited to alevins and fry and must be carried out in a closed system for at least 2 d. Treatment-related mortality may be significant under some conditions. Finally, the long-term permanence of tetracycline marks is still somewhat uncertain, although the best evidence suggests that marks in internal structures in large fish remain detectable for at least 5 years.

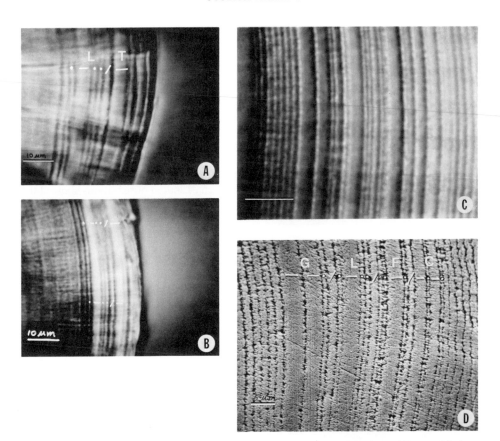

FIGURE 13.—Temperature marks produced in lake trout otoliths by varying the width of C-zone. Marks are coded in Morse Code: narrow C-zones = dots, wide C-zones = dashes, and intermediate-width C-zones = spaces. (A) "LT" in a sagitta. Photomicrograph is from a video monitor; bar = 10 μm. (B) "LT" in a lapillus. Photomicrograph is from a video monitor; bar = 10 μm. (C) "GLFC" in a sagitta. Photomicrograph is from a video monitor; bar = 10 μm. (D) Same mark as in C. Scanning electron micrograph was taken in the secondary electron mode; bar = 5 μm.

(3) Both immersion and injection experiments with chloride salts of terbium and europium produced generally negative responses, because of precipitate formation and detection difficulties. In contrast, immersion and injection treatments with strontium chloride produced strontium-rich bands in the otoliths. The bands were detected and identified in SEM examination with both a back-scattered electron detector and an energy dispersive X-ray spectrometer. Figure 15 illustrates strontium marks and the time course of exposure, uptake, and incorporation. The first band was due to a midday-to-midday 24-h immersion in a solution of 300 mg/L. The bright areas, which indicate high strontium content, correspond to portions of C-zones; the dark line, indicating lower strontium content and deeper etching, is an M-zone. The difference in the widths of the two C-zone por-

FIGURE 14.—Otolith mark (arrow) produced by exposure of a lake trout fingerling to acetazolamide (100 mg/kg) during a diel temperature cycle. Photomicrograph is from a video monitor; bar = 10 μm.

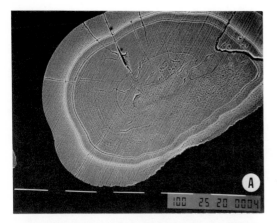

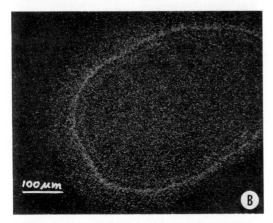

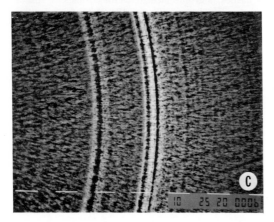

tions may be ascribed to a slight lag in uptake and then a residual deposition of strontium after the fish were transferred out of the strontium-enriched water. A similar response is indicated in the 52-h strontium exposure represented in the second band (midday to midafternoon).

A chemical marking system could employ variable-duration exposures and intervening times to produce a unique mark pattern in otoliths. The concentrations of strontium used here did not cause any mortality or other detectable detrimental effects. Limited observations showed that embryos did not take up much of the chemical; however, more work is needed to confirm this finding. The technique could be used to mass-mark alevins and fry, but the fish must be maintained in a closed system during treatment. The technique has also been applied successfully to a variety of larval and juvenile marine fishes (Brothers, G. D. Johnson and P. Keener, unpublished). The marks produced should be permanent, but preparation and analytical procedures are time-consuming and require highly specialized and expensive instrumentation.

Conclusions

Natural otolith marks, both chemical and structural, have great potential for identification of stock, as well as in detailed studied of life histories and tracking of fish movements. Although results of the studies presented here demonstrate the feasibility of several possible marking procedures to be used for the early life stages of lake trout and other fishes, thermal marking seems to be the most promising for use in large-scale stocking programs.

Temperature manipulations can be used to produce a very large variety of unique and easily visible marks in the otoliths of salmonid embryos, alevins, and fingerlings (see Mosegaard 1986 for some calculations of the number of marks possible). I expect that thermal marking is appropriate for all salmonids, and probably for many other fishes, on the basis of observations of thermal lability of otolith microstructure (Brothers, unpublished). Species-specific procedures vary due to thermal tolerances and response characteristics

FIGURE 15.—Otolith marks produced by two immersion exposures—one of 24 h, the other of 52 h—of a lake trout to strontium chloride during diel temperature cycles. Scanning electron micrographs were taken in the back-scattered electron mode. (A) The two bright bands indicate areas enriched with an element (strontium) of relatively higher atomic mass. Bars = 100 μm. (B) Verification of the presence of strontium: the scanning electron micrograph of an elemental map was produced by tuning the energy-dispersive X-ray spectrometer for strontium emmisions. Bar = 100 μm. (C) Detail of the otolith; strontium enrichment is most apparent in the C-zones. The time course of the experiment can be determined. Bars = 10 μm.

of different species. The technique is easily applied to mass-marking in either open or closed rearing systems. Procedures for preparation and viewing of the mark are the easiest and simplest of all the marking techniques examined.

Tetracycline marking of alevins and fry is possible, but it requires maintenance of fish in a closed system. Increased mortality may result from treatment conditions. The marks are weakly expressed, are of limited variety, and require more specialized viewing conditions. Long-term persistence of the marks is uncertain. Immersion marking of marine species produces much better results.

Acetazolamide marking by injection produces a mark qualitatively similar to those induced by temperature manipulations, but does not offer the precise control of the latter method.

Immersion treatment of salmonids, and probably all fishes, with strontium (as strontium chloride) will produce a discrete mark after 1–2 d of exposure in a closed system. The mark is a band or bands of strontium-enriched material detectable with an appropriately equipped SEM. Preparation and viewing are relatively difficult and expensive.

Acknowledgments

The lake trout research reported in this paper was supported by the Great Lakes Fishery Commission. The Tunison Laboratory of Fish Nutrition (U.S. Fish and Wildlife Service) generously provided space, facilities, and personnel that were essential for the successful completion of this research. Charles Hall gave up valuable fishing time to review the manuscript.

References

Bagenal, T. B., F. J. H. Mackerth, and J. Heron. 1976. The distinction between brown trout and sea trout by the strontium content of their scales. Journal of Fish Biology 5:555–557.

Balon, E. K. 1980. Early ontogeny of the lake charr, *Salvelinus (Cristivomer) namaycush*. Pages 485–562 *in* E. K. Balon, editor. Charrs: salmonid fishes of the genus *Salvelinus*. Dr. W. Junk, The Hague.

Beckman, D. W., C. A. Wilson, F. Lorica, and J. M. Dean. 1990. Variability in incorporation of calcein as a fluorescent marker in fish otoliths. American Fisheries Society Symposium 7:547–549.

Behrens Yamada, S., and T. J. Mulligan. 1982. Strontium marking of hatchery reared coho salmon, *Oncorhynchus kisutch* Walbaum, identification of adults. Journal of Fish Biology 20:5–10.

Behrens Yamada, S., and T. J. Mulligan. 1990. Screen-

ing of elements for the chemical marking of hatchery salmon. American Fisheries Society Symposium 7:550–561.

Behrens Yamada, S., T. J. Mulligan, and S. J. Fairchild. 1979. Strontium marking of hatchery-reared coho salmon (*Oncorhynchus kisutch*, Walbaum). Journal of Fish Biology 14:267–275.

Belanger, S. E., D. S. Cherry, J. J. Ney, and D. K. Whitehurst. 1987. Differentiation of freshwater versus saltwater striped bass by elemental scale analysis. Transactions of the American Fisheries Society 116:594–600.

Bergstedt, R. A., R. L. Eshenroder, C. Bowen II, J. G. Seelye, and J. C. Locke. 1990. Mass-marking of otoliths of lake trout sac fry by temperature manipulation. American Fisheries Society Symposium 7:216–223.

Bilton, H. T. 1986. Marking chum salmon fry with oxytetracycline. North American Journal of Fisheries Management 6:126–128.

Blackler, R. W. 1974. Recent advances in otolith studies. Pages 67–90 *in* F. R. Harden Jones, editor. Sea fisheries research. Wiley, New York.

Brothers, E. B. 1981. What can otolith microstructure tell us about daily and subdaily events in the early life history of fish? Rapports et Procès-Verbaux des Réunions Counseil International pour l'Exploration de la Mer 128:393–394.

Brothers, E. B. 1984. Otolith studies. American Society of Ichthyologists and Herpetologists Special Publication 1:50–57.

Brothers, E. B. 1985. Methodological approaches to the examination of otoliths in aging studies. Pages 319–330 *in* R. C. Summerfelt and G. E. Hall, editors. Age and growth of fish. Iowa State University Press, Ames.

Calaprice, J. R. 1971. X-ray spectrometric and multivariate analysis of sockeye salmon (*Oncorhynchus nerka*) from different geographic regions. Journal of the Fisheries Research Board of Canada 28:369–377.

Calaprice, J. R., and F. P. Calaprice. 1970. Marking animals with microtags of chemical elements for identification by X-ray spectroscopy. Journal of the Fisheries Research Board of Canada 27:317–330.

Calaprice, J. R., H. M. McShefferey, and L. A. Lapi, 1971. Radioisotope X-ray fluorescence spectrometry in aquatic biology: a review. Journal of the Fisheries Research Board of Canada 28:1583–1594.

Campana, S. E. 1983. Calcium deposition and check formation during periods of stress in coho salmon, *Oncorhynchus kisutch*. Comparative Biochemistry and Physiology A, Comparative Physiology 75:215–220.

Campana, S. E., and J. D. Neilson. 1982. Daily growth increments in otoliths of starry flounder (*Platichthys stellatus*) and the influence of some environmental variables in their production. Canadian Journal of Fisheries and Aquatic Sciences 39:937–942.

Campana, S. E., and J. D. Neilson. 1985. Microstructure of fish otoliths. Canadian Journal of Fisheries and Aquatic Sciences 42:1014–1032.

Casselman, J. M., J. J. Collins, E. J. Crossman, P. E. Ihssen, and G. R. Spangler. 1981. Lake whitefish (*Coregonus clupeaformis*) stocks of the waters of Lake Huron. Canadian Journal of Fisheries and Aquatic Sciences 38:1772–1789.

Choate, J. 1964. Use of tetracycline drugs to mark advanced fry and fingerling brook trout (*Salvelinus fontinalis*). Transactions of the American Fisheries Society 93:309–311.

Dabrowski, K., and K. Tsukamoto. 1986. Tetracycline tagging in coregonid embryos and larvae. Journal of Fish Biology 29:691–698.

Dannevig, E. H. 1956. Chemical composition of the zones in cod otoliths. Journal du Conseil, Conseil International pour l'Exploration de la Mer 21:156–159.

Degans, E. T., W. G. Deuser, and R. L. Haedrich. 1969. Molecular structure and composition of fish otoliths. Marine Biology 2:105–113.

Elrod, J. H., and C. P. Schneider. 1986. Evaluation of coded wire tags for marking lake trout. North American Journal of Fisheries Management 6:272–276.

Garland, D. E. 1987. Recopilación de antecedentes sobre estudios de edad y crecimiento en peces basados en la microstructura de sus otolitos. Pages 151–166 in P. Arana, editor. Manejo y desarrollo pesquero. Universidad Católica de Valparaíso, Escuela de Ciencias del Mar, Valparaíso, Chile.

Gauldie, R. W., D. A. Fournier, D. E. Dunlop, and G. Coote. 1986. Atomic emission and proton microprobe studies of the ion content of otoliths of chinook salmon aimed at recovering the temperature life history of individuals. Comparative Biochemistry and Physiology A, Comparative Physiology 84:607–616.

Gauldie, R. W., E. J. Graynoth, and J. Illingworth. 1980. The relationship of the iron content of some fish otoliths to temperature. Comparative Biochemistry and Physiology A, Comparative Physiology 66:19–24.

Gauldie, R. W., and A. Nathan. 1977. Iron content of the otolith of tarakihi (Teleostei: Cheilodactylidae). New Zealand Journal of Marine and Freshwater Research 11:179–191.

Gjosaeter, J., P. Dayaratne, O. A. Bergstad, H. Gjosaeter, M. I. Sousa, and I. M. Beck. 1984. Aging tropical fish by growth rings in the otoliths. FAO (Food and Agriculture Organization of the United Nations) Fisheries Circular 776.

Gleason, T.R., and C. Recksiek. 1990. Preliminary field verification of daily growth increments in the lapillar otoliths of juvenile cunners. American Fisheries Society Symposium 7:562–565.

Gutierrez, E., and B. Morales-Nin. 1986. Time series analysis of daily growth in *Dicentrarchus labrax* L. otoliths. Journal of Experimental Marine Biology and Ecology 103:163–179.

Hettler, W. F. 1984. Marking otoliths by immersion of marine fish larvae in tetracycline. Transactions of the American Fisheries Society 113:370–373.

Hurley, G. V., P. H. Odense, R. K. O'Dor, and E. G. Dawe. 1985. Strontium labeling for verifying daily growth increments in the statolith of the short-finned squid (*Illex illecebrosus*). Canadian Journal of Fisheries and Aquatic Sciences 42:380–383.

Ihssen, P. E., H. E. Booke, J. M. Casselman, J. M. McGlade, N. R. Payne, and F. M. Utter. 1981. Stock identification: materials and methods. Canadian Journal of Fisheries and Aquatic Sciences 38:1838–1855.

Jensen, A. C., and K. B. Cumming. 1967. Use of lead compounds and tetracycline to mark scales and otoliths of marine fishes. Progressive Fish Culturist 29:166–167.

Kariyama, M., M. Akai, and S. Nishijima. 1969. Three-color fluorescent labelling method for calcified tissues in a reptile, *Caiman crocodilus*. Archives of Oral Biology 14:1349–1350.

Klar, G. T., and N. C. Parker. 1986. Marking fingerling striped bass and blue tilapia with coded wire tags and microtaggants. North American Journal of Fisheries Management 6:439–444.

Kobayashi, S., R. Yuki, T. Furui, and T. Kosugiyama. 1964. Calcification in fish and shellfish—I. Tetracycline labelling patterns on scale, centrum and otolith in young goldfish. Bulletin of the Japanese Society for Scientific Fisheries 30:6–13.

Koenings, J. P., J. Lipton, and P. McKay. 1986. Quantitative determination of oxytetracycline uptake and release by juvenile sockeye salmon. Transactions of the American Fisheries Society 115:621–629.

Lanzing, W. J. R., and J. S. Hynd. 1966. Tetracycline distribution in body tissues of marine fishes. Australian Journal of Science 29:117–178.

Lapi, L. A., and T. J. Mulligan. 1981. Stock identification using a microanalytic technique to measure elements present in the freshwater growth region of scales. Canadian Journal of Fisheries and Aquatic Sciences 38:744–751.

McKern, J. L., H. F. Horton, and K. V. Koski. 1974. Development of steelhead trout (*Salmo gairdneri*) otoliths and their use for age analysis and for separating summer from winter races and wild from hatchery stock. Journal of the Fisheries Research Board of Canada 31:1420–1426.

Messieh, S. N. 1972. Use of otoliths in identifying herring stocks in the southern Gulf of Saint Lawrence and adjacent waters. Journal of the Fisheries Research Board of Canada 29:1113–1118.

Miller, W. P. 1963. Neutron activation analysis of stable dysprosium biologically deposited in the bone of chinook salmon fingerlings. Master's thesis. University of Washington, Seattle.

Moreau, G., and C. Barbeau. 1979. Différenciation de populations anadromes et dulcicoles de Grands Corégones (*Coregonus clupeaformis*) par la composition minérale de leurs écailles. Journal of the

Fisheries Research Board of Canada 36: 1439–1444.

Moreau, G., C. Barbeau, J. J. Frenette, J. Saint-Onge, and M. Simoneau. 1983. Zinc, manganese, and strontium in opercula and scales of brook trout (*Salvelinus fontinalis*) as indicators of lake acidification. Canadian Journal of Fisheries and Aquatic Sciences 40:1685–1691.

Mosegaard, H. 1986. Growth in salmonid otoliths. Doctoral dissertation. University of Uppsala, Uppsala, Sweden.

Mugiya, Y. 1977. Effect of acetazolamide on the otolith growth of goldfish. Bulletin of the Japanese Society for Scientific Fisheries 43:1053–1058.

Mugiya, Y., and J. Muramatsu. 1982. Time-marking methods for scanning electron microscopy in goldfish otoliths. Bulletin of the Japanese Society for Scientific Fisheries 48:1225–1232.

Mugiya, Y., N. Watabe, J. Yamada, J. M. Dean, D. G. Dunkleberger, and M. Shimizu. 1981. Diurnal rhythm in otolith formation in the goldfish, *Carassius auratus*. Comparative Biochemistry and Physiology A, Comparative Physiology 68:659–662.

Mulligan, T. J., L. Lap, R. Kieser, S. Behrens Yamada, and D. L. Duewer. 1983. Salmon stock identification based on elemental composition of vertebrae. Canadian Journal of Fisheries and Aquatic Sciences 40:215–229.

Mulligan, T. J., F. D. Martin, R. A. Smucker, and D. A. Wright. 1987. A method of stock identification based on the elemental composition of striped bass *Morone saxatilis* (Walbaum) otoliths. Journal of Experimental Marine Biology and Ecology 114:241–248.

Muncy, R. J., and A. P. D'Silva. 1981. Marking walleye eggs and fry. Transactions of the American Fisheries Society 110:300–305.

Muncy, R. J., N. C. Parker, and H. A. Poston. 1990. Inorganic chemical marks induced in fish. American Fisheries Society Symposium 7:541–546.

Neilson, J. D., and G. H. Geen. 1982. Otoliths of chinook salmon (*Oncorhynchus kisutch*): daily growth increments and factors influencing their production. Canadian Journal of Fisheries and Aquatic Sciences 39:1340–1347.

Neilson, J. D., and G. H. Green. 1985. Effects of feeding regimes and diel temperature cycles on otolith increment formation in juvenile chinook salmon, *Oncorhynchus tshawytscha*. U.S. National Marine Fisheries Service Fishery Bulletin 83:91–101.

Neilson, J. D., G. H. Geen, and D. Bottom. 1985a. Estuarine growth of juvenile salmon (*Oncorhynchus tshawytscha*) as inferred from otolith microstructure. Canadian Journal of Fisheries and Aquatic Sciences 42:899–908.

Neilson, J. D., G. H. Geen, and B. Chan. 1985b. Variability in dimensions of salmonid otolith nuclei: implications for stock identification and microstructure interpretation. U.S. National Marine Fisheries Service Fishery Bulletin 83:81–89.

Odense, P. H., and V. H. Logan. 1974. Marking Atlantic salmon (*Salmo salar*) with oxytetracycline. Journal of the Fisheries Research Board of Canada 31:348–350.

Ophel I. L., and J. M. Judd. 1968. Making fish with stable strontium. Journal of the Fisheries Research Board of Canada 25:1333–1337.

Papadopoulou, C., G. D. Kanias, and E. Moraitopoulou-Kassimati. 1978. Zinc content in otoliths of mackerel from the Aegean. Marine Pollution Bulletin 9:106–108.

Papadopoulou, C., G. D. Kanias, and E. Moraitopoulou-Kassimati. 1980. Trace element content in fish otoliths in relation to age and size. Marine Pollution Bulletin 11:68–72.

Radtke, R. L., and T. E. Targett. 1984. Rhythmic structural and chemical patterns in otoliths of the Antarctic fish *Notothenia larseni:* their implication to age determination. Polar Biology 3:203–210.

Rahn, B. A., and S. M. Perrin. 1970. Calcein blue as a fluorescent label in bone. Experimentia 26:519–520.

Rosa, H. C., and P. Re. 1985. Influence of exogenous factors on the formation of daily microgrowth increments in otoliths of *Tilapia mariae* (Boulanger, 1899) juveniles. Cybium 9:341–357.

Schmitt, P. D. 1984. Marking growth increments in otoliths of larval and juvenile fish by immersion in tetracycline to examine the rate of increment formation. U.S. National Marine Fisheries Service Fishery Bulletin 82:237–242.

Schneider, R. L., and S. V. Smith. 1982. Skeletal Sr content and density in *Porites* spp. in relation to environmental factors. Marine Biology 66:121–131.

Suzuki, H. K., and A. Mathews. 1966. Two-color fluorescent labelling of mineralizing tissues with tetracycline and 2, 4-bis [N, N'-di (carbomethyl) aminomethyl] fluorescein. Stain Technology 41:57–60.

Tanaka, K., Y. Mugiya, and J. Yamada. 1981. Effects of photoperiod and feeding on daily growth patterns in otoliths of juvenile *Tilapia nilotica*. U.S. National Marine Fisheries Service Fishery Bulletin 79:459–466.

Taubert, B. D., and D. W. Coble. 1977. Daily rings in otoliths of three species of *Lepomis* and *Tilapia mossambica*. Journal of the Fisheries Research Board of Canada 34:332–340.

Tsukamoto, K. 1985. Mass marking of ayu eggs and larvae by tetracycline-tagging of otoliths. Bulletin of the Japanese Society for Scientific Fisheries 51:903–911.

Van Coillie, R., A. Rousseau, and G. Van Coillie. 1974. Composition minérale des écailles de *Catostomus commersoni* issu de deux mileaux différents: étude par microscopie électronique analytique. Journal of the Fisheries Research Board of Canada 31:63–66.

Victor, B. 1982. Daily otolith increments and recruitment in two coral-reef wrasses, *Thalassoma bifasciatum* and *Halichoeres bivittatus*. Marine Biology 71:203–208.

Volk, E. C., S. L. Schroder, and K. L. Fresh. 1990. Inducement of unique otolith banding patterns as a practical means to mass-mark juvenile Pacific salmon. American Fisheries Society Symposium 7:203–215.

Volk, E. C., R. C. Wissmar, C. A. Simensted, and D. M. Eggers. 1984. Relationship between otolith microstructure and the growth of juvenile chum salmon (*Oncorhynchus keta*) under different prey rations. Canadian Journal of Fisheries and Aquatic Sciences 41:126–133.

Weber, D. D., and G. J. Ridgway. 1962. The deposition of tetracycline drugs in bones and scales of fish and its possible use for marking. Progressive Fish Culturist 24:150–155.

Weber, D. D., and G. J. Ridgway. 1967. Marking Pacific salmon with tetracycline antibiotics. Journal of the Fisheries Research Board of Canada 24:849–865.

Wild, A., and T. J. Foreman. 1980. The relationship between otolith increments and time for yellowfin and skipjack tuna marked with tetracycline. InterAmerican Tropical Tuna Commission Bulletin 17:509–560.

Wilson, C. A., D. W. Beckman, and J. M. Dean. 1987. Calcein as a fluorescent marker of otoliths of larval and juvenile fish. Transactions of the American Fisheries Society 116:668–670.

Wilson, K. H., and P. A. Larkin. 1980. Daily growth increments in the otoliths of juvenile sockeye salmon (*Oncorhynchus nerka*). Canadian Journal of Fisheries and Aquatic Sciences 37:1495–1498.

Yoklavich, M. M., and G. W. Boehlert. 1987. Daily growth increments in otoliths of juvenile black rockfish, *Sebastes melanops:* an evaluation of autoradiography as a new method of validation. U.S. National Marine Fisheries Service Fishery Bulletin 85:826–832.

American Fisheries Society Symposium 7:203–215, 1990
© Copyright by the American Fisheries Society 1990

Inducement of Unique Otolith Banding Patterns as a Practical Means to Mass-Mark Juvenile Pacific Salmon

Eric C. Volk, Steven L. Schroder, and Kurt L. Fresh

Washington Department of Fisheries, Room 115 General Administration Building
Olympia, Washington 98504, USA

Abstract.—Recent research in our laboratory has shown that simple environmental manipulations during incubation and rearing periods can produce specific banding patterns in the otolith microstructure of juvenile salmonids. In embryonic chum salmon *Oncorhynchus keta*, optically dense bands were produced by sudden drops in water temperature. In chum salmon alevins, optically dense otolith zones corresponded with exposures to cool water as little as 1.7°C below ambient temperature. There was a positive relationship between interband spacing in embryo otoliths, or the width of optically dense zones in alevin otoliths, and the number of temperature units accumulated during exposure to cool water. By exposing salmonid embryos or alevins to regularly repeating thermal cycles, we created unique patterns of optically dense bands or zones on the otoliths. Alternating cycles of feeding and starvation imposed on juvenile fall chinook salmon *O. tschawytscha* also induced identifiable otolith patterns; periods of starvation produced zones of relatively low optical density compared to the rest of the otolith. By means of these techniques, we have successfully marked up to 350,000 hatchery salmonids en masse. The main challenges to implementation of this marking technique on a production scale are the logistics of marking large numbers of fish and the processing and analysis of very many otoliths when adult fish return.

A host of fishery evaluation and allocation issues depend upon our ability to identify specific groups of fish, and an important concern in fisheries management is to identify the origins of captured and harvested fish. Economic considerations, concerns for fish health, and the sheer numbers to be marked dictate the development of a procedure that can mark millions of hatchery-incubated juvenile fishes simultaneously. There are many ways to approach this objective, and one of the most promising is the inducement of specific marks or patterns in the otoliths of juvenile fishes by means of environmental manipulations.

A mass-marking system should produce unique and identifiable marks or patterns that persist throughout the life of the fish; the marks should be easy to apply to a large number of juvenile fish simultaneously, and treatments should not impose undue stress. The embryonic and alevin life history stages are particularly well suited to mass-marking of hatchery-produced salmonids because huge numbers of fish are concentrated in a small area. Otolith marking is one of the few viable techniques for marking these life history stages because the otoliths form well before hatching in salmonids.

There is ample evidence for the influence of environmental factors on otolith microstructure, and a number of investigators have successfully induced patterns in the otoliths of fishes (e.g., Taubert and Coble 1977; Brothers 1981; Tanaka et al. 1981; Campana and Nielson 1982; Victor 1982). The object of the foregoing studies was not to release marked fish, but to study otolith growth or validate the periodicity of otolith increments. However, Brothers (1985) has demonstrated that environmentally induced otolith marks can serve as a means to mark hatchery-reared lake trout alevins *Salvelinus namaycush* scheduled for release.

In this study, we present results from our research on Pacific salmon. We successfully induced a number of unique and obvious marks in the otoliths of chum salmon *Oncorhynchus keta* at the embryo and alevin stages and in the otoliths of fingerling fall chinook salmon *O. tschawytscha*. We also report on our initial efforts to implement this mass-marking technique on a production scale, and present a review of our techniques for rapid preparation of otolith sections and for electronic data acquisition. These components are important in the successful implementation of a mass-marking scheme based on otoliths.

Methods

Thermal Manipulation Experiments

We performed a series of thermal manipulation experiments on incubating chum salmon eggs and

TABLE 1.—Water temperatures, dissolved oxygen concentrations, and water-flow regimes experienced by chum salmon embryos and alevins during thermal marking. Numbers in parentheses represent the standard deviation around the mean of each measured variable. Numbers in quotations refer to the approximate temperature of each water source.

Variable	Water sources			
	Well	"7°C"	"5°C"	"3°C"
Temperature (°C)				
Mean	8.8	7.1	5.0	3.4
	(0.11)	(0.20)	(0.22)	(0.30)
Range	8.7–8.9	6.5–7.4	4.5–5.4	2.8–4.0
Dissolved oxygen (mg/L)				
Mean	9.9	10.5	11.0	11.0
	(0.68)	(0.57)	(0.46)	(0.52)
Range	9.2–11.6	9.7–12.0	10.1–11.4	9.9–12.4
Water flow (L/min)				
Mean	6.7	6.1	6.1	6.7
	(1.31)	(1.34)	(0.93)	(1.24)
Range	4.9–8.1	3.8–7.7	4.7–7.2	5.3–8.9

alevins. The purpose of these experiments was to examine the effect of changes in water temperature on otolith banding patterns in hatchery-reared chum salmon. Experimental eggs were selected from eggs taken at the Washington Department of Fisheries' George Adams salmon hatchery. Eggs were fertilized and subsequently held in wire-mesh baskets in a deep trough until eyed. After eggs were shocked and dead ones were culled, approximately 2,000 eggs were placed in each of 12 Heath incubator trays. At this point, eggs had accumulated 260 temperature units (1 temperature unit is 1°C above 0°C for 24 h) and treatment groups of embryos began to be subjected to thermal manipulations. At this stage of development, primordial otolith elements were beginning to coalesce.

The main water supply for the George Adams facility is from a well and its temperature remained fairly constant (8.7–8.9°C) over the 68-d experimental period. The well water was pumped into a head box equipped with three water-chilling units, which cooled the incoming water to about 3.4°C. This chilled water was then mixed with incoming well water to create two additional water sources of approximately 5 and 7°C, respectively. Each of the four distinct water sources supplied a different stack of Heath trays. The mean and range for each thermal regime are shown in Table 1. Water flows were monitored every 3 d, and dissolved oxygen weekly (Table 1). Although other physical factors were not monitored, we had no reason to believe that unmanipulated environmental conditions were in any way unusual during the experimental periods.

All treatment groups of embryos or alevins were moved on a predetermined and repetitive schedule between Heath tray stacks supplied with relatively warm and relatively cool water. A single thermal cycle, involving a standard exposure to warm water and a sudden shift to cooler water for a variable period, was repeated as many times as possible within the experimental time period. The treatment schedules for each group are shown in Table 2. For embryos, exposure to the relatively warm water that served as a control always lasted 2 d in each repeating cycle. Exposure to one of the coolwater sources had either the same 2-d duration or was planned to make the temperature-unit accumulations for warm and cool water approximately equal in each thermal cycle. For the embryonic groups, regular temperature shifts were repeated until hatching, when the newly hatched alevins were returned to control water until their yolk material was resorbed. At this point, 200 fish from each treatment group were transferred to compartmentalized deep troughs, reared for 14–21 d while they were fed ad libitum, and then killed and preserved in 95% ethanol.

The six groups of embryos used to produce alevins for thermal manipulation were allowed to hatch in the control water source. Each group was then subjected to a specific cycle of thermal alterations between this relatively warm water and cool water (Table 2). Exposure to warm control water always lasted 5 d in each cycle. Exposure to one of the coolwater sources lasted either 5 d or an appropriate number of days to make temperature-unit accumulations during the

TABLE 2.—Thermal treatment schedules experienced by each experimental group of chum salmon embryos or alevins. Values shown in the water-temperature column are approximate source temperatures from Table 1. Treatments 1–6 involved groups of embryos, treatments 7–12 involved groups of alevins, and the treatment-13 group served as a control.

| | Thermal treatment | | | | |
| | Warm phase | | Cool phase | | |
Treatment	Number of days	Water temperature (°C)	Number of days	Water temperature (°C)	Number of repetitions
1	2	8.8	2	7	7
2	2	8.8	3	7	6
3	2	8.8	2	5	7
4	2	8.8	4	5	5
5	2	8.8	2	3	9
6	2	8.8	6	3	5
7	5	8.8	5	7	4
8	5	8.8	5	5	4
9	5	8.8	5	3	4
10	5	8.8	6	7	3
11	5	8.8	9	5	3
12	5	8.8	15	3	2
13	All	8.8			

warm and cool phases approximately equal. These cycles were repeated until the fish had resorbed their yolk material, at which time they were also transferred to deep troughs, fed ad libitum for 2 weeks, killed and preserved. In all experimental groups, mortality was monitored.

Feeding Manipulation Experiments

The effect of food deprivation on otolith banding patterns was also examined. In this case, fall chinook salmon originating from eggs taken at the George Adams hatchery and incubated at the University of Washington's Big Beef Creek hatchery were used. These fish were incubated and reared in accordance with normal hatchery practices. After yolk absorption, 200 fish were selected at random for each of four treatment groups and one control group. Each group was reared outside in 265-L plastic tanks. The fish were fed a maximum ration of a standard hatchery diet (Biodiet®) and food was delivered eight times daily by automatic feeders.

All groups subjected to food deprivation had food withheld for 5-d intervals in a repeating cycle with 5 d of food, for five or six cycles. Fish in control groups were fed throughout. Although all of these fish were maintained in experimental enclosures throughout the 112-d experimental period, alternating cycles of feeding and starvation were only imposed for half that time on any one group. Two groups had treatments during the first 56 d, and two groups had treatments during the second half of the experiment. Among the groups receiving experimental treatments at the same time, food-deprivation intervals were staggered by 5 d. Feeding schedules for these fish are shown in Table 3. At the end of the experimental period, fish were killed and preserved in 95% ethanol. Mortality was monitored throughout the experiment.

Otolith Marking on a Large Scale

Dungeness hatchery coho salmon.—In the fall of 1986, we uniquely marked the entire production of coho salmon *O. kisutch* from the Washington

TABLE 3.—Treatment cycles, numbers of repetitions, numbers of low-density otolith zones produced, and starting times of the feeding and starvation cycles used to mark the otoliths of juvenile fall chinook salmon.

Treatment	Number of cycles	Number of low-density zones	Start day	N
1 5 d no food, 5 d food	5	5	21	10
2 5 d food, 5 d no food	5	5	21	10
3 5 d no food, 5 d food	6	5	51	10
4 5 d food, 5 d no food	6	5	51	10
5 Continuous food	1	0	1	10

TABLE 4.—Thermal schedules, numbers of cycles, and water temperatures used to mark the otoliths of juvenile coho salmon at the Dungeness hatchery and juvenile chum salmon at the Skagit hatchery. The numbers in parentheses represent the standard deviations around the mean.

Population	Treatment	Number of cycles	Thermal regime (°C)	
			Cold	Warm
Dungeness coho salmon	3 d warm, 3 d cold	4	3.7 (1.3)	7.4 (1.2)
	18 d cold	1	4.5 (0.9)	
	3 d warm, 3 d cold	4	6.2 (0.7)	8.5 (0.9)
Skagit chum salmon	3 d warm, 3 d cold	9	6.6 (0.52)	7.9 (0.44)

Department of Fisheries' Dungeness hatchery. The otoliths of approximately 350,000 coho salmon embryos and alevins were marked by means of periodic changes in their thermal environment.

After eggs were shocked, they were divided equally among 32 Heath trays grouped into four stacks of eight. The tray stacks were insulated with 5-cm foam blocks. The main water source for the Dungeness hatchery is a variable-temperature stream, which is typically cold. Thermal data for this water source are presented in Table 4. Because the water was cold, it was necessary to create a source of relatively warm water by pumping stream water into a head box equipped with three water heaters, which elevated the temperature of incoming water approximately 3.5°C. However, significant temperature fluctuations were unavoidable due to the water source. The Heath-tray stacks were plumbed such that by turning a valve, two stacks would receive the ambient-temperature water while the other two received heated water from the head box. Every 3 d, the water source for each stack was reversed, in an attempt to create unique, thermally induced otolith marks in all fish (Table 4). These thermal changes were conducted throughout embryo incubation, discontinued briefly while the eggs hatched, and then resumed several days after hatching, until the alevins had resorbed their yolk materials. After the experiments, fish were put in ponds as usual. Some fish were killed periodically and preserved in 95% ethanol to assess the success of the marking procedure. Water temperature was monitored throughout the experiment with recording thermometers (Table 4). Flow rates and dissolved oxygen were always within safe incubation limits. Mortality was also monitored.

Skagit chum salmon.—We sought to mark 95,000 juvenile chum salmon alevins with thermal manipulations to induce specific otolith banding patterns. The water source for the Skagit hatchery is fairly constant and cold, so heaters were once again employed to create two constant-temperature water sources, which differed by approximately 1.3°C (Table 4). An arrangement similar to that at Dungeness allowed delivery of warm or cool water to specific groups of incubating fish. Fish received nine cycles of warm and cool water until absorption of their yolk material. Treatment schedules are shown in Table 4. At the conclusion of marking, several hundred fish were killed and preserved in 95% ethanol. Water temperature was monitored with recording thermometers throughout the experiment. Flow rates and dissolved oxygen were always within safe incubation limits.

Otolith Analyses

All otoliths (we worked exclusively with sagittae) were processed in accordance with a standard procedure. Specimens were dissected from the fish and placed, medial surface down, on glass plates in groups of up to 10. All otoliths processed together were taken from fish whose size range was less than 5 mm, to ensure similarly sized otoliths. A rubber mold was then placed around the array and polyester resin poured in. After the resin hardened, otoliths lay at the bottom surface of the hard, clear polyester plug. Otoliths were lapped by hand on 600-grit sandpaper on a Buehler Ecomet 3® lapping machine until the primordia of the otoliths were visible. The otolith block was then hand-polished on the same machine with Texmet® cloth and 1.0-µm alumina. Polished plugs were then mounted with epoxy on glass slides previously lapped to a uniform thickness. The slides with the bonded plugs were mounted on a Buehler Isomet® diamond saw and all but 0.5 mm of the otolith resin block was removed. Slides were then placed on a Logitech® LP-30 precision lapping and polishing machine, and lapped to a uniform thickness of between 50 and 150 µm; thickness depended upon the specimens being prepared. Initial lapping was done with 5.0-µm alumina, and specimens received their final polish on the LP-30 with Texmet cloth and 1.0-μm alumina abrasive.

Data were collected from otolith sections by means of a Biosonics Optical Pattern Recognition System®. This system employs a microscope, video camera, frame grabber, digitizing board, and microcomputer to gather, store, and analyze otolith band data.

We established a standard protocol for the measurement of all otolith specimens. At 100× magnification, we produced a reference line, which connected the rostrum of the otolith with its opposite side through the center of the otolith core. At 200×, we subjectively chose the point on the reference line that fell closest to the middle of the primordial core. This represented the beginning point for radial measurements. At a 75° angle to the reference line, in the anterodorsal quadrant of the otolith, a radius was drawn with the computer mouse, and otolith bands or zones were either recognized automatically by the machine or scored manually as desired. Band or zone number and interband widths (IBW) or zone widths were then stored in the computer.

In arriving at a figure for mean IBW, all discrete measurements from all otoliths in a group were combined. For alevins, the beginning and end of each optically dense zone were scored on the digitizer so that zone number and width were recorded. As previously described, widths of all discrete zones were combined to determine the mean within each treatment group.

Results

Thermal Manipulation Experiments with Chum Salmon Embryos

Cyclic shifts of relatively warm and cool incubation water effectively induced obvious and unique incremental banding patterns in the otoliths of chum salmon embryos. Changes of less than 2°C were sufficient to produce a system of optically very dense bands (Figure 1).

These patterns were characterized by the number of optically dense bands in the area formed before hatching and by the space between each band (Table 5). The number of dense otolith bands closely corresponded to the number of thermal-shift cycles experienced by the fish, and there was a positive relationship between the number of temperature units accumulated during each thermal cycle (x) and the mean IBW (y, μm): $y =$ $0.56x + 5.50$; $r = 0.89$. Mean IBWs differed ($P < 0.05$) among the various groups except between those in treatments 4 and 2.

Dissolved oxygen levels and water flow always remained within the safe range for incubating chum salmon. Aside from incubation at reduced egg densities, there was no reason to believe that experimental fish were in any way different from those incubated according to normal hatchery procedures. Mortality never exceeded 1.5% in any group.

Thermal Manipulation Experiments with Chum Salmon Alevins

Regular shifts between relatively warm and cool water also produced distinctive effects in the otoliths of chum salmon alevins (Figure 2). In contrast to embryos, in which thermal shifts produced patterns of discrete and very dense bands, similar treatment of alevins produced a system of broader otolith zones, alternating between regions of higher and lower optical density. The ratio of otolith increments, counted from the last optically dense zone to the edge of the otoliths, and the number of days between the last cold exposure and the sacrifice of the fish, was not different from unity, which suggested that optically dense zones corresponded to periods of exposure to cold water. (The daily periodicity of otolith increments during nonexperimental bouts was verified in control fish, for which the ratio of posthatching otolith increments to the number of days since hatching was not significantly different from 1.0 [$P < 0.05$].)

The numbers of optically dense (coolwater) and less dense (warmwater) zones and their respective mean widths for each of the alevin treatment groups are shown in Table 6. During the first thermal shift cycle at the outset of these experiments, our chiller system failed, and water temperatures were elevated for 5 d. The general effect of this mishap was to eliminate the first optically dense zone in treatments 7–10, because rising temperatures encompassed almost the entire first nominal exposure to cool water. For treatments 11 and 12, the first optically dense zone was truncated by the initially high water temperatures at the beginning of the first exposure to cool water. This effect is shown in Figure 2C, D, where the first optically dense zone is clearly narrower than the ensuing ones. We have eliminated effects of this first coolwater exposure on otoliths from data analysis in all groups. Thus, for the purpose of discussion, treatments 7–9 exhibited three op-

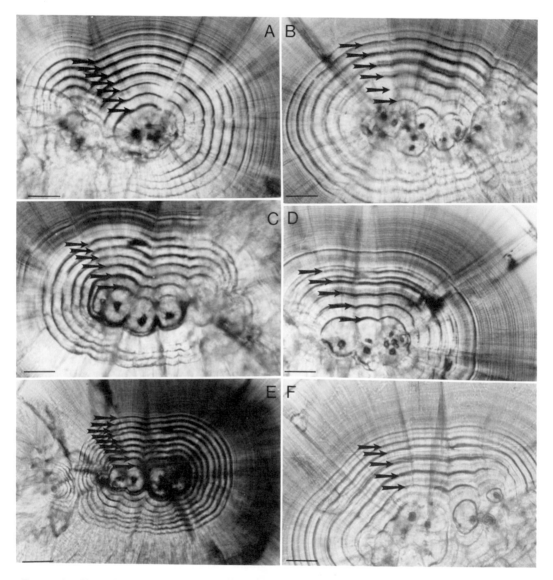

FIGURE 1.—Photomicrographs of otoliths from juvenile chum salmon that had been thermally manipulated as embryos in one of six temperature regimes. Arrows mark optically dense zones induced by sudden thermal changes. Scale bars = 100 μm. Temperatures in quotation marks below are approximate.

(A) Treatment-1 fish 51 d after hatching; its thermal cycle was 2 d at 8.8°C and 2 d at "7°C."
(B) Treatment-2 fish 51 d after hatching; its thermal cycle was 2 d at 8.8°C and 3 d at "7°C."
(C) Treatment-3 fish 56 d after hatching; its thermal cycle was 2 d at 8.8°C and 2 d at "5°C."
(D) Treatment-4 fish 56 d after hatching; its thermal cycle was 2 d at 8.8°C and 4 d at "5°C."
(E) Treatment-5 fish 54 d after hatching; its thermal cycle was 2 d at 8.8°C and 2 d at "3°C."
(F) Treatment-6 fish 49 d after hatching; its thermal cycle was 2 d at 8.8°C and 6 d at "3°C."

tically dense zones, treatments 10 and 11 showed two dense zones, and treatment 12 had only one dense zone. The number of dense zones corresponded to the number of coolwater exposures experienced by each group (Table 6). Mean widths of optically dense zones produced during coolwater phases differed among alevins in different treatment groups, except between fish in treatments 10 and 11 ($P < 0.05$). In contrast, the mean widths of zones produced during warmwater phases did not differ from one another, except in the case of treatments 8 and 10 ($P < 0.05$; Table

TABLE 5.—Numbers of otolith bands produced, mean distances between bands, thermal-cycle repetitions, and numbers of temperature units accumulated during one thermal cycle for each group of chum salmon marked as embryos. The values in parentheses represent the standard deviations around the mean interband widths (IBW).

Treatment	Mean IBW (μm)	Mean band number	Cycle repetitions	N	Temperature units accrued per cycle
1	12.3 (1.44)	7.0	7	20	32.2
2	15.6 (2.47)	6.0	6	20	39.3
3	10.9 (1.30)	7.0	7	20	28.0
4	15.9 (2.54)	5.0	5	20	38.0
5	9.7 (1.97)	9.6	9	20	24.8
6	20.0 (6.94)	5.6	5	20	38.4

6). There was a positive relationship between the mean width (y, μm) of zones due to cool water in alevin treatment groups and the number of temperature units (x) accumulated during bouts of cold ($y = 0.25x + 1.99$; $r = 0.82$).

The magnitude of the temperature change between relatively warm and relatively cool water appeared (subjectively) to affect the optical density of the otolith zone produced by cool water. When the temperature change was about 6°C, the otolith zone produced by cool water was very dense, and increment definition was poor. When the change was only 2°C, these zones were less dense and increments were clearer (Figure 2A, B).

Feeding Experiments with Juvenile Fall Chinook Salmon

Alternating 5-d periods of feeding and starvation also had an effect upon otolith microstructure of juvenile fall chinook salmon reared at Big Beef Creek. This feeding regime produced areas of high and low optical density within a well-defined otolith region (Figure 3). It was difficult to verify (through periodic sacrifices) the correspondence between starvation events and otolith zones of high or low optical density, because we suspected a lag between cause and effect. However, the presence of these anomalous zones of low optical density in otoliths of experimental fish subjected to periods of food deprivation suggests that starvation intervals were the cause (Figure 3).

For groups 1 and 2, five low-density zones corresponded to the five food deprivation bouts to which the fish were subjected. However, for groups 3 and 4, one less low-density zone than expected was recorded (Table 3). The mean widths of the first high- and low-density otolith zones corresponding to the first 10-d cycle of feeding and starvation are shown in Table 7 for each treatment. All experienced the same repeating 10-d cycle, and mean widths of the resulting zones were not significantly different ($P > 0.05$).

To determine if food-deprivation patterns were distinguishable between replicate groups whose treatments were staggered by 5 d, we compared the mean distance from the otolith focus to the start of the first low-density zone in groups 1 and 2. Although ranges overlapped substantially, mean placement of the first low-density zone along the otolith radius was 273.72 μm (SD = 12.53) for treatment 1 and 285.51 μm (SD = 12.03) for treatment 2. These values represent a significant difference ($P < 0.05$).

Mortality never exceeded 2.5% in these groups.

Implementation

With the results from the foregoing experiments, we devised simple plans for marking the otoliths of 350,000 Dungeness hatchery coho salmon and 95,000 Skagit hatchery chum salmon by thermal manipulations. Because our goal was to create a single unique pattern in each case, the thermal treatments were not elaborate. Treatment schedules and thermal data are shown in Table 4. The correspondence between the number of optically dense zones and the number of coldwater bouts is clear for the Skagit chum salmon (Figure 4). We also successfully induced identifiable marks on the otoliths of Dungeness coho salmon. However, the more variable thermal regime at the Dungeness facility produced some spurious otolith marks as a result of unscheduled temperature drops, which made the correspondence between otolith patterns and environmental changes less obvious.

Discussion

It is clear that obvious and unique band patterns can be induced in the otoliths of chum salmon embryos through regular, periodic shifts of less than 2°C in water temperature. Although in most treatments there was a precise correspondence between the number of optically dense bands

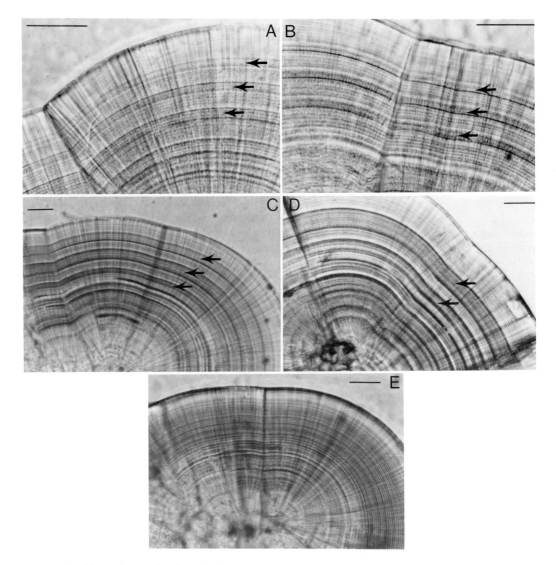

FIGURE 2.—Photomicrographs of otoliths from juvenile chum salmon that had been thermally manipulated as alevins in one of five temperature regimes. Arrows mark optically dense zones induced by exposures to relatively cool water. Scale bars = 100 μm. Temperatures in quotation marks below are approximate.

    (A) Treatment-7 fish 51 d after hatching; its thermal cycle was 5 d at 8.8°C and 5 d at "7°C."
    (B) Treatment-9 fish 56 d after hatching; its thermal cycle was 5 d at 8.8°C and 5 d at "3°C."
    (C) Treatment-11 fish 55 d after hatching; its thermal cycle was 5 d at 8.8°C and 9 d at "5°C."
    (D) Treatment-12 fish 62 d after hatching; its thermal cycle was 5 d at 8.8°C and 15 d at "3°C."
    (E) Control fish 60 d after hatching; its thermal regime was a constant 8.8°C.

produced in embryonic otoliths and the number of thermal cycles experienced by the fish, more than 50% of the samples from treatments 5 and 6 showed band counts one in excess of that expected from the number of cycle repetitions. Our explanation is that the otolith check corresponding to the time of hatching was counted as a thermally induced, dense band. The two can look

very similar, and these thermal cycles were repeated until hatching had commenced, so possible confusion is understandable. In this case, fish in treatments 5 and 6 were distinguishable from all others on the basis of interband width.

    There was also a positive relationship between temperature units accumulated during each cycle and the mean interband width for fish in treat-

TABLE 6.—Numbers of optically dense otolith zones created during coolwater exposures and of less dense zones created by warmwater exposures of chum salmon alevins. Mean zone widths and numbers of temperature units accumulated during zone formation are also shown. The numbers in parentheses represent the standard deviations around the mean zone widths.

Treatment	Number of zones	Mean zone width (μm)	Number of temperature units accumulated	N
		Warmwater phase		
7	4	14.14 (2.00)	44.0	20
8	4	14.92 (2.85)	44.0	20
9	4	13.39 (2.66)	44.0	20
10	3	12.75 (1.67)	44.0	20
11	3	13.06 (1.74)	44.0	20
12	3	14.28 (1.77)	44.0	20
		Coolwater phase		
7	3	6.44 (1.49)	17.0	20
8	3	8.48 (2.01)	25.0	20
9	3	9.81 (1.19)	35.5	20
10	3	11.21 (2.13)	42.6	20
11	2	11.21 (1.54)	44.0	20
12	1	17.82 (2.83)	44.0	20

ments 1–6 (Table 5). Fish in treatment 6, which experienced 6 d of very low temperatures, showed a greater interband width for the same

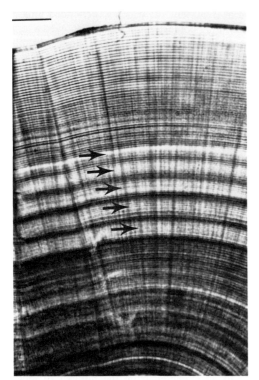

FIGURE 3.—Photomicrograph of an otolith from a juvenile fall chinook salmon that had been subjected to alternating 5-d bouts of starvation and feeding as a fingerling. Arrows mark zones corresponding to periods of food deprivation. Scale bar = 100 μm.

number of temperature units accumulated than did fish in treatments 2 and 4, which had shorter exposures to less extreme temperatures.

It has been suggested by Brothers (1985) that optically dense otolith bands or zones, which result from greater deposition of organic matrix in the calcified otolith, are formed during periods of depressed or declining water temperatures. In our chum salmon embryos, it seemed that dense bands were produced by the sudden change from relatively warm to relatively cool water, because the thickness of the bands did not vary with the length of exposure to cool water. Furthermore, fish in treatment 1, which were subjected to a cycle of two warm and two cool days, showed the characteristic dense bands each followed by three less-dense bands (Figure 5). If otolith increments were produced daily during this time, this configuration supports the notion that these dense bands were recorded during the 1-d event of thermal change. Although we cannot unequivocally dem-

TABLE 7.—Mean widths of the first high- and low-optical-density otolith zones in fall chinook fingerlings caused by 5 d of feeding and 5 d of food deprivation. The treatment numbers correspond to those described in Table 5. The values in parentheses represent one standard deviation around the mean.

Treatment	Mean width of first otolith zone produced by feeding cycle, μm	N
1	24.5 (3.78)	10
2	24.5 (3.78)	10
3	25.1 (2.66)	10
4	24.1 (2.45)	10

FIGURE 4.—(A) Otolith from a juvenile Skagit hatchery chum salmon that had been subjected as an alevin to a thermal cycle of 3 d warm and 3 d cool water. Arrows mark optically dense zones corresponding to periods of coolwater exposure. Scale bar = 100 μm.

(B) Otolith from a juvenile Dungeness hatchery coho salmon that had been subjected as both embryo and alevin to a thermal cycle of 3 d warm and 3 d cool water. Scale bar = 200 μm.

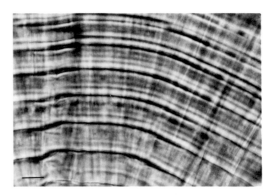

FIGURE 5.—Optically dense bands in an otolith from a 51-d-old (posthatch) chum salmon that had been subjected as an embryo to a thermal cycle of 2 d at 8.8°C and 2 d at about 7°C. Scale bar = 25 μm.

onstrate that the sudden rise in temperature was not responsible for the production of these dense bands, results from our alevin studies and those of Brothers (1985) suggest that this explanation is unlikely.

Alevin chum salmon subjected to cycles of thermal change also exhibited distinctive otolith effects. In contrast to otoliths of control fish, otoliths of experimental fish exhibited a distinctive series of zones characterized by greater optical density. These higher-density zones appeared to correspond to periods of exposure to cool water, because there was a close correspondence between the number of otolith increments counted from the end of the last dense zone to the otolith edge, and the number of days elapsed between the last coldwater exposure and the sacrifice of the fish. As we expected, the width of optically dense otolith zones bore a positive relationship to the number of temperature units accu-

mulated during coldwater exposure. Thus, the lack of significantly different mean widths of the dense zone of fish in treatments 10 and 11 was expected, because fish in these treatments accumulated a similar number of temperature units during coldwater exposure. However, as before, fish in treatment 12, which experienced long periods at the coldest temperatures, showed a larger mean width of the dense zone than fish in other treatments that had accumulated the same number of temperature units during exposure to less extreme temperatures (Table 6).

Subjection of juvenile fall chinook fingerlings to periodic starvation also produced distinctive otolith patterns. The patterns were characterized by alternating zones of lesser and greater optical density. Individual otolith increments were generally apparent within each of these zones. Direct verification of the correspondence between zone density and feeding or starvation bouts was difficult due to the time elapsed between environmental events and otolith effects. However, it seems logical that because the zones of low optical density were not present in otoliths of control fish, periodic food withdrawal was probably responsible for their production.

The number of low-density zones within the otoliths always corresponded precisely with the number of starvation episodes experienced by the fish in treatments 1 and 2. Fish from treatments 3 and 4, however, exhibited one less zone than would be expected from the number of feeding cycles they were subjected to (Table 3). The last bout of starvation took place just before sacrifice in these groups, which argues for a possible lag between the initiation of food withdrawal and its effect on the otolith. Other evidence for this phenomenon comes from comparison of the number of posttreatment otolith increments from fish in groups 1 and 2. For fish in group 1, which ended the last foodless period 55 d before sacrifice, there was an average of 47.1 increments between the last low-density zone and the otolith edge. For fish in group 2, which had 50 d between the last starvation period and sacrifice, this value was 42.4. If these increments were laid down daily, which is suggested by the mean difference of 4.7 increments, then a lag of 7–8 d between food withdrawal and effect on the otolith is suggested.

We attempted to distinguish otolith patterns from fish in replicate (staggered) treatments, which had been exposed to the same schedule of feeding and starvation, but for which schedules had commenced 5 d apart. Starting on different days would alter the placement of the pattern in the otolith. Although mean values for the distance from the otolith focus to the beginning of the pattern differed between the two groups, there was substantial overlap among individuals (Table 7). This overlap suggests that feeding regimes must be more distinctive between the groups if otolith patterns are to be reliably distinguished on an individual basis.

Although successful, these feeding experiments were done in very small tanks that supported only 200 fish. Under these conditions, we were somewhat confident that all fish received approximately equal rations of food. However, in a raceway environment, equal access to food may not be the case. We have looked at otoliths from fish that were subjected to similar starvation and feeding bouts, but were reared under production conditions in a hatchery. Patterns were often evident, but they were variable and not always clear (unpublished data). Furthermore, in our study, fish went from maximum rations to total lack of food. During normal hatchery rearing, fish are not always fed maximum rations, and we wonder if the degree of contrast in otolith patterns could be influenced by this feeding regime. Certainly, the more that growth rates are affected by feeding manipulations, the more chance there is that a recognizable otolith pattern will be produced.

All of our efforts to mass-mark hatchery populations have involved the use of embryos and alevins and manipulation of the temperature of their water. While conceptually simple, the central challenge of this technique is to adapt the physical plant of a hatchery to allow such manipulations. Aside from logistics, the main considerations in this regard are the volume of chilled or heated water required to mark fish and the temperature differential between that water and water at ambient temperature. These two factors largely determine the cost of marking fish by this method.

It is theoretically possible to provide any volume of thermally altered water, given large enough heating or chilling units. However, our initial attempts to implement this technique have shown that hatchery populations can be divided into subgroups that receive their thermally induced marks on different days. Thus, thermally altered water need only be supplied to part of the population on any given day. Because the otolith patterns are based upon the number of optically dense bands or zones and their interband spacing or width, the same pattern is still produced in all otoliths, but simply displaced in the otolith very slightly. Naturally, this strategy of water use can be employed to greater or lesser degree by subdividing the population to be marked. Such a tactic allows substantial savings of thermally altered water.

Another important consideration is the normal water temperature in a hatchery. We have found that when the water source has a constant temperature, temperature changes of less than 2°C during incubation are sufficient to induce recognizable marks in fish otoliths. However, where the hatchery water source varies in temperature, a temperature differential can be maintained between warm and cool water, but temperature of both will fluctuate according to that of the incoming water. This fluctuation can create spurious effects or "noise" in the otolith patterns as a result of natural thermal events in the source water. The contrast in thermally induced marks produced by constant- and variable-temperature water sources can be seen by comparing figures 4A and B. At the Skagit hatchery, where we marked 95,000 juvenile chum salmon, incubation water temperature was constant. Examination of a Skagit chum salmon otolith shows very clear optically dense zones corresponding to a shift in water temperature of about 3.5°C every 3 d. In contrast, at the Dungeness hatchery, where we marked 350,000 juvenile coho salmon, the water source was a variable-temperature stream. The otoliths from these fish showed a more confusing pattern because the correspondence between otolith marks and thermal events was less obvious. In this particular case, the problem was not debilitating because our goal was simply to identify these fish among any other coho salmon. However, it seems that subtle differences between patterns produced with a variable-temperature water supply may be lost in the "noise." This confusion may be less evident if the magnitude of the thermal change used for marking overshadows the normal fluctuations of the variable-temperature water source.

If mass-marking of otoliths is to be useful, specimen-preparation and data-collection systems must be able to accommodate large sample sizes efficiently. There is a variety of methods for production of quality otolith sections, but most are slow and tedious, and the application of otolith-marking procedures in a large and complex fishery entails the processing of thousands of otoliths a year. Some sort of automated system, such as that outlined here, is essential to such an endeavor. With our current system of preparation, we are able to process hundreds of otoliths per day. However, it is important to remember that significant limitations are inherent in mass-preparation systems because morphological variations, even among similarly sized otoliths, mean that sections of samples mounted together are not precisely comparable. This variation results in a measurable rejection rate among specimens prepared together, the magnitude of which is proportional to the size variation of the otoliths mounted together (our unpublished data). Regarding mark recognition, this practical consideration dictates that otolith patterns should be made as obvious as possible. Although many subtle differences in incremental banding patterns may be produced, it may not be cost-effective to retrieve them in large-scale programs.

As with otolith preparation, the ease of interpretation of otolith marks increases with the simplicity of patterns. In our mass-marking projects, we created a limited number of patterns at each hatchery. These patterns could be distinguished by eye. However, when several similar otolith patterns are being used, an electronic scanning device with a microscope, video system, and computer must become part of the deciphering process.

It is clear from the results of this study that large groups of juvenile salmonids can be effectively marked by induction of specific banding patterns in their otoliths, via simple environmental manipulations. The use of water-temperature changes to produce these patterns in embryos and alevins is advantageous because the marks are universally distributed within a population, and

large numbers of fish in early developmental stages can be treated simultaneously in a small space. Such changes are also apparently stress-free. Perhaps the one drawback to thermally induced otolith marking is the initial cost of the heating and chilling equipment, and the adaptation of facilities to this equipment. We are just beginning to implement these marking procedures on a grand scale, and our goal is to make heating and chilling units modular so that they can be used in a number of different locations.

The use of feeding manipulations for mass-marking of older salmonids also holds promise as a practical marking method; however, its drawbacks include a more complex physiological interaction between environmental event and otolith effect, and the difficulty of obtaining an even distribution of the effects of food withdrawal among all fish in a population. The attribute that makes feeding manipulations particularly attractive for further investigation is cost-effectiveness.

The extent to which otolith marking is applied to fishery problems in the future may depend upon our ability to solve technical problems related to the handling and processing of very large sample sizes. Our proposed projects in the next 5 years involve the analysis of 10,000–20,000 otoliths per year. Their preparation cannot be done manually or individually without a large technical force constantly at work. As a result, the caveats associated with our mass-processing techniques will partly dictate our practical approach to otolith-pattern induction. In short, patterns will have to be sufficiently obvious and distinguishable to allow room for variability in otolith preparation.

There is little question that the future of fish marking will involve a variety of mass-marking procedures. Because of economic and biological reasons, otolith marking will be an important tool. Otolith-marking procedures are already employed by the Washington Department of Fisheries to evaluate other tagging methods, as well as various cultural practices and release strategies. On the horizon, it is easy to see the direct application of these techniques to mixed-stock fishery questions.

Acknowledgments

We greatly appreciate the help of Gene Sanborn, who was largely responsible for the design, construction, and maintenance of our water-cooling and heating systems. His expertise was integral to the success of their effort. Thanks also go to Mark Carr, who assisted in physical plant operations and helped with dissections. Finally, we thank all the hatchery personnel at the George Adams, Dungeness, and Skagit facilities for their cooperation and help in completing this work.

References

Brothers, E. B. 1981. What can otolith microstructure tell us about daily and sub-daily events in the early life history of fish? Rapports et Procès-Verbaux de Réunions, Conseil International pour l'Exploration de la Mer 178:393–394.

Brothers, E. B. 1985. Otolith marking techniques for the early life history stages of lake trout. Great Lakes Fishery Commission, Research Completion Report, Ann Arbor, Michigan.

Campana, S. E., and J. D. Nielson. 1982. Daily growth increments in otoliths of starry flounder (*Platichthys stellatus*) and the influence of some environmental variables in their production. Canadian Journal of Fisheries and Aquatic Sciences 39:937–942.

Tanaka, K., Y. Mugiya, and J. Yamada. 1981. Effects of photoperiod and feeding on daily growth patterns in otoliths of juvenile *Tilapia nilotica*. U.S. National Marine Fisheries Service Fishery Bulletin 79:459–466.

Taubert, B. D., and D. W. Coble. 1977. Daily rings in otoliths of three species of *Lepomis* and *Tilapia mossambica*. Journal of the Fisheries Research Board of Canada 34:332–340.

Victor, B. 1982. Daily otolith increments and recruitment in two coral reef wrasses, *Thalassoma bifasciatum* and *Halichoeres bivittatus*. Marine Biology 71:203–208.

American Fisheries Society Symposium 7:216–223, 1990

Mass-Marking of Otoliths of Lake Trout Sac Fry by Temperature Manipulation

ROGER A. BERGSTEDT

Hammond Bay Biological Station, 11188 Ray Road, Millersburg, Michigan 49759, USA

RANDY L. ESHENRODER

Great Lakes Fishery Commission, 1451 Green Road, Ann Arbor, Michigan 48105, USA

CHARLES BOWEN II

National Fisheries Center–Great Lakes, 1451 Green Road, Ann Arbor, Michigan 48105, USA

JAMES G. SEELYE AND JEFFREY C. LOCKE

Hammond Bay Biological Station

Abstract.—The otoliths of 676,000 sac fry of lake trout *Salvelinus namaycush* in 1986, and of 1,100,000 in 1987, were marked by daily manipulation of water temperature. The fish were stocked into Lake Huron in the spring. Otolith marks consisted of groups of daily growth rings accentuated into recognizable patterns by steadily raising and lowering the temperature about 10°C (from a base of 1–4°C) over 14 h. In 1987, groups of marked and control fish were held for 6 months. The otoliths were removed from samples of the fish, embedded in epoxy, thin-sectioned by grinding in the sagittal plane, etched, and viewed by using a combination of a compound microscope (400–1000×) and a video enhancement system. One or more readable otolith sections were obtained from 39 of a sample of 40 fish. Three independent readers examined 41 otoliths for marks and correctly classified the otoliths, with accuracies of 85, 98, and 100%, as being from marked or unmarked fish. The exact number of rings in a recognizable pattern sometimes differed from the number of temperature cycles to which the fish were exposed. Counts of daily rings within groups of six rings varied less than counts within groups of three rings.

During the 1940s and 1950s, lake trout *Salvelinus namaycush* became severely depleted in Lake Superior, and probably became extinct or neared extinction in the other Great Lakes. The sequence of events leading to the declines was reviewed in detail for Lake Huron by Berst and Spangler (1973), for Lake Ontario by Christie (1973), for Lake Erie by Hartman (1973), for Lake Superior by Lawrie and Rahrer (1973), and for Lake Michigan by Wells and McLain (1973). The declines were generally attributed to a combination of fishery exploitation and lethal attacks by the sea lamprey *Petromyzon marinus*. Control of the sea lamprey with a selective lamprey larvicide (Pearce et al. 1980; Smith and Tibbles 1980) and the later stocking of more than 100 million yearling lake trout (Great Lakes Fishery Commission 1983) did not result in appreciable reproduction except in Lake Superior, where remnant populations contributed to the recovery (Eshenroder et al. 1984).

Inadequate use of appropriate spawning sites has been hypothesized as a factor that contributed to the failure to restore natural lake trout reproduction (Eshenroder et al. 1984). Lake trout were routinely stocked as yearlings at historic spawning sites, but the fish apparently did not become imprinted on those sites, and returned to spawn in lower proportions than native fish (Krueger et al. 1986). Binkowski (1984) and Foster (1984) suggested that the stocking of earlier life stages might be more effective in imprinting lake trout on potential spawning sites. However, an evaluation of this hypothesis requires a marking method that differentiates between fish stocked as fry and fish stocked as yearlings, as well as between stocked and naturally produced young.

Because the methods currently used for marking lake trout in the Great Lakes (fin clips and coded wire tags) are most applicable to fingerlings or larger fish, Brothers (1990, this volume) tested in the laboratory a technique for placing thermal marks on the otoliths of very young lake trout. We used this technique to mark large numbers of lake trout sac fry stocked on an offshore reef in Lake Huron during 1986–1987. Here we describe the

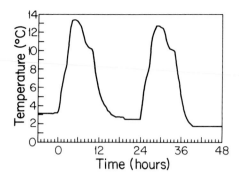

FIGURE 1.—Water temperature during two temperature cycles used to mark otoliths of lake trout sac fry, 30–31 March 1987. Heating cycles began at about 0800 hours each day.

large-scale implementation of an inexpensive and effective method (Brothers 1990) for thermally marking the otoliths of large numbers of young fish by exposure to diel temperature cycles, describe and evaluate preparation of the otoliths for decoding, and evaluate the readability of the mark.

Methods

Marking.—We subjected three groups of lake trout sac fry to diel temperature cycles, as suggested by Brothers (1990), to produce identifiable marks on their otoliths. Otoliths of lake trout of the Lake Superior strain were marked during spring in 1986 and 1987; otoliths of lake trout of the Seneca Lake strain were marked only in spring 1987. In both years, control groups of lake trout of the Lake Superior strain were held under identical conditions. All lake trout were obtained from the Iron River (Wisconsin) National Fish Hatchery as eyed eggs and incubated at the Hammond Bay Biological Station in eight 16-tray Heath[1] incubators. Each incubator was supplied with filtered Lake Huron water at ambient temperatures at the rate of 7 L/min. Fry density was about 10,000 per tray; we estimated density from total number of eggs and number of trays used.

During marking of the fry, each of four incubators was supplied with water, at a rate of 4.0 L/min, from a 1,900-L mixing tank in place of the normal supply. The mixing tank was supplied with a constant flow (16 L/min) of water at ambient temperature and two potential flows (each 2.1 L/min) of heated water (58°C) controlled by solenoid valves. Water was heated in a propane-fired

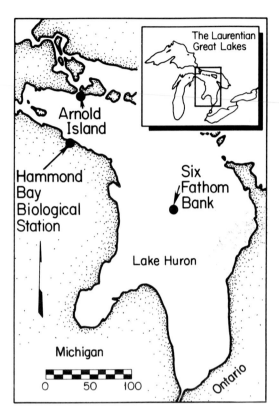

FIGURE 2.—Stocking sites for lake trout sac fry with thermally marked otoliths stocked during 1986–1987. The scale is in kilometers.

water heater. The solenoid valves were opened and closed by timers at the same times each day to increase and decrease the temperature over the ensuing 14 h. During each temperature cycle, one solenoid valve was on for the first 3 h, both for the next 4 h, and one for the next 3 h. The water in the mixing tank then returned to ambient temperature in about 4 h, and remained so until the start of the next cycle. The large volume of water in the mixing tank resulted in relatively even changes in temperature (Figure 1). For all marks applied, the mean ambient temperature at the start of the cycle and the mean rise in temperature were 2.4°C (range, 1.1–4.4°C) and 10.6°C (9.4–12.8°C), respectively. Control fish were held at ambient temperatures.

Marks consisted of one or two groups of accentuated daily rings that formed recognizable patterns. Sac fry were exposed to the described temperature cycle for a number of consecutive days to produce a corresponding number of accentuated daily rings (i.e., a group of rings) on the

---

[1]Reference to trade names does not imply endorsement by the U.S. Fish and Wildlife Service.

otoliths. If more than one group of rings was desired, a space between them was produced by holding the sac fry at ambient temperatures for 9–13 d.

Lake trout of the Lake Superior strain stocked in 1986 were marked with a group of five rings followed by a group of two (the mark was represented by the notation 5,2). On 23 April 1986, 570,000 sac fry marked 5,2 were released on Six Fathom Bank in Lake Huron (Figure 2). Lake trout of the Lake Superior strain stocked in 1987 were marked with a group of three accentuated daily rings followed by a group of six (mark = 3,6). However, because heated water could be supplied to only four incubators at a time, the sac fry were marked in two lots. On 14 April 1987, 1,100,000 sac fry marked 3,6 were released on Six Fathom Bank (Figure 2). Lake trout of the Seneca Lake strain stocked in 1987 were marked with a single group of four accentuated daily rings (mark = 4). On 13 March 1987, 106,000 sac fry marked 4 were released near Arnold Island in Lake Huron (Figure 2).

Reference collections from each marked group were taken 2–4 d after marking was completed. In addition, control fish and fish marked 3,6 were held at the Hammond Bay Biological Station for 6 months after they were marked. All lake trout killed for the reference collections were preserved and stored in 95% ethanol until the otoliths were removed.

During spring 1988, we conducted an experiment to measure the effect of the marking procedure on survival. Four groups of 500 sac fry each were exposed to nine consecutive diel temperature cycles, as described previously. Four groups of 500 sac fry each were also held as controls. The mean numbers dead in control and marked groups after 90 d were compared by analysis of variance (ANOVA).

Sectioning of otoliths.—The otoliths (sagittae) were located and removed from the preserved lake trout with the aid of a dissecting microscope (12–50×) and dried in air. Beyond this point, the method of preparation varied with the size of the otolith.

Each of the small otoliths taken from sac fry was placed individually, sulcus side up, in a drop of thin-section epoxy (Hillquest, Seattle, Washington) on a glass slide. Slides were then placed in an oven at 55°C for 5–10 min to remove air bubbles. Upon removal from the oven, the otoliths were reexamined for proper orientation and repositioned if necessary. Any remaining air bub-

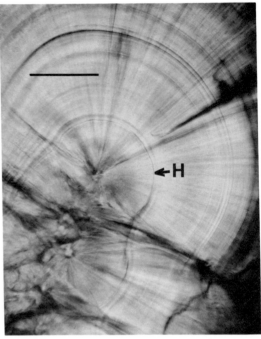

FIGURE 3.—Section of an otolith taken from an unmarked fingerling lake trout (control) 6 months after the marking period. H = hatching mark. Bar = 200 μm.

bles were removed with a needle. Setting of the epoxy required overnight storage at room temperature. Small otoliths were ground to the midsagittal plane with 600-grit silicon carbide sandpaper, followed by 1,000-grit silicon carbide grinding abrasive (Bruce Bar, Bruce Products, Howell, Michigan) on a glass plate. It was not difficult to locate the midsagittal plane. The first area of continuous growth surrounding the otolith core structure (Geffen 1983; Brothers 1990) was relatively featureless and the hatching mark formed a well-defined reference mark (Figure 3). We periodically checked each otolith microscopically (40–400×) during the grinding process and continued to remove material until the hatching mark was visible. If it was visible, any thermal marks laid down outside it were also visible. After the otoliths were ground, they were polished with aluminum oxide polishing compound on a wet felt lap and etched with 1% (0.03 N) HCl for 15–30 s to enhance microstructural features. The weak acid accented the surface contours by removing material from the proteinaceous or dark portion of the growth increment (Pannella 1980; Mugiya et al. 1981). A small camel's-hair brush was used to expose the reactive surface to the acid, and also to remove any gas bubbles that blocked the field of view. Etching time was best regulated by observation of the process under a dissecting microscope; continuous observation was particularly important for small otoliths because it helped prevent over-etching and loss of microstructural features.

Larger otoliths (from fingerling lake trout held for 6 months after marking) were more opaque and required grinding on both sides. Caps from 1.5-mL microcentrifuge test tubes (Bio-Rad Laboratories) were used as molds for embedding the otoliths. The caps were tapered and slightly wider at the open end; this shape helped secure the preparation while the first side of the otolith was ground. The caps were filled with thin-section epoxy and an otolith was positioned sulcus side down in each. The epoxy was then heated as previously described. After setting overnight, the epoxy plugs were pried from the caps and placed on a hard flat surface with the otolith end down. The caps were forced partly over them so that the embedded otolith remained beyond the cap lip. The cap was then used to grip the epoxy plug during grinding. Otoliths were ground to the midsagittal plane on the first side with 600-grit silicon carbide sandpaper, followed by 1,000-grit silicon carbide grinding abrasive, on a glass plate. The epoxy plugs were periodically checked microscopically (40–400×) to ensure that they were ground precisely to the midsagittal plane. After the first side of the otolith was ground, the flat surface was washed with 95% ethanol, polished, dried in air, and bonded to a glass slide with a drop of cyanoacrylate glue. Epoxy resin overlying the otolith was removed with 400-grit sand paper before the second side was ground, polished, and etched as described for small otoliths.

To evaluate the success rate of our sectioning technique, we attempted to remove and section the otoliths from a sample of 40 lake trout of the Lake Superior strain (20 control fish and 20 marked 3,6) 6 months after marking. Sections were permanently mounted and later evaluated for mark recognition.

Mark recognition.—The otolith sections were wetted with mineral or immersion oil and viewed without a coverslip by using a compound microscope and transmitted light. The magnification required to detect a thermal mark on an otolith varied from 400× for the best sections to 1,000× for most other sections. Counting the number of accentuated daily rings within a group of rings always required 1,000×. The camera port on the microscope was fitted with a high-resolution video camera that enhanced both resolution and contrast. The camera was helpful in counting individual rings and resolving uncertainties about some of the marks.

In an evaluation of the readability of the marks, three readers independently examined 41 permanently mounted otolith sections. Two of the readers had previous experience in aging fish from otoliths; the third, an experienced histologist, did not. The test sample consisted of 19 otolith sections from unmarked control fish and 22 from fish marked 3,6 (1 each from 18 fish and 2 each from 2 fish). The slides were viewed in a random sequence; readers knew that the marked fish had been marked 3,6. The reader was asked to judge whether each fish was marked or unmarked, and count the number of rings evident in each group of rings. Because the reading of each otolith had two possible outcomes (correct or incorrect), the proportion of fish correctly identified as marked or unmarked by the three readers and the 95% confidence intervals were calculated by use of the binomial distribution.

Results and Discussion

Marking

The sac fry showed neither unusual behavior nor increased mortality during the marking process. Although the rise in temperature was as much as 12.8°C during a temperature cycle, the sac fry never seemed distressed. Daily estimates of mortality were not attempted, but the numbers of sac fry lost in 1986 and 1987 were low from the start of marking through stocking, and no increases in mortality associated with marking were noted. In 1988, the mean number dead after 90 d in four lots of 500 control fry was 10.5 (SD = 4.65) and in four lots of 500 marked fry was 13.8 (SD = 6.34). The number dead in the marked groups was not significantly different (ANOVA, $P = 0.44$) from the number dead in the control groups. Examination of a more delayed mortality caused by the temperature changes was not within the scope of this study but should be investigated.

The marking process was inexpensive relative to the number of fish that could be marked. The flow of heated water was controlled by washing-machine solenoid valves that were, in turn, controlled by ordinary appliance timers. Because no adjustments were required after a series of temperature cycles began (Figure 1), personnel costs were low because only periodic checks of the equipment were required. The largest cost was that of the propane used to heat the water. If efficiency were 100%, each cycle (Figure 1) required 8.2 kg of propane to supply heated water to four 16-tray incubators, or about 74 kg for the 3,6 mark. Our apparatus was used to mark about 650,000 fry at a time, so these costs were low.

Sectioning of Otoliths

We successfully prepared sections of one or both otoliths from 39 of 40 fingerling lake trout. We believe that we could do even better with additional practice.

Because thermal marks were not equally clear and well-defined in all sectors of an otolith, it was critical that the section be kept in the sagittal plane. On most otoliths, daily growth increments, including thermal marks, were better defined on the posterior portion. In some, only a limited sector toward the posterior end of the otolith contained enough detail to allow interpretation. If that end of the otolith were tilted downward during grinding, the area where the thermal marks were defined best could be lost before the hatching mark was revealed.

The sectioning of otoliths marked in an early life stage and later recovered from fish of a larger size is a difficult but manageable task. Although we sectioned only otoliths taken from fingerlings, we are confident that the technique would also yield reliable results with those from larger fish. Because of the small size of the otoliths at marking, the greatest challenge was to obtain a section through the small area that contained the marks. If the absence of marks is also of interest, it is absolutely critical that the section be made in an appropriate plane. Otherwise, marks cannot be seen, and the fish would be misclassified as unmarked. Use of the hatching mark to establish the proper plane consistently resulted in sections that were suitable for determining the presence or absence of marks. Volk et al. (1984) used the otolith core as a guide for establishing sections through the centers of otoliths. Either the core or hatching mark can serve well as a reference mark. However, if marking were done before hatching, the otolith core would of course have to be used as the guide.

The process of grinding, polishing, and mounting otoliths by hand was very exacting and severely limited the number of sections that could be prepared each day. One person could process no more than 25 otoliths in a day if both sides were to be ground. With experience, greater productivity should be possible, and elimination of

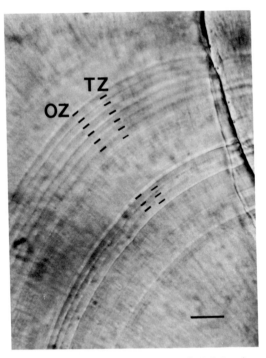

FIGURE 4.—Section of an otolith marked 3,6, taken from a fingerling lake trout 6 months after the marking period. OZ = opaque zone; TZ = translucent zone. Bar = 20 μm.

acid etching might further accelerate the sectioning process without reducing readability (E. Brothers, EFS Consultants, personal communication). The number of otoliths that can be sectioned per day might also be increased by grinding multiple otoliths of similar size at the same time (E. C. Volk, Washington Department of Fisheries, personal communication).

Mark Recognition

Daily marks on the otoliths consisted of a broad translucent zone, followed by a narrow opaque zone. Differences in the transparency of these zones gave the appearance of hills (translucent zones) and valleys (opaque zones). The temperature cycles increased both the width and the contrast of the daily increments. The most visible features of daily marks were the narrow opaque zones (Figure 4) that Brothers (1990) associated with the end of each temperature cycle (thermal night). He also believed that the fluctuations in temperature accentuated the contrast between these elements. A series of temperature cycles produced a regular pattern that was difficult to mistake for one produced by random events (Figure 4). The broader translucent zones were formed during the periods of elevated temperature and were wider than normal daily increments at ambient temperatures.

When we counted rings, we believed it was best to concentrate on the opaque zones because the inner edge of the first translucent zone was often poorly defined (see the inner edge of the six-ring group in Figure 4). However, a distinct opaque zone preceding the first translucent zone was not always due to the marking process. If a distinct random mark (as in Figure 3) preceded the marking process by a day, the illusion of an additional mark could be created (as on the inner edge of the three-ring group in Figure 4).

Otoliths from control fish were not featureless. Some of the daily rings on otoliths from control fish were as distinct as those accentuated by the temperature cycles (Figure 3). These usually appeared in random patterns, but in some cases a series of two or three clearly defined rings (Figure 3) could be mistaken for thermal marks. The distinctness of natural daily rings tended to increase with distance from the otolith core. Confusion between distinct natural daily rings and thermally accentuated rings on lake trout otoliths would be minimized if marking occurred before or soon after hatching.

Otolith sections from reference collections of sac fry taken 2–4 d after marking failed to show all the rings from the most recent mark. The outer rings appeared to be crowded near the edge and were difficult to count (Figure 5). However, all unsectioned otoliths examined fresh at the time the reference collections of sac fry were taken clearly showed the presence of the most recent mark. We are not sure of the reason for this discrepancy. Because the marks were clearly present in fresh specimens 2–4 d after marking and in specimens killed 6 months later, the marks must have been present when the reference collections were taken. The outer marks were either obscured by diffraction at the edge of the otolith, or etching and sectioning affected the thin outer margin. Rings should be visible shortly after marking is completed, but to minimize the possibility of losing information near the edge, we advise holding fish for at least a week after marking before any reference collections are taken. Because of this problem, our evaluations were confined to collections taken in 1987, 6 months after marking.

Independent reading of a blind sample of 41 otolith sections from fish held 6 months after

FIGURE 5.—Section of an otolith marked 3,6, taken from a lake trout sac fry 2 d after the marking period. E = otolith edge. Bar = 20 μm.

marking indicated that thermally marked fish could be separated from control fish with reasonable certainty. Of 41 sections examined, all were properly classified by the first reader, 40 (98%) by the second, and 35 (85%) by the third. Overall, two marked fish were misclassified as control fish and five control fish as marked fish. The 95% confidence intervals for the proportions of otolith sections that would be classified correctly by the three readers in similar samples of the same size were 93–100, 92–100, and 73–97%.

Our success in classifying marked and control fish suggested that, if otoliths are properly sectioned and if the information is retained on the otoliths of adults, we could identify thermally marked adult lake trout. The three readers together correctly identified 94% of the otoliths examined, and the two experienced readers together correctly identified 99%. This observation further suggests that a reader with experience should be able to identify marked fish with almost no errors.

Counts of individual rings within groups of rings frequently differed from the number of temperature cycles to which the fish had been exposed. None of the counts for the six-ring groups differed

TABLE 1.—Distribution of the number of rings counted (percentages in parentheses) by three independent readers on 22 otoliths marked with groups of three and six daily rings accentuated by temperature manipulations. Because readers 1 and 3 misclassified a marked fish as unmarked and readers 2 and 3 did not assign a count to a three-mark group, row totals are less than 22.

Reader	Number of rings counted in three-mark group					Number of rings counted in six-mark group		
	0	1	2	3	4	5	6	7
1		1 (5)	9 (43)	11 (52)			19 (90)	2 (10)
2	2 (10)	1 (5)	7 (33)	9 (43)	2 (10)	2 (9)	19 (86)	1 (5)
3			4 (20)	15 (75)	1 (5)		19 (90)	2 (10)
Total	2 (3)	2 (3)	20 (32)	35 (56)	3 (5)	2 (3)	57 (89)	5 (8)

from the expected value by more than one ring (Table 1). For the three readers combined, 89% of the six-ring groups were judged as containing six rings. The variability in counts was greater for the three-ring groups (Table 1), for which counts ranged from zero (three missed rings) to four (one extra ring). For the three readers combined, only 56% of the three-ring groups were judged to contain three rings. The readers were aware of the expected counts, and variation in a blind test would probably have been greater.

The variability in our counts of individual rings within the three-ring and six-ring groups suggests two conclusions: (1) with the marking procedure we used, counts often varied by at least one from the count expected; and (2) it was less difficult to count the rings in the six-ring group than in the three-ring group. Because the thermally induced marks were usually strongly defined, we believe that the errors were largely caused by uncertainty about whether naturally occurring rings were also part of the mark. Because the regular pattern established in a group of six consecutive rings made it easier to judge whether a ring was part of the pattern, we believe that five or more rings should be grouped together. Control of lighting and ambient temperatures (not done in our study) might also reduce the number of naturally occurring distinct rings and thereby reduce variability of the counts.

Thermal marking of otoliths, as we approached it, would have its greatest utility in situations where a large number of marks were not required. Otherwise, the variability in ring counts could result in confusion between marks. Because we had few groups of fish to separate, it was possible to use marks that were very different and not easily confused. Many important issues concerning contributions of hatchery and wild fish could also be settled with use of a small number of distinct marks. With more precise temperature control, more elaborate marks are possible (Brothers 1990).

Acknowledgements

We thank John Hudson for his assistance in sectioning the otoliths and Michael Mac for reading the otoliths. This paper is contribution 708 of the National Fisheries Center–Great Lakes.

References

Berst, A. H., and G. R. Spangler. 1973. Lake Huron—the ecology of the fish community and man's effects on it. Great Lakes Fishery Commission Technical Report 21.

Binkowski, F. P. 1984. Stocking practices. Great Lakes Fishery Commission Technical Report 40:10–14.

Brothers, E. B. 1990. Otolith marking. American Fisheries Society Symposium 7:183–202.

Christie, W. J. 1973. A review of the changes in the fish species composition of Lake Ontario. Great Lakes Fishery Commission Technical Report 23.

Eshenroder, R. L., T. P. Poe, and C. H. Olver, editors. 1984. Strategies for rehabilitation of lake trout in the Great Lakes: proceedings of a conference on lake trout research. Great Lakes Fishery Commission Technical Report 40.

Foster, N. R. 1984. Physiology and behavior. Great Lakes Fishery Commission Technical Report 40:22–27.

Geffen, A. J. 1983. The deposition of otolith rings in Atlantic salmon, Salmo salar L., embryos. Journal of Fish Biology 23:467–474.

Great Lakes Fishery Commission. 1983. Summary of management and research. Great Lakes Fishery Commission Annual Report 1981:20–32.

Hartman, W. L. 1973. Effects of exploitation, environmental changes, and new species on the fish habitats and resources of Lake Erie. Great Lakes Fishery Commission Technical Report 22.

Krueger, C. C., B. L. Swanson, and J. H. Selgeby. 1986. Evaluation of hatchery-reared lake trout for reestablishment of populations in the Apostle Islands Region of Lake Superior, 1960–84. Pages 93–107 in R. H. Stroud, editor. Fish culture in fisheries management. American Fisheries Society, Fish Culture and Fisheries Management sections, Bethesda, Maryland.

Lawrie, A. H., and J. F. Rahrer. 1973. Lake Superior—a case history of the lake and its fisheries. Great Lakes Fishery Commission Technical Report 19.

Mugiya, Y., N. Watabe, J. Yamada, J. M. Dean, G. G. Dunkelberger, and M. Shimuzu. 1981. Diurnal rhythm in otolith formation in the goldfish, *Carassius auratus*. Journal of Comparative Biochemistry and Physiology A, Comparative Physiology 78:289–293

Pannella, G. 1980. Growth patterns in fish sagittae. Pages 519–560 in D. C. Rhoads and R. A. Lutz, editors. Skeletal growth of aquatic organisms: biological records of environmental change. Plenum, New York.

Pearce, W. A., R. A. Braem, S. M. Dustin, and J. J. Tibbles. 1980. Sea lamprey (*Petromyzon marinus*) in the lower Great Lakes. Canadian Journal of Fisheries and Aquatic Sciences 37:1802–1810.

Smith, B. R., and J. J. Tibbles. 1980. Sea lamprey (*Petromyzon marinus*) in Lakes Huron, Michigan, and Superior: history of invasion and control, 1936–78. Canadian Journal of Fisheries and Aquatic Sciences 37:1780–1801.

Volk, E. C., R. C. Wissmar, C. A. Simenstad, and D. E. Eggers. 1984. Relationship between otolith microstructure and the growth of juvenile chum salmon (*Oncorhynchus keta*) under different prey rations. Canadian Journal of Fisheries and Aquatic Sciences 41:126–133.

Wells, L., and A. L. McLain. 1973. Lake Michigan—man's effects on native fish stocks and other biota. Great Lakes Fishery Commission Technical Report 20.

American Fisheries Society Symposium 7:224–231, 1990

Use of the Brain Parasite *Myxobolus neurobius* in Separating Mixed Stocks of Sockeye Salmon

ADAM MOLES, PATRICIA ROUNDS, AND CHRISTINE KONDZELA

National Marine Fisheries Service, Alaska Fisheries Science Center, Auke Bay Laboratory
Post Office Box 210155, Auke Bay, Alaska 99821, USA

Abstract.—Prevalence of the myxosporidian brain parasite *Myxobolus neurobius* is being used to separate international stocks of sockeye salmon *Oncorhynchus nerka* in southeastern Alaska. Distribution studies have shown that sockeye salmon from southeastern Alaska are mostly parasitized, whereas those of Canadian origin mostly lack the parasite. Parasite prevalence is stable from year to year. Parasite presence, in combination with genetic loci detectable with electrophoretic techniques and characteristics of scales could add precision to stock-separation estimates and can be used to detect shifts in stock abundance during a single season.

Commercial net fisheries in southeastern Alaska harvest sockeye salmon *Oncorhynchus nerka* from the short island and mainland streams of Alaska, and intercept fish of Canadian origin bound for the transboundary rivers of British Columbia. Interception of Canadian fish as they migrate through U.S. territorial waters has generated concern for stock separation based on country of origin. Over 100 streams producing sockeye salmon exist throughout southeastern Alaska, and have been estimated to contribute 69% of the 1–2 million sockeye salmon harvested annually (McPherson and McGregor 1986). Several methods of stock separation have been used to assess the stock contributions to the mixed-stock fisheries, the most common being examination of scale characteristics. This method assigns probable origin based on distances between circuli on the scale.

Parasites have been successfully used as biological indicators to differentiate between populations of freshwater fish (Black 1981; Dick and Belosevic 1981; Frimeth 1987), Pacific herring *Clupea harengus pallasi* (Arthur and Arai 1980), and Pacific salmon *Oncorhynchus* spp. (Uzmann et al. 1957; Margolis 1963; Konovalov 1971). Parasites have been used to separate seven stocks with precision in the Fraser River (Bailey and Margolis 1987). Parasites picked up during early freshwater life were used to separate Asian and North American stocks of marine sockeye salmon (Margolis 1963). The plerocercoids of *Triaenophorus* were unique to Alaskan stocks of sockeye salmon, whereas the nematode *Truttaedacnitus* was limited to Asian sockeye salmon. Compared to artificial markers, parasites are less expensive

to analyze, do not involve mark-and-recapture techniques, require smaller sample sizes, and are stable from year to year.

Parasites have shown particular potential for determining the nation of origin of sockeye salmon in the northern British Columbia–southeast Alaska mixed-stock fishery. Research has focused on the sporozoan brain parasite *Myxobolus neurobius* (10 μm), which infects the fish in fresh water and remains throughout the period of marine residence. Initial sampling of streams indicated that sockeye salmon in the largely coastal streams in southeastern Alaska were parasitized, whereas sockeye salmon from the interior Canadian river systems in British Columbia generally lacked the parasite (Leo Margolis, Department of Fisheries and Oceans, personal communication).

If the principal Alaskan stocks are all parasitized at stable levels, parasite presence could serve as a useful tag for separating salmon stocks in the mixed fisheries in southeastern Alaska. Determining the parasite prevalence in Alaskan fishery management areas 101 to 111 (Figure 1) would provide information on the relative contributions of Alaskan and Canadian stocks to the fishery. During the season, fisheries managers could get such information on a weekly basis, because the techniques, equipment, and labor required for parasite analysis are simple compared to scale and electrophoretic analyses.

Despite the increasingly important role of *M. neurobius* in stock separation, its basic biology is not well known. Dana (1982) described the morphology of the vegetative and spore stages in detail, and proposed a life cycle based on transmission experiments and periodic examination of a natural sockeye salmon population. He suggested that *M. neurobius* (like many other Myx-osporea) requires only one host and that spores are released from deteriorating carcasses of spawners, which are ingested by salmon fry the following spring. Dana (1982) also proposed that, once ingested, the immature spores migrate to the brain of the fish, where they mature and remain until the host's death. Its presence throughout the life of the fish suggested that *Myxobolus* would make a useful tag.

We undertook the present study to examine in detail the distribution of *Myxobolus* in southeastern Alaska and to better understand those factors that bear on the use of the parasite as a tag. We report on the distribution of the parasite, its year-to-year stability, and the initial results related to its use as a tag for stock separation.

Methods

Alaska Department of Fish and Game (ADFG) and National Marine Fisheries Service (NMFS) personnel sampled sockeye salmon in 45 spawning streams in southeastern Alaska (Figure 1) in 1986 and 1987 to establish the baseline distribution of *Myxobolus*. Locations were chosen to cover a wide geographical range; particular attention was paid to streams with sizable runs of sockeye salmon. Heads of 50–100 returning spawners were collected from each stream, often with matching scale and tissue samples for electrophoretic analyses, frozen, and sent to the NMFS laboratory at Auke Bay, Alaska.

In the laboratory, brains were removed and examined for parasite presence or absence by means of a variation of the pepsin-digest method developed for the detection of whirling disease (Canadian Department of Fisheries and Oceans 1984). A pepsin digest in a horizontal shaker bath was followed by centrifugation at 1,200× gravity for 5 min. A slide made from the pellet was examined with a phase-contrast microscope (250× magnification) for the presence of spores. Any sample with 1–10 spores was rerun to confirm the presence of spores; the presence of a few spores could have been the result of cross-contamination. The presence of at least 10 spores in 300 fields was required for a positive sample. Levels of infection (i.e., heavy, light) in each fish were not estimated. Care was taken to avoid cross-contamination, by use of disposable supplies for each sample. Sex, length, weight, and age data were collected for each fish, and matching samples were taken for scale and electrophoretic determinations of stock identity.

In 1986 and 1987, sockeye salmon heads collected from fish in the commercial gill-net fisheries were examined weekly for the presence of *Myxobolus* to assess use of the parasite as an inseason stock-separation tag. Each week in 1986, approximately 300 matched parasite, scale, and tissue samples were taken from commercial fishery samples in areas 104 and 106, as well as from fishery samples in areas 108 and 111 (Figure 1). In 1987, only areas 106-41 and 111 were sampled. Nine thousand heads were examined over the 2-year period.

The utility of the parasite as a stock-separation tool for sockeye salmon was examined by comparing estimates based on scale characteristics with those based on parasite prevalences. The percentage of each stock's contribution to a given week's fishery (based on scale analysis) was multiplied by the estimated amount of parasitism of that stock, determined from baseline distribution data. For example, scale characters were used to estimate that 7% of the sockeye salmon in the area-104 seine fishery in week 34 were of McDonald Lake origin. Parasitism for McDonald Lake fish was 100%. Therefore 7% of the fish in the fishery should be parasitized. The calculation was repeated for the other three stocks present in the fishery to give a total estimated number of parasitized fish for that week. This estimated level of parasitism in the fishery based on scale analysis was compared with the actual level of parasitism.

Results and Discussion

Baseline Distribution and Characterization

Parasite distribution was a function of the characteristics of the freshwater rearing environment. Nearly 3,500 sockeye salmon from 45 baseline sites in southeastern Alaska were sampled for the presence of *Myxobolus* over a 2-year period (from 1986 to 1987). Thirteen of the locations were sampled both years. Prevalence ranged from 0 to 100%; in half of the streams, fish populations were 90 to 100% parasitized (Figure 1).

In the following lakes, more than 85% of fish were parasitized: Speel, Crescent, Windfall, Luck, Salmon Bay, Situk, Thoms, Red Bay, Karta, Helm, Kutlaku, Kook, Kegan, McDonald, Hugh Smith, Petersburg, Upper Sarkar, Auke, Redoubt, Klawock, Sitkoh, and Leask (Table 1). These waters are all nonglacial, single-lake systems that are typically within a few kilometers of the ocean. Spawning occurs in one to several streams flowing into a nursery lake, or in the lake. If a second nursery lake is present, the stream

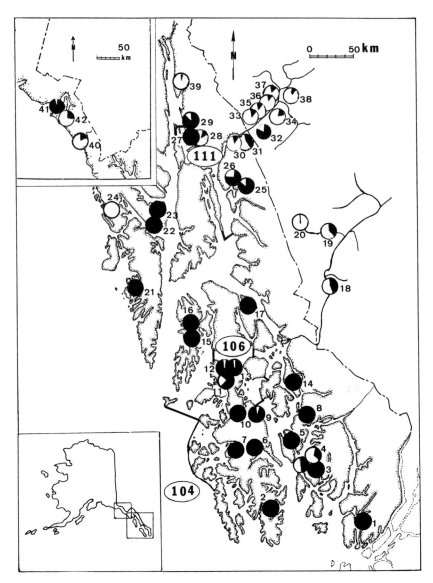

FIGURE 1.—Distribution of the brain parasite *Myxobolus neurobius* in returning adult sockeye salmon in southeastern Alaska. The dark portion of each circle corresponds to the percentage of fish parasitized. Each uncircled number corresponds to a stream listed in Table 1. Circled numbers are fishery areas.

between it and the first lake is short, usually less than a kilometer long (ADFG 1986). The streams that produce the vast majority of sockeye salmon in southeastern Alaska are of this coastal variety, in which fish are highly parasitized. Scale annuli show that young sockeye salmon grow in these nursery lakes for 2 years before migrating to sea; small proportions may migrate in their first summer or as 3-year-olds.

At some sites, returning adults were less than 30% parasitized; these sites were Ford Arm Lake,

Chutine Lake, Lace River, East River, Taku River, Steep Creek, and the Old Situk River (Table 1). Some of these rivers were sampled at multiple sites. The majority of sites sampled in this group were streams or side channels of large rivers; the only lake rearing systems containing fish with this low level of prevalence were Chutine and Ford Arm. These sites are all in northern southeastern Alaska (as are several lakes with heavily parasitized fish). Sites having fish with intermediate levels of parasitism (30–85%) were the Naha River, Stikine River,

TABLE 1.—Prevalence of *Myxobolus neurobius* in adult sockeye salmon returning to southeastern Alaska streams in 1986 and 1987. Map locations correspond to numbered areas on Figure 1.

Map location	Stock	Year	Sample size	Proportion of fish infected
1	Hugh Smith Lake	1986	10	1.00
		1987	127	0.99
2	Kegan Lake	1986	50	1.00
3	Leask Lake	1986	41	1.00
4	Naha River	1986	79	0.48
	Heckman Lake	1987	89	0.33
5	Helm Lake	1986	50	1.00
		1987	50	1.00
6	Karta Lake	1986	10	0.80
		1987	99	0.99
7	Klawock Lake	1987	50	0.99
8	McDonald Lake	1986	50	0.98
		1987	97	1.00
	Yes Bay	1987	10	0.90
9	Luck Lake	1987	49	0.95
10	Sarkar Lake	1986	44	0.98
		1987	42	1.00
11	Sutter Creek	1986	36	0.72
12	Red Bay Lake	1986	53	0.96
13	Salmon Bay Lake	1986	41	0.95
	Shaul Creek	1986	10	1.00
14	Thoms Lake	1986	50	1.00
		1987	60	0.98
15	Alecks Lake	1986	45	0.98
		1987	100	0.98
16	Kutlaku Lake	1986	42	1.00
		1987	50	1.00
17	Petersburg Lake	1986	50	1.00
18	Stikine River	1986	50	0.46
19	Chutine River	1986	50	0.40
20	Chutine Lake	1986	64	0.02
21	Redoubt Lake	1986	56	1.00
22	Sitkoh Lake	1987	100	1.00
23	Kook Lake	1987	80	1.00
24	Ford Arm Lake	1987	68	0.00
25	Crescent Lake	1987	100	0.87
26	Speel Lake	1986	100	0.85
		1987	100	0.74
27	Auke Creek	1986	32	1.00
		1987	99	1.00
28	Steep Creek	1986	28	0.00
		1987	50	0.16
29	Windfall Lake	1987	50	0.90
30	Yehring Creek	1986	89	0.12
		1987	60	0.13
31	Fish Creek	1986	20	0.40
32	South Fork Slough	1986	59	0.78
33	Tuskwa Slough	1986	60	0.13
		1987	52	0.15
34	Coffee Slough	1986	39	0.00
		1987	33	0.18
35	Shutahini Slough	1986	128	0.13
36	Chum Salmon Slough	1987	97	0.10
37	Honakta Slough	1986	59	0.19
38	Nakina River	1986	53	0.15
39	Lace River	1986	60	0.05
40	East Alsek River	1987	91	0.11
41	Situk River	1987	50	0.96
42	Old Situk River	1987	45	0.27

Chutine River, Sutter Creek, and two side channels on the Taku River. The Stikine and Taku Rivers are large and have both lake-type nursery areas and riverine habitats.

The distribution of parasitism was a function of habitat, apart from some glacial lakes and systems with high washout rates for carcasses. Except for the complex Stikine system, locations with glacial

TABLE 2.—Year-to-year variability in prevalence of *Myxobolus neurobius* in adult sockeye salmon. Map locations correspond to numbered areas on Figure 1.

Map location	Stock	Percentage of fish parasitized				
		1982[a]	1983[a]	1984[a]	1986	1987
5	Helm Lake		100		100	100
16	Kutlaku Lake				100	100
2	Hugh Smith Lake	93	95		100	99
10	Sarkar Lake (Upper)				98	100
14	Thoms Lake		99		100	98
1	Kegan Lake	100	88		100	
8	McDonald Lake	93	94		98	100
6	Karta River	97	94		80	99
15	Alecks Lake				98	98
13	Salmon Bay Lake	100	99		96	
25	Crescent Lake		98			87
26	Speel Lake		90		85	74
9	Luck Lake		98			95
4	Naha River		34	39	48	
	Heckman Lake					33
18	Stikine River		34		46	
19	Chutine River			47	40	
34	Coffee Slough				33	18
28	Steep Creek				0	16
33	Tuskwa Slough				13	15
30	Yehring Creek				12	13
20	Chutine Lake			9	2	

[a] Pacific Salmon Commission (1987).

turbidity had sockeye salmon populations with infection rates less than 20%. The Taku River channel habitats, Steep Creek Lake, Chutine Lake, and the Lace River are all glacially turbid and had fish populations with low levels of parasitism. The presence of a nonglacial nursery lake in a system favored high parasite levels among the fish. The heavily parasitized fish in nonglacial nursery lake systems represented over two-thirds of the estimated sockeye salmon escapement from southeastern Alaska. South Fork Slough, a lake-like habitat, had fish with the highest level of parasitism among Taku River sites. Carcass washout could also be a factor in reducing the presence of spores in a system. Although hard to quantify in river habitats, carcass washout should be greater in habitats with high water velocity and less depth (i.e., channel habitats) than in off-channel habitats and nursery-lake systems. Ford Arm Lake (0% parasitism), which in all other respects resembles a nursery lake habitat in which fish are highly parasitized, is estimated to have a carcass-washout rate of about 50% of returning adults (L. Shaul, ADFG, personal communication). Spawning in Steep Creek (0–16% parasitism) occurs in clear riffles, and carcasses are washed out to nearby glacial Mendenhall Lake, where the fry also grow.

Interannual Stability

Because juvenile sockeye salmon, particularly in a lake, share the same environment, they have an equal risk of parasitism. Parasite prevalence from year to year was relatively stable for the fourteen locations sampled in both 1986 and 1987 (Table 2). A *t*-test performed on prevalence data from 1986 and 1987 resulted in no significant difference between those years. This result means that individual stocks do not require annual re-sampling, as is the case with other stock-separation methods. Comparison of 1986–1987 levels with 1982–1985 levels (Table 2) supports the idea that prevalence is very stable, especially in systems in which 95% or more of the fish are parasitized. In the more complex river systems that support several stocks, the logistics of duplicating sampling sites, and timing with respect to stock migration, make year-to-year comparisons more difficult. In such systems, spawning areas may not be in rearing areas. In river systems where there is a lot of annual variation in conditions (e.g., the Taku and Stikine rivers), there is probably corresponding variation in degree of carcass washout and migration of fry to other places to grow.

Stability of parasitism was great in locations having high or low proportions of parasitized fish. Stability in systems having intermediate proportions of parasitized fish (30–85%) showed the widest variation. These areas are complex riverine habitats with multiple stocks. The resulting percentages of parasitized fish were probably a mixture of highly parasitized and lightly parasitized stocks. Factors affecting the relative numbers of fish of different stocks that mix together will affect the percentage of parasitized fish in the entire system.

Stock Separation

Distribution studies showed that most of the southeastern Alaska sockeye salmon stocks were heavily parasitized with *Myxobolus neurobius*. In contrast, relatively few Canadian stocks were parasitized with the brain spore (Pacific Salmon Commission 1987). The primary contributing streams, particularly in southern southeastern Alaska, all had heavily parasitized fish. The proportion of parasitized sockeye salmon in a mixed-stock fishery sample could be interpreted as the maximum contribution from Alaskan stocks.

Parasite prevalence in sockeye salmon taken in the area-111 gill-net drift fishery ranged from 17 to 87% in 1986. The stock groups that constituted the area-111 fishery, as identified by scale-pattern analysis, were mainstem Taku River, Little Trapper Lake, Tatsamenie Lake, Crescent Lake, Speel Lake, and Kuthai Lake (McGregor and Walls 1987). Parasite prevalence was 85% among fish in Crescent and Speel lakes and 100% among fish in Kuthai Lake in the Canadian reaches of the Taku River system. Fish in the mainstem Taku River showed a 13% prevalence, and the other stocks all were generally free of the spore.

The percentage of fish that each of these six stocks contributed weekly to the area-111 fishery has been estimated by linear discriminate function analysis of scale patterns (McGregor and Walls 1987). If the weekly contribution of each stock is multiplied by the percentage of fish parasitized in that stock, the figure should agree with the proportion of fish that were parasitized in that week's sample. For area 111, the agreement between scale-pattern analysis and parasite prevalence in 1986 is within 5% (Figure 2a). In Figure 2, a column showing actual percentage represents the weekly level of parasitism and a column showing predicted percentage represents the percentage of the fish that should have been parasitized based on scale characters.

A similar analysis was performed on data from areas 106 and 104. There were five major stocks contributing to the area-106-41 gill-net drift fish-

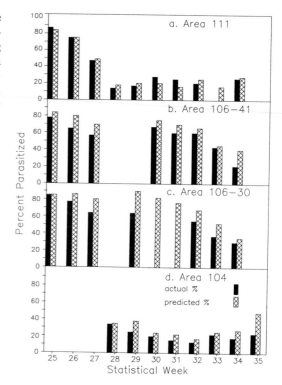

FIGURE 2.—Prevalence of *Myxobolus neurobius* in the brains of sockeye salmon in the 1986 southeastern Alaska commercial sockeye salmon fishery areas (subareas for area 106) and the predicted level of parasitism based on stock estimates from scale-pattern analysis.

ery: Alaska I, Alaska II, Nass–Skeena, Tahltan, and non-Tahltan Stikine (Jensen et al. 1988). The Alaska I stock was composed of fish from the myriad coastal and island streams throughout southern southeastern Alaska, and the Alaska II stock was composed primarily of fish from McDonald Lake.

Almost all of the populations of sockeye salmon in southeastern Alaska are heavily parasitized. The only major stock in the area-106-41 mixed fishery that was not heavily parasitized was that from the Naha River. The three Canadian stocks had less than 2% prevalence of the parasite (Pacific Salmon Commission 1987). The percentage of fish parasitized should have been the same as the percentage of fish of Alaskan stocks, as determined by scale-pattern analysis. If scale-pattern and parasite data were both drawn from the same stocks, both bars in Figure 2 should be the same height. The percentage of fish with *Myxobolus* in 1986 was consistently 5–15% lower than the values predicted by scale analysis (Figure 2b). Both bars should estimate the maximum contribution of U.S. fish to the fishery. A similar comparison of parasite prevalence in and scale patterns of fish from areas 106-30 and 104 in 1986 also resulted in underestimates based on percentage of parasitism (Figure 2c, d). Rigorous statistical analysis of matched scale, electrophoretic, and parasite data is needed to explain the discrepancies.

The differences in stock estimates based on scale characters and parasite prevalence could be accounted for by overestimation of the Alaskan contribution, based on scale characters, or by the possibility that the commercial fishery is more heavily exploiting stocks of sockeye salmon with low levels of parasitism. Systems like the Naha River may contribute more heavily to the fishery than to the escapement figures.

It is clear, however, that the vast majority of southeastern Alaska sockeye salmon are infected with the parasite and that the parasite's distribution is a function of characteristics in the freshwater rearing environment. All of the factors that are correlated with parasite distribution are related to the extent of the exposure of the host to the parasite. The less variable the habitat and the longer the fry are exposed, the greater the prevalence of the parasite within a system. Fish that share the same lake environment also share the same risk of parasitism.

Myxobolus shows great promise as the subject of a low-cost method of estimating the national origin of sockeye salmon in the mixed-stock fishery. It may prove to have particular utility in separating stocks such as the Stikine and McDonald, which have similar scale characters but much different rates of parasitism (33 versus 100%). Used together, scale characters and parasites may allow good separation of some stocks, because each method results in some error through misclassification.

Myxobolus also shows excellent promise as a tool for stock separation in southeastern Alaska. The level of parasitism in a given system is stable over time; stability eliminates the need for constant baseline sampling that is required for other methods. Only a small sample size is needed because data (presence or absence) are easy to collect and analyze. The heads are available as a by-product of processing, and a single investigator can analyze over 100 samples per day. The precision of the method could be improved even more by the addition of other parasites to the analyses.

Acknowledgments

We thank Andrew McGregor, Jan Weller, Keith Pahlke, Karl Hoffmeister, and the port sampling crews of the Commercial Fisheries Division of the Alaska Department of Fish and Game for collecting the thousands of samples needed to complete this study. We also thank Glen Oliver, Andrew McGregor, and Kathleen Jensen of ADFG for their assistance in all phases of this work.

References

ADFG (Alaska Department of Fish and Game). 1986. Catalog of waters important for spawning, rearing or migration of anadromous fishes. ADFG, Southeastern Region Resource Management Region I, Juneau.

Arthur, J. R., and H. P. Arai. 1980. Studies on the parasites of Pacific herring (*Clupea harengus pallasi* Valenciennes): survey results. Canadian Journal of Zoology 58:64–70.

Bailey, R. E., and L. Margolis. 1987. Comparison of parasite fauna of juvenile sockeye salmon (*Oncorhynchus nerka*) from southern British Columbian and Washington State lakes. Canadian Journal of Zoology 65:420–431.

Black, G. A. 1981. Metazoan parasites as indicators of movements of anadromous brook charr (*Salvelinus fontinalis*) to sea. Canadian Journal of Zoology 59:1892–1896.

Canadian Department of Fisheries and Oceans. 1984. Fish health protection regulations: manual of compliance. Canadian Special Publication of Fisheries and Aquatic Sciences 31 (revised).

Dana, D. 1982. The biology of transmission of *Myxobolus neurobius* Schuberg and Schroder, 1905, a myxosporean parasite of salmonid fishes. Master's thesis. Simon Fraser University, Burnham, Canada.

Dick, T. A., and M. Belosevic. 1981. Parasites of Arctic charr *Salvelinus alpinus* (Linnaeus) and their use in separating sea-run and non-migrating charr. Journal of Fish Biology 18:339–347.

Frimeth, J. P. 1987. Potential use of certain parasites of brook charr (*Salvelinus fontinalis*) as biological indicators in the Tabusintac River, New Brunswick, Canada. Canadian Journal of Zoology 65:1989–1995.

Jensen, K. A., G. T. Oliver, and I. R. Frank. 1988. Scale pattern analysis of sockeye salmon stock compositions in Alaska's districts 106 and 108 and Canada's Stikine River fisheries, 1986. Alaska Department of Fish and Game, Division of Commercial Fisheries, Technical Fisheries Report 89–02, Juneau.

Konovalov, S. M. 1971. Differentiation of local populations of sockeye salmon *Oncorhynchus nerka* (Walbaum). University of Washington Publications in Fisheries New Series 6.

Margolis, L. 1963. Parasites as indicators of the geographical origin of sockeye salmon, *Oncorhynchus nerka* (Walbaum), occurring in the North Pacific Ocean and adjacent seas. International North Pacific Fisheries Commission Bulletin 11:101–156.

McGregor, A. J., and S. L. Walls. 1987. Separation of principal Taku River and Port Snettisham sockeye salmon (*Oncorhynchus nerka*) stocks in southeastern Alaska and Canadian fisheries of 1986 based on scale pattern analysis. Alaska Department of Fish and Game, Commercial Fisheries Division, Technical Data Report 213, Juneau.

McPherson, S. A., and A. J. McGregor. 1986. Abundance, age, sex, and size of sockeye salmon (*Oncorhynchus nerka* Walbaum) catches and escapements in southeastern Alaska in 1985. Alaska Department of Fish and Game, Technical Data Report 188, Juneau.

Pacific Salmon Commission. 1987. Stock identification of sockeye salmon using biological markers. Pacific Salmon Commission, Northern Boundary and Transboundary River Technical Committees, TC-NB/TR-8701, Vancouver, Canada.

Uzmann, J. R., R. H. Lander, and M. N. Hesselholt. 1957. Parasitological methods for identification and abundance estimates of downstream migrant races of salmon. Proceedings of the Alaska Science Conference 8:93–94.

American Fisheries Society Symposium 7:232–236, 1990

Washington Department of Fisheries' Mobile Tagging Units: Construction and Operation

GERALD C. SCHURMAN AND DANIEL A. THOMPSON

Washington Department of Fisheries, 115 General Administration Building
Olympia, Washington 98504, USA

Abstract.—The Washington Department of Fisheries (WDF) has been using the coded wire tag extensively since the late 1960s. To expedite its rapidly growing program, WDF developed the self-contained mobile tagging unit in 1972. Currently, WDF operates five mobile tagging units and tags approximately 8 million fish annually. Another 1–2 million fish are marked in the trailers during other fish identification projects, including freeze-branding and fin-marking. In the spring of 1986, WDF completed construction of two mobile tagging units. Optimal operational design was incorporated into the new units, which utilize the most advanced coded-wire-tagging technology. These trailers were designed and constructed mostly by WDF employees, thus keeping costs down and allowing flexibility during construction. Materials for construction and supplies for operation cost approximately US$30,000 per unit, not including the coded-wire-tagging equipment (quality control device, injector, power supply). The WDF tagging system separates tagging from marking within the trailer and operates with a crew of 10 and 1 supervisor. More than 30,000 fish can be tagged and fin-clipped during one 8-h shift, and up to 100,000 fish/d can be fin-marked by a crew of 16. The highly efficient operation of the units results in a high-quality marking–tagging operation, with minimum stress to the fish.

Before 1985, the Washington Department of Fisheries (WDF) operated three mobile tagging units and marked 5.5 to 6 million Pacific salmon *Oncorhynchus* spp. and steelhead *O. mykiss* (formerly *Salmo gairdneri*) annually. This number includes all fish tagged with coded-wires, fin-marked, and freeze-branded. The signing of the Pacific Salmon Treaty Act and the development of the Lower Snake River Compensation Plan expanded WDF's annual marking requirements to more than 10 million fish after 1985. To accommodate this increase, funds were provided for the construction of two new mobile tagging units in the initial implementation budgets of both programs. The WDF has been operating mobile units since 1982, and was thus able to design an optimum facility that used the most recent coded-wire-tagging technology and allowed flexibility in adaptation to other marking procedures. The goal was to design a trailer that was durable and virtually maintenance-free, one that could be constructed primarily by WDF employees; thus costs were reduced during the construction process. The trailer had to function in such a way as to keep stress on fish to a minimum during marking operations, while allowing the crew to operate at maximum speed and efficiency.

Construction and Operation

The mobile tagging unit is a trailer, used mainly to tag salmonids with coded wire tags. (Figures 1 and 2). The unit is completely self-contained, with its own sinks, tables, troughs, anesthetic system, plumbing, pumps, and electrical system. The mobile unit can be moved to and operated at any fish-rearing station in the state. The trailer is set up next to the rearing pond that contains the fish to be tagged.

The mobile unit receives 220-V, single-phase electricity from an outlet that is specially provided at the tagging locations. A power cord, made of 8-gauge, four-strand wire with heavy weatherproof plugs on each end, attaches to the trailer and power source. The unit's electrical system supplies power to the lighting, heating, air-conditioning, and pumping systems, and also to various 110-V receptacles. The lighting system consists of two parallel rows of 1.17-m double-lamp fluorescent fixtures running the entire length of the trailer. The fixtures are lightweight and moisture-resistant. This system provides excellent light for the meticulous work being performed. Two 2,000-W heaters, as well as an air-conditioning unit, provide a comfortable working environment. The 110-V receptacles are used for the power-pack converters, which convert 110-V alternating current to 12-V direct current required to operate the coded-wire-tag injectors. The 3.4-kW water pump (Flygt model B2070) operates on 220-V single-phase electricity and is completely submersible. This pump provides 750L/min (at 207 kN/m^2 of pressure) to the intricate plumbing system within the trailer.

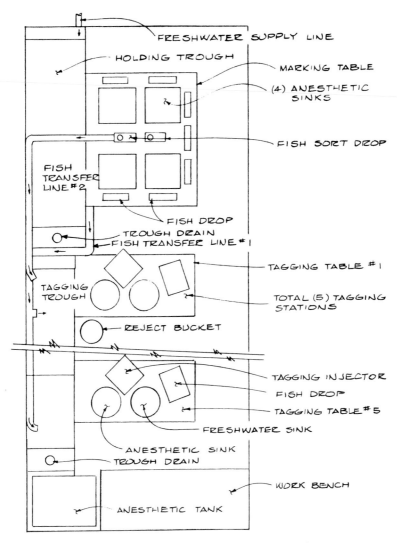

FRESHWATER SUPPLY LINE

HOLDING TROUGH

MARKING TABLE

(4) ANESTHETIC SINKS

FISH SORT DROP

FISH TRANSFER LINE #2

FISH DROP

TROUGH DRAIN

FISH TRANSFER LINE #1

TAGGING TABLE #1

TAGGING TROUGH

TOTAL (5) TAGGING STATIONS

REJECT BUCKET

TAGGING INJECTOR

FISH DROP

TAGGING TABLE #5

FRESHWATER SINK

ANESTHETIC SINK

TROUGH DRAIN

WORK BENCH

ANESTHETIC TANK

FIGURE 1.—Floor plan for mobile tagging unit.

Three sets of steps are required for the mobile unit, two for the entrance doors and one for the fish door. The steps were made of aluminum and were designed to be lightweight and stable. They break down into easy-to-handle components that are stored inside during transport.

The primary goal in the design and construction of the new units was to improve upon the older trailers in maintenance and overall operation, yet keep the marking system presently used and preferred by WDF. The older trailers required constant maintenance. Because the working area inside the trailers is subject to constant wetness and humidity, trailers required yearly scraping and painting. Metal and wooden parts rusted and

deteriorated, and required periodic replacement and repairs. The nonskid floor paint peeled in places, creating slick spots when the floor was wet.

The new trailer shells were constructed by the Wells Cargo Company of Ogden, Utah. The shells were 9.75 m long and 2.44 m wide, with a steel chassis and vertical body frame covered with aluminum siding. The entire trailer shell was insulated, as was the trailer floor. The primary objective for the interior of the shell was to eliminate maintenance. The floor consisted of two layers of 19-mm marine plywood covered with 3-mm aluminum treadplate welded to be 100% watertight. This type of floor eliminated the need

FIGURE 2.—Washington Department of Fisheries mobile tagging unit.

for nonskid paint and thus improved safety. The walls were covered with white Kemlite, a plastic-coated plywood paneling that is durable, scratch-resistant, and waterproof. The ceiling panels were covered with a white vinyl material that does not need painting.

During construction by WDF, all work tables were built with 19-mm marine plywood, covered with formica, and sealed with varnish (Figure 3). The prefabricated storage cabinets, purchased from a local source, were built to be strong enough to withstand the stress of extra weight and vibration during transport. They were covered inside and out with formica and do not need painting. All metal components, materials, and fasteners used in construction were aluminum or stainless steel, and will not rust.

The system used in WDF's units separates the fin-marking area from the tagging area (Figure 1). The main holding trough is in the front area of the trailer (Figure 3). This aluminum trough, 3.04 m long × 61 cm wide × 46 cm deep, is where fish to be marked are initially held. The holding trough is loaded with fish from the outside via an access door immediately adjacent to the trough. Fish are netted from the holding trough and anesthetized, and their adipose fins are clipped by crew members working at the marking table (Figure 3). Clipped fish are dropped into a 10-cm-diameter transfer line (Figures 1, 3, and 4), which carries them back to the tagging trough (Figures 1–4). This trough is 5.79 m long × 51 cm wide × 33 cm deep, and is used to hold fish for tagging. It was designed with five separate screened holding areas that hold fish for the individual tagging stations; fish are diverted to them from the transfer line coming from the marking table (Figure 4). In the tagging area, fish are dipped from the trough

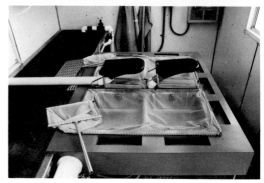

FIGURE 3.—Upper, front work area with tagging tables, sort lines, and fish transfer line. Lower, marking table and sorters.

and reanesthetized in the tagging table's anesthetic sink. When the fish are fully anesthetized, they are moved to a freshwater sink and the tag is injected (Figures 1–3). While fish from the freshwater sink are being tagged, another netful of fish is being anesthetized in the anesthetic sink. With the two-sink method, the machine operator is always working with anesthetized fish and does not have to wait while fish are being anesthetized. After tagging, the fish are passed through the quality control device (under each tagging table; Figure 5), which directs tagged fish through a 10-cm-diameter transfer line to a hatchery pond. Attached to this return line, on the outside of the trailer, is a 6.10-m section of 10-cm-diameter heavy-duty radial-flex hose, which allows the fish to be routed in any direction. Fish without a tag are diverted into a reject bucket (Figure 5) located at the tagging table and are retagged. The two-area system allows the supervisor maximum flexibility in handling the day-to-day operations of the unit. Flexibility is needed to control tagging quality and speed of production at each tagging location.

FIGURE 4.—Tagging trough, anesthetic cooling system, and fish transfer line.

FIGURE 5.—Upper, position of quality control device and reject bucket. Lower, rear work area and anesthetic tank.

The anesthetic system in the new units has been greatly improved. All WDF mobile units use recirculating anesthetic systems. The anesthetic solution is pumped from a main tank located at the rear of the trailer (Figure 5) to all the work stations through 3-cm-diameter polyvinyl chloride (PVC) pipe. The anesthetic returns via gravity flow to the main tank in a 5-cm-diameter PVC drainline. The new units were built with a 265-L main anesthetic tank. The anesthetic solution is pumped, with a 74.5-W submersible pump, through a cooling system to the work stations. The cooling system consists of 30.48 m of 3-cm-diameter aluminum pipe running under a false bottom built into the tagging trough (Figure 4). Water supplied to the trough circulates freely around the pipe and, by convection, keeps the anesthetic solution to within 1°C of the ambient water temperature over a 4-h period.

Sorting fish is an important function in the mobile units. Fish are sorted to remove those that are not to be tagged and also to separate fish into different size groups. Fish that are not to be tagged include fish with major injuries or deformities, fish of a different species and, in some cases, fish that are simply too small or too large to be tagged. Sorting fish by size and routing them to separate machines for tagging is routine in the mobile units. This sorting not only ensures that nearly 100% of the fish in the sample group will be tagged, but also improves the quality of tag placement and improves overall tag retention.

The design of the marking table and the system is optimal for sorting. Fish to be sorted are simply dropped into a sorter located at the middle of the marking table (Figure 3). Via a 5-cm-diameter PVC transfer line, they are sent to a tagging machine that is set up to tag fish of that specific size (Figure 3). The new trailers were built with more sorting capability than the older ones. A second sorter was installed in the center of the marking table, and fish can now be sorted and sent to three separate tagging machines. An added function of the second sorter is that it can be set up to allow fish to be sorted directly from the marking table into a 10-cm-diameter transfer line and back to the hatchery pond. This line is separate from the tagged-fish return line, so sorted fish can be sent wherever desired. An important factor in the sorting process is that our system places no more stress on the fish than the normal fin-marking procedure, and does not interfere with the flow of fish through the trailer. Because the sorting process generally slows down the marking crew, the marking table in the new trailers was built to accommodate one more person.

The mobile units were primarily designed and built to accommodate coded-wire tagging and related equipment. The WDF also uses the units for other fish-marking projects, mainly fin-clipping and freeze-branding. When fin-clipping is done, a crew of 16 is used. Two crew members work at each tagging table, using the quality control device's fish drop to return the fish to the pond, and six crew members work at the marking table. The numbers of fish clipped daily can vary greatly, depending on the size of the fish and the fin or combination of fins to be clipped. Each worker is expected to mark at least 5,000 fish/d.

The WDF has contracted to freeze-brand salmon and steelhead in support of the Lower Snake River Compensation Plan. Freeze-branding can also be used concurrently with fin-clipping and tagging with coded wire tags. The WDF mobile units can easily be set up for freeze-branding only, or for freeze-branding, fin-clipping, and tagging. When only branding is done, the trailer can accommodate a crew of eight, who use eight branding pots. Approximately 40,000 fish can be branded during an 8-h day at a cost of US$47/1,000 fish marked. (Costs are in 1990 US$ except for Table 1.) If the same fish is to be freeze-branded, fin-clipped, and tagged with coded wire tags, a second portable anesthetic sink must be added at each tagging station; the sink can be installed in minutes. A crew of 16 can brand, mark, and tag at least 25,000 fish/d. This costs $145/1,000 fish marked.

With a crew of five fin-markers, five machine operators, and one supervisor, the average daily coded-wire-tag output expected in the new mobile units ranges from 30,000 to 35,000 fish/d. Based on 30,000 fish/d, the cost is $98/1,000 fish tagged. These costs are currently used by WDF for budgeting purposes. The costs for constructing a completely outfitted mobile unit are given in Table 1; they do not include the salaries of WDF employees who worked on the project. Six person-weeks were used in planning, designing, and ordering

TABLE 1.—Costs for a mobile tagging unit (1986 US$).

Item	Cost
Construction costs	
Trailer shell	$17,500.00
Troughs (holding, tagging, anesthetic)	3,285.00
Main pump	2,102.00
Anesthetic pump	90.00
Cabinets	433.00
Electrical	3,000.00
Polyvinyl chloride pipe and fittings	1,200.00
Miscellaneous plumbing, valves, hose, fittings	450.00
Water hose (heavy-duty supply and outlet fittings and couplers)	450.00
Tables (sinks, lumber, formica)	1,200.00
Aluminum (straps, trim, bracing, sorters, net frames, fish-door steps, legs)	500.00
Netting (dipnets, rejects buckets, marking-table sinks)	425.00
Subtotal	$30,665.00
Supplies	
Scissors (12 Noyes, 12 Kneebent)	$ 1,200.00
Miscellaneous (hand tools, hardware, hand calculator, first-aid kit, scales, hand counters, trailer jack, jack stands, aprons)	800.00
Subtotal	$ 2,000.00
Tagging equipment	
Mark IV tagging injectors (five at $11,400 each, with quality control device)	$57,000.00
Grand total	$89,665.00

components and materials. A further 15 person-weeks were required for actual construction.

Summary

In its 16 years of experience operating mobile tagging units, WDF has produced a state-of-the-art mobile tagging unit incorporating flexibility and optimum design and construction considerations. During the last 2 years of continuous operation, the trailer has remained virtually maintenance-free, and its flexibility in other uses has been highly successful. The anesthetic, heat-exchange, and lighting systems are efficient. Cleaning and disinfecting the trailers has been made easier by the use of plastic-coated walls and aluminum floors.

Acknowledgments

We express our appreciation to Steve Phelps, Brett DeMond, and Kent Dimmitt for their involvement in our paper.

American Fisheries Society Symposium 7:237–243, 1990

Effects of Time and Fish Size on Coded Wire Tag Loss from Chinook and Coho Salmon

H. LEE BLANKENSHIP

Washington Department of Fisheries, 115 General Administration Building
Olympia, Washington 98504, USA

Abstract.—Coded wire tag (CWT) loss by salmonids has been treated and reported inconsistently by fishery agencies on the Pacific coast of the USA and Canada. Inconsistent treatment of tag-loss estimation can affect final estimates of the CWT groups. This report provides information on CWT loss rates, factors influencing tag loss, length of time over which tag loss occurs, and frequency of naturally occurring adipose fin loss. Four groups of chinook salmon *Oncorhynchus tshawytscha* and four of coho salmon *O. kisutch* were tagged and held for up to 293 d after tagging, to estimate the level of CWT loss. The average size of the fish ranged from 0.9 to 7.6 g. Final tag-loss rates ranged from 1.1 to 5.3%. No significant tag loss was observed in any of the groups later than 29 d after tagging. With tagging-crew experience level and average fish size (2.1 g) held constant, the effects of half-length (0.5-mm-long) CWTs and standard-length (1.1-mm-long) CWTs on tag loss were compared, and no significant difference was found. However, when experience of the tagging crew and tag length (1.1 mm) were held constant, fish sizes (mean, 1.6 and 1.1 g) had a significant effect on tag loss. The mean tag loss for the smaller fish (1.1 g) was 1.8%; the larger fish (1.6 g) had a mean tag loss of 1.1%. Returning adult coho salmon were monitored at four Puget Sound hatcheries for naturally missing adipose fins, the frequency of which was 0.5%. Observations on naturally reared coho salmon smolts, and returning adults from several Puget Sound streams, showed that 0.06% of fish were missing adipose fins.

The coded wire tag (CWT; Jefferts et al. 1963) has been used as a major stock identification tool for chinook salmon *Oncorhynchus tshawytscha* and coho salmon *O. kisutch* by many fishery agencies in the Pacific Northwest, including the Washington Department of Fisheries (WDF). Successful use of the CWT requires associating a visually detectable secondary mark with each tagged fish. The external mark of choice has been the surgical removal (i.e., clipping) of the adipose fin. Because some tag loss is inevitable subsequent to the implantation of CWT, recapture rates must be adjusted to reflect the rate of tag loss. This adjustment is usually accomplished by determination of the number of fish that lack an adipose fin and that do not carry a CWT. It is assumed that each such fish lost a tag, and CWT release figures are adjusted accordingly. However, salmon can also lose adipose fins naturally. When a fish loses an adipose fin naturally, it looks just like a fin-clipped fish that has lost its CWT.

Since 1973, WDF has produced yearly progress reports that document the estimated catch of groups of tagged fish that were recovered in that particular year's fisheries. In the 1974 report, a 15% tag-loss rate was routinely used to adjust the release figure for each tagged group. At the time of release, tag loss from juveniles was estimated to average about 5%, but 15% of adults returning to the hatchery without an adipose fin were tagless. It was assumed that 15% represented the true rate of tag loss, and that tag loss took place after release from the hatchery but before recruitment to the fishery. The corresponding adjustment of release figures artificially inflated the estimated catch of tagged fish. In a similar case, the Oregon Department of Fish and Wildlife (ODFW) estimated the number of salmonids caught with each tag code by allocating the estimate of fish missing an adipose fin to each code in proportion to that code's frequency in the sample (Johnson 1985). In essence, ODFW also equalized tag loss among all codes recovered and also credited to those codes fish that had lost adipose fins naturally.

Conversely, if tag loss is used to adjust release figures before loss has stopped, an artificially low catch of tagged fish can be estimated. Before 1982, WDF tag-loss checks were not performed at all, or only within 2 d after tagging, and about half of the tag-loss estimates presently reported by Canada Department of Fisheries and Oceans are made within 1 d after tagging (Johnson 1987).

The investigations described here were conducted (1) to estimate the number of coho salmon naturally missing adipose fins, (2) to measure the

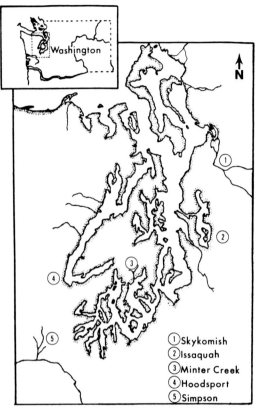

FIGURE 1.—Study area where coded wire tag loss and fish naturally missing adipose fins were observed.

rate of CWT loss from chinook and coho salmon, (3) to determine the effects of fish size at tagging and of tag size on the rates of CWT loss, and (4) to determine the length of time over which CWT loss occurs. The results of these studies should provide a basis for appropriate adjustment of release data for tagged fish. This adjustment factor can be

consistently applied for more accurate estimates of tag-recovery rates in fisheries.

Methods

Tag loss.—Fish in eight test groups (approximately 10,000 fish each) were tagged and observed for tag loss at Minter Creek Hatchery (Figure 1) to provide information on tag loss as a function of tag length and fish size. Four groups were coho salmon and four groups were chinook salmon. Coho salmon in three groups were tagged with standard-length tags (1.1 mm) and averaged 2.2, 4.1, and 7.6 g/fish. Coho salmon in the fourth group averaged 0.9 g/fish, and were tagged with half-length tags (0.5 mm) (Table 1). Standard-length tags CWTs were used on chinook salmon in one group, in which fish averaged 2.1 g/fish. Half-length tags were implanted in chinook salmon in three groups; these fish averaged 1.1, 1.6, and 2.1 g per fish.

Standard WDF tagging procedures, recommended by the Pacific Marine Fisheries Commission (PMFC 1983), were used. Fish were tagged in a mobile trailer, then passed through a quality control device to confirm the presence of a tag. Because of their small size, the coho salmon given half-length CWTs were tagged by highly experienced personnel. The other three lots of coho salmon and all four lots of chinook salmon were tagged by a "typical" crew consisting of local residents with some previous experience with coded wire tagging. The tagging crews were not informed of the study design.

Coho salmon were sampled for tag loss three or four times during the first week, then once a week for the next 3 weeks, and once every other week thereafter until release, which was about 6 months after tagging. Two permanent WDF staff members conducted all sampling for tag loss throughout the study.

TABLE 1.—Coded wire tag (CWT) loss rates from different salmon groups. Half-length CWTs were 0.5 mm long; standard-length CWTs were 1.1 mm.

Salmon species	Average fish weight (g)	Length of CWT	Number tagged	Tag loss (%)[a]	Number of days from tagging to release
Coho	0.9	Half	9,034	5.33	293
Coho	2.2	Standard	10,168	1.45	293
Coho	4.1	Standard	10,230	5.13	185
Coho	7.6	Standard	10,855	1.65	121
Chinook	2.1	Standard	9,847	1.96	201
Chinook	2.1	Half	10,272	1.48	198
Chinook	1.6	Half	10,279	1.13	212
Chinook	1.1	Half	10,545	1.84	215

[a]The tag loss shown represents the mean computed from samples of 3,000 fish. Samples were taken 4 weeks after tagging and every 2 weeks thereafter until release.

Each time coho salmon were sampled for tag loss, the group was crowded into a small, confined area of a standard concrete raceway. Approximately 3,000 fish were dipnetted from the confined area and placed in a holding pen. From the holding pen, sublots of approximately 50 fish were netted and placed in an anesthetic solution (1 g/20 L) of tricaine (MS-222) until each worker had checked a total of 1,500 fish. To check for the presence of a CWT, each fish was passed through a Northwest Marine Technology magnetic field detector. When a fish did not trigger an audible signal from the detector, it was designated "no tag" and put in a holding bucket of fresh water. After staff had each checked 1,500 fish, they exchanged buckets containing the "no-tag" fish and repeated the process to verify the absence of tags. To assure magnetization of any tags that might have been present, each "no-tag" fish was passed three times, with different orientation, through the magnetic field of a large horseshoe magnet. Chinook salmon were checked in the same manner, but more frequently: four times the first week, three times the second week, twice the third and fourth weeks, once each week for the fifth through the eighth week, and then once every 3 weeks until release.

Fish size and tag size were included as variables in the studies of chinook salmon. The typical crew that tagged the two size-groups of chinook salmon spent the first day tagging fish that were not included in the study groups. This extra day was added because WDF supervisors have observed that tag loss for a tagging crew usually is highest the first day, probably because of the crew's unfamiliarity with the tagging operation, and because of initial adjustments in equipment. For the first 4 h on the second day of tagging, the crew tagged fish averaging 1.1 g. Fish averaging 1.6 g were tagged for 4 h in the afternoon. On the third day of tagging, the order of size-groups was reversed, and the larger fish were tagged first. Fish in the two groups of chinook averaging 2.1 g/fish were tagged with the half-length CWT and standard-length CWTs and were treated in the same manner.

Fish naturally missing adipose fins.—Juvenile coho and chinook salmon were examined for lack of adipose fins in both hatchery and natural environments. During hatchery tagging operations (1978–1986 broods), tagging crews were asked to record the occurrence of fish naturally missing adipose fins. The same observations were made during trapping and tagging of wild coho salmon

smolts on several small streams entering Puget Sound and Grays Harbor on the Washington coast (Figure 1), for the same broods.

Four Puget Sound and Hood Canal hatcheries (Figure 1) were monitored in 1979 for returning adult coho salmon (1976 brood) naturally missing their adipose fins. The hatcheries (Minter Creek, Issaquah, Skykomish, and Hoodsport) were selected because they had not released fish with clipped fins from that brood year. Similarly, observations on native adult fish returning to their natural environment were made at the fish passage system at Sunset Falls on the Skykomish River (Figure 1), where coho salmon returned in 1978. Simpson hatchery on the Washington coast (Figure 1) was also included because an abnormally high occurrence of juveniles naturally missing their adipose fins had been noted previously by a WDF tagging supervisor.

Two permanent staff members from the WDF coded-tag recovery laboratory were responsible for all sampling of adults, taking snouts from adult fish that were missing an adipose fin. Fish with a questionable mark (i.e., partial adipose fin) were treated as marked fish, in accordance with the procedure for sport and commercial fishery sampling. All snouts taken from fish without adipose fins were checked for tags at the WDF tag-recovery laboratory. Verification of absence of a CWT was made by passing the snouts through an X-ray machine.

Results

Tag Loss

The eight tag-loss study groups (Table 1) were observed during periods ranging from 121 to 293 d. Rates of tag loss ranged from 1.13 to 5.33%. The percentage of fish without tags rose sharply (Figures 2, 3) during the first 1–4 weeks after tagging, and then became stable. Chi-square trend tests with one degree of freedom (Armitage 1973) did not indicate any significant increase in total loss for any of the tag groups after 29 d (Table 2). Groups tagged with half-length tags showed no significant change after 17 d.

Mean tag losses for samples taken from day 29 until the end of testing for the chinook salmon group averaging 2.1 g/fish were 1.96% for the full-length tag and 1.48% for the half-length tag; the difference between loss rates was not significant ($\chi^2 = 2.2$, $P > 0.05$). In contrast, losses of half-length tags differed significantly between 1.6-

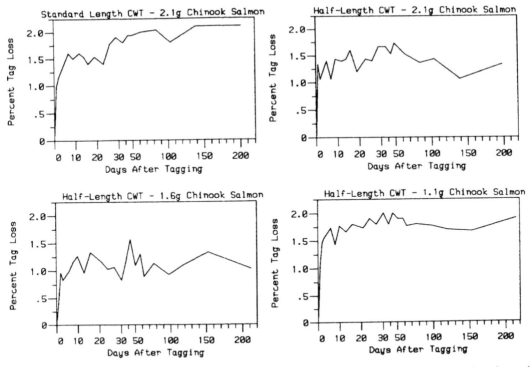

FIGURE 2.—Cumulative percentages of tag loss by coded-wire-tagged (CWT) chinook salmon. Tag size and average weight of the fish varied at the time of tagging. Fish were sampled chronologically after tagging.

g chinook salmon (1.13%) and 1.1-g fish (1.84%; $\chi^2 = 5.03$, $P < 0.05$).

The three coho salmon groups tagged with standard-length tags by three separate crews had mean tag-loss rates of 1.62% for 7.6-g fish, 5.13% for 4.1-g fish, and 1.45% for 2.2-g fish. Half-length tags placed in 0.9-g coho salmon by an experienced crew were lost at a rate of 5.33% over 17 d.

Fish Naturally Missing Adipose Fins

Among juveniles of the 1978–1986 hatchery broods, an average 0.13% of coho salmon (annual range, 0.05–0.23%) and 0.03% of chinook salmon (0.01–0.07%) naturally lacked adipose fins (Table 3). Tagging personnel noted that adipose fins were only partially missing at times, and that some of the fins had recently been lost because some wounds appeared fresh. Coho salmon were observed at the hatchery 3–12 months prior to release, and the chinook salmon were mostly observed 1–2 months prior to release. Observations on naturally reared juvenile coho salmon showed an average percentage of 0.06% without adipose fins in the 1978, 1979, 1982, and 1983 broods (Table 3). No missing adipose fins were

marked by fresh or new wounds, among naturally reared fish.

Among adult hatchery coho salmon returning in the fall of 1979, an average 0.95% naturally lacked adipose fins (Table 4). Exclusion of Simpson hatchery, known for its high incidence of naturally missing fins, lowers the rate to 0.52%. Only 3-year-old fish were included in the observations (i.e., jacks—precocious males—were identified by scale analysis and CWTs and excluded from the analysis).

Issaquah and Hoodsport hatcheries each received one stray 3-year-old coho salmon that had an adipose fin mark and contained a CWT, while Minter Creek had 11 such fish. The tagged adult (code 63-16/50) found at Hoodsport hatchery was from a group that had been planted in a nearby stream system from another hatchery. All fish from that release lot had been tagged so no adjustment was made to the observed figures from Hoodsport. The 12 fish with CWT (code 5-34/4) found at Minter Creek and Issaquah were released from a saltwater rearing pen. Twenty percent of fish in the net-pen group had been released with tags, so the numbers of unmarked adults returns

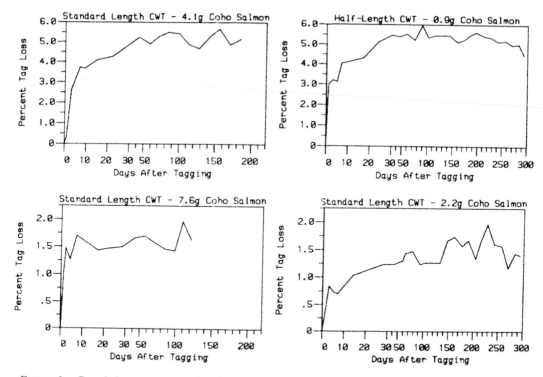

FIGURE 3.—Cumulative percentages of tag loss by coded-wire-tagged (CWT) coho salmon. Tag size and average weight of the fish varied at the time of tagging. Fish were sampled chronologically after tagging.

to Minter Creek and Issaquah hatcheries were adjusted downward by four for each CWT recovered.

Of native adult coho salmon returning in 1978 to Sunset Falls on the Skykomish River, 0.04% (8 of 20,388) were naturally missing their adipose fins.

Discussion

None of the eight lots of fish in this study showed a significant change in the rate of tag loss after 29 d, and the loss from four groups essentially stopped within 17 d. Morley and associates (R. B. Morley, Canada Department of Fisheries and Oceans, unpublished data) recently collected data that support these results. They used the standard-length CWT on three lots of chum salmon O. keta that averaged 2.2 g/fish, and found that tag losses stabilized at 2.5% in an average of 23 d. However, when they used the standard-length tag on five lots of chum salmon that averaged only 0.46 g/fish, tag loss did not stabilize until an average of 98 d after tagging, when it had

TABLE 2.—Length of time (days) from tagging until tag loss for different salmon groups. There was no significant change shown by a chi-square trend test. Half-length CWTs were 0.5 mm; standard-length CWTs were 1.1 mm.

Salmon species	Average weight (g)	CWT	Number of days from tagging to release	Number of days from tagging to last day of significance[a]	Significance level (df = 1)
Coho	0.9	Half	293	17	0.55
Coho	2.2	Standard	293	28	2.77
Coho	4.1	Standard	185	29	0.18
Coho	7.6	Standard	121	28	0.23
Chinook	2.1	Standard	201	29	1.66
Chinook	2.1	Half	198	16	0.07
Chinook	1.6	Half	212	16	0.08
Chinook	1.1	Hlaf	215	10	0.02

[a]Tag-loss estimates were made three times per week for 4 weeks after tagging and every 2 weeks thereafter until release to determine when tag loss stabilized.

TABLE 3.—Wild and hatchery juvenile coho and chinook salmon naturally missing the adipose fin.

Brood year	Coho salmon				Chinook salmon	
	Hatchery		Wild		Hatchery	
	Sample size	Percent of fish without adipose fin	Sample size	Percent of fish without adipose fin	Sample size	Percent of fish without adipose fin
1978	1,584,363	0.07	81,851	0.06	1,910,648	0.03
1979	374,872	0.05	160,980	0.06	2,357,352	0.02
1980	1,769,136	0.27			1,571,896	0.01
1981	1,837,703	0.06			2,118,485	0.02
1982	1,807,108	0.17	86,043	0.07	2,160,401	0.07
1983	2,420,601	0.11	70,514	0.05	2,534,583	0.02
1984	1,869,764	0.11			4,971,039	0.02
1985	1,839,242	0.09			5,317,846	0.06
1986	1,642,374	0.23			5,363,942	0.02
Total	15,145,163	0.13	399,388	0.06	28,306,192	0.03

reached 38.6%. Similarly, when they used the standard-length tag on chum salmon averaging 0.85 g/fish, tag loss continued for an average of 48 d before it stabilized at 16%.

Intuitively, standard-length CWTs do not seem as suitable as half-length CWTs for tagging salmonids as small as or smaller than 0.1 g/fish because of high tag loss and possible physical damage, as reported by Morrison and Zajac (1987). The smallest salmon on which I successfully used the half-length CWT averaged 0.9 g/fish. Thrower and Smoker (1984), however, reported using half-length CWTs on pink salmon O. gorbuscha as small as 0.25/g fish with reasonable tag loss (4%).

Each of the two lots of chinook salmon, tagged with half-length CWTs by similar tagging crews had low tag loss rates; although the larger (1.6-g) fish had a slightly lower tag-loss rate than the smaller (1.1-g) fish. However, when fish as large as 2.1 g were tagged by the same crew, no significant difference in tag-loss rate was noted between half-length and full-length CWTs. Consequently, with fish larger than 2.1 g, I recommend use of full-length CWTs because of the

TABLE 4.—Hatchery coho salmon adults naturally missing the adipose fin, 1976 brood.

Hatchery	Number of adults sampled[a]	Percent of fish without adipose fin
Minter Creek	7,248	0.79
Issaquah	1,614	0.25
Skykomish	5,598	0.18
Simpson	5,477	2.12
Hoodsport	289	1.73
Total	20,226	0.95

[a]Issaquah and Hoodsport each had one stray adult with a coded wire tag, and Minter Creek had 11. Expansion figures were applied to these recoveries and subtracted from the total number of adults sampled to correct for known strays.

limited numbers of codes available for half-length CWTs. For fish smaller than 2.1 g, I recommend half-length CWTs to avoid excessive and prolonged tag loss.

Adult hatchery coho salmon had a higher incidence of naturally missing adipose fins (0.52%) than hatchery juveniles (0.13%). This may be an artifact of sampling juveniles only halfway through their hatchery rearing period. Many losses of adipose fins from juveniles were recent, and seemed to be the result of a dynamic process that probably continued through the rest of the rearing period. I postulate that adipose fins were bitten off during feeding periods at the hatchery. Perhaps fins were mistaken for food, or fish became aggressive due to crowding. This postulate is supported by the observed rate of adipose fin loss among stream-reared wildstock coho salmon, which is approximately one-tenth that among hatchery coho salmon. Further support for this hypothesis is provided by the observation that low (0.03%) rate of the missing fins among hatchery chinook salmon juveniles, which were typically reared at lower densities and for half as long compared with coho salmon. In addition, net-fishery data from 1980 showed that 0.12% (35 of 29,503) of fish were naturally missing their adipose fins in the Puget Sound chum salmon fishery, the highest incidence occurring in the vicinity of chum salmon hatcheries (WDF, unpublished data).

Regardless of how or when fish lose their adipose fins naturally, it is necessary to differentiate actual CWT loss and naturally occurring loss of the adipose fin. The necessity for making this distinction becomes clear if one imagines a hatchery release group of 2 million fish in which a natural adipose fin loss rate is 0.5%. This rate would result in 10,000 fish naturally without adipose fins. If 50,000 fish are released with tags, and there is an actual tag-loss rate of 5% (resulting in 2,500 actual tag losses), upon return the tag loss would seem to be 25% (10,000 natural plus 2,500 actual losses) instead of the actual 5% (2,500 losses).

The easiest way to account for tag loss is to adjust the release figures downward to reflect the actual number of released, marked fish expected to retain CWTs. Results from experimental groups suggest that a final level of tag loss can be ascertained by waiting 29 d after tagging. Some tag loss may occur after this time, but for practical purposes, estimates made after 29 d provide a reasonable measure of tag loss when a CWT of the proper length is applied to a salmon of suitable size.

Acknowledgments

I offer my special thanks to JoAnn Lincoln and Jeff McGowan for their dedication and perserverance in conducting the numerous tag retention checks.

References

Armitage, P. 1973. Statistical methods in medical research. Wiley, New York.

Jefferts, K. B., P. K. Bergman, and H. F. Fiscus. 1963. A coded wire identification system for macro-organisms. Nature (London) 198:460–462.

Johnson, J. K. 1985. 1983 Pacific salmonid coded wire recoveries. Pacific Marine Fisheries Commission, Portland, Oregon.

Johnson, J. K. 1987. Pacific salmonid coded wire tag releases through 1986. Pacific Marine Fisheries Commission, Portland, Oregon.

Morrison, J., and D. Zajac. 1987. Histologic effect of coded wire tagging in chum salmon. North American Journal of Fisheries Management 7:439–440.

PMFC (Pacific Marine Fisheries Commission). 1983. Coded-wire tagging procedures for Pacific salmonids. PMFC, Portland, Oregon.

Thrower, F. P., and W. W. Smoker. 1984. First adult return of pink salmon tagged as emergents with binary-coded wires. Transactions of the American Fisheries Society 113:803–804.

American Fisheries Society Symposium 7:244–252, 1990

Performance of Half-Length Coded Wire Tags in a Pink Salmon Hatchery Marking Program

LARRY PELTZ

Alaska Department of Fish and Game
Division of Fisheries Rehabilitation, Enhancement, and Development
Box 669, Cordova, Alaska, 99574, USA

JACK MILLER

Alaska Department of Fish and Game
Division of Fisheries Rehabilitation, Enhancement, and Development
333 Raspberry Road, Anchorage, Alaska 99518, USA

Abstract.—Over 200,000 emergent fry (0.2 g) of pink salmon *Oncorhynchus gorbuscha* were tagged with half-length coded wire tags at each of three Prince William Sound (Alaska) hatcheries in 1986. The tags were recovered in 1987 in the commercial purse-seine fishery and in the brood-stock returns to each hatchery. The adjusted tag-retention rate from fry release to adult return was excellent (83.6–86.3%) at Armin F. Koernig (AFK) and Esther hatcheries, but poor (48.6%) at Cannery Creek Hatchery. The low tag-retention rate was attributable to shallow tag placement. Comparison of proportions of marked fry among the fish at release and of adults at return revealed that more tagged adults than anticipated returned to AFK Hatchery, slightly fewer than anticipated returned to Esther Hatchery, and far fewer than anticipated returned to Cannery Creek Hatchery. The major problems identified in this marking program were tag retention, differential mortality of tagged fry, and wide variation in proportions of marked fry in release groups with the same tag code. The most important criterion for designing a large-scale marking program appears to be maintenance of a constant proportion of marked fish among all release groups. On the basis of our results, we conclude that half-length coded wire tags can be used to estimate return proportions from pink salmon hatchery releases numbering in the hundreds of millions.

The first hatchery for pink salmon *Oncorhyncus gorbuscha* on Prince William Sound, Alaska, began operation in 1975. Currently, four hatcheries with a combined capacity of 600 million pink salmon eggs are in operation. These are expected to produce a yearly return of approximately 25 million adult pink salmon. Historic catch records indicate that an additional 5–25 million pink salmon can be expected annually from wild stock production. Both hatchery and wild stocks are harvested in mixed-stock purse-seine fisheries in seven of the eight commercial fishing districts in Prince William Sound. Consequently, to preserve wild stocks of pink salmon, which usually are less abundant than hatchery stocks, it is necessary to identify and quantify differences between wild and hatchery stocks.

Coded wire tags (Jefferts et al. 1963) have been used to mark salmon smolts and fingerlings for over 20 years. More recently, half-length coded wire tags have been successfully used on smaller fish (Opdycke and Zajac 1981; Thrower and Smoker 1984), but have not been used to evaluate large-scale releases of hundreds of millions pink salmon. Large numbers of fish must be tagged to ensure that an adequate number of tags are recovered in the commercial fisheries to identify and quantify differences between hatchery and wild stocks. Tagging hundreds of thousands of pink salmon fry requires a large investment of money in personnel and equipment.

We designed this study in part to test the practicality of tagging large numbers of pink salmon with half-length coded wire tags. We determined rates of tag application, overnight and long-term tag retention, tag placement, changes in the ratio of tagged to untagged fish in the time between release and return, and numbers of tags recovered from the commercial fishery.

Methods

Tag application.—Application of half-length coded wire tags was conducted at Armin F. Koernig Hatchery (AFK), Esther Hatchery, and Cannery Creek Hatchery (Figure 1) in spring 1986. Over 200,000 tags were applied at each of

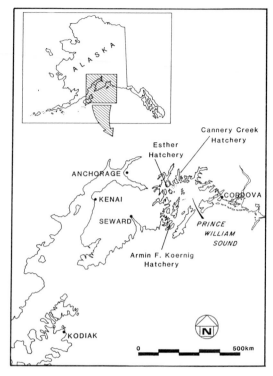

FIGURE 1.—Map of Prince William Sound, Alaska, showing pink salmon hatchery locations.

these hatcheries; there were two tag codes per hatchery. The Esther and AFK hatcheries each applied one lot of approximately 150,000 tags and a second lot of approximately 50,000 tags. Cannery Creek Hatchery applied two lots of approximately 100,000 tags. Each hatchery divided their fry into groups for early and late release. Tag code A was used to represent the early releases and code B represented the late releases. Northwest Marine Technology model MK-2 tagging machines and quality control devices were used for the tagging. Crew size and work schedule differed among facilities. Crews at Esther and Cannery Creek hatcheries consisted of two people who clipped adipose fins (to externally identify tagged fish) and one person who operated the tagging machine. The AFK Hatchery crew consisted of one person who clipped adipose fins and one person who did the tagging. The goal of tag application at each hatchery was to disperse the tagged fish equally among lots, resulting in approximately equal proportions of marked fish in each hatchery's release lots.

Each daily lot of tagged fish was held overnight (8–24 h) and 100–200 fry were randomly collected

from that lot the following day to evaluate the overnight tag-retention rate. Approximately 20–40 fry that had been marked by each fin-clipper were randomly collected daily to evaluate the quality of the adipose fin clips. An imaginary line close to the base of the adipose fin was used as a grading criterion, and the clip was not counted as valid if the distal portion of the clipped fin base crossed the line. The number of marked fish was the total number of fish tagged multiplied by both the estimated percentage that retained the tag overnight, and the estimated percentage with a valid adipose fin clip (a marked fry possessed an adipose fin clip and a wire tag).

All pink salmon fry released from AFK and Esther hatcheries were held and fed in saltwater net-pens for 18–41 d before they were released. A concerted effort was made to add the same proportion of marked fry into each net-pen relative to the unmarked fry. All emergent fry released from Cannery Creek were collected in two raceways and released every 3–4 d. Due to short holding periods and a later-than-expected emigration, a greater proportion of marks were inadvertently applied early in the emergence period. Some marked fry were held for up to 10 d during the late stages of emergence to obtain a better distribution of tags.

Tag placement.—Tag placement in the fry was quantified with a preservation and clarification technique adapted from Hanken and Wassersug (1981). Approximately 150 randomly selected fry from each hatchery were preserved in a 5% solution of neutral buffered formalin for 1 month. Next, the samples were rinsed for 4 d in several changes of tap water. The samples were then soaked in a 0.5% solution of potassium hydroxide (KOH) for 48 h to clarify the tissues that would otherwise obscure the view of the tag. Then the samples were soaked in a series of dilutions of 0.5% KOH and glycerin for 48 h. The dilutions began at 25% glycerin–75% KOH (0.5%) and increased to 50, 75, and finally 100% glycerin.

Snouts of cleared fry were observed with the aid of a dissecting microscope to evaluate the position of the tags in the snouts. A computer program was written to create a file that could be read, analyzed, summarized, and graphed by the software programs LOTUS 1-2-3 or SYMPHONY. Eight observations—X and Y coordinates of each end of the tag from both dorsal and lateral views—were made for each fry and plotted on a grid.

TABLE 1.—Total numbers of fry and numbers of fry marked with half-length coded wire tags released from three Prince William Sound pink salmon hatcheries in 1986.

	Hatchery					
	Cannery Creek		Armin F. Koernig		Esther	
Variable	Code A	Code B	Code A	Code B	Code A	Code B
Total number of fry	19,536,042	33,385,800	77,467,000	35,510,000	23,491,205	11,164,795
Number of marked fry	100,509	109,653	155,303	48,235	154,065	57,021
Number of release lots	5	10	7	4	3	1
Ratio, total:marked						
Mean	194	304	499	736	152	196
Range	106–257	145–781	409–640	537–898	134–193	

Tag recovery.—Commercial catches of adult pink salmon were sampled to find fish without adipose fins in July and August 1987 at selected processing plants in Cordova and Seward (Figure 1). At each plant, one person scanned the fish on the sorting line as they were pumped from the tenders. One person at each hatchery scanned the brood stock at the spawning rack to find pink salmon without adipose fins. The goal at each hatchery was to recover 200 marked brood fish. At least 25% of fish at Esther Hatchery and 50% at AFK and Cannery Creek hatcheries were scanned daily. Heads from all fish without adipose fins were sent to the Alaska Department of Fish and Game's coded wire tag recovery laboratory in Juneau for tag extraction and decoding.

Chi-square comparisons (Fleiss 1981) of the tag-code ratio (number of recoveries with code B divided by number of recoveries with code A) in the brood-stock returns and in the commercial fishery were made at each hatchery for each tag code. The coefficient of variation was calculated for the ratio of numbers of marked to total fish in the brood-stock returns, for each tag code at each hatchery; Poisson rates were assumed (Snedecor and Cochran 1967).

Results

Tagging

Each of the three hatcheries exceeded the goal of releasing 200,000 pink salmon fry with a valid wire tag and a valid fin clip (Table 1). The ratios of total to marked fry were most consistent among release lots at Esther Hatchery (Figure 2). The three release lots were placed in three feeding pens so it was easy to maintain approximately equal proportions of marked and unmarked fish in each pen. At AFK Hatchery, seven net-pens were used to release 11 lots of pink salmon fry. Marked fry were added to the pens in proportions as equal as possible, but fewer marked fry were included in

the later releases. The proportions of marked fry released from Cannery Creek Hatchery fluctuated widely. Cannery Creek Hatchery did not have a saltwater rearing program, so the fry lots were released volitionally. Fry emergence was later than normal, and a greater proportion of marked fish were released during the early stages of the emergence.

The overnight tag-retention rates exceeded 95% at all three hatcheries (Table 2). The percentage of valid fin clips was similar among the three hatcheries. After the number of marked fish was adjusted to reflect those retaining tags overnight and the percentage of those with valid clips, all three hatcheries released approximately the same percentage of validly marked fish. Marking rates were higher for crews with three workers, but the numbers of valid marks applied per person-hour of work were essentially equal.

Tag Placement

There was considerable variation in tag placement in the fish snouts at the three hatcheries (Figure 3). The tag depth in fish from AFK Hatchery was deeper and the tag placement more variable than in fish from either Cannery Creek or Esther hatcheries. Most of the variability was along the dorsoventral axis.

Tag Recovery

Over 2.7 million pink salmon from the Prince William Sound commercial fishery were examined for marks in 1987 (Table 3). Half-length coded wire tags were found in 69.1% of the fish with missing adipose fins; 1,178 tags had Esther Hatchery codes, 1,060 had AFK Hatchery codes, and 268 had Cannery Creek Hatchery codes.

Examination of brood-stock returns at all three hatcheries resulted in 716 tag recoveries (Table 4). The percentage of fish with missing adipose fins and valid tags was higher at AFK Hatchery and

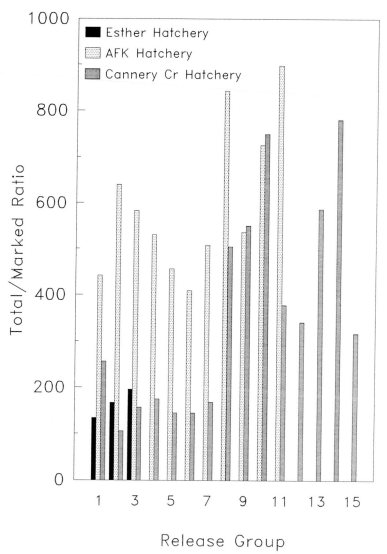

FIGURE 2.—Number of unmarked pink salmon fry released for each marked fry at three Prince William Sound hatcheries in 1986. AFK is Armin F. Koernig.

Esther Hatchery than at Cannery Creek Hatchery.

Comparisons of the ratios of the two tag codes from each facility at release and return revealed differences in the marine survival rates between lots released early (tag code A) and late (B) (Table 5). The lots of fish released late at Cannery Creek Hatchery were 1.7 times more abundant than the lots of fish released early both among brood stock and in the commercial fisheries. At AFK Hatchery, the lots of fish released early were 4.2 times more abundant than the fish released late among the brood stock, but only 1.9 times more abundant

in the commercial fishery. At Esther Hatchery, the lots of fish released early were 2.0 times more abundant than the fish released late among the brood stock, but only 1.4 times more abundant in the commercial fishery.

Table 6 shows ratios of marked pink salmon in fry releases and brood-stock returns during the study.

Discussion

Potential sources of error related to the tagging procedure include differential mortality, loss of tags, regeneration of clipped adipose fins, and

TABLE 2.—Tag-retention rates, percentages of validly fin-clipped and marked (tagged and clipped) pink salmon fry, and rates of tag application at three Prince William Sound hatcheries in 1986.

	Hatchery		
Variable	Cannery Creek	Armin F. Koernig	Esther
Percentage of tags retained overnight	95.3%	97.5%	97.6%
Percentage of valid clips	93.3%	91.8%	92.2%
Percentage of validly marked fish	88.9%	89.5%	89.9%
Number of validly marked fish per hour	643	415	622
Number of workers per shift	3	2	3
Number of valid marks per worker-hour	214	208	207

increased straying of adults because of olfactory damage attributable to tagging (Zajac 1985). Other complications that could affect data analysis are nonrandom distribution of marks in the population (Ricker 1975), fish that naturally lack an adipose fin, the presence of wild fish among returning brood stock, and error in determination of proportion of marked fish among the original hatchery releases.

In our study, differential mortality may have occurred at Cannery Creek Hatchery. Marked fry were held up to 10 d in fresh water during the later stages of emergence, and some of them died. Holding emergent pink salmon fry in fresh water without feed for more than a few days may weaken the fry and decrease the potential of survival. A second source of differential mortality is increased susceptibility of newly marked pink salmon fry to predation (Parker et al. 1963). No saltwater rearing occurred at Cannery Creek Hatchery, and it is possible that the physical effects of marking, together with the extended freshwater holding, made the fry more vulnerable once they left the hatchery. Differential mortality did not appear to be a problem at either AFK or Esther hatcheries, where all marked fry were held and fed in saltwater net-pens for a period of time prior to release. As suggested by Martin et al. (1981), differential survival of marked fry should decrease with increased rearing time.

The percentages of returning brood stock with a clipped adipose fin but no tag were indicators of the tag-loss rate. At Cannery Creek Hatchery, only 42% of the fish without an adipose fin contained a tag. The most consistent tag placement occurred among fry released from Cannery Creek

ARMIN F. KOERNIG HATCHERY

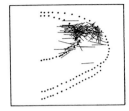

CANNERY CREEK HATCHERY

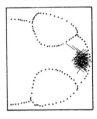
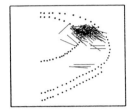

ESTHER HATCHERY

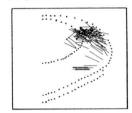

FIGURE 3.—Dorsal (left) and lateral (right) views of half-length coded wire tag placements in random samples of tagged pink salmon fry at three Prince William Sound hatcheries.

Hatchery (Figure 3). The tags appeared to be placed more shallowly in the snouts of fry at Cannery Creek Hatchery than at either of the other two hatcheries. The shallow tag placement seemed to allow a larger proportion of tags to be shed. This conclusion is supported by similar observations at Kitoi Bay Hatchery, where improved tag retention was noted as depth of tag placement increased (T. Joyce, Alaska Department of Fish and Game, unpublished data).

Regeneration of clipped adipose fins apparently did not occur. Discounting of marks based on fin-clip grading has not been widely practiced, although it was suggested by Moberly et al. (1977). We discounted marks in this project; however, the grading may have been too severe. Evidence of this stringent grading occurred at AFK Hatchery, where 8.2% of the marks were discounted based on the quality of fin clips, but there was an overall gain of 4.1% (Table 6) in the

TABLE 3.—Recovery of half-length coded wire tags in 1987 from pink salmon released from three Prince William Sound pink salmon hatcheries in 1986.

Commercial fishing district	Number of fish			Number of tagged fish recovery from					
				Cannery Creek Hatchery		Armin F. Koernig Hatchery		Esther Hatchery	
	Examined	Missing adipose fin[a]	Marked[b]	Code A	Code B	Code A	Code B	Code A	Code B
Eastern	323,458	97	19	1	1	2	4	0	9
Northern	214,970	348	199	37	71	0	2	15	74
Coghill	79,565	51	27	1	6	0	1	4	15
Northwest	44,203	37	28	0	2	0	8	6	12
Southwest	1,137,357	2,036	1,601	24	37	103	738	153	546
Montague	18,715	15	9	2	2	1	2	1	1
Southeast	132,581	63	2	0	0	1	1	0	0
Other[c]	789,585	980	623	35	49	40	157	132	210
Total	2,740,434	3,627	2,508	100	168	147	913	311	867

[a] Fish without an adipose fin.
[b] Fin-clipped fish containing a valid half-length coded wire tag.
[c] Includes samples from tenders with fish from more than one district and samples from hatchery sales.

proportion of marked fish found among the brood-stock fish. All tag-recovery personnel were instructed to consider a fish with any piece of adipose fin missing as a marked fish. Total regeneration of the adipose fin is necessary for a poor fin clip to be missed on a returning adult salmon. Fin-clip grading is necessary for quality control, but its use to discount marked fish may be questionable.

Increased straying of adults because of olfactory damage could cause a loss of marked fish and an increase in the ratio of total to marked fish between fry release and adult return. The AFK Hatchery was the most likely location of the three hatcheries for olfactory damage to occur, as suggested by tag placement data (Figure 3). The decrease in the ratio of total to marked fish at AFK Hatchery between fry release and adult return (Table 6) indicates that straying of marked fish was minimal. Olfactory damage caused by tag application has been documented (Morrison and Zajac 1987), but the effects on homing are unknown (Zajac 1985). Our results suggest that minimal straying was caused by the half-length coded wire tags.

Nonrandom distribution of marked fish seems to have been a serious problem at two of the hatcheries. The AFK Hatchery released 11 lots of

fish, with diverse proportions of marked fish; Cannery Creek Hatchery released 15 lots. It is highly likely that fish in different release lots had different survival rates. Apparently fish in some of the lots with high proportions of marked fish released from AFK Hatchery survived at rates higher than the rest of the release lots. The opposite probably occurred at Cannery Creek Hatchery, where it seems that fish in lots with lower marking proportions survived better, because large numbers of marks seem to have been lost.

The occurrence of fish that naturally lack adipose fins in the brood-stock returns could also lower apparent tag-retention rates. The rate of occurrence of such fish in wild or hatchery-reared salmonid populations is not well documented and it may vary dramatically among years or fish populations. Examination of Prince William Sound pink salmon at processing plants in 1985 showed that an estimated 1 per 2,400 fish naturally lacked adipose fins (S. Sharr, Alaska Department of Fish and Game, unpublished data). When this rate is extrapolated to the brood-stock returns at the three hatcheries, the adjusted tag-retention rates are 48.6, 83.6, and 86.3% for brood stock

TABLE 4.—Recovery of half-length coded wire tags from pink salmon brood stock at three Prince William Sound hatcheries in 1987.

Hatchery	Number of fish					Number of tags	
	Total	Examined	Missing adipose fin[a]	Marked[b]	Percentage with valid tags	Code A	Code B
Cannery Creek	136,231	108,577	333	140	42.0%	51	89
Armin F. Koernig	202,914	100,438	286	204	71.3%	189	14
Esther	242,957	69,286	460	372	80.9%	313	58

[a] Fish without an adipose fin.
[b] Fish without an adipose fin and containing a half-length coded wire tag.

TABLE 5.—Tag-code ratios of pink salmon from three Prince William Sound hatcheries, at release, among brood-stock returns, and in the commercial fishery.

Variable	Cannery Creek Hatchery			Armin F. Koernig Hatchery			Esther Hatchery		
	Code A	Code B	B/A	Code A	Code B	B/A	Code A	Code B	B/A
Fry hatchery release	100,509	109,653	1.09	155,303	48,235	0.31	154,065	57,021	0.37
Adult brood-stock recoveries	51	89	1.75	189	14	0.07	313	58	0.19
Total commercial fishery recoveries	100	168	1.68	913	147	0.16	932	246	0.26
Combined fishery and brood-stock recoveries after data expansion	1,017	1,714	1.69	9,083	1,429	0.16	9,727	2,501	0.26

from Cannery Creek, AFK, and Esther hatcheries, respectively.

Wild pink salmon in brood-stock returns can also decrease the proportion of marked fish and make it appear that marked fish have been lost. Neither AFK nor Esther hatcheries have a substantial amount of spawning habitat in the hatchery terminal area. Conversely, the Cannery Creek Hatchery brood-stock collection area provides approximately 10,000 m² of suitable spawning habitat for pink salmon that do not enter the hatchery raceways. Substantial numbers of adult pink salmon could be produced from the natural spawning and artificially decrease the proportion of marked fish. In addition, the large number of adult pink salmon (0.5–3.0 million) that return to hatchery terminal areas may attract fish returning to other parts of Prince William Sound. Only one tagged fish was found that had returned to the wrong hatchery. If straying among hatchery stocks is indicative of straying by in wild fish, the level of brood-stock dilution attributable to straying is minimal.

An error in the determination of the original proportion of marked fry could occur if either the number of marked fish or the estimated number of total fish were inaccurate. The marked fish were counted by the tagging machine, and a major error in counting the number of marked fish was unlikely. The procedure for estimating the total number of fish at all three hatcheries was nearly identical. All fry were passed through Northwest Marine Technology model FC-1 fry counters as they emerged from the incubators. A small error may have been associated with this enumeration, but the amount of error should have been consistent among hatcheries.

The ratios of total to marked fish at release and return should be similar; any differences should be attributable to a source. These ratios indicated a small proportional gain in marked brood stock at AFK Hatchery (i.e., a small decrease in the ratio), a modest loss of marked brood fish at Esther Hatchery, and a large loss of marked brood stock at Cannery Creek Hatchery (Table 6). The coefficients of variation for the inverse ratio (marked

TABLE 6.—Ratios of marked pink salmon from three Prince William Sound hatcheries, in fry releases in 1986 and in brood-stock returns in 1987.

Ratio or statistic	Hatchery					
	Cannery Creek		Armin F. Koernig		Esther	
	Code A	Code B	Code A	Code B	Code A	Code B
Total:marked fish						
At release	194	304	499	736	152	196
In brood-stock	570	893	479	705	180	230
Change	193%	194%	−5.0%	−4.2%	18.4%	17.3%
Marked:total fish						
In brood stock	0.0018	0.0011	0.0021	0.0014	0.0056	0.0043
SE	0.0001	0.0001	0.0001	0.0001	0.0003	0.0002
Coefficient of variation[a]	7.2%	9.2%	6.9%	8.4%	5.1%	5.8%

[a] SE = $100 \cdot$ SE/mean.

to total fish) in the brood stock were smaller than the difference in tag-code ratios at release and return at Esther and Cannery Creek hatcheries. These discrepancies indicate that sources other than random variation in the brood-stock sampling are at least partially responsible for the difference in ratios of total to marked fish at release and return. The change in the ratio at Esther Hatchery can be attributed mainly to the tag-retention rate. The excessive decline in the ratio of total to marked fish at Cannery Creek Hatchery was attributable to a combination of sources, which included differential mortality of marked fry, differential survival among lots of fry, a poor tag-retention rate, nonrandom distribution of marks, and the presence of wild pink salmon in the brood stock returns. At AFK Hatchery, the coefficient of variation for the ratio of marked to total fish in the brood stock was larger than the difference in tag-code ratios at release and return. This discrepancy indicates that random variation in the brood-stock sampling could be totally responsible for the difference in tag-code ratios at release and return. However, the adjusted tag loss rate of 16.4%, together with the observed rate of tag gain of 4.6%, yields an estimated net gain of 21% in the ratio of marked to total fish. In addition to random variation in the brood-stock sampling, excessive discounting of the adipose-fin marks and better survival of fish in release lots with the lower ratios of total to marked fish may have contributed to the gain in marked fish.

Chi-square analysis of the returns of fish with different tag codes in the Cannery Creek Hatchery brood stock and in the commercial fishery showed no significant difference ($P > 0.5$; Table 5). Comparison of the combined fishery and brood-stock recovery tag-code ratio to the tag code ratio at release reveals that fry released later (code B) survived 1.6 times better than those released earlier (code A). Chi-square analyses of data from both AFK and Esther hatcheries showed significant differences ($P < 0.05$) between the tag code returns in the brood stock and the commercial fishery (Table 5). Because the tag-code returns were statistically different, combining the hatchery brood-stock and commercial fishery recoveries after expansion (all recoveries were expanded to include the total return) provides the best comparison of survival of fish of different tag codes. At AFK Hatchery, fish released early (code A) survived 1.9 times better than fish released later (code B). Similarly, at Esther Hatch-

ery, fish released early (code A) survived 1.4 times better than those released later (code B).

Recommendations for Future Use

Half-length coded wire tags can be used successfully to evaluate the results of large-scale pink salmon hatchery production. Tag recoveries in the commercial fishery and in the hatchery brood stock were sufficient to determine pink salmon hatchery contributions to the Prince William Sound commercial fishery (L. R. Peltz and H. Geiger, Alaska Department of Fish and Game, unpublished data). An additional benefit is the capability to determine survival differences between fish with different tag codes. The ability to evaluate the success of various release strategies, with different tag codes, is invaluable. For example, adjusting release strategies to increase survival at the marine stage from the present average of 5% to just 6% would increase the return to the Prince William Sound area fisheries and hatcheries by 5 million pink salmon.

Results from this study point to several recommendations pertinent to large-scale pink salmon HLCWT marking programs.

(1) Marking is most suitable at hatcheries with rearing programs. Rearing increases the potential for maintaining equal proportions of marked fish in all release groups. In addition, rearing for any length of time before release allows marked fry to recuperate from the tagging process and reduces the potential for differential survival of unmarked and marked fry.

(2) Tag placement should be monitored closely. The tag should be placed as deeply in the nose as possible without causing irreparable damage to the fish. Quality should not be sacrificed for quantity. Dramatic increases in tagging rates may signify a decline in the quality of tag placement.

(3) Adipose-fin clips should be monitored closely to maintain good quality control. Discounting of marks based on grading of adipose fin clips is of questionable value, but should be used until the utility of this practice can be tested further.

(4) Equal proportions of marked fish should be maintained in all release groups. This consideration may be the most important in any plan for a large-scale marking program.

Acknowledgments

We thank Kit Rawson, Mike Kaill, and Tim McDaniel for their inputs into the project design. Dave Gaither did an excellent job training the tagging crews. The tagging crews at Cannery Creek, Esther, and Armin F. Koernig hatcheries tagged over 600,000 fish. The cooperation and funding provided by Prince William Sound Regional Aquaculture Corporation, operator of Esther and AFK Hatcheries, is greatly appreciated. Sam Sharr of the Commercial Fisheries Division, Alaska Department of Fish and Game (ADFG), and his crew collected the tag-recovery data from adult fish in the commercial fishery. Karen Crandall and her crew at the ADFG coded wire tag recovery laboratory excised the tags from the heads and provided us with the decoded data. Hal Geiger provided assistance with the statistical analysis. Carol Schneiderhan and Mary Ann McKean assisted us with the figures, and Bill Hauser and Tim McDaniel reviewed the manuscript. Keith Pratt and Bill Hauser assisted with the graphics.

References

Fleiss, J. L. 1981. Statistical methods for rates and proportions. Wiley, New York.

Hanken, J., and R. Wassersug. 1981. The visible skeleton. Functional Photography (July–August): 22–26.

Jefferts, K. B., P. K. Bergman, and H. F. Fiscus. 1963. Coded wire identification system for macro-organisms. Nature (London) 198:460–462.

Martin, R. M., W. R. Heard, and A. C. Wertheimer. 1981. Short-term rearing of pink salmon (*Oncorhynchus gorbuscha*) fry: effect on survival and biomass of returning adults. Canadian Journal of Fisheries and Aquatic Sciences 38:554–558.

Moberly, S. A., R. Miller, K. Crandall, and S. Bates. 1977. Mark–tag manual for salmon. Alaska Department of Fish and Game, Juneau.

Morrison J., and D. Zajac. 1987. Histologic effect of coded wire tagging in chum salmon. North American Journal of Fisheries Management 7:439–441.

Opdycke, J. D., and D. P. Zajac. 1981. Evaluation of half-length binary-coded wire tag application in juvenile chum salmon. Progressive Fish-Culturist 43:48.

Parker, R. R., E. C. Black, and P. A. Larkin. 1963. Some aspects of fish-marking mortality. International Commission for the Northwest Atlantic Fisheries, Special Publication 4:117–122.

Ricker, W. E. 1975. Computation and interpretation of biological statistics of fish populations. Fisheries Research Board of Canada Bulletin 191.

Snedecor, G. W., and W. G. Cochran. 1967. Statistical methods. Iowa State University Press, Ames.

Thrower, R. P., and W. W. Smoker. 1984. First adult return of pink salmon tagged as emergents with binary-coded wires. Transactions of the American Fisheries Society 113:803–804.

Zajac, D. P. 1985. A cursory evaluation of the effects of coded wire tagging upon salmonids. U.S. Fish and Wildlife Service, Fisheries Assistance Office, Olympia, Washington.

American Fisheries Society Symposium 7:253–258, 1990
© Copyright by the American Fisheries Society 1990

Retention Rates of Half-Length Coded Wire Tags Implanted in Emergent Pink Salmon

W. Michael Kaill[1] and Kit Rawson[2]

Alaska Department of Fish and Game
Division of Fisheries Rehabilitation, Enhancement, and Development
333 Raspberry Road, Anchorage, Alaska 99518, USA

Timothy Joyce

Alaska Department of Fish and Game
Division of Fisheries Rehabilitation, Enhancement, and Development
Kitoi Bay Hatchery, 211 Mission Road, Kodiak Alaska 99615, USA

Abstract.—We evaluated the use of half-length coded wire tags to mark emergent pink salmon *Oncorhynchus gorbuscha* at a production hatchery in Alaska. Fish in rearing cages and hatchery releases were used to estimate short-term and long-term tag loss, respectively. Fish tested for tag retention had had their adipose fins clipped. Short-term estimates of tag retention from net-pen holding experiments ranged from 93 to 100%. Tag lots for long-term tag-retention experiments were about 100,000 each year. Tagged pink salmon were recovered at the hatchery spawning rack and in the fishery by visual recognition of the adipose fin clip. Estimated long-term tag retention was about 75, 50, 65, and 84% for recovery years 1983–1986. Reasons for apparently inconsistent long-term tag retention may include human error and naturally missing adipose fins. Better results might be obtained from reduced dependence on fish with adipose fin clips for tag recovery.

Coded wire tags (Jefferts et al. 1963) have been used to tag fingerling-sized or larger salmonids, usually chinook salmon *Oncorhynchus tshawytscha* and coho salmon *O. kisutch* weighing 5 g or more (Eames 1983). Our goal was to test for the first time the feasibility of marking production lots of emergent (<0.25 g) pink salmon *O. gorbuscha* with a 0.5-mm or half-length version of the standard 1-mm-long coded wire tag (Opdycke and Zajac 1981; Thrower and Smoker 1984). We also evaluated the feasibility of detecting the tags with an electronic device to eliminate the costs of visual recovery as well as the problems caused by fish naturally missing adipose fins (adipose fins are clipped from wire-tagged fish for subsequent recognition). For coded-wire-tag evaluation to be cost-effective at a high level of production, large numbers of fry must be inexpensively tagged and recovered.

Methods

Tagging and fin clipping.—Implantation of half-length coded wire tags in newly emergent pink

salmon at Kitoi hatchery followed standard procedures for full-length tags (Moberley et al. 1977), except as noted below. In 1982, initial tagging was done on fish that had been reared for 15 d because we thought that a mean weight of 0.32 g was necessary for effective tagging. However, size variability within this first group caused difficulties. Fish in a second group were tagged immediately after emergence. These smaller fish (mean weight, 0.26 g) presented no difficulties for either adipose fin clipping or tagging. The homogeneous size of fish in this group made tag placement easier because a single head mold was appropriate for all fish. Another advantage of using emergent fish was that we did not need to use salt water in the tagging process and were able to avoid its corrosive effect on metal equipment.

Tagging was accomplished in two 6-h shifts per day for 6 d/week throughout the period of tagging. During each shift, two people, a clipper and a tagger, worked together. The clipper kept the operation supplied with untagged fish, anesthesized the fish, and clipped the adipose fin. The tagger operated the tagging machine. At the midpoint of each shift the two people switched jobs.

Accurate records were kept of tagging rates. Tagged fish were kept in fresh water overnight. The next day, dead fish were picked out and counted, and 100 of the live fish were run through

[1]Present address: Marine Research Co. of Alaska, Box 22210, Juneau, Alaska 99802, USA.

[2]Present address: Tulalip Fisheries Department, 3901 Totem Beach Road, Marysville, Washington 98720, USA.

TABLE 1.—Numbers of pink salmon tagged with coded wire tags and released at Kitoi Bay Hatchery, 1982–1986.

Year	Number of release groups	Number released (1,000s)	Nominal number tagged (1,000s)	Valid number tagged (1,000s)	Ratio, tagged:untagged
1982	3	87,583.9	102.2	93.9	1.07×10^{-3}
1983	2	71,793.1	93.2	88.0	1.23×10^{-3}
1984	2	87,065.6	89.5	84.6	0.97×10^{-3}
1985	2	75,100.8	116.3	112.3	1.49×10^{-3}
1986	2	97,773.4	89.7	89.3	0.91×10^{-3}

the tagging machine's quality control device to check for the presence of a tag. For each tagger's shift (i.e., for each 3-h period) the number of valid tags implanted per hour (VTPH) was computed:

$$VTPH = (TT - M) \times TR/H;$$

TT = total number of tags inserted (as read from the machine);

M = number of dead fish recovered the following morning;

TR = percentage of tags retained in the sample of 100 run through the quality control device; and

H = number of hours spent tagging (typically 3 h).

Table 1 summarizes the tagging history for 1982–1986.

Tag retention and tagging mortality.—A sample of approximately 1,500 fry was held from 20 to 40 d at Kitoi during each year of tagging except 1986, when 500 fish were held. Holding time depended on the hatchery schedule for fry emergence and release. Holding pens measured 1 m × 1 m × 1 m. Fish were marked with adipose fin clips and injected with half-length coded wire tags. Each day, dead fish were removed and counted. At the end of the holding period, all remaining fish were checked for the presence of a tag with the quality control device.

In 1985 and 1986, pink salmon were also held in saltwater pens for up to 240 d to determine tag retention and adipose fin regeneration. All groups were checked for missing adipose fins. The adipose fin was clipped on all tagged fish in order to determine if this mark could be used to help estimate tag loss.

If a salmon is missing its adipose fin and no tag is found, the first assumption often made is that the fish has lost its tag. This assumption can be misleading in determination of tag loss, because even a small number of fish that naturally lack

adipose fins in the population can bias this kind of data. We retained the heads of all returning adults found without adipose fins. The heads were sent to the Alaska Department of Fish and Game's coded wire tag recovery laboratory in Juneau, where they were examined for tags. In 1986, the tagging crew examined each pink salmon fry prior to fin marking to determine if the adipose fin was naturally missing.

A tagged fish might be missed during the recovery process if its adipose fin were intact because the fin had not been clipped or had regenerated. To check for the former, the crew taking eggs at Kitoi hatchery ran spot checks on 1,000 returning fish with adipose fins. The fish were put through the tubular electronic tag detector on several occasions during the egg-taking period in 1983, 1984, and 1985. In 1983 and 1984, about 6,500 and 4,200 fish, respectively, were run through the tubular detector at a fish processor in Kodiak.

To check for regeneration of clipped adipose fins, fish held in pens for monitoring of short-term tag retention were examined for regrowth of adipose fins during tag-detection tests. No adipose fin regeneration was observed.

Tag recovery.—Fish returning to the Kitoi Bay Hatchery are either caught in local fisheries or are used as brood fish for eggs. There is no large-scale straying of fish from the facility to other locations, nor is there straying from other systems into the group returning to Kitoi Bay.

Recovery of marked fish took place in local fisheries as well as from the brood-stock fish. In 1983 and 1984, fish were recovered from salmon tenders in the local fisheries. These fish were screened for tags by visual examination for missing adipose fins. In 1983, 1984, and 1985, fish also were screened electronically at Alaska Pacific Seafoods, a processor in Kodiak.

Electronic screening was accomplished with a metal cylinder (about 1.5 m long, with a 10- or 15-cm inside diameter) called a tubular detector. Whole fish or heads are passed through the detector. When one of them contains a magnetized particle, the resulting distortion of the magnetic field is detected, and a beep or other signal is triggered.

Results

Tag Application

Tagging rates started out at a few hundred per hour in 1982 as the crew, experienced in fin clipping but not tagging, learned the techniques.

TABLE 2.—Short term (40-d) mortality of fin-clipped and tagged pink salmon fry at Kitoi Bay Hatchery in 1982. HLCWT = half-length coded wire tag.

Treatment group	Number held	Number dead	Estimated mortality (95% confidence interval)
Reared fry			
HLCWT and adipose fin clip	1,500	14	0.95% (0.55–1.69%)
Adipose fin clip	1,005	10	1.00% (0.51–1.89%)
Emergent fry			
HLCWT and adipose fin clip	1,500	2	0.13% (0.02–0.54%)
Adipose fin clip	1,011	3	0.30% (0.08–0.94%)
Left ventral fin clip	1,030	5	0.48% (0.18–1.20%)

By the end of the season, a maximum hourly rates of 800 valid tags was attained; the average rate was 500 valid tags per hour and the minimum was 200. These levels were maintained through the study.

We used the average valid tag rate of 500/h (for a crew of two people), the cost of purchasing tags in 1983, and labor and food costs at Kitoi hatchery in 1983 to estimate the cost of applying one tag. This cost was $0.13, exclusive of the cost of tagging equipment (personnel, $0.08; food for workers, $0.01; miscellaneous supplies, $0.005; purchase of tags, $0.035).

Short-Term Tag Retention and Mortality

Tag-retention rates for fish fin-clipped, tagged, and held in saltwater rearing pens at Kitoi averaged 93% (95% confidence interval, 91–94%) in 1982 and 97% (96–98%) in 1983. A 1983 group that

was only tagged and not fin-clipped also had a 97% tag-retention rate. Short-term mortality during the 40-d holding period was 1% or less in 1982 and 1983 (Table 2).

In 1985, fish were tested with the tubular tag detector in April when they were loaded into the pens, in August, and again on October 29 when the experiment was terminated. Fish without tags were not removed from the pens. Tag retention was never less than 95%. In 1986, when the tag injector was set for deeper injections, the pens were loaded in April; fish were checked in August, and checked again by the tag laboratory in December when the experiment was terminated. In all cases during 1986, the fish examined had 100% tag retention. There was no adipose fin regeneration.

Long-Term Tag Retention

From the Kitoi releases, we estimate long-term tag retention from the fraction of adults that had missing adipose fins and that also had tags. We assumed that the incidence of fish that naturally lacked adipose fins was so low that all returning adults with no adipose fins were from the tagging operation of the previous season. For 5 years of tagging and recovery, the apparent tag-retention rates computed this way ranged from 49 to 84% (Table 3). These levels of long-term retention were much lower than we had expected. We suspected that we had underestimated the number of fish that naturally lacked adipose fins, so a specific effort was made in 1986 to estimate the number of such fish. Out of 90,000 fish examined, four were missing adipose fins.

TABLE 3.—Recovery of pink salmon tagged with coded wires at or near Kitoi Bay Hatchery. All tags were recovered the year after the juvenile pink salmon were released.

Year tagged	Location	Total examined	Fish recovered Number with adipose fin clip	Number with tag	Apparent tag retention (95% confidence interval)
1982	Fishery	17,556	24	18	75% (58–92%)
1982	Egg-take	132,466	235	173	74% (68–79%)
1983	Fishery	46,136	31	16	52% (34–69%)
1983	Fishery	4,184	[a]	4	
1983	Egg-take	139,881	90	44	49% (39–59%)
1984	Fishery	66,400	[a]	57	
1984	Egg-take	131,676	163	104	64% (56–71%)
1985	Egg-take	137,590	271	227	84% (79–88%)
1986	Egg-take	140,277	102	58	57% (47–66%)

[a] Tags were detected with a tubular detector.

Tag Recovery

Attempts were made to use the tubular detector on the tender boats, but the motion and shock transmitted through the steel decks caused the detectors to produce an excessive number of false positive readings.

Discussion

In this project we investigated the half-length coded wire tag as an evaluation tool for production hatcheries releasing 100 million fry annually. To serve this function, the technique must be feasible and affordable from two points of view. First, tags must be implanted such that tag retention is high and tag-caused mortality is low in both fresh water and the ocean. Second, tags must be recovered effectively from fisheries involving millions of returning adult salmon. Tagged fish can be detected visually by the absence of adipose fins or electronically. We found some difficulties with use of the adipose fin mark as a means of identifying tagged fish.

The data suggest that, in the short term, half-length coded wire tags are effective for emergent pink salmon. Short-term tag retention was consistently high, and those were no apparent detrimental effect on the fish. In the long term, however, a major portion of returning adults did not seem to retain the tags, even though the short-term retention was greater than 95%. The poor long-term retention could have been caused by several factors.

Naturally Missing Adipose Fins

At the hatchery, one-sixth to one-half of the tags were apparently lost between the time juveniles were released and the time adults returned (Table 3). Such low levels of long-term tag retention, if observed consistently, would imply a serious limitation on the usefulness of half-length wire tags for pink salmon. We believe that fish that naturally lack adipose fins account for part of this apparent tag loss.

In many tagging studies, a large proportion of the fish released, from 5 to 50%, is marked. Kitoi Bay Hatchery releases up to 100 million fry per year. Of that number, we tag 100,000 or fewer fry. We are able to examine large numbers of returning adults, but the small proportion of fish tagged (0.1% or less) may result in problems with estimates of long-term tag retention.

In a 1979 study of chum salmon *O. keta* returning to the Alaska Department of Fish and Game

hatchery at Beaver Falls, G. Freitag (Southern Southeast Regional Aquaculture Association, Ketchikan, Alaska, personal communication) observed enough fish that naturally lacked adipose fins (0.7 fish/1,000) to make him question his evaluation.

In 1986, we looked for emergent pink salmon at Kitoi Bay Hatchery that naturally lacked adipose fins, and found 4 out of 90,000, or 0.044/1,000. We did not attempt to measure loss of adipose fins from environmental causes. If this rate of naturally missing adipose fins is compared to the rate at which pink salmon were marked (1/1,000) it is evident that an error of 5% or more is possible. If the rate of missing adipose fins were as high as 0.5/1,000, there could be as much as a 50% depression of the apparent tag-retention rate if it were computed, as we did, by comparison of the total number of adipose-fin-clipped adults with the number of those that had tags.

Our result and the foregoing observations have implications for tagging studies of pink salmon. If adipose fin clips are to be used to locate fish tagged with coded wires, rates for the natural occurrence of fish without adipose fins need to be estimated for each population of fish being tagged, for each year of tagging.

In 1985 and 1986, we held samples of about 1,000 tagged fry in saltwater pens at Kitoi Bay Hatchery to monitor actual tag loss. The 1985 experiment ended on October 29 when the fish were inadvertently killed by unseasonably heavy flows of cold fresh water from a nearby stream. Tag-retention rates remained at 95% or higher throughout the holding period. In 1986, the fish were held until December, and until they were lost to similar causes, tag retention was 100%. These experiments suggest that properly tagged fish retain half-length tags, and our results seem to be consistent with those of Blankenship (1990, this volume). His findings with chinook and coho salmon of at least 0.9 g tagged with half-length coded wires showed that no significant tag loss occurred after 4 weeks.

We were not able to use fish taken from production lots for the 1985 and 1986 tag-loss experiments. The result is that tagging was done with precision that may not be typical for day-to-day operations.

Visual Detection of Adipose Fin Marks

In 1983 and 1984, preliminary experiments were conducted in which 1,000 fish previously checked for adipose fin marks were passed through a

TABLE 4.—Ratios of tagged to untagged pink salmon at the time coded wire tags were inserted, and 1 year later when adults returned.

Year tagged	Ratio, tagged:untagged			Method of recovery
	Release[a]	Recovery[a]	Percent difference	
1982	1.07	1.03	−4	Fishery: hand recovery
	1.07	1.31	+22	Egg-take
	1.07	1.08	+1	Processor: tubular detector
1983	1.23	0.35	−72	Fishery: hand recovery
	1.23	0.96	−22	Fishery: tubular detector
	1.23	0.31	−75	Egg-take
	1.23	0.96	−22	Processor: tubular detector
1984	0.97	0.86	−27	Fishery: tubular detector
	0.97	0.79	−18	Egg-take
1985	1.49	1.65	+16	Egg-take
1986	0.91	0.51	−50	Egg-take

[a]Proportions shown are 1,000 times actual proportions.

tubular detector. In 1983, one pink salmon was detected in this way (1/1,000). This fish had an adipose fin clip and half-length tag. In 1984 no marked fish were found among the 1,000 examined. Adults visually checked for the presence of an adipose fin clip were reexamined at day's end during the 1983 egg-take. Out of a day's production of 6,500 fish spawned, as many as 2 marked fish were found, an error rate of 0.31/1,000.

Variable Quality Control

Over the course of the project, we became increasingly aware of the importance of deep and consistent placement of tags. It is important to work with the tagging crew to emphasize concern for proper tag placement. The crew's tendency is to emphasize speed. Often the taggers compete among themselves to validly tag the most fish during a work shift. Experienced personnel can attain tagging speeds of up to 850 pink salmon fry per hour, with 99–100% overnight tag retention. If a tagger is capable of tagging at a rate of 800/h, we feel that it is advisable to regulate such a speed to about 600/h, to assure high quality of tag placement. Attempts to go faster often result in poor tag placement and low retention. In the case of coded wire tags, it is necessary to establish quality control first, and then speed. Inexperienced taggers cannot be expected to average more than 300 valid tags per hour in their first season. Even slower rates can be expected during the first few weeks.

It is not clear how much physical damage is done to a fish injected with a half-length coded wire. Although more study is necessary in this area, damage seems minimal if high quality control is maintained with respect to tag placement. Work on disease issues (Zajac 1985) and effects of coded wire tags on magnetic orientation for navigation (Quinn and Groot 1983) suggest that the tags are not serious problems. Morrison and Zajac (1987) found olfactory nerve damage in properly tagged chum salmon. Such damage could effect the homing ability of returning adults. Based on lack of obvious straying to Kitoi Bay Hatchery, we found no evidence of impaired homing ability in this study.

Comparison of the ratios of tagged to untagged pink salmon at time of release to that at recovery is presented in Table 4. In some cases, the difference in the proportion of tagged fish between release and recovery is equivalent to the error rate found in our spot checks. This error rate was caused by error in visual recovery of fish with adipose fin clips, as well as by fish that naturally lacked adipose fins. The magnitude of these effects seen in our data, as well as reported in the literature, suggests that further refinement of technique is necessary in order to address these sources of error.

Recommendations

Pink salmon are usually released in large numbers, which creates a need to mark large numbers of fry to maintain reasonable proportions of marked fish in the release. Large numbers of adults must recovered in a short period of time, and then screened for marks. In future work with half-length coded-wire tagging of pink and chum salmon, researchers should explore the possibility of investing project resources in the production of a higher proportion of tagged fish for automatic electronic tag detection, and in a careful, moderately paced tagging routine by well-trained technicians. Elimination of the adipose fin mark as an indicator of wire tags HLCWTs may eliminate what may be the two greatest sources of apparent tag loss: naturally missing adipose fins, and error in mark detection. Development of standard operating procedures and quality control should be a priority. Some of the cost of instituting these changes could be offset by elimination of adipose fin marks and the cost of hand recovery of marks from returning adults.

Early in the project, informal investigations, such as dissection of newly tagged fry, were conducted in the hatchery to determine the best placement for tags. For best tag retention, we believe that the tag should be injected as deeply as possible, compared to equivalent settings for smolt-sized fish. Experience has shown that tags properly placed in the snout of a pink salmon fry are not visible to the unaided eye. If tags are visible, the depth is too shallow and poor long-term retention may result.

With the support of industry we can install electronic tag detectors that automatically remove tagged fish from conveyors in fish-processing plants. We hope that manufacturers will develop tags with stronger magnetic fields, and sensitive, easy-to-deploy detectors that are not triggered by physical shock or small metal particles. Much work needs to be done before such ideas become reality, but the initial steps have been taken.

References

Blankenship, H. L. 1990. Effects of time and fish size on coded wire tag loss from chinook and coho salmon. American Fisheries Society Symposium 7:237–243.

Eames, J. E. 1983. An evaluation of four tags suitable for marking juvenile salmon. Transactions of the American Fisheries Society 112:464–468.

Jefferts, K. B., P. K. Bergman, and H. F. Fiscus. 1963. Coded wire identification system for macro-organisms. Nature (London) 198:460–462.

Moberley, S. A., R. Miller, K. Crandall, and S. Bates. 1977. Mark-tag manual for salmon. Alaska Department of Fish and Game, Division of Fisheries Rehabilitation Enhancement and Development, Juneau.

Morrison, J., and D. Zajac. 1987. Histological effect of coded-wire tagging on chum salmon. North American Journal of Fisheries Management 7:439–441.

Opdycke, J. D., and D. P. Zajac. 1981. Evaluation of half-length binary-coded wire tag application in juvenile chum salmon. Progressive Fish-Culturist 43:48.

Quinn, T. P., and C. Groot. 1983. Orientation of chum salmon (Oncorhynchus keta) after internal and external magnetic field alteration. Canadian Journal of Fisheries and Aquatic Sciences 40:1598–1606.

Thrower, F. D., and W. W. Smoker. 1984. First adult return of pink salmon tagged as emergents with binary coded wires. Transactions of the American Fisheries Society 113:803–804.

Zajac, D. P. 1985. A cursory evaluation of the effect of coded wire tagging upon salmonids. U.S. Fish and Wildlife Service, Fisheries Assistance Office, Olympia, Washington.

American Fisheries Society Symposium 7:259–261, 1990
© Copyright by the American Fisheries Society 1990

Effects on Survival of Trapping and Coded Wire Tagging Coho Salmon Smolts

H. Lee Blankenship and Patrick R. Hanratty

Washington Department of Fisheries, 115 General Administration Building
Olympia, Washington 98504, USA

Abstract.—The effects of trapping and tagging on the survival of migrating coho salmon smolts *Oncorhynchus kisutch* were tested. Fish were trapped by means of a temporary V-shaped weir of small-mesh screened panels, which channeled migrating smolts into live boxes. Fish were then tagged with coded wire tags. The effects were measured over three brood years with hatchery-reared coho salmon that were planted above the weir (test group) and below the weir (control group). Over three brood years, survival of the test groups averaged 84% of that of the control groups.

Several types of traps and weirs have been used in research on, and management of, wild stocks of juvenile Pacific salmon *Oncorhynchus* spp. A commonly used weir employed by all the state fishery management agencies and Canada on the Pacific coast of North America for capture of coho salmon *O. kisutch* is a V-shaped fence weir of small-mesh screen panels, described by Armstrong and Argue (1975). Since 1975, the Washington Department of Fisheries (WDF) has trapped and tagged over 1 million wild coho salmon with coded wire tags (CWTs; Jefferts et al. 1963). The primary purpose of this activity is to estimate survival and evaluate contributions of wild coho salmon stocks to fisheries. A basic question that has been raised about the resulting data concerns the effects of the trapping and tagging on the coho salmon smolts. Wedemeyer et al. (1980) have shown that stress, scale loss, exposure to anesthetic, etc., can reduce survival of salmon smolts. Tagging agencies have recommended that smolting salmon not be tagged (Pacific Marine Fisheries Commission 1983). However, an adequate method has not been found to capture sufficient numbers of wild coho salmon for tagging prior to smoltification. This study was designed to quantify the effects of trapping and tagging on survival of coho salmon smolts.

Methods

Minter Creek, located on Carr Inlet in southern Puget Sound, was chosen as the study site (Figure 1). Salo and Bayliff (1958) described Minter Creek as representative of numerous small streams that contribute in large measure to the production of wild coho salmon in Puget Sound. Minter Creek drains a watershed of 22 km² and has a moderate

overall gradient of about 1.3%. Stream flows range from a high of 28 m³/s to a low of 0.5 m³/s. Minter Creek is typical of many streams in which fish were previously trapped by WDF, except that a WDF salmon hatchery is located on the creek just above high tide.

Each year, approximately 1,000 adult coho salmon are passed upstream of the hatchery rack for natural spawning. In order to differentiate our study fish from the naturally reared fish, a freeze brand was applied in February, approximately 2 months prior to smoltification, to a control and a test group of juvenile coho salmon reared at Minter Creek Hatchery. In a test group of 25,000 juveniles, each fish was branded with an "S" on the left side, just below the dorsal fin. In the control group of 25,000 each fish was branded with an "S" on the right side, just below the dorsal fin; in addition, its adipose fin was clipped and it was tagged with a CWT. Tagging was conducted according to standard procedures recommended by the Pacific Marine Fisheries Commission (1983). Prior to planting, both groups were checked for brand, tag retention, or both. Each year, both groups were reared in adjacent cement raceway ponds. The study was planned to last four consecutive brood years, 1980–1983.

For this study, we made three assumptions. First, we assumed that the effects of trapping and tagging were similar for hatchery-reared coho salmon smolts and naturally reared smolts. Second, we assumed that there was no differential survival after 3 months between test fish, which were branded, and control fish, which were branded, fin-clipped, and tagged. Third, we assumed that all tagged fish were equally susceptible

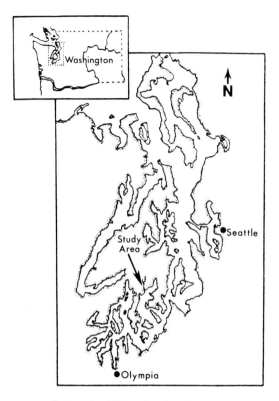

FIGURE 1.—Minter Creek study area.

FIGURE 2.—Minter Creek V-shaped fence weir.

to capture in the fisheries that were sampled for the occurrence of CWTs.

From the last week in April until the last week in May, approximately equal numbers of control and test fish were planted on a schedule coinciding with the anticipated migration pattern of the naturally reared fish in Minter Creek. Fish were planted daily, Tuesday through Saturday. The fish were dipped out of the raceway into a 20-L bucket of water, and then counted as they were slowly poured into a 600-L tank truck, which transported them 5 km to the release site. The control fish (tagged) were planted immediately downstream of the weir. The test fish (brands only) were planted 1 km upstream of the weir.

During the third week in April, a V-shaped fence weir was installed on Minter Creek 2 km above high tide. The weir consisted of seven 180-cm-long panels, two connecting trap boxes measuring 120 cm × 180 cm, and a holding box of 120 cm × 180 cm (Figure 2). Daily maintenance of the weir included the transfer of fish from the two trap boxes to a holding box to help reduce stress from overcrowding and excessive current.

Trapped fish were tagged every Tuesday, Thursday, and Saturday, from the last week in April to the first week in June. Each tagging day, the fish were transferred by dip net to a portable trough, and then groups of approximately 100 fish were anesthetized in a solution (1 g/20 L) of tricaine (MS-222). The trapped smolts were sorted to differentiate between naturally reared fish (no brand) and hatchery-reared fish (brand on left side). The test fish were tagged and, after recovery from anesthesia, released immediately downstream of the weir. A sample group was retained for 24 h to determine immediate tag loss and mortality.

Results and Discussion

Although the study was planned for four consecutive brood years, a family of otters wiped out most of the 1982 brood; this loss left three broods on which to report.

Data on fish released, adjusted for tag loss, and on observed recovery rates for fish tagged with CWTs in each brood are given in Table 1. Observed recoveries include all of those recorded from the sport and commercial fisheries and those noted at hatchery racks, as reported by WDF and British Columbia (coastwide coded wire tag database, University of Washington, unpublished) during 1983, 1984, and 1986.

TABLE 1.—Released and recovered coho salmon with coded wire tags.

Release group	Tag code	Number of tagged and released fish	Fish recovered	
			Number	%
1980 brood				
Above weir	62-25-53	17,926	586	3.27
Below weir	63-25-09	22,365	901	4.03
1981 brood				
Above weir	63-25-60	20,117	326[a]	1.62[a]
Below weir	63-25-58	22,424	432	1.93
1983 brood				
Above weir	63-34-27	21,090	813	3.85
Below weir R-1	63-24-54	7,766	358	4.61
Below weir R-2	63-28-55	7,772	318	4.09
Below weir R-3	63-28-56	7,780	340	4.37
R-1, R-2, R-3	Combined	23,318	1,016	4.36

[a]This value was corrected by six recoveries because 150 wild smolts were inadvertently tagged with this code.

Comparisons of test and control groups were made with observed rather than expanded data for recovered fish. If survivors of all groups had the same spatiotemporal distribution in the fisheries, it would probably be best to draw inferences from the data at hand (i.e., the observed values). In such a case, adjusting these data through expansion would probably increase the error variance relative to the estimated variance components of the design factors, thus decreasing the power of the tests for factor effects.

The difference in survival between the control groups planted below the weir and the test groups planted above the weir was tested with a two-sample test for binomial proportions (Zar 1984) for the three broods. The one-tailed null hypothesis was rejected for each of the brood years at the 0.01 level.

The test-group smolts planted above the weir survived at rates that were 81, 84, and 88% (average, 84%) as high as those of the controls for the 1980, 1981, and 1983 broods, respectively. Supporting evidence for this range of relative survival, and for the assumption that the effects of trapping and tagging are the same for naturally reared smolts and hatchery-reared smolts, was

given by Lister et al. (1981), who used fence weirs and tagging procedures like ours. The authors reported that survival of trapped and tagged native coho salmon smolts was 84%, relative to smolts not trapped and tagged, by the time fish returned to the spawning grounds. To adjust for the 84% relative survival of trapped and tagged smolts, we suggest that a 1.19 factor be applied to survival data for wild coho salmon smolts trapped and tagged as migrants, so that survival of smolts can be directly compared with survival of pre-smolts tagged with CWTs in hatcheries.

Acknowledgments

We express our appreciation to Jeff McGowan, John Sneva, and Dan Thompson for their involvement in planting of fish, tagging, trap installation, and maintenance.

References

Armstrong, R. W., and A. W. Argue. 1975. Trapping and coded-wire tagging of wild coho and chinook juveniles from the Cowichan River system. Fisheries and Environment Canada, PAC/T-77-14, Vancouver, Canada.

Jefferts, K. B., P. K. Bergman, and H. F. Fiscus. 1963. A coded-wire identification system for macro-organisms. Nature (London) 198:460–462.

Lister, D. B., L. M. Thorson, and I. Wallace. 1981. Chinook and coho salmon escapements and coded-wire tag returns to the Cowichan–Koksilah river system, 1976–1979. Canadian Manuscript Report of Fisheries and Aquatic Sciences 1608.

Pacific Marine Fisheries Commission (PMFC). 1983. Coded-wire tagging procedures for Pacific salmonids. PFMC, Portland, Oregon.

Salo, E. O., and W. H. Bayliff. 1958. Artificial and natural production of silver salmon at Minter Creek, Washington. Washington Department of Fisheries Research Bulletin 4.

Wedemeyer, G. A., R. L. Saunders, and W. C. Clarke. 1980. Environmental factors affecting smoltification and early marine survival of anadromous salmonids. U.S. National Marine Fisheries Service Marine Fisheries Review 42(6):1–14.

Zar, J. H. 1984. Biostatistical analysis, 2nd edition. Prentice-Hall, Englewood, Cliffs, New Jersey.

American Fisheries Society Symposium 7:262–266, 1990

Coded Wire Tag Retention by, and Tagging Mortality of, Striped Bass Reared at the Hudson River Hatchery

DENNIS J. DUNNING AND QUENTIN E. ROSS

New York Power Authority, 123 Main Street, White Plains, New York 10601, USA

BRUCE R. FRIEDMANN

EA Engineering, Science, and Technology, Inc.
The Maple Building, 3 Washington Center, Newburgh, New York 12550, USA

KENNETH L. MARCELLUS

Consolidated Edison Company of New York, Inc.
4 Irving Place, New York, New York 10003, USA

Abstract.—Coded wire tags were implanted in 1,370,426 juvenile striped bass *Morone saxatilis* produced at a commercial hatchery on the Hudson River between 1983 and 1987. Three tag locations were used: nasal cartilage, cheek muscle, and musculature of the nape. The tag retention rate, 8–10 weeks after tagging, was significantly different among body locations. Most tag loss occurred within 2 weeks after marking. Tags implanted vertically in the cheek and in the nape were retained by most fish. The retention rate for tags implanted vertically in the cheek ranged from 86 to 94%, and the lowest retention rate was associated with the highest peak tagging rate (417 fish/h). The mortality rate, 48 h after tagging, of fish tagged vertically in the cheek was less than 1% during 1985 and 1987 and about 3% during 1986. The significantly higher rate observed during 1986 was associated with the greatest number of fish tagged at the Hudson River hatchery in a year.

Striped bass *Morone saxatilis,* with median total lengths of at least 77 mm, have been stocked into the Hudson River since 1983. Stocking has been part of an effort to evaluate supplementation of wild stocks and mitigation of mortality associated with impingement of fish on the intake screens of four power plants. Stocking will continue on an annual basis through 1990. Striped bass for stocking are reared at a commercial hatchery in Verplanck, New York, on the Hudson River.

At the hatchery, striped bass are tagged prior to release so that they can be distinguished from naturally spawned fish. All of the striped bass produced at the Hudson River hatchery during a given year are marked, because the cost of tagging hatchery fish is considerably less than the cost of recapturing them in the Hudson River (Dunning and Ross 1986), and because large numbers of tagged fish are required for statistically robust estimates of the contribution of hatchery fish to the Hudson River stock (Heimbuch et al. 1990, this volume).

In this paper, we present data on the retention rates of coded wire tags implanted in juvenile striped bass, and we include data on mortality within 72 h of tagging.

Methods

Tagging

Coded wire tags were implanted in 1,370,426 striped bass between 1983 and 1987, at a commercial hatchery on the Hudson River. We tested three tag locations (nasal cartilage, cheek muscle and musculature of the nape) and two tag types (color-coded and binary-coded wire tags). The location and type of tag used were not the same in all years and were changed as new data were obtained. The tagging and evaluation procedures are described below by year. All fish lengths are given as total lengths.

1983.— In 1983, striped bass of approximately 65–100 mm were selected for tagging. All fish were anesthetized with quinaldine; then the second dorsal fin of each fish was clipped and a color-coded wire tag was implanted in its nasal cartilage (snout) with a Smith-Root tagging machine. The presence of a tag was verified with a magnetic-field detector. After they were marked, fish were given a prophylactic bath of salt (1%) and Neo-terramycin (20 mg/L) and allowed to recover for up to 1 h in a flow-through tank.

On September 20, 1983, 480 hatchery fish were tagged as described and held for examination of tag retention. Checks were performed 2, 4, 8, and

10 weeks after tagging. Fish without tags were not removed after each check, and thus accumulated throughout the 10-week examination period.

1984.—Binary-coded, 1-mm-long stainless steel wire tags, produced by Northwest Marine Technology, Inc. (NMT), were implanted approximately 2 mm deep with NMT tag injectors in 1984. The presence of tags was determined with NMT quality control devices. Flow-through water was supplied, and an attempt was made to keep water temperatures at or below 25°C to reduce the stress associated with tagging. Water temperatures ranged from 19 to 27°C, and most readings were less than 25°C.

Fish were not fed for at least 12 h before they were marked. Fish were anesthetized with quinaldine, fin-clipped, tagged, checked with a quality control device, and piped with water to tanks supplied with oxygenated water, where they could recover. Fish with no apparent tag were transported to separate holding tanks for subsequent rechecking or retagging. Only fish 65 mm and longer were tagged. Three tag locations were used during 1984: horizontally in the cheek (the cephalic portion of the adductor mandibularis) ventral to the eye; vertically in the cheek, but at a slight posterior angel, posterior to the eye; and in the musculature of the nape immediately posterior to the head. Mortality resulting from tags implanted in the cheek was evaluated for a random sample of 10% of each tagging machine's production per shift; the sample contained up to 100 fish per machine when one or two machines operated, and up to 50 fish per machine when three or four machines operated. Approximately equal numbers of unmarked fish were held as controls. Tagged and control fish were held for 72 h. All tagged fish that were alive after 72 h were held for determination of tag retention at 2-week intervals through the next 10 weeks. Fish without tags were not removed after each check. Mortality resulting from tags implanted in the nape was not evaluated.

1985–1987.—Binary-coded, 1-mm-long stainless steel wire tags were used in 1985–1987 as in 1984. Tags were inserted vertically in the cheek, as in 1984. Only fish 65 mm or longer were tagged.

Mortality due to tagging was evaluated with a random sample of 10% of each marking day's output (a sample of up to 50 fish/d), which were held in the hatchery for 48 h. An equal number of unmarked fish was held from each marking day to serve as controls. All surviving marked fish were combined into 2-week cohorts and held for tag retention studies.

All marked fish held for tag-retention studies were checked for the presence of a wire tag at 2-week intervals for up to 10 weeks. Checks were made with a quality control device or with a magnetic-field detector. All fish that retained a tag after each check were held for further study, whereas those fish with no apparent tags were removed.

Each year, 1983–1987, personnel were trained and certified to tag fish. Potential taggers were provided with and asked to read a set of standard operating procedures for marking striped bass at the Hudson River hatchery. Taggers also read literature prepared by NMT with regard to the tagging apparatus. Next, the tagging process was demonstrated. Then each potential tagger marked 100 fish, which were retained for 1 h. When 95% of the test fish retained tags and survived, a person was certified. If the survival rate for fish in the entire hatchery on any given day fell below 95%, attempts were made to determine and correct the source of the problems.

Data Analysis

During 1983 and 1984, the tag-retention rate was determined for a single cohort at 2-week intervals after tagging, and was calculated as the number of fish with tags divided by the total number of fish alive at the end of that interval. Because fish without tags were not removed from the sample, the proportion of fish with tags at the last time interval was the best estimate of the final tag-retention rate. The 2-week-interval estimates are not multiplicative.

From 1985 through 1987 the retention rate at each 2-week interval was calculated as

$$R_i = (\sum T_{ij} / \sum N_{ij}) \times 100;$$

T is the number of fish with tags, i is the 2-week period after tagging, j is the cohort of fish tagged, and N is the number of fish alive at the time that fish were checked for tags. Fish without tags were removed from the sample once they were identified. Thus, each 2-week interval provides an independent estimate of the retention rate. Therefore, the overall retention rate was

$$R = [\Pi(R_i / 100)] \times 100.$$

The tagging mortality was calculated as the number of fish dead divided by the number of fish tagged, corrected for mortality of control fish. The correction was made by subtracting mortality of control fish from the unadjusted mortality. Data were excluded from analysis when mortality of

TABLE 1.—Retention of coded wire tags by striped bass, measured at 2-week intervals after tagging during 1983 and 1984. Sample sizes are in parentheses.

Year	Tag placement	Tag-retention rate, %[a]				
		0–2 weeks	3–4 weeks	5–6 weeks	7–8 weeks	9–10 weeks
1983	Snout	98.5 (469)	63.0 (468)	[b]	63.0 (446)	65.0 (331)
1984	Horizontal in cheek	28.0 (585)	30.4 (690)	22.2 (676)	30.7 (670)	29.8 (661)
	Vertical in cheek	83.7 (2,553)	82.7 (1,547)	85.8 (2,075)	87.0 (1,915)	86.8 (760)
	Nape	99.4 (173)	93.8 (161)	[b]	99.3 (153)	

[a] Fish were kept from one sampling interval to the next, regardless of whether they retained a tag or not.
[b] No data available.

control fish was either greater than 25% or more than twice that of tagged fish.

The peak tagging rate was calculated as the maximum number of fish tagged per day, divided by the number of tagging machines used and the number of hours during which tagging was conducted.

Differences in tag retention and tagging-induced mortality between years were tested by constructing a 95% confidence interval around the rates for the year in which the peak tagging rate was highest. Calculations were in accordance with procedures described by Steel and Torrie (1960). Values outside the 95% confidence interval were considered different.

Results

Retention

During 1983, the retention rate of color-coded wire tags implanted in the snouts of juvenile striped bass was high during the first 2 weeks after tagging (98.5%), but declined abruptly during the third and fourth weeks after tagging (to 63.0%). It remained relatively constant from that point through the end of the 10-week period of examination (Table 1).

In 1984, coded wire tags were retained significantly better $(P < 0.05)$ when implanted in the nape (99.3%) than when placed vertically in the cheek (86.8%). Both these placements resulted in significantly higher retention $(P < 0.05)$ than horizontal cheek placements (29.8%). After the first 2 weeks, the loss rate remained relatively consistent for all of the tag locations (Table 1).

During 1985 and 1986, there was no significant difference $(P < 0.05)$ in the overall retention of coded wire tags implanted vertically in the cheek (Table 2). Tag loss in both years decreased noticeably after the second posttagging week after tagging.

During 1987, tag-retention rate (85.3%) was significantly lower $(P < 0.05)$ than that observed during 1985 (93.5%) and 1986 (92.5%), and similar

TABLE 2.—Retention of coded wire tags inserted vertically into the cheek of striped bass, 1985–1987. Sample sizes are in parentheses.

Year	Tag-retention rate, %[a]					
	0–2 weeks	3–4 weeks	5–6 weeks	7–8 weeks	9–10 weeks	Overall
1985	94.8 (1,845)	98.8 (1,763)	99.8 (1,701)	100.0 (1,564)	100.0 (523)	93.5
1986	95.0 (1,023)	99.1 (911)	99.2 (840)	99.1 (712)	[b]	92.5
1987	93.3[c] (2,491)	[b]	[b]	[b]	91.3 (383)	85.3

[a] Only fish that retained tags were kept from one sampling interval to the next.
[b] No data available.
[c] Forty-eight hour retention.

TABLE 3.—Mortality of striped bass immediately and 24, 48 and 72 h after insertion of coded wire tags, 1983–1987. Numbers of fish tagged are in parentheses.

Year	Mortality, %			
	Initial	24 h[a]	48 h[a]	72 h[a]
1983	1.7 (62,617)	[b]	[b]	[b]
1984[c]	2.8 (155,871)	5.5 (4,650)	18.2 (5,400)	21.1 (4,650)
1985	0.7 (288,433)	0.1 (1,898)	0.3 (1,898)	0.3 (1,898)
1986	0.8 (535,966)	[b]	3.6 (1,475)	[b]
1987	0.1 (327,539)	[b]	0.3 (2,550)	[b]

[a]Adjusted for mortality of control fish.
[b]No data available.
[c]Values for tags implanted vertically and horizontally in the cheek only.

to that observed during 1984 (86.8%), when the same tag location and orientation were used.

Mortality

There were significant differences ($P < 0.05$) in mortality immediately after tagging when values were compared among years from 1983 through 1987 and from 1985 through 1987 (Table 3). The lowest rate was observed during 1987 (0.1%), and the highest was observed during 1984 (2.8%). There also were significant differences ($P < 0.05$) in mortality 48 h after tagging when values were compared among years 1984–1987 and 1985–1987. The lowest values were observed during 1985 and 1987 (0.3%), and the highest value was observed during 1984 (18.2%).

Tagging Rate

The peak tagging rate increased each year from 1984 through 1987 (170, 340, 383, and 417 fish tagged per hour per machine). The tagging rate was not determined during 1983. The peak tagging rate did not increase consistently as the total number of fish tagged increased (Table 3).

From 1985 through 1987, when all tags were implanted vertically in the cheek, there was no apparent relationship between the peak tagging rate and mortality immediately after tagging or 48 h later. There was also no relationship between the total number of fish tagged and mortality immediately after tagging. However, there was a high positive correlation ($r = 0.99$) between the number of fish tagged and mortality 48 h after tagging.

From 1985 through 1987, there was a high negative correlation ($r = -0.89$) between the peak tagging rate and the retention of tags 10 weeks after tagging, but no apparent relationship between the total number of fish tagged and the retention of tags 10 weeks after tagging.

Discussion

The relatively low retention rate we observed for tags implanted in the snout, and the relatively high retention rate for tags implanted vertically in the cheek, indicate that the location of the tag strongly affects retention by juvenile striped bass. These results are consistent with those reported by Klar and Parker (1986) for juvenile striped bass, and by Williamson (1987) and Fletcher et al. (1987) for largemouth bass *Micropterus salmoides*, which are morphologically similar to striped bass. The majority of tag losses from juvenile striped bass occurred within the first 2 weeks after tagging which is consistent with the observations made by Fletcher et al. (1987) on largemouth bass.

We consistently observed loss of tags implanted in the cheek of juvenile striped bass from 1985 through 1987 (unweighted average, 9.6%), whereas no loss was observed by Klar and Parker (1986). This difference may be attributable to the extended tagging season and the high tagging rate at the Hudson River hatchery. Klar and Parker (1986) tagged a maximum of 2,000 striped bass per year, but between 288,000 and 536,000 were tagged annually at the Hudson River hatchery from 1985 through 1987. It is possible that fatigue and boredom of the taggers were more of a problem at the Hudson River hatchery, where the tagging season lasts about 45 d, than for Klar and Parker (1986), whose tagging was conducted over several days, despite the formal program to train and certify taggers at the Hudson River hatchery.

A significant decline in retention of tags 10 weeks after tagging at the Hudson River hatchery, from 1985 (93.5%) through 1987 (85.3%), was associated with a 23% increase in the peak tagging rate. The hourly tagging rate for Klar and Parker (1986) was the same order of magnitude as the peak tagging rate at the Hudson River hatchery from 1985 through 1987 (N. C. Parker, U.S. Fish and Wildlife Service, personal communication). Because the tagging rate for Klar and Parker was approximate, we cannot discount the possibility that the tagging rate at the Hudson River hatchery was actually higher and contributed to the difference in the retention between the two programs.

The retention of coded wire tags implanted in the nape was relatively high, and similar to that of tags implanted vertically in the cheek. This is not surprising because tags are implanted in muscle in both these locations, whereas tags are implanted in cartilage in the snout. However, we do not have a satisfactory explanation for the significantly lower retention rate observed for coded wire tags implanted horizontally in the cheek compared with those implanted vertically. One obvious difference between the two orientations is that tags implanted horizontally run perpendicular to the muscle striation, whereas those implanted vertically run parallel to the muscle striation. Nonetheless, it appears that orientation as well as location strongly affects retention.

Mortality associated with coded wire tags implanted vertically in the cheek (1985–1987) occurred up to 72 h after tagging, albeit at a relatively low level. The mortality rate 48 h after tagging was less than 1% in each year except 1986. The significantly higher mortality rate during 1986

was associated with the greatest number of fish tagged at the Hudson River hatchery in a year, suggesting that tagger fatigue and boredom increases directly with the number of fish to be tagged.

Acknowledgments

Funding for this project was provided by Central Hudson Gas and Electric Corporation, Consolidated Edison Company of New York, the New York Power Authority, Niagara Mohawk Power Corporation, and Orange and Rockland Utilities under terms of the Hudson River Cooling Tower Settlement Agreement. This project was developed in cooperation with the New York State Department of Environmental Conservation. Tagging was conducted by EA Science and Technology.

References

Dunning, D. J., and Q. E. Ross. 1986. Parameters for assessing Hudson River striped bass stocking. Pages 391–397 in R. Stroud, editor. Fish Culture in Fisheries Management. American Fisheries Society, Fish Culture Section and Fisheries Management Section, Bethesda, Maryland.

Fletcher, D. H., F. Haw, and P. K. Bergman. 1987. Retention of coded wire tags implanted into cheek musculature of largemouth bass. North American Journal of Fisheries Management 7:436–439.

Heimbuch, D. G., D. J. Dunning, H. Wilson, and Q. E. Ross. 1990. Sample-size determination for mark–recapture experiments: Hudson River case study. American Fisheries Society Symposium 7:684–690.

Klar, G. T., and N. C. Parker. 1986. Marking fingerling striped bass and blue tilapia with coded wire tags and microtaggants. North American Journal of Fisheries Management 6:439–444.

Steel, R. G., and J. H. Torrie. 1960. Principles and procedures of statistics. McGraw-Hill, New York.

Williamson, J. H. 1987. Evaluation of wire nose tags for marking largemouth bass. Progressive Fish-Culturist 49:156–158.

American Fisheries Society Symposium 7:267–271, 1990

Magnetic Tag Detection Efficiency for Hudson River Striped Bass

MARK T. MATTSON

Normandeau Associates, Inc., 25 Nashua Road, Bedford, New Hampshire 03102, USA

BRUCE R. FRIEDMAN

EA Engineering, Science, and Technology, Inc.
The Maple Building, 3 Washington Center, Newburgh, New York 12550, USA

DENNIS J. DUNNING AND QUENTIN E. ROSS

New York Power Authority, 123 Main Street, White Plains, New York 10601, USA

Abstract.—The objective of this study was to quantify tag detection efficiency in recovery efforts for striped bass *Morone saxatilis* from the Hudson River hatchery. Two types of detection errors were observed, type I or failure to detect a tag when one was present (false negative); and type II or detection of a tag when none was present (false positive). Since 1983, more than 1.3 million hatchery-reared fingerling striped bass have been released in the Hudson River. Each of these fish was tagged with an internal magnetic coded wire tag. More than 64,000 striped bass have been caught in four annual winter–spring trawling periods since 1984. During trawling, all fish were examined for magnetic tags with a metal detector. All fish registering a positive response on the detector were taken to the laboratory for removal and reading of the magnetic tags. As a test for type-I errors, a random sample of 3,805 striped bass was first passed through the detector used in the field and then through a tube-shaped detector with 100% efficiency for confirmation. Two out of 26 tagged fish (7.7%) escaped detection with the detector used in the field. In tests for type-II errors, this detector falsely indicated that magnetic tags were present in 12 (3.4%) of the 348 fish that gave a positive response. Fishhooks (1.7%), metal particles (0.3%), other agency tags (0.3%), and false signals due to boat motion (1.1%) were four sources of false positive results identified in this study. Removal and reading of magnetic tags in the laboratory was, therefore, an important part of this program, because it permitted verification of the tag origin and the hatchery cohort, as well as recognition of sources of error due to false positive results.

Type-I and type-II errors are typically discussed in the context of statistical hypothesis testing (Sokal and Rohlf 1981). A type-I error arises if we reject a true null hypothesis, and a type-II error results if we accept a false null hypothesis. In the context of a mark–recapture study of fish populations, type-I and type-II errors can affect the fundamental assumption that marks are recognized and reported (Cormack 1968; Ricker 1975; Seber 1982). In this paper, we examine type-I and -II errors in the detection of coded wire tags in stripped bass *Morone saxatilis*.

Magnetic coded wire tags are 1-mm × 0.25-mm pieces of wire that, when embedded in the head musculature or cartilage of fish, permit detection and identification of marked fish (Wydoski and Emery 1983). Notches etched in the wire provide the necessary data in binary form. The coded tag usually must be removed from the fish before the code can be read. In the field, fish are monitored with a metal detector designed by the tag manufacturer to detect the tags. Each fish causing a positive signal by the field detector may be taken

to the laboratory for extraction and reading of the tag. (Although metal detectors respond to changes in local magnetic fields, "field detector" in this paper means one used during field sampling.)

Type-I errors can arise if it is assumed that all fish that elicit a positive signal from the field detector have a wire tag. The null hypothesis is that all fish tagged with a wire will be detected if examined with a tag detector. Nondetection can occur if the tag is shed, or if the field detector is not sensitive enough. Type-II errors can occur under tests of this same null hypothesis if fishhooks or other metallic particles are present and elicit a positive response from the detector. We also considered detection of striped bass with coded wires from hatcheries outside of the Hudson River to be type-II errors, because the intention of the stocking program was to determine the proportion of fish reared at the Hudson River hatchery among the striped bass population. False positive responses may also be obtained if the sensitivity of the detector is set too high during periods of excessive boat motion or vibration of the boat due to wind, waves, and rough weather.

In the laboratory, type-II errors can be corrected by removal and reading of tags from fish that tested positive with the field detector. Type-I errors, however, must be evaluated by other methods. Tag shedding can be evaluated if batches of tagged fish are held and periodically monitored for the proportion of fish retaining tags. For striped bass reared at the Hudson River hatchery, the 10-week shedding rates for tags inserted at a 45° angle from the vertical ranged from 6.5 to 14.7% in four annual hatchery programs between 1984 and 1987 (Dunning et al. 1990, this volume).

The objective of this study was to quantify type-I and type-II error rates for detection of coded wire tags in recovery efforts for striped bass from the Hudson River hatchery. During four annual sampling periods since 1984, we observed variability in the sensitivity of the field detectors used to monitor fish for the presence of tags. We therefore employed a more sensitive metal detector on randomly selected days during field sampling to determine the proportion of tags not detected by the field detector. The type-I error rate was quantified and compared to the type-II error rate, which was due to the presence of fishhooks and other metallic particles that triggered the field detector.

Methods

Since 1983, more than 1.3 million fingerling striped bass with a median total length (TL) greater than 76 mm have been stocked into the Hudson River from a hatchery at Verplanck, New York (67 km upriver from the southern tip of Manhattan in New York, New York). The number of hatchery-reared striped bass stocked in each year was 61,357 in 1983, 147,153 in 1984, 284,578 in 1985, 529,563 in 1986, and 324,579 in 1987. To facilitate evaluation, all striped bass produced at the hatchery were tagged with coded wires prior to release. Tags were placed at an angle of 45° from the vertical, in the cephalic portion of the adductor mandibularis (cheek) muscle, posterior to the eye. Wire tags were implanted with a standard injector (Northwest Marine Technology, Inc., Shaw Island, Washington). A random sample of tagged fish from each batch was retained for 10 weeks to evaluate tag shedding (Dunning et al. 1990).

Otter-trawl and Scottish-seine sampling was conducted during the winter–spring periods of 1984, 1985–1986, 1986–1987 and 1987–1988 to capture primarily age-1 and older striped bass and to examine these fish for coded wire tags. All striped bass caught were passed through the detection field of a V-shaped field detector. This detector was designed by the tag manufacturer (Northwest Marine Technology, Inc.) to detect tags in fish under the weather and operating conditions experienced in field recapture efforts. The detector emits a high-pitched tone or beep when metallic objects such as wire tags are moved slowly through the magnetic field, which is located approximately midway between the sides of the detector.

Standardized written procedures were followed when fish were tested for tags with the field detector. The device was placed in a stable, vertical position on the boat, away from most magnetic disturbances. The detector was adjusted to the highest possible sensitivity without continually emitting a false beep. Before and after each fish was monitored, a wire tag embedded at a 45° angle in a wand was passed through the detection field to verify the sensitivity and stability of the detector. Fish were removed one at a time from a holding tank and held around the body in the vicinity of the dorsal fins by wet, gloved hands. With a vertical, sweeping motion, each fish's head was passed through the detection field of the detector. Each fish was passed three or four times through the detection field. Striped bass that tested positive were called "suspected" hatchery recaptures, and were frozen and taken to the laboratory for verification.

In the laboratory, suspected hatchery recaptures were first checked with a field detector and then beheaded. The cheek muscle of fish at least 125 mm TL or the head of smaller fish was digested in sodium hydroxide (NaOH) to dissolve most tissue, cartilage, and bony parts. The wire tag was then removed and read. If no tag was found, the remainder of the fish was cut into major body parts, and each part was digested in NaOH and examined for a tag or other metallic objects.

An extremely sensitive and stable tube-shaped magnetic tag detector was also used in the field on randomly selected days to evaluate the detection efficiency of the field detector. This tubular aluminum detector was approximately 1.3 m long, had an inside diameter of 100 mm, and was mounted on a set of four legs, which held the device at a 30° angle with respect to the boat deck. Magnetic shielding surrounded the entire tube and excluded local magnetic disturbances. This device

TABLE 1.—Recoveries of coded wire tags from striped bass reared at the Verplanck (New York) hatchery, stocked in the Hudson River, and recaptured during four annual sampling periods.

Statistic	Annual sampling period			
	1984	1985–1986	1986–1987	1987–1988
Sampling dates	9 Apr–7 Jun	11 Nov–12 May	21 Dec–8 May	9 Nov–22 Apr
Number examined for tags[a]	1,620	20,820	14,136	28,192
Number of suspected recaptures[b]	0	3	99	246
Number of verified recaptures[c]	0	2	95	238[d]
Number verified by hatchery cohort (number stocked)				
1983 (61,357)	0	0	0	0
1984 (147,153)	0	0	5	4
1985 (284,578)		2	52	82
1986 (529,563)			38	127
1987 (324,579)				25

[a]Number of fish examined with a magnetic tag detector in the field.

[b]Number of fish giving positive response from the detector in the field.

[c]Number of suspected recaptured fish with Verplanck hatchery tags recovered upon laboratory examination.

[d]Not included were one fish recaptured with a coded wire tag from the U.S. Fish and Wildlife Service Manning Hatchery in Cedarville, Maryland, and one fish for which the recovered tag was lost before it was read.

was supplied with running water, and has been typically used in large-scale tag-recovery laboratories for Pacific salmonids. In 1986–1987 and 1987–1988, about 10% of the striped bass caught were first tested for wire tags with the field detector and then with the more sensitive tube detector. Based on the manufacturers' specifications and daily calibration checks with a wire-imbedded wand, we considered the tube detector to be 100% efficient for detecting fish with wire tags.

For striped bass that were tested for tags with both types of detectors, the type-I error rate (nondetection) was calculated as the number of fish with tags that escaped detection with the field detector, divided by the number of fish with tags that were detected by the tube detector. The type-II (false positive) error rate was calculated as the number of fish testing positive for tags with the field detector that did not have tags from the Hudson River hatchery when examined in the laboratory, divided by the total number of fish testing positive. Both type-I and type-II errors were expressed as percentages.

Results

In 1986–1987, the type-I (nondetection) error rate of the field detector was determined by testing each of 2,138 striped bass for wire tags first with the field detector and then with the more sensitive tube-shaped detector. Two of 15 tagged fish escaped detection with the field detector, resulting in an error rate of 13.3%. In 1987–1988, 1,667 fish were checked first with the field detector and then with the tube detector; all 11 tagged

fish were detected by the field unit. The overall type-I error rate for wire tags was 7.7% (2/26).

In the 1984 trawling period, none of the 1,620 fish caught and tested with the field detector were positive for tags (Table 1). Three fish tested positive for tags in 1985–1986 among 20,820 fish caught, but laboratory examination revealed a fishhook in the buccal cavity of 1 fish and tags from the Hudson River hatchery in two fish (Table 2). The type-II error rate in 1985–1986 was 33.3%, based on the small sample of three fish. In 1986–1987, 14,136 striped bass were tested; 99 tested positive with the field detector. However, three fish had no tag and one fish had a fishhook when examined in the laboratory; the remaining 94 fish had Hudson River hatchery tags. In 1986–1987, the type-II error rate was 4.0%. In 1987–1988, 28,192 striped bass were tested with the field detector and 246 fish tested positive (Table 1). However, five fish had fishhooks, one had a 5-mm metal shaving embedded in the gill arch, one had a tag from another hatchery, and one had a tag that was lost before it could be read (Table 2); the remaining 238 fish had Hudson River hatchery tags. If the one fish for which the tag was lost was from the Hudson River hatchery, the type-II error rate was 2.8% in 1987–1988. The type-II error rate from false positive tag detection was 3.4% over all four sampling periods.

Discussion

The overall type-I error rate from failure to detect coded wire tags in Hudson River hatchery striped bass was within the range of the reported 10-week shedding rates for such tags. Nondetec-

TABLE 2.—Sources of false positive detection of coded wire tags in the 1985–1986, 1986–1987, and 1987–1988 annual sampling periods for Hudson River hatchery striped bass.[a]

Source of error	Percent and number (in parentheses) of false positive recaptures (type-II error) in			
	1985–1986 ($N = 3$)	1986–1987 ($N = 99$)	1987–1988 ($N = 246$)	Total ($N = 348$)
Fishhook	0.0 (0)	1.0 (1)	2.0 (5)	1.7 (6)
Metal particles	0.0 (0)	0.0 (0)	0.4 (1)	0.3 (1)
Other agency tag[b]	0.0 (0)	0.0 (0)	0.4 (1)	0.3 (1)
No tags	33.3 (1)	3.0 (3)	0.0 (0)	1.1 (4)
Total	33.3 (1)	4.0 (4)	2.8 (7)	3.4 (12)

[a] N is the total number of fish caught that elicited a positive response from the detector in the field.

[b] One fish of 337 mm total length was recaptured in 1987–1988 with a coded wire tag indicating it was released in Chesapeake Bay from the U.S. Fish and Wildlife Service's Manning Hatchery at Cedarville, Maryland.

tion of tags is, therefore, equal in magnitude to tag shedding as the major source of type-I error in the hatchery striped bass program for the Hudson River. Procedural changes may have been responsible for the zero type-I error achieved in 1987–1988. In 1986–1987, the field detector was set on a portable table on the stern deck of the sampling vessel, near the holding tank. In 1987–1988, monitoring was conducted in the wheelhouse on a table bolted to the wall. This change in location and the stability of the attached table may have permitted the field detector to be operated at higher sensitivity. The relative sensitivity of the field detectors to vibrations, boat motion, and local magnetic fields indicates the need to determine empirically, and to standardize, the optimum procedures for placement, calibration, and use.

The observed overall type-II error rate of 3.4% shows the need for laboratory examination of all striped bass eliciting a positive response from the field detector. One fish was recaptured with another agency's tag during this study, and we considered this a type-II error for detection of Hudson River hatchery striped bass. Had we not read the tag, we would have assumed that this fish originated in the Hudson River hatchery. The discovery of fishhooks as the primary source (6/12 or 50%) of false positive results suggests that analysis of type-II error rates based on fish size may be useful, because hook-and-line fishing may selectively affect larger fish. The absence of tags in 33% (4/12) of the striped bass giving false positive results suggests that a second important source of type-II error may be high-sensitivity adjustment of the detector, or a changing sensitivity threshold due to boat motion or vibration. Our changes in the placement of the detector, described in the preceding paragraph, may have

reduced false positive detection from 3% in 1986–1987 to 0% in 1987–1988. The presence of metallic particles was the least important source of type-II error in this otter-trawl and Scottish-seine sampling program. However, false positive detection due to metal particles may be more common for fish recovered from other sources, such as cooling-water-intake screening devices at power generating stations (unpublished data).

Acknowledgments

Funding for this project was provided by Central Hudson Gas and Electric Corporation, Consolidated Edison Company of New York, Inc., the New York Power Authority, Niagara Mohawk Power Corporation, and Orange and Rockland Utilities under terms of the Hudson River Cooling Tower Settlement Agreement. This project was developed in cooperation with the New York State Department of Environmental Conservation. Field work was conducted by Normandeau Associates, Inc., and verification of coded wire tags in the laboratory was conducted by EA Engineering, Science, and Technology, Inc. We are grateful for the constructive comments provided by the editor and two anonymous reviewers.

References

Cormack, R. M. 1968. The statistics of capture–recapture methods. Oceanography and Marine and Biology: An Annual Review 6:455–506.

Dunning, D. J., Q. E. Ross, B. R. Friedman, and K. L. Marcellus. 1990. Coded wire tag retention by, and tagging mortality of, striped bass reared at the Hudson River hatchery. American Fisheries Society Symposium 7:262–266.

Ricker, W. E. 1975. Computation and interpretation of biological statistics of fish populations. Fisheries Research Board of Canada Bulletin 191.

Seber, G. A. F. 1982. The estimation of animal abundance, 2nd edition. Griffin, London.

Sokal, R. R., and F. J. Rohlf. 1981. Biometry, 2nd edition. Freeman, San Francisco.

Wydoski, R., and L. Emery. 1983. Tagging and marking. Pages 215–328 *in* L. A. Nielsen and D. L. Johnson, editors. Fisheries techniques. American Fisheries Society, Bethesda, Maryland.

American Fisheries Society Symposium 7:272–280, 1990

Insertion and Detection of Magnetic Microwire Tags in Atlantic Herring

J. A. MORRISON

Department of Agriculture and Fisheries for Scotland Marine Laboratory
Victoria Road, Aberdeen, Scotland, United Kingdom

Abstract.—Magnetic binary-coded microwire tags were used in a tagging experiment, started in 1983, carried out on Atlantic herring *Clupea harengus harengus* in the northern part of the North Sea. Monitoring for the presence of microtags was carried out in 1984 and 1985 with a prototype mechanised tag-detection system. Petersen estimates of the stock in 1983, derived from the tag-return data, agreed generally with acoustic estimates made in 1983 and with virtual population analysis (VPA) of the stock in 1983, derived from later VPA assessments. Estimates from the tagging experiment had a high variance, however, because good recruitment in 1984 and 1985 caused low densities of tagged fish. The results suggest that the tag–recapture method can be used successfully for herring stock estimation, but is not the most suitable method for a stock undergoing a rapid expansion.

The decline of the stocks of Atlantic herring *Clupea harengus harengus* in the North Sea during the late 1960s and the early 1970s has been well documented, but it has been less well understood that the assessment of these stocks has been difficult during the closure of the herring fisheries, which extended from 1977 to 1983. At the time of the closure, estimates of the biomass of spawning herring in the northern and central North Sea were made with data derived from commercial catches and surveys by research vessels. The method was to regress the series of annual estimates of stock biomass, which were generated from catch-based virtual population analysis (VPA), against indices of larval abundance derived from annual surveys coordinated by the International Council for the Exploration of the Sea (ICES) and carried out in autumn in the North Sea. The regression was used to estimate the spawning stock biomass in the most recent year from the appropriate value of the larval index.

This procedure was followed because the VPA stock biomass estimates for the most recent years of an assessment are highly dependent on the correctness of the assumptions made about the fishing mortality rate in the most recent year, and are thus subject to considerable potential error. During the period of closure, VPA assessments could not be updated realistically because of the lack of catch data. Larval indices could thus be used only as very approximate indicators of the biomass of spawning stock, because of uncertainty about the validity of the relationship existing between stock biomass and the larval index at the historically low levels of stock biomass that had then been reached.

Scientists from ICES member countries consequently began looking at other methods of stock assessment, including tagging and acoustic estimation (Anonymous 1980). One method that was proposed was a mark–recepture experiment with a system of magnetic microtags developed by Northwest Marine Technology, Inc. (Shaw Island, Washington). Coded microwires seemed to have considerable advantages over tagging with traditional internal tags because their small size would reduce tagging damage to "full" (spawning) herring. Accordingly, an ad hoc ICES planning group was set up to review the technical problems in applying this system to the tagging of Atlantic herring. In 1979, proposals were made to carry out a major tagging experiment in statistical subarea IV and division VIa, to estimate the sizes of Atlantic herring stocks in these areas, to estimate recruitment to these stocks, to examine their interrelationships, and to estimate natural mortality.

Although the general idea of a microtagging experiment under ICES auspices gained some support in principle, support was not sufficient to allow the original proposals to go ahead. However, so that the technique could be evaluated, Scotland, Norway, and the Netherlands agreed to cooperate, and a field trial of the technique was planned to start in Shetland waters in June–July 1983. By that time, valuable experience in use of the tagging equipment and a 15-cm-diameter tu-

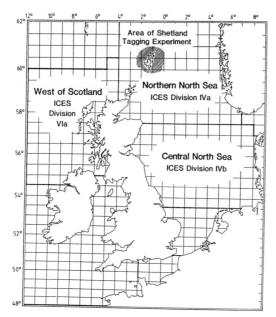

FIGURE 1.—Location of operation, microtagging experiment in the Shetland area, 1983–1985.

bular tag detector had been gained, first during an acoustic survey of North Sea herring, with RV *Tridens* (Corten 1980), and later in pilot tagging and monitoring experiments on Atlantic herring in the Firth of Clyde (Morrison 1982). In the latter experiments, Atlantic herring were tagged on board chartered ring-net vessels (Strange 1977); screening of microtags with a 22-cm-diameter tubular detector, was carried out at processing factories ashore. From the experience gained, it was recognised that a mechanized screening device was essential for any large-scale tag-monitoring program. By 1983 a prototype screening device had already been tested successfully in Canada, and thus it was possible to go ahead with the planning for the Shetland experiment.

Methods

Tagging, 1983.—Tagging was carried out from a 40-m Scottish purse-seine vessel chartered for a 5-week period during June and July 1983. This charter was jointly financed by the Marine Laboratory Aberdeen and the Dutch Government, and scientific staff were provided by the Aberdeen, Bergen, and Ijmuiden fisheries laboratories. The vessel operated in the Shetland area in the northern part of division IVa (Figure 1) and fish were tagged over as wide an area as possible, including places in which the commercial fleets were not

operating. Atlantic herring were caught in purse seines of between 130 and 170 m depth, and after capture were brailed into the vessel's refrigerated seawater tanks, which were fitted with nets and used as holding facilities. The brails were fitted with canvas liners to minimize damage to the fish during transfer. Microtag injectors supplied by Northwest Marine Technology and operating on the ship's 24-V power supply were set up on top of the tagging tanks. Magnetizers were fixed to the ship's rail, as described by Morrison (1982), and a water supply was connected to each outlet funnel to help wash the tagged fish down a 7.5-cm-diameter plastic pipe and into the sea. Atlantic herring were removed from the holding cages in the refrigerated seawater tanks by hand-held net and initially were transferred to the tagging tanks before they were tagged. This system was discontinued to avoid unnecessary handling. Thereafter, fish were removed directly from the net cages and tagged. No anaesthetics were used during the tagging operation. All fish were inspected for damage, and all fish that had more than very slight scaling were discarded. The layout of the tagging vessel and details of the tagging operation are shown in Figure 2.

Tagging with magnetic microtags was carried out in a fashion identical to that described by Morrison (1982); each fish was tagged in the musculature of the dorsum immediately posterior to the skull. In addition, every 50th fish tagged was checked after magnetization to ensure that the tagging system was operating correctly. To estimate the number of fish released that were carrying a correctly magnetized tag, a sample of approximately 50 fish was taken from each injector on each tagging occasion, and the fish were subsequently checked with a magnetic field detector for the presence of magnetized tags. The proportion of fish containing a correctly magnetized tag was then used as a correction factor to estimate the effective number of fish released with tags. Tagging rates were very variable, and largely dependent on the condition of the fish and the consequent amount of selection required. However, in good weather conditions, and with three tag injectors, it was possible to tag between 3,000 and 4,000 Atlantic herring in 1 d.

Tagging mortality.—To provide estimates of the initial mortality due to handling and tagging, three experiments were conducted. Atlantic herring were tagged and magnetized as usual and then liberated into one of the unnetted refrigerated seawater tanks aboard the purse seiner.

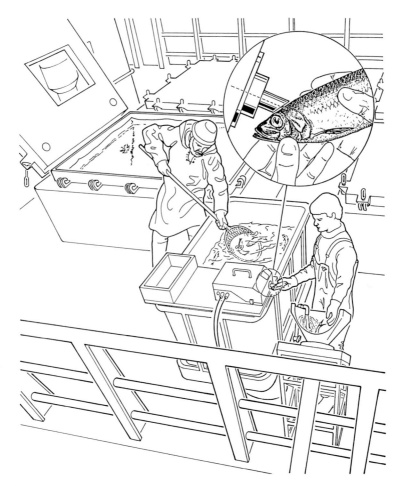

FIGURE 2.—Layout and details of the on-board tagging system for Atlantic herring.

These tanks were of 60–90 m³ capacity and were provided with pumped seawater at ambient sea temperature (9–12°C). The duration of each of the experiments was determined by operational requirements of the vessel concerned and, for this reason, each experiment ran for a different number of days.

Tag shedding and longer-term mortality.—In initial experiments with the herring microtagging apparatus, the tag-injector needle occasionally struck a scale and drove it into the muscle tissue along with the microtag. In these cases, the microtags were not securely lodged and were liable to be lost. To ascertain whether such tag loss was a major problem, 80 Atlantic herring were tagged, checked with a magnetic field detector to ensure that a tag was present, and held in an aquarium at the Marine Laboratory, Aberdeen, for 6 months.

Tag recovery, 1984 and 1985.—There were major difficulties in sampling large quantities of Atlantic herring from commercial catches at Shetland, because most of the catch was transshipped to processing ships from eastern European nations. To overcome this problem, a 40-m Scottish purse-seine vessel was specifically chartered in both 1984 and 1985 as a catcher vessel to catch Atlantic herring for monitoring.

The automatic tag-detection system was first set up on the quayside near a fish-meal plant at Bressay in Shetland at the beginning of June 1984 (Morrison and MacDonald 1986), and used in the same location in 1985. Throughout the course of the 1984 purse-seine charter, all catches made by the Scottish vessel were discharged through this detector. The Norwegian government chartered a Norwegian purse-seine vessel for 2 weeks in 1984

TABLE 1.—Details of vessels employed in tagging or monitoring in the Shetland area, 1983–1985.

Year	Vessel	Operational period	Operational role
1983	FV *Research*	6 Jun–14 Jul	Tagging
1984	FV *Vigilant*	11 Jun–14 Jul	Monitoring and tagging
	FV *Klaring*	16 Jun–2 Jul	Monitoring
	RV *Tridens*	2–12 Jul	Monitoring
1985	FV *Vigilant*	10 Jun–3 Aug	Monitoring and tagging

to assist in the monitoring operations. On this vessel, catches were carried by conveyor belt to the 22-cm-diameter tubular detector described by Morrison (1982). The same year the Dutch government made RV *Tridens* available for a short period to monitor catches for tags in areas around Shetland and also elsewhere in the northern North Sea. Catches were monitored aboard *Tridens* with an existing conveyor belt system coupled to a 15-cm-diameter tubular detector. In 1985, however, only the shore-based tag-detection system was used. Details of the tagging and catcher vessels and their periods of operation are shown in Table 1.

At the fish-meal plant, Atlantic herring were brailed from the seawater tanks aboard the purse seiner into a hopper on the pier, from which they were conveyed through the detector and onto another conveyor belt running to the fish-meal factory. The detection system consisted of an aluminium box section about 2 m long containing detection coils and a two-position deflecting paddle. When a tagged fish on the conveyor belt passed through the coil system, the coils were energized and the paddle activated, thus deflecting a batch of fish (including the tagged one) into a suitable container. Each fish was then passed through the detector again until the tagged individual had been isolated. This arrangement is illustrated in Figure 3. Up to 35 tonnes of fish per hour could be passed through this detection system.

Tag detection.—The tag-detection system used at Shetland was tested on each occasion. An aluminium rod containing a magnetized tag was dropped onto the conveyor belt and checked for consistent detection and deflection over at least five trials. On five occasions in 1984 and on seven occasions in 1985, more extensive tests of efficiency were carried out. In these tests, single tagged fish were thrown into the hopper of the detector when it was full of fish, and the ability of the detector to separate out the tagged fish was

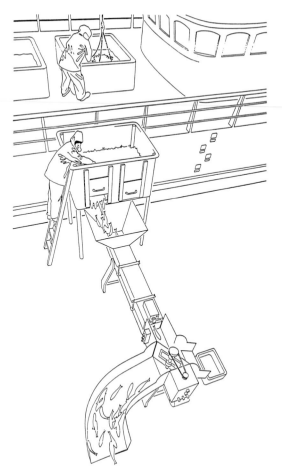

FIGURE 3.—Layout of automatic tag recovery system.

noted in 10 successive trials. In each trial, the tagged fish was initially swept out of the detector along with approximately 20–30 kg of fish. The flow of fish to the hopper was then shut off, the hopper was allowed to empty, and the fish that had been ejected were poured slowly back into the hopper. If a tag was detected, a further small batch of fish was ejected. These fish were placed individually, head first, on the conveyor belt, and passed into the detection tunnel. In this way the tagged fish was finally isolated from the batch. Any trial in which the tagged fish was not isolated was deemed to be a "failure." Fish that passed through the tunnel in such a way that the tag they contained lay almost at right angles to the longitudinal axis of the tunnel seemed to give a relatively weak signal, and this orientation was considered to be the cause of occasional failure to activate the deflecting gate.

Biological samples.—Atlantic herring samples of approximately 25 kg were taken on each occasion that fish were caught. These fish were measured and checked for maturity, and their otoliths were sampled as necessary to establish the age and racial composition of the tagged population. Similar sampling was also carried out with respect to the monitoring operations.

Petersen estimates of stock size.—Adjusted Petersen estimates were based on Chapman's (1951) formula:

$$N = \frac{(M + 1)(C + 1)}{(R + 1)}$$

N = number of individuals in the population in 1983;

M = number of individuals tagged in 1983 after allowance for tag loss and tagging mortality;

C = number of individuals monitored for tags in 1984 and 1985;

R = number of tagged individuals recovered in 1984 and 1985.

Confidence limits on the Petersen estimates were based on Chapman's (1951) formula:

$$V(N) = \frac{N^2(C - R)}{(C + 1)(R + 2)}.$$

Results

Number of Fish Tagged at Shetland

During each tagging operation, sampling was carried out to assess the proportion of the total number of fish tagged that would subsequently activate a magnetic field detector. This proportion was approximately 95%. When corrected for this factor, the total number of tagged Atlantic herring released at Shetland in 1983 was approximately 48,000. Availability of herring shoals at that time restricted tagging operations to the east side of Shetland.

Limited tagging operations with microwires were also carried out at Shetland in 1984 and 1985, but these will not be considered further in this paper.

Tag Monitoring, 1984 and 1985

In 1984, 1,550 tonnes of Atlantic herring were monitored in total; 48 fish with microtags were recovered. Of these recoveries, 32 were from 1983

releases. A high proportion of the 1983 tags were taken to the west of Shetland.

In 1985, 1,259 tonnes of Atlantic herring were monitored, and 48 microtags and one Norwegian internal tag were recovered. Of these tags, 28 were from 1983 releases.

Short-Term Mortality

Three experiments were carried out aboard different purse-seine vessels to estimate short-term mortality due to tagging. In the first of these experiments, 206 Atlantic herring were held for 13 d, and only 6 died (3%); in the second, 100 fish were held for 6 d, and only 4 died (4%); in the third, 102 fish were held for 3 d, and none died. These experiments demonstrated that initial tagging mortality is very low, although they did not give precise estimates of mortality.

Tag Shedding and Long-Term Mortality

In the experiment carried out at the Marine Laboratory Aberdeen, 75 of 80 Atlantic herring tagged (93%) still retained their tags at the end of a 6-month period. Eight months later, when the experiment was terminated, no further tags had been shed.

Efficiency of the Automatic Tag-Detection System

On the basis of tests carried out at Bressay in 1984 and 1985, an overall value of 80% detection efficiency was used in all subsequent calculations. Actual detection rates were 80% in 8 of the 12 tests, 70% in 2 tests, and 90% and 100% in 1 test each.

Population Assessments, 1983 Tagging Data

The basic data required for population estimation are summarized in Table 2. Densities of tagged fish from the 1983 release are also shown for each year class in June and July of 1984 and 1985 in Table 3.

In 1984, a new year class (that of 1981), carrying no tags, entered the fishery. These fish constituted 51 and 44%, respectively, of the total number of fish monitored in June and July (Figure 4). By 1985, a further untagged year class had been recruited to the population, and the proportion of fish that could not contain tags from the 1983 tagging constituted 82 and 59% of the total monitored in June and July, respectively. This recruitment caused a very considerable dilution of the

TABLE 2.—Summary of data required for population estimates of Atlantic herring, based on the 1983 release of fish tagged with coded microwires in the Shetland area.

Measure	Year class			
	1980	1979	pre-1979	Total
1983 tagging				
Thousands of fish tagged	6.8	12.4	29.1	48.3
1984 monitoring				
Thousands of fish monitored				
Jun	1,525.5	653.7	432.9	2,612.1
Jul	686.4	324.0	442.4	1,452.8
Number of tags recovered				
Jun	5	3	5	13
Jul	3	5	11	19
1985 monitoring				
Thousands of fish monitored				
Jun	215.2	99.9	62.0	377.1
Jul	683.9	459.5	576.4	1,719.8
Number of tags recovered				
Jun	0	0	0	0
Jul	1	6	21	28

tagged population and consequently very much reduced the chances of tag recovery.

In 1984, density of tagged fish was considerably lower in June than in July (Table 3). In 1985, no tags were recovered at all in June, despite the monitoring of nearly 350 tonnes of Atlantic herring caught to the east and west of Shetland, whereas tags were recovered in both these areas in July.

Stock sizes in 1983 were calculated for year classes 1980 and older, by means of the data from Table 2. These calculations were carried out with the assumption that 15% of tags were lost due to tagging mortality and tag loss combined. Although about 5% higher than the value shown by the relevant experiments, this value was used in order to take into account additional estimated mortality, such as that due to loss of newly tagged fish to predation by gannets. For the purpose of this assessment, the tag recovery rate of 80%, estimated

TABLE 3.—Densities of Atlantic herring (as coded microwire tags recovered per million fish monitored) that were tagged in 1983 and recaptured in 1984 and 1985.

Sampling		Year class			Weighted mean of 1980
Year	Month	1980	1979	pre-1979	and older year classes
1984	Jun	3.00	4.59	11.55	4.81
	Jul	4.37	15.43	24.86	13.07
1985	Jun	0	0	0	0
	Jul	1.46	13.06	36.43	16.27

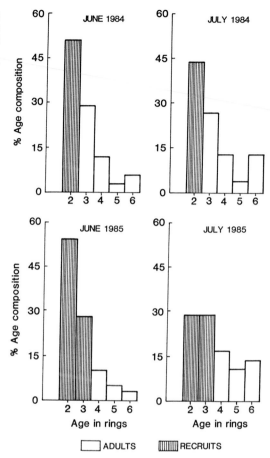

FIGURE 4.—Age composition of Atlantic herring catches monitored for tags, 1984 and 1985. Age is in otolith rings.

from the detection system, was adopted. Estimates of stock size in numbers, based on the foregoing variables, are shown in Table 4, along with 95% confidence limits on the Petersen estimates.

Population Data from Other Sources

Estimates of Atlantic herring population size in ICES division IVa, and divisions IVa and IVb combined, during July 1983 are given in Table 4. The VPA estimates were derived from stock estimates on 1 January 1983, adjusted to account for one-third of the annual fishing mortality and one-half of the annual natural mortality estimated for that year.

Since 1984, successive VPA assessments of the division-IVa stock, carried out by the ICES Herring Assessment Working Group for the area south of 62°N, have resulted in increased estimates of the size of this stock in 1983 (of year

TABLE 4.—Estimates of the size of the 1983 Atlantic herring population in the northern North Sea, based on Petersen mark–recapture estimates from fish tagged with coded microwires in 1983 and recovered in 1984 and 1985, and on acoustic surveys and virtual population analyses conducted in statistical divisions IVa and IVb during July 1983. Data are in millions of fish; 95% confidence half-intervals of Petersen estimates are in parentheses.

Estimate	Year class			Total
	1980	1979	pre-1979	
Petersen				
1984 recovery				
Jun and Jul	1,162 (670)	937 (540)	1,031 (440)	3,130 (966)
Jul	794 (648)	488 (346)	730 (364)	2,012 (820)
1985 recovery				
Jun and Jul	2,599 (3,002)[a]	655 (414)	585 (222)	3,839 (3,039)[a]
Jul	1,977 (2,282)[a]	538 (340)	528 (200)	3,043 (2,316)[a]
Acoustic survey in Jul 1983[b]	571	300	431	1,302
Virtual population analysis in Jul 1983[c]				
Division IVa	988	392	413	1,793
Division IVa and IVb	1,705	653	569	2,927

[a]Inefficient estimate based on return of only one tag.
[b]Anonymous (1984).
[c]Anonymous (1986).

class 1980 and older). The most recent estimate is now very different from the 1983 acoustic estimate.

Two sets of comparisons are shown in Table 4. Firstly, estimates derived from tag-return data for June and July are compared with VPA estimates for IVa and IVb combined. This comparison was done on the basis of suggestions in Anonymous (1985) and Saville and Morrison (1985) that Atlantic herring from division IVb, as well as from division IVa, were present in the Shetland area in June every year. These comparisons agreed well in 1984, but less well in 1985. Secondly, estimates derived from tag-return data for July are compared with VPA estimates of the stock in division IVa. These comparisons again agreed well in 1984 but poorly in 1985. This latter comparison was considered to be valid because it was likely that by July the mixed herring stock believed to be present in Shetland waters in June would have separated into its spawning components and dispersed toward spawning grounds (Wood 1937). Thus, Atlantic herring still in Shetland waters were likely to be part of the division-IVa spawning stock. As previously mentioned, the density of fish with 1983 tags at Shetland was considerably lower in June than in July in both 1984 and 1985.

Discussion

Initially, it was proposed that for an assumed division-IVa stock size of 500,000 tonnes, the liberation of 50,000 tagged fish would yield approximately 100 tags per 1,000 tonnes monitored in the ensuing year. It was calculated that this number would be sufficient to estimate stock abundance within reasonable error bounds. However, even though subsequent VPA estimates of the stock in 1983 were below this level, and almost 50,000 tagged fish were released in 1983, there were substantially fewer fish recovered than had been expected. There was a variety of reasons for the low number recovered. Firstly, in both 1984 and 1985 large-scale recruitment tended to dilute the pool of tagged fish. Secondly, in both years the nonrecruit component of the sampled population contained a much lower density of tagged fish in June than in July, suggesting the presence in June of an additional stock component that was not part of the tagged population. Tag recoveries per tonne in July (when corrected for recruitment) were close to those anticipated.

In terms of a simple Petersen estimate, the presence of this additional stock component in the recapture sample invalidates the assumptions of a closed population. However, it was suggested earlier that the tag-return rates over the entire sampling period might be used to provide a Petersen estimate for divisions IVa and IVb combined. There are at present, however, no data with which to estimate what proportion of the division-IVb stock migrates into the Orkney–Shetland area, and whether this proportion is constant from year to year. Therefore, although there is some level of agreement between the Petersen and VPA estimates for the combined IVa and IVb populations, these data should be viewed with caution when the validity of the microtagging method is evaluated.

It seems likely, however, that a valid comparison can be made between Petersen estimates from data on tag recoveries in July and the VPA estimates for division IVa. With the data from the fish recaptured in 1984, the Petersen estimates encompass the value of the VPA estimate for the same area. The estimated confidence limits are quite high, however, and may also be underestimates of the true confidence limits because they contain no allowances for variance in estimates of detector efficiency, mortality due to tagging, or tag loss. Nevertheless, the Petersen estimate of the Atlantic herring stock in division IVa, for fish of year class 1980 and older in 1983 (based on July 1984 recapture data) is within 12% of the most recent estimate derived from VPA (Anonymous 1986).

The comparable estimate based on the July 1985 recapture data is well above the VPA estimate. However, because almost 60% of the estimated stock in numbers is made up of recruit fish of year class 1980, and because only one tag was recovered from fish belonging to this cohort in 1985, the estimate cannot be considered to be a statistically significant one. If the same data are used and a Petersen estimate calculated with the total number of fish sampled in the relevant cohorts and the total number of tags recovered, a stock estimate of fish of the 1980 year class and older in 1983 is 1,961 million plus or minus 645 million fish, which is within 9% of the most recent VPA estimate.

Conclusions

The main objective of the Shetland experiment undertaken by Scotland, Norway, and the Netherlands was to evaluate the use of magnetic microtagging in herring stock assessment and, in so doing, to provide an independent estimate of the size of the population in division IVa. The results suggest that, in general terms, the first objective was met. Although there is still a need to obtain better estimates of tag loss and tagging mortality, the available evidence indicates that these factors do not seriously hinder the further use of this method. In relation to the tag detection system used, there are still obvious areas of improvement in the design of this prototype system, but the level of detection estimated (approximately 80%) is perfectly acceptable for the type of experiment described.

With respect to the second objective of the experiment, however, the expanding population in division IVa and the rather complicated stock composition of the population in the Shetland area have meant that estimates derived from this experiment must be viewed cautiously. Because of the potential level of sampling error caused by low density of tagged fish, the estimates of stock size do not provide a useful alternative to existing assessments, although it is possible that the variance of existing estimates from VPA and acoustic surveys is also high. What can be concluded, however, is that the estimates do not conflict with existing estimates from other sources, and to this extent the results of this experiment provide independent confirmation of the validity of recent assessments for this area.

As a result of its technical advantages, magnetic microtagging has considerable potential for stock assessment of Atlantic herring, and perhaps of other pelagic species in appropriate situations. In view of the recent increase in the North Sea herring stock, however, a greatly increased scale of operations would now be needed for successful assessment with this method. The enhanced operation would entail the release of at least 100,000 tagged fish per year and annual monitor of at least several thousand tonnes of fish.

The microtagging technique seems ideally suited to a situation in which stock sizes are low and stable with little recruitment, conditions that prevailed in the North Sea in the late 1970s. These conditions are those most unsuitable for acoustic and larval surveys, which are most appropriate when stock sizes are large. Under the former circumstances, research effort would need to be shifted from other existing programs.

In conclusion, the equipment used in these experiments constitutes an important breakthrough in technology, and its use should be seriously considered for estimation of the sizes of stocks in cases where the logistics are not so formidable.

Acknowledgments

I thank members of the scientific staffs of the Bergen, Ijmuiden, and Aberdeen laboratories, whose hard work in difficult conditions contributed to the success of this experiment. Thanks are also due to the skippers and crews of the charter and research vessels employed in the tagging and monitoring operations and to the managers and employees of Herring Byproducts Limited at Bressay, who all made vital contributions. R. S. Bailey and R. Jones of the Marine Laboratory, Aberdeen, made useful suggestions on the text, and A. Rice, also of the Marine Laboratory, prepared the illustrations. Finally, the advice and prompt technical back-up provided by Northwest Marine Technology made possible the successful completion of a difficult technical operation.

References

Anonymous. 1980. Report of the Herring Assessment Working Group for the area north of 62°N. International Council for the Exploration of the Sea, Co-operative Research Report 96, Copenhagen.

Anonymous. 1985. Report of the Herring Assessment Working Group for the area south of 62°N. International Council for the Exploration of the Sea, C.M. 1985/Assessment 12, Copenhagen.

Anonymous. 1986. Report of the Herring Assessment Working Group for the area south of 62°N. International Council for the Exploration of the Sea, C.M. 1986/Assessment 19, Copenhagen.

Chapman, D. G. 1951. Some properties of the hypergeometric distribution with applications to zoological sample census. University of California Statistical Publications 1:131–160.

Corten, A. 1980. Experiments on the detection of magnetic wire tags in herring catches at sea. International Council for the Exploration of the Sea, C.M. 1980/H:36, Copenhagen.

Morrison, J. A. 1982. A preliminary evaluation of tagging and detection methods of coded microwire tags in herring tagging experiments. International Council for the Exploration of the Sea, C.M. 1982/H:48, Copenhagen.

Morrison, J. A., and W. S. MacDonald. 1986. Further evaluation of tagging and detection of microwire tags in herring and the potential of this technique in herring stock assessment. International Council for the Exploration of the Sea, C.M. 1986/H:13, Copenhagen.

Saville, A., and J. A. Morrison. 1985. Evaluation of tagging and detection using microwire tags in herring and the potential of this technique in herring stock assessment. International Council for the Exploration of the Sea, C.M. 1985/H:26, Copenhagen.

Strange, E. S. 1977. An introduction to commercial fishing gear and methods used in Scotland. Scottish Fisheries Information Pamphlet 1, Aberdeen.

Wood, H. 1937. Movements of herring in the northern North Sea. Scientific Investigations of Fishery Board for Scotland III, Edinburgh.

American Fisheries Society Symposium 7:281–285, 1990

Head Mold Design for Coded Wire Tagging of
Selected Spiny-Rayed Fingerling Fishes

S. Bradford Cook,[1] William T. Davin, Jr., and Roy C. Heidinger

Fisheries Research Laboratory and Department of Zoology, Southern Illinois University
Carbondale, Illinois 62901-6511, USA

Abstract.—Binary-coded magnetic wire tags inserted into the nasal capsule have been used extensively to mark salmonids. Recently, attempts have been made to use coded wire tags to mark various spiny-rayed fishes. Due to morphological and behavioral differences, high tag-retention rates are not obtained for spiny-rayed fished when the standard salmonid head mold is used. We designed head molds for the purpose of tagging channel catfish *Ictalurus punctatus,* bluegills *Lepomis macrochirus,* walleyes *Stizostedion vitreum,* and largemouth bass *Micropterus salmoides.* Two full-head mold designs were tested in this study. Construction of a closed head mold, which completely surrounded a fish's head, required little finishing work, but tag retention by walleyes and largemouth bass used with that mold was only 50–85%. An open head mold, in which deep lateral notches were ground, was used on all four species of fish, and tag retention ranged from 91 to 100%. The higher tag-retention rate realized with the open head mold was due in part to its capacity to accept fish of a wider range of sizes. Proper fit of the mold is essential for correct implantation of the tag; correct implantation, in turn, assures higher retention rates.

The coded-wire-tag system described by Jefferts et al. (1963) for the marking of macroorganisms has been successfully applied on a large scale to salmonids (Ebel 1974; Argue et al. 1979). Binary-coded, full-length magnetic wire tags are internal tags consisting of a small wire (1 mm × 0.25 mm) that is implanted into the fish's nasal cartilage or bone (Bergman et al. 1968). Notches cut into the tag carry a binary code; this system allows for 262,000 different codes. Recently, the microtagging technique has been used on some nonsalmonid species, with varying results. Gibbard and Colura (1980) tagged red drums *Sciaenops ocellata* of 50-mm total length (TL) in the nasal cartilage, but only 27% of tags were retained at the end of 1 year. In another study, largemouth bass *Micropterus salmoides,* averaging 92 mm TL, were tagged in the nasal cartilage; at the end of 16 months, 59% of the tags were retained, and the tags had no effects on growth or survival (Williamson 1983). A smaller amount of nasal cartilage compared to salmon (Gibbard and Colura 1980) and poor head-mold sizing (Williamson 1983) have been suggested as reasons for wire-tag loss from nonsalmonids.

Coded wire tags are implanted with a tagging machine equipped with a head mold, which holds and aligns the fish during tagging. Head molds

used for tagging salmonids are based on a half-head mold (Jenkinson and Bilton 1981). This design is effective because salmonids open their mouths when anesthetized, and thus the oral cavity can be used for alignment. Unlike salmonids, many fish do not open their mouths when anesthetized. This characteristic makes the half-head mold unsuitable for tagging spiny-rayed fishes, because proper alignment of the fish in the mold becomes difficult and as a result tag placement varies.

This study was undertaken to design and evaluate suitable full-head molds for coded wire tagging of channel catfish *Ictalurus punctatus,* bluegills *Lepomis macrochirus,* walleyes *Stizostedion vitreum,* and largemouth bass. Two types of full-head molds were tested in this study to determine the appropriateness of the molds for selected spiny-rayed fishes.

Methods and Results

Fishes used in this investigation were obtained locally. Before tagging, the fish were acclimated in 210-L aquaria with recirculated water, and received a prophylactic treatment of formalin (25 mg/L). They were fed daily a commercially prepared feed containing tetracycline. Northwest Marine Technology, Inc. (Shaw Island, Washington) furnished the Mark II microtagging unit used for this study. The system consisted of a tag injector, a quality control device, and a battery-operated magnetic field detector. Each mold con-

---

[1]Present address: Florida Game and Fresh Water Fish Commission, 3900 Drane Field Road, Lakeland, Florida 33811, USA.

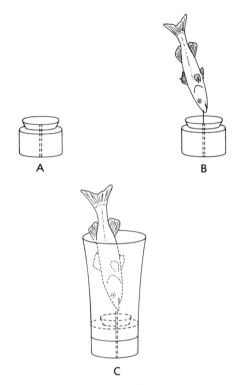

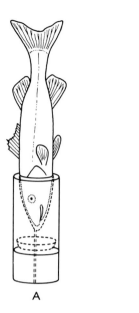

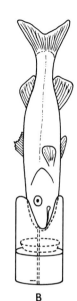

FIGURE 2.—Comparison of the two types of molds used. (A) Side view of closed-form mold. (B) Side view of open-form mold.

FIGURE 1.—Construction of full-head molds. (A) Reusable metal base. (B) Placement of the fish on the needle. (C) The tape cup into which resin is poured around the fish.

sisted of two parts, a reusable metal casting base (Figure 1A), and a fiberglass form.

Before the mold was constructed, the head of a representative fish of the species for which the mold was to be made was dissected and an appropriate area (cartilage or bone) was located for tag implantation. A freshly killed fish was impaled on the needle so that the previously located target area was penetrated and the most anterior part of the fish was approximately 5 mm from the casting base. To prevent the needle from glancing off the nasal bone during tagging, the needle was inserted slightly from the side and downward (Figure 1B). Masking tape (5 cm wide) was wrapped around the base so it encircled the fish (Figure 1C) and served as a temporary form. Resin was poured into the form until it reached the posterior end of the operculum, and was allowed to harden thoroughly (3–5 h), after which the fish and tape were removed.

Two types of mold designs were used in this study. The principal difference between them was the amount of material ground away with a

Dremel Moto-Tool.® The first type (closed form; Figure 2A) was made with polyester resin and ground out slightly so that the eyes of the fish did not contact the mold. The second type (open form; Figure 2B) had an open, blunt, V-shaped notch cut into both sides in the area of the fish's eyes (Heidinger and Cook 1988).

All fish were anesthetized with tricaine (MS-222, 50 mg/L) or quinaldine (1.5 mg/L) before they were tagged. The tagging unit then triggered a protrusible needle, which inserted the tag into the targeted nasal tissue. Tagged fish were passed through a quality control device, which magnetized the tag, separated out untagged fish, and counted the tagged fish.

Tagging with the Closed-Form Mold

Two species, walleyes and largemouth bass, were tagged with the closed-form mold. The walleyes were used to obtain preliminary estimates of the size range of fish that could be tagged effectively with a selected mold, to compare the effects of tagging on growth and survival, and to evaluate tag retention. Only tag retention and mold compatibility were measured for the largemouth bass.

Preliminary estimates of mold compatibility were obtained by tagging 25 juvenile walleyes, 5 each measuring 70, 76, 82, 86, and 94 mm TL, with a mold constructed from an 81-mm walleye.

Immediately following tagging, the fish were X-rayed to assess the tag position. Radiographs indicated consistent tag placement within the target area of all the 82-mm fish. The 76- and 86-mm walleyes were satisfactorily tagged; however, some of the tags, 60 and 80%, respectively, were slightly outside the target area. None of the tags completely penetrated the skin of the 94-mm fish, whereas the tags in the 70-mm fish had all passed through the nasal cartilage and lodged in the roof of the mouth. Tag placement was unacceptable when the tagged more than 5 mm shorter or longer than the 81-mm fish for which the mold was made. To verify this finding and to test larger molds, a second group of 25 walleyes (mean TL, 94 mm) was tagged. The fish were anesthetized, sorted into three groups (I, 76–82 mm; II, 83–93 mm; III, 94–114 mm), and tagged with a mold constructed from an 81-mm, an 87-mm, and a 100-mm walleye, respectively. The tagged walleyes were held in a 210-L aquarium for 2 weeks, and then X-rayed to determine retention and position of the tags.

No tag loss or mortality occurred among the 25 walleyes held for 2 weeks. Radiographs indicated that the tag placement varied as much as 3 mm around the target area, although all tags were in the nasal capsule. No statistically significant relationship was found between tag placement and the size of the fish for any mold size (one-way analyses of variance, $P > 0.05$).

To determine the effects of tagging on growth and survival, and to check short-term tag-retention rates, fingerling walleyes were separated into two size-groups: A, 81 walleyes with a mean TL of 85 mm; and B, 70 walleyes with a mean TL of 106 mm. One-half of each group was tagged (41 and 35 fish, respectively) and the other half received only a fin clip. Churchill (1963) has shown that fin clipping had no significant effect on the survival or growth rate of fingerling walleyes. On the basis of previous findings, group-A fish were tagged with molds made from walleyes within ±5 mm of their TL; group-B fish were tagged with molds from fish within ±10 mm of their TL. All the fish were held for 48 h in two 210-L aquaria, checked for tag retention at 24 and 48 h, and then stocked in two 0.3-hectare ponds, one group per pond. The ponds had no other fish and had been fertilized to augment plankton growth. The walleyes were left in the ponds for 44 d, and then the ponds were drained and the fish were harvested. The fish were measured and checked with the magnetic field detector, although Williamson (1987) did not consider this method of detection to be as reliable as X-ray analysis.

No significant difference in growth or survival was found when the fin-clipped walleyes and the tagged walleyes were compared within groups A or B after 44 d in the ponds (one-way analysis of variance, $P > 0.05$). There was, however, some variation in tag retention between groups A and B (Table 1). The tag-retention rate in group A was 85% after 44 d; however, 66% of those losses occurred within the first 24 h. After 44 d, in tag retention group B was 71%; no loss was recorded during the initial 48-h holding period.

Three groups of largemouth bass were tagged to assess tag-retention rates and mold compatibility (Table 1). Group A consisted of 50 fish that had a mean TL of 75 mm; they were all tagged with the aid of a mold constructed from a 77-mm largemouth bass. The 23 fish in group B had a mean TL of 85 mm and were tagged with an 81-mm mold. The 56 fish in group C had a mean TL of 108 mm and were tagged with the aid of a 110-mm mold. All fish were held for 9 months in 210-L aquaria with recirculated water, and were checked weekly for tag loss. Tag-retention rates varied by group: they were 50%, 83, and 77% for groups A, B, and C, respectively. In all three groups, 90% of the tag loss occurred within the first 14 d. Growth and survival were not evaluated because no untagged control groups were used.

Tagging with the Open-Form Mold

Based upon the results of the tagging with the closed form, we selected a limited size range of 10 mm for all size-classes of fish to be tagged with a given mold (Heidinger and Cook 1988). Either 100 or 101 fish from each size-class were tagged. For each species and size-class tagged, a control group of untagged fish (not fin-clipped) was handled in a similar manner.

Fish were maintained in 210-L aquaria for 2 weeks, checked for the presence of tags, and transferred into drainable 0.3-hectare ponds that contained forage of aquatic macroinvertebrates and fathead minnows Pimephales promelas. The fish were checked for the presence of tags with a magnetic field detector, measured, and counted at 3-month intervals. At the conclusion of the study, all ponds were drained, and the fish were again checked. Growth comparisons between tagged fish and control fish were analyzed by Student's t-test.

TABLE 1.—Survival of, and retention of nasally inserted coded wire tags by, fingerling fish used with open and closed head molds. Values for control fish are in parentheses.

Species and mold type	Group	Study duration (months)	Length of fish for head mold (mm)	Sample size	Mean total length Initial (mm)	Final (mm)	Survival (%)	Tag retention (%)
Channel catfish								
Open[a]	A	6	96	100(100)	89(89)	141(144)	62(67)	96
	B	9	156	101(100)	147(146)	178(181)	67(73)	91
	C	6	174	100(100)	171(170)	206(204)	58(61)	99
Bluegill								
Open[a]	A	6	56	100(100)	51(49)	58(63)[b]	100(100)	95
	B	9	89	101(100)	85(85)	126(124)	100(100)	93
	C	6	126	100(100)	119(120)	142(140)	100(100)	98
Walleye								
Open[a]	A	6	58	100(100)	51(51)	99(97)	71(78)	96
	B	6	78	100(100)	73(73)	123(133)[b]	94(91)	100
Closed	A	1.5[c]	81	41(40)	85(84)	102(101)	100(97)	85
	B	1.5[c]	100	35(35)	106(106)	119(121)	99(97)	71
Largemouth bass								
Open[a]	A	6	78	100(100)	72(72)	100(116)[b]	100(98)	100
	B	6	100	100(100)	95(96)	122(129)[b]	99(100)	95
Closed	A	9	77	50	75	167		50
	B	9	81	23	85	169		83
	C	9[d]	110	56	108	159		77

[a] These data are found in Heidinger and Cook (1988).
[b] Mean length of tagged fish was significantly smaller than that of controls ($P < 0.05$).
[c] This trial lasted for 44 d.
[d] Twenty-three percent of the fish had scoliosis at the start of the experiment.

Three size-classes of channel catfish (group A, mean TL = 89 mm; group B, mean TL = 147 mm; and group C, mean TL = 171 mm) were tagged with open-form molds constructed from fish of 96, 156, and 174 mm TL, respectively (Table 1). Tag-retention was 99% for group C, 96% for group A, and 91% for group B. These fish were held for 6 months, 6 months, and 9 months, respectively. There were no significant effects on survival and growth of fish in any of the three size-classes (Heidinger and Cook 1988).

Bluegills (group A, mean TL = 51 mm; group B, mean TL = 85 mm; group C, mean TL = 119 mm) were tagged with molds made from 56-, 89-, and 126-mm bluegills (Table 1). Tag-retentions after 6, 9, and 6 months were 95, 93, and 98% for groups A, B, and C, respectively. Control fish grew significantly more than tagged fish in group A ($P \le 0.05$), but not in the other two groups. No fish died in the three groups (Heidinger and Cook 1988).

Two groups of walleye fingerlings (group A, mean TL = 51 mm; group B, mean TL = 73 mm) were tagged with the aid of head molds constructed from walleyes of 58 and 78 mm (Table 1). After 6 months, tag retention was 96% in group A and 100% in group B. Control fish grew significantly better than tagged fish in group B, but not in group A. Neither group of walleyes exhibited

significant differences in survival between control and tagged fish (Heidinger and Cook 1988).

Two size-groups of largemouth bass (group A, mean TL = 72 mm; group B, mean TL = 95 mm) were tagged with the aid of head molds constructed from 78- and 100-mm largemouth bass (Table 1). Six months after tagging, fish in group A exhibited 100% tag retention, and fish in group B had 95% tag retention. In both groups, control fish gained significantly greater lengths than tagged fish; however, survival was 98% or better in all cases (Heidinger and Cook 1988).

Discussion

Because a magnetic field detector is not considered to be as reliable as X-ray analysis (Williamson 1987), fish that tested negative with the detector were again run through the quality control device in the laboratory. In the field (pond studies) only the field detector was used, and because this device does not tend to give false positives but does occasionally give false negatives, the tag-retention rates may be slightly underestimated.

A properly fitted head mold is the most important factor that affects tag placement and loss rate. Although less grinding and preparation was required to construct the closed-form mold, it did not consistently yield satisfactory results. The restricted mold orifice reduced the size range of fish that could be tagged effectively; consequently, tag-retention rates decreased when fish of a size outside the optimal range for a mold were tagged. It appears that the optimal range of fish that could be effectively accommodated by the closed form was ±3 mm of the "model" fish for walleyes less than 80 mm TL, ±5 mm for 81–110-mm walleyes, and ±5 mm for largemouth bass less than 110 mm TL. In addition allowing less than 90% tag retention, the closed form would be unsuitable for large-scale tagging operations because of the time factor involved in precise grading of fishes and the number of mold changes required.

Fishes tagged when the open form was used retained tags at rates greater than 90%, which would be acceptable for most research needs. This type of mold tolerated a wider range of fish sizes (10 mm) than the closed form (±3 mm), which reduced the amount of grading as well as the number of mold changes. A less restricted throat on this mold also facilitated rapid placement of the fish into the mold.

Tagging of these spiny-rayed fishes (channel catfish, bluegills, walleyes, and largemouth bass) with either form of full-head mold did not significantly alter the survival of the fingerlings. Significant reductions in growth were found in bluegills, walleye, and largemouth bass tagged with the open mold; however, absolute differences from control fish were small. Crumpton (1985) found that largemouth bass tagged in the nasal cartilage attained total lengths that were not significantly different ($P > 0.05$) from those of control fish. Although care must be taken in grinding the molds to ensure that no mechanical damage occurs to the fish during tagging, we do not believe that tagging with the full-head molds would result in any long-term detrimental effects on growth or survival of these fishes.

References

Argue, A. W., L. M. Patterson, and R. W. Armstrong. 1979. Trapping and coded-wire tagging of wild coho, chinook and steelhead juveniles from the Cowichon–Koksilah river system, 1976. Canada Fisheries and Marine Service Technical Report 850.

Bergman, P. K., K. B. Jefferts, H. F. Fiscus, and R. C. Hager. 1968. A preliminary evaluation of an implanted coded wire fish tag. Washington Department of Fisheries, Fisheries Research Papers 3:63–84.

Churchill, W. S. 1963. The effect of fin removal on survival, growth, and vulnerability to capture of stocked walleye fingerlings. Transactions of the American Fisheries Society 92:298–300.

Crumpton, J. E. 1985. Effects of micromagnetic wire tags on the growth and survival of fingerling largemouth bass. Proceedings of the Annual Conference Southeastern Association of Fish and Wildlife Agencies 37:391–394.

Ebel, W. J. 1974. Marking fishes and invertebrates. III. Coded wire tags useful in automatic recovery of chinook salmon and steelhead trout. U.S. National Marine Fisheries Service. Marine Fisheries Review 36(7):10–13.

Gibbard, G. L., and R. L. Colura. 1980. Retention and movement of magnetic nose tags in juvenile red drum. Annual Proceedings of the Texas Chapter, American Fisheries Society 3:22–29. (Texas Parks and Wildlife Department, Austin.)

Heidinger, R. C., and S. B. Cook. 1988. Use of coded wire tags for marking fingerling fishes. North American Journal of Fisheries Management 8:268–272.

Jefferts, K. B., P. K. Bergman, and H. F. Fiscus. 1963. A coded wire identification system for macro-organisms. Nature (London) 198:460–462.

Jenkinson, D. W., and H. T. Bilton. 1981. Additional guidelines to marking and coded wire tagging of juvenile salmon. Canadian Technical Report of Fisheries and Aquatic Sciences 1051.

Williamson, J. H. 1983. Retention of wire nasal tags by largemouth bass. Annual Proceedings of the Texas Chapter, American Fisheries Society 6:21. (Texas Parks and Wildlife Department, Austin.)

Williamson, J. H. 1987. Evaluation of wire nose tags for marking largemouth bass. Progressive Fish-Culturist 49:156–158.

American Fisheries Society Symposium 7:286–292, 1990

Tag Retention, Survival, and Growth of Red Drum Fingerlings Marked with Coded Wire Tags

Britt W. Bumguardner, Robert L. Colura, and
Anthony F. Maciorowski[1]

*Texas Parks and Wildlife Department, Perry R. Bass Marine Fisheries Research Station
Star Route Box 385, Palacios, Texas 77465, USA*

Gary C. Matlock

Texas Parks and Wildlife Department, 4200 Smith School Road, Austin, Texas 78744, USA

Abstract.—Retention of coded wire tags by, and survival and growth of, fingerling red drums *Sciaenops ocellatus* were monitored in cages in Matagorda Bay, Texas, and in saltwater ponds near Palacios, Texas. Tags were implanted in the adductor mandibularis muscle of 2,100 56-d-old pond-cultured fish. Fingerlings were retained in the laboratory for 24 h before they were stocked in cages or ponds. The mean tag retention rate after 24 h was 67.2%. One hundred twenty tagged fish were then placed in six 19-L circular cages (20 fish/cage). Three cages were placed in Matagorda Bay and three in 0.1-hectare saltwater ponds. Equal numbers of untagged fish were selected as controls and treated identically. Tag retention by, and survival of, caged fish were monitored daily for 7 d, and averaged 86 and 81%, respectively. Survival of control fish was 86%. Three 0.1-hectare ponds were each stocked with 500 tagged red drums, and three 0.1-hectare ponds with 500 untagged fish. The ponds were drained after 23 d and the fish were transferred to six 0.2-hectare ponds for an additional 91 d. Mean size of tagged fish held in ponds after 23 d and average survival of tagged fish after 114 d were significantly less than those of control fish. Cumulative tag loss was 33% 24 h after tagging, 43% after 8 d, and 54% after 23 d, when tag retention stabilized at approximately 46%. The proportion of initially tagged fish surviving and retaining tags was 7% after 114 d.

Red drum fingerlings *Sciaenops ocellatus* have been stocked by the Texas Parks and Wildlife Department (TPWD) since 1975 (Matlock 1986). However, quantification of their contribution to wild stocks and to the catches of anglers has been difficult. The fish, which are typically less than 40 mm total length when stocked (Matlock 1986), are frequently released in spring and summer when no red drums as small as 100 mm occur naturally in marine bays. Survival of stocked fish can be monitored by analysis of length frequencies for only about 1 year (Dailey and McEachron 1986; Matlock et al. 1986; McEachron and Green 1986). Although red drum fingerlings are also stocked in the fall months, the presence of similarly sized wild fish during the fall precludes the use of length frequencies to identify hatchery fish.

Tagging of hatchery fish has been attempted, but with little success. Only 10 of 5,942 hatchery-reared red drums (mean length, 152 mm) tagged with monel jaw tags on the opercula were recaptured (Matlock et al. 1984). Three fingerlings

tagged in the snout with coded wire microtags were recaptured from over 38,000 40-120-mm red drums released in St. Charles Bay, Texas (Matlock et al. 1986). The low recapture rate was probably due to tag loss. Gibbard and Colura (1980) found that only 27% of coded wire tags placed in snouts of 50-mm red drums were retained after 1 year in saltwater ponds.

Tag placement may affect retention. Coded wire tags implanted in the cheek musculature of striped bass *Morone saxatilis* and largemouth bass *Micropterus salmoides* were retained better than tags placed in snouts (Klar and Parker 1986; Fletcher et al. 1987; Williamson 1987).

In this investigation we estimated retention of microtags placed in the cheek musculature of juvenile red drums. The effects of tagging on growth and survival were also examined.

Methods

The study was conducted at the TPWD Perry R. Bass Marine Fisheries Research Station (MFRS) located near Palacios, Texas. Red drum larvae produced by photoperiod- and temperature-induced spawning (McCarty et al. 1986) were

---

[1]Present address: Battelle Columbus Division, 505 King Avenue, Columbus, Ohio 43210-2693, USA.

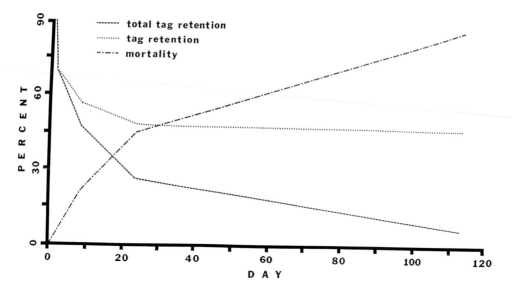

FIGURE 1.—Percent mortality of, and tag retention and total tag retention by red drum fingerlings 1, 8, 23, and 114 d after they were tagged with coded wires. Tag retention refers to survivors; total tag retention refers to all fish.

obtained from the Gulf Coast Conservation Association—Central Power and Light Marine Development Center on 21 May 1987. Larvae were stocked into two 0.2-hectare ponds previously prepared according to general methods described by Porter and Maciorowski (1984). Fish were harvested 23 June 1987, and 10,000 individuals (25

mm mean TL) were stocked into a 0.2-hectare pond; their mean length was 25 mm (all fish lengths in this paper are total lengths). Fish were fed 0.5–1 kg of trout feed (Murry Elevators, Murry, Utah) daily until 3 July 1987, when the fish were harvested. The surviving fish (7,100) were placed in a 9,500-L circular fiberglass tank. Tank

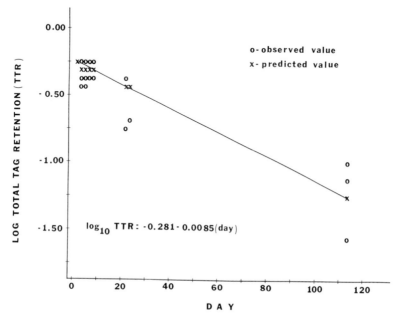

FIGURE 2.—Total retention (all fish) of coded wire tags by red drum fingerlings over time. Predicted values were derived from the regression equation.

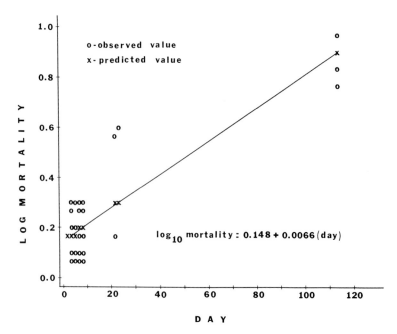

FIGURE 3.—Mortality of tagged red drum fingerlings over time. Predicted values were derived from the regression equation.

water was recirculated through a biological filter system, and water temperature was reduced to 20°C with a 746-W water chiller (Frigid Units, Toledo, Ohio). Fish were fed chopped shrimp and trout feed ad libitum three times daily. Mortality was negligible between 3 and 13 July 1987.

About 2,500 fish were transferred to a 3.0 × 0.6 × 0.6-m tank on 13 July 1987. Groups of approximately 50 fish each were dipnetted from the tank and placed in a 190-L tank, anesthetized with a commercial fish calmer (Hypno®, Jungle Laboratories, Cibolo, Texas) and tagged. Tagged fish were placed in a second 3.0 × 0.6 × 0.6-m tank for recovery.

The tagging procedure consisted of placing a coded wire tag in the adductor mandibularis, a small muscle posterior to the eye, with a model MK2A tagging unit (Northwest Marine Technology, Shaw Island, Washington) modified to use an interrupted-cycle mode. A side mold, constructed of plastic, was used to orient fish for tag placement. Tagged fish were passed through the magnetic field of a Northwest Marine Technology quality control device to magnetize the tags. Control fish were not subjected to the handling associated with tag implantation. Fish ($N = 100$) averaged ($\pm$ SD) 52 $\pm$ 5 mm and 1.16 $\pm$ 0.38 g at tagging; 2,124 fish were tagged at this time and

subsequently monitored for tag retention, survival, and growth in cages and ponds.

Tag retention approximately 24 h after tagging was determined by checks of 220 fish with a Northwest Marine Technology field sampling detector. Seven-day survival and tag retention was determined by holding fingerlings in 19-L cages constructed from plastic buckets and plastic screen material. Twenty tagged fish were placed randomly in each of six cages on 14 July 1987. Three cages were placed in Matagorda Bay near the MFRS pump intake. The remaining cages were placed in a 0.1-hectare pond at the MFRS. Tag retention was confirmed by passing each fish through the detector before placing it in the cage. Three control cages with 20 untagged fish each were similarly placed at the bay and pond sites. Due to limited pond availability, cages containing tagged fish were placed in a pond stocked with tagged fish, and cages holding untagged fish were placed in a pond stocked with untagged fish. This arrangement was used to prevent accidental release of tagged or untagged fish in a pond containing fish subjected to the opposite treatment. Tag retention was monitored daily with the detector. Fish were fed approximately 7 g of trout feed and their survival was determined daily by visual inspection. At the end of the study all tagged fish

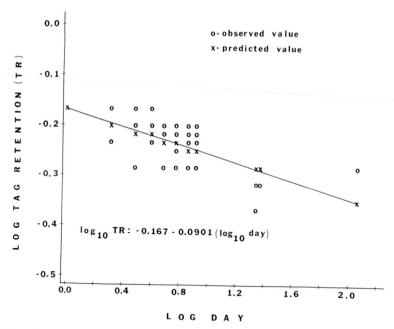

FIGURE 4.—Retention (survivors only) of coded wire tags by red drum fingerlings over time. Predicted values were derived from the regression equation.

from cages were preserved for later X-ray examination.

Long-term (114-d) tag retention and growth and survival of tagged fish were estimated by rearing tagged fish in culture ponds. Three 0.1-hectare ponds were stocked on 14 July 1987 with 500 tagged fish (mean ± SD, 52 ± 5 mm; $N = 100$) each. Tag presence was not reconfirmed with the detector before fish were stocked. An additional three 0.1-hectare ponds received 500 untagged fish (52 ± 5 mm) each. Fish in each pond were fed 0.45 kg trout feed daily. Survivors were harvested 23 d after stocking, and checked with the detector for tag retention. Total length (nearest mm) and

weight (nearest 0.01 g) were determined for 30 fish from each pond at harvest. Surviving red drums harvested from the six 0.1-hectare ponds were transferred to six 0.2-hectare ponds as a group.

Fish in each 0.2-hectare pond received 0.45 kg of trout feed daily for the first 58 d and 0.9 kg daily for the last 33 d. Fish were harvested 91 d after stocking and tag retention determined by examination of all fish with the detector. At harvest all fish that seemed to have lost their tags were preserved in 10% formalin for X-ray examination. Fifty-six of 108 fish that retained tags were also preserved. Preserved fish from both cage and pond studies were X-rayed. The number of fish

TABLE 1.—Mean (± SD) lengths and weights and survival of, and retention of coded wire tags by, red drum fingerlings transferred from 0.1-hectare to 0.2-hectare ponds and held for 91 d. F- and t-statistics are based on 178 and 4 df; the asterisk denotes a significant difference between tagged and untagged fish ($P \leq 0.05$). Survival was calculated from the original 500 fish stocked in 0.1-hectare ponds. Tag retention is the percentage of surviving fish with tags. Total tag retention is the proportion of initially tagged fish surviving and retaining tags.

Variable	Tagged fish			Untagged fish			Statistic
	Pond 8	Pond 10	Pond 12	Pond 7	Pond 9	Pond 11	
Total length (mm)	219±22	230±21	229±19	209±23	216±23	244±21	$F = 0.23$
Weight (g)	106±30	120±34	121±31	89±31	97±30	122±31	$F = 0.03$
Survival (%)	23	18	6	57	38	25	$t = 4.325*$
Tag retention (%)	44.4	44.6	51.7				
Total tag retention (%)	10.2	8.0	2.0				

BUMGUARDNER ET AL.

TABLE 2.—Mean (± SD) lengths and weights and survival of, and retention of coded wire tags by, red drum fingerlings held 23 d in 0.1-hectare ponds. F- and t-statistics are based on 179 and 4 df; asterisks denote significant differences between tagged and untagged fish ($P \leq 0.001$). Means along a row with a letter in common are not significantly different ($P > 0.05$). Tag retention is the percentage of surviving fish with tags; total tag retention is the proportion of initially tagged fish surviving and retaining tags.

Variable	Tagged fish			Untagged fish			Statistic
	Pond 1	Pond 3	Pond 6	Pond 2	Pond 4	Pond 5	
Total length (mm)	75±8 xy	69±73 z	71±12 yz	83±10 w	80±11 wx	78±8 wx	$F = 10.59^*$
Weight (g)	3.9±1.3 yz	3.2±1.0 z	3.6±2.1 yz	5.3±1.9 x	5.1±2.1 x	4.5±1.4 y	$F = 7.61^*$
Survival (%)	83	44	41	80	35	35	$t = 0.24$
Tag retention (%)	48.6	43.4	47.6				
Total tag retention (%)	39.8	18.8	19.2				

that showed tags upon X-ray examination was compared to the number of fish that showed tags upon examination with the detector.

Survival of tagged and untagged red drums in cages at bay and pond sites was compared by two-way analysis of variance (ANOVA; Sokal and Rohlf 1981). Tag retention by caged fish at bay and pond sites, and survival of tagged and untagged fish at harvest from 0.1- and 0.2-hectare ponds, were compared with t-tests ($P = 0.05$). Mean length and weight of fish harvested from 0.1- and 0.2-hectare ponds were compared by one-way ANOVA and analysis of covariance (Sokal and Rohlf 1981), respectively. Linear regression analysis was used to estimate mortality, tag retention by live fish (i.e., percentage of surviving fish with tags), and total tag retention by all tagged fish (i.e., proportion of initially tagged fish that survived and retained tags). Death was treated as a tag loss for estimating total tag retention. To calculate comparable total retention rates, stocking densities of fish in cages were adjusted for initial (24-h) mortality due to tagging (2.1%) and tag loss (32.7%) by dividing stocking density by 0.659. Stocking densities of fish in ponds were adjusted for initial mortality due to

tagging (2.1%) by dividing density by 0.979. Statistical analyses were performed with the Statistical Analysis System (SAS 1985).

Results

Tag loss within 24 h of tagging was 32.7% and mortality of tagged fish was 2.1%. Mean (± SD) percentages of tagged and untagged fish surviving after 7 d in cages were 81 ± 11 and 86 ± 6%, respectively. Mean tag retention of surviving fish in cages (adjusted for initial tag loss and mortality) was 57.9 ± 4%; total tag retention was 45.6 ± 5.2%. Tagged and untagged fish held in ponds for 23 d had mean survival rates of 56 ± 23% and 50 ± 26%, respectively; adjusted tag retention was 46.5 ± 2.8%, and total tag retention was 25.9 ± 12.0%. Mean percentages of tagged and untagged fish surviving after 114 d in ponds were 16 ± 9% and 40 ± 16%, respectively. Tag retention was 46.9 ± 4.2%, and total tag retention (± SD) was 7.0 ± 3.7 after 114 d, for fish held in ponds (Figure 1). Coefficients of determination for the best linear regression models were $r^2 = 0.84$ for total tag retention (Figure 2), 0.70 for mortality (Figure 3), and 0.59 for tag retention (Figure 4).

TABLE 3.—Survival of, and retention of coded wire tags by, red drum fingerlings held in cages for 7 d. F- and t-statistics are based on 11 and 4 df; none are significant ($P > 0.05$). Tag retention is the percentage of surviving fish with tags; total tag retention is the proportion of initially tagged fish surviving and retaining tags. Columns for cages 1–3 show results of three replicates.

Variable	Tagged fish						Untagged fish						Statistic
	Bay			Pond			Bay			Pond			
	Cage 1	Cage 2	Cage 3	Cage 1	Cage 2	Cage 3	Cage 1	Cage 2	Cage 3	Cage 1	Cage 2	Cage 3	
Survival (%)	75	70	85	90	95	70	85	80	80	85	95	90	$F = 1.12$
Tag retention (%)	86.7	92.9	76.5	77.8	89.5	92.9							$t = 0.20$
Total tag retention (%)	46.1	56.0	42.8	42.8	42.8	42.8							$t = 1.39$

The field sampling detector detected tags in small fish more accurately than in larger fish. Visual examination of X-ray exposures of 97 preserved 52-mm red drum fingerlings from cages showed that 88 of the fish retained tags. The detector indicated that 87 of the 97 fish retained tags when fish were examined both before and after preservation. Examination of X-ray exposures from 186 fish of 226 mm held for 114 d in ponds showed that tags were present in 76 fish. The detector, however, indicated that only 56 and 67 fish retained tags before and after preservation, respectively. Examination of X-ray exposures indicated that approximately 25% of the tags in caged fish and 51% of the tags in pond-held fish were not in the expected locations.

Tagging significantly reduced survival at 114 d (Table 1) and initially repressed growth of tagged fish (Table 2). Tagged red drum fingerlings in ponds 3 and 6 were significantly shorter and lighter than untagged fish after 23 d in ponds (Table 2), but no differences were found after 114 d (Table 1). Survival of tagged and untagged red drum fingerlings in cages was not significantly different for fish held at bay and pond sites, nor was tag retention different for fish held at either site (Table 3).

Discussion

Implanting coded wire tags in the cheek musculature of small red drums is not a practical marking method. The high initial (24-h) tag loss of 32.7% was probably due to the difficulty of placing tags correctly in the cheek musculature of small fish. This conclusion is supported by the high percentage (25%) of tags at locations other than the intended site of implantation (adductor mandibularis) after 7 d, as determined by X-ray examination. Fletcher et al. (1987) stated that the majority of wire tag loss in largemouth bass occurred soon after tagging; this result was similar to our findings with red drums. Loss of tags due to shedding apparently stopped after 23 d; thereafter, death was the major factor affecting the number of tagged fish present. Although survival of tagged fish at 114 d was significantly less than that of untagged fish, all mortality was not necessarily related to tagging, as evidenced by death of up to 75% of the untagged fish after 114 d. The apparent increase in survival in pond 9 (from 35 to 38%) was due to an error in counting while fish were transferred from 0.1-hectare to 0.2-hectare ponds.

Coded wire tags have been successfully implanted in several fish species at various body sites (Leary and Murphy 1975; Gibbard and Colura 1980; Krieger 1982; Crumpton 1985; Klar and Parker 1986; Fletcher et al. 1987; and Williamson 1987). However, most studies have been conducted on fish longer than 100 mm. Gibbard and Colura (1980) reported 27% tag retention at 12 months for 50-mm fish tagged in the snout. The 7% total tag-retention rate of red drum fingerlings we tagged in the cheek musculature was approximately one-fourth that reported by Gibbard and Colura (1980). This difference may be partially due to the small amount of tissue present in the adductor mandibularis of a red drum less than 50 mm. Correct tag placement is difficult in small fish; tag retention and survival may improve if larger red drums are tagged.

Apparent discrepancies between the detector and X-ray methods of detecting tags point to the difficulties of using microtagged red drums in large-scale stockings. Reliance on the detector would effectively reduce detection of tags in larger fish (>220 mm). This reduction, in turn, would cause the proportion of tagged fish in the population to be underestimated.

Acknowledgments

We acknowledge the assistance of Maxine Kubecka and the staff of the Palacios Veterinary Clinic in procurement and interpretation of X-ray exposures, and the Marine Development Center staff for help with fish tagging. This study was conducted with partial funding from the U.S. Fish and Wildlife Service, Federal Aid in Fish Restoration 15.605 (Project F-36-R).

References

Crumpton, J. E. 1985. Effects of micromagnetic wire tags on growth and survival of fingerling largemouth bass. Proceedings of the Annual Conference Southeastern Association of Fish and Wildlife Agencies 37:391–394.

Dailey, J. A., and L. W. McEachron. 1986. Survival of unmarked red drum stocked into two Texas bays. Texas Parks and Wildlife Department, Coastal Fisheries Branch, Management Data Series 116, Austin.

Fletcher, D. H., F. Haw, and P. K. Bergman. 1987. Retention of coded wire tags implanted into cheek musculature of largemouth bass. North American Journal of Fisheries Management 7:436–439.

Gibbard, G. L., and R. L. Colura. 1980. Retention and movement of magnetic nose tags in juvenile red drum. Annual Proceedings of the Texas Chapter, American Fisheries Society 3:22–29. (Texas Parks and Wildlife Department, Austin.)

Klar, G. T., and N. C. Parker. 1986. Marking fingerling striped bass and blue tilapia with coded wire tags and microtaggants. North American Journal of Fisheries Management 6:439–444.

Krieger, K. J. 1982. Tagging herring with coded-wire tags. U.S. National Marine Fisheries Service Marine Fisheries Review. 44(3):18–21.

Leary, D. F., and G. I. Murphy. 1975. A successful method for tagging the small, fragile engraulid, *Stolephorus purpureus*. Transactions of the American Fisheries Society 104:53–55.

Matlock, G. C. 1986. A summary of 10 years of stocking fishes into Texas bays. Texas Parks and Wildlife Department, Coastal Fisheries Branch, Management Data Series 104, Austin.

Matlock, G. C., B. T. Hysmith, and R. L. Colura. 1984. Returns of tagged red drum stocked into Matagorda Bay, Texas. Texas Parks and Wildlife Department, Coastal Fisheries Branch, Management Data Series 63, Austin.

Matlock, G. C., R. J. Kemp, Jr., and T. J. Heffernan. 1986. Stocking as a management tool for a red drum fishery, a preliminary evaluation. Texas Parks and Wildlife Department, Coastal Fisheries Branch, Management Data Series 75, Austin.

McCarty, C. E., J. E. Geiger, L. N. Sturmer, B. A. Gregg, and W. P. Rutledge. 1986. Marine finfish culture in Texas: a model for the future. Pages 249–262 *in* R. Stroud, editor. Fish culture in fisheries management. American Fisheries Society, Fish Culture Section and Fisheries Management Section, Bethesda, Maryland.

McEachron, L. W., and W. W. Green. 1986. Trends in relative abundance and size of selected finfish in Texas bays: November 1975–June 1984. Texas Parks and Wildlife Department, Coastal Fisheries Branch, Management Data Series 91, Austin.

Porter, C. W., and A. F. Maciorowski. 1984. Spotted seatrout fingerling production in saltwater ponds. Journal of the World Mariculture Society 15:222–232.

SAS. 1985. SAS/STAT guide for personal computers, version 6 edition. SAS Institute, Cary, North Carolina.

Sokal, R. R., and F. J. Rohlf. 1981. Biometry, 2nd edition. Freeman, San Francisco.

Williamson, J. H. 1987. Evaluation of wire nosetags for marking largemouth bass. Progressive Fish-Culturist 49:156–158.

American Fisheries Society Symposium 7:293–303, 1990
© Copyright by the American Fisheries Society 1990

Evaluation of Marking Techniques for Juvenile and Adult White Sturgeons Reared in Captivity

C. E. BORDNER[1]

Aquaculture and Fisheries Program, University of California, Davis
Davis, California 95616, USA

S. I. DOROSHOV

Department of Animal Science, University of California, Davis

D. E. HINTON

School of Veterinary Medicine, University of California, Davis

R. E. PIPKIN

Department of Animal Science, University of California, Davis

R. B. FRIDLEY[2]

Aquaculture and Fisheries Program, University of California, Davis

FRANK HAW

Northwest Marine Technology, Inc., Shaw Island, Washington 98286, USA

Abstract.—Hatchery-produced, 5-month-old, 39-g fingerlings of white sturgeon *Acipenser trans-montanus* were marked by combinations of wire microtag implants and amputation of one of the four barbels. Six control and treatment groups were mixed and raised in tanks with artificial feeding for 6 months. Marking treatments did not affect fingerling survival (96–100%) or growth (3.0–3.6%/d). All microtags were retained when implanted under the first dorsal scute or 4–5 mm deep in the snout cartilage. Shallow (2–3 mm) implantation in the snout resulted in 40% microtag loss. Amputated barbels did not regenerate, which allowed identification of fish after the 6-month rearing period. Radiographic and histological examinations of tagging sites indicated that there was very little disturbance of the tissue. Domestically raised white sturgeon brood stock, 3–6 years old, were marked by subcutaneous ink tattoos on the abdomen near the pelvic fins. Marks that identified the stock origin, sex, and individual identity were seen clearly after 1 year, and enabled us to make observations on growth rates of individual fish. These observations indicated that ripe males ceased growth during the winter season.

Recent progress in the culture of North American species of sturgeons *Acipenser* spp. and paddlefish *Polyodon spathula* necessitates reliable marking techniques to identify fish and evaluate their performance. Marking is important for monitoring both domestic brood stocks and hatchery-produced fish that are reared either in natural or commercial aquacultural environments. Efficacious sturgeon tags should be permanently retained, easily detected, and sturdy enough to withstand handling of the fish, and they should have minimal effects on survival and growth of the fish. The longevity of sturgeons and the extended

period of rapid growth between birth and sexual maturity preclude efficient use of conventional external tags (Belyaeva 1963).

The need for reliable marking techniques in large-scale production of fingerling sturgeons is so critical that some investigators have experimented with radioactive tags (Shekhanova 1955). Fin clipping of juvenile fish is unreliable due to regeneration of soft tissue and a high incidence of fin erosion on unmarked sturgeons (Milstein 1957). Implantable magnetic coded wire microtags, developed by Jefferts et al. (1963) and Bergman et al. (1968), may offer a means of marking sturgeon fingerlings; these tags have been used successfully with a wide variety of fish (Leary and Murphy 1975; Gibbard and Colura

---

[1]Deceased.
[2]To whom correspondence should be addressed.

1980; Klar and Parker 1986; Fletcher et al. 1987) and other aquatic animals (Schwartz 1981; Joule 1983; Wickins et al. 1986; Montfrans et al. 1986). These microtags, which carry information in the form of notches, have been particularly useful in evaluation of the survival and return of hatchery-produced juveniles.

Development of domestic sturgeon brood stocks requires the ability to identify individual fish. Marking of fish selected for the potential brood stocks is usually carried out when fish are 2–3 years old, after sex has been determined by gonadal biopsy. External mechanical tags, transmitters, or subcutaneously injected dyes are not fully appropriate because of tag losses, or because of difficulty in reading or deciphering information at the moment the fish is captured.

The objective of the work reported here was to evaluate marking techniques for juvenile and adult white sturgeons *Acipenser transmontanus* during growth of fingerlings and culture of domestic brood stock. We examined the use of coded wire microtags in combination with amputation of barbels for fingerlings, and ink tattoo marks for adult fish raised in tanks.

Methods

Marking of Juveniles

Sibling (one hatchery mating) 5-month-old white sturgeon fingerlings, obtained from wild brood stock in the Sacramento River, were divided randomly into six experimental groups of 31–34 animals each. The fish were obtained as fry from a commercial hatchery (Arrowhead Fishery, Red Bluff, California), and raised in the laboratory for 3 months. Ten fish were randomly sampled to determine initial standard length and body weight. After segregation, the groups of fish were kept in identical 1.22-m-diameter, 270-L circular fiberglass tanks supplied with aeration and a continuous flow of well water. The temperature was maintained between 18 and 21°C, and photoperiod matched ambient conditions. The fish were fed a commercial trout diet (Silvercup brand, Sterling H. Nelson and Sons, Murray, Utah), by means of automatic feeders (Sweeney Enterprises, Bourne, Texas), at an initial rate of 2% of the body weight per day; pellet size and feeding rate were adjusted periodically for growth of the fish. Culture methods followed those outlined by Conte et al. (1988).

Treatments were selected to compare combinations of tagging methods and tagging sites. Three groups were implanted with stainless-steel wire microtags inserted by a Mark IV tag injector (Northwest Marine Technology, Shaw Island, Washington). Tags measuring 0.254 mm × 2.0 mm were injected by experienced personnel from the tag-manufacturing company. Each of the three groups of fish received tags at a different anatomical location: either deep (3–4 mm) or shallow (1–2 mm) sites in the dorsomedial snout cartilage, or under the first dorsal scute. Tags were injected medially at a downward angle without the use of a head mold (Figure 1). Snout tagging was accomplished with the snout facing the injector; for scute tagging, the injector needle was inserted under the posterior edge of the first dorsal scute in an effort to deposit the tag near the center of the cartilage surrounded by the scute. One control group received no injection; another group received an injection in the snout with the tagging needle, but no implant (sham control). Immediately after implantation, presence of the tag was verified with a magnetic field detector.

Barbel amputation was accomplished by severing either the left or rightmost barbel close to the snout, with the aid of a magnifying glass and fine scissors (Figure 2). In two treatments (control and scute-tagged fish) barbels were left intact. These fish were also compared to fish in the untreated control group. After treatment, and subsequent to all other handling procedures, the fish were treated prophylactically with 8 mg/L nitrofurazone for 1 h under static conditions. Microtag retention was verified with the detector 5 d after marking. After 20 d, the fish were again examined with the detector and moved into two 1.83-m-diameter tanks of 1,600 L each. On day 107, the fish were transferred to two 3.66-m-diameter tanks of 6,400 L each. The fish were weighed and examined for tag retention again on days 69, 131, and 180, when the experiment was concluded. Before each sampling period, the fish fasted for 24 h, and then were anesthetized by immersion for 5 min in 90 mg/L tricaine (MS-222, Argent Chemical, Redland, Washington). Three white sturgeons from each treatment group except the untagged group and the sham implant group (Table 1) were killed on days 69 and 180, and heads (snout tags) and transversely sectioned regions containing the first dorsal scute (scute tags) were removed and placed in Davidson's fixative (Humason 1979). Samples of normal and amputated barbels were also preserved. After 24 h of fixation, tissues were rinsed three times in 50% ethanol and stored in 70% ethanol until radiographic and histologic analyses were performed. Radiography was per-

FIGURE 1.—Orientation of a fingerling white sturgeon with respect to the tag-injection needle during implantation of wire tags in snout cartilage. Also visible is the first dorsal scute.

formed at the Department of Radiology, School of Veterinary Medicine, University of California, Davis. After radiography, tissues were trimmed and blocks were oriented to approximate the site of tag implantation. By use of the radiographs, we were able to locate individual wire tags. To prevent interference with subsequent microtomy, tags were removed and histologic analysis was performed on the remaining tissue, which included the implantation track as well as tissue immediately adjacent to the tag. Tissue blocks were dehydrated by passage through a graded series of ethanol solutions, cleared by passage

FIGURE 2.—Fingerling white sturgeon positioned for amputation of a barbel.

through xylenes, and infiltrated with paraffin. Paraffin blocks were sectioned to 5-μm thickness, and slides were stained with hematoxylin and eosin. Histologic interpretations represent a consensus of microscopic observations by two histopathologists who provided independent analyses of tissues.

One-way analyses of variance (ANOVAs) were used to compare the weights of fish in each group on days 69, 131, and 180. Duncan's multiple-range test was used to compare mean weights at each sampling. A two-way ANOVA was used to examine the combined effects and possible interactions of tagging and barbel amputation. Contrast analyses were performed on day-180 weights to compare all tagged fish against all untagged fish, to separate the effects of removal of different barbels, and to compare all operated-upon fish against intact ones and sham controls against tagged fish. Tag retention was evaluated by chi-square analysis of the proportion of each group retaining the tags at each sampling. Statistical analyses were performed on a microcomputer with SAS software (SAS Institute, Cary, North Carolina). Statistical procedures conformed to those recommended by Steele and Torrie (1980).

Marking of Brood Stock

Marking of brood stock was accomplished with a Spaulding Special Electric Tattoo Marker (110

TABLE 1.—Growth and survival of white sturgeon fingerlings subjected to different marking treatments. Weight gain and survival are from day 0. Mean body weights at a given time did not differ significantly among the six fish groups (P > 0.05).

Microtag placement (barbel removed)	Time (d)	Mean body weight, g (SE)	Weight gain (%/d)	Survival (%)	N
	0	38.8 (4.7)			10
No tag (none)	69	148.7 (7.2)	4.11	100	33
	131	225.9 (11.0)	3.68	97	32
	180	263.1 (16.4)	3.21	97	32
Sham inplant (left)	69	144.2 (4.9)	3.94	100	33
	131	216.0 (10.3)	3.49	100	33
	180	261.1 (16.0)	3.18	100	33
No tag (right)	69	147.0 (5.0)	4.04	100	34
	131	236.7 (10.0)	3.89	100	31
	180	288.1 (13.6)	3.57	100	31
Deep snout (right)	69	143.9 (5.4)	3.93	100	33
	131	210.2 (11.5)	3.37	100	30
	180	245.3 (17.0)	2.96	97	29
Dorsal scute (none)	69	142.2 (6.6)	3.86	100	31
	131	227.1 (12.4)	3.70	96	25
	180	275.7 (20.5)	3.39	96	25
Shallow snout (left)	69	139.3 (6.2)	3.75	100	33
	131	214.7 (13.0)	3.46	100	28
	180	251.2 (19.2)	3.04	100	28

V), model SSMC-1 (Spaulding-Rogers, Voorhees-ville, New York). This inexpensive unit has parts that can be easily removed for cleaning and replacement, and is operated by fingertip pressure on the switch. Tattooing was achieved by rapid repeated penetration of the skin by a four-point needle bar, to a depth of approximately 0.3 mm, which allowed the black ink to travel from the reservoir into the subcutaneous layer. Each needle bar retained effective sharpness for marking 40–50 fish.

For marking, adult white sturgeons were anesthetized as described previously and placed ventral side up in a hooded stretcher (Figure 3). The ventral side of the white sturgeon provided a flat white area with good contrast for writing with black tattoo ink. Tattoos (letters and numbers up to 3 cm in height) were drawn immediately proximal to the base of a pelvic fin (the right fin of females and the left fin of males). In most cases, the application was repeated two or three times over the same tattoo symbol to ensure good penetration and resolution. The animal was then placed back in fresh water; recovery was rapid

and no ill effects were visible. The entire procedure took about 3 min.

In October–November 1986, we marked several hundred white sturgeons maintained as domestic brood stock at the university and at several commercial farms in northern California. Fish were 3–6 years old and ranged in weight from 5 to 27 kg. These brood stock fish were raised in outdoor freshwater tanks or raceways and fed commercial trout diets. At marking, fish were sexed by gonadal biopsy. To determine the utility of this marking technique, we reared mixed groups from five broodstock colonies (four commercial farms and university hatchery) in a communal tank 8 m in diameter and 1.2 m deep. The tank was stocked with 20 males and 15 females in November 1986. All fish were sampled in May 1987 and again in November 1987 to examine individual growth and sexual development. Body weights at each sampling were plotted as histograms to indicate the shapes of the population curves. Monthly instantaneous growth rates, ($\log_{10}$ final weight $-$ $\log_{10}$ initial weight)/6, were calculated for individual fish for each of the 6-month growing periods.

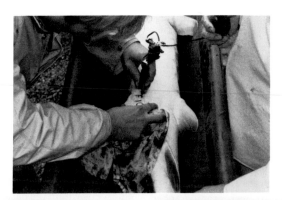

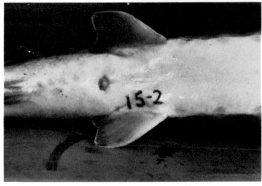 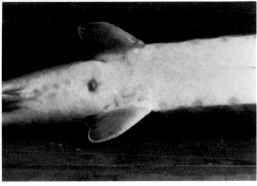

FIGURE 3.—Top, tattooing of white sturgeon brood stock; bottom left, fresh tattoo mark; bottom right, the same tattoo after a 12-month growing season.

Comparisons of growth of males and females were made with t tests. During this 1-year growout period, water temperature ranged seasonally from 12 to 20°C. Fish were fed commercial trout diet at a rate of 0.3% of body weight per day. Food was delivered every 2 h by an automatic feeder installed near the tank wall.

Results

Marking of Juveniles

Fingerling white sturgeons that were marked by microtag implantation and barbel amputation recovered fully a few minutes after anesthesia. Amputation of barbels was not accompanied by bleeding.

No mortality was observed in any treatment during the 69-d posttagging period. Later, only three stunted fish were removed (one each from groups with no tag, tag in the dorsal scute, and tag shallowly implanted in the snout). Stunting of a few individuals is usually observed during the growout of white sturgeons and apparently was not related to the tagging treatments.

The fish grew well during the experiment, exhibiting overall linear increases in body weight and reaching 245–288 g after the 6-month rearing period (Table 1). The individual weights remained normally distributed in each treatment; coefficients of variation (100·SD/mean) were 20–40%. Gains in body weight ranged from 3.0 to 3.6%/d in different treatments. There were no significant differences in body weight among the control and treatment groups at days 69, 131, or 180 ($P <$ 0.05). Contrast analyses indicated no significant differences among any combinations of treatment groups. Data on survival and weight gain are shown in Table 1.

All of the wire microtags were retained through day 180 when implantation was under the first dorsal scute or in the deep snout cartilage. The retention of microtags implanted shallowly in the snout was not as successful: 9% of fish had lost the tags after 5 d, 33.3% after 69 d, and 39.4% after 180 d (Table 2). On day 69, several fish with shallow snout or scute tags failed to elicit a positive response from the detector. These fish

TABLE 2.—Retention of wire microtags by juvenile white sturgeons. Retention was significantly poorer for shallow snout placements than for other placements ($P < 0.05$) at day 69 and thereafter.

Time (d)	% Microtag retention after placement in		
	Deep snout	Dorsal scute	Shallow snout
5	100	100	90.9
20	100	100	87.9
69	100	100	66.7
131	100	100	60.6
180	100	100	60.6

(two from each treatment) were radiographed along with the fish killed for tag site histology. The radiographs revealed that all of the fish given shallow snout tags that tested negative with the detector had indeed lost their tags, whereas those tagged under the scute had retained the tags but had failed to trigger the detector. After fixation and radiography, these sites were reexamined with the detector, and this time sections of white sturgeons with scute tags triggered the device. At later samplings, fish that tested negative on the first pass through the detector were isolated and reexamined in different orientations to the detector. In all cases, scute or deep snout tags elicited positive responses. Figure 4 shows the results of radiography of wire-tagged fish.

Barbel removal was not associated with any ill effects. The barbel stump healed rapidly and formed a slightly bulbous tip; regeneration did not occur during the 6-month experiment. Gross examination of barbel stumps indicated no unusually deleterious response.

Tissues from wire injection sites showed no sign of alteration under a dissecting microscope. Skin surface and the cut surface of tissue blocks resembled elements of the contralateral tissues. Similarly, histopathologic analysis revealed little alteration. No morphological evidence of reaction to a foreign body was seen. In sections through the deep snout (Figures 5 and 6), tracks or passageways for implantation within the head cartilage were visible. Within these sections, variable amounts of areolar connective tissue were seen. On occasion, mononuclear cell aggregates resembling cells within the basal layer of epidermis were present (Figure 5). Skin changes near implantation tracks included increased pigmentation of dermis (not shown in figures) and mild infiltration of inflammatory cells in subdermal areolar connective tissue of the perichondrium (Figure 7).

Marking of Brood Stock

The tattoo marks used for the identification of captive brood stock were well preserved after 6 months of growth. After 1 year of growout, the ink started to fade (Figure 3). However, the marks were clearly visible and no single identification was lost. All fish were retattooed to facilitate further monitoring of their performance.

The mean body weight of fish in the mixed group raised in the communal tank increased from 9.9 to 13.6 kg (37.4%/year). Histograms of weight distributions revealed that population growth patterns involved asymmetrical changes—e.g., from positively to negatively skewed distributions—as the fish grew (Figure 8). Reconstruction of growth

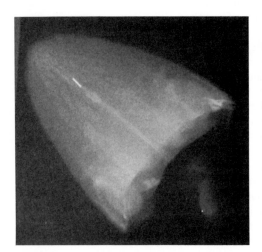
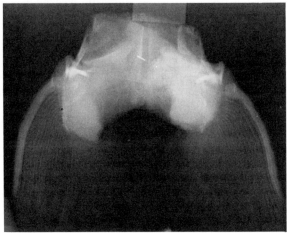

FIGURE 4.—Radiographs of white sturgeon sections containing a wire tag deep in the snout cartilage (left) and beneath the first dorsal scute (right). The tags (2 mm long) appear clearly as white lines.

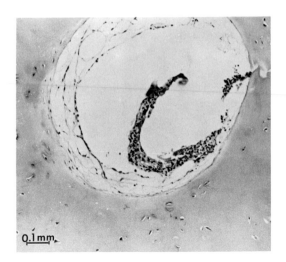

FIGURE 5.—Cross section of a microtag implantation track in the chondrocranium of a white sturgeon. Areolar connective tissue loosely fills the opening in cartilage. Cells resemble those of the basal layer of the epidermis; they represent a mild form of response by the host (if inflammatory) or may indicate translocation from the skin. The surrounding cartilage adjacent to the track shows a slight alteration in staining properties when compared to other areas.

rates for each marked individual explained these growth patterns. Statistical analysis of data shown in Figure 9 revealed that individual growth was affected by two factors: sex of the fish and season.

During the winter season (November–May) males did not grow, whereas females gained weight at a mean rate of 4.2%/month ($P < 0.001$). During the summer (May–November) both sexes grew at similar rates of 2.5–3.0%/month. An analysis of gametogenesis in individual fish indicated that growth was associated with the stage of sexual maturation. Males, especially in group E (university broodstock, 1982 year class) had fully mature testes containing ripe sperm during the winter season, whereas all females in the tank population were immature. Our preliminary interpretation is that the ripe males decreased food intake during the winter reproductive season, and females gained an advantage in growth by consuming the excess food.

Discussion

Marking of Juveniles

The absence of significant effects of wire tag implantation or barbel amputation on the growth or survival of white sturgeons suggests that both these techniques may be used with confidence for the identification of hatchery-reared fingerlings. There is no reason to believe that implanted wire microtags affect the survival or growth of white sturgeon fingerlings released in the natural environment; however, amputation of barbels, which play no role in intensive growout systems, may potentially affect feeding behavior in the natural

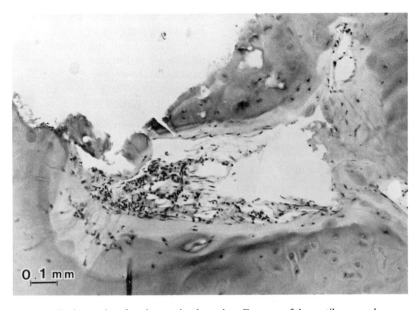

FIGURE 6.—Intracartilaginous site of a microtag implantation. Fracture of the cartilage may have accompanied the implant, or more likely when the tissue block was trimmed prior to processing.

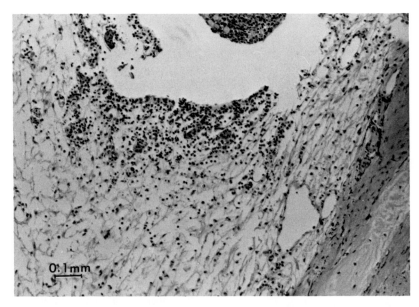

FIGURE 7.—Dermis of a white sturgeon head near site of a microtag implant. Note the mild cellular infiltration in the areolar connective tissue of the perichondrium.

environment. Barbels of white sturgeons are important for tactile reception and taste recognition (Detlaf et al. 1981; Buddington and Christofferson 1985). Therefore, although the four sturgeon appendages allow sixteen possible barbel-clipping schemes, the removal of one or more barbels may affect the fitness of the fish. Yet, barbel clipping can be used successfully in captivity—for example, to identify the families in selective breeding programs. Pavlov et al. (1970) observed no effect on feeding by captive juvenile *Huso huso* after complete extirpation of barbels. Our preliminary findings suggest that amputation of barbels may

be a useful technique, not only for sturgeons, but for other barbelled fish as well. Investigations on barbel-stump histology are continuing.

Complete retention of scute or deep snout-tags by white sturgeons compares favorably with results obtained with other species, although the time period of our study was shorter. Tag retention in other species has varied from over 98% for Atlantic salmon *Salmo salar* (Isaksson and Bergman 1978), and over 97% for lake trout *Salvelinus namayeush* (Elrod and Schneider 1986), to as low as 27% for red drums *Sciaenops ocellatus* (Gibbard and Colura 1980). Tag retention is influenced

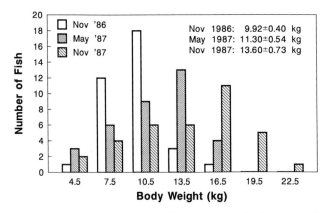

FIGURE 8.—Weight distributions for 35 brood-stock white sturgeons raised communally. Fish were sampled at 6-month intervals. The overall mean weights and standard errors at each sampling data are given at the upper right.

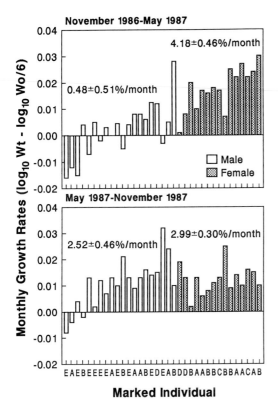

FIGURE 9.—Monthly growth rates, $(\log_{10}$ final weight $-\log_{10}$ initial weight$)/6$, of individual male and female white sturgeons raised communally during the winter season (top), and summer season (bottom). Values above the bars are means with standard errors. Individual fish are indicated by letters along the abscissa, which refer to different farm stocks.

by skill of the tagger and also by the type of tissue into which the tag is implanted (Klar 1985; Williamson 1987). Retention rates of at least 95% are desired for effective use of these tags. In white sturgeons, the large amount of nearly homogeneous cartilage at the tagging sites probably increased tag retention.

The most effective site of tag implantation may vary among species. In salmonids, the connective tissue and cartilage of the snout seem to be the best sites of tag placement (Wydoski and Emery 1983). Other locations, such as the body or cheek musculature, are more suitable for other species. Some information is available on the acute and chronic effects of wire tag implantation on the histology of implantation sites in other species (Bergman et al. 1968; Morrison and Zajac 1987).

The virtual lack of tissue reaction to the tag implant contrasts with reactions associated with external tags. The tag is implanted relatively deep within the cartilage, obviating surface abrasion and irritation after the initial trauma of implantation. This factor may be partially responsible for the minimal superficial tissue response in white sturgeons. We did not examine wire tags after their removal from implantation sites during histological analysis. Subsequent investigations may involve scanning electron microscopy of individual tags to determine tissue responses at tag surfaces. By removal of tags before our morphological analysis, we may have removed host tissue that responded. The absence of appreciable response in areas adjacent to tracks and implantation sites, however, leads us to believe that adherent tissue that might have been removed would have contained little additional evidence of alteration. Dimensions of the track (Figure 5) and the wire diameter permitted little additional space for tissue reaction beyond that already observed. The act of implantation may translocate cells and tissue components from superficial to deep sites. The areolar connective tissue shown in Figure 5 is identical in morphology to the perichondrium shown in Figure 6. Likewise, cells shown in Figure 5 closely resemble cells in basal regions of the epidermis.

The duration of the present study was sufficient to observe the acute, subacute, and chronic effects of implantation, but was insufficient to evaluate protracted effects. Acute and chronic toxicity was shown to be minimal, but it is entirely possible that over many years the implant may result in localized tissue reactions possibly deleterious to the host. Long-term tissue response is related to the component metals used in implants and tags (see Waalkes et al. 1987 for discussion). Cadmium, chromium, and nickel are carcinogenic. It may be possible to minimize undesired long-term effects by careful selection of metals. Type-302 stainless steel, used in these wire tags because of its magnetic properties, has been recommended for internal use in humans (Bardos 1977). Few studies of metals known to be carcinogenic in mammals have been conducted with fish, however.

The loss of tags from the shallow snout implantation site was probably due to the shallow placement of the tags, although it is possible that they were actively expelled. The migration and expulsion of relatively much heavier and larger transmitter-capsule implants has been studied in channel catfish *Ictalurus punctatus* (Marty and Sommerfelt 1986). Heavier capsules were expelled more readily than lighter ones, and the tendency for expulsion was greater in gravid females than in males or spent females. The wire tag has an extremely light weight, and it can be implanted by injection instead of surgically, which undoubtedly contribute to its improved retention relative to other tags. The initial difficulty in detection of wire tags implanted deep in the snout or under the scute was a result of the rapid growth of the fish. The tags had apparently not moved or migrated, but the increased mass of the fish resulted in a greater distance between tag and detector, making the orientation of the tag more critical for detection. Once the proper orientation was determined, no further detection problems were encountered.

Use of coded wire microtags, with or without external comarkers such as clipped barbels, appears to be a most suitable marking technique for juvenile white sturgeons. This technique may be of critical importance for studies on sturgeon populations and hatchery enhancement.

Marking of Brood Stock

Adult wild white sturgeons have been marked with various external tags and exhibited low tag retention as well as tag losses at handling (Galbreath 1985). Our experience in the use of conventional tags for rapidly growing cultured sturgeons was not satisfactory: disc, anchor, and spaghetti tags had low retention rates, and sometimes affected the health of the fish. In addition to external tags, removal of dorsal scutes and subcutaneous injection of dyes have been suggested (Belyaeva 1963). Those methods seem to offer little advantage because they are hazardous to the fish and unreliable for identification.

Tattooing of adult white sturgeons solved many problems encountered in work with large domestically raised fish. Much of the white sturgeon's body is covered with subdermal bony scutes of varying sizes. Three areas that seem to have a minimal number of scutes are the soft, fleshy areas at the bases of the pelvic and pectoral fins, and the ventral part of the snout, anterior to the mouth. Despite earlier work in which sturgeon tattoos were placed on the ventral snout region (Cochnauer et al. 1985), we selected alternative sites to avoid injury to the nerves and sensory receptors in this specialized tissue. The base of the pelvic fin was an ideal location because of its large area and the ease with which a tattoo mark in this location could be read while a fish was netted from a tank. Anticipated advantages were realized in subsequent culture conditions. Our

populations of white sturgeons raised in communal tanks showed that important new information can be obtained by means of this simple and inexpensive technique.

In conclusion, each of the marking techniques evaluated in this report was useful under particular conditions. Implantation of coded wire microtags in cartilage of fingerling white sturgeons was an efficient means of identifying large numbers and different groups of fish. Each tag will probably remain in place throughout the life of the fish, and can supply positive, detailed information. Amputation of one barbel provided effective, if limited, identification of fingerlings, and may provide an alternative to fin clipping as a secondary comarker in some situations. Tattooing is an appealing alternative for larger fish subject to inspection at least annually. Although not permanent, tattoos are inexpensive, durable, and easy to detect. The absence of deleterious effects of all of these methods confirms that, under the appropriate circumstances, identification of white sturgeons may be readily achieved.

Acknowledgments

We thank Bill Walsh, Brendan Moore (Aquaculture and Fisheries Program, University of California, Davis), Boyd Bentley, Joel VanEenennaam, Frank Chapman, Kevin Kroll (Department of Animal Science, University of California, Davis), Peter K. Bergman, and Richard D. Fralick (Northwest Marine Technology, Inc.) for their input and assistance in this research, and Arrowhead Fishery, Red Bluff, California, for the donation of white sturgeons used in this study. The use of trade names in this article does not imply endorsement by the authors.

References

Bardos, D. I. 1977. Stainless steels in medical devices. Pages 42/1–42/10 *in* D. Peckner and I. M. Bernstein, editors. Handbook of stainless steels. McGraw-Hill, New York.

Belyaeva, V. N. 1963. [Experimental tagging of sturgeon juveniles raised at Kizansky fish hatchery.] Pages 44–46 *in* E. N. Pavlovsky, editor. [Sturgeon management in the USSR.] Academy of Science, Moscow. (In Russian.)

Bergman, P. K., K. B. Jefferts, H. F. Fiscus, and R. C. Hagar. 1968. A preliminary evaluation of an implanted, coded wire fish tag. Washington Department of Fisheries, Fisheries Research Paper 3:63–84.

Buddington, R. K., and J. D. Christofferson. 1985. Digestive and feeding characteristics of the chondrosteans. Pages 31–42 *in* F. P. Binkowski and S. I.

Doroshov, editors. North American sturgeons: biology and aquaculture potential. Dr. W. Junk, Dordrecht, The Netherlands.

Cochnauer, T. G., J. R. Lukens, and F. E. Partridge. 1985. Status of white sturgeon, *Acipenser transmontanus*, in Idaho. Pages 127–134 *in* F. P. Binkowski and S. I. Doroshov, editors. North American sturgeons: biology and aquaculture potential. Dr. W. Junk, Dordrecht, The Netherlands.

Conte, F. S., S. I. Doroshov, P. B. Lutes, and E. M. Strange. 1988. Sturgeon hatchery manual for the white sturgeon (*Acipenser transmontanus* Richardson), with application to other North American Acipenseridae. University of California, Division of Agriculture and Natural Resources, Publication 3322, Davis.

Detlaf, T. A., A. S. Ginzburg, and O. I. Schmalganzen. 1981. [Development of acipenserid fish]. Nauka, Moscow. (In Russian.)

Elrod, J. H., and C. P. Schneider. 1986. Evaluation of coded wire tags for marking lake trout. North American Journal of Fisheries Management 6:264–271.

Fletcher, D. H., F. Haw, and P. K. Bergman. 1987. Retention of coded wire tags implanted into cheek musculature of largemouth bass. North American Journal of Fisheries Management 7:436–439.

Galbreath, J. L. 1985. Status, life history, and management of Columbia River white sturgeon, *Acipenser transmontanus*. Pages 119–126 *in* F. P. Binkowski and S. I. Doroshov, editors. North American sturgeon: biology and aquaculture potential. Dr. W. Junk, Dordrecht, The Netherlands.

Gibbard, G. L., and R. L. Colura. 1980. Retention and movement of magnetic nose tags in juvenile red drum. Annual Proceedings of the Texas Chapter, American Fisheries Society 3:22–29. (Texas Parks and Wildlife Department, Austin.)

Humason, G. L. 1979. Animal tissue techniques, 4th edition. Freeman, San Francisco.

Isaksson, A., and P. K. Bergman. 1978. An evaluation of two tagging methods and survival rates of different age and treatment groups of hatchery-reared Atlantic salmon smolts. Journal of Agricultural Research in Iceland 10(1):74–99.

Jefferts, K. B., P. K. Bergman, and H. F. Fiscus. 1963. A coded wire identification system for macro-organisms. Nature (London) 198:460–462.

Joule, B. J. 1983. An effective method for tagging marine polychaetes. Canadian Journal of Fisheries and Aquatic Sciences 40:540–541.

Klar, G. T. 1985. Coded wire tagging techniques for striped bass. U.S. Fish and Wildlife Service, Research Information Bulletin 8-520, Washington, D.C.

Klar, G. T., and N. C. Parker. 1986. Marking fingerling striped bass and blue tilapia with coded wire tags and microtaggants. North American Journal of Fisheries Management 6:439–444.

Leary, L. F., and G. I. Murphy. 1975. A successful method for tagging the small, fragile engraulid, *Stolephorus purpureus*. Transactions of the American Fisheries Society 104:53–55

Marty, G. D., and R. C. Summerfelt. 1986. Pathways and mechanisms for expulsion of surgically implanted dummy transmitters from channel catfish. Transactions of the American Fisheries Society 115:577–589.

Milstein, V. V. 1957. [Artificial propagation of sturgeon.] Pishepromizdat, Moscow. (In Russian.)

Montfrans, J., J. Capelli, R. J. Orth, and C. H. Ryer. 1986. Use of microwire tags for tagging juvenile blue crabs (*Callinectes sapidus* Rathbun). Journal of Crustacean Biology 6:370–376.

Morrison, J., and D. Zajac. 1987. Histologic effects of coded wire tagging in chum salmon. North American Journal of Fisheries Management 7:315–330.

Pavlov, D. S., Y. N. Sbikin, and I. K. Popova. 1970. [Role of sensory organs in feeding behavior of juvenile sturgeon.] Zoologischeskii Zhurnal 159:872–880. (In Russian.)

Schwartz, F. J. 1981. A long term internal tag for sea turtles. Northeast Gulf Science 5:87–93.

Shekhanova, I. A., 1955. [Use of radioactive phosphorus (P^{32}) for marking juvenile sturgeon.] Rybnoe Khozyaistvo 11:44–56. (In Russian.)

Steele, R. G. D., and J. H. Torrie. 1980. Principles and procedures of statistics, 2nd edition. McGraw-Hill, New York.

Waalkes, M. P., S. Rehm, K. S. Kasprzak, and H. J. Issaq. 1987. Inflammatory, proliferative, and neoplastic lesions at the site of metallic identification ear tags in Wistar [Crl:(WI)BR] rats. Cancer Research 47:2445–2450.

Wickins, J. F., T. W. Beard, and E. Jones. 1986. Microtagging cultured lobsters, *Homarus gammarus* (L.), for stock enhancement trials. Aquaculture and Fisheries Management 17:259–265.

Williamson, J. H. 1987. Evaluation of wire nose tags for marking largemouth bass. Progressive Fish-Culturist 49:156–158.

Wydoski, R., and L. Emery. 1983. Tagging and marking. Pages 215–237 *in* L. A. Nielsen and D. L. Johnson, editors. Fisheries techniques. American Fisheries Society, Bethesda, Maryland.

American Fisheries Society Symposium 7:304–310, 1990

Evaluation of Coded Microwire Tags Inserted in Legs of Small Juvenile American Lobsters

JAY S. KROUSE AND GLENN E. NUTTING

Maine Department of Marine Resources, West Boothbay Harbor, Maine 04575, USA

Abstract.—We conducted laboratory experiments and field studies to evaluate the suitability of coded microwire tags for mark–recapture studies of juvenile American lobsters *Homarus americanus*. In laboratory tests, microwire tags were implanted in the propodus of the walking legs of 99 small lobsters (12–22 mm carapace length [CL]). This site was selected so the tag of a recaptured lobster could be easily and safely excised from the amputated leg tip for subsequent decoding. In three tests, tag retention ranged from 52 to 86% after the first molt. Among animals that started the next cycle with tags, 62–100% retained tags through the second molt; of these, 88–100% kept tags through the third molt. Tag loss was greater for smaller lobsters and inversely related to the time elapsed between tag application and ecdysis. Tagged and control (untagged) lobsters had similar growth increments, frequencies of molt, and survival. Of 603 lobsters (12–24 mm CL) tagged and released in the wild from 1985 through 1987, 72 (12%) were recaptured, of which 30 had molted. Ten recaptured lobsters at large for at least 1 year were estimated to have undergone an average of 3.4 molts. Results indicate that microwire tagging, with perhaps some slight modifications to improve tag retention, is a convenient and efficient means of labelling small juvenile lobsters for studies of growth and movement.

Along the Maine coast, the American lobster *Homarus americanus* is an intensively fished species (86% annual exploitation rate, Thomas 1973), which supports the state's most valuable commercial fishery (US$54.5 million ex-vessel value in 1987). Concern for the scientific management of the resource has prompted studies of American lobster movement, mortality, and growth. These studies have involved tagging considerable numbers of lobsters, primarily large juveniles and adults (>60 mm carapace length, CL), with persistent external tags designed to remain attached through ecdysis (see reviews by Krouse 1980; Stasko 1980). More recently, investigators have recognized the importance of a thorough understanding of the life processes of small juvenile lobsters. Accordingly, they have developed miniaturized persistent back tags for American lobsters as small as 20–25 mm CL (Bernstein and Campbell 1983; J. S. Krouse and G. E. Nutting, Maine Department of Marine Resources, unpublished data). Although tag retention was high (>87%) for lobsters held under laboratory conditions, our observations on lobsters smaller than 30 mm CL that were released and recaptured in the wild, were less favorable. Return rates were lower for smaller lobsters, indicating a proportionately greater tag loss, lower survival, or both. Entanglement of tags with the substrate, leading to tag loss, might be more prevalent among cryptic juveniles. Also, firm implantation of the tags' anchors was more difficult with lobsters less than 30 mm CL.

The injected binary-coded microwire tag first described by Jefferts et al. (1963) and successfully applied to crabs (Tutmark et al. 1967; van Montfrans et al. 1986), shrimp (West and Chew 1968; Prentice and Rensel 1977), and lobsters (Ennis 1972; Wickins and Beard 1984; Wickins et al. 1986) seems to be suitable for marking small juvenile American lobsters for multiple release–recapture studies. The tag is biologically inert, should be retained through numerous molts, can be located in tissue not likely to be consumed, and should not alter survival, growth, or behavior patterns.

The objectives of our study were to evaluate, on the basis of laboratory experiments, the retention of microwire tags inserted in leg tips of small American lobsters (12–24 mm CL), and to determine any influence the tags might have on growth and survival. In addition, limited field releases of tagged animals were made to assess the feasibility and practicality of this tagging technique for multiple release–recapture studies. In order that recaptured lobsters could be easily identified after detection and then released with minimal trauma, we implanted the microtags in the propodus of the pereiopods. In this location the tag can be quickly recovered without any deleterious effects.

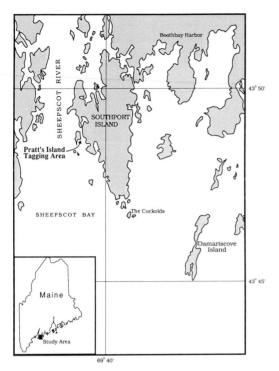

FIGURE 1.—Location of the intertidal study area at Pratt's Island near Boothbay Harbor, Maine.

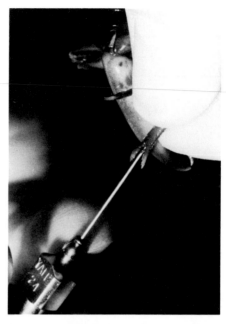

FIGURE 2.—Injection of microwire tag into the propodus of an American lobster's second pereiopod.

Methods

Test animals and study area.—Juvenile American lobsters were collected by hand during periods of low spring tides at Pratt's Island beach southwest of Boothbay Harbor, Maine, near the mouth of the Sheepscot River estuary (Figure 1). This intertidal area is characterized by areas of gradually sloping coarse sand–shell beach interspersed with bedrock ledges covered with rocks of assorted sizes and brown algae *Fucus* spp. Each captured lobster was carefully placed in a 18.9-L bucket with copious seaweed to minimize injurious intraspecific interactions. Prior to tagging, which usually occurred a few hours after capture, lobsters were refrigerated (4–8°C) in moist seaweed. Before tag insertion, carapace length was measured to the nearest millimeter, and each animal's sex and claw condition were recorded. After tagging, the lobsters were immediately assigned to one of the laboratory experiments or released in the vicinity of original capture at Pratt's Island.

Tagging technique.—The binary-coded microwire tags were 0.25 mm in diameter and 1 mm long, and were fabricated of biologically nonreactive type-302 stainless steel (Northwest Marine

Technology, Shaw Island, Washington). A 24-gauge hypodermic needle attached to a brass syringe with a push-rod ejector was used to implant tags through the articulating membrane of the dactylus into the propodus of the second pereiopod (Figure 2). This location was selected to facilitate excision of the tags for identification without measurable harm to the recaptured animal, and to minimize the possibility that humans would ingest unrecovered tags. A lobster was held by hand in an inverted position with the targeted leg secured between the tagger's thumb and index finger while the other hand directed the needle into the slender propodus. After tag insertion, proper placement of the tag was checked visually with the aid of transmitted light. The lobster was moved through the posts of a horseshoe magnet to magnetize the tag. Next, to ensure that the tag had been properly magnetized, the tagged animal was passed over a magnetic sensing unit, which emitted an audible signal upon exposure to a magnetized tag.

Laboratory study.—In four 38.8-L aquaria, test animals were held singly in 0.7-L plastic food containers (eight per tank). Each container had plastic mesh side and bottom panels to allow water circulation. Air was supplied to each aquarium to create a flow of water through the undergravel filter system. Fresh, temperature-condi-

TABLE 1.—Mean sizes of American lobsters tagged with coded microwires, and of untagged controls, in four laboratory experiments, 1984–1987.

	Date		Group treatment	N	Mean (SE) carapace length at tagging, mm
Experiment	Initiated	Ended			
1	30 Aug 1984	30 Sep 1985	Tagged	28	16.8 (0.5)
2	8 Aug 1985	31 Jan 1985	Control	8	18.3 (0.8)
			Tagged	22	16.5 (0.5)
3	16 Jun 1986	30 Jun 1987	Control	8	16.8 (1.3)
			Tagged	24	18.4 (0.7)
4	16 Jun 1987	31 May 1988	Control	7	16.4 (0.8)
			Tagged	25	17.4 (0.3)

tioned sea water was used to replace about one-third of the test water every 2 months. Temperatures, which ranged from about 15 to 21°C, were maintained below ambient levels during summer months by cool tap water circulated through titanium tubing immersed in each tank. During the remainder of the year, water of ambient temperature was used. Salinity and pH were monitored periodically. Lobsters were fed, usually biweekly, a diet of minced mussels *Mytilus edulis,* clams *Mya arenaria,* and snails *Littorina littorea* supplemented with brown seaweed. Any uneaten food was removed a day after feeding. Observations were made daily for molting, mortality, and any unusual behavior. Newly molted lobsters were allowed to consume their cast shells and were not measured for at least a week after ecdysis so that the chances of autotomy (spontaneous limb loss) were reduced. Tagged lobsters were checked visually and with the magnetic detector for tag loss. Details of the four experiments are presented in Table 1.

Analysis of covariance was used to compare the growth rates of tagged and untagged (control) groups. Comparisons of the size distributions of lobsters in each treatment, and of those animals that retained and lost tags, were done with unpaired *t*-tests. Molt-period data were evaluated with the same test. Changes in tag retention with succeeding molts (I–III) were analyzed with a chi-square test for differences in probabilities (Conover 1980).

Field releases.—Juvenile lobsters (12–24 mm CL) were tagged and released at our intertidal sampling area in the Sheepscot River at about monthly intervals, April–November, from 1985 to 1987. Because there were few batches of individually coded tags (57 in all), it was necessary to use tags with the same code several times throughout the season. Although this situation sometimes impeded a positive identification, we were usually able to resolve the problem by reference to the lobster's size, sex, and claw condition, as well as to a secondary mark made by removal of the distal margin of a uropod corresponding to the year of tagging. Such marks have been shown not to affect the growth of juvenile lobsters (Bernstein and Campbell 1983).

From monthly research catches, all hand-collected American lobsters ranging from 12 to 40 mm CL (the largest size possibly attained since tagging) were carefully screened for tags. Each lobster was passed across the magnetic tag detector several times. An audible beep indicated the presence of a tag. Upon detecting a recaptured animal, we severed the tagged propodus with a

TABLE 2.—Retention of microwire tags by American lobsters through one to three molts, under laboratory conditions in four experiments. Lobsters in experiment 4a were retagged following tag loss in experiment 4. Data are number of lobsters that survived through the molt indicated *and* retained tags; in parentheses, these animals are expressed as percentages of all survivors that began the molt cycle with tags. Starting sample sizes are given in Table 1.

	Experiment					
Molt	1	2	3	4	4a	Total
I	24 (85.7%)	3 (75.0%)	15 (62.5%)	13 (52.0%)	7 (70.0%)	62 (68.1%)
II	18 (94.7%)		12 (80.0%)	8 (61.5%)	6 (100%)	44 (83.0%)
III	2 (100%)		7 (87.5%)	3 (100%)		12 (92.3%)

TABLE 3.—Molt-period data for American lobsters tagged with coded microwires, and their untagged controls, in laboratory experiments. In experiment 4a, lobsters were retagged after losing tags in experiment 4. Data are means±SE (N).

Experiment	Group	Days from tagging to first molt	Days from first to second molt	Days from second to third molt
1	Tag kept	47.2±6.6 (24)	109.3±12.0 (18)	115.5±24.7 (6)
	Tag lost	22.0±10.5 (4)		
3	Control	35.4±10.6 (8)	109.5±10.8 (6)	82.5±6.9 (6)
	Tag kept	54.3±10.7 (16)	110.5±12.2 (11)	108.0±11.0 (6)
	Tag lost	46.0±14.8 (8)		
4	Control	13.0±6.7 (7)	83.4±12.1 (7)	
	Tag kept	23.6±11.8 (13)	90.2±13.5 (6)	
	Tag lost	17.8±9.2 (12)		
4a	Tag kept	57.6±9.17 (7)	123.7±11.6 (6)	

scalpel at the joint next to the carpus. The tag was then excised, soaked in 0.1 N hydrochloric acid, and decoded under a dissecting microscope.

Results and Discussion

Laboratory Study

Tag retention.—Among the tagged aquarium lobsters that survived through the first molt, 68% retained their tags (Table 2). Among those that began the next molt cycle with tags and survived through the second molt, tag retention improved to 83%; from the second through the third molts, tag retention was 92% among the 13 survivors. The improvement from the first to the later molts was significant ($P < 0.05$).

Tag loss varied widely among experiments (Table 2) for reasons that were not readily apparent. The same tagging technique was used for all tests, the size distribution of lobsters tagged (Table 1) was similar among experiments ($P > 0.05$), and there were no apparent relationships between tag loss and the date each test began or the intermolt period,.

Tagging success was influenced by temporal proximity to ecdysis, tag placement, and lobster size. Lobsters that lost tags generally had their first molt sooner after tagging (18–46 d) than did those that retained tags (24–54 d; $P < 0.01$; Table 3). Thus, the likelihood of tag loss seemed to be inversely related to the time elapsed between tag application and the first molt. Perhaps as the time of ecdysis approached, the developing cuticle occupied a larger volume beneath the exoskeleton and made it difficult to position the injected tag completely under the membrane that would become the new outer shell. Any tag not implanted totally beneath the cuticle without altering the membrane's integrity might be lost at molting. We have observed a few recently molted lobsters with either partially protruding tags or scarred leg tips indicative of tag loss.

The importance of proper tag placement was

TABLE 4.—Mean carapace lengths of American lobsters that retained and lost microwire tags after their first molts in laboratory experiments.

Experiment	Lobsters retaining tags		Lobsters losing tags	
	N	Mean (SE) length, mm	N	Mean (SE) length, mm
1	24	17.2 (0.6)	4	15.0 (0.9)
3	15	19.5 (0.7)	9	17.3 (1.2)
4	13	17.7 (0.4)	12	17.0 (0.3)

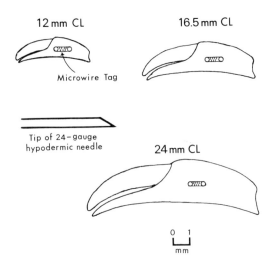

FIGURE 3.—Sizes of propoduses from American lobsters with carapace lengths (CL) of 12, 16.5, and 24 mm in relation to sizes of the microwire tag and applicator needle.

reflected, we believe, by the marked increase in tag retention for molts subsequent to the first molt (Table 2). We think that most tags not correctly positioned were lost at the first molt, and that subsequent high retention rates were due to well-placed tags.

Size of lobsters also seemed to be related to tag loss (Table 4). In all tests retention was less for smaller animals ($P < 0.05$). This might be explained by the diminutive tagging site of 12–14-mm lobsters and the difficulty of inserting the hypodermic needle and tag into the internal confines of the cuticle (Figure 3). Wickins and Beard (1984) also reported reduced tag retention for European lobsters *H. gammarus* close to 12 mm CL.

We noted that American lobsters tagged just prior to ecdysis not only had a higher rate of tag loss, they also were more prone to autotomize the pereiopod selected for tagging. Likewise, Bailey and Dufour (1987) reported that snow crabs *Chionoecetes opilio* tagged immediately before the molt were at risk for appendage autotomy or damage to the soft limb developing beneath the exoskeleton.

Survival.—Although 116 of 122 lobsters in all tests (95%) died within 13 months (mean, 7.9 months), none of these deaths appeared to be related to the injected tag because death rates were similar in both the control and tagged groups. Two animals were cannibalized, but the remaining deaths were unexplained. They may

have been associated with some toxic material in the experimental tanks. Some of the materials in question, such as hot glue and black plexiglass, were replaced. Further work will be necessary to identify the lethal agent.

Tagged animals did not display any adverse responses to the tags. No signs of infection were observed at the tagging sites, nor did any behavior patterns such as feeding and locomotion seem to be affected. Moreover, Wickins et al. (1986) found no microtag-associated mortality of *H. gammarus,* some of which had been tagged for more than 3 years.

Growth.—Comparisons of molt frequency and size increment at molting for tagged and untagged American lobsters revealed similar growth rates for both groups. Analysis of covariance indicated no statistical differences ($P > 0.05$) between the coefficients of the linear regressions of postmolt CL or premolt CL for the control and tagged lobsters. Accordingly, the regressions were combined for each test (Figure 4). The molt occurrence averaged 2.7 for each group. Similarly, Prentice and Rensel (1977) reported normal growth for spot prawns *Pandalus platyceros* with microtags injected into the thoracic sinus. Van Montfrans et al. (1986) found that the size increment at molting was unaffected by tagging for juvenile blue crabs *Callinectes sapidus,* although the frequency of molts may have been lower for tagged crabs.

Field Study

Of 603 juvenile American lobsters (12–24 mm CL) tagged and released in the Sheepscot River from 1985 through 1987, we recaptured by hand 72 (12%). The recaptured lobsters had a mean CL of 23.4 mm (15–34 mm) and had been at large an average of 10.1 months. Thirty recaptured animals had molted. On the basis of percentage increase in size, we estimated that 12 lobsters had molted once since release, 6 twice, 4 thrice, 2 four times, and 3 five times (three recaptured animals were excluded for lack of positive identification). Overall size increases ranged from 5 to 77% (mean increase, 30.6%) and the mean number of molts was 2.2 per animal. Ten recaptured lobsters at large at least 1 year were estimated to have molted an average of 3.4 times.

Recommendations

Our field and laboratory studies have indicated that the coded microwire tag can be applied successfully to juvenile American lobsters as

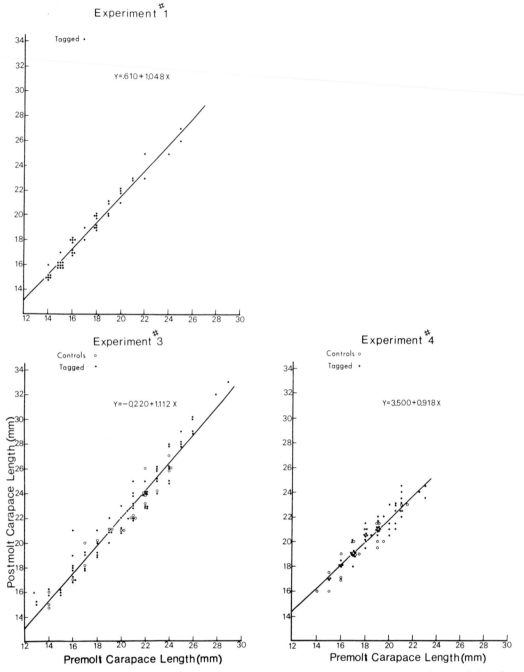

FIGURE 4.—Relations of postmolt (Y) to premolt (X) carapace lengths for tagged and untagged (control) American lobsters in laboratory experiments.

small as 12 mm CL to investigate movement, growth, and possibly mortality rates. By implanting the tag in the leg tip, the tag of a recaptured lobster can be easily detected, excised, and decoded without the need of a costly X-ray reader.

However, we emphasize that further work is required to determine if amputation of the propodus segment affects growth.

Although levels of tag retention in this study were sufficiently high we believe improvements could be made by (1) ensuring proper orientation of the tag during implantation (tagging with the aid of magnification would be helpful), (2) not tagging lobsters just prior to ecdysis, and, possibly, (3) using the shorter 0.5-mm-long tag now available.

For those studies that require only knowledge of whether or not a tag is present, and not of its identity, investigators should consider implanting the microtag beneath the base of the fifth pereiopod. Wickens et al. (1986) reported that juvenile *H. gammarus* (9–15 mm CL) tagged in this manner retained 85-100% of the tags as some lobsters attained sizes of 90–102 mm CL after 24–29 molts. Considering the high rate of tag retention and the ease of tagging large groups of juveniles with this technique, this marking system appears well suited for evaluating the contribution of hatchery-produced lobsters to wild stocks.

Acknowledgments

We express our appreciation to Sally Adams, Jill Benedict, and Fran Pierce for their help with the laboratory experiments and assistance with the data analysis. We are grateful to Northwest Marine Technology Company for furnishing tags and technical advice.

References

Bailey, R. F. J., and R. Dufour. 1987. Field use of an injected ferromagnetic tag on the snow crab (*Chionoecetes opilio* O. Fab.) Journal du Conseil, Conseil International pour l'Exploration de la Mer 43:237–244.

Bernstein, B. B., and A. Campbell. 1983. Contribution to the development of methodology for sampling and tagging small juvenile lobsters *Homarus americanus*. Canadian Manuscript Report of Fisheries and Aquatic Sciences 1741.

Conover, W. J. 1980. Practical nonparametric statistics, 2nd edition. Wiley, New York.

Ennis, G. P. 1972. Growth per moult of tagged lobsters (*Homarus americanus*) in Bonavista Bay, Newfoundland. Journal of the Fisheries Research Board of Canada 29:143–148.

Jefferts, K. B., P. K. Bergman, and H. F. Fiscus. 1963. A coded wire identification system for macro-organisms. Nature (London) 198:460–462.

Krouse, J. S. 1980. Summary of lobster, *Homarus americanus*, tagging studies in American waters (1898-1978). Canadian Technical Report of Fisheries and Aquatic Sciences 932:135–140.

Prentice, E. F., and J. E. Rensel. 1977. Tag retention of the spot prawn, *Pandalus platyceros*, injected with coded wire tags. Journal of the Fisheries Research Board of Canada 34:2199–2203.

Stasko, A. B. 1980. Tagging and lobster movements in Canada. Canadian Technical Report of Fisheries and Aquatic Sciences 932:141–150.

Thomas, J. C. 1973. An analysis of the commercial lobster (*Homarus americanus*) fishery along the coast of Maine, August 1966 through December 1970. NOAA (National Oceanic and Atmospheric Administration) Technical Report NMFS (National Marine Fisheries Service) SSRF (Special Scientific Report Fisheries) 667.

Tutmark, G. J., W. Q. B. West, and K. K. Chew. 1967. Preliminary study on the use of Bergman–Jefferts coded tags on crabs. Proceedings of the National Shellfisheries Association 57:24–26.

van Montfrans, J., J. Capelli, R. J. Orth, and C. H. Ryer. 1986. Use of microwire tags for tagging juvenile blue crabs (*Callinectes sapidus* Rathbun). Journal of Crustacean Biology 6:370–376.

West, W. Q. B., and K. K. Chew. 1968. Application of the Bergman–Jefferts tag on the spot shrimp, *Pandalus platyceros* Brandt. Proceedings of the National Shellfisheries Association 58:93–100.

Wickins, J. F., and T. W. Beard. 1984. Micro-tagging juvenile lobsters (*Homarus gammarus* L.)—preliminary results. International Council for the Exploration of the Sea C.M. 1984/K:7.

Wickins, J. F., T. W. Beard, and E. Jones. 1986. Microtagging cultured lobsters, *Homarus gammarus* (L.), for stock enhancement trials. Aquaculture and Fisheries Management 17:259–265.

American Fisheries Society Symposium 7:311–315, 1990
© Copyright by the American Fisheries Society 1990

Visible Implanted Fish Tag

FRANK HAW,[1] PETER K. BERGMAN, AND RICHARD D. FRALICK

Northwest Marine Technology, Shaw Island, Washington 98286, USA

RAYMOND M. BUCKLEY AND H. LEE BLANKENSHIP

Washington State Department of Fisheries, 115 General Administration Building
Olympia, Washington 98504, USA

Abstract.—After we observed that coded wire tags and fragments of polypropoylene implanted into tissues of some perciform and salmonid fishes remained externally visible, we implanted alphanumerically coded tags into periocular transparent tissue of 119 rainbow trout *Oncorhynchus mykiss* (formerly *Salmo gairdneri*), 149–280 mm long, that were under observation from 22 to 44 weeks. Tags, made from polyester and diazo film, were 0.08–0.18 mm thick, 0.6–1.3 mm wide, and 1.5–4.0 mm long. Five tags were shed, but none later than 4–7 weeks after implantation. Larger alphanumerically coded laminated implants were more easily read externally. Most tissue reactions to the implants were mild to moderate; more severe responses may have resulted from concentrations of exposed laminating adhesive. The visible implanted tag seems to offer a viable alternative to percutaneous devices in some fishes. Work continues on tag design, development of an efficient injector, and application of the method to other fishes.

Tagging methods often raise a variety of unanswered questions regarding rates of tag loss and effects on growth, migration, and mortality of fish. Commonly at fault is the percutaneous attachment, which permanently penetrates through a defect in the skin and is used for external fish tags. Various failure modes and mechanisms have been described in association with percutaneous devices in animals (Hall et al. 1984; von Recum 1984). An improved externally visible fish tag is needed, and we have conducted research to develop such a system.

Our initial attempts to develop a new system focused on improving the compatibility of external tags with fish tissue by use of test materials and smaller tags. Despite the favorable external appearance of some of these new tags, histological analysis indicated that the tag's percutaneous attachment soon became inflamed, infected, or both. Chronic lesions at tag attachment sites on Atlantic salmon *Salmo salar* have been analyzed histopathologically by Roberts et al. (1973a, 1973b, 1973c) and by Morgan and Roberts (1976).

Influenced by our success with internal coded wire tags (Jefferts et al. 1963; Bergman et al. 1968; Fletcher et al. 1987), and anticipating the development of coded wire tags capable of identifying individual specimens, we experimented with shallow implantations that would facilitate benign

surgical recovery, so that it would be unnecessary to kill the fish to recover the wire tag. We injected tags into the scalps of small (35–72 mm) pumpkinseeds *Lepomis gibbosus*, above the eyes and parallel to the long body axis, but slightly off center. These implanted coded wire tags have remained clearly visible through the tissue for 19 months (Figure 1). Visual location of the implanted tags and the small size of the required incision simplified surgical recovery with little apparent detrimental effect on the host.

We then attempted similar procedures on small (42–60 mm) slenderhead darters *Percina phoxocephala*. The scalps of these small fish are thin, so the preopercle area was selected for implantation. Although the implanted tags were highly visible, the cylindrical shape and thickness of the wire appeared incompatible with the fish's anatomy, and shedding occurred over a period of a few weeks. Flat, rectangular surrogate tags were then fashioned from 2.0 (metric) blue monofilament polypropylene suture, shaved on two sides with a scalpel and cut into segments about 1 mm long. With a modified insulin syringe, these tags were injected into (42–70 mm) slenderhead darters and orangethroat darters *Etheostoma spectabile*. Retained implants remain highly visible after 13 months. Thus, implanted tags with an externally legible code appeared to offer an alternative to existing external markers for certain fishes.

Our attention then turned to salmonids. Although coded wire tags injected into scalps of adult

---

[1]Present address: Northwest Marine Technology, 2401 Bristol Court, S.W., Olympia, Washington 98502, USA.

311

FIGURE 1.—Coded wire tag (white line at arrow) implanted into the scalp of a pumpkinseed.

FIGURE 2.—Visible tag being implanted into transparent tissue of a rainbow trout.

chinook salmon *Oncorhynchus tshawytscha* and coho salmon *O. kisutch* were invisible, they were clearly visible when implanted into transparent tissue (adipose eyelid) posterior to the eye. Tags injected into a few such specimens, maintained in a marine aquarium for longer than a year, have shown no signs of rejection or visual occlusion. We had observed that similar tissue was present in representatives of various salmonid genera (*Oncorhynchus, Salmo, Salvelinus*, and *Prosopium*). We therefore made plans to test a visible implanted tag in domestic rainbow trout *Oncorhynchus mykiss* (formerly *Salmo gairdneri*).

Methods

Large blank tags.—Blank (uncoded) tags were made to test retention, visibility, and compatibility with the tissue of the fish. With a transparent adhesive, red or silver pigment was laminated inside clear polyester sheets, 0.08 or 0.05 mm thick, with a finished thickness of 0.18 mm (red) and 0.13 mm (silver). Rectangular tags (1.3 mm wide × 4.0 mm long) with rounded ends were cut from this material.

Tag injectors were made from 0.2-mm-thick stainless steel. The hand-operated devices cut holes in tissue that were slightly larger than the tags. Tags were implanted upon slight withdrawal of the injector tip that had cut an accommodating fissure.

Live rainbow trout were obtained from Nisqually Trout Farm, Lacey, Washington. In April

1987, we tagged 20 fish (239–280 mm total length TL; mean, 257 mm). The fish were deeply anesthetized with MS-222 (tricaine) and we tagged 10 with red and 10 with silver tags. Tags were injected dorsoventrally (Figure 2) at a point behind, but even with, the upper margin of the eye, so as to rest behind the middle of the eye adjacent to the hard, solidly pigmented underlying tissue. The tagged fish, as well as others described herein, were confined and maintained in two rectangular covered pens (0.8 m × 2.2 m × 0.7 m deep) immersed in raceways containing production rainbow trout.

These tagged fish were examined 27 times, under anesthesia, during the ensuing 44 weeks. Six fish were killed and subjected to a standard histological examination 5, 16, and 40 weeks after they were tagged.

Small coded tags.—Alphanumeric coding seemed to be practical for the new tags. To determine its visibility, and the stability and compatibility of various tag materials, shapes, and sizes with the tissue of the fish, we implanted and examined various prototype tags. In one such test, we compared square- and round-ended rectangular alphanumerically coded tags implanted in 57 rainbow trout (158–226 mm; 197 mm, mean total length) on July 22. These smaller laminated

FIGURE 3.—A 2.5-mm-long, alphanumerically coded tag implanted 29 weeks previously in a 185-mm rainbow trout.

tags (2.5 mm × 0.9 mm × 0.13 mm thick), constructed and injected as previously described, bore photographically produced 0.7-mm-high, three-digit silver alphanumerics surrounded by solid black (Figure 3). Close-up photographs of 50 implanted tags were taken on August 13. The tagged fish were examined 16 times during the ensuing 29 weeks. Fish were killed for closer examination 2, 3, and 17 weeks after tagging.

Smaller and thinner tags.—Tagging on September 4 involved 42 fish even smaller (149–172 mm TL; mean 158 mm) and smaller tags. The rectangular tags (0.6 mm × 1.5 mm long), bearing 0.5-mm-high alphanumerics, were of two thicknesses and constructions. Half of the fish were implanted with tags 0.08 mm thick, and half with tags 0.13 mm thick. The thinner material was unlaminated diazo film, and transparent alphanumerics were surrounded by solid black. The other was the same as that described in the previous test. These fish were examined 10 times during the ensuing 22 weeks. Fish were killed for closer examination 10, 18, and 22 weeks after they were tagged.

Results

Large Blank Tags

The transparent tissue appeared to easily accommodate large implanted tags. Exposed tacky adhesive was noticeable along the edges of the tags. An examination after 4 d revealed that one silver tag had been shed and that entry wounds were open and ragged. Eight days after tagging, healing was well under way, and it appeared

relatively complete at the next examination on day 15. Apart from the five sacrificed fish and two escapees, all of the fish have survived with no additional tag loss. All tags seemed to remain in the original position.

Thirteen weeks after tagging, one of the red tags seemed to be abnormal. At week 16, the condition appeared better, but the fish was sacrificed for a histological examination. An acute inflammatory reaction, which could have been in the process of resolution, was discovered, but the tag itself appeared to be in its original condition.

Histological analysis indicated that the targeted transparent tissue was relatively acellular, contained few vessels, and was primarily stromal. Despite its designation (adipose eyelid), it contained little fat. Apart from the reaction described above, tissue responses to the implants were mild, ranging from essentially none to the formation of a thin fibrohistiocytic envelope surrounding the tag and accompanied by a mild chronic inflammatory infiltrate.

These uncoded tags remained highly visible for the 44 weeks of observation. The silver tag coloration is likely to enhance the visibility of the alphanumeric coding more than the red. In contrast to the clearer tissue in the few tagged chinook salmon, coho salmon, Atlantic salmon, and steelhead (anadromous rainbow trout) fingerlings maintained in our marine aquarium, a thin layer of melanophores normally overlies the rainbow trout tags.

Different reactions to subcutaneous implants have been observed elsewhere. Within about 6 months, melanophores became concentrated directly over, and obscured, diazo film tags implanted into musculature anterior to the dorsal fin of small poeciliid fishes (Heugel et al. 1977). After 3–4 weeks, pigmentation also obscured vinyl tags implanted beneath the previously white belly skin of rainbow trout (Butler 1957, 1962).

Small Coded Tags

A sharper and more precise injector associated with small tags reduced tissue damage, but the exposed and tacky laminating adhesive was more concentrated within the smaller tag perimeters. Three tags were shed between weeks 2 and 6, and none thereafter. The shed tags were from fish that were shorter than the mean length of fish in the group. Two of the three lost tags were square-ended and one round-ended, but no conclusion can be drawn about the difference between shapes. An analysis of the photographs taken before tags were shedding was also inconclusive,

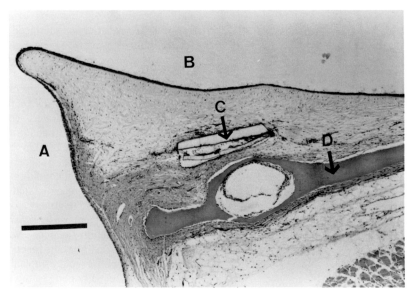

FIGURE 4.—Coronal-plane photomicrograph of an implanted laminated tag, showing a moderately intense reaction, in the adipose eyelid of a rainbow trout. The eye would occupy the space to the left (A). A, anterior edge; B, lateral edge; C, tag; D, bone. Bar = 0.5 mm.

although square-ended tags appeared somewhat closer to the anterior margin of the target.

Two fish in this experiment were unaccounted for; otherwise no mortality occurred during the 29 weeks the fish were under observation. The tissue surrounding only one of these tags—a square-ended one implanted for 2 weeks—was histologically examined. Little tissue reaction was noted.

Smaller and Thinner Tags

Only one small, thin tag (laminated) was shed by fish during 22 weeks of observation. This shedding occurred between weeks 4 and 7; we had been dissatisfied at tagging with this relatively shallow implant. No fish died during the test.

Small size, the lack of a reflective surface to highlight the alphanumerics, and normal tissue pigmentation—further exacerbated by poor natural light conditions—rendered the unlaminated diazo film tags difficult to read, at best. Laminated tags were far more easily read, but magnification, artificial light, or both, were advantageous even so.

Smaller laminated tags drew in and concentrated the exposed adhesive perimeter. This adhesive may have caused the more severe tissue reactions observed among the five specimens that were histologically examined 18 and 22 weeks after implantation. Two reactions were mild, two were moderately intense fibrohistocitic reactions with focal chronic inflammation (Figures 4 and 5),

and one was an intense acute and chronic inflammation associated with fragments of detached adhesive. Reactions to the three unlaminated tags examined ranged from mild to moderate.

Discussion

Although important questions remain unanswered, the visible implanted tags seem viable and more suitable than percutaneously attached tags for some smaller fishes. Large fish appear to be even better adapted for tagging with visible implants. We continue work on tag design, development of a more efficient hand-operated injector, and the use of the method on other fishes.

Existing fisheries literature is a poor source of information on the presence of transparent tissue that might be implantation sites for visible implants. Because transparent tissues can be altered by preservatives and freezing, it is best to search for them on live or fresh specimens. Our limited observations of the ocular areas of fishes suggest that extensive areas of clear tissue occur among the Elopidae, Albulidae, Clupeidae, Engraulidae, Osmeridae, and Carangidae, and on *Scomber* spp. in the Scombridae. Other parts of the body, such as adipose fins in the Ictaluridae, as well as those body parts previously noted for darters and pumpkinseeds, seem to offer potential sites for visible implants in an even wider variety of fishes.

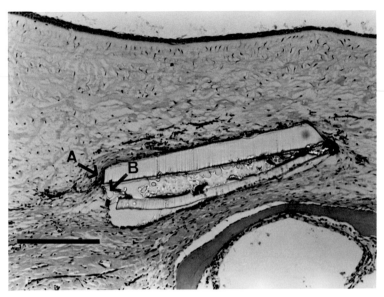

FIGURE 5.—Enlargement of Figure 4, showing details of tissue reactions and tag. A, fibrotic and chronic inflammatory cellular response; B, laminating adhesive. Bar = 0.2 mm.

Recent observations at Norfolk National Fish Hatchery, Arkansas, indicated that rainbow trout at that location have thinner layers of transparent tissue, providing more difficult targets for implants, than the specimens involved in our present work. We can offer no explanation for this difference.

Acknowledgments

Thomas Horbett (Department of Chemical Engineering, University of Washington) furnished us with key reference material and advice regarding problems with percutaneous devices and biocompatible materials. Dennis F. Peck (Black Hills Pathology, Ltd, Olympia, Washington) took special interest in the experiment, conducted the histological analysis, and provided the photomicrograph and its interpretation, mentioned in the text. Paul James, seeking a means to identify individual live specimens at the Oklahoma State University Wildlife Research Unit, collected the darters.

References

Bergman, P. K., K. B. Jefferts, H. F. Fiscus, and R. C. Hager. 1968. A preliminary evaluation of an implanted coded wire fish tag. Washington Department of Fisheries, Fisheries Research Paper 3:63–84.

Butler, R. L. 1957. The development of a vinyl plastic subcutaneous tag for trout. California Fish and Game 43:201–212.

Butler, R. L. 1962. Recognition and return of trout tags by California anglers. California Fish and Game 48:5–18.

Fletcher, D. H., F. Haw, and P. K. Bergman. 1987. Retention of coded wire tags implanted into cheek musculature of largemouth bass. North American Journal of Fisheries Management 7:436–439.

Hall, C. W., P. A. Cox, and S. R. McFarland. 1984. Some factors that influence prolonged interfacial continuity. Journal of Biomedical Materials Research 18:383–393.

Heugel, B. R., G. R. Joswiak, and W. S. Moore. 1977. Subcutaneous diazo film tag for small fishes. Progressive Fish-Culturist 39:98–99.

Jefferts, K. B., P. K. Bergman, and H. F. Fiscus. 1963. A coded wire identification system for macroorganisms. Nature (London) 198:460–462.

Morgan R. I. G., and R. J. Roberts. 1976. The histopathology of salmon tagging. IV. The effect of severe exercise on the induced tagging lesion in salmon parr at two temperatures. Journal of Fish Biology 8:289–292.

Roberts, R. J., A. McQueen, W. M. Shearer, and H. Young. 1973a. The histopathology of salmon tagging. I. The tagging lesion in newly tagged parr. Journal of Fish Biology 5:497–503.

Roberts, R. J., A. McQueen, W. M. Shearer, and H. Young. 1973b. The histopathology of salmon tagging. II. The chronic tagging lesion in returning adult fish. Journal of Fish Biology 5:615–619.

Roberts, R. J., A. McQueen, W. M. Shearer, and H. Young. 1973c. The histopathology of salmon tagging. III. Secondary infections associated with tagging. Journal of Fish Biology 5:621–623.

von Recum, A. F. 1984. Applications and failure modes of percutaneous devices: a review. Journal of Biomedical Materials Research 18:323–336.

American Fisheries Society Symposium 7:317–322, 1990

ELECTRONIC TAGS

Feasibility of Using Implantable Passive Integrated Transponder (PIT) Tags in Salmonids

Earl F. Prentice, Thomas A. Flagg, and Clinton S. McCutcheon

Northwest Fisheries Center, National Marine Fisheries Service
2725 Montlake Boulevard East, Seattle, Washington 98112, USA

Abstract.—The technical and biological feasibility of using passive integrated transponder (PIT) tags for tagging salmonids has been evaluated by the U.S. National Marine Fisheries Service. Each tag is 12.0 mm long by 2.1 mm in diameter and is coded with one of 34×10^9 codes. When energized at 400 kHz, the tag transmits a return signal at 40 to 50 kHz. The tag can be detected in situ at a distance up to 18 cm, which eliminates the need to anesthetize, handle, or restrain fish during data gathering. The tag's longevity is estimated at 10 or more years. The body cavity of juvenile and adult salmonids was found to be an acceptable site for implantation. The PIT tag did not adversely affect growth or survival in laboratory and field tests. Swim-chamber tests showed no significant effect of the tag on respiratory rate, tail-beat frequency, stamina, or survival of juvenile salmonids. Tag retention within the body cavity was nearly 100% for salmonids ranging in size from 50 to 800 mm, fork length. Previously PIT-tagged salmon that were hand-stripped of sperm and eggs showed high tag retention and no adverse effects of the tag.

The ability to recognize individuals or groups within a population is important in fisheries research, and many types of tags and marks have been developed to aid biologists in such recognition (Rounsefell 1963; Farmer 1981). No single technique has been totally satisfactory from a biological or technical standpoint. In 1983, the National Marine Fisheries Service (NMFS) began a study to evaluate the technical and biological feasibility of implanting passive integrated transponder (PIT) tags in salmonids. This paper describes how the tags operate, and it discusses their biological acceptability in salmonids tested under laboratory and field conditions. Details of the tests can be found in Prentice et al. (1984, 1985, 1986, 1987). We present additional information on PIT tagging and on monitoring systems elsewhere in this volume (Prentice et al. 1990a, 1990b).

PIT Tags in Operation

The PIT tag consists of an antenna coil that has about 1,200 wraps of a specially coated copper wire 0.0254 mm in diameter. The antenna coil is bonded to a pad and an integrated circuit chip. The electronic components of the tag are encapsulated in a glass tube 12.0 mm long by 2.1 mm in diameter (Figure 1).

The passive tag relies on an external source of energy to operate. The excitation energy comes from a tuned loop that is part of the tag interrogation system. The system transmits an alternating current via the loop at 400 kHz to establish a magnetic field. When a tag enters the magnetic field for as little as 25 ms, an induction current is established in the transponder antenna coil. The induced current energizes the transponder's integrated circuit. The circuit divides the fundamental frequency by 8 and 10, which results in a frequency shift between 40 and 50 kH giving mark–space coding. The signal from the tag is received by the loop of the interrogation system and passed through a filter to separate it from the 400-kHz excitation signal. The tag signal is then passed through amplifiers and filters, where it is decoded.

Each tag is programmed at the factory with one of about 34×10^9 unique code combinations. The identification code consists of a preamble and a 40-bit code arranged as five 8-bit bytes, each with a parity bit. One reading error per byte can be detected and self-corrected, but more than one error may result in a bad code identification. For a complete description of coding and decoding, see Housley (1979). The data received by the interrogation system are changed via tag-reader cards to 10 ASCII-encoded hexidecimal charac-

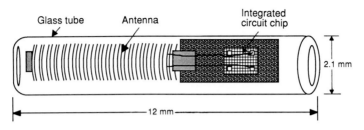

FIGURE 1.—Diagram of a typical PIT tag.

ters for transmission to a computer via an RS232 port, or they are stored within the internal memory of the interrogation system. A portable tag interrogation system (Prentice et al. 1990a) or a fixed tag monitor system (Prentice et al. 1990b) is used to detect and display tag code information. The rate of data transfer is 4,000 bits/s. The range for tag detection varies with the monitoring equipment used—up to 7.6 cm for hand-held tag interrogators, and about 18 cm for fixed full-loop interrogators. The tag can be read easily through soft and hard tissue, seawater, fresh water, glass, and plastic, and with difficulty through metal. Extreme cold or heat (−90 to 60°C) does not appreciably affect detection or reading of the tag. Successful monitoring can take place when the tag is moving at velocities up to 3.6 m/s. Because of the passive nature of the tag, an operational life of 10 years or more is expected.

The operator requires no special permits to use the interrogation system other than what is required by the Federal Communications Commission (FCC) in the USA, or its equivalent elsewhere, to operate low-powered transmitting devices. These permits pertain only to permanent specialized monitoring systems (e.g., at hydroelectric dams), and not to the hand-held interrogation system, which has already been certified by the FCC. No special certification is required to operate tag monitoring equipment.

Biological Suitability

Fish tags should not alter growth, survival, behavior, or reproduction; they should be retained; and they should have a long functional life. Laboratory tests were conducted to examine these factors for PIT tags implanted in salmonids. Juvenile and adult chinook salmon *Oncorhynchus tshawytscha*, sockeye salmon *O. nerka*, steelhead *O. mykiss* (formerly *Salmo gairdneri*) and Atlantic salmon *Salmo salar* were used in the studies. The fish ranged in fork length from about 55 to over

800 mm. Tags were injected into the body cavity with a modified hypodermic syringe and a 12-gauge needle (Prentice et al. 1986, 1987, 1990a).

Tag Retention, Fish Survival, and Fish Growth

Tag retention and operational longevity were investigated during 1986 and 1987. Two groups of 300 juvenile fall chinook salmon were established, one with and the other without PIT tags. The two groups were maintained in fresh water until smolted and then transferred to seawater and held in separate cages. Observations on growth (fork length), survival, and tag retention and operation were made on eight occasions. After 570 d, tag operation was 100% and retention was 98%. Growth of tagged fish was slightly depressed from day 1 to 20, after which it was approximately equal for the control and tagged fish until day 409 (Figure 2). Up to day 409, survival for the control and tagged fish was 89.7% and 86.7%, respectively. Afterward, comparisons of growth and survival were confounded by the appearance in the control group of many precocious males subject to high mortality. At the termination of the study (570 d), survival was 36.4% for control fish and 76.5% for tagged fish.

Also in 1986, we conducted a study to determine the minimum size at which juvenile fall chinook salmon can be successfully PIT-tagged. Approximately 200 fish in each of four size-groups were tagged and held separately (Table 1). Fish ranged in fork length from 56 to 120 mm at the time of tagging. Two water sources—well water and stream water—were used to test the effects of stream water containing fish pathogens on tag wounds and tag retention. Tag retention was excellent for all test groups (99–100%). Healing usually occurred within 14 d, regardless of the water source. Survival of PIT-tagged fish ranged from 97.0 to 100% in the well-water groups and

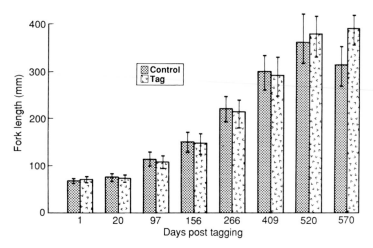

FIGURE 2.—Comparison of growth for 200 PIT-tagged and 200 untagged (control) fall chinook salmon. Columns show means; bars represent standard deviations.

from 95.0 to 98.0% in the stream-water groups. We found no association between survival and fish size, water source, or presence of the tag.

In 1987, using two sizes of fingerlings and one size of smolts, we studied the effect of fish size on tag retention, growth, and survival for sockeye salmon (Table 2). The minimum and maximum fork length at the time the test groups were established ranged from 55 to 107 mm. Each size-group was divided into a control (fish handled but not tagged) and a tagged group. Survival, never below 96.5%, was uniformly high for all test and control groups, and did not differ between them. Fingerlings as well as smolts exhibited high tag retention. The results of this and the previous experiment with fall chinook salmon are in general agreement. Both experiments indicate that

the PIT tag can be injected into juvenile salmonids without jeopardizing growth or survival.

To further determine if the PIT tag compromised juvenile salmonids, we conducted a series of field tests on outmigrating yearling chinook salmon, underyearling chinook salmon, and steelhead collected at hydroelectric dams. We compared the survival of PIT-tagged fish with that of control fish (handled but not tagged), coded-wire-tagged fish (CWT), CWT plus freeze-branded fish, and freeze-branded fish. The tests were conducted at Lower Granite Dam on the Snake River and McNary Dam on the Columbia River. Fish from all five treatments were combined in a common holding cage, where they received a continuous supply of untreated river water and examined daily for mortalities. No measurable differ-

TABLE 1.—Growth, survival, and PIT-tag retention for juvenile fall chinook salmon in well and stream water.

Treatment[a] and test group (G)	Number of fish	Test period (days)	Mean (SD) fork length, mm		Survival (%)	PIT-tag retention (%)
			Starting	Ending		
Control–well	202	135	77 (5)	125 (8)	100.0	
Control–stream	200	135	77 (5)	126 (8)	99.0	
PIT-tagged						
Well–G1	201	139	66 (3)	121 (6)	99.5	100.0
Well–G2	200	135	78 (5)	131 (8)	100.0	100.0
Well–G3	201	134	84 (5)	129 (8)	100.0	100.0
Well–G4	200	137	99 (6)	138 (9)	97.0	100.0
Stream–1	200	139	66 (3)	122 (7)	95.0	99.0
Stream–2	200	135	77 (5)	127 (7)	98.0	100.0
Stream–3	203	134	85 (5)	130 (8)	95.0	100.0
Stream–4	202	137	100 (6)	135 (9)	98.0	100.0

[a]The artesian well water had a constant temperature of 10°C and was pathogen-free; the stream water came from Big Beef Creek, had an ambient temperature of 9.3°–14.4°C, and contained pathogens.

TABLE 2.—Growth, survival, and PIT-tag retention for juvenile sockeye salmon.

Fish group and treatment	Number of fish per group	Mean (SD) fork length, mm		Survival (%)	PIT-tag retention (%)
		Starting	Ending		
Small presmolts					
Control	200	67 (4.8)	135 (9.6)	99.5	
Tagged	200	68 (4.1)	137 (10.7)	99.5	100
Large presmolts					
Control	200	82 (6.8)	134 (10.1)	98.5	
Tagged	200	83 (6.5)	130 (10.2)	99.0	98.5
Smolts					
Control	200	96 (4.0)	143 (9.3)	97.0	
Tagged	200	99 (3.8)	144 (9.2)	96.5	100

ence in survival was noted among the groups at the end of 14 d (Table 3).

Wound Healing and Tissue Response to the Tag

The insertion of a PIT tag or other foreign body into a fish is a trauma that may provoke such host reactions as inflammation, melanomacrophage aggregation, encapsulation, and rejection. In 1986, we examined the responses of fall chinook salmon (3.7 g, average weight) to PIT tags. The fish were held in tanks 1.2 m in diameter and supplied with constant 10°C well water. Wound healing was documented visually and histologically. PIT-tagging procedures followed the methods described by Prentice et al. (1990a), and all tags were placed in the body cavity.

Random samples of 10 fish were taken from the population ($N = 161$) on days 22, 30, and 45 post-tagging and examined histologically. Tissues were embedded in paraffin, sectioned at 6-μm thicknesses, and stained with hematoxylin and eosin. By day 22, the injection wound consisted of granulation tissue that had replaced the dermis and underlying muscle damaged during injection

of the tag, and the peritoneum and epidermis had regenerated. By days 30 and 45, the injection site was difficult to locate histologically and complete healing of the wound had occurred.

No host reaction to the tag was observed in any fish. Neither melanomacrophage accumulations nor tissue adhesions were noted, which indicates that the fish did not recognize the tag as a foreign body. The glass-encapsulated tag appears to be biologically inert.

Tagging wounds on the remaining fish were visually evaluated between days 14 and 45. The tag wound had closed on all fish by day 14, when a scar was noticeable on most of them. By day 30, the epidermal pigmentation appeared normal. These observations support the histological evidence and indicate that complete healing occurs within 2 weeks.

Tag retention was 100% during this study. We found most of the tags near the abdominal musculature posterior to the pyloric caeca close to the spleen, but some were between the midgut and the pyloric caeca (Table 4). No tissues adhered to the tags. We have preliminary evidence, however, that some tags may become encapsulated with tissue

TABLE 3.—Percent survival at day 14 post-tagging of fish tagged or marked with PIT tags and other devices at Columbia River dams.

Dam (year)	Species	Control	Tag or mark			
			PIT tag	Freeze brand	Coded wire tag (CWT)	CWT plus freeze brand
Lower Granite	Yearling chinook salmon	95	98	96	97	99
(1986)	Steelhead	100	99	100	99	97
McNary (1985)	Age-0 chinook salmon	96	87	94	92	93
	Yearling chinook salmon	86	83	86	80	89
McNary (1986)	Steelhead	89	87	93	91	94
	Age-0 chinook salmon	64	65	59	68	66

TABLE 4.—PIT-tag locations within the body cavities of juvenile fall chinook salmon over time. Data are percentages of fish examined.

	Days after tagging					
Tag location	14	15	16	23	28	36	39
Near abdominal musculature[a]	91.7	100	100	100	92.8	100	100
Elsewhere[b]	8.3	0	0	0	7.2	0	0

[a]Often embedded posteriorly among the pyloric caeca near the spleen or in the adjacent adipose tissue.
[b]Generally between the midgut and air bladder or between the liver and pyloric caeca. Tags were never found between the pyloric caeca and midgut. All fish retained their tags and none of the tags protruded through the abdominal wall.

after a year or so in the host. The uniform locations noted during this study confirm that a reliable implantation technique has been developed.

Effects of PIT Tagging on Swimming Performance

We evaluated the effects of the PIT tag on swimming performance in 726 juvenile chinook salmon and steelhead (Prentice et al. 1986, 1987). The fish were tested in a modified Blaska respirometer-stamina chamber described by Smith and Newcomb (1970). Experimental fish of several sizes were obtained from a hatchery or collected as in-river migrants (Table 5). We recorded all tests on videotape, which we examined at slow speed to determine swimming stamina, stride efficiency (tail beats per minute), and respiratory rate. Fish were held up to 14 d following the test and monitored for survival.

Neither the act of tagging nor the presence of the PIT tag compromised swimming stamina, stride efficiency, or respiratory rate of juvenile salmonids. Moreover, post-test survival was not affected by the PIT tag, and tag retention was 100% (Table 5). After 14 d, all tagged fish were killed and examined for signs of tissue reaction to the tags. No adverse tissue reactions or tag migrations within the peritoneal cavity were noted. On the basis of these tests, we conclude that the PIT tag

TABLE 5.—Swimming performance, post-test survival, and tag retention of PIT-tagged and control juvenile salmonids.

Species (mean length) and test group[a]	Stride efficiency: mean (SD) tail beats/min[b]	Respiratory rate: mean (SD) opercular beats/min[c]	Swimming stamina: mean (SD) body lengths/s[d]	Post-test[e]	
				Survival (%)	Tag retention (%)
Laboratory tests					
Steelhead (83 mm)					
Control	94.3 (22.4)	143.1 (21.2)	5.2 (0.7)	100	100
PIT-tagged	95.8 (22.7)	143.7 (20.8)	5.0 (0.7)	100	100
Steelhead (112 mm)					
Control	95.5 (16.6)	148.8 (16.8)	4.7 (0.6)	100	100
PIT-tagged	99.8 (17.1)	147.7 (15.4)	4.6 (0.5)	100	100
Steelhead (171 mm)					
Control	122.5 (18.4)	135.5 (21.3)	3.1 (0.3)	100	100
PIT-tagged	125.6 (18.4)	135.6 (17.3)	3.1 (0.3)	100	100
Fall chinook salmon (67 mm)					
Control	122.9 (37.9)	137.2 (27.5)	5.5 (0.4)	100	100
PIT-tagged	125.1 (37.3)	136.8 (22.6)	5.4 (0.4)	100	100
Fall chinook salmon (89 mm)					
Control	124.4 (29.7)	130.8 (15.3)	4.7 (0.6)	100	100
PIT-tagged	124.4 (28.9)	130.7 (17.2)	4.6 (0.6)	100	100
Field tests					
Steelhead (201 mm)					
Control	129.1 (20.9)	145.7 (19.3)	2.9 (0.5)	70.0	100
PIT-tagged	125.8 (17.8)	145.7 (16.3)	2.8 (0.8)	70.0	100
Yearling chinook salmon (137 mm)					
Control	131.8 (23.8)	125.0 (7.5)	3.2 (0.7)	63.6	100
PIT-tagged	124.8 (25.1)	114.3 (16.1)	3.4 (0.8)	56.7	100
Age-0 chinook salmon (111 mm)					
Control	129.6 (35.6)		5.2 (1.2)	26.7	100
PIT-tagged	125.3 (33.0)		5.2 (1.4)	30.0	100

[a]Laboratory fish were reared and tested at the Big Beef Creek facility near Seabeck, Washington: 96–144 PIT-tagged and 32–48 control fish were tested for each group. Migrant fish were collected and tested at the McNary Dam juvenile fish collection facility near Umatilla, Oregon; 30 PIT-tagged and control fish were tested for each group.
[b]Tail-beat rate required to maintain a swimming speed of 1 body length/s.
[c]Respiratory rate is number of opercular beats per minute until impingement on a screen barrier.
[d]Swimming stamina is fatigue level (time to impingement) in body lengths per second.
[e]Post-test holding period was 14 d for laboratory tests and 5 d for field tests.

should not compromise swimming performance of juvenile salmonids during downstream migration.

Effects on Maturing Fish

We investigated whether morphological and physiological changes during maturation altered the response of salmonids to PIT tags. We used 21 male and 60 female maturing Atlantic salmon *Salmo salar* ranging in weight from 2,500 to 10,000 g and in length from 61 to 80 cm. We PIT-tagged the fish according to the method of Prentice et al. (1990a), and examined them several times prior to spawning to determine wound condition, tag retention, readiness to spawn, and general condition. Eggs were collected by hand stripping.

No adverse tissue reaction was noted. All tag wounds were closed and healing by the third day following tagging. No infection or discoloration appeared in the area of the tag. All 21 males matured, and milt was collected from each fish. Tag retention was 100% for the males. Forty-eight females were spawned. Tag retention was 83% for the spawned females and 100% for the nonspawners. Four tags were passed during the first egg stripping and four more during the second through fourth strippings. When a tag was passed, it was easily seen among the eggs. The presence of tags did not appear to adversely affect egg quality or survival.

Future Applications

The PIT tag is the first generation of sophisticated identification systems to take advantage of the computer age. Its use is not limited to salmon—prawns and crabs also have been tagged, and the tag is applicable to any animal that can accept and retain it. Advantages and applications of the PIT tag include individual identification of brood stock, serial measurements (e.g., growth) of individuals, reduction in number of replicates, and combination of treatments. It can also be used in behavioral studies to monitor animal movements automatically or through capture–recapture methods. Pelagic animals might be monitored by means of trawl nets equipped with PIT-tag detectors mounted in an open cod-end or a specially constructed purse seine, whereas movement of benthic animals might be monitored with a grid system or underwater sled. Innovation is the key to future use of the PIT tag.

Acknowledgments

Support for this project came from electrical ratepayers through the Bonneville Power Administration. We also extend appreciation to the personnel of the Fish Passage Center, U.S. Fish and Wildlife Service (Fisheries Assistance Office, Vancouver, Washington), U.S. Army Corps of Engineers at McNary and Lower Granite dams, and the NMFS staff at McNary Dam for assistance in this study.

References

Framer A. S. D. 1981. A review of crustacean marking methods with particular reference to penaeid shrimp. Kuwait Bulletin of Marine Science 2:167–183.

Housley, T. 1979. Data communications and teleprocessing systems. Prentice-Hall, Englewood Cliffs, New Jersey.

Prentice, E. F., T. A. Flagg, and C. S. McCutcheon. 1987. A study to determine the biological feasibility of a new fish tagging system. Report (contract DE-A179-83BP11982, project 83-19) to Bonneville Power Administration, Portland, Oregon.

Prentice, E. F., T. A. Flagg, C. S. McCutcheon, D. F. Brastow, and D. C. Cross. 1990a. Equipment, methods, and an automated data-entry station for PIT tagging. American Fisheries Society Symposium 7:335–340.

Prentice, E. F., T. A. Flagg, C. S. McCutcheon, and D. F. Brastow. 1990b. PIT-tag monitoring systems for hydroelectric dams and fish hatcheries. American Fisheries Society Symposium 7:323–334.

Prentice, E. F., D. L. Park, T. A. Flagg, and C. S. McCutcheon. 1986. A study to determine the biological feasibility of a new fish tagging system. Report (contract DE-A179-83BP11982, project 83-19) to Bonneville Power Administration, Portland, Oregon.

Prentice, E. F., D. L. Park, and C. W. Sims. 1984. A study to determine the biological feasibility of a new fish tagging system. Report (contract DE-A179-83BP11982, project 83-19) to Bonneville Power Administration, Portland, Oregon.

Prentice, E. F., C. W. Sims, and D. L. Park. 1985. A study to determine the biological feasibility of a new fish tagging system. Report (contract DE-A179-83BP11982, project 83-19) to Bonneville Power Administration, Portland, Oregon.

Rounsefell, G. A. 1963. Marking fish and invertebrates. U.S. Fish and Wildlife Service Bureau of Commercial Fisheries Fishery Leaflet 545:1–12.

Smith, L. S., and T. W. Newcomb. 1970. A modified version of the Blazka respirometer and exercise chamber for large fish. Journal of the Fisheries Research Board of Canada 27:1321–1324.

American Fisheries Society Symposium 7:323–334, 1990

PIT-Tag Monitoring Systems for Hydroelectric Dams and Fish Hatcheries

EARL F. PRENTICE, THOMAS A. FLAGG, CLINTON S. MCCUTCHEON, AND DAVID F. BRASTOW

Northwest Fisheries Center, National Marine Fisheries Service
2725 Montlake Boulevard East, Seattle, Washington 98112, USA

Abstract.—Juvenile salmonids implanted with passive integrated transponder (PIT) tags can be monitored remotely as they are released from fish hatcheries or as they pass through specially designed facilities at hydroelectric dams. We have also designed and tested a system that monitors PIT-tagged adult salmonids. The systems record the individual PIT-tag code, time, date, and location of detection. Interrogation systems at dams can monitor fish traveling up to 3.7 m/s and provide tag detection efficiency above 95% and reading accuracy (correct code identification) above 99.0%. The information collected at each dam is automatically transferred to a central data base for storage and processing. The system used to monitor hatchery releases can process over 20,000 fish/h (at a ratio of 1:4 tagged to untagged) with a 93%, or higher, PIT-tag detection efficiency and a reading accuracy above 99.0%.

Salmonids in the Columbia River basin implanted with passive integrated transponder (PIT) tags can be interrogated remotely by means of a computer-based PIT-tag monitoring system. Details on the tag, how it operates, and its biological and technical suitability have been presented by Prentice et al. (1984, 1985, 1986, 1987) and are reviewed elsewhere in this volume (Prentice et al. 1990a, 1990b). The PIT tag, available from Destron-Identification Devices, Inc. (D-IDI)[1], consists of an integrated circuit and a coil (antenna) encapsulated together in a glass tube. The integrated circuit is factory-programmed with a unique code (a 10-digit hexadecimal number displayed in an alphanumeric format—e.g., 7F7131000) which is automatically transmitted whenever the circuit is energized. The tag is energized and read when the fish passes through the loop antennas of the monitoring system. Individual code, time, date, and location of detection are recorded for each PIT tag interrogated by the monitoring system. The system can passively monitor juvenile PIT-tagged salmonids as they are released from fish hatcheries or as they pass downstream through specially designed facilities at hydroelectric dams. A system to passively monitor adult salmon also has been designed.

In this paper, we describe and evaluate the PIT-tag monitoring systems we designed for hydroelectric dams and fish hatcheries. All elec-

tronic components of the monitoring system are commercially produced by D-IDI.

Systems at Hydroelectric Dams

Most outmigrating salmonids in the Columbia River basin encounter hydroelectric dams that impede migration and increase mortality (Figure 1). Several of these dams include collection and diversion facilities for passing migrants around the turbines to increase fish survival. A typical juvenile collection–diversion facility consists of traveling screens that divert fish from the dam's turbine intakes into gatewells and then into a series of conduits leading to a wet separator (Figure 2). The separator reduces the volume of water and removes debris. Fish are then diverted to a raceway for later transport downstream via truck or barge, directly to a barge for transportation downstream, or back into the river below the dam.

Monitoring systems for PIT-tagged juvenile salmonids have been installed at three Columbia Basin dams that have collection–diversion facilities. The systems are positioned so that all the fish exiting the wet separator are passively interrogated for PIT tags. The prototype was installed at the wet separator at McNary Dam in 1985 (Prentice et al. 1986) and modified in 1986 (Figure 3) (Prentice et al. 1987). Subsequent systems were installed at Lower Granite Dam on the Snake River in 1986 (Figure 4) and at Little Goose Dam in 1987 (Figure 5).

---

[1]Reference to trade names does not imply endorsement by the National Marine Fisheries Service.

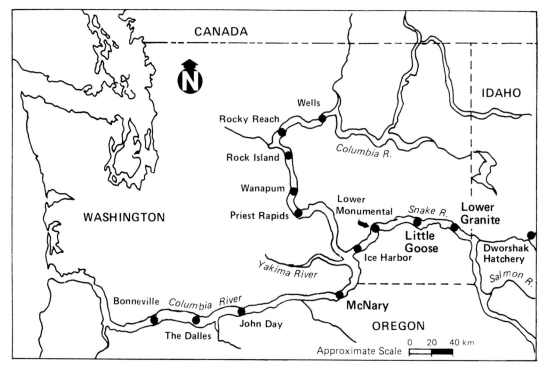

FIGURE 1.—Hydroelectric dams on the Columbia and Snake rivers. Those dams with PIT-tag monitoring systems appear in bold print.

In 1987, a prototype system for monitoring PIT tags in adult salmonids was installed at Lower Granite Dam at the entrance to an existing fish trap (Figure 6) (Prentice et al. 1987). All adult fish passing over this dam are trapped for biological sampling. Fish entering the trap pass over one of two false weirs, down a pipe 31 cm in diameter, through a coded wire tag (CWT) detector, and finally through a PIT-tag monitoring system.

We evaluated the efficiency and accuracy of all these systems by passing a set of known tags through their monitors at various times during the field season. Over 3 years, they detected more than 95% of the tags and correctly read more than 99% of the codes (Prentice et al. 1986, 1987). Two minor equipment problems that reduced efficiency and accuracy were corrected in 1988.

The PIT-tag monitoring systems consist of several components (Table 1) interconnected by shielded cable as shown schematically in Figure 7. A dual loop antenna assembly (DLAA) comprises a waterproof aluminum radio frequency (RF) shield housing, two transmitting and receiving loop antennas wrapped around a nonmetallic pipe or flume, and two loop tuners (LT). The number and size of DLAAs at each of the dams vary

(Table 1). The DLAAs were constructed by NMFS personnel according to the specifications of the manufacturer and were modified for each application and location.

For each loop antenna of the DLAA, the number of wire wraps varies with the cross-sectional area of the pipe or flume and functions as a 400-kHz exciter coil and tag sensor. The loop antennas are wrapped in opposite directions and energized with opposite polarities to reduce radiated RF signals generated by the system. One LT is attached to each loop antenna to tune it to the correct 400-kHz energizing signal and aid in reducing RF emissions. Each LT is connected to the dual exciter (DE), which energizes the loop antennas and receives and amplifies the returning signal from a tag. The DE consists of two independent circuit boards (one for each loop antenna), connectors for signal input and output, and a tuning system and meter to tune the DLAA to 400 kHz for maximum efficiency. The maximum operational distance between the loop tuner and the DE is 6.1 m. Power for the DE is supplied by a dual power supply (DPS) that converts 110-V AC power to a variable DC voltage. Each DPS independently powers one of two DE circuit boards.

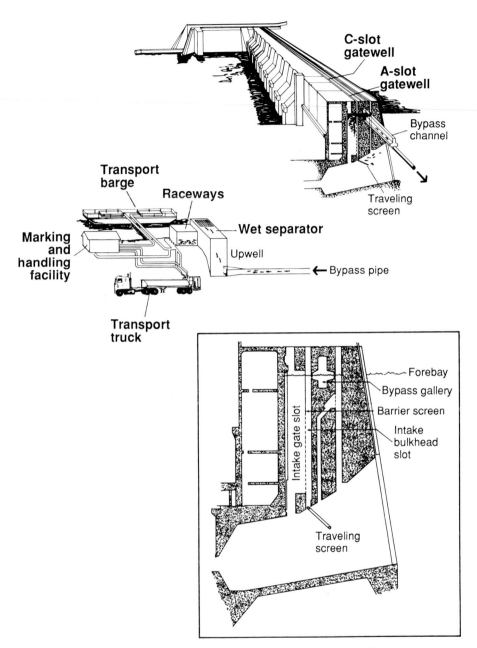

FIGURE 2.—Side view of a hydroelectric dam showing a fish collection–diversion system.

Power levels are controlled by switches within the DPS, and a dual filter between the DE and the DPS reduces RF signal interference.

A PIT tag is energized as it passes through the electromagnetic field of the loop antenna, which causes it to emit a coded low-frequency (40–50 kHz) signal. PIT-tag signals are amplified at the DE and are then sent to a standard (STD) bus controller for processing. The maximum distance between the DE and the controller is 61 m. One controller can process signals for as many as three DEs. During tag interrogation, the controller demodulates and decodes the amplified tag return signal from the DE. In addition to decoding the 10-digit tag code number, the controller produces a 2-digit check sum of the tag code (the code's

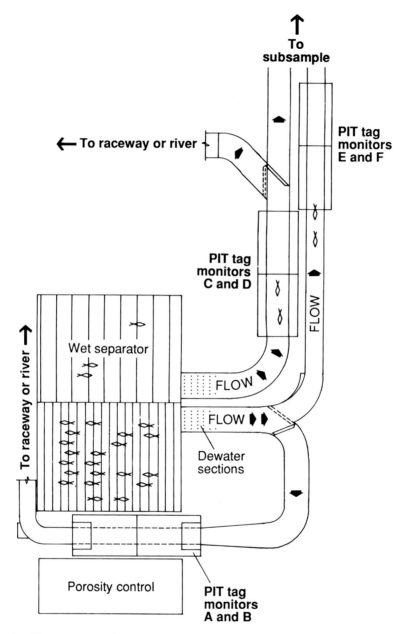

FIGURE 3.—Wet separator and PIT-tag monitoring system for juvenile salmonids at McNary Dam.

hexidecimal sum), a 2-digit origin code (loop antenna identification number), a 2-digit system code (controller number), and the date and time of day (hour, minute, and second). The time of day and date are generated hourly by the controller, even in the absence of PIT tags. All information is transferred independently from the controller via separate standard RS232 ports to a printer, and via a multiport to a computer compatible with a MicroSoft Disk Operational System (MS-DOS). Each DLAA, and its supporting electronics, can operate as an independent system to provide backup in the event of an electronic problem. The multiport controls the simultaneous transmission of information from one or more controllers to the computer. Furthermore, buffers within the con-

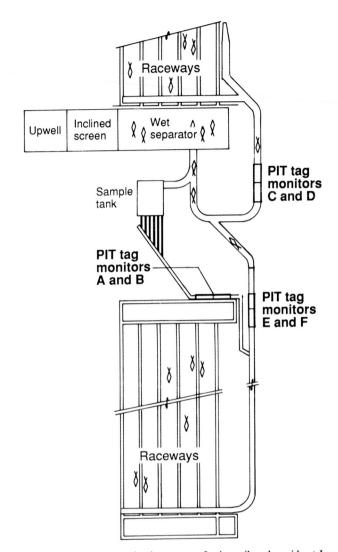

FIGURE 4.—Wet separator and PIT-tag monitoring system for juvenile salmonids at Lower Granite Dam.

troller, multiport, printer, and computer protect the system from becoming overloaded with information.

The system is designed to interrogate, decode, and process tag code information at rates in excess of one tag code per second (average), with peak rates of 10 codes/s for a maximum duration of 1 s. Signal interference can occur if two or more tags are present at the same time in the excitation field of the loop antenna. This situation may prevent either tag from being read. If a tag remains in the fringe reading range of the loop antenna for several seconds, an incorrect reading may occur. The DLAA, DE, LT, and dual filter are designed to operate in exposed conditions at temperatures of −20 to 50°C and at humidities of 0 to 100%. However, the controller, power supply, multiport, printer, and computer must operate in a protected environment.

Data Collection and Transfer

The computer in the PIT-tag monitoring system enables data to be stored in a specific format (ASC II) on electronic media and to be transferred via telephone lines. DoubleDos software allows concurrent operation of a PIT-tag monitoring program and a communication program (ProComm) that can send data to a central data-processing site. A program developed by the NMFS formats data received from the monitor controllers, and it

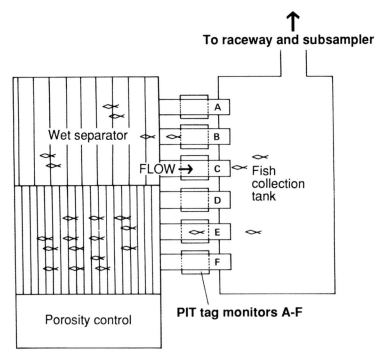

FIGURE 5.—Wet separator and PIT-tag monitoring system for juvenile salmonids at Little Goose Dam.

creates new files at 0000 hours every day. The title of the file and the time, date, and location of the monitoring system begin each entry. Hourly date–time stamps and tag-code information are added as tag codes come in.

At this time, PIT-tag monitoring sites at hydroelectric projects are queried daily for the previous day's files. The data-collection computer at the dam is accessed via telephone by a computer operator who transfers the files to a centralized computer in Seattle, Washington. A file from each of the monitoring sites is stored and edited for errors and system operation, and a processed file is generated. The processed file is available to users by 1200 hours on the day the file is received.

Hatchery Release Monitors

In some studies, there are waiting periods between tagging and release when tags are rejected or deaths occur. In such situations, it is important to identify the code of every PIT-tagged fish at the time of release so that losses during the waiting period are accounted for. Prentice et al. (1986, 1987) described a PIT-tag system for monitoring releases in hatchery raceways, which was tested at Dworshak National Fish Hatchery (DNFH) in 1986 (Figure 8). The monitoring systems at the dams and at hatcheries differed primarily in the size and number of their DLAAs and supporting electronic units (Table 1). At DNFH, there were four DLAAs, each consisting of a pipe 10.2 cm in diameter and 61.0 cm long. The four DLAAs were fitted to a raceway discharge so that all fish, tagged and nontagged, passed through the four DLAAs. The tag interrogation, decoding, and recording rate was about 20,000 fish/h (tagged and untagged combined) at a ratio of one tagged to four untagged fish. The tag-detection efficiency of this system was 93%, and the reading accuracy was over 99%.

Field Studies

Several of the PIT-tag monitoring systems for juvenile salmonids were evaluated in a series of field tests conducted in 1985 and 1986. We determined the tag-reading efficiency of monitors at Lower Granite and McNary dams for migrating yearling chinook salmon *Oncorhynchus tshawytscha*, underyearling chinook salmon, and steelhead *O. mykiss* (formerly *Salmo gairdneri*). In each test, PIT-tagged fish were released into a wet separator upstream from the tag monitors (Table 2). Tag-detection efficiency ranged from 95 to 100%, and tag-reading accuracy (correct code recognition) exceeded 99%. The monitoring equipment remained active up to 7 months with-

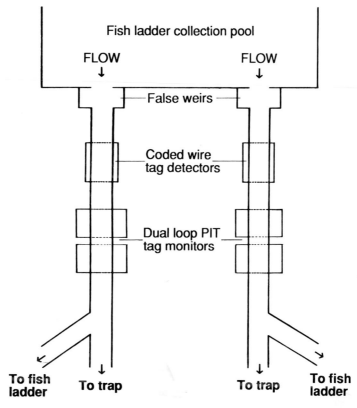

FIGURE 6.—Fish trap and PIT-tag monitoring system for adult salmonids at Lower Granite Dam.

TABLE 1.—Components required for PIT-tag systems used to monitor juvenile and adult salmonids at dams and hatchery raceways.[a]

Location	Monitor type	Size of DLAA (cm)	Number of DLAAs	Number of components required per DLAA				Number of components required per location			
				LT[b]	DE[b]	Filter[b]	DPS[b]	Con-troller[b]	Prin-ter	Multi-port[c]	Com-puter
McNary Dam	Juvenile	15 × 46 × 122	4	2	1	1	1	3	2	1	1
		15 × 31 × 122	2	2	1	1	1				
		15 dia × 22	1	2	1	1	1				
Little Goose Dam	Juvenile	10 dia × 61	6	2	1	1	1	2	2	1	1
Lower Granite Dam	Juvenile	15 × 46 × 122	4	2	1	1	1	2	2	1	1
		25 dia × 122	2	2	1	1	1				
Lower Granite Dam	Adult	31 dia × 122	4	2	1	1	1	2	2	1	1
DNFH	Juvenile	10 dia × 61	4	2	1	1	1	2	2	1	1
NMFS	Juvenile (pump system)	15 dia × 122	2	2	1	1	1	1	1	1	1

[a]Abbreviations used: DLAA = dual loop antenna assembly; LT = loop tuner; DE = dual exciter; DPS = dual power supply; dia = diameter; NMFS = National Marine Fisheries Service; DNFH = Dworshak National Fish Hatchery.

[b]Model numbers for PIT-tag monitoring equipment from Destron-Identification Devices, Inc.: LT = 800-0069-01; DE = 800-0026-00; DPS = 800-0027-00; Filter = 761-0050-00; Controller = 800-0028-00.

[c]Model number and source of multiport: Multiport model 528-H from Bay Technical Associates, Bay Saint Louis, Mississippi.

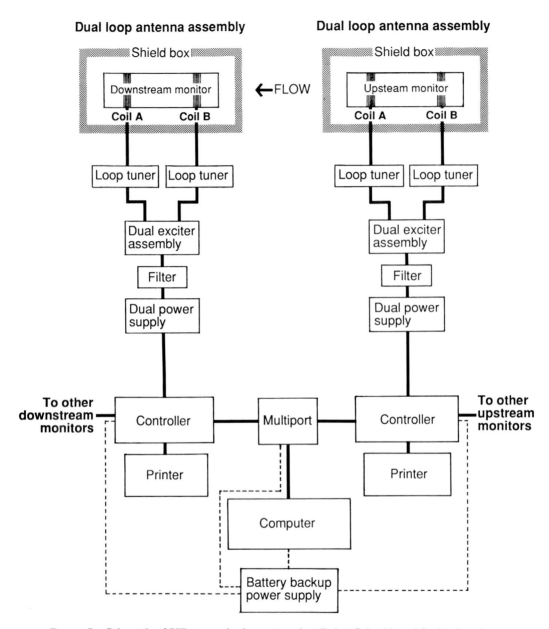

FIGURE 7.—Schematic of PIT-tag monitoring systems installed at Columbia and Snake river dams.

out major problems and proved to be reliable under field conditions.

To further evaluate the PIT-tag system, we compared it with freeze branding, a traditional marking method for juvenile salmonids. The migrations of juvenile salmonids in the Columbia Basin have been studied annually since 1964 (Raymond 1974). Usually, groups of fish are marked (either at the hatchery or in-river), released, and then sampled at collector dams—e.g., McNary,

Little Goose, and Lower Granite. Freeze branding has been the traditional method used to identify these groups of fish (Park and Ebel 1974). At the dams, freeze-brand and PIT-tag data are acquired in fundamentally different ways. The PIT-tag detectors are deployed to interrogate all fish in the migrant bypass system. To obtain freeze-brand data, a subsample from the bypass population is examined for marks, which are then used in extrapolations to estimate the number of a partic-

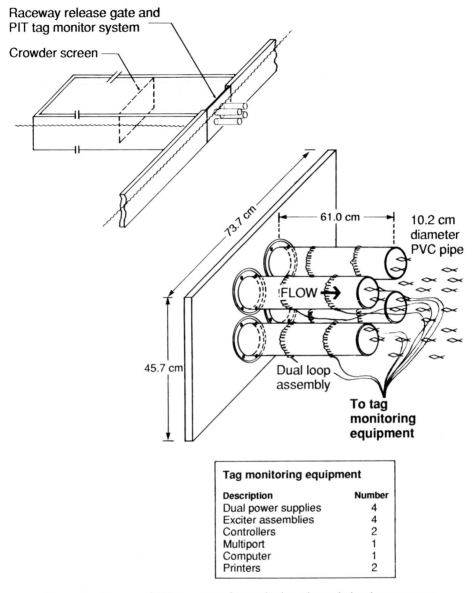

Raceway release gate and PIT tag monitor system

Crowder screen

73.7 cm

61.0 cm

10.2 cm diameter PVC pipe

FLOW →

45.7 cm

Dual loop assembly

To tag monitoring equipment

Tag monitoring equipment	
Description	**Number**
Dual power supplies	4
Exciter assemblies	4
Controllers	2
Multiport	1
Computer	1
Printers	2

FIGURE 8.—Diagram of PIT-tag system for monitoring releases in hatchery raceways.

ular marked group in the entire bypass system (Giorgi and Sims 1987). The methods also differ notably in the time required for data recovery. Detection of PIT tags are known to the second, whereas brands are pooled over a 24-h period and processed once a day.

Another drawback of the freeze-brand method is the amount of physical handling of many unmarked as well as marked individuals required to gather data. Because branded fish make up only a small portion of the outmigrants, hundreds of thousands of salmonids must be handled each year at the collector dams to obtain freeze-brand code information. The PIT-tag system alleviates this added stress on migrant salmonids.

For certain studies, the use of PIT tags in lieu of brands has the potential to produce statistically and biologically comparable results with a 90 to 95% reduction in the number of fish treated. In 1985 and 1986, we compared the collection ratios of freeze-branded and PIT-tagged chinook salmon and steelhead. The test groups were released into

TABLE 2.—Results of field tests that measured the ability of monitors to detect PIT tags in juvenile chinook salmon and steelhead at hydroelectric dams.

Year	Fish	Number of fish released	Tags detected (%)
	Lower Granite Dam		
1986	Yearling chinook salmon	340	98.5
1986	Steelhead	480	98.1
Subtotal		820	98.3
	McNary Dam		
1985	Yearling chinook salmon	584	97.9
1985	Age–0 chinook salmon	260	95.4
1986	Yearling chinook salmon	480	96.5
1986	Steelhead	480	96.0
1986	Age-0 chinook salmon	480	99.0
Subtotal		2,284	97.2
	Both dams		
Total		3,104	97.5

the reservoir of McNary Dam or released from Dworshak National Fish Hatchery and monitored at the juvenile fish collection facilities at McNary and Lower Granite dams (Table 3). Detection rates for PIT tags were as high as or higher than those for freeze-branded fish. Generally, PIT-tagged fish at McNary Dam were recovered in a

greater proportion than their freeze-branded counterparts (Table 3). Also, over a 5-d period, recoveries of serial releases of PIT-tagged and freeze-branded chinook salmon smolts at McNary Dam indicated that the PIT tag provided data with less statistical variation (Table 3).

The discrepancy in recovery data between PIT-tagged and branded fish suggests a bias may be associated with the recovery process. It may be an anomaly of the sampling mechanism or of the brand reading and transcription process. Personnel of the NMFS are conducting research to identify the source of this error.

Future PIT-Tag Monitoring Systems

We are now evaluating a PIT-tag monitoring system that processes fish as they are pumped from fish hatchery raceways to transport trucks or barges. The system consists of two DLAAs, each 15 cm in diameter and 152 cm long, attached to the intake of a fish pump (Figure 9). The electronic components of the system are the same as for the PIT-tag monitoring systems previously described. Tag interrogation, decoding, and recording rate are being evaluated for different pumping rates and ratios of tagged to untagged fish.

A disadvantage of the PIT-tag monitoring systems is its range of detection, which is limited to a radius of about 18 cm. Future efforts will be directed at increasing this range. With an expanded detection system, it would be possible to

TABLE 3.—Detection of PIT-tagged and freeze-branded chinook salmon and steelhead released into the Columbia River system in 1985 and 1986.

Species (year)	Treatment	Number released	Number of groups	Tag detection[a]			
				Lower Granite Dam		McNary Dam	
				Observed	%	Observed	% (SD)
		Releases from Dworshak Hatchery					
Yearling chinook salmon (1986)	Branded	40,675	1	4,659	11.5	3,402	8.9
	PIT-tagged	2,450	1	464	18.9	264	10.8
Steelhead (1986)	Branded	35,025	1	7,061	20.2	389	1.1
	PIT-tagged	2,424	1	928	38.1	45	1.8
		Releases into McNary Reservoir					
Age-0 chinook salmon (1985)	Branded	4,400	5			758	19.0 (9.0)
	PIT-tagged	400	5			64	16.0 (4.0)
Age-0 chinook salmon (1986)	Branded	5,000	5			1,371	27.4 (3.7)
	PIT-tagged	500	5			142	28.4 (1.6)
Yearling chinook salmon (1986)	Branded	5,000	5			2,101	39.6 (9.9)
	PIT-tagged	500	5			318	63.6 (2.5)

[a]Detection of freeze-branded fish is based on actual number of fish observed expanded by the prevailing sample rate, whereas the detection of PIT-tagged fish is based upon actual number of fish observed.

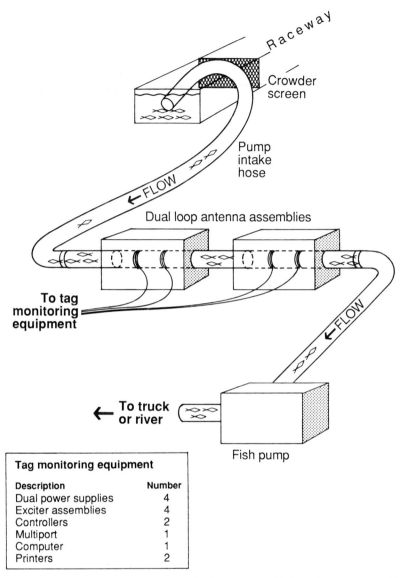

Tag monitoring equipment	
Description	**Number**
Dual power supplies	4
Exciter assemblies	4
Controllers	2
Multiport	1
Computer	1
Printers	2

FIGURE 9.—Diagram of experimental system for monitoring PIT-tagged fish as they are pumped from raceways to release or transport points.

interrogate all the adult salmonids that migrate through fish ladders at hydroelectric dams.

Acknowledgments

We thank the Bonneville Power Administration for funding this project and the U.S. Army Corps of Engineers for allowing us to install and evaluate the PIT-tag monitoring systems at their facilities. We also thank Richard Frazier, Phillip Weitz, and their NMFS staffs for engineering services and installation of the PIT-tag monitoring systems.

Reference

Giorgi, A. E., and C. Sims. 1987. Estimating the daily passage of juvenile salmonids at McNary Dam on the Columbia River. North American Journal of Fisheries Management 7:215–222.

Park, D. L., and W. J. Ebel. 1974. Marking fishes and invertebrates. II. Brand size and configuration in relation to long-term retention on steelhead trout and chinook salmon. U.S. National Marine Fisheries Service Marine Fisheries Review 36(7):16.

Prentice, E. P., T. A. Flagg, and C. S. McCutcheon. 1987. A study to determine the biological feasibility of a new fish tagging system. Report (contract DE-A179-83BP11982, project 83-19) to Bonneville Power Administration, Portland, Oregon.

Prentice, E. P., T. A. Flagg, C. S. McCutcheon, D. S. Brastow, and D. C. Cross. 1990a. Equipment, methods, and an automated data-entry station for PIT tagging. American Fisheries Society Symposium 7:335–340.

Prentice, E. P., T. A. Flagg, and C. S. McCutcheon. 1990b. Feasibility of using implantable passive integrated (PIT) tags in salmonids. American Fisheries Society Symposium 7:317–322.

Prentice, E. P., D. L. Park, T. A. Flagg, and C. S. McCutcheon. 1986. A study to determine the biological feasibility of a new fish tagging system. Report (contract DE-A179-83BP11982, project 83-19) to Bonneville Power Administration, Portland, Oregon.

Prentice, E. P., D. L. Park, and C. W. Sims. 1984. A study to determine the biological feasibility of a new fish tagging system. Report (contract DE-A179-83BP11982, project 83-19) to Bonneville Power Administration, Portland, Oregon.

Prentice, E. P., C. W. Sims, and D. L. Park. 1985. A study to determine the biological feasibility of a new fish tagging system. Report (contract DE-A179-83BP11982, project 83-19) to Bonneville Power Administration, Portland, Oregon.

Raymond, H. L. 1974. Marking fishes and invertebrates. I. State of the art of fish branding. U.S. National Marine Fisheries Service Marine Fisheries Review 36(7):1–9.

American Fisheries Society Symposium 7:335–340, 1990

Equipment, Methods, and an Automated Data-Entry Station for PIT Tagging

EARL F. PRENTICE, THOMAS A. FLAGG, CLINTON S. MCCUTCHEON,
DAVID F. BRASTOW, AND DAVID C. CROSS

Northwest Fisheries Center, National Marine Fisheries Service
2725 Montlake Boulevard East, Seattle, Washington 98112, USA

Abstract.—A miniaturized PIT (passive integrated transponder) tag, developed by private industry, was adapted for tagging juvenile salmonids by the U.S. National Marine Fisheries Service. The PIT tag enables individual fish to be identified by a unique 10-digit alphanumeric code read in situ with a detector–decoding system. The body cavity proved to be a satisfactory site for tag implantation, and techniques were tested that gave up to 100% tag retention and high fish survival. Hand-held and stationary automatic tag injectors were developed that allowed tagging of 150–300 fish/h. A computer-based system was designed to automatically record PIT-tag code and associated information, such as fish length and weight. A discussion of tagging technique and required equipment is presented.

The National Marine Fisheries Service (NMFS) has evaluated the technical and biological feasibility of tagging juvenile and adult salmonids with a miniaturized passive integrated transponder (PIT). This PIT tag, developed by Destron-Identification Devices, Inc. (D-IDI)[1,2], consists of an integrated microchip bonded to an antenna coil. The electronic components of the tag are encapsulated in a glass tube about 12 mm long × 2 mm in diameter. Each tag is programed at the factory with one of about 34×10^9 unique (10-digit alphanumeric) code combinations (e.g., 7F7E2136A1). Having no power of its own, the tag is energized by a 400-kHz external signal that enables the tag to transmit a unique 40–50-kHz signal to the interrogation equipment, where the code is immediately processed (decoded), displayed, and (optionally) stored on a computer. The PIT tag system allows for passive (in situ) collection of a tag code from an individual fish without handling the fish.

The body cavity was found to be a satisfactory implant site for the PIT tag, and no detrimental effects of the tag on salmonids (*Oncorhynchus* spp. and *Salmo* spp.) have been observed (Prentice et al. 1984, 1985, 1986, 1987). In 1986, fish-handling and PIT-tagging guidelines were developed for juvenile and adult salmonids, and a computer-based system was developed to automatically record tag code, fish length, and fish

weight. Discussions and descriptions of the tagging equipment, tagging technique, and data-recording equipment are presented in this paper. Additional information on the biological evaluation of the PIT tag, monitoring systems, and uses in migration studies are presented elsewhere in this volume (Prentice et al. 1990a, 1990b).

Tagging Equipment

Two tag-injection systems have been developed: a manual (hand-held) tag injector and a stationary semiautomated system for rapidly tagging large numbers of fish (Prentice et al. 1986, 1987). The tag is implanted with a 12-gauge hypodermic needle in both systems.

The hand-held system consists of a modified plastic syringe that may vary in size from 5 to 10 cm³ depending on the operator's preference (Figure 1). A hole drilled in the end of the barrel prevents the operator from injecting air into the fish during tagging. A rod attached to the end of the syringe plunger pushes the PIT tag through the needle as the plunger is depressed. This rod, when fully extended, should reach only to the end of the needle, which should be at least 25 mm long. Each tag must be manually inserted into the needle. This procedure is satisfactory for small numbers of fish but becomes ineffective as numbers increase.

We developed a semiautomated injection system for rapidly tagging fish. This bench-mounted unit operates entirely on bottled compressed gas (e.g., carbon dioxide) and incorporates a gas-activated ramrod to push the PIT tag through the 12-gauge needle (Figure 2). The PIT tags for use in

---

[1]Reference to trade names does not imply endorsement by the National Marine Fisheries Service.

[2]Destron-Identification Devices, Inc., Boulder, Colorado.

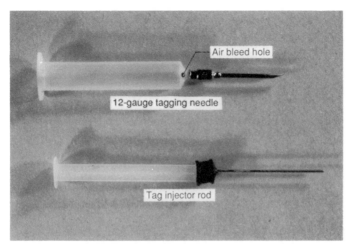

FIGURE 1.—Hand-held PIT-tagging syringe developed by the National Marine Fisheries Service.

the injector are contained in a removable clip that allows them to gravity-feed into the breech of the tagging machine. Each clip is loaded with about 150 tags. A sliding weight at the top of the clip helps deliver tags to the breech of the injector. A foot-operated switch activates the gas-ram. After injecting the tag, the plunger retracts, which allows a new tag to drop into position for the next implantation. Adjustments to the speed of the gas-ram operation and length of gas-ram exten-

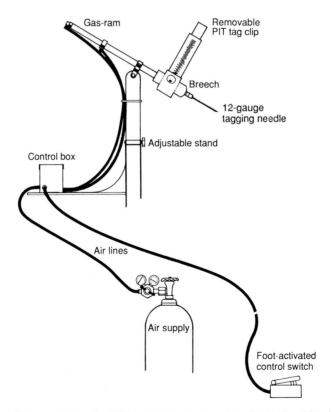

FIGURE 2.—Diagram of the semiautomatic PIT tag injector developed by the National Marine Fisheries Service.

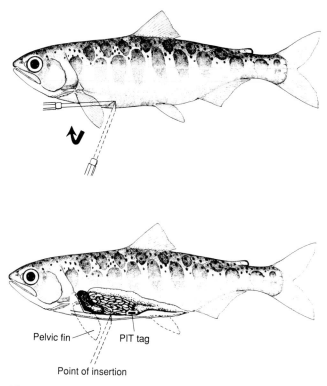

FIGURE 3.—Illustrated implantation of PIT tags, showing point of insertion through the body wall musculature, needle injection angle to implantation site, and position of the implanted PIT tag within body cavity.

sion can be made. The tagging rate with the semiautomated injector—over 300 fish/h—is more than double that of the hand-held unit.

Tagging Methods

The following methods were developed for inserting the PIT tag into the body cavity of juvenile and adult salmonids. The needle is inserted anteroventrally and the tag implanted posterior to the pyloric caeca in the area of the pelvic girdle. The following fish-handling and PIT-tagging guidelines have been developed: the fish should be in good health with no signs of disease; feeding should be suspended 2 d prior to tagging; all fish should be anesthetized for tagging; and fish should be fed a post-tagging maintenance ration for 3 d so the gut does not expand and force the tag through the unhealed needle wound.

The point of needle entry for PIT tagging depends on fish size. For salmonids less than 200 g, we recommend a point just posterior to the pectoral fins alongside the midventral line (Figure 3). On larger fish, the point should be anterior to the pelvic girdle, again just off the midventral line. For all fish, the bevel of the needle

should face away from the body, and the needle should be held at a 20–45° angle, depending on fish size (less angle for smaller fish). Needle pressure should be just enough to penetrate the body wall. Once the needle passes through the body wall musculature, the syringe angle should be decreased until the barrel parallels the body wall. The needle is then inserted farther until its point is posterior to the pyloric caeca near the pelvic girdle, and the tag is implanted. Correctly implanted PIT tags have up to 100% retention and little measurable effect on fish survival (Prentice et al. 1986, 1987).

Tagging equipment is disinfected with 60–90% ethanol periodically during tagging and when moved from site to site. The semiautomatic injector is cleaned after 300–400 taggings by running an ethanol-soaked pipe cleaner through the needle and breech to remove accumulated fish mucus and scales. Care should be taken to keep the tag clip dry; otherwise, surface tension will interfere with tag delivery.

No appreciable host tissue response or infection resulting from tagging procedures has been observed in salmonids. Tagging wound condition,

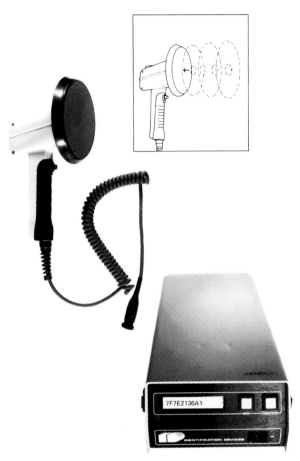

FIGURE 4.—Portable (battery-powered) PIT tag detector–decoder; the detector receives pulses from a tag, which the decoder displays in alphanumeric code.

tag placement within the body cavity, and histological effects on the tag have been examined for groups of juvenile fall chinook salmon *O. tshawytscha* and sockeye salmon *O. nerka* up to 19 cm long (Prentice et al. 1986, 1987). The needle puncture wound appeared to close almost immediately, and as early as 2 weeks after tagging, little visible evidence of external trauma remained at the injection site. Most of the tags stayed in place posterior to the pyloric caeca. Normal scar tissue usually replaced the dermis and underlying muscle tissue within 3 weeks. Complete healing usually occurred 4–6 weeks after tagging.

Computer-Interfaced PIT-Tagging System

After tagging, tag presence and code are verified with a detector–decoding system. The system can be portable, consisting of a battery-powered, hand-held scanner, or it can be interfaced with a computer. The portable detector–decoder[3] displays the tag number (code) on a liquid crystal display (LCD) screen and can store over 1,000 code combinations for subsequent retrieval via a computer (Figure 4). The portable unit is useful for field applications when tagging and interrogating small numbers of fish. A stationary, computer-interfaced system is used when tagging larger numbers of fish.

An integrated system for PIT-tagging fish and recording tag code, fish length and weight, and written comments was assembled and tested under field conditions (Prentice et al. 1987). This computerized system makes it possible to maintain individual records for large numbers of fish. It

---

[3]Model numbers for Destron-Identification Devices, Inc., equipment: D-IDI Loop Detector FS-5102; D-IDI PIT tag Decoder/Detector HS-5101.

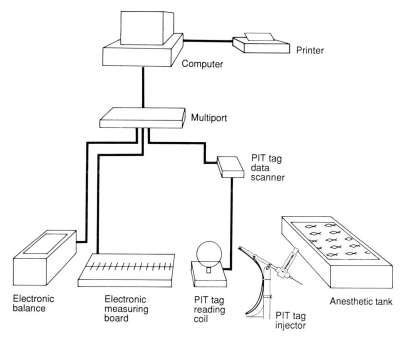

FIGURE 5.—Diagram of a PIT-tagging station linked to a computer.

consists of several commercially available components (Figure 5). A portable PIT tag decoder is connected to a tabletop loop detector measuring 150 × 150 mm. This detector system is used to interrogate, decode, and transmit the tag code through a multiport[4] to a computer compatible with MicroSoft-Disk Operation System (MS-DOS) and a printer for storage. Length (±1 mm) and comments (custom configured) are documented with a digitizing board[5] and weight (±0.2 g) via an electronic balance[6]. Information from the digitizing board also is routed through the multiport to the computer and printer.

An important part of the system is the computer program that controls information flow to and from the computer. The program[7] is written in Turbo Pascal, and the computer files are in ASCII (text) format. These programs are menu-driven and allow custom configuration that can be accessed at any time during program operation (e.g.,

length or weight may be optional or mandatory selections). In addition, a menu selection is available to allow length, weight, and additional comments to be recorded for fish without PIT tags.

The procedure for using the system requires several steps. First a fish is removed from an anesthetic bath and injected with a PIT tag. The fish is then manually passed through the tag-detection loop. The tag code appears on the computer screen (all information displayed on the computer screen is in an expanded format for ease of reading), an audible tone is emitted by the data scanner, and a light appears on the scanning loop. The operator then places the fish on a protective plexiglass cover fitted over the digitizing board, with the fish's head positioned against a stop that acts as a zero reference point. An electronic stylus is activated at the point where the length measurement is to be taken. Length information is displayed under the PIT-tag code on the computer screen. Next, the fish may be weighed on the electronic balance, placed in a recovery tank, or returned to the rearing area. Weight information appears on the computer screen beneath the tag code and the length information. Comments may be keyed into the file at any time via the digitizer board or computer. All information is entered automatically into the computer, and a printed hard copy is made when the next PIT-tagged fish

[4]Model 525-H from Bay Technical Associates, Bay Saint Louis, Mississippi.

[5]Model 23120-9 with kit 23064-03 from CalComp Corporation, Anaheim, California.

[6]Model FY3000 from A & D Engineering Inc., Milpitas, California.

[7]The program was written by David Brastow of the Northwest Fisheries Center, National Marine Fisheries Service, Seattle, Washington.

transmits its tag code. The tagging and documentation rate for this system, when used with an automatic PIT-tag injector, is in excess of 300 fish/h.

Summary

PIT tags last throughout the life cycle of their hosts. They can be detected and decoded in living fish in fresh and salt water, and they eliminate the need to anesthetize, handle, restrain, or kill fish during data retrieval. Used with computer stations, they allow repeated identification and measurements of individuals within a population. The automated weighing and measuring station can be run with or independent of PIT-tag analyses and should have broad application in fisheries management.

Acknowledgments

We thank the Bonneville Power Administration for funding this project (83-319).

References

Prentice, E. P., T. A. Flagg, and C. S. McCutcheon. 1987. A study to determine the biological feasibility of a new fish tagging system. Report (contract DE-A179-83BP11982, project 83-19) to Bonneville Power Administration, Portland, Oregon.

Prentice, E. P., T. A. Flagg, and C. S. McCutcheon. 1990a. Feasibility of using implantable passive integrated (PIT) tags in salmonids. American Fisheries Society Symposium 7:317–322.

Prentice, E. P., T. A. Flagg, C. S. McCutcheon, and D. F. Brastow. 1990b. PIT tag monitoring systems for hydroelectric dams and fish hatcheries. American Fisheries Society Symposium 7:323–334.

Prentice, E. P., D. L. Park, T. A. Flagg, and C. S. McCutcheon. 1986. A study to determine the biological feasibility of a new fish tagging system. Report (contract DE-A179-83BP11982, project 83-19) to Bonneville Power Administration, Portland, Oregon.

Prentice, E. P., D. L. Park, and C. W. Sims. 1984. A study to determine the biological feasibility of a new fish tagging system. Report (contract DE-A179-83BP11982, project 83-19) to Bonneville Power Administration, Portland, Oregon.

Prentice, E. P., C. W. Sims, and D. L. Park. 1985. A study to determine the biological feasibility of a new fish tagging system. Report (contract DE-A179-83BP11982, project 83-19) to Bonneville Power Administration, Portland, Oregon.

American Fisheries Society Symposium 7:341–345, 1990
© Copyright by the American Fisheries Society 1990

Use of PIT Tags to Individually Identify Striped Bass and Red Drum Brood Stocks

WALLACE E. JENKINS AND THEODORE I. J. SMITH

South Carolina Wildlife and Marine Resources Department
Post Office Box 12559, Charleston, South Carolina 29412, USA

Abstract.—Captive brood fish of striped bass *Morone saxatilis* and red drum *Sciaenops ocellatus* were implanted with passive integrated transponder (PIT) tags for long-term individual identification and monitoring of growth, survival, and spawning success. Tags were implanted in both species by intramuscular injection posterior to the dorsal fin. Striped bass brood fish ranged from 1.6 to 6.6 kg and red drums from 4.2 to 11.7 kg at tagging. Injection wounds on all fish healed quickly, and no behavioral changes or health problems were traced to tagging. Both groups of fish continued to grow and spawn after tagging and no deaths were caused by the procedure. Gross dissections of fish that died as a result of hatchery operations revealed little tissue trauma associated with the tag. Tags implanted in striped bass in September 1985 could still be read after 2 years. Tag retention rate was 97% for striped bass after 2 years and 100% for red drums after 10.5 months. This study indicates that PIT tags can be useful in individually identifying captive brood fish.

In recent years, interest has grown in commercially raising striped bass *Morone saxatilis* and red drums *Sciaenops ocellatus* for food. At present, wild brood stocks are used to support hatchery operations, which means the source is unpredictable as well as controversial. Eventually, aquaculturists must develop domesticated brood fish to meet the needs of private hatcheries.

Controlled spawning of captive striped bass and red drum has been achieved (Roberts et al. 1978; Arnold et al. 1979; Smith and Jenkins 1988; Henderson-Arzapalo and Colura 1987; Smith 1988). Rearing success remains low, however, because of diseases, poor water quality, equipment failure, and human error. Hatchery managers are especially troubled by the lack of a reliable technique for identifying individual brood fish. Individual identification is valuable because it allows managers to compile breeding histories and select particular spawners. It promotes efficient use of rearing facilities, because mixed populations can be held together. And it lessens stress on brood fish by reducing the times they are handled during hatchery work.

In the past, individual brood fish have been identified by means of colored threads placed in the dorsal fin, various forms of mutilation (e.g., branding, fin clipping), and external tags (Bayless 1972; Wydoski and Emery 1983). Each of these methods has advantages and disadvantages, yet none is well suited for long-term identification of captive brood fish. Work by Prentice et al. (1984) has demonstrated that passive integrated transponder (PIT) tags can be used to monitor long-term survival of hatchery-produced salmon released in the wild. Our work evaluated the use of PIT tags for long-term individual identification of captive striped bass and red drum brood stock.

Methods

Thirty-nine striped bass were used in the study. These fish were progeny of wild brood stock and had been reared in earthen ponds for about 1 month. In 1982 and 1983, the small juveniles, averaging 25 mm in total length (TL), were received at the South Carolina Wildlife and Marine Resources Department (SCWMRD), Marine Resources Research Institute, for use in development of cultured brood stock. The 1982 fish ($N = 25$) were reared in two cylindrical tanks, each 3.7 m in diameter and 0.9 m deep, in an indoor controlled environment chamber where photoperiod and temperature were manipulated (Smith and Jenkins 1988). Water in these indoor tanks was recirculated after passage through a biological filter. The 1983 fish ($N = 14$) were reared outdoors in a cylindrical fiberglass tank 6.1 m in diameter and 1.5 m deep. This tank also received recirculated water processed through a biological filter.

In contrast to the striped bass, the red drums were captured as adults by anglers who donated the fish to SCWMRD for use in spawning and stock-enhancement experiments. The red drums ($N = 21$) were kept in three 3.7-m-diameter × 0.9-m-deep cylindrical fiberglass tanks in an environmental control room in July 1987. These tanks received recirculated salt water (29–32‰ salinity) processed through a biological filter. To expedite spawning, the photoperiod and temperature were manipulated as described by Roberts et al. (1978) to simulate an annual cycle in only 150 d. Spawning was induced at a photoperiod of 12 h light and a temperature of 27°C.

Water quality variables were monitored weekly in all systems and maintained within acceptable limits by the addition of new water as necessary. During the study, fish were fed once to several times daily depending on their size and condition, and on water temperature. Striped bass were fed 38%-protein, pelleted trout rations and chopped squid, supplemented with live topminnows *Fundulus* spp. in the months prior to and after spawning. The wild red drums were fed a variety of chopped fish, including striped mullet *Mugil cephalus,* Atlantic menhaden *Brevoortia tyrannus,* spot *Leiostomus xanthurus,* and Atlantic croaker *Micropogonias undulatus.*

Tagging method and location.—Individual fish of each species received a PIT tag (Identification Devices, Inc., Westminster, Colorado). The PIT tag is an electronic tag, approximately 10 mm long × 2.1 mm in diameter, that can be coded with one of 35×10^9 unique codes. The tag can be automatically detected and decoded in situ, eliminating the need to kill fish during data retrieval (Prentice et al. 1984).

Prior to tagging, fish were anesthetized in a bath containing 100 mg of MS-222 (tricaine) per liter of culture water. Each fish was then weighed to the nearest gram and its total length measured to the nearest millimeter. The sex was recorded for mature fish. Betadine was applied to the injection site for the first group of striped bass, but subsequently this procedure was discontinued. The tag was implanted immediately posterior to the dorsal fin by intramuscular injection with a 12-gauge hypodermic needle. The tag was injected at approximately a 45° anterior angle (Figure 1) to a depth of 18 mm. This site was chosen because the scales are small and easily penetrated, the tag is not apt to migrate once imbedded within the musculature, and the risk of damage to internal organs and reproductive capacity is slight. Tags were read with a hand-held detector (Handwand Reader model 3100, Identification Devices) before and immediately after implantation to verify that each tag was functional.

Individuals from the 1982 group of striped bass were tagged on four occasions (Table 1). Four

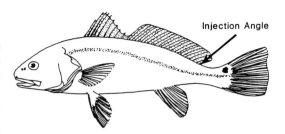

FIGURE 1.—Schematic diagram of a red drum showing site and angle of injection of PIT tags.

striped bass were randomly selected and tagged in September 1985. They were placed in tanks with untagged fish ($N = 21$) and observed for 6 months for differences in behavior, growth, and spawning success. After this 6-month period, the control group plus the 1983 group of striped bass were all PIT-tagged during 5 months in 1986.

Eighteen of the red drums were tagged in July 1987. Eight tagged fish were placed in tank 1 and 10 in tank 2. Three untagged fish were used as a control and placed in tank 3. These three fish were subsequently tagged in November 1987.

Striped bass received polyethylene-encased tags marketed prior to 1986, whereas red drums received more recent models encased in glass (model IX 0007A, BioSonics, Seattle, Washington). Striped bass ranged from 1.6 to 6.6 kg at time of tagging (Table 1). Red drums were larger, ranging from 4.2 to 11.7 kg.

Results

Behavior

Visual observations indicated that tagged and untagged fish behaved the same. Typically, fish handled during sampling did not feed for several days, whether they were tagged or not; both groups resumed normal feeding at about the same time. Visibility in the culture tanks was excellent, yet tagged and untagged fish could not be distinguished, either immediately after tagging or during long-term culture. During sampling sessions, tagged fish could only be identified by using the electronic tag reader.

Survival

The four striped bass tagged in 1985 all survived for 6 months. One female in this group died after spawning in February 1986, but this death can be attributed to stress from repeated handling and strip-spawning, because 5 of the 21 untagged

TABLE 1.—Tagging and growth data for striped bass and red drum brood stock implanted with PIT tags.

	Tagging data			Sample data			
Date	Number of fish	Mean weight (range) of fish (kg)	Mean total length (range) (mm)	Date	Mean weight (range) (kg)	Tag retention rate (%)	Duration (months)
			Striped bass				
Sep 1985[a]	4	3.10 (2.55–3.65)	618 (590–650)	Apr 1987	5.67 (4.54–6.99)	100	19[b]
Feb 1986[a]	5	4.14 (2.55–5.01)	655 (561–724)	Apr 1987	5.43 (4.71–6.12)	80	14
Mar 1986[a]	8	3.99 (2.67–5.21)	661 (600–725)	Apr 1987	6.25 (4.94–6.62)	100	13[c]
Jul 1986[a]	8	4.99 (3.77–6.66)	708 (660–788)	Apr 1987	6.86 (5.21–8.89)	100	9[d]
Jul 1986[e]	14	2.07 (1.66–2.96)	524 (476–614)	Apr 1987	3.11 (2.18–4.31)	100	9[d]
			Red drum				
Jul 1987	8 (tank 1)	5.90 (4.65–8.85)	846 (795–964)	Jan 1988	8.30 (6.93–10.23)	100	6[f]
Jul 1987	10 (tank 2)	6.03 (4.21–11.68)	813 (710–1,055)	Jan 1988	8.52 (6.31–13.29)	100	6[f]
Nov 1987	3 (tank 3)	9.17 (8.41–9.49)	935 (920–960)	Jan 1988	8.90 (8.18–9.32)	100	3[g]

[a] Fish obtained in 1982.
[b] Tags still detected in remaining fish in December 1987, 27.5 months after tagging.
[c] Tags still detected in remaining fish in April 1988, 25 months after tagging.
[d] Tags still detected in remaining fish in April 1988, 21 months after tagging.
[e] Fish obtained in 1983.
[f] Tags still detected in fish in May 1988, 10.5 months after tagging.
[g] Tags still detected in fish in May 1988, 7 months after tagging.

striped bass handled during 1986 spawning season also died. All other striped bass survived to the end of the study in April 1987.

Red drums were induced to spawn in tanks by manipulation of temperature and photoperiod, so they were handled very little during hatchery operations. After being cultured for 11 months, and induced to spawn repeatedly, their survival was 100%.

No mortalities could be attributed to the tagging procedure or to the tag. Striped bass that died during hatchery operations were dissected. No adverse tissue response was noted in the one tagged fish that died.

Growth

Five groups of striped bass were tagged from September 1985 to July 1986 and monitored until April 1987 (Table 1). All fish grew throughout the observation period. Preliminary data indicate that growth of tagged and untagged fish was similar. From September 1985 through March 1986, the untagged controls increased from a mean weight of 3.1 to 4.6 kg, while the mean weight of two tagged fish increased from 3.5 to 4.7 kg.

Three groups of red drums were tagged between July and November of 1987. As with the striped bass, the tagged and untagged red drums appeared to grow similarly during the first half of the observation period (Figure 2). However, given the preliminary nature of the data and the small sample sizes, it is not possible to definitely compare growth rates.

Spawning Success

Spawning success among striped bass is highly variable in hatcheries when naturally ripe fish are collected from the wild. In the laboratory, spawning success with cultured brood stock can be influenced by environmental conditions, diet, and the experience of hatchery personnel in determining time of ovulation. The cultured striped bass in this study received a diet that may not have been ideal for maturation, and the techniques employed to hasten maturation were experimental. Consequently, the results of our spawning trials are even more variable than those encountered among wild fish. In 1986, 6 of 11 females were in a mature condition, but only 4 fish spawned. Three of the spawning fish were untagged, while one was tagged. The tagged fish produced 112,401 eggs per kilogram of body weight, while the three untagged fish produced a mean of 115,284 eggs/kg. In 1986, all males, tagged and untagged, were in running-ripe condition.

From days 52 to 137, the red drums in tanks 1 and 3 were spawning. From days 52 to 112, the six tagged red drum females in tank 1 produced 62.1 million eggs, while the two untagged females in tank 3 produced 17.6 million eggs. These egg-production levels are equivalent to 20,000 and 15,000 eggs per kilogram of body weight per day

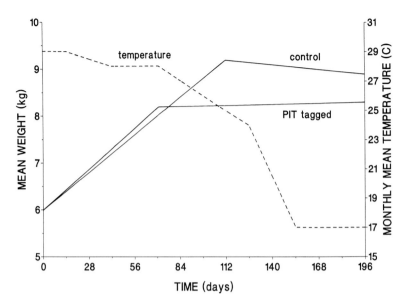

FIGURE 2.—Growth of 10 red drums PIT-tagged on July 13, 1987 (day 0), and of 3 untagged (control) fish in a controlled environmental system. The control fish were tagged on day 112. Broken line indicates water temperature (°C).

for the tagged and untagged females, respectively. On day 112, the red drums in tank 3 were weighed and tagged. The newly tagged fish continued to spawn until day 137, during which time they produced 6.9 million eggs (14,400 eggs per kilogram per day), compared to 8.6 million eggs (6,800 eggs per kilogram per day) for the tagged fish in tank 1. The continuation of spawning in spite of handling and tagging on day 112 indicates that neither the tagging procedure nor the PIT tags impaired reproductive ability of the red drums.

The tagged fish in tank 2 did not spawn, possibly because only one small male (6.3 kg) was in the tank. Red drum females respond to a spawning cue from the males—a drumming sound produced by muscles acting on the swim bladder. This behavior was not noted in tank 2 prior to day 112. On day 112, a large male (14.1 kg) was removed from tank 1 and placed in tank 2, but no spawning occurred subsequently. Egg production decreased in tank 1 after removal of this large male, however. During May–June 1988, successful spawning occurred in all tanks containing tagged red drums.

Tag Readability and Retention

In December 1987, the only remaining striped bass from the first group tagged died. At this time (27.5 months), its tag could still be detected. The tags implanted in the striped bass during the

spring and summer of 1986 were still functioning in samples taken in April 1988 (maximum of 25 months). The tag in one striped bass tagged in February 1986 could not be detected in a May 1986 sample. Because all the other fish in the tank had functioning tags, it was assumed that the tag malfunctioned or had not been retained. In 1987 and 1988 samples, all fish contained a functional tag. Thus, retention rate for striped bass was 97.4% The untagged fish was subsequently retagged.

All tags implanted in red drums in July were still functioning during the last sample in June 1988 (up to 10.5 months; Table 1).

Summary and Conclusion

Use of PIT tags allowed us to collect information on growth, maturation, spawning for individual fish, and to establish long-term trends within our brood stocks of striped bass and red drums. The tags also promoted more efficient use of our limited research rearing facilities.

Because PIT tags are internal, the potential for infection associated with external tags is less, and there are no projections apt to be bitten by other fish. As a result, stresses associated with external tagging methods are reduced or eliminated with PIT tags.

We found no differences in behavior or spawning response between tagged and untagged fish.

Preliminary data suggest that growth of tagged and untagged fish does not differ substantially. Retention rates were 100% for red drums tagged up to 10.5 months and 97.4% for striped bass tagged up to 28 months. This tag appears to offer a degree of reliability and versatility not available with other tags. We expect PIT tags to become increasingly important in hatchery operations and in stocking programs involving commercially valuable or endangered species (Smith et al. 1990, this volume).

Acknowledgments

This paper is contribution 247 of the South Carolina Marine Resources Center. Research support was provided by the Sea Grant College Program under contract NA85AA-D-SG-121, and by the state of South Carolina. Reference to trade names does not imply endorsement. We thank Ray Haggerty for his assistance throughout these studies. Linda K. Greene typed the manuscript and Karen Swanson prepared the figures.

References

Arnold, C. R., W. H. Bailey, T. D. Williams, A. Johnson, and J. L. Lasswell. 1979. Laboratory spawning and larval rearing of red drum and southern flounder. Proceedings of the Annual Conference Southeastern Association of Fish and Wildlife Agencies 31:437–440.

Bayless, J. D. 1972. Artificial propagation and hybridization of striped bass, *Morone saxatilis* (Walbaum). South Carolina Wildlife and Marine Resources Department, Columbia.

Henderson-Arzapalo, A., and R. L. Colura. 1987. Laboratory maturation and induced spawning of striped bass. Progressive Fish-Culturist 49:60–63.

Prentice, E. P., C. W. Sims, and D. L. Park. 1984. A study to determine the biological feasibility of a new fish tagging system. Report (Contract DE-A-179-83BP11982, Project 83-19) to Bonneville Power Administration, Portland, Oregon.

Roberts, D. E., B. V. Harpster, and G. E. Henderson. 1978. Conditioning and induced spawning of the red drum (*Sciaenops ocellata*) under varied conditions of photoperiod and temperature. Proceedings of the World Mariculture Society 9:311–332.

Smith, T. I. J. 1988. Aquaculture of striped bass (*Morone saxatilis*) and its hybrids in North America. Aquaculture Magazine 14(1):40–49.

Smith, T. I. J., W. E. Jenkins. 1988. Culture and controlled spawning of striped bass (*Morone saxatilis*) to produce striped bass and striped bass × white bass (*M. chrysops*) hybrids. Proceedings of the Annual Conference Southeastern Association of Fish and Wildlife Agencies 40:152–162.

Smith, T. I. J., S. D. Lamprecht, and J. W. Hall. 1990. Evaluation of tagging techniques for shortnose sturgeon and Atlantic sturgeon. American Fisheries Society Symposium 7:134–141.

Wydoski, R., and L. Emery. 1983. Tagging and marking. Pages 215–237 in L. A. Nielsen and D. L. Johnson, editors. Fishery techniques. American Fisheries Society, Bethesda, Maryland.

American Fisheries Society Symposium 7:346–352, 1990

Relative Success of Telemetry Studies in Michigan

JAMES S. DIANA, DAVID F. CLAPP, AND ELIZABETH M. HAY-CHMIELEWSKI

School of Natural Resources, University of Michigan, Ann Arbor, Michigan 48109, USA

GARY SCHNICKE, DEL SILER, AND WILLIAM ZIEGLER

Department of Natural Resources, Post Office Box 300, Crystal Falls, Michigan 49920, USA

RICHARD D. CLARK, JR.

Department of Natural Resources, Institute for Fisheries Research, University Museum Annex
Ann Arbor, Michigan 48109, USA

Abstract.—We have used telemetry in several recent studies of fish behavior in Michigan. Our work involved both radio and ultrasonic systems with a variety of species in a wide range of aquatic habitats. Although radiotelemetry was simpler to use than ultrasonic telemetry, it did not prove reliable in large, deep lakes. Work with radio tags on lake sturgeons *Acipenser fulvescens* and chinook salmon *Oncorhynchus tshawytscha* was largely unsuccessful because the fish regularly moved deeper than 10 m, where signals faded. Ultrasonic telemetry eliminated this problem for lake sturgeons and allowed year-round observation. Radiotelemetry was used successfully to study northern pike *Esox lucius* and largemouth bass *Micropterus salmoides* in small, shallow lakes, and the behavior of brown trout *Salmo trutta* in small rivers was successfully evaluated with radio transmitters, although short battery life was a problem. Similar work on northern pike in rivers proved less satisfactory because the fish moved extensively. Programable radio receivers facilitated searches for fish, particularly in rivers. Ultrasonic transmitters that emitted coded signals over one channel were more convenient than multichannel systems.

Methods of studying fish movements and behavior have evolved from use of simple systems involving traps or physical tags to use of complex systems involving telemetry or acoustics. Physical tagging studies commonly involve large numbers of animals over wide geographic areas; they generate release and recapture data but provide little information on behavior. Trapping studies are somewhat more limited in numbers of fish treated and geographic range; they provide even less information on behavior. At the other extreme are telemetry studies, which produce much behavioral information for a few fish. All three methods are based on similar assumptions: that the fish are typical, and that behavior and survival are not altered by handling or the presence of a device on the body. All three have been used successfully to study fish movements under different behavioral and habitat constraints.

Telemetry systems available for use in fish behavior studies can be separated into radio and ultrasonic categories. Radio transmitters commonly use wavelengths from 40 to 200 MHz. The signal is transmitted through air and picked up by an antenna. Reception range is related to battery size, current drain, and depth of the fish in the water. Obstructions generally do not completely block signals but may reduce range. Separate fish are usually monitored on different channels.

Ultrasonic transmitters use wavelengths of 20–300 kHz. The signal is transmitted through water and picked up by a hydrophone. The range is proportional to battery size and current drain. Although water depth itself does not reduce signal detection, thermal stratification can weaken the signal, and physical obstacles (islands, vegetation) can block it. Separate fish are usually tracked by a combination of different channels and different pulse frequencies for each transmitter. A detailed description of ultrasonic telemetry has been provided by Stasko and Pincock (1977).

Since 1979, we have conducted telemetry studies on six species in Michigan: Lake sturgeon *Acipenser fulvescens*, chinook salmon *Oncorhynchus tshawytscha*, largemouth bass *Micropterus salmoides*, brown trout *Salmo trutta*, walleye *Stizostedion vitreum*, and northern pike *Esox lucius*. Their habitats consisted of inland streams, inland lakes, and the Great Lakes. This paper evaluates the successes and failures of our efforts, and it describes behaviors observed for largemouth bass, lake sturgeons, and brown trout.

Methods

Lake sturgeons were tracked from 1985 to 1987 in Black Lake, a 4,101-hectare lake in Cheboygan County. The purpose of this work was to evaluate habitat preference and behavior of the lake sturgeon, which is listed as a threatened species in Michigan. Four fish, ranging in size from 20 to 40 kg, were implanted with radio transmitters (AVM Instrument Company model SM2, weight 30 g in air) during spring 1985, and eight fish were implanted with ultrasonic transmitters (Sonotronics [ST] UCTT-85, weight 20 g in air) during spring 1986 (Hay-Chmielewski 1987). The fish were collected from the Black River and Black Lake by dip net or gill net, surgically implanted with transmitters as described by Crossman (1977), and released immediately. The radio-tagged fish were tracked for 1–33 d with an AVM model LA 12 receiver, and ultrasonic-tagged fish were tracked for 78–245 d with an ST model USR-5 receiver. Tracking of fish in the river and lake was conducted during periods of open water and ice cover. Daily locations were made when possible, supplemented by hourly locations during July 1986 and January 1987.

Chinook salmon were tracked during July 1985 in northeastern Lake Huron near Rogers City. Their depth distribution was evaluated with temperature-sensitive transmitters, and this distribution was compared to prey distribution. Three fish, ranging in size from 2 to 10 kg, were collected by angling, surgically implanted with radio transmitters (Custom Telemetry and Consulting, CTC, weight 35 g in air), and released immediately. Tracking with an AVM model LA 12 radio receiver was unsuccessful: fish were never relocated after 1 d.

Largemouth bass were tracked in Third Sister Lake, a 4-hectare lake in Washtenaw County, during 1980–1981. Their onshore–offshore movements were followed, and seasonal changes in their temperature preferences were related to vertical migrations and growth (Diana 1984). Four fish, ranging in size from 800 to 2,000 g, were collected by angling, surgically implanted with radio transmitters (AVM model SM2, weight 30 g in air), and released. They were tracked for 30–278 d with an AVM model LA 12 receiver. Tracking was conducted during periods of open water and ice cover. The only measurements made were of daily horizontal displacement (the linear distance between locations on successive days).

Walleyes were tracked from April 1987 to January 1988 in Chicagon Lake, Iron County. This study, like the one described later for northern pike, was designed to evaluate habitat selection and behavior, and to detect relationships between angling techniques and fish behavior. Three fish were collected by angling and surgically implanted with radio transmitters (Advanced Telemetry Systems, ATS). They were held overnight in a mild tetracycline solution (Winter 1983), then released. They were tracked for 150–270 d with an ATS Challenger 200 Scanning Receiver. Tracking was done during seasons of open water and ice cover, and daily displacements were measured.

Brown trout were studied in the South Branch of the Au Sable River, Crawford County, from 1986 to 1988 (Clapp 1988). This study's objective was to determine whether trophy brown trout were residents or transients in a catch-and-release area. Thirteen fish were collected by electroshocking and surgically implanted with radio transmitters (CTC). Tracking was conducted on eight of these fish for 52–250 d with a radio receiver (ATS Challenger 200). Fish were located every other day; on intervening days, one fish was tracked continually to determine its diel activity cycles.

Northern pike were tracked from March to October 1986 in the Iron River, Iron County. Three fish were captured by angling and surgically implanted with ATS radio transmitters. They were held overnight in a mild tetracycline solution, then released. The fish were followed for 180–240 d in open water during summer and winter. Only daily locations were recorded.

Fish locations for all studies were determined by triangulation on shore landmarks with a siting compass (Diana et al. 1977). Locations were plotted on maps to determine daily displacement, which was measured only from diurnal sitings of fish. Lake sturgeons were also sampled on an hourly basis through 24-h periods, and brown trout were often followed continuously. Associated habitat variables (depth, current velocity, substrate type, distance from shore) were measured daily at location sites.

Results and Discussion

Comparisons of Telemetry Systems

Signal strength and detection distance depended on battery size and current drain on the transmitter, which in turn set the transmitter size and signal duration. The study habitat, and in

TABLE 1.—Duration of contact (means and ranges) for six fish species tracked with radiotelemetry in Michigan lakes or rivers.

Species	System	Number of fish tracked	Mean (range) contact duration, days
Lakes			
Lake sturgeon	Radio	4	25 (1–33)
	Ultrasonic	8	204[a] (78[b]–245[c])
Chinook salmon	Radio	3	0 (0–1)
Largemouth bass	Radio	4	171[a] (30–278)
Walleye	Radio	3	210[b] (150[d]–270)
Rivers			
Brown trout	Radio	8	149[a] (52–250[c])
Northern pike	Radio	3	220 (180–240)

[a]Mean includes any fish whose transmitter either malfunctioned or was still functioning when tracking ended.
[b]One transmitter malfunctioned.
[c]Tracking discontinued while transmitter still functioning.
[d]Fish was caught by angler.

particular the water depth, affected the success of each telemetry system. Radiotelemetry studies of chinook salmon and lake sturgeons in large, deep lakes were unsuccessful (Table 1). The signals from the AVM and CTC radio transmitters faded significantly with depth, and at a depth of 10 m the signal was only detectable within 30 m. Rarely could the researchers get this close to a specific fish in a large lake, so regular contact was not maintained for more than a few days. Lake sturgeons were successfully tracked as long as 245 d in the same lake with ultrasonic telemetry, which indicates that the poor radio transmission, and not fish movement, was the probable cause of failure.

In small, shallow lakes, radiotelemetry proved effective. Largemouth bass were successfully tracked in waters up to 20 m deep, although they preferred the shallow, littoral habitat. The tagged specimens were tracked for 30–278 d. Signal strength was generally sufficient to allow transmitters to be detected from any point in the lake. Certain species were tracked by radiotelemetry in large, deep lakes when appropriate transmitters were used. Walleyes were tracked successfully in a deeper lake with ATS radio transmitters. These transmitters gave slightly greater distance detection at depth—their range was 300 m at a depth of 10 m, compared to a 100-m range at the same depth for AVM or CTC transmitters. It is unknown whether the walleyes moved deeper in the water column, but they were often found over deep water. They also showed occasional daily displacements up to several kilometers.

Detection distance as a function of transmitter depth also depended on battery size and current drain. Batteries of 600 mA-h or better, which allowed a current drain of 0.2 mA/d, gave good transmission in water up to 10 m deep and had a long transmitter life (up to 270 d). Smaller batteries and current drains limited transmission distance and duration. Because of these trade-offs, few dealers offer standard fish transmitters, but rather each transmitter type is custom-made with batteries and antenna types (whip or contained) as options.

Radiotelemetry was successful in small rivers, where brown trout were followed up to 250 d, in spite of daily movements that occasionally reached 1,600 m. However, in another river, northern pike were more difficult to locate. The variable movements of northern pike, including forays into marshes and floodings, made these fish more difficult to locate than brown trout, which remained in the main channel. In both cases, however, the fish were followed for extended periods. Northern pike movements (up to 4 km/d) were extreme in spring, when contact tended to be lost; but contact was regained later in the summer for all three fish.

Radio signals weakened with depth, and detection distance at a depth of 3 m ranged from 200 to 1,000 m. Radio signals along rivers were obstructed by islands, terrestrial vegetation, and river meanders. Thus, signal strength through the air was not consistent for all locations.

Ultrasonic signals were independent of depth and often could be detected over distances up to 1.6 km. However, these signals can be strongly attenuated by the thermocline in stratified lakes (Stasko and Pincock 1977) or by warm water (at an isothermal 26°C in Black Lake, range was only 400 m, 600 m less than at cooler temperatures), as well as by islands, vegetation, or other physical obstructions (Diana et al. 1977). Such obstructions were not common in Black Lake, and therefore only temperature affected signal detection.

Ease of location also differed considerably among telemetry systems. Radio transmitters required less effort to monitor in general, because one did not have to be stationary and lower a hydrophone into the water to receive a signal. This situation was particularly advantageous for winter tracking. However, good signal detection for radios often required use of large aerial yagi antennas (about 3.5 m long), which were cumbersome. Smaller, hand-held yagis gave variable results.

Both radio and ultrasonic systems proved awkward when channels had to be changed manually to find fish on different frequencies. Use of a programable scanning radio receiver (ATS Challenger 200) expedited this search. However, most radio transmitters drifted from 2 to 4 kHz over time. Drift of frequency, and tracking slightly off frequency, reduced detection distance. This was especially a problem with manually tuned receivers (AVM LA 12) or with fish that were not regularly followed.

Search time also was shortened by uniquely coded signals for ultrasonic transmitters (ST model UCTT-85) and receivers (ST model USR-5). This system allowed unique identification of each fish by the output pattern of the transmitter, thus all fish could be monitored on one channel.

The choice of transmitter size involves a trade-off between detection distance and transmitter life (Stasko and Pincock 1977). To increase battery strength, one must increase the size of the transmitter. Stronger batteries (and larger current drains) can yield both longer transmitter life and better distance reception. However, large transmitters can disturb fish behavior. Generally, transmitters should be kept to a size less than 3% of a fish's body weight (Winter 1983; Summerfelt and Mosier 1984). For our studies, we used large fish (0.8–42 kg), so transmitter size was not a significant problem.

Survival following surgical implantation was variable. While contact was maintained, no mortality of largemouth bass, northern pike, walleyes, or lake sturgeons was attributable to surgery (Diana et al. 1977; Hay-Chmielewski 1987). Chinook salmon, however, were very difficult to implant. After capture by angling, these fish often appeared exhausted and did not recover from anoxia. After collection, only a few specimens were healthy enough for implantation. Brown trout surgery also was variable. Fourteen fish collected in water less than 15°C survived well (± 80%), whereas four fish implanted at water temperature near 25°C died. Also, incisions that cut through fin supports of brown trout apparently slowed recovery, compared to incisions through the ventral surface.

Fish Behavior

Largemouth bass in Third Sister Lake showed extensive summer movements with no indication

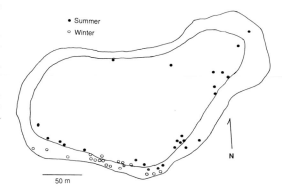

FIGURE 1.—Locations of a 44-cm largemouth bass tracked in Third Sister Lake from 8 July 1980 to 13 April 1981. Solid circles indicate open-water locations; open circles show ice-cover locations. Inner line is the 2-m depth contour.

of a limited home range (for example, see data for fish in Figure 1). In comparison, winter activity was extremely limited. Habitat selection indicated that the four subjects preferred shallow water (<2 m) in or near vegetation (Figure 2a). However, in midsummer when surface temperature exceeded 25°C, largemouth bass appeared to move farther from vegetation over deeper and cooler water (chi-square = 14.6, $N = 25$, $P < 0.01$, Figure 2b). This lake has a steeply sloping bottom, and locations in vegetation were less than 2 m in depth, whereas locations farther than 5 m from vegetation generally had depths greater than 3.5 m (Figure 2b).

The behavior noted for largemouth bass agreed, more or less, with that noted in other studies. The fish were active mainly in the summer and avoided areas with temperatures above the optimum for growth (27°C; Niimi and Beamish 1974). Warden and Lorio (1975) found similar large summer ranges and limited winter movement for largemouth bass in a Missouri reservoir. However, Winter (1977) found limited home ranges for largemouth bass from a small Minnesota lake in the summer. Ostensibly these differences involved variation among the lakes. Habitat selection was similar during the three studies, and the relationship between distance from shore and surface temperatures was similar for this study and that of Warden and Lorio (1975).

Lake sturgeons moved extensively in Black Lake and showed no indication of a limited home range. The fish were continually active, apparently foraging along the bottom. Daily displace-

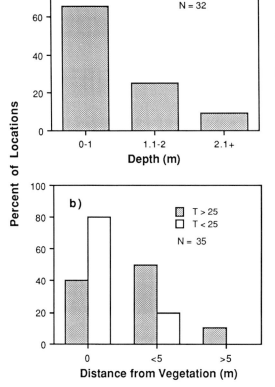

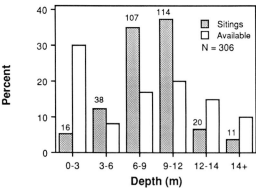

FIGURE 3.—Percent frequency of locations at various depths, and the frequency of those depths in the lake, for the lake sturgeons tracked in Black Lake, Michigan. N is number of locations.

FIGURE 2.—(a) Percent of locations at each depth for four largemouth bass tracked in Third Sister Lake during all seasons. (b) Percent of locations at each distance from vegetation for three largemouth bass tracked in Third Sister Lake during summer and fall. N is number of locations; T is temperature (°C).

ure 4). The fish showed continual, sporadic activity with no indication of diurnal or nocturnal peaks. Depth and substrate selection by lake sturgeons probably represented use of areas with preferred food types (Harkness and Dymond 1961; Hay-Chmielewski 1987). Habitat selection was generally similar to that seen in studies by Williams (1951) and Harkness and Dymond (1961).

Large brown trout in the South Branch of the Au Sable River also showed variable movements without sharply defined home ranges (Figure 5). Average daily displacements ranged from 100 to 400 m from May to August, and from 200 to more than 1,000 m from August to January. Movements during spring–summer were generally less than

ment ranged from 18 to 7,000 m, and was strongly correlated with water temperature ($y = 167.51 + 75.54x$, $r^2 = 0.89$; y = daily displacement in m and x = temperature in °C). Daily displacement was also correlated with individual body length ($y = 7732 - 101.05L + 0.38L^2$, $r^2 = 0.95$; y = daily displacement in m and L = total length in cm). Lake sturgeons selected depths in 10 m (summer) or 7 m (winter) commonly, and avoided shallow sites (Figure 3). They generally used areas with mucky bottoms (75% of locations), which was also the most common bottom type. They apparently avoided marl substrates, found in a shallow part of the lake. The fish positively selected areas with sloped bottoms, compared to the flat, middle section of the lake (chi-square = 200, $P < 0.01$, $N = 306$).

Hourly linear displacements were much larger for lake sturgeons in summer than in winter (Fig-

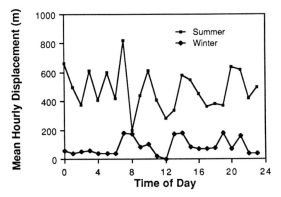

FIGURE 4.—Mean hourly displacements at each time of day (hours from midnight) for eight lake sturgeons tracked during July 1986 (rectangles) and January 1987 (diamonds) in Black Lake.

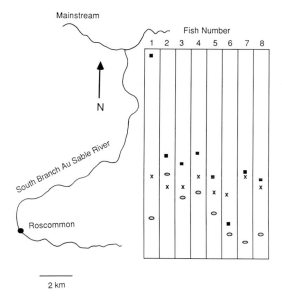

FIGURE 5.—Map of the study area and the areas frequented by eight brown trout in the Au Sable River, Michigan. Downstream extreme of movement (square), upstream extreme (oval), and point of release (×) are given for each fish. Fish 1, 4, and 7 were tracked from 15 June 1987 to 10 March 1988, fish 2 from 12 May to 26 June 1986, fish 3 from 12 May to 15 August 1986, fish 5 from 5 November 1986 to 24 January 1987, fish 6 from 5 November 1986 to 11 January 1987, and fish 8 from 5 November 1986 to 14 February 1987.

those during fall–winter, suggesting that spawning-related movements were extensive.

Spring–summer movements occurred almost exclusively at night and were generally of two types. A brown trout might move out from an area of resting cover near dusk, show periodic activity (sometimes venturing as far as 1,600 m from this cover) throughout the night, and then return near dawn. Sometimes, a fish selected a new area of cover the following dawn. Brown trout generally changed locations about once every 2 d, although one fish stayed at a single site for 47 d without moving to a new daytime resting site. The variation in distance moved was related primarily to volume discharge, temperature, and day length ($r^2 = 0.30$, multiple regression, $P < 0.05$).

Brown trout selected sites in summer that had significantly slower velocities, sandier and siltier bottoms, and closer proximity to heavy cover than did randomly chosen sites in the river (t-test, $P = 0.05$). This selection was probably for resting rather than feeding habitat, because habitat use was monitored during the day when these trout are inactive.

The behavior of trophy brown trout differed considerably from the territoriality described by Bachman (1984), who relied on visual observations in studying small brown trout in a Pennsylvania stream. Jenkins (1969), also relying on visual observations, noted that brown trout populations included territorial residents as well as nomadic fish. In a tagging experiment, Shetter (1967) found that large brown trout in the Au Sable system moved extensively, whereas small fish remained local residents. Thus, the behavior of brown trout seems related to their size. Larger brown trout become piscivorous (Alexander 1977; Stauffer 1977), so differences in behavior probably involve changes in foraging strategies.

Conclusions

Telemetry systems were useful in establishing behavior of free-ranging fish over long periods in a variety of habitats. Major problems occurred in the use of radiotelemetry for fish in deep water, where ultrasonic systems worked better. Marked differences were seen in the daily movements of individuals of the same species, especially walleyes and brown trout.

Acknowledgments

Funds for these studies were provided by the Michigan Department of Natural Resources and the Michigan Fly Fishing Club. Field assistance was provided by A. Abramson, K. Gardiner, P. Larmour, R. Lockwood, B. Merrill, F. Simonis, and the fisheries crew from District 5. Some of this study is a result of work sponsored by the Michigan Sea Grant College Program, under grant NA 86AA-D-SG043, project R/GLF-18, from the National Oceanic and Atmospheric Administration, U.S. Department of Commerce, and by funds from the state of Michigan.

References

Alexander, G. R. 1977. Consumption of small trout by large predatory brown trout in the North Branch of the Au Sable River, Michigan. Michigan Department of Natural Resources, Research Report 1855, Ann Arbor.

Bachman, R. A. 1984. Foraging behavior of free-ranging wild and hatchery brown trout in a stream. Transactions of the American Fisheries Society 113:1–32.

Clapp, D. 1988. Long-range movements, habitat choice, and activity cycles of trophy brown trout *Salmo trutta* in the South Branch of the Au Sable River, Michigan. Master's thesis. University of Michigan, Ann Arbor.

Crossman, E. J. 1977. Displacement and home range movements of muskellunge determined by ultrasonic tracking. Environmental Biology of Fishes 1:145–158.

Diana, J. S. 1984. The growth of largemouth bass, *Micropterus salmoides* (Lacepede), under constant and fluctuating temperatures. Journal of Fish Biology 24:165–172.

Diana, J. S., W. C. Mackay, and M. Ehrman. 1977. Daily movements and habitat preference of northern pike (*Esox lucius*) in Lac Ste. Anne, Alberta. Transactions of the American Fisheries Society 106:560–565.

Harkness, W. J. K., and J. R. Dymond. 1961. The lake sturgeon: the history of its fishery and problems of conservation. Ontario Department of Lands and Forests, Fish and Wildlife Branch, Toronto.

Hay-Chmielewski, E. M. 1987. Habitat preferences and movement patterns of the lake sturgeon (*Acipenser fluvescens*) in Black Lake, Michigan. Michigan Department of Natural Resources, Fisheries Research Report 1949, Ann Arbor.

Jenkins, T. M. 1969. Social structure, position choice, and microdistribution of two trout species (*Salmo trutta* and *Salmo gairdneri*) resident in mountain streams. Animal Behavior Monographs 2:56–123.

Niimi, A. J., and F. W. H. Beamish. 1974. Bioenergetics and growth of largemouth bass (*Micropterus salmoides*) in relation to body weight and temperature. Canadian Journal of Zoology 52:447–456.

Shetter, D. S. 1967. Observations on movements of wild trout in two Michigan stream drainages. Michigan Department of Natural Resources, Research Report 1743, Ann Arbor.

Stasko, A. B., and D. G. Pincock. 1977. Review of underwater biotelemetry, with emphasis on ultrasonic techniques. Journal of the Fisheries Research Board of Canada 34:1261–1285.

Stauffer, T. E. 1977. A comparison of the diet and growth of brown trout (*Salmo trutta*) from the South Branch and the Main Stream, Au Sable River, Michigan. Michigan Department of Natural Resources, Research Report 1845, Ann Arbor.

Summerfelt, R. C., and D. Mosier. 1984. Transintestinal expulsion of surgically implanted dummy transmitters by channel catfish. Transactions of the American Fisheries Society 113:760–766.

Warden, R. L., and W. L. Lorio. 1975. Movements of largemouth bass (*Micropterus salmoides*) in impounded waters as determined by underwater telemetry. Transactions of the American Fisheries Society 104:696–702.

Williams, J. E. 1951. The lake sturgeon, Michigan's largest fish. Michigan Department of Conservation, Report 1297, Ann Arbor.

Winter, J. D. 1977. Summer home range movements and habitat use by four largemouth bass in Mary Lake, Minnesota. Transactions of the American Fisheries Society 106:323–330.

Winter, J. D. 1983. Underwater biotelemetry. Pages 371–395 *in* L. A. Nielsen and D. L. Johnson, editors. Fisheries techniques. American Fisheries Society, Bethesda, Maryland.

American Fisheries Society Symposium 7:353–356, 1990

Effects of Dummy Ultrasonic Transmitters on Juvenile Coho Salmon[1]

MARY L. MOSER,[2] ALAN F. OLSON, AND THOMAS P. QUINN

School of Fisheries WH-10, University of Washington, Seattle, Washington 98195, USA

Abstract.—The miniaturization of ultrasonic transmitters has made it possible to track small fish. We investigated the feasibility of using internal transmitters with small salmonids by testing the effects of dummy transmitters on the behavior and swimming performance of juvenile coho salmon *Oncorhynchus kisutch*. We inserted cylindrical dummy tags resembling two sizes of available transmitters (33 × 9 and 17 × 7 mm) into the stomachs of 28 juvenile coho salmon (150–221 mm, total length). Of the 14 fish to receive the smaller tags, all recovered within 5 h, and they exhibited no significant reductions in swimming performance or alterations in behavior compared to controls. The larger tags were regurgitated by 4 of 14 fish, and those fish retaining the tags took significantly longer to recover their buoyancy than did controls. Later, however, the larger tags did not alter feeding and aggressive behavior, nor did they impair swimming ability. We recommend the smallest available transmitters for tracking studies of small salmonids. If larger tags must be used, fish should be allowed at least 36 h to recover from tag insertion before they are monitored.

Ultrasonic transmitters have been used in a variety of studies of individual fish movements (McCleave and LaBar 1972; Dodson et al. 1973; Tytler et al. 1978; Quinn and terHart 1987). Such studies often require a trade-off between larger transmitters that broadcast longer and stronger, and smaller ones that minimize effects on fish behavior and swimming performance. As ultrasonic transmitters are further miniaturized, smaller fish can be tracked. Tests of dummy ultrasonic transmitters in Atlantic salmon *Salmo salar* (McCleave and Stred 1975; Fried et al. 1976) indicated that telemetry was feasible for juvenile salmonids as small as 200 mm total length (TL); and stomach implantation of radio transmitters had little effect on yearling chinook salmon *Oncorhynchus tshawytscha* from 149 to 195 mm in fork length (Stuehrenberg et al. 1986). In the present study we evaluated the effects of two sizes of dummy transmitters on juvenile coho salmon *Oncorhynchus kisutch* as small as 150 mm TL. We compared the feeding behavior, aggressive interactions (Chapman 1962), buoyancy compensation, and swimming performance of tagged and control fish.

Methods

Juvenile coho salmon (150–221 mm TL; 72.0 g mean weight) were obtained from the University of Washington's Seward Park hatchery in November 1987. We selected this size range to include sizes of coho salmon smolts from the Chehalis River, Washington. (Wild smolts from Scatter Creek, a tributary to the Chehalis River, tracked by us in spring 1988 averaged 193 mm TL).

All fish were anesthetized (MS-222, 50 mg/L) and freeze-branded (Everest and Edmundson 1967) for later identification. The fish were fed ad libitum while being held at least 10 d at 12.5°C. By the end of this acclimation period, all freeze brands were readable, all fish were alive, and approximately 75% of the fish were feeding regularly.

The large dummy tags had dimensions similar to those of transmitters available from Vemco, Ltd., Nova Scotia; and the small tags resembled a prototype designed by the staff of the Ministry of Agriculture, Fisheries, and Food, Lowestoft, England (Table 1). The dummies consisted of plastic tubing weighted with nails and coated with epoxy cement. The fish were anesthetized, weighed, and measured. Dummy tags were gently pushed down the esophagi of submerged fish with a smooth wooden rod, and sham insertions were performed on controls. For each replicate, two fish from each treatment group (large tags, small tags, and controls) were placed in a 61.5-L glass aquarium for behavioral observations. The six fish had unique brands to permit individual identification

---

[1]Contribution 766 from the School of Fisheries, University of Washington.
[2]Present address: Zoology Department, Box 7617, North Carolina State University, Raleigh, North Carolina, 27695-7617, USA.

TABLE 1.—Characteristics of dummy ultrasonic transmitters inserted into juvenile coho salmon 150–221 mm in total length.

Tag size	Length (mm)	Diameter (mm)	Weight in air (g)	Weight in water (g)	% of mean fish weight
Large	33.0	9.0	4.6	2.5	4.5–14.5
Small	17.0	7.0	1.5	0.85	1.4–4.7

but, to avoid bias, the observers were unaware of treatment assignments. The experiment was replicated seven times.

After tag insertion, the vertical position and general behavior of each fish were recorded for 5 min every 30 min during the first hour after tag insertion, every hour for the next 4 h, and every day for up to 4 d. Recovery was designated when the fish left the tank bottom. Observations were terminated if a fish regurgitated its tag. Twenty-four hours after tag insertion, the fish were fed and individual feeding activity was recorded. All these fish were killed no later than 4 d after tag insertion, and the presence or absence of food in the gut was noted, along with tag position. Retention of dummy transmitters by an additional six fish from each tagged group was monitored for 4 weeks after insertion. The effect of tags on recovery time was tested with the Wilcoxon rank sum test, and the effects on feeding and aggressive behaviors were tested with the binomial test (Zar 1984).

We evaluated swimming performance using three modified Blaska respirometer–stamina chambers (Smith and Newcomb 1970). Fish were tested either 1 or 3 d after tag insertion to determine the influence of recovery time on swimming performance. At least five fish from each treatment group and recovery time were tested. Individual fish were allowed to acclimate for 30 min at current velocities of less than 0.5 body length (BL)/s. The velocity was then increased to 1.5 BL/s and fish were exercised for 15 min. After

each completed 15-min interval, the velocity was increased by 0.5 BL/s until the fish tired and could no longer swim against the current. Critical swimming speed (U_{crit}) was calculated following Beamish (1978):

$$U_{crit} = U_{max} + (T/T_i) \, U_i;$$

U_{max} = highest velocity maintained for the prescribed period (BL/s);
U_i = velocity increment (BL/s);
T = time the fish swam at fatigue velocity (min);
T_i = time interval (min).

The effects of tag treatment and recovery time on U_{crit} were determined by analysis of variance (Zar 1984).

After noting high variation in swimming performance among individuals within treatment groups, we ran additional swimming tests to evaluate individual differences. We determined swimming performance for at least five fish from each treatment group, both before and 3 d after tag insertion, and then made treatment comparisons using paired t-tests (Zar 1984).

Results

Four of the 14 fish tagged with large dummy transmitters regurgitated the tags (Table 2) within 4 h of tag insertion. Fish that regurgitated were all less than 180 mm TL. Fish retaining the large dummy tags took significantly longer to recover than did controls ($P \leq 0.05$). The two smallest fish

TABLE 2.—Observations on regurgitation of dummy ultrasonic tags, feeding activity, aggressive behavior, and recovery time for experimental and control coho salmon.

Tag size	Number of fish that			Mean (SD) time to recovery (h)
	Regurgitated tag	Fed	Showed aggression	
Large	4 of 14	6 of 10[a]	5 of 10[a]	6.28 (9.90)[a,b]
Small	0 of 14	13 of 14	7 of 14	1.80 (1.16)
Control		11 of 14	8 of 14	0.89 (1.08)[c]

[a] Only fish that retained their tags were considered.
[b] Significantly longer than controls; $P \leq 0.05$.
[c] Recovery from sham transmitter insertions.

TABLE 3.—Mean critical swimming speeds U_{crit} (SD) of experimental and control coho salmon tested either 1 or 3 d after tag insertion. Additional fish were tested before and 3 d after tag insertion. BL is body lengths.

Tag size	Recovery time (d)	Mean U_{crit} (BL/s)	Mean U_{crit} (BL/s)	
			Before tag insertion	3 d after tag insertion
Control	1	3.26 (0.09) $N = 7$		
Small	1	3.06 (0.14) $N = 5$		
Large	1	2.94 (0.38) $N = 5$		
Control	3	3.72 (0.10) $N = 5$		
Small	3	3.48 (0.16) $N = 6$	3.19 (0.29) $N = 5$	2.90 (0.12) $N = 5$
Large	3	3.14 (0.18) $N = 5$	3.35 (0.12) $N = 6$	3.14 (0.13) $N = 6$

tagged with the large transmitters, 163 and 165 mm TL, required 24 and 26 h to recover, respectively. Their long recovery times resulted in high overall variance (Table 2). Feeding and aggressive behavior of fish that retained the large tags were not significantly different from that of controls ($P > 0.05$). No small tags were regurgitated. Recovery time, feeding activity, and incidence of aggressive encounters for fish with small tags were not significantly different from those of controls ($P > 0.05$).

All large tags and 12 of the 14 small tags were found posteriorly in the stomachs of dissected fish. Two small tags had passed into the intestine. Food was found anterior and posterior to tags of both sizes.

The exact dates of tag loss could not be determined for fish held 4 weeks after tagging. However, all 12 fish retained the dummies for at least 1 week, and 6 fish held them for the entire period. Four of the recovered tags were small and two were large.

Mean critical swimming speeds of tagged fish were not lower than those of controls ($P > 0.05$, Table 3). Fish tested before and after tag insertion also showed no significant change in U_{crit} ($P > 0.05$). Both control and experimental fish that were allowed to recover for 3 d had higher mean U_{crit} than did fish tested 1 d after tag insertion ($P < 0.05$). Interaction of treatment and recovery time was not significant ($P > 0.05$).

Discussion

We found that small transmitters (17×7 mm) did not significantly alter behavior or swimming performance of juvenile coho salmon between 150 and 221 mm TL. Fish bearing the small tags fed as frequently as controls and displayed agonistic behavior typical of healthy coho salmon. Even the smallest individuals tagged with small dummies recovered within 4 h and were swimming normally 24 h after tag insertion. The small tags represented up to 4.7% of the body weight of the fish tested. Similarly, dummy transmitters representing up to 5% of the body weight did not significantly reduce swimming performance of Atlantic salmon (McCleave and Stred 1975) or chinook salmon smolts (Stuehrenberg et al. 1986).

The large dummy transmitters represented 4.5–14.5% of the weight of tested fish. These tags did not significantly affect swimming performance, feeding, or aggressive behavior. However, four of the smallest fish tested (TL < 180 mm) regurgitated the large tags, and two fish under 170 mm required over 24 h to recover their buoyancy. The larger tags were equivalent to 9.5% of the mean body weight for fish smaller than 180 mm TL. These results led us to conclude that the 33×9-mm tags are appropriate for coho salmon over 180 mm TL. Although salmonid smolts typically are smaller than 180 mm, individuals of this size have been reported. Smolts of steelhead *Oncorhynchus mykiss* (formerly *Salmo gairdneri*) from Lyons and Wells hatcheries on the Columbia River averaged over 190 mm in fork length (Fish Passage Center 1986), and chinook salmon smolts collected for radio tracking in the Columbia River ranged from 150 to 205 mm FL (Stuehrenberg et al. 1986).

For tags weighing up to 5% of a fish's body weight, we suggest a minimum recovery period of 4 h to allow for buoyancy compensation and possible regurgitation. Fried et al. (1976) recommended an 8-h recovery period, based on dummy tag experiments with Atlantic salmon smolts (>200 mm). For tags exceeding 9% of the body weight we suggest a recovery period of 36 h.

Dummy-tagged juvenile salmonids unable to gulp air cannot regain their buoyancy after tagging (Fried et al. 1976). After field tagging, we suggest holding fish in shallow small-meshed enclosures allowing the fish easy access to the surface. Such holding pens would also permit recovery of regurgitated tags. In our experience, fish that retain transmitters of both tested sizes for 4 h will retain them at least 1 week more.

Our dissections revealed that neither large nor small transmitters prevented food passage and that small tags may be excreted after 1 week. The ability of some smolts to eliminate small transmitters could constrain study designs.

Acknowledgments

We thank Ernest Brannon (University of Washington) and the Seward Park hatchery staff for providing the fish used in this study. We also thank Thomas Flagg (National Marine Fisheries Service) and Lynwood Smith (University of Washington) for loaning the swimming stamina chambers. This project was funded by the Washington Department of Fisheries and the University of Washington's College of Ocean and Fishery Sciences Select Program.

References

Beamish, F. W. H. 1978. Swimming capacity. Pages 101–172 in W. S. Hoar and D. J. Randall, editors. Fish physiology, volume 7. Academic Press, New York.

Chapman, D. W. 1962. Aggressive behavior in juvenile coho salmon as a cause of emigration. Journal of the Fisheries Research Board of Canada 19:1047–1080.

Dodson, J. J., W. C. Leggett, and R. A. Jones. 1973. The behavior of adult American shad (*Alosa sapidissima*) during migration from salt to fresh water as observed by ultrasonic tracking techniques. Journal of the Fisheries Research Board of Canada 29:1445–1449.

Everest, F. H., and E. H. Edmundson. 1967. Cold branding for field use in marking juvenile salmonids. Progressive Fish-Culturist 29:175–176.

Fish Passage Center. 1986. Smolt monitoring program: estimation of survival. U.S. Department of Energy, Bonneville Power Administration, and Division of Fish and Wildlife, Project 80-1, Portland, Oregon.

Fried, S. M., J. D. McCleave, and K. A. Stred. 1976. Buoyancy compensation by Atlantic salmon (*Salmo salar*) smolts tagged internally with dummy telemetry transmitters. Journal of the Fisheries Research Board of Canada 33:1377–1380.

McCleave, J. D., and G. W. LaBar. 1972. Further ultrasonic tracking and tagging studies of homing cutthroat trout (*Salmo clarki*) in Yellowstone Lake. Transactions of the American Fisheries Society 101:44–54.

McCleave, J. D., and K. A. Stred. 1975. Effect of dummy telemetry transmitters on stamina of Atlantic salmon (*Salmo salar*) smolts. Journal of the Fisheries Research Board of Canada 32:559–563.

Quinn, T. P., and B. A. terHart. 1987. Movements of adult sockeye salmon (*Oncorhynchus nerka*) in British Columbia coastal waters in relation to temperature and salinity stratification: ultrasonic telemetry results. Canadian Special Publication of Fisheries and Aquatic Sciences 96:61–77.

Smith, L. S., and T. W. Newcomb. 1970. A modified version of the Blazka respirometer and exercise chamber for large fish. Journal of the Fisheries Research Board of Canada 27:1321–1324.

Stuehrenberg, L. C., A. E. Giorgi, C. W. Sims, J. Ramonda-Powell, and J. Wilson. 1986. Juvenile radio-tag study: lower Granite Dam. National Marine Fisheries Service, Northwest and Alaska Fisheries Center, Coastal Zone and Estuarine Studies Division, Annual Report 85-35, Seattle, Washington.

Tytler, P., J. E. Thorpe, and W. M. Shearer. 1978. Ultrasonic tracking of the movements of Atlantic salmon smolts (*Salmo salar* L) in the estuaries of two Scottish rivers. Journal of Fish Biology 12:575–586.

Zar, J. H. 1984. Biostatistical analysis. Prentice-Hall, Englewood Cliffs, New Jersey.

American Fisheries Society Symposium 7:357–363, 1990
© Copyright by the American Fisheries Society 1990

Use of Radiotelemetry to Estimate Survival of Saugers Passed through Turbines and Spillbays at Dams

FORREST W. OLSON

CH2M Hill, Post Office Box 91500, Bellevue, Washington 98009, USA

E. SUE KUEHL

BioSonics, Incorporated, 4520 Union Bay Place NE, Seattle, Washington 98105, USA

KERBY W. BURTON AND JAMES S. SIGG

Lewis & Associates, Incorporated, Post Office Box 1383, Portsmouth, Ohio 45662, USA

Abstract.—Of 48 live, radio-tagged saugers *Stizostedion canadenses* passed through a 23-MW bulb turbine at Greenup Dam on the Ohio River, 41 (85.4%) survived as determined by their behavior in the tailwaters of the dam. The status of the other seven fish (14.6%) was classified as "undetermined." Of 41 live, radio-tagged saugers passed through a spillbay at the dam, 17 fish (41.5%) survived, and the status of 24 fish (58.5%) was undetermined. Results of supplemental experiments defining the limitations of radiotelemetry techniques at the site demonstrated that lack of evidence of survival did not necessarily indicate mortality. Therefore, these results provided only a lower-limit estimate of survival. The difficulties in using radiotelemetry to distinguish between live and dead fish are compounded by the tendency of live saugers to sound and dead ones to sink in slow water.

The Vanceburg Hydroelectric Generating Station Number 1 is located at the U.S. Army Corps of Engineers Greenup Locks and Dam on the Ohio River approximately 216 km east of Cincinnati, Ohio. The generating station began operation in 1982. Several studies were conducted in 1986 and 1987 to determine impacts of the new facility on fisheries and to provide information for mitigation planning. Among these was a study to estimate the survival rate of adult saugers *Stizostedion canadenses* passed through bulb-turbine blades at the generating station and through open tainter gates at the spill dam. The study design stipulated the use of radiotelemetry to determine percent survival or mortality of the test fish after they passed through the respective conduits. Saugers were selected because they are an important Ohio River game fish and are relatively abundant near the dam during spring and late autumn. Saugers also have large stomachs, which makes them good subjects for gastric tags.

The generating station consists of three horizontal-axis bulb turbines with 6-m-diameter adjustable Kaplan runners with fixed-guide vanes. Generating capacity at the rated head (8 m) is approximately 23 MW per unit. The maximum flow through each turbine at maximum efficiency and head is approximately 337 m³/s.

The spill dam consists of 10 gate piers and 9 nonsubmersible tainter gates. Each gate is 30.5 m wide and, when closed, rests on a gate sill at elevation 146.3 m. The upstream pool is maintained at an elevation of 157 m, and the minimum downstream pool elevation is 148 m. Water passing through the gates enters a dentated stilling basin.

Methods

Experimental approach.—The experimental approach to assessing survival of adult fish at the dam consisted of releasing radio-tagged fish into a turbine intake and immediately upstream of an open tainter gate so they would pass quickly through the respective conduits. Directional radio antennas and receivers below the dam were used to monitor fish movements. Live and killed fish were similarly released with the expectation that differences in movement between the live and dead fish could be discerned, thereby providing criteria upon which to estimate percent survival of the fish released alive. Estimates of mortality were limited to the time the fish remained trackable.

The original study design called for fieldwork in the spring when saugers are abundant in the tailwater of Greenup Dam; however, high flows prevented us from getting more than one day's data in April 1987. The fieldwork was postponed

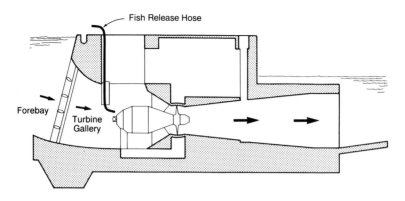

FIGURE 1.—Cross section of the generating station at Greenup Dam showing where test fish were released into turbines.

until October 1987. Summer testing was specifically avoided because high water temperatures and low dissolved oxygen levels would be close to the tolerance limits for saugers. The stress of handling and tagging fish under these adverse conditions could have biased the results.

Environmental conditions monitored during the testing periods included river and dam gate discharge, turbine discharge, water temperature, dissolved oxygen, conductivity and turbidity.

Radiotelemetry equipment.—Radio transmitters (tags) were manufactured by Custom Telemetry and Consulting of Athens, Georgia. Each tag weighed 3.5 to 4.2 g in air and measured approximately 10 by 35 mm. Expected battery life was approximately 5 d following activation with a magnetic switch. Transmission frequencies ranged from 49.0 to 50.2 MHz. Three pulse rates were used.

Radio signals were monitored with three receivers manufactured by two companies: Custom Electronics of Urbana, Illinois, and Smith-Root of Vancouver, Washington. Directional hand-held antennas were manufactured by AF Antronics of White Health, Illinois. Specifications for the radiotelemetry equipment are available through the authors.

Test fish collection and tagging.—The test fish were collected by electrofishing at night in the 3-km reach below the dam. Saugers less than about 200 mm total length were not collected because they were found to be too small for the radio tags. Test fish were held at the generating station in 750-L plastic tanks supplied with a continuous flow of river water pumped from the turbine gatewell. Tagged fish were placed in an-other flow-through tank and observed up to 48 h for tag regurgitation, stress, and mortality.

Fish were individually anesthetized, weighed, and measured prior to tagging. Fish were tagged internally by placing the transmitter in the mouth and then pushing it into the stomach with a plunger. A small rectangular piece of compressed sponge was attached to the side of each tag, and a short piece of surgical tubing was stretched over the tag and part of the sponge. The tag was inserted with the exposed end of the sponge pointing toward the mouth. Once inside the fish's stomach, the wetted sponge expanded into a wedge that checked tag regurgitation. This technique has been used successfully on juvenile salmonids by staff of the National Marine Fisheries Service in Seattle, Washington (K. Liscom, personal communication).

External marks and tags on each fish facilitated identification of radio-tag frequencies. During the April tests, fish were marked by clipping fins in unique combinations. During the October tests, small plastic or metal clips with engraved numbers were attached to the fish's operculum.

Test fish release procedures.—Test fish were released into the intake of turbine unit (Ohio shore) via a 10-cm-diameter, 20-m-long "Kanoflex" hose with a smooth interior. The discharge end of the hose was attached to a turbine gallery fyke-net frame used in previous studies at the site (Olson et al. 1988). The frame was lowered into the gatewell slot with a gantry crane so that the discharge end of the hose extended 4.6 m below the intake ceiling (Figure 1). At this location the intake water velocity was estimated to be 1.5–2.4 m/s during the fish releases. Turbine operation

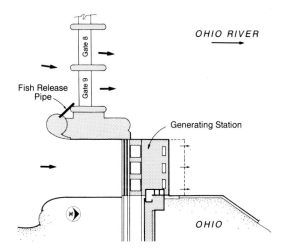

FIGURE 2.—Plan of Greenup Dam generating station and east tainter gates showing location of dam gates where test fish were released.

was normal for the river conditions at the time, and no special turbine adjustments were made for the study.

At the spillbay, the test fish release system consisted of a 15-m-long section of 10-cm-diameter plastic pipe reinforced on the outside with angle iron. One end of the pipe was secured to the deck of the guide cell adjacent to dam gate 9 (Figure 2). The discharge end of the pipe was lowered into position with a rope-and-pulley apparatus attached to the gate pier. The discharge end of the pipe was positioned approximately 2 m above the gate sill, 6 m above the gate opening, and 2 m from the gate pier wall. During all test releases, gate 9 was opened 0.6 m to produce a discharge of approximately 220 m³/s. No other gates were open during testing.

Test fish were individually dropped headfirst into the hose or pipe and then flushed through with flow provided by a fire hose. The nozzle of the hose was removed to minimize turbulence in the pipe.

Radio tracking.—To monitor the movement of test fish that passed through the turbine, receiver stations were set up on the Ohio shore and on the powerhouse deck (Figure 3). The shore stations were approximately 30 m from the water's edge, which put the antennas at a higher elevation and increased tracking range. Other stations on shore and on the powerhouse deck were used occasionally to obtain additional bearings or to improve reception.

To monitor test fish released through the dam

gate, receiver stations were established near the downstream (west) pier of dam gate 9 and on the northwest corner of the powerhouse deck (Figure 3). An attempt was made to detect signals from a shore station 90 m below the powerhouse, but the powerhouse apparently blocked the signals.

At least two receiver stations operated during each fish release. Communication radios were used to coordinate tracking and to collect information from all stations. Directional bearings were radioed to one person designated as the central recorder. Each receiver operator also kept notes on signal strength, signal variations, accuracy of bearing measurements, location of signals relative to turbine discharge backroll or slack water, and other pertinent observations.

Results

Environmental Conditions

Table 1 summarizes river conditions and operating conditions at the generating station for the April and October 1987 test dates. On April 30, 1987, the river flow was high at 5,239 m³/s with a total head available for power generation of only 4 m (compared to a maximum of 10 m). Only 224 m³/s passed through each of the three turbines. During the October 19–24 testing period, the river was relatively low and clear. No flow passed through the dam gates in October except during the gate-passage test on October 24, and the turbines operated at reduced flows. Only two of the three turbines operated during most of the October testing period. Water conductivity was 300–325 μS/cm in October compared to 200 μS/cm in April. On October 21, a weather front entered the area, bringing high winds and a sudden drop in air temperature. Interference due to atmospheric conditions caused a significant reduction in receiving range and terminated tracking that day.

Tag Regurgitation and Handling Stress

Of the 164 test and control fish tagged during the study, five died and five showed visible signs of stress during the 1–2-d holding period. The five stressed fish were killed and released as killed test fish.

During the spring test, 8 of 34 tags were regurgitated within 20 h of tag insertion. Five of the eight regurgitators were larger than 290 mm. On the basis of this observation, we did not tag live fish larger than 300 mm during the autumn tests. In October we inserted 130 tags, 7 of which were regurgitated in the first overnight holding period;

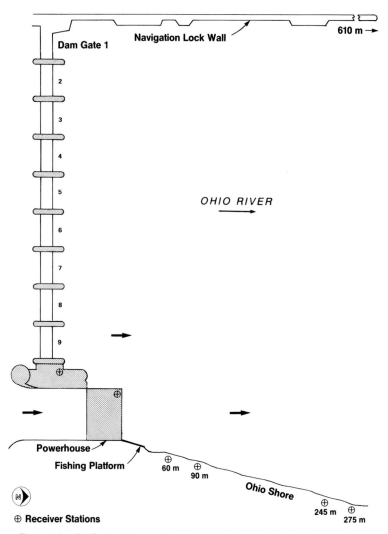

FIGURE 3.—Radio-tracking stations and tailwater area below Greenup Dam.

again we noted that larger fish were more apt to regurgitate tags. In October we also held 29 tagged fish as controls for up to 7 d to determine the occurrence of regurgitation after the first overnight period. Eight of the 29 tags were regurgitated after the first overnight period. We concluded that regurgitation was possible at any time during testing. Of primary concern was the possibility that the sudden pressure drop experienced by fish passing through the turbine might cause rapid expansion of their swim air bladder and encourage tag expulsion. We had no way to test this in the field.

Survival Detection Criteria

During the April field test, four of the five dead fish sent through the turbine were not detected by

the receivers. The one detected fish was observed only briefly, in the upwell area of the turbine discharge. In contrast, all 10 living fish were detected by at least one receiver. The average tracking time for fish released live was 14 min. Most of the live fish moved rapidly downstream. Because the tailwater elevation was relatively high during the April test, we concluded that the dead fish quickly sank to the bottom beyond the range of the radio receivers, and that the live fish apparently stayed near the surface within tracking range.

During the initial release of test fish in October, we again observed that dead fish were detected briefly, if at all, in the turbine discharge upwell. The water apparently was still too deep to permit

TABLE 1.—Ohio River conditions and operational characteristics of the Vanceburg Generating Station during radiotelemetry test with saugers in 1987.

Variable	Apr 30	Oct 19	Oct 20	Oct 21	Oct 22	Oct 23	Oct 24
Total river discharge (m³/s)	5,239	515	635	405	405	523	409
Downstream gauge height (m)	9.2	4.1	4.3	4.1	4.0	4.1	4.0
Powerhouse discharge (m³/s)	671	515	635	405	405	302	409
Turbines operating	1, 2, 3	1, 3	1, 2, 3	1, 3	1, 3	2, 3	2, 3
Water temperature (°C)	13.7	16.5	16.0	16.0	13.7	14.4	14.0
Dissolved oxygen (mg/L)	10.5	8.0	8.0	8.0	8.2	8.0	8.1
Conductivity μS/cm	200	325	300	320	325	300	
Secchi depth (m)	0.30	1.75			2.00	2.00	

tracking. As noted in the spring tests, however, most of the live test fish were trackable. We conducted several experiments to determine if our inability to track dead fish could be used as a criterion to determine survival.

To test the relationship between fish depth and radio-tracking range, live radio-tagged fish were tethered to inflated balloons with various cord lengths, then released into the frontroll of the turbine discharge. A fish tethered to a 1-m cord moved downstream past the end of the lock wall in about 10 min. The signal was lost when the fish was approximately 365 m from the receiver. Another fish was attached to a balloon on a 2.5-m cord with a 113-g lead weight added near the fish to keep the radio transmitter at a depth of approximately 2.5 m. Maximum tracking range for this fish when released in the tailrace was only about 120 m. This observation, coupled with the sharp decrease in transmitter range with depth (Winter 1983), supported our hypothesis that fish at the 4.5- to 6-m depth of the tailwater probably were untrackable. This particular experimental fish swam toward shore into a riprap area where the radio signal was lost when the fish was approximately 60 m from the receiver. An angler then retrieved the tethered fish. Apparently we lost the signal because the fish was behind or under a rock.

To test whether live fish could act dead, we tossed three radio-tagged live fish into the frontroll of the turbine discharge. Two of them remained trackable for several minutes and moved rapidly downstream in a manner similar to patterns observed for most of the live turbine-released fish; however, the third was trackable for only 3 s after it reached the water. Because this fish was alive and vigorous when released, we concluded that it sounded immediately.

Two live radio-tagged fish were released into the slack water downstream of gate 9 to determine whether the test fish tended to sound when encountering slack or slow-moving water. Depth at this location was approximately 6 m. The radio receiver was positioned on the powerhouse deck about 30 m from where the fish entered the water. After the fish reached the water, signals from them were received for 2 and 20 s. Apparently they sounded quickly.

From the results of these experiments, we concluded that signals received downstream from the turbine discharge provided evidence that a fish had not sunk or sounded and therefore was alive. However, our results also indicated that some live fish sounded and thereby became indistinguishable from dead fish. With this information, plus the data obtained during preliminary test releases, we developed the following criteria to identify fish that survived passage through the turbine or dam gate: (1) upstream movement, (2) cross-current movement, (3) rapid downstream movement, and (4) signal reception from areas of deep, nonupwelling water (indicating surface orientation of the fish). Fish that did not exhibit one or more of the above behaviors were classified as undetermined. They might have died, sounded, regurgitated their tag during turbine passage, or remained deep when exiting the draft tube or stilling basin.

Turbine Tests

Sixty-six fish were passed through the turbine unit during the spring and autumn field tests. Of these, 48 were alive upon release and 18 were killed just before release. Seventeen of the 18 killed fish were undetected or were detected briefly (less than 2 min) in the upwelling of the turbine discharge, and the other washed into shallow water near shore just below the up-

TABLE 2.—Radio-tracking time and survival status of live saugers passed through a bulb turbine at the Vanceburg Generating Station at Greenup Dam, 1987

Minutes tracked	Date	Number of fish tagged	Survival status	
			Survived	Undetermined
0–3	Apr 30	2	1	1
	Oct 19–22	16	10	6
4–10	Apr 30	2	2	0
	Oct 19–22	19	19	0
11–20	Apr 30	3	3	0
	Oct 19–22	2	2	0
>20	Apr 30	3	3	0
	Oct 19–22	1	1	0
Total		48	41	7
Percent			85.4	14.6

TABLE 3.—Radio-tracking time and survival status of live saugers passed through a tainter gate at Greenup Dam on October 23 and 24, 1987.

Minutes tracked	Number of fish tagged	Survival status	
		Survived	Undetermined
0–2	24	1	23
3–6	13	12	1
7–9	3	3	0
>9	1	1	0
Total	41	17	24
Percent		41.5	58.5

welling. Of the 48 live fish released in the turbine intake, 41 survived passage as indicated by their behavior in the tailwater (Table 2). The status of the remaining seven fish was unknown. As previously noted, we concluded that lack of survival evidence did not provide conclusive evidence of death. Therefore, the 85.4% of the test fish identified as surviving represents only an estimate of minimum immediate survival of fish that passed through the turbine. Mortality after the tracking period could not be assessed.

Dam Gate Tests

Forty-one live fish and nine dead fish equipped with radio tags were released through a tainter gate at the dam. The results of this test were inconclusive because of our inability to distinguish clearly between movements of live and dead fish. Of the 41 live fish, only 17 demonstrated movements in the tailwater characteristic of live fish (Table 3). Both the live and sacrificed fish tended to remain in the turbulent water of the stilling basin immediately downstream of the gate. Survival of saugers passing through the gate could have been as low as 41.5%. We believe, however, that most live fish probably sounded as soon as they exited the stilling basin and were untrackable with the radio receivers. Because only one of nine gates was open at the dam during the test period, the water velocity decreased abruptly as the discharge encountered the expanse of slack water below the dam. Earlier experiments at the site demonstrated that saugers tend to sound quickly in slow water and disappear from transmitter reception range.

Discussion and Conclusion

Radiotelemetry is valuable in estimating fish survival only if the fish killed can be distinguished from those that survive. Our test results indicate that distinction is often difficult to make at dams on large rivers, particularly when proof of survival depends on consistent tracking patterns. We demonstrated that although sacrificed fish did not show tracking patterns like those of live fish, live fish occasionally behaved as though they were dead. Therefore, the survival rates obtained in this study only represent lower limit estimates. Based on our experimental results, we believe that some of the fish whose survival status was classified as undetermined actually survived.

We estimated the minimum short-term survival for saugers passed through bulb turbines to be 85.4%. This is considerably lower than an estimate of short-term plus extended survival for juvenile salmonids that went through similar turbines. In a mark–recapture study of individually branded coho salmon *Oncorhynchus kisutch* and steelhead *O. mykiss* (formerly *Salmo gairdneri*) 685,000 smolts passed through bulb turbines at Rock Island Dam on the Columbia River in Washington. Researchers estimated that 94.3% of the coho salmon and 96.1% of the steelhead survived (Olson and Kaczynski 1980; Anderson 1984). In a radiotelemetry study similar to ours, survival of Atlantic salmon *Salmo salar* was estimated at 98% for smolts passed through bulb turbines at Essex Dam on the Merrimack River (Knapp et al. 1982). Although the Greenup Dam study used saugers as test fish in a warmwater environment, we find it difficult to believe that the sauger, a relatively hardy fish, would survive turbine passage at a lower rate than salmonid smolts. The saugers (average length 253 mm) used in this study were approximately the same size as the Atlantic salmon (range, 225–350 mm) used in the

Essex study, but larger than the coho salmon (average length, 117 mm) and steelhead (average 166 mm) used in the Rock Island study. Although smaller fish sometimes survive passage through turbines better than larger fish (Eicher Associates 1985), the size differences in these studies seem too minor to account for the differences in mortality estimates.

Studies of salmonid survival at dam gates also are available for comparison. At McNary Dam on the Columbia River, which has a 27-m head, estimated survival of chinook salmon smolts *Oncorhynchus tshawytscha* passed through the spill gates was 98% (Schoeneman et al. 1961). Survival of coho salmon smolts passed through a turbulent and highly dentated stilling basin at Rocky Reach Dam on the Columbia River was estimated at 99.6% (Heinle and Olson 1981). Based on the results of these studies, it appears very unlikely that the estimate of 41.5% minimum survival at Greenup Dam represents the true value.

This study documented some limitations of using radiotelemetry to estimate survival at dams. The depth of the Greenup Dam tailwater, coupled with the tendency of saugers to sound in slow water, reduced the tracking range and contributed to the problem of distinguishing live from dead fish. The tailwater depth decreased from approximately 12 to 6 m between spring and autumn, but a corresponding increase in water conductivity from 200 to 325 µS/cm may have offset any improvement in tracking range (Winter 1983). Atmospheric conditions (e.g., dramatic changes in weather) can also affect the radio tracking range (Fred Ore, AF Antronics, White Heath, Illinois, personal communication). On a day of sudden high winds and a substantial drop in temperature during this study, radio receiving range in open air was reduced to several meters rather than the usual several hundred meters.

If test fish regurgitated their radio tags when exposed to the sudden pressure drop at the turbine runner, this would have biased our results. Although surgical implantation of tags in the peritoneal cavity would eliminate the possibility of regurgitation, sudden expansion of the fish's swim bladder during pressure drops might cause the incision to rupture, again biasing the results. This problem could be avoided if sufficient time were

allowed for the incision to heal, but larger tags would be needed to accommodate the longer-life batteries.

The fish might also regurgitate tags as a fright response to turbines. We became aware of this possibility one morning when we found several tags in a tank that an otter had entered the previous night.

In summary, we found that radiotelemetry can provide conservative estimates of fish survival at dams. Site-specific and behavioral factors that affect tag retention and signal reception can lower the quality of the estimates, and should be accounted for. Radiotelemetry did not provide information on long-term survival, because live fish sounded or left the study area soon after being detected.

References

Anderson, D. R. 1984. Review of analysis and inference procedures for the bulb turbine survival experiments conducted at Rock Island Dam, 1979. Report to Chelan County Public Utility District, Wenatchee, Washington.

Eicher Associates, Inc. 1985. Proceedings: turbine passage workshop. Report to Electric Power Research Institute, Research Project 1745-18, Portland, Oregon.

Heinle, D. R., and F. W. Olson. 1981. Survival of juvenile coho salmon passing through the spillway at Rocky Reach Dam. Report to Chelan County Public Utility District, Wenatchee, Washington.

Knapp, W. E., B. Kynard, and S. P. Gloss. 1982. Potential effects of Kaplan, Ossberger, and bulb turbines on anadromous fishes of the northeast United States. U.S. Fish and Wildlife Service, Final Report. Newton Corner, Massachusetts.

Olson, F. W., and V. W. Kaczynski. 1980. Survival of downstream migrant coho salmon and steelhead trout through bulb turbines. Chelan County Public Utility District, Wenatchee, Washington.

Olson, F. W., J. F. Palmisano, G. E. Johnson, and W. R. Ross. 1988. Fish population and entrainment studies for the Vanceburg Hydroelectric Generating Station No. 1. Report to City of Vanceburg, Kentucky.

Schoeneman, D. E., R. Pressely, and C. O. Junge, Jr. 1961. Mortality of downstream migrant salmon at McNary Dam. Transactions of the American Fisheries Society 90:58–72.

Winter, J. D. 1983. Underwater biotelemetry. Pages 371–395 in L.A. Nielsen, and D.L. Johnson, editors. Fisheries techniques. American Fisheries Society, Bethesda, Maryland.

American Fisheries Society Symposium 7:364–369, 1990

Radio Transmitters Used to Study Salmon in Glacial Rivers

John H. Eiler

Alaska Fisheries Science Center, Auke Bay Laboratory, National Marine Fisheries Service
Post Office Box 210155, Auke Bay, Alaska 99821, USA

Abstract.—Radiotelemetry was used to study adult sockeye salmon *Oncorhynchus nerka* and coho salmon *O. kisutch* in large glacial rivers in southeastern Alaska and northwestern British Columbia. Approximately 600 adult salmon were tagged with radio transmitters from 1984 through 1987. Motion-sensitive transmitters equipped with mortality sensors were useful for determining the activity of radio-tagged fish. High-frequency transmitters (150–151 MHz) had greater signal range in shallow water but could not be received at depths greater than 6.0 m. Low-frequency transmitters (30–31 MHz) had a signal range of about 2.0 km at depths of 15.0 m.

Many glacial rivers in Alaska are important producers of Pacific salmon *Oncorhynchus* spp. However, their turbid nature makes it difficult to obtain visual information on salmon populations. Radiotelemetry provides a method for studying salmon in glacial systems because it can monitor movements of radio-tagged fish even under extremely turbid conditions.

The Taku River is a large, glacial river in northwestern British Columbia and southeastern Alaska (Figure 1). It drains a remote watershed of over 16,000 km², where access is mostly by aircraft. Water visibility for most of the drainage is extremely poor because of turbidity. Five species of Pacific salmon return to this river to spawn.

From 1984–1987, radiotelemetry was used to follow the movements of adult salmon returning to the Taku River drainage. Part of the research involved developing telemetry equipment suitable for studying adult salmon in large, glacial rivers. This report describes the radio transmitters used during the study, compares the suitability of high- and low-frequency transmitters, and evaluates transmitters equipped with motion and mortality sensors.

Methods

Salmon were captured with fish wheels (Meehan 1961) near the mouth of the Taku River. Sockeye salmon *O. nerka* and coho salmon *O. kisutch* were tagged with two-stage radio transmitters (Figure 2) made by Advanced Telemetry Systems (Isanti, Minnesota)[1] and Lotek Engineering (Aurora, Ontario, Canada). Transmitters pro-

duced by other vendors were used in 1984, but were unsuitable for this study. The fish were held in a tagging cradle submerged in a trough of fresh water. Anesthesia was not used during the study. The transmitter was inserted through the mouth and placed in the stomach of the fish. The standard transmitter weighed approximately 20 g, was 6.5 cm long × 2.0 cm in diameter, and had an external transmitting antenna 30.0 cm long that extended out of the fish's mouth. In 1986, larger transmitters 8.0 cm in length were used. The tagging procedure took approximately 2 min to complete. Tagged fish were released into quiet water out of the main current.

Transmitter frequencies were spaced 10 kHz apart within the 30–31 MHz range used in this study. Different pulse rates were used to identify transmitters placed on the same frequency. Typically, two transmitters placed on one frequency had rates of 60 and 90 pulses/min. In 1986, three transmitters were placed on each frequency at rates of 60, 80, and 100 pulses/min. The lifespan of the transmitters was approximately 4 months.

The transmitters used from 1985 to 1987 were equipped with motion and mortality sensors. The motion sensor, a small mercury switch sensitive to movement (Figure 3), inserted additional pulses into the base pulse rate each time the transmitter moved. The mortality sensor was activated if the motion sensor was not triggered for over 6 h. Transmitters in the mortality mode had a pulse rate substantially faster than the base pulse rate— e.g., rates of 60 and 90 pulses/min changed to 150 and 180 pulses/min. The transmitters reverted to the base pulse rate if the motion sensor was activated.

We located salmon tagged with radio transmitters every week (up to 15 weeks) using a Cessna

---

[1]Reference to trade names does not imply endorsement of the products by the National Marine Fisheries Service.

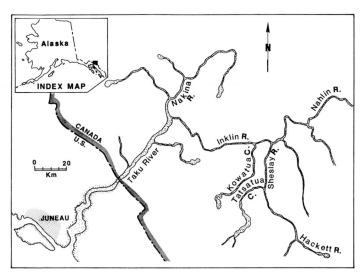

FIGURE 1.—Map of Taku River drainage. Steep Creek and Mendenhall Lake are in the shaded area around Juneau.

185 floatplane that had quarter-wavelength whip antennas mounted on each side. The antennas were coupled with a switch box that allowed reception of transmitter signals from either or both sides of the plane. An Advanced Telemetry Systems programmable scanning receiver was used to detect the presence of the transmitters. Aerial surveys were flown at a height of approximately 300 m. Fish location was determined to the

nearest kilometer. The activity mode of the transmitter was recorded as active, inactive, or mortality (inactive for over 6 h). During the fall, we made precise locations (within 10 m) of radiotagged salmon on spawning areas using helicopters equipped with quarter-wavelength whip antennas. We used directional loop antennas to locate radio-tagged salmon during foot surveys of spawning areas.

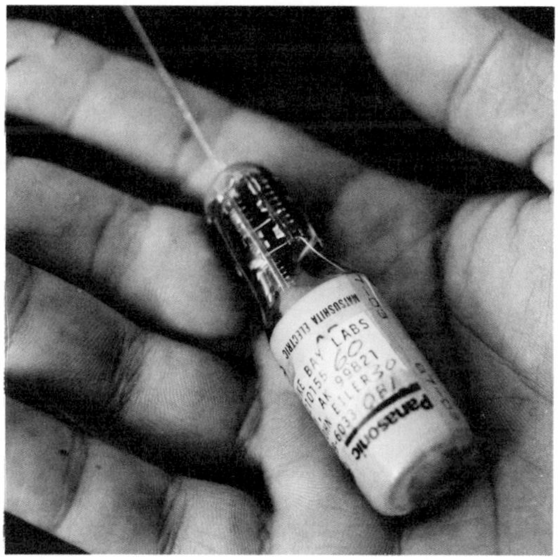

In 1985, coho salmon were tagged with motion-sensitive transmitters in Steep Creek, a small stream near Juneau, Alaska (Figure 1). Steep Creek is clear, shallow, and easy to survey on foot, which makes it ideal for observing tagged fish. Radio-tagged salmon in Steep Creek were located daily. Visual observations were made of the fish, and information was collected on the response of the motion sensor to actual movements. Movements were classified into four basic categories: holding, swimming (slow, moderate, fast), reverse-direction turns (escape behavior), and spawning.

An evaluation of high- and low-frequency transmitters at different depths was conducted at Mendenhall Lake, a small (390-hectare) glacial lake near Juneau, Alaska (Figure 1). Mendenhall Lake is isothermic and has a low, uniform conductivity of 30 μS/cm. Low- (30–31 MHz) and high-(150–151 MHz) frequency transmitters made by Advanced Telemetry Systems were placed in salmon carcasses attached to a horizontal structure and lowered to various depths in the lake. All

FIGURE 2.—Standard radio transmitter used to study salmon on the Taku River. It measures 6.5 × 2.0 cm, weighs approximately 20 g, and has an antenna about 30.0 cm long which extends from the fish's mouth.

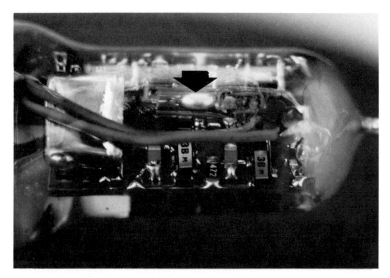

FIGURE 3.—Radio transmitter equipped with a motion sensor consisting of a mercury switch (indicated by arrow).

the transmitters had the same power rating, battery type, transmitting antenna, and expected lifespan. Differences in power output between transmitters of the same frequency range were minor, not exceeding 10%. A transect was established between the horizontal structure and a point 5.0 km away. We attempted to receive transmitter signals at 0.5-km intervals along the transect while in a helicopter hovering 300 m above the ground. Two-element H-antennas and quarter-wavelength whip antennas were used to receive signals from high- and low-frequency transmitters, respectively. Tests were conducted for transmitters at depths of 0, 1.5, 3.0, 4.5, 6.0, 9.0, 12.0, and 15.0 m. Signal reception was determined by one person to standardize the results.

Results and Discussion

Transmitter Suitability

We tagged 398 sockeye salmon and 186 coho salmon with radio transmitters during this study (Table 1). Most of the fish responded well to tagging. Sustained upriver movement was observed for 330 sockeye salmon and 125 coho salmon tagged in the Taku River. Radio-tagged fish caught in the riverine gill-net fishery or recovered on spawning areas were in good condition. No internal damage caused by the transmitter was observed. The 20 sockeye salmon and 44 coho salmon tagged in Steep Creek exhibited no adverse effects attributable to the transmitters or the tagging procedure. The length of the fish tagged

during the study ranged from 405 to 670 mm for sockeye salmon and 405 to 725 mm for coho salmon.

Sixty-eight of the sockeye salmon and 61 of the coho salmon were not detected upriver. These fish died after being tagged (because of predation or handling), or were lost as they traveled upstream, or moved out of the river into Taku Inlet. There is evidence for all three possibilities. The large transmitters (8.0 cm long) used in 1986 were difficult to insert, particularly into smaller fish, and in several cases they ruptured the stomach

TABLE 1.—Numbers of adult salmon tagged with radio transmitters in southeastern Alaska and northwestern British Columbia, 1984–1987.

Year	Location	Number of	
		Sockeye salmon	Coho salmon
1984	Taku River	93	17
	Steep Creek	20	12
1985	Taku River	3	
	Steep Creek		32[a]
1986	Taku River	282[b,c]	71[b,c]
1987	Taku River		54[c]
Total		398	186

[a]Transmitters were equipped with motion sensors.
[b]Transmitters measured 8.0 cm, 1.5 cm longer than those used with other fish.
[c]Transmitters were equipped with motion sensors and mortality sensors.

lining. Harbor seals *Phoca vitulina* and other predators were frequently observed in the vicinity of the tagging site, and tagged fish were susceptible to predation. Several fish tagged with radio transmitters were recovered in the commercial fishery in Taku Inlet.

Large numbers of adult salmon return to the Taku River (1987 Transboundary River Technical Committee, Pacific Salmon Commission, Vancouver, British Columbia, unpublished report) which makes it necessary to tag relatively large numbers of fish to obtain representative samples. Placing several transmitters on the same frequency with different pulse rates allowed us to increase the number of fish that could be tagged and identified with radio transmitters.

Transmitters placed on the same frequency and spaced 30 pulses/min apart (60 and 90 pulses/min) were readily distinguishable by ear. However, it was difficult to separate transmitters with rates only 20 pulses/min apart, especially at faster rates (e.g., 80 and 100 pulses/min). The difficulty was compounded during aerial surveys when time was limited and large numbers of fish were being tracked.

Motion and Mortality Sensors

Thirty-two coho salmon were tagged with motion-sensitive transmitters in Steep Creek, and we collected detailed information on the response of the motion sensor to different types of movement. Holding or slow-swimming movements were not sufficient to trigger the motion sensor. Moderate swimming resulted in active readings approximately half of the time. Fast swimming, reverse-direction turns, and spawning activity consistently triggered the motion sensor. The signal produced by spawning was distinctive and easily distinguished from other types of activity, making it possible to document spawning activity even under turbid conditions.

Motion-sensitive transmitters made it possible to verify spawning within specific areas. Burger et al. (1985) used movement data to identify spawning areas for chinook salmon *O. tshawytscha* in the Kenai River, Alaska. Fish were assumed to be spawning when movement was restricted to a localized area after upriver movement had stopped. However, radio-tagged sockeye salmon and coho salmon in the Taku River remained in localized areas for prolonged periods before moving to spawning grounds, which made movement patterns a poor indicator of spawning grounds.

Motion sensors also helped us find radio-tagged fish on the spawning grounds, because the fish were often inactive until approached.

In the Taku River, 407 salmon were tagged with transmitters equipped with motion and mortality sensors. Tracking results showed that the mortality sensors were reliable. Only one transmitter was known to register mortality while the fish remained alive, and it later reverted to the base pulse rate. Twenty-five fish whose transmitters reported mortality were recovered on spawning areas, where all 25 were dead. The general locations of other transmitters registering mortality indicated the sensors were accurately reporting dead fish, even though the transmitters were not recovered.

Motion sensors have been used extensively to study terrestrial animals (Garshelis et al. 1981) but less often to study fish. Luke et al. (1979) developed a transmitter that monitored activity with electrodes placed in the axial or opercular muscles of fish, and transmitters with motion sensors similar to ours have been used to study fish in Norway (Dalen 1977). However, most studies of fish activity have focused on determining changes in location by continuously monitoring movements over time (Mackay and Craig 1983). This technique has limited application in remote areas where time constraints exist or the fish are restricted to localized sites. Sockeye salmon spend varying periods in holding areas before moving on to spawning grounds (Foerster 1968). We observed this pattern not only for sockeye salmon but for coho salmon in the Taku River. Motion-sensitive transmitters were extremely useful in these situations; without a method to determine activity, it was usually impossible to determine whether stationary transmitters represented dead fish or restricted movements in holding areas.

We found it difficult to evaluate activity using only the information provided by motion sensors due to the activity patterns shown by adult salmon. Radio-tagged coho salmon holding in small ponds in Steep Creek were relatively inactive for long periods before moving on to spawning areas, as were radio-tagged salmon on the Taku River. Mortality sensors gave us a quick and efficient means of monitoring the condition of radio-tagged fish, and were particularly useful during aerial surveys and when large numbers of fish were being tracked.

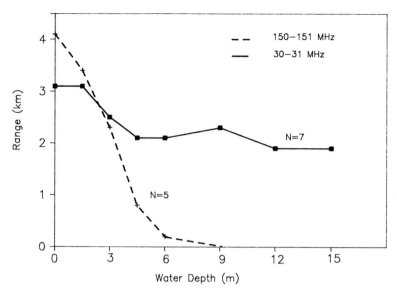

FIGURE 4.—Signal range of high- (150–151 MHz) and low-frequency (30–31 MHz) transmitters at different water depths in a glacial lake in southeastern Alaska.

Comparison of High- and Low-Frequency Transmitters

Signal range, measured as a radial distance from transmitter to receiver, was determined for seven low- and five high-frequency transmitters at different depths (Figure 4). The high-frequency transmitters had greater range at shallow depths (0 m and 1.5 m) than the low-frequency transmitters. However, the ranges for low- and high-frequency transmitters were similar at a depth of 3.0 m, and low-frequency transmitters had greater range in deeper water (1.9 km at a depth of 15 m). The range for high-frequency transmitters decreased substantially at 4.5 m, and there was no reception from depths greater than 6.0 m. Estimated range for radio-tagged salmon in the Taku River was approximately 2–3 km, which agreed with the results of our depth-evaluation tests. The results also were similar to those reported by Burger (1980), and by Mackay and Craig (1983).

It is difficult to predict the ranges of high- and low-frequency transmitters because of the number of variables that may affect those ranges. Signal transmission and reception can be affected not only by depth, but by the water conductivity (Winter et al. 1978, Velle et al. 1979), external noise (Winter 1983), the type of receiving antenna used (Knight 1978; Gilmer et al. 1981), and differences among equipment operators (Kolz and Johnson 1981). Early in our study, we found that

there are also substantial differences between transmitters made by different vendors.

Our tests at Mendenhall Lake attempted to standardize major variables and look specifically at the effect of depth on the range of high- and low-frequency transmitters in an aquatic system with low conductivity. One source of bias that was not standardized was the efficiency of the receiving antennas. High- and low-frequency antennas commonly used on aerial surveys are selected for their aerodynamic properties, not for their receiving capabilities. Low-frequency antennas suitable for aerial tracking are much less efficient than high-frequency antennas.

The results of our depth evaluations indicate that high-frequency transmitters are preferable to low-frequency transmitters in shallow rivers with low conductivity. Here, the high-frequency tags have greater range and can be detected by smaller (and more directional) antennas. Lower frequencies, however, are more suitable for use in deeper rivers and lakes, where their signal range is greater. The physical characteristics of the aquatic system must be carefully considered before a frequency range is selected.

Conclusions

Radiotelemetry was effective in studying adult sockeye salmon and coho salmon in the Taku River. Most salmon responded well to the tagging procedure. The standard transmitter (6.5 cm in length) used during the study did not appear to have any adverse effects on the salmon, although some problems were experienced with the larger transmitters (8.0 cm in length).

Motion-sensitive transmitters were useful in determining activity of radio-tagged salmon. Transmitters equipped with mortality sensors were reliable indicators of the condition of the fish, and were particularly useful in aerial surveys of large numbers of radio-tagged salmon.

Pronounced differences in signal range were observed for high- and low-frequency transmitters at different depths. High-frequency transmitters had greater range in shallow water (<3.0 m), but could not be received at depths greater than 6.0 m. Low-frequency transmitters had a range of at least 2.0 km at depths down to 15.0 m.

Acknowledgments

Many people assisted on this project. In particular, I wish to thank B. D. Nelson, R. F. Bradshaw, J. R. Greiner, J. M. Lorenz, R. P. Stone, and J. J. Pella, who assisted with fieldwork. H. R. Carlson, J. H. Helle, and J. C. Olsen critically reviewed the manuscript and assisted with the project. E. L. Landingham and R. F. Bradshaw produced the figures.

References

Burger, C. V. 1980. Comparison between high and low frequency radio transmitters in the Kenai River, Alaska. Underwater Telemetry Newsletter 10(1):9–10.

Burger, C. V., R. L. Wilmot, and D. B. Wangaard. 1985. Comparison of spawning areas and times for two runs of chinook salmon (*Oncorhynchus tshawytscha*) in the Kenai River, Alaska. Canadian Journal of Fisheries and Aquatic Sciences 42:693–700.

Dalen, J. 1977. Underwater telemetry work in Norway. Underwater Telemetry Newsletter 7(1):5–9.

Foerster, R. E. 1968. The sockeye salmon. Roger Duhamel, Queen's Printer, Ottawa, Canada.

Garshelis, D. L., H. B. Quigley, C. R. Villarrubia, and M. R. Pelton. 1981. Assessment of telemetric motion sensors for studies of activity. Canadian Journal of Zoology 60:1800–1805.

Gilmer, D. S., L. M. Cowardin, R. L. Duval, L. M. Mechlin, C. W. Shaiffer, and V. B. Kuechle. 1981. Procedures for the use of aircraft in wildlife biotelemetry studies. U.S. Fish and Wildlife Service Resource Publication 140.

Knight, A. E. 1978. Biotelemetry tracking under the ice cover of Lake Winnipesaukee, New Hampshire. Biotelemetry and Patient Monitoring 5(1):38.

Kolz, A. L., and R. E. Johnson. 1981. The human hearing response to pulsed-audio tones: implications for wildlife telemetry design. Pages 27–34 *in* F. M. Long, editor. Proceedings of the third international conference on wildlife biotelemetry. University of Wyoming, Laramie.

Luke, D. M., D. G. Pincock, P. D. Sayre, and A. L. Weatherly. 1979. A system for the telemetry of activity-related information from free swimming fish. Pages 77–85 *in* F. M. Long, editor. Proceedings of the second international conference on wildlife biotelemetry. University of Wyoming, Laramie.

Mackay, W. C., and J. F. Craig. 1983. A comparison of four systems for studying the activities of pike, *Esox lucius* L. and perch, *Perca fluviatilis* L. and *P. flavescens*. Pages 22–30 *in* D. G. Pincock, editor. Proceedings of the fourth international conference on wildlife biotelemetry. Applied Microelectronics Institute and Technical University of Nova Scotia, Halifax, Canada.

Meehan, W. R. 1961. Use of a fishwheel in salmon research and management. Transactions of the American Fisheries Society 90:490–494.

Velle, J. I., J. E. Lindsay, R. W. Weeks, and F. M. Long. 1979. An investigation of the loss mechanisms encountered in propagation from a submerged fish telemetry transmitter. Pages 228–237 *in* F. M. Long, editor. Proceedings of the second international conference on wildlife biotelemetry. University of Wyoming, Laramie.

Winter, J. D. 1983. Underwater biotelemetry. Pages 371–395 *in* L. A. Nielsen and D. L. Johnson, editors. Fisheries techniques. American Fisheries Society, Bethesda, Maryland.

Winter, J. D., V. B. Kuechle, D. B. Siniff, and J. R. Tester. 1978. Equipment and methods for radio tracking freshwater fish. Minnesota Agricultural Experiment Station Miscellaneous Report 152.

American Fisheries Society Symposium 7:370–374, 1990

Pulse-Coded Radio Tags for Fish Identification

LOWELL STUEHRENBERG, ALBERT GIORGI, AND CHUCK BARTLETT

U.S. National Marine Fisheries Service, Northwest Fisheries Center
2725 Montlake Boulevard East, Seattle, Washington 98112, USA

Abstract.—The radio tag is an effective tool for in situ research on fish behavior, but its application has been limited by the number of tagged animals that can be exposed to a treatment. This paper describes a pulse-coded radiotelemetry system that greatly decreases this limitation. We have used the system in researching adult and juvenile salmonids *Oncorhynchus* spp. in the Columbia River basin. For juvenile fish, we used a miniaturized tag to assess the relationship between the amount of water spilled at Lower Granite Dam on the Snake River and the proportion of downstream-migrating juvenile salmonids passing over the dam's spillway. For adults, we used a larger tag to identify sources of chinook salmon *O. tshawytscha* losses as they migrated between hydroelectric dams.

Radiotelemetry can provide detailed, in situ information about fish behavior. However, research applications have been limited by the small number of animals observable at any one time. This shortcoming resulted from the limited number of unique codes available with traditionally designed radio-tag systems, coupled with the inability of monitoring systems to discriminate among tagged individuals that simultaneously transmitted on the same tracking frequency. The National Marine Fisheries Service has now developed a radiotelemetry system that substantially reduces these limitations. The first part of the system is a tag that can be encoded with one of 600 unique codes. The second component is a receiver that can identify and record up to 144 uniquely coded tags as they simultaneously occupy a single detection field.

These improved capabilities not only increase the number of fish trackable at any time, they also allow broader applications. With the new system, it is possible to release groups of tagged migrating salmon and detect sufficient numbers at strategically located monitor sites to generate statistically sound recovery estimates. The purpose of this paper is to describe the new telemetry system, provide examples of its performance, recommend appropriate applications, and identify its limitations.

Pulse-Coded System

The tag, with modifications, can be used on adult or juvenile salmonids. Tags operate on crystal-controlled frequencies assigned (in the USA) by the Federal Communication Commission and spaced 10 kHz apart between 30.17 and 30.25 MHz. Tags for adult fish have miniature integrated circuit components, whereas the tags for juveniles have hybrid chip circuitry. The adult tag is encased in a polystyrene capsule 87 mm long × 19 mm in diameter; it weighs 28 g in water. The juvenile tag is 26 × 9 × 6 mm and weighs 2.9 g; its circuitry is coated with Humiseal[1], and the entire tag is dipped in a 50:50 mixture of beeswax and paraffin. Each type of tag has a 127-mm external whip antenna and an exposed ground band to maximize transmission range. Each is placed in the gut with the antenna protruding from the mouth. Range from surface-oriented fish to a surface-monitoring system is 3.2 km for the adult and 1.6 km for the juvenile tags.

The individual code for a tag is produced by breaking the radio pulse into two parts—power and code pulses—and then setting the time between the parts differently for each tag on an assigned channel (Figure 1). Battery life for all tags is similar because of approximately equal pulse lengths and pulse rates. At a pulse rate of 600 ms, tag life is 1 month for adults and 7 d for juveniles.

For tracking system includes a broad-band radio receiver, a decoder–microprocessor, and two data-output devices (paper and magnetic tape). The receiver simultaneously monitors all channels and sets an electrical latch when a channel is active. The decoder–microprocessor scans the channels until finding a latch. It then inputs data for a period of two coded pulses, checks pulse length and sequence, measures the code period,

---

[1]Reference to trade names does not imply endorsement by the National Marine Fisheries Service.

PULSE SEQUENCE

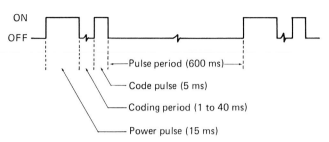

FIGURE 1.—Pulse sequence for the National Marine Fisheries Service's pulse-coded radio tag. Each 0.5 ms within the coding period (1–40 ms) has a code value of 1.

and outputs the data to the recording devices. The paper record provides a real-time quality control. The digital cassette record provides the permanent storage medium for computer entry and subsequent data analyses. Data from each recorded pulse includes the pulse code, frequency, date, time, monitor identifier, and antenna number. If tags remain on the monitor for extended periods, there are provisions in the monitor to reduce the number of records stored and turn off the paper recorder.

Tag Effects on Host

Adults

Recovery time reported for fish tagged at multiple release sites (Monan and Liscom 1973) indicate that adult salmonids take approximately 24 h to adjust to the tag. Subsequent behavior patterns appear normal. For example, Liscom and Stuehrenberg (1983) observed similar entrance patterns at dam fish ladders for tagged and untagged adult salmonids. Adults may regurgitate tags, however. Chinook salmon *Oncorhynchus tshawytscha* have a low regurgitation rate of less than 1%, but the rate for adult steelhead *O. mykiss* (formerly *Salmo gairdneri*) is approximately 5%.

Juveniles

The miniaturized version of the pulse-coded tag has been used increasingly to investigate migratory behavior of anadromous yearling chinook salmon (Giorgi et al. 1985, 1988; Stuehrenberg et al. 1986) and juvenile steelhead, (K. L. Liscom, National Marine Fisheries Service, personal communication). For yearling chinook salmon, we have examined the effects of the tag on swimming stamina, mortality, buoyancy control, and regurgitation rates. Those investigations were reported

in detail by Stuehrenberg et al. (1986) and Giorgi et al. (1988). A synopsis of that work is presented here.

Radio-tagged chinook salmon (149–195 mm fork length, FL) showed a slight, but not significant, decrease in swimming stamina relative to controls. Mean values of critical swimming speed (Beamish 1978) were 4.04 and 4.43 body lengths/s for tagged and control groups ($N = 13$) (Stuehrenberg et al. 1986). These values fall within ranges observed for migrants in the Snake River (Swan et al. 1987).

Tag regurgitation averaged 2.7% and always occurred within 4 h of tagging. We recommend holding the fish for 10–24 h before releasing them, so tag regurgitation should be of little concern. Generally, we found that fish larger than 150 mm FL could accommodate the tag.

Results from two bioassays indicated that tag-related mortality averaged 3.7% for the 24 h after tagging. During the same period, none of the controls died; but for the 24–48-h period, tagged and control groups had the same mortality, ranging from 0.7 to 1.6%. These data indicate that tag-related mortality typically occurs within 24 h following tag insertion. Consequently, when such mortality might bias the results of a field study, we recommend a holding period of 24 h.

Of the biological responses we examined, buoyancy compensation was the most difficult to interpret. The tag weight in proportion to body weight averaged 5.3% and ranged from 2.4 to 7.8%. The miniaturized tag did interfere with the fish's ability to adjust swim bladder volume. We determined this by measuring changes in bladder volume through time, using the Cartesian Diver method described by Saunders (1965). We found that tagged fish had difficulty reestablishing the

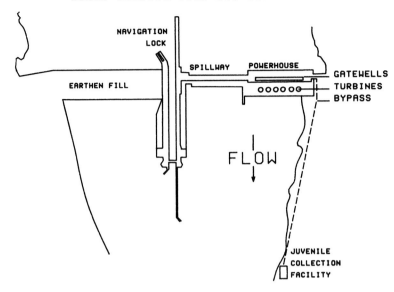

FIGURE 2.—Lower Granite Lock and Dam, where radio tags were used to determine the effect of different spill rates on spillway passage for juvenile chinook salmon. Receivers were located at the spillway, along the face of the dam, and within the turbine intakes, gatewells, and bypass flume.

gas volumes needed for normal buoyancy (Stuehrenberg et al. 1986, Giorgi et al. 1988). None of the 73 tagged fish we observed, however, were unable to swim. Nevertheless, buoyancy problems of tagged fish could be an important consideration for some studies.

Pulse-Coded Tag Study Designs

Studies of Juveniles

In 1985 and 1986, we conducted a series of radiotelemetry studies with yearling chinook salmon at Lower Granite Dam on the Snake River (Figure 2) (Stuehrenberg et al. 1986; Giorgi et al. 1988). Our primary objective was to assess the feasibility of using miniaturized radio tags to estimate the effectiveness of spill as a passage alternative at that site.

Monitors were positioned at the various passageways used by the yearlings: turbine intakes, spillway, gatewells, and fingerling bypass system. Additionally, three sets of monitors were located 1.4, 3.2, and 6.1 km downstream from the dam.

Two types of antennas were used. Underwater antennas were suspended in all gatewells, along the face of the dam in front of the powerhouse, in each spill opening, and in the fingerling bypass system. Three-element beam antennas were used at the downstream transect sites and the powerhouse tailrace. The powerhouse and spillway antennas were ganged together with line amplifiers. Each amplifier boosted the signal to a level equal to the signal lost in the line between underwater antennas. This equalized tag signals at the monitor for radio-tagged smolts at both ends of the powerhouse or spillway antennas.

Two groups ($N = 100$) of radio-tagged yearling chinook salmon (>155 mm FL) were released 6 km upstream from the dam. When the first group arrived at the dam, the spill level was held constant for 3 d at 20% of the river flow. When the fish from the second group arrived, spill was held constant for 3 d at 40%. Radio-tag signal recoveries from the various monitors were used to estimate the proportion of each test group that passed through the spillway.

Tag recovery data from the downstream monitors indicated that not all tagged fish were detected as they passed the dam. We corrected for this by developing a statistical model to estimate passage rates. Using the model, we estimated the spillway passage rate at 40.5% (95% confidence interval, CI = 28.7–52.3%) and at 60.6% (95% CI = 46.8–74.4%) when 20 and 40% of the river flow, respectively, went through the spillway.

We also tested whether radio tags could be used to estimate fish survival through turbines and

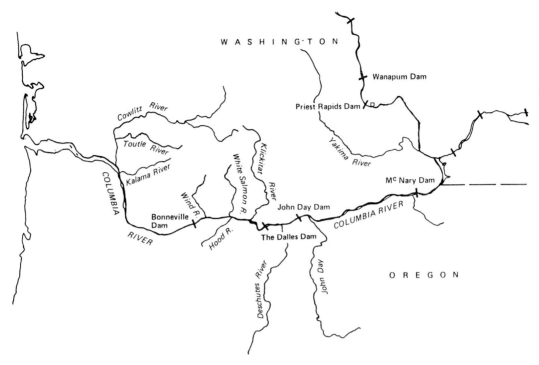

FIGURE 3.—Map of the Columbia River including the area between Bonneville and McNary dams where previously unreported losses of adult salmon were documented with radio tags.

spillways at hydroelectric dams on the Snake River. We implanted the tags in dead as well as live fish, released the fish together into the tail-race, and recorded their travel times to the downstream transect monitors (the designated recovery sites for survival studies). We could not discriminate between live and dead fish on the basis of their transit times. Some of the dead fish, by drifting, reached the downstream monitors at the same rate as live fish. In a river such as the Snake, where flows are considerable, it is unlikely that radio tags can be used to estimate mortality.

For studies of juvenile salmonids, we feel that the best use of the pulse-coding system is in situations where a large group must be tested in a short time. The system's greatest limitation is in the effect of tags on fish buoyancy and, possibly, on vertical distribution.

Studies of Adults

We have also employed pulse-coded tags to study migratory behavior of adult salmonids. In 1982, the tag was used to identify sources of adult chinook salmon losses as the fish migrated between hydroelectric dams (Liscom and Stuehrenberg 1983). In that study, the number of radio-

tagged fish exposed to dams and fisheries was varied over time so that it remained directly proportional to historical runs counted at Bonneville Dam (U.S. Army Corps of Engineers 1981). We released 286 radio-tagged fish in weekly proportions that corresponded to the historical temporal distribution (e.g., if 26% of the historical run occurred during week 3, 26% of the tagged fish were released during week 3).

Monitors at the exits of fish ladders at Bonneville, The Dalles, John Day, and McNary dams (Figure 3) recorded the passage time between locations. In addition, fish movement in the 232 km of reservoirs between the dams was surveyed twice daily from an airplane.

Data from this study of 286 fish revealed three unreported areas of adult loss. The largest was an 18% loss in the impoundments between the dams, where we suspect fish are spawning. A 16% loss consisted of fish that entered tributaries of pools behind Bonneville and The Dalles dams. A 7% loss involved radio-tagged fish that, after being released into the fish ladder at Bonneville Dam, exited into the forebay and then fell back through the turbines. Several of these fish entered the hatchery directly below the dam, leading us to believe they initially overshot the hatchery, were then counted over Bonneville Dam, and subsequently fell back to their targeted spawning area. The 12% lost to the fishery (net and sport) was in the range expected by fisheries managers.

For studies of adult salmonids, we feel that the pulse-coding system's greatest application is in documenting the fates of individuals in a tagged population.

References

Beamish, F. W. H. 1978. Swimming capacity. Pages 101–187 in W. S. Hoar and D. J. Randall, editors. Fish physiology, volume 7. Academic Press, New York.

Giorgi, A. E., L. C. Stuehrenberg, D. R. Miller, and C. W. Sims. 1985. Smolt passage behavior and flow-net relationship in the forebay of John Day Dam. U.S. National Marine Fisheries Service, Coastal Zone and Estuarine Studies Division, Final Report, Seattle, Washington.

Giorgi, A. E., L. C. Stuehrenberg, and J. Wilson. 1988. Juvenile radio-tag study: Lower Granite Dam, 1985–86. U.S. National Marine Fisheries Service, Coastal Zone and Estuarine Studies Division, Final Report, Seattle, Washington.

Liscom, K. L., and L. C. Stuehrenberg, 1983. Radio tracking studies of "Upriver Bright" fall chinook salmon between Bonneville and McNary dams, 1982. U.S. National Marine Fisheries Service, Coastal Zone and Estuarine Studies Division, Final Report, Seattle, Washington.

Monan, G. E., and K. L. Liscom. 1973. Radio tracking of adult spring chinook salmon below Bonneville and The Dalles dams, 1972. U.S. National Marine Fisheries Service, Coastal Zone and Estuarine Studies Division, Final Report, Seattle, Washington.

Saunders, R. L. 1965. Adjustment of buoyancy in young Atlantic salmon and brook trout by changes in swim bladder volume. Journal of the Fisheries Research Board of Canada 27:1331–1336.

Stuehrenberg, L. C., A. E. Giorgi, C. W. Sims, J. Ramonda-Powell, and J. Wilson. 1986. Juvenile radio-tag study: Lower Granite Dam. U.S. National Marine Fisheries Service, Coastal Zone and Estuarine Studies Division, Annual Report, Seattle, Washington.

Swan, G. W., A. E. Giorgi, T. Coley, and W. T. Normon. 1987. Testing fish guiding efficiency of submersible traveling screens at Little Goose Dam; Is it affected by smoltification levels in yearling chinook salmon? U.S. National Marine Fisheries Service, Coastal Zone and Estuarine Studies Division, Final Report, Seattle, Washington.

U.S. Corps of Engineers. 1981. Annual fish passage report, Columbia and Snake Rivers. North Pacific Division, Portland, Oregon.

American Fisheries Society Symposium 7:375–379, 1990
© Copyright by the American Fisheries Society 1990

Use of Ultrasonic Transmitters to Track Demersal Rockfish Movements on Shallow Rocky Reefs

KATHLEEN R. MATTHEWS,[1] THOMAS P. QUINN, AND BRUCE S. MILLER

Fisheries Research Institute, School of Fisheries
University of Washington WH-10
Seattle, Washington 98195, USA

Abstract.—Ultrasonic transmitters were used to monitor the movements of copper rockfish *Sebastes caurinus* and quillback rockfish *S. maliger* on shallow rocky reefs in Puget Sound, Washington. In tests with aquarium specimens, we found that rockfish routinely regurgitated dummy transmitters (48 mm long × 15 mm wide) placed in their stomachs, so we resorted to external tags. We harnessed each transmitter to a rectangular piece of plastic, which we attached to the fish with Petersen discs just beneath the dorsal fins. No tags were shed during the 3-month study. Tag size limited us to rockfish at least 27 cm in total length. Reception of the 50–69-kHz tags varied with habitat. Transmitters on fish associated with low-relief rocky bottoms having dense stands of bull kelp *Nereocystis leutkeana* were detectable for 1 km in open water, but transmitters on fish associated with high-relief rocky bottoms, regardless of kelp cover, had reduced signal strengths. If a tagged fish was underneath rocks (verified by scuba observation), the signal could only be detected for 50–100 m. If a rockfish was associated with, but not under, high-relief rock, the signal could be detected for 200–500 m. Because these rockfish were sedentary and did not make quick movements, searchers could locate tagged individuals under rocks. Ultrasonic tracking provided information on nocturnal movements, movements during times of strong current (7.4 km/hr), and return routes followed after experimental displacement.

This paper describes an application of ultrasonic telemetry to rockfish *Sebastes* spp. on shallow rocky reefs. Ultrasonic tracking has provided information on the behavior of various fish in their natural habitats (Stasko and Pincock 1977; Hawkins and Urquhart 1983), but apparently it has not been tried with demersal species on rocky reefs. We wanted to use transmitters on shallow rocky reefs in Puget Sound to study home ranges of rockfish and movements of the fish after experimental displacement.

To use telemetry with rockfish, we were faced with several problems. There are no published reports on transmitter attachment methods for rockfish or any demersal rocky reef fish, so we had to test and modify attachment methods. We were concerned about reception in our study area, which is characterized by dense stands of kelp and high-relief rock substrate. In shallow marine waters, signal transmission can be blocked by bottom relief or vegetation (Stasko and Pincock 1977), and researchers have suggested that dense vegetation can significantly reduce signal range for ultrasonic transmitters (Ziebell 1973; Winter

1983). Dense macrophytes and suspended algae were reported to reduce signal strength from several hundred meters to a few meters (Winter 1983). The objectives of this paper are to describe the attachment method finally used and others that were evaluated, to report reception ranges in rocky reef areas, and to provide details on ultrasonic tracking of shallow-water marine fish.

Methods

Tag attachment.—Several attachment methods were evaluated before the final method was chosen. Originally, we intended to place transmitters (48 mm long by 15 mm wide, 18 g in air, 8.3 g in water) in the stomachs of rockfish, which reportedly is less traumatic than external attachment or surgical implantation (Mellas and Haynes 1985). We also preferred internal transmitters because, unlike external tags, they would not be lost or fouled when rockfish wedged under rocks and vegetation. Three quillback rockfish *S. maliger* and two copper rockfish *S. caurinus* were held in aquaria for tests with dummy transmitters, which were made of nylon rods similar to transmitters in size and weight. The dummies were coated with glycerin and pushed through the esophagus into the stomach with a smooth plastic rod. No regurgitation occurred during the first 15–20 min, but

[1]Present address: Department of Fisheries and Oceans, Pacific Biological Station, Nanaimo, British Columbia V9R 5K6, Canada.

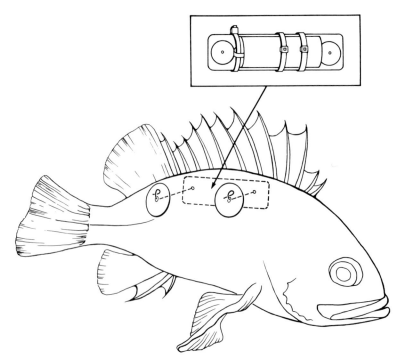

FIGURE 1.—Technique for attaching a harnessed transmitter to a rockfish by means of Petersen discs.

all the dummy transmitters were regurgitated within 24 h.

We then tried to attach external tags to the caudal peduncle of the rockfish, using a method effective for red drums *Sciaenops ocellatus* (Carr and Chaney 1977). In this method, the transmitter is attached to an inelastic pull-tie harness that is wrapped around the caudal peduncle. However, even when the pull ties were cinched tight, the tag always slipped off, probably because of the thickness of the caudal peduncle in relation to the tail.

We attempted to attach transmitters by threading inelastic pull ties through the dorsal musculature, as has been done on tuna (Holland et al. 1985). In this method, the nylon loop embedded in the transmitter is used as an anchor for one pull tie, and another tie is wrapped around the tag and through the body to prevent the transmitter from dangling. The two pull ties are inserted through the musculature and cinched tight. The method failed. Not only did it mutilate rockfish musculature, but there was not enough space on the fish to secure the tag.

Finally, we tried an external "sidesaddle harness" similar to that described by Haynes and Gray (1979) and Mellas and Haynes (1985). To assemble the harness, the transmitter was first attached with three inelastic pull ties to a rectan-

gular piece of flexible plastic (70 mm long × 20 mm wide × 1 mm thick) that had two holes drilled through opposite ends (Figure 1). One tie was pulled through the embedded transmitter loop, then wrapped around the plastic rectangle, and the remaining two ties were cinched down around the transmitter and plastic piece. Petersen discs were then lined up over the holes on the plastic, and pins were pushed through the muscle just beneath the dorsal fins. On the opposite site of the fish, the pins were run through two more Petersen discs, snipped of extra length, and twisted tight. The pins prevented the transmitter from sliding off the plastic as well as securing it to the fish. Before the harness was assembled, the transmitter was activated by connecting the two exposed wires, soldered, and epoxied.

In the field, rockfish were captured underwater with a fish anesthetic and a hand-held net, and slowly brought to the boat for tag attachment (Matthews and Reavis 1990, this volume). After attaching the tag, we returned the fish to the bottom in a net bag and observed them for signs of stress or inability to swim. They revived in 5–10 min.

The transmitters (Vemco V3-1 Lo pingers, Vemco Limited, Nova Scotia) operated at one of four crystal-controlled frequencies—50, 60, 65.54,

TABLE 1.—Summary of tracking data for copper and quillback rockfish tagged with ultrasonic transmitters in central Puget Sound, 1987.

Tag number	Rockfish species	Total length (cm)	Tracking interval	Tracking duration (d)[a]	Comment
6325	Copper	28	Jul 21–Aug 11	22	Captured and released at Orchard Rocks
6326	Quillback	30	Aug 28–Sep 18	22	Captured and released at Orchard Rocks
6324	Copper	33	Jul 24–Aug 14	22	Captured and released at Bainbridge
6330	Copper	30	Aug 2–Aug 20	19	Captured and released at Bainbridge
6322	Quillback	38	Jul 30–Aug 15	17	Moved 500 m
6323	Copper	27.5	Aug 31–Sep 20	21	Moved 500 m
6328	Copper	34	Jul 29–Aug 17	20	Moved 500 m
6329	Quillback	38	Aug 27–Sep 16	21	Moved 500 m
6331	Quillback	37	Jul 29–Aug 27	30	Moved 500 m
6332	Copper	29.5	Aug 28–Sep 22	26	Moved 500 m
6333	Copper	27.5	Jul 31–Aug 26	27	Moved 500 m

[a]Tracking ended when transmitter failed, so tracking duration equals transmitter life.

or 69 kHz—corresponding to preset channels on the Vemco VR 60 receiver. For this project we had three tags on each of the four frequencies. Individual tags were differentiated by their pulse period, which was decoded and displayed by the receiver. Each tag had an expected battery life of 40 d.

To locate the transmitter, we used a Vemco V-10 directional hydrophone bolted to a 2.4-m section of 5.0-cm-diameter galvanized steel pipe. The hydrophone cable was run through the pipe and attached to the receiver on the boat. The pipe was attached to a bracket placed amidships on our research vessel, a 5.8-m Glasply boat. An elbow and a 0.3-m section of pipe was added to the larger pipe section as a handle, which let us rotate the hydrophone a full 360°. The hydrophone was approximately 2 m below the water's surface. Lines secured the pipe fore and aft to prevent it from bending while the boat was under way.

We located transmitters by searching the study area and rotating the hydrophone 360°. We immediately knew which direction to steer the boat because the signal was strongest when the hydrophone was pointed towards the transmitter. As we approached the transmitter, signal intensity increased until the hydrophone was directly over the tagged fish, at which time the signal strength did not change regardless of hydrophone direction. We then took Loran C readings, depth soundings, and visual compass bearings on four charted features (buoys, piers, etc.) in the study area. For each fish, we accumulated a series of locations charted on a map, collected almost daily for the life of transmitter's battery.

Study area.—The study was conducted in Rich Passage on the southeastern side of Bainbridge Island in central Puget Sound. This is an extensive hard-bottom region with high- and low-relief reefs. The high-relief rocky reef, Orchard Rocks, is approximately 600 m offshore, and during the summer it has an overstory of bull kelp *Nereocystis leutkeana*. It is an offshore pinnacle with boulders and ledges that rise as much as 5 m off the bottom. The entire reef measures approximately 5 hectares. Nearby, extending along most of the southern shoreline of Bainbridge Island, is a low-relief area with an extensive kelp bed. It has a featureless flat bottom. Depths encountered during ultrasonic tracking ranged from 3 to 30 m. The water temperature during this study ranged from 11 to 13°C.

From July 21 through September 22, 1987, we tagged and periodically located 11 rockfish ranging from 27 to 38 cm, total length (Table 1). Up to seven tags were tracked in the study area at one time. Two copper rockfish were captured, tagged, released, and monitored at their original point of capture on the low-relief reef. One quillback rockfish and one copper rockfish were captured, tagged, released, and monitored at their original point of capture on the high-relief reef. Data were collected on locations during days, nights, and rapid tidal currents. Four copper and three quillback rockfish were captured at Orchard Rocks, tagged, and released at the Bainbridge Island low-relief reef, 500 m from the capture site. These displaced fish were monitored as they returned to their home sites.

Results and Discussion

Although the transmitters were supposed to function for 40 d, they ranged in life from 17 to 30 d (Table 1). Six of the displaced fish were followed until they returned to their home sites, but one fish could not be located after 17 d. Two fish (tags

6331 and 6333) were recaptured immediately after the signal could no longer be detected, and we confirmed that the transmitters had failed. The transmitters appeared normal with no signs that water had penetrated the epoxy seal and corroded the wires. During scuba surveys we resighted several more tagged fish that we had been unable to locate with ultrasonic telemetry, and again it appeared the transmitters had prematurely failed. These visual sightings were proof that the fish had not left the area or died.

Reception of the 50–69-kHz transmitters varied with the habitat. Transmitters on fish associated with the low-relief reef gave signals detectable up to 1 km away, even under dense kelp cover. Transmitters on fish associated with high-relief areas of the reef had the poorest reception. If a fish was underneath rock (verified by scuba observation), the signal was only detectable for 50–100 m. If it was near but not under rock, the signal was detectable for 200–500 m.

Our technique for attaching tags externally to demersal rockfish was effective but could be improved. For future studies, we recommend adding a thin sheet of neoprene between the fish and plastic plate to reduce abrasion. A smaller tag might also be tolerated better by the fish. When the tagged rockfish were first released, they appeared stressed and had trouble maneuvering; however, within a few minutes they swam normally. During our scuba surveys (Matthews and Reavis 1990) we saw tagged fish behaving normally. After the transmitter batteries died, we speared several fish and inspected them for tag-related injuries. There were signs of abrasion but the fish had been behaving normally and had food in their stomachs. One tagged fish was recaptured in January after wearing a tag since August. The fish appeared healthy and was swimming normally, although the pin holes had become enlarged. We removed the tag, reoutfitted the fish with a modified anchor tag, and released it for further observation. More evidence that tags did not impair swimming ability came from comparative information gathered in a companion study (Matthews and Reavis 1990). We found that displaced rockfish tagged with a modified anchor tag returned to their home sites as quickly as did those with transmitters. For these demersal species, whose movements do not depend on swimming speed and endurance, it appeared that external attachment was acceptable.

Because the rockfish we studied are slow, we could follow them even when high-relief rocks reduced signal reception, although displaced fish were sometimes difficult to locate. For researchers studying more mobile fish that associate with high-relief rock, the reduced reception could present a problem. For example, black rockfish *S. melanops* are considerably more mobile than copper and quillback rockfish, and they may move up to 80 km (Culver 1987). Black rockfish, however, tend to hover over the tops of rocks (Moulton 1977) rather than under them, which would facilitate reception.

Another reception problem arose when seven transmitters were used simultaneously in the study area. Transmitters in the same area but on different frequencies sometimes interfered with each other. Specifically, the signal from one frequency sometimes "bled" onto the channel being monitored. This confounded our searches when we mistakenly believed we were receiving a signal from a particular transmitter. However, with experience we came to recognize signal bleeds.

Even in areas of dense bull kelp, reception was good and presented no problems. This is contrary to Ziebell's (1973) report of significant signal loss when catfish were tracked in dense vegetation. The bull kelp was so thick, incidentally, that it entangled our boat, propeller, and hydrophone support. In addition to the surface canopy of bull kelp, there was a dense understory of *Pterygophora californica* and *Agarum fimbriatum*, which did not hinder reception. Reception might be reduced around giant kelp *Macrocystis pyrifera*, which can form denser stands and has more vertical structure.

Telemetry provided information not available from other sources. In our scuba study of tagged rockfish movements (Matthews and Reavis 1990), we were not able to collect any information on movements at night or during periods of strong current; Rich Passage tidal currents up to 7 km/h are common, and they make scuba observation impractical and unsafe. Our displacement experiments with anchor-tagged fish proved that the fish returned, but they provided no information on what routes were taken. Ultrasonic telemetry led to a detailed understanding of rockfish movement.

Among the major drawbacks to using ultrasonic transmitters are the small sample size, the cost of transmitters, and the prolonged observation required. Also, researchers have to assume that the fish sampled are displaying representative behavior. By combining ultrasonic tagging with a companion study using anchor tags (Matthews and Reavis 1990), in which we tagged 435 rockfish and made 588 resightings, we were able to overcome some of the limitations of each tagging technique.

Acknowledgments

We thank the Washington Department of Fisheries (WDF), specifically Greg Bargmann, Cyreis Schmitt, and Wayne Palsson, for generous support of this project, which included the use of the WDF boat *Research 3*. Robert Reavis and Vince Macurdy assisted with all phases of the fieldwork, and their help is greatly appreciated. Also assisting with field operations were Alan Olson and Amy Unthank. The University of Washington Graduate School Research Fund provided funds to purchase the ultrasonic tracking equipment. This is contribution 765 from the University of Washington Fisheries Research Institute.

References

Carr, W. E. S., and T. B. Chaney. 1977. Harness for an attachment of an ultrasonic transmitter to the red drum, *Sciaenops ocellata*. U.S. National Marine Fisheries Service Fishery Bulletin 74:998–1000.

Culver, B. N. 1987. Results from tagging black rockfish (*Sebastes melanops*) off the Washington and northern Oregon coast. Pages 231–240 *in* Proceedings of the international rockfish symposium. University of Alaska, Sea Grant Report 87-2, Fairbanks.

Hawkins, A. D., and G. G. Urquhart. 1983. Tracking fish at sea. Pages 103–166 *in* A. G. MacDonald and I. G. Priede, editors. Experimental biology at sea. Academic Press, London.

Haynes, J. M., and R. H. Gray. 1979. Effects of external and internal radio transmitter attachment on movement of adult chinook salmon. Pages 115–128 *in* Proceedings of the second international conference on wildlife biotelemetry. University of Wyoming, Laramie.

Holland, K., R. Brill, S. Ferguson, R. Chang, and R. Yost. 1985. A small vessel technique for tracking pelagic fish. U.S. National Marine Fisheries Service Marine Fisheries Review 47(4):26–32.

Matthews, K. R., and R. H. Reavis. 1990. Underwater tagging and visual recapture as a technique for studying movement patterns of rockfish. American Fisheries Society Symposium 7:168–172.

Mellas, E. J., and J. M. Haynes. 1985. Swimming performance and behavior of rainbow trout (*Salmo gairderi*) and white perch (*Morone americana*): effects of attaching telemetry transmitters. Canadian Journal of Fisheries and Aquatic Sciences 42:488–493.

Moulton, L. L. 1977. An ecological analysis of fishes inhabiting the rocky nearshore regions of northern Puget Sound, Washington. Doctoral dissertation. University of Washington, Seattle.

Stasko, A. B., and D. G. Pincock. 1977. Review of underwater biotelemetry, with an emphasis on ultrasonic techniques. Journal of the Fisheries Research Board of Canada 34:1261–1285.

Winter, J. D. 1983. Underwater biotelemetry. Pages 371–395 *in* L. A. Nielsen and D. L. Johnson, editors. Fisheries techniques. American Fisheries Society, Bethesda, Maryland.

Ziebell, C. D. 1973. Ultrasonic transmitters for tracking channel catfish. Progressive Fish-Culturist 35:28–32.

American Fisheries Society Symposium 7:380–383, 1990

Use of Staple Sutures to Close Surgical Incisions for Transmitter Implants

Donald G. Mortensen

Alaska Fisheries Science Center, Auke Bay Laboratory, U.S. National Marine Fisheries Service
Post Office Box 210155, Auke Bay, Alaska 99821, USA

Abstract.—Dummy ultrasonic transmitters were implanted in juvenile chinook salmon *Oncorhynchus tschawytscha* through an incision along the ventral midline, which was closed with staple sutures. Other juveniles underwent the same operation but did not receive implants. The fish were held and observed in a seawater net pen for 15 d, and were then killed and examined. Closure with surgical staples reduced handling time relative to needle-and-thread closure by 50%. All fish swam normally 6 h after surgery, and they displayed normal feeding responses 24 h after surgery. The staple sutures held well in 48 of the 50 operations. The incisions healed, and only one fish showed signs of internal infection. The results indicated that staple sutures can be used effectively with small salmon, and that implantation of transmitters in the visceral cavity is relatively safe for chinook salmon.

For over 20 years, researchers have surgically implanted transmitters in various fish (Stasko and Pincock 1977). Hart and Summerfelt (1975) provided a detailed description of a relatively simple surgical method for implanting ultrasonic transmitters in the visceral cavity of flathead catfish *Pylodictis olivaris*. Their method has since been successfully used to implant transmitters in channel catfish *Ictalurus punctatus* (Summerfelt and Mosier 1984), rainbow trout *Oncorhynchus mykiss* (formerly *Salmo gairdneri*; Chisholm and Hubert 1985), and muskellunge *Esox masquinongy*) (Crossman 1977; Miller and Menzel 1986). The method involves anesthetizing the fish, making an incision along the linea alba, inserting a small transmitter in the visceral cavity, and closing the incision with needle-and-thread sutures.

In 1986, a modification of this method was used to implant dummy ultrasonic transmitters in the visceral cavity of juvenile chinook salmon *Oncorhynchus tschawytscha*. The objective of this study was to determine the effectiveness of surgical staple sutures, and to document the short-term biological effects and mortality associated with the various steps of the surgical procedure.

The modified surgical method was tested for possible use in studies of the residence time and migration routes of juvenile chinook salmon in the marine waters of southeast Alaska. Stomach inserts were ruled out for these studies because of concerns that the transmitters would interfere with feeding behavior in the relatively small fish. External attachment of transmitters also was not considered, on the basis of a study by Lewis and

Muntz (1984) that indicated the chronic drag from transmiters externally attached to rainbow trout led to decreased swimming performance. Another study of rainbow trout, by Mellas and Haynes (1985), showed that drag from external transmitters caused exhaustion sooner than did surgical implants.

Methods

The juvenile chinook salmon used in this experiment came from a population cultured in seawater net pens at our laboratory's research station at Little Port Walter, Alaska. They were divided into four groups of 25 each, according to the degree of handling received (Table 1). The fish in group 1 served as controls. They were randomly selected and transferred by dip net from the culture net pen to a saltwater net pen 7 m × 7 m × 7 m, where they received no further handling.

The fish in group 2 were dipped from the culture net pen and anesthetized in a solution of tricaine (MS-222) until loss of equilibrium occurred. Each fish was then marked with a uniquely numbered anchor tag (Figure 1A), placed in a recovery tank of fresh seawater until equilibrium was regained (about 3 min), and transferred to the saltwater net pen with the group-1 fish.

The fish in group 3 were anesthetized and marked with uniquely numbered anchor tags. Each fish was placed upside down in a tagging cradle in a shallow recovery tank. The abdomen was above the water line, but the gills were immersed in fresh seawater. A scalpel was used to make a 1.5–2.0-cm incision along the linea alba

TABLE 1.—Treatment groups ($N = 25$) for testing the effects of surgical implantation techniques on juvenile chinook salmon.

Group	Fork length (cm): mean (SD)	Weight (g): mean (SD)	Treatment
1	28.4 (1.1)	361 (2.4)	Control
2	28.2 (1.3)	346 (3.3)	Anesthetic and anchor tag
3	28.1 (1.2)	351 (1.7)	Anesthetic, anchor tag, incision, and staple sutures
4	27.9 (1.7)	349 (3.1)	Anesthetic, anchor tag, incision, dummy transmitter, and staple sutures

anterior to the base of the pelvic fins (Figure 1B). The outside of the incision, as well as the peritoneal cavity, was flooded with a 3% solution of providone iodine. Four or five long-leg staple sutures were used to close the incision, instead of the usual needle-and-thread sutures (Figure 1D). All surgical instruments, including the staples, were disinfected in a 5% solution of providone iodine prior to surgery. Each fish was removed from the tagging cradle and allowed to right itself before being released into the holding pen.

Each fish in group 4 received the same treatment as the group 3 fish, but a disinfected dummy transmitter was implanted in the visceral cavity before the incision was stapled shut (Figure 1C). Dummy transmitters were fabricated from fiberglass tubing 8.5 mm in diameter and 38 mm long. Each dummy was filled with lead shot and fiberglass resin so that it weighed about 5.14 g in air, or about 1.5% of the fish's body weight. The ends of the dummy transmitters were dipped in fast-setting epoxy to eliminate sharp edges and to simulate the sealing caps of actual transmitters. Figure 1E shows the incision sutured shut with staples. The staples remain in the fish and eventually become imbedded in the flesh.

A diver entered the holding pen 6 h after completion of the surgery to retrieve any fish that might have died and to observe fish behavior. These observations continued daily for 15 d, after which the study was terminated and the fish were killed. Each of the fish that had undergone surgery was examined for evidence of healing, suture failure, retention of the dummy transmitter, gross signs of internal infection, and the presence of food in the stomach.

Results

Implantation surgery took an average of 3 min; practice lessened the time required. Postoperative holding times ranged from 1 to 5 min. Six hours after the last fish was placed in the holding pen, none of the fish exhibited abnormal swimming behavior. Pelletized food was presented to the salmon 24 h after surgery, and they fed aggressively. There were no mortalities during the study.

At the end of the study, all fish had food in their stomachs. The fish from groups 1 and 2 appeared normal. One fish in group 3 was missing four of its five suture staples, and its incision was partly open. There was no apparent internal infection. The incisions of the remaining 24 fish in group 3 were tightly closed by the staples, and healing had begun. One fish in group 4 was missing three of its original five sutures, and its incision was partly open; however, the dummy transmitter remained within the visceral cavity. Closer examination of the visceral cavity revealed large amounts of mucus, necrotic serosal tissue, and hemorrhagic areas on both the body wall and intestinal mass adjacent to the dummy transmitter. The remaining 24 salmon in group 4 had tightly closed incisions that were beginning to heal, and their visceral cavities showed no internal infection or injury.

Discussion

Handling stress should be kept to a minimum in fish-tagging procedures. Hart and Summerfelt (1975) noted that transmitters could be implanted in channel catfish in 15 min. Marty and Summerfelt (1986) were able to complete their surgeries, also on channel catfish, in 6 min. Researchers who required 8 min, on the average, to implant ultrasonic transmitters in wild juvenile chinook found that suturing with needle and thread was the most time-consuming portion of the procedure (D. Mortensen, unpublished data). The staple sutures used in the present study reduced surgery time at least by 50%, which subjected fish to less stress

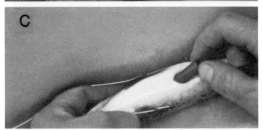

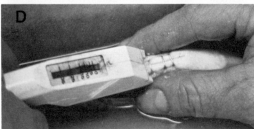

FIGURE 1.—Procedures for evaluating staple sutures used in surgeries on juvenile chinook salmon. (A) Fish from groups 2, 3, and 4 were marked with a uniquely number anchor tag. (B) Fish from groups 3 and 4 were incised for 1.5–2.0 cm along the linea alba. (C) Dummy transmitters were implanted in the visceral cavities of fish in group 4. (D) The incisions of fish in groups 3 and 4 were closed with staple sutures. (E) The staples formed occlusive seals.

from handling and anesthesia, and may have shortened recovery time.

Surgical implantation of transmitters in fish was thought to be suited only to long-term studies because of certain postoperative effects (Stasko and Pincock 1977). For example, fish recovering from implantation surgery have responded to the added weight of the transmitter by becoming negatively buoyant. Buoyancy compensation was accomplished in 3 h by 85% of Atlantic salmon smolts *Salmo salar*, whose transmitter weight averaged 4.4% of their body weight (Fried et al. 1976). Buoyancy compensation by juvenile chinook salmon with dummy transmitters averaging 1% of their body weight should take no longer. Blood loss from the surgical procedure should not be a problem, given the absence of major arteries and veins in the linea alba. Assuming the staple sutures close the incision tightly, postoperative stress from fluid loss should be negligible. Therefore, the major postoperative effect to overcome is the handling associated with the tagging process. In this study, the reappearance of normal feeding and swimming behavior 24 h after surgery indicated that postoperative trauma was not extensive and that staple sutures have merit. The failure of the sutures in two salmon was probably caused by faulty suture placement. The infection noted in one salmon from group 4 probably resulted from the open incision, which allowed bacteria to enter the visceral cavity, and was exacerbated by the dummy transmitter.

Wild juvenile chinook salmon implanted with ultrasonic transmitters have been tracked successfully during 15-d studies in Alaska (D. Mortensen, unpublished data). Throughout the studies, the fish were active and swam almost constantly, with occasional high-speed bursts. However, transmitter expulsion may limit how long juvenile chinook salmon can be tracked. Expulsion within 1–3 months of transmitter implantation has been noted for channel catfish (Summerfelt and Mosier 1984; Marty and Summerfelt 1986) and rainbow trout (Chisholm and Hubert 1985). The problem of transmitter expulsion should be resolved before ultrasonic transmitters are used in long-term tracking studies of juvenile chinook salmon.

References

Chisholm, I. M., and W. A. Hubert. 1985. Expulsion of dummy transmitters by rainbow trout. Transactions of the American Fisheries Society 114:766–767.

Crossman, E. J. 1977. Displacement and home range of muskellunge determined by ultrasonic tracking. Environmental Biology of Fishes 1:145–158.

Fried, S. M., J. D. McCleave, and K. A. Stred. 1976. Buoyancy compensation by Atlantic salmon (*Salmo salar*) smolts tagged internally with dummy telemetry transmitters. Journal of the Fisheries Research Board of Canada 33:1377–1380.

Hart, L. G., and R. C. Summerfelt. 1975. Surgical procedures for implanting ultrasonic transmitters into flathead catfish (*Pylodictis olivaris*). Transactions of the American Fisheries Society 114:56–59.

Lewis, A. E., and W. R. A. Muntz. 1984. The effects of external ultrasonic tagging on the swimming performance of rainbow trout, *Salmo gairdneri* Richardson. Journal of Fish Biology 25:577–585.

Marty, G. D., and R. C. Summerfelt. 1986. Pathways and mechanisms for expulsion of surgically implanted dummy transmitters from channel catfish.

Transactions of the American Fisheries Society 115:577–589.

Mellas, E. J., and J. M. Haynes. 1985. Swimming performance and behavior of rainbow trout (*Salmo gairdneri*) and white perch (*Morone americana*): effects of attaching telemetry transmitters. Canadian Journal of Fisheries and Aquatic Sciences 42:488–493.

Miller, M. L., and B. W. Menzel. 1986. Movements, homing, and home range of muskellunge, *Esox masquinongy*, in West Okoboji Lake, Iowa. Environmental Biology of Fishes 16:243–255.

Stasko, A. B., and D. G. Pincock. 1977. Review of underwater biotelemetry, with emphasis on ultrasonic techniques. Journal of the Fisheries Research Board of Canada 34:1261–1285.

Summerfelt, R. C., and D. Mosier. 1984. Transintestinal expulsion of surgically implanted dummy transmitters by channel catfish. Transactions of the American Fisheries Society 113:760–766.

American Fisheries Society Symposium 7:384–389, 1990
© Copyright by the American Fisheries Society 1990

Monitoring the Nearshore Movement of Red King Crabs under Sea Ice with Ultrasonic Tags

PAUL C. RUSANOWSKI[1]

NORTEC, A Division of ERT, 750 West Second Avenue, Anchorage, Alaska 99501, USA

E. LINWOOD SMITH AND MARK COCHRAN

E. L. Smith and Associates, 3030 North Longhorn Drive, Tucson, Arizona 85749, USA

Abstract.—Nearshore movements of red king crabs *Paralithodes camtschaticus* were tracked during 1986–1988 near Nome, Alaska, to assess the potential effects of placer gold dredging on the animals' use of winter habitat. External ultrasonic tags were secured to their carapaces and the animals were tracked with a digital receiver and directional hydrophone. Tags were coded by pulse intervals for individual recognition. After release, tagged red king crabs were relocated at intervals of 15 min to 4 h during daylight hours for periods ranging up to 10 d.

The Nome offshore placer project is the realization of an offshore gold-mining operation initiated in 1962 by Shell Oil Company. Western Gold Exploration and Mining Company Limited Partnership (WestGold) began dredging offshore for placer gold in October 1985 in the vicinity of Nome, Alaska (Figure 1).

The relatively small population of red king crabs *Paralithodes camtschaticus* in Norton Sound is most concentrated near Nome (Wolotira et al. 1977, 1979; Powell et al. 1983). A seaward (southwesterly) migration consisting at least of adult males occurs in early summer, and a return northeastward migration occurs in late fall (Powell et al. 1983). In general these crabs inhabit shallow, nearshore waters (less than 30 m deep), as do juvenile and spawning red king crabs in other regions of their range (Jewett and Powell 1981). In the winter, the Norton Sound population near Nome supports a native through-the-ice subsistence fishery (Ellanna 1983; Powell et al. 1983). Since the mid-1980s, it has also supported a small winter commercial fishery.

The Alaska Department of Fish and Game trapped and Floy wire-tagged red king crabs in Norton Sound from 1982 to 1986. However, most of the recovered tags came from animals taken in the offshore commercial fishery conducted during the summer. Information on red king crabs in nearshore areas was limited to word-of-mouth accounts from village elders and town residents, and to the occasional recapture of a tagged animal.

An extensive program was begun in 1986 to document the effects of WestGold's offshore mining operation on benthic winter habitat used by red king crabs. As part of the program, the pre- and postmining distribution and abundance of red king crabs within the mining leases were monitored to determine whether local commercial and subsistence fisheries were affected by the gold dredging. We contributed to the project by using ultrasonic tags to monitor nearshore movements of red king crabs in their winter habitat.

Methods

From 1986 to 1988, we tracked red king crabs during periods of ice cover and open water in the nearshore area of Norton Sound near Nome, Alaska (Figure 1). Various equipment and techniques were tried, but only those that proved most successful are described here in detail.

Red king crabs were captured in baited, cone-shaped crab pots. All the crabs captured in each pot were released except for a single mature male, which was tagged, returned to the pot so that tracking equipment could be calibrated and tested, and then released.

Specifications of the ultrasonic tag selected for use are shown in Table 1. Receiving equipment consisted of a digital receiver with a range of 10–110 kHz and a directional hydrophone. The manufacturer modified the receiver before the second year of the study by increasing preamplifier gain and installing an impedance-matching transformer to enhance receiver sensitivity. External speaker amplifiers and stereo headphones were used with the receiver. Hydrophone direction was determined with a hand-held Silva com-

---

[1]Present address: WestGold, 184 East 53rd Avenue, Anchorage, Alaska 99518, USA.

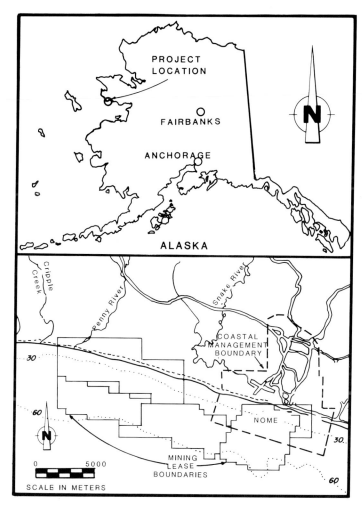

FIGURE 1.—Map of the study area, with details on mining lease boundaries. The 30- and 60-m depth contours are shown as dotted lines.

pass. The system cost approximately US$1,500, plus $225 for each additional tag.

Ultrasonic tags were bound to the back of the red king crab carapace with two nylon-coated wires. One end of each wire was looped through a livestock castration ring that had been placed around the base of each of the animal's third legs, and the other end was tied to the tag (Figure 2). A small hole was drilled through the end cap of the tag to facilitate attachment of the wire. Lead sleeves were clamp onto the wires to strengthen the attachment. Wire length was adjusted to prevent the tag from slipping over the posterior of the carapace. To secure the ultrasonic tags in this manner, it was necessary to use mature red king crabs with a carapace diameter larger than 100 mm.

At each station, we fixed our position with a Micrologic ML-8000 Loran C receiver and navigator-positioning system, and we determined a bearing to the sonic tag with the directional hydrophone and hand-held compass.

We relocated tagged crabs by monitoring at a series of listening stations, where we used a 20-cm-diameter power auger to cut a hole in the ice for the hydrophone. The support vehicle (a snowmachine) was then moved to a second location several hundred meters away and the procedure was repeated. Tag position was computed by plotting bearing intercepts on a lattice chart.

Results and Discussion

Sixteen red king crabs were tagged and released from 1986 through 1988. Twelve of them were

TABLE 1.—Specifications high-power coded ultra-sonic tags (CHP-85-1)[a] used to track red king crabs in Norton Sound.

Characteristic	Specification
Frequency	75 kHz nominal
Life	1 year, shipped operating
Range	3,000 m in seawater
Size	16 mm in diameter, 112 mm long
Construction	Epoxy-sealed, polyvinyl chloride case; predrilled cap ends for attachment
Weight	29 g in air, 10 g in water
Code	Up to 18 pulses and pauses

[a]Sonotronics, Tucson, Arizona.

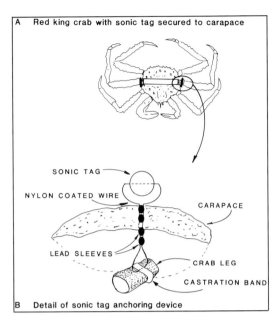

FIGURE 2.—Location and attachment technique for ultrasonic tags used on red king crabs.

tracked for periods ranging from 4 h to 10 d. Three animals escaped from crab pots prior to tracking and were not relocated. Only one tag failure occurred, caused by a faulty internal battery connection damaged during shipment.

Early efforts to monitor nearshore daily and short-term movements of red king crabs under ice met with very limited success. Improvements in equipment and techniques were made during subsequent open-water and under-ice monitoring efforts. The final configuration of equipment and methods, which proved highly successful, has been described in the previous section. The improvements were in three areas: tag attachment, signal reception, and relocation.

Ultrasonic Tag Attachment

Adult male red king crabs range from 90 to 140+ mm in carapace length and nearly the same in width. Numerous spines cover the carapace, making it a poor surface for tag attachment. Attempts to glue the sonic tag to the spiny carapace with Sea-poxy putty, and to secure the tag with baling wire looped around the bases of the crab's third legs, were unsuccessful. Tags were successfully secured to red king crabs, however, when mason's twine was used instead of baling wire. The twine was attached to the base of the third leg by looping it through a stockman's rubber castration ring, which had been applied with the help of castration ring pliers that opened the heavy device so it could be passed over the leg. This procedure was repeated on the opposite leg, and the twine was tied to each castration ring in such a manner that the tag stayed on the dorsal side of the carapace without hindering mobility of the crab. This method resulted in at least a 2-month retention of tags. However, the twine eventually frayed where it came in contact with the shell. For the most effective attachment (al-ready described in the Methods section), we substituted 18-kg-test, nylon-coated, stainless steel fishing leader for the mason's twine.

Signal Reception

Problems with signal reception were not anticipated because the high-power tags we selected had a range of at least 3,000 m in seawater, and tracking was to be conducted at depths of 8–20 m. However, under rough and heavily ridged sea ice in 1986, detection distances varied and were less than 500 m. We attempted to improve this range by lowering the hydrophone from immediately below ice to a depth of 9 m, but were unsuccessful. In the intervening open-water season, the receiver was modified by the manufacturer to increase sensitivity. In contrast to 1986, the nearshore ice in 1987 was frozen into a flat expanse approximately 1.8 km wide × 11 km long in the study area. Signal reception problems were not experienced in the 1987 season, and tags were occasionally detected at distances of 4,000 m. In 1988, ice conditions were similar to those of 1986, and signal ranges were limited to less than 1 km. The temperature gradient is the primary factor affecting signal range, abetted by salinity, pressure, and the presence of ice crystals or air bubbles in the water column. The roughness of

TABLE 2.—Tracking data for red king crabs under sea ice near Nome, Alaska, in 1987.

Tag code	Tracking duration	Total distance traveled (m)	Net distance moved (m)
258	9 d	9,994	950
2327	6 d	3,512	500[a]
249	5 d	1,233	10
2534	33 h	1,435	335
2228	30 h	2,576	360
2345	24 h	1,463	825

[a]Recovered by a local subsistence fisherman 49 d after release, approximately 2,000 m from the release point.

surface ice also affects range, possibly in conjunction with one or more of the above factors.

As a means of identifying individual crabs when more than one tag is used, a unique pulsed code is built into each tag. For example, the code 23 would be represented by continuous repetition of the following sequence: 2 pulses, pause, 3 pulses, long pause. Pulsed codes of 2 to 4 digits were used in this study. Simple low-count codes of 2 or 3 digits proved most effective. The larger and higher code-counts significantly increased the time required to distinguish a code from others and to locate the direction of maximum signal strength. To expedite relocation and code identification, only two animals were released and tracked in any one location.

Relocation

We tested several relocation methods before finding one that was reliable in winter arctic conditions. The simplest system, relying on a compass and bearing shots on local landmarks, proved unsuitable because of a shortage of recognizable landmarks. A Motorola mini-ranger was highly reliable, but it required an elaborate setup of shore stations and logistics to support the navigation network. The approach that proved most effective was to use a Loran C unit, which was portable, required no shore stations, and ran on a single 12-V battery. We made additional adjustments in our relocation method in response to winter weather. Headphones proved awkward and extremely cold to the ears; so, when signal strengths were moderate to strong, we traded the headphones for a small amplifier–speaker. Also, we employed a simple, easily adjusted compass

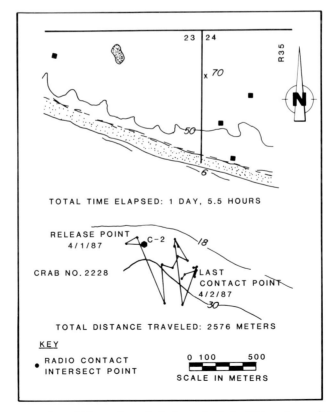

FIGURE 3.—Movements of a red king crab tracked ultrasonically for 30 h under sea ice, April 1–2, 1987.

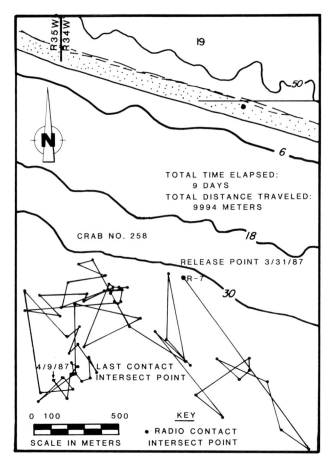

FIGURE 4.—Movements of a red king crab tracked ultrasonically for 9 d under sea ice, March 31–April 9, 1987.

for bearing movements, because it could be used even while we wore heavy gloves.

Crab Movements

The results of our 1987 winter tracking project are shown in Table 2. Seven red king crabs were tagged and released during a 10-d field program. One of the animals immediately moved out of range of our receiving equipment. Representative movements are shown in Figures 3 and 4 for two of the remaining six animals. The animals either quickly moved out of range of our equipment, or they appeared to wander haphazardly in the vicinity of the release point. Animal 2327 was captured by a subsistence fisherman 49 d after release, when it had only moved a net distance of approximately 2,000 m. We also observed localized meandering by animals tagged in the winter of 1988. In contrast to these under-ice movements, the animals tagged in early June just prior

to migration offshore showed movements of 2–4 km within 1–2 d of their release. They always moved in an offshore direction towards known summer concentration areas.

Summary

Equipment and technique modifications resulted in the development of an effective ultrasonic tracking method for red king crabs under sea ice. The method also was suitable for tracking animals in open water. It relied on readily available tracking equipment and ultrasonic tags, positioning with Loran C, and attachment of tags by means of nylon-coated wire. Red king crabs were tracked as long as 10 d, and tagged animals were recovered in subsistence and commercial fisheries up to 2 months after release. Ice conditions appeared to indicate water properties that can limit the range of transmitter signals. In years of rough, heavily ridged ice (1986, 1988), tags were detected at ranges of less than 1,000 m; but in 1987, when relatively smooth, ridge-free ice covered the study area, detection ranges in excess of 3,000 m were recorded.

Acknowledgments

We acknowledge the support of WestGold, which funded the work and authorized publication of this paper. We also thank Charles Lean of the Alaska Department of Fish and Game for generously donating his time and assistance on numerous occasions.

References

Ellanna, L. J. 1983. Nome: resource uses in a middle-size regional center of northwestern Alaska. Pages 82–120 in Resource use and socioeconomic systems: case studies of fishing and hunting in Alaskan communities. Alaska Department of Fish and Game, Subsistence Division, Technical Paper 61, Juneau.

Jewett, S. C., and G. C. Powell. 1981. Nearshore movement of king crab. Alaska Seas and Coasts 9(3):6–8.

Powell, G. C., R. Peterson, and L. Schwarz. 1983. The red king crab, Paralithodes camtschatica (Tilesius) in Norton Sound, Alaska: history of biological research and resource utilization through 1982. Alaska Department of Fish and Game, Information Leaflet 222, Juneau.

Wolotira, R. J., T. M. Sample, and M. Morin. 1977. Demersal fish and shellfish resources of Norton Sound, the southeastern Chukchi Sea, and adjacent waters in the baseline year 1976. U.S. National Marine Fisheries Service, Northwest and Alaska Fisheries Center, Seattle, Washington.

Wolotira, R. J., T. M. Sample, and M. Morin. 1979. Baseline studies of fish and shellfish resources of Norton Sound and the southeastern Chukchi Sea. Pages 258–572 in Environmental assessment of the alaskan continental shelf, final reports of principal investigators, volume 6. National Oceanic and Atmospheric Administration, Outer Continental Shelf Environmental Assessment Program, Boulder, Colorado.

American Fisheries Society Symposium 7:390–394, 1990

Evaluation of Pressure-Sensitive Radio Transmitters Used for Monitoring Depth Selection by Trout in Lotic Systems

Thomas H. Williams[1] and Robert G. White

U.S. Fish and Wildlife Service, Montana Cooperative Fishery Research Unit[2]
Montana State University, Bozeman, Montana 59717, USA

Abstract.—Pressure-sensitive radio transmitters were field-tested to determine their effectiveness in monitoring depth selection by brown trout *Salmo trutta* and rainbow trout *Oncorhynchus mykiss* in a large river. The surgically implanted transmitters had a mean length of 70.9 mm and an average weight of 9.6 g in water. Eight transmitters were implanted; of these, six were located and monitored, and four were recovered. Substantial electronic drift, detected by recalibrating the transmitters recovered after 21 d, indicated that the depth data were unreliable. The transmitters must be redesigned to reduce size, to overcome problems of electronic drift, and to increase pulse repetition in relation to change in depth if they are to be of value in similar behavioral studies.

Radiotelemetry is commonly used with fish in lotic systems to determine migration behavior (McCleave et al. 1978; Gray and Haynes 1979) and habitat selection (Chisholm et al. 1987; Hurley et al. 1987). Temperature-sensing transmitter have been used to monitor depth selection by steelhead *Oncorhynchus mykiss* (anadromous rainbow trout; formerly *Salmo gairdneri*) in thermally stratified lentic environments (Haynes et al. 1986). For fish of rivers and streams, where there is no thermal stratification, radiotelemetry relies on pressure-sensitive transmitters to indicate depth. Such transmitters vary pulse repetition in response to changes in hydrostatic pressure at different depths. Haynes (1978) applied this technology to monitor swimming depths of migrating adult chinook salmon *Oncorhynchus tshawytscha* in relation to supersaturated levels of dissolved gas. Because of high power requirements, the transmitters were large and could be implanted only in fish weighing more than 5.0 kg. In the present study, we have taken advantage of smaller pressure-sensitive transmitters to estimate depths selected by resident brown trout *Salmo trutta* and rainbow trout in relation to changes in dissolved gas levels in the Bighorn River, Montana.

Methods

Our pressure-sensitive radio transmitters were designed and built by Custom Telemetry and Consulting (CTC) of Athens, Georgia.[3] They were tested in the Bighorn River in August 1987. The cylindrical transmitters had a mean length of 70.9 mm, a mean diameter of 20.3 mm, and an average weight of 9.6 g in water (Table 1). They were designed to be temperature-compensating, and their frequencies were in the 30-MHz range in order to operate in the highly conductive Bighorn River water (mean conductivity was 858 µS/cm in 1987).

Prior to implantation, transmitters were operated for 3 d and recalibrated to correct for electronic drift caused by battery power drain (Haynes 1978). Calibration tests were conducted in Afterbay Reservoir below Yellowtail Dam, where water quality and temperature resembled conditions in the Bighorn River. Each transmitter was suspended on a metered line at depths of 0, 1, 2, and 3 m, and pulse repetitions were measured. Transmitters recovered from trout after 21 d were recalibrated by the same methods to determine if electronic drift had occurred.

Eight transmitters were surgically implanted in the abdomens of four brown trout and four rainbow trout ranging in weight from 0.98 to 1.83 kg and in total length from 44.7 to 53.1 cm (Table 2). We followed the general procedures of Hart and Summerfelt (1975). Erythromycin (Erythro-200, 25 mg/kg) was injected into the dorsal sinus, and

---

[1] Present address: BioSystems Analysis, Inc., 3152 Paradise Drive, Building 39, Tiburon, California 94920, USA.

[2] The unit is jointly supported by Montana State University, the Montana Department of Fish, Wildlife, and Parks, and the U.S. Fish and Wildlife Service.

[3] Reference to trade names or manufacturers does not imply U.S. Government endorsement of commercial products.

TABLE 1.—Specifications of pressure-sensitive radio transmitters implanted in brown trout and rainbow trout in the Bighorn River, Montana.

Character	Description
Length (mean)	70.9 mm
Diameter (mean)	20.3 mm
Weight in air	33.6–35.0 g
Weight in water	9.3–10.0 g
Battery type	3.5-V lithium
Sensor	Cantilever silicon with pleated stainless steel diaphragm
Frequency	30.071–30.276 MHz
Theoretical life span	28–35 d

tetracycline (Polyotic) was applied to the incision to reduce the risk of infection. Transmitter weight in water did not exceed 1.25% of fish weight out of water, as recommended by Winter et al. (1978). Trout were retained in river holding cages for 12 h before release.

After release, tagged trout were located and monitored with a 124-cm^2 loop antenna and two different receivers. A programable scanning receiver and pulse counter (model 2000 Challenger Programable Scanner and Pulse Decoder) manufacturer by Advanced Telemetry Systems, Inc. (ATS) of Isanti, Minnesota, were used to locate fish (reception distance, approximately 400 m). When within about 40 m of the signal, we switched to a receiver and pulse counter manufactured by CTC (model CTC-AR-12 and Pulse Counter) to record pulse repetitions. Pulse repetition was measured from the leading edge of one pulse to the leading edge of the succeeding pulse. Pulse repetition time decreased with increases in hydrostatic pressure. The CTC unit measured pulse repetitions in thousandths of a second, and the ATS unit measured them in tenths of a second.

A linear regression was used to determine the relationship between pulse repetition and water depth for each transmitter. To determine depth resolution, we used unlimited simultaneous discrimination intervals (Lieberman et al. 1967, 1971). This method accounts for both the uncertainty in the observed pulse repetition and the uncertainty in the relationship between pulse repetition and water depth. Methods that do not take both uncertainties into account are not appropriate for measurements from precalibrated instruments. Transmitter calibration was done by regressing pulse repetition on known depths. We then converted field pulse repetitions to confidence intervals for various depths, using this regression equation along with estimates of both types of uncertainty. A computer program, designed by Milo Adkison and Dan Gustafson of Montana State University, performed the regression analysis.

Results and Discussion

The pressure-sensitive radio transmitters were ineffective in determining depth selection by trout. Accurate determination of depth was not possible because of insufficient change in pulse repetition with depth as well as electronic drift.

Lack of sufficient change in pulse repetition with depth resulted in a low calculated regression slope. Because the regression slope is used to interpret depth change, and because the discrimination interval influences depth resolution of the transmitter, depth resolution was not sufficient to accurately predict the depth of tagged fish. Pulse repetition typically changed about 0.030 s over a depth range of 0 to 3 m (Figure 1). The maximum observed change in pulse repetition was 0.048 s (Table 3). This deficiency was exacerbated by the relatively broad confidence intervals associated with the regression line (Figure 1). A steeper regression slope with better correlation between pulse repetition and depth would decrease the width of the discrimination interval and improve depth determination.

TABLE 2.—Lengths and weights of brown trout and rainbow trout implanted with pressure-sensitive radio transmitters, August 1987, Bighorn River, Montana.

Transmitter frequency (MHz)	Trout species	Total length (cm)	Weight (kg)	Transmitter weight in water (g)	Percent of trout weight[a]
30.071	Brown	51.3	1.09	10.0	0.92
30.088	Rainbow	50.8	1.31	9.6	0.73
30.097	Rainbow	52.6	1.83	9.5	0.52
30.110	Brown	51.8	0.99	9.3	0.94
30.125	Brown	51.1	1.00	9.8	1.00
30.138	Brown	53.1	1.72	9.3	0.54
30.258	Rainbow	46.2	1.13	9.5	0.84
30.276	Rainbow	44.7	0.98	10.0	1.02

[a] The transmitter weight in water as a percentage of the trout weight in air.

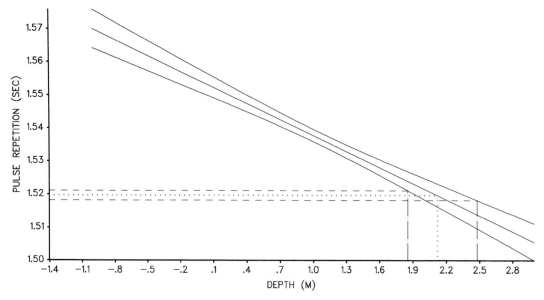

FIGURE 1.—Unlimited simultaneous discrimination intervals for a pressure-sensitive radio transmitter operating at a frequency of 30.097 MHz before implantation, Bighorn River, Montana. Solid lines are regressions of pulse repetition on depth and confidence-interval bands. Horizontal dashed lines are confidence intervals for mean pulse repetition given observed repetitions. Dotted lines give point estimates of depth based on regressions. Vertical dashed lines give discrimination intervals for depth given repetitions.

Electronic drift introduced more error into our calculations of fish depth. During field operation, the relationship between depth and pulse repetition changed and became weaker. Correlations between pulse repetition and depth were reduced for all four recovered transmitters, and regression slopes were flatter for three of them—30.097, 30.125, and 30.276 (Table 3).

To determine depth resolution of recovered transmitters that showed electronic drift, we

calculated the extreme low and high depth values at a particular pulse repetition for calibrations made before implantation and after recovery (Figure 2). Depth resolution of an individual transmitter at any given pulse repetition increases with proximity to the mean calibration depth, because the width of the confidence region of the regression increases as one extrapolates beyond the data. Consequently, we recommend that calibration be done at all experimental

TABLE 3.—Slopes and pulse-repetition values for regressions of pulse repetition on depth for pressure-sensitive radio transmitters implanted in brown trout and rainbow trout, August 1987, Bighorn River, Montana.

| Transmitter frequency | Slope | Repetition intercept | Pulse repetition (s) at depth | | | | r^2 |
			0 m	1 m	2 m	3 m	
30.071[a]	0.00083	1.554	1.554	1.555	1.555	1.556	0.6658
30.071[b]	−0.00132	1.539	1.539	1.537	1.536	1.535	0.1801
30.088[a,c]	−0.01196	1.298	1.298	1.287	1.275	1.263	0.9914
30.097[a]	−0.01612	1.551	1.551	1.535	1.519	1.503	0.9818
30.097[b]	−0.01402	1.528	1.528	1.513	1.498	1.482	0.9438
30.110[a]	−0.01616	1.629	1.629	1.613	1.597	1.581	0.9928
30.125[a]	−0.01274	1.582	1.582	1.569	1.557	1.544	0.9580
30.125[b]	−0.01027	1.526	1.526	1.515	1.505	1.495	0.4044
30.138[a]	−0.00838	1.548	1.548	1.539	1.531	1.523	0.9720
30.258[a]	−0.00917	1.527	1.527	1.518	1.509	1.500	0.9278
30.276[a]	−0.01472	1.551	1.551	1.536	1.521	1.507	0.9729
30.276[b]	−0.01235	1.542	1.542	1.530	1.518	1.505	0.3348

[a]Preimplantation calibration.
[b]Postimplantation calibration for recovered transmitters.
[c]Transmitter 30.088 was designed with a faster pulse repetition.

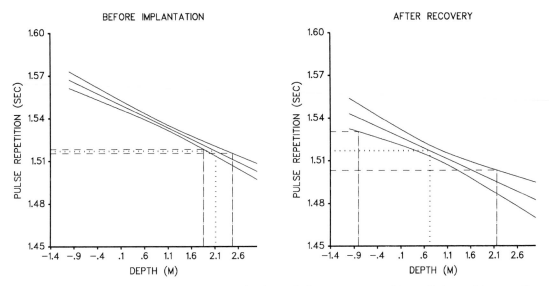

FIGURE 2.—Unlimited simultaneous discrimination intervals for a pressure-sensitive radio transmitter operating at a frequency of 30.097 MHz before implantation and after recovery from the Bighorn River. Solid lines are regressions of pulse repetition on depth and confidence-interval bands. Horizontal dashed lines are confidence intervals for mean pulse repetition given observed repetition. Dotted lines give point estimates of depth based on regressions. Vertical dashed lines give discrimination intervals for depth given repetition.

depths. The best depth resolution during calibration before implantation was 22 cm, whereas the best depth resolution for a recovered transmitter was 3.02 m.

The transmitters need to be modified so that pulse repetition changes markedly with change in depth. Also, the transmitters should be smaller and less subject to electronic drift. These problems must be corrected before this type of transmitter can have practical field value for studies of resident salmonids or similar-sized fish of lotic habitats.

An increase in change of pulse repetition with depth should correct several problems. It should result in steeper regression lines which, in turn, will enhance the receiver's capability to discriminate specific pulse repetitions and provide more accurate determination of fish depth. Another benefit should be the ability to use a stopwatch for monitoring depth. This would provide a backup method if a receiver or pulse counter malfunctioned. With short-lived (21-d) tags, this is particularly important. Reduction in size should make the transmitter more useful for studies of resident trout.

The problems encountered with the pressure-sensitive transmitters appear to be correctable. Pressure-sensitive radio transmitters have great potential value for monitoring depth selection by fish. Further refinement of the transmitters promises to provide a valuable tool for ecologists.

Acknowledgments

We thank George Liknes, Cal Sprague, and Jim Brammer for field assistance; Dennis Christenson for his cooperation; Milo Adkison, Dan Gustafson, and Martin Hamilton for statistical advice; and Vicki Bokum, D.V.M., for implanting the transmitters. This work was funded by the U.S. Department of Interior, Bureau of Reclamation.

References

Chisholm, I. M., W. A. Hubert, and T. A. Wesche. 1987. Winter stream conditions and use of habitat by brook trout in high-elevation Wyoming streams. Transactions of the American Fisheries Society 116:176–184.

Gray, R. H., and J. M. Haynes. 1979. Spawning migration of adult chinook salmon (*Oncorhynchus tshawytscha*) carrying external and internal radio transmitters. Journal of the Fisheries Research Board of Canada 36:1060–1064.

Hart, L. G., and R. C. Summerfelt. 1975. Surgical procedures for implanting ultrasonic transmitters into flathead catfish (*Pylodictis olivaris*). Transactions of the American Fisheries Society 104:56–59.

Haynes, J. M. 1978. Movement and habitat studies of chinook salmon and white sturgeon. Battele, Pacific Northwest Laboratory, Richland, Washington.

Haynes, J. M., D. C. Nettles, K. M. Parnell, M. P. Voiland, R. A. Olsen, and J. D. Winter. 1986. Movements of rainbow steelhead trout (*Salmo gairdneri*) in Lake Ontario and a hypothesis for the influence of spring thermal structure. Journal of Great Lakes Research 12:304–313.

Hurley, S. T., W. A. Hubert, and J. G. Nickum. 1987. Habitat and movements of shovelnose strugeons in the upper Mississippi River. Transactions of the American Fisheries Society 116:655–662.

McCleave, J. D., J. H. Power, and S. A. Rommel, Jr. 1978. Use of radio telemetry for studying upriver migration of adult Atlantic salmon (*Salmo salar*). Journal of Fish Biology 12:549–558.

Lieberman, G. J., R. G. Miller, Jr., and M. A. Hamilton. 1967. Unlimited simultaneous discrimination intervals in regression. Biometrika 54:133–145.

Lieberman, G. J., R. G. Miller, Jr., and M. A. Hamilton. 1971. Errata to Biometrika (1967) 54:133–145. Biometrika 58:687.

Winter, J. D., V. B. Kuechle, D. B. Siniff, and J. R. Tester. 1978. Equipment and methods for radio tracking freshwater fish. University of Minnesota, Agricultural Experiment Station, Miscellaneous Report 152, St. Paul.

American Fisheries Society Symposium 7:395–406, 1990

GENETIC MARKS

Genetic Marking of an Alaskan Pink Salmon Population, with an Evaluation of the Mark and the Marking Process

S. Lane[1] and A. J. McGregor[2]

Juneau Center for Fisheries and Ocean Sciences, University of Alaska–Fairbanks
11120 Glacier Highway, Juneau, Alaska 99801 USA

S. G. Taylor

U.S. National Marine Fisheries Service
Auke Bay Laboratory, Auke Bay, Alaska 99821 USA

A. J. Gharrett[3]

Juneau Center for Fisheries and Ocean Sciences, University of Alaska–Fairbanks
and
U.S. National Marine Fisheries Service
Auke Bay Laboratory, Auke Bay, Alaska 99821 USA

Abstract.—Genetic marks were bred into even- and odd-year runs of pink salmon *Oncorhynchus gorbuscha* in a southeastern Alaskan stream. Two infrequent but naturally occurring alleles at the duplicated *Mdh-3,4* loci for malate dehydrogenase were used to selectively mark the late portion of the odd-year, upstream run in Auke Creek. Marking was accomplished by electrophoretically screening 3,906 returning adults in 1979, and by spawning 390 that had the *70* but not the *130* allele. The frequencies of the *70* and *130* alleles in the hatchery-incubated fish were changed from 0.055 and 0.046 to 0.508 and 0.000, respectively. Of 178,219 genetically marked fish released in spring 1980, 60,000 were fin-clipped so they could be distinguished from wild pink salmon produced in Auke Creek when both returned in 1981. Electrophoretic examination of 1,048 fin-marked adults returning in fall 1981 showed that frequencies of the marker alleles did not differ from those of the parents or those of the marked fry that emigrated in 1980. Similarly, no changes were detected at other loci. In fall 1981, genetically marked fish made up about 47.6% of the late, upstream run. The marked fish spawned in Auke Creek, thereby marking the wild population. The frequencies of the marker alleles have remained stable in this run for four generations (1981–1987). Similar results have been obtained for genetically marked even-year runs.

Most fisheries for Pacific salmon *Oncorhynchus* spp. exploit mixtures of populations (stocks). Effective management of the fisheries requires optimal escapement to all spawning grounds, which means managers need to know the various origins of the exploited populations (Larkin 1981). Such knowledge can be obtained only if the fish carry markers indicating those origins.

Methods for marking juveniles include micro-wire tagging and fin excision. Such marks may be used to determine the origin of individual fish as well as to estimate the contribution of marked stocks to mixed fisheries. However, they often increase mortality or affect homing behavior (Ricker 1976; Morrison and Zajac 1987), and they must be repeated for each brood.

Where it is sufficient to estimate the composition of mixtures, as opposed to determining the destination of each individual, biological markers may be useful. Biological markers reflect the genetic composition and environmental experience of a population. Environmentally determined biological markers include the pattern of scale circuli (e.g., Marshall et al. 1987) and the species and frequency of occurrence of parasites (e.g., Moles et al. 1990, this volume). The environments experienced by populations must be

---

[1]Present address: Taku Smokeries, 5422 Shaune Drive C-1, Juneau, Alaska 99801, USA.

[2]Present address: Alaska Department of Fish and Game, Commercial Fisheries Division, Post Office Box 20, Douglas, Alaska 99824, USA.

[3]To whom correspondence should be addressed.

sufficiently different to produce resolvable differences in the biological markers. Variation in environmental conditions can also cause year-to-year changes in some of these characters for a population.

Another kind of biological marker is a genetic marker. Differences in the frequencies of alleles often develop between populations as a result of natural selection, random drift, or both (Wright 1943, 1951). Such differences may be used to estimate the compositions of mixed fisheries (Pella and Milner 1987). At present, the most commonly used genetic markers are protein variants detected by electrophoresis.

If natural differences in allelic frequency profiles do not exist, genetic markers may be produced by intentionally changing the frequencies of alleles in one or more populations. Concerns that arise with the concept of genetic marking include the efficacy of the marking process, the phenotypic effect of the marker, and the ability of researchers to distinguish between marked and unmarked populations. Practical and theoretical guidelines for genetic marking are presented elsewhere in this proceedings (Gharrett and Seeb 1990, this volume).

Biological questions about genetic marking require empirical answers. One question asks whether it is practical to genetically mark an anadromous stock. This can be answered by producing a marked stock in a salmon hatchery. Another question asks whether marking reduces the fitness of the stock. Because intentional genetic marking of a population alters the natural genetic composition, it is possible that the marker or genes linked with the marker may alter the fitness of the population. Defenders of marking, usually argue that the protein markers used occur naturally at low levels in the population; therefore, increases in their frequencies are unlikely to substantially decrease the fitness of that population. Fitness may also decrease if genetic variability declines or inbreeding results during the intense selection necessary to mark the population. Loss of genetic variability can be reduced by using large numbers of breeders. The possibility that genetic marking might change fitness should be considered whenever marking is done. The effect of marking on fitness can be evaluated by monitoring the persistence of the marker in the population. If the marker or genes associated with it are deleterious, a decrease in the frequency of the marker over time would be expected.

To address questions involving the practicality of genetic marking and persistence of the marks, we bred genetic markers into two populations of pink salmon *O. gorbuscha* in southeastern Alaska. The life history of pink salmon is unique among salmonids. The fry migrate to sea soon after absorbing their yolk sacs. In the northern Pacific Ocean, almost invariably they mature and return to spawn and die in their second year. As a result of the rigid two-year life cycle, no genetic exchange occurs between even- and odd-year runs. Allelic frequency differences between pink salmon populations returning to the same streams in even and odd years are greater than differences among populations from different geographic regions in the same year (Aspinwall 1974; Johnson 1979; McGregor 1983). Because the allelic frequency profiles of even- and odd-year pink salmon differ, the appropriate loci for genetic marking also differ.

We evaluated our genetic marking experiments in three ways. First, if the marker allele was selectively neutral, its frequency should not change during the life cycle. Therefore, allelic frequencies at the marker locus were monitored through the life cycle of the marked hatchery fish.

Second, because many of the genetically marked fish would spawn naturally in the stream when they returned, we expected wild populations within the stream to become genetically marked. The frequency of the marker should then persist in those subpopulations, unless gene flow or selection altered it.

Finally, the selection required to mark hatchery fish can cause changes in allelic frequencies at other loci as a result of chromosomal linkage, gametic disequilibrium, or both (Hedrick et al. 1978). Allelic frequency changes caused by nonrandom associations among loci should be minimized by the use of a sufficient number of breeders. To test this hypothesis, we looked for changes in allelic frequencies at protein-coding loci other than those used for genetic marking, and we tested for gametic disequilibrium.

Methods

Study Site

Auke Creek is 18 km north of Juneau, Alaska, where it flows about 350 m from Auke Lake into Auke Bay. The National Marine Fisheries Service operates a salmon hatchery and a weir for juvenile and adult fish at the mouth of Auke Creek. The stream has natural runs of both even- and odd-year pink salmon. Within each year class, five identifiable subpopulations are separated temporally, spatially, or both: early run intertidal, early run upstream, late run intertidal, late run upstream, and Lake Creek, the last in a tributary to Auke Lake (Taylor 1980). The early run returns from late July to late August; the late run during September. The appearance of late-run fish is preceded by a sharp decline in the number of early-run fish and is clearly indicated by the arrival of incompletely mature, or "sea bright," fish. Between 1971 and 1980, the early run averaged about 2,500 fish and the late run about 7,000, although there was large variation among years. Beginning in 1976, the hatchery's stocks of pink salmon were derived from late-run upstream spawners. Hatchery releases of pink salmon ceased in 1983, but outmigrant and returning fish are still counted at the weir.

Adult Collection and Culturing Procedures

Adults returning to the weir were screened for the appropriate marker allele. A skeletal muscle sample was taken from each fish; the fish was uniquely tagged and held until the sample had been analyzed electrophoretically. Fish not used for breeding were released. Fish selected as genetically marked brood stock were held in wire cages in the stream until they were sexually mature, at which time they were killed and their gametes were extracted immediately. Typically, eggs from five females were pooled and fertilized with the milts of five males, which were added sequentially before eggs and milt were thoroughly mixed.

Fertilized eggs were incubated in gravel incubators (Bailey and Taylor 1974) until larval development was complete. Fry were allowed to emigrate at will the following spring. A sample of genetically marked hatchery fry was marked externally by excision of the adipose and left pelvic fins.

Electrophoresis

Protein electrophoresis was used to determine the genetic compositions of samples. Eye, muscle, liver, and heart tissues were sampled from individuals that had been used for brood stock for genetic marking, as well as from other adults. Tissue samples not analyzed immediately were stored at $-20°C$. Fry were frozen in water. Electrophoresis was conducted on tissue extracts as described by Utter et al. (1974).

Enzymes, International Union of Biochemistry numbers (IUBNC 1984), and abbreviations for protein-coding loci reported in this paper are: Aconitate hydratase (4.2.1.3) Aco-4, adenosine deaminase (3.5.4.4) Ada-2, aspartate aminotransferase (2.6.1.1) Aat-3, glycerol-3-phosphate dehydrogenase (1.1.1.8) G3p-1, malate dehydrogenase (1.1.1.37) Mdh-3,4, malic enzyme (1.1.1.40) Me-1, phosphoglucomutase (5.4.2.2) Pgm-2, 6-phosphogluconate dehydrogenase (1.1.1.44) Pgd, and two peptidase loci, dipeptidase (3.4.13.11) with L-leucyl-L-leucine as substrate Pep(Ll-1), and proline dipeptidase (3.4.13.9) with L-phenylalanyl-L-proline as a substrate Pep(Pp-2). Tissue specificity of enzymes and optimum buffer conditions were described by Gharrett and Thomason (1987). Other electrophoretically detectable loci had too little variability for statistical comparisons in the present study. Alleles at a locus are designated by electrophoretic mobility of the proteins they code relative to that of the protein coded by the most common allele (mobility 100).

Allelic frequencies were estimated directly from observed phenotypic frequencies for all loci except Mdh-3,4. For Mdh-3,4, which are isoloci with common alleles of identical electrophoretic mobility (Allendorf and Thorgaard 1984), we assumed, for our allelic frequency estimates, that all the variability was expressed at one locus and that the other locus was fixed. This assumption was made because we could not determine at which locus a variant allele existed. In addition, we were unable to determine the number of doses of the 70 allele(s) with the buffer initially used to resolve these loci (pH 6.1 amine–citrate buffer; Clayton and Tretiak 1972). We estimated allelic frequencies from the frequency of individuals homozygous for the 100 allele(s). For this, we assumed Hardy–Weinberg and gametic equilibria, which are probably good assumptions for natural populations. After the frequencies of the 70 allele(s) had been altered, however, gametic equilibrium no longer existed between Mdh-3 and Mdh-4 because artificial selection created a surplus of 70/100 gametes and a deficit of 100/100 and of 70/70 gametes. For the first several generations following genetic marking, where genetically marked and unmarked fish interbred, the assumption of diploidy was much better than the assumption of tetraploidy for estimating allelic frequencies from the frequency of the homozygous 100 individuals. The assumption also works well for allelic frequencies less than about 0.2 (diploid estimate) when variability occurs at both loci and equilibria existed. The subsequent use of a different buffer system (Markert and Faulhaber

TABLE 1.—Stability of pink salmon alleles at loci used for genetic marking. Frequencies of *Mdh-3,4* alleles are presented for odd-year fish and of *G3p-1* alleles for even-year fish in Fish Creek, which is approximately 8 km from Auke Creek. Frequencies were estimated for *Mdh-3,4* by assuming all the variability was expressed at one locus. Homogeneity was tested with log-likelihood (*G*) ratios (Sokal and Rohlf 1981).

| Locus and year | N | Frequency of *Mdh-3,4* allele | | | Frequency of *G3p-1* allele | | G | df | P |
		100	70	130	100	200			
Odd-year runs									
Mdh-3,4							7.16	4	0.13
1969[a]	75	0.893	0.060	0.047					
1971[a]	64	0.844	0.125	0.031					
1979	259	0.898	0.056	0.046					
Even-year runs									
G3p-1							1.82	1	0.18
1970[a]	81				0.710	0.290			
1978	274				0.763	0.237			

[a]From Aspinwall (1974).

1965) enabled us to unequivocally score the different *70* allele phenotypes.

Genetic Marking

Marker alleles were chosen for the odd- and even-year runs based on allelic frequency data collected by McGregor (1983) and Lane (1984) and on the marking effort possible (number of fish screened per fish spawned; Gharrett and Seeb 1990). Because the intensive selection involved in genetic marking can increase inbreeding if the brood stock is small or there are unequal numbers of males and females, the goal of the screening process was to identify about 200 males and 200 females with genotypes appropriate for genetic marking. A brood stock of this size and composition has an effective population size of about 400. Choice of marker alleles was limited to loci expressed in skeletal muscle from which samples could be taken easily without killing the fish. Only loci that were easy and inexpensive to stain, and that had strong activity, were considered.

Odd-year run.—For the odd-year marking effort, we expected to obtain a brood stock of 400 males and females by screening about 4,000 returning adults. With this effort, an allele present in the population at a frequency of about 0.05 could be increased to a frequency of about 0.50 (Gharrett and Seeb 1990). The *Mdh-3,4 70* allele was chosen for the odd-year mark because (1) the frequencies had remained stable in nearby Fish Creek for six generations (Table 1); (2) the frequency of the *Mdh-3,4 100* allele was similar (*G* = 22.091; 21 df; *P* = 0.39) among 22 subpopulations from Auke Creek and 11 nearby streams sampled in 1979, and frequencies of the *70* and *130* alleles

were low in all subpopulations, never exceeding 0.11 (McGregor 1983); and, (3) multiple malate dehydrogenase loci were expressed, and it was hypothesized that, if any potentially deleterious effects resulted from altering the genetic composition, the additional activity might buffer the effect. In addition to increasing the frequency of the *70* allele, we removed the *130* allele from the marked population.

Screening for the marker alleles was begun when no fish from the early run remained in Auke Creek. Screening lasted from 13 to 27 September 1979, so the marking effort involved fish returning in the latter part of the late run. Of the 3,906 fish screened, 407 carried the *70* allele but not the *130* allele. The marking effort was one fish spawned for every 10 screened.

Between 1 April and 7 May 1980, 178,219 genetically marked pink salmon fry from the 1979 brood were released from Auke Creek Hatchery. In addition, 74,047 naturally produced fry (weir count) emigrated from Auke Creek. Because returning genetically marked hatchery fish would be externally indistinguishable from returning stream-spawned fish, 60,000 hatchery fry were fin-clipped. A sample of the hatchery fry was taken for electrophoretic analysis.

In 1981, 412 of the fin-clipped adult pink salmon were spawned in the hatchery, and their progeny were released in 1982. None of these fry were fin-clipped. The hatchery fry were sampled for electrophoretic analysis.

Even-year run.—The *G3p-1 200* allele was chosen to mark the even-year run. This locus was strongly expressed in muscle, and its allelic frequencies were similar among 10 subpopulations in four nearby streams examined in 1978 and 1980 (G = 14.91; 9 df; P = 0.09) (Lane 1984). Allelic frequencies at the *G3p-1* locus had been stable in Fish Creek over five generations (Table 1). Between 6 and 20 September 1980, we screened 7,710 returning adults. We used 396 fish to breed the genetic mark into the hatchery stock. One fish was spawned for every 19.5 screened. Hatchery fry were sampled for genetic analysis.

Genetically marked pink salmon from these spawners were released between 25 March and 7 May 1981. Of 175,827 fry released, 85,747 were marked with clipped adipose and left pectoral or left pelvic fins so they could be distinguished from stream spawners when they returned in 1982. Samples were taken from hatchery fry for electrophoretic analysis. Nongenetically marked controls incubated in the hatchery died as a result of incubator failure. To provide a reference, 27,026 of the 111,416 wild fry that passed the weir in spring 1981 were marked with clipped adipose and right pelvic fins.

In 1982, 91 of the fin-clipped returns were spawned in the hatchery, and their progeny (39,914 fry) were released in the spring of 1983. These fry were also sampled for electrophoretic analysis.

Analysis.—Homogeneity among allelic frequencies was tested with log-likelihood (G) ratios (Sokal and Rohlf 1981). Tests were not done if the expected number of observations for an allele in a collection was less than four. The significance level of multiple simultaneous tests was adjusted as described by Cooper (1968).

Contributions of genetically marked fish to mixtures of marked and unmarked returns were estimated from the proportion

$$x = (p_m - p_2)/(p_1 - p_2);$$

p_m is the frequency of the marker allele in the mixture, p_1 the allelic frequency of the marked fish, p_2 the allelic frequency of the unmarked fish, and x the proportional contribution of the genetically marked fish to the mixture (Wallace 1981). The asymptotic variance of the estimator, which we obtained from a Taylor series expansion, is

$$V(x) = [1/(p_1 - p_2)]^2 V(p_m)$$
$$+ [(p_m - p_2)/(p_1 - p_2)^2]^2 V(p_1)$$
$$+ [(p_m - p_1)/(p_1 - p_2)^2]^2 V(p_2);$$

$V(p_i) = p_i(1 - p_i)/2N_i$ and N_i is the size of the sample from which p_i were estimated.

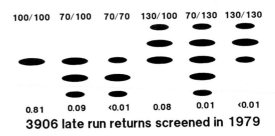

100/100 70/100 70/70 130/100 70/130 130/130

0.81 0.09 <0.01 0.08 0.01 <0.01

3906 late run returns screened in 1979

0.97 0.03

390 hatchery spawners chosen

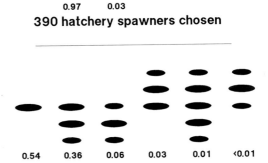

0.54 0.36 0.06 0.03 0.01 <0.01

stream spawners in 1981

FIGURE 1.—Strategy used to genetically mark the late upstream run of pink salmon in Auke Creek in 1979. Electrophoretic patterns of proteins coded by the *100*, *70*, and *130* alleles of the *Mdh-3,4* locus are shown as resolved in a tris–borate–EDTA buffer (Markert and Faulhaber 1965). The patterns reflect the relative mobilities of the proteins. All patterns depicted have two additional proteins coded by *100* alleles from the second of the isoloci. Estimated phenotypic frequencies are given beneath the patterns for each phase of marking. Electrophoretic patterns after 1981 reflect the interbreeding of genetically marked fish with unmarked stream spawners.

Departures from Hardy–Weinberg equilibrium expectations were tested with chi-square goodness-of-fit tests. Gametic (linkage) equilibrium was estimated in odd-year pink salmon and tested according to Hill (1974).

Results

Odd-Year Run

The 1979 Auke Creek Hatchery brood stock was genetically marked by increasing the frequency of the *Mdh-3,4 70* allele from 0.055 to 0.508 and decreasing the frequency of the *130* allele from 0.046 to 0.000 (Figure 1). The survival

TABLE 2.—Releases and returns (numbers of fish) and survivals (percentages) of wild and genetically marked hatchery pink salmon from Auke Creek. Year-class refers to the year the parents of the outmigrating fry and subsequently returning adults spawned in the wild or were bred in the hatchery. Estimates for survival of genetically marked fry were based on the proportion of late run returns that were genetically marked.

Fish group (survival)	Year-class			
	1979	1980	1981	1982
Outmigrating fry				
Wild	74,047	111,416	164,413	169,552
Fin-clipped	0	27,026	0	0
Hatchery	178,219	175,827	134,843	39,914
Fin-clipped	60,000	85,700	0	
Returning adults				
Early run	3,425	6,391	13,802	4,178
Late run	11,025	4,262	6,244	1,093[a]
Total (marked				
and unmarked)	14,450	10,653	24,827	5,271
(Survival)	(5.7%)	(3.7%)	(8.3%)	(2.5%)
Fin-clipped hatchery	1,469	365		
(Survival)	(2.4%)	(0.4%)		
Fin-clipped wild		367		
(Survival)		(1.4%)		
Estimated survival,				
unclipped fish				
Wild (early plus late)	12.4%	11.2%		
Hatchery (late)	3.8%	0.5%		

[a]Distinction between runs was difficult because of a relatively large early run and a small, early returning late run.

of the offspring of these breeders from fertilized egg to eyed embryo, from eyed embryo to outmigrating fry, and from fertilized egg to outmigrating fry was 80, 73, and 59%, respectively. In September 1981, 1,469 fin-clipped adults from the 1979 year-class returned to the Auke Creek weir, for an outmigrant-to-adult return rate of 2.45% (Table 2).

Electrophoretic data were obtained for 1,048 of the returning fin-clipped adults in 1979. Comparisons of allelic frequencies at Mdh-3,4 among the 1979 hatchery spawners, the fry sampled from the hatchery release in 1980, and the fin-clipped returns from these in 1981 (Table 3) indicated that no detectable changes occurred during the life cycle of the first generation of odd-year marked fish ($P = 0.31$). Likewise, the change in the frequency of the 70 allele between the 1981 hatchery spawners and resulting fry in 1982 was not significant ($P = 0.053$). Because these fry were not fin-clipped at release, it was not possible to compare their allelic frequencies with the known hatchery-produced genetically marked returns in 1983.

Protein-coding loci used to compare allelic frequencies of the marked stock at different life history stages were Aat-3, Ada-2, G3p-1, Pep(Ll-1), Me-1, Pgm-2, and Pep(Pp-2) (Table 4). The comparison included 1979 adults before marking, 1979 hatchery brood stock, hatchery fry from the 1979 brood stock, 1981 hatchery brood stock, and fry from the 1981 hatchery brood stock. The G value, summed over all loci, although large, was not significant ($P = 0.05$). A single locus, Pep(Pp-2), contributed much of the heterogeneity to this test, and this was also not significant when multiple simultaneous testing was taken into account (Cooper 1968); however, the significance level correction reduces the test's power to detect actual differences. The Pep(Pp-2) locus was the most variable locus we examined, and it was often difficult to interpret some of the banding patterns on gels. In addition, the heterogeneity came from a single sample, the 1979 brood stock, which failed Hardy–Weinberg expectations. The departures may reflect processes related to the genetic marking experiment, but it is also possible that inexperience in reading gel patterns at the beginning of the project was responsible.

Among the loci not involved in the odd-year genetic marking experiment, four deviations from Hardy–Weinberg expectations were observed for the 23 tests made on the 1979 and 1981 brood stock and hatchery fry sampled in 1980 and 1982 (Table 4). None of these departures from Hardy–Weinberg expectations is significant if adjustment is made for multiple testing (Cooper 1968); but this many departures is more than one would expect from random chance. Only one of the 165 tests for gametic equilibrium was significant at the 0.05 level, which is fewer than would be expected by random chance alone.

Adults without clipped fins that returned in 1981 showed a large increase in the frequency of the Mdh-3,4 70 allele relative to that existing before genetic marking. The increase indicated that genetically marked fish made a substantial contribution to the run. We estimated that the genetically marked fish contributed $47.6 \pm 3.5\%$ (mean ± SD) of the late-run return (Table 5). Using this rate of return and counts of outmigrating fry and returning adults, we estimated the adult return rates for 1979 year-class fish to be 3.8 and 12.4%, respectively, for genetically marked and wild pink salmon (Table 2). In contrast, the frequency of the Mdh-3,4 70 allele for the 1983 adult return was not significantly different from the 1981 returns (Table 6), indicating relatively low survival of offspring

TABLE 3.—Stability of marker alleles during the life cycle of genetically marked pink salmon. *Mdh-3,4 100* frequencies were estimated by assuming all the variability was expressed at one of the loci. Homogeneity was tested with log-likelihood (G) ratios (Sokal and Rohlf 1981).

Generation and sample	N	Frequency of Mdh-3,4 allele			Frequency of G3p-1 allele		G	df	P
		100	*70*	*130*	*100*	*200*			
Odd-year runs; *Mdh-3,4* locus									
First marked generation							2.35	2	0.31
1979 hatchery brood stock[a]	390	0.492	0.508	0.000					
1980 hatchery fry[b]	658	0.516	0.484	0.000					
1981 clipped[c] returns	1,048	0.490	0.510	0.000					
Second marked generation							3.73	1	0.05
1981 hatchery brood stock	412	0.470	0.530	0.000					
1982 hatchery fry[b]	649	0.513	0.487	0.000					
Even-year runs; *G3p-1* locus									
First marked generation							4.72	2	0.09
1980 hatchery brood stock	396				0.124	0.876			
1981 hatchery fry	695				0.109	0.891			
1982 clipped[c] returns	202				0.148	0.852			
Second marked generation							3.88	2	0.14
1982 hatchery brood stock	91				0.193	0.808			
1983 hatchery fry	383				0.223	0.777			
1984 clipped[c] returns	74				0.155	0.845			

[a]Allelic frequencies were estimated from the frequency of *100* homozygote among screened fish.
[b]Allelic frequencies were estimated from the frequency of *100* homozygotes.
[c]Adipose and paired fin clips.

from the 1981 hatchery brood stock. Apparently the second generation of marking had little effect. If genetically marked hatchery fish had contributed significantly to the 1983 adult return, the frequency of the *Mdh-3,4 70* allele would have increased from the level observed for 1981 returns. Because we did not fin-clip the outmigrant hatchery fry in 1982, there were no fin-clipped returns in 1983, so the gametic contribution of hatchery-spawned fish to the total run could not be estimated.

The fate of the marker allele in late upstream spawners of the odd-year run has been followed each generation since marking. No significant changes in the frequencies of alleles at *Mdh-3,4* were seen over the first four generations ($P = 0.85$; Table 6), although a trend toward decreasing frequency of the *70* allele was apparent.

Even-Year Run

The immediate result of genetic marking in the 1980 Auke Creek late run was to increase the frequency of *G3p-1 200* from 0.199 in the screened fish to 0.876 in the hatchery brood stock. The survival of the offspring of the 1980 breeders from fertilized egg to eyed embryo, from eyed embryo to migrating fry, and from fertilized egg to migrating fry was 78, 66, and 52%, respectively.

In 1982, 365 fin-clipped hatchery fish and 367 fin-clipped wild fish returned to the Auke Creek

weir at rates of 0.4 and 1.4%, respectively (Table 2). These return rates were considerably less than the overall return of 11.2% estimated for wild fish. According to counts made at the weir in 1980 and 1982 (Table 2), the ratio of early to late run fish changed from 1 to 1.5 in 1979 to 3.0 to 1 in 1980, which represented nearly a fivefold difference in the survival of the two runs over a single generation. This change reflected the magnitude of naturally occurring fluctuations in abundance. We could not estimate the return rate for late-run wild fish that did not have fin clips, but probably it was near the rate observed for wild fin-clipped fish—about 1.4%.

Comparisons of allelic frequencies of *G3p-1* among the 1980 hatchery spawners, the hatchery fry produced from this brood stock and released in 1982, and the 1982 fin-clipped adult returns showed no significant differences ($P = 0.09$; Table 3) during the life cycle of the first generation of even-year marked fish. Likewise, no differences were detected among the 1982 hatchery brood stock, the hatchery fry produced from these brood stock and released in 1983, and 1984 fin-clipped hatchery returns ($P = 0.14$; Table 3).

For the even-year run, the loci at which different life history stages were compared after marking were *Aat-3, Aco-4, Ada-2, Pep(Ll-1), Me-1, Pgd,* and *Pep(Pp-2)*. No heterogeneity was observed among the 1980 adults before selection for mark-

TABLE 4.—Allelic frequencies at protein-coding loci other than the loci used for marking Auke Creek pink salmon. These data were gathered during different phases of the genetic-marking process, from premarking to two generations postmarking. Test of homogeneity and G statistics (df) are given for each locus for each brood-year type. Total G statistics (df) are given for each marking experiment. A denotes adults before selection for marking, B means brood stock, and F is fry.

| | Aat-3 | | | Aco-4 | | | Ada-2 | | | | G3p-1 | | | Pep(Ll-1) | | | |
| | | Allele | | | Allele | | | Allele | | | | Allele | | | Allele | | |
Sample or statistic	N	100	85	N	100	85	N	100	87	130	N	100	200	N	100	85	1150
Odd-year runs																	
1979 A	96	0.781	0.219				98	0.883	0.117	0.000	98	0.883	0.117	98	0.765	0.235	0.000
1979 B	359	0.751	0.249[a]								364	0.875	0.125	254	0.783	0.217	0.000
1980 F	456	0.730	0.270				258	0.857	0.143	0.000	408	0.879	0.121[b]	200	0.772	0.228	0.000
1981 B	360	0.722	0.278				412	0.854	0.146	0.000[b]	412	0.880	0.120	407	0.733	0.267	0.000
1982 F	393	0.762	0.238				396	0.847	0.153	0.000	395	0.896	0.104	385	0.739	0.261	0.000
G (df)		5.67 (4)						1.74 (3)				1.97 (4)			5.99 (4)		
Even-year runs																	
1980 A	63	0.659	0.341	98	0.944	0.056	98	0.959	0.041	0.000				89	0.854	0.056	0.090
1980 B	396	0.712	0.288[b]	397	0.951	0.049	397	0.956	0.044	0.000				387	0.861	0.041	0.098
1981 F	345	0.744	0.256	595	0.956	0.044	200	0.938	0.058	0.005							
1982 B				90	0.967	0.033	90	0.956	0.028	0.017				90	0.867	0.056	0.078
G (df)		4.47 (2)			1.45 (3)			2.15 (3)							1.86 (4)		

[a] Hardy–Weinberg proportions not observed ($P < 0.01$).

ing, or among the selected 1980 and the 1982 breeders ($P = 0.53$; Table 4). Three of 14 tests for Hardy–Weinberg expectations failed, which was more than one would expect at random.

Genetically marked fish contributed $12.5 \pm 2.4\%$ (mean $\pm$ SD) to the mixture of non-fin-clipped fish returning to the Auke Creek weir in September 1982, and they contributed virtually nothing ($0.0 \pm 6.9\%$) to the returns in 1984 (Table 5). The resultant strength of the *G3p-1 200* marker in the late upstream run of Auke Creek was approximately 0.25, compared to an initial frequency of 0.199. No change was observed between the years 1982 and 1984 ($P = 0.28$; Table 6).

Discussion

The questions we have addressed asked if genetically marking a pink salmon stock is feasible, and if the process measurably reduces the fitness of the marked stock. Because the possibility of reduced fitness affects considerations of feasibility, we will consider our evaluations of the markers first.

A primary concern in altering allelic frequencies to mark fish populations is whether or not increasing the frequencies of genotypes having the desired marker decreases the fitness of the marked population, either because the marker alleles themselves are deleterious or because they are linked to deleterious alleles. Selection acts on genotypes (phenotypes); however, as a result of most selection regimes, allelic frequencies also change. For some statistical tests of selection at a locus, changes in allelic frequency are more sensitive and more testable than changes in genotypic frequencies (Gharrett and Seeb 1990). The frequency of the marker alleles, *Mdh-3,4 70* and *G3p-1 200*, did not change during the life cycle of the marked hatchery populations, nor did that of the *Mdh-3,4 70* allele among the stream spawners in four generations after marking. Given the sample size (1,048) examined for the 1981 returning fin-clipped adults, it can be concluded that if selection occurred against genotype(s) with the *Mdh-3,4 70* marker allele, the selection coefficient was less than 0.15 (Gharrett and Seeb 1990).

Genetic marking involves relatively intense selection for relatively infrequent alleles. Unless many individuals are taken for brood stock, it is possible that other loci may be missampled. The departures from Hardy–Weinberg expectations we observed in excess of random chance may reflect missampling; however, several of the departures occurred at *Pep(Pp-2)*, which we initially had trouble scoring. The absence of gametic disequilibrium between loci other than the marker loci is indicative of random selection of other loci during marking.

TABLE 4.—Extended.

Sample or statistic	N	Me-1 Allele 100	130	70	N	Pgd Allele 100	90	N	Pgm-2 Allele 100	150	N	Pep(Pp-2) Allele 100	93	109	Total G (df), P-value
							Odd-year runs								
1979 A	97	0.946	0.054	0.000				98	0.964	0.036	97	0.680	0.113	0.206	
1979 B	364	0.964	0.034	0.001				364	0.975	0.025	361	0.697	0.090	0.213[b]	
1980 F	400	0.951	0.049	0.000				408	0.976	0.025					
1981 B	412	0.970	0.030	0.000				412	0.973	0.027	412	0.672	0.148	0.180	
1982 F	398	0.974	0.026	0.000				398	0.961	0.039	396	0.685	0.130	0.186	
G (df)		8.52 (4)							3.88 (4)			14.37 (6)			42.14 (29) P = 0.05
							Even-year runs								
1980 A	98	0.816	0.184	0.000	91	0.950	0.049				98	0.546	0.286	0.168	
1980 B	396	0.796	0.205	0.000	394	0.950	0.049				398	0.518	0.308	0.175[b]	
1981 F	588	0.816	0.184	0.000	595	0.959	0.041				593	0.559	0.280	0.161[b]	
1982 B	87	0.862	0.138	0.000	90	0.944	0.056				89	0.511	0.264	0.224	
G (df)		4.65 (3)				1.28 (3)						6.97 (6)			22.83 (24) P = 0.53

[b] Hardy–Weinberg proportions not observed ($P < 0.05$).

Intense selection for infrequent alleles might also select for relatives, who would be more likely than unrelated fish to carry the rare allele. As a result, the effective population size (N_e) would be less than that predicted from the numbers of males and females used. The only way to compensate for such a reduction is to increase the number of spawners. We used about 200 males and 200 females to produce an estimated N_e of about 400, which far exceeded the recommendations of other authors, such as the 60 of Ryman

and Staahl (1980). In addition, the apparent gametic equilibrium indicated that loci other than the marker loci were randomly selected, and reflected more than a few familial lineages.

The most important test of our genetic marker was its persistence in the stream spawners over several generations. One objective of genetic marking is to permanently mark a population. If either natural selection or immigration (gene flow) changes the frequency of the allelic marker, the marker becomes of limited use. The frequency of

TABLE 5.—Estimates of contributions of genetically marked Auke Creek Hatchery fish to the late upstream run. Equations for estimates of contribution and variance are in the text.

Generation and sample	N	Frequency of Mdh-3,4 allele 100	70	130	Frequency of G3p-1 allele 100	200	Proportion of genetically marked fish (SD)
		Odd-year runs; Mdh-3,4 locus					
First marked generation							
1980 late wild fry[a]	100	0.910	0.020	0.070			
1981 fin-clipped[b] returns	1,048	0.490	0.510	0.000			
1981 non-fin-clipped returns[c]	392	0.727	0.253	0.020			47.6% (3.5%)
		Even-year runs; G3p-1 locus					
First marked generation							
1981 late wild fry	396				0.846	0.154	
1982 fin-clipped[b] returns	202				0.148	0.852	
1982 non-fin-clipped returns[c]	635				0.759	0.241	12.5% (2.4%)
Second marked generation							
1983 late wild fry	175				0.720	0.280	
1984 fin-clipped[b] returns	74				0.155	0.845	
1984 non-fin-clipped returns[c]	528				0.740	0.260	0% (6.9%)

[a] Allelic frequencies were estimated from the frequency of 100 homozygotes.
[b] Adipose and paired fin clips.
[c] Mix of wild and genetically marked adults.

TABLE 6.—Stability of frequencies of marker alleles in the upstream late run of pink salmon in Auke Creek after genetic marking. Fish returns were derived from stream-spawned and genetically marked fish. *Mdh-3,4 70* was the allelic marker for the odd-year run and *G3p-1 200* for the even-year run. Homogeneity was tested with log-likelihood (*G*) ratios (Sokal and Rohlf 1981).

| Locus and year | N | Frequency of *Mdh-3,4* allele | | | Frequency of *G3p-1* allele | | G | df | P |
		100	70	130	100	200			
Odd-year runs									
Mdh-3,4							2.70	6	0.85
1981	392	0.727	0.253	0.020					
1983	95	0.711	0.268	0.021					
1985	149	0.735	0.245	0.020					
1987	110	0.777	0.205	0.018					
Even-year runs									
G3p-1							1.16	1	0.28
1982	635				0.759	0.241			
1984	528				0.740	0.260			

REPEAT OF PREVIOUS page

either natural selection or immigration (gene flow) changes the frequency of the allelic marker, the marker becomes of limited use. The frequency of the *Mdh-3,4 70* marker has not changed detectably since the first generation returned. Although no statistically significant change was observed among these collections, a 16% decrease was observed between 1985 and 1987 (Table 6). This change may reflect a relatively large early run and very poor late run (Table 2) rather than changes associated with fitness. The early run was quite large and its late portion extended into September, whereas the late run ended in mid-September. As a result, there was no clear distinction between the two runs in 1987.

Relative survival of different subpopulations and experimental groups of Auke Creek pink salmon was difficult to estimate. Outmigrating fry from early and late runs were indistinguishable; and the ratio of returning wild adults in the early and late runs changed nearly fivefold during the experiment, the late run predominating from 1979 to 1981 and the early run thereafter. Returns of wild Auke Creek pink salmon were similar to those of other streams in the region during the years of the marking experiment. Because of the huge natural fluctuations in survival, it was impossible to attribute the survival of hatchery-produced fish to genetic or non-genetic causes.

In one fin-clipping experiment, we compared the survival of genetically marked and wild fish from the 1980 year-class (Table 2). The results showed that outmigrating wild fry had a survival of 1.4%, compared to 0.4% for outmigrating hatchery fry. These results are similar to those of Bailey et al. (1976), who observed survivals of

1.35 and 0.8%, respectively, for wild and hatchery fry. The rate of return for the fin-clipped wild fish (1.4%) was also much lower than the total estimated for wild fish (11.2%). Bailey et al. (1976) obtained an egg-to-outmigrant survival of 74%, which is much higher than the 50 to 60% we observed. The major difference appeared in the survival from eyed egg to fry, which was 95% for Bailey et al. (1976) and about 75% for us. Although we did not determine the cause of incubator mortality, probably the survivors were stressed prior to emergence; the incubator that held randomly selected controls for the 1980 brood stock malfunctioned, resulting in complete loss; and the experimental incubator used for the 1982 brood stock had substantial losses. Some of the egg mortality may have resulted from handling the females during screening (Lane 1984). Such handling results in bruised and ruptured eggs.

The results of these marking experiments indicate that, although there are no genetic reasons against genetic marking, there may be some non-genetic practical considerations. One such consideration involves the species being marked. For species that return in large numbers, such as pink salmon, the ratio of fish screened to fish spawned can be very large, and a large N_e can be maintained. For species with small returns that limit the number of eggs available to hatcheries, it may be necessary to screen only the males. In this case, the ratio of males screened to males spawned must be greater than when both sexes are screened, and care must be taken to maintain a large N_e. For spawning broods of unequal numbers of males and females,

$$N_e = 4N_m N_f/(N_f + N_m);$$

N_m is the number of males and N_f the number of

females used (e.g., Falconer 1981). This strategy may also be preferable for species with large returns, because handling the females during the screening process reduces egg survival.

Another consideration involves the strategy used to introduce the mark into the population. If a mark is bred into an anadromous fish population by releasing a large marked component of hatchery-raised fish to return and spawn with an unmarked wild component (as might be done to enhance existing runs by fry planting), the success of the marking program will largely depend on marine survival of the hatchery-produced fish. If survival of this group is high, the mark will become strongly established in the run; but if survival is poor, the mark will be weak. In contrast, when a typical ocean-ranching operation is established, eggs are taken from a nearby source, raised to fry or smolt in a hatchery, and released to return to the hatchery at maturity. High marine survival is not critical for establishing the mark because brood stock and existing unmarked stock do not mix. If marine survival is poor, the run size will be poor, but the frequency of a selectively neutral marker allele will not be affected.

Finally, production losses that result from small decreases in fitness might be acceptable—both biologically and economically—if the genetic mark enables stocks to be managed more efficiently. Better management means that optimum escapement can more often be achieved, and optimum escapement generally results in increased production.

Acknowledgments

We appreciate the technical assistance of A. Ricci, M. Thomason, S. Shirley, and C. Smoot. We are grateful to the many individuals who assisted with the sampling and marking efforts. The study was done in cooperation with the National Marine Fisheries Service Auke Bay Laboratory and the Alaska Territorial Sportsmen. D. Campton and an anonymous reviewer made many suggestions that we incorporated. This work is a result of research sponsored by the Alaska Sea Grant College Program under grant NA81AA-D-00009, project R/02-05, and by the University of Alaska.

References

Allendorf, F. W., and G. H. Thorgaard. 1984. Tetraploidy and the evolution of salmonid fishes. Pages 1–53 in B. J. Turner, editor. Evolutionary genetics of fishes. Plenum, New York.

Aspinwall, N. 1974. Genetic analysis of North American populations of pink salmon, Oncorhynchus gorbuscha, possible evidence for the neutral mutation–random drift hypothesis. Evolution 28:295–305.

Bailey, J. E., J. J. Pella, and S. G. Taylor. 1976. Production of fry and adults of the 1972 brood of pink salmon, Oncorhynchus gorbuscha, from gravel incubators and natural spawning at Auke Creek, Alaska. U.S. National Marine Fisheries Service Fishery Bulletin 74:961–971.

Bailey, J. E., and S. G. Taylor. 1974. Salmon fry production in a gravel incubator hatchery, Auke Creek, Alaska, 1971–1972. NOAA (National Oceanic and Atmospheric Administration) Technical Memorandum NMFS (National Marine Fisheries Service) ABFL-3, Auke Bay, Alaska.

Clayton, J. W., and D. N. Tretiak. 1972. Amine-citrate buffer for pH control of starch gel electrophoresis. Journal of the Fisheries Research Board of Canada 29:1169–1172.

Cooper, D. W. 1968. The significance level in multiple tests made simultaneously. Heredity 23:614–617.

Falconer, D. S. 1981. Introduction to quantitative genetics, 2nd edition. Longman, New York.

Gharrett, A. J., and J. E. Seeb. 1990. Practical and theoretical guidelines for genetically marking fish populations. American Fisheries Society Symposium 7:407–417.

Gharrett, A. J., and M. A. Thomason. 1987. Genetic changes in pink salmon (Oncorhynchus gorbuscha) following their introduction into the Great Lakes. Canadian Journal of Fisheries and Aquatic Sciences 44:787–792.

Hedrick, P., S. Jain, L. Holden. 1978. Multilocus systems in evolution. Evolutionary Biology 11:101–108.

Hill, W. G. 1974. Estimation of linkage disequilibrium in randomly mating populations. Heredity 33:229–239.

IUBNC (International Union of Biochemistry, Nomenclature Committee). 1984. Enzyme nomenclature 1984. Academic Press, Orlando, Florida.

Johnson, K. R. 1979. Genetic variation in populations of pink salmon (Oncorhynchus gorbuscha) from Kodiak Island, Alaska. Master's thesis. University of Washington, Seattle.

Lane, S. 1984. The implementation and evaluation of a genetic mark in a hatchery stock of pink salmon (Oncorhynchus gorbuscha) in southeast Alaska. Master's thesis. University of Alaska, Juneau.

Larkin, P. A. 1981. A perspective on population genetics and salmon management. Canadian Journal of Fisheries and Aquatic Sciences 38:1469–1475.

Markert, C. L., and I. Faulhaber. 1965. Lactate dehydrogenase isozyme patterns of fish. Journal of Experimental Zoology 159:319–332.

Marshall, S., and nine coauthors. 1987. Application of scale pattern analysis to the management of Alaska's sockeye salmon. Canadian Special Publication of Fisheries and Aquatic Sciences 96:307–326.

McGregor, A. J. 1983. A biochemical genetic analysis of northern southeast Alaskan pink salmon (Oncorhynchus gorbuscha). Master's thesis. University of Alaska, Juneau.

Moles, A., P. Rounds, and C. Kondzela. 1990. Use of the brain parasite *Myxobolus neurobius* in separating mixed stocks of sockeye salmon. American Fisheries Society Symposium 7:224–231.

Morrison, J., and D. Zajac. 1987. Histologic effect of coded wire tagging in chum salmon. North American Journal of Fisheries Management 7:439–441.

Pella, J. J., and G. B. Milner. 1987. Use of genetic marks in stock composition analysis. Pages 247–276 *in* N. Ryman and F. Utter, editors. Population genetics & fishery management. University of Washington Press, Seattle.

Ricker, W. E. 1976. Review of the rate of growth and mortality of Pacific salmon in salt water, and noncatch mortality caused by fishing. Journal of the Fisheries Research Board of Canada 33:1483–1524.

Ryman, N., and G. Ståhl. 1980. Genetic changes in hatchery stocks of brown trout (*Salmo trutta*). Canadian Journal of Fisheries and Aquatic Sciences 37:82–87.

Sokal, R. R., and F. J. Rohlf. 1981. Biometry, 2nd edition. Freeman, San Francisco.

Taylor, S. G. 1980. Marine survival of pink salmon fry from early and late spawners. Transactions of the American Fisheries Society 109:79–82.

Utter, F. M., H. O. Hodgins, F. W. Allendorf. 1974. Biochemical genetic studies in fishes: potentialities and limitations. Pages 213–238 *in* D. C. Malins and J. R. Sargent, editors. Biochemical and biophysical perspectives in marine biology, volume 1. Academic Press, San Francisco.

Wallace, B. 1981. Basic population genetics. Columbia University Press, New York.

Wright, S. 1943. Isolation by distance. Genetics 28:114–138.

Wright, S. 1951. The genetical structure of populations. Annals of Eugenics (London) 15:323–354.

American Fisheries Society Symposium 7:407–417, 1990

Practical and Theoretical Guidelines for Genetically Marking Fish Populations

A. J. Gharrett

Juneau Center for Fisheries and Ocean Sciences University of Alaska–Fairbanks
11120 Glacier Highway, Juneau, Alaska 99801, USA, and
U.S. National Marine Fisheries Service Auke Bay Laboratory, Auke Bay, Alaska 99821, USA

J. E. Seeb

Fisheries Research Laboratory, Department of Zoology
Southern Illinois University at Carbondale, Carbondale, Illinois 62901, USA

Abstract.—Researchers should consider several biological and genetic factors before choosing marker alleles and altering their frequencies to mark a fish population. Information on the range and time of spawning and the sizes of the target population and the populations from which it is to be discriminated are needed to determine the utility of a mark. Life history information is needed to determine the extent of follow-up marking necessary. The most efficient marking strategy depends on the genetic variability within and among populations, as well as on the resources available to mark the population and subsequently to detect the mark in mixtures. Selection for single allele markers can produce optimum genetic marks. Genetic variability, which contributes to the long-term success of a population, can be sustained by selecting a relatively large brood stock. The marked population should be monitored over several generations to assure that the mark is not lost through natural selection or diluted by immigration from unmarked populations.

Fish marking is used to elucidate the biology of species as well as to provide information necessary for managing fisheries. In general, marking enables researchers to identify individuals or populations in a mixture. Marks include fin clips, brands, dyes, and internal and external tags (see reviews in this symposium).

Biological marks that occur naturally can also be used in analyses of mixed populations. Examples include scale patterns reflecting the environmental experience of the fish (e.g., Marshall et al. 1987), incidence of parasites (Moles et al. 1990, this volume), and genetic differences among isolated populations (e.g., Schweigart et al. 1977; Grant et al. 1980; Murphy et al. 1983). Genetic marks are usually detected in fish by means of protein electrophoresis.

If distinctive genetic differences among populations do not exist naturally, they may be bred into one or more of the populations. That process, known as intentional genetic marking, is the focus of this paper. Strong selection can produce uniquely marked lots of fish (Reisenbichler and McIntyre 1977; Gharrett and Shirley 1985). Alternatively, if the objective is to estimate compositions of mixtures, rather than to identify origins of individuals in the mixture, differences may be produced among populations simply by altering allelic frequency profiles (e.g., Seeb et al. 1986,

1990, this volume; Lane et al. 1990, this volume). A major advantage of genetic markers is that they are heritable. In theory, once a population has been marked, the marker will be inherited from generation to generation, provided it is not maladaptive.

A major goal of genetic marking, as with any marking technology, is to produce the best mark for the available marking and recovery resources. For microwire tagging, fin clipping, and use of external tags, viability of the marked fish and tag retention are essential to success. For genetic marking, the long-term viability of marked fish is subject to additional concerns over matters such as the extent of inbreeding incurred by the selection process and the selective neutrality of the allelic markers. A major consideration is the quality of a genetic mark: the extent to which a marked population can be detected in a mixture of marked and unmarked populations. Genetic marks are based on altered frequencies of gene expression, so the presence of a marked population cannot always be discerned in small samples from a mixture.

When a genetic marking program is planned, several questions must be addressed, including how to choose a marker allele and how to evaluate its performance. The answers require biological and genetic information for both the population that will be marked and the populations from which it is to be discriminated. In this paper, we provide quantitative guidelines for choosing appropriate marker alleles, and we describe how to examine the selective neutrality of a marker allele.

Premarking Considerations

Biological Factors

To use genetic marking, researchers commonly need to know the distribution and abundance of the stocks involved, and where and when they spawn. The marked population must be large enough to be detectable at the time and place it is to be sampled. The relative and absolute sizes of populations are important factors in deciding the potential utility of a mark. (The detectability of genetically marked fish in a mixture is considered later.)

Genetic marking requires some means, such as a weir or a hatchery, for separating fish with particular genotypes from a population. Marking can occur where no fish previously existed or where populations are already present. Where no population exists, a marked population can be established by transplantation of previously screened gametes, embryos, or juveniles; where populations already are present, allelic frequencies in one or more populations can be altered by selective breeding.

Marking subpopulations requires knowledge of how their spawning groups are separated in time or space. In addition, when marking a previously existing population, one must take into account the dilution of the marker allele by natural production of that population (Lane et al. 1990).

Most species of fish have overlapping generations and require follow-up genetic marking in successive years throughout the life span of the longest-lived individuals. For a transplanted population, genetic marking must be done each year until marked fish that can be used as brood stock return. Marking an existing stock also requires marking in successive years; however, after the first returns of marked fish, only fish from year-classes that have not been marked need to be screened, provided the year-classes are readily distinguishable. The dilution of marks added to an existing population may require additional marking effort.

Genetic Factors

Intentional genetic marking is necessary only when discrimination among populations by means of physical, biological, or genetic characteristics is not possible or practical. Genetic marking uses selective breeding to alter frequencies of alleles in the marked population so it can be distinguished from unmarked populations. In this process, three factors must be considered: baseline genetic infor-

mation, effects of radically altered allelic frequencies on the population, and effective population sizes.

For allelic markers to be useful, they must differ in frequency from those of other populations. To ensure that the marked population will indeed be distinct from other stocks, baseline information on allelic frequencies must first be obtained from all populations that could potentially contribute to a mixture. Such baseline information is also needed to choose the marker alleles.

The genetic composition of a genetically marked population differs from naturally occurring frequencies. Although it is usually presumed that artificial changes in the natural frequencies of the marker alleles will have little affect on the fitness of the marked stock, the marked population should be monitored for several generations to determine if this presumption is true. Natural selection could eliminate fish that have maladaptive marker alleles or maladaptive hitchhiking alleles closely linked to the markers. Also, monitoring is necessary because the representation of marked fish in a population could decrease over generations if there were an influx of unmarked fish.

Inbreeding is a potential problem in genetic marking programs. The selective breeding process greatly reduces the effective population size (N_e), because the intensive selection for the marker allele results in a limited number of breeders. Also, if the genetic mark is relatively rare in a population, the probability of selecting related individuals for brood stock is increased. In order to maintain genetic variability, which may be essential to the long-term survival of the marked population, the size of the brood stock should be as large as possible. Using a large number of spawners reduces the chance that deleterious hitchhiking alleles will be enhanced in the marked population. If 200 spawners (100 males and 100 females) are used to found the marked population and subsequently maintain it, the change in the inbreeding coefficient will be only 0.0025 per generation. More than 20 generations would be required to reach a moderate inbreeding coefficient F of 0.05. If different numbers of males and females are used, $N_e = 4N_m N_f/(N_m + N_f)$, N_m and N_f being the numbers of males and females, respectively. For species with overlapping generations, N_e is approximately the size of the brood in one year times the average age of breeders (Falconer 1981). Inbreeding resulting from the selection of related individuals will also be reduced by increasing the number of breeders, which should increase the representation of more families among the brood stock. However, it is not possible to quantify how selection of relatives decreases N_e, or how increasing the brood stock compensates for the decrease.

Genetic Marking

Marking Strategy

The objective of genetic marking is to produce a population distinct from other stocks. After it has been determined that genetic marking is appropriate for a particular application, and after baseline information has been obtained for the stocks involved, the marking strategy can be selected. That strategy depends on the genetic variability among the populations as well as within the population to be marked. A useful marker allele cannot be abundant in populations other than the marked population.

Genetic marking is accomplished by breeding only those individuals carrying the marker allele or alleles. How strongly a population is marked depends on the marking effort, which we define as the ratio of fish screened to fish spawned (bred), and on the marking strategy. We compared five strategies: (I) selecting for a single allele (1 locus); (II) simultaneously selecting for two alleles, one at each of two loci, when the premarking frequencies of those alleles—p at locus 1 and q at locus 2—are equal $(+/+, p = q)$; (III) selecting for an allele at one locus and against another allele at a second locus when the premarking frequencies of the two alleles are equal $(+/-, p = q)$; (IV) selecting for an allele at one locus and against an allele at a second locus when the frequency of the allele at the second locus is 0.10 $(+/-, q = 0.10)$; and (V) selecting for an allele at one locus and against an allele at a second locus when the frequency of the second allele is 0.20 $(+/-, q = 0.20)$.

For strategies involving positive selection (I and II), increased marking effort results in the progressive elimination of genotypes that have the fewest marker alleles, until only those homozygous for markers remain. For example, when selection is for the single allele A', the AA genotype is eliminated first, then AA'. If the initial frequency of the A' allele is higher than 0.30 or 0.40, alleles other than the marker alleles may be eliminated by relatively low marking efforts (Figure 1, top). If the initial frequency is less than 0.025, the effort may involve removing only homozygous AA types. For intermediate initial frequencies, an inflection occurs at some level of marking effort that indicates the exhaustion of AA genotypes and the beginning of removal of AA' types (Figure 1, top).

Selection for two alleles (II) is similar to selection for a single allele except that there are nine possible genotypes. When, as a result of increased marking effort, two different genotypes possessing the same number of marker alleles A' and B' (e. g., $AA'BB$ and $AABB'$) become the remaining genotypes with the fewest marker alleles, they are eliminated simultaneously at the same rate. The $A'A'BB$ and $AAB'B'$ types are removed before $AA'BB'$ types to accelerate approach to gametic phase equilibrium. The number of inflections in the curves that describe the postmarking frequencies as a function of marking effort reflect the progressive exhaustion of different genotypic classes (Figure 1, bottom).

Selection for an allele at one locus and against alleles at a second locus (III, IV, and V) is accomplished by first eliminating the undesired allele, and then following the strategy for selecting a single allele.

Detectability

Efforts to use natural genetic differences among populations to detect distinct stocks in, and to estimate stock compositions of, mixed fisheries have focused on the development of algorithms for multilocus variability. Verification of the algorithms has been conducted principally by computer simulation with existing data sets (e.g., Fournier et al. 1984; Pella and Milner 1987). Although these algorithms can resolve some mixtures into component stocks, they are inappropriate for testing hypotheses and can be used only indirectly (by simulation) to examine the discriminating power resulting from allelic frequency differences between populations. To date, a systematic quantitative treatment of discrimination based on allelic differences has not been completed for even one locus; such analyses for multilocus differences between populations are significantly more complex.

Intentional genetic marking is necessary only when discrimination among populations is not otherwise possible. Intentional marking does allow a more tractable analytical model, however: because all the stocks are initially similar, a mixture consists only of marked and unmarked fish and changes in allelic frequencies are needed at only one or a few loci to uniquely mark the stock. Analyses of such a model are pertinent to the naturally occurring allelic frequency differences exploited by multilocus stock separation algorithms.

The minimum proportion of a genetically marked population that can be detected in a

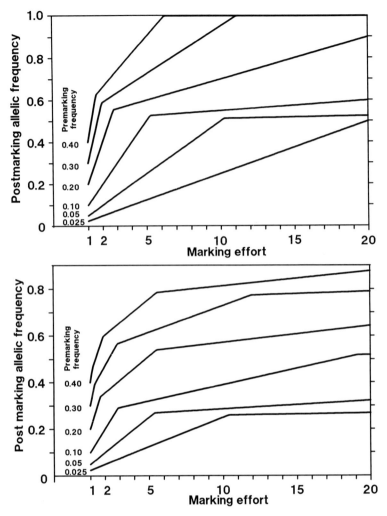

FIGURE 1.—Marker allele frequencies attainable as a function of the marking effort and premarking allelic frequency. Above: the results of marking at a single locus (Gharrett et al. 1983). Below: the results of simultaneously selecting for alleles at two loci when their premarking frequencies (p_0, q_0) are the same. Marking effort is the number of fish screened per fish spawned.

mixture with other populations depends on the genetic composition of the other populations, the "strength" of the mark (resulting from the marking effort and the marking strategy used), the size of the sample, and the criterion of selection.

We used a chi-square goodness-of-fit test to compare premarking allelic or genotypic frequencies with those observed in a mixture in which the unmarked populations were assumed to have allelic frequencies equal to the premarking frequencies of the marked population. This assumption is often valid (e.g., Lane et al. 1990), particularly in the geographical area near the donor population.

We defined the minimum detectable proportion of the population in the mixture as that proportion

detectable with a power of 90%, using the goodness-of-fit test conducted at a 10% level. The probability that any given proportion of the genetically marked population is detected in the mixture is determined from the power curve of the alternative hypothesis (presence of the marked population) for a particular sample size (N) from the mixture (see Appendix).

We examined the five strategies described above and present the curves that show the greatest discrimination with tests based on either allele or genotype frequencies at one or two loci. In many instances, especially when only small samples of a mixture were assayed, the genotypic model was inappropriate because one or more of the genotypes

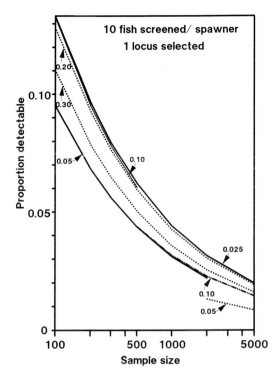

10 fish screened/ spawner
1 locus selected

FIGURE 2.—The minimum proportion of a genetically marked population that can be detected with a probability of 0.90 in a mixture in which allelic frequencies of unmarked fish equal the premarking frequencies. The minimum detectable proportion is shown, for various premarking allele frequencies (numbers along curves), as a function of the size of the sample from a mixture. The frequency of the marker allele is the result of a marking effort of 10 fish screened per fish spawned. A single locus was marked. Shown are the lesser of the values calculated from allelic frequencies (solid line) and genotypic frequencies (broken line).

occurred too infrequently to meet assumptions of asymptoticity. For all tests, the smallest expected class size was at least five. In some instances, the allelic frequency difference was more discriminating. The sample sizes used here, $N = 300$ and 1,000, approximate the minimum generally necessary to resolve a 5–10% contribution by marked fish with the maximum reasonable recovery effort. An N of 300 fish per statistical week has been used for stock composition estimates of sockeye salmon *Oncorhynchus nerka* in southeastern Alaska (Gharrett, unpublished data). An example of the relationship between recovery effort, N, and detectability, m, is shown for strategy (I), given a marking effort of 10 fish screened for each fish spawned (Figure 2).

Relationships between marking effort and detectability indicate that the premarking frequencies in the marked population (Figure 3), the

recovery effort (Figure 3), and the marking effort (Figures 3 and 4) must all be considered when the optimum marking strategy is chosen (Figure 4).

The optimum composites of the five strategies differed for samples of 300 fish (Figure 5, top) and 1,000 fish (Figure 5, bottom), because more genotypic or allelic classes exceeded the minimum of five for the larger sample size, which permitted more discriminating tests. The sawtooth nature of the composite curves indicated that even more discriminating strategies existed for some intervals.

Evaluation

One attractive aspect of genetic markers is that they are heritable. If genetic marking does not reduce the fitness of the population, and if no large numbers of unmarked fish stray into the marked population, the population will remain marked from generation to generation. To confirm the persistence of the marker, the frequency of marker alleles must be monitored over time. Tests for changes resulting from gene flow are similar to those used to detect the marked population in a mixture.

Natural selection can alter allelic frequencies in different ways. In particular, it can alter the frequency of only the homozygous genotype for an allele (complete dominance), or of any genotype carrying the allele (partial or no dominance), or of both homozygous genotypes (overdominance) (Falconer 1981). If marker alleles or closely linked alleles are deleterious, they will reduce the fitness of the population, and their frequencies will decrease. Because the frequencies are highest just after marking, the largest decreases for deleterious alleles should occur in the first generation.

Because of the potential loss in production from reduced fitness and the difficulty in resolving selection, tests for selection need to be conservative; that is, the power of the test is much less important than an indication that the marker frequencies are decreasing. Therefore, to investigate selection against the marker allele, we used a chi-square goodness-of-fit test (Appendix). This test compared the frequencies of marker alleles or genotypes to calculated postselection frequencies to determine the minimum detectable selection coefficients for three selection schemes (complete dominance, no dominance, and overdominance) at a single locus (Figure 6). The overdominance curves Figure 6 are based on a premarking equilibrium frequency of 0.10 for the marker allele. Note that even when 1,000 individuals were sampled, a selection coefficient had to be nearly 0.15 before it became noticeable.

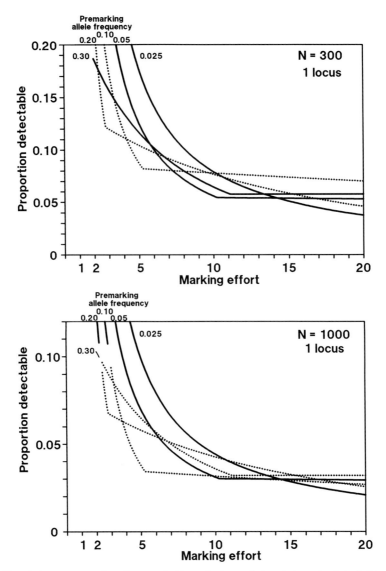

FIGURE 3.—The minimum proportion of a genetically marked population that can be detected in a mixed sample of 300 or 1,000 fish with a probability of 0.90, shown for various premarking allele frequencies as a function of the marking effort (number of fish screened per fish spawned) used to mark a single locus. Shown are the lesser of the values calculated from allelic frequencies (solid line) and genotypic frequencies (broken line).

Discussion

The success and utility of a genetic marker bred into a population is directly related to the effort expended in the marking and recovery efforts (Figures 2–4). The best marking strategy for a marking effort of five fish screened per fish spawned and a recovery sample of 300 fish is quite different from the best strategy for a marking effort of 10 and a recovery of 1,000 fish (Figure 5).

The baseline genetic information acquired prior to marking is needed for identifying alleles that will uniquely mark the population and for choos-

ing the best marking strategy given the resources available. Choice of marker alleles is limited by the genetic variability in the populations of interest. Populations or species with more polymorphic loci are more likely to have alleles that make good genetic markers.

The marking strategy selected can also increase detectability. Selection for a relatively low-frequency allele at a single locus is often the best strategy (Figure 5), because a single locus produces the largest allelic frequency difference between the marked and unmarked populations. In

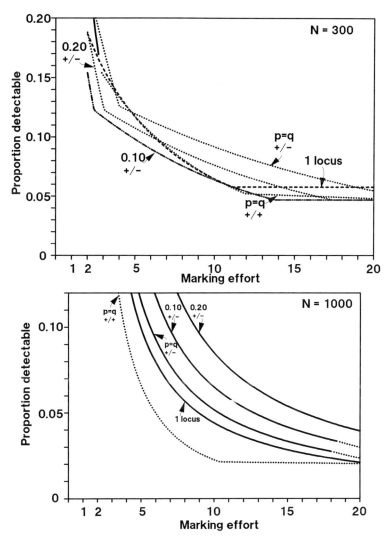

FIGURE 4.—The minimum proportion of a genetically marked population that can be detected in a mixed sample of 300 or 1,000 fish with a probability of 0.90, shown for various marking strategies as a function of the marking effort (number of fish screened per fish spawned) used to mark a single locus. All five marking strategies in the figure select for an allele at one locus with a premarking frequency of 0.30 (above) or 0.025 (below): *1 locus* selects only for that locus; *0.20 +/−* simultaneously selects at a second locus against an allele with a premarking frequency of 0.20; *p=q +/+* simultaneously selects at a second locus for an allele at frequency 0.30 (above) or 0.025 (below); *0.10 +/−* simultaneously selects at a second locus against an allele at a frequency 0.10; and *p=q +/−* simultaneously selects at a second locus against an allele at frequency 0.30 (above) or 0.025 (below). Shown are the lesser of the values calculated from allelic frequencies (solid line) and genotypic frequencies (broken line).

contrast, small changes at several loci generate a large number of genotypes at low frequencies not amenable to the model described here.

As marking effort is increased, progressively less frequent alleles become the best markers. For example, when the marking effort exceeds 15 fish screened per fish spawned, an allele with a frequency of about 0.025 is appropriate; but with a marking effort of 5, an allele with a frequency of about 0.10 is good. Alleles with intermediate frequencies are appropriate for marking efforts between 5 and 15.

Practical considerations may determine which marking strategy is most appropriate. Allelic markers must be chosen from among the alleles occurring naturally in the population, and those alleles may not be at frequencies that would best mark the population for a given marking effort. In

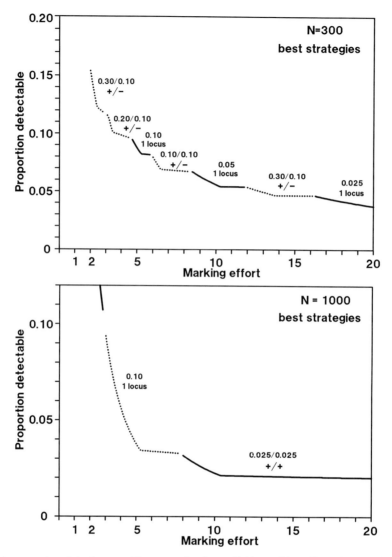

FIGURE 5.—A composite of the best marking strategies for available marking effort when marked fish must be detected in mixed samples of 300 fish (above) or 1,000 fish (below). Seven marking strategies are involved: *0.30/0.10 +/−* represents selection at one locus for an allele with a premarking frequency of 0.30 and at a second locus against an allele with a premarking frequency of 0.10; *0.20/0.10 +/−* denotes selection at one locus for an allele with a premarking frequency of 0.20 and at a second locus against an allele with a premarking frequency of 0.10; *0.10/0.10 +/−* is selection at one locus for an allele with a premarking frequency of 0.10 and at a second locus against an allele with a premarking frequency of 0.10; *0.025/0.025 +/+* is selection at one locus for an allele with a premarking frequency of 0.025 and at a second locus against an allele with a premarking frequency of 0.025; *0.10 1 locus* is selection at a single locus for an allele with a premarking frequency of 0.10; *0.05 1 locus* is selection at a single locus for an allele with a premarking frequency of 0.05; and *0.025 1 locus* is selection at a single locus for an allele with a premarking frequency of 0.025. Shown are the lesser of the values calculated from allelic frequencies (solid line) and genotypic frequencies (broken line).

addition, in order to detect some allelic markers, fish must be sacrificed to obtain tissues in which the allele is expressed. In many instances, it is desirable to keep the fish alive to spawn at a later time. For some applications it may not be possible

to screen both sexes because many or all of the females may be needed for production.

For example, in genetically marking a population of chum salmon *Oncorhynchus keta,* Seeb et al. (1986) were limited to male spawners because

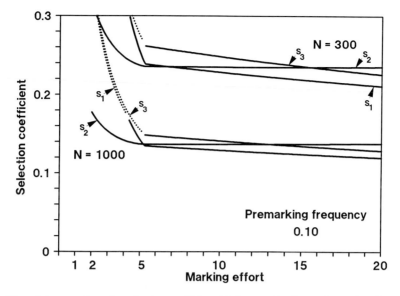

FIGURE 6.—The minimum value of a selection coefficient (s) that can be detected as a function of the marking effort used to mark a population at one locus that has an initial marker allele frequency of 0.10. Selection coefficients were estimated from a chi-square goodness-of-fit test ($P_\alpha = 0.10$). The three selection regimes depicted are s_1, complete dominance (selection against the homozygote for the marker allele); s_2, no dominance (selection against the marker allele); and s_3, overdominance (heterozygote superiority) for samples of 300 and 1,000. For s_3, the ratio of selection coefficients is the ratio of premarking allele frequencies, 0.9/0.1. Shown are the lesser of the values calculated from allelic frequencies (solid line) and genotypic frequencies (broken line).

all the females were needed for production. They had to determine the genotypes of the males without killing them (to ensure sufficient fish for spawning), which meant their choice of marker alleles was limited to loci expressed in muscle that could be sampled nonlethally. Because no single allele had an acceptable premarking frequency, they chose an alternative marking scheme that screened potential sires for two alleles with premarking frequencies of about 0.01 and 0.10 (a +/+ marking strategy). Each year for 5 years, they marked fish over a long spawning season of several weeks. The marking effort was approximately 15 males screened per male spawned. Spawning the selected males with unselected females involved a marking effort of about 8 for the less frequent allele and about 3 for the more frequent allele. Selecting for an allelic marker in only one sex reduces the marking effort by one-half for lower-frequency alleles and by more than one-half for the more frequent alleles. Selecting for alleles at two loci increased the chances of including unrelated individuals in the brood stock, thereby reducing the inbreeding effect that may result from selecting for rare alleles.

After a population has been genetically marked, it should be monitored to ensure the persistence of the marker in the population. The frequency of

the marker allele would be expected to decrease if it were not selectively neutral or if there were immigration of unmarked fish into the population. It is not a simple matter to distinguish changes in frequencies caused by natural selection from those caused by gene flow. Regardless of the cause, decreases in the frequency of the marker alleles over time diminishes their utility.

Tests for selection against genetic markers are quite weak. Using a large sample of 1,000, we found that a value of approximately 0.15 is minimally detectable for the selection coefficients s_1 (complete dominance), s_2 (no dominance), and s_3 (overdominance) (Figure 6). A selection coefficient of 0.15, which may not be detectable, can result in a 10% decrease in the fitness of a population. Only substantial decreases in fitness are noticeable. On the other hand, advantages gained in managing the population, such as increased control over spawning escapement, that increase production of the marked population should more than compensate for small decreases in fitness that might result from genetic marking.

Acknowledgment

We greatly appreciate the advice and help of J. Pella. J. Olsen, J. A. Gharrett, and J. Helle provided helpful comments on the manuscript. Support for J. E. Seeb came from Washington Sea Grant Program NA86AA-D-SG044, project R/A-50. Support for A. J. Gharrett came from the National Marine Fisheries Service Auke Bay Laboratory and from the Alaska Sea Grant College Program under grant NA81AA-D-00009 (project R/02-05), and the University of Alaska.

References

Bishop, Y. M. M., S. F. Fienberg, and P. W. Holland. 1977. Discrete multivariate analysis: theory and practice. MIT Press, Cambridge, Massachusetts.

Falconer, D. S. 1981. Introduction to quantitative genetics. Longman, New York.

Fournier, D. A., T. D. Beacham, B. E. Riddell, and C. A. Busack. 1984. Estimating stock composition in mixed stock fisheries using morphometric, meristic, and electrophoretic characteristics. Canadian Journal of Fisheries and Aquatic Sciences 41:400–408.

Gharrett, A. J., S. Lane, A. J. McGregor, and S. G. Taylor. 1983. The genetic mark: its implementation and evaluation. Pages 387–402 in Proceedings of the second North Pacific aquaculture symposium. Tokai University, Shimizu, Japan.

Gharrett, A. J., and S. M. Shirley. 1985. A genetic examination of spawning methodology in a salmon hatchery. Aquaculture 47:245–256.

Grant, W. S., G. B. Milner, P. Krasnowski, and F. M. Utter. 1980. Use of biochemical variants for identification of sockeye salmon (*Oncorhynchus nerka*) stocks in Cook Inlet, Alaska. Canadian Journal of Fisheries and Aquatic Sciences 37:1236–1247.

Lane, S., A. J. McGregor, S. G. Taylor, and A. J. Gharrett. 1990. Genetic marking of an Alaskan pink salmon population, with an evaluation of the mark and the marking process. American Fisheries Society Symposium 7:395–406.

Marshall, S., and nine coauthors. 1987. Application of scale pattern analysis to the management of Alaska's sockeye salmon. Canadian Special Publication of Fisheries and Aquatic Sciences 96:307–326.

Moles, A., P. Rounds, and C. Kondzela. 1990. Use of the brain parasite *Myxobolus neurobius* in separating mixed stocks of sockeye salmon. American Fisheries Society Symposium 7:224–231.

Murphy, B. R., L. A. Nielsen, and B. J. Turner. 1983. Use of genetic tags to evaluate stocking success for reservoir walleyes. Transactions of the American Fisheries Society 112:457–463.

Pella, J. J., and G. B. Milner. 1987. Use of genetic marks in stock composition analysis. Pages 247–276 in N. Ryman and F. M. Utter, editors. Population genetics & fishery management. University of Washington Press, Seattle.

Reisenbichler, R. R., and J. D. McIntyre. 1977. Genetic differences in growth and survival of juvenile hatchery and wild steelhead trout, *Salmo gairdneri*. Journal of the Fisheries Research Board of Canada 34:123–128.

Schweigert, J. F., J. J. Ward, and J. W. Clayton. 1977. Effects of fry and fingerling introductions on walleye (*Stizostedion vitreum vitreum*) production in West Blue Lake, Manitoba. Journal of the Fisheries Research Board of Canada 34:2142–2150.

Seeb, J. E., L. W. Seeb, and F. M. Utter. 1986. Use of genetic marks to assess stock dynamics and management programs for chum salmon. Transactions of the American Fisheries Society 115:448–454.

Seeb, L. W., J. E. Seeb, R. L. Allen, and W. K. Hershberger. 1990. Evaluation of adult returns of genetically marked chum salmon, with suggested future applications. American Fisheries Society Symposium 7:418–425.

Appendix: Selection Coefficients and Power of Detection

A chi-square goodness-of-fit statistic is computed as

$$X^2 = N\sum_i (f_i - fob_i)^2/f_i;$$

f_i is the expected allelic, gametic, or genotypic frequency, fob_i is the observed frequency, and N is the number of alleles, gametes, or genotypes.

Selection Measurements

Three selection regimes are considered: complete dominance with selection against the marker allele, no dominance with selection proportional to the number of marker alleles in a genotype, and overdominance (heterozygote superiority).

For natural selection, f_i is the frequency in the marked population and fob_i is the postselection frequency. For complete dominance with selection (s_1) against the marker allele, the postselection genotypic frequencies will be $p^2/(1 - s_1q^2)$, $2pq/(1 - s_1q^2)$, and $(1 - s_1)q^2/(1 - s_1q^2)$ for genotypes AA, AA', and $A'A'$, respectively. Postselection allelic frequencies can be derived from the genotypic frequencies. For a type I error of $P_\alpha = 0.10$,

$$s_1 > (2.706/2Npq^3)^{1/2}/[1 + q^2(2.706/2Npq^3)^{1/2}]$$

for allele frequency comparisons, and

$$s_1 > [4.605/Npq^2(1 + q)]^{1/2}/\{1 + q^2[4.605/Npq^2(1 + q)]^{1/2}\}$$

for genotypic comparisons.

For no dominance, postselection frequencies are $p^2/(1 - qs_2)$, $2pq(1 - s_2/2)/(1-qs_2)$, and $q^2(1 - s_2)/(1 - qs_2)$. For a type I error of $P_\alpha = 0.10$,

$$s_2 > [2.706(2)/Npq]^{1/2}/\{1 + q[2.706(2)/Npq]^{1/2}\}$$

for allele frequency comparisons, and

$$s_2 > [4.605(2)/Npq]^{1/2}/\{1 + q[4.605(2)/Npq]^{1/2}\}$$

for genotypic comparisons.

Overdominance (heterozygote superiority) is one explanation for maintenance of variability in a population. If the naturally occurring frequencies reflect equilibrium frequencies, the ratio of the frequencies equals the ratio of the reciprocal of the selection coefficients against the homozygous types. For the premarking frequencies, p and q, let $q/p = r$. Then the postselection genotypic frequencies are $p^2(1 - rs_3)/(1 - rs_3p^2 - s_3q^2)$, $2pq/(1 - rs_3p^2 - s_3q^2)$, and $q^2(1 - s_3)/(1 - rs_3p^2 - s_3q^2)$. For a type I error of $P_\alpha = 0.10$,

$$s_3 > [2.706/2Npq(pr - q)^2]^{1/2}/\{1 + (p^2r + q^2)[2.706/2Npq(pr - q)^2]^{1/2}\}$$

and

$$s_3 > \{4.605/N[p^2r^2 + q^2 - (p^2r + q^2)^2]\}^{1/2}/\{1 + (p^2r + q^2) \cdot [4.605/N(p^2r^2 + q^2 - [p^2r + q^2]^2)]^{1/2}\}$$

for allelic and genotypic comparisons, respectively.

Detectability in a Mixture

In a mixture comprising a genetically marked population and unmarked populations for which the frequencies equal the premarking frequencies of the marked stock,

$$p\text{mix}_i = m(p\text{m}_i) + (1 - m)p_i = m(p\text{m}_i - p_i) + p_i;$$

$p\text{mix}_i$ are the frequencies (allelic or genotypic) observed in the mixture, $p\text{m}_i$ the frequencies in the marked population, p_i the frequencies in the other contributors to the mixture (presumed to equal the premarking frequencies), and m the proportional contribution of the marked population to the mixture. To determine the detectability of the marked population in the mixture, the power of the test must be used. The power is computed from the probability density function for the goodness-of-fit statistic when the alternative hypothesis is true; it is the area under the density function for values greater than the critical value of the test. For the chi-square goodness-of-fit test described above, $f\text{obs}_i = p\text{mix}_i$ and $f_i = p_i$. The distribution of the alternative hypothesis is described by a noncentral chi-square distribution with a noncentrality parameter:

$$\lambda = \sum_i (p\text{mix}_i - p_i)/p_i + (N - 1)\sum_i (p\text{mix}_i - p_i)^2/p_i$$

or

$$\lambda = m\sum_i (p\text{m}_i - p_i)/p_i + m^2 (N - 1)\sum_i (p\text{m}_i - p_i)^2/p_i;$$

i indexes the genotypes or the alleles or gamete types, and N is the number of individuals in the sample for genotypes or twice the number of individuals for alleles or allele pairs (Bishop et al. 1977). For a given critical value of the goodness-of-fit test ($P_\alpha = 0.10$ here) and the degrees of freedom in the test, a noncentrality parameter can be obtained for which the power of the test is 90%. For that noncentrality parameter value and the chosen sample sizes (N), the proportion (m) of the marked population in the mixture can be determined from the quadratic equation above. This is the minimum proportion for which detection of the marked population is 90% certain.

American Fisheries Society Symposium 7:418–425, 1990
© Copyright by the American Fisheries Society 1990

Evaluation of Adult Returns of Genetically Marked Chum Salmon, with Suggested Future Applications

L. W. Seeb and J. E. Seeb

Fisheries Research Laboratory and Department of Zoology, Southern Illinois University
Carbondale, Illinois 62901, USA

R. L. Allen

Washington Department of Fisheries, Room 115 General Administration Building
Olympia, Washington 98504, USA

W. K. Hershberger

School of Fisheries, University of Washington, Seattle, Washington 98195, USA

Abstract.—Chum salmon *Oncorhynchus keta* from a tributary of Puget Sound, Washington, were genetically marked at the aspartate aminotransferase (*Aat1,2*) and phosphogluconate dehydrogenase (*Pgdh*) loci and released from gravel incubation boxes (egg boxes) over a 5-year period (1976–1980). We compared egg-box and natural production by monitoring allozyme frequencies in returning adults from 1979 to 1984. Egg-box production accounted for 0.20 to 0.91 of all fish returning to the stream. This study demonstrated the utility of genetically marking, on a production scale, fish that are too small for other tags, and it showed the benefit of mark transmission to offspring. It is possible to identify genetically marked salmon stocks in mixed-stock aggregations by standard methods now widely used by countries of the North Pacific rim.

The use of genetic marks, typically allozyme variants, to identify stocks of fish is increasingly common in fisheries and aquaculture studies. These allozyme variants, detected through protein electrophoresis, are generally inherited according to simple Mendelian principles and are permanent throughout the life of the organism (Utter et al. 1987a).

Genetic marks find application in at least three types of studies (Seeb et al. 1990). First, naturally occurring allozyme variation can be used to identify particular stocks in mixed-stock aggregations (Grant et al. 1980; Beacham et al. 1985; Williamson et al. 1986; Millar 1987; Pella and Milner 1987; Utter et al. 1987b). Second, stocks with naturally occurring, distinct allele frequencies can be introduced into an experimental situation or a population of resident fish (Schweigert et al. 1977; Sato et al. 1982; Murphy et al. 1983). Third, fish with particular genetic marks can be bred to increase mark frequency, thereby producing a uniquely identifiable population. The latter category includes both experimental studies (Reisenbichler and McIntyre 1977; Gharrett and Shirley 1985) and studies of hatchery production (Taggart and Ferguson 1984; Seeb et al. 1986; Gharrett and Seeb 1990, this volume; Lane et al. 1990, this volume; and reviewed in Utter and Seeb 1990, this

volume). Intentional genetic marking of a stock of chum salmon *Oncorhynchus keta* is the focus of this paper.

The life history of chum salmon makes them particularly suitable for genetic marking. Unlike coho salmon *O. kisutch* and chinook salmon *O. tshawytscha,* whose fry remain and grow for some time in fresh water, chum salmon migrate to seawater almost immediately upon hatching. Their small size makes such marking techniques as microwire tagging and fin clipping difficult if not potentially harmful. Morrison and Zajac (1987) recently reported histological evidence showing that half-length coded wire tags substantially damaged one of the olfactory nerves in approximately half the chum salmon examined. With genetic marking, the parents are screened, but no handling of their small progeny is required.

In a previous study, we genetically marked a population of chum salmon at Kennedy Creek, a small tributary of Totten Inlet in southern Puget Sound, Washington (Seeb et al. 1986). We marked all eggs placed in streamside gravel incubation boxes (e.g., Bams 1972, 1974) from 1976 to 1980. Male spawners were selected for expression of the *115* allele, *Aat1,2(115),* for aspartate aminotransferase (AAT, enzyme number 2.6.1.1: IUBNC 1984) and for expression of the *90* allele, *Pgdh(90),* for phosphogluconate dehydrogenase (PGDH, 1.1.1.44). The respective frequencies were increased from approximately 0.12 and 0.01 in the unselected fish to 0.25 and 0.11 among the progeny of the selected males. Our objectives were to (1) conduct a mark–recapture experiment to estimate the total abundance of young chum salmon produced in Totten Inlet, (2) evaluate the relative contributions of wild and stocked fish to subsequent spawning runs, and (3) evaluate the biological and management implications of genetic marking in general.

Seeb et al. (1986) reported results of objective (1). In the present paper, we concentrate on objectives (2) and (3). We use data from 5 years of marking and 6 years of monitoring the allozyme phenotypes of returning adults to estimate the contribution of the gravel incubation boxes (egg boxes) from each brood year to the total Kennedy Creek run. We also suggest possible applications of genetic marking in future studies.

Methods

Our methods were described by Seeb et al. (1986). The selection process for genetic marking was performed during the years 1976–1980. The Washington Department of Fisheries constructed a weir and trap across the stream, which stopped upstream migration. All females were either used for the egg-box take or passed upstream for natural spawning; only males were available for selection. Each day, males were removed from the trap, anesthetized, and tagged, and their muscles were biopsied for electrophoresis. After electrophoresis, males were sorted and those with the desired *Aat1,2* and *Pgdh* phenotypes were used to fertilize the eggs. The other males were passed upstream or discarded. Beginning in 1979, with the first returns of genetically marked fish, we took scale samples from returning adults for age determination.

In 1981, when the genetic marking had been completed, we focused on monitoring the adult returns of both sexes. Because there was no need to keep the fish alive for a selection process, heart rather than skeletal muscle was chosen for electrophoresis. (In chum salmon, *Aat1,2* isozymes in the heart are much less subject to degradation during storage than are those in skeletal muscle.) Hearts for electrophoretic analysis and scale samples for age determination were collected from returning adult salmon each year from 1981 to 1984. Samples were collected from freshly killed individuals immediately after spawning, and num-

bers were assigned to all samples so that the electrophoretic phenotype and age could be matched for each individual. The hearts were frozen as soon as possible and remained frozen until analysis. The age of each fish was determined from the scale samples by the Washington Department of Fisheries.

All tissue samples were electrophoretically analyzed for genetic variation at the *Aat1,2* and *Pgdh* loci on an amine–citrate gel, pH 6.1 (Clayton and Tretiak 1972). The *Aat1,2* loci are isoloci, both encoding identical allele products. We could not assign AAT alleles to individual loci, so we treated the isolocus pair as a single, tetrasomic locus. The phenotype of each individual was based on the number of observed doses (0–4) of the variant allele (Allendorf et al. 1975). The dosages were determined by their relative intensities, which are easily distinguished in heart or freshly biopsied muscle tissue of chum salmon. The genetic basis of these phenotypes has been previously documented through inheritance tests (Seeb and Seeb 1986). We were able to score the nine possible *Aat1,2* genotypes only as five possible phenotypes corresponding to the number of doses of the variant allele. For example, we could not differentiate a *100/115/100/100* individual from a *100/100/100/115* individual. Both were assigned a phenotype reflecting one dose of the variant allele. At the *Pgdh* locus, we scored three possible genotypes, giving us 15 possible combinations of *Aat1,2* and *Pgdh* phenotypes.

Phenotype frequencies of wild and egg-box juveniles (Seeb et al. 1986) from each brood year were used to estimate the proportion of each group in the returning adult population; we used the admixture analysis described by Millar (1987). Maximum-likelihood estimates of the contribution of each stock were calculated from the EM algorithm (Dempster et al. 1977); variances were calculated from the asymptotic covariance matrix (Milner et al. 1981). Separate analyses were conducted for each genetically marked brood, because the frequencies varied between years as a result of the marking procedure.

Estimates of naturally spawned eggs were made by the Washington Department of Fisheries and were based on spawning-ground counts and fecundity estimates from hatchery females. To estimate wild fry production, we assumed a 10% egg-to-fry survival, the average value the Washington Department of Fisheries found in gravel nests in Kennedy Creek. Outmigrating fry were trapped and counted to estimate egg-box release.

TABLE 1.—Allelic frequency estimates for adult chum salmon returning to Kennedy Creek. The 1976–1980 data are from males screened for genetic marking; the 1981–1984 data are from marked males and females.

Year of return	N	Frequency by locus and allele			
		Aat1,2		Pgdh	
		100	115	100	90
1976	1,407	0.879	0.121	0.994	0.006
1977	257	0.881	0.119	0.998	0.002
1978	1,081	0.871	0.129	0.996	0.004
1979	1,174	0.868	0.132	0.992	0.008
1980	2,745	0.836	0.164	0.979	0.021
1981	522	0.819	0.181	0.980	0.020
1982	612	0.783	0.217	0.922	0.078
1983	454	0.763	0.237	0.912	0.088
1984	704	0.800	0.200	0.943	0.057

A ratio of success for wild-fry production and egg-box release was calculated as follows. (1) The ratio (A) of the estimated wild-fry production to the estimated egg-box release was calculated. (2) A second ratio (B) of the estimated contribution of adults from natural spawning to the estimated contribution of adults from the egg boxes was calculated. (3) The index of relative success of wild compared to egg-box fish was calculated as B/A. The ratios would be equal and the index would equal 1.0 if both groups performed equally.

Results

The allele and phenotype frequencies of returning adults were monitored for 9 years, 1976–1984 (Table 1). During the first 5 years, the monitoring was in conjunction with the actual genetic marking. From 1981 to 1984, the returning adult frequencies were monitored without further selection. The first returns from the marking program began in 1979, at which time age determination from scales also began.

Chum salmon in Puget Sound have overlapping generations; the majority of fish return as 3- and 4-year-olds, and a few return as 5-year-olds. Originally, we planned to sample approximately 200 fish from both of the 3-year-old and 4-year-old age-groups from each year to minimize the variance around the maximum-likelihood estimates. However, during the study, the distribution among age-classes occasionally shifted dramatically. Of the the 653 fish sampled in 1980, only 24 were 4-year-olds. In 1984, an unusually high percentage of the fish were 5-year-olds (10%). This was the only year in which the sample of 5-year-olds was large enough for admixture analysis.

Maximum-likelihood estimates of the relative contribution of the egg box and wild fish were calculated from the phenotype frequencies (Table 2) for returning adults of each age-class, and from the baseline wild and egg-box juvenile frequencies reported by Seeb et al. (1986). Significant differences between baseline phenotype frequencies allowed us to distinguish the egg-box from the wild fish in each year (Table 3, Figure 1); the precision of our estimates of proportional contribution depended on the sample size.

We compared the relative success of the egg-box program in different years, using estimates of the number of fry produced in the stream, the number of fry released from the egg boxes, and the number of adults that returned from each (Table 4). Indexes of relative success (Table 4) ranged from 1.1 to 8.0, indicating that wild fry often were more successful than enhancement fry in terms of fry-to-adult survival.

The 3-year-olds that returned in 1984 were the progeny of the first brood from nonselective mating, after 5 years of screening for particular males. The selection process ended in 1980, and all fish from eggs spawned in 1981 were the result of nonselective hatchery matings and wild spawnings. The phenotype frequencies (Table 2) of this group reflect the lack of selection. A χ^2 goodness-of-fit test to Hardy–Weinberg expected values was performed by pooling the rare phenotypes. No significant difference from the expected distribution was observed ($\chi^2 = 1.988$, df = 5).

Discussion

Evaluation of the Study

Intentional genetic marking of chum salmon in Kennedy Creek was performed for 5 years, 1976–1980, a period longer than the average life cycle of chum salmon in western Washington. Over that period, close to 17 million eggs were genetically marked and placed in streamside

TABLE 2.—Frequencies of the 15 possible combinations of *Aat1,2* and *Pgdh* phenotypes among returning Kennedy Creek chum salmon adults for which age data also were available. Sample sizes are given in parentheses.

Allele combination		Frequency by year of return and age-group											
		1979	1980		1981		1982		1983		1984		
Aat1,2 locus	*Pgdh* locus	Age 3 (106)	Age 3 (629)	Age 4 (24)	Age 3 (48)	Age 4 (462)	Age 3 (297)	Age 4 (299)	Age 3 (44)	Age 4 (379)	Age 3 (352)	Age 4 (240)	Age 5 (66)
100/100/100/100	*100/100*	0.415	0.469	0.500	0.333	0.454	0.297	0.314	0.159	0.214	0.460	0.229	0.258
100/100/100/115	*100/100*	0.377	0.385	0.125	0.479	0.363	0.336	0.401	0.410	0.380	0.389	0.383	0.333
100/100/115/115	*100/100*	0.132	0.102	0.042	0.021	0.120	0.158	0.124	0.182	0.198	0.088	0.179	0.076
100/115/115/115	*100/100*	0.047	0.014	0.333	0.063	0.028	0.047	0.020	0.136	0.029	0.014	0.042	0.076
115/115/115/115	*100/100*	0.000	0.000	0.000	0.000	0.002	0.000	0.003	0.000	0.003	0.000	0.004	0.000
100/100/100/100	*100/90*	0.018	0.016	0.000	0.083	0.008	0.064	0.064	0.045	0.098	0.020	0.054	0.121
100/100/100/115	*100/90*	0.001	0.025	0.000	0.021	0.018	0.087	0.047	0.045	0.068	0.020	0.071	0.091
100/100/115/115	*100/90*	0.000	0.005	0.000	0.000	0.007	0.013	0.027	0.023	0.010	0.008	0.008	0.045
100/115/115/115	*100/90*	0.000	0.000	0.000	0.000	0.000	0.003	0.000	0.000	0.000	0.000	0.013	0.000
115/115/115/115	*100/90*	0.000	0.000	0.000	0.000	0.000	0.000	0.000	0.000	0.000	0.000	0.000	0.000
100/100/100/100	*90/90*	0.000	0.000	0.000	0.000	0.000	0.000	0.000	0.000	0.000	0.000	0.008	0.000
100/100/100/115	*90/90*	0.000	0.000	0.000	0.000	0.000	0.003	0.000	0.000	0.000	0.000	0.004	0.000
100/100/115/115	*90/90*	0.000	0.000	0.000	0.000	0.000	0.000	0.000	0.000	0.000	0.000	0.004	0.000
100/115/115/115	*90/90*	0.000	0.000	0.000	0.000	0.000	0.000	0.000	0.000	0.000	0.000	0.000	0.000
115/115/115/115	*90/90*	0.000	0.000	0.000	0.000	0.000	0.000	0.000	0.000	0.000	0.000	0.000	0.000

gravel incubation boxes. The marked fish had elevated frequencies for relatively rare alleles at *Aat1,2* and *Pgdh*. They also had unique distributions of *Aat1,2–Pgdh* phenotype combinations that were characterized by high proportions of the rarer classes compared to an unselected stock. The shifted allele frequencies and unique phenotype distributions represented identifiable stocks as defined statistically by Millar (1987).

A long-term consideration is the continuing identifiability of the Kennedy Creek population relative to other known populations of chum salmon. At present, the allelic frequencies generated by our marking program distinguish the Kennedy Creek population from all other known

populations of chum salmon in the north Pacific (e.g., Okazaki 1982; Beacham et al. 1985). The Washington Department of Fisheries is evaluating how precisely the contribution of the Kennedy Creek stock can be estimated from admixtures of numerous chum salmon stocks in Washington.

The frequencies of the *Aat1,2(115)* and *Pgdh(90)* alleles, originally estimated at 0.121 and 0.006, will likely stabilize at approximately 0.22 and 0.06, respectively. These figures will fluctuate because the actual magnitude of the shift in allelic frequency varied during the marking years. In the 1984 returns, for example, the overall frequency of *Pgdh(90)* was 0.057 ($N = 704$), but frequencies were 0.024 ($N = 372$) for 3-year-olds, 0.089 ($N =$

TABLE 3.—Maximum-likelihood estimates of the proportions of chum salmon adults of five brood years originating from naturally spawning (wild) and hatchery (egg-box) fish.

Brood and year of return	Age	N	Estimated contribution		Standard deviation
			Wild	Egg-box	
1976 brood					
1979	3	106	0.67	0.33	0.110
1980	4	26	[a]	[a]	
1977 brood					
1980	3	768	0.80	0.20	0.043
1981	4	462	0.74	0.26	0.052
1978 brood					
1981	3	48	0.50	0.50	0.176
1982	4	303	0.32	0.68	0.069
1979 brood					
1982	3	308	0.23	0.77	0.062
1983	4	379	0.09	0.91	0.056
1984	5	66	0.09	0.91	0.124
1980 brood					
1983	3	44	0.21	0.79	0.156
1984	4	260	0.39	0.61	0.069

[a] Sample size too small to make estimate.

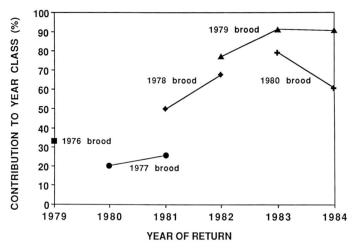

FIGURE 1.—Estimated contribution of egg-box chum salmon to the returning adult Kennedy Creek population during 1979–1984. Results for five brood years are shown.

251) for 4-year-olds, and 0.123 ($N = 66$) for 5-year-olds. Similar variation was observed for *Aat1,2(115)*, which had an overall 1984 frequency of 0.200 and year-class frequencies of 0.163, 0.250, and 0.233.

As a consequence of these variations, future frequencies will depend on the relative contributions of year-classes. Annual fluctuations in frequency will decline as year-classes mix and generations overlap. These results emphasize the need to continually monitor a genetically marked stock once the marking process is completed. Both phenotype and allele frequencies should be followed to assess the permanency and stability of

the genetic markers. Correlated changes at other loci also are potentially informative (Lane et al. 1990).

We can now evaluate one or our original study objectives—to determine the relative contributions of wild and egg-box fry to subsequent spawning runs. The egg boxes were installed to improve embryo survival rates, thereby providing surplus eggs for transplantation to other sites and surplus adults for harvest by commercial fishermen. The egg-box fry performed poorly in the first year, 1976, when very high egg-to-fry mortality rates occurred because of siltation in the egg boxes. Subsequently, settling boxes were in-

TABLE 4.—Comparison of the wild-fry production, egg-box release, and relative returns of chum salmon to Kennedy Creek for five brood years.

	Fry			Returning adults		
Brood	Estimated wild-fry production (1,000s)[a]	Estimated egg-box release (1,000s)[a]	Ratio of wild-fry production to egg-box release (A)	Age at return	Ratio of estimated wild contribution to estimated egg-box contribution (B)[b]	Success of wild-fry production relative to egg-box release (B/A)
1976	781	3,144	0.25	3	2.0	8.00
1977	863	327	2.64	3	4.0	1.52
				4	2.8	1.06
1978	702	2,198	0.32	3	1.0	3.12
				4	0.5	1.56
1979	148	1,576	0.09	3	0.3	3.33
				4	0.1	1.11
				5	0.1	1.11
1980	450	3,826	0.12	3	0.3	2.50
				4	0.6	5.00

[a] Provided by Washington State Department of Fisheries.
[b] From Table 3.

stalled to keep silt from entering the egg boxes. However, the wild chum salmon fry continued to produce adult returns at a greater rate than the egg-box fry (Table 4). The egg box program was discontinued partially based on these findings.

The level of inbreeding must be considered in the evaluation of any genetic-marking program. Seeb et al. (1986) calculated inbreeding coefficients for Kennedy Creek for the years 1976–1980. The coefficients had magnitudes of approximately 10^{-3}, well below the threshold of 10^{-1} at which measurable effects first become identifiable. These calculations do not take into account that individuals with rare alleles may be more closely related than individuals selected at random. However, the use of two independent marks, as in this study, reduces this potential source of inbreeding.

Future Directions of Genetic Marking

Large-scale genetic-marking programs have been successfully implemented for chum salmon and pink salmon *O. gorbuscha* (Seeb et al. 1986; Lane et al. 1990). We anticipate that future studies involving a genetic mark will rely on intentional marking rather than on stocks with natural differences in allelic frequencies, because the latter are few. Moreover, stocks with natural genetic marks may also differ in such characteristics as growth and development. With intentional genetic marking, only the selected allozyme and tightly linked loci differ. The majority of the genetic background is, presumably, unchanged.

Admixture analysis—Genetic marking has the potential to increase the precision of admixture analyses of fish populations. The most important limiting factor in the application of admixture analysis is the lack of variation among stocks of fish. Often, populations from large geographic regions have similar gene frequencies and cannot be resolved into their components. Various authors (Millar 1987; Pella and Milner 1987) advocated combining stocks into larger groupings either before or after the composition analysis. Genetic marking of particular target stocks could create the variation needed to make the stocks identifiable in a statistical sense.

Alternatively, increasing the number of loci in the baseline with the hope of finding further discriminating loci has been suggested as a way of increasing the identifiability of contributing stocks (Pella and Milner 1987). However, this would require resampling of all contributing stocks, which could be difficult for salmonid species, some of whose spawning stocks are highly inaccessible because of geographic barriers or political boundaries. Possibly a genetic mark could be chosen from loci within the existing database, which would make resampling the entire database unnecessary. In that case, only the genetically marked populations would need to be resampled.

Genetic marking and other genetic manipulations—Many potentially advantageous genetic manipulations are now being considered and implemented by fisheries managers and aquaculturists. They include production of transgenics (e.g., Chourrout et al. 1986), all-female populations (Donaldson 1986), and diploid and triploid hybrids (Seeb et al. 1988), to name a few. Genetic marking could be used to evaluate these types of projects.

An important step in the commercial production of transgenics will be the qualitative confirmation that transferred genes controlling quantitative traits such as growth have been stably incorporated into the host genome. The fusion of the gene controlling the quantitative trait to an allozyme marker could provide such confirmation. Expression of the allozyme marker in the host and its progeny would indicate that the qualitative gene had been stably incorporated.

Also, the first step in producing families of all-female genotypes, a popular procedure in the culture of rainbow trout *Oncorhynchus mykiss* (formerly *Salmo gairdneri*), is the production of F_1 diploid gynogens (Bye and Lincoln 1986). Diploid gynogenesis also produces individuals homozygous for many allozyme markers suitable for genetic marking (Allendorf et al. 1986; Seeb and Seeb 1986), thus providing a reservoir of marks to choose from when producing the all-female F_2 generation.

Finally, interspecies hybrids may be impossible to identify by means of meristic or morphological characters (Leary et al. 1983; Campton and Utter 1985), but they are often identifiable by their allozyme phenotypes (e.g., Magee and Philipp 1982). Triploid hybrids will likely possess one or more allozyme marks, and allozyme analysis may play a central role in studies of these organisms.

Acknowledgments

The assistance of many individuals from the Washington Department of Fisheries was vital for the completion of this project; we especially thank Kevin Bauersfeld, Tom Burns, and John Sneva. Financial support was provided by the Washington Department of Fisheries; J.E.S. was supported by the Washington Sea Grant Program through project RA/50.

References

Allendorf, F. W., J. E. Seeb, K. L. Knudsen, G. H. Thorgaard, and R. F. Leary. 1986. Gene-centromere mapping of 25 loci in rainbow trout. Journal of Heredity 77:307–312.

Allendorf, F. W., F. M. Utter, and B. P. May. 1975. Gene duplication within the family Salmonidae: II. Detection and determination of the genetic control of duplicate loci through inheritance studies and the examination of populations. Pages 415–432 in C. L. Markert, editor. Isozymes, volume 4. Genetics and evolution. Academic Press, New York.

Bams, R. A. 1972. A quantitative evaluation of survival to the adult stage and other characteristics of pink salmon (Oncorhynchus gorbuscha) produced by a revised hatchery method which simulates optimal natural conditions. Journal of the Fisheries Research Board of Canada 29:1151–1167.

Bams, R. A. 1974. Gravel incubators: a second evaluation on pink salmon, Oncorhynchus gorbuscha, including adult returns. Journal of the Fisheries Research Board of Canada 31:1379–1385.

Beacham, T. D., R. E. Withler, and A. P. Gould. 1985. Biochemical genetic stock identification of chum salmon (Oncorhynchus keta) in southern British Columbia. Canadian Journal of Fisheries and Aquatic Sciences 42:437–448.

Bye, V. J., and R. F. Lincoln. 1986. Commercial methods for the control of sexual maturation in rainbow trout (Salmo gairdneri R.). Aquaculture 57:299–310.

Campton, D. E., and F. M. Utter. 1985. Natural hybridization between steelhead trout (Salmo gairdneri) and coastal cutthroat trout (Salmo clarki clarki) in two Puget Sound streams. Canadian Journal of Fisheries and Aquatic Sciences 42:110–119.

Chourrout, D., R. Guyomard, and L. Houdebine. 1986. High efficiency gene transfer in rainbow trout (Salmo gairdneri Rich.) by microinjection into egg cytoplasm. Aquaculture 51:143–150.

Clayton, J. W., and D. N. Tretiak. 1972. Amine-citrate buffers for pH control in starch gel electrophoresis. Journal of the Fisheries Research Board of Canada 29:1169–1172.

Dempster, A. P., N. M. Laird, and D. B. Rubin. 1977. Maximum likelihood estimation from incomplete data via the EM algorithm. Journal of the Royal Statistical Society B: Methodological 39:1–38.

Donaldson, E. M. 1986. The integrated development and application of controlled reproduction techniques in Pacific salmonid aquaculture. Fish Physiology and Biochemistry 2:9–24.

Gharrett, A. J., and J. E. Seeb. 1990. Practical and theoretical guidelines for genetically marking fish populations. American Fisheries Society Symposium 7:407–417.

Gharrett, A. J. and S. M. Shirley. 1985. A genetic examination of spawning methodology in a salmon hatchery. Aquaculture 47:245–256.

Grant, W. S., G. B. Milner, P. Krasnowski, and F.M. Utter. 1980. Use of biochemical genetic variants for identification of sockeye salmon (Oncorhynchus nerka) stocks in Cook Inlet, Alaska. Canadian Journal of Fisheries and Aquatic Sciences 37:1236–1247.

IUBNC (International Union of Biochemistry, Nomenclature Committee). 1984. Enzyme nomenclature 1984. Academic Press, Orlando, Florida.

Lane, S., A. J. McGregor, S. G. Taylor, and A. J. Gharrett. 1990. Genetic marking of an Alaskan pink salmon population, with an evaluation of the mark and the marking process. American Fisheries Society Symposium 7:395–406.

Leary, R. F., F. W. Allendorf, K. L. Knudsen. 1983. Consistently high meristic counts in natural hybrids between brook trout and bull trout. Systematic Zoology 32:369–376.

Magee, S. M., and D. P. Philipp. 1982. Biochemical genetic analyses of the grass carp female × bighead carp male F_1 hybrid and the parental species. Transactions of the American Fisheries Society 111:593–602.

Millar, R. B. 1987. Maximum likelihood estimation of mixed stock fishery composition. Canadian Journal of Fisheries and Aquatic Sciences 44:583–590.

Milner, G. B., D. J. Teel, F. M. Utter, and C. L. Burley. 1981. Columbia River stock identification study: validation of genetic method. Annual report of research (FY80). National Oceanic and Atmospheric Administration, Northwest and Alaska Fisheries Center, Seattle, Washington.

Morrison, J., and D. Zajac. 1987. Histologic effect of coded wire tagging in chum salmon. North American Journal of Fisheries Management 7:439–441.

Murphy, B. R., L. A. Nielsen, and B. J. Turner. 1983. Use of genetic tags to evaluate stocking success for reservoir walleyes. Transactions of the American Fisheries Society 112:457–463.

Okazaki, T. 1982. Genetic study on population structure in chum salmon (Oncorhynchus keta). Bulletin Far Seas Research Laboratory (Shimizu) 19:25–116.

Pella, J. J., and G. B. Milner. 1987. Use of genetic marks in stock composition analysis. Pages 247–276 in N. Ryman and F. Utter, editors. Population genetics and fishery management. University of Washington Press, Seattle.

Reisenbichler, R. R., and J. D. McIntyre. 1977. Genetic differences in growth and survival of juvenile hatchery and wild steelhead trout, Salmo gairdneri. Journal of the Fisheries Research Board of Canada 34:123–128.

Sato, R., K. Naka, and R. Ishida. 1982. Possible application of isozyme as a genetic marker for examination of fish stocking. Bulletin of the National Research Institute of Aquaculture 3:11–19.

Schweigert, J. G., F. J. Ward, and J. W. Clayton. 1977. Effects of fry and fingerling introductions on walleye (Stizostedion vitreum vitreum) production in West Blue Lake, Manitoba. Journal of the Fisheries Research Board of Canada 34:2142–2150.

Seeb, J. E., and L. W. Seeb. 1986. Gene mapping of isozyme loci in chum salmon. Journal of Heredity 77:399–402.

Seeb, J. E., L. W. Seeb, and F. M. Utter. 1986. Use of genetic marks to assess stock dynamics and management programs for chum salmon. Transactions of the American Fisheries Society 115:448–454.

Seeb, J. E., G. H. Thorgaard, and F. M. Utter. 1988. Survival and allozyme expression in diploid and triploid hybrids between chum, chinook, and coho salmon. Aquaculture 72:31–48.

Seeb, L. W., J. E. Seeb, and A. J. Gharrett. 1990. Genetic marking of fish populations. Pages 223–239 in D. H. Whitmore, editor. Electrophoretic and isoelectric focusing techniques in fisheries management. CRC Press, Boca Raton, Florida.

Taggart, J. B., and A. Ferguson. 1984. An electrophoretically-detectable genetic tag for hatchery-reared brown trout (*Salmo trutta* L.). Aquaculture 41:119–130.

Utter, F., P. Aebersold, and G. Winans. 1987a. Interpreting genetic variation detected by electrophoresis. Pages 21–45 in N. Ryman and F. Utter, editors. Population genetics and fishery management. University of Washington Press, Seattle.

Utter, F., D. Teel, G. Milner, and D. McIsaac. 1987b. Genetic estimates of stock compositions of 1983 chinook salmon, *Oncorhynchus tshawytscha,* harvests off the Washington coast and the Columbia River. U.S. National Marine Fisheries Service Fishery Bulletin 85:13–23.

Utter, F. M., and J. E. Seeb. 1990. Genetic marking of fishes: overview focusing on protein variation. American Fisheries Society Symposium 7: 426–438.

Williamson, J. H., G. J. Carmichael, M. E. Schmidt, and D. C. Morizot. 1986. New biochemical markers for largemouth bass. Transactions of the American Fisheries Society 115:460–465.

American Fisheries Society Symposium 7:426–438, 1990

Genetic Marking of Fishes:
Overview Focusing on Protein Variation

FRED M. UTTER

National Marine Fisheries Service
2725 Montlake Boulevard East, Seattle, Washington 98112, USA

JAMES E. SEEB

Fisheries Research Laboratory and Department of Zoology
Southern Illinois University, Carbondale, Illinois 62901, USA

Abstract.—Most marks commonly used to identify fish are restricted to one generation. An exception to this are genetic marks such as variant proteins encoded by different alleles and detected by electrophoresis which are completely heritable. Selective breeding of distinct genotypes (i.e., genetic marking) has created identifiable groups and even populations. Such genetically marked populations maintain the long-term performance potential of the parent population if distinguishing genotypes have negligible effect on fitness and if there is an adequate effective number of breeding individuals from the parent stock in the marked population. Genetic marking has been used with salmon *Oncorhynchus* spp. to estimate hatchery contributions to a population of spawning adults and to detect straying between populations. Similar genetic marking projects are feasible for any cultured population. Cultured marine species are particularly suitable for genetic marking because the generally reduced genetic differences among populations compared with freshwater and anadromous species make marked populations easier to identify. Genetic marking can also be extended to wild populations if potential breeders can be screened and only those of appropriate genotype are permitted to spawn. Much additional genetic variation other than at protein-coding genes is potentially useful for genetic marking, including variation in mitochondrial DNA and genetic variants that can be detected immunologically.

Many kinds of distinguishing marks can be used to identify fish. The value of most of them, however, is limited by the inability of individuals to transmit their distinguishing mark to subsequent generations. Marks limited to the immediate generation cannot provide long-term information, for instance, on the relative contribution of a stock to a spawning population composed of multiple stocks. Only heritable characteristics can provide such information.

Especially valuable to fish researchers are the heritable characters based on genes that code for proteins, and that are detected by electrophoresis. The usefulness of these characters lies in their purely genetic basis. Information about many different gene loci can be obtained and interpreted with relative ease (Utter et al. 1987). Since the mid-1960s, the advantages of genetic marks have resulted in a proliferation of studies involving electrophoretically detected differences in protein-coding genes within and among populations of all types of living organisms (see Lewontin and Hubby 1966 for a classical early study, and more recent reviews of the scope of such studies by Hamrick et al. 1979 and Nevo et al. 1984). A broad, systematic sample of fish that represent diverse environments has now been examined electrophoretically (Shaklee 1983; Nevo et al. 1984; Gyllensten 1985).

This paper focuses on the use of variation in protein-coding genes as genetic marks for cultured fish. Particular emphasis is placed on genetic marking—the process of breeding selected individuals to produce offspring that are genetically distinguishable (Allendorf and Utter 1979; Seeb et al. 1986). Pertinent studies involving genetic marks and genetic marking are reviewed, and future directions for genetic marking with nonprotein-coding genes are considered.

Genetic Marks

Alleles and Genotypes of Protein-Coding Loci

Detailed methods for detecting genetic variants of proteins by electrophoretic separation and histochemical staining are amply described in various sources (e.g., Shaklee and Keenan 1986; Aebersold et al. 1987), as are the bases for genetic interpretation of the observed variation (Utter 1987; Utter et al. 1987). Therefore, we will present

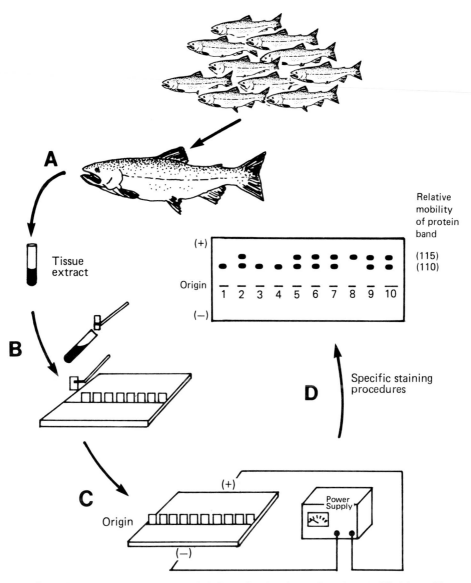

FIGURE 1.—Steps for obtaining allelic genotypic information by electrophoresis. (Modified from Gharrett and Utter 1982.)

only a brief outline of procedures and genetic interpretations of electrophoretically detected protein variations.

The procedures leading from the intact organism to the genetically interpretable electrophoretic pattern are outlined in Figure 1. Crude protein extracted from a small fragment of tissue, such as muscle or liver (Figure 1A), is absorbed by filter paper and placed in a starch gel (Figure 1B). Samples from many individuals (we regularly run up to 50) can be placed side by side in a single

gel. When an electric current is applied to the gel (Figure 1C), allelically encoded proteins (i.e., genetically distinct forms encoded by the same locus) often migrate different distances because of their different net charges.

In one system of nomenclature (Allendorf and Utter 1979), alleles are designated according to the relative mobility of the proteins they encode. The number *100* is assigned to the allele responsible for the most commonly observed protein. An allele would be designated *145* if it encoded a protein that migrated 45% farther than the one encoded by allele *100*. Related proteins migrating different distances are readily identified by staining procedures that are specific for a particular protein class (Figure 1D). Specificity in staining, especially for enzymatic proteins, permits identification of both the activity and the exact location of a particular protein from a complex mixture of proteins in each extract.

Patterns similar to those observed in Figure 1D permit determination of genotypes (i.e., designation of presumptive genes encoding the observed character) as well as the frequencies of genotypes and alleles in samples. Such one- and two-banded patterns typify expressions of monomeric proteins (active proteins consisting of single subunits) such as the enzyme phosphoglucomutase (PGM). The reader is referred to Allendorf and Utter (1979) and Utter et al. (1987) for guidelines for genetic interpretations of more complex electrophoretic patterns and for criteria to assure that a particular genetic interpretation is valid.

If the genetic basis of the electrophoretic patterns like those of Figure 1D has been verified, the pattern for each individual reflects the contribution of two allelic doses, one inherited from each parent. The patterns for individuals 1, 3, and 4 may be interpreted as direct products of the same allele, say *Pgm(100)*, which was inherited from each parent, and which resulted in the genotype *100/100* for the PGM locus and in the production of only one type of protein molecule to form the band designated 100 on Figure 1D. These individuals are homozygous for the *Pgm(100)* allele. Similarly, individual 8 is homozygous for the *Pgm(115)* allele. This individual has a genotype of *115/115* and produces only one type of protein molecule, which forms the band designated 115 on Figure 1D. The remaining individuals, with two-banded expressions, inherited different PGM alleles from each parent. They are heterozygous, their genotypes are *100/115*, and they produce both types of protein molecules. The genotype frequencies of this sample of 10 individuals, then, are *100/100*, 3; *100/115*, 6; *115/115*, 1. There are 20 alleles whose frequencies can be counted directly from the 10 genotypes. There are six *100* alleles from the three *100/100* genotypes and six more from the six *100/115* genotypes for a total of 12; similarly, there are eight *115* alleles altogether. The allele frequencies, then, are 12/20 = 0.6 *100* and 8/20 = 0.4 *115*. Knowledge of genotypes, alleles, and their frequencies is fundamentally important in the use of protein-coding genes as marks to identify individuals and populations.

Monitoring Migration and Transplants through Natural Marks

Any populations of a species that are distinguishable from other populations by significant differences in their allele frequencies at one or more detectable loci are genetically marked. Isolated conspecific groups frequently are genetically marked relative to one another. Some instances are examined below in which such differences have been used to monitor the presence of genetically marked groups in areas where they migrated or were transplanted.

Once the genotypes of a region's populations are sufficiently distinguished, frequency distributions can be used to obtain estimates of the relative abundance of contributing populations in areas of intermingling. An overview of this application of genetically marked populations was given by Milner et al. (1985), and the procedure for obtaining estimates is outlined in Figure 2. This procedure has been used with populations of chinook salmon *Oncorhynchus tshawytscha* in the northwestern USA and British Columbia. Currently, sets of allelic data representing more than 20 polymorphic loci and 100 distinguishable populations are being used in the management of these fisheries (D. Teel, U.S. National Marine Fisheries Service, and J. Shaklee, Washington Department of Fisheries, personal communications). Similar procedures are being applied in investigations of mixed populations of other species of Pacific salmon in this region and in Alaska, including chum salmon *O. keta* (Beacham et al. 1985a), Pink salmon *O. gorbuscha* (Beacham et al. 1985b), sockeye salmon *O. nerka* (Utter et al. 1984), and coho salmon *O. kisutch* (G. Winans, U.S. National Marine Fisheries Service, personal communication).

Existing differences have been effectively used to monitor the effects of stock transplants. Transplanted populations of walleye *Stizostedion vitreum* marked by frequency differences for malate dehydrogenase (MDH) alleles were monitored in two studies. Fixed allelic differences (i.e., no alleles shared by stocked and native populations) permitted Schweigert et al. (1977) to directly measure contributions of stocked larvae and juveniles introduced into West Blue Lake, Manitoba. Murphy et al. (1983) observed differences in frequencies of alleles common to both stocked and

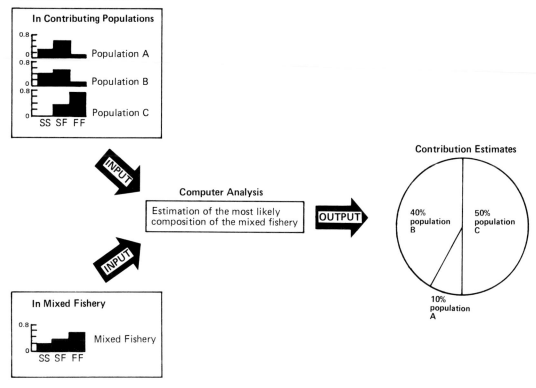

FIGURE 2.—Schematic of procedure used to estimate the contributions of separate stocks in a mixed fishery, based on known genotype frequencies for the stocks and the mixture (from Milner et al. 1985). Histograms indicate genotype frequencies for two alleles (S and F) in three populations, and in the mixture to which each contributes. These data are used to obtain estimates of individual contributions to the mixture (see Pella and Milner 1987).

naturally spawned populations in Claytor Lake, Virginia. Shifts in allele frequencies that resulted from the presence of stocked fish were used to estimate the proportion of stocked fish.

Differences in allele frequencies have been used in several instances to monitor the consequences of transplanting salmonid populations; the consequences have varied considerably (Table 1). The complexities of adapting to new environments are reflected in the different abilities of transplanted groups to become established, and in both the presence and the absence of introgression within and between species.

Genetic Marking

Basic Ideas

Genetic marking—accomplished by breeding individuals of specified genotypes to produce identifiable progeny—is a logical extension of procedures that use natural genetic marks. The idea became feasible when the existence of widespread allelic variation was demonstrated by electrophoresis (see Moav et al. 1976; Utter et al. 1976; Gharrett et al. 1983). In addition to some generally appealing features, genetic marking has distinct attributes that make it possible to obtain information not otherwise available (Table 2). For example, the technique is valuable because it eliminates the need to handle or deform individuals to impose a permanent and readily detected mark that persists for a lifetime. The uniqueness of genetic marking, of course, lies in the heritable nature of the mark. No other marking process also marks subsequent generations.

Screening Process

The process of genetic marking can be divided into three stages: prescreening, identification of individuals with appropriate genotypes for use as breeders (screening), and the actual breeding. Prescreening allows suitable marker loci to be identified in the parent population before the

TABLE 1.—Examples of transplanted salmonid populations monitored through natural differences in their allele frequencies.

Species	Observation	Reference
Rainbow trout[a]	Life history differences between native and transplanted populations	Utter et al. (1976)
	Persistence of native populations in spite of transplants	Allendorf et al. (1980); Wishard et al. (1984)
	Introgression of transplanted and native populations	Campton and Johnson (1985)
Other trout native to western North America	Varying degrees of introgression of transplanted rainbow trout with native populations of cutthroat,[b] Gila,[c] and Apache trout[d]	Allendorf and Phelps (1981); Busack and Gall (1981); Loudenslager et al. (1986)
Chum salmon	Persistence of run timing and allele frequencies of transplanted population	Okazaki (1978)
	Returns, based on numbers of fish released, much lower for transplanted than native fish	Altukhov and Salmenkova (1987)

[a]*Oncorhynchus mykiss* (formerly *Salmo gairdneri*).
[b]*Oncorhynchus clarki* (formerly *Salmo clarki*).
[c]*Oncorhynchus gilae* (formerly *Salmo gilae*).
[d]*Oncorhynchus apache* (formerly *Salmo apache*).

hectic time of spawning. Prescreening of 50 to 100 individuals usually provides adequate estimates of genotype frequencies of potential marker loci; alleles not detected by such screenings are probably too rare to be useful as genetic markers. Expression of loci in tissues that can be biopsied (e.g., skeletal muscle, eye fluid, blood) is essential if fish screened for selection of' spawners must remain alive up to the time of spawning.

Suitable spawners may be identified from biopsied tissues or from tissues of sacrificed fish. Use of biopsied tissue is more labor-intensive and limits the number of loci that can be screened. On the other hand, it permits sampling from brood stocks with small numbers of spawning individuals, and may substantially extend the collection period for gametes. Biopsied individuals receive physical tags so that fish whose genotypes have been selected for breeding can be identified. A number of biopsy procedures that minimize mortality have been described (e.g., Crawford et al. 1977; Vuorinen and Piironen 1984; Carmichael et al. 1986; Seeb et al. 1986).

Suitable breeders may be identified from tissues of sacrificed fish for populations having a surplus of spawners. In this case, the screening process must occur within the time limits of in vitro gamete viability (although procedures presently exist for extended viability of both sperm and eggs well beyond 48 h: e.g., Stoss 1983).

Genetic marking can proceed once individuals having suitable genotypes for use as spawners are identified. Some applications of genetic marking are considered in the following sections.

Experimental Applications

The three studies summarized below are good examples of straight-forward applications of genetic marking. They were conducted in confined or restricted locations over short periods. In two of them, progeny classification could be based on the parental genotypes; in two of them, there was little or no mortality from fertilization to termination of the experiment.

Genetically marked steelhead (anadromous rainbow trout) were used to compare the perfor-

TABLE 2.—Comparison of the characteristics of marks used to identify fish.

Mark	Handling of individuals	Effect on fitness	Permanence of mark in individual	Genetic component expressed in descendants	Minimum size requirement
Coded wire nose tag	Yes	Small	High	No	Yes
Otolith marking	No	None	100%	No	No
Fin clip	Yes	Large	Variable	No	Yes
Visible external tag	Yes	Large	Variable	No	Yes
Scale analysis	No	None	Variable	Low	Yes
Genetic marking	No	Negligible	100%	Yes	No

TABLE 3.—Scheme for estimating the proportion of eggs fertilized by male chum salmon of different dominance states (α, $\beta1$, $\beta2$) in multiple matings with individual females. Examples of three possible results are given. Note that estimated contributions of $\beta1$ and $\beta2$ males are twice their observed frequencies in progeny because heterozygous parents would contribute the variant allele (B) to only half of their progeny.

	Breeding scheme				Possible results						
	Genotype				Observed frequencies				Estimated contribution of males		
			Males		Locus 1		Locus 2				
Locus	Females	α	$\beta1$	$\beta2$	AA	AB	AA	AB	α	$\beta1$	$\beta2$
1	AA	AA	AB	AA	1.00	0.00	1.00	0.00	1.00	0.00	0.00
2	AA	AA	AA	AB	0.75	0.25	1.00	0.00	0.50	0.50	0.00
					0.90	0.10	0.90	0.10	0.60	0.20	0.20

mances of progeny of hatchery fish and wild fish from the Deschutes River, Oregon, in creek and pond environments (Reisenbichler and McIntyre 1977). Fish were placed in their respective environments before or shortly after hatching, which precluded mechanical marking. These environments were either barren or depopulated, and downstream migration was blocked by weirs. The breeding scheme involved gamete pools from parents that were homozygous for one of two alleles of a lactate dehydrogenase locus (Ldh-4). Breedings created nine lots of each of three possible Ldh-4 genotypes in pure hatchery (HH), pure wild (WW), and hatcher–wild (HW) progeny. The lots, arranged in three sets, permitted identification within each set of individual HH, HW, or WW progeny. Researchers could also use the lots to compare relative performances of different Ldh-4 genotypes in different habitats. Performance was better for hatchery fish in pond environments and for wild fish in stream environments; no differences were detected among Ldh-4 genotypes.

In the second study, genetic marking was used to measure the relative spawning contribution of male chum salmon that occupied different positions in the dominance hierarchy (Schroder 1982). Males of different dominance states and genotypes (determined from muscle biopsies) were introduced into an experimental spawning chamber containing a mature female of known genotype. The spawning behavior was recorded on videotape. All eggs were collected for subsequent rearing after each of six spawning sequences. It was possible to estimate the spawning contribution of dominant and satellite males by examining the genotypes of the progeny. The breeding scheme and possible results are presented in Table 3. The experimental data indicated that one or more satellite males contributed to five of the six mating sequences, in which they accounted for an average 28% of the progeny.

In a similar study measurements were made of the individual contributions of multiple males that fertilized single egg lots of pink salmon *Oncorhynchus gorbuscha* (Gharrett and Shirley 1985). Screening of over 500 individuals produced seven males and eight females with genotypic differences at four loci, which permitted identification of paternity in multiple fertilizations. Sperm from four males was added either sequentially (in four different sequences) or simultaneously to individual egg lots. Highly significant differences were observed in the contributions of individual males in both types of tests. The authors recommended the use of single male and female crosses in hatchery situations where brood stocks are limited; this procedure would eliminate disproportionate representation of males that have more competitive sperm, thereby reducing variance in family size and optimizing the effective population size (discussed below).

Changing Allele Frequencies in Populations

Genetic marking of a population is a more extensive venture than marking of an experimental group. Screening and marking of more individuals are required, particularly if multiple year-classes are involved. If the mark is intended as a permanent feature of the population, the marking process and the mark itself are elevated in importance because considerations extend beyond the short term to future generations. In general, the goal in marking a population is to create an identifiable group that retains the evolutionary fitness of the parent stock. To meet this goal, the mark must have a negligible effect on individual performance and the marked population must include an adequate sampling of genes over all other loci from the parent stock.

Equivalent performance of selected genotypes. —The possibility that genetic marking might select deleterious alleles is a complex issue that is heatedly debated and difficult or impossible to resolve. There is no question that genetic marking involves different genotypes, and that varied genotypic fitnesses are major elements in evolutionary processes. The uncertainty arises because much genetic variation is largely or entirely unaffected by selection; that is, variation can be neutral, or nearly so (see discussions in Hartl 1980). An ideal genetic mark should fall into this category. There is good evidence that allelic differences for genes encoding *some* proteins in *some* fishes cause differences in fitness (e.g., Powers et al. 1983; Mork et al. 1984). Alternate interpretations seem more compelling for other reports of variable fitness of allelically encoded proteins in fishes (e.g., Tsuyuki and Williscroft 1977; Chilcote et al. 1986). However, empirical data from diverse animal species support an assumption of neutrality for most protein-coding alleles (Chakraborty et al. 1980; Ihssen et al. 1981). Because neutrality can never be proved, some guidelines for identifying appropriate allelic variants for genetic marking are offered here.

(1) Be cautious of variants if good evidence for selection exists for their protein class in related organisms. Hemoglobins, tranferrins, esterases, and null alleles for any protein-coding locus are notable examples.

(2) Seek variants that occur widely and in diverse environments.

(3) Conversely, be careful of rare alleles or those with substantial frequencies in restricted environments.

(4) Monitor the performance of comparable marked and unmarked individuals and populations, and—when feasible—design experiments to test the relative fitness of genotypes involved in the marking process (see Reisenbichler and McIntyre 1977, summarized above).

Adequate sampling of parent population.—An adequate sampling of genes to create a genetically marked population requires sufficient breeders of both sexes to minimize both inbreeding depression and loss of rare alleles (see discussions in Ryman and Ståhl 1980; Allendorf and Ryman 1987). The degree of inbreeding is a function of the effective population size (N_e); the latter is defined by

$$N_e = (4N_m N_f)/(N_m + N_f);$$

N_m and N_f are the respective numbers of males and females participating in breeding. The N_e is strongly affected by the number of the sex used

least in a breeding program. Inbreeding depression (ΔF) is related to N_e by

$$\Delta F = \frac{1}{2} N_e,$$

and is only slightly affected by further increase in N_e beyond 60. Because N_e is also affected by the variance in family size (see Gharrett and Shirley 1985, summarized above), efforts should be made to equalize the contributions of individual matings. Loss of rare alleles, which is related only to the total number of parents, affects the capability of the population to adapt to changing environments. Ryman and Stahl (1980) noted that an N_e of 100 individuals results in a reduction in average heterozygosity (a reflection of inbreeding) of only 0.5%, but it reduces the number of alleles per locus (a reflection of loss of rare alleles) by 50%. The above considerations led Allendorf and Ryman (1987) to postulate 25 individuals of each sex (with equal contributions from individual matings) as the absolute minimum for establishing a new population from a parent group; they recommended a much greater number to minimize inbreeding and maintain established populations over successive generations.

An adequate sampling of genes in the marked population also depends on the frequency of a marker allele in the parent population and the desired change of this frequency in the marked population. These considerations dictate the screening effort required for the marking process. The degree of change in a single generation of genetic marking depends further upon the availability of male and female gametes. Gharrett et al. (1983) calculated the effort required to change the frequency of an allele when equal numbers of both sexes are screened and used as breeders (Figure 3). Under the conditions presumed in Figure 3, screening 600 individuals to identify 100 breeders would shift the allele frequency from 0.10 in the parent population to 0.53 in the marked population. Under the same conditions, screening 2,000 individuals would change a parent population frequency of 0.01 to 0.20. A limiting threshold allele frequency of around 0.025 in the parent population is indicated for practical use in genetic marking under the conditions of Figure 3. Although other conditions will change this threshold, it is evident that a marker allele with a low frequency in the parent population has limited application if the marked population is to approach the performance capabilities and evolutionary potential of the parent population.

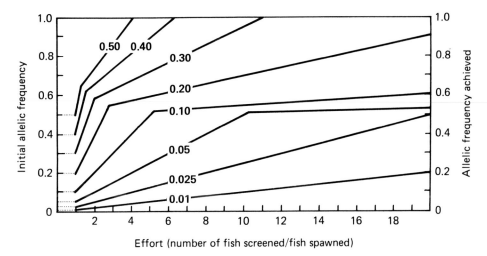

FIGURE 3.— Marking effort (number of fish screened for each breeding fish) required to change allele frequencies from eight different baseline frequencies. The following conditions are presumed: equal numbers of both sexes will be used as breeders; and the priority for spawning genotypes (based on a predetermined number of individuals to be screened) will be (1) homozygotes for the marker allele, (2) heterozygotes for the marker allele, and (3) nonbearers of the marker allele. The inflection points on lines for baseline frequencies of 0.05 and higher represent a screening effort sufficient to exclude individuals that lack the selected allele from brood stock (i.e., only priority-1 and -2 individuals will be bred). Beyond the inflection point, the change in allele frequency requires greater effort because more heterozygous individuals are screened than are actually used. (Modified from Gharrett et al. 1983.)

Examples of Population Changes

A segment of a chum salmon population in south Puget Sound, Washington, was genetically marked every year for 5 years (Seeb et al. 1986). The purposes of this study were (1) to estimate the regional population of juvenile chum salmon through dilution of the mark in samples from an adjacent saltwater area (Totten Inlet), and (2) to identify contributions of marked fish to the total spawning population of their natal stream (Kennedy Creek) upon their return as spawning adults. To maximize the number of eggs that could be taken, only males were screened as prospective parents of the marked group. The number of spawners used during each of the 5 years averaged 203 screened males and 1,323 unscreened females, which resulted in average N_e and ΔF values of 695 and 0.001, respectively. Variant alleles for two protein systems were chosen for marking to assure that an adequate number of males would be available for spawning. One protein-coding system, Aat-1,2, consisted of a set of isoloci for the enzyme aspartate aminotransferase (see Allendorf and Thorgaard 1984; Utter et al. 1987) for which phenotypes had to be determined from the combined contribution of two loci (i.e., four allelic doses); phenotypes with two or more doses of the 115 allele were used. The second

protein system—6-phosphogluconate dehydrogenase—was encoded by a single locus, Pgdh; genotypes with at least one dose of the 90 allele were used. The average allele-frequency changes between the unmarked and marked populations were from 0.12 to 0.25 for the Aat-1,2(115) allele and from 0.01 to 0.11 for the Pgdh(90) allele. These differences were sufficient to allow adequate estimates of marked fish in population mixtures by maximum-likelihood procedures (see Milner et al. 1985). Estimates of the total juvenile population in Totten Inlet, ranging from 2.2 to 4.3 million, indicated that an expansion of the procedure could be effectively used to estimate the juvenile production of the entire south Puget Sound region. The estimated contributions of marked fish to the spawning population of Kennedy Creek ranged from 6 to 29% of the entire run (Seeb et al. 1990, this volume).

Late runs of pink salmon returning to Auke Creek, near Juneau, Alaska, were genetically marked for four consecutive years (Gharrett 1985; Lane et al. 1990, this volume). Even- and odd-year runs represented isolated gene pools because of the rigid 2-year life cycle of pink salmon in this area. Screening of 3,906 odd-year adults for genotypes with a variant malate dehydrogenase (MDH) allele, Mdh-3,4(70), identified 390 males and females that served as the brood stock for genetic marking in 1979. (Although *Mdh-3,4* are isoloci, the variant allele was limited to only one of the loci, i.e., no more than two doses of the variant allele were observed). Random mating of the selected brood stock increased the frequency of the *70* allele from 0.054 in the parent population to 0.510 in the genetically marked adults returning in 1981 (identified by fin clips imposed on juveniles prior to release). Similarly, screening of 7,380 even-year adults for genotypes of an alpha-glycerophosphate dehydrogenase (AGP) allele, *Agp(200)*, identified a brood stock of 396 individuals, mating of which increased the frequency of the *Agp(200)* allele from 0.199 in the parent population to 0.852 in the fin-clipped adults that returned in 1982. Second-generation genetically marked fish were randomly bred from returning fin-clipped adults of both even- and odd-year runs; and the new allele frequencies persisted in the adults produced from these crosses. Monitored returns from the first and second generations indicated no straying of genetically marked fish to adjacent streams. Likewise, no straying was detected in the Auke Creek drainage for the gentically marked odd-year run. There was evidence of straying in the drainage, however, for marked fish of the even-year run: Allele-frequency shifts towards that of the marked population occurred in late runs both upstream and downstream from the hatchery.

The chum and pink salmon studies had instructive similarities and differences. In both studies, sufficient individuals were screened in a given year to identify enough breeding parents for an N_e well in excess of 100. Thus the goal of adequately representing the parent population in the founder population was approached. The alleles selected for marking appeared to be suitable with regard to fitness: they occurred over broad geographic ranges, and their frequencies varied according to expectations of neutrality (Aspinwall 1974; Okazaki 1982). The marked populations of both species provided a means to measure straying, over several generations, among populations of their respective regions.

Differences in the two studies highlight the flexibility inherent in genetic-marking projects. Equal numbers of both sexes were selected for marking pink salmon. Consequently, the allele-frequency differences between parent and founder populations were greater than was possible for the chum salmon study, in which selected males were bred with randomly chosen females. This greater difference in allele frequency gave greater "power" to the mark (see Gharrett et al. 1983)—the presence of marked pink salmon or their descendants was more readily detected in situations where marked and unmarked populations had intermingled or interbred. Nevertheless, the allelic differences produced in chum salmon were sufficient to achieve the study's goals.

These studies also demonstrated the general applicability of genetic marking in ocean farming. The principles can be applied to any artificially propagated group in which distinction is desired. Indeed, genetic marking can be extended to naturally propagated groups, provided the individuals that make up the spawning population can be selected through screening (see Hedgecock et al. 1976).

The ability to genetically mark a population gives biologists, managers, and growers a powerful tool for identifying marked populations in areas where they have intermingled or interbred. Mandatory genetic marking has been suggested for all hatchery stocks of Atlantic salmon *Salmo salar* (Ståhl 1987) as a means of improving their detection in population mixtures (a major current concern; see Saunders 1966), and as a means of monitoring intentional or unintentional introductions. For example, it has been estimated that roughly 10% of the Atlantic salmon harvested in Norwegian fisheries are escapees from net-pen culture (K. Hindar, Norsk Biotech, personal communication). Such escapement is viewed as a serious threat to wild Norwegian populations because it disrupts adapted gene pools and introduces diseases (see Ståhl 1987).

Mandatory marking of hatchery stocks could be readily and generally implemented. It could be straightforwardly carried out with an existing population (as in the pink salmon study), and would be even more readily achieved when a new stock is established (as in the chum salmon study). Such an identifiable "brand" would also be valuable to aquaculturists who seek protection in harvests where their fish predominate.

Our discussion of genetic marks and marking has been limited to anadromous and freshwater species, particularly salmonids, because studies to date have focused on them. Marine fishes that are amenable to culture also are promising subjects for genetic marking because they have a much lower degree of observed genetic divergence than freshwater and anadromous species (Gyllensten 1985). Marked populations of a cultured marine species would be conspicuous in a particular region amidst a relatively homogeneous background of local aggregates maintained by natural reproduction. Adequate numbers of marked individuals would permit estimates of population sizes and migration patterns similar to those obtained in the chum salmon study. Genetically marked populations would be especially valuable in evaluating restocking programs for marine species.

Other Genetic Marks

Although protein-coding genes are presently the most useful source of genetic marks for fish, the number of protein loci that can be electrophoretically screened for genetic variation (in excess of 100) still represents substantially less than 1% of a species' DNA. A vast reservoir of genetic markers undoubtedly exists within the remaining segment of the genome. Some of this variation is considered below. Much more of it will undoubtedly become available for genetic marks and for genetic marking as the pace of development in molecular biology continues to accelerate.

Morphological marks.—Morphological markers inherited in a simple Mendelian pattern have the obvious advantage of immediate visibility. Although this attribute recommends them as genetic marks, morphological marks have never been widely used. There are justifiable reasons for this neglect. First, Mendelian morphological markers are uncommon relative to protein markers—the studies described above would not have been possible with morphological markers, because none exist in the populations considered. A second shortcoming of morphological marks is their dominant or recessive expression; i.e., heterozygous individuals usually cannot be identified. In contrast, genotypes for protein-coding loci often are directly interpretable from electrophoretic phenotypes through codominant allelic expression (see Utter et al. 1987). Finally, individuals with marked morphological phenotypes often have lower fitness than wild types, whereas appropriate electrophoretic markers are quasineutral. One example is albinism in rainbow trout (Bridges and von Limbach 1972), a recessive genetic mark that has been used experimentally (e.g., Parsons and Thorgaard 1985).

Direct examination of DNA.—Though the value of protein variants as genetic markers has been clearly demonstrated, they remain only indirect reflections of the actual genetic variation.

Direct information concerning genetic variation must come from examination of DNA, the substance of genes. A widely used approach for studying DNA polymorphism is to examine DNA fragments that have been enzymatically cleaved at specific restriction sites identified by sequences of four to six bases. The procedures involved are much more laborious than those used to collect data for protein polymorphism. Nevertheless, they provide access to an enormous reservoir of genetic information that cannot be approached by protein analysis.

The cytoplasmic DNA of mitochondria (mtDNA) accounts for about 1% of all an individual's genes; the remaining 99% (including protein-coding loci detected by electrophoresis) reside in the nucleus. For several reasons, this small fraction of the genome has been intensively investigated in recent years (see Ferris and Berg 1987). The maternal inheritance of mtDNA permits maternal lineages to be followed. Mutation rates of mtDNA are up to 10 times higher than those of protein-coding nuclear genes in vertebrates. Mitochondria can be isolated and mtDNA purified with relative ease. Different maternal lineages can be recognized when purified mtDNA is enzymatically cleaved and restriction-site polymorphisms are revealed by differences in fragment size. Similar patterns of relationships among populations and species have been indicated by the relatively few comparable data sets involving protein-coding genes and mtDNA lineages in fishes (e.g., see Berg and Ferris 1984; Bermingham and Avise 1986). Distinctions among conspecific populations based on mtDNA tend to be more qualitative, and maternal inheritance has some distinct advantages—in particular, it can be used to determine direction of hybrid crosses (Avise and Saunders 1984). On the other hand, mtDNA cannot yield some types of information available from diploid genotypes of nuclear genes, such as heterozygosity and "Hardy–Weinberg" and gametic-phase equilibria.

An advantage of mtDNA in genetic marking is that screening would be limited to females. The concerns about the founder population, including adequate representation of genes from the parent population, still apply, however, as they do for any genetic mark. Several generations of backcrossing to males from the parent population would be required if one or very few females with a rare mtDNA type were the basis of the founder population. Dividing the eggs of one female with a rare mtDNA variant among 200 males would still produce a founder population that had an N_e of approximately 4 without subsequent backcrossing.

Immunological variation.—The use of immunological procedures to identify genetic marks in fish was largely abandoned when it became obvious that protein electrophoresis was a vastly more efficient procedure for finding and applying such variation (Hodgins 1972). It is time to reconsider at least limited use of immunological procedures to detect genetic marks and to produce markers. Some fishes have too little protein variation for use in genetic marking (Gyllensten et al. 1985; Seeb et al. 1987). In other vertebrates, some species with little variation among protein-coding loci show substantial immunogenic variation. In the American buffalo *Bison bison*, for example, only one polymorphic protein-coding locus (Sartore et al. 1969; Baccus et al. 1983) but nine polymorphic blood-group loci have been identified (Stormont et al. 1986). The technical problems outlined by Hodgins (1972) relating to immunological procedures and genetic marks for fish appear to be resolvable without great difficulty. In the absence of more readily detected sources of genetic variation, heritable antigenic variations should be sought for genetic marks and used in genetic marking.

References

Aebersold, P. B., G. A. Winans, D. J. Teel, G. B. Milner, and F. M. Utter. 1987. Manual for starch gel electrophoresis: A method for detection of genetic variation. NOAA (National Oceanic and Atmospheric Administration) Technical Report NMFS (National Marine Fisheries Service) 61.

Allendorf, F. W., D. M. Espeland, D. T. Scow, and S. Phelps. 1980. Coexistence of native and introduced rainbow trout in the Kootnenai River drainage. Proceedings of the Montana Academy of Sciences 39:28–36.

Allendorf, F. W., and S. R. Phelps. 1981. Isozymes and the preservation of genetic variation in salmonid fishes. Ecological Bulletin (Stockholm) 34:37–52.

Allendorf, F. W., and N. Ryman. 1987. Genetic management of hatchery stocks. Pages 141–160 *in* N. Ryman and F. Utter, editors. Population genetics & fisheries management. University of Washington Press, Seattle.

Allendorf, F. W., and G. H. Thorgaard. 1984. Tetraploidy and the evolution of salmonid fishes. Pages 1–53 *in* B. Turner, editor. Evolutionary genetics of fishes. Plenum, New York.

Allendorf, F. W., and F. M. Utter. 1979. Population genetics. Pages 407–454 *in* W. S. Hoar, D. J. Randall, and J. R. Brett, editors. Fish physiology, volume 8. Academic Press, New York.

Altukhov, Y. P., and E. A. Salmenkova. 1987. Stock transfer relative to the natural organization, man-
agement and conservation of fish populations. Pages 333–343 *in* N. Ryman and F. Utter, editors. Population genetics & fisheries management. University of Washington Press, Seattle.

Aspinwall, N. 1974. Genetic analysis of duplicate malate dehydrogenase loci in the pink salmon, *Oncorhynchus gorbuscha*. Genetics 76:65–72.

Avise, J., and N. Saunders. 1984. Hybridization and introgression among species of sunfish (*Lepomis*): analysis by mitochondrial DNA and allozyme markers. Genetics 108:237–255.

Baccus, R., N. Ryman, M. Smith, C. Reuterwall, and D. Cameron. 1983. Genetic variability and differentiation of large grazing mammals. Journal of Mammalogy 64:109–120.

Beacham, T. D., R. E. Withler, and A. P. Gould. 1985a. Biochemical genetic stock identification of chum salmon (*Oncorhynchus keta*) in southern British Columbia. Canadian Journal of Fisheries and Aquatic Sciences 42:437–448.

Beacham T. D., R. E. Withler, and A. P. Gould. 1985b. Biochemical genetic stock identification of pink salmon (*Oncorhynchus gorbuscha*) in southern British Columbia. Canadian Journal of Fisheries and Aquatic Sciences 42:1474–1483.

Berg, W. J., and S. D. Ferris. 1984. Restriction endonuclease analysis of salmonid mitochondrial DNA. Canadian Journal of Fisheries and Aquatic Sciences 41:1041–1047.

Bermingham, E., and J. C. Avise. 1986. Molecular zoogeography of freshwater fishes in the southeastern United States. Genetics 113:939–965.

Bridges, W. R., and B. von Limbach. 1972. Inheritance of albinism in rainbow trout. Journal of Heredity 63:152–153.

Busack, C. A., and G. A. E. Gall. 1981. Introgressive hybridization in populations of Paiute cutthroat trout (*Salmo clarki seleniris*). Canadian Journal of Fisheries and Aquatic Sciences 38:939–951.

Campton, D. E., and J. M. Johnston. 1985. Electrophoretic evidence for a genetic admixture of native and nonnative rainbow trout in the Yakima River, Washington. Transactions of the American Fisheries Society 114:782–793.

Carmichael, G. J., J. H. Williamson, M. E. Schmidt, and D. C. Morizot. 1986. Genetic marker identification in largemouth bass with electrophoresis of low-risk tissues. Transactions of the American Fisheries Society 115:455–459.

Chakraborty, R., P. A. Fuerst, and M. Nei. 1980. Statistical studies on protein polymorphism in natural populations. III. Distribution of allele frequencies and the number of alleles per locus. Genetics 94:1039–1063.

Chilcote, M. W., S. A. Leider, and J. J. Loch. 1986. Differential reproductive success of hatchery and wild summer-run steelhead under natural conditions. Transactions of the American Fisheries Society 115:726–735.

Crawford, B. A., S. A. Leider, and J. M. Tipping. 1977. Technique for rapidly taking samples of skeletal muscle from live adult steelhead trout. Progressive Fish-Culturist 39:125.

Ferris, S. D., and W. J. Berg. 1987. The utility of mitochondrial DNA in fish genetics and fishery management. Pages 277–299 in N. Ryman and F. Utter, editors. Population genetics and fisheries management. University of Washington Press, Seattle.

Gharrett, A. J. 1985. Genetic interaction of Auke Creek Hatchery pink salmon with natural spawning stocks in Auke Creek. University of Alaska, Report SFS UAJ-8509, Juneau.

Gharrett, A. J., S. Lane, A. J. McGregor, and S. G. Taylor. 1983. The genetic mark: it's implementation and evaluation. Pages 387–342 in A. Nagai, editor. Proceedings of the second north Pacific aquaculture symposium. Tokai University, Tokyo.

Gharrett, A. J., and S. M. Shirley. 1985. A genetic examination of spawning methodology in a salmon hatchery. Aquaculture 47:245–256.

Gharrett, A. J., and F. M. Utter. 1982. Scientists detect genetic differences. Sea Grant Today 12(2):3–4.

Gyllensten, U. 1985. The genetic structure of fish: differences in the intraspecific distribution of biochemical genetic variation between marine, anadromous, and freshwater species. Journal of Fish Biology 26:691–699.

Gyllensten, U., N. Ryman, and Ståhl. 1985. Monomorphism of allozymes in perch (*Perca fluviatilis* L.). Hereditas 102:57–61.

Hamrick, J. L., Y. B. Linehart, and J. B. Mitton. 1979. Relationships between life history characteristics and electrophoretically detectable genetic variation in plants. Annual Review of Ecology and Systematics 10:173–200.

Hartl, D. 1980. Principles of population genetics. Sinauer Associates, Sunderland, Massachusetts.

Hedgecock, D., R. A. Shleser, and K. Nelson. 1976. Applications of biochemical genetics to aquaculture. Journal of the Fisheries Research Board of Canada 33:1108–1119.

Hodgins, H. O. 1972. Serological and biochemical studies in racial identification of fishes. Pages 199–208 in R. Simon and P. Larkin, editors. The stock concept in Pacific salmon. H. R. MacMillan Lectures in Fisheries, University of British Columbia, Vancouver, Canada.

Ihssen, P. E., H. E. Booke, J. M. Casselman, J. M. McGlade, N. R. Payne, and F. M. Utter. 1981. Stock identification: materials and methods. Canadian Journal of Fisheries and Aquatic Sciences 38:1838–1855.

Lane, S., A. J. McGregor, S. G. Taylor, and A. J. Gharrett. 1990. Genetic marking of an Alaskan pink salmon population, with an evaluation of the mark and the marking process. American Fisheries Society Symposium 7:395–406.

Lewontin, R. C., and J. Hubby. 1966. A molecular approach to the study of genic heterozygosity in natural populations. II. Amount of variation and degree of heterozygosity in natural populations of *Drosophila pseudoobscura*. Genetics 54:595–609.

Loudenslager, E. J., J. N. Rinne, G. A. E. Gall, and R. E. David. 1986. Biochemical genetic studies of native Arizona and New Mexico trout. Southwestern Naturalist 31:221–234.

Milner, G. B., D. J. Teel, F. M. Utter, and G. A. Winans. 1985. A genetic method of stock identification in mixed populations of Pacific salmon, *Oncorhynchus* spp. U.S. National Marine Fisheries Service Marine Fisheries Review 47(1):1–8.

Moav, R., T. Brody, G. W. Wohlfarth, and G. Hulata. 1976. Applications of electrophoretic genetic markers to fish breeding. I. Advantages and methods. Aquaculture 9:217–228.

Mork, J., R. Giskeodegard, and G. Sundnes. 1984. The hemoglobin polymorphism in Atlantic cod (*Gadus morhua* L.): genotypic differences in somatic growth and in maturing age in natural populations. Pages 721–732 in E. Dahl, D. S. Danielssen, E. Moksness, and P. Solemdal, editors. The propagation of cod *Gadus morhua* L. Institute of Marine Research, Flødevigen Biological Station, Flødevigen rapporlserie 1, Arendal, Norway.

Murphy, R. R., L. A. Nielsen, and B. J. Turner. 1983. Use of genetic tags to evaluate stocking success for reservoir walleyes. Transactions of the American Fisheries Society 112:457–463.

Nevo, E., A. Beiles, and R. Ben-Shlomo. 1984. The evolutionary significance of genetic diversity: ecological, demographic and life history correlates. Lecture Notes in Biomathematics 53:12–213.

Okazaki, T. 1978. Genetic differences of two chum salmon (*Oncorhynchus keta*) populations returning to the Tokachi River. Bulletin Far Seas Fisheries Research Laboratory (Shimizu) 16:121–128.

Okazaki, T. 1982. Genetic study on population structure in chum salmon (*Oncorhynchus keta*). Bulletin Far Seas Fisheries Research Laboratory (Shimizu) 19:25–116.

Parsons, J. E., and G. H. Thorgaard. 1985. Production of androgenetic diploid rainbow trout. Journal of Heredity 76:177–181.

Pella, J. J., and G. B. Milner. 1987. Use of genetic marks in stock composition analysis. Pages 247–276 in N. Ryman and F. Utter, editors. Population genetics & fisheries management. University of Washington Press, Seattle.

Powers, D. A., L. DiMichele, and A. R. Place. 1983. The use of enzyme kinetics to predict differences in cellular metabolism, developmental rate, and swimming performance between Ldh-B genotypes of the fish *Fundulus heteroclitus*. Genetics and Evolution 10:147–170.

Reisenbichler R. R., and J. D. McIntyre. 1977. Genetic differences in growth and survival of juvenile hatchery and wild steelhead trout, *Salmo gairdneri*. Journal of the Fisheries Research Board of Canada 34:123–128.

Ryman, N., and G. Ståhl. 1980. Genetic changes in hatchery stocks of brown trout (*Salmo trutta*). Canadian Journal of Fisheries and Aquatic Sciences 37:82–87.

Sartore, G., C. Stormont, B. G. Morris, and A. A. Grunder. 1969. Multiple electrophoretic forms of carbonic anhydrase in red cells of domestic cattle *(Bos taurus)* and American buffalo *(Bison bison)*. Genetics 161:823–831.

Saunders, R. L. 1966. Some biological aspects of Greenland salmon fishery. Atlantic Salmon Journal 1966:17–23.

Schroder, S. L. 1982. The influence of intrasexual competition on the distribution of chum salmon in an experimental stream. Pages 275–285 *in* E. L. Brannon and E. O. Salo, editors. Salmon and trout migratory behavior symposium. University of Washington, College of Fisheries, Seattle.

Schweigert, J. G., F. J. Ward, and J. W. Clayton. 1977. Effects of fry and fingerling introductions on walleye *(Stizostedion vitreum vitreum)* production in West Blue Lake, Manitoba. Journal of the Fisheries Research Board of Canada 34:2142–2150.

Seeb, J. E., L. W. Seeb, D. W. Oates, and F. M. Utter. 1987. Genetic variation and postglacial dispersal of populations of northern pike in North America. Canadian Journal of Fisheries and Aquatic Sciences 44:556–561.

Seeb, J. E., L. W. Seeb, and F. M. Utter. 1986. Use of genetic marks to assess stock dynamics and management programs for chum salmon. Transactions of the American Fisheries Society 115:448–454.

Seeb, L. W., J. E. Seeb, R. L. Allen, and W. K. Hershberger. 1990. Evaluation of adult returns of genetically marked chum salmon, with suggested future applications. American Fisheries Society Symposium 7:418–425.

Shaklee, J. B. 1983. The utilization of isozymes as gene markers in fisheries management and conservation. Isozymes: Current Topics in Biological and Medical Research 11:213–247.

Shaklee, J. B., and C. P. Keenan. 1986. A practical laboratory guide to the techniques and methodology of electrophoresis and its application to fish fillet identification. Australia Commonwealth Scientific and Industrial Research Organization, Marine Laboratories, Report 177, Melbourne.

Ståhl, G. 1987. Genetic population structure of Atlantic salmon. Pages 121–140 *in* N. Ryman & F. Utter, editors. Population genetics and fisheries management. University of Washington Press, Seattle.

Stormont, C. J., B. G. Morris, Y. Suzuki, and J. Dodd. 1986. Blood typing beefalo cattle. Pages 359–364 *in* G. E. Dickerson and R. K. Johnson, editors. Proceedings 3rd world congress on genetics applied to livestock production, volume 9. University of Nebraska, Agricultural Communications, Lincoln.

Stoss, J. 1983. Fish gamete preservation and spermatozoan physiology. Pages 305–350 *in* W. S. Hoar, D. J. Randall, and E. M. Donaldson. Fish physiology, volume 9. Part B. Academic Press, New York.

Tsuyuki, H., and S. N. Williscroft. 1977. Swimming stamina differences between genotypically distinct forms of rainbow *(Salmo gairdneri)* and steelhead trout. Journal of the Fisheries Research Board of Canada 34:996–1003.

Utter, F. M. 1987. Protein electrophoresis and stock identification in fishes. NOAA (National Oceanic and Atmospheric Administration) Technical Memorandum NMFS (National Marine Fisheries Service) SEFC-199:62–103. Southeast Fisheries Center, Miami.

Utter, F. M., P. Aebersold, J. Helle, and G. Winans. 1984. Genetic characterization of populations in the southeastern range of sockeye salmon. Pages 17–32 *in* J. Walton and D. Houston, editors. Proceedings of the Olympic wild fish conference. Peninsula College, Fisheries Technology Program, Port Angeles, Washington.

Utter, F., P. Aebersold, and G. Winans. 1987. Interpreting genetic variation detected by electrophoresis. Pages 21–45 *in* N. Ryman and F. Utter, editors. Population genetics & fisheries management. University of Washington Press, Seattle.

Utter, F. M., F. W. Allendorf, and B. May. 1976. The use of protein variation in the management of salmonid populations. Transactions of the North American Wildlife and Natural Resources Conference 41:373–384.

Vuorinen, J., and J. Piironen. 1984. Electrophoretic identification of Atlantic salmon *(Salmo salar)*, brown trout *(S. trutta)* and their hybrids. Canadian Journal of Fisheries and Aquatic Sciences 41:1834–1837.

Wishard, L. N., J. E. Seeb, F. M. Utter, and D. Stefan. 1984. A genetic investigation of suspected redband trout populations. Copeia 1984:120–132.

American Fisheries Society Symposium 7:439–458, 1990

Gametic Disequilibrium Analysis as a Means of Identifying Mixtures of Salmon Populations

ROBIN S. WAPLES

Northwest Fisheries Center, National Marine Fisheries Service
2725 Montlake Boulevard East, Seattle, Washington 98112, USA

PETER E. SMOUSE[1]

Departments of Human Genetics and Biology, University of Michigan
Ann Arbor, Michigan 48109, USA

Abstract.—We evaluated the power of a multilocus test for gametic disequilibrium (the nonrandom association of alleles at different gene loci) to detect mixtures of gene pools. Monte Carlo methods were used to simulate mixtures drawn from pairs of populations with allele-frequency differences corresponding to those found in an extensive data base for chinook salmon *Oncorhynchus tshawytscha*. Results were as follows. (1) For sample sizes of $N = 100$ drawn from a 50:50 mixture of two populations, the power to detect a mixture was limited (20–30%) if allele-frequency differences in the source populations were typical of those for populations from geographically adjacent areas; but the power was generally high (80–100%) for populations separated by greater distances. In the first case, larger samples increased the probability of detecting a mixture. (2) Unequal mixture proportions reduced the power of the test, and mixtures were difficult to detect if one population contributed less than 10%. (3) In some cases, the test retained appreciable power to detect disequilibria in F_1 progeny from a mixture, but the mixture event would be difficult to detect in F_2 and later generations unless initial genetic differences between contributing populations were large. (4) Disequilibria caused by genetic drift were important considerations except when the effective number of breeders was at least twice as large as the sample size. Analysis of a set of electrophoretic data for chinook salmon suggested that the high levels of gametic disequilibrium observed in hatchery (but not in wild) populations were caused by a mixture of stocks, a small effective number of breeders, or both.

Anadromous salmonids *Oncorhynchus* spp. of northwestern North America spend one or more years at sea before entering fresh water to spawn. Because these fish have strong tendencies to return to their stream of origin (Scheer 1939; Harden Jones 1968), each breeding population potentially represents an independent evolutionary unit (Ricker 1972). Indeed, the large body of electrophoretic data for salmonids gathered in the past two decades demonstrates a considerable degree of genetic heterogeneity among populations for all of the anadromous salmonid species found in northwestern North America (reviewed by Utter et al. 1989). Nevertheless, the fidelity of spawning adults to their natal stream is not absolute; some straying of individuals from wild populations occurs, primarily between contiguous localities (e.g., Ricker 1972). Elevated rates of straying have been reported for hatchery fish released away from the rearing site (Ricker 1972;

Lister et al. 1981), particularly for individuals released in areas not previously colonized by the species (Withler 1982).

These two observations—genetic heterogeneity among wild populations, and increased potential for genetic mixing resulting from enhancement practices—present a dilemma for fisheries managers who seek to rebuild depleted salmonid populations. Evidence suggests that many of the physiological, behavioral, and life history differences among wild populations are genetically based, presumably the result of many generations of adaptation to local environmental conditions (e.g., Helle 1981 and references therein). Wild gene pools may thus represent irreplaceable resources that merit conservation.

Many wild populations are endangered, however, primarily because of human activities (overfishing, destruction of habitat, blockage of migratory routes by hydroelectric dams, pollution), and large-scale supplementation programs are necessary to maintain them even at present levels (Wahle and Pearson 1984; Fraidenburg and Lin-

---
[1]Present address: Center for Theoretical and Applied Genetics, Rutgers University, New Brunswick, New Jersey 08903-0231, USA.

coln 1985). These enhancement programs, which involve increased hatchery production and off-site release, inevitably increase the potential for genetic interactions between hatchery and wild fish. To the extent that the interactions compromise coadapted gene pools in wild populations, they may adversely affect the genetic health of the species.

Answers to the following questions are important in dealing with the potential effects of hatchery enhancement. (1) Given that artificial propagation and transport of stocks have occurred on a broad scale for decades (Mathews 1980), sometimes without detailed records, is it possible to determine which putatively "wild" populations are less genetically "pure" than expected? (2) Do hatchery stocks being considered for enhancement programs show evidence of heterogeneous genetic composition? Some hatchery lines have been initiated from multiple brood stocks, and transfer of eggs, juveniles, and adults among hatcheries is a common practice. Because of uncertainty about the relative success of contributing gene pools, the genetic status of some hatchery stocks is presently uncertain. (3) Can the genetic consequences of proposed, large-scale enhancement programs be monitored effectively? This last question is particularly important in view of the current Columbia River Basin Fish and Wildlife Plan (NWPPC 1987), which advocates doubling the run size for salmonids in the Columbia River basin through increased hatchery production and supplementation efforts.

Traditional marking methods are not well suited to address these questions. Physical tags are valuable to monitor the movement of individual fish but provide no direct information about the genetic consequences of straying migrants. Such migrants are relevant to the evolutionary future of a population only if they produce progeny that contribute gametes to subsequent generations. Therefore, the use of genetic marks is necessary to assess the extent of this gametic contribution. The problem would be simple if each population contained unique (and therefore diagnostic) alleles, but salmonid populations typically are characterized by different frequencies of the same suite of alleles. If allele frequencies in the source populations are known and differ sufficiently from one another, estimates of mixture proportions may be fairly accurate (Glass and Li 1953). Even in the absence of genetic information for source populations, a mixture of gene pools in a blind sample may be detected by the presence of a statistically significant deficiency of heterozygotes (Wahlund effect) relative to binomial (Hardy–Weinberg) expectations. This test, however, has limited utility. Its interpretation is ambiguous, because a variety of other factors can cause heterozygote deficiency within a single gene pool; it is not a powerful test of alternative hypotheses (Lewontin and Cockerham 1959; Fairbairn and Roff 1980); and a single generation of random mating eliminates heterozygote deficiencies due to the Wahlund effect. Analyses based on allele frequencies or heterozygote deficiencies are more useful if data for several gene loci are available (Weir and Cockerham 1984; Long 1986; Smouse and Long 1988), but these approaches suffer from the same limitations because they consider each locus individually.

More information about the gene pools represented in a sample can be obtained by a cross-locus or gametic disequilibrium approach—that is, by considering the correlations between alleles at different gene loci. In a population in gametic equilibrium, the expected frequency of gametes containing allele I at locus 1 and allele J at locus 2 is the product of the frequencies of the two alleles. In such a population, alleles I and J are uncorrelated in the sense that an individual's genotype at locus 1 provides no information about its genotype at locus 2. Departures from gametic equilibrium can be caused by natural selection, lack of recombination (physical linkage), sampling errors in a finite population (genetic drift), and population subdivision. In a mixture of gene pools with different frequencies of I and J, observed gametic frequencies will depart from the independence expectations, resulting in gametic disequilibrium. These disequilibria constitute a two-locus Wahlund effect (Sinnock 1975), but the power of gametic disequilibrium analysis is greater than the combined power of the single-locus tests for heterozygote deficiency. Furthermore, whereas the single-locus Wahlund effect disappears after one generation of random mating, gametic disequilibria decay at a rate equal to the recombination rate between loci (50% per generation for unlinked loci) and may be detectable for a number of generations after an episode of genetic exchange (Nei and Li 1973).

Multilocus gametic disequilibrium analysis has proved to be a powerful means of detecting gene flow between populations (Smouse and Neel 1977; Smouse et al. 1983). To date, the relatively few applications of disequilibrium analysis to salmonids (e.g., Ryman et al. 1979; Campton and

Johnston 1985; Campton and Utter 1985) have only considered disequilibrium between individual pairs of loci. Brown (1975) has shown, however, that large sample sizes are required to demonstrate statistically even moderately large amounts of disequilibrium between a single pair of loci. In our analysis, we used an overall measure of disequilibrium that combines information for all possible pairs of loci, thus considerably increasing the power of the test. Although such an approach is not appropriate for the analysis of gametic disequilibrium resulting from natural selection or physical linkage (because strong disequilibria between certain pairs of loci might be obscured in a combined test), it is useful in the present context because disequilibria caused by a mixture of gene pools results from a breeding history common to all loci.

The practical utility of gametic disequilibrium analysis may be limited by the amount of genetic information available for the species of interest (number of polymorphic loci resolved, magnitude of differences between populations, number of individuals sampled). In this study, we created artificial mixtures containing the amount of genetic information typically found in mixtures of salmonid populations to address the following questions. (1) How large must allele-frequency differences between populations be before a mixture can be identified as such? (2) How small a mixture fraction can be detected, and how large must sample sizes be for the test to have reasonable power? (3) How far back in time does our power of resolution extend (given that mixture disequilibrium decays with random mating following the initial event)?

To illustrate the procedure for gametic disequilibrium analysis, we have used allozyme frequencies found in a series of hatchery and wild populations of chinook salmon *Oncorhynchus tshawytscha*.

Methods

Gametic Disequilibrium

Consider a pair of genetic loci, A and B, each with two codominant alleles, and a pair of source populations, 1 and 2, with different allele frequencies for these two loci. Let P_{1A} and $1 - P_{1A}$ be allele frequencies for locus A in population 1, and let P_{2A} and $1 - P_{2A}$ be frequencies for locus A in population 2, with equivalent expressions for locus B. If these two populations contribute to a mixed population in proportions m and $(1 - m)$,

respectively, a two-locus gametic disequilibrium is generated in the mix (Nei and Li 1973):

$$D^{(0)} = m(1 - m)(P_{1A} - P_{2A})(P_{1B} - P_{2B}). \quad (1a)$$

A correlational form is constructed by standardization to the mean allele frequencies observed in the mix:

$$\rho^{(0)} = \frac{m(1 - m)(P_{1A} - P_{2A})(P_{1B} - P_{2B})}{[\overline{P}_A(1 - \overline{P}_A)\overline{P}_B(1 - \overline{P}_B)]^{1/2}}; \quad (1b)$$

$\overline{P}_A = mP_{1A} + (1 - m)P_{2A}$ and $\overline{P}_B = mP_{1B} + (1 - m)P_{2B}$. The size of the initial disequilibrium depends both on the magnitude of the genetic difference between the two source populations and on the mixture fractions, and the maximum disequilibrium is generated by a 50:50 mix. It can also be shown that the square of this initial cross-locus disequilibrium is of the same magnitude as the single-locus Wahlund effects, gauged by the single-locus F-statistics (see Smouse 1982):

$$F_A^{(0)} = \frac{m(1 - m)(P_{1A} - P_{2A})^2}{\overline{P}_A(1 - \overline{P}_A)}$$

and

$$F_B^{(0)} = \frac{m(1 - m)(P_{1B} - P_{2B})^2}{\overline{P}_B(1 - \overline{P}_B)}. \quad (2)$$

Unlike heterozygote deficiencies due to the Wahlund effect, which disappear after a single generation of random mating, the gametic disequilibrium decays only gradually following random mating. For unlinked loci in the tth generation, we expect

$$D^{(t)} = D^{(0)}[\tfrac{1}{2}]^t \quad (3a)$$

as well as

$$\rho^{(t)} = \rho^{(0)}[\tfrac{1}{2}]^t. \quad (3b)$$

The population mixture leaves a characteristic signature on the population in the form of gametic disequilibrium, but the strength of that signature declines steadily over time. When more than two loci are involved, one simply defines equation (1b) for each pair of the K loci in the set, and then constructs a $K \times K$ correlation matrix of pairwise disequilibrium values.

Test for Gametic Disequilibrium

The quantities in equations (1–3) are defined in terms of population parameters. We need to be able to estimate these parameters for a sample from a possible mixture of gene pools. The strategy we shall use was described more fully by Smouse and Neel (1977) and Smouse et al. (1983). In brief, we envisage the same two populations and a set of K two-allele loci. For the A locus, we define an indicator variable Y_A, which takes one of three values,

$$Y_A = \begin{cases} 1 & \text{if genotype is } AA \\ \frac{1}{2} & \text{if genotype is } Aa \quad (4) \\ 0 & \text{if genotype is } aa. \end{cases}$$

We then define a similar variable for each of the other loci. For a K-locus genotype, we have a vector of indicator variables, with one variable for each locus. For example, the genotype (AA, Bb, cc, DD, Ee, ff, GG, ...) is represented by the vector $\mathbf{Y} = (1, \frac{1}{2}, 0, 1, \frac{1}{2}, 0, 1, \dots)^T$. Extension to multiple alleles at each locus is straightforward (Smouse and Neel 1977).

The next order of business is to estimate the variances (S_{kk}) and covariances (S_{jk}) of these K indicator variables. These estimates take the form ($\overline{P}_k$ is the mean allele frequency at locus k)

$$S_{kk} = \frac{\sum\limits_{n=1}^{N} (Y_{kn} - \overline{P}_k)^2}{(N-1)}$$

and

$$S_{jk} = \frac{\sum\limits_{n=1}^{N} (Y_{jn} - \overline{P}_j)(Y_{kn} - \overline{P}_k)}{(N-1)}. \quad (5a)$$

The estimated covariance matrix for these $\mathbf{Y}$-vectors is thus

$$\mathbf{S} = \begin{bmatrix} S_{AA} & S_{AB} & S_{AC} & \dots & S_{AK} \\ S_{AB} & S_{BB} & S_{BC} & \dots & S_{BK} \\ S_{AC} & S_{BC} & S_{CC} & \dots & S_{CK} \\ & & & & \\ S_{AK} & S_{BK} & S_{CK} & \dots & S_{KK} \end{bmatrix} \quad (5b)$$

The diagonal elements correspond to the F-statistics defined in equation (2). The sum of these diagonal elements leads to a composite test of single-locus departures from random mating (Smouse et al. 1983). Here, we are more interested in the off-diagonal terms (i.e., the covariances S_{AB}, $S_{AC,}$ etc.). The estimated covariance matrix is converted into its estimated correlation matrix equivalent (Smouse and Neel 1977) as

$$\mathbf{R} = \begin{bmatrix} 1 & r_{AB} & r_{AC} & \dots & r_{AK} \\ r_{AB} & 1 & r_{BC} & \dots & r_{BK} \\ r_{AC} & r_{BC} & 1 & \dots & r_{CK} \\ & & & & \\ r_{AK} & r_{BK} & r_{CK} & \dots & 1 \end{bmatrix} \quad (6)$$

the element r_{jk} has expectation ρ_{jk}, which takes the form of equation (1b) for a mixture of two populations in Hardy–Weinberg equilibrium.

Under the null hypothesis of gametic equilibrium, the correlation matrix is strictly diagonal in expectation (i.e., we expect all off-diagonal elements to be zero). The test of departure from the null expectation is

$$\lambda = -M\log_e[det\mathbf{R}] \approx \chi^2_{K(K-1)/2}; \quad (7)$$

"det" indicates the determinant, and $M = N - \frac{3}{2} - (K+1)/3$ (Smouse and Neel 1977). Although equation (7) provides our chi-square test criterion, it conveys little sense of the magnitude of the disequilibrium contained within $\mathbf{R}$. We can, however, obtain a measure of "effective correlation" by constructing a matrix $\mathbf{R}_e$ of the form

$$\mathbf{R}_e = \begin{bmatrix} 1 & r_e & r_e & \dots & r_e \\ r_e & 1 & r_e & \dots & r_e \\ r_e & r_e & 1 & \dots & r_e \\ & & & & \\ r_e & r_e & r_e & \dots & 1 \end{bmatrix} \quad (8)$$

and setting r_e to the value that makes [det $\mathbf{R}_e$] = det $\mathbf{R}$. For a mixture of two populations in Hardy–Weinberg equilibrium, the r_e value is the same as that obtained from equation (1b) under the assumption that, within each population, allele frequencies are the same at all loci. Both r_e and λ are computed with the aid of a computer program (PANMIX), which is available from the authors on request. The convertibility of equation (6) into equation (8) makes the use of "average, replicate loci" a reasonable substitute for more complicated allele-frequency arrays.

Genetic Distance

The power of gametic disequilibrium analysis to detect mixtures of gene pools is determined by the amount of genetic information contained in a mixture, which is a function of the number of individuals and loci sampled, the mixture proportions, the magnitude of the allele-frequency differences between contributing populations at each locus, and the number of generations of interbreeding since the mixture event. Even if temporal effects are ignored (for the moment), there are too many other variables for an exhaustive evaluation. Nevertheless, we can elucidate the essential features of the problem with a genetic distance measure that summarizes the genetic divergence between two populations.

Consider K loci, each with two alleles. A measure of the degree of genetic separation between two populations contributing to a mixture is

$$\Delta = m(1 - m) \sum_{k=1}^{K} \frac{(P_{1k} - P_{2k})^2}{\bar{P}_k(1 - \bar{P}_k)}; \qquad (9)$$

P_{1k} and P_{2k} are frequencies of one allele at the kth locus in populations 1 and 2, respectively, $\bar{P}_k$ is the corresponding mean allele frequency at the kth locus in the mixture, and m and $(1 - m)$ are the mixture fractions of the two populations (Smouse 1982). (The right side of equation (9) is equivalent to Wright's F_{ST} between two populations.)

As we pointed out earlier, admixture converts this genetic disparity between two populations into gametic disequilibrium within the mixture (Nei and Li 1973), as shown in equations (1a) and (1b). In the present context, we wanted to know how the power of the test was affected by sample size, given allele frequency differences typical of those found between real salmonid populations. Accordingly, we obtained estimates ($\hat{\Delta}$) of Δ for a large number of pairwise comparisons of chinook salmon populations, using allele frequencies observed in samples and setting $m = (1 - m) = 0.50$. (We relaxed the 50:50 mixture requirement later.) We were thus able to quantify the amount of genetic information potentially available for gametic disequilibrium analysis in mixtures of salmonid gene pools. Within broad limits of sample size (N), mixture fractions, and allele frequency, $2N\hat{\Delta}$ is distributed approximately as chi-square with K degrees of freedom (Smouse 1982).

Chinook Salmon Data Base

We focused on a large data base for chinook salmon compiled over the last decade by means of protein electrophoresis (described by Aebersold et al. 1987). The data included genotypic frequencies for well over 100 populations and as many as 100 loci (National Marine Fisheries Service and Washington Department of Fisheries, unpublished data). In the present study, allele frequencies in samples from 137 populations were examined for 21 polymorphic loci: aconitate hydratase (*Ah-4*), adenosine deaminase (*Ada-1*, *Ada-2*), al-

cohol dehydrogenase (*Adh*), aspartate aminotransferase (*Aat-3*, *Aat-4*), glucose-6-phosphate isomerase (*Gpi-1*, *Gpi-3*), glutathione reductase (*Gr*), hydroxyacylglutathione hydrolase (*Hagh*), isocitrate dehydrogenase (*Idh-2*), lactate dehydrogenase (*Ldh-4*, *Ldh-5*), mannose-6-phosphate isomerase (*Mpi*), peptidases (glycyl-leucine [*Dpep-1*], leucylglycyl-glycine [*Tapep-1*], leucyltyrosine [*Pep-lt*], phenylalanyl-proline [*Pdpep-2*]), phosphogluconate dehydrogenase (*Pgdh*), phosphoglycerate kinase (*Pgk-2*), and superoxide dismutase (*Sod-1*). The loci surveyed included all those with at least a moderate degree of genetic variability in some populations. We omitted only the isoloci (pairs of duplicated loci sharing alleles with identical electrophoretic mobility; Allendorf and Thorgaard 1984; Waples 1988) and certain other loci for which not all possible genotypes could be resolved. In computing $\hat{\Delta}$ for pairs of populations, we assumed equal mixture fractions, and we ignored loci not scored in one population and loci with frequency of the same allele greater than 0.98 in both populations. (Essentially monomorphic loci provide negligible resolution.)

The populations represented three major geographic regions: California, Oregon–Washington–Idaho, and Canada (Figure 1). Within these regions, the populations were clustered in smaller groups that corresponded to management units for certain mixed-stock fishery analyses (see WDF 1988). In general, management groups included populations from a restricted geographic area. Two types of pairwise comparisons were made. For *within-group* comparisons, $\hat{\Delta}$ was computed for all pairs of populations within the same management group. The *between-group* comparisons included all pairs of populations in different management groups but within the same region. For both types of comparisons, results were tabulated separately for each of the three regions.

Evaluating the Mixtures

Analysis of the artificial mixtures created by computer simulation involved three steps. First, parametric allele frequencies were chosen for the two populations contributing to the mixture. Next, a mixture of the two populations was made with predetermined proportions, and a sample was taken for analysis. Finally, the PANMIX program was used to compute the test statistic (given by equation 7), which was compared with the critical chi-square value (probability of a type-I error $\alpha = 0.05$) for $K(K - 1)/2$ degrees of

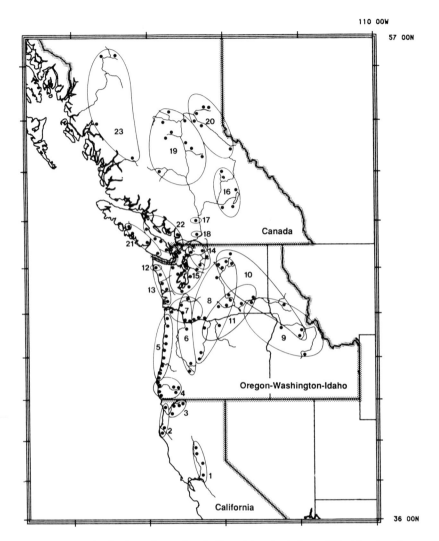

FIGURE 1.—Geographic distribution of samples (indicated by dots) from 137 chinook salmon populations. Samples are clustered into 23 groups (indicated by ellipses) corresponding to management units for some mixed-stock fisheries. Groups are organized into three major regions: California, Oregon–Washington–Idaho, and Canada. Some groups overlap geographically because they contain stocks that return to spawn at different times (SP = spring run; SU = summer run; F = fall run). California region: (1) Sacramento River; (2) California coast; (3) Klamath River. Oregon–Washington–Idaho region: (4) southern Oregon coast; (5) northern Oregon coast; (6) lower Columbia River (SP); (7) lower Columbia River (F); (8) upper Columbia River (SP); (9) Snake River (SP); (10) upper Columbia River and Snake River (SU); (11) upper Columbia River and Snake River (F); (12) northern Washington coast (SP, SU); (13) Washington coast (F); (14) northern Puget Sound (SP); (15) Puget Sound (SU, F). Canada region: (16) Thompson River; (17) lower Fraser River (SP); (18) lower Fraser River (F); (19) mid-Fraser River (SP, SU); (20) upper Fraser River (SP, SU); (21) western Vancouver Island; (22) Georgia Strait; (23) central British Columbia.

freedom. For each set of initial conditions, 500 replicate mixtures were analyzed, and the percentage yielding statistically significant evidence of gametic disequilibrium was recorded. Mixtures with $\Delta = 0$ (two populations with identical allele frequencies at each locus) were also analyzed to

evaluate the type-I error rate (probability of rejecting the null hypothesis when it is true).

Sample allele frequencies from the chinook salmon data base were used to determine parametric allele frequencies for the source populations in the simulations. For a comparison of interest, the total observed $\hat{\Delta}$ was divided by the number of loci used to find the average $\hat{\Delta}$ per locus. Allele frequencies for the source populations were then chosen to yield a genetic distance of $\Delta = \hat{\Delta}$ per locus. In most simulations, all loci were given the same average allele frequency ($\overline{P}_1 = \overline{P}_2 \ldots = \overline{P}_K$); P_{1k} and P_{2k} were chosen to satisfy the required Δ per locus. Given our conversion of the **R** matrix from equation (6) into a single "effective correlation" (r_e), we can always find a $\overline{P}$ to represent any particular array. Simulations using actual allele frequencies observed in salmon populations (different at each locus) yielded results similar to those using a single average $\overline{P}$ (see Results). Although analysis of any number of alleles per locus is possible with PAN-MIX, we considered only diallelic loci. This was appropriate in the present context, because few loci had a third allele with frequency greater than 0.05 in any chinook salmon population.

Sample in same generation as mixture.— The first mixtures analyzed were designed to model populations that had commingled but not interbred. Thus, the sample of size N for analysis consisted of mN individuals from population 1 and $(1 - m)N$ individuals from population 2. For each population, the K-locus genotype of the individuals contributing to the mix was determined by a series of K random numbers. Probabilities of the three possible genotypes at each locus were set (independently at each locus) to the corresponding expectations from an assumed Hardy–Weinberg equilibrium. This was equivalent to sampling from a randomly mating population of infinite size and ignored any disequilibrium due to genetic drift.

Initially, all simulations were done with a total sample from the mix of $N = 100$ and $m = (1 - m) = 0.5$ (50 fish from each population). These simulations ignored the likelihood that a sample taken from a population with a 50:50 mix will not contain these mixture proportions exactly, resulting in $m(1 - m) \neq 0.25$. However, for samples of size $N = 100$, both m and $(1 - m)$ will be in the interval 0.4–0.6 about 95% of the time, so ignoring this source of sampling error had little effect on the term $m(1 - m)$. This would not be the case, however, if the actual mixture fractions were very unequal. Therefore, in simulations that evaluated the effects of sample size and mixture fraction, the source population for each individual in the sample was determined by a random variable in such

a way that the individual came from population 1 with probability m and from population 2 with probability $(1 - m)$.

Sampling subsequent generations.—We also evaluated the probability of detecting a mixture event if the gene pools hybridized and progeny were sampled, in which case disequilibrium should decay according to equations (3a) and (3b). This entailed considering two additional parameters: the effective number of breeders (N_b) for each generation, and the number of generations elapsed between the mixture event and the sample. A further complication, the pattern of overlapping generations found in most species of salmonids, was not modeled explicitly, but we evaluate and discuss its importance in the Results section.

The temporal simulations were carried out as follows. In generation 0, a mixture was created by sampling individuals from two populations, as described below. From this mixture, two independent random samples were taken, each by sampling without replacement. The first sample, of N individuals, was used for analysis by PANMIX; the second, numbering N_b, was the breeding population for the next generation. The size of the mixture in the initial generation (and the number of progeny in later generations) was N_t, the larger of N and N_b. This permitted a sample of more individuals than actually bred, as might occur in sampling juveniles of fecund species such as salmon. The breeders were arbitrarily divided into $N_b/2$ males and $N_b/2$ females and were allowed to pair randomly without respect to population of origin. In each generation, N_t offspring were produced, one each from N_t pairwise matings of males and females chosen randomly with replacement from the breeding population. Under these conditions, each male and female had an equal opportunity to contribute gametes to the next generation, and the distribution of progeny number per parent was binomial. The breeders thus behaved as an "ideal" population (Kimura and Ohta 1971), with effective number = census number = N_b. In each of 3–5 generations following the mixture event, N individuals were selected for genetic analysis.

The temporal analysis also permitted evaluation of the importance of gametic disequilibrium arising from a force quite apart from any mixture of gene pools—genetic drift (Ohta and Kimura 1969; Hill 1981). Because the magnitude of disequilibrium due to drift is larger for small populations, we also needed to consider the background levels

TABLE 1.—Mean estimates of genetic distance ($\hat{\Delta}$) and number of informative loci used for pairwise comparisons of chinook salmon populations in three major regions (California, Oregon–Washington–Idaho, Canada). For each region, data were compiled for comparisons of populations within and between management groups.

Comparison	California: 3 groups 13 populations	Oregon–Washington– Idaho: 12 groups 86 populations	Canada: 8 groups 38 populations
Within groups			
Mean $\hat{\Delta}$	0.07	0.19	0.21
Range of $\hat{\Delta}$	0.02–0.28	0.01–1.35	0.03–0.55
Mean number of loci	5.5	8.2	9.5
Between groups			
Mean $\hat{\Delta}$	0.50	0.46	0.48
Range of $\hat{\Delta}$	0.15–1.1	0.03–1.85	0.05–1.2
Mean number of loci	6.8	8.6	10.3

of disequilibrium to be expected in each of the source populations prior to mixing. In the previous analyses, genotypes of individuals for the initial mix were chosen multinomially (and independently) at each locus, which is equivalent to sampling from two infinitely large source populations, each in gametic equilibrium. In the temporal analysis, the initial samples (N_b breeders chosen from each of two source populations) were allowed to breed for one generation before mixing, and N_t progeny from the two populations were chosen as the mixture in generation 0. The initial mixture, therefore, contained gametic disequilibrium from one generation of genetic drift in each population, as well as that from a mixture of two gene pools.

Results

The genetic distance values for the pairwise comparisons of the chinook salmon populations are summarized in Table 1. Two points are worth noting. First, mean $\hat{\Delta}$ was considerably larger for comparisons between than within management groups for each of the three regions. Second, the absolute magnitude of mean $\hat{\Delta}$ in each category was similar in the three regions, except that within-group genetic differences were smaller in California than in the other two regions, and the number of informative loci increased from California to Oregon–Washington–Idaho to Canada. Figure 2 shows the distribution of $\hat{\Delta}$ values for the three regions; in each region, a considerable range is apparent for both the within- and between-group categories.

Test under the Null Hypothesis

The first simulated mixtures we analyzed were designed to evaluate properties of the test statistic under the null hypothesis (no disequilibrium ex-cept that due to sampling error in choosing the N individuals for analysis). Accordingly, these mixtures were created by sampling from two populations with identical allele frequencies (i.e., $\Delta = 0$). Using $N = 100$, we evaluated the true α of the test for allele frequencies in the range 0.5–0.975. The number of loci used was allowed to vary from 5 to 15, the approximate range found in the real data set, and the nominal α level was set at 0.05. Results of these simulations (Figure 3) demonstrated that, in spite of the sensitivity of some disequilibrium measures to allele frequency (Hedrick 1987), probability of a type-I error for the multilocus test was very close to the nominal α level unless $\bar{P} > 0.95$. This result was independent of the number of loci used.

Mixture Event in Current Generation

To evaluate power of the test (probability of rejecting the null hypothesis when it is false) under conditions suitable for its use, we analyzed 50:50 mixtures from pairs of populations having the same Δ values and number of informative loci as the average $\hat{\Delta}$ for between- and within-group comparison in each of the three regions (see Table 1). When the mean number of loci in Table 1 was intermediate between two integers, we ran simulations using both the higher and lower number of loci and averaged the results (Figure 4). The power was much greater for between-group than for within-group comparisons. Typical mixtures of populations in the same management group in the Oregon–Washington–Idaho and Canada regions could be detected about 20–30% of the time, whereas power for typical between-group comparisons was 80–100%. The situation was similar but even more extreme for California: gametic disequilibrium analysis had little power to detect within-group mixtures, but a typical mixture of

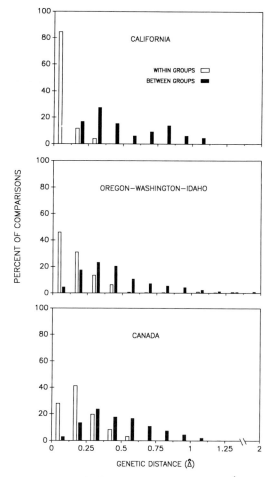

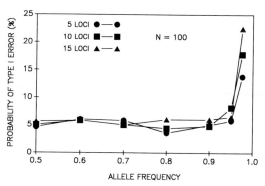

FIGURE 3.—Probability of a significant test of gametic disequilibrium under the null hypothesis (no disequilibrium except that due to error in choosing a finite sample). Total sample size (N) was 100 individuals; parametric allele frequencies were the same at all loci. Each data point represents 500 replicate simulations for each allele frequency and number of loci.

FIGURE 2.—Distribution of genetic distance ($\hat{\Delta}$) values for comparisons of chinook salmon populations in three major regions. Data were tabulated for comparisons within the same management group and between different management groups.

populations from different management groups was virtually certain to be detected. Further, power for within-group comparisons increased considerably—much more than did the true α level—for allele frequencies above 0.8. Clearly, it is important to consider the allele frequencies that actually occur in chinook salmon populations.

Accordingly, we identified two comparisons with $\hat{\Delta}$ and number of loci similar to the means for within- and between-group comparisons shown in Table 1. We used the Oregon–Washington–Idaho region, because it contained the most populations and was intermediate between the other regions in number of informative loci. Upper Columbia River spring run stocks (group 8 on Figure 1) were selected because fish from these populations (Car-

TABLE 2.—Observed frequencies of the common allele in samples from three chinook salmon populations used in comparison 1 (Carson–Klickitat) and comparison 2 (Carson–John Day). Loci with no data for one population or with allele frequency greater than 0.98 in both populations were omitted. $\hat{\Delta}$ for comparison 1 is typical of between-management-group differences in the Oregon–Washington–Idaho area; comparison 2 is typical of within-group differences. For each comparison, $\overline{P}_k$ is the unweighted mean frequency of the two samples; $\hat{\Delta}$ is a measure of genetic distance.

Locus	Reference population Carson	Comparison 1 $\hat{\Delta} = 0.47$, 9 loci		Comparison 2 $\hat{\Delta} = 0.19$, 7 loci	
		Klickitat	$\overline{P}_k$	John Day	$\overline{P}_k$
Ada-1	0.970	0.980	0.975	1.000	0.985
Ah-4	1.000	0.930	0.965		
Gr	0.990	0.760	0.875		
Hagh	0.878	0.830	0.854	0.992	0.935
Mpi	0.889	0.730	0.810	0.882	0.886
Pep-lt	0.980	0.929	0.954	1.000	0.990
Pgk-2	0.870	0.430	0.650	0.717	0.794
Sod-1	0.830	0.690	0.760	0.700	0.765
Tapep-1	0.890	0.950	0.920	0.992	0.941
Mean			0.863		0.899

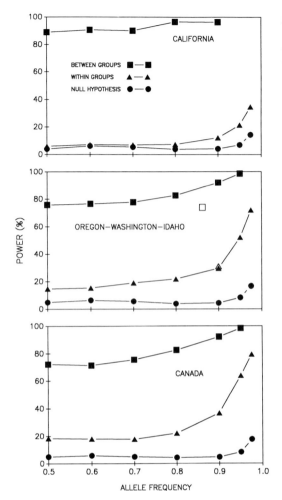

FIGURE 4.—Probability of successfully identifying a typical mixture of two chinook salmon populations from the same (within) or different (between) management groups in the three major areas. Results shown are from 500 replicate simulations of each parameter set for $N = 100$ and equal mixture proportions. Filled squares and triangles are for simulations using Δ (genetic distance) values equal to observed $\hat{\Delta}$ values given in Table 2; filled circles are for simulations with $\Delta = 0$ (equivalent to sampling from a single population). In each case, all loci were given the same mean allele frequency. Results of simulations using actual allele frequencies (different at each locus) for two comparisons in Table 3 are plotted at the overall mean allele frequency for the comparison. For comparison 1 (open square), Δ was typical of between-group comparisons; for comparison 2 (open triangle), Δ was typical of within-group comparisons.

son hatchery in particular) have been distributed widely throughout the Columbia River Basin (Howell et al. 1985). The Carson–John Day comparison ($\hat{\Delta} = 0.19$, 7 loci) was typical of within-

group comparisons for the region, and the Carson–Klickitat pair ($\hat{\Delta} = 0.47$, 9 loci) was about as divergent genetically as a typical between-group comparison, even though they are in the same management group. Observed allele frequencies in samples from these populations are presented in Table 2. Results of simulations using these frequencies as parameters are shown as open symbols on Figure 4 (middle plot). Their abscissa values are the means of the $\bar{P}_k$ (mean frequency of the common allele over all loci), which was 0.90 for the Carson–John Day comparison and 0.86 for Carson–Klickitat. For the former comparison, power was nearly identical to that observed in simulations using a single average allele frequency (different in the two populations) at every locus; power was slightly reduced (74% versus about 85%) for the Carson–Klickitat pair when actual allele frequencies were used. We also evaluated the null hypothesis under conditions of varying allele frequencies per locus by setting allele frequencies in both populations equal to $\bar{P}_k$ for the kth locus (see Table 2). For the Carson–Klickitat comparison, the percentage of significant tests under these conditions (6.8%) was close to the nominal α level, whereas the type-I error rate was somewhat elevated (11.8%) for the Carson–John Day pair.

The slightly reduced power for the Carson–Klickitat pair in simulations using actual allele frequencies can be attributed to two factors. First, this comparison involved nine loci, whereas power for the between-group comparisons in Oregon–Washington–Idaho was determined by averaging results for simulations with eight and nine loci (mean number of loci used, 8.6; Table 1). For the same total Δ, power was reduced in simulations using more loci because the test statistic had more degrees of freedom. Second, most of the $\hat{\Delta}$ for Carson–Klickitat was due to differences at just two loci (*Gr* and *Pgk-2*); in such a case, the approximation based on a single average allele frequency at all loci is not entirely accurate. Nevertheless, the "average, replicate loci" approach yielded results qualitatively similar to those obtained with frequencies that differed at each locus, and the advantages gained by reducing the number of variables was considerable. In subsequent analyses, therefore, we used a single allele frequency (different in each population) for all loci. We chose $\bar{P}_k = 0.8$, a frequency that is high enough to be similar to the mean observed in most salmonid populations but low enough to avoid inflated probability of a type-I error. (The 11.8% rejection rate under the null hypothesis for the Carson–John Day comparison

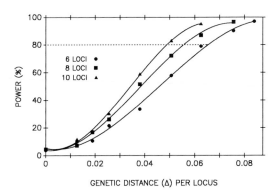

FIGURE 5.—Power to detect a mixture of two populations as a function of genetic distance (Δ) between them. Numbers of loci used (6, 8, and 10) are typical for comparisons of chinook salmon populations in California, Oregon–Washington–Idaho, and Canada regions, respectively. Results shown are from 500 replicate simulations of each parameter set for $N = 100$, mean allele frequency $\bar{P}_k = 0.8$ for every locus, and population proportions $m = (1 - m) = 0.5$.

apparently was due to inclusion of two loci, *Ada-1* and *Pep-lt*, with an average allele frequency greater than 0.98.)

This tactic allowed us to address an issue mentioned earlier (Figure 2): in all three regions, there was considerable variance in $\hat{\Delta}$ values for individual comparisons, both within and between groups. So far, however, we have only considered "average" or "typical" comparisons. To be more specific, we needed a description of power as a continuous function of Δ and the number of loci used. We obtained this by setting $\bar{P}_k = 0.8$ for all loci and evaluating power as a function of Δ per locus for simulations with six loci (typical of California), eight loci (Oregon–Washington–Idaho), and 10 loci (Canada).

As expected, for a given Δ per locus, power increased with the number of loci used (Figure 5). By drawing horizontal lines through the curves and then vertical lines from the points of intersections to the abscissa, it is possible to find Δ per locus required for any desired power. We defined five categories of power (0–20%, 20–40%, 40–60%, 60–80%, 80–100%), found the associated range of Δ-per-locus values for each curve, and plotted the percentage of comparisons from each region that fell in each power category (Figure 6). Because the variability in number of loci within regions was not accounted for, these results should be considered approximations. Nevertheless, Figure 6 gives a more complete picture than does Figure 4 of the likelihood of

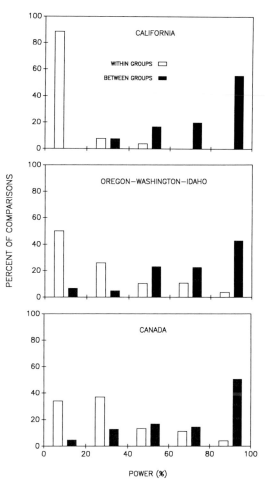

FIGURE 6.—Percentage of within-management-group and between-management-group comparisons of chinook salmon populations falling in each power category for the three major regions. Genetic distance (Δ) values for each power category were obtained from Figure 5; these were used to order $\hat{\Delta}$ values (Figure 2) for pairwise comparisons of samples from 137 populations.

detecting a 50:50 mixture of two chinook salmon populations by gametic disequilibrium analysis. Although considering "typical" between- and within-group comparisons is useful, it is clear that some populations in the same management group were very different genetically, while some in different management groups were quite similar.

Sample size.—All of the above presume a sample of 100 individuals from the mixture, which is a common sample size for chinook salmon and other salmonid populations. In many situations, however, larger samples may be available or smaller ones may be necessary. To evaluate the effect of sample size, we again considered "typi-

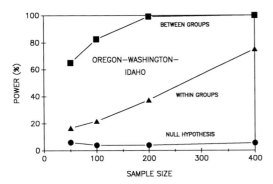

FIGURE 7.—Power to detect mixtures of two chinook salmon populations as a function of sample size. Results shown are from 500 replicate simulations of each parameter set for eight loci, mean allele frequency $\bar{P}_k = 0.8$ for every locus, equal mixture proportions, and genetic distance (Δ) equal to the mean for the appropriate comparison in the Oregon–Washington–Idaho region (Table 1).

cal'' between- and within-group comparisons for the Oregon–Washington–Idaho region. We set all $\bar{P}_k = 0.8$, assumed equal mixture proportions, and varied N from 50 to 400. In these simulations (Figure 7), use of samples as small as 50 caused little change in power for the within-group comparisons, but the probability of correctly identifying a mixture of two populations from different management groups decreased sharply (from about 90% to about 60%). Therefore, we recommend that at least 100 individuals be sampled whenever possible. Increasing N as high as 400 had little effect on between-group comparisons,

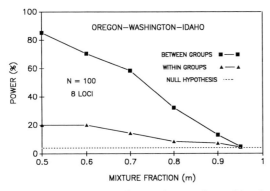

FIGURE 8.—Power to detect mixtures of two chinook salmon populations as a function of mixture fraction (m) of the more common population. Results shown are from 500 replicate simulations of each parameter set for eight loci, mean allele frequency $\bar{P}_k = 0.8$ for every locus, and allele frequency differences typical of those for the appropriate comparison in the Oregon–Washington–Idaho region (Table 1).

because power in this case was already high; however, large sample sizes greatly increase the probability of detecting mixtures of populations that are less distinct genetically, such as those within the same management group.

Unequal mixture proportions.—We next relaxed another assumption—that the sample was a mixture of two populations in equal proportions. Once again, we used typical within- and between-group comparisons for the Oregon–Washington–Idaho region to evaluate power as mixture fractions depart from equality. With sample size fixed at $N = 100$, power declined rapidly as the proportional contribution of the second population decreased (Figure 8). With $(1 - m) \leq 0.1$, the test had essentially no power, even for mixtures of populations with typical between-group differences in allele frequency. To a limited extent, this decline in power can be offset by taking larger samples. However, if the second population makes a small contribution to the mixture (less than 10%), even very large sample sizes ($N = 400$) do not appreciably increase the power to detect a mixture.

Decay of Disequilibrium over Time

All results described to this point were obtained by analyzing samples taken in the same generation as the mixture event. If random interbreeding among individuals from the two source populations occurs after the mixture, single-locus heterozygote deficiencies are not expected in the progeny, but any gametic disequilibrium arising from the mixture itself will decay at a rate of 0.5 per generation and may be detectable for some time. Before evaluating power of the test to detect disequilibrium in progeny, we need to examine the importance of a new parameter, N_b, the effective number of breeders each generation.

The previous analyses relied on samples whose genotypes were drawn independently at each locus by multinomial sampling from populations with known allele frequencies. This is equivalent to drawing a sample from a population in gametic equilibrium with $N_b = \infty$. In a finite population, some multilocus gametes will be produced at frequencies greater than the product of the respective allele frequencies, while others will occur less frequently (and some gametes may not be possible at all, given the limited number of multilocus genotypes in the breeders). The result is gametic disequilibrium due to genetic drift (finite sampling), and the power to detect it depends on the parameters N

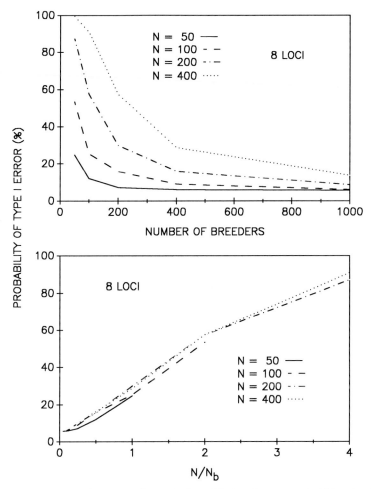

FIGURE 9.—Probability of a significant test of gametic disequilibrium due to genetic drift under the null hypothesis (no allele frequency differences between two source populations). As shown by the lower graph, the probability of a significant test depends primarily on the ratio of progeny sample size (N) to the effective number of breeders (N_b) that produce the progeny, rather than on actual values of N and N_b. Results shown are from 500 replicate simulations for eight loci and mean allele frequency $\bar{P}_k = 0.8$ at each locus. In the lower graph, values of N_b used were 50, 100, 200, 400, and 1,000. Simulations were run for three generations after the mixture, and the mean percentage of significant tests was calculated over generations 1–3.

and N_b. As an illustration, consider a breeding population with $N_b = 2$, and let the two-locus genotypes of the male and female be $AABb$ and $AaBB$, respectively. Possible gametes are AB and Ab from the male and AB and aB from the female. No ab gametes are produced, and only four of the nine possible two-locus genotypes can occur in progeny from this pair: $AABB$, $AABb$, $AaBB$, $AaBb$. The result is gametic disequilibrium in the progeny because of departures from multilocus panmixia in the parents. The magnitude of the disequilibrium in the progeny is inversely related to the number of parents (N_b), but the probability of

detecting the disequilibrium statistically is an increasing function of the number of progeny sampled (N).

Population size.—We reevaluated the null hypothesis that the sample was taken from a single population by analyzing mixtures, using $\bar{P}_k = 0.8$ for eight loci and various values of N and N_b (Figure 9). Simulations were run for three generations after mixing occurred, and the mean percentage of significant tests was calculated over generations 1–3. It was not entirely clear whether the populations had reached a steady-state level of drift disequilibrium after this time, but levels of

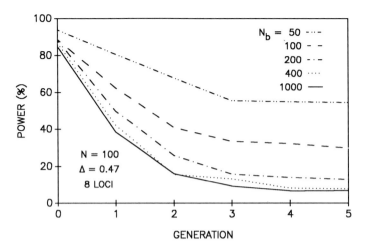

FIGURE 10.—Power to detect gametic disequilibrium in samples ($N = 100$) of progeny produced by random hybridization of two populations in generation 0. Initial mixture proportions are assumed to be equal. Results are for 500 replicate simulations for eight loci, mean allele frequency $\overline{P}_k = 0.8$, total genetic distance (Δ) = 0.47, and a range of number of breeders (N_b) per generation. Residual power at generation 5 is primarily due to disequilibrium due to finite population size.

disequilibrium did not, in general, increase substantially over generations 1–3. The importance of considering the parameter N_b is evident in the upper panel of Figure 9. The lower panel indicates that, for N between 100 and 400, the probability of a type-I error is primarily determined by the ratio N/N_b, not by the absolute values of N or N_b. Waples (1989) and Waples and Teel (in press) found the probability of a significant test statistic similarly depended on the ratio of N/N_b when

they compared allele frequencies in temporally spaced samples. In the test for gametic disequilibrium, probability of a type-I error for $\alpha = 0.05$ can be very large (30% or greater) if $N \geq N_b$, and is moderately inflated unless $N \leq N_b/4$.

The effects of N_b are also evident in Figure 10, which shows how power declines with time after the mixture event for a typical between-group comparison in the Oregon–Washington–Idaho region. With sample size held constant at $N = 100$, power de-

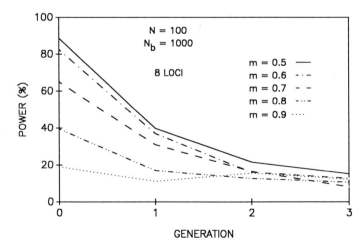

FIGURE 11.—Power to detect gametic disequilibrium in samples ($N = 100$) of progeny produced by random hybridization of two populations combined in generation 0 in various proportions [m, $(1 - m)$]. Results are for 500 replicate simulations for eight loci, mean allele frequency $\overline{P}_k = 0.8$, number of breeders per generation $N_b = 1,000$, and allele frequency differences typical of those for between-group comparisons in the Oregon–Washington–Idaho region.

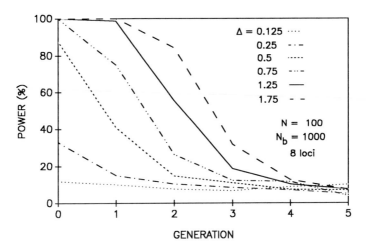

FIGURE 12.—Power to detect gametic disequilibrium in samples ($N = 100$) of progeny produced by random hybridization in generation 0 of two populations separated by a range of genetic distance (Δ) values. Results are for 500 replicate simulations for eight loci, mean allele frequencies $\overline{P}_k = 0.8$, equal mixture proportions, and $N_b = 1,000$.

clined rapidly over the initial generations, approaching an asymptotic value equal to the type-I error rate for the appropriate N_b (see Figure 9).

For simulations depicted in Figure 11, we set $N = 100$ and $N_b = 1,000$ (thus effectively negating any disequilibrium due to genetic drift) and monitored power over three generations as a function of mixture fraction, again using the typical between-group comparison for Oregon–Washington–Idaho as a model. For these conditions, and even with equal mixture fractions, the test had essentially no power to detect disequilibrium in progeny two generations removed from the mixture event. However, for $m = 0.3–0.7$, the test retained about a 40–50% chance of detecting disequilibrium in F_1 progeny.

Of course, higher levels of disequilibrium will persist longer if genetic differences between populations are larger than those used in Figure 11 ($\Delta = 0.47$ for equal mixture fractions), and about 30–40% of all within-region comparisons of chinook salmon populations yield $\hat{\Delta}$ values greater than 0.5 (Figure 2). Figure 12 shows power over time for a range of Δ values. If Δ is as large as 1.75, a mixture can be identified about half the time even after three generations of random mating. However, with eight loci, this corresponds to Δ per locus = 0.22; for intermediate allele frequencies, this requires allele-frequency differences in the two populations of about 0.45–0.5 at each locus.

Overlapping Generations

The results obtained thus far have been based on a model with discrete generations. However, with the exception of the pink salmon *Oncorhynchus gorbuscha*, which has a rigid 2-year life cycle, Pacific salmon return to spawn at a variety of ages. Chinook salmon, for example, may spawn at any of several ages (e.g., 3, 4, 5). A typical spawning population is thus a mixture drawn from several different (albeit interrelated) gene pools, each separated from the others by one or more years. To the extent that allele frequencies within a population differ over time, gametic disequilibrium will be produced in each spawning population. We attempted to evaluate the magnitude of this background disequilibrium and its possible effect on our ability to detect mixtures involving other populations.

TABLE 3.—Mean genetic distance (Δ) per locus for comparisons of allele frequencies of spawners returning in different years to the same population. (Based on simulations described in Waples 1990).

Number of spawners	Δ per locus[a]
25	0.012
50	0.006
100	0.003
200	0.0016
500	0.0006
1,000	0.0003

[a] Mean for comparisons of spawning populations 1 and 2 years apart in the same stream.

Waples (1990) conducted a series of computer simulations of temporal variability in chinook salmon populations due to genetic drift. Each spawning population was chosen by sampling progeny from populations that bred 3, 4, and 5 years previously. We computed Δ values from the variance of the difference in allele frequencies in breeding populations in different years [mean $(P_{1k} - P_{2k})^2$ is the numerator of equation (3)], which was recorded for the simulations. Mean $(P_{1k} - P_{2k})^2$ for a spawning population was computed as the average for the three pairwise comparisons of population allele frequencies 3, 4, and 5 years earlier. We obtained Δ by dividing this number by $\overline{P}(1 - \overline{P})$—i.e., the denominator of equation (3), $\overline{P}$ being the mean allele frequency observed in the simulations. The Δ values shown in Table 3 are for simulations with 50% of the spawners returning at age 4 and 25% each at ages 3 and 5, but similar results were obtained when the percentage returning at age 4 was allowed to vary over the range 30–70%.

As expected, the Δ-per-locus values were inversely related to the number of spawners each year, but, even for very small N_b, Δ was less for a comparison of different years' spawning populations than for most comparisons of populations in the same management group. For example, with $N_b = 100$, mean Δ per locus for between-year comparisons was 0.003, so total Δ for eight loci was only about 0.024. Even allowing for the possibility that the temporally caused disequilibrium might be slightly higher than indicated by the Δ values in Table 3 (because a typical chinook salmon spawning population may be a mixture of three year classes, not just two), it was apparent that for $N_b \geq 100$, the interbreeding of parents of different ages did not produce significant levels of background disequilibrium. Mixing of year-classes produced larger disequilibrium if N_b was less than 100; when the number of breeders was this small, however, the disequilibrium produced by genetic drift became much more important. Therefore, background disequilibrium due to mixing age-classes did not appear to be an important source of noise in our analyses.

The pattern of overlapping year classes can, however, be expected to affect the rate of decay of disequilibrium. Consider an initial mixture of two populations in generation 0, and assume random interbreeding to produce progeny that return at ages 3, 4, and 5. Not only will gametic disequilibrium caused by the mixture event decay after the episode of random mating, it will also be diluted as it is spread over three subsequent spawning populations. Therefore, for a species with an age structure like that of chinook salmon, a single-mixture event will probably not produce detectable, long-lasting disequilibrium in the extended population.

Selection of Loci

Although we avoided extreme allele frequencies in most of the simulations, we included loci in the computation of Δ if $P_k \leq 0.98$ in at least one of the two samples. This was done because even minor differences in allele frequency correspond to reasonably large Δ values if allele frequencies are high; such loci thus potentially contain some information regarding mixture of gene pools. However, because maximum power for these loci is somewhat limited and the type-I error rate increases considerably if $\overline{P}_k$ is very high, we recomputed $\hat{\Delta}$ for the pairwise comparisons, omitting loci with $(P_{1k} + P_{2k})/2 > 0.95$. On the average, this reduced the number of loci per comparison by about two and typically reduced $\hat{\Delta}$ by about 0.02–0.07. According to the curves in Figure 5, this had little effect on power. However, extreme allele frequencies were avoided in this way, thus reducing the probability of a type-I error. Because this was achieved with no apparent sacrifice in power, we recommend omitting loci with $\overline{P} > 0.95$ in tests for gametic disequilibrium.

Example

Use of the test for gametic disequilibrium can be illustrated with an example from actual data. Waples and Teel (in press) discussed temporal differences in allele frequency found in 21 coastal populations of chinook salmon from Oregon and California that were sampled in two different years. In three California hatchery and nine Oregon wild populations, the temporal changes in allele frequency were small enough that they could be explained satisfactorily by stochastic forces alone (genetic drift and sampling error). For the nine hatchery stocks from Oregon, however, over 35% of all loci showed statistically significant changes in allele frequency between 1981 and 1985. It is difficult to explain these changes by drift unless N_b was at least as small as 50; and it is necessary to postulate very large selection coefficients to explain them by natural selection. Therefore, we examined the possibility that mixtures of fish from other gene pools caused allele-frequency changes in the later samples.

TABLE 4.—Significance levels of tests for multilocus gametic disequilibrium in 1981 and 1985 samples from nine Oregon hatchery stocks of chinook salmon. Overall single locus excesses (+) or deficiencies (−) of heterozygotes are indicated for those samples with significant gametic disequilibrium. Significance levels are also given for multilocus contingency chi-square tests[a] of equality of allele frequencies across years. Blank spaces indicate $P > 0.05$; S denotes spring run, F is fall run.

	Gametic disequilibrium		Allele frequency
Hatchery	1981	1985	
Cedar Creek	0.05 (+)	0.01 (+)	0.01
Cole Rivers (S)			
Cole Rivers (F)			0.01
Elk River			0.01
Fall Creek	0.01 (+)		0.01
Rock Creek	0.05 (−)	0.01 (−)	0.01
Salmon		0.01 (−)	0.01
Trask (S)		0.05 (+)	0.01
Trask (F)	0.01 (+)	0.01 (+)	0.01

[a] Waples and Teel (in press).

Table 4 presents the results of tests of overall heterozygote deficiency (Wahlund effect) and multilocus gametic disequilibrium for the 18 samples from the 9 Oregon hatcheries. In most of these hatchery stocks there were substantial departures from random assortment of alleles at different gene loci. Four of the 1981 samples and five of the 1985 samples showed statistically significant disequilibrium (9/18 = 50% of tests significant), and four of the latter samples were significant at $\alpha = 0.01$. In contrast, only one of the 23 samples from California hatchery and Oregon wild populations showed significant ($P < 0.05$) gametic disequilibrium.

What do these results tell us about possible mixtures of gene pools? If the allele-frequency changes were caused by admixture in the interim between the two samples, we would expect disequilibrium in the later (1985) samples; this was observed. That disequilibrium also was observed in the earlier samples may indicate that the phenomenon is recurrent. If admixture occurred in the same generation as the samples (as in transfer of eggs or juveniles), we would also expect heterozygote deficiencies due to the Wahlund effect. In fact, none of the departures from Hardy–Weinberg proportions were statistically significant; for those samples with significant gametic disequilibrium, at least as many had heterozygote excesses as deficiencies (Table 4). As mentioned earlier, the Hardy–Weinberg test is not very powerful, but the occurrence of some heterozygote excesses suggests that the disequilibria in general were not due to an actual mixture of fish from multiple gene pools at the time of sampling. The observed heterozygote excesses can be explained satisfactorily by sampling error, but it is worth noting that they could also be the result of unequal allele frequencies in the two sexes in the previous generation. In any case, it seems unlikely that all the disequilibria were due to transfer of eggs that produced the juveniles sampled, although this explanation cannot be ruled out for those stocks with heterozygote deficiencies.

Transfer of eggs or fish prior to 1985 (the year-class of the later sample), or straying of fish from other hatchery or wild populations, might be responsible for the observed disequilibrium. No Wahlund effect would be expected because in this scenario, the sample was at least one generation removed from the admixture event. If this scenario is correct, the mixture proportions must have been nearly equal and the contributing stocks divergent genetically; if not, we would not expect power to be as great as observed (50% of tests were significant) after even a single generation of mating (see Figure 11). Hatchery records give little evidence, however, for recent imports of eggs or fish into most of these stocks (Waples and Teel, in press).

Results depicted in Figures 9–11 suggest another explanation for the results—that the disequilibria were due to genetic drift attributable to a small effective number of breeders. The curve for $N = 100$ in Figure 9 is appropriate for this example because $\overline{N} = 95$ and $\overline{K} = 7.7$ for the Oregon hatchery stocks. A 50% rate of test significance is about what can be expected under the null hypothesis (a single gene pool) if N_b was as small as 50. Waples and Teel (in press) showed that the shifts in allele frequency observed in these Oregon hatchery stocks also can be explained by genetic drift only if N_b was 50 or less. This number of effective breeders seems small for hatcheries at which the returning adults may number in the hundreds, but the breeding protocols followed in some hatcheries, in combination with a high variance in family size, might cause N_b to be considerably less than the number of mature adults (Simon et al. 1986; Waples and Teel, in press).

The present analysis cannot rule out other explanations for the disequilibrium, such as natural or artificial selection, but both the allele-frequency and gametic disequilibrium data are consistent with the proposition that fish from different gene pools were introduced (or strayed) into some of these Oregon hatchery stocks prior to the 1985 sample. However, it is necessary to postulate a large proportion of introduced fish from relatively divergent stocks to explain the disequilibrium by admixture alone. It seems likely that a low number of effective breeders also was a factor.

Discussion

Results described here give us a perspective for evaluating the potential usefulness of gametic disequilibrium analysis in identifying samples that represent a mixture of gene pools. In addition, they provide a basis for answering some of the more specific questions pertaining to genetic interactions between salmonid populations. We can summarize as follows.

Properties of the Test Statistic

(1) Over the range of conditions examined here (sample size, 50–400; number of loci, 5–15), the probability of a statistically significant test due to sampling error alone is very close to the desired level ($\alpha = 0.05$) unless allele frequencies are near 0 or 1. However, the multilocus test is very sensitive to the effects of genetic drift, and gametic disequilibrium due to a finite number of breeders can seriously inflate the probability of a significant test unless sample size is substantially smaller than N_b. This presents something of a dilemma, because samples of fewer than 100 individuals will seldom provide meaningful information regarding mixture events unless genetic differences between populations are very large. Some idea of the size of the breeding population is needed to interpret test results, particularly if sample size is large ($N > 100$).

Given that samples of at least 100 are required in most cases for reasonable power, and that the test is strongly affected by genetic drift unless $N < N_b/2$, multilocus gametic disequilibrium analysis involving as many as eight loci is not well suited to evaluating mixtures of populations with $N_b < 200$. If Δ per locus is large, however, smaller samples can be used, thus reducing the noise created by disequilibrium due to drift.

(2) Mixtures in which one of the populations contributes less than 10% cannot be detected successfully unless Δ is large and large sample sizes are available.

(3) With moderately large genetic differences between populations (e.g., total $\Delta \approx 0.5$), the test retains some power to detect mixtures after one generation of random mating, provided the initial

mixture proportions [m, $(1 - m)$] do not differ by more than a factor of about 2. Little power remains two generations after the mixture event unless initial genetic differences were large.

Relevance to Salmonids

The results provide answers to some of the questions posed earlier about mixtures of salmon gene pools.

(1) Considerable heterogeneity exists in the degree of genetic differentiation between pairs of potentially interbreeding populations, and the power of the test to detect these mixtures varies accordingly. Some pairs of populations are so similar genetically that a mixture cannot reliably be identified even under ideal conditions (similar mixture proportions, large sample size, mixture event in current generation). Others are divergent enough that considerable power remains even under less favorable conditions. If populations that might be contributing to a suspected mixture can be identified and characterized genetically, results described here provide a context for interpreting the test statistic for gametic disequilibrium. This type of scenario might occur, for example, if eggs (or fry) had been exchanged between hatcheries but their genetic contribution to succeeding generations was uncertain, or if fish were transplanted outside the hatchery and their genetic interactions with resident fish were to be monitored.

(2) In the absence of information about potential source populations, our results of comparisons within and between management groups in the three regions permit a rough approximation of the probability of successfully detecting gametic disequilibrium if a mixture of two populations exists. Such information might be valuable to hatchery managers, for example, when transfer records are missing or incomplete, or when it is desirable to identify and preserve wild gene pools that have not been substantially affected by strays or transplants of uncertain origin.

(3) Given that mixtures to which one population contributes less than 10% are difficult to detect unless Δ is very large, and given that geographically adjacent populations of chinook salmon tend to be genetically similar (Utter et al. 1989), incidental straying between populations will be difficult to detect by gametic disequilibrium analysis.

(4) Under certain conditions (Δ typical of between-group comparisons, subequal mixture frac-

tions), the test retains appreciable power to detect a mixing event in assessments of F_1 progeny derived from random interbreeding of the source populations.

(5) The importance of the parameter N_b indicates that tests of gametic disequilibrium should be interpreted with caution, particularly in those instances in which a reasonable estimate of the effective number of breeders is not available. However, the disequilibrium due to drift can be used to advantage under certain circumstances. In many populations, for example, it may be possible to rule out recent admixture from other gene pools, but it is often very difficult to tell whether juvenile samples have been randomly drawn from the population as a whole, or instead represent the progeny of one or a few families. In such a scenario, the magnitude and significance level of the disequilibrium would provide information regarding the number of breeders responsible for the sample.

Although the above analyses were designed with specific reference to chinook salmon, the results are also relevant for other salmonid species. Steelhead *Oncorhynchus mykiss* (formerly *Salmo gairdneri*) and coastal cutthroat trout *O.* (formerly *S.*) *clarki clarki* have levels of heterozygosity similar to or higher than those of chinook salmon, and there is evidence of significant genetic heterogeneity among some of their populations (e.g., Parkinson 1984). Chum salmon *O. keta*, pink salmon, and sockeye salmon *O. nerka* all have slightly lower levels of heterozygosity but substantial genetic differences between at least some regions (e.g., Utter et al. 1980). The usefulness of gametic disequilibrium analysis to detect mixtures in these species must, therefore, be evaluated case by case. Currently available data (e.g., Wehrhahn and Powell 1987) suggest that the level of genetic variability in coho salmon *O. kisutch* may be too small for the techniques described here to be of much use.

Results obtained here, particularly those depicted in Figures 5 and 12, should also be useful to those interested in using gametic disequilibrium analysis for other taxa where mixtures of gene pools are a possibility.

Acknowledgments

We thank the National Marine Fisheries Service, Seattle, and the Washington Department of Fisheries for use of unpublished data. The manuscript benefited from comments by Donald Campton, Stevan Phelps, Gary Winans, and an anonymous reviewer. Support by the National Research Council (postdoctoral fellowship to R.S.W.), the National Institutes of Health (grant NIH-R01-GM32589 to P.E.S.), the NJAES (grant 32102 to P.E.S.), and the USDA is gratefully acknowledged.

References

Aebersold, P. B., G. A. Winans, D. J. Teel, G. B. Milner, and F. M. Utter. 1987. Manual for starch gel electrophoresis: a method for the detection of genetic variation. NOAA (National Oceanic and Atmospheric Administration) Technical Report NMFS (National Marine Fisheries Service) 61:1–19.

Allendorf, F. W., and G. H. Thorgaard. 1984. Tetraploidy and the evolution of salmonid fishes. Pages 1–53 *in* B. Turner, editor. Evolutionary genetics of fishes. Plenum, New York.

Brown, A. H. D. 1975. Sample sizes required to detect linkage disequilibrium between two or three loci. Theoretical Population Biology 8:184–201.

Campton, D. E., and J. M. Johnston. 1985. Electrophoretic evidence for a genetic admixture of native and nonnative rainbow trout in the Yakima River, Washington. Transactions American Fisheries Society 114:782–793.

Campton, D. E., and F. M. Utter. 1985. Natural hybridization between steelhead trout (*Salmo gairdneri*) and coastal cutthroat trout (*Salmo clarki clarki*) in two Puget Sound streams. Canadian Journal of Fisheries and Aquatic Sciences 42:110–119.

Fairbairn, D. J., and D. A. Roff. 1980. Testing genetic models of isozyme variability without breeding data: can we depend on the χ^2? Canadian Journal of Fisheries and Aquatic Sciences 37:1149–1159.

Fraidenburg, M. E., and R. H. Lincoln. 1985. Wild chinook salmon management: an international conservation challenge. North American Journal of Fisheries Management 5:311–329.

Glass, B., and C. C. Li. 1953. The dynamics of racial intermixture—an analysis based on the American Negro. American Journal of Human Genetics 5:1–20.

Harden Jones, F. R. 1968. Fish migration. Edward Arnold, London.

Hedrick, P. 1987. Gametic disequilibrium measures: proceed with caution. Genetics 117:331–341.

Helle, J. H. 1981. Significance of the stock concept in artificial propagation of salmonids in Alaska. Canadian Journal of Fisheries and Aquatic Sciences 38:1665–1671.

Hill, W. G. 1981. Estimation of effective population size from data on linkage disequilibrium. Genetical Research (Cambridge) 38:209–216.

Howell, P., K. Jones, D. Scarnecchia, L. Lavoy, W. Kendra, and D. Ortmann. 1985. Stock assessment of Columbia River anadromous salmonids, volume 1. Bonneville Power Administration, Portland, Oregon.

Kimura, M., and T. Ohta. 1971. Theoretical aspects of population genetics. Princeton University Press, Princeton, New Jersey.

Lewontin, R. C., and C. C. Cockerham. 1959. The goodness-of-fit test for detecting natural selection in randomly mating populations. Evolution 13:561–564.

Lister, D. B., D. G. Hickey, and I. Wallace. 1981. Review of the effects of enhancement strategies on the homing, straying, and survival of Pacific salmonids, volume 1. Canada Department of Fisheries and Oceans, Vancouver.

Long, J. C. 1986. The allelic correlation structure of Gainj and Kalam speaking people. I. The estimation and interpretation of Wright's F-statistics. Genetics 112:629–647.

Mathews, S. B. 1980. Trends in Puget Sound and Columbia River coho salmon. Pages 133–145 in W. J. McNeil and D. C. Himsworth, editors. Salmonid ecosystems of the north Pacific. Oregon State University Press, Corvallis.

Nei, M., and W.-H. Li. 1973. Linkage disequilibrium in subdivided populations. Genetics 75:213–219.

NWPPC (Northwest Power Planning Council). 1987. Columbia River basin fish and wildlife program. Portland, Oregon.

Ohta, T., and M. Kimura. 1969. Linkage disequilibrium due to random genetic drift. Genetical Research 13:47–55.

Parkinson, E. A. 1984. Genetic variation in populations of steelhead trout (Salmo gairdneri) in British Columbia. Canadian Journal of Fisheries and Aquatic Sciences 41:1412–1420.

Ricker, W. E. 1972. Hereditary and environmental factors affecting certain salmonid populations. Pages 27–160 in R. C. Simon and P. A. Larkin, editors. The stock concept of Pacific salmon. H. R. MacMillan Lectures in Fisheries, University of British Columbia, Vancouver, Canada.

Ryman, N., F. W. Allendorf, and G. Ståhl. 1979. Reproductive isolation with little genetic divergence in sympatric populations of brown trout (Salmo trutta). Genetics 92:247–262.

Scheer, T. T. 1939. Homing instinct in salmon. Quarterly Review of Biology 14:408–430.

Simon, R. C., J. D. McIntyre, and A. R. Hemingsen. 1986. Family size and effective population size in a hatchery stock of coho salmon (Oncorhynchus kisutch). Canadian Journal of Fisheries and Aquatic Sciences 43:2434–2442.

Sinnock, P. 1975. The Wahlund effect for the two locus model. American Naturalist 109:565–570.

Smouse, P. E. 1982. Genetic architecture of swidden agricultural tribes from the South American rainforests. Pages 139–178 in M. Crawford and J. Mielke, editors. Current developments in anthropological genetics, volume 2. Plenum, New York.

Smouse, P. E., and J. C. Long. 1988. A comparative F-statistics analysis of the Yanomama of lowland South America and the Gainj and Kalam of highland New Guinea. Pages 32–46 in B. S. Weir, G. Eisen, M. M. Goodman, and G. Namkoong, editors. Proceedings II international conference on quantitative genetics. Sinauer Associates, Sunderland, Massachusetts.

Smouse, P. E., and J. V. Neel. 1977. Multivariate analysis of gametic disequilibrium in the Yanomama. Genetics 85:733–752.

Smouse, P. E., J. V. Neel, and W. Liu. 1983. Multiple-locus departures from panmictic equilibrium within and between village gene pools of Amerindian tribes at different stages of acculturation. Genetics 104:133–153.

Utter, F. M., D. Campton, S. Grant, G. B. Milner, J. Seeb, and L. Wishard. 1980. Population structures of indigenous salmonid species of the Pacific Northwest. Pages 285–304 in W. J. McNeil and D. C. Himsworth, editors. Salmonid ecosystems of the North Pacific. Oregon State University Press, Corvallis.

Utter, F., G. Milner, G. Ståhl, and D. Teel. 1989. Genetic population structure of chinook salmon, Oncorhynchus tshawytscha, in the Pacific Northwest. U.S. National Marine Fisheries Service Fishery Bulletin 87:239–264.

Wahle, R. J., and R. E. Pearson. 1984. History of artificial propagation of coho salmon, Oncorhynchus kisutch, in the mid-Columbia River system. U.S. National Marine Fisheries Service Marine Fisheries Review 46(3):34–43.

Waples, R. S. 1988. Estimation of allele frequencies at isoloci. Genetics 118:371–384.

Waples, R. S. 1989. Temporal variation in allele frequencies: testing the right hypothesis. Evolution 43:1236–1251.

Waples, R. S. 1990. Temporal changes of allele frequency in Pacific salmon: implications for mixed-stock frequency analysis. Canadian Journal of Fisheries and Aquatic Sciences 47:968–976.

Waples, R. S., and D. J. Teel. In press. Conservation genetics of Pacific salmon. I. Temporal changes in allele frequency. Conservation Biology.

WDF (Washington Department of Fisheries). 1988. Analysis of the 1987 Washington May troll chinook fishery. WDF, GSI Summary Report 88-2, Olympia.

Wehrhahn, C. F., and R. Powell. 1987. Electrophoretic variation, regional differences, and gene flow in the coho salmon (Oncorhynchus kisutch) of southern British Columbia. Canadian Journal of Fisheries and Aquatic Sciences 44:822–831.

Weir, B. S., and C. C. Cockerham. 1984. Estimating F-statistics for the analysis of population structure. Evolution 38:1358–1370.

Withler, F. C. 1982. Transplanting Pacific salmon. Canadian Technical Report of Fisheries and Aquatic Sciences 1079.

American Fisheries Society Symposium 7:459–469, 1990

Genetic Differentiation of Pacific Herring Based on Enzyme Electrophoresis and Mitochondrial DNA Analysis

J. F. Schweigert and R. E. Withler

Department of Fisheries and Oceans, Biological Sciences Branch
Pacific Biological Station, Nanaimo, British Columbia V9R 5K6, Canada

Abstract.—Starch-gel electrophoresis of proteins and restriction endonuclease analysis of mitochondrial DNA (mtDNA) were used to investigate genetic relationships among samples of Pacific herring *Clupea harengus pallasi* from five areas (three regions) along the coast of British Columbia. Electrophoretic data for seven polymorphic loci revealed little variability among the three regions, or among three areas within one region. There was significant heterogeneity in allelic frequencies at one or more loci between years in three of the five areas sampled. The 11 restriction enzymes used in the mtDNA analysis identified 13 different mtDNA phenotypes among 31 female Pacific herring sampled from five areas. Each of six mtDNA phenotypes that occurred more than once was sampled from at least two regions, and two of them were present in samples from all three regions. The study provided little evidence of genetic differentiation among Pacific herring sampled from geographically defined stocks in coastal waters of British Columbia. These results are consistent with previous biochemical genetic surveys of both Atlantic herring *C. h. harengus* and Pacific herring.

A genetically accurate description of stock structure is crucial to the success of any fisheries management scheme. The demise of the least productive stocks as a result of consistent exploitation of mixed stocks has been well documented (Ricker 1973; Sinclair et al. 1985). A universally acceptable definition of a herring stock, however, has been particularly elusive (Iles and Sinclair 1982; Smith and Jamieson 1986).

Pacific herring *Clupea harengus pallasi* spawn adhesive demersal eggs in the intertidal zone in numerous discrete locations, which may range in size from less than 1 hectare to hundreds of hectares, all within an area of several square kilometers (Haegele and Schweigert 1985). In addition to this spatial segregation, temporal variation in spawning may occur within a localized area (Sinclair and Tremblay 1984; Haegele and Schweigert 1985).

Although there is some evidence that Pacific herring home to geographic regions encompassing an amalgam of spawning beaches, the level of straying among regions is sometimes as great as 30% (Hourston 1982; Wheeler and Winters 1984). Thus, there is clearly the opportunity for substantial gene flow among individual spawning beaches and over larger geographic areas. Previous genetic surveys of both Pacific herring and Atlantic herring *Clupea harengus harengus* have revealed little or no differentiation of putative stocks (Andersson et al. 1981; Kornfield et al. 1982; Grant 1984; Grant and Utter 1984; Ryman et al. 1984; Kornfield and

Bogdanowicz 1987). Conversely, morphological and meristic differentiation of herring from discrete geographic regions has been interpreted as evidence for the existence of genetically distinct stocks (Tester 1937, 1949; Schweigert 1981, 1990, this volume; Meng and Stocker 1984; King 1985). Much of the morphological differentiation, however, may be environmentally mediated.

Most of the foregoing genetic and morphometric studies were conducted on spatial scales far coarser than those required for fisheries management. The purpose of this study was to investigate genetic differentiation of Pacific herring in British Columbia as reflected by variation in allelic frequencies among statistical areas, which constitute an intermediate spatial scale consistent with fisheries management needs. Differentiation among areas was compared with variability between years within areas. In addition, during one year we used restriction endonuclease analysis of mitochondrial DNA (mtDNA) to assess the potential of this technique for future stock-identification efforts.

Methods

Starch-gel electrophoresis of enzymes.—Pacific herring for electrophoretic analysis were collected in 1985 and 1986 from commercial sac-roe fisheries. They were taken along the British Columbia coast at 12 locations in five statistical areas encompassing three distinct geographic regions (area 3 on the north coast; areas 14, 15, and 17 in

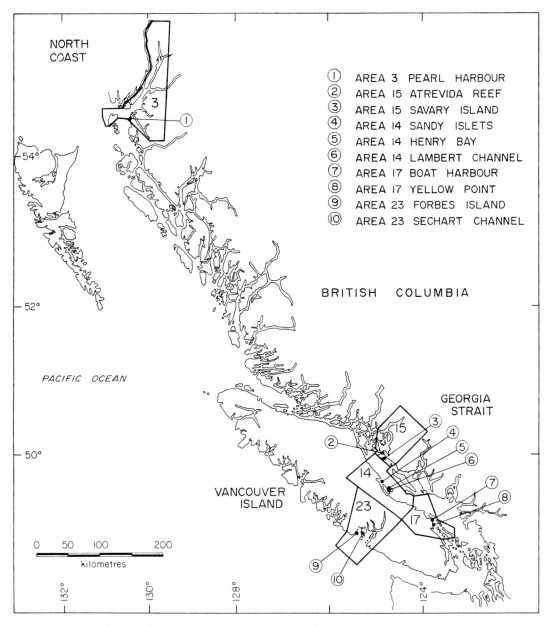

FIGURE 1.—Distribution of British Columbia samples of Pacific herring from which electrophoretic data were obtained. The areas are statistical boundaries used in reporting catch and other biological data.

Georgia Strait; and area 23 on the west coast of Vancouver Island; Figure 1). White muscle tissues were collected from approximately 100 Pacific herring at each location, and the tissue samples were stored at −20°C prior to analysis. Horizontal starch-gel electrophoresis (Utter et al. 1974) was used to measure genetic variation at seven of the polymorphic loci described by Grant and Utter (1984). In designating protein loci and alleles, we followed the system proposed by Allendorf and Utter (1979) and were consistent with the interpretation of Grant (1981) and Grant and Utter (1984). The seven loci examined and the buffer systems used are given in Table 1.

Allelic frequencies for each locus were determined by summing the numbers of each allele and dividing by the total number of alleles. The fre-

TABLE 1.—Enzymes and loci examined for protein variation in Pacific herring. Buffer systems used were (1) the amine–citrate buffer described by Clayton and Tretiak (1972), (2) a tris–citric acid–lithium hydroxide–boric acid buffer described by Ridgway et al. (1970), and (3) a tris–boric acid–EDTA buffer described by Markert and Faulhaber (1965). Parenthetic enzyme numbers are those of IUBNC (1984).

Enzyme	Locus	Buffer
Adenosine deaminase (3.5.4.4)	*Ada*	1, 3
Glyceraldehyde-3-phosphate dehydrogenase (1.2.1.12)	*Gap-1*	1
Isocitrate dehydrogenase (1.1.1.42)	*Idh-2*	1
Malate dehydrogenase (1.1.1.37)	*Mdh-4*	1
Malic enzyme (1.1.1.40)	*Me-2*	1
Phosphoglucose (glucose-6-phosphate) isomerase (5.3.1.9)	*Pgi*	2, 3
Phosphoglucomutase (5.4.2.2)	*Pgm-1*	2, 3

quency of genotypes at each locus in each sample was tested for departures from Hardy–Weinberg equilibrium by chi square tests. We tested the equality of allelic frequencies among three geographic regions (Georgia Strait, west coast of Vancouver Island, and north coast of British Columbia), among three areas within Georgia Strait, and between years for samples taken from Georgia Strait, using the likelihood-ratio statistic (G-test) (Sokal and Rohlf 1969). Any alleles that occurred at frequencies of less than 0.025 were combined with those of the next-rarest allele prior to analysis. An approximate F-ratio statistic (G-statistic summed over loci per degrees of freedom) was used to test the relative magnitude of these sources of variation (Grant and Utter 1984). We used the genetic distance of Nei (1978) to calculate genetic divergence among stocks for the seven loci of Table 1; and we constructed a cluster dendrogram from the matrix of interstock genetic distances with the unweighted pair-group method with arithmetic averages (UPGMA), as described by Sneath and Sokal (1973). Total genic diversity was partitioned into temporal and spatial components as described by Chakraborty et al. (1982). All statistical analyses were performed with the BIOSYS package of FORTRAN programs (Swofford and Selander 1981) on a VAX 11-780 minicomputer.

Restriction enzyme digestion of mtDNA. —Pacific herring for restriction endonuclease analysis were collected in 1984 from five statistical areas (areas 14, 15, and 17 in Georgia Strait; area 23 on the west coast of Vancouver Island; and area 3 on the north coast of British Columbia; Figure 1). Ovaries were removed from five to nine fish at each location and were kept on ice prior to mtDNA extraction and cesium chloride purification (Lansman et al. 1981). The restriction enzymes used for digestion recognized either six-base nucleotide sequences (*Eco* RI, *Bam* HI, *Hind* III, *Bgl* I, *Bgl* II, *Kpn* I, *Pst* I, *Sma* I, and *Xba* I) or six-base multiple sequences (*Hinc* II and *Hae* II).

Separation of restriction fragments was carried out by horizontal agarose electrophoresis with 0.7% gels. The mtDNA fragments were visualized with ethidium bromide and photographed under ultraviolet light. Fragment sizes were estimated by comparison with the migration distances of phage lambda DNA fragments produced by digestion with *Hind* III or *Hind* III and *Eco* RI.

The maximum-likelihood method of Nei and Tajima (1983) was used to estimate levels of nucleotide substitution between mtDNA phenotypes. The pairwise estimates of percentage nucleotide substitution were used to generate a dendrogram of the genetic relationships based on UPGMA clustering (Sneath and Sokal 1973).

Results

Electrophoretic Analysis of Enzymes

Of 84 comparisons of observed genotypic frequencies in the Pacific herring samples with frequencies expected under Hardy–Weinberg equilibrium, six (7%) deviated significantly ($P < 0.05$). These six cases, resulting from heterozygote or homozygote deficiencies, occurred at *Ada*, *Idh-2*, and *Me-2*.

Estimates of total heterogeneity in the sample indicated significant differences at four loci—*Ada*, *Gap-1*, *Idh-2*, and *Me-2*. The overall level of heterogeneity in allelic frequencies between 1985 and 1986 was significant because of annual variability in allelic frequencies at one or two loci in each of three of the five areas sampled (Tables 2, 3). Significant annual variability occurred in one or more areas at *Ada*, *Gap-1*, *Idh-2*, and *Me-2*.

Allelic frequencies at *Ada* and *Gap-1* differed significantly among the three areas within Georgia Strait, in spite of the annual variation observed at each of these loci within areas (Table 3). Significant regional variability among Georgia Strait, the west coast of Vancouver Island, and the north coast of British Columbia occurred only at *Gap-1* (Tables 2, 3). The frequency of the *Gap-1*$^{-100}$ allele was lower in northern British Columbia waters (area 3) than in any of the more southerly areas (Table 2). In addition, the frequency of the *Gap-1*$^{-100}$ allele was lower in the two northern

TABLE 2.—Allelic frequencies observed at seven polymorphic loci (Table 1) over 2 years for Pacific herring from five areas in British Columbia.

Stock,[a] Year	Ada					Gap-1			Idh-2				
	N	100	91	80	115	85	N	-100	-50	N	100	88	106	116	82
Area 3														
Pearl Hr, 1985	90	0.539	0.372	0.067	0.022	0.000	100	0.735	0.265	100	0.120	0.395	0.475	0.010	0.000
Pearl Hr, 1986	99	0.551	0.439	0.010	0.000	0.000	92	0.712	0.288	100	0.110	0.460	0.410	0.020	0.000
Pooled	189	0.545	0.407	0.037	0.010	0.000	192	0.724	0.276	200	0.115	0.428	0.442	0.015	0.000
Area 14														
Sandy Is, 1985	95	0.621	0.363	0.011	0.005	0.000	99	0.788	0.212	98	0.173	0.332	0.480	0.015	0.000
Henry B, 1985	100	0.540	0.435	0.015	0.010	0.000	102	0.755	0.245	103	0.136	0.311	0.549	0.005	0.000
Lambert C, 1986	196	0.503	0.452	0.033	0.013	0.000	197	0.827	0.173	199	0.113	0.430	0.452	0.003	0.003
Pooled	391	0.541	0.426	0.023	0.010	0.000	398	0.799	0.201	400	0.134	0.375	0.484	0.006	0.001
Area 15														
Atrevida R, 1985	49	0.480	0.398	0.061	0.000	0.061	59	0.669	0.331	60	0.142	0.392	0.458	0.008	0.000
Savary I, 1986	224	0.536	0.391	0.067	0.007	0.000	235	0.817	0.183	241	0.127	0.365	0.500	0.008	0.000
Pooled	273	0.526	0.392	0.066	0.006	0.011	294	0.787	0.213	301	0.130	0.370	0.492	0.008	0.000
Area 17														
Yellow Pt, 1985	71	0.528	0.430	0.014	0.007	0.021	93	0.806	0.194	100	0.105	0.360	0.525	0.005	0.005
Boat Hr, 1985	97	0.582	0.351	0.046	0.015	0.005	102	0.858	0.142	102	0.147	0.328	0.495	0.015	0.015
Boat Hr, 1986	83	0.536	0.434	0.012	0.018	0.000	93	0.892	0.108	93	0.075	0.425	0.484	0.011	0.005
Pooled	251	0.552	0.401	0.026	0.014	0.008	288	0.852	0.148	295	0.110	0.369	0.502	0.010	0.008
Georgia Strait[b]	915	0.539	0.409	0.037	0.010	0.005	980	0.811	0.189	996	0.126	0.372	0.491	0.008	0.003
Area 23														
Forbes I, 1985	86	0.523	0.384	0.070	0.000	0.023	99	0.803	0.197	99	0.101	0.399	0.475	0.020	0.005
Sechart C, 1986	98	0.566	0.403	0.020	0.010	0.000	99	0.818	0.182	100	0.130	0.470	0.390	0.010	0.000
Pooled	184	0.546	0.394	0.043	0.005	0.011	198	0.811	0.190	199	0.116	0.435	0.432	0.015	0.002

Stock,[a] year	Mdh-4				Me-2			
	N	100	70	130	N	100	90	40	110
Area 3								
Pearl Hr, 1985	100	0.940	0.060	0.000	99	0.874	0.086	0.030	0.010
Pearl Hr, 1986	100	0.935	0.065	0.000	92	0.804	0.076	0.092	0.027
Pooled	200	0.938	0.063	0.000	191	0.840	0.081	0.060	0.018
Area 14								
Sandy Is, 1985	99	0.944	0.056	0.000	99	0.884	0.071	0.040	0.005
Henry B, 1985	102	0.946	0.054	0.000	102	0.824	0.083	0.064	0.029
Lambert C, 1986	200	0.957	0.043	0.000	197	0.876	0.056	0.030	0.038
Pooled	401	0.951	0.049	0.000	398	0.865	0.067	0.041	0.027
Area 15								
Atrevida R, 1985	99	0.939	0.061	0.000	99	0.813	0.096	0.061	0.030
Savary I, 1986	242	0.946	0.054	0.000	242	0.849	0.083	0.041	0.027
Pooled	341	0.994	0.056	0.000	341	0.839	0.087	0.047	0.028
Area 17								
Yellow Pt, 1985	100	0.950	0.050	0.000	100	0.750	0.150	0.040	0.060
Boat Hr, 1985	102	0.931	0.069	0.000	101	0.837	0.094	0.054	0.015
Boat Hr, 1986	94	0.947	0.053	0.000	94	0.824	0.090	0.069	0.016
Pooled	296	0.943	0.058	0.000	295	0.803	0.112	0.054	0.031
Georgia Strait[b]	1,038	0.946	0.054	0.000	1034	0.839	0.086	0.047	0.028
Area 23								
Forbes I, 1985	97	0.923	0.077	0.000	100	0.840	0.060	0.040	0.060
Sechart C, 1986	100	0.915	0.075	0.010	99	0.874	0.051	0.040	0.035
Pooled	197	0.919	0.076	0.005	199	0.857	0.056	0.040	0.048

[a] Abbreviations: B, Bay; C, Channel; Hr, Harbor; I, Island; Is, Islet; Pt, Point; R, Reef.
[b] Areas 14, 15, and 17 pooled.

areas within Georgia Strait (areas 14 and 15) than in the more southerly area 17 (Table 2).

Heterogeneity in allelic frequencies was not significantly greater among regions than among areas within Georgia Strait ($F = 1.17$; df = 26, 26; $P > 0.05$). Similarly, heterogeneity among areas within Georgia Strait was not significantly greater

than that between years within these areas ($F = 0.91$; $df = 26, 39$; $P > 0.05$).

Our hierarchical gene-diversity analysis (Chakraborty et al. 1982) was consistent with these results (Table 4). Virtually all of the variation in allelic frequencies (>99%) occurred within samples, so was common to all areas. Variation among

TABLE 2.—Continued.

Stock,[a] year	Pgi							Pgm-1			
	N	100	145	45	180	190	55	N	75	100	65
Area 3											
Pearl Hr, 1985	99	0.823	0.177	0.000	0.000	0.000	0.000	99	0.768	0.232	0.000
Pearl Hr, 1986	100	0.780	0.210	0.010	0.000	0.000	0.000	100	0.765	0.235	0.000
Pooled	199	0.801	0.194	0.005	0.000	0.000	0.000	199	0.766	0.234	0.000
Area 14											
Sandy Is, 1985	99	0.788	0.197	0.010	0.000	0.000	0.005	98	0.755	0.240	0.005
Henry B, 1985	102	0.809	0.186	0.000	0.000	0.000	0.005	101	0.683	0.317	0.000
Lambert C, 1986	200	0.810	0.185	0.005	0.000	0.000	0.000	194	0.737	0.263	0.000
Pooled	401	0.804	0.188	0.005	0.000	0.000	0.003	393	0.728	0.271	0.001
Area 15											
Atrevida R, 1985	100	0.780	0.220	0.000	0.000	0.000	0.000	100	0.720	0.280	0.000
Savary I, 1986	242	0.783	0.202	0.024	0.000	0.000	0.000	239	0.736	0.262	0.002
Pooled	342	0.782	0.207	0.010	0.000	0.000	0.000	339	0.731	0.267	0.001
Area 17											
Yellow Pt, 1985	100	0.750	0.230	0.010	0.005	0.005	0.000	99	0.763	0.237	0.000
Boat Hr, 1985	102	0.804	0.191	0.000	0.000	0.005	0.000	102	0.755	0.240	0.005
Boat Hr, 1986	95	0.795	0.200	0.005	0.000	0.000	0.000	94	0.681	0.319	0.000
Pooled	297	0.783	0.207	0.005	0.002	0.003	0.000	295	0.734	0.264	0.000
Georgia Strait[b]	1,040	0.791	0.200	0.007	0.000	0.001	0.001	925	0.728	0.271	0.001
Area 23											
Forbes I, 1985	100	0.785	0.215	0.000	0.000	0.000	0.000	98	0.745	0.255	0.000
Sechart C, 1986	100	0.705	0.285	0.010	0.000	0.000	0.000	100	0.750	0.250	0.000
Pooled	200	0.745	0.250	0.005	0.000	0.000	0.000	198	0.748	0.252	0.000

samples within years, among years within areas, and among areas within regions, were all of similar magnitude (~0.25% of the total variation), whereas regional differentiation was even less (Table 4).

Pairwise genetic distances were low, ranging from 0.0 to a maximum of 0.009 between the Atrevida Reef sample and the 1986 Boat Harbour sample. The dendrogram based on genetic distances also reflected the considerable annual variability in allelic frequencies and a lack of geographic patterns (Figure 2). In 1985, there was no significant separation between the Georgia Strait samples and samples from the north coast and west coast of Vancouver Island. The 1986 sam-

ples from the latter two regions differed from other samples because of higher $Idh-2^{88}$ frequencies (Table 2). The 1985 Atrevida Reef sample was characterized by an unusually low $Gap-1^{-100}$ frequency for a southern population. This may have resulted from sampling error due to the small number of individuals scored for this locus.

Restriction Enzyme Analysis of mtDNA

The fragments resulting from restriction enzyme digestion provided well-resolved electrophoretic patterns. Restriction fragment-length polymorphisms (RFLPs) were detected among

TABLE 3.—Results of G-tests of heterogeneity of allelic frequencies at seven loci, among and within Pacific herring stocks sampled along the coast of British Columbia. Asterisks denote $P < 0.05^*$ or $P < 0.01^{**}$.

Sources of variation	Ada		Gap-1		Idh-2		Mdh-4		Me-2		Pgi		Pgm		Total		Stan-dardized measure[a]
	df	G	df	G	df	G	df	G	df	G	df	G	df	G	df	G	
Among regions	4	0.29	2	14.61**	6	10.41	2	3.19	6	11.82	4	5.60	2	2.43	26	48.35**	1.86
Within Georgia Strait	4	11.36*	2	9.67**	6	5.94	2	0.63	6	10.76	4	2.96	2	0.07	26	41.39*	1.59
Between years	10	26.90**	5	17.55**	15	26.62*	5	0.47	15	23.23	10	14.35	5	3.89	65	113.01**	1.74
Area 3	2	14.84**	1	0.50	3	2.71	1	0.00	3	8.54*	2	3.55	1	0.00	13	29.79**	2.29
Area 14	2	7.26*	1	3.59	3	12.39**	1	0.41	3	6.92	2	0.06	1	0.25	13	30.88**	2.38
Area 15	2	0.20	1	10.56**	3	0.69	1	0.03	3	1.62	2	5.05	1	0.12	13	18.27	1.41
Area 17	2	1.55	1	3.18	3	5.93	1	0.01	3	4.57	2	0.09	1	3.52	13	18.85	1.45
Area 23	2	3.05	1	0.07	3	4.90	1	0.02	3	1.58	2	5.60	1	0.00	13	15.22	1.17
Total	18	38.55**	9	41.83**	27	42.97*	9	4.29	27	45.81*	14	22.91	9	6.39	117	202.75**	1.73

[a]Approximate F-ratio based on the sum of G-statistics over loci divided by degrees of freedom.

TABLE 4.—Hierarchical analysis of relative gene diversity for Pacific herring samples from three regions of British Columbia in 1985 and 1986. H_S and H_T are Hardy–Weinberg expectations of heterozygosity in the subpopulation and total population, respectively.

Estimates of relative gene diversity (G)	Ada	Gap-1	Idh-2	Mdh-4	Me-2	Pgi	Pgm	Mean	SE
$H_S H_T$ (within samples)	0.9995	0.9794	0.9956	0.9995	0.9957	0.9986	1.0000	0.9955	0.0028
$G_{SY(T)}$ (between samples within years)	0.0050	0.0006	0.0008	0.0	0.0109	0.0003	0.0027	0.0029	0.0015
$G_{YA(T)}$ (between years within areas)	0.0[a]	0.0112	0.0047	0.0[a]	0.0[a]	0.007	0.0[a]	0.0024	0.0016
$G_{AR(T)}$ (between areas within regions)	0.0001	0.0108	0.0[a]	0.0	0.0028	0.0[a]	0.0[a]	0.0020	0.0015
G_{RT} (between regions)	0.0[a]	0.0[a]	0.0021	0.0012	0.0[a]	0.0015	0.0010	0.0008	0.0003

[a] Negative variance components were assumed to equal zero in calculations of means and standard errors.

Pacific herring mtDNA with 9 of the 11 endonucleases used (Table 5). All of the RFLPs could be accounted for by the gain or loss of cut sites, and each cut-site alteration appeared to result from a single nucleotide substitution. There was no evidence of mtDNA size variation among the samples due to DNA additions or deletions. The sum of fragment sizes for all three *Bgl* II restriction patterns was approximately 21,000 base pairs (bp), which exceeded the estimated size of the mitochondrial genome based on restriction by all other enzymes by about 4,000 bp. We assumed that this discrepancy was due to the partial digestion of a fragment shared by all three *Bgl* II patterns (4,250 bp) into two smaller fragments (possibly the 2,700 bp and 1,600 bp fragments). Thus, the mtDNA sizes estimated from digestion by *Bgl* II were the sums of fragment sizes exclusive of the 4,250 bp fragment (Table 5). The mean mtDNA genome size, calculated as the average of all digestion pattern sums (Table 5), was 16,730 bp+261 bp (SD).

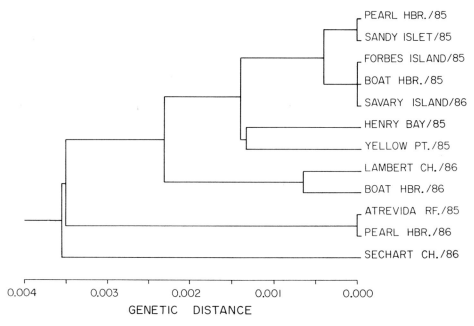

FIGURE 2.—Clustering dendrogram from the unweighted pair-group method with arithmetic averages, produced from Nei's (1978) genetic distance coefficient, for seven electrophoretic loci of Pacific herring in British Columbia. Numbers following the names refer to sampling years.

TABLE 5.—Sizes of fragment base pairs (bp) produced by restriction enzyme digestion of Pacific herring mitochondrial DNA. *Hinc* II and *Hae* II are six-base multiple nucleotide sequences; the others are six-base simple sequences. Fragment totals for digestion patterns are at the bottom of each column.

Bgl I			*Bgl* II			*Hinc* II		*Sma* I		*Hae* II	*Hind* III	
A	B	C	A	B	C	A	B	A	B	A	A	B
8,100	8,100				9,500		5,000	5,800		4,450		12,750
		6,700	6,300	6,300		3,950	3,950	3,850	3,850	2,630	7,600	
5,650	5,650	5,650	4,250[a]	4,250[a]	4,250[a]	2,870[b]	2,870[b]	3,550	3,550	2,300	5,100	
2,400		2,400	3,350			2,630			3,150	2,280	3,340	3,340
	1,380			2,880		2,320			2,550	2,000	890	890
		1,300	2,700	2,700	2,700	740	740	2,050	2,050	1,620		
	1,030		1,900	1,900	1,900	630	630	1,450	1,450	1,400		
			1,600	1,600	1,600	500	500					
			1,030	1,030	1,030	430	430					
				510								
16,150	16,160	16,050	16,880	16,920	16,730	16,940	16,990	16,700	16,600	16,680	16,930	16,980

Eco RI	*Pst* I				*Bam* HI		*Kpn* I		*Xba* I		
A	A	B	C	D	A	B	A	B	A	B	C
9,500	17,000				17,000			16,800	5,000	5,000	5,000
4,100			13,000			8,900	14,800				4,900
3,150		11,500		11,500		7,900	1,920		4,700	4,700	4,700
		5,300							2,790	2,790	
				4,600					2,100[c]	2,100	2,100
			3,800							2,000	
				800							
16,750	17,000	16,800	16,800	16,900	17,000	16,800	16,720	16,800	16,690	16,590	16,700

[a]The *Bgl* II 4,250-bp digestion fragment was assumed to remain after partial digestion into the 2,700-bp and 1,600-bp bands, and was therefore excluded from the estimate of total genome size.

[b]There were assumed to be two *Hinc* II digestion fragments 2,870 bp in size; therefore, this fragment size was included twice in the estimate of total genome size.

[c]There were assumed to be two *Xba* I fragments 2,100 bp in size in the A digestion pattern, one of which resulted from the digestion of the 4,900-bp fragment of the C pattern into the 2,790-bp and 2,100-bp fragments. In pattern B, one of the 2,100-bp fragments was further digested into a 2,000-bp (and, presumably, a 100-bp) fragment.

Thirteen mtDNA restriction phenotypes were present among the 31 Pacific herring samples. Their distributions among the five statistical areas is shown in Table 6. Depending on phenotype, the 11 restriction enzymes recognized between 44 and 49 cut sites, and sampled approximately 300 bp (1.8%) of the mitochondrial genome. Two of the three most common phenotypes were present in all three major geographic sampling regions— Georgia Strait, west coast of Vancouver Island,

TABLE 6.—Distribution of mitochondrial DNA digestion patterns among five samples of Pacific herring from British Columbia.

Mitochondrial DNA phenotype	Digestion pattern[a]	Georgia Strait			West coast Vancouver Island	North coast
		Area 14	Area 15	Area 17		
1	AAABAAAAA		1	2	1	2
2	AAAAAAAAA	3		1	1	
3	BAAAAAAAA			1	1	3
4	CAABAAAAA	1	1		1	
5	BABACAAAA	1	1			1
6	CAABBAAAB			1		1
7	BBABAAAAA			1		
8	AABAAAAAA		1			
9	CCBDAAABB		1			
10	AAAAABBAA		1			
11	AABACAAAA					1
12	AABCAAAAA					1
13	AABAAAAAB				1	
Total		5	6	6	5	9

[a]Letters represent digestion patterns from Table 5 for *Bgl* II, *Bgl* I, *Hind* III, *Pst* I, *Xba* I, *Hinc* II, *Bam* HI, *Sma* I, *Kpn* I.

TABLE 7.—Pairwise estimates of percentage nucleotide sequence divergence for mitochondrial DNA phenotypes of Pacific herring (above the dashed diagonal; SDs are in parentheses) and minimum numbers of mutational events (below the diagonal).

Pattern	Mitochondrial DNA digestion patterns												
	1	2	3	4	5	6	7	8	9	10	11	12	13
1	–	0.187 (0.187)	0.373 (0.266)	0.187 (0.187)	0.771 (0.394)	0.569 (0.333)	0.365 (0.260)	0.381 (0.272)	1.144 (0.482)	0.568 (0.333)	0.582 (0.341)	0.569 (0.333)	0.582 (0.341)
2	1	–	0.187 (0.187)	0.381 (0.272)	0.582 (0.341)	0.771 (0.394)	0.556 (0.326)	0.192 (0.192)	1.358 (0.532)	0.381 (0.272)	0.390 (0.279)	0.381 (0.272)	0.390 (0.279)
3	2	1	–	0.569 (0.333)	0.381 (0.272)	0.959 (0.440)	0.365 (0.260)	0.381 (0.272)	1.543 (0.568)	0.568 (0.333)	0.582 (0.341)	0.569 (0.333)	0.582 (0.341)
4	1	2	3	–	0.981 (0.451)	0.381 (0.272)	0.556 (0.326)	0.582 (0.341)	0.959 (0.440)	0.771 (0.393)	0.789 (0.403)	0.771 (0.394)	0.789 (0.403)
5	4	3	2	5	–	1.390 (0.545)	0.754 (0.385)	0.390 (0.279)	1.580 (0.582)	0.980 (0.450)	0.196 (0.196)	0.582 (0.341)	0.596 (0.350)
6	3	4	5	2	7	–	0.937 (0.430)	0.981 (0.451)	0.959 (0.440)	1.170 (0.491)	1.200 (0.505)	1.171 (0.493)	0.789 (0.403)
7	2	3	2	3	4	5	–	0.754 (0.385)	1.508 (0.555)	0.936 (0.429)	0.959 (0.440)	0.937 (0.430)	0.959 (0.440)
8	2	1	2	3	2	5	4	–	1.171 (0.493)	0.581 (0.341)	0.196 (0.196)	0.192 (0.192)	0.196 (0.196)
9	6	7	8	5	7	5	8	6	–	1.766 (0.616)	1.390 (0.545)	1.358 (0.532)	0.981 (0.451)
10	3	2	3	4	5	6	5	3	9	–	0.789 (0.403)	0.771 (0.393)	0.789 (0.403)
11	3	2	3	4	1	6	5	1	7	4	–	0.390 (0.279)	0.399 (0.286)
12	3	2	3	4	3	6	5	1	7	4	2	–	0.390 (0.279)
13	3	2	3	4	3	4	5	1	4	4	2	2	–

and north coast of British Columbia—despite the small numbers of samples analyzed from each area (Table 6). Unique phenotypes (those sampled only once) occurred in all three regions in proportion to sample size (four from Georgia Strait, two from the north coast, and one from the west coast of Vancouver Island).

Estimates of the pairwise percentage nucleotide substitution among the 13 mtDNA phenotypes (Table 7) ranged from 0.187±0.187 (SD) to 1.766±0.616 (mean, 0.718±0.380). They were used to construct a dendrogram of genetic relationships (Figure 3). Three of the phenotypes (6, 9, and 10) were quite distinct from each other, and from the remaining 10 phenotypes. Two of these three were unique phenotypes sampled from Georgia Strait, and the third was sampled once from Georgia Strait and once from the north coast (Table 6). The 10 remaining phenotypes formed two major clusters—one contained the three most common phenotypes (1, 2, and 3), and the other was composed primarily of unique phenotypes (Figure 3). Pacific herring samples from all three major geographic regions were included in both clusters.

Discussion

The genetic data collected in this study support earlier findings of little genetic differentiation

among Pacific herring in coastal North American waters south of the Gulf of Alaska (Utter et al. 1984; Grant and Utter 1984). Significant differences among geographic regions of British Columbia occurred only at the $Gap-1$ locus, evidently as the result of a previously documented north–south cline of increasing $Gap-1^{-100}$ frequency (Grant 1981). The $Gap-1$ cline and variability at Ada also resulted in a weak differentiation among the areas sampled within Georgia Strait. However, because allelic frequencies at these two loci differed between years in some areas, stock differentiation based on allozyme frequencies may have been more apparent than real. Mitochondrial DNA analysis revealed that common mitochondrial phenotypes were shared not only among areas within Georgia Strait, but also among the three larger geographic regions. Because of the great diversity of mtDNA phenotypes discovered in Pacific herring, we conclude that sample sizes were much too small to determine if any of the "unique" phenotypes actually were geographically restricted.

Even if Pacific herring stocks defined on a geographic basis at spawning time are genetically discrete populations, the failure to find great genetic differentiation among them is not surprising, given the tremendously large size of spawning

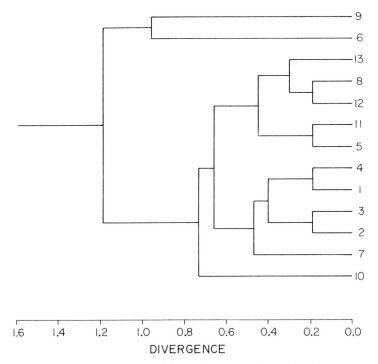

FIGURE 3.—Clustering dendrogram from the unweighted pair-group method with arithmetic averages for mtDNA phenotypes, calculated from the estimated percentage nucleotide sequence divergences presented in Table 7 for Pacific herring in British Columbia.

populations and the relatively short evolutionary time available for population subdivision. Grant and Utter (1984) postulated that any differentiation of herring populations of British Columbia has occurred since the retreat of the last Pleistocene glaciers, 10,000–15,000 years ago. Of greater importance, however, both electrophoretic loci that indicated some degree of stock structure also showed significant annual variability within stocks. Kornfield et al. (1982) documented significant annual variability in allelic frequencies for geographically defined stocks of western Atlantic herring, which they attributed to sampling error.

The significant year-class variation we observed could be the result of sampling variance, or the variability in allelic frequencies could be due either to genetic drift or natural selection. Because herring population sizes are large and individuals are reproductively active over several years, genetic drift seems an unlikely source of the considerable variability observed, unless levels of recruitment vary tremendously among years. It is not possible to determine the effect, if any, of selection on allelic frequencies among Pacific herring. If gene flow among herring in

distinct geographic regions of British Columbia is as great as seems possible, selection must be invoked to explain the persistent cline of Gap-1 allelic frequencies. However, annually variable selective forces that operate within the confines of a statistical area are extremely difficult to identify.

Our analysis of gene diversity also indicated little geographic differentiation. Most of the variation in allelic frequencies was common to all samples, as was demonstrated recently by Grant (1986) for Atlantic as well as Pacific herring. Variation between years was comparable to variation between samples collected within the same statistical area in a single year. This indicates that differentiation occurs on a much finer geographic scale than that defined by statistical area; moreover, area-specific samples in successive years may have been taken from different spawning populations. Hence, some of the temporal variability we observed may have resulted from the failure to accurately resample populations.

The mitochondrial genome evolves more rapidly than the nuclear genome, at least in primates (Brown et al. 1979; Brown 1983), and analyses of mtDNA variability have revealed strong geographic patterns of genetic divergence in mammals (Lansman et al. 1983; Avise 1986) and in freshwater fishes (Avise and Smith 1974; Bermingham and Avise 1986). However, for the Pacific herring sampled in this study, restriction enzyme digestion of mtDNA provided no evidence of genetic differentiation among fish collected from Georgia Strait, the west coast of Vancouver Island, and the north coast of British Columbia. Of the 13 mtDNA phenotypes identified from 31 fish, seven were sampled only once. In no case was a phenotype that occurred more than once confined to a single geographic area. These results are similar to those obtained for Atlantic herring by Kornfield and Bogdanowicz (1987), who found many unique mtDNA phenotypes among samples collected along the coast of Maine and New Brunswick, where common phenotypes tended not to be site-specific. The average level of nucleotide substitution among their Atlantic herring phenotypes (1.66%) was greater than that among Pacific herring phenotypes (0.72%). However, their estimate for Atlantic herring was based solely on results from seven restriction enzymes that produced polymorphic digests, whereas our Pacific herring estimates were based on results from nine polymorphic and two monomorphic enzyme digestion patterns.

Given the large size of spawning populations of Pacific herring, the widespread occurrence of ancestral mtDNA phenotypes and their derivatives is to be expected (Avise 1986; Neigel and Avise 1986). Larger sample sizes of herring from each locality than were used in this study, and perhaps a survey of a greater proportion of the mitochondrial genome, will be required to properly assess the distribution of mtDNA variability within and among geographic areas.

The determination of whatever genetic population structure exists among Pacific herring in British Columbia waters is important to the successful management of this species. Thus, in spite of the accumulating evidence for a lack of genetic differentiation over broad geographic regions, we concur with the recommendation of Kornfield and Bogdanowicz (1987) that the management of herring be continued on the basis that individual spawning groups may represent semidiscrete genetic units. As such, they should be preserved as sources of genetic variability that may be important to the long-term viability of the herring resource.

Acknowledgments

We are grateful to R. W. Armstrong, Fisheries Branch, for coordinating the charter vessel program through which Pacific herring samples for electrophoresis and mtDNA analysis were obtained. The electrophoretic analyses were performed by Helix Biotech Ltd., Richmond, British Columbia, under contract to the Department of Fisheries and Oceans. The restriction endonuclease analysis of mitochondrial DNA was performed by David Banfield and W. Kelly Thomas, Department of Biological Sciences, Simon Fraser University, Burnaby, British Columbia, under contract to the Department of Fisheries and Oceans.

References

Allendorf, F. W., and F. M. Utter. 1979. Population genetics of fish. Pages 407–454 in W. S. Hoar, D. S. Randall, and J. R. Brett, editors. Fish physiology, volume 8. Academic Press, New York.

Andersson, L., N. Ryman, R. Rosenberg, and G. Ståhl. 1981. Genetic variability in Atlantic herring (Clupea harengus harengus): description of protein loci and population data. Hereditas 95:69–78.

Avise, J. D. 1986. Mitochondrial DNA and the evolutionary genetics of higher animals. Philosophical Transactions of the Royal Society of London, B: Biological Sciences 312:325–342.

Avise, J. C., and M. H. Smith. 1974. Biochemical genetics of sunfish. I. Geographic variation and subspecies intergradation in the bluegill, Lepomis macrochirus. Evolution 28:42–56.

Bermingham, E. T., and J. C. Avise. 1986. Molecular zoogeography of freshwater fishes in the southeastern United States. Genetics 113:939–966.

Brown, W. M. 1983. Evolution of animal mitochondrial DNA. Pages 62–88 in M. Nei and R. K. Koehn, editors. Evolution of genes and proteins. Sinauer Associates, Sunderland, Massachusetts.

Brown, W. M., M. George, and A. C. Wilson. 1979. Rapid evolution of animal mitochondrial DNA. Proceedings of the National Academy of Sciences of the USA 76:1967–1971.

Chakraborty, R., M. Haag, N. Ryman, and G. Ståhl. 1982. Hierarchical gene diversity analysis and its application to brown trout population data. Hereditas 27:17–22.

Clayton, J. W., and D. N. Tretiak. 1972. Amine-citrate buffers for pH control in starch gel electrophoresis. Journal of the Fisheries Research Board of Canada 29:1169–1172.

Grant, W. S. 1981. Biochemical genetic variation, population structure, and evolution of Atlantic and Pacific herring. Doctoral dissertation. University of Washington, Seattle.

Grant, W. S. 1984. Biochemical population genetics of Atlantic herring, Clupea harengus. Copeia 1984:357–364.

Grant, W. S. 1986. Biochemical genetic divergence between Atlantic, Clupea harengus, and Pacific, C. pallasi, herring. Copeia 1986:714–719.

Grant, W. S., and F. M. Utter. 1984. Biochemical population genetics of Pacific herring (Clupea pal-

lasi). Canadian Journal of Fisheries and Aquatic Sciences 41:856–864.

Haegele, C. W., and J. F. Schweigert. 1985. Distribution and characteristics of herring spawning grounds and description of spawning behavior. Canadian Journal of Fisheries and Aquatic Sciences 42 (Supplement 1):39–55.

Hourston, A. S. 1982. Homing by Canada's west coast herring to management units and divisions as indicated by tag recoveries. Canadian Journal of Fisheries and Aquatic Sciences 39:1414–1422.

Iles, T. D., and M. Sinclair. 1982. Atlantic herring: stock discreteness and abundance. Science (Washington, D.C.) 215:627–633.

IUBNC (International Union of Biochemistry, Nomenclature Committee). 1984. Enzyme nomenclature 1984. Academic Press, Orlando, Florida.

King, D. P. F. 1985. Morphological and meristic differences among spawning aggregations of north-east Atlantic herring *Clupea harengus* L. Journal of Fish Biology 26:591–607.

Kornfield, I., and S. M. Bogdanowicz. 1987. Differentiation of mitochondrial DNA in Atlantic herring, *Clupea harengus.* U.S. National Marine Fisheries Service Fishery Bulletin 85:561–568.

Kornfield, I., B. D. Sidell, and P. S. Gagnon. 1982. Stock definition in Atlantic herring *(Clupea harengus harengus):* genetic evidence for discrete fall and spring spawning populations. Canadian Journal of Fisheries and Aquatic Sciences 39:1610–1621.

Lansman, R. A., J. C. Avise, C. F. Aquadro, J. F. Shapiro, and S. W. Daniel. 1983. Extensive genetic variation in mitochondrial DNAs among geographic populations of the deer mouse, *Peromyscus maniculatus.* Evolution 37:1–16.

Lansman, R. A., R. O. Shade, J. F. Shapiro, and J. C. Avise. 1981. The use of restriction endonucleases to measure mitochondrial DNA sequence relatedness in natural populations. III. Techniques and potential applications. Journal of Molecular Evolution 17:214–226.

Markert, C. L., and I. Faulhaber. 1965. Lactate dehydrogenase isozyme patterns of fish. Journal of Experimental Zoology 159:319–322.

Meng, H. J., and M. Stocker. 1984. An evaluation of morphometrics and meristics for stock separation of Pacific herring *(Clupea harengus pallasi).* Canadian Journal of Fisheries and Aquatic Sciences 41:414–422.

Nei, M. 1978. Estimation of average heterozygosity and genetic distance from a small number of individuals. Genetics 89:583–590.

Nei, M., and F. Tajima. 1983. Maximum likelihood estimation of the number of nucleotide substitutions from restriction site data. Genetics 105:207–215.

Neigel, J. E., and J. C. Avise. 1986. Phylogenetic relationships of mitochondrial DNA under various demographic models of speciation. Pages 515–534 *in* S. Karlin and E. Nevo, editors. Evolutionary processes and theory. Academic Press, New York.

Ricker, W. E. 1973. Two mechanisms that make it impossible to maintain peak period yields from stocks of Pacific salmon and other fishes. Journal of the Fisheries Research Board of Canada 30:1275–1286.

Ridgway, G. J., S. W. Sherburne, and R. D. Lewis. 1970. Polymorphism in the esterases of Atlantic herring. Transactions of the American Fisheries Society 99:147–151.

Ryman, N., U. Lagercrantz, L. Andersson, R. Chakraborty, and R. Rosenberg. 1984. Lack of correspondence between genetic and morphologic variability patterns in Atlantic herring *(Clupea harengus).* Heredity 53:687–704.

Schweigert, J. F. 1981. Pattern recognition of morphometric and meristic characters as a basis for herring stock identification. Canadian Technical Report of Fisheries and Aquatic Sciences 1021.

Schweigert, J. F. 1990. Comparison of morphometric and meristic data against truss networks for describing Pacific herring stocks. American Fisheries Society Symposium 7:47–62.

Sinclair, M., V. C. Anthony, T. D. Iles, and R. N. O'Boyle. 1985. Stock assessment problems in Atlantic herring *(Clupea harengus)* in the northwest Atlantic. Canadian Journal of Fisheries and Aquatic Sciences 42:888–898.

Sinclair, M., and M. J. Tremblay. 1984. Timing of spawning of Atlantic herring *(Clupea harengus harengus)* populations and the match–mismatch theory. Canadian Journal of Fisheries and Aquatic Sciences 41:1055–1065.

Smith, P. J., and A. Jamieson. 1986. Stock discreteness in herrings: a conceptual revolution. Fisheries Research (Amsterdam)4:223–234.

Sneath, P. H., and R. R. Sokal. 1973. Numerical taxonomy. Freeman, San Francisco.

Sokal, R. R., and F. J. Rohlf. 1969. Biometry. Freeman, San Francisco.

Swofford, D. L., and R. B. Selander. 1981. BIOSYS-1: a FORTRAN program for the comprehensive analysis of electrophoretic data in population genetics and systematics. Journal of Heredity 72:281–283.

Tester, A. L. 1937. Populations of herring *(Clupea pallasi)* in the coastal waters of British Columbia. Journal of the Biological Board of Canada 3:108–144.

Tester, A. L. 1949. Populations of herring along the west coast of Vancouver Island on the basis of mean vertebral number, with a critique of the method. Journal of the Fisheries Research Board of Canada 7:403–420.

Utter, F. M., H. O. Hodgins, and F. W. Allendorf. 1974. Biochemical genetic studies of fishes: potentialities and limitations. Pages 213–237 *in* D. Malins, editor. Biochemical and biophysical perspectives in marine biology, volume 1. Academic Press, San Francisco.

Wheeler, J. P., and G. H. Winters. 1984. Homing of Atlantic herring *(Clupea harengus harengus)* in Newfoundland waters as indicated by tagging data. Canadian Journal of Fisheries and Aquatic Sciences 41:108–117.

American Fisheries Society Symposium 7:470–474, 1990

Mitochondrial DNA Variability in Brook Trout Populations from Western Maryland

J. M. Quattro[1] and R. P. Morgan II

University of Maryland, Center for Environmental and Estuarine Studies
Appalachian Environmental Laboratory, Gunter Hall, Frostburg, Maryland 21532, USA

R. W. Chapman

The Johns Hopkins University, Chesapeake Bay Institute, Shady Side, Maryland 20764, USA

Abstract.—Restriction endonuclease analysis of purified mitochondrial DNA (mtDNA) was used to determine levels of sequence heterogeneity within and among local populations of brook trout *Salvelinus fontinalis* from western Maryland. Cleavage of the brook trout mtDNA with five restriction endonucleases produced 14 reproducible fragments. The brook trout mtDNA genome was estimated to be 16,540 ± 99 (SD) base pairs long. Eighty-four nucleotides, or approximately 0.51% of the brook trout mtDNA genome, were sampled in 143 individuals. Our results provided no genetic basis for differentiating populations within each of the two drainages studied. The restriction endonuclease *Pst* I, however, produced a fixed difference in polymorphic fragment patterns for brook trout of the Ohio River drainage and the Atlantic slope drainage, and may have value as a genetic mark. Divergence in mtDNA sequence between these allopatric populations was estimated at 0.6% ± 0.6; thus it appears they had a common female ancestor between 200,000 and 600,000 years ago.

The brook trout *Salvelinus fontinalis* is the native trout of the Appalachians, although its major distribution is throughout eastern Canada and Hudson Bay drainages to New England, the Great Lakes drainage, and the upper Mississippi basin (Scott and Crossman 1973; Hendricks 1980). Its historical distribution has been radically altered by salmonids introduced from western North America. The anadromous and freshwater forms of this species are taxonomically indistinct (Hendricks 1980).

The original distribution of brook trout in western Maryland are poorly known. Brook trout prefer first- and second-order streams with high gradients, but we estimate that only 5–10% of Maryland waters theoretically capable of supporting these fish now have self-sustaining populations. Siltation from farming, lumbering, and urbanization, as well as the loss of riparian vegetation, led to the eradication of many of the state's brook trout populations. In the western portion of the state, changes in water quality caused primarily by drainage from coal mining and agricultural runoff also eliminated stocks. Now, by our conservative estimate, the native brook trout of Maryland consist of 60 fragmented populations.

Two or three of these may represent the remains of coastal and perhaps anadromous populations, because Maryland was thought to be of the southernmost historical limit of anadromous salmonids on the Atlantic coast (Scott and Crossman 1973).

The identification of discrete stocks is of primary importance to the management and rehabilitation of both recreational and commercial fisheries (Ihssen et al. 1981; Larkin 1981). Ignorance of the composition of a stock can result in dramatic changes in its biological features (Altukhov 1981). There is no perfect technique for identifying discrete stocks of fishes, but restriction analysis of mitochondrial DNA (mtDNA) closely approximates an ideal method for the quantification of genetic differences among conspecific populations (Ferris and Berg 1987). Several recent studies have demonstrated sufficient levels of intra- and interspecific variation in mtDNA in fishes to warrant its use as a stock-specific mark, although others have failed (e.g., Avise et al. 1984; Avise and Saunders 1984; Berg and Ferris 1984; Graves et al. 1984; Gyllensten et al. 1985; Wilson et al. 1985; Bermingham and Avise 1986; Birt et al. 1986; Thomas et al. 1986; Avise and Vrijenhoek 1987; Gyllensten and Wilson 1987).

Unique properties of mtDNA make it an appealing tool with which to study the population structure of organisms at or below the species level. In contrast to nuclear DNA, the circular,

---

[1]Present address: Center for Theoretical and Applied Genetics, Cook College/Rutgers University, Post Office Box 231, New Brunswick, New Jersey 08903, USA.

duplex, mtDNA molecule appears to be maternally transmitted and does not recombine during meiotic divisions (Brown et al. 1979; Avise and Vrijenhoek 1987). In mammalian species, it evolves much more rapidly than homologous, single-copy, nuclear DNA (Brown et al. 1979), and it lacks such complicating features found in nuclear DNA as spacer sequences, intervening sequences, and classes of repetitive DNA (Brown 1983).

The purpose of this study was to evaluate the utility of mtDNA markers for identifying discrete populations of brook trout in western Maryland. The initial study focused on brook trout along the eastern continental divide because of their vulnerability to coal mining and lumbering, as well as to increasing recreation and land-use practices. The number of populations in this area was estimated at 22 including some of the largest continuous populations of brook trout in Maryland, both in the Ohio and in the Potomac river drainages. Levels of intrapopulational mtDNA heterogeneity were also determined. In gathering this baseline information, we hoped to provide insight into the status of brook trout populations in western Maryland, and to help hatchery managers select appropriate brood stock for restoration projects.

Methods

Using a Coffelt model BP-3 backpack electroshocker, we collected 143 gravid female brook trout in the fall of 1985 and 1987 from 10 sampling sites in western Maryland (Figure 1). Three samples came from the Ohio River drainage and the remaining seven from the Atlantic slope drainage. Sampling areas were selected based on a Maryland Department of Natural Resources survey of "natural" trout populations (Hughes et al. 1980), as well as on a review of stocking records of brook trout in Allegany and Garrett counties. Samples were immediately packed in wet ice and returned to the laboratory, where we isolated mtDNA from brook trout eggs by means of the rapid technique as outlined by Chapman and Powers (1984).

Purified mtDNA from each specimen was digested with five restriction endonucleases according to manufacturer's instructions (Bethesda Research Laboratories). The endonucleases (and their recognition sites) were *Eco* RI (5'-GAATTC-3'), *Hind* III (5'-AAGCTT-3'), *Pst* I (5'-CTG-CAG-3'), *Sst* I (5'-GAGCTC-3'), and *Xba* I (5'-TCTAGA-3'); the letters stand for base sequences (A, adenine; C, cytosine; G, guanine; T, thymine). Restriction fragments produced by digestion were electrophoretically separated in 0.8% and

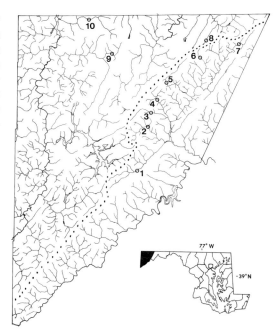

FIGURE 1.—Sample locations of brook trout in western Maryland. The eastern continental divide is indicated as a dotted line. The 10 sampling stations are designated by numbers: 1—Crabtree Creek, 2—Middle Fork, 3—Monroe Run, 4—Big Run, 5—Poplar Lick, 6—Blue Lick, 7—Upper Savage River, 8—Two Mile Creek, 9—Little Bear Creek, and 10—Mill Run.

1.0% agarose gels. We used a tris–borate buffer system and a one-kilobase DNA ladder (Bethesda Research Laboratories) as a molecular weight standard. After electrophoresis, gels were immersed in a solution of ethidium bromide (0.5 µg/mL) in TBE buffer (89 mM tris, 89 mM boric acid, 2.5 mM EDTA, pH 8.3; Maniatis et al. 1982), and the fragments were visualized by transillumination at 340 nm. Gels were photographed under ultraviolet illumination with a Polaroid MP-4 camera equipped with a Kodak 23A orange filter.

Migration distances of restriction fragments were consistently measured from the photographs as the distance from the sample origins to the middle of the bands. Molecular weights of restriction fragments were estimated by regression equations describing the standard DNA markers (molecular weight versus migration distance; $\log_{10}$ scales). No attempt was made to score restriction fragments smaller than 500 base pairs. Fragments of coincident mobility were assumed to be homologous. Each unique single-endonuclease digestion profile was designated by an upper case letter (Table 1). Each fish was assigned a five-letter

TABLE 1.—Restriction endonuclease digestion profiles for brook trout from western Maryland, which indicate two genotypes, designated A and B.

Restriction endonuclease	Digestion profile (base pairs)	Genotype designation
Eco RI	8 , 495	
	8 , 173	
Sum	16 , 668	A
Hind III	5 , 079	
	3 , 735	
	2 , 962	
	2 , 394	
	1 , 807	
Sum	15 , 977	A
Pst I-A	12 , 819	
	3 , 704	
Sum	16 , 523	A
Pst I-B	16 , 540	
Sum	16 , 540	B
Sst I	16 , 540	
Sum	16 , 540	A
Xba I	5 , 748	
	4 , 760	
	3 , 739	
	2 , 181	
Sum	16 , 428	A

composite code describing its five observed digestion phenotypes (Table 2). Since our restriction enzymes recognized only unique hexanucleotide sequences, we pooled data and calculated pairwise estimates of percent sequence divergence using the comparative fragment method of Nei and Li (1979).

Results

Our procedures yielded enough purified mtDNA from the ovarian tissue of an average

TABLE 2.—Composite mitochondrial DNA genotypes observed in populations of brook trout from western Maryland.

Drainage	Composite genotype[a]	Number of fragments scored	Population (sample size)
Atlantic slope	AAAAA	14	Big Run (23)
			Blue Lick (21)
			Crabtree Creek (17)
			Middle Fork (13)
			Monroe Run (8)
			Poplar Lick (28)
			Upper Savage (2)
Ohio River	AABAA	13	Little Bear (20)
			Mill Run (5)
			Two Mile (6)

[a] Letters refer to observed digestion profiles for the following restriction endonucleases, respectively: Eco RI, Hind III, Pst I, Sst I, Xba I.

brook trout (approximately 15 cm) for six or seven restriction endonuclease digestions. Although this yield may not be sufficient for an extensive restriction endonuclease survey, it is more than adequate when several diagnostic restriction enzymes are to be used in a qualitative survey. We found that by using the rapid isolation protocol of Chapman and Powers (1984), we needed only 20–25 unfertilized mature eggs per restriction digest. That amount can be stripped safely from individual females in the field or at the hatchery.

The five restriction enzymes used in the subsequent analysis produced 14 detectable restriction fragments (Table 1). Our techniques were not sensitive enough to detect additions, deletions, or rearrangements in the brook trout mtDNA genome.

The size of the brook trout mtDNA molecule, estimated from the average of the sums of the fragment sizes produced by the three enzymes that cleaved the brook trout mtDNA molecule more than once, was approximately $16,540 \pm 99$ (SD) base pairs in length. (Hind III was not included in this estimated because of inadequate resolution of fragments less than 500 base pairs.) We sampled 84 nucleotides (14 fragments × 6 bases) per individual mtDNA, which represented slightly more than 0.5% of the brook trout mitochondrial genome.

Of the five restriction endonucleases used in this study, only Pst I produced a polymorphic restriction fragment profile. We attributed the difference between the Pst I-A and Pst I-B digestion phenotypes to the loss or gain of a single restriction site (confirmed by a Pst I–Eco RI double digestion), which we assumed to result from a single nucleotide substitution. None of the other four restriction endonucleases revealed any inter- or intrapopulational mtDNA heterogeneity.

Composite mtDNA genotypes indicated there were two mtDNA clones among the 143 brook trout assayed in this study (Table 2). Sequence divergence between the two clones was estimated to be approximately 0.6%. These clones concorded with the major drainage patterns of western Maryland. Digestion profile Pst I-A was indicative of all brook trout collected in the Atlantic slope drainage, and the Pst I-B digestion profile marked all brook trout collected in the Ohio River drainage.

Discussion

We assayed 10 putative populations of brook trout for genetic heterogeneity, using restriction endonuclease analysis of mtDNA. Our primary goals were to identify mtDNA fragment polymorphisms that would distinguish natural stocks of brook trout along the continental divide of western Maryland, and to measure the level of genetic heterogeneity within stream systems. Such data should be useful in the management and restoration of brook trout populations in western Maryland, and should provide analytical protocols for use in eastern Maryland where remmant anadromous and Piedmont populations occur. Because of different selection regimes on each side of the divide, it is important to develop genetically appropriate stocks specific for either the Ohio River or Atlantic slope drainages (Meffe 1986).

Composite genotypes of fragment patterns allowed us to reduce the 10 populations to two distinct matriarchal assemblages, or mtDNA clonal lineages, that fell on either side of the major geographic feature of the area—the eastern continental divide. No intra- or interpopulational sequence divergence was detected within a drainage system by the analysis of mtDNA, perhaps because only a few restriction endonucleases were employed on a large number of fish from a small area.

There are at least two possible explanations for this low level of genetic variation. Small population size, coupled with migration among populations, may not allow for the retention of variation. Alternatively, sequence heterogeneity within populations may exist but was not revealed by the restriction endonucleases we used. Standard electrophoretic data also could be used to assess variation in brook trout, but interpretation of isozyme banding patterns would be difficult because of tetraploidy (Stoneking et al. 1981).

The estimate of percent sequence divergence for eastern- and western-slope brook trout was low (0.6% ± 0.6), indicating that the fish had a common female ancestor between 200,000 and 600,000 years ago. This figure is based on assumptions that brook trout mtDNA mutates at a rate similar to that of mammalian species—approximately 2% per million years (Brown et al. 1979)—and that all accumulated sequence divergence occurred after the common lineage split. An extensive survey of southeastern fishes (Bermingham and Avise 1986) provided estimates of average mtDNA divergence times ranging from 105,000 to 4,115,000 years, depending on the species and the mtDNA clones compared.

The size of the brook trout mtDNA obtained in this study (16,540 ± 99 base pairs) agreed with the estimate of 16,700 base pairs for mtDNAs of chinook salmon *Oncorhynchus tshawytscha*, rainbow trout *O. mykiss* (formerly *Salmo gairdneri*), brown trout *Salmo trutta*, and brook trout (Berg and Ferris 1984); 16,500 base pairs for mtDNAs of steelhead (anadromous rainbow trout) and cutthroat trout *O. clarki* (formerly *Salmo*) (Wilson et al. 1985); and 16,700 base pairs for Atlantic salmon *Salmo salar* (Birt et al. 1986). Identical *Eco* RI, *Hind* III and *Xba* I digestion profiles were observed for a brook trout from a California hatchery by Berg and Ferris (1984). However, their observed *Pst* I digestion profile was perhaps three and four restriction site losses different from the *Pst* I-A and *Pst* I-B digestion phenotypes observed in this study. Similarly, *Pst* I-B and *Xba* I digestion profiles identical to ours were observed for three Newfoundland brook trout by Gyllensten and Wilson (1987). However, their *Hind* III digestion profile diverged from the phenotype observed in this study, although the absolute number of mutational events relating the two patterns is unclear.

Results of this study, when coupled with the work of Berg and Ferris (1984) and Gyllensten and Wilson (1987), demonstrate extensive levels of mtDNA restriction site polymorphism among populations of brook trout across broad geographic areas. Although the studies overlapped by only four restriction endonucleases, they identified four distinct brook trout clones in North America. Only two restriction endonucleases, *Pst* I and *Hind* III, were needed to unambiguously identify these four stocks. Levels of mtDNA variation uncovered in these three studies suggest that mtDNA markers have a role in the management of natural brook trout stocks in North America.

It is common management practice to augment or replace depleted natural stocks of brook trout with hatchery strains. However, little regard is given to the potential genetic impact of hatchery-reared fish on natural populations. By using fixed mtDNA markers in hatchery fish as "genetic brands," useful information on the impact of introduced fish on natural populations could be collected (Berg and Ferris 1984), especially if more restriction endonucleases were used to develop a comprehensive picture of the mtDNA genotype. Careful consideration of the geographic patterns of mtDNA genotypes as they relate to geological events (Bermingham and Avise 1986) may also provide important data regarding the basis of the genetic structure of natural populations. Taken together, this new information can lead to sound management decisions that preserve genetic variation in brook trout.

Acknowledgments

We thank D. R. Kenney, P. G. Piavis, and L. F. Vukovich for their assistance in field collections. This manuscript was submitted in partial fulfillment of the requirements for the degree of Masters of Science in Fisheries Management, Frostburg State University. This is contribution 1912 from the Appalachian Environmental Laboratory, University of Maryland Center for Environmental and Estuarine Studies.

References

Altukhov, Y. P. 1981. The stock concept from the viewpoint of population genetics. Canadian Journal of Fisheries and Aquatic Sciences 38:1523–1538.

Avise, J. C., E. Bermingham, L. G. Kessler, and N. C. Saunders. 1984. Characterization of mitochondrial DNA variability in a hybrid swarm between subspecies of bluegill sunfish (*Lepomis macrochirus*). Evolution 38:931–941.

Avise, J. C., and N. C. Saunders. 1984. Hybridization and introgression among species of sunfish (*Lepomis*): analysis by mitochondrial DNA and allozyme markers. Genetics 108:237–255.

Avise, J. C., and R. C. Vrijenhoek. 1987. Mode of inheritance and variation of mitochondrial DNA in hybridogenetic fishes of the genus *Poeciliopsis*. Molecular Biology and Evolution 4:514–525.

Berg, W. J., and S. D. Ferris. 1984. Restriction endonuclease analysis of salmonid mitochondrial DNA. Canadian Journal of Fisheries and Aquatic Sciences 41:1041–1047.

Bermingham, E., and J. C. Avise. 1986. Molecular zoogeography of freshwater fishes in the southeastern United States. Genetics 113:939–965.

Birt, T. P., J. M. Green, and W. S. Davidson. 1986. Analysis of mitochondrial DNA in allopatric anadromous and nonanadromous Atlantic salmon, *Salmo salar*. Canadian Journal of Zoology 64:118–120.

Brown, W. M. 1983. Evolution of animal mitochondrial DNA. Pages 62–88 *in* M. Nei and R. K. Koehn, editors. Evolution of genes and proteins. Sinauer Associates, Sunderland, Massachusetts.

Brown, W. M., M. George, Jr., and A. C. Wilson. 1979. Rapid evolution of animal mitochondrial DNA. Proceedings of the National Academy of Sciences of the USA 76:1967–1971.

Chapman, R. W., and D. A. Powers. 1984. A method for the rapid isolation of mtDNA from fishes. University of Maryland Sea Grant Program, Technical Report UM-SG-TS-84-05, College Park.

Ferris, S. D., and W. J. Berg. 1987. The utility of mitochondrial DNA in fish genetics and fishery management. Pages 277–299 *in* N. Ryman and F. Utter, editors. Population genetics and fishery management. University of Washington Press, Seattle.

Graves, J. E., S. D. Ferris, and A. E. Dizon. 1984. Close genetic similarity of Atlantic and Pacific skipjack tuna (*Katsuwonus pelamis*) demonstrated with restriction endonuclease analysis of mitochondrial DNA. Marine Biology 79:315–319.

Gyllensten, U., R. F. Leary, F. W. Allendorf, and A. C. Wilson. 1985. Introgression between two cutthroat trout subspecies with substantial karyotypic, nuclear, and mitochondrial genomic divergence. Genetics 11:905–915.

Gyllensten, U., and A. C. Wilson. 1987. Mitochondrial DNA of salmonids: Inter- and intraspecific variability detected with restriction enzymes. Pages 301–317 *in* N. Ryman and F. Utter, editors. Population genetics and fishery management. University of Washington Press, Seattle.

Hendricks, M. L. 1980. *Salvelinus fontinalis* (Mitchill), brook trout. Page 114 *in* D. S. Lee, C. R. Gilbert, C. H. Hocutt, R. E. Jenkins, D. E. McAllister, and J. R. Stauffer, editors. Atlas of North American freshwater fishes. North Carolina State Museum of Natural History, Raleigh.

Hughes, D. M., H. J. Stinefelt, and S. E. Rivers. 1980. Survey and inventory of natural trout waters. Maryland Department of Natural Resources, Federal Aid in Fish Restoration, Project F-26-R, Final Report, Annapolis.

Ihssen, P. E., H. E. Booke, J. M. Casselman, J. M. McGlade, N. R. Payne, and F. M. Utter. 1981. Stock identification: materials and methods. Canadian Journal of Fisheries and Aquatic Sciences 38:1838–1855.

Larkin, P. A. 1981. A perspective on population genetics and salmon management. Canadian Journal of Fisheries and Aquatic Sciences 38:1469–1475.

Maniatis, T., E. F. Fritsch, and J. Sambrook. 1982. Molecular cloning: a laboratory manual. Cold Spring Harbor Laboratory, Cold Spring Harbor, New York.

Meffe, G. K. 1986. Conservation genetics and the management of endangered fishes. Fisheries (Bethesda) 11(1):14–23.

Nei, M., and W.-H. Li. 1979. Mathematical model for studying genetic variation in terms of restriction endonucleases. Proceedings of the National Academy of Sciences of the USA 76:5269–5273.

Scott, W. B., and E. J. Crossman. 1973. Freshwater fishes of Canada. Fisheries Research Board of Canada Bulletin 184.

Stoneking, M., D. J. Wagner, and A. C. Hilderrand. 1981. Genetic evidence suggesting subspecific differences between northern and southern populations of brook trout (*Salvelinus fontinalis*). Copeia 1981:810–819.

Thomas, W. K., R. E. Withler, and A. T. Beckenbach. 1986. Analysis of Pacific salmonid evolution. Canadian Journal of Zoology 64:1058–1064.

Wilson, G. M., W. K. Thomas, and A. T. Beckenbach. 1985. Intra- and interspecific mitochondrial DNA sequence divergence in *Salmo*: rainbow, steelhead, and cutthroat trout. Canadian Journal of Zoology 63:2088–2094.

American Fisheries Society Symposium 7:475–491, 1990

Restriction Endonuclease Analysis of Striped Bass Mitochondrial DNA: The Atlantic Coastal Migratory Stock

Isaac I. Wirgin,[1,2] Peter Silverstein, and Joseph Grossfield[3]

Department of Biology, The City College of the City University of New York
138th Street at Convent Avenue, New York, New York 10031, USA

Abstract.—We investigated mitochondrial DNA (mtDNA) diversity in striped bass *Morone saxatilis* of the Atlantic coastal migratory stock. Our goals were to identify distinct Mendelian populations and to determine if the ancestry of individual fish collected in coastal waters could unequivocally be assigned to geographic origin based on mtDNA genotypes. We examined fish from several Hudson River sites in New York, five Chesapeake Bay spawning locales in Maryland and Virginia, and the Roanoke River in North Carolina. Subadult striped bass were also collected from Holyoke, Massachusetts, on the Connecticut River. In addition, a coastal collection of adults was obtained from Montauk Point, Long Island, New York. Highly purified mtDNA was prepared by cesium chloride density-gradient ultracentrifugation and digested with a battery of 12 restriction endonucleases, including four-, five-, and six-base-cutting enzymes. Mitochondrial DNA fragments were visualized by either autoradiography of end-labeled sequences or Southern blot analysis. Mitochondrial DNA sequence diversity ($p = 0.0004$) was among the lowest reported for any animal species. Five mtDNA "major length variants" were identified, three of them common within the populations examined. Significant differences in the frequencies of these major-length genotypes were found between composite Hudson River and Chesapeake Bay spawning aggregations and between Chesapeake Bay and Roanoke River collections. No difference in the frequencies of these characters was observed between Hudson and Roanoke river fish. "Minor length variants" and base substitutions, which were detected only by the use of four-base-cutting enzymes, were found within the Chesapeake Bay and the Roanoke River groups. These innate markers permitted identification of ancestry of individual Chesapeake and Roanoke fish. One minor length variant was found in moderate frequency at two upper Chesapeake Bay spawning locales. Differences in the frequencies of these characters allowed discrimination between upper and lower Chesapeake Bay fish. Heteroplasmy for length variation was observed in approximately 20% of striped bass contributing to the coastal stock.

The endemic range of striped bass *Morone saxatilis* extends along the Atlantic coast from the St. Lawrence River in Canada (Magnin and Beaulieu 1967) to the St. Johns River in northern Florida (McLane 1958). At the limits of this range (southern North Carolina to Florida and in the Canadian Maritime provinces), striped bass are believed to prefer rivers, and they exhibit only localized mobility within natal streams (Beaulieu 1962; Dudley et al. 1977). Within the center of the Atlantic range, from northern North Carolina to at least Maine (Merriman 1941), adult striped bass make seasonal feeding migrations. These coastal migrants go north in the spring and south in the fall. Most fish in the coastal migration are over 2 years of age (Kohlenstein 1981), and at times a preponderance of females has been observed (Schaefer 1968; Oviatt 1977).

Historically, this coastal group of striped bass has supported lucrative commercial and recreational fisheries within spawning rivers and at multiple coastal sites (Boreman 1986). Although these fisheries have been subject to dramatic population fluctuations (Koo 1970; Van Winkle et al. 1979), the past decade has witnessed a particularly severe decline in landings (Boreman and Austin 1985). Relying on estuarine returns of striped bass tagged in coastal waters, Merriman (1941) theorized that the coastal fishery was supported by migrants spawned in several major estuarine systems, including various tributaries of Chesapeake Bay and the Hudson, Roanoke, and Delaware rivers. Today, it is believed that the Hudson and Chesapeake are the predominant contributors to the coastal fishery (Waldman et al. 1988), and that the relative contributions of these two spawning systems vary from year to year, depending on relative year-class strengths. It has

---

[1]Present address: Institute of Environmental Medicine, New York University Medical Center, Long Meadow Road, Tuxedo, New York 10987, USA.
[2]To whom reprint requests should be addressed.
[3]Deceased.

been demonstrated that the Chesapeake and Hudson systems experience asynchronous year-class strengths (Young 1984; Boreman and Austin 1985).

The Atlantic coastal striped bass fishery is managed currently as a single stock. Hypothesized stock differentiation suggests that this is inappropriate. A detailed knowledge of population partitioning would permit management on the basis of individual spawning systems rather than at the species level. Such an approach to management could be based on the discovery of innate tags that identify population ancestry of individual fish taken in the mixed fishery. Given adequate sampling, these innate tags would permit an estimate of the relative contribution of individual spawning systems to the coastal fishery within any spatial or temporal window. The tags might also reveal the ancestry of striped bass collected outside the traditional spatial or temporal confines of the Atlantic coastal stock, such as the overwintering contingents in the upper Bay of Fundy in Nova Scotia (Rulifson 1987) or Narragansett Bay in Rhode Island (O'Brien 1977).

Attempts to identify stocks of striped bass have drawn on mark–recapture methods (Merriman 1941; Clark 1968), meristic (Lewis 1957; Raney 1957) and morphometric analysis (Lund 1957), and isoelectric focusing of eye lens proteins (Fabrizio 1987a). (For a review of striped bass stock-identification studies, see Waldman et al. 1988.) With the exception of the initial effort by Morgan et al. (1973), genetic studies of polymorphism at protein loci have been unable to detect sufficient genetic variability to distinguish hypothesized populations (Hitron 1974; Otto 1975; Grove et al. 1976; Sidell et al. 1980; Wirgin 1987). Only small differences could be found among Hudson, Chesapeake, and Roanoke fish, or among fish of the various tributaries of Chesapeake Bay. Freshwater populations of striped bass are also extremely monomorphic (Rogier et al. 1985). In an isoelectric focusing study of striped bass, Fabrizio (1987a, 1987b) did not find differences in the genetically determined primary structure of eye lens proteins; however, variation in the concentration of individual eye lens proteins did permit identification of individual stocks. During research reported in this paper, we examined whether genetic variability in mitochondrial DNA (mtDNA) can indicate stock differentiation, and whether such variation produces markers that can be used to identify the ancestry of individual fish.

Mitochondrial DNA is a double-stranded, circular molecule found in multiple copies within the cells of eukaryotic organisms. It is relatively small, measuring 16,000–19,500 base pairs (bp). Usually all copies of mtDNA within a single individual are identical (homoplasmic); however, some invertebrate and lower vertebrate species have exhibited mtDNA heterogeneity (heteroplasmy) within the cellular population of a single individual (Densmore et al. 1985; Harrison et al. 1985; Wirgin et al. 1989). The mitochondrial DNA molecule of conspecific individuals may be screened for base-sequence differences by the use of restriction enzymes. These enzymes recognize and cleave specific four- to eight-base-pair sequences in the mtDNA molecule to produce mtDNA fragments. The number and size of these fragments may be quantified electrophoretically. Base-sequence diversity in mtDNA is evidenced by differences in the number of mtDNA fragments visualized following electrophoretic separation. These differences probably reflect single-point mutations in the base sequence recognized by the restriction enzymes. Diversity in the overall length of the mtDNA molecule (which consists of repeated mtDNA sequences) has been demonstrated in species of parthenogenetic lizards (Densmore et al. 1985; Moritz and Brown 1987), newts (Wallis 1987), and some fish (Bermingham et al. 1986; Wirgin et al. 1989), but the mechanism whereby these length variants are generated has yet to be demonstrated.

Several characteristics of mtDNA make it a potentially distinctive genetic mark. Mitochondrial DNA is believed to be maternally inherited (Lansman et al. 1983; Gyllensten et al. 1985); thus, barring further mutations, all descendants of a single ancestral female exhibit identical mitochondrial genomes. In this absence of recombination, it is easy to assign maternal ancestry of individual fish from gel banding patterns. Change of base sequences in the mitochondrial genome of primates is 5 to 10 times more rapid than that of nuclear DNA (Brown et al. 1979). As a result, mtDNA may be an excellent tool for exploring and quantifying genetic distances between populations that have recently become reproductively isolated, that have recently gone through population bottlenecks, or that have experienced common selective pressures on the products of the nuclear genome.

High levels of mtDNA sequence differentiation have been observed in populations of some freshwater fishes (Avise et al. 1984; Bermingham and

TABLE 1.—Striped bass samples subjected to mitochondrial DNA analysis.

Sample site	Date	N	Sex[a]	Standard length (mm)
Chesapeake Bay, Maryland–Virginia				
Chesapeake and Delaware Canal	May 8, 1984	13	13 M	230–281
Upper Bay (Aberdeen)	May 10, 1983	10	10 M	271–417
Choptank River	May 9, 1983	1	1 M	270
Choptank River	Apr 18, 1984	13	13 M	245–319
Rappahannock	Apr 25, 1984	13	13 M	216–394
James River	Apr 26, 1984	12	12 M	205–344
Roanoke River, North Carolina				
Weldon	June 1, 1983	9	9 M	369–479
Weldon	May 17, 1984	14	14 M	279–355
Hudson River, New York–New Jersey				
Upper Hudson				
Claverack	Jun 3, 1984	6	6 M	429–525
Troy	July 15, 1984	9		213–410
Mid-Hudson				
Croton Bay	May 21, 1983	2	1 M, 1 F	490–639
Croton Bay	Mar 21, 1983	5	2 M	320–330
Cornwall	Jun 5, 1983	4	4 M	450–539
Cornwall	May 31, 1984	6	5 M, 1 F	392–552
Danskammer	Jun 27, 1983	2		207–210
Lower Hudson				
Hoboken	Mar 3, 1983	15	4 M	114–212
Hoboken	Mar 17, 1983	1	1 F	306
Westway	Dec 17, 1983	14	6 M, 1 F	185–290
George Washington Bridge	Mar 3, 1983	5	3 M	205–269
Upper Manhattan	Jun 20, 1983	4		137–220
Upper Manhattan	Jul 18, 1984	5	1 M	162–255
Connecticut River, Massachusetts				
Holyoke	Jun 22, 1983	13	3 M	205–255
Coastal collection, New York				
Montauk	Nov 8, 1984	12	Unknown	501–850[b]
Montauk	Nov 11, 1984	12	Unknown	546–838[b]

[a] Sex was not determined for all fish, so the sum of males (M) and females (F) may not equal the total number of fish for a site.
[b] Total length (mm).

Avise 1986). In general, anadromous, catadromous, and marine fish have shown much lower levels of mtDNA diversity (Graves et al. 1984; Avise et al. 1986, 1987) and a lack of geographic cohesion of mtDNA genotypes (Kornfield and Bogdanowicz 1987). We have examined populations of striped bass in the southeastern USA and have found significant differences in the frequency of mtDNA genotypes between Gulf of Mexico and Atlantic populations (Wirgin et al. 1989). Based on the presence of unique mtDNA sequences, we concluded that the maternal descendants of an original Gulf population still existed in the Apalachicola River system in northwest Florida.

Our objective in the study reported here was to determine if the mtDNA of striped bass from Atlantic coast populations has sufficient diversity to serve as an effective tool in stock discrimination. We asked whether statistically significant stocks of striped bass could be identified, and

whether mtDNA genotypes permitted the identification of the ancestry of individuals from different spawning systems. Using a variety of restriction enzymes, we also compared the rate and mode of mtDNA evolution in striped bass with patterns observed in other organisms.

Methods

Collections.—Table 1 lists the dates and sites of striped bass collections. To ensure accurate designation of all sampled fish, we participated in all collections. Except for the Montauk coastal collection, adult striped bass were sampled at breeding times in the major spawning rivers. This was done to ensure they were native to the rivers in which they were collected. An exception was the sample collected at Troy, New York, in July. Small young-of-the-year and yearling striped bass obtained at various Hudson sites were assumed to be of Hudson ancestry because Chesapeake stocks that young do not make coastal migrations (Kohlenstein 1981). All fish (with the exception of Montauk samples) were transported alive to our laboratory and maintained in holding tanks until processed. Standard lengths and sex were recorded where possible.

mtDNA purification and characterization.—Mitochondria were obtained from freshly killed striped bass with the exception of Montauk samples. Depending on tissue availability, we used liver, muscle, gonad, or brain as our sources of mtDNA. Often, we conducted separate mtDNA preparations from two tissues to compare their forms of mtDNA. Given the protein monomorphism exhibited by all striped bass populations previously investigated, we felt that our best strategy for uncovering polymorphic markers was to generate and score a large number of mtDNA fragments. Each fragment was viewed as, potentially, a separate discriminatory genetic marker. The more fragments scored, the greater the percentage of the mitochondrial genome screened, and the greater the possibility of uncovering base substitutions and small DNA insertions or deletions. This dictated the use of several four- and five-base-cutting restriction enzymes, which, because of their shorter recognition sites, are more likely than six-base-cutters to produce large numbers of DNA fragments. The many small fragments (≤ 500 bp) produced by these enzymes can only be visualized if highly purified mtDNA has been used. We purified mtDNA through multiple rounds of cesium chloride density-gradient ultracentrifugation as outlined by Wirgin (1987).

In summary, following tissue homogenization, multiple rounds of differential centrifugation were used to isolate a mitochondrial pellet. This pellet was purified by application to a preformed sucrose gradient (1.5 and 1.0 M in a 10-mM tris, 1-mM EDTA, pH 8.0 buffer) which was spun for 1 h on an ultracentrifuge. The mitochondrial fraction was then lysed with sodium dodecyl sulfate (SDS), which released mtDNA into the supernatant. Mitochondrial DNA was then separated from any nuclear DNA contamination by three rounds of cesium chloride (CsCl) density-gradient ultracentrifugation. Altogether, 72 h of ultracentrifugation were required per sample. This was followed by butanol extraction to remove ethidium bromide, overnight dialysis to remove CsCl, alcohol precipitation (100% and 70%) to concentrate the mtDNA and remove any residual salts, and resuspension of the dried mtDNA pellet in TEN buffer (10 mM tris, 1 mM EDTA, 1 mM NaCl; pH 7.5). Although our yields of mtDNA were quite low (<1 μg of mtDNA/g of tissue), they still provided sufficient purified mtDNA to visualize via radioactive end-labeling of the products of at least 20 restriction digests.

Digest conditions followed the manufactures' recommendations (Bethesda Research Laboratories Incorporated, International Biotechnologies Incorporated, and New England Biolabs), except that digest times were extended to 4–6 h. Approximately 2–10 ng of mtDNA were digested with 5–10 units of restriction enzyme in each reaction. Digest conditions were empirically optimized for each restriction enzyme to reduce the possibility of partial digests. We used 12 restriction enzymes to characterize mtDNA genotypes. Not all enzymes were used in screening all striped bass; however, a minimum of six restriction digests were performed on each sample.

We end-labeled mtDNA fragments with a cocktail of all four α-^{32}P-labeled nucleotides (New England Nuclear) for 1 h, using the Klenow fragment of DNA polymerase I (International Biotechnologies Incorporated) and following the manufacturer's recommendations. Fragments obtained from most digests were aliquoted and separated on both agarose (1.0%–1.8%) and polyacrylamide gels (4.0%–12%) to visualize and score a range of fragments of widely different molecular sizes. For all four- and five-base-cutter digests, at least two and sometimes three gels were used to separate a range of DNA fragments of widely different sizes. The DNA fragments were visualized by autoradiography with and without intensifying screens. Intensifying screens reduce the exposure time of autoradiographs, albeit at a loss of resolution. Molecular sizes of all striped bass mtDNA fragments were determined by comparison with digested phage or plasmid DNA size standards.

Preparations of mtDNA from 10 small Hudson River fish (<11 cm total length) did not prove suitable for end-labeling of fragments. Gonzalez-Villasenor et al. (1986) previously reported on the use of Southern blot analysis (Southern 1975) with cloned mtDNA from mummichogs *Fundulus heteroclitus* as a probe to visualize mtDNA from various fish species. In this procedure, mtDNA fragments produced by restriction enzyme digestion are separated electrophoretically and then transferred by capillary action to a nitrocellulose filter. The immobilized mtDNA fragments are then visualized by hybridization to a probe of cloned mtDNA that has been radioactively labeled by a procedure called nick-translation. We adopted a variation of this procedure to visualize the mtDNA fragments from these small fish. Rather than make our probe using cloned mtDNA, we used mtDNA from a single striped bass that we knew from end-labeling experiments to be ultrapure. This approach did not provide a limitless supply of mtDNA, but the mtDNA isolated from the liver or eggs of a single fish provided enough mtDNA to make probes sufficient for 50–100 hybridizations. Less than 20 ng of mtDNA were used to make each probe.

The mtDNA from these small fish was digested only with four restriction enzymes—*Xba* I, *Hind* III, *Ava* II, and *Taq* I—and electrophoresed on 1.0–1.2% agarose gels. We did not electrophorese this mtDNA on polyacrylamide gels. The mtDNA was denatured and neutralized, and fragments were transferred to a nitrocellulose filter (Schleicher & Schuell, BA 85) by capillary action for 16–20 h, according to the methods of Southern (1975) and Maniatis et al. (1982). Highly purified striped bass mtDNA was radiolabeled by nick-translation in the presence of deoxycytidine 5'-triphosphate, tetra(triethylammonium) salt, (α-^{32}P), specific activity 3,000 curies/mmol (New England Nuclear) (Rigby et al. 1977). Filters were hybridized at 65°C for 16–20 h with these radiolabeled probes in the presence of dextran sulfate (modified procedure of Wahl et al. 1979). Filters were initially washed in three baths of 2.0× SSC (3 M NaCl, 0.3 M sodium citrate, pH 7.0) and 0.1% SDS for 15 min at room temperature. The final high-stringency wash was in 0.1× SSC and 0.1% SDS at 65°C for 30 min to remove nonspecifically bound probe molecules. Filters were air-dried and subjected to autoradiography with intensifying screens for 1–6 d.

Our initial measure of genetic similarity between mtDNAs from different individuals was the proportion of shared fragments in their digestion profiles, as revealed electrophoretically (Upholt 1977). We then quantified the extent of mtDNA diversity between stocks by estimating the proportion of nucleotides (p) in the mtDNA genome that differed for the most divergent individuals. We estimated this value directly from restriction site differences between individuals, using the methodology of Nei and Li (1979) and making certain assumptions about the random distribution and frequency of restriction sites in the mitochondrial genome and its mode of mutation.

Results

The mean sizes and numbers of striped bass mtDNA fragments obtained for the various restriction enzymes are presented in Table 2. The mean size of fragments was based on all digests done with a particular restriction enzyme. Mean mtDNA genomic size for striped bass was based on all digests done on all individuals used in this study. The mean mtDNA size for the coastal stock, based on all digests (>1,200) performed on the three spawning stocks, was 17,860 base pairs (bp). In total, 155 fragments were generated for most fish, representing 730 bp. This constituted approximately 4.1% of the entire striped bass mtDNA nucleotide sequence. Four-base-cutters produced a mean of 26 fragments, five-base-cutters 21 fragments, and six-base-cutters 5 fragments. For the purpose of stock identification, our strategy was to use several four- and five-base-cutters to produce a large number of fragments, which would enable us to screen a sizable proportion of the mitochondrial genome with a limited number of restriction enzymes.

Base Substitutions

Despite the use of several four- and five-base-cutters, base substitutions were extremely rare among the populations compared. None of the six-base-cutters revealed any base sequence divergence. The five-base-cutters *Ava* II and *Hinf* I also failed to reveal any nucleotide sequence divergence, despite generating a large number of mtDNA fragments (42). The four-base-cutter *Rsa* I revealed the greatest amount of base sequence heterogeneity among the individuals sampled. Four rare genotypes were observed, each in a single fish. These rare *Rsa* I genotypes were all different from the variant *Rsa* I genotype we previously reported for two fish from the Apalachicola system (Wirgin et al. 1989). All other striped bass displayed the common *Rsa* I genotype. Three of these unique *Rsa* I genotypes were found in the Rappahannock River; the fourth was observed in a single fish from the Roanoke River. The *Taq* I endonuclease also generated a rare restriction genotype observed in only one fish from the Roanoke River. These five rare genotypes uniquely identify these individuals, all of which were from populations at the southern limits of the coastal migratory stock. The level of striped bass mtDNA sequence variation was $p = 0.0004$, one of the lowest reported for any animal species. It is also the same maximum level of

TABLE 2.—Mean sizes[a] (numbers of base pairs) of mitochondrial DNA fragments in restriction endonuclease digests of tissues from Atlantic coastal striped bass. "Poly" means polymorphic.

Fragment	Ava I	Ava II	Bgl II	Eco RI	Hinf I	Hind III	Mbo I	Pvu II	Rsa I	Sac I	Taq I	Xba I
										Restriction endonuclease		
1	3,952	3,715	7,737	6,330	Poly	7,971	1,861	Poly	2,816	11,277	Poly	5,635
2	3,645	3,126	Poly	5,104	1,900	6,033	1,604	5,608	Poly	Poly	1,702	3,535
3	2,329	Poly	2,792	2,791	1,643	2,296	1,360	1,328	2,136	1,276	1,602	3,456
4	2,126	1,926	1,947	Poly	1,429	Poly	1,262		1,974		1,381	3,233
5	1,396	1,651			1,257		1,163		1,204		1,135	1,703
6	1,357	1,184			1,068		1,092		1,104		1,085	130
7	1,216	1,146			868		1,023		838		924	105
8	Poly	693			625		925		703		840	
9	417	653			603		865		498		806	
10	398	515			560		809		459		776	
11	261	383			533		735		431		735	
12		222			509		690		405		687	
13		208			495		642		366		627	
14					483		526		346		597	
15					460		410		330		509	
16					428		381		317		492	
17					391		349		303		452[b]	
18					361		336		254		325	
19					348		321		222		257	
20					323		298		206		213	
21					222		266		172		204	
22					195		213		103		196	
23					167		148		96		187	
24					117		115		71		118	
25					111		98				115	
26					97		90					
27					92		70					
28					59		63					
29					41		47					

[a]Mean sizes are based on all restriction digests by an enzyme for all fish.
[b]Polymorphic fragment.

mtDNA diversity that we reported for striped bass populations in the southeastern USA (Wirgin et al. 1989).

Length Polymorphisms

Major length variation.—We define major length variation as differences of more than 100 bp in the size of the mtDNA molecule between individual fish. Such differences are not large when compared to differences of several thousand base pairs reported for a few individual newts and parthenogenetic lizards (Wallis 1987; Moritz and Brown 1987). We observed five different lengths of the mtDNA molecule in Atlantic coastal striped bass populations. Each length differed by approximately 100 bp from the next-smallest size. Therefore, the largest genotype was more than 500 bp larger than the smallest. As expected, all restriction enzymes revealed this length polymorphism (see Figures 1–3 for *Hind* III, *Ava* II, and *Ava* I patterns). When mtDNA from several different tissues from a single fish were examined, all exhibited identical length genotypes. For digests with most restriction enzymes, length variation

was indicated by differences in the electrophoretic mobility of a single mtDNA fragment. All other fragments compared between individual fish showed identical electrophoretic mobility. The difference in molecular size (bp) among the polymorphic fragments generated for all single digests was consistent for all enzymes used. The *Mbo* I and *Xba* I restriction digest patterns did not display major length variation as differences in the mobility of single fragments (Figure 4). For *Mbo* I, each length variant appeared as comigrating bands (two or more mtDNA fragments with identical electrophoretic mobility because of similar molecular size). For fish representative of each different length genotype, different sequential fragments (fragments 2–6) generated by *Mbo* I appeared as doublets. We have demonstrated for southeastern striped bass that length variation in fragments generated by *Xba* I appears as differences in the number of copies (6–10) of a 130-bp fragment, which we resolved on 10% polyacrylamide gels (Wirgin et al. 1989). This observation suggests that major length variation in striped bass results from multiple duplications of the

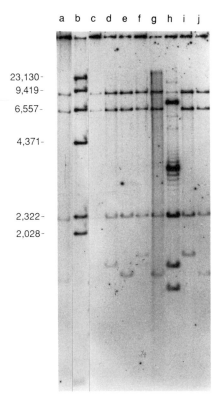

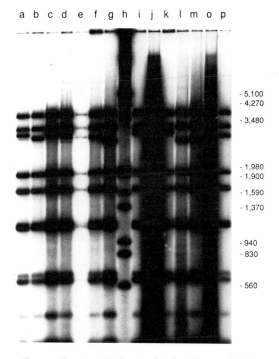

FIGURE 1.—*Hind* III digest of striped bass mitochondrial DNA electrophoresed on a vertical 1.0% agarose gel made in tris–borate–EDTA buffer. We used the Klenow fragment of DNA polymerase I to end-label fragments with a cocktail of all four ^{32}P-radionucleotides. Major length variation can be seen in the size of the smallest fragment. Four different-length genotypes can be seen on this gel. Lane a—genotype *E*; lane b—molecular weight standard λ digested with *Hind* III; lane c—genotype not discernible from this gel; lane d—genotype *C*; lane e—genotype *D*; lane f—genotype *B* (heteroplasmic for *C*); lane g—genotype *D*; lane h—white perch *Morone americana* mtDNA digested with *Hind* III; lane i—genotype *B*; lane j—genotype *D*. Numbers along the left edge are base-pair lengths of DNA standards.

FIGURE 2.—*Ava* II digest of striped bass mitochondrial DNA on a 1.4% vertical agarose gel. Major length variation can be seen in the third-largest mtDNA fragment. All other fragments show identical electrophoretic mobility, and therefore share the same molecular size. Lane d exhibits heteroplasmy for four different length genotypes; heteroplasmy for a single extra band can be seen in lane a. Lane a—genotype *B* (heteroplasmic for *C*); lane b—genotype *C*; lane c—genotype *B*; lane d—genotype *A-1* (heteroplasmic for *B*, *C*, and *D*); lane e—genotype *C*; lane f—genotype *C*; lane g—genotype *D*; lane h—molecular weight standard λ digested with *Hind* III and *Eco* RI; lane i—genotype *B* lane j—genotype not discernible from this gel; lane k—genotype *B*; lane l—genotype *D*; lane m—genotype *D*; lane o—genotype not discernible from this gel; lane p—genotype *C*.

130-bp fragment. However, in the present study we did not run fragments generated by *Xba* I on the requisite 10% polyacrylamide gels to visualize the difference in copy number.

Because all major length polymorphisms are recognizable as single fragments, all the additions and deletions responsible for this size variation seem to be confined to a single region of the mtDNA molecule. We have mapped this major length variation to a single site in the striped bass mtDNA molecule (Wirgin et al. 1989). All fragments displaying this variation, with the excep-

tion of *Xba* I products, were visualized on agarose gels. The different classes of size variants were most easily scored and molecular size determinations were most accurate when we used four- or five-base-cutters. When we used certain six-base-cutters, such as *Pvu* II, the polymorphic fragment was quite large (10,000 bp), sometimes making it difficult to accurately size the variants. The larger the size of the mtDNA fragment, the more difficult it is to recognize differences in molecular size among individuals.

Three of the length genotypes (*B*, *C*, and *D*) were common among the populations surveyed and two were rare (Table 3). No single length

1 2 3 4 5 6 7 8 9

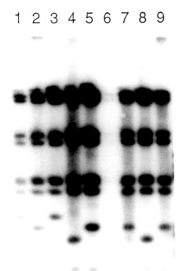

a b c d e f g h i j k l

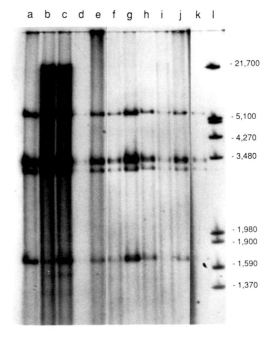

- 21,700

- 5,100

- 4,270

- 3,480

- 1,980
- 1,900
- 1,590

- 1,370

FIGURE 3.—*Ava* I digest of striped bass mitochondrial DNA run on 1.1% agarose gel. Fragments of mtDNA were visualized by Southern blot analysis. Ultrapure mtDNA used in making this probe was isolated from a single striped bass. Major length variation can be seen in the smallest mtDNA fragment visualized on this gel. Lane 1—genotype *B*; lane 2—genotype *C*; lane 3—genotype *B*; lane 4—genotype *D*; lane 5—genotype *C*; lane 6—genotype *C*; lane 7—genotype *C*; lane 8—genotype *D*; lane 9–genotype *C*.

FIGURE 4.—*Xba* I digest of striped bass mitochondrial DNA electrophoresed on a vertical 1.0% agarose gel. Although representatives of all three common major length genotypes are shown, no differences can be seen in the molecular size of any of these DNA fragments. We have previously shown that length variation in fragments generated by *Xba* I can be seen as multiple copies of a 130-base-pair fragment that was best visualized on 10% polyacrylamide gels. In addition, unlike southeastern striped bass populations, *Xba* I recognition site gain or loss was not observed among representatives of the Atlantic coastal migratory stock. Lane l contains a molecular weight standard: λ DNA digested with *Hind* III and *Eco* RI.

genotype was fixed in any population, although the frequencies of length variants did differ among the individual rivers and overall systems surveyed. The largest of the three common genotypes (*B*) was observed in 36% of Hudson River striped bass. This genotype was noted in two (3%) Chesapeake Bay striped bass, one from the Choptank River and one from the Chesapeake and Delaware (C & D) Canal. Because of minor size variation in a *Taq* I fragment, however, the sample from the Choptank River was unequivocally distinguishable from all Hudson River striped bass. Therefore, all Hudson River striped bass possessing the *B* genotype were distinguishable from all Chesapeake Bay striped bass except for the single specimen from the C & D Canal. It is possible that our limited spatial and temporal sampling in the Chesapeake underrepresented the frequency of this genotype among Chesapeake populations. For the Choptank River and C & D Canal, our sample was almost entirely composed of 1982 year-class males whose mtDNA genotypic frequencies may not have been entirely representative of these spawning systems. It is also quite possible that the subadult males did not

demonstrate the same degree of spawning system fidelity as might be observed in older fish or in females.

In addition, two (3%) Hudson River striped bass displayed a rare large-length genotype (*A-1*) also noted in a single Chesapeake sample from the Rappahannock River. The single Rappahannock striped bass bearing this genotype was uniquely identifiable because it had the rare aforementioned *Rsa* I base substitution, which was not observed in the two Hudson River fish. By scoring all major length variants, we could separate 36% of Hudson River striped bass from all Chesapeake samples except for the single specimen from the C & D Canal.

At the systems level, when we compared major length frequencies of the Chesapeake with those

TABLE 3.—Frequency of major mitochondrial DNA (mtDNA) length variants in samples of Atlantic coastal striped bass. Data are number of fish (% of whole collection).

Sample	mtDNA gentype				
	A-1	B	C	D	E
Roanoke River					
1983		3 (33%)	3 (33%)	3 (33%)	
1984		1 (7%)	4 (29%)	9 (64%)	
James River			6 (50%)	6 (50%)	
Rappahannock River	1 (8%)		6 (46%)	6 (46%)	
Choptank River		1 (7%)	8 (57%)	4 (29%)	
Chesapeake and		1 (8%)	8 (61%)	4 (29%)	1 (7%)
Delaware Canal			8 (61%)	4 (31%)	
Upper Chesapeake Bay			8 (80%)	2 (20%)	
Lower Hudson River	2 (5%)	17 (38%)	13 (30%)	12 (27%)	
Mid-Hudson River		5 (26%)	10 (53%)	4 (21%)	
Upper Hudson River		6 (40%)	3 (20%)	6 (40%)	
Connecticut River	1 (8%)	3 (22%)	1 (8%)	8 (62%)	
Montauk		5 (21%)	17 (71%)	2 (8%)	

of the Hudson, highly significant differences were observed ($\chi^2 = 22.58$, $P < 0.01$, df = 2). In addition, the Chesapeake total differed significantly from that observed in the Roanoke ($\chi^2 = 8.33$, $0.05 < P < 0.01$, df = 2). However, significant differences were not observed when we compared Hudson to Roanoke totals, which suggested closer relatedness for striped bass inhabiting these two systems ($\chi^2 = 4.92$, df = 2). In many cases, cell size for this analysis was small.

When we divided the Chesapeake system into two major components—upper bay (either C & D Canal and upper Bay–Aberdeen or C & D Canal, upper Bay–Aberdeen, and Choptank River) and lower bay (James River and Rappahannock River)—we found that these Chesapeake subsystems, considered independently, differed significantly from Hudson totals. The magnitude of this difference was far greater when we compared only upper Chesapeake Bay (using either pooling) to Hudson River totals. This observation suggested a closer relatedness between Hudson River and lower Chesapeake Bay stocks. When we compared upper (C & D Canal and upper Bay-–Aberdeen, Choptank River) and lower Chesapeake Bay frequencies, significant differences in major length variants were not observed ($\chi^2 = 3.94$, df = 2). Similarly, significant differences were not found when we compared upper and lower Chesapeake Bay frequencies independently to that observed for the Roanoke total. Neither were significant differences observed among the various tributaries of Chesapeake Bay. Similarly, Hudson stock proved homogeneous regarding this length characterization, even though we sampled almost the entire length of the river's striped bass distribution.

Minor size variation.—We define minor size variation as differences in the electrophoretic mobility of small mtDNA fragments (<500 bp) between individuals, which are resolved best on polyacrylamide gels. In this study, one four-base-cutter, *Taq* I, revealed minor size variation in two small fragments visualized on 6% polyacrylamide gels (Figure 5). Differences in the sizes of both of these minor lengths variants were easily recognized and scored, and were useful in stock discrimination. Ten striped bass, of upper or middle Chesapeake origin, displayed a unique fragment for this enzyme (fragment 17). Altogether, about 25% of our entire Chesapeake collection exhibited this *Taq* I genotype, and the percentage was much higher in some Chesapeake systems. We did not find these minor size variants in either lower Chesapeake Bay tributaries—the James or Rappahannock rivers—nor in our Hudson or Roanoke collections. Minor size variation was not seen in southeastern striped bass (Wirgin et al. 1989). One fish from our coastal collection at Montauk did exhibit this unique *Taq* I variant.

The electrophoretic mobility of this polymorphic fragment indicated that it was 16 bp smaller than the common fragment of 452 bp. In addition, one fish of Chesapeake ancestry exhibited electrophoretic variation in two other small *Taq* I-generated fragments, which measured 204 and 196 bp. These minor size variants appeared to differ by five and one bp, respectively, from the common genotypes for these fragments. Despite the small size differences, these variant *Taq* I fragments were identifiable on appropriate polyacrylamide gel concentrations. The striped bass exhibiting the unique minor size variants represented all three major length genotypes. All of these

a b c d e f g h i j k l m

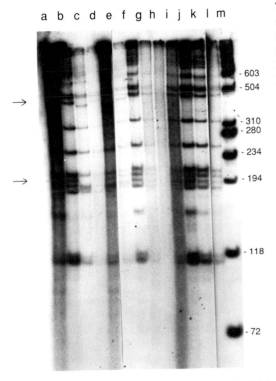

- 603
- 504

- 310
- 280

- 234

- 194

- 118

- 72

FIGURE 5.—*Taq* I digest of striped bass mitochondrial DNA run on 6% polyacrylamide gel. This gel concentration was used to resolve striped bass mtDNA fragments ranging from approximately 100 to 600 base pairs (bp). Micro-length variants unique to several tributaries of the Chesapeake Bay are visible. The common *Taq* I genotype, which is fixed in the Roanoke and Hudson river populations, has a 452-bp fragment not found in approximately 20% of Chesapeake Bay fish. The variant genotype, represented by a 436-bp fragment, can be seen in lanes c, d, and e. The common genotype may be seen in all other lanes. Additional size variation in some of these same Chesapeake samples may be seen in a 200-bp fragment in lanes c and d. Lane m contains a molecular weight standard: φX-174 digested with *Hae* III.

minor size variants were unique to the Chesapeake system and, therefore, they allowed unequivocal identification of ancestry of their carriers in our collection. Given the high frequency of these minor size variants in our Choptank sample (6 of 10 fish), this marker should prove quite effective in identifying Choptank fish in a mixed sample. However, these observations were based on small sample sizes collected primarily in a single season.

Heteroplasmy

Heteroplasmy refers to the expression of more than one mtDNA genotype in a single individual.

The minority form of the molecule must be present at a concentration of at least 1–5% of the majority form to achieve the threshold of detectability by current DNA technologies (Avise and Lansman 1983). We have observed a substantial level of heteroplasmy for the major length variants in our study. Approximately 20% of all the coastal striped bass we sampled proved heteroplasmic. These individuals, in addition to exhibiting the normal complement of fragments for any given enzyme digest, also presented one or more extra bands of DNA. The extra heteroplasmic band was always adjacent to the DNA fragment that revealed major length variation among the populations surveyed. In all cases, samples that we called heteroplasmic displayed this pattern for all restriction enzymes used, with the exception of *Mbo* I and *Xba* I digests patterns, which showed comigrating bands for length variants. In addition, several samples showed more than one heteroplasmic genotype. One striped bass from Montauk exhibited four different length genotypes (Figure 2). We have also seen heteroplasmy for four length genotypes in a single fish from our southeastern collection. In all cases, the molecular size of the extra heteroplasmic band(s) corresponded to that of other known major length variants among striped bass surveyed, although not necessarily from the same population. There did not appear to be a correlation between size of the heteroplasmic band (length genotype) and its relative DNA concentration as determined densitometrically. In other words, the heteroplasmic band with the lower DNA concentration could be either smaller or larger in molecular size than the DNA fragment that had the higher DNA concentration. Heteroplasmy was found in all three major spawning systems and the coastal Montauk collection. We have not used heteroplasmy as a tool in stock discrimination. It is notable that most examples of heteroplasmy in the literature involve length variants rather than base substitutions which suggests that, in these organisms, the rate of generation of length variants far exceeds that of base substitutions or, alternatively, that the rate of fixation of new base substitutions is much more rapid than that of length variants.

Ancestry of Connecticut River Fish

Length frequencies of mtDNA in the Connecticut River differed significantly from those reported for the Chesapeake total, which probably precludes a recent Chesapeake origin for, or cur-

rent exchange with, Connecticut River striped bass. With the exception of the James and Rappahannock rivers in lower Chesapeake Bay, all other single sites listed for the Chesapeake showed highly significant differences in mtDNA length frequencies when compared to the Connecticut River total ($\chi^2 = 14.11$, $0.05 < P < 0.01$, df = 2). The *Taq* I minor-length genotype present in about 25% of all Chesapeake Bay fish was absent from all Connecticut River fish. However, both the Hudson and Roanoke fish apparently share common genotypic frequencies with those observed in the Connecticut. Both the Hudson and Connecticut rivers share a rare-length genotype, which would suggest exchange between these two systems. Because of the absence of a unique Connecticut River mtDNA genotype, the sharing of a rare genotype, and similar major length genotypic frequencies, we postulate current gene flow with the Hudson River.

Discussion

Stock Identification

Mendelian populations are identified by significant differences in genotypic frequencies among sites sampled. Significant differences in mtDNA major-length genotypes have permitted us to differentiate between Hudson River and Chesapeake Bay stocks of striped bass. In addition, significant differences in mtDNA lengths were shown for Chesapeake and Roanoke populations. This is the first demonstration of genetic differences between these heretofore hypothesized Mendelian populations. We did not observe significant differences between Roanoke and Hudson samples. This similarity between Hudson and Roanoke samples is difficult to explain. Meristic and morphometric studies both indicated significant differences between these two stocks (Raney 1957; Lund 1957). By additionally considering information from minor length variants and rare base substitutions (generated exclusively through the use of four-base-cutting enzymes), we were able to differentiate between an upper Chesapeake Bay population and a James–Rappahannock population. These results confirmed the separation of Chesapeake stocks on the north–south cline proposed by Raney (1957) and Raney and deSylva (1953), which was based on meristic characters. Raney (1957) also suggested a closer relatedness between stocks of the Hudson and James rivers than between Hudson and upper Chesapeake Bay stocks. Based on the frequencies of major length

variants and an absence of minor size variants in either system, we also found a reduced amount of genetic divergence between Hudson and lower Chesapeake stocks. Our observation of genetic homogeneity of the Hudson River stock was expected, given the lack of potential geographic isolation within the Hudson, and it confirmed the results of protein electrophoresis work by Hitron (1974). It refuted earlier suggestions of population subdivisions in the Hudson proposed by Raney and DeSylva (1953).

mtDNA as an Innate Genetic Tag to Identify Individuals

One pivotal question concerns the utility of mtDNA as a tag to unambiguously identify individuals of unknown ancestry in the coastal fishery. Such a tag would permit an estimate of the relative contribution of the different spawning populations to the coastal fishery on a spatial or temporal basis. The rarity of detected base substitutions currently limits the applicability of mtDNA restriction site polymorphisms as an effective tool in stock discrimination. The *Rsa* I and *Taq* I digests did permit the detection of three unique variants in the Rappahannock River (21%) and two unique variants in the Roanoke River (9%). The striped bass with these base substitutions may be uniquely assigned as to river of ancestry. Base sequence diversity was not observed among any Hudson River striped bass.

In a mixed fishery, major length variation permits identification of a moderate percentage (35%) of Hudson River fish, those exhibiting the *B* length genotype, if we have not underrepresented the frequency of this genotype in our Chesapeake collection. Minor size variants were observed only in Chesapeake Bay. In total, almost 25% of the Chesapeake Bay striped bass examined exhibited unique *Taq* I minor size variants. The *Taq* I genotypes were found in 36% of upper Bay (Aberdeen) and 60% of Choptank River samples. Thus, these genetic tags could serve to identify a substantial percentage of representatives of these systems in a mixed fishery. No minor size variants were detected in the two lower Chesapeake Bay rivers sampled—the James and Rappahannock rivers. Scoring *Taq* I microvariants could aid in distinguishing upper Bay and Choptank River striped bass from those of the lower Chesapeake Bay. Surprisingly, we also failed to observe this genotype in any of our samples from the C & D Canal, possibly because of an age and sex bias in

our collection. Also, both Hudson and Roanoke River striped bass were monomorphic for the common *Taq* I minor size variant. Thus, this genotype could be used to uniquely identify individual Chesapeake fish from all striped bass of the Hudson and Roanoke rivers.

The minor size variants were observed in three different *Taq* I fragments. Despite the use of several other four- and five-base-cutters, no other restriction enzyme yielded fragments displaying these small differences. This is not unexpected, however. Unless the sequence containing the small added or deleted DNA tracts occurs in a small restriction fragment (10–500 bp), there is little chance of detecting it. The ability to detect small DNA additions or deletions depends on the size of the fragment being scored. The smaller the DNA fragment, the more sensitive is the procedure for revealing minor size variants. It is possible that if a battery of additional four-base-cutting restriction enzymes were used to screen the mtDNA genome of these fish, more base substitutions or minor size variants would be found. We have retained sufficient mtDNA from almost all of our striped bass samples to conduct further analyses with such a battery of four-base-cutting enzymes. Our study of American shad *Alosa sapidissima*, by drawing primarily on four- and five-base-cutting restriction enzymes such as *Hinf* I and *Dde* I, has revealed extensive minor size polymorphisms in small mtDNA fragments resolved on polyacrylamide gels (K. Nolan et al., City College of New York, unpublished data).

Minor size variants were observed in representatives of all three common classes of major length variants. A distribution of minor variants among many classes of major length variants in lizards was also reported by Densmore et al. (1985). The presence of these minor variants in all common classes of major length variants suggests that the event(s) resulting in these small additions or deletions preceded the generation of the various major size classes of mtDNA. However, the absence of *Taq* I minor variants from Hudson River striped bass indicates that the event(s) occurred following the isolation of Hudson and Chesapeake stocks. An absence of heteroplasmy for these minor length variants suggests that their generation is infrequent. That individuals showing heteroplasmy for major length variants do not exhibit heteroplasmy for these minor length variants is further evidence that these DNA additions or deletions should map to a distinct site in the mtDNA molecule.

Recent studies on human mtDNA by Singh et al. (1987) and Vigilant et al. (1988) have raised an interesting possibility concerning the molecular basis of striped bass minor size variants. These workers also observed slight differences in the electrophoretic mobility of small fragments (<500 bp) generated by *Taq* I and *Hinf* I from human mtDNA electrophoresed on polyacrylamide gels. The fragments did not show altered electrophoretic mobility on conventional agarose gels. DNA sequencing of these variant fragments from a modest number of individuals revealed single base substitutions outside of restriction enzyme recognition sites. A single thymine–cytosine transition was found near the center of the variable fragment. This altered the fragment's electrophoretic mobility, probably by affecting its curvature (Vigilant et al. 1988). For two reasons, this raises the possibility that if more four-base-cutting enzymes were used in stock-identification studies, more minor variants would be found. First, a greater percentage of the mitochondrial genome would be scored for mutations in enzyme recognition sites. Second, polyacrylamide gel electrophoresis of small mtDNA fragments might permit detection of base substitutions outside of enzyme recognition sites.

The Mode of mtDNA Evolution in Striped Bass

Almost all population surveys of vertebrate mtDNA have recorded substantial levels of base substitutions (Brown 1983; Avise and Lansman 1983; Brown 1985; Birley and Croft 1986). Mean p values between conspecific individuals in comparisons of higher primates have ranged between 0.3 and 4% (Avise and Lansman 1983). We have calculated a mean p value for all conspecific comparisons of mtDNA reported for fish of 3.3% (Wirgin 1987)—almost two orders of magnitude greater than that observed in striped bass (0.04%). Previously, it was thought that mtDNA genomic size was very stable, with the exception of the small additions or deletions found among humans (Cann et al. 1984; Wrischnik et al. 1987) and domestic cattle (Hauswirth et al. 1984). The question now arises as to whether our observations concerning mtDNA evolution in striped bass reflect a unique condition, or whether they represent a different mode or tempo of mtDNA change in lower vertebrates.

Levels of mtDNA sequence divergence detected among southeastern U.S. populations of

sunfishes *Lepomis* spp. and bowfins *Amia calva* are comparable with those reported among primate species (Avise et al. 1984; Avise and Saunders 1984; Bermingham and Avise 1986). We have observed substantial levels of mtDNA sequence diversity among individuals in small collections of all the other *Morone* species as well as in congenetic *Morone* comparisons (Wirgin 1987). It seems unlikely that a deceleration of the rate of mtDNA change is universal for all lower vertebrate taxa or even for lower percoid fishes.

When Chesapeake contributions to the coastal stock were high (>80%; Berggren and Lieberman 1978), 85–90% of the fish tagged at coastal sites were females (Schaefer 1968; Holland and Yelverton 1973; Oviatt 1977). Analysis of recapture data on striped bass tagged in the Potomac River has demonstrated that few young Chesapeake males leave the bay (Kohlenstein 1981). Given this sex-biased vagility of striped bass females in the coastal fishery, the possibility that females might stray into foreign spawning systems is increased. Birky et al. (1983) demonstrated that geographically specific differentiation of mtDNA genotypes may be disrupted by female-specific mobility. Thus, greater vagility of females would be responsible for the lack of geographically delineated mtDNA sequence diversity observed in striped bass. This hypothesis, however, does not account for the extreme mtDNA monomorphism found within all spawning stocks.

The older literature (Merriman 1941; Raney 1952; Koo 1970), as well as observations on the current situation (Boreman and Austin 1985) emphasized that striped bass making up the coastal migratory stock experience population fluctuations. Historically, this stock has been supported by recruitment from unusually large year-classes produced at irregular intervals. It may be hypothesized that individual striped bass populations periodically undergo population bottlenecks of sufficient intensity to drastically reduce intrapopulation genetic polymorphism. The maternal inheritance of the mtDNA molecule halves the effective population size, thereby accentuating this effect. As these large year-classes age, the number of reproducing individuals declines dramatically. In addition, males dominate the spawning grounds. During the 1974–1975 period of relatively large abundance of striped bass in the Chesapeake, Wilson et al. (1976) reported a male-to-female ratio of approximately 4:1 at spawning time in the Potomac River. Currently, with population sizes in the Chesapeake reduced, the ratio

and actual numbers of spawning females in the various Chesapeake tributaries may be greatly diminished. Therefore, it is possible that a very small number of unusually old females are now, and at times historically have been, responsible for year-class production in many or most Chesapeake spawning rivers. It is conceivable that, presently and in the distant past, small population size has dramatically reduced the pool of mtDNA diversity within individual populations. However, to reduce the level of intraspecific mtDNA polymorphism, this same effect should have occurred over the entire range of the striped bass in the recent evolutionary past. In the Hudson system, striped bass show an equally depauperate level of intrasystem mtDNA polymorphism. Given the immense size of the Hudson system, and current estimates of its population, however, we doubt that its population of spawning females has crashed recently to levels sufficient to extinguish mtDNA diversity.

Despite the publication of several studies exploring the extent of structural protein diversity among striped bass populations (Otto 1975; Grove et al. 1976; Sidell et al. 1980; Rogier et al. 1985; Fabrizio 1987a, 1987b; Wirgin 1987), little or no such diversity has been found. Relying on data obtained from avian genera, Kessler and Avise (1985) suggested that a parallel paucity of both mtDNA base sequence divergence and protein variation provides strong evidence of recent common ancestry. In the case of striped bass, two independent genomes—contrasting greatly in inter- and intragenic structure, employing autonomous replication and transcription systems, and differing in repair ability and functional constraints—still displayed concordant low levels of genetic polymorphism. This evidence suggests that either speciation of striped bass was fairly recent, or that all current striped bass evolved from a mitochondrially monomorphic population in the not-too-distant evolutionary past. Our detection of mtDNA sequence diversity exclusively in southern stocks (Rappahannock and Roanoke rivers) suggests that these populations may have served as maternal "seed" populations for current stocks.

In summary, our observations on striped bass mtDNA change reflect two independent phenomena. First, given the frequency of length variants and heteroplasmy for length polymorphisms, the occurrence of addition or deletion events in the striped bass genome must be very high; and it may reflect unique molecular mechanisms whereby these variants are generated at an unusually high level. Second, the lack of mtDNA sequence divergence can only be attributed to the recent life history patterns of striped bass populations. This observation is supported by the moderate levels of sequence diversity we have observed in all other *Morone* species.

Utility of mtDNA Polymorphisms in Striped Bass Stock Identification

Because of its mode of inheritance, mitochondrial DNA has several limitations for identification of striped bass stocks. First, only information concerning the maternal ancestry of a fish is provided. Knowledge is not generated on the possible introgression of paternally derived genes into the nuclear genome of a fish displaying a unique mtDNA genotype. For instance, we have found unique mtDNA genotypes in striped bass from the Apalachicola system, a Gulf of Mexico tributary, which suggests the continued existence of a maternal lineage of Gulf ancestry. Given the introduction of Atlantic fish into the Apalachicola system by transplantation beginning 20 years ago, and a generation time of 5–6 years, it is possible that a fish displaying a unique Gulf mtDNA genotype may have been the product of several crosses with Atlantic fish. As a result, a fish displaying a Gulf mtDNA genotype may have only a small percentage of Gulf nuclear genes. Second, because the mitochondrial genome contains only a very small fraction of the number of genes found in the nuclear genome, a knowledge of the mitochondrial genotype of an individual tells us little concerning its overall genetic composition. Third, the number of spawning striped bass females in many major spawning systems may decline to very low levels, a condition that can extinguish mtDNA polymorphisms. Fourth, straying by females between spawning populations is particularly disruptive to established discontinuities in mtDNA genotypic frequencies.

Mitochondrial DNA polymorphisms can be put to immediate use in monitoring hatchery-reared individuals to be released in the wild. During two recent spawning seasons (1987–1988), the mtDNA genotypes of all female brood stock collected from the Apalachicola River drainage were cataloged. Progeny from these fish are being hatchery-reared and retained for future use as brood stock. Thus, in the future there will be an available pool of striped bass brood stock of Apalachicola ancestry with various known mtDNA genotypes. The genetically marked progeny of these females can be used for differential-stocking strategies and to evaluate performance of Gulf versus Atlantic striped bass in Gulf of Mexico drainages.

Recommended Approaches for Future Stock-Identification Studies Using mtDNA

Several different approaches are available to prepare mtDNA and to visualize the digested mtDNA restriction fragments (Lansman et al. 1981; Chapman and Powers 1984). These techniques produce mtDNA of different purity, which determines the method of mtDNA fragment visualization. Generally, the more purified the mtDNA produced, the more costly, technically demanding, and time-consuming is the purification procedure. We prepared highly purified mtDNA and visualized digested mtDNA fragments by end-labeling with ^{32}P-radionuclides. In experienced hands, this approach yields sufficient mtDNA from most tissues from small fish (down to 5–8 cm) to do many digests, and it also permits the detection of very small mtDNA fragments (<25 bp). Is it necessary to use this approach? In our opinion, it depends on the life history of the species, size of the specimens, and degree of resolution desired. For a genetically monomorphic species, such as striped bass, highly purified mtDNA pemits the use of four- and five-base-cutting restriction enzymes that detect population-specific markers unidentifiable with ethidium bromide visualization of semipurified mtDNA. Furthermore, when only very small fish are available, end-labeling still permits the visualization of the products of many digests. For other species, we recommend an initial survey of highly purified mtDNA from approximately 20 individuals from each spawning system. Once polymorphic markers have been identified, large numbers of individuals can be screened by means of the Southern blot analysis or by ethidium bromide staining of rapidly purified mtDNA. The choice of technique greatly depends on tissue availability and on the molecular size of the polymorphic marker(s).

Acknowledgments

This work was supported by grants from the Hudson River Foundation, the Sport Fishing Institute, and, in part, by grant 666281 from the PSC-CUNY Research Award Program of the City University of New York. The following individuals were invaluable in sample acquisition: William Dey, Bruce Friedman, Bill Kriete, Everett Nack,

Harley Spier, and John Waldman. We thank John Waldman and several anonymous reviewers for their criticisms of early drafts of this manuscript.

References

Avise, J. C., E. Bermingham, L. G. Kessler, and N. C. Saunders. 1984. Characterization of mitochondrial DNA variability in a hybrid swarm between subspecies of bluegill sunfish (*Lepomis macrochirus*). Evolution 38:931–941.

Avise, J. C., G. S. Helfman, N. C. Saunders, and L. S. Hales. 1986. Mitochondrial DNA differentiation in North Atlantic eels: population genetic consequences of an unusual life history pattern. Proceedings of the National Academy of Sciences of the USA 83:4350–4354.

Avise, J. C., and R. A. Lansman. 1983. Polymorphism of mitochondrial DNA in populations of higher animals. Pages 147–161 *in* M. Nei and R. K. Koehn, editors. Evolution of genes and proteins. Sinauer Associates, Sunderland, Massachusetts.

Avise, J. C., C. A. Reeb, and N. C. Saunders. 1987. Geographic population structure and species differences in mitochondrial DNA of mouthbrooding marine catfishes (Ariidae) and demersal spawning toadfishes (Batrachoidae). Evolution 41:991–1002.

Avise, J. C., and N. C. Saunders. 1984. Hybridization and introgression among species of sunfish (*Lepomis*): analysis by mitochondrial and allozyme markers. Genetics 108:237–255.

Beaulieu, G. 1962. Results of tagging striped bass in the Saint Lawrence River during the period 1945–1960. Naturaliste canadien 89:217–236.

Berggren, T. J., and J. T. Lieberman. 1978. Relative contribution of Hudson, Chesapeake, and Roanoke striped bass, *Morone saxatilis*, stocks to the Atlantic coast fishery. U.S. National Marine Fisheries Service Fishery Bulletin 76:335–345.

Bermingham, E., and J. C. Avise. 1986. Molecular zoogeography of freshwater fishes in the southeastern United States. Genetics 113:939–965.

Bermingham, E., T. Lamb, and J. C. Avise. 1986. Size polymorphism and heteroplasmy in the mitochondrial DNA of lower vertebrates. Journal of Heredity 77:249–252.

Birky, C. W., T. Maruyama, and P. Fuerst. 1983. An approach to population and evolutionary genetic theory for genes in mitochondria and chloroplasts, and some results. Genetics 103:513–527.

Birley, A. J., and J. H. Croft. 1986. Mitochondrial DNAs and phylogenetic relationships. Pages 108–137 *in* S. K. Dutta, editor. DNA systematics. CRC Press, Boca Raton, Florida.

Boreman, J. 1986. Striped bass research and management—how far have we come in 25 years. Pages 25–31 *in* A. L. Pacheco, editor. Recreation fisheries and the environment. Past, present and future. U.S. National Marine Fisheries Service, Sandy Hook Laboratory, Technical Series Report 32, Highlands, New Jersey.

Boreman, J., and H. M. Austin. 1985. Production and harvest of anadromous striped bass stocks along the Atlantic coast. Transactions of the American Fisheries Society 114:3–7.

Brown, W. M. 1983. Evolution of animal mitochondrial DNA. Pages 62–68 *in* M. Nei and R. K. Koehn, editors. Evolution of genes and proteins. Sinauer Associates, Sunderland, Massachusetts.

Brown, W. M. 1985. The mitochondrial genome of animals. Pages 95–130 *in* R. J. MacIntyre, editor. Molecular evolutionary genetics. Plenum, New York.

Brown, W. M., M. George, Jr., and A. C. Wilson. 1979. Rapid evolution of animal mitochondrial DNA. Proceedings of the National Academy of Sciences of the USA 76:1967–1971.

Cann, R. L., W. M. Brown, and A. C. Wilson. 1984. Polymorphic sites and the mechanism of evolution in human mitochondrial DNA. Genetics 106:479–499.

Chapman, R. W., and D. A. Powers. 1984. A method for the rapid isolation of mitochondrial DNA from fishes. University of Maryland Sea Grant Program, Technical Report, UM-SG-TS-84-05, College Park.

Clark, J. R. 1968. Seasonal movements of striped bass contingents of Long Island Sound and the New York Bight. Transactions of the American Fisheries Society 97:320–343.

Densmore, L. D., J. W. Wright, and W. M. Brown. 1985. Length variation and heteroplasmy are frequent in mitochondrial DNA from parthenogenetic and bisexual lizards (Genus *Cnemidophorus*). Genetics 110:689–707.

Dudley, R. G., A. W. Mullis, and J. W. Terrell. 1977. Movements of adult striped bass (*Morone saxatilis*) in the Savannah River, Georgia. Transactions of the American Fisheries Society 106:314–322.

Fabrizio, M. C. 1987a. Contribution of Chesapeake Bay and Hudson River stocks of striped bass to Rhode Island coastal waters as estimated by isoelectric focusing of eye lens proteins. Transactions of the American Fisheries Society 116:588–593.

Fabrizio, M. C. 1987b. Growth-invariant discrimination and classification of striped bass by morphometrics and electrophoretic methods. Transactions of the American Fisheries Society 116:728–736.

Gonzalez-Villasenor, L. I., A. M. Burkhoff, V. Corces, and D. A. Powers. 1986. Characterization of cloned mitochondrial DNA from the teleost *Fundulus heteroclitus* and its usefulness as an interspecies hybridization probe. Canadian Journal of Fisheries and Aquatic Sciences 43:1866–1872.

Graves, J. E., S. D. Ferris, and A. E. Dizon. 1984. Close genetic similarity of Atlantic and Pacific skipjack tuna (*Katsuwonus pelamis*) demonstrated with restriction endonuclease analysis of mitochondrial DNA. Marine Biology 79:315–319.

Grove, T. L., T. J. Berggren, and D. A. Powers. 1976. The use of innate tags to segregate spawning stocks of striped bass, *Morone saxatilis*. Estaurine Processes 1:166–176.

Gyllensten, U., D. Wharton, and A. C. Wilson. 1985. Maternal inheritance of mitochondrial DNA during backcrossing of two species of mice. Journal of Heredity 76:321–324.

Harrison, R. G., D. M. Rand, and W. C. Wheeler. 1985. Mitochondrial DNA size variation within individual crickets. Science (Washington, D.C.) 228:1446–1447.

Hauswith, W. W., M. J. Van de Walle, P. H. Lapis, and P. D. Olivio. 1984. Heterogeneous mitochondrial DNA D-loop sequences in bovine tissue. Cell 37:1001–1007.

Hitron, J. W. 1974. Serum transferrin phenotypes in striped bass, *Morone saxatilis*, from the Hudson River. Chesapeake Science 15:246–247.

Holland, B. F., Jr., and G. F. Yelverton. 1973. Distribution and biological studies of anadromous fishes off-shore North Carolina. North Carolina Department of Natural and Economic Resources, Division of Commercial and Sport Fisheries, Special Scientific Report 24, Raleigh.

Kessler, L. G., and J. C. Avise. 1985. A comparative description of mitochondrial DNA differentiation in selected avian and other vertebrate genera. Molecular Biology and Evolution 3:109–125.

Kohlenstein, L. C. 1981. On the proportion of the Chesapeake Bay stock of striped bass that migrates into the coastal fishery. Transactions of the American Fisheries Society 110:168–179.

Koo, T. S. 1970. The striped bass fishery in the Atlantic states. Chesapeake Science 11:73–93.

Kornfield, I., and S. M. Bogdanowicz. 1987. Differentiation of mitochondrial DNA in Atlantic herring, *Clupea harengus*. U.S. National Marine Fisheries Service Fishery Bulletin 85:561–568.

Lansman, R. A., J. C. Avise, and M. D. Huettel. 1983. Critical experimental test of the possibility of "paternal leakage" of mitochondrial DNA. Proceedings of the National Academy of Sciences of the USA. 80:1969–1971.

Lansman, R. A., R. O. Shade, J. F. Shapira, and J. C. Avise. 1981. The use of restriction endonucleases to measure mitochondrial DNA sequence relatedness in natural populations. III. Techniques and potential applications. Journal of Molecular Evolution 17:214–226.

Lewis, R. M. 1957. Comparative study of populations of the striped bass. U.S. Fish and Wildlife Service Special Scientific Report—Fisheries 204.

Lund, W. A., Jr. 1957. Morphometic study of the striped bass, *Roccus saxatilis*. U.S. Fish and Wildlife Service Special Scientific Report—Fisheries 216.

Magnin, E., and G. Beaulieu. 1967. Le bar, *Roccus saxatilis* (Walbaum), du Fleuve Saint-Laurent. Naturaliste canadien 94:539–555.

Maniatis, T., E. F. Fritsch, and J. Sambrook. 1982. Molecular cloning. A laboratory manual. Cold Spring Harbor Laboratory, Cold Spring Harbor, New York.

McLane, W. M. 1958. Striped bass investigations. Florida Game and Fresh Water Fish Commission, Federal Aid in Fish Restoration, Project F-4-R, Completion Report, Tallahassee.

Merriman, D. 1941. Studies on the striped bass (*Roccus saxatilis*) of the Atlantic coast. U.S. Fish and Wildlife Service Fishery Bulletin 50:1–77.

Morgan, R. P., T. S. Y. Koo, and G. E. Krantz. 1973. Electrophoretic determination of populations of the striped bass, *Morone saxatilis*, in the Upper Chesapeake Bay. Transactions of the American Fisheries Society 102:21–32.

Moritz, C., and W. M. Brown. 1987. Tandem duplications in animal mitochondrial DNAs: variation in incidence and gene content among lizards. Proceedings of the National Academy of Sciences of the USA. 84:7183–7187.

Nei, M., and W.-H. Li. 1979. Mathematical model for studying genetic variation in terms of restriction endonucleases. Proceedings of the National Academy of Sciences of the USA 76:5269–5273.

O'Brien, J. F. 1977. Investigations of the striped bass, *Morone saxatilis* (Walbaum), overwintering in the upper Pettaquamscutt River estuary. Master's thesis. University of Rhode Island, Kingston.

Otto, R. S. 1975. Isozyme systems of the striped bass and congeneric percichthyid fishes. Doctoral dissertation. University of Maine, Orono.

Oviatt, C. A. 1977. Menhaden, sport fish and fisherman. University of Rhode Island, Marine Technical Report 60, Kingston.

Raney, E. C. 1952. The life history of the striped bass, *Roccus saxatilis* (Walbaum). Bulletin of the Bingham Oceanographic Collection, Yale University 14(1):5–97.

Raney, E. C. 1957. Subpopulations of the striped bass *Roccus saxatilis* (Walbaum), in tributaries of Chesapeake Bay. U.S. Fish and Wildlife Service Special Scientific Report—Fisheries 208:85–107.

Raney, E. C., and D. P. De Sylva. 1953. Racial investigations of the striped bass, *Roccus saxatilis* (Walbaum). Journal of Wildlife Management 17:495–509.

Rigby, P. W. J., M. Dieckmann, C. Rhodes, and P. Berg. 1977. Labeling deoxyribonucleic acid to high specific activity *in vitro* by nick translation with DNA polymerase I. Journal of Molecular Biology 113:237–251.

Rogier, C. G., J. J. Ney, and B. J. Turner. 1985. Electrophoretic analysis of genetic variability in a landlocked striped bass population. Transactions of the American Fisheries Society 114:244–249.

Rulifson, R. A., S. A. McKenna, and M. L. Gallagher. 1987. Tagging studies of striped bass and river herring in Upper Bay of Fundy, Nova Scotia. North Carolina Department of Natural Resources and Community Development, Division of Marine Fisheries, Completion Report AFC-28-1, Morehead City.

Schaefer, R. H. 1968. Sex composition of striped bass from the Long Island surf. New York Fish and Game Journal 15:117–118.

Sidell, B. D., R. G. Otto, D. A. Powers, M. Karweit, and J. Smith. 1980. Apparent genetic homogeneity of spawning striped bass in the Upper Chesapeake Bay. Transactions of the American Fisheries Society 109:99–107.

Singh, G., N. Neckelmann, and D. C. Wallace. 1987. Conformational mutations in human mitochondrial DNA. Nature (London) 329:270–272.

Southern, E. M. 1975. Detection of specific sequences among DNA fragments separated by gel electrophoresis. Journal of Molecular Biology 98:503–517.

Upholt, W. B. 1977. Estimation of DNA sequence divergence from comparisons of restriction endonuclease digests. Nucleic Acids Research 4:1257–1265.

Van Winkle, W., B. L. Kirk, and B. W. Rust. 1979. Periodicities in Atlantic coast striped bass (*Morone saxatilis*) commercial fisheries data. Journal of the Fisheries Research Board of Canada 36:54–62.

Vigilant, L., M. Stoneking, and A. C. Wilson. 1988. Conformation mutation in human mtDNA detected by direct sequencing of enzymatically amplified DNA. Nucleic Acids Research 16:5945–5955.

Wahl, G. M., M. Stern, and G. R. Stark. 1979. Efficient transfer of large DNA fragments from agarose gels to diazobenzyloxymethyl-paper and rapid hybridization by using dextran sulfate. Proceedings of the National Academy of Sciences of the USA 76:3683–3687.

Waldman, J. R., J. Grossfield, and I. Wirgin. 1988. A review of stock discrimination techniques for striped bass. North American Journal of Fisheries Management 8:410–425.

Wallis, G. P. 1987. Mitochondrial DNA insertion polymorphism and germ line heteroplasmy in the *Triturus cristatus* complex. Journal of Heredity 58:229–238.

Wilson, J. S., R. P. Morgan, P. W. Jones, H. R. Lunsford Jr., J. Lawson, and J. Murphy. 1976. Potomac River fishery study—striped bass spawning stock assessment. Final report 1975. University of Maryland, Chesapeake Biological Laboratory, UM-CEES 76–14, Solomons.

Wirgin, I. I. 1987. Molecular evolution in the fish genus *Morone*. Doctoral dissertation. City University of New York, New York.

Wirgin, I. I., R. Proenca, and J. Grossfield. 1989. Mitochondrial DNA diversity among populations of striped bass in the southeastern United States. Canadian Journal of Zoology 67:891–907.

Wrischnik, L. A., R. G. Higuchi, M. Stoneking, H. A. Erlich, N. Arnheim, and A. C. Wilson. 1987. Length mutations in human mitochondrial DNA: direct sequencing of enzymatically amplified DNA. Nucleic Acids Research 15:529–542.

Young, B. H. 1984. A study of the striped bass in the marine district of New York State IV. New York Department of Environmental Conservation, Division of Marine Resources, Completion Report AFC-12, Albany.

American Fisheries Society Symposium 7:492–498, 1990
© Copyright by the American Fisheries Society 1990

Technique for Determining Mitochondrial DNA Markers in Blood Samples from Walleyes

NEIL BILLINGTON AND PAUL D. N. HEBERT

Department of Biological Sciences, Great Lakes Institute, University of Windsor
Windsor, Ontario N9B 3P4, Canada

Abstract.—Recent studies of freshwater fish indicate that base sequence divergence in mitochondrial DNA (mtDNA) may be useful for stock identification, and that polymorphisms of mtDNA restriction fragments may serve as genetic tags in introduced fish. Most mtDNA studies have been limited by the need to kill fish to obtain tissue samples. In this paper we describe a nonlethal technique that involves probing the mtDNA in total DNA extracts from blood samples collected from live fish. We tested this technique on walleyes *Stizostedion vitreum* from Lake Simcoe, Ontario. We found that 0.5-mL blood samples provided a reliable source of DNA, and that mtDNA fragments could be readily resolved with a ^{32}P-labeled walleye mtDNA probe. This nonlethal technique for detecting mtDNA markers makes it possible for fisheries managers to use such markers with stocked fish.

Fisheries for walleyes *Stizostedion vitreum* exist in the Great Lakes region of North America and across Canada. Declining yields in the last 20 years have made stock manipulation and identification increasingly important components of management programs. Discrimination of walleye spawning stocks has relied almost entirely on tagging studies (Ferguson and Derkson 1971; Wolfert and Van Meter 1978; Bodaly 1980). Allozyme studies have generally failed to reveal substantial differentiation in local gene frequencies, and attempts to differentiate walleye stocks by morphological or physiological criteria have also proven ineffective (Colby and Nepszy 1981). Recent studies of freshwater fish indicate that base sequence divergence in mitochondrial DNA (mtDNA) may be useful for stock identification (Avise et al. 1984; Berg and Ferris 1984; Wilson et al. 1985, 1987; Bermingham and Avise 1986; Thomas et al. 1986; Grewe 1987; Gyllensten and Wilson 1987; Billington and Hebert 1988). Furthermore, known mtDNA restriction fragment patterns may serve as genetic tags in introduced fish (Ferris and Berg 1987; Grewe 1987; Billington and Hebert 1988).

The study of mtDNA variation is based on digestion of the molecule by restriction endonucleases and separation of the resulting fragments according to molecular weight by gel electrophoresis. Variability in fragment patterns is interpreted as genetic variation at the nucleotide level (Ferris and Berg 1987). Examination of intraspecific mtDNA variation has several advantages over traditional methods of investigating genetic

diversity among conspecific individuals (reviewed by Avise et al. 1987; Ferris and Berg 1987). Animal mtDNA is a circular molecule approximately 17,000 base pairs (bp) in length and many copies of it occur in each cell, making it relatively easy to purify. The rate of mtDNA nucleotide substitutions among higher vertebrates is approximately 5–10 times that of nuclear DNA (Moritz et al. 1987), which further enhances the resolution capability of mitochondrial DNA studies. Finally, the strict maternal inheritance of mtDNA (Hutchinson et al. 1974; Gyllensten et al. 1985) may prove particularly useful in the context of fish population studies, especially in philopatric species such as walleye (Billington and Hebert 1988).

Currently, there is only one technical disadvantage to this technique. Studies of polymorphs rely either on mtDNA purified by ultracentrifugation procedures, or on hybridization of a probe with the mtDNA present in total DNA extracted from soft-tissue samples such as gonads, liver, or heart (Gonzalez-Villasenor et al. 1986; Ferris and Berg 1987). In both cases the tissue donors are killed except when eggs and sperm are collected from live fish during the spawning season.

The purpose of this paper is to describe a technique that permits analysis of the mtDNA genotypes of live fish. We extracted total DNA from blood samples, digested it with restriction endonucleases, and visualized mtDNA restriction fragment patterns by hybridization with ^{32}P-labeled mtDNA. A similar approach has been used successfully in meadow voles *Microtus pennsyl-*

vanicus (Plante et al. 1987) and humans (Kan et al. 1977; Denaro et al. 1981; Johnson et al. 1983).

Methods

Twenty-four walleyes were collected from Lake Simcoe, Ontario, on October 14, 1987. We drew small (0.5-mL) blood samples prior to killing the fish and removing their livers. These samples were returned on ice to the laboratory. Mitochondrial DNA was extracted from the liver of each fish and purified by the protocol of Billington and Hebert (1988). Blood samples were stored at 4°C until processed to extract total DNA.

Collection of blood samples from fish.—Small (100–500-μl) blood samples were collected from the sinus venosus (Wingo and Muncy 1984) with a 25-mm-long, 23-gauge hypodermic needle attached to a 3-mL syringe. The syringe and needle were rinsed with 0.5 M EDTA (pH 8.0) immediately before use. The EDTA served as an anticoagulant (Hesser 1960) and as an inhibitor of nonspecific nuclease activity. The blood was immediately transferred to 1 mL of 1 × SSC (standard sodium citrate: 150 mM NaCl; 15 mM sodium citrate) in a 1.5-mL microcentrifuge tube. Blood samples could then be stored at 4°C for at least a month without loss or degradation of DNA.

Extraction of total DNA.—We extracted total DNA from blood cells by a protocol modified from Plante et al. (1987). Blood cells were pelleted in a microcentrifuge for 30 s at 16,000 × gravity, following which the supernatant was discarded. The cells were hemolyzed by rapid suspension in 750 μL of sterile double-distilled water, then immediately made isotonic by adding 250 μL of 5 × SSC. The red cell ghosts and white blood cells were pelleted by centrifugation for 2 min and the supernatant was discarded. The pellets were resuspended in 400 μL of 0.2 M sodium acetate (pH 5.2), lysed by treatment with 15 μL of 20% SDS (sodium dodecyl sulfate) for 5 min at room temperature, and incubated for 1–6 h at 65°C with 25 μL proteinase K (10 mg/mL).

The resulting slurry was extracted once with phenol, twice with phenol–chloroform–isoamyl alcohol (25:24:1), and once with chloroform–isoamyl alcohol (24:1), the DNA was recovered by ethanol precipitation (Maniatis et al. 1982). Pellets were then resuspended in 100 μL of 1 mM TE (1 mM tris-HCl, pH 7.6; 0.1 mM EDTA, pH 8.0) and stored at 4°C. Total DNA samples stored in this manner were stable for fish for at least 3 months (N. Billington, unpublished data) and for

the crustacean *Daphnia* sp. for up to 2 years (Stanton 1988).

Yields of up to 300 ng of total DNA were obtained from the 0.5-mL blood samples. Yields were improved by back-extracting the first phenol phase with 300 μL TE as recommended by Maniatis et al. (1982).

Restriction analysis of mtDNA.—A 15-μL aliquot of each total DNA extract (containing approximately 30–45 ng DNA per aliquot) was separately digested with 3–6 units of the restriction endonucleases *Hind* III (A/AGCTT), *Nco* I (C/CTAGG), *Sca* I (AGT/ACT), and *Eco* RI (G/AATTC), plus *Sal* I (G/TCGAC) double digest).[1] Ten units of RNase T1 were added to each sample, and the total volume was adjusted to 30 μL with 1 × reaction buffer. Digests were run overnight (>16 h) under buffer and temperature conditions recommended by the suppliers (Bethesda Research Laboratories). After a minimum of 12 h, 5 μL of stop-dye buffer (7 M urea; 50% sucrose; 1 mM EDTA; 0.1% bromophenol blue) were added to terminate digestion and to act as a loading buffer. The samples were then loaded onto horizontal submarine agarose gels (0.7% in 1 × TAE: 40 mM tris-acetate; 1 mM EDTA) and electrophoresed overnight at 20 V. Twenty-five nanograms of phage lambda-DNA digested with *Hind* III were loaded in one lane of each gel as a size standard.

After electrophoresis was completed, the DNA was stained with ethidium bromide and photographed under ultraviolet illumination. The gel was transferred to a glass dish, the DNA was denatured with 1.5 M NaCl and 0.5 M NaOH, and the gel was neutralized with 1 M tris-HCl, pH 8.0, and 1.5 M NaCl. We then immobilized DNA fragments on a nitrocellulose filter (S&S BA85, Mandel) by Southern transfer (Southern 1975; Maniatis et al. 1982), using a 20 × SSC transfer buffer for 16 h at room temperature.

Preparation of ^{32}P-labeled probe.—A walleye mtDNA probe with a specific activity of at least 10^8 disintegrations·min^{-1}·μg^{-1} was prepared with [α-^{32}P]dCTP and [α-^{32}P]dATP by nick-translation (Rigby et al. 1977; Maniatis et al. 1982). The reaction mixture contained 5 μL (0.5 μg) mtDNA, 4 μL sterile double-distilled water, 2 μL buffer (500 mM tris-HCl, pH 7.6; 50 mM MgCl$_2$; 100 mM β-mercaptoethanol; 100 μg bovine serum albumin/mL), 1 μL each of unlabeled dGTP and dTTP

---

[1]A = adenine, C = cytosine, G = guanine, T = thymine.

(2 mM stock solutions), 2.5 μL each of the ^{32}P-labeled nucleotide stocks (Amersham; 111 TBq/mmol, 370 MBq/mL), 1 μL DNase (100 ng/mL), and 1 μL DNA polymerase I (10 units). The mixture was incubated at 15°C for 1 h, after which the reaction was terminated by the addition of 20 μL of stop buffer (0.2 M EDTA, pH 8.0; 1% SDS; 40 mg blue dextran/mL, 0.1 mg bromophenol blue/mL). The nick-translated DNA was separated from unincorporated nucleotides by means of Sephadex G-50 columns (Maniatis et al. 1982). Previous experience, with *Daphnia* mtDNA probes, had shown that 1 μg of probe prepared in this fashion was sufficient to hybridize with at least 400 blotted digestions (Stanton 1988). In order to probe the lambda standard, we used the same procedure to label 500 ng of lambda-DNA.

Using a protocol similar to that described by Gonzalez-Villasenor et al. (1986), we have cloned walleye mtDNA into the phage vector EMBL-3. Rather than growing up the cloned probe, however, we found it more convenient to use walleye mtDNA purified by density-gradient centrifugation as a hybridization probe. Walleye mtDNA purified from a Lake Simcoe fish by the protocol of Billington and Hebert (1988) was used to prepare the ^{32}P-labeled probe in all hybridizations reported here.

Hybridization.—The nitrocellulose filter containing the walleye mtDNA restriction fragments was sealed in a plastic bag. The filter was prehybridized with 15–20 mL of hybridization buffer (5 × SSC; 0.5% SDS; 5 × Denhardt's solution) and 1 mL of salmon sperm solution (10 mg salmon sperm DNA/mL [Sigma D-1626] in 1 mM TE, sheared by sonication, denatured by boiling for 10 min, and quenched on ice). The buffer was introduced by cutting the corner of the bag and injecting the solution with a pipette, then resealing the bag. Prehybridization was carried out at 68°C for 16 h. The walleye and lambda probes were denatured by boiling for 10 min, quenched on ice for at least 30 min, and then added to a tube containing 700 μL of hybridization buffer. The contents of this tube were mixed gently and injected into a cut corner of the plastic bag. After as many air bubbles as possible were expressed, the bag was resealed and inverted several times to allow the contents to mix thoroughly. The bag was sealed in a larger plastic bag, then incubated, flat, at 68°C for a minimum of 16 h, during which time it was periodically agitated.

After the incubation period, the liquid contents of the bag were drained into a radioactive waste

container and the filter was washed twice in 500 mL of a 2 × SSC:0.1% SDS solution for 15 min at room temperature. This was followed by a 1-h wash in 500 mL of a second posthybridization buffer (1 × SSC:0.1% SDS) at 65°C, and a final 1-h wash in 500 mL of 0.2 × SSC:0.1% SDS buffer, also at 65°C. The filter was then dried between two pieces of Whatman filter paper and wrapped in cling-film wrap. Autoradiographs were prepared by exposing the filter to X-ray film (Fugi-RX), and by using Dupont Cronex intensifying screens at −70°C for 48–96 h.

Results

We readily visualized mtDNA fragment patterns from five restriction endonucleases by probing the Southern blot of walleye total DNA obtained from blood samples with a ^{32}P-labeled walleye mtDNA probe. Polymorphic fragment patterns were seen for *Sca* I (Figure 1) and *Nco* I, whereas *Eco* RI, *Hind* III, and *Sal* I patterns were monomorphic. These fragment patterns were identical to those obtained when walleye mtDNA samples, purified by the density-gradient centrifugation method from liver tissue (Billington and Hebert 1988), were cut with the same endonucleases (Table 1). However, the data obtained with the blood-sample technique were collected in 2 weeks, whereas it took us 5 weeks to obtain the same amount of information by the purified mtDNA method.

One fish had a mtDNA molecule 0.3 kilobases (kb) larger than that normally found in other walleyes. This length polymorphism was resolved by both methods. For *Sca* I, the length increase occurred in the large "A" fragment (lane 7 in Figure 1.). The length increase also mapped to the large *Hind* III, large *Nco* I, and large *Sal* I fragments. The precise location of this 0.3-kb increase has not been determined, but the above fragments span the area of the walleye mtDNA molecule that includes the D-loop, where other length variants of walleye mtDNA have been recorded (Billington and Hebert 1988).

Discussion

Mitochondrial DNA markers offer exceptional promise as tools in fishery management, both in stock identification and as genetic tags that persist for many generations. To exploit this potential, it is essential that mtDNA markers be detectable in live fish. This crucial advance has now been made. We visualized restriction fragments of mtDNA in total DNA extractions, obtained from

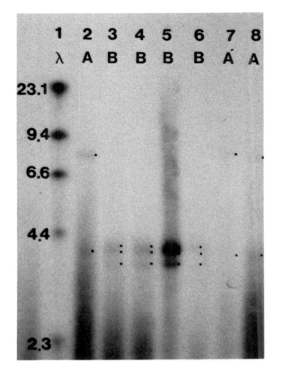

FIGURE 1.—Autoradiograph of a Southern blot of total DNA from seven Lake Simcoe walleyes (lanes 2–8), cut with the endonuclease *Sca* I and probed with ^{32}P-labeled walleye mtDNA. Letters A and B refer to the *Sca* I fragment patterns obtained for walleye mitochondrial DNA (Billington and Hebert 1988). Note length variant A' in lane 7. Lambda-DNA digested with endonuclease *Hind* III and probed with a ^{32}P-labeled lambda probe is included in lane 1 as a size standard. Numbers at the left are fragment sizes in kilobase pairs.

method of end-labeling. A comparison of the techniques is presented in Table 2.

Various tissues were considered as possible nonlethal sources of total DNA. In earlier experiments, eggs, fin, muscle, and skin tissue were used as sources of total DNA, in addition to blood. No DNA could be extracted from fin or skin tissues, although egg and muscle tissue proved satisfactory sources of total DNA for probing (N. Billington, unpublished data). In the context of sampling live fish, small blood samples obtained with a syringe and hypodermic needle were much easier to collect than were muscle samples, which required a biopsy (e.g., Uthe 1971; McAndrew 1981). In walleyes, blood samples may be taken without an anaesthetic (Wingo and Muncy 1984; N, Billington, personal observation). Furthermore, yields of total DNA were several times greater from 0.1–0.5 mL of blood than from 0.3–0.5 g of fresh or frozen muscle tissue (N. Billington, unpublished data). Blood samples have another advantage for field collection of data. They may be kept at room temperature for a few days or stored at 4°C for at least a month (Plante et al. 1987; N. Billington, unpublished data), whereas muscle tissue samples need to be powdered in liquid nitrogen or stored frozen.

Mitochondrial DNA markers promise to be a valuable tool for fisheries management. The unique combinations of mtDNA restriction fragment patterns found in walleyes could be used as genetic markers in introduced fish (Billington and Hebert 1988). These mtDNA markers will be passed on by females to their offspring because of the maternal inheritance of mtDNA. By employing the blood-sample technique to screen fish, researchers could selectively breed those female walleyes with distinctive mtDNA fragment patterns in hatcheries to produce mitochondrially marked lines. Alternatively, fish to be used in transfer programs could be screened for mtDNA

small (0.5-mL) blood samples collected from live walleyes, by hybridization with a ^{32}P-labeled walleye mtDNA probe. Mitochondrial DNA fragment patterns obtained by this method were identical to those obtained by the more time-consuming

TABLE 1.—Comparison of mitochondrial DNA (mtDNA) fragment patterns A and B (as defined by Billington and Hebert 1988) obtained for 24 Lake Simcoe walleyes by probing total DNA from blood samples and end-labeling purified mtDNA from liver samples. A' is a length polymorphism approximately 300 base pairs longer than the A pattern.

	Method		
Endonuclease	Probing	End-labeling	Notes
Hind III	23 A, 1 A'	23 A, 1 A'	Monomorphic
Eco RI plus *Sal* I	23 A, 1 A'	23 A, 1 A'	Monomorphic, double digest
Nco I	6 A, 1 A', 17 B	6 A, 1 A', 17 B	Polymorphic
Sca I	6 A, 1 A', 17 B	6 A, 1 A', 17 B	Polymorphic

TABLE 2.—Comparison of three techniques for visualizing patterns of mitochondrial DNA (mtDNA) restriction fragments: (1) visualization of purified mtDNA fragments by end-labeling or ethidium bromide staining (e.g., Ferris and Berg 1987; Billington and Hebert 1988), (2) probing mtDNA fragments in digests of total DNA extracted from soft tissue samples (e.g., Gonzalez-Villasenor et al. 1986; Ferris and Berg 1987), and (3) probing mtDNA fragments in digests of total DNA extracted from blood samples (current study).

Feature	End-labeling or staining of purified mtDNA	mtDNA probes of total DNA	
		Soft tissues	Blood
Tissue requirements	Liver, heart, gonads	Liver, heart, gonads	Blood
Amount of tissue	1–5 g	0.1–1 g	0.5 mL
Effect on specimen	Usually death	Usually death	Little trauma
Special requirements	Ultracentrifuge	mtDNA probe	mtDNA probe
Number of restriction enzymes	>30	6–12	6–12
Number of samples per run	15–30	24–400	24–400
Laboratory time (six restriction enzyme analyses)	350 h	170 h	150 h
Laboratory costs per sample (ratio)	7	3	3
Level of resolution	20–17,000 base pairs	500–17,000 base pairs	500–17,000 base pairs
Scientific interpretation	Excellent	Intermediate	Intermediate
Fisheries applications	Detection of markers for use in stock identification	Visualization of known markers, surveys	Visualization of known markers, nonlethal surveys

markers so that only those of a known clone would be introduced to other water bodies. The fate of those fish in the wild, and of their offspring, could be monitored by subsequent determination of mtDNA fragment patterns. Such sampling could easily be performed as part of the annual surveys of fish populations already undertaken by management personnel.

Mitochondrially marked fish would be more useful for stocking programs than are fish with variant allozyme markers. Although allozyme markers are useful for following single cohorts of introduced fish (Ward and Clayton 1975; Schweigert et al. 1977; Murphy et al. 1983), they soon lose value because genotypic frequencies tend to return to Hardy–Weinberg equilibrium after a single generation. Furthermore, current physical methods of marking fish, such as fin clips or artificial tags, often are inefficient and expensive, and do not allow fish to be monitored for more than a single generation.

The blood-sample technique may also be suitable for interspecific comparisons of variation in fish mtDNA. For example, Gonzalez-Villasenor et al. (1986) used a cloned mtDNA probe from the mummichog *Fundulus heteroclitus* as an interspecific hybridization probe. They were able to hybridize the *Fundulus* probe and complementary mtDNA fragments obtained from members of six fish families (Centrarchidae, Sciaenidae, Salmonidae, Cichlidae, Percichthyidae, Ictaluridae). We anticipate that the homology in DNA sequences

should be sufficient to allow walleye mtDNA to be used to probe related species, such as sauger *Stizostedion canadense,* zander *S. lucioperca,* yellow perch *Perca flavescens,* and European perch *P. fluviatilis.*

In summary, we have demonstrated that walleye mtDNA fragment patterns in total DNA digests, obtained from small blood samples collected from live fish, can be vizualized by hybridization with a ^{32}P-labeled walleye mtDNA probe. This method allows mtDNA markers for up to six diagnostic restriction endonucleases to be determined easily in fish used in stocking programs. It also allows mtDNA variation in large numbers of live fish to be surveyed routinely with a limited number of endonucleases in order to examine differences among stocks. For example, nine mitochondrial clones can be resolved in 10 Great Lakes walleye populations, based on restriction patterns for six endonucleases (Billington and Hebert 1988). Thus, it should be possible with the probing technique to determine the relative proportions of these clones in additional walleye populations without the need to kill fish. Because most of the techniques are relatively simple, it should be possible for many fisheries laboratories to perform the total DNA extractions, digests, and Southern blotting with only minimal expenditures for equipment or staff training. Only the hybridization stage with the radioactively labeled ^{32}P-probes would need to be conducted in a specialized laboratory. Large-scale implementation of mtDNA markers in fisheries management should be practicable almost immediately with the techniques we have described.

Acknowledgments

We thank David Stanton for developing the DNA hybridization system at Windsor. Lake Simcoe walleyes were collected with the assistance of Ontario Ministry of Natural Resources personnel at the Sibbald Point Lake Simcoe Fisheries Assessment Unit. Drew Bodaly, Peter Grewe, Steven Schwartz, David Stanton, and Bob Ward commented on earlier drafts of the manuscript. We also thank Gary Winans for his suggestions on improving the manuscript. This work was funded by Fisheries and Oceans Canada, Great Lakes Fishery Commission, and NSERC Natural Sciences and Engineering Research Council (Operating and Strategic) grants to P.D.N.H.

References

Avise, J. C., E. Bermingham, L. G. Kessler, and N. C. Saunders. 1984. Characterization of mitochondrial DNA variability in a hybrid swarm between subspecies of bluegill sunfish (*Lepomis macrochirus*). Evolution 38:931–941.

Avise, J. C., and seven coauthors. 1987. Intraspecific phylogeography: the mitochondrial DNA bridge between population genetics and systematics. Annual Review of Ecology and Systematics 18:489–522.

Berg, W. J., and S. D. Ferris. 1984. Restriction endonuclease analysis of salmonid mitochondrial DNA. Canadian Journal of Fisheries and Aquatic Sciences 41:1041–1047.

Bermingham, E., and J. C. Avise. 1986. Molecular zoogeography of freshwater fishes in the southeastern United States. Genetics 113:939–965.

Billington, N., and P. D. N. Hebert. 1988. Mitochondrial DNA variation in Great Lakes walleye (*Stizostedion vitreum*) populations. Canadian Journal of Fisheries and Aquatic Sciences 45:643–654.

Bodaly, R. A. 1980. Pre- and post-spawning movements of walleye, *Stizostedion vitreum*, in Southern Indian Lake, Manitoba. Canadian Technical Report of Fisheries and Aquatic Sciences 931.

Colby, P. J., and S. J. Nepszy. 1981. Variation among stocks of walleye (*Stizostedion vitreum vitreum*): management implications. Canadian Journal of Fisheries and Aquatic Sciences 38:1814–1831.

Denaro, M., and six coauthors. 1981. Ethnic variation in *Hpa* I endonuclease cleavage patterns of human mitochondrial DNA. Proceedings of the National Academy of Sciences of the USA 78:5788–5772.

Ferguson, R. G., and A. J. Derksen. 1971. Migrations of adult and juvenile walleyes (*Stizostedion vitreum vitreum*) in southern Lake Huron, Lake St. Clair, Lake Erie, and connecting waters. Journal of the Fisheries Research Board of Canada 28:1133–1142.

Ferris, S. D., and W. J. Berg. 1987. The utility of mitochondrial DNA in fish genetics and fishery management. Pages 277–299 *in* N. Ryman and F. Utter, editors. Population genetics & fishery management. University of Washington Press, Seattle.

Gonzalez-Villasenor, L. I., A. M. Burkhoff, V. Corces, and D. A. Powers. 1986. Characterization of cloned mitochondrial DNA from the teleost *Fundulus heteroclitus* and its usefulness as an interspecies hybridization probe. Canadian Journal of Fisheries and Aquatic Sciences 43:1866–1872.

Grewe, P. M. 1987. Divergence of mitochondrial DNA among the subfamily Salmoninae with special reference to the genus *Salvelinus*. Masters thesis. University of Windsor, Windsor, Canada.

Gyllensten, U., D. Wharton, and A. C. Wilson. 1985. Maternal inheritance of mitochondrial DNA during backcrossing of two species of mice. Journal of Heredity 76:321–324.

Gyllensten, U., and A. C. Wilson. 1987. Mitochondrial DNA of salmonids: inter- and intraspecific variability detected with restriction enzymes. Pages 301–317 *in* N. Ryman and F. Utter, editors. Population genetics & fishery management. University of Washington Press, Seattle.

Hesser, E. F. 1960. Methods for routine fish hematology. Progressive Fish-Culturist 22:164–171.

Hutchinson, C. A., III, J. E. Newbold, S. S. Potter, and M. H. Edgell. 1974. Maternal inheritance of mammalian mitochondrial DNA. Nature (London) 251:536–538.

Johnson, M. J., D. C. Wallace, S. D. Ferris, M. C. Rattazzi, and L. L. Cavalli-Sforza. 1983. Radiation of human mitochondrial DNA types analysed by restriction endonuclease cleavage patterns. Journal of Molecular Evolution 19:255–271.

Kan, Y. K., A. M. Dozy, R. Trecartin, and D. Todd. 1977. Identification of a nondeletion effect in α-Thalassemia. New England Journal of Medicine 297:1081–1084.

Maniatis, T., E. F. Fritsch, and J. Sambrook. 1982. Molecular cloning: a laboratory manual. Cold Spring Harbor Laboratory, Cold Spring, New York.

McAndrew, B. J. 1981. Muscle biopsy technique for fish stock management. Veterinary Record 108:516.

Moritz, C., T. E. Dowling, and W. M. Brown. 1987. Evolution of animal mitochondrial DNA: relevance for population biology and systematics. Annual Review of Ecology and Systematics 18:269–92.

Murphy, B. R., L. A. Neilsen, and B. J. Turner. 1983. Use of genetic tags to evaluate stocking success for reservoir walleyes. Transactions of the American Fisheries Society 112:457–463.

Plante, Y., P. T. Boag, and B. N. White. 1987. Nondestructive sampling of mitochondrial DNA from voles (*Microtus*). Canadian Journal of Zoology 65:175–180.

Rigby, P. J., M. Dieckmann, C. Rhodes, and P. Berg. 1977. Labeling deoxyribonucleic acid to high specific activity *in vitro* by nick translation with DNA polymerase I. Journal of Molecular Biology 113:237–251.

Schweigert, J. F., F. J. Ward, and J. W. Clayton. 1977. Effects of fry and fingerling introductions on walleye (*Stizostedion vitreum vitreum*) production in West Blue Lake, Manitoba. Journal of the Fisheries Research Board of Canada 34:2142–2150.

Southern, E. 1975. Detection of specific sequences among DNA fragments separated by gel electrophoresis. Journal of Molecular Biology 98:503–517.

Stanton, D. J. 1988. Evolution of asexuality in *Daphnia pulex:* implications of mitochondrial DNA analysis. Doctoral dissertation. University of Windsor, Windsor, Canada.

Thomas, W. K., R. E. Whithler, and A. T. Beckenbach. 1986. Mitochondrial DNA analysis of Pacific salmonid evolution. Canadian Journal of Zoology 64:1058–1064.

Uthe, J. F. 1971. A simple field technique for obtaining small samples of muscle from living fish. Journal of the Fisheries Research Board of Canada 28:1203–1204.

Ward, F. C., and J. W. Clayton. 1975. Initial effects of fry introduction on year-class strengths of West Blue lake walleye *Stizostedion vitreum vitreum* (Mitchill) using fry marked with distinctive malate dehydrogenase isozyme phenotypes as an identifying mark. Internationale Vereinigung für Theoretische und Angewandte Limnologie Verhandlungen 20:2442–2451.

Wilson, G., W. K. Thomas, and A. T. Beckenbach. 1985. Intra- and interspecific mitochondrial DNA sequence divergence in *Salmo:* rainbow, steelhead, and cutthroat trouts. Canadian Journal of Zoology 63:2088–2094.

Wilson, G. M., W. K. Thomas, and A. T. Beckenbach. 1987. Mitochondrial DNA analysis of Pacific Northwest populations of *Oncorhynchus tshawytscha*. Canadian Journal of Fisheries and Aquatic Sciences 44:1301–1305.

Wingo, W. M., and R. J. Muncy. 1984. Sampling walleye blood. Progressive Fish-Culturist 46:53–55.

Wolfert, D. R., and H. D. Van Meter. 1978. Movements of walleyes tagged in eastern Lake Erie. New York Fish and Game Journal. 25:16–22.

American Fisheries Society Symposium 7:499–513, 1990
© Copyright by the American Fisheries Society 1990

Genetic Marking of Fish by Use of Variability in Chromosomes and Nuclear DNA

Ruth B. Phillips

Department of Biological Sciences, University of Wisconsin-Milwaukee
Post Office Box 413, Milwaukee, Wisconsin 53201, USA

Peter E. Ihssen

Fisheries Research Branch, Ontario Ministry of Natural Resources
Box 5000, Maple, Ontario L6A 1S9, Canada

Abstract.—Chromosome and nuclear DNA polymorphisms within a species can be used as genetic markers by methods described in this paper. Chromosome polymorphisms found in fish include variations in number, structure, and banding patterns. Nuclear DNA polymorphisms include variations in size and sequence of restriction fragments obtained from single-copy or multiple-copy genes. The relative advantages of using different genetic markers—including variation in chromosomes, allozymes, nuclear DNA, and mitochondrial DNA—involve availability, cost, and expertise.

Genetic variations in chromosomes and DNA sequences have potential as markers to differentiate fish populations and produce marked stocks. Until recently, only variations in protein allozymes were used for this purpose (reviewed in Utter et al. 1987). The use of chromosome markers has been limited because they have been described for few species, they require examination of dividing cells, and they are more time-consuming to analyze than are allozyme markers (reviewed by Thorgaard and Allen 1987). The technology for DNA markers, which have been investigated only recently in fish, is more time-consuming and costly than that required for protein markers (Allendorf et al. 1987). The use of DNA markers for stock identification in fish has been limited to differences in mitochondrial DNA (reviewed in Ferris and Berg 1987), partly because very few nuclear DNA sequences have been examined.

In this paper we describe some potential chromosome and nuclear DNA markers and the methods by which they can be detected. In the future, chromosome markers probably will be detected by flow cytometry or by analysis of DNA sequences. Because DNA sequence analysis should yield the most sensitive markers, we also examine how the methods involved can be simplified so that they are quicker and more cost-effective.

Chromosome Markers

Chromosome Polymorphisms

Variations in chromosome number.—The most common type of chromosome number variation in fish results from Robertsonian translocations. In these mitotic events, which do not change the amount of genetic material, either two uniarmed chromosomes fuse to form a biarmed chromosome, or a biarmed chromosome splits to form two uniarmed chromosomes. Such translocations have been reported for many species (reviewed in Gold 1979; Sola et al. 1981; Thorgaard 1983b; Hartley 1987) and especially for salmonids (Allendorf and Thorgaard 1984) (Table 1).

The most extensive population studies on Robertsonian translocations in salmonids to date have been done on rainbow trout (Thorgaard 1983a) and pink salmon (Phillips and Kapuscinski 1987, 1988). The chromosome number in rainbow trout varies between 58 and 64. The variation shows a geographic pattern; the majority of west coast populations either are homogeneous or vary in only one chromosome pair (e.g., 58, 59, and 60 in a given population). In the case of pink salmon, which have a 2-year life cycle, all fish spawned in even-numbered years that have been examined to date ($N = 22$) have a chromosome number of 52, whereas odd-year fish ($N = 76$) have numbers varying between 52 an 54, the majority of individuals in most populations having 53 or 54. In the fish with 53 chromosomes, one member of the chromosome pair with the nucleolar organizer region has split to form two chromosomes; in the fish with 54 chromosomes, this splitting has occurred in both members of the chromosome pair. In some of the odd-year populations, an additional rearrangement—an inversion—has occurred, so that there are two distinct types of karyotypes

TABLE 1.—Intraspecific Robertsonian polymorphisms in salmonid species.

Species	2n[a]	NF[b]	References
Salmo			
Altantic salmon *S. salar*	54–56	72	Roberts (1968, 1970)
		74	Barsienne (1981)
	56–58	74	Hartley and Horne (1984)
Brown trout *S. trutta*	77–82	102	Zenzes and Voiculescu (1975)
Oncorhynchus			
Pink salmon *O. gorbuscha*	52–54	104	Gorshkov and Gorshkova (1981)
	52–53	104	Phillips and Kapuscinski (1987)
	52–54	104	Phillips and Kapuscinski (1988)
Sockeye salmon *O. nerka*	56–58	104	Fukuoka (1972), Gorshkov and Gorshkova (1978), Thorgaard (1978)
Cutthroat trout *O. clarki*	64–68	104	Gold and Gall (1977), Loudenslager and Thorgaard (1979)
Rainbow trout *O. mykiss*	58–64	104	Ohno et al. (1965), Gold (1977), Thorgaard (1983a), Hartley and Horne (1982)
Salvelinus			
Arctic char *S. alpinus*	78–80	98	Viktorovsky (1978), Vasilev (1975)
Iwana *S. leucomaenis*	84–86	100	Ueda and Ojima (1983b)
Dolly Varden *S. malma*	82	98	Ueda and Ojima (1983b)
	76–78	96	Viktorovsky (1978)
	78–80	98	Viktorovsky (1978)
Coregonus			
Common whitefish *C. laveretus*	80–82	102	Viktorovsky and Ermolenko (1981)
Broad whitefish *C. nasus*	58–60	92	Viktorovsky and Ermolenko (1981)

[a] 2n is the diploid chromosome number.
[b] NF represents the number of chromosome arms.

with 53 and 54 chromosomes (Figure 1). Thus several odd-year populations can be differentiated on the basis of this polymorphism (Figure 2).

Variation in chromosome number in natural populations also occurs in the form of spontaneous polyploids, especially triploids. Induction of triploidy could be used to mark fish stocks, but usually it is employed to obtain sterile fish. This topic has been reviewed recently (Thorgaard 1983b; Thorgaard and Allen 1987).

Variations in chromosome structure.—Chromosome structural changes include duplications, deficiencies, translocations, and inversions. Many of these rearrangements could be detected with chromosome banding techniques, such as Giemsa (G) or replication banding, that produce species-specific banding patterns and allow identification of each chromosome pair (Comings 1978); but these techniques are not routinely used on fish chromosomes. Intraspecific chromosome rearrangements, such as inversions that change the number of chromosome arms or arm ratios, can be detected without chromosome banding and have been reported in several species, including the goodeid *Ilyodon fucidens* (Turner et al. 1985),

Atlantic salmon (Roberts 1970; R. B. Phillips and S. E. Hartley, unpublished), and chum salmon *Oncorhynchus keta* (Kulikova 1971). In the case of the inversion in pink salmon, described above, the nucleolar organizer region (NOR) is located within the inverted section so that the chromosome with the inversion can be identified by means of banding techniques specific for NORs.

Small duplications or inversions can cause changes in the size of the short arms in submetacentrics, which are detectable without G-banding. This type of variation has been widely reported in many fish species (reviewed by Sola et al. 1981).

Variations in chromosome banding patterns.— Apart from the species-specific banding patterns produced by G-banding or replication banding, several banding techniques can be used to reveal intraspecific polymorphisms in vertebrate chromosomes. Variations in the constitutive heterochromatin—the highly repetitive DNA usually found in tandem blocks of varying sizes near centromeres and telomeres—can be detected by C-banding, by various fluorescent banding techniques such as Q-banding (reviewed by Comings 1978; Sumner 1982), and by digestion with specific

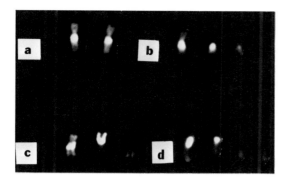

FIGURE 1.—Partial karyotypes of the seventh-largest chromosome pair from pink salmon with different chromosome numbers, stained with CMA3 to reveal the nucleolar organizer regions (NORs). Represented are fish with, (a) 52 chromosomes, (b) 53 chromosomes without an inversion, (c) 53 chromosomes with inversion of the NOR to the short arm, and (d) 54 chromosomes with inversion of the NOR (d).

restriction enzymes followed with Giemsa staining (Miller et al. 1983). Many fluorescent dyes are specific for DNA of a particular base sequence (reviewed by Schweizer 1981) and usually reveal a subset of the repetitive DNA in lower vertebrates (Schmid 1980; Phillips and Hartley 1988). Variations in the size and location of the NORs, which are the sites of the multiple-copy ribosomal RNA genes, can be detected by silver staining (Howell and Black 1980) and by chromomyocin A3 (CMA3) staining, which stains NORs regardless of transcriptional activity (Amemiya and Gold 1986).

Chromosome banding methods that detect intraspecific variation in the constitutive heterochromatin have not been used routinely in studies of fish chromosomes. A summary of C-banded karyotypes from the published literature through 1984 (Gold et al. 1986) listed only 30 fish species. Large C bands were found primarily in salmonids. However, several laboratories are now routinely using C-banding, and karyotypes with prominent C bands have been reported in blenniids (Garcia et al. 1987), scorpaenids (Thode et al. 1985), gobiids (Giles et al. 1985), and prochilodontids (Feldberg et al. 1987). Population studies have revealed stock-specific variations in heterochromatin in several *Salvelinus* species including iwana (Ueda and Ojima 1983a, 1983b) lake trout *S. namaycush* (Phillips et al. 1989b and Arctic char (Pleyte et al. 1989).

Quinacrine stains a subset of C bands in several salmonid species (see Table 2), and intraspecific

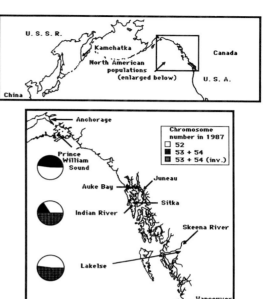

FIGURE 2.—Locations in the Pacific basin where pink salmon with variable chromosome counts were collected. Populations sampled in 1986 and 1987 are shown in the lower map. All individuals sampled in 1986 had 52 chromosomes. Pie charts show the proportions of individuals in 1987 found with 52 chromosomes, 53 or 54 chromosomes with no inversion, and 53 or 54 chromosomes with the inversion (inv) of the chromosome containing the nucleolar organizer region.

variation occurs in the presence of some of these Q bands (presence–absence or +/− variants; see Figure 3). In Atlantic salmon, 10–13 of the C bands also stained with quinacrine in North American fish from New Brunswick and Maine, but not in European fish from Scotland (Phillips and Hartley 1988). In Arctic char, the average number of Q bands in Scottish fish was 9.9, compared with 3.2 in fish from Northeast Territories (Pleyte et al. 1988). Variation in the size and presence of Q bands on specific metacentric chromosomes was found to be heritable in lake trout (Phillips and Ihssen 1985b, 1986a), and stock differences in the frequencies of +/− variants on several chromosomes were found in lake trout (Phillips and Ihssen 1986b; Phillips et al. 1989b). The largest difference was between the Seneca Lake, New York, stock and the western Great Lakes stocks. Genetic distances based on the

TABLE 2.—Intraspecific heterochromatin polymorphisms in salmonids.

Species	C bands	Q bands[a]	References
Salmo			
Atlantic salmon	+	+	Hartley and Horne (1984), Phillips and Hartley (1988)
Brown trout	+	−	Zenses and Volisecu (1975), Phillips and Hartley (1988)
Oncorhynchus			
Pink salmon	+	−	Phillips (unpublished)
Chum salmon	+	+	Phillips (unpublished)
Coho salmon[b]	+	+	Phillips (unpublished)
Sockeye salmon[c]	+	[c]	Thorgaard (1978)
Chinook salmon[d]	+	+	Phillips et al. (1985)
Rainbow trout	+	+	Thorgaard (1976), Phillips and Hartley (1988)
Salvelinus			
Arctic char	+	+	Pleyte et al. (1989)
Brook trout[e]	+	+	Phillips and Zajicek (1982)
Iwana	+	+	Ueda and Ojima (1983a)
Dolly Varden	+	+	Abe and Muramoto (1974)
Lake trout	+	+	Phillips and Ihssen (1986a), Phillips et al. (1988b)

[a]Quinacrine stains a subset of C bands in each species.
[b]*O. kisutch.*
[c]Not examined for Q bands.
[d]*O. tshawytscha.*
[e]*S. fontinalis.*

frequencies of these chromosome markers were similar to those calculated on the basis of allozyme frequencies for these stocks (Ihssen et al. 1988).

Intraspecific variation in the size, number, and chromosomal location of NORs revealed by silver staining and CMA3 staining has been found in several fish species. The CMA3 stains the NORs, regardless of transcriptional activity, in many fish and amphibian species (Schmid 1982; Amemiya and Gold 1986), whereas silver stains only the active NORs. The most extensive studies have been done on cyprinids (Gold 1984; Gold and Amemiya 1986) and on salmonids (Phillips and Ihssen 1985a; Phillips et al. 1986b, 1988, 1989a). Intraspecific variation in the total number and chromosomal location of NORs has been found in members of the genus *Salvelinus* (Table 3 and Figure 4). Although the total number and chromosomal location of NORs are constant for different cells of a given individual, the number of NORs varies from 4 to 8 among individual brook trout, from 3 to 12, among lake trout, and from 2 to 8 among Arctic char. Population surveys of lake trout and arctic char have revealed stock-specific differences in the average number of NORs per diploid genome (Table 4).

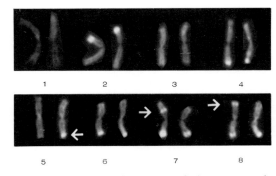

FIGURE 3.—Partial karyotype of the metacentric chromosomes stained with quinacrine, showing Q-band polymorphisms in lake trout. Arrows indicate presence–absence variants.

Methods for Detecting Chromosome Markers in Fish

Methods for preparing chromosome slides and staining.—Chromosome methods for fish have been reviewed recently (Thorgaard and Disney 1990) and are summarized in diagram form in Figure 5. Some of the methods are noninvasive, but others kill the fish (Kligerman and Bloom 1977). Methods that do not kill the fish include culturing blood, (Hartley and Horne 1985), regenerating tissue from clipped fins, and genotyping the parents by examining offspring in the embryonic stage. Both haploid and diploid embryos can be examined and parental genotypes can be deduced in this manner. Methods requiring death of the fish include preparations from kidney and embryonic tissue. As shown in Figure 5, the first step is to obtain either mitotically active cells such as those found in embryonic, gill, or kidney tissues, or to culture active cells from blood or other tissues. Next cells are treated with colchicine to produce condensed chromosomes, exposed to a hypotonic solution to swell the cells to enhance spreading, and then fixed in Carnoy's fixative. A suspension of fixed cells is dropped onto slides, and the slides are stained. Finally, chromosomes are examined under the microscope and photographs are made of the karyotypes.

Although variations in chromosome number can be detected without specific banding techniques, detection of many of the polymorphisms requires specific staining of chromosome slides. Variations in the total amount of constitutive heterochromatin are detected by C-banding. Subsets of this repetitive DNA may be stained with various fluorescent dyes, such as quinacrine and DAPI for DNA sequences rich in adenosine and

TABLE 3.—Number of nucleolar organizer regions (NORs) per diploid genome in species of *Salvelinus*. Numbers indicate the range of NORs found in different individuals of a given species.

Species (N)	Chromosomal location of NORs				Total NORs
	Acrocentric short arms	Acrocentric telomeres	Submetacentric telomeres	Metacentric telomeres	
Bull trout[a] (12)	2				2
Brook trout (20)	2–8[b]	0–2		0–1	4–10
Lake trout (91)	2–6[b]	2–4[b]		0–2	4–12
Arctic char (60)	0–2	0–2	2[b]	0–3	2–6
Dolly varden (12)			2		2

[a] *S. confluentus.*
[b] Indicates major NOR sites in species with NORs at more than one chromosomal location.

thymine and with mithromycin and CMA3 for sequences rich in guanine and cytosine (Phillips and Hartley 1988). Banding with specific DNA restriction enzymes can also reveal specific subsets of heterochromatin in fish (Lloyd and Thorgaard 1988; Hartley, personal communication). To detect the nucleolar organizer regions, slides may be stained with silver (Howell and Black 1980) or CMA3 (Amemiya and Gold 1986).

Although many chromosome polymorphisms respond to stains and are easily scored directly from slides, for a detailed analysis it may be desirable to photograph cells and prepare karyotypes. For a quantitative analysis of band numbers and sizes, an image analyzer system can be used either with negatives projected onto a microfiche screen or with photographs. Several programs are available for analysis of karyometric features (Green and Bogart 1980), and we have written a computer program specifically for chromosome-band analysis (Phillips et al. 1989b).

Genotyping parents for production of a marked stock.—In genotyping parent fish for producing a marked stock, methods that do not kill the fish are necessary. Blood culture has obvious advantages, but seasonal variation in the success of blood culture can affect the efficiency of this method (Hartley, Phillips, and Thorgaard, unpublished observations). Another method involves clipping fins of the parent fish and making a direct chromosome preparation from the regenerating tissue a few days later.

When it is difficult to sample tissues from the parents, crosses can be made in various combinations, and desired crosses can be selected by genotyping some of the embryos midway through development. Eggs do not take up much room in hatching trays, and embryos can be genotyped

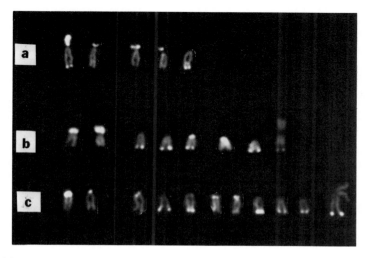

FIGURE 4.—Partial karyotypes from three lake trout stained with CMA3, showing intraspecific variation in the number of chromosomes with CMA3 bands. Fish represented were from (a) Lewis Lake, Wyoming, (b) Seneca Lake, New York, and (c) Killala Lake, Ontario.

TABLE 4.—Intraspecific variation in the mean number of nucleolar organizer regions (NORs) per diploid genome in lake trout and Arctic char.[a]

		Chromosomal location of NORs				
Stock	N	Acrocentric short arms	Acrocentric telomeres	Submetacentric telomeres	Metacentric telomeres	Total NORs
Lake trout						
Wyoming (from Lake Michigan)	32	2.1	1.8		0.5	4.4 ± 1.8
Marquette (south Lake Superior)	45	3.5	2.8	0.2	0.9	7.4 ± 2.2
Michipicoten Island (north Lake Superior)	14	5.5	3.7	0.2	1.0	10.4 ± 2.1
Arctic char						
Northwest Territories	20		0.2	2.0	0.1	2.3 ± 0.6
Labrador	20		1.1	2.0	0.3	3.4 ± 1.0
Scotland[b]	20	0.6	1.2	1.6	2.5	5.9 ± 1.4

[a]Data on lake trout are from Phillips et al. (1989b); data on Arctic char are from Phillips et al. (1988).
[b]Pooled sample including benthic and pelagic fish from Loch Rannoch, Scotland.

before the eye-up stage, so this method is an alternative to direct genotyping of parents. Also, it allows gametes to be collected in the field and crosses to made up to 3 days later in the laboratory. In some cases, both haploid and diploid embryos can be produced from each female, and the genotype of the female parents can be determined directly from the haploid embryos. Haploid embryos survive until after the eye-up stage, and chromosome preparations can be made from them at the same time as they are from diploid embryos. *Genotyping fish from natural populations.*—When fish are sampled from natural populations, it is often easier to obtain tissue samples in the field than to bring many fish to the laboratory. Because chromosome methods involve working with living tissues, this can present a problem for field studies. Although drawing blood is not difficult, the blood must be transported back to the laboratory quickly, on ice, to obtain good results. If fish can be brought back in buckets, there are alternatives to the blood-culture technique. For example, the fish can be injected with colchicine, and their kidney or gill tissues can be sampled a few hours later (gill samples do not require sacrifice of the fish). If fish can be held for a few days, their fins can be clipped and regenerating tissue will become available. Probably the easiest method is to collect gametes (eggs and milt) from spawning fish directly into sealable plastic bags and ship them on ice back to the laboratory, where various matings can be made and embryos can be genotyped a few weeks later.

Future methods.—A major drawback of chromosome markers is the necessity of working with living tissue. If chromosome markers could be identified in cells that are not actively dividing, or

in blood or other tissues that have not been cultured, they would be more practical. Both of these approaches may be possible in the future.

Most of the polymorphisms involving differences in chromosome banding patterns could be detected by analysis of nuclear DNA, whose polymorphisms involve either variations in the repetitive DNA (C and Q bands) or in the ribosomal RNA genes (NORs). Because both nuclear polymorphisms consist of DNA sequences present in multiple copies, their identification should be uncomplicated. When DNA probes specific for nuclear polymorphisms become available, the polymorphisms could be detected by dot blotting or by sandwich hybridization. (discussed subsequently).

In some cases karyotypic features can be determined on interphase cells, which is advantageous because it is quicker to fix cells and prepare slides than to extract DNA from a large number of samples. The ploidy of cells, for example, has been determined from smears containing interphase cells. This can be done by determining the DNA content of nuclei from slides stained with dyes specific for DNA (reviewed in Thorgaard 1983b), and by nucleolar counts after slides are stained with silver nitrate. The second method is applicable only to species with NORs on a single chromosome pair, which is the case for most fish species including salmonids of the genera *Oncorhynchus* and *Salmo*. We have successfully identified triploids from several trout and salmon fingerlings by removing small pieces of gill tissue and placing them directly into small vials of fixative; slides are prepared later from the fixed tissue, and the cell smears are stained with silver nitrate (Phillips et al. 1986a). Most triploid cells have three nucleoli and most diploid cells have

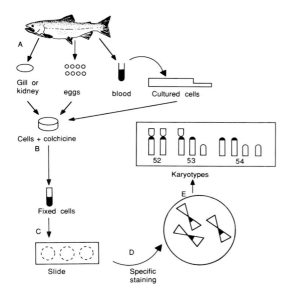

FIGURE 5.—Methods for detection of chromosome polymorphisms. (**A**) Tissue is sampled and treated to obtain dividing cells. Gill, kidney, and embryonic tissues normally divide rapidly, whereas blood must be cultured for a few days to obtain dividing cells; in both cases, mitosis is blocked with colchicine for a few hours to obtain shortened chromosomes. (**B**) After colchicine treatment and hypotonic treatment to swell the cells to enhance spreading, the cells are fixed by the addition of Carnoy's fixative. (**C**) Slides are prepared from fixed cells and (**D**) stained with either Giemsa or specific stains to reveal chromosome bands. (**E**) The preparations are observed under the microscope and polymorphisms are recorded. Photographs may be made for a permanent record of karyotypes.

two; thus they are easy to differentiate. Intraspecific variation in the number of chromosomes with NORs in lake trout can be detected by this method, but only in young fish because nucleoli tend to fuse in the cells of older fish (Phillips, unpublished). As specific DNA probes become available, in situ hybridization with cell nuclei may become an attractive method (discussed later).

A second drawback of chromosome markers is the time and expertise needed to analyze slides. The best markers are those that can be scored quickly from slides. The chromosome number polymorphism in west coast pink salmon is an example. The specific chromosome pair involved in the polymorphism can be identified by silver staining or CMA3 staining. In fish with 52 chromosomes the silver- or CMA3-stained bands are found near the middle of two large metacentric chromosomes. When one of these chromosomes

splits, two much smaller acrocentric chromosomes are produced, one of which has the silver band at the centromere end (see Figure 1).

The technique of flow cytometry allows many karyotypic features to be determined rapidly for dividing cells from blood and other tissues without culturing them. Flow cytometry has been used successfully to detect triploid fish (Thorgaard et al. 1982) and could be used to identify chromosome markers in fish cells. For identification of triploid cells, a fluorescent dye that stains DNA is added to the cell suspension; as single cells flow by the laser beam, the fluorescence is measured and the amount of DNA per cell is determined. In flow-cytometry studies of human chromosome polymorphisms (reviewed by Carrano et al. 1983 and Green et al. 1984), cells are disrupted and single chromosomes flow by the laser beam, where the fluorescence of each chromosome is measured. This produces a flow karyotype in which each chromosome can be identified and minor variations in size and staining properties of specific chromosomes can be detected. For example, structural and numerical chromosome aberrations were accurately detected recently in a blind study of amniocyte cultures (Gray et al. 1988). Robertsonian rearrangements found in fish populations have chromosomes with altered sizes and would be easy to identify with flow cytometry. Polymorphic bands could also be identified with these techniques.

Nuclear DNA Markers

Types of Nuclear DNA Markers

Tandemly repetitive DNA (satellite DNA).—Most eukaryotes contain substantial amounts of repetitive DNA in their genomes, some of it organized in tandem arrays and other types dispersed throughout the genome. The tandem repeats are usually short sequences of 10–300 base pairs present in 10^5–10^6 copies, often in blocks at the centromeres or telomeres of chromosomes. Usually they are not transcribed, and frequently they have a base composition (percentage of guanine plus cytosine) different from the rest of the DNA, in which case they are called satellite DNA. Such DNA tends to evolve rapidly, so that major differences in the amounts and in base composition are found between closely related species or even between populations of a single species (Brutlag 1980).

The C-banding technique detects this repetitive DNA at the cytological level, where differences in base composition result in differences in fluorescent banding of DNA subsets. Our cytogenetic studies on salmonids suggest that intraspecific differences occur in the amount and sequence of this tandemly repetitive DNA. Because tandemly repetitive DNA is present in large numbers of copies, it should be visible as sharp bands when digests of genomic DNA are electrophoresed on agarose gels. Intraspecific differences in the amount of this DNA has been found (B. Turner, P. Grewe, S. Hartley, and R. Phillips, unpublished observations). Molecular studies are needed to characterize tandemly repetitive DNA and to determine the extent of intraspecific variation.

Dispersed repetitive DNA ("fingerprints").— Sequences of repetitive DNA dispersed throughout the genome occur in most vertebrates. Recently, Jeffreys et al. (1983) isolated probes specific for a core sequence of 10–15 base pairs common to many dispersed repetitive DNAs. These probes detected a hypervariable "minisatellite" DNA, which is so polymorphic in human populations that each individual has a unique "fingerprint." The DNA probes that detect this minisatellite DNA have been used to identify intraspecific variations in other vertebrates including fishes (Wetton et al. 1987; Castelli et al. 1990, this volume). Because this DNA is hypervariable, it can be used to identify specific individuals and should be useful in studies of breeding structure. It could be used as a mark to identify fish of a specially constructed stock (reviewed by Castelli et al. 1990).

Moderately repetitive DNA: the rRNA genes.— Moderately repetitive DNA sequences include several multigene families that code for proteins or RNAs needed in large amounts in the cell. One of these multigene families consists of the ribosomal RNA (rRNA) genes. There are four types of rRNAs in higher organisms: 5S, 5.8S, 18S, and 28S (S is the sedimentation coefficient, 1×10^{-13}s). Genes (rDNA) for three of these—5.8S, 18S, and 28S—occur as multiple copies of a repeating unit in higher organisms (Figure 6a). Each cell has many such copies, which are tandemly arranged in huge blocks situated on one or more chromosomes and can be visualized cytologically as nucleolar organizer regions (Hsu et al. 1975). The rDNA repeating unit is especially suited for use in phylogenetic comparisons, because it contains both highly conserved coding regions and nontranscribed regions that evolve more rapidly than single-copy genes. Correction of sequences occurs in the clusters of ribosomal

RNA cistrons, which tends to maximize differences between geographically isolated populations or closely related species (reviewed by Gerbi 1985). Stock-specific differences have been found in the nontranscribed spacer region (NTS) in many organisms including mice (Suzuki et al. 1986), drosophila (Williams et al. 1985), and tree frogs (Romano and Vaughn 1986). The large number of copies of these genes simplifies their detection. For example, there are over 2,000 copies in several trout species (Popodi et al. 1985). Using different restriction enzymes, we have identified several polymorphisms in the NTS in lake trout (Figures 6b, 6c, and 7), which have different frequencies in different stocks.

Single-copy DNA: coding and noncoding sequences.—Differences in the sequences of single-copy DNA can also be used as genetic markers. The simplest method of detecting sequence differences is by digesting genomic DNA with specific restriction enzymes and looking for differences in the size of restriction fragments obtained. Restriction-fragment-length polymorphisms (RFLPs) can be examined for any genes or DNA sequences for which specific DNA probes are available (Cooper and Schmidtke 1986). Few genes in fish have been cloned, so relatively few probes are available for studying single-gene polymorphisms. Researchers recently identified more than 500 RFLPs in humans by screening random genomic clones (Schumm et al. 1988). This same approach could be used with fish. For example, gene "libraries" that represent all the DNA from a single fish (cut into small pieces and cloned into bacteriophage) are now available for several species. One could use random clones from gene libraries for a given species as probes or simply synthesize short oligonucleotides and use them as probes to look for RFLPs.

Methods for Detecting Nuclear DNA Markers

Standard methods.—Standard methods involve the following steps (Figure 8). First, DNA is extracted from a convenient tissue (liver, blood, sperm, etc.), usually by phenol extraction. Then the DNA samples are analyzed by the Southern blotting technique (Southern 1975). In this procedure, the DNA samples are digested with a specific restriction enzyme and electrophoresed on agarose gels. The separated DNA fragments are transferred to a nitrocellulose or a nylon filter, after which they are hybridized with a specific radioactive DNA probe. This probe can be spe-

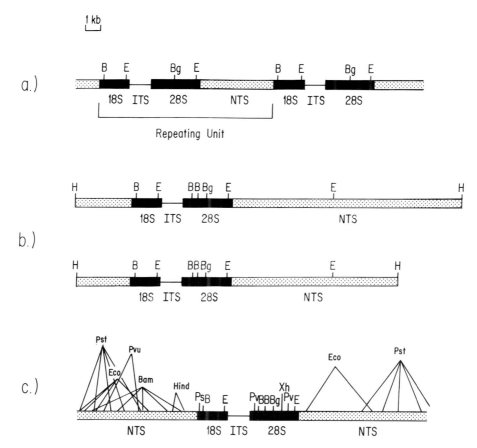

FIGURE 6.—(a) Diagram of the repeating unit that contains the 5.8S, 18S, and 28S ribosomal RNA genes (rDNA) in vertebrates. The 18S and 28S coding regions (shown in black), along with the ITS (internal transcribed spacer containing the 5.8S coding region), are transcribed as a single unit. The NTS (nontranscribed spacer region) varies between species in sequence and length. The Bam HI (B), Eco RI (E) and Bgl II (Bg) restriction sites shown in the coding regions are common to all vertebrates.

(b) Restriction map of the rDNA in lake trout, from Popodi et al. (1985), showing the repeating unit with the 18S and 28S coding regions, the internal transcribed region (ITS), and the variable-length nontranscribed spacer region (NTS; two of the length variants obtained when the unit is digested with Hind III [H] are shown here). The restriction sites in the coding regions are common to all fish examined to date.

(c) Restriction map of the rDNA in lake trout, showing the variable restriction sites found in the nontranscribed spacer region (NTS) with the following enzymes: Hind III, Bam HI, Eco RI, Pvu I, and Pst I. (Abbreviations used for restriction sites in the coding regions are: B = Bam HI, H = Hind III, E = Eco RI, Ps = Pst I, Pv = Pvu I, and Xh = Xho I). Note the large number of variable sites in the NTS "upstream" (left) of the 18S coding regions.

cific for a known single-copy gene such as the hemoglobin gene, a multicopy gene such as the rRNA genes, or any other DNA sequence. Then autoradiography is carried out and the DNA fragments that hybridize with the probe are detected on X-ray film. The specific fragment patterns obtained from different individuals are analyzed and the genotypes are determined.

Amplification of specific sequences by polymerase chain reaction.—Specific nuclear sequences of interest can be quickly amplified by using the polymerase chain reaction (PCR; Saiki et al. 1985), if the sequence of the flanking region is known. The amplified DNA can be analyzed by direct sequencing or by restriction digestion, agarose electrophoresis, and examination of ethidium bromide stained gels.

Oligonucleotide probes and dot blotting.—When restriction fragment polymorphisms in a given gene have been identified and the sequence of the portion of the gene detected is known, specific oligonucleotide probes that bind only with

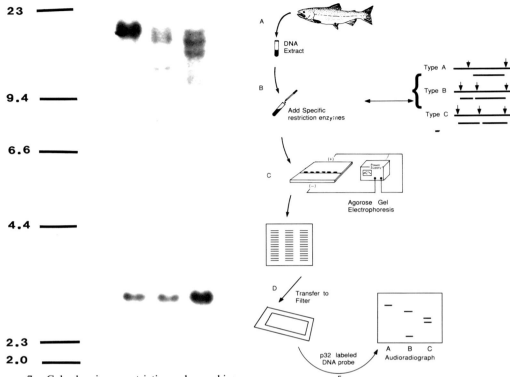

FIGURE 7.—Gels showing a restriction polymorphism in the nontranscribed spacer region (NTS) of the rRNA genes of lake trout; scale at left is in kilobases (kb). DNA from three individuals was digested with Bam HI, electrophoresed through agarose, and hybridized with a probe to 28S rDNA. Each individual has a different phenotype for the large variable band (15–22 kb) that spans the NTS, and all three are identical for the small band (3 kb) that spans the coding region. Left pattern has band A, middle pattern has bands A, B, and D, and right pattern has bands A, B, and C. The last two patterns represent three different sizes of the repeating unit.

DNA from a given genotype can be synthesized. For example, two different oligonucleotide probes have been synthesized, one complementary to the DNA of normal hemoglobin and the other to the DNA of sickle-cell hemoglobin. Genomic DNA from different individuals can be applied to nitrocellulose paper in a dot-blot format and tested to see if they hybridize with these probes (Rabin and Dattagupta 1987). Other variations on this technique involve amplification of the specific DNA sequences to be studied by using PCR and hybridization in solution with an oligonucleotide probe. Few fish genes have been sequenced to date, but these methods may be useful in the future.

Sandwich hybridization.—Another simplified technique, which eliminates the need for agarose

FIGURE 8.—Methods for detection of nuclear DNA polymorphisms. (**A**) The tissue sample (liver, blood, sperm, etc.) is collected and homogenized, and DNA is extracted. (**B**) The DNA sample is digested with a specific restriction enzyme, and (**C**) the DNA fragments are separated by agarose gel electrophoresis. (**D**) DNA fragments are transferred to a nitrocellulose or nylon filter. (**E**) A ^{32}P-labeled probe for the specific gene is added to the filter, and the DNA fragments that hybridize with the probe are detected by autoradiography. In this example, DNA of type A has two restriction sites that yield one large fragment; DNA of type B has an additional restriction site, to the left of center, that yields two fragments of unequal size; and DNA of type C has an additional restriction site, further to the right, that yields two medium-sized fragments.

electrophoresis and Southern transfer, is "sandwich" hybridization. This technique was developed for identification of specific DNA sequences in connection with the diagnosis of human genetic disease (Woodhead et al. 1986) and does not require sequencing of the gene in question. The method involves a double ("sandwich") hybridization in which the DNA sample to be identified is tested to determine whether it will form a link between a solid support (a resin) and a labeled DNA probe (Figure 9). First, a region of DNA is

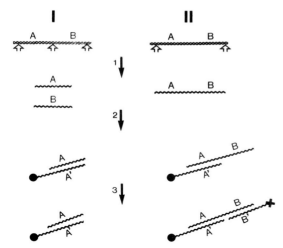

FIGURE 9.—"Sandwich hybridization" method for detecting differences between the DNA of fish stock I (on the left) and stock II (on the right). In the DNA region shown, individuals of stock I have a restriction site (indicated by the middle white arrow) that is missing in the DNA of individuals of stock II. In step 1, digestion with the appropriate restriction enzyme divides the DNA of stock I into two small pieces, A and B, whereas the DNA of stock II remains in one piece. In step 2, the DNA is mixed with resin-bound DNA (A'), which is complementary to A and binds to the fragment A of stock I and the intact fragment AB of stock II. In step 3, the labeled probe B' (which will hybridize with B) is added and a "sandwich" hybridization is formed with the DNA from stock II but not with the DNA from stock I.

identified where individuals of one stock lack a restriction site present in the DNA of individuals of the other stock. Digestion with the appropriate enzyme divides the DNA fragment of stock I into two small pieces, A and B, but leaves this DNA fragment intact in stock II. After digestion, the entire DNA sample is mixed with resin-bound DNA (A') which is complementary to A and thus will bind the A fragment of stock I and the intact AB fragment of stock II. Following this, a labeled DNA probe B' that will hybridize with the B region is added. This probe can only bind to the AB fragment of stock II. Thus, only DNA from a fish of stock II will be labeled in the resin-bound fraction.

The resin-bound DNA (A') can be prepared ahead of time and kept in the refrigerator. Although the labeled DNA probe B' usually is a radioactive probe, this technique can be integrated with various nonradioactive detection systems and can be automated to process many samples. The DNA extracted from fresh or frozen samples can be used. Because the procedure involves a small region of DNA, a crude extraction method able to yield small pieces of DNA is adequate. To obtain fragments A' and B' for the assay, subcloning is necessary, but this can be done by using standard conditions of digestion, ligation, transformation, and selection (Maniatis et al. 1982).

In situ hybridization methods.—DNA sequences can be detected with specific DNA probes and hybridized to chromosomes in metaphase cells or to cell nuclei in interphase cells. This technique has the advantage of eliminating the need for DNA extraction; if interphase cells are used, tissue can be fixed immediately in the field and brought back to the laboratory, where slides can be prepared when convenient. If biotin-labeled DNA is used as a probe, the need for working with radioactive isotopes is eliminated. For example, in situ hybridization with fluorescent, biotin-labeled rDNA probes has been used to detect human chromosomes in mouse–human hybrid DNA (Pinkel et al. 1986). The percentage of human DNA in the cells was accurately determined from cell nuclei. This technique could be applied to studies of intraspecific differences in satellite DNA or rDNA. Recently, in situ hybridization has been combined with dual-beam flow cytometry to detect satellite DNA sequences in nuclei from mouse cells in suspension (Trask et al. 1985).

Pulsed field-gradient electrophoresis.—Pulsed field-gradient electrophoresis (PFG) is a new technique (Schwartz and Cantor 1984; Carle and Olsen 1984) that allows resolution of much larger DNA fragments (up to 9 million base pairs) on agarose gels (reviewed by Van Ommen and Verkerk 1986). This technique, which separates yeast chromosomes, is being used to study large linkage complexes such as the HLA genes (Hardy et al. 1986). By using an enzyme that does not affect the repeating unit, researchers have cut out whole nucleolar organizer regions from the DNA of African clawed toads *Xenopus* sp. (Reeder, personal communication). This technique will allow analysis of the copy number and molecular structure of the rDNA at each NOR. It could also be used to analyze the composition and number of copies of tandemly repetitive DNA on different chromosomes. Analyses of multicopy genes by this technique would probably reveal both stock-specific and individually specific markers, which could be identified on the same gel.

Discussion

Two other types of genetic markers based on protein variability and variability in mitochondrial DNA are used routinely in fish biology. The utility of protein allozyme variation detected by electrophoresis has been reviewed previously (Allendorf et al. 1987) and is considered elsewhere in this volume. The standard methodology used for detecting allozyme variation has been presented by Utter et al. (1987). The major advantages of this method are technical simplicity and relatively low cost. The major disadvantage is the lack of appropriate markers for some species. The utility of mitochrondial DNA as a genetic marker in fish has been reviewed recently (Ferris and Berg 1987). Analysis of mitochondrial DNA has proved useful in stock identification for several fishes (Avise et al. 1984; Wilson et al. 1985; Grewe and Hebert 1987; Gyllensten and Wilson 1987). Mitochondrial DNA markers differ principally from nuclear DNA markers in being maternally inherited. This characteristic makes them useful for the analysis of hybrid swarms and in marking stocks, because all offspring resemble the female parent.

The techniques for detecting mitochondrial markers are similar to those for nuclear markers (Ferris and Berg 1987). There are two basic alternatives. In the first method, the mitochondrial DNA is separated from the nuclear DNA by ultracentrifugation and rigorously purified prior to restriction analysis, which is performed as shown in steps B and C of Figure 8. Restriction fragments are visualized with ethidium bromide staining or end-labeling, and specific fragment patterns are analyzed as described earlier. The second method, involving Southern blotting, is identical to that shown in Figure 8. In this case, genomic DNA (both the mitochondrial and nuclear forms) is isolated, digested with restriction enzymes, electrophoresed on agarose gels, transferred to a filter, and hybridized with a probe consisting of labeled mitochondrial DNA. Filters can be washed and reprobed several times with different DNA probes, so both nuclear and mitochondrial DNA markers can be examined in preparations from the same individuals. This is especially advantageous when hybrids are analyzed and in species with relatively few genetic markers. Certain tissues such as sperm have very little mitochondrial DNA (mtDNA), but there is sufficient mtDNA in fish blood so that if its total DNA is extracted, both mtDNA and nuclear DNA markers can be examined (N. Billington, University of Windsor, personal communication).

In selecting genetic tags for a particular purpose, investigators need to consider the frequency of markers, the sensitivity of markers, and the ease and cost of genotyping fish. The methodology for allozyme markers is the simplest and least expensive (Utter et al. 1987), so such markers are preferable when suitable ones exist for the species or populations involved. Mitochondrial DNA is maternally inherited, so fewer individuals are needed to mark a new stock; but hybrids between marked and unmarked stocks can not be detected with mitochondrial DNA alone. Nuclear DNA markers are very specific and can be used to detect hybrids, but detection of DNA markers involves lengthier and costlier procedures than does detection of allozyme or chromosome markers. Although the cost of detecting chromosome markers is about the same as that for allozymes, lack of expertise in cytogenetics has limited the application of this technology. At present, chromosome markers would probably be favored when they are present in species with little allozyme variation, and when expertise and funds are not available for DNA analysis. Development of simpler and less expensive methods (such as sandwich hybridization and in situ hybridization) to identify chromosome and DNA polymorphisms will result in much wider application of these markers.

References

Abe, S., and J. I. Muramoto. 1974. Differential staining of chromosomes of two salmonid species, *Salvelinus leucomaenis* (Pallas) and *Salvelinus malma* (Walbaum). Proceedings of Japan Academy 50:507–511.

Allendorf, F. W., and G. H. Thorgaard. 1984. Tetraploidy and the evolution of salmonid fishes. Pages 1–53 *in* B. J. Turner, editor. Evolution of fishes. Plenum, New York.

Allendorf, F., N. Ryman, and F. Utter. 1987. Genetics and fishery management: past, present and future. Pages 1–19 *in* Ryman and Utter (1987).

Amemiya, C. T., and J. R. Gold. 1986. Chromomycin A₃ stains nucleolar organizer regions of fish chromosomes. Copeia 1986:226–231.

Avise, J. C., E. Bermingham, L. G. Kessler, and N. C. Saunders. 1984. Characterization of mitochondrial DNA variability in a hybrid swarm between subspecies of bluegill sunfish *(Lepomis macrochirus)*. Evolution 38:931–941.

Barsienne, J. V. 1981. Intercellular polymorphism of chromosome sets in the Atlantic salmon. Tsitologiya 23:1053–1059. (In Russian.)

Brutlag, D. V. 1980. Molecular arrangement and evolution of heterochromatic DNA. Annual Review of Genetics 14:121–144.

Carle, G. F., and M. V. Olson. 1984. Separation of chromsomal DNA molecules from yeast by orthogonal field alteration gel electrophoresis. Nucleic Acids Research 12:5647–5664.

Carrano, A. V., J. W. Gray, R. G. Langlois, and L.-C. Yu. 1983. Flow cytogenetics: methodogy and application. Pages 195–209 in J. D. Rowley. Chromosomes and cancer. Academic Press, New York.

Castelli, M., J.-C. Philippart, G. Vassart, and M. Georges. 1990. DNA fingerprinting in fish: a new generation of genetic markers. American Fisheries Society Symposium 7:514–520.

Comings, D. E. 1978. Mechanisms of chromosome banding and implications for chromosome structure. Annual Review of Genetics 12:25–46.

Cooper, D. N., and J. Schmidtke. 1986. Diagnosis of genetic disease using recombinant DNA. Human Genetics 73:1–11.

Feldberg, E., L. A. C. Bertollio, L. F. de A. Toledo, F. Foresti, O. M. Filho, and A. F. dos Santos. 1987. Biological aspects of Amazonian fishes. IX. Cytogenetic studies in two species of the genus Semaprochilodus (Pisces, Prochilodontiae). Genome 29:1–4.

Ferris, S. D., and W. J. Berg. 1987. The utility of mitochondrial DNA in fish genetics and fishery management. Pages 277–299 in Ryman and Utter (1987).

Fukuoka, H. 1972. Chromosomes of the sockeye salmon (Oncorhynchus nerka). Japanese Journal of Genetics 47:459–464.

Garcia, E., M. C. Alvarez, and G. Thode. 1987. Chromosome relationships in the genus Blennius (Blenniidae, Perciformes) C banding patterns suggest two karyoevolutional pathways. Genetica 72:27–36.

Gerbi, S. A. 1985. Evolution of ribosomal DNA. Pages 419–518 in R. J. Macintyre, editor. Molecular evolutionary genetics. Plenum, New York.

Giles, V., G. Thode, and M. C. Alvarez. 1985. A new Robertsonian fusion in the multiple chromosome polymorphism of a mediterranean population of Gobius paganellus (Gobiidae, Perciformes). Heredity 55:255–260.

Gold, J. R. 1977. Systematics of western North American trout (Salmo) with notes on the redband trout of Sheepheaven Creek, California. Canadian Journal of Zoology 55:1858–1873.

Gold, J. R. 1979. Cytogenetics. Pages 353–405 in W. S. Hoar, D. J. Randall and J. R. Brett, editors. Fish physiology, volume 8. Academic Press, New York.

Gold, J. R. 1984. Silver staining and heteromorphism of chromosomal nucleolus organizer regions in North American cyprinid fishes. Copeia 1984:133–139.

Gold, J. R., and C. T. Amemiya. 1986. Cytogenetic studies in North American minnows (Cyprinidae). XII. Patterns of chromosomal nucleolus organizer region variation among 14 species. Canadian Journal of Zoology 64:1869–1877.

Gold, J. R., C. T. Amemiya, and T. R. Ellison. 1986. Chromosomal heterochromatin differentiation in North American cyprinid fishes. Cytologia 51:557–566.

Gold, J. R., and Gall. 1977. Chromosome cytology in the cutthroat trout series Salmo clarki (Salmonoidea). Cytologia 42:377–382.

Gorshkov, S. A., and G. V. Gorshkova. 1978. Chromosome sets in seasonal races of Oncorhynchus nerka in the Azabachye Lake (Kamchatka). Zoologicheskii Zhurnal 57:1382–1387.

Gorshkov, S. A., and G. V. Gorshkova. 1981. Chromosome polymorphism of the pink salmon Oncorhynchus gorbuscha (Walb). Tsitologiya 23:954–960.

Gray, J. W., and eight coauthors. 1988. Application of flow karyotyping in prenatal detection of chromosome aberrations. American Journal of Human Genetics 42:49–59.

Green, D. M., Bogart, J. P. 1980. An interactive microcomputer based karyotype analysis system for phylogenetic cytotaxonomy. Computers in Biology and Medicine 10:219–227.

Green, D. K., and six coauthors. 1984. Karyotyping and identification of human chromosome polymorphisms by single fluorochrome flow cytometry. Human Genetics 66:143–146.

Grewe, P. M., and P. D. N. Hebert. 1987. Mitochondrial DNA diversity among brood stocks of the lake trout (Salvelinus namaycush). Great Lakes Fishery Commission, Research Completion Report, Ann Arbor, Michigan.

Gyllensten, U., and A. C. Wilson. 1987. Mitochondrial DNA of salmonids: inter- and intraspecific variability detected with restriction enzymes. Pages 301–317 in Ryman and Utter (1987).

Hardy, D. A., J. I. Bell, E. O. Long, T. Lindsten, and H. O. McDevitt. 1986. Mapping of the class II region of the human major histocompatibility complex by pulsed-field gel electrophoresis. Nature (London) 323:453–455.

Hartley, S. E. 1987. The chromosomes of salmonid fishes. Biological Reviews of the Cambridge Philosophical Society 62:197–214.

Hartley, S. E., and M. T. Horne. 1982. Chromosome polymorphism in the rainbow trout (Salmo gairdneri Richardson). Chromosoma (Berlin) 87:461–468.

Hartley, S. E., and M. T. Horne. 1984. Chromosome polymorphism and constitutive heterochromatin in Atlantic salmon, Salmo salar. Chromosoma (Berlin) 89:377–380.

Hartley, S. E., and M. T. Horne. 1985. Cytogenetic techniques in fish genetics. Journal of Fish Biology 26:575–582.

Howell, W. M., and D. A. Black. 1980. Controlled silver staining of nucleolus organizer regions with a protective colloidal developer: a 1-step method. Experentia (Basal) 36:1014–1045.

Hsu, T. C., S. E. Spirito, and M. L. Pardue. 1975. Distribution of 18S and 28S ribosomal genes in mammalian genomes. Chromosoma 53:25–36.

Ihssen, P. E., J. C. Casselman, G. W. Martin, and R. B. Phillips. 1988. Biochemical genetic differentiation of lake trout *(Salvelinus namaycush)* stocks of the Great Lakes region. Canadian Journal of Fisheries and Aquatic Sciences 45:1018–1029.

Jeffreys, A. J., V. Wilson, and S. L. Thein. 1983. Hypervariable minisatellite regions in human DNA. Nature (London) 314:67–73.

Kligerman, A. D., and S. E. Bloom. 1977. Rapid chromosome preparations from solid tissues of fishes. Journal of Fisheries Research Board Canada 34:266–269.

Kulikova, N. I. 1971. Intraspecific variability of karyotypes of the chum salmon *(Oncorhynchus keta* (Walb)). Journal of Ichthyology 11:977–983.

Lloyd, M. A., and G. H. Thorgaard. 1988. Restriction endonuclease banding of rainbow trout chromosomes. Chromosoma (Berlin) 96:171–177.

Loudenslager, E. J., and G. H. Thorgaard. 1979. Karyotypic and evolutionary relationships of the Yellowstone *(Salmo clarki bouvieri)* and west-slope *(S. c. lewisi)* cutthroat trout. Journal of the Fisheries Research Board of Canada 36:630–635.

Maniatis, T., E. F. Fritsch, and J. Sambrook. 1982. Molecular cloning. Cold Spring Harbor Laboratory, Cold Spring Harbor, New York.

Miller, D. A., Y. Choi, and O. J. Miller. 1983. Chromosome localization of highly repetitive human DNAs and amplified ribosomal DNA with restriction enzymes. Science (Washington, D.C.) 219:395–397.

Ohno, S., C. Stenius, E. Faisst, and M. T. Zenzes. 1965. Post-zygotic chromosomal rearrangements in rainbow trout *(Salmo irideus* Gibbins). Cytogenetics 4:117–129.

Phillips, R. B., and S. E., Hartley. 1988. Fluorescent banding patterns of the chromosomes of the genus *Salmo.* Genome 30:193–197.

Phillips, R. B., and P. E. Ihssen. 1985a. Chromosome banding in salmonid fishes: nucleolar organizer regions in *Salmo* and *Salvelinus.* Canadian Journal of Genetics and Cytology 27:433–440.

Phillips. R. B., and P. E. Ihssen. 1985b. Identification of sex chromosomes in lake trout *(Salvelinus namaycush).* Cytogenetics and Cell Genetics 39:14–18.

Phillips, R. B., and P. E. Ihssen. 1986a. Inheritance of Q band chromosomal polymorphisms in lake trout. Journal of Heredity 77:93–97.

Phillips, R. B., and P. E. Ihssen. 1986b. Stock structure of lake trout from the Great Lakes region as determined by chromosome and isozyme polymorphisms. Great Lakes Fishery Commission, Research Completion Report, Ann Arbor, Michigan.

Phillips, R. B., and A. R. Kapuscinski. 1987. Robertsonian polymorphism in pink salmon *(Oncorhynchus gorbuscha)* involving the NOR region. Cytogenetics and Cell Genetics 44:148–152.

Phillips, R. B., and A. R. Kapuscinski. 1988. High frequency of translocation heterozygotes in odd year stocks of pink salmon *(Oncorhynchus gorbuscha).* Cytogenetics and Cell Genetics 48:178–182.

Phillips, R. B., K. A. Pleyte, and S. E. Hartley. 1988. Stock differences in the number and chromosomal position of the nucleolar organizer regions (NORs)

in Arctic char *(Salvelinus alpinus).* Cytogenetics and Cell Genetics 48:9–12.

Phillips, R. B., K. A. Pleyte, and P. E. Ihssen. 1989a. Patterns of chromosomal nucleolar organizer region (NOR) variation in fishes of the genus *Salvelinus.* Copeia 1989:47–53.

Phillips, R. B., and K. D. Zajicek. 1982. Q band chromosomal banding polymorphisms in lake trout *(Salvelinus namaycush).* Genetics 101:222–234.

Phillips, R. B., K. D. Zajicek, and P. E. Ihssen. 1989b. Population differences in chromosome-banding polymorphisms in lake trout. Transactions of the American Fisheries Society 118:64–73.

Phillips, R. B., K. D. Zajicek, P. E. Ihssen, and O. Johnson. 1986a. Application of silver staining to the identification of triploid fish cells. Aquaculture 54:313–319.

Phillips, R. B., K. D. Zajicek, and F. M. Utter. 1985. Q-band chromosomal polymorphisms in chinook salmon *(Oncorhynchus tshawyscha)* Copeia 1985:273–278.

Phillips, R. B., K. D. Zajicek, and F. M. Utter. 1986b. Chromosome banding in salmonid fishes: nucleolar organizers in *Oncorhynchus.* Canadian Journal of Genetics and Cytology 28:502–510.

Pinkel, D., T. Straume, and J. W. Gray. 1986. Cytogenetic analysis using quantitative, high-sensitivity, fluorescence hybridization. Proceedings of the National Academy of Sciences of the USA 83:2934–2938.

Pleyte, K. A., R. B. Phillips, and S. E. Hartley. 1989. Q band chromosomal polymorphisms in Arctic char *(Salvelinus alpinus).* Genome 32:129–133.

Popodi, E. M., D. Greve, R. B. Phillips, and P. J. Wejksnora. 1985. The ribosomal RNA genes in three salmonid species. Biochemical Genetics 23:997–1010.

Rabin, D., and N. Duttagupta. 1987. A simple DNA diagnostic method for human genetic disorders. Human Genetics 75:120–122.

Roberts, F. L. 1968. Chromosomal polymorphisms in North American land locked *Salmo salar.* Canadian Journal of Genetics and Cytology 10:865–875.

Roberts, R. L. 1970. Atlantic salmon *(Salmo salar)* chromosomes and speciation. Transactions of the American Fisheries Society 99:105–111.

Romano, P. R., and J. C. Vaughn. 1986. Restriction endonuclease mapping of ribosomal RNA genes: sequence divergence and the origin of the tetraploid tree frog *Hyla versicolor.* Biochemical Genetics 24:329–347.

Ryman, N. and Utter, F., editors. 1987. Population genetics & fishery management. University of Washington Press, Seattle.

Saiki, R. K., and six coauthors. 1985. Enzymatic amplification of beta globin genomic sequences and restriction site analysis for diagnosis of sickle cell anemia. Science (Washington, D.C.) 230:1350–1354.

Schmid, M. 1980. Chromosome banding in Amphibia. IV. Differentiation of GC- and AT-rich chromosome regions in Anura. Chromosoma (Berlin) 77:83–103.

Schmid, M. 1982. Chromosome banding in Amphibia. VII. Analysis of the structure and variability of NORs in Anura. Chromosoma 87:327–344.

Schumm, J. W., and twelve coauthors. 1988. Identification of more than 500 RFLP's by screening random genomic clones. American Journal of Human Genetics 42:143–159.

Schwartz, D. C., and C. R. Cantor. 1984. Separation of yeast chromosome-sized DNAs by pulsed field gradient gel electrophoresis. Cell 37:67–75.

Schweizer, D. 1981. Counterstain-enhanced chromosome banding. Human Genetics 57:1–14.

Sola, L., S. Cataudella, and E. Capanna. 1981. New developments in vertebrate cytotaxonomy 3. Karyology of bony fishes—a review. Genetica 54:285–328.

Southern, E. M. 1975. Detection of specific sequences among DNA fragments separated by gel electrophoresis. Journal of Molecular Biology 98:503.

Sumner, A. T. 1982. The nature and mechanisms of chromosome banding. Cancer Genetics and Cytogenetics 6:59–87.

Suzuki, H., and nine coauthors. 1986. Evolutionary implication of heterogeneity of the nontranscribed spacer region of ribosomal DNA repeating units in various subspecies of Mus musculus. Molecular Biology and Evolution 3:126–137.

Thode, G., M. C. Alvarez, E. Garcia, and V. Giles. 1985. Variation in C banding pattern and DNA values in two scorpion fishes (Scorpaena porcus and S. notata, Teleostei). Genetica 689:69–74.

Thorgaard, G. H. 1976. Robertsonian polymorphism and constitutive heterochromatin distribution in chromosomes of the rainbow trout (Salmo gairdneri). Cytogenetics and Cell Genetics 17:174–184.

Thorgaard, G. H. 1978. Sex chromosomes in the sockeye salmon: a Y-autosome fusion. Canadian Journal of Genetics and Cytology 29:349–354.

Thorgaard, G. H. 1983a. Chromosomal differences among rainbow trout populations. Copeia 1983:650–662.

Thorgaard, G. H. 1983b. Chromosome set manipulation and sex control in fish. Pages 405–534 in W. S. Hoar, D. J. Randall, and E. M. Donaldson, editors. Fish physiology, volume 9. Part B. Academic Press, New York.

Thorgaard, G. H., and S. K. Allen. 1987. Chromosome manipulation and markers in fishery management. Pages 319–332 in Ryman and Utter (1987).

Thorgaard, G. H., and J. E. Disney. 1990. Chromosome preparation and analysis. Pages 171–190 in C. B. Schreck and P. B. Moyle, editors. Methods for fish biology. American Fisheries Society, Bethesda, Maryland.

Thorgaard, G. H., P. S. Rabinovitch, M. W. Shen, G. A. E. Gall, J. Propp, and F. M. Utter. 1982. Triploid rainbow trout identified by flow cytometry. Aquaculture 29:305–309.

Trask, B., G. van den Engh, J. Landegent, N. J. in de Wal, and M. van der Ploeg. 1985. Detection of DNA sequences in nuclei in suspension by in situ hybridization and dual beam flow cytometry. Science (Washington, D.C.) 230:1401–1403.

Turner, B. J., T. A. Grudzen, K. P. Adkinsson, and R. A. Worrell. 1985. Extensive chromosomal divergence within a single river basin in the goodeid fish, Ilyodon furcidens. Evolution 39:122–134.

Ueda, T., and Y. Ojima. 1983a. Geographic and chromosomal polymorphisms in the Iwana (Salvelinus leucomaenis). Proceedings of the Japan Academy Series B, Physical and Biological Sciences 59(10):259–262.

Ueda, T., and Y. Ojima. 1983b. Karyotypes with C banding patterns of two species in the genus Salvelinus of the family Salmonidae. Proceeding of the Japan Academy Series B, Physical and Biological Sciences 59(10):343–346.

Utter, F., P. Abersold, and G. Winans. 1987. Interpreting genetic variation detected by electrophoresis. Pages 21–46 in Ryman and Utter (1987).

Van Ommen, G. J. B., and J. M. H. Verkerk. 1986. Restriction analysis of chromosomal DNA in a size range up to two million base pairs by pulsed field gradient electrophoresis. Pages 113–133 in K. E. Davies, editor. Human genetic diseases: a practical approach. IRL press, Washington, D.C.

Vasilev, V. P. 1975. Karyotypes of various forms of Arctic char (Salvelinus alpinus) from the waters of Kamchatka. Problems of Ichthyology 15(3):417–429. (In Russian.)

Viktorovsky, R. M. 1978. The evolution of karyotypes in chars of the genus Salvelinus. Tsitologiya 20:576–579. (In Russian.)

Viktorovsky, R. M., and L. N. Ermolenko. 1981. The chromosomal complement of Coregonus nasu, and C. lavaretus and the problem of the coregonus karyotype divergence. Tsitologia 24:797–780. (In Russian.)

Wetton, J. H., R. E. Carter, D. T. Parkin, and D. Walters. 1987. Demographic study of a wild house sparrow population by DNA fingerprinting. Nature (London) 327:147–152.

Williams, S. M., R. DeSalle, and C. Strobeck. 1985. Homogenization of geographical variants at the nontranscribed spacer of rDNA in Drosophila mercatorum. Molecular Biology and Evolution 2:338–346.

Wilson, G. M., W. K. Thomas, and A. T. Beckenbach. 1985. Intra- and inter-specific mitochondrial DNA sequence divergence in Salmo: rainbow, steelhead, and cutthroat trouts. Canadian Journal of Zoology 63:2088–2094.

Woodhead, J. L., R. Fallon, H. Figueiredo, J. Langdale, and A. D. B. Malcolm. 1986. Alternative methods of gene diagnosis. Pages 51–64 in K. E. Davis, editor. Human genetic diseases: a practical approach. IRL Press, Washington, D.C.

Zenzes, M. T., and I. Voiculescu. 1975. C banding patterns in Salmo trutta, a species of tetraploid origin. Genetica 45:531–536.

American Fisheries Society Symposium 7:514–520, 1990

DNA Fingerprinting in Fish:
A New Generation of Genetic Markers

MANOLA CASTELLI AND JEAN-CLAUDE PHILIPPART

Laboratoire de Démographie des Poissons et de Pisciculture Expérimentale
Service d'Ethologie et Aquarium, Faculté des Sciences, Université de Liège
Quai Van Beneden 22, 4020 Liège, Belgique

GILBERT VASSART

Institut de Recherche Interdisciplinaire et Unité de Génétique Clinique et Moléculaire
Hôpital Erasme, Faculté de Médecine, Université Libre de Bruxelles
Route de Lennik 808, 1070 Bruxelles, Belgique

MICHEL GEORGES

Laboratoire de Génétique, Faculté de Médecine Vétérinaire, Université de Liège
Rue des Vétérinaires 45, 1070 Bruxelles, Belgique

Abstract.—The recently discovered "families of hypervariable minisatellites" have proved to be a valuable source of genetic markers in humans and other animal species. Several members of those families can be visualized simultaneously by Southern blot hybridization to yield individual-specific "DNA fingerprints." These fingerprints, also called "bar codes," are efficient genetic tools with a wide variety of applications, such as individual identification, paternity testing, and linkage analysis. We have demonstrated that similar, individual-specific DNA fingerprints can be obtained from a fish, the common barbel *Barbus barbus*, by using four probes known to reveal hypervariability in humans and other species: wild-type M13 bacteriophage, Jeffreys' core sequence, the human α-globin hypervariable region, and a mouse probe related to part of the *Drosophila* Per gene. These DNA fingerprints may provide fish biologists with a powerful tool in ecological, behavioral, and population studies.

Dispersed in the genome of higher eukaryotes are several thousand segments of DNA, known as "minisatellites," that do not code for any protein (Wyman and White 1980; Bell et al. 1982; Proudfoot et al. 1982; Capon et al. 1983; Jeffreys et al. 1985b; Stoker et al. 1985; Jarman et al. 1986; Knott et al. 1986). These minisatellites are characterized by short, tandem repeats or motifs. A significant proportion of the minisatellites show genetic polymorphism, because of variation in the number of repeated motifs (Figure 1). Heterozygosities exceeding 90% are not uncommon at individual loci (Jeffreys et al. 1985b). Consequently, "hypervariable minisatellites" are a potential source of valuable genetic markers.

Jeffreys et al. (1985b) found that the majority of these sequences belong to "families of hypervariable minisatellites" whose members are dispersed throughout the genome. They observed that, when a repeated motif is used as a probe under low-stringency conditions, several hundred related sequences can be recognized per genome. Nakamura et al. (1987) developed a strategy to isolate over 250 human hypervariable minisatel-lites, or VNTR (variable number of tandem repetitions), as markers. It is anticipated that soon a nearly complete map of the human genome, based entirely on these kinds of markers, will be available.

Different members of the families of hypervariable minisatellites may be simultaneously visualised when different probes are used in Southern blot hybridizations. Restriction patterns for minisatellites show variability so extensive that it can be used to distinguish between closely related individuals, which explains why researchers have called the restriction patterns "DNA fingerprints" and "DNA bar-codes" (Jeffreys et al. 1985c; Ali et al. 1986; Jarman et al. 1986; Vassart et al. 1987) (Figure 2). These bar-codes have proved to be very efficient genetic tools in a variety of applications, including forensic science (Gill et al. 1985; 1987), paternity testing (Jeffreys et al. 1985a), linkage analysis (Jeffreys et al. 1986), determination of twin zygosity (Hill and Jeffreys 1985), and identification of post-transplant cell populations (Thein et al. 1986).

Similar DNA fingerprints have now been established for a variety of animal species, including

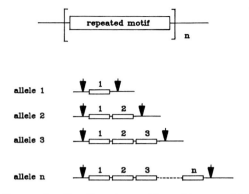

FIGURE 1.—Genetic polymorphism of hypervariable minisatellites is based on variation in the number of tandem repetitions (VNTR). Consequently, these polymorphisms are called VNTR markers.

domestic animals (Burke and Bruford 1987; Jeffreys et al. 1987; Jeffreys and Morton 1987; Wetton et al. 1987; Georges et al. 1988). They are expected to play an important role in breeding strategies.

In this paper we show that families of hypervariable minisatellites also exist in fish. We obtained individual-specific DNA fingerprints from the common barbel *Barbus barbus* by using four probes known to detect families of hypervariable minisatellite sequences in other species, including humans.

Biology and Genetics of *Barbus barbus*

The common barbel is a large cyprinid (maximum length, 90–100 cm) found in European running waters belonging to the "barbel zone" (Huet 1949). There, it often makes up more than 50% of the ichthyomass and plays a major ecological role. Its importance to commercial fishing is rather low except in certain eastern regions; but it is a valuable species for sportfishing because of its size, vigor, and abundance.

The common barbel is a gregarious, rheophilic, and benthic cyprinid that exhibits nocturnal habits and mainly feeds on aquatic invertebrates. Spawning occurs in May–June when temperatures reach 14–18°C. It is a lithophilous spawner (Balon 1975), laying its eggs on gravelly bottoms in fast-flowing waters. Detailed investigations of the biology and population dynamics of barbels in Belgium were reported by Philippart (1987).

Few data on the genetics of the common barbel are available. Recently, a team from the University of Montpellier, France, began a comparative study of the enzyme polymorphisms of wild barbels from geographically isolated river systems (the Rhône River in France and the Ourthe River in Belgium). The first results indicated a very low level of variability between the two populations— the only polymorphic enzyme system identified was glucose phosphate isomerase. Other barbel proteins, such as transferrins, are known to be highly polymorphic (Stratil et al. 1983).

Methods

Animals.—We used a single family of common barbels consisting of two parents and 16 full siblings. These hatchery-reared barbels were obtained from the experimental fish-culture station of the University of Liège (Philippart et al. 1987).

DNA extraction.—A piece of fin was cut off and homogenized in the following buffer: 8 M urea; 0.3 M NaCl; 10 mM tris–HCl, pH 8.5; 2% sodium dodecyl sulfate (SDS); 10 mM EDTA. The mixture was extracted at least three times with phenol–chloroform (1:1). After two ether extractions, the DNA was precipitated with 2 volumes of ethanol in 0.1 M NaCl and resuspended in TE buffer (10 mM tris-HCl, pH 7.5; 1 mM EDTA).

Restriction enzyme digestion.—The following enzymes were used: *Hae* III, *Hinf* I and *Mbo* I (Amersham, Boehringer, Pharmacia). Ten micrograms of DNA were digested by 75 units of enzyme under the manufacturer's recommended conditions.

Gel electrophoresis and Southern blotting.—The restricted DNA was run in a 1% agarose gel in TAE buffer (tris-acetate 40 mM; 1 mM EDTA) at constant voltage of 3 V/cm for 24 h with continuous buffer recirculation. The gels were stained with ethidium bromide and photographed under ultraviolet light, soaked in 0.5 M NaOH and 1.5 M NaCl for two 30-min periods, and equilibrated in 1 M NH_4 acetate, 0.03 M NaOH, again for two 30-min periods. We then blotted the gels overnight onto nitrocellulose membranes (Schleicher and Schuell, BA85), using the same solution as our transfer medium. After transfer, filters were washed in 3×SSC (20×SSC = 3 M NaCl, 0.3 M Na citrate) and baked at 80°C for 2 h.

Probes.—Four different probes were used: wild type M13mp9 phage (Vassart et al. 1987), plasmid pα3'HVR64 containing the human α-globin 3' hypervariable region (Reeders et al. 1985), plasmid pSP64.2.5EI containing a mouse sequence related to part of the *Drosophila* Per gene (Shin et al. 1985), and plasmid pUCJ containing approxi-

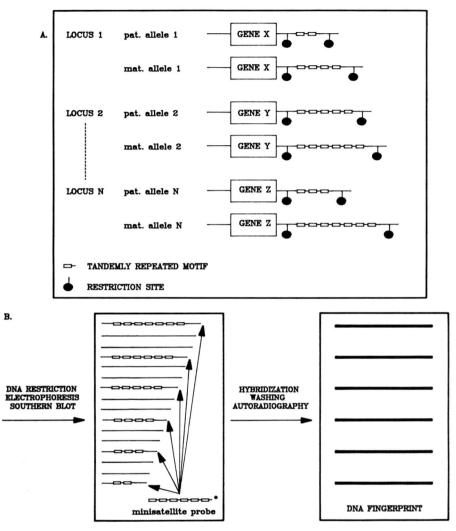

FIGURE 2.—Production of DNA fingerprints. (A) Genomic DNA extracted from nucleated cells is digested with specific-restriction endonucleases that cut the DNA into short, well-defined nucleotide sequences. (B) Endonuclease-digested DNA fragments are separated by size in a gel. After gel electrophoresis, the DNA is transferred to a nitrocellulose sheet (by Southern blotting) and hybridized to a ^{32}P-labeled probe that associates the fragments with paternal (pat.) or maternal (mat.) alleles. An individual-specific DNA fingerprint is revealed after washing and autoradiography.

mately 25 tandem copies of Jeffreys' core sequence (Georges et al. 1988).

Hybridization and washing.—The filters were prehybridized for at least 2 h in 35% formamide, 6×SSC, 5 mM EDTA, and 0.25% dried skimmed milk at 42°C, and were hybridized for at least 12 h at 42°C after addition of the probe to a final concentration of about 5×10^5 counts/min per milliliter. The filters were washed 1 h at 65°C in

2×SSC and 0.1% SDS, and were autoradiographed at −80°C with intensifying screens.

Results

Figure 3 illustrates restriction patterns obtained with the four probes for common barbel DNA. As can be seen, DNA fingerprints allow each of the 16 full siblings to be distinguished from the others.

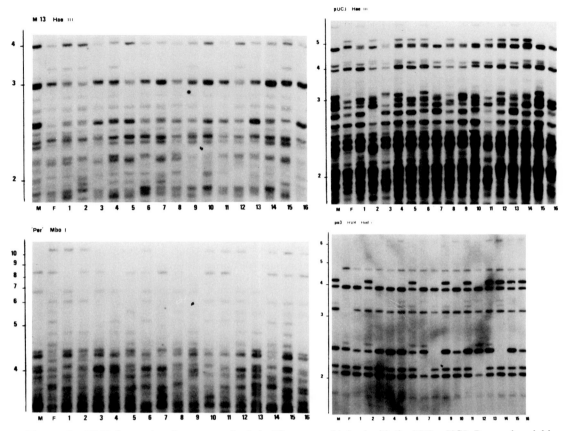

FIGURE 3.—DNA fingerprints for common barbels. They were obtained with the M13, pUCJ, Per, and α-globin probes. M and F correspond to male and female parents; individuals 1 to 16 are full siblings. The restriction enzymes used are indicated alongside the probe symbols. The ordinate scale marks DNA fragment sizes in kilobases.

For every offspring, each band could always be traced back to at least one of the parents, which indicates somatic and germ-line stability of the minisatellite sequences. Out of approximately 1,100 fragments studied in the offspring, only one exception to this rule was found: individual 10 presented a new band of 3.9 kilobases with the pUCJ probe on *Hinf* I-restricted DNA. This new band probably represents a neomutation.

When pooling the results obtained with the three enzymes (*Hae* III, *Hinf* I and *Mbo* I), we distinguished a mean of 8, 18.5, 23, and 22.5 differently segregating bands within the fingerprint of each parent with the M13, pUCJ, α-globin and Per probes respectively. Segregation analysis of these bands indicated that the parents were heterozygous for at least 77% of the fragments. Assuming that this heterozygosity corresponds to $2q(1 - q)/[2q(1 - q) + q^2] = (2 - 2q)/(2 - q)$, the mean allelic frequency, q, of the explored alleles equals 0.37.

Using this value, one can predict the probability of shared bands for unrelated individuals as $x = q^2 + 2q(1 - q) = 0.60$. This value is very close to the 0.56 value we found for the two common barbel parents (to our knowledge unrelated), which shared 41 bands from a mean of 72 per individual.

The segregation ratios of 56 bands (31 maternal, 25 paternal) present in the heterozygous state in the fingerprint of only one of the two parents were determined for the 16 offspring. Only eight of these bands (four paternal and four maternal) showed a segregation ratio that deviated significantly ($0.01 < P < 0.05$) from the expected 8:8 value. One band was transmitted to three offspring, four bands were transmitted to four offspring, and three bands to 12 offspring. These results could point towards non-Mendelian behavior for some of the systems explored, but more data are needed to confirm or invalidate this possibility. Moreover, the frequency distribution

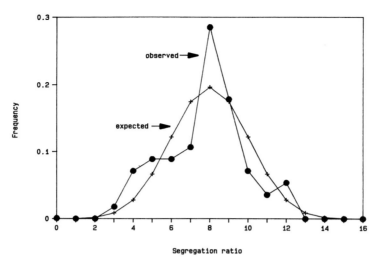

Segregation ratio

FIGURE 4.—Comparison of the observed segregation ratios for common barbels and the expected binomial distribution of those ratios. The segregation ratios of 56 bands (31 maternal and 25 paternal), which were present in the heterozygous state in the fingerprint of only one of the two parents, were determined for the 16 offspring. The absence of significant deviation from the expected binomial distribution as shown by the one-sample runs test (Siegel 1956; seven runs were scored for this sample, the critical values being 4 and 13 respectively) indicated a simple Mendelian behavior.

of the observed segregation ratios shows absence of significant deviation from the expected binomial distribution (Figure 4).

Preliminary linkage studies indicate the absence of close linkage between the majority of minisatellite loci explored. Clustering of all those sequences at a single or restricted number of loci can thus be excluded.

Discussion

We demonstrated that DNA fingerprinting can be extended to fish biology. The high degree of variation in the minisatellites explored, as well as the considerable number of scorable systems make these bar-codes a suitable tool for a variety of applications based on polymorphic genetic systems.

We are currently using these markers to exclude paternal genetic contribution in gynogenesis and to monitor the degree of heterozygosity in individuals produced by gynogenesis (induced parthenogenesis with all-maternal inheritance; Chourrout 1987). Moreover, studying the genetic bar-codes of individuals produced by gynogenesis should allow us to estimate minisatellite–centromere distances, according to Allendorf et al. (1986) for gene–centromere mapping of enzymatic loci in rainbow trout, and to test if the minisatellites detectable by fingerprints are telomeric. Lethal mutations in genes, possibly playing an important role in development, may be identified by

experimenting with fish whose gynogenetic offspring are deficient in one of the two expected homozygote genotypes for a minisatellite marker. Similar studies could, of course, be carried out on androgenetic animals (induced parthenogenetic individuals with all-paternal inheritance; see Chourrout 1987).

Many other applications of this methodology are possible. In aquaculture, DNA fingerprints could be used to identify, by linkage analysis, genomic segments carrying "quantitative trait loci" that play important roles in determining economically important traits such as growth, fertility, and disease resistance. The use of DNA fingerprints for individual identification and paternity testing opens a whole are up to ecological and behavioral studies, as has already been illustrated for a variety of bird species (Wetton et al. 1987; Burke and Bruford 1987). For a mean probability of band sharing $x = 0.60$, the probability that two unrelated individuals would share the same fingerprints with the four probes used in this study equals $0.60^{72} = 10^{-16}$. As for paternity testing, the expected number of obliged paternal fragments can be estimated at ±17.7. From a mean of 72 resolvable fragments, only 28.8 (40%) will be unique to the father, and 22.2 of these will be present in the heterozygous state (77%), of which half will be transmitted to the offspring. The 6.6 bands present in the homozygous state will, of course, all be transmitted. In consequence, the total number of obliged paternal fragments amounts to ± 17.7. The probability that an unrelated individual would have each of these fragments in his offspring is $0.60^{17.7} = 10^{-4}$, which also equals the probability of missing a wrong paternity.

At this stage, the use of DNA fingerprints to identify and manage fish stocks is questionable. Indeed, the variation observed is so high that this tool may be unusable to discriminate among fish populations. Extensive studies are required to determine whether population-specific bands can be recognized in the fingerprints. However, after identifying families of hypervariable minisatellites in fish, we should be able to isolate individual members of these families, as in humans, and use them as locus-specific markers. The study of one locus at a time should greatly simplify genetic analysis and allow researchers to identify and trace specific alleles. These systems should provide valuable stock-specific markers.

Acknowledgments

This work has been supported by the Institut pour l'Encouragement de la Recherche Scientifique dans l'Industrie et l'Agriculture (IRSIA), by Actions Concertées (Ministère de la Politique Scientifique), by the Fonds National de la Recherche Scientifique, and by the Association Recherche Biomédicale et Diagnostic. M. Castelli is a fellow of the IRSIA.

References

Ali, S., C. R. Müller, and J. T. Epplen. 1986. DNA fingerprinting by oligonucleotide probes specific for simple repeats. Human Genetics 74:239–243.

Allendorf, F. W., J. E. Seeb, K. L. Knudsen, G. H. Thorgaard, and R. F. Leary. 1986. Gene–centromere mapping of 25 loci in rainbow trout. Journal of Heredity 77:307–312.

Balon, E. K. 1975. Reproductive guilds of fishes: a proposal and definition. Journal of the Fisheries Research Board of Canada 32:821–864.

Bell, G. I., M. J. Selby, and W. J. Rutter. 1982. The highly polymorphic region near the human insulin gene is composed of simple tandemly repeating sequences. Nature (London) 295:31–35.

Burke, T., and M. W. Bruford. 1987. DNA fingerprinting in birds. Nature (London) 327:149–152.

Capon, D. J., E. Y. Chen, A. D. Levinson, P. H. Seeburg, and D. V. Goeddel. 1983. Complete nucleotide sequence of the T24 human bladder carcinoma oncogene and its normal homologue. Nature (London) 302:33–37.

Chourrout, D. 1987. Genetic manipulations in fish: review of methods. Pages 111–126 in K. Tiews,

editor. Proceedings, world symposium on selection, hybridization, and genetic engineering in aquaculture, volume 2. Heenemann Verlagsgesellschaft, Berlin.

Georges, M., A. S. Lequarré, M. Castelli, R. Hanset, and G. Vassart. 1988. DNA fingerprinting in domestic animals using four different minisatellite probes. Cytogenetics and Cell Genetics 47:127–131.

Gill, P., A. J. Jeffreys, and D. J. Werret. 1985. Forensic application of DNA "fingerprints." Nature (London) 318:577–579.

Gill, P., J. E. Lygo, S. J. Fowler, and D. J. Werret. 1987. An evaluation of DNA fingerprinting for forensic purposes. Electrophoresis 8:38–44.

Hill, A. V. S., and A. J. Jeffreys. 1985. Use of minisatellite DNA probes for determination of twin zygosity at birth. Lancet 1985(1):1394–1395.

Huet, M. 1949. Aperçu des relations entre la pente et les populations piscicoles. Schweizerische Zeitschrift für Hydrologie 11:332–351.

Jarman, A. P., R. D. Nicholls, D. J. Weatherall, J. B. Clegg, and D. R. Higgs. 1986. Molecular characterisation of a hypervariable region downstream of the human α-globin gene cluster. EMBO (European Molecular Biology Organization) Journal 5:1857–1863.

Jeffreys, A. J., J. F. Y. Brookfield, and R. Semeonoff. 1985a. Positive identification of an immigration test-case using human DNA fingerprints. Nature (London) 317:818–819.

Jeffreys, A. J., and D. B. Morton. 1987. DNA fingerprints of dogs and cats. Animals Genetics 18:1–15.

Jeffreys, A. J., V. Wilson, R. Kelly, B. A. Taylor, and G. Bulfield. 1987. Mouse DNA "fingerprints": analysis of chromosome localization and germ-line stability of hypervariable loci in recombinant inbred strains. Nucleic Acids Research 15:2823–2836.

Jeffreys, A. J., V. Wilson, and S. L. Thein. 1985b. Hypervariable "minisatellite" regions in human DNA. Nature (London) 314:67–73.

Jeffreys, A. J., V. Wilson, and S. L. Thein. 1985c. Individual–specific "fingerprints" of human DNA. Nature (London) 316:76–79.

Jeffreys, A. J., V. Wilson, S. L. Thein, D. J. Weatherall, and B. A. J. Ponder. 1986. DNA "fingerprints" and segregation analysis of multiple markers in human pedigrees. American Journal of Human Genetics 39:11–24.

Knott, T. J., S. C. Wallis, R. J. Pease, L. M. Powell, and J. Scott. 1986. A hypervariable region 3′ to the human apolipoprotein B gene. Nucleic Acids Research 14:9215–9216.

Nakamura, Y., and ten coauthors. 1987. Variable number of tandem repeat (VNTR) markers for human gene mapping. Science (Washington, D.C.) 235:1616–1622.

Philippart, J. C. 1987. Démographie, conservation et restauration du barbeau fluviatile, Barbus barbus (Linné) (Teleostei, Cyprinidae) dans la Meuse et ses affluents. Quinze années de recherches. Annales de la Societe Royale Zoologique de Belgique 117:49–62.

Philippart, J. C., P. Poncin, and C. Mélard. 1987. La domestication du barbeau fluviatile, *Barbus barbus* (L.) (Cyprinidae) en vue de la production massive contrôlée d'alevins pour le repeuplement des rivières. Résultats et problèmes. Pages 227–238 *in* K. Tiews, editor. Proceedings, world symposium on selection, hybridization and genetic engineering in aquaculture, volume 1. Heenemann Verlagsgesellschaft, Berlin.

Proudfoot, N. J., A. Gil, and T. Maniatis. 1982. The structure of the human zeta-globin gene and a closely linked, nearly identical pseudogene. Cell 31:553–563.

Reeders, S. T., and seven coauthors. 1985. A highly polymorphic DNA marker linked to adult polycystic kidney disease on chromosome 16. Nature (London) 317:542–544.

Shin, H. S., T. A. Bargiello, B. T. Clark, F. R. Jackson, and M. W. Young. 1985. An unusual coding sequence from a *Drosophila* clock gene is conserved in vertebrates. Nature (London) 317:445–448.

Siegel, S. 1956. The one-sample runs test. Pages 52–58 *in* H. F. Harlow, editor. Nonparametric statistics for the behavioural sciences. McGraw-Hill International Student Editions, Tokyo.

Stoker, N. G., K. S. E. Cheah, J. R. Griffin, F. M. Pope, and E. Solomon. 1985. A highly polymorphic region 3′ to the human Type II collagen gene. Nucleic Acids Research 13:4613–4622.

Stratil, A., P. Bobak, V. Tomasek, and M. Valenta. 1983. Transferrins of *Barbus barbus, Barbus meridionalis petenyi* and their hybrids. Genetic polymorphism, heterogeneity and partial characterization. Comparative Biochemistry and Physiology B, Comparative Biochemistry 76:845–850.

Thein, S. L., A. J. Jeffreys, and H. A. Blacklock. 1986. Identification of posttransplant cell population by DNA fingerprint analysis. Lancet 1986(2):37.

Vassart, G., M. Georges, R. Monsieur, H. Brocas, A. S. Lequarré, and D. Christophe. 1987. A sequence in M13 phage detects hypervariable minisatellites in human and animal DNA. Science (Washington, D.C.) 235:683–684.

Wetton, J. H., R. E. Carter, D. T. Parkin, and D. Walters. 1987. Demographic study of a wild house sparrow population by DNA fingerprinting. Nature (London) 327:147–149.

Wyman, A., and R. White. 1980. A highly polymorphic locus in human DNA. Proceedings of the National Academy of Sciences of the USA 77:6754–6758.

American Fisheries Society Symposium 7:521–540, 1990

Genetic Markers Identified by Immunogenetic Methods

W. C. Davis and R. A. Larsen

Department of Veterinary Microbiology and Pathology, College of Veterinary Medicine
Washington State University, Pullman, Washington 99164, USA

M. L. Monaghan

Department of Farm Animal Clinical Studies, Faculty of Veterinary Medicine
University College Dublin, Ballsbridge Dublin 4, Ireland

Abstract.—Extensive investigations of higher vertebrates have established the leukocyte as an ideal cell for use in immunogenetic studies. It is readily isolated from blood, and it has a large number of cell membrane molecules that are coded for by genes randomly distributed on different chromosomes. Monoclonal antibody technology and flow microfluorimetry allow such molecules to be identified and characterized and their allelic variants to be distinguished. In this report, we describe the use of these technologies to identify and analyze leukocyte antigens and antigens of the major histocompatibility complex in large food animals, and we discuss how these technologies can be used to identify genetic markers in fish.

The possibility of using antibodies to identify genetic markers was first recognized at the turn of the 20th century with the discovery of naturally occurring antibodies to the A, B, and O blood-group molecules (antigens) (Klein 1986). Sets of alleles were found that coded for a molecule bearing the A or B blood-group antigens (recognized by anti-A and anti-B antibodies, respectively) and for a molecule bearing neither A nor B antigens (anti-A and anti-B negative). The different antigenic phenotypes were identified with a simple hemagglutination assay as A^+, B^+, AB^+, or AB^- (the O blood group). Subsequent research revealed that antigenic variation is not unique and that specific antibodies could be obtained by immunizing laboratory animals (e.g., rabbits, mice, rats, goats, and sheep) with allelic products. The findings established that any molecule has the potential to be a genetic marker if two criteria are met: (1) Two or more alleles must encode antigenically distinct forms of the molecule, and (2) immunization with the antigenic variants must elicit the formation of antibodies that distinguish each form of the antigen. This in turn led to the development of immunogenetics in the 1930s, and to the wide application of immunological techniques to problems in biomedical and basic research (Klein 1986). More recently, the use of immunogenetic techniques to study genetic markers has been greatly extended by three discoveries: (1) Antibody-secreting cells (B lymphocytes) are programed to produce a single type of antibody (monoclonal antibody) specific for a single

antigenic determinant, (2) cells producing monoclonal antibodies or expressing antigenic markers can be immortalized (Kohler and Milstein 1975; Jonak and Kennett 1984; Langone and Van Vunakis 1986; Jonak et al. 1988), and (3) genes coding for antigens of interest can be isolated, transferred, and expressed in tissue culture-adapted cell lines (Graham and Van Der Eb 1973; Kavathas and Herzenberg 1983; Lalor and Herzenberg 1987; Figure 1).

These discoveries have afforded the means to use antibodies in new areas of immunogenetic research. For example, it is now possible to identify and purify antigenic molecules with monoclonal antibodies that recognize antigenic determinants conserved on all allelic forms of a molecule. Subsequently, the purified molecules can be used to produce monoclonal antibodies that react with allele-specific antigenic determinants (Figure 2). It is also possible to use monoclonal antibodies to identify and map the chromosomes that bear genes encoding antigenic protein molecules (Figure 3). A specific antigen can be purified to homogeneity, by means of monoclonal antibody immunoaffinity-chromatography, and then partially or completely sequenced. Then, DNA analogs of the natural gene (based on the amino acid composition of the molecule) can be constructed and used to locate the gene on the chromosome. The ability to immortalize cells, and to transfer and express genes encoding membrane molecules, has provided a way to establish cell lines for genetic studies (Figure 1).

Monoclonal Antibody and Transfection Technology

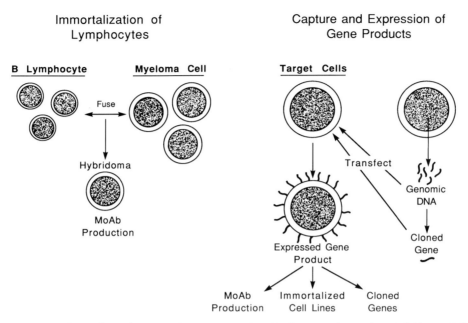

FIGURE 1.—Methods available for producing monoclonal antibodies (MoAbs) and establishing cell lines that express antigenic markers. Cell lines producing monoclonal antibodies can be prepared by fusing antibody-producing B lymphocytes with tissue culture-adapted myeloma cells (derived from a malignant B lymphocyte), and by transferring (transfecting) B lymphocytes with genomic DNA obtained from certain types of malignant lymphoid cells. Transfection technology can also be used to immortalize lymphoid cells bearing antigenic markers, and to accomplish cross-species transfer of genes coding for antigenic membrane molecules.

We are using these advances in technology to study the genetic mechanisms governing the expression of the immune response and resistance to disease in food animals (Figures 1–3; Davis et al. 1984, 1987; Davis 1985). In this report, we describe the strategies we have devised to develop immune reagents for our research, and discuss how similar approaches might be used to develop reagents for use in immunogenetic studies in fish.

Development of Monoclonal Antibodies

As shown in Figure 4, antibody-producing B lymphocytes are immortalized through fusion with a tissue culture-adapted line of myeloma cells (Kohler and Milstein 1975; Langone and Van Vunakis 1986). Mice are immunized once or several times with an antigen and then, 3 d before the spleen is taken for fusion, they receive a final injection of antigen intravenously. Nucleated spleen cells are separated from erythrocytes and fused with myeloma cells. The preparation of fused cells is distributed in 96-well culture plates.

After 24 h of culture, medium containing HAT (hypoxanthine, aminopterin, and thymidine) is added to the primary cultures to block proliferation of unfused myeloma cells. The myeloma cell lines used as fusion partners have an induced defect in the enzyme hypoxanthine guanine phosphoribosyl transferase, which is essential for synthesis of DNA by the salvage pathway (Langone and Van Vunakis 1986). Aminopterin blocks the normal pathway of nucleotide synthesis, causing unfused myeloma cells to die. Hybrids containing the unaltered enzyme are able to use the salvage pathway of DNA synthesis and proliferate in the presence of aminopterin. Unfused B lymphocytes die in culture without specific treatment because of the absence of essential growth factors. At 8 to 12 d, tissue culture supernatants are collected and tested for the presence of antibodies. Fused cells (hybridomas) producing antibodies are expanded and then cryopreserved. Supernatants are collected at the time of freezing and tested further to verify specificity. Hybridomas producing mono-

Identification of Polymorphic Antigenic Determinants

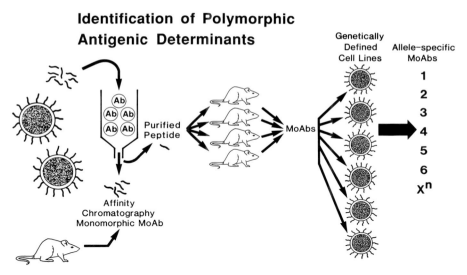

FIGURE 2.—Flow diagram to illustrate how monoclonal antibodies (MoAbs) to conserved antigenic determinants present on all allelic variants of a molecule can be used to produce monoclonal antibodies to polymorphic antigenic determinants (i.e., antigenic determinants that define each allelic form of the molecule). Monoclonal antibodies can be bound to a solid support matrix and used (by means of immunoaffinity) to purify the allelic forms of the molecule. The purified molecules can then be used to immunize mice to produce allomorph-specific antibodies. Primary cultures of hybridomas can be screened against genetically defined cell lines or preparations of cells to identify hybrids that produce allomorph-specific monoclonal antibodies.

clonal antibodies of interest are then cloned, and stocks of the antibody are made for research.

Although the methodology is straightforward, we had to resolve four problems before we could make efficient use of it to produce monoclonal antibodies to leukocyte antigens (Figure 5). Three problems concerned the strategies of producing monoclonal antibodies and assaying for speci-

Chromosome Mapping

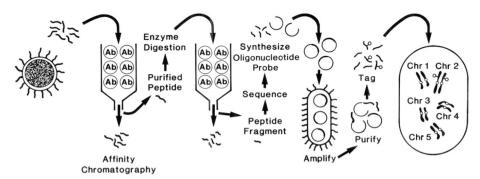

FIGURE 3.—Flow diagram to illustrate how monoclonal antibodies can be used to facilitate development of DNA probes for mapping the relative positions of genes that encode membrane molecules. An antigenic molecule is identified with a monoclonal antibody (Ab) and subsequently purified by immunoaffinity chromatography. Following purification, it is either sequenced directly or is enzymatically degraded and subjected to another round of immunoaffinity purification. The amino acid sequence of the purified antigenic fragment is determined. DNA analog probes (based on the amino acid sequence of a portion of the molecule) are then prepared and used to identify the native gene or, as indicated here, they are amplified directly, labeled, and used to localize the position of the gene on the chromosome (Chr). The procedure provides a means to identify antibody-defined markers encoded by genes on different chromosomes, and to follow the patterns of inheritance of such genetically marked chromosomes.

Cell Fusion

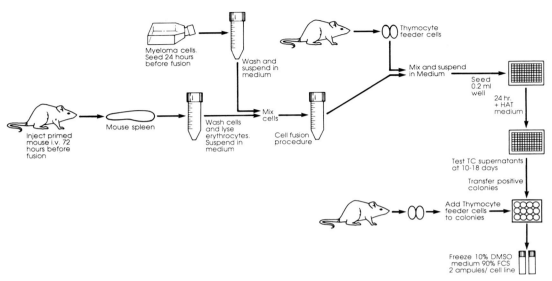

FIGURE 4.—Flow diagram detailing the basic steps for producing monoclonal antibodies. Mice are first immunized with crude or purified preparations of antigen. Following two or more rounds of immunization, the mice are injected intravenously (i.v.) with the antigen. Three days later, the mice are killed and lymphocyte preparations are prepared from the spleen. Following washing and removal of contaminating erythrocytes, polyethylene glycol is used as a reagent to fuse the lymphocytes with tissue culture-adapted mouse myeloma cells. The fused cells are washed, mixed with thymocytes (filler–feeder cells), and distributed at high dilution into 96-well tissue culture plates. A selective growth medium containing HAT (hypoxanthine, aminopterin, and thymidine) is added 24 h later to block the growth of unfused myeloma cells. At 8 to 12 d, tissue culture (TC) supernatants are collected and tested for the presence of specific antibody (see Figures 6–8). Hybrids producing antibodies of interest are expanded and cryopreserved in dimethyl sulfoxide (DMSO) and serum. Supernatants are saved for later evaluation. Following evaluation, hybrids that produce monoclonal antibodies of immediate interest are cloned, and preparations of antibody are prepared for use.

ficity, and the fourth concerned the survival and outgrowth of hybridomas in primary cultures.

Our first problem was to determine how to increase the yield of monoclonal antibodies for the different molecules in leukocyte membranes. Under the usual conditions of immunization, hybridomas yield a heterogeneous set of antibodies, some of which identify the same molecules. The frequency of hybridomas that produce antibodies specific for the same molecule depends on the immunogenicity (the capacity to elicit an immune response) of the molecule, and on the consequent proliferation of B lymphocyte clones producing the monoclonal antibodies. The more cells there are in a given clone, the greater the probability of multiple hybrids. Moreover, dominance of a few immunogenic molecules can lead to the generation of hybrids that produce the same spectrum of monoclonal antibodies in successive fusions.

Our second problem was to devise a way to establish phylogenetic relations of the monoclonal

Identification
of
Genetic Markers

Leukocytes

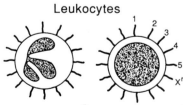

MoAbs $1,2,3,\ldots X^n$ Antigens $1,2,3,\ldots X^n$

FIGURE 5.—Diagrammatic sketch that emphasizes the potential of leukocytes as a source of antigenic molecules for immunological and genetic investigations. Individual subpopulations of leukocytes (lymphocytes, monocytes or macrophages, and granulocytes) bear molecules with particular activities and molecules common to all leukocytes. It is technically possible to prepare monoclonal antibodies specific for each variant and invariant antigenic determinant on each type of molecule present in the cell membrane.

antibody-defined molecules in different species (Davis et al. 1987). There is limited information on the composition of the lymphoid system in different species, and on variations in the distribution and expression of related molecules. For example, the discovery that class-II molecules encoded by the major histocompatibility complex (Deeg et al. 1982; Accolla et al. 1983; Lunney et al. 1983; Burger et al. 1984; Iwaki et al. 1984; Pescovitz et al. 1984, 1985; Lunney and Pescovitz 1988) as well as other lymphocyte markers (Thy-1 molecules: Reif and Allan 1964; Acton et al. 1974; Williams 1976; Thierfelder 1977; Dalchau and Fabre 1979; Ades et al. 1980; McKenzie and Fabre 1981; Foon et al. 1984; and CD4 molecules: Moscicki et al. 1983; Wood et al. 1983; Jefferies et al. 1985) vary in expression or distribution in different species has shown that even well-characterized molecules may be difficult to identify.

Our third problem was to determine which assay would allow us to establish the specificity of the monoclonal antibodies in primary cultures, thereby avoiding repeated selection of antibodies of the same specificity.

Our fourth problem was to define the culture conditions that maximized survival and outgrowth of newly formed hybrids.

To resolve the first two problems, we took advantage of new techniques that permit analysis of the fine specificity of antibodies and, in addition, exploit the full potential of mice to produce antibodies of interest. Several studies had shown that phylogenetically conserved antigenic determinants on major histocompatibility-complex and leukocyte-differentiation molecules are occasionally recognized by monoclonal antibodies prepared for a single species (Davis et al. 1987). Such cross-reactivity had demonstrated phylogenetic relations among molecules derived from different species, and it had established that related molecules have a similar tissue distribution. These findings suggested to us that it should be possible to (1) identify sets of monoclonal antibodies that could be employed across species to define the major histocompatibility complex and establish its role in immune regulation, (2) identify monoclonal antibodies that could be used to compare the expression and function of homologous leukocyte-differentiation molecules in multiple species, and (3) identify cross-reactive monoclonal antibodies to major histocompatibility-complex and leukocyte-differentiation molecules, which could be useful in developing monoclonal antibodies to polymorphic antigenic determinants in two or more species. The exploitation of specificity to both species-restricted and conserved antigenic determinants (i.e., conserved antigenic determinants on homologous molecules) seemed the simplest solution to the problem of working with outbred species. The development of an extensive set of cross-reactive monoclonal antibodies would minimize the need for producing comparable sets of monoclonal antibodies for each species. It would also allow the development of sets of reagents for distinguishing molecules used as genetic markers in two or more species.

To resolve the third problem, we evaluated the potential of various assays in screening for monoclonal antibodies and in genetic studies.

To resolve the fourth problem, we examined the factors that affect the growth of newly formed hybridomas, and we devised improvements in culture methods.

Strategies for Preparation and Analysis of Monoclonal Antibodies

Production of Cross-reactive Monoclonal Antibodies

When we examined existing monoclonal antibodies selected on the basis of their pattern of reactivity in a given species, we found that most of them had a narrow specificity for determinants present on related molecules in one or a few species. Only occasionally did a monoclonal antibody exhibit cross-reactivity (i.e., specificity for a highly conserved determinant). We obtained similar results when we immunized mice with cells from a single species (Davis et al. 1984). Analysis of various strategies of immunization, however, showed us that the frequency of cross-reactive monoclonal antibodies could be augmented. Hyperimmunization with leukocytes from multiple species resulted in the expansion of B cell clonotypes that produced cross-reactive monoclonal antibodies (Table 1). Moreover, by altering immunization procedures, we were able to expand B cell clonotypes that produced monoclonal antibodies specific for molecules of special interest. In addition, our studies demonstrated how detection of cross-reactive monoclonal antibodies could be improved by employing immunization procedures that excluded the species used for primary screening. When the intent was to identify antibodies reactive with antigens in distantly related species, such as conserved determinants on leukocyte-differentiation antigens, we excluded closely related species from the immuni-

TABLE 1.—Comparison of the effect of different regimens of immunization with leukocytes from one or more species of mammals on the yields of hybridomas that produce monoclonal antibodies to conserved antigenic determinants.

Cell fusion[a] (number of species used to immunize mice)		Number of cell lines selected for further evaluation	Number of monoclonal antibodies reactive with peripheral blood from one or more species					
			Cow	Goat	Sheep	Pig	Horse	Human
B[b]	(1)	35/550	35	4	3	3	4	1
PT[b]	(1)	97/1,206	5	5	5	97	4	5
HT[c]	(1)	110/930[d]	1	1	1		110	
H[b]	(6)	73/890	73	10	8	9	14	15
TH[c]	(4)	120/1,190	120	14	13	12	9	12

[a]The acronyms denote origin of the cell line producing the monoclonal antibodies and the regimen of immunization used to generate the cell lines: B = immunized with bovine peripheral blood mononuclear cells (PBM); HT = immunized with equine thymocytes; PT = immunized with porcine thymocytes; H = immunized with equine, canine, caprine, rat, and rabbit PBM; TH = immunized with bovine, caprine, equine, and rat thymocytes.
[b]Monoclonal antibodies identified by complement-mediated antibody cytotoxicity.
[c]Monoclonal antibodies identified by enzyme-linked immunosorbent assay.
[d]Number of hybridomas producing antibody per number of wells in 96-well culture plates containing one or more hybridomas.

zation process. In studies conducted thus far, we have used various regimens of immunization and cells from two or more species to assess the spectrum of monoclonal antibody specificities produced.

Methods of Assay

Our efforts to identify and characterize monoclonal antibodies specific for surfaced-expressed proteins demonstrated that few methods exist that provide adequate resolution, especially for comparisons of the specificity of multiple monoclonal antibodies in primary cultures. Our initial attempts to use complement-mediated killing (microcytotoxicity assay) and enzyme-linked immunosorbent assays (ELISA) (Figures 6, 7) revealed that many of the antibodies of interest could not be readily detected (Davis et al. 1984; Davis et al. 1986). The extent of killing of target cells in the microcytotoxicity assay varied with the quantitative expression of the antigen detected, thus providing little indication of whether an antigen was expressed by a subpopulation or by all leukocytes. In our initial studies, we identified numerous cytotoxic monoclonal antibodies that killed between 10 and 80% of peripheral blood mononuclear cells. Subsequently we discovered that many of the detected antigens were actually present on all peripheral-blood mononuclear cells. Using the enzyme-linked immunosorbent assay, we found that many of the molecules of interest were not always detectable after fixation of peripheral blood mononuclear cells with formaldehyde or glutaraldehyde, and that the interaction of the antibody with the molecule failed to yield a consistent pattern of reactivity. Some of

the latter difficulties with the enzyme-linked immunosorbent assay were attributable to variations in the concentration of the antigen-bearing cells rather than to a lack of sensitivity of the assay. The problems we encountered emphasized the

Microcytotoxicity Assay

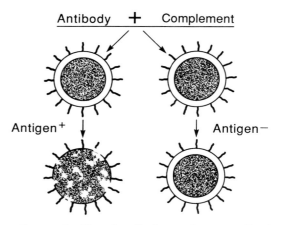

FIGURE 6.—Use of antibody-complement-mediated cytoxicity to identify cells that express a given membrane molecule. Complement-fixing antibodies bound to the cell membranes serve as a site for attachment and activation of complement (a set of serum proteins). Activation of complement at the cell surface leads to the lysis of antigen[+] cells. Dead cells are distinguished from live cells by vital dyes or by phase and fluorescence microscopy. Such complement-mediated lysis can be used to study the distribution of antigens on different cell types and to analyze the patterns of inheritance of antibody-defined genetic markers.

Enzyme Linked Immunosorbent Assay

Flow Microfluorimetry

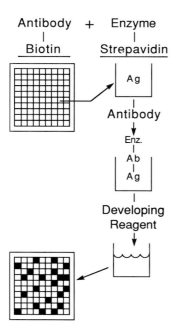

FIGURE 7.—Use of the enzyme-linked immunosorbent assay (ELISA) to study antigens. Crude or purified preparations of antigenic molecules are bound to 96-well microtiter test plates and then reacted with a set of specific antibodies. After washing, the plates are reacted with a second antibody that is coupled to an indicator enzyme (e.g., horseradish peroxidase-linked goat anti-mouse antibody). Then a chromogenic substrate is added, and enzymatic degradation of the chromogen leads to the development of a colored product in antigen-positive wells. The ELISA can be used in studies with any cell membrane-bound or soluble antigenic molecule that binds to the plastic in the test plates.

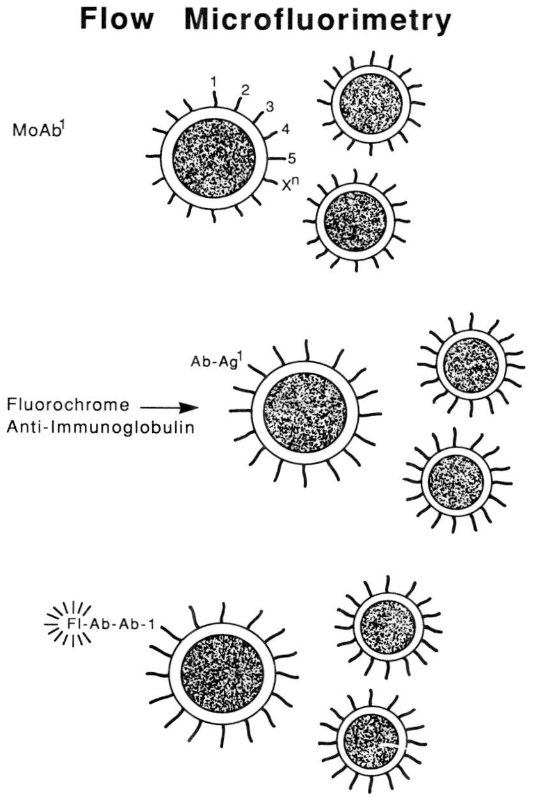

FIGURE 8.—Use of antibody coupled directly or indirectly with a fluorochrome such as fluorescein (which emits green fluorescence) or phycoerythrin (which emits red fluorescence) to identify antigen+ (Ag) cells by fluorescence microscopy or flow microfluorimetry. In the indirect assay, cells are reacted with a monoclonal antibody (MoAb) and then with one or two second-step antibodies specific for mouse antibodies, such as fluorescein-coupled or phycoerythrin-coupled goat anti-mouse antibody (or both). When used together, the second-step reagents react with different classes of mouse antibody (Davis et al. 1987). Positive cells exhibit a ring of fluorescence (Fl) under the fluorescent microscope. Fluorochrome-labeled antibodies can be used to analyze the patterns of expression of membrane molecules and to determine whether any two molecules are expressed on the same or different subpopulations of leukocytes.

need for an assay technique that readily distinguished the patterns of reactivity of monoclonal antibodies with large and small populations of cells in peripheral-blood mononuclear cells, and that also distinguished mononuclear cells from granulocytes.

Flow microfluorimetry (Figure 8) gave us the needed resolution (Herzenberg and Herzenberg 1978; Kung et al. 1981; Landay et al. 1983; Lanier et al. 1983a, 1983b; Davis et al. 1984, 1987; Lanier and Loken 1984; Davis 1985; Lewin et al. 1985). A flow microfluorimeter (cytometer) is an automated instrument interfaced with a computer. It

allowed us to distinguish different populations of cells on the basis of cell size, light-scattering properties, and fluorescence. As illustrated in Figure 9, we used single- and dual-parameter analyses to distinguish between granulocytes and mononuclear cells, and to estimate cell size, the percentage of cells expressing a given membrane molecule, and the level of antigen expression

Dual Parameter Anaysis

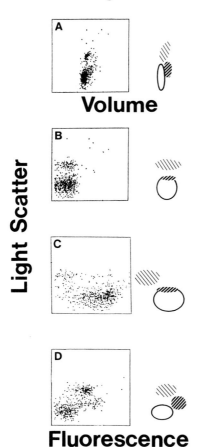

Volume

Light Scatter

Fluorescence

FIGURE 9.—Representative dot-plot profiles of cells analyzed with the Becton Dickinson flow microfluorimeter (FACS™ Analyzer). They illustrate how flow cytometry is used to distinguish different types of leukocytes on the basis of cell size, light-scattering properties, and fluorescence intensity (natural autofluorescence and fluorescence imparted by the attachment of a fluorochrome-conjugated antibody). The profiles were obtained with caprine peripheral-blood leukocytes. The dot plots represent two-dimensional comparisons of two characteristics of the cells. The abscissa and ordinate are divided into 255 channels each. The diagrams alongside each dot-plot profile depict the populations of labeled and unlabeled cells. In **A**, light-scattering properties of leukocytes (wide-angle scatter) are plotted against cell volume to discriminate among granulocytes (uppermost dot profiles), mononuclear lymphocytes and monocytes (lower right dot profiles), and lymphocytes (lower left dot profiles). In **B** through **D**, granulocytes (upper dot profiles) are distinguished from mononuclear cells (lower dot profiles) on the basis of light-scattering properties, and antigen-positive cells (labeled with fluoresceinated anti-

(based on the intensity of fluorescence of labeled cells). Variations in the pattern and level of antigen expression on given molecules in populations of cells permitted the identification of monoclonal antibodies that recognized antigenic determinants on the same or different molecules (single fluorescence analysis, Figure 9). Dual-fluorescence analysis provided the means to determine whether any two molecules were expressed on single or multiple lineages of cells (Figure 10).

Recent studies have shown that dual-fluorescence flow microfluorimetry can also be used to simultaneously examine the specificity of a monoclonal antibody for related molecules expressed on leukocytes from two species. This is accomplished by internally labeling one population of cells with a red fluorochrome (Hydroethidine, Polysciences; Bucana et al. 1986). The labeled cells are then mixed with an equal number of unlabeled cells and reacted with a monoclonal antibody and fluorescein-coupled goat anti-mouse immunoglobulin. As illustrated in Figure 11, staining with Hydroethidine resolves the populations of cells in one graphical dimension (red fluorescence), and the fluoresceinated goat anti-mouse immunoglobulin resolves antigen-positive cells on the other (green fluorescence).

Methods of Preparation

Factors affecting the outgrowth of primary hybridomas.—Although previous studies had established the value of cross-reactive monoclonal antibodies, most of the cross-reactive antibodies in use had been identified by serendipity. Few attempts had been made to develop strategies to optimize production and analysis of cross-reactive monoclonal antibodies (Davis et al. 1983,

← body) are distinguished from antigen-negative cells based on differences in fluorescence. The light-scattering and fluorescence characteristics of unlabeled cells are shown in B. In C and D, dots in line with those in B represent unlabeled cells, whereas dots to the right represent labeled cells. Variations in the profiles of labeled cells are caused by different concentrations of fluoresceinated goat anti-mouse immunoglobulin on antigen-positive cells. The variations in C and D illustrate that patterns of labeling can be used to establish whether monoclonal antibodies recognize the same or different membrane molecules. Profile C is of cells labeled with a monoclonal antibody that recognizes an antigenic molecule present on the majority of mononuclear cells. Profile D is of cells labeled with an antibody that recognizes a molecule expressed only on granulocytes (upper set of dot profiles) and monocytes (lower right set of dot profiles).

Dual Color Analysis

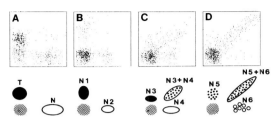

FIGURE 10.—Representative dot-plot profiles of bovine peripheral blood mononuclear cells (PBM). They illustrate how flow microfluorimetry can be used to determine whether two separate membrane molecules are present on the same or different subpopulations of leukocytes. The dot plots represent a comparison of labeling with a red (phycoerythrin) fluorochrome (ordinate) and with a green (fluorescein) fluorchrome (abscissa). Leukocytes were reacted with two monoclonal antibodies specific for different membrane molecules, and were then reacted with a phycoerythrin-coupled antibody and a fluorescein-coupled antibody. The diagrams below each dot-plot profile depict the populations of labeled and unlabeled cells. Profile **A** is a profile of cells reacted with anti-T-lymphocyte and anti-N-lymphocyte monoclonal antibodies; it shows that T- and N-lymphocytes can be distinguished by molecules expressed exclusively on them. Profiles **B** through **C** are of lymphocytes reacted with combinations of antibodies that recognize molecules specific to different N-lymphocytes. The antibodies used in B distinguish subpopulations of N-lymphocytes that express either one or the other N-lymphocyte antigen. The antibodies used in C and D distinguish subpopulations that express one or both N-lymphocyte antigens.

Dual Color Analysis

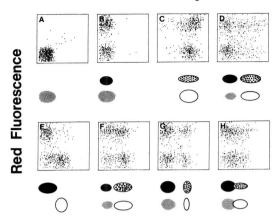

Green Fluorescence

FIGURE 11.—Representative dot-plot profiles that illustrate how flow microfluorimetry can be used to determine if a monoclonal antibody recognizes an antigenic determinant on homologous molecules from two species. The profiles are of cultured caprine and ovine peripheral blood mononuclear cells treated with some combination of Hydroethidine (a red fluorochrome taken up by living cells), monoclonal antibodies, and fluorescein-coupled goat anti-mouse antibody. The diagrams below each dot-plot profile depict the populations of labeled and unlabeled cells. Profile **A** is of mixed, untreated cells. Profile **B** is a mixture of untreated caprine (lower dot profiles) and Hydroethidine-treated ovine (upper-left quadrant) peripheral blood mononuclear cells. Profiles **C** through **H** are of the same cell mixture reacted with different monoclonal antibodies and fluorescein-coupled goat anti-mouse antibody. Note that in each set of profiles, it is possible to determine whether a monoclonal antibody reacts with an antigenic determinant expressed on homologous molecules in one species (plot E) or both species (plots C, D, F, G, and H).

1987). Consequently, our first endeavors to use monoclonal antibody technology focused on maximizing the yield of hybrids obtained in each fusion. Because previous studies by others had revealed low frequencies of the B lymphocyte clones that produced antibody to many of the antigens of interest, it was essential to optimize conditions for fusion and outgrowth of myeloma cells for these B cells. We also found considerable variation in the number of hybrids reported for successive fusions, with yields varying from 300 to 1,000 hybrids from the fusion of 10^8 nucleated spleen cells with SP2/0 or X63.853 myeloma cells (Langone and Van Vunakis 1986). In our initial studies, we confirmed that variation in outgrowth was in part attributable to the presence or absence of culture medium supplements and filler cells (i.e., thymocytes or peritoneal macrophages). Subsequently, we discovered that the major factor affecting outgrowth was the absence of growth factors critical to survival of hybrids during the

first 24 h following fusion. When a B cell mitogen (STM, Ribi Immunochemical Research Corporation) was added to the medium and addition of HAT was delayed for 24 h, hybrid yield increased to 2,000–3000 hybrids per 5×10^7 spleen cells. We found that the effect of the mitogen on outgrowth was consistent and that similar yields of hybrids could be obtained from fusion of freshly prepared and cryopreserved spleen cells (Table 2). These findings provided us with the flexibility and productive potential needed to develop monoclonal antibodies for use in outbred species.

Monoclonal Antibodies Reactive with Leukocyte Membrane Molecules

In our initial studies, we were interested in obtaining monoclonal antibodies to all membrane

TABLE 2.—Comparison of the effect of a B lymphocyte mitogen (STM) on the outgrowth of hybridomas made with fresh and cryopreserved murine spleen cells.

Fusion (number of spleen cells in fusion)	Control: thymocytes only		STM + thymocytes	
	Number of wells with colonies (96-well plates per set)	Number of colonies per plate (colonies per well)	Average number of wells per plate with one or more colonies (number of 96-well plates)	Average number of colonies per plate (colonies per well)
AMG[a] (5 × 10⁷)	65 (1)	145 (1–11)	96 (9)	439 (1–15)
CACTB[a] (5 × 10⁷)	79 (1)	173 (1–13)	90 (9)	251 (1–16)
BAQ[b] (5 × 10⁷)	25 (1)	33 (1–3)	90 (9)	249 (1–11)
BLV[b] (5 × 10⁷)	57 (1)	113 (1–27)	80 (9)	287 (1–16)

[a]Fresh spleen cells were used in the fusion. The letter designations refer to the name of the fusions.
[b]Cryopreserved spleen cells were used in the fusion.

antigens that might prove useful in the analysis of the bovine immune system, as well as in genetic studies (Davis et al. 1984). The microcytotoxicity and enzyme-linked immunosorbent assays were used to obtain the first sets of monoclonal antibodies. Flow microfluorimetry was employed to characterize subsequent sets of monoclonal antibodies (Davis 1985; Davis et al. 1987). The monoclonal antibodies listed in Table 3 were selected from a large series of monoclonal antibodies currently under investigation; they illustrate how we have used flow microfluorimetry and patterns of cross-reactivity to distinguish monoclonal antibodies that recognize antigenic determinants on homologous molecules in different species, and also to distinguish molecules for potential use as genetic markers.

Identification and Characterization of Monoclonal Antibodies Specific for Gene Products of the Major Histocompatibility Complex

The major histocompatibility complex (generically referred to as the MHC) consists of a set of closely linked genes that are centrally involved in regulation and expression of the vertebrate immune response. The antigenic molecules encoded by class-I and class-II genes are highly polymorphic and are the main molecules associated with regulating immune recognition. Thus they are important for use in genetic marking and analysis of the immune response. To identify antibodies that clearly reacted with antigenic determinants encoded by the major histocompatibility-complex (bovine lymphocyte antigen [BoLA]) class-I and -II gene products in cattle, we screened bovine peripheral blood mononuclear cells with sets of monoclonal antibodies specific for either mouse or human MHC molecules (referred to as H-2 in the mouse and HLA in humans) by either micro-

cytotoxicity or flow microfluorimetry. In addition, we screened our initial sets of monoclonal antibodies for cross-reactivity with mouse and human lymphocytes. Monoclonal antibodies found to cross-react were then analyzed by flow microfluorimetry to determine the distribution of the antigens on peripheral blood mononuclear cells in the different species, and also to establish the patterns of labeling obtained with known anti-class-I and anti-class-II monoclonal antibodies (Table 3). Analysis of anti-H-2 class-I monoclonal antibodies yielded three monoclonal antibodies (anti-CA4CA, -CA8PA, -CA48PA) reactive with human and bovine cells. Examination of available anti-HLA class-I and class-II monoclonal antibodies, reported to cross-react with bovine cells, yielded one anti-class-II monoclonal antibody (H4; Chatterjee et al. 1982; Lewin et al. 1985). Analysis of our antibodies yielded sets of monoclonal antibodies that reacted with both mouse lymph node cells and human peripheral blood mononuclear cells. The screening of the monoclonal antibodies positive for both mouse and human cells on 29 inbred strains of mice of known H-2 genotype revealed one monoclonal antibody (anti-H58A) that reacted with an antigenic determinant on class-I molecules coded for by H-$2K^{f,p,pv}$ and H-$2D^{w16}$ alleles, and three monoclonal antibodies (anti-H42A, -TH21A, and -TH81A) that reacted with determinants on class-II molecules coded for by H-2 haplotypes H-2^{w3} (anti-H42A and -TH21A) and H-$2^{b,f,r,s,u,v,w4,w6,w7,w17,w22}$ *(anti-TH81A).*

Comparative studies of the staining patterns obtained with these monoclonal antibodies established the presence of consistent patterns that could be used to identify other monoclonal antibodies with specificity for either class-I or -II molecules. As shown in Table 3, this finding permitted us to identify sets of monoclonal antibodies reactive with cells from two or more species. Biochemical studies confirmed that the monoclonal antibodies recognized molecules with the characteristic molecular weights of class-I and -II molecules in each species tested (Davis et al. 1987).

As shown in Figure 4 and Table 3, the monoclonal antibodies characterized by these procedures have proved exceptionally useful in the identification of class-I and -II molecules in several species of higher vertebrates. In addition, they have afforded us the means to extend our studies on the phylogenetic origin and relation of alleles in different species, and the reagents needed to develop monoclonal antibodies to polymorphic determinants.

Conservation of major histocompatibility antigenic determinants.—The screening of our present panel of monoclonal antibodies has shown that any monoclonal antibody-defined determinant on class-I molecules may vary in expression between species. Some determinants may be absent or in low frequency in one species and in high frequency in other species, as exemplified by antigenic determinant H58A. This determinant behaves as an allele-specific determinant in mice, where it is encoded by a few related *H-2K* alleles and one *H-2D* allele. In pigs and cats, it behaves as a polymorphic determinant. In all other species examined, it behaves as a monomorphic determinant (unpublished observations).

In contrast, analysis of the monoclonal antibody-defined class-II molecules has thus far revealed polymorphism only for the three determinants present on class-II molecules in mice and other species (H42A, TH21A, and TH81A). These determinants behave as polymorphic determinants in mice, and as monomorphic determinants (i.e., they are expressed on most of the allelic variants in a given species) in all other species examined.

Identification of polymorphic determinants by microcytotoxicity and flow microfluorimetry.—We have used two methods to screen for polymorphic determinants: microcytotoxicity and flow microfluorimetry. The microcytotoxicity assay has been used when some information has been available on alloantisera-defined polymorphic antigenic determinants. Comparison of distribution and patterns of inheritance has allowed us to demonstrate that most monoclonal antibody-defined determinants differ from alloantisera-defined determinants. In addition, the microcytotoxicity assay has been used to determine if immunoaffinity-purified molecules prepared with monoclonal

antibodies to monomorphic determinants can be used to produce monoclonal antibodies to polymorphic determinants. This assay has permitted us to test primary cultures from fused cells against animals of known genotype to detect monoclonal antibodies that react with allelic products coded for by one or several major histocompatibility-complex genes.

Using these techniques in studies with cattle, we have established that class-I molecules (immunoaffinity-purified with monoclonal antibodies) can be used as immunogens to produce monoclonal antibodies that define such polymorphic determinants as HELP1F, SAG4A, SAG7A, and SAG8A (Table 3). However, the assays have proven less useful for screening for polymorphic determinants in species whose antigenic variation is poorly known. Large numbers of animals of unknown lineage must be tested to establish whether antigenic variation exists. Because of logistics, this has been a difficult problem to resolve; but preliminary studies with flow microfluorimetry, indicate that it can be minimized. Because monoclonal antibodies specific for a conserved determinant on a molecule define the frequency of cells expressing the molecule, it is possible to mix cells from two or more animals and still determine the concentration of the antigen-positive cells. Thus sets of mixed cells can be used to screen for monoclonal antibodies to polymorphic determinants. New monoclonal antibodies reactive with monomorphic determinants on a given molecule will react with the whole population of antigen-positive cells, whereas the monoclonal antibodies to polymorphic determinants will react only with a fraction of the cells. This method was used recently to establish that the class-I antigen recognized by anti-H58A is polymorphic in the cat (Davis et al., unpublished), which indicates the technique can be used to identify monoclonal antibody-defined polymorphic determinants on other membrane molecules.

Identification and Characterization of Monoclonal Antibodies Reactive with Leukocyte-Differentiation Antigens

When the preceding method was used to identify leukocyte-differentiation molecules (i.e., molecules that are expressed on one or more lineages of mature leukocytes) in cattle, it revealed that few monoclonal antibodies specific for antigenic determinants on human leukocytes react with leukocytes from other species. Consequently, a

TABLE 3.—Monoclonal antibodies selected for use in immunogenetic and immunologic studies in ruminants and other mammalian species.[a] Ig = immunoglobulin.

Monoclonal antibodies	Ig isotype	Flow-cytometric analysis of the patterns of monoclonal antibody reactivity with antigenic determinants present on peripheral-blood leukocytes from different species								Predominant specificity
		Cattle	Sheep	Goat	Cape buffalo	Water buffalo	Pig	Horse	Human	
Anti-H58A*	IgG$_{2A}$	+	+	+	+	+	+ (P)	+	+	MHC-class I
Anti-HELP1F	IgM	+ (P)	+ (P)	NT	+	+	−	−	−	MHC-class I
Anti-SAG4A	IgG$_{2A}$	+ (P)	NT	NT	NT	NT	NT	NT	NT	MHC-class I
Anti-SAG7A	IgG$_{2A}$	+ (P)	NT	NT	NT	NT	NT	NT	NT	MHC-class I
Anti-SAG8A	IgG$_{2A}$	+ (P)	NT	NT	NT	NT	NT	NT	NT	MHC-class I
Anti-CA4C-A*	IgM	+ (P)	+ (P)	−	+	NT	+ (P)	−	+ (P)	MHC-class I
Anti-CA8P-A*	IgG$_{2A}$	+ (P)	−	−	+ (P)	NT	+ (P)	−	+ (P)	MHC-class I
Anti-CA10PB*	IgG$_{2A}$	+ (P)	−	−	−	NT	−	−	−	MHC-class I
Anti-CA48PA1*	IgG$_{2A}$	+ (P)	+ (P)	+ (P)	+	NT	+ (P)	+ (P)	+ (P)	MHC-class I
Anti-H42A*	IgG$_{2A}$	+	+	+	+	+	+	+	+	MHC-class II
Anti-TH14B	IgG$_{2A}$	+	+	+	+	+	+	+	+	MHC-class II
Anti-TH21A*	IgG$_{2B}$	+ (P)	+	+	+	+	+	+	+	MHC-class II
Anti-TH81A5*	IgG$_{2A}$	+ (P)	+	+	+	+	+	+	+ (P)	MHC-class II
Anti-CAT82A	IgG$_1$	+	+	+	NT	NT	+	+	NT	MHC-class II
Anti-BAT83A	IgG$_1$	+	+	+	NT	NT	NT	NT	NT	MHC-class II
Anti-BAGB44A	IgG$_1$	+	+	+	NT	NT	NT	+	NT	MHC-class II
Anti-TH97A	IgG$_{2A}$	+(Thy)	+(Thy)	+(Thy)	NT	NT	−	−	NT	BoT1 (CD1-like)
Anti-B26A4	IgM	+	−	+	+	+	−	−	−	BoT2 (BoCD2, SRBCr)
Anti-CH61A	IgG$_1$	+	−	−	+	+	−	−	−	BoT2 (BoCD2, SRBCr)
Anti-CH128A	IgG$_1$	+	−	−	+	+	−	−	−	BoT2 (BoCD2, SRBCr)
Anti-CH132A	IgM	+	−	+	+	−	−	−	−	BoT2 (BoCD2, SRBCr)
Anti-CH134A	IgG$_1$	+	−	−	+	+	−	−	−	BoT2 (BoCD2, SRBCr)
Anti-BAQ95A	IgG$_1$	+	−	+	NT	+	NT	NT	−	BoT2 (BoCD2, SRBCr)
Anti-CACT98A	IgM	+	−	+	NT	+	−	−	−	BoT2 (BoCD2, SRBCr)
Anti-BAT18A	IgG$_1$	+	−	+	NT	+	NT	NT	NT	(CapCD2, SRBCr)
Anti-BAT42A	IgG$_1$	+	−	+	NT	+	NT	NT	NT	(CapCD2, SRBCr)
Anti-BAT76A	IgG$_{2A}$	+	−	+	NT	+	NT	NT	NT	(CapCD2, SRBCr)
Anti-BAQ82A	IgM	+	+	−	NT	+	NT	NT	−	Pan T (Bo6, CD6 ?)
Anti-BAQ91A	IgG$_{2B}$	+	−	−	NT	+	NT	NT	−	Pan T (Bo6, CD6 ?)
Anti-IL A11	IgG$_{2A}$	+	−	−	+	+	NT	NT	NT	BoT4 (BoCD4)
Anti-IL A12	IgG$_{2A}$	+	−	−	−	−	NT	NT	NT	BoT4 (BoCD4)
Anti-SBUT4	IgG$_{2A}$	−	+	+	NT	NT	−	−	−	SBUT4 (OvCD4)
Anti-CACT33A	IgM	+ (P)	−	−	NT	−	−	−	−	BoT4 (BoCD4)
Anti-CACT83A	IgM	+	−	−	NT	−	−	−	−	BoT4 (BoCD4)
Anti-CACT87A	IgM	+	−	−	NT	−	−	−	−	BoT4 (BoCD4)
Anti-CACT91C	IgM	+	−	−	NT	−	−	−	−	BoT4 (BoCD4)
Anti-CACT93A	IgM	+	−	−	NT	−	−	−	−	BoT4 (BoCD4)
Anti-CACT138A	IgG$_1$	+	−	−	NT	+	−	−	−	BoT4 (BoCD4)
Anti-GC1A	IgG$_{2A}$	−	+	+	NT	NT	−	−	−	CapCD4
Anti-GC17A	IgM	+	+	+	NT	−	NT	NT	NT	CapCD4
Anti-GC50A	IgM	+	+	+	NT	+	NT	NT	NT	CapCD4
Anti-IL A17	IgG$_1$	+	−	−	−	+	−	−	−	BoT8 (BoCD8)
Anti-SBUT8	IgG$_{2A}$	−	+	+	NT	NT	−	NT	NT	SBUT8 (OvCD8)
Anti-CACT80C	IgG$_1$	+	+	+	NT	+	−	−	NT	BoT8 (BoCD8)
Anti-CACT85A	IgG$_1$	+	−	−	NT	−	−	−	−	BoT8 (BoCD8)
Anti-CACT88C	IgG$_3$	+	−	+	NT	−	−	−	−	BoT8 (BoCD8)
Anti-CACT130A	IgG$_3$	+	−	−	NT	−	−	−	−	BoT8 (BoCD8)
Anti-BAQ111A	IgM	+	−	+ (P)	NT	+	NT	NT	−	BoT8 (BoCD8)
Anti-TH82D1	IgG$_1$	+ (P)	+	+	+	NT	−	+ (P)	+ (P)	BoT8 (BoCD8)
Anti-BAT82A	IgG$_1$	+	+	+	NT	−	−	NT	NT	CapCD8
Anti-BAGB25A	IgM	−	−	+	NT	NT	NT	NT	NT	CapCD8
Anti-BAT63A	IgM	+	+	+	NT	+	+	NT	NT	T subpop
Anti-PIg45A	IgG$_{2B}$	+	+	+	+	+	+	+/−	+	IgM
Anti-BIg715A	IgG$_1$	+	NT	NT	NT	NT	NT	NT	NT	IgG1
Anti-BIg623A	IgG$_3$	+	NT	NT	NT	NT	NT	NT	NT	IgG2
Anti-BIg312D3	IgG$_1$	+	NT	NT	NT	NT	NT	NT	NT	IgA
Anti-BIg501E	IgG$_1$	+	+	+	NT	+	NT	NT	NT	"Lambda light chain"
Anti-BIg43A	IgG$_1$	+	+	+	+	+	NT	NT	NT	"Kappa light chain"
Anti-BAS3A	IgG$_1$	+	+	+	NT	+	NT	NT	NT	B
Anti-BAS9A	IgM	+	+	+	NT	+	NT	NT	−	B
Anti-BAQ44A	IgM	+	+	+	NT	+	NT	NT	−	B
Anti-BAQ155A	IgG$_1$	+	+	+	NT	+	NT	NT	NT	B subpop
Anti-BAS21A	IgG$_1$	+	+	+	NT	−	NT	NT	NT	B, T

TABLE 3.—Continued.

Monoclonal antibodies	Ig isotype	Flow-cytometric analysis of the patterns of monoclonal antibody reactivity with antigenic determinants present on peripheral-blood leukocytes from different species								Predominant specificity
		Cattle	Sheep	Goat	Cape buffalo	Water buffalo	Pig	Horse	Human	
Anti-BAG32A	IgG$_1$	−	+	+	NT	NT	NT	NT	NT	B, T
Anti-BAG36A	IgG$_1$	+	+	+	NT	−	NT	NT	NT	B, T
Anti-GS5A	IgG$_1$	+	+	+	+	−	NT	NT	NT	B, T
Anti-GX18A	IgG$_1$	+	+	+	+	−	NT	NT	NT	B, T
Anti-TH90A	IgM	+	+	+	+	+	−	+	+	B, M
Anti-TH92A	IgM	+	+	+	+	+	+	−	+	B, M
Anti-HUH43A	IgM	+	+	+	+	+	NT	+	+	subpop PBM (SRBCr+)
Anti-E4C	IgM	+	+	+	+	+	+	+	+	subpop PBM (SRBCr+)
Anti-E40A	IgM	+	+	+	+	+	+	+	+	subpop PBM (SRBCr+)
Anti-H18A2	IgM	+	+	+	+	+	+	+	+	subpop PBM (SRBCr+)
Anti-TH71A	IgM	+	+	+	+	+	NT	+	+	subpop PBM (SRBCr+)
Anti-B7A1	IgM	+	+	+	+	+	−	−	−	N
Anti-BAQ4A	IgG$_1$	+	+	+	NT	+	NT	NT	−	N
Anti-BAQ89A	IgG$_1$	+	+	+	NT	+	NT	NT	NT	N subpop
Anti-BAQ90A	IgG$_3$	+	+	+	NT	+	NT	NT	NT	N subpop
Anti-BAQ102A	IgM	+	+	+	NT	+	NT	NT	−	N
Anti-BAQ128A	IgG$_1$	+	+	+	NT	+	NT	NT	NT	N subpop
Anti-BAQ136A	IgM	+	+	+	NT	+	NT	NT	+	N subpop
Anti-BAQ159A	IgG$_1$	+	+	+	NT	+	NT	NT	NT	N subpop
Anti-CACTB6A	IgM	+	+	+	NT	+	NT	NT	NT	N subpop
Anti-CACTB14A	IgG$_1$	+	+	+	NT	+	NT	NT	NT	N subpop
Anti-CACTB22-1A	IgG$_1$	+	+	+	NT	+	NT	NT	NT	N subpop
Anti-CACTB31A	IgG$_{2B}$	+	+	+	NT	+	NT	NT	NT	N subpop
Anti-CACTB32A	IgG$_1$	+	+	+	NT	+	NT	NT	NT	N subpop
Anti-DH16A	IgM	+	+	+	+	+	−	+	+	T subpop, N subpop
Anti-BAGB3A	IgG$_1$	−	+	+	NT	NT	+	NT	NT	G, N, B, T subpop
Anti-BAGB7A	IgG$_1$	−	+	+	NT	NT	−	NT	NT	T, B, N subpop
Anti-BAGB17A	IgG$_1$	−	+	+	NT	NT	−	NT	NT	T, B, N
Anti-BAT31A	IgG$_1$	+	+	+	NT	+	+	NT	NT	G, M, T, B, N
Anti-BAT75A	IgG$_1$	+	−	+	NT	+	−	NT	NT	G, M, T, B, N
Anti-GX24A	IgG$_1$	+	+	+	+	NT	−	−	−	G, PBM subpop
Anti-DH59B	IgG$_1$	+	+	+	+	+	+	+	+	G, M
Anti-BAG18A	IgM	+	+	+	NT	NT	NT	NT	NT	G, PBM subpop

[a]The acronyms denote origin of the hybridoma cell line producing the monoclonal antibody and the regimen of immunization used to generate the cell line: B = immunized with bovine peripheral blood mononuclear cells; HT = immunized with equine thymocytes; PT = immunized with porcine thymocytes; H = immunized with equine, canine, caprine, rat, and rabbit peripheral blood mononuclear cells; TH = immunized with bovine, caprine, equine, and rat thymocytes. GS, GX = immunized with caprine peripheral blood mononuclear cells; CH, DH, HUH = immunized with bovine, caprine, porcine, equine, canine, mink, and human peripheral blood mononuclear cells; PIg = immunized with bovine and porcine immunoglobulin M (IgM); BIg = immunized with bovine IgG$_1$ and IgG$_2$; RH = immunized with rat lymphocytes and equine fibroblasts infected with equine infectious anemia virus; CA = CBA (H-2^a) mice immunized with B10.A (H-2^a) lymphocytes; BAQ, BAG, BAGB, BAS = immunized with human T lymphocyte tumor (JM) cells, bovine, caprine, ovine, porcine, and equine peripheral-blood mononuclear cells; CACT and CACTB = immunized with bovine concanavalin A-stimulated peripheral blood mononuclear cells; BAT, CER = immunized with caprine, ovine, bovine, and elk concanavalin A-stimulated lymphocytes and caprine, ovine, and bovine thymocytes; CAT = immunized with feline, canine, and ferret peripheral-blood mononuclear cells and thymocytes. PBM = peripheral-blood mononuclear cells; MHC = major histocompatibility complex; T = T lymphocyte; CD = cluster of differentiation, a terminology established to assign a common sequential designation for monoclonal antigens defined by their antibodies on human leukocytes; BoCD2 and CapCD2 = pan T sheep red blood cell receptor; BoCD4, CapCD4, and OvDC4 = T$_{helper}$-lymphocytes; BoCD8, CapCD8, and OvCD8 = T$_{cytotoxic/suppressor}$-lymphocytes; B = B lymphocyte; N = NonT/NonB lymphocyte; G = granulocyte; M = monocyte; NT = not tested; P = antigen polymorphic; − = negative; * = reacts with mouse leukocytes; SRBCr =sheep red blood cell receptor; PBM = peripheral blood mononuclear cells.

series of fusions had to be performed to obtain the first set of monoclonal antibodies cross-reactive with subpopulations of leukocytes in cattle and one or more other species. Leukocytes from cattle were used in the primary screening of culture supernatants. Supernatants from positive cultures were later screened for reactivity with cells from other species. Monoclonal antibodies specific for major histocompatibility-complex class-I and -II molecules, and a monoclonal antibody specific for a conserved determinant on bovine immunoglobulin M (IgM), were used to provide flow-microfluirimetric profiles of cells that expressed known molecules. A sheep red blood cell (SRBC) rosette-inhibition assay was used to identify monoclonal antibodies specific for the bovine homologue of the sheep red blood cell receptor (SRBCr, a molecule present on all T cells that is now reffered to as cluster-of-differentiation molecule 2 [CD2] in humans) (Grewal and Babiuk 1978; Howard et al. 1981; Davis 1985; Davis et al. 1987). In some studies, immunohistochemistry was performed on fresh-frozen tissues to verify specificity (Davis et al. 1988).

Flow-microfluorimetric analysis of supernatants from primary cultures of hybridomas yielded a large set of hybrids that produce monoclonal antibodies capable of detecting molecules on a few or all leukocytes. After cloning the relevant cell lines, we prepared monoclonal antibody from each cell line for investigation. Monoclonal antibodies that yielded similar patterns of labeling were grouped and examined by single- and dual-flow microfluorimetry to determine if they detected the same or different membrane molecules. Further analysis revealed that a set of six monoclonal antibodies reacted with the sheep red blood cell receptor (as defined by their capacity to inhibit sheep red blood cell rosetting with bovine lymphocytes) (Table 3). Comparative dual-fluorescence studies with these monoclonal antibodies allowed us to identify four additional anti-sheep red blood cell receptor antibodies (anti-CACT98A, -BAT18A, -BAT42A, -BAT76A), three antibodies (anti-BAQ82A, -BAQ91A, 2CACT141A) reactive with a different antigen present on all T cells, and two sets of monoclonal antibodies that reacted with subpopulations of T cells. Comparison of the flow-microfluorimetric profiles obtained with monoclonal antibodies known to react with the bovine homologues of CD4 (anti-IL A11, -IL A12; Baldwin et al. 1986) and CD8 (anti-IL A17; Ellis et al. 1986) established that eight of the monoclonal antibodies were specific for CD4 and seven for CD8. Comparative dual-flow microfluorimetry with surface immunoglobulin M^+ ($sIgM^+$) cells permitted us to identify monoclonal antibodies specific for immunoglobulin lambda and kappa chains, and monoclonal antibodies specific for molecules predominantly expressed on B lymphocytes. Comparative dual-flow microfluorimetry also permitted us to identify molecules present on two or more lineages of leukocytes and molecules on a unique lineage of cells that do not express known T, B, or monocyte markers—nonT/nonB (N) lymphocytes (we used anti-B7A1). Further dual-fluorescence studies revealed a set of monoclonal antibodies (anti-H18A2, -TH71A, -HUH43A, -E4C, -E40A) that reacted with a small population of leukocytes with the CD2 (sheep red blood cell receptor) that does not appear to be of T lymphocyte lineage. Single-fluorescence analysis of prep-

arations of leukocytes containing granulocytes, monocytes, and lymphocytes permitted us to identify monoclonal antibodies reactive with granulocytes or monocytes or both; we could also identify monoclonal antibodies reactive with granulocytes and subpopulations of lymphocytes.

Following the development of an extensive panel of monoclonal antibodies known to react with bovine leukocytes, we undertook studies to determine which monoclonal antibodies recognized conserved antigenic determinants in other species. As summarized in Table 3, the studies revealed that several monoclonal antibodies obtained with different immunization procedures reacted with cells from other species. The studies also established that homologous molecules, especially in ruminants, exhibit similar or identical patterns of expression. This finding allowed us to establish that cross-reactive monoclonal antibodies can be used as standards to simultaneously screen primary hybridoma culture supernatants against cells from one or two species to distinguish monoclonal antibodies specific for new molecules. As noted in Figure 12 and Table 3, the use of such monoclonal antibodies (prepared in our laborabory and anti-SBUT4 and -SBUT8 prepared by Maddox et al. 1985) facilitated identification of new monoclonal antibodies specific for the goat homologues of CD2 (anti-BAT18A, -BAT42A, -BAT76A, -CACT98A), CD4 (anti-GC1A, -GC17A, -GC50A), CD8 (anti-BAGB25A, -BAT82A, -CACT80C), major histocompatibility complex class-II molecules (anti-BAGB44A, -BAT83A, -CAT82A), and monoclonal antibodies specific for additional molecules (anti-BAGB3A, -BAGB7A, -BAGB17A, -BAT63A, -BAT70A, -BAT75A). Comparative studies with bovine, caprine, and ovine leukocytes permitted us to demonstrate that the composition of peripheral blood leukocytes in these species is similar, and to establish that the unique population of N lymphocytes, first defined in cattle (with anti-B7A1: Davis et al. 1987), is common in ruminants.

Additional but more limited studies indicate that the monoclonal antibodies we generated recognize homologous molecules in both wild and domestic ruminants. As shown in Table 3, this has facilitated analysis of the lymphoid system in species (e.g., the Cape and water buffaloes) for which there has been little opportunity to produce monoclonal antibodies. As we have shown, the extent of cross-reactivity is such that immunological investigations can now be conducted in these

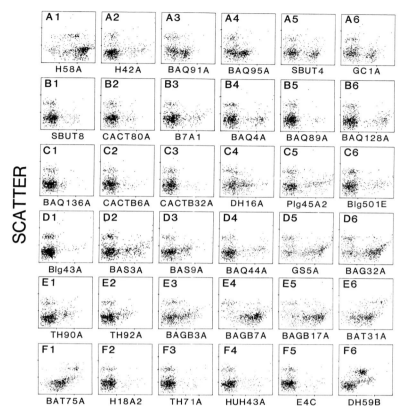

FIGURE 12.—Representative dot-plot profiles of caprine leukocytes reacted with monoclonal antibodies that define membrane molecules in two or more species. The profiles illustrate how flow microfluorimetry has facilitated identification of an extensive set of monoclonal antibodies that clearly recognize different membrane molecules. Compare A5 and A6 (both monoclonal antibodies recognize the T_{helper}-cell antigen CapCD4) and B1 and B2 (both monoclonal antibodies recognize the $T_{cytotoxic/suppressor}$-cell antigen CapCD8) with the other profiles. Monoclonal antibodies that recognize the same molecule yield identical dot-plot profiles. Monoclonal antibodies that recognize different molecules exhibit unique patterns of expression. The monoclonal antibodies listed in Table 3 that recognize the same molecule were identified by comparing their patterns of labeling.

species without the need to develop extensive sets of new monoclonal antibodies.

Strategies for Identification and Use of Leukocyte Membrane Molecules in Immunogenetic and Immunologic Studies in Fish

Identification of Genetic Markers and MoAb Reagents

It is evident that advances in flow microfluorimetry and monoclonal antibody technology have afforded the means to address the major problems encountered in developing immune reagents for studies in any outbred species. Monoclonal antibody technology provides a highly efficient tech-

nique for developing antibodies specific to monomorphic and polymorphic antigenic determinants on any given protein. Flow microfluorimetry provides a way to identify cell types on the basis of size, light-scattering properties, and expression of one or more membrane antigens. We have been able to exploit both advances to develop monoclonal antibodies for use in genetic and immunologic investigations. By using a dual strategy in the development of monoclonal antibodies, we have been able to take advantage of the phylogentic conservation of some antigenic determinants to develop cross-reactive monoclonal antibodies that identify honologous molecules in two or more species. As summarized in Table 3, the develop-

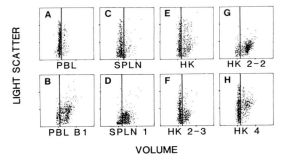

FIGURE 13.—Selected dot-plot profiles of leukocytes separated by density, illustrating the capacity of a flow microfuorimeter to resolve populations of fish leukocytes on the basis of size and light-scattering characteristics. Profile **A** is of unseparated peripheral blood leukocytes (PBL) on the right and erythrocytes on the left; **B** is a profile of partially purified lymphocytes from PBL. Profile **C** is of unseparated leukocytes from spleen (SPLN), and **D** is of partially purified lymphocytes from SPLN. Profile **E** is of u nseparated leukocytes from head kidney (HK); **F** is a profile of partially purified lymphocytes from HK; **G** is a profile of partially purified lymphoblasts from HK; and **H** is a profile of melanocytes separated from HK lymphocytes.

ment of a set of cross-reactive monoclonal antibodies to major histocompatibility-complex class-I and -II molecules has permitted us to identify homologous molecules in multiple species, and to establish the specificity of monoclonal antibodies that recognize antigenic determinants present on major histocompatibility-complex molecules present in one or only a few species. The development of a set of monoclonal antibodies comprising antibodies that recognize antigenic determinants on leukocyte-differentiation molecules in one or more species has permitted us to define the immune system in cattle, goats, and sheep, and to begin elucidating the evolution of the lymphoid system in mammals. As noted in Table 3, ongoing studies have established that many of the monoclonal antibodies react with relevant molecules in Cape and water buffaloes. We can predict that cross-reactivity will extend to other species.

In addition, we are now in a position to begin developing monoclonal antibodies for immunogenetic studies. Each of the antigenic molecules we have identified has the potential to serve as a genetic marker to map chromosomes and identify linkage groups that contain genes that regulate the expression of desirable traits, such as resistance to disease, rate of growth, and milk production. Studies with class-I antigens have shown that immunoaffinity-purified molecules can be used as

immunogens to produce monoclonal antibodies to polymorphic antigenic determinants. In addition, ongoing efforts to develop monoclonal antibodies specific for monomorphic determinants has serendipitously yielded monoclonal antibodies to polymorphic determinants on some molecules of interest. The latter finding has provided support for our contention that it will be possible to produce monoclonal antibodies that recognize polymorphic determinants on molecules identifiable with monoclonal antibodies.

Preliminary studies suggest that similar strategies can be used to develop monoclonal antibodies for use in genetic and immunologic investigations in fish. As shown in Figure 13, analysis of preparations of leukocytes taken from rainbow trout *Oncorhynchus mykiss* (formerly *Salmo gairdneri*), has shown that cell populations can be distinguished on the basis of size and light-scattering properties, as in higher vertebrates. Thus it will be possible to use flow microfluorimetry to determine whether a given monoclonal antibody-defined molecule is expressed on one or more lineages of cells. When appropriate, Hydroethidine can be used to mark populations of cells from different lymphoid organs for comparative studies of antigen distribution and expression. In addition, the dye can be used to mark populations of cells so that cells from two individuals (of the same or different species) can be examined simultaneously to determine if a given monoclonal antibody detects polymorphic or phylogenetically conserved antigenic determinants. When a panel of monoclonal antibodies has been developed, they can be used in multiple ways to characterize that immune system in fish and develop a battery of monoclonal antibodies for genetic studies. An intensive effort should yield sets of monoclonal antibodies equivalent to those now available for research in mammals.

Choice of Assay for Immunogenetic Studies

To this point, emphasis has been placed on the description of strategies that have proved effective in the identification of molecules suitable for use in functional and immunogenetic studies. It is apparent, however, that the potential of immunogenetic techniques rests not only on reagents and markers, but also on the types of assays available for detection and analysis of polymorphisms. The choice of assay for immunogenetic studies depends on the types and source of antigens selected for use as genetic markers. It appears that the needs differ for immunogenetic studies of mammals and fish. We have focused on the identification of polymorphic markers on leukocytes because of the ready accessibility of leukocytes for repeated sampling and because they permit the use of a single assay system (either the enzyme-linked immunosorbent or the microcytotoxicity assay) to simultaneously monitor allelic variation in multiple genes. For our studies, the microcytotoxicity assay is the method of choice. Automated equipment and associated computer software have been developed to facilitate data acquisition and analysis.

For immunogenetic studies of fish, the enzyme-linked immunosorbent assay may prove to be more appropriate. One of the constraints of working with the microcytotoxicity assay is the need to use live cells. Samples must be obtained under sterile conditions and used before the cells deteriorate. This does not present a problem when animals can be brought to the testing center. When blood must be shipped over long distances, however, the quality of the samples can deteriorate. When this happens, the reliability of the microcytotoxicity assay is diminished. The enzyme-linked immunosorbent assay offers the possibility of using either live or dead cells. Thus, with fish, it would be possible to obtain and fix blood samples for later testing. The cells could be either fixed as a free cell suspension following isolation from erythrocytes or fixed directly to microtiter plates (Figure 14). Alternatively, crude antigen extracts could be prepared from preparations of cells lysed in nonionic detergents for use in a modified enzyme-linked assay technique, "dot blotting," which involves the binding of solubilized proteins to nitrocellulose paper. The antigens can be preserved in the lysate until used. Each of these modes of antigen preservation affords a means of preparing a stock of stable antigen for multiple sampling and analysis.

An added advantage of enzyme-linked assays is that they can take advantage of other immunogenetic markers, such as those currently detected by electrophoresis and enzymatic activity (Lundstrom 1987). Polymorphic differences in molecules that affect electrophoretic mobility are, on many occasions, accompanied by changes in antigenic composition. Such differences can be detected with monoclonal antibodies.

Acknowledgments

The excellent technical assistance of M. J. Hamilton during the course of the studies is acknowledged. The trout cells used to illustrate the potential of flow microfluorimetry in fish research were prepared by A. Greenlee in the laboratory of J. A. Congleton, Department of Fish and Wildlife Resources, University of Idaho, Moscow, Idaho.

The studies reported here were supported in part by USDA–SEA (U.S. Department of Agriculture–Science and Education Administration)

Choice of Assay
Fish Immunogenetics

ELISA

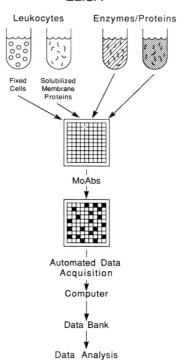

FIGURE 14.—Use of enzyme-linked immunosorbent assay (ELISA) to determine genetic phenotypes of fish polymorphic for one or several antibody-defined genetic markers. Whole leukocytes, leukocyte or tissue extracts, or perparations of serum proteins can be bound to the microtiter test plates and reacted with a panel of monoclonal antibodies (MoAbs). Plates showing positive reactions can be recorded with an automated ELISA reader, and the data can be accumulated in a computer for later analysis. The ELISA offers a flexible assay for analysis of multiple types of antigenic molecules. The only limiting criteria are that the antigens bind consistently with the plastic used in the test plates, and that the antigenic determinants not be blocked or denatured following attachment to the plastic.

Special Research grants 82-CRSR-2-2045, 83-CRSR-2-2281, 84-CRSR-2-2457, 86-CRSR-2-2913, and 86-CRCR-1-2241; USDA Animal Disease Research Unit; Formula Funding Public Law 95-113; the Agricultural Research Center Project 3073; Carnation Research Farm; the American Cancer Society; and the Washington State Technology Center, International Program Development Office of Washington State University. R. A. Larsen was supported by National Science Award AI 07025 from the National Institute of Allergy and Infectious Diseases. M. L. Monaghan was supported by a fellowship from University College, Dublin, and by Norden Animal Health (Irl.) Ltd.

References

Accolla, R. S., D. Birnbaum, and M. Pierres. 1983. The importance of cross-reactions between species: mouse allo-anti-Ia monoclonal antibodies as a powerful tool to define human Ia subsets. Human Immunology 8:75–82.

Acton, R. T., R. J. Morris, and A. F. Williams. 1974. Estimation of the amount and tissue distribution of the rat Thy-l antigen. European Journal of Immunology 4:598–602.

Ades, E. W., R. K. Zwerner, R. T. Acton, and C. M. Balch. 1980. Isolation and partial characterization of the human homologue of Thy-l. Journal of Experimental Medicine 151:400–406.

Baldwin, C. L., A. J. Teale, J. Naessens, B. M. Goddeeris, N. D. MacHugh, and W. I. Morrison. 1986. Characterization of a subset of bovine T lymphocytes that express BoT4 by monoclonal antibodies and function: similarity to lymphocytes defined by human T4 and murine L3T4. Journal of Immunology 136:4385–4392.

Bucana, C., I. Saiki, and R. Nayar. 1986. Uptake and accumulation of vital dye hydroethidine in neoplastic cells. Journal of Histochemistry and Cytochemistry 34:1109–1115.

Burger, R., I. Scher, S. O. Sharrow, and E. M. Shevach. 1984. Non-activated guinea pig T cells and thymocytes express Ia antigens: FACS analysis with alloantibodies and monoclonal antibodies. Immunology 51:93–102.

Chatterjee, S., D. Bernoco, and R. Billings. 1982. Treatment with anti-Ia and anti-blast/monocyte monoclonal antibodies can prolong skin allograft survival in nonhuman primates. Hybridoma 1:369–377.

Dalchau, R., and J. W. Fabre. 1979. Identification and unusual tissue distribution of the canine and human homologues of Thy-l (theta). Journal of Experimental Medicine 149:576–591.

Davis, W. C. 1985. The use of monoclonal antibodies to define the bovine lymphocyte antigen system (BoLA), leukocyte differentiation antigens, and other polymorphic antigens. Pages 119–143 in W. C. Davis, J. N. Shelton, and C. W. Weems, editors. Characterization of the bovine immune system and the genes regulating expression of immunity with particular reference to their role in disease resistance. Washington State University, Department of Veterinary Microbiology and Pathology, Pullman.

Davis, W. C., J. A. Ellis, N. D. MacHugh, and C. L. Baldwin. 1988. Bovine pan T-cell monoclonal antibodies reactive with a molecule similar to CD2. Immunology 63:165–167.

Davis, W. C., and six coauthors. 1987. The development and analysis of species specific and cross reactive monoclonal antibodies to leukocyte differentiation antigens and antigens of the major histocompatibility complex for use in the study of the immune system in cattle and other species. Veterinary Immunology and Immunopathology 15: 337–376.

Davis, W. C., T. C. McGuire, and L. E. Perryman. 1983. Biomedical and biological application of monoclonal antibody technology in developing countries. Periodicum Biologorum 85:259–282.

Davis, W. C., L. E. Perryman, and T. C. McGuire. 1984. The identification and analysis of major functional populations of differentiated cells. Pages 121–150 in N. J. Stern and H. R. Gamble, editors. Hybridoma technology in agricultural and veterinary research. Roman and Allanheld, Totowa, New Jersey.

Davis W. C., L. E. Perryman, and T. C. McGuire. 1986. Construction of a library of monoclonal antibodies for the analysis of the major histocompatibility gene complex and the immune system of ruminants. Pages 88–115 in W. I. Morrison, editor. The ruminant immune system in health and disease. Cambridge University Press, Cambridge, England.

Deeg, H. J., and eight coauthors. 1982. Unusual distribution of Ia-like antigens on canine lymphocytes. Immunogenetics 16:445–457.

Ellis, J. A., and six coauthors. 1986. Characterization by a monoclonal antibody and functional analysis of a subset of bovine T lymphocytes that express BoT8, a molecule analogous to human CD8. Immunology 58:351–358.

Foon, K. A., R. H. Neubauer, C. J. Wikstrand, R. W. Schroff, H. Rabin, and R. C. Seeger. 1984. A monoclonal antibody recognizing human Thy-1: distribution on human and non-human primate haematopoietic cells. Journal of Immunogenetics (Oxford) 11:233–243.

Graham, F. L., and A. J. Van Der Eb. 1973. Transformation of rat cells by DNA of human adenovirus 5. Virology 54:536–539.

Grewal, A. S., and L. A. Babiuk. 1978. Bovine T lymphocytes: an improved technique of F rosette formation. Journal of Immunological Methods 24:355–361.

Herzenberg, L. A., and L. A. Herzenberg. 1978. Analysis and separation using the fluorescence activated cell sorter (FACS). Pages 22.1–22.21 in D. W. Weir, editor. Handbook of experimental immunology. Blackwell Scientific Publications, Oxford, England.

Howard, F. D., J. A. Ledbetter, J. Wong, C. P. Bieber, E. B. Stinson, and L. A. Herzenberg. 1981. A human T lymphocyte differentiation marker defined by monoclonal antibodies that block E-rosette formation. Journal of Immunology 126:2117–2122.

Iwaki, Y., P. I. Terasaki, T. Kinukawa, T. H. Thai, T. Rott, and R. Billing. 1983. Crossreactivity between human and canine Ia antigens using a mouse monoclonal antibody (CIA). Transplantation (Baltimore) 36:189–191.

Jefferies, W. A., J. R. Green, and A. F. Williams. 1985. Authentic T helper CD4 (W3/25) antigen on rat peritoneal macrophages. Journal of Experimental Medicine 162:117–127.

Jonak, J. L., and R. H. Kennett. 1984. Methods for transfection of human DNA into primary mouse lymphocytes and NIH/3T3 mouse fibroblasts. Pages 418–422 in R. H. Kennett, K. B. Bechtol, and T. J. Mckearn, editors. Monoclonal antibodies and functional cell lines: progress and applications. Plenum, New York.

Jonak, Z. L., J. A. Owen, and P. Machy. 1988. Strategies for the immortalization of B lymphocytes. Pages 163–191 in C. A. K. Borrebaek, editor. In vitro immunization in hybridoma technology. Elseveir Science Publishers Amsterdam.

Kavathas, P., and L. A. Herzenberg. 1983. Stable transformation of mouse L cells for human membrane T-cell differentiation antigens, HLA and beta₂-microglobulin: selection by fluorescence-activated cell sorting. Proceedings of the National Academy of Sciences of the USA 80:825–829.

Klein, J. 1986. Natural history of the major histocompatibility complex. Wiley, New York.

Kohler, G., and C. Milstein. 1975. Continuous culture of fused cells secreting antibody of predefined specificity. Nature (London) 256:495–497.

Kung, P. C., and six coauthors. 1981. Creating a useful panel of anti-T cell monoclonal antibodies. International Journal of Immunopharmacology 3:175–181.

Lalor, P. A., and L. A. Herzenberg. 1987. Transfection of genes encoding lymphocyte differentiation antigens: applications in veterinary immunology. Veterinary Immunology and Immunopathology 17:291–302.

Landay, A., L. Gartland, and A. T. Clement. 1983. Characterization of a phenotypically distinct subpopulations of Leu-2+ cells that suppresses T cell proliferative responses. Journal of Immunology 131:2757–2761.

Langone, J. I., and H. Van Vunakis, editors. 1986. Immunochemical techniques, part 1. Hybridoma technology and monoclonal antibodies. Methods in Enzymology 121:1–947.

Lanier, L. L., E. G. Engleman, P. Gatenby, G. F. Babcock, N. L. Warner, and L. A. Herzenberg. 1983a. Correlation of functional properties of human lymphoid cell subsets in surface marker phenotypes using multiparameter analysis and flow cytometry. Immunological Reviews 74:143–160.

Lanier, L. L., A. M. Le, J. H. Phillips, N. L. Warner, and J. F. Babcock. 1983b. Subpopulations of natural killer cells defined by expression of the Leu-7

(HNK-1) and Leu-11 (NKP-15) antigens. Journal of Immunology 131:1789–1796.

Lanier, L. L., and M. R. Loken. 1984. Human lymphocyte subpopulations identified by using three color immunofluorescence and flow cytometry analysis: correlation of Leu-2, Leu-3, Leu-7, Leu-8 and Leu-11 cell surface antigen expression. Journal of Immunology 132:151–156.

Lewin, H. A., W. C. Davis, and D. Bernoco. 1985. Monoclonal antibodies that distinguish bovine T and B lymphocytes. Veterinary Immunology and Immunopathology 9:87–102.

Lunney, J. K., B. A. Osborne, S. O. Sharrow, C. Devaux, M. Pierres, and D. H. Sachs. 1983. Sharing of Ia antigens between species. IV. Interspecies cross reactivity of monoclonal antibodies directed against polymorphic mouse Ia determinants. Journal of Immunology 130:2786–2793.

Lunney, J. K., and M. D. Pescovitz. 1988. Differentiation antigens of swine lymphoid tissues. Pages 421–454 in M. Miyasaka and Z. Trnka, editors. Differentiation antigens in lymphopoietic tissues. Marcel Dekker, New York.

Lundstrom, R. C. 1987. The potential use of monoclonal antibodies for identification of fish stocks. Pages 137–148 in H. E. Kumpf, R. N. Vaught, C. B. Grimes, A. G. Johnson, and E. L. Nakamura, editors. Proceedings of the stock identification workshop. NOAA (National Oceanic and Atmospheric Administration) Technical Memorandum NMFS (National Marine Fisheries Services) SEFC (Southeast Fisheries Center) 199, Miami.

Maddox, J. F., C. R. Mackay, and M. R. Brandon. 1985. Surface antigens, SBU-T4 and SBU-T8, of sheep T lymphocyte subsets defined by monoclonal antibodies. Immunology 55:739–748.

McKenzie, J. L., and J. W. Fabre. 1981. Human Thy-1: unusual localization and possible functional significance in lymphoid tissues. Journal of Immunology 126:843–850.

Moscicki, R. A., E. P. Amento, S. M. Krane, J. T. Kurnick, and R. R. Colvin. 1983. Modulation of surface antigens of a human monocyte cell line, U937, during incubation with T lymphocyte-conditioned medium: detection of T4 antigen and its presence on normal blood monocytes. Journal of Immunology 131:743–748.

Pescovitz, M. D., J. K. Lunney, and D. H. Sachs. 1984. Preparation and characterization of monoclonal antibodies reactive with porcine PBL. Journal of Immunology 133:368–375.

Pescovitz, M. D., J. K. Lunney, and D. H. Sachs. 1985. Murine anti-swine T4 and T8 monoclonal antibodies: distribution and effects on proliferative and cytotoxic T cells. Journal of Immunology 134: 37–44.

Reif, A. E., and J. M. V. Allen. 1964. The AKR thymic antigen and its distribution in leukemias and nervous tissue. Journal of Experimental Medicine 120:413–433.

Thierfelder, S. 1977. Haemopoietic stem cells of rats but not of mice express Thy-1.1 alloantigen. Nature (London) 269:691–693.

Williams, A. F. 1976. Many cells in rat bone marrow have cell surface Thy-l antigen. European Journal of Immunology 6:526–528.

Wood, G. S., N. L. Warner, and R. A. Warnke. 1983. Anti-Leu3/T4 antibodies react with cells of monocyte/macrophage and Langerhans lineage. Journal of Immunology 131:212–216.

American Fisheries Society Symposium 7:541–546, 1990

CHEMICAL MARKS

Inorganic Chemical Marks Induced in Fish

ROBERT J. MUNCY[1]

U.S. Fish and Wildlife Service
Mississippi Cooperative Fish and Wildlife Research Unit
Post Office Drawer BX, Mississippi State, Mississippi 39762, USA

NICK C. PARKER[2]

U.S. Fish and Wildlife Service
Southeastern Fish Culture Laboratory
Route 3, Box 86, Marion, Alabama 36756, USA

HUGH A. POSTON

U.S. Fish and Wildlife Service
Tunison Laboratory of Fish Nutrition, Cortland, New York 13045, USA

Abstract.—Researchers have attempted to mark batches of fish internally through feeding, immersion, and injections with stains, dyes, radioactive isotopes, rare earth compounds, metallic elements, and fluorescent compounds. In general, immersion in dyes and stains produced short-term marks detectable for days rather than years. Injections of dyes and stains provided marks detectable for longer periods, but they subjected fish to greater stresses during handling and marking. Rare earth salts administered in fish feeds were deposited in the body structure at a level of a 1 μg/g or less, and detection by neutron activation analysis was difficult after 6 months. When oxytetracycline and calcein were injected or fed to fish, they produced marks that remained detectable after several years in both captive and wild individuals.

Internal marking with chemicals has been considered for rapidly marking large numbers of fish of various sizes without handling them individually (Emery and Wydoski 1987). Research on fish nutrition, physiology, and toxicology has increased knowledge of the uptake and internal metabolism of chemical compounds tested as internal markers. In general, when fish are fed or immersed in solutions of chemicals, metabolically active compounds are taken up more rapidly, reach higher concentrations in the body, and are dispersed and excreted faster than are metabolically inactive compounds. Fish metabolism and growth can dissipate or dilute chemicals not strongly bonded within more stable systems such as bone. Direct injection of less reactive metallic compounds (such as rare earth salts), liquid latex,

and pigments bypasses metabolic barriers but exposes individual fish to increased handling stress (Hoss 1967; Harrell 1985). Batch-marking precludes the recognition of individual fish (Wydoski and Emery 1983).

Recently improved radioisotope techniques (Heydorn 1984) and detection instruments (Willard et al. 1981; Skoog 1985) have expanded research capabilities to include more of the 12 characteristics listed by Everhart et al. (1975) for the "ideal mark." Radioactive iron and cesium (Scott 1961, 1962) and cerium and cobalt (Hoss 1967) have been used to mark fish; however, human health and safety concerns and adverse public reaction toward radioactivity (Everhart et al. 1975) have constrained the use of radioactive isotopes in field studies. Nonradioactive rare earth compounds of dysprosium were tested first by Miller (1963) as markers for fingerling chinook salmon *Oncorhynchus tshawytscha*. Subsequent studies addressed the uptake, retention, and detection potentials and problems of rare earth compounds administered to various fish species

---

[1]Present address: Route 3, Box 299, Moneta, Virginia 24121, USA.

[2]Present address: Texas Cooperative Fish and Wildlife Research Unit, Department of Range and Wildlife Sciences, Texas Tech University, Lubbock, Texas 79409-2125, USA.

by immersion, injection, and feeding (Babb et al. 1967; Shibuya 1979; Michibata 1981; Michibata and Hori 1981; Muncy and D'Silva 1981; Zak 1984; Kato 1985). Incorporation of visible and fluorescent chemicals into new bone has sparked continued interest in marking fish by immersion, injection, and feeding of such substances as lead versenate (Fry et al. 1960; Jensen and Cummings 1967), tetracycline compounds (Weber and Ridgway 1962; Bilton 1986; Koenings et al. 1986; Babaluk and Campbell 1987), and calcein (Wilson et al. 1987).

Guillou and de La Noüe (1987) suggested that hatchery-reared fish released in the wild be identified by their levels of specific elements and by statistical techniques that detect combinations of elements absorbed from hatchery diets. Microanalytical and statistical techniques were applied to identify different elements or combinations of compounds known as chemoprints (Calaprice et al. 1971) in wild fish stocks (Lapi and Mulligan 1981; Mulligan et al. 1983; Schroeder 1983; Wiener et al. 1984; Guillou and de La Noüe 1987). Behrens Yamada et al. (1987) cautioned that stock and treatment effects, as well as influences of most elements, have not been well studied.

Application and Detection Techniques

Researchers have tested dyes and stains, rare earth elements, heavy metals, calcein, and tetracycline compounds for incorporation into fish eggs (Muncy and D'Silva 1981), larval fish (Jessop 1973; Hettler 1984; Behrens Yamada and Mulligan 1987), and juvenile fish (Ennis and Ziebell 1965; Zak 1984; Koenings et al. 1986; Babaluk and Campbell 1987; Wilson et al. 1987). Dyes and stains such as Bismark brown Y, natonal fast blue 8GXM, and hydrated chromium oxide have been used mainly for short-term marking because they fade over time (Kelly 1967; Jessop 1973; Everhart et al. 1975; Wydoski and Emery 1983). Fluorescent compounds of tetracycline and calcein bind to alkaline earth metals in fish tissues (Wilson et al. 1987) and to bones during periods of growth. Alkaline earth metals, when bonded to recently formed bone, reduced the dispersal of fluorescent compounds (Koenings et al. 1986).

Externally visible dyes and stains can be detected without instruments, but they tend to increase predation on marked fish. Injections of metallic cadmium sulfide, mercuric sulfide (Hansen and Stauffer 1964), and chromium oxide (Kelly 1967) lasted more than 1 year, whereas most dyes and stains dispersed more rapidly. Fluorescent marks can be examined under ultraviolet light (Hettler 1984; Bilton 1986; Wilson et al. 1987) and detected at levels of less than 0.2 μg/g by fluorometric techniques (Koenings et al. 1986).

Detecting low concentrations of nonvisible chemical compounds requires sophisticated support facilities and trained operators. Rare earth elements have been analyzed in fish samples by means of neutron activation analysis, X-ray-excited optical luminescence, dye laser techniques, resonance ionization spectroscopy, and atomic absorption spectroscopy. X-ray fluorescense spectroscopy has proved effective in detecting small (μg/g) concentrations of strontium (Behrens Yamada and Mulligan 1987). Lapi and Mulligan (1981) were able to identify the freshwater lakes from which sockeye salmon *Oncorhynchus nerka* originated by the chemical composition of the scales as revealed by X-ray scanning electron microscopy.

Retention of Internal Chemical marks

Injection of dyes and stains (Kelly 1967) usually has resulted in short-term marks (Laird and Stott 1978; Wydoski and Emery 1983). In comparative tests summarized by Table 1, rare earth compounds lasted longer at higher concentrations when injected than when administered by immersion or feeding techniques (Miller 1963; Michibata and Hori 1981; Zak 1984). Injection of metallic compounds (Fry et al. 1960; Hansen and Stauffer 1964) produced intensive color patterns recognizable for up to 4 years. Injection of tetracycline (Smith 1984; Babaluk and Campbell 1987) provided internal marks recognizable longer than 2 years.

Researchers have successfully marked small fish by immersing them in strontium (Behrens Yamada and Mulligan 1987), tetracycline (Hettler 1984), and calcein solutions (Wilson et al. 1987), but postimmersion exposure of small fish to sunlight may reduce detectable levels of oxytetracycline (Lorson and Mudrak 1987). Eggs of walleyes *Stizostedion vitreum* immersed in europium and terbium salts retained up to 5 μg/g, and sac fry hatched from marked eggs contained from 5 to 50 ng/g (Muncy and D'Silva 1981).

When salts of the rare earth elements terbium and europium were fed to striped bass *Morone saxatilis* for 84 d, they accumulated in scales at levels (detectable by neutron activation analysis) below 1 μg/g. The levels corresponded to dietary concentrations and duration of feeding. Relative concentrations of rare earth elements decreased

TABLE 1.—Chemical methods and detection techniques used for marking fish internally. Application techniques: In, injection; F, feeding; Im, immersion. Detection techniques: V, visual; Uv, ultraviolet; NAA, neutron activation analysis; AAS, atomic absorption spectroscopy; XEOL, X-ray-excited optical luminescence; XFS, X-ray fluorescence spectroscopy; DL, dye laser; Fl, fluorometric; EM, electron microscopy; RIS, resonance ionization spectroscopy. The time required to apply a mark and the period of detectability for a specific concentration are given when applicable.

Chemical	Application		Detection		Amount detected ($\mu g/g$)	Reference
	Method	Time	Method	Time		
Dyes and stains	Im	3 h	V	6 d		Jessop (1973)
	F	?	V	77 d		Moser et al. (1986)
	In		V	1 year		Kelly (1967)
Rare earth elements	Im	0.5 h	XEOL	21 d	5	Muncy and D'Silva (1981)
	Im	0.5 h	DL	10 d	0.0002	McWilliams[a]
	Im, F	30 d	AAS	30 d	2,000	Zak (1984)
	F	84 d	NAA	1.5 year	0.6	Muncy et al. (1988)
	F	84 d	RIS	1.5 year	0.1	Coutant (1990)
	F	40 d	NAA	2 years	0.1	Kato (1985)
	In		NAA	2 years	1	Michibata and Hori (1981)
Tetracycline	Im	2 h	V, Uv	8 d		Hettler (1984)
	F	14 d	Uv	2 years		Bilton (1986)
	F	40 d	Fl	1 year	0.6	Koenings et al. (1986)
	In		Uv	2 years		Babaluk and Campbell (1987)
Calcein	Im	2 h	Uv	27 d		Wilson et al. (1987)
Pollutants	F		EM		?	Lapi and Mulligan (1981)
	F		AAS		0.05	Wiener et al. (1984)
Lead	In		V	2 years		Fry et al. (1960)
Cadmium	In		V	4 years		Hansen and Stauffer (1964)
Mercury	In		V	4 years		Hansen and Stauffer (1964)
Cobalt	Im	1 d	NAA	36 d	?	Hoss (1967)
Strontium	F	42 d	AAS	42 d	200	Guillou and de la Noüe (1987)
	F	80 d	XFS	75 d	1	Behrens Yamada et al. (1987)
	Im	49 d	XFS	169 d	1	Behrens Yamada and Mulligan (1987)
Manganese	F	60 d	XFS	75 d	1	Behrens Yamada et al. (1987)
Natural mixtures	N		EM		1	Calaprice et al. (1971)
	N		XFS		1	Calaprice et al. (1971)
	N		XFS		1	Mulligan et al. (1983)

[a] Unpublished. McWilliams, R. H. 1986. Natural lakes investigations. Iowa Conservation Commission, Federal Aid Fish Restoration, Project F-95-R, Annual Report, Des Moines.

during a 400-d period after termination of the labeled diets probably because of dilution within the scale mass by growth (Muncy et al. 1988). Dilution of rare earth elements in fish scales was also reported by Michibata (1981), Zak (1984), and Kato (1985). Feeding of oxytetracycline (OTC) to fry of rainbow trout *Oncorhynchus mykiss* (Trojnar 1973) and chum salmon *O. keta* (Bilton 1986) produced concentrated fluorescent bands still visible under ultraviolet light 1 and 2 years later.

Koenings et al. (1986) demonstrated that concentrations of OTC in juvenile sockeye salmon *O. nerka*, detected by fluorometric procedures, declined in muscle and other soft tissues soon after treatment stopped but remained stable in bones; also, uptake of OTC by bones increased over the period of feeding. Moser et al. (1986) found little uptake of stains or tetracycline from one feeding. Improvements in technique allowed Koenings et al. (1986) to measure the uptake, loss, and retention of fluorescent compounds at concentrations as small as 0.17 µg/g. The percentages of fish marked by different tetracycline formulations and concentrations, treatment techniques, and lengths of treatment have been determined by several investigators, including Weber and Ridgway (1962), Scidmore and Olson (1969), Trojnar (1973), Odense and Logan (1974), Hettler (1984), Bilton (1986), Babaluk and Campbell (1987), Lorson and Mudrak (1987). The use of tetracycline to validate age marks for fish would require increased levels if the OTC marks were to be located visually (Smith 1984; Babaluk and Campbell 1987).

Conclusions

The experience of Muncy and D'Silva (1981) and Muncy et al. (1988), plus this review of the published literature on marking fishes with chemicals, indicates that application of chemical markers continues to be limited by inadequate field-detection techniques. Compounds such as rare earth salts are not normally abundant in natural environments of fish and may not be readily absorbed at levels above 5 µg/g (Zak 1984; Kato 1985; Muncy et al. 1988). Nutritional studies on mammals (Hutcheson et al. 1979; Kennelly et al. 1980) have demonstrated low absorption of lanthanide markers from the digestive tract. Compounds that are both readily available and absorbed, such as strontium (Behrens Yamada et al. 1987; Guillou and de La Noüe 1987), can be masked by background levels taken up by fish from some environments.

Lapi and Mulligan (1981) used microanalytical techniques to distinguish sockeye salmon by differences in the chemical composition of their scales and to classify mixed stocks of these fish according to their natal lakes. Fluorescent compounds (OTC and calcein) bonded to particular sites in bone during active growth appear to have better potential now for field studies of fish because detection techniques are more readily available and less expensive than those used for rare earth elements. Identification of fish marked with fluorescent compounds may be complicated by the presence of hatchery-reared fish fed OTC for disease control and the presence of natural fluorescent compounds in study areas.

Growth of the aquaculture industry has increased the need to distinguish farm-raised from wild stocks, both to ensure that marketed fish are legal and to provide quality control; these additional incentives may stimulate the development

of improved detection techniques for fluorescent compounds (Parker 1988). Analyses of soft tissues and bones of fish from highly variable natural and aquaculture environments (Behrens Yamada et al. 1987; Coutant 1990, this volume) present many challenges and opportunities for improving instrumentation (Skoog 1985) to detect chemical fingerprints. To obtain the maximum value of tools and techniques available, researchers must understand the basis of analytical techniques and be aware of improvements that lower detection limits, mask background levels, remove interferences, and determine the metabolic pathways of uptake and incorporation of chemicals used to mark fish internally.

References

Babaluk, J. A., and J. S. Campbell. 1987. Preliminary results of tetracycline labelling for validating annual growth increments in opercula of walleyes. North American Journal of Fisheries Management 7:138–141.

Babb, L., W. P. Miller, W. E. Wilson, Jr., G. L. Woodruff, and A. J. Novotny. 1967. Neutron activation analysis of stable dysprosium biologically deposited in the bone of chinook salmon fingerlings. Transactions of the American Nuclear Society 19:449.

Behrens Yamada, S., and T. J. Mulligan. 1987. Marking nonfeeding salmonid fry with dissolved strontium. Canadian Journal of Fisheries and Aquatic Sciences 44:1502–1506.

Behrens Yamada, S., T. J. Mulligan, and D. Fournier. 1987. Role of environment and stock on the elemental composition of sockeye salmon (Oncorhynchus nerka) vertebrae. Canadian Journal of Fisheries and Aquatic Sciences 44:1206–1212.

Bilton, H. T. 1986. Marking chum salmon fry vertebrae with oxytetracycline. North American Journal of Fisheries Management 6:126–128.

Calaprice, J. R. 1971. X-ray spectrometric and multivariate analysis of sockeye salmon (Oncorhynchus nerka) from different geographic regions. Journal of the Fisheries Research Board of Canada 28:369–377.

Calaprice, J. R., H. M. McSheffrey, and L. A. Lapi. 1971. Radioisotope X-ray spectrometry in aquatic biology: a review. Journal of the Fisheries Research Board of Canada 28:1583–1594.

Coutant, C. C. 1990. Microchemical analysis of fish hard parts for reconstructing habitat use: practice and promise. American Fisheries Society Symposium 7:574–580.

Emery, L., and R. Wydoski. 1987. Marking and tagging of aquatic animals: an indexed bibliography. U.S. Fish and Wildlife Service Resource Publication 165.

Ennis, E. J., and C. D. Ziebell. 1965. A negative response in an attempt to mark chum salmon by feeding dyed pellets. Progressive Fish-Culturist 27:146.

Everhart, W. H., A. W. Eipper, and W. D. Youngs. 1975. Principles of fishery science. Cornell University Press, Ithaca, New York.

Fry, F. E. J., D. Cucin, J. C. Kennedy, and A. Papson. 1960. The use of lead versenate to place a time mark on fish scales. Transactions of the American Fisheries Society 89:149–153.

Guillou, A., and J. de La Noüe. 1987. Use of strontium as a nutritional marker for farm-reared brook trout. Progressive Fish-Culturist 49:34–39.

Hansen, M. J., and T. M. Stauffer. 1964. Cadmium sulfide and mercuric sulfide for marking sea lamprey larvae. Transactions of the American Fisheries Society 93:21–26.

Harrell, R. M. 1985. Experimental marking techniques for young-of-year, hatchery-reared striped bass. Proceedings of the Annual Conference Southeastern Association of Fish and Wildlife Agencies 37:295–303.

Hettler, W. F. 1984. Marking otoliths by immersion of marine fish larvae in tetracycline. Transactions of the American Fisheries Society 113:370–373.

Heydorn, K. 1984. Neutron activation analysis for clinical trace element research, volumes 1 and 2. CRC, Boca Raton, Florida.

Hoss, D. E. 1967. Marking post-larval paralichthid flounders with radioactive elements. Transactions of the American Fisheries Society 96:151–156.

Hutcheson, D. P., B. Venugopal, D. H. Gray, and T. Luckey. 1979. Lanthanide markers in a single sample for nutrient studies in humans. Journal of Nutrition 109:702–707.

Jensen, A. C., and K. B. Cummings. 1967. Use of lead compounds and tetracycline to mark scales and otoliths of marine fishes. Progressive Fish-Culturist 29:166–167.

Jessop, B. M. 1973. Marking alewife fry with biological stains. Progressive Fish-Culturist 35:90–93.

Kato, M. 1985. Recent information on europium marking techniques for chum salmon. NOAA (National Oceanic and Atmospheric Administration) Technical Report NMFS (National Marine Fisheries Service) 27:67–73.

Kelly, W. H. 1967. Marking freshwater and a marine fish by injected dyes. Transactions of the American Fisheries Society 96:163–175.

Kennelly, J. J., F. X. Aherne, and M. J. Apps. 1980. Dysprosium as an inert marker for swine digestibility studies. Canadian Journal of Animal Science 60:441–446.

Koenings, J. P., J. Lipton, and P. McKay. 1986. Quantitative determination of oxytetracycline uptake and release by juvenile sockeye salmon. Transactions of the American Fisheries Society 115:621–629.

Laird, L. M., and B. Stott. 1978. Marking and tagging. IBP (International Biological Programme) Handbook 3:84–100.

Lapi, L. A., and T. J. Mulligan. 1981. Salmon stock identification using a microanalytic technique to measure elements present in the freshwater growth region of scales. Canadian Journal of Fisheries and Aquatic Sciences 38:744–751.

Lorson, R. D., and V. A. Mudrak. 1987. Use of tetracycline to mark otoliths of American shad fry. North American Journal of Fisheries Management 7:453–455.

Michibata, H. 1981. Labeling fish with an activable element through their diet. Canadian Journal of Fisheries and Aquatic Sciences 38:1281–1282.

Michibata, H., and R. Hori. 1981. Labeling fish with an activable element. Canadian Journal of Fisheries and Aquatic Sciences 38:133–136.

Miller, W. P. 1963. Neutron activation analysis of stable dysprosium biologically deposited in the bone of chinook salmon fingerlings. Master's thesis. University of Washington, Seattle.

Moser, M., J. Sakanari, and G. Cailliet. 1986. Vital stain in bait as a tag. North American Journal of Fisheries Management 6:600–601.

Mulligan, T. J., L. Lapi, R. Kieser, S. B. Yamada, and D. L. Duewer. 1983. Salmon stock identification based on elemental composition of vertebrae. Canadian Journal of Fisheries and Aquatic Sciences 40:215–229.

Muncy, R. J., and A. P. D'Silva. 1981. Marking walleye eggs and fry. Transactions of the American Fisheries Society 110:300–305.

Muncy, R. J., N. C. Parker, and H. A. Poston. 1988. Marking striped bass with rare earth elements. Proceedings of the Annual Conference Southeastern Association of Fish and Wildlife Agencies 41:244–250.

Odense, P. H., and V. H. Logan. 1974. Marking Atlantic salmon, Salmo salar, with oxytetracycline. Journal of the Fisheries Research Board of Canada 31:348–350.

Parker, N. C. 1988. Aquaculture—natural resource managers ally? Transactions of the North American Wildlife and Natural Resources Conference 53:584–593.

Schroeder, G. L. 1983. Stable isotope ratios as naturally occurring tracers in the aquaculture food web. Aquaculture 30:203–210.

Scidmore, W. J., and D. E. Olson. 1969. Marking walleye fingerlings with oxytetracycline antibiotic. Progressive Fish-Culturist 31:213–216.

Scott, D. P. 1961. Radioactive iron as a fish mark. Journal of the Fisheries Research Board of Canada 18:383–391.

Scott, D. P. 1962. Radioactive caesium as a fish and lamprey mark. Journal of the Fisheries Research Board of Canada 19:149–157.

Shibuya, M. 1979. Non-destructive activation analysis for Eu in fish materials. Radioisotopes 28:64–68. (In Japanese.)

Skoog, D. A. 1985. Principles of instrumental analysis, 3rd edition. Saunders, Philadelphia.

Smith, S. E. 1984. Timing of vertebral-band deposition in tetracycline-injected leopard sharks. Transactions of the American Fisheries Society 113:308–313.

Trojnar, J. R. 1973. Marking rainbow trout fry with tetracycline. Progressive Fish-Culturist 35:52–54.

Weber, D., and G. Ridgway. 1962. The deposition of tetracycline drugs in bones and scales of fish and its possible use for marking. Progressive Fish-Culturist 24:150–155.

Wiener, J. G., G. A. Jackson, T. W. May, and B. P. Cole. 1984. Longitudinal distribution of trace elements (As, Cd, Cr, Hg, Pb, and Se) in fishes and sediments in the upper Mississippi River. Pages 139–170 in J. G. Wiener, R. V. Anderson, and D. R. McConville, editors. Contaminants in the upper Mississippi River. Butterworth, Stoneham, Massachusetts.

Willard, H. H., L. L. Merritt, Jr., J. A. Dean, and F. A. Settle, Jr. 1981. Instrumental methods of analysis, 6th edition. Wadsworth, Belmont, California.

Wilson, C. A., D. W. Beckman, and J. M. Dean. 1987. Calcein as a fluorescent marker of otoliths of larval and juvenile fish. Transactions of the American Fisheries Society 116:668–670.

Wydoski, R., and L. Emery. 1983. Tagging and marking. Pages 215–237 in L. A. Nielsen and D. L. Johnson, editors. Fisheries techniques. American Fisheries Society, Bethesda, Maryland.

Zak, M. A. 1984. Mass-marking American shad and Atlantic salmon with the rare earth element, samarium. Master's thesis. Pennsylvania State University, State College.

American Fisheries Society Symposium 7:547–549, 1990

Variability in Incorporation of Calcein as a Fluorescent Marker in Fish Otoliths

DANIEL W. BECKMAN AND CHARLES A. WILSON

Coastal Fisheries Institute, Center for Wetland Resources
Louisiana State University, Baton Rouge, Louisiana 70803, USA

FE LORICA

Department of Marine Sciences, Center for Wetland Resources
Louisiana State University

JOHN MARK DEAN

Belle W. Baruch Institute, University of South Carolina
Columbia, South Carolina 29208, USA

Abstract.—We determined the effect of calcein concentration and immersion time on the incorporation of a fluorescent mark in the otoliths of sciaenid fish, and we evaluated the utility of calcein immersion for marking otoliths of laboratory-reared red drums *Sciaenops ocellatus*. A fluorescent calcein mark was incorporated into the otoliths of all wild spot *Leiostomus xanthurus* and spotted seatrout *Cynoscion nebulosus* immersed for 2 or 4 h at concentrations of 100 or 200 mg calcein/L of estuarine water. However, there were inconsistencies in the incorporation of fluorescent calcein marks in otoliths of laboratory-reared red drums immersed at the same dosages and in the times required to produce a mark in wild fish.

Larval and juvenile fish that are immersed in tetracycline solutions incorporate a fluorescent mark in their otoliths (Hettler 1984; Schmitt 1984; Tsukamoto 1985). These marks can be used to validate age-determination techniques and to identify the fish later. Wilson et al. (1987) immersed juvenile sciaenid fish in a calcein solution; the resulting incorporation of a fluorescent mark in the otoliths indicated that calcein could provide an alternative to tetracycline marking. Work still is needed to determine the effects of factors such as fish size, calcein dosage, and chemical properties of the immersion water on the incorporation of calcein. In the study reported here, we determined the effects of different dosage levels and immersion times on the incorporation of calcein marks in otoliths of spot *Leiostomus xanthurus* and spotted seatrout *Cynoscion nebulosus,* and we evaluated the utility of calcein immersion for marking otoliths of laboratory-reared larval and juvenile red drums *Sciaenops ocellatus.*

Methods

Juvenile spot ($N = 13$) were captured by seine from estuarine waters in southwestern Louisiana on June 20, 1986. Fish were immersed for 2 h at one of three concentrations of calcein in water obtained from the site of capture (15‰ salinity;

Table 1). Following immersion, fish were transferred to submerged wire-mesh cages in the area of capture (Wilson et al. 1987) and held for 5 d.

In a second set of experiments, spotted seatrout (40–100 mm) were captured along the north shore of Lake Pontchartrain, Louisiana, in a bag seine. On September 19, 1986, and October 5, 1987, fish were captured and immersed in calcein solutions for 2 h at 100 mg/L and 4 h at 200 mg/L, respectively. Water for immersions was obtained from the site of capture. After immersion, fish were held in translucent enclosures for 7 and 10 d, respectively.

In a third experiment, laboratory-reared red drums were immersed in a calcein solution of 200 mg/L prepared with distilled tap water and Instant Ocean. Immersion time and salinity were varied (Table 1). Fish were held from 7 to 9 d following immersion.

In all experiments, the treated fish were killed and preserved in 95% ethanol until their otoliths (sagittae) were removed. Otoliths were embedded, sectioned through the core in the transverse plane, and examined under ultraviolet light (as described by Wilson et al. 1987) to detect fluorescent marks resulting from incorporation of calcein.

TABLE 1.—Production of fluorescent marks in the otoliths of wild and laboratory-reared sciaenids at different calcein concentrations, immersion times, and salinities. Fish lengths are standard lengths.

| Species | N | Immersion bath | | | | Number of surviving fish with fluorescent mark | |
		Calcein concentration (mg/L)	Time (h)	Salinity (‰)	Temperature (°C)	Present	Absent
Spot	5	100	2	15	22	5	0
(wild; 79–114 mm)	3	10	2	15	22	2	1
	5	1	2	15	22	0	5
Spotted seatrout	17	100	2	1	32	4	0
(wild; 40–100 mm)	8	200	4	1–5	25	4	0
Red drum (laboratory-reared;	13	200	3	15	22	0	13
treatment 1: 21–35 d old;	4	200	3	5	17	4	0
treatments 2–4: 40–198 mm)	3	200	4	15	22	1	2
	3	200	4	5	22	2	1

Results and Discussion

All spot ($N = 5$) immersed in the 100-mg/L calcein solution in estuarine water exhibited a distinct fluorescent mark in otolith sections, which corresponded to calcein marking. Of the three spot immersed in the 10-mg/L solution, a faint fluorescent mark was observed near the sulcus acousticus in otolith sections from two fish, but there was no detectable mark in the third. None ($N = 5$) of the spot immersed in the 1-mg/L solution exhibited detectable fluorescent marks in otoliths. No spot died or showed adverse effects during immersion at any dosage.

All otoliths from surviving spotted seatrout had a fluorescent mark corresponding to the date of immersion. The sagittae from those immersed in the 200-mg/L solution for 4 h ($N = 4$) had much brighter fluorescent marks than the fish immersed at 100 mg/L for 2 h ($N = 4$). The poor survival noted in both the experiments was attributed to cannibalism (Tucker 1988).

None of the 13 surviving laboratory-reared larval red drums (treatment 1) and only one of three 86–117-mm juveniles (treatment 3) immersed in calcein (200 mg/L, 15‰ salinity) exhibited fluorescent marks in otolith sections. Calcein dosages and immersion times were greater than or comparable with those that produced a mark in otoliths of wild spot and spotted seatrout. The times and dosages were also greater than those that produced fluorescent marks in otoliths of wild red drums and Atlantic croakers Micropogonias undulatus (Wilson et al. 1987). In contrast, immersion in 200-mg/L calcein solutions at 5‰ salinity for 3 h (treatment 2, $N = 4$) and 4 h (treatment 4, $N = 3$) produced fluorescent marks in juvenile red drum otoliths for all but one of our fish—an individual in the 4-h treatment.

There was no indication that incorporation of calcein varied with species collected in the wild. All fish immersed in 100–200-mg/L final concentrations of calcein made up in habitat water had fluorescent marks in their otoliths, both in our experiments and in those of Wilson et al. (1987). Because of a limited availability of fish for experimental treatments, we were unable to define factors that control deposition of calcein in otoliths; however, the only inconsistency in the incorporation of calcein marks (at the same calcein concentrations) was between field-and laboratory-reared red drums (Wilson et al. 1987). These inconsistencies could have been caused by metabolic variability between laboratory-reared and wild fish, or by the artificial seawater we used. Temperature, pH, and salinity were similar in field and laboratory studies. Hettler (1984) observed that tetracycline was incorporated into some but not all spot and pinfish Lagodon rhomboides at certain tetracycline concentrations. All our red drums were successfully marked in the field at dosages lower than those that created no marks in laboratory-reared red drums, which suggests the need for further study of the effects of water quality, laboratory handling and treatment, and physiology on calcein marking.

References

Hettler, W. F. 1984. Marking otoliths by immersion of marine fish larvae in tetracycline. Transactions of the American Fisheries Society 113:370–373.

Schmitt, P. D. 1984. Marking growth increments in otoliths of larval and juvenile fish by immersion in tetracycline to examine the rate of increment formation. U.S. National Marine Fisheries Service Fishery Bulletin 82:237–242.

Tsukamoto, K. 1985. Mass marking of ayu eggs and larvae by tetracycline-tagging of otoliths. Bulletin of the Japanese Society of Scientific Fisheries 51:903–911.

Tucker, J. W., Jr. 1988. Growth of juvenile spotted seatrout on dry feeds. Progressive Fish-Culturist 50:39–41.

Wilson, C. A., D. W. Beckman, and J. M. Dean. 1987. Calcein as a fluorescent marker of otoliths of larval and juvenile fish. Transactions of the American Fisheries Society 116:668–670.

American Fisheries Society Symposium 7:550–561, 1990

Screening of Elements for the Chemical Marking of Hatchery Salmon

SYLVIA BEHRENS YAMADA[1] AND TIMOTHY J. MULLIGAN

Department of Fisheries and Oceans, Fisheries Research Branch
Pacific Biological Station, Nanaimo, British Columbia V9R 5K6, Canada

Abstract.—Chemical marking is gaining acceptance as a fish-marking technique. To be useful as a chemical mark, an element must be (1) rare in the environment and in fish tissue, (2) taken up and retained in fish tissue, (3) nontoxic, and (4) measurable by standard analytical techniques. We screened 19 elements as potential chemical marks for young salmon by adding them to the food or ambient water. For an open-water system, we found that it is more economical to induce a chemical mark via food than via water. We induced permanent marks in young sockeye salmon *Oncorhynchus nerka* by adding manganese or strontium (10,000 μg/g) and rubidium (5,000 μg/g) to their food for 40 d. The manganese and strontium marks were detected 2 years after treatment; the rubidium mark was detectable for at least 6 months. Strontium, manganese, and rubidium can be recommended as practical chemical markers, but bromine, cesium, and barium need further testing. As a fish grows, the concentration of a chemical mark decreases in a predictable manner. However, because calcified tissues such as scales and vertebrae grow concentrically, they reflect the chemical history of the growing fish. By analyzing the appropriate regions of scales and vertebrae, it is possible to distinguish adult fish that were marked as fingerlings.

Biological studies often call for the release and subsequent recapture of marked animals. Ideally, a mark should not affect an animal adversely in the wild and should be readily distinguishable on recapture. Easily recognized marks such as fin clips, toe clips, dyes, and attached numbered tags have inherent drawbacks in that they may reduce survival by interfering with routine activities such as locomotion, feeding, reproduction, and avoidance of predators. Radioisotope marks bypass this difficulty, but they are unacceptable if a species is used for human consumption. An alternative technique with none of these disadvantages is nonradioactive chemical marking—the introduction of a biologically rare element into an animal's tissue.

Chemical marking has been achieved by peritoneal injections, by adding the element to the food, and (in the case of aquatic species) by adding it to the ambient water. When a rare element is introduced via food or ambient water, the animals do not have to be handled during the marking process. This feature is most important for animals that are unduly stressed by handling or are too small or numerous to mark by other techniques.

To be useful as a chemical mark, an element must be (1) rare in the environment and animal tissue, (2) taken up and retained by animal tissue,

(3) nontoxic, and (4) measurable by standard analytical techniques. Rubidium, strontium, and manganese, as well as the rare earth elements europium and samarium, meet these criteria. Rubidium, which is chemically similar to potassium, has been used (as an RbCl food additive) to mark Mexican bean beetles *Epilachna varivestis* (Shepard and Wadill 1976) and cabbage loopers *Trichoplusia ni* (Stimmann et al. 1973). Diets with Rb concentrations as high as 14,000 μg/g did not significantly alter mating, fecundity, fertility, longevity, or response to sex pheromones of cabbage loopers, and a rubidium mark was detected in Mexican bean beetles 34 d after marking.

Stable strontium, which replaces calcium in calcareous tissue (Behrens Yamada et al. 1979; Spangenberg 1979), has been used to mark goldfish, *Carassius auratus*, (Ophel and Judd 1968), salmon *Oncorhynchus* spp. (Behrens Yamada et al. 1979, 1987; Behrens Yamada and Mulligan 1982, 1987), and brook trout *Salvelinus fontinalis* (Guillon and de la Noüe 1987). In all these studies, the strontium mark was induced via the diet. Ophel and Judd (1968) found that, 1 year after marking, single scales from marked goldfish contained 10 times as much strontium as those from control fish. Guillon and de la Noüe (1987) found that farm-reared brook trout could be distinguished from wild stocks by feeding them 200-μg/g concentrations of strontium in their diet. This technique was devised to discourage poach-

---

[1]Present address: Zoology Department, Oregon State University, Corvallis, Oregon 97331-2914, USA.

ing on farm-reared stocks. Studies from our laboratory have demonstrated that strontium marking is a simple and inexpensive technique for permanently marking young salmon. Smolts of coho salmon *Oncorhynchus kisutch* were fed a diet to which 10,000 μg/g of strontium was added for 60 d. Prior to their seaward migration, the treated smolts contained 31 times as much strontium in their vertebrae as did the control fish. The strontium mark was still detectable in samples of whole vertebrae from returning "jacks" (precocious males) 6 months later (Behrens Yamada et al. 1979) and in scale and vertebral cores from adults 18 months later (Behrens Yamada and Mulligan 1982).

A strontium mark also can be induced in non-feeding alevins by adding $SrCl_2$ to the rearing water. The addition of strontium (1 μg/mL) to the rearing waters of nonfeeding sockeye salmon *Oncorhynchus nerka* for 49 d resulted in a 10-fold increase in vertebral strontium concentrations. As the sockeye salmon grew, the induced mark became diluted in a predictable manner but was still detectable after 169 d (Behrens Yamada and Mulligan 1987).

Other elements also provide useful chemical marks. Behrens Yamada et al. (1987) found that manganese and strontium work equally well as chemical markers for young sockeye salmon. An addition of 10,000 μg of strontium or manganese per gram of food resulted in a 33-fold increase of either element in sockeye salmon vertebrae. Manganese at this concentration may be slightly toxic to salmon, because Mn-treated fish grew more slowly than control or Sr-treated fish during the treatment period. Shibuya (1979) added the rare earth element europium to the diet of salmon fry. The europium mark was still detectable 3 years later when the fish returned to their home stream. Another rare earth element, samarium, was induced in medaka *Oryzias latipes* and goldfish via the diet (Michibata 1981) and peritoneal injections (Michibata and Hori 1981). These investigators found that samarium was retained in internal organs for 1 year after the labeled diet was eaten, and that the mark could still be detected 2 years after the peritoneal injection.

Behrens Yamada and Mulligan (1987) addressed the effect of stock on chemical marking of salmon. Although a stock effect was observed for zinc uptake, it did not appear that minor stock effects could ever mask an induced chemical mark.

The aims of the present study were (1) to screen elements for use as chemical markers for salmon, (2) to determine the most cost-effective mechanism for chemical tag uptake into biological tissue; and (3) to determine the best time during salmon development for inducing a chemical mark.

We selected the elements lanthanum, europium, cerium, and chromium for their rarity in biological systems; manganese, aluminum, lanthanum, and zinc for their apparent ability to be concentrated at least 1,000 times in the scales of pink salmon *Oncorhynchus gorbuscha* (Klokov and Frolenko 1970); strontium and rubidium for their lack of toxicity; and barium and zinc for their importance in distinguishing sockeye salmon populations from different lakes (Lapi and Mulligan 1981).

Chemical markers are absorbed across the gut and gill epithelia. Some elements such as strontium appear to be taken up more readily through the gills (Ophel and Judd 1967; Schiffman 1961), whereas others such as zinc and manganese are taken up more readily through the gut (Hoss 1964; Pentreath 1973; Willis and Sunda 1984). We compared the relative merits of treating the diet and the water by adding Sr, La, and Ba to both. To test the effect of age on chemical marking, we fed Mn- and Sr-enriched food to members of the same year-class as fry and again as 1-year-old fish.

Methods

Experimental design.—Chemical marking experiments were set up at the beginning of June in 1974, 1975, and 1976, and were monitored for the subsequent 2 years (Table 1). During each year, 200 hatchery-raised sockeye salmon fry were introduced into 35-L tanks (flushing rate, 2 L/min). Treatments were replicated five times and were assigned randomly within the five blocks. The bottoms of all tanks except those designated as "no Al" were fitted with aluminum screens.

Daily rations of experimental diets were weighed into styrofoam cups and placed next to their assigned tanks. Both cups and tanks were color-coded to distinguish the various treatments. Fish were fed five times daily at the rate of 3% of their body weight per day. Food and water treatments lasted for 40 d, after which fish were raised according to standard hatchery practice. In our study of the time required for mark induction, we fed one cohort of sockeye salmon Sr or Mn at 10,000 μg/g of feed either as fry or as yearlings.

TABLE 1.—Experimental treatments for the chemical marking of salmon at Rosewall Creek. Each treatment was replicated in five tanks.

Element	Concentration in food (μg/g)					Concentration in water (μg/mL)	
	100	1,000	4,300	5,000	10,000	0.1	1.0
Strontium		×			×	×	×
Manganese		×			×		
Rubidium				×			
Bromine		×			×		
Copper		×					
Zinc	×	×			×		
Chromium	×	×					
Cobalt		×					
Nickel		×					
Yttrium		×					
Cesium				×			
Tin		×					
Iodine			×				
Barium	×	×			×	×	×
Magnesium		×					
Europium	×	×			×		
Lanthanum	×	×			×	×	×
Cerium	×	×			×		
Aluminum	Presence or absence of aluminum grid in bottom of tank						

Preparation of experimental diets.—Experimental diets were prepared by adding the chloride of the desired cation (except for Br, Sn, and I, for which NaBr or SnI$_4$ were used) to Oregon Moist Starter Mash (Moore Clark, La Connor, Washington). The appropriate amount of the compound of the test element was dissolved in 400 mL of distilled water and added to 1 kg of starter mash. The mixture was blended for 6 min, frozen, chopped, and forced through a commercial meat grinder. The spaghetti-like strings of food leaving the meat grinder were frozen, crumbled by hand, sieved, and refrozen. Food particles were dried in a fume hood at room temperature for 3 h until the finished produce resembled Oregon moist pellets in size, color, and water content. Control food was prepared in a similar fashion. All experimental diets were then frozen until needed.

Addition of chemicals to the rearing water.—Stock solutions of Ba, La, or Sr were added to the water-supply hoses of the experimental rearing tanks through a multichannel, peristaltic metering pump. The flow rates in the water supply hoses were adjusted to 2 L/min ± 5% and those in the metering pump channels to 1 mL/min ± 5%. By adjusting the concentration of the chosen element in the stock solution, we achieved the desired concentration in the rearing tank.

Sampling.—Mortalities were recorded daily. Length and weight measurements were taken periodically, and behavioral observations were made to determine whether any of the chemicals adversely affected the fish. Fish were sampled for later chemical analysis immediately after treatment and at 6 months, 1 year, and 2 years after treatment. For the first sampling period, five whole fish (exclusive of the gut) were pooled from each tank to make one sample. Vertebrae of fish from the same tank were pooled to make subsequent samples.

Chemical analysis.—Several techniques were used to analyze the experimental elements in fish tissue (Table 2). Most samples were examined by X-ray fluorescence spectrometry, either with an X-ray tube operated at 35 kV or an americium source. A detailed description of this procedure can be found in Mulligan et al. (1983). Other elements were examined in a commercial laboratory, where plasma spectrometry was used for Ba and Mg and graphite furnace atomic absorption spectrometry for Al. Of the rare earth elements, only one europium sample was analyzed by neutron activation, by W. R. Schell of the University of Washington. The strength of an induced chemical mark was expressed as a ratio of the concentration of the element in treated tissue to that in control tissue (Tables 2, 3).

Results

Mortality

Fish survived well in all but one of the treatments. Lanthanum, administered at 1 μg/mL in the ambient water, killed all the fish within 24 h,

TABLE 2.—Elements screened for the chemical marking of hatchery salmon. A question mark indicates that statistically significant amounts of the induced chemical could have been present in salmon tissues at the designated time; time and cost constraints prevented analysis of those samples.

Analytical technique	Element	Detection limit μg/g	Strength of induced chemical mark (multiple of control value)			
			Just after treatment: whole fish	6 months after treatment: vertebrae	1 year after treatment: vertebrae	2 years after treatment: vertebrae
X-ray fluorescence	Strontium	0.6	40×	4×		1.5×
spectrometry,	Manganese	8.0	43×	9×		2×
35 kV,	Rubidium	3.0	261×	6×	?	?
molybdenum tube,	Bromine	4.0	44×	?	?	?
molybdenum filter	Copper	2.0	2×	1×		
	Zinc	2.0	2×	1×		
	Chromium	2.0	Detected	Not detected		
	Cobalt	2.0	Detected	Not detected		
	Nickel	2.0	Detected	Not detected		
	Yttrium	5.0	Not detected			
X-ray fluorescence	Cesium	?	Detected	Detected	?	?
spectrometry	Tin	?	Not detected			
Americium source	Iodine	?	Not detected			
Plasma	Barium	0.2	64×	?	?	?
spectrometry	Magnesium	0.1	1×			
Neutron activation	Europium				Detected	?
Not easily analyzed	Lanthanum Cerium					
Graphite furnace atomic absorption spectrometry	Aluminum	0.5	Not detected			

probably because La ions at this concentration interfered with the function of the gill epithelium. Lanthanum at 0.1 μg/mL in the water and at 10,000 μg/g in the food did not result in reduced survival or growth.

Fish fed a diet with additional zinc (10,000 μg/g) failed to grow and became "pin heads." This food may have been distasteful, because control fish offered pellets containing 10,000 μg Zn/g soon learned to reject them.

Behavior

Fish in all but two of the treatments grew and behaved normally. Bromine at 10,000 μg/g and cesium at 5,000 μg/g in the food caused reversible behavioral changes. Bromine-treated fish exhibited classical symptoms of bromism: sluggishness and reduced irritability. They weighed 5% more than control fish and displayed distended abdomens. Control fish were startled by a tapping sound on the tank and they schooled in response to shadows overhead, but bromine-treated fish showed no responses to these stimuli. Three days after treatment ceased, however, these fish behaved normally.

Cesium-treated fish were noted for their hyperactivity and brilliant blue-green coloration. They

were extremely responsive to acoustic and visual stimulation. They weighed 36% less than control fish, which indicated that energy may have been channeled into movement rather than growth. As with bromide, all these effects were reversed once cesium treatment stopped.

Screening of Elements

All of the 19 elements we chose met the first two criteria for useful elemental tags: they were relatively rare in fish tissues and were nontoxic, at least at some concentrations. Ten of the elements had to be eliminated as potential chemical markers because they were not easily detected by the methods available to us (Table 2). (See the Appendix for a comparison of detection techniques.) Chromium, Co, Ni, Y, I, and Sn were barely detectable (or were undetectable) in fish tissue by X-ray fluorescence spectrometry, and Al was not detectable by graphite furnace atomic absorption spectrometry. The rare earth elements Eu, Ce, and La must be analyzed by neutron activation. One series of bone, muscle, and scale samples from fish fed Eu at 10,000 μg/g still contained this element 1 year after treatment. (Because neutron activation requires the use of a nuclear reactor,

TABLE 3.—Concentrations of induced elements in whole-fish samples just after treatment. Concentrations are given in μg/g ± 95% confidence limit of the mean. The strength of the induced chemical mark is calculated by dividing the concentration of the element in treated tissue by that in control tissue.

Element	Additional concentration in the food	Concentration in whole-fish tissue	Strength of induced mark
Strontium	0 (1975)	50±5	
	0 (1976)	53±5	
	1,000 (1975)	348±27	7×
	1,000 (1976)	335±65	6×
	10,000 (1975)	1,997±150	40×
	10,000 (1976)	2,233±100	42×
Manganese	0	11±2	
	1,000	56±10	5×
	10,000	476±90	43×
Rubidium	0	32±6	
	5,000	8,362±100	261×
Bromine	0	32±7	
	1,000	796±12	25×
	10,000	1,416±100	44×
Copper	0	27±29	
	1,000	46±11	1.7×
Zinc	0	174±9	
	100	266±24	1.5×
	1,000	322±30	1.8×
Chromium	0	Not detected	
	100	Not detected	
	1,000	Detected in 2 of 7 samples	
Cobalt	0	Not detected	
	1,000	53±6	
Nickel	0	Not detected	
	1,000	24±9	
Cesium	0	Not detected	
	5,000	Detected	
Barium	0	0.17±0.09	
	1,000	10.9±3.7	64×
Magnesium	0	963±211	
	1,000	1,070±100	1×

we could not afford to analyze any more of the rare earth fish samples.)

Fish showed a wide variety of responses in elemental uptake and retention (Table 2). Magnesium was not taken up at all by the fish tissue, whereas Cu and Zn were taken up but not retained 6 months after treatment. Of the 19 elements screened, only 6 (Sr, Mn, Rb, Br, Cs, and Ba) showed promise as practical chemical tags (Tables 2 and 3).

Experimental fish took up Cs and retained it for at least 6 months. The Cs concentrations in treated fish tissue fell into the lower detection limits for our X-ray fluorescent spectrometer, and we recommend that a more sensitive technique such as atomic absorption be employed for Cs analysis. Bromine and Ba showed great promise as chemical tags in that a mere 1,000 μg/g addition

of either element to the food resulted in 25-fold and 64-fold increases, respectively, in whole-fish tissue. Time and cost restraints prevented us from analyzing subsequent samples.

The addition of 5,000 μg Rb/g to the food resulted in a 260-fold increase in the Rb concentration of whole fish immediately after treatment, and the addition of 10,000 μg of Sr or Mn per gram of feed resulted in a 40-fold increase of either element (Table 3). Six months after cessation of treatment, the strength of the induced chemical marks of Rb, Sr, and Mn in vertebra samples was of the same order of magnitude (Table 2).

In general, the greater the concentration of an element in the food, the greater was its concentration in biological tissue. A 10-fold increase in the concentration of Sr or Mn in the food resulted in a 6- to 8-fold increase in the strength of the

TABLE 4.—Calculation of the induced strontium mark in whole-fish tissue just after treatment. The induced Sr mark is calculated by subtracting the Sr concentration of the control from that of the Sr treatments. The expected Sr mark for a dual treatment is derived by adding observed Sr marks for the respective single treatments. Error terms represent 95% confidence limits of the mean.

Additional Sr concentration in the food (μg/g)	Additional Sr concentration in the water (μg/mL)	Sr concentration in fish tissue (μg/g)	Induced Sr mark (μg/g)	Expected Sr mark in dual treatment (μg/g)
0	0	53±5	0	
0	0.1	90±4	37±4	
0	1.0	393±2	340±3	
1,000	0	335±65	282±65	
1,000	0.1	373±32	320±32	319±65
1,000	1.0	649±61	596±61	622±65

induced chemical mark (Table 3). Bromine and Zn, however, showed biological regulation such that a 10-fold increase in concentration in the food resulted in less than a 2-fold increase in the chemical mark (Table 3).

Mechanism of Chemical Tag Uptake

Of the three elements (Sr, Ba, and La) added to the rearing water of young sockeye salmon, only Sr was observed by our X-ray fluorescence spectrometer. The greater the Sr concentration in the ambient water, the greater was the Sr concentration in fish tissue (Table 4). The uptake of Sr via food and via water was completely additive in that the observed Sr concentration in the mark from a dual food–water treatment agreed with the sum of concentrations in marks from the respective single treatments (Table 4).

The costs of inducing Sr marks via food and via water are compared in Table 5. For our open-water system with a 6% flushing rate per minute, it was 200 times more expensive to induce a Sr tag via water than via food.

Time for Chemical Marking

Marking salmon just before they go to sea should assure the strongest chemical mark at recapture. This generalization is supported by our experiment in which sockeye salmon from the same cohort were marked either as fry or as yearlings. The treatment consisted of an addition of 10,000 μg/g Sr or Mn in the food for 40 d. As 2-year-olds, the fish marked as yearlings contained twice as much Sr or Mn in their vertebrae as did those marked as fry (Table 6).

Discussion

Of the 19 elements screened, only 6 showed potential as practical chemical marks. Of these six, Sr, Mn, and Rb hold great promise, and Br, Cs, and Ba need more testing. The remaining elements either were not concentrated and retained in fish tissue, or they were not easily observed by the techniques we used.

The mode of action of most chemical markers is simply to replace an abundant element in biolog-

TABLE 5.—Cost comparison (1990 U.S. dollars) of inducing a strontium mark via food and via water.

Treatment	Application rate	Sr concentration in whole-fish tissue (μg/g ± 95% confidence interval)	Amount of SrCl$_2$ crystal used to treat five tanks of fish for 40 d (g)	Cost of SrCl$_2$
Sr in water	1.0 μg/mL; 6% flushing rate per minute	393±2	1,756.8	$115.00
Sr in food	1,000 μg/g; 3% body weight per day	335±65	7.6	$ 0.50
Sr in food	10,000 μg/g; 3% body weight per day	2,233±100	76.0	$ 5.00

TABLE 6.—Concentration and location of induced strontium or manganese in vertebrae of the same cohort of sockeye salmon sampled in June 1977. Some fish were marked as fingerlings and sampled 23 months after cessation of treatment; others were marked as yearlings and sampled 15 months after cessation of treatment. Concentrations are given in μg/g $\pm$ 95% confidence limit of the mean. The strength of the induced chemical mark is the concentration of the element in treated tissue divided by that in control tissue.

Element	Time since end of treatment (months)	Additional concentration in the food	Concentration in vertebrae	Strength of induced mark	Location of induced mark in vertebral cross section
Sr		0	322±41		
	23	1,000	406±30	1.3×	Near center
	23	10,000	495±42	1.5×	Near center
	15	10,000	996±317	3.1×	Halfway to periphery
Mn		0	21.7±4.0		
	23	1,000	28.1±2.7	1.3×	Near center
	23	10,000	43.8±7.5	2.0×	Near center
	15	10,000	90.0±27.9	4.1×	Halfway to periphery

ical tissue with a rare one. Strontium replaces Ca in the inorganic crystalline matrix of bone, Rb and Cs replace K in the intracellular fluid, and Br replaces Cl in the extracellular fluid (Gray 1960; Bowen 1966; Eisenberg 1973). The actions of Mn and Ba are more complex. Manganese, a catalyst for many biochemical reactions, plays an important metabolic role in bone formation and has a structural role in the organic component of bone (Bowen 1966; Guggenheim and Gaster 1973). Barium acts like a biological inhibitor by chelating essential metabolites (Davies 1972). Strontium, Mn, and Ba tend to concentrate predominantly in bony tissue, whereas Rb, Cs, and Br tend to concentrate predominantly in soft tissue.

An important consideration in the choice of an elemental marker is that the element must be benign at concentrations that induce a detectable chemical mark. This criterion was met for elements delivered via food at the following concentrations: Sr and Mn at 10,000 μg/g, Rb at 5,000 μg/g, and Br and Ba at 1,000 μg/g. Both Br at 10,000 and Cs at 5,000 μg/g resulted in reversible behavioral changes—Br-treated fish were sluggish, and Cs-treated fish were hyperactive. Both Rb and Cs resemble K in chemical properties. Because K and Cl ions play important roles in determining the electrochemical equilibrium of cell membranes, any substitution with ions of different sizes or membrane permeabilities (such as Cs and Br) can greatly alter nervous irritability (Potts and Parry 1964). Cesium increases excitability by depolarizing nerve cell membranes, and Br decreases excitability by hyperpolarizing them. Fish fed Rb at 5,000 μg/g showed no abnormal behavior, which indicates that Rb, with its smaller atomic radius, is a better biological substitute for K than Cs and consequently a better

chemical marker. The behavioral changes observed with the Br and Cs treatments might be ameliorated by use of lower concentrations of these elements in the food. Because Br uptake seemed to reach a maximum when concentrations in the food were between 1,000 and 10,000 μg/g, a practical dosage for Br might fall between 2,000 and 5,000 μg/g.

The strength of the induced chemical mark increased with the concentration of the element in the treatment. This generalization held for the three elements that were added at more than one concentration: Sr in food and water, Mn and Br in food. Toxic effects and biological saturation were demonstrated only with the highest Br treatment, indicating that it may be possible to induce stronger Sr and Mn marks by increasing the food concentration of these elements above 10,000 μg/g. The marginal increase in the chemical mark, however, may not outweigh the increased cost of the chemical and the risk of toxic effects.

The two mechanisms of elemental uptake—via food and via water—appear to be completely additive for Sr. No interference or synergism was observed between waterborne (0.1 and 1.0 μg/mL) and dietary (1,000 μg/g) doses of Sr given simultaneously. Our results indicate that, at these concentrations, the gill and gut epithelia act independently in taking up Sr from the environment. Unless one uses a closed water system, as did Ophel and Judd (1967) to induce an Sr mark in goldfish, it is more cost-effective to induce a chemical mark via food.

Two processes account for the decrease of an induced chemical mark: elimination and dilution by growth. Elimination appears to be important for elements concentrated in soft tissue, such as Rb, but not for elements concentrated in bony tissue, such as Sr and Mn. In two separate experiments, dilution by growth totally explained the reduction of a Sr mark in coho salmon (Behrens Yamada et al. 1979; Behrens Yamada and Mulligan 1987). It appears that once the bone matrix is laid down, it remains relatively inert metabolically.

Concentrically growing bony tissues such as vertebrae and scales afford unique sampling opportunities. Because they develop much like annual rings on a tree, they reflect the chemical history of the growing fish. By taking vertebrae or scale cores (the central portion) of a certain diameter, one can reconstruct the chemical composition of those tissues at a former time. This was demonstrated by Behrens Yamada and Mulligan (1982) for coho salmon marked with Sr as smolts, prior to release into the ocean. When the fish returned as spawning adults 18 months later, samples were taken of whole vertebrae and of the central cores of individual vertebrae and scales. Strontium-marked adults could be distinguished from controls by examination of the cores but not by study of the whole vertebrae.

The best time for inducing a chemical mark in fish depends on several considerations. Ideally, one would want to induce as strong a mark as possible just prior to release, and to recover the fish before the mark became undetectable. These requirements can be relaxed when bone-seeking elemental markers are used, such as Sr (and possibly Mn and Ba), and when vertebral or scale cores can be taken. Then, it may actually be advantageous to mark fish as young as possible in order to economize on chemical costs. Fish marked at a smaller size display the ring of the chemical marker closer to the center of scales and vertebrae than do fish marked at a larger size (Table 6). This feature assures that the chemical mark will be included even in slightly off-center core samples.

Rubidium promises to become an excellent marker for fish; it already has been successfully applied to bean beetles in the field (Shepard and Waddil, 1976). After treatment, whole fish samples contained 260 times more Rb than did the controls. The strength of the Rb mark 6 months after treatment was 6× greater in vertebrae and 8× greater in muscle than in control fish. Even though Rb is concentrated predominantly in soft tissue, enough of it may still be deposited in the matrix of the bone that the coring technique can be used with older fish. This possibility needs further investigation.

Applications

Of the 19 elements tested, we recommend Sr, Mn, and Rb as chemical markers for young salmon. These elements are totally benign, are retained by fish tissue, and are easily analyzed with X-ray fluorescence or atomic absorption spectrometry. We recommend that these elements be administered in food in their chloride crystal form at concentrations of 10,000 $\mu g/g$ for Sr or Mn and at 5,000 $\mu g/g$ for Rb.

A previous field study has demonstrated that Sr is an excellent long-term marker (Behrens Yamada et al. 1979; Behrens Yamada and Mulligan 1982). Because Sr replaces Ca in bony tissue, it is laid down in the concentrically growing rings of scales and vertebrae. Even though a Sr mark in whole vertebrae is diluted as a growing fish adds unmarked bone, the Sr mark can still be detected in scale and vertebrae cores from adult salmon.

Nonfeeding salmon embryos can also be marked by adding Sr to their rearing water (Behrens Yamada and Mulligan 1987). The induced Sr mark has been detected 169 d after treatment and it should still be distinguishable in adults by the coring method. The technique of marking incubating salmon, combined with core sampling, may be a practical way to mark sockeye, pink, and chum salmon O. keta that leave the hatchery right after incubation.

Manganese and Rb might show promise equal to that of Sr once field tests with these elements are conducted. Because both Mn and Rb are retained in bone, the coring technique should be applicable with these elements. The uptake and retention patterns for Mn mirror those for Sr. The only evidence from our previous work for a possible toxic effect with Mn was observed when three of four stocks of young sockeye salmon showed a reversible stunting effect after being fed an additional 10,000 μg Mn/g for 60 d (Behrens Yamada et al. 1987). No such stunting effect was observed in this study after 40 d of treatment. Preliminary results for Mn-treated chinook salmon O. tshawytscha indicate that Mn may have a much more pronounced stunting effect on this species (T. J. Mulligan, personal observation).

In the search for more elemental tags for salmon, it may prove profitable to test whether two elements can be combined into a dual tag. If they can be, three elements could be combined two at a time to produce six distinct marks. Another way to get more marks would be to induce a Sr mark at more than one time. By using an electron microprobe on scale samples, one should be able to detect the Sr ring induced at each treatment time. This technique has been successfully demonstrated for squid cuttlebone (Hurley et al. 1985).

Both the Oregon Department of Fish and Wildlife and the Canadian Department of Fisheries and Oceans are planning to use an induced strontium mark to distinguish wild from hatchery-produced salmonids. R. Ewing and M. Smith of the Oregon Department of Fish and Wildlife will be marking all the hatchery chinook salmon in the Willamette River system by adding strontium to their diet. The object of this study is to determine the extent to which wild fish are being produced in the river.

T. J. Mulligan and L. Lapi of Fisheries and Oceans, Canada, want to determine why coded-wire-tagged salmon are returning to their hatcheries of origin at rates 20% lower than predicted. These low return rates have been observed for six hatcheries over periods ranging from 5 to 14 years. Possible reasons for the low return rates are (1) loss of coded wire tags, (2) straying of unmarked fish from another system, (3) higher mortality among coded-wire-tagged fish, and (4) higher straying rates by coded-wire-tagged fish. Multiple tags are needed to evaluate the relative importance of each of these four possibilities. All the fish in a hatchery need to be marked to distinguish them from wild fish of the same system and to identify strays from other system. This will be accomplished by adding strontium to all the food of the hatchery fish. Fish marked with coded wire tags and adipose fin clips will also contain a manganese mark. Any returning fish with both strontium and manganese marks will have come from the hatchery and will have originally been marked with a coded wire tag. In this way, the loss rates for coded wire-tags and natural adipose fins can be determined.

Acknowledgments

It is a pleasure to thank G. Johnston, R. Humphreys, and their staff at Rosewall Creek Hatchery for invaluable help in rearing the experimental fish. S. Fairchild provided technical assistance in all phases of the research, D. Marshal fitted the raw X-ray data, and J. Valenter and C. Allen typed the manuscript. R. J. Aruney, L. Curtis, R. Kendall, G. R. Sauer, T. D. Schowalter, N. C. Parker, and an anonymous reviewer made suggestions for improving our manuscript.

References

Behrens Yamada, S., and T. J. Mulligan. 1982. Strontium marking of hatchery-reared coho salmon, *Oncorhynchus kisutch* Walbaum, identification of adults. Journal of Fish Biology 20:5–9.

Behrens Yamada, S., and T. J. Mulligan. 1987. Marking nonfeeding salmonid fry with dissolved strontium. Canadian Journal of Fisheries and Aquatic Sciences 44:1502–1506.

Behrens Yamada, S., T. J. Mulligan, S. J. Fairchild. 1979. Strontium marking of hatchery-reared coho salmon (*Oncorhynchus kisutch* Walbaum). Journal of Fish Biology 14:267–275.

Behrens Yamada, S., T. J. Mulligan, and D. Fournier. 1987. The role of environment and stock on the elemental composition of sockeye salmon (*Oncorhynchus nerka*) vertebrae. Canadian Journal of Fisheries and Aquatic Sciences 44:1206–1212.

Bowen, H. J. M. 1966. Trace elements in biochemistry. Academic Press, London.

Davies, I. J. T. 1972. The clinical significance of essential biological metals. Heinemann Medical Books, London.

Eisenberg, E. 1973. The biological metabolism of strontium. Pages 435–442 *in* I. Zipken, editor. Biological mineralization. Wiley, New York.

Gray, C. H. 1960. Laboratory of toxic agents. Royal Institute of Chemistry, London.

Guggenheim, K., and D. Gaster. 1973. The role of Mn, Cu, and Zn on the physiology of bones and teeth. Pages 443–462 *in* I. Zipkin, editor, Biological mineralization. Wiley, New York.

Guillou, A., and J. de la Noüe. 1987. Use of strontium as a nutritional marker for farm-reared brook trout. Progressive Fish-Culturist 49:34–39.

Heydorn, K. 1984. Neutron activation analysis for clinical trace element research, volumes 1 and 2. CRC, Boca Raton, Florida.

Hoss, D. E. 1964. Accumulation of zinc-65 by flounder of the genus *Paralichthys*. Transactions of the American Fisheries Society 93:364–368.

Hurley, G. V., P. H. Odense, R. K. O'Dor, and E. G. Dawe. 1985. Strontium labelling for verifying daily growth increments in the statolith of the short-finned squid (*Illex illecebrosus*). Canadian Journal of Fisheries and Aquatic Sciences 42:380–383.

Klokov, V. K., and L. A. Frolenko. 1970. Elementary chemical composition of scales of pink salmon. Izvestiya Tikhookeanskogo Nauchno-Issledovatel'skogo Instituta Rybnogo Khozyaistva i Okeanorgrafii 71:159–168. Translated from Russian: Fisheries Research Board of Canada Translation Series 2576, 1973, Ottawa.

Lapi, L. A., and T. J. Mulligan. 1981. Salmon stock identification using a microanalytical technique to measure elements present in the freshwater growth region of scales. Canadian Journal of Fisheries and Aquatic Sciences 38:744–751.

Michibata, H. 1981. Labeling fish with an activable element through their diet. Canadian Journal of Fisheries and Aquatic Sciences 38:1281–1282.

Michibata, H., and R. Hori. 1981. Labeling fish with an activable element. Canadian Journal of Fisheries and Aquatic Sciences 38:133–136.

Mulligan, T. J., L. Lapi, R. Kieser, S. B. Yamada, and D. L. Duewer. 1983. Salmon stock identification based on elemental composition of vertebrae. Canadian Journal of Fisheries and Aquatic Sciences 40:215–229.

Ophel, I. L., and J. M. Judd. 1967. Experimental studies of radiostrontium accumulation by freshwater fish from food and water. Pages 859–865 in B. Aberg and F. P. Hungate, editors. Radioecological concentration processes. Pergamon Press, Oxford, England.

Ophel, I. L., and J. M. Judd. 1968. Marking fish with stable strontium. Journal of the Fisheries Research Board of Canada 25:1333–1337.

Pentreath, R. J. 1973. The accumulation and retention of zinc-65 and manganese-54 by the plaice, *Pleuronectes platessa* L. Journal of Experimental Marine Biology and Ecology 12:1–18.

Potts, W. T. W., and G. Parry. 1964. Osmotic and ionic regulation in animals. Pergamon Press, Oxford, England.

Schiffman, R. H. 1961. The uptake of strontium from diet and water by rainbow trout. U.S. Atomic Energy Commission Research and Development Report HW-72107.

Shepard, M., and V. H. Waddil. 1976. Rubidium as a marker for Mexican bean beetles, *Epilachna varivestis* (Coleoptera:Coccinellidae). Canadian Entomologist 108:337–339.

Shibuya, M. 1979. Practical techniques of activation analysis to agricultural and biological samples. VIII.7. Nondestructive activation analysis for europium in fish materials. Radioisotopes 28:62–66. (In Japanese.)

Skoog, D. A. 1985. Principles of instrumental analysis, 3rd edition. Saunders, Philadelphia.

Spangenberg, D. B. 1979. Statolith synthesis and ephyra development in *Aurelia* metamorphosing in strontium and low in calcium. Scanning Electron Microscopy 2:433–438.

Stimmann, M. W., W. W. Wolf, and W. L. Berry. 1973. Cabbage loopers: biological effects of rubidium in the larval diet. Journal of Economic Entomology 66:324–326.

Willard, H. H., L. L. Merritt, Jr., and J. A. Dean. 1981. Instrumental methods of analysis, 6th edition. Wadsworth, New York.

Willis, J. N., and W. G. Sunda. 1984. Relative contributions of food and water in the accumulation of zinc by two species of marine fish. Marine Biology 80:273–279.

Appendix: Comparison of Analytic Techniques

Modern elemental analysis employs instrumental procedures almost exclusively. Wet-chemical processing is limited by the specific premeasurement processes used to increase the concentration of a desired element or to eliminate undesirable interference or matrix effects in the subsequent instrumental analysis. (For a comparison of analytical techniques, see Willard et al. 1981; Heydorn 1984; Skoog 1985). The instrumental techniques are all similar in that they cause individual atoms of the elements of interest to emit some form of characteristic radiation that can be detected and measured by electronic means. The analytic techniques popular today differ in the type of radiation that is excited and subsequently measured.

Of the many techniques available, four or five are by far the most popular for elemental analyses of inorganic elements. These are (1) atomic absorption spectrometry, (2a) inductively coupled plasma optical emission spectrometry or (2b) inductively coupled plasma mass spectroscopy, (3) X-ray fluorescence spectrometry, and (4) neutron activation analysis. Each technique is capable of multielement analysis. Sometimes this is accomplished in one simultaneous analysis; at other times elements are analyzed sequentially. Simultaneous elemental analysis techniques are popular because they save time and involve less sample preparation, but they can require expensive instruments that need complex maintenance. No single technique is superior for all elements in either performance or ease of operation. As always, the choice involves compromise. We shall briefly describe each instrumental technique.

Atomic absorption spectrometry is a relatively old method that has the advantages of simplicity and ease of equipment maintenance. Atoms of the element of interest are caused to emit visible light radiation in a special hollow cathode lamp. This light is passed through a plasma, often in the form of a flame, that contains an aspirated solution of the sample. The atoms of interest in the plasma absorb this element-specific radiation and subsequently reradiate it. The initial radiation from the lamp is focused on the detector positioned on the opposite side of the plasma; because the reradiated light is emitted in all directions, there is a measured decrease in the intensity of the original light when the element is present in the plasma. Drawbacks of this method are a relatively limited concentration range for detection of the element of interest and the need to dilute or concentrate the original sample to accommodate this range. Most atomic absorption systems employ single-element analysis, so sequential measurements of the same sample must be made when multielement analysis is desired.

Inductively coupled plasma optical emission spectroscopy uses electronic means to create an intensely hot plasma into which the compound for analysis is injected. The resulting light that results is spectroscopically examined for wavelengths characteristic of the elements being analyzed. When one uses mass instead of optical spectrometry for analysis, the individual atoms are detected in a magnetic particle spectrograph. Both techniques are capable of simultaneous multielement analysis and both involve complex instruments that require special handling. Compared to atomic absorption spectrometry, a much larger range of elemental concentrations can be examined without dilution, and analysis time is relatively short (typically, one sample every 2 min). Disadvantages include multielement interferences and high equipment costs.

X-ray fluorescence spectrometry, another older instrumental technique, is capable of simultaneous multielement analysis. Its lower limit of detection is not as low as most other techniques unless special preprocessing is employed. If preprocessing is not required, very little sample preparation is necessary. The sample need not be dissolved, as it must be for the previous techniques, but some sort of sampling grinding is normally used to ensure homogeneity. An X-ray generating tube is normally used to fluoresce characteristic X-rays from the elements in the sample, which are subsequently analyzed spectroscopically. Often more than one X-ray tube is necessary to cover all the elements of interest and to perform sequential analyses. The equipment is reliable and relatively robust, requiring little maintenance. Matrix effects and interelemental interferences are among the principal disadvantages.

Neutron activation analysis uses a source of neutrons (normally from a nuclear reactor) to cause the sample to become radioactive. The gamma radiation emitted by the unstable radionuclides is spectroscopically analyzed to obtain elemental identification. Often a decay period of several days to several weeks must elapse before gamma ray analysis. Many times, no sample preparation is required. Almost no matrix effects are present and multielement analysis is standard. Disadvantages include the need for access to a nuclear reactor, interelemental interferences, and

TABLE A.1.—The periodic chart of the elements is roughly reproduced for elements heavier than Ar. Under each element is a column of numbers representing instrumental methods particularly sensitive for analysis of that element. The numbers correspond to (1) atomic absorption, (2) inductively coupled plasma optical emission, (3) X-ray fluorescence, and (4) neutron activation.

K	Ca	Sc	Ti	V	Cr	Mn	Fe	Co	Ni	Cu	Zn	Ga	Ge	As	Se	Br	Kr
1	1	1	1	1		1						1					
2	2		2			2	2							2	2		
		3		3	3			3	3	3	3						
		4			4			4						4		4	

Rb	Sr	Y	Zr	Nb	Mo	Tc	Ru	Rh	Pd	Ag	Cd	In	Sn	Sb	Te	I	Xe
1	1				1			1				1					
		2	2	2	2					2	2		2	2			
3	3				3												
					4									4			

Cs	Ba	La	Hf	Ta	W	Re	Os	Ir	Pt	Au	Hg	Tl	Pb	Bi	Po	At	Rn
1					1							1					
			2	2	2		2			2			2	2			
	3	3			3												
4		4	4	4	4												

Ce	Pr	Nd	Pm	Sm	Eu	Gd	Tb	Dy	Ho	Er	Tm	Yb	Lu
					1		1	1	1				
2	2	2		2	2	2	2	2	2	2	2		2
3		3											
4		4		4	4	4	4		4		4	4	4

Th	Pa	U	Np	Pu	Am	Cm	Bk	Cf	Es	Fm	Jd	No	Lr
2													
3		3											
4		4											

compromises required to perform simultaneous multielement analysis.

Because analytic chemistry is a complex subject, it would be foolish to list "typical" costs and detection limits. This information, while important, depends particularly on the elements to be analyzed, their concentration in the sample, and the nature and amount of the sample. When the elements have been artificially induced into tissue, only those few are of interest. The number and identification of these elements are the most important determinant of the best analytic technique. For example, Sr can be detected easily in bone by X-ray fluorescence, whereas the rare earth element Eu, which is also easily incorporated into bone, is difficult to measure with this technique (Table A.1). In contrast, neutron activation analysis is very sensitive for many of the rare earths, but is not a good choice for Sr. The best solution may be to choose several elements for potential use and then consult a competent analytic chemist for help in selecting the best detection technique.

American Fisheries Society Symposium 7:562–565, 1990
© Copyright by the American Fisheries Society 1990

Preliminary Field Verification of Daily Growth Increments in the Lapillar Otoliths of Juvenile Cunners

Timothy R. Gleason[1] and Conrad Recksiek

Department of Fisheries, Animal and Veterinary Science
University of Rhode Island, Kingston, Rhode Island 02881, USA

Abstract.—Seventy-five field-captured juvenile cunners *Tautogolabrus adspersus* were immersed for 2 h in a solution of tetracycline hydrochloride (500 mg/L) and monovalent salts nearly isotonic to seawater. After immersion, the 64 surviving cunners were released on an artificial reef in Narragansett Bay, Rhode Island. Eight tetracycline-marked juvenile cunners were recaptured 13–19 d after release. Ultraviolet and bright-field microscopy confirmed the presence of daily growth increments in the lapillar otoliths of five of these recaptured fish.

The presence of daily growth increments in fish otoliths was first described by Pannella (1971). Since that time, daily otolith increments have been used for aging larval and juvenile fish of numerous species (Campana and Neilson 1985). The microstructural increments in otoliths also have been used to determine growth (Townsend and Graham 1981; Wilson and Larkin 1982), mortality (Crecco et al. 1983), and age at settlement (Victor 1982, 1986; Brothers et al. 1983).

Beamish and McFarlane (1983) stressed the need to validate aging techniques for all stages of a species' life history, but validation studies still are uncommon. Assessments of the periodicity of otolith increment formation generally involve laboratory rearing of known-age fish (Brothers et al. 1976; Neilson and Geen 1982; Wilson and Larkin 1982; Jones 1986; Jones and Brothers 1987), but some investigators have used marks on the otolith as time checks. Victor (1982), for example, used both stress and supplemental feeding to produce detectable marks on the otoliths of tropical labrids in the field. For large juveniles and adult fish, the injection of tetracycline has proved an effective means of temporally marking otoliths (Wild and Foreman 1980; Campana and Neilson 1982). Tetracycline chelates to calcium, and the resulting compound exhibits a gold fluorescence when viewed with ultraviolet light (Milch et al. 1957). Larvae and small juveniles are generally too small to inject. Several investigators have immersed larval and juvenile fish in a solution of tetracycline to mark otoliths (Campana and Neilson 1982;

Hettler 1984; Schmitt 1984; Tsukamoto 1985; Dabrowski and Tsukamoto 1986). The latter studies have been confined primarily to the laboratory (Hettler 1984; Tsukamoto 1985; Dabrowski and Tsukamoto 1986).

We report here our efforts to mark juvenile cunners *Tautoglabrus adspersus* by tetracycline immersion to investigate the periodicity of increment formation in their otoliths. The marked fish were released to and recovered from the ocean. Although tetracycline immersion is practical for marking small (<20 mm) marine fishes (Hettler 1984), the identification of recaptured young fish has not been based on chemical marks alone. We undertook this study not only to validate daily formation of otolith increment, but also to assess the feasibility of using mass marking with tetracycline to identify experimentally released juvenile cunners in Narragansett Bay.

Methods

Seventy-five juvenile cunners, 15–40 mm standard length (SL), were captured between 1800 and 1900 hours at Bass Rock (41°24′28″N, 71°27′28″W), Narragansett Bay, on 1 September 1987. Fish were captured with a suction device by two scuba divers and transported to the laboratory in a covered and aerated 35-L cooler containing seawater from the dive site.

All fish remained in the cooler overnight and for one full day. They were fed live *Artemia salina* once daily while in the laboratory. At 0625 hours on 3 September, the fish were transferred by dip nets to a bucket containing 10 L of a monovalent salt solution: sodium chloride (23.9 g/L), sodium sulfate (4.0 g/L), potassium chloride (0.7 g/L), and sodium bicarbonate (0.2 g/L) (Spotte et al. 1984); the solution also contained a 500-mg/L concentra-

[1]Present address: Science Applications International Corporation, at the Environmental Research Laboratory, U.S. Environmental Protection Agency, 27 Tarzwell Drive, Narragansett, Rhode Island 02882, USA.

tion of tetracycline hydrochloride.[2] (We determined in earlier tests with concentrations of 250 mg/L and 500 mg/L of tetracycline that the latter was most effective, and that it marked 100% of the cunners exposed for 2 h.) Aeration was provided throughout the exposure period. The solution had been allowed to equilibrate to ambient seawater temperature (22°C) overnight. After 2 h the surviving fish were removed from the tetracycline solution and divided among three nylon mesh (1-mm) cages, 30.5 × 16.5 × 8.75 cm, in flow-through seawater aquaria. The fish were kept in the cages overnight.

At 0630 hours the next morning, 4 September, the 64 fish were transferred in their cages to a cooler containing ambient-temperature seawater and transported to an artificial reef constructed for the study. There, two scuba divers released them from their cages.

The artificial reef was constructed near an abandoned jetty at South Ferry, Narragansett, Rhode Island (41°29′30″N, 71°25′15″W), 4 d before the fish were released. It consisted of stacked, algae-covered rocks at a mean low-water depth of 3 m. It was approximately 1.5 m wide, 2.0 m long, and 0.3 m high.

Ten juvenile cunners were captured at the artificial reef on 16 September, 12 d after release (13 d after tetracycline marking). Two cunners were captured 6 d later, 22 September (19 d after tetracycline marking). Because of poor visibility and heavy silting at the site, no more attempts were made to recapture fish.

Captured fish were killed with tricaine (MS-222), measured to the nearest 0.1 mm SL, and frozen until otoliths were removed. The sagittae and lapilli were removed, rinsed in water to remove all tissue, and mounted in thermoplastic on glass slides. Sagittae were mounted sulcus-side down. Lapilli were oriented with what appeared to be a small knob facing up.

All otoliths required further preparation to expose the increments. They were ground to the center by hand on 600-grit lapping paper and then polished with 0.3-μm-grit carborundum powder. Deionized water was used as a lubricant for grinding. Fluorescence was detected with a Zeiss compound research microscope equipped with epiplan objectives, a 35-mm camera attachment, and an ultraviolet (UV) light source with excitation filter bands from 450 to 490 nm and barrier

[2]Tetracycline hydrochloride, T-3383, Sigma Chemical Company, St. Louis, Missouri.

filters at 520 nm on up. To count the increments, we used either a compound microscope and bright-field illumination or photographs taken with the fluorescent microscope at 400 and 1,000× magnification.

Results and Discussion

Sixty-four (85%) of the cunners survived the 2-h tetracycline exposure. Immersion in a solution of modified artificial seawater and tetracycline hydrochloride was an effective means of placing a discrete temporal marker on the otoliths of juvenile cunners (Figure 1A). The mortality observed during marking was believed to have resulted from handling stress, in particular from the transfer of fish just prior to tetracycline exposure. In later laboratory experiments that minimized handling stress, 100% survival was achieved (our unpublished data). Fluorescent microscopy revealed that 8 of the 12 juvenile cunners captured at the artificial reef had been marked with tetracycline. Seven of the 10 fish captured on 16 September and one of the two fish captured 22 September were marked. Over the 19-d period, 12.5% of the 64 fish released 4 September were recaptured.

Sagittae were difficult to prepare, often cracking prior to or during grinding. In addition, increments in the sagittae were, in general, difficult to resolve. The lapilli were much simpler to prepare and interpret. Victor (1986) reported similar findings for some tropical labrids. We were able to read lapilli from five of the eight recaptured cunners. Faulty preparation of samples was responsible for our inability to read otoliths from the other three.

Examination of otoliths with ultraviolet and bright-field microscopy indicated a daily periodicity for increment formation (Table 1; Figure 1). Often a mark corresponding to the fluorescent mark was observed with bright-field illumination (Figure 1B); presumably it had been caused by stress. The presence of the stress-induced mark facilitated counting with bright-field illumination. In three fish, the number of increments deposited outside the fluorescent ring matched the number of days elapsed since exposure to tetracycline (Table 1). In the other two samples, the increments numbered either one less or one more than the number of days (Table 1). We may have caused this variation in the number of increments by undergrinding or overgrinding the otoliths during preparation.

The results of this study suggest there are daily growth increments on the lapillar otoliths of juvenile cunners in the wild. This agrees with a

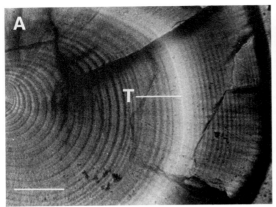

FIGURE 1.—Left lapillus, ground and polished, from a cunner (21.4 mm standard length) recaptured 22 September, 19 d after it had been marked with tetracycline and 18 d since it had been released in Narragansett Bay. The otolith increments for this fish have been tabulated in Table 1. (**A**) Otolith with both ultraviolet light and bright-field illumination. The fluorescent ring, T, caused by tetracycline marking is visible, as are the increments from the center to the edge. (Scale bar = 0.025 mm.) (**B**) Same view and magnification as (**A**) with bright-field illumination only. A check mark, C, coincident with the tetracycline mark is visible.

laboratory study in which we found a daily periodicity in otolith increment formation for 19 juvenile cunners marked with tetracycline and held for 3–62 d (our unpublished data). To ascertain absolute ages of juvenile cunners, the time of first increment formation and the rate of increment deposition in fertilized eggs and larvae must be determined for the lapillar otolith. The relatively rapid development of cunner eggs and larvae, described by Kuntz and Radcliffe (1917), suggests that otolith development occurs during the first few days following fertilization. Victor (1986) indicated that, in the tropics, labrid otoliths appear within 1–2 d of fertilization.

In summary, we mass-marked juvenile cunners in a monovalent salt solution with 500 mg tetracycline/L. By releasing and recapturing tetracy-

cline-labeled cunners on an artificial reef, we were able to show a daily pattern (on average) for lapillar otolith increment formation in juveniles under natural conditions.

Because of our work with cunners, it now appears practical to use mass-marking techniques in field studies of the early life history of at least some marine fishes. Tsukamoto (1985) and Dabrowski and Tsukamoto (1986) have successfully mass marked eggs and larvae, yet they concentrated on, respectively, diadromous and freshwater species. As yet they have not reported results from beyond the laboratory, though Tsukamoto's (1985) suggestion that the techniques may be applied to study early life stages of ayu *Plecoglossus altivelis* in the sea appears practical.

With cunners, and probably with other reef fish of temperate as well as tropical waters, it is feasible to mass-mark and release recently metamorphosed juvenile fish. Dabrowski and Tsukamoto (1986) successfully placed multiple marks in otoliths, which has opened the way to new coding methods. With the capability of rearing and chemically marking batches of coded juveniles, the prospect of defining and executing experiments in recruitment processes becomes a reality.

TABLE 1.—Otolith increment data for five recaptured juvenile cunners that had been released on an artificial reef in Narragansett Bay, Rhode Island, on 4 September 1987, 1 d after they had been immersed in tetracycline. The number of increments for each fish was determined according to the criteria outlined by Victor (1986).

Standard length at recapture (mm)	Days since marking	Increments formed since marking	Total increments
20.7	13	13	48
25.7	13	13	66
18.5	13	12	52
19.9	13	13	60
21.4[a]	19	20	57

[a] The left lapillus from this specimen is illustrated in Figure 1.

Acknowledgments

This work was supported in part by the Department of Fisheries, Animal and Veterinary Science, and by the Rhode Island Experiment Station, College of Resource Development, University of Rhode Island. This paper is contribution 2426 of the Rhode Island Experiment Station. The authors extend special thanks to the following persons and organizations for their generous support and assistance: Xu Liu-Xiong, Cynthia Bearse, David Bengston, Walter Berry, Ann Durbin, Edward Durbin, Einar Hjorleifsson, Cornelia Mueller, Sheila Polofsky, Paul Selvitelli, and Philip Stuart-Sharkey and the Graduate School of Oceanography Diving Program, the Environmental Research Laboratory of the Environmental Protection Agency at Narragansett, and Science Applications International Corporation of Narragansett and Newport, Rhode Island.

References

Beamish, R. J., and G. A. McFarlane. 1983. The forgotten requirement for age validation in fisheries biology. Transactions of the American Fisheries Society 112:735–743.

Brothers, E. B., C. P. Mathews, and R. Lasker. 1976. Daily growth increments in otoliths from larval and adult fishes. U.S. National Marine Fisheries Service Fishery Bulletin 74:1–8.

Brothers, E. B., D. M. Williams, and P. F. Sale. 1983. Length of larval life in twelve families of fishes at "One Tree Lagoon," Great Barrier Reef, Australia. Marine Biology 76:319–324.

Campana, S. E., and J. D. Neilson. 1982. Daily growth increments in otoliths of starry flounder (*Platichthys stellatus*) and the influence of some environmental variables in their production. Canadian Journal of Fisheries and Aquatic Sciences 39:937–942.

Campana, S. E., and J. D. Neilson. 1985. Microstructure of fish otoliths. Canadian Journal of Fisheries and Aquatic Sciences 42:1014–1032.

Crecco, V., T. Savoy, and L. Gunn. 1983. Daily mortality rates of larval and juvenile American shad (*Alosa sapidissima*) in the Connecticut River with changes in year class strength. Canadian Journal of Fisheries and Aquatic Sciences 40:1719–1728.

Dabrowski, K., and K. Tsukamoto. 1986. Tetracycline tagging in coregonid embryos and larvae. Journal of Fish Biology 29:691–698.

Hettler, W. F. 1984. Marking otoliths by immersion of marine fish larvae in tetracycline. Transactions of the American Fisheries Society 113:370–373.

Jones, C. 1986. Determining age of larval fish with the otolith increment technique. U.S. National Marine Fisheries Service Fishery Bulletin 84:91–103.

Jones, C., and E. B. Brothers. 1987. Validation of the otolith increment aging technique for striped bass, *Morone saxatilis*, larvae reared under suboptimal feeding conditions. U.S. National Marine Fisheries Service Fishery Bulletin 85:171–178.

Kuntz, A., and L. Radcliffe. 1917. Notes on the embryology and larval development of twelve teleostean fishes. U.S. Bureau of Fisheries Bulletin 35:89–134.

Milch, R. A., D. P. Rall, and J. E. Tobie. 1957. Bone localization of the tetracyclines. Journal of the National Cancer Institute 19:87–93.

Neilson, J. D., and G. H. Geen. 1982. Otoliths of chinook salmon (*Oncorhynchus tshawytscha*): daily growth increments and factors influencing their production. Canadian Journal of Fisheries and Aquatic Sciences 39:1340–1347.

Pannella, G. 1971. Fish otoliths: daily growth layers and periodical patterns. Science (Washington, D.C.) 173:1124–1127.

Schmitt, P. D. 1984. Marking growth increments in otoliths of larval and juvenile fish by immersion in tetracycline to examine the rate of increment formation. U.S. National Marine Fisheries Service Fishery Bulletin 82:237–241.

Spotte, S., G. Adams, and P. M. Bubucis. 1984. GP2 medium is an artificial seawater for culture or maintenance of marine organisms. Mystic Marinelife Aquarium, Sea Research Foundation Contribution 47:229–240. (Mystic, Connecticut.)

Townsend, D. W., and J. J. Graham. 1981. Growth and age structure of larval Atlantic herring, *Clupea harengus harengus*, in the Sheepscot River Estuary, Maine, as determined by daily growth increments in otoliths. U.S. National Marine Fisheries Service Fishery Bulletin 79:123–130.

Tsukamoto, K. 1985. Mass-marking of ayu eggs and larvae by tetracycline-tagging of otoliths. Bulletin of the Japanese Society of Scientific Fisheries 51:903–911.

Victor, B. C. 1982. Daily otolith increments and recruitment in two coral-reef wrasses, *Thalassoma bifasciatum* and *Halichoeres bivittatus*. Marine Biology 71:203–208.

Victor, B. C. 1986. Duration of the planktonic larval stage of one hundred species of Pacific and Atlantic wrasses (family Labridae). Marine Biology 90:317–326.

Wild, A., and T. J. Foreman. 1980. The relationship between otolith increments and time for yellowfin and skipjack tuna marked with tetracycline. Inter-American Tropical Tuna Commission Bulletin 17:509–541.

Wilson, K. H., and P. A. Larkin. 1982. Relationship between thickness of daily growth increments in sagittae and change in body weight of sockeye salmon (*Oncorhynchus nerka*) fry. Canadian Journal of Fisheries and Aquatic Sciences 39:1335–1339.

American Fisheries Society Symposium 7:566–571, 1990

Distinguishing Populations of Herring by Chemometry of Fatty Acids

Otto Grahl-Nielsen and Kjell Arne Ulvund

Department of Chemistry, University of Bergen
Allégt. 41, N-5007 Bergen, Norway

Abstract.—Three populations of herring *Clupea harengus harengus* were distinguished by a rapid and simple chemometric method. The three populations consisted of local spring spawners from a Norwegian fjord, Atlantoscandic fish that spawn in spring along the Norwegian coast, and autumn spawners from the North Sea. All were sampled on their spawning grounds. Heart tissue and bone from the upper jaw were subjected to methanolysis followed by gas-chromatography of the resulting fatty-acid methyl esters. When we applied principal component data analysis to representative fatty acids—9 from the bone and 10 from the heart samples—we were able to distinguish the three populations of herring. The heart samples, in which only 12% of the fatty acids were bound in neutral lipids, gave somewhat better separation than the bone samples, in which 47% of the fatty acids were bound in the neutral lipids.

Approximately 200,000 tonnes of herring *Clupea harengus harengus*[1] are harvested each year from the Skagerrak–Kattegat area by the bordering countries. The herring belong to different populations, some spawning in the fall in the North Sea and others spawning in Kattegat in the spring. Contributions may also come from Baltic populations. Because of fishery regulatory problems, it is desirable to obtain knowledge of how much each population contributes to the total number of herring in the area.

Herring studies now in progress need new marking techniques to investigate fish origins. Most physical or electronic tagging equipment is costly or fairly large, which makes inborn genetic or chemical marks preferable. Good results have been obtained by using fatty acid profiles as chemical marks. For example, eggs of Atlantic cod *Gadus morhua* have been distinguished from those of haddock *Melanogrammus aeglefinus* by their fatty acids (Knutsen et al. 1985), as have five species of trematode parasites *Gyrodactyle* spp. in the ratfish (Chimaeridae) (Berland et al. 1990). It has also been possible to distinguish different populations of the harp seal *Phoca groenlandica* (Grahl-Nielsen and Tvedt 1988) by use of the fatty acid profile in tissues from their eye lens and jawbone.

In the present investigation, our objective was to determine whether the composition of fatty acids could be used as a chemical mark for identifying various herring populations. It has been shown that the composition of fatty acids in deposited fat of herring depends on diet (Linko et al. 1985). However, the composition of the fatty acids in the phospholipids apparently is not influenced by the diet; rather, it is genetically controlled. Thus our question became: Are there detectable differences among populations of herring in Skagerrak–Kattegat with respect to the composition of their fatty acids in tissues that contain low amounts of storage fat?

To answer this question, we relied on a chemometric method of determining the fatty acid profile in tissues (Kvalheim et al. 1983; Grahl-Nielsen and Barnung 1985; Barnung and Grahl-Nielsen 1987; Ulvund and Grahl-Nielsen 1988; Hove et al. 1989). Because many samples had to be examined, the analytical method selected was as simple as possible. The common method of fatty acid analysis involves the extraction of lipids and the separation of the various lipid fractions before the final analysis for fatty acids. Our less-laborious method was based on direct methanolysis of all fatty acids present in the tissue sample, followed by gas chromatography; it sacrificed information about the fatty acids in the various classes of lipids.

The polar phospholipids were the target of our study. Because our method included all fatty acids, even those present in triacyl glycerols, we preferred tissues low in deposited fat. No systematic investigation of the distribution of the various lipid classes in different tissues of herring has

---

[1]This subspecies is known as Atlantic herring is North America, but "Atlantic" often refers to a particular stock of the subspecies in European terminology. In this paper, "herring" refers to *C. h. harengus*.

been carried out; therefore, our choice of tissue had to be based on expectations. We selected the heart and bone from the upper jaw as our sources. Samples of herring from the various populations in the Skagerrak–Kattegat area were not available, so we tested the method on herring from an Atlantoscandic population, a North Sea population, and a local population from a Norwegian fjord.

Methods

The heart and jawbone tissues to be analyzed were dissected from the herring, carefully separated from other types of tissue, rinsed in seawater, dried, and transferred to 15-mL vials, one bone or one heart sample in each vial. Approximately 2 mL of anhydrous methanol containing 2 N HCl was added, and the vials were securely closed with teflon-lined screw caps. They were heated for 15 h at 100° C. After the methanol was evaporated by N_2 gas down to about 0.5 mL, 0.5 mL of water was added and the mixture was extracted twice by 2 mL of hexane. One microliter of the combined extracts was chromatographed on a 30-m×0.32-mm fused silica column (J&W Scientific Inc.) with 50% cyanopropylmethyl–50% methylphenylpolysiloxane (0.25 μm thick) as the stationary phase and helium as the mobile phase. The column was mounted in a Hewlett-Packard 5890 gas chromatograph with splitless injection and flame ionization detection. The sample was injected at 130° C, and the oven temperature was then raised 3° C/min to 210° C, where it was left isothermal. The detector output was coupled to a VG Multichrome laboratory data system for storage and treatment of the chromatograms. The areas under the peaks were integrated.

The peaks were identified by comparison with a chromatogram of a standard mixture of fatty acid methyl esters. Chromatograms with the identified peaks are shown in Figure 1. Representative peaks were selected from the chromatograms: 9 peaks for the bone samples and 10 peaks for the heart samples. The selected peaks are marked in Figure 1. The relative amounts, expressed in area units, of the selected peaks were subjected to principal component analysis by the program SIMCA (Kyalheim and Karstang 1987). To eliminate differences in injected amounts, the values for each sample were normalized by expressing them relative to the mean. To level out the more than 50-fold difference between the fatty acids present in largest and smallest amounts, the data were autoscaled. Thereafter, the samples were

placed in a 9-dimensional space (10-dimensional for the heart samples); that is, there was one coordinate for each of the fatty acids. Two new coordinates, the principal components, were generated in this space in the directions of the largest and second-largest variation of the samples. In this manner, the dimensionality was reduced from 9 and 10 to 2. The relation between the samples was then visualized by a projection of the samples on the plane made up of the two principal components.

Results and Discussion

The principal component (PC) analyses were carried out on herring populations two at a time—North Sea versus Norwegian Sea (Atlantoscandic), North Sea versus Trondheimfjord, and Norwegian Sea versus Trondheimfjord populations. The resulting three PC plots for the bone samples are shown in Figure 2, and the three PC plots for the heart samples are shown in Figure 3.

With the exception of the North Sea population versus the Norwegian Sea population for bone tissue, there was complete separation in all cases. Overall, the heart tissue gave better results than the bone tissue. The figures show that differences among populations were mainly along the first PC. This indicated that the differences involved major systematic variation among the samples.

The composition of fatty acids was different in the two tissues, as indicated by the gas chromatograms in Figure 1. Moreover, when we used an internal strandard, the fatty acid 17:0, we found that the fatty acids constituted 0.3% of the wet weight of the bone tissue and 1.1% of the heart tissue. By extracting the lipids from the tissues and fractionating the extract into neutral lipids, free fatty acids, and polar lipids, we were able to identify large differences between the two tissues. In the bone tissue, 47% of the fatty acids were bound in neutral lipids, 31% were bound in polar lipids, and 22% were present as free fatty acids. The corresponding figures for the heart tissue were 11% in neutral lipids, 70% in polar lipids, and 19% as free fatty acids. Because the composition of polar membrane lipids is thought to better reflect genotype than does the composition of the neutral lipids (Linko et al. 1985), the above percentages indicate that heart is the tissue of choice in distinguishing different populations of herring.

This preliminary investigation suggests that chemometric analysis of fatty acids has the potential to be used to distinguish different populations

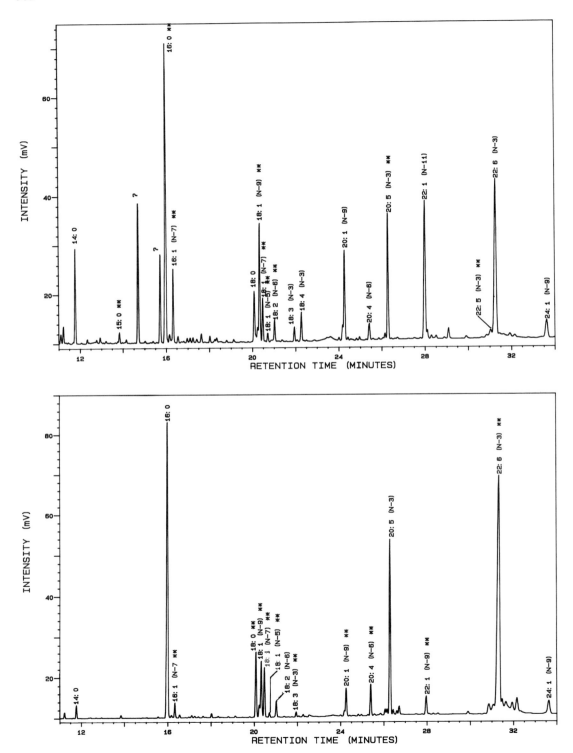

FIGURE 1.—Typical gas chromatograms of bone (top) and heart (bottom) tissue from herring. The peaks marked with asterisks (9 for bone and 10 for heart tissue) were used for multivariate computations.

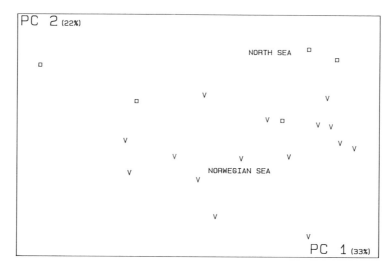

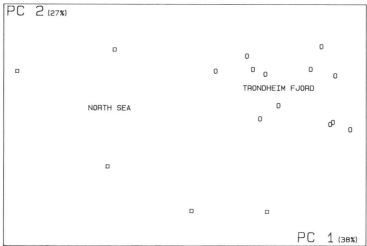

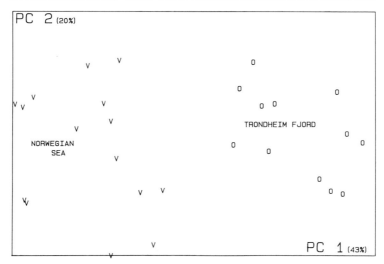

FIGURE 2.—Principal component (PC) plots based on bone tissue of herring. Each symbol represents one herring: □ from the North Sea, V from the Norwegian Sea (Atlantoscandic), and 0 from Trondheimfjord. The proportion, in percent, of the total variance between the samples accounted for by the respective PCs is indicated on the axes.

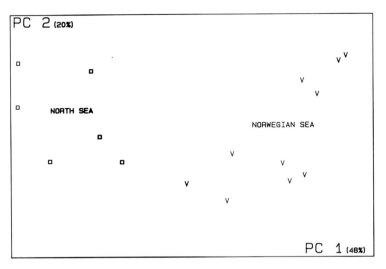

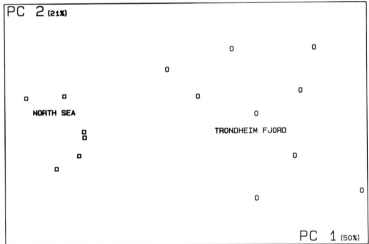

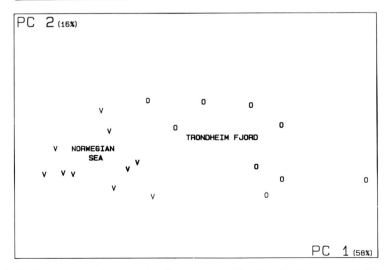

FIGURE 3.—Principal component plots based on heart tissue of herring. Symbols are as in Figure 2.

of herring. The method is simple and can be carried out rapidly on a large number of samples. When a gas chromatograph and a computer are at hand, the cost of the analyses is minor. In these respects, the method compares favorably with genetic marks. The possible applications of the method remain to be determined. Investigations of several herring populations from Norwegian fjords, from the Atlantic, from the North Sea, and from Skagerrak–Kattegat are in progress. The influence of diet on fatty acid profiles in different tissues will be studied by sampling herring from a population in a Norwegian fjord over time.

Acknowledgments

We thank A. Johannessen, I. Røttingen, and A. Aglen for supplying samples of herring, and A. Johannessen and E. Bakken for helpful discussions and marine biological advice.

References

Barnung, T., and O. Grahl-Nielsen. 1987. The fatty acid profile in cog eggs and larvae. Developmental variations and responses to oil pollution. Sarsia 72:415–417.

Berland, B., G. A. Bristow, and O. Grahl-Nielsen. 1990. Chemotaxonomy of *Gyrocotyle* species, parasites of chimaerid fishes (*Holocephali*), by chemometry of their fatty acids. Marine Biology 105:185–189.

Grahl-Nielsen, O., and E. Tvedt. 1988. Selinvasjon på norskekysten. Hvor kommer selen fra? Naturen 1988:42–45.

Grahl-Nielsen, O., and T. Barnung. 1985. Variations in the fatty acid profile of marine animals caused by environmental and developmental changes. Marine Environmental Research 17:218–221.

Hove, H. T., O. Grahl-Nielsen, and A. Rogstad. 1989. Assay for dinoflagellate toxins in mussels by gas chromatography and principal component analysis. Analytica Chimica Acta 22:35–42.

Knutsen, H., E. Moksnes, and N. B. Vogt. 1985. Distinguishing between one-day-old cod (*Gadus morhua*) and haddock (*Melanogrammus aeglefinus*) eggs by gas chromatography and SIMCA pattern recognition. Canadian Journal of Fisheries and Aquatic Sciences 42:1823–1826.

Kvalheim, O. M., and T. V. Karstang. 1987. A general-purpose program for multivariate data analysis. Chemometrics and Intelligent Laboratory Systems 2:235–237.

Kvalheim, O. M., K. Øygard, and O. Grahl-Nielsen. 1983. SIMCA multivariate data analysis of blue mussel components in environmental pollution studies. Analytica Chimica Acta 150:145–152.

Linko, R. R., J. K. Kaitaranta, and R. Vuorela. 1985. Comparison of the fatty acids in Baltic herring and available plankton feed. Comparative Biochemistry and Physiology B, Comparative Biochemistry 82:699–705.

Ulvund, K. A., and O. Grahl-Nielsen. 1988. Fatty acid composition in eggs of Atlantic cod (*Gadus morhua*). Canadian Journal of Fisheries and Aquatic Sciences 45:898–901.

American Fisheries Society Symposium 7:572–573, 1990

Tagging at the FBI, Present and Future

Lynn D. Lasswell III

Federal Bureau of Investigation
Chemistry/Toxicology Unit, FBI Laboratory
Washington, D.C. 20535, USA

Abstract.—The FBI Laboratory frequently examines evidence that has been chemically tagged. The primary method of our laboratory's analyses is mass spectrometry. The gas chromatograph–mass spectrometer and the triple-stage quadrupole allow us to detect tagging compounds at the nanogram level. Typical cases encountered in the laboratory include bank robbery and theft from interstate shipments. Bank robberies frequently involve a device that, when triggered, disperses a dye and tear-gas mixture on the robber, who literally tags himself. Analysis of trace quantities of these compounds can often link a suspect to a bank robbery. In theft cases, a soluble marker added to petroleum products allows them to be identified later.

Tagging plays an important part in forensic science. Edmond Locard (1928) of Lyons, France, observed in the early 1900s that whenever two objects come in contact, as in a violent crime, there is an exchange of material between them. Locard's speciality was dust, but this principle is the basis for numerous examinations conducted in the FBI Laboratory, such as the analysis and comparison of hairs on a suspect's clothing that are similar to the victim's hair and the hairs on a victim's clothing that are similar to the suspect's hair. This type of exchange tagging may also involve fibers, blood, semen, glass, pieces of metals, soil, tool marks, and chemicals.

In the Chemistry/Toxicology Unit of the FBI Laboratory, we deal in the detection, comparison, and identification of various chemical substances. An example of one of our examinations, which again relies on the exchange principle, involves tagged bank robbers. In this case the violent crime is the robbery and the material exchanged is a packet of exploding money. This bank security device looks exactly like a bundle of currency; however, it contains a small package of electronics and chemicals. Most of these packets are actually made from real Federal Reserve notes with their centers punched out. The chemicals contained in the packets include a red dye, methylaminoanthraquinone (MAAQ), tear gas (CS or CN), and an igniter and fuel to vaporize and disperse the dye and tear gas. The purpose of the device is to make the robber abandon the bag of money. As an added benefit to us, the robber is usually tagged with both MAAQ and tear gas. In addition to the robber's clothes, the getaway car and residence may also be tagged.

Typically, the evidence we examine in the laboratory consists of money passed to someone else or found in the suspect's possession, or of clothing, upholstery from the getaway vehicle, or items from a residence such as grout from a kitchen sink.

The FBI Laboratory uses several techniques to screen for MAAQ, including color analysis, solubility tests, and thin-layer chromatography, but our identification technique is the gas chromatograph–mass spectrometer (GC-MS). Using it, we can detect amounts of MAAQ as low as 1 ng when operating in our most sensitive mode—negative chemical ionization (Martz et al 1983). If a trace of the dye can be seen, it can be identified.

Another tag examined in our laboratory involves fuels stolen from interstate shipments. When fuel theft is anticipated, we furnish our field investigators with a readily detectable but unusual chemical, such as a halogenated aromatic compound, which they use to tag the fuel for subsequent identification. With our GC-NS technique, we can detect most sensitive halogenated aromatic compounds at concentrations as low as one part per 10^9. This detection threshold enables us to tag 100,000,000 gallons of fuel with 1 gallon of our marker when properly mixed.

In the future, we will be making improvements on our basic GC-MS technique. One of these is mass spectrometry–mass spectrometry (MS–MS). We are already using a triple-stage quadrupole in the MS–MS mode for many of our analyses, which allows us to reach even lower levels of detection. In combination with this technique, we intend to introduce liquid samples into our GC–MSs, probably by supercritical fluid chromatography. This will increase the number of substances we can analyze by allowing us to use nonvolatile and heat-sensitive samples.

References

Locard, E. 1928. Dust and its analysis. Police Journal 1:177–192.

Martz, R. M., D. J. Reutter, and L. D. Lasswell III. 1983. A comparison of ionization techniques for gas chromotography/mass spectroscopy analysis of dye and lachrymator residues for exploding bank security devices. Journal of Forensic Sciences 23:200–207.

Safferstein, R., editor. 1988. Forsenic science handbook, volume 2. Prentice-Hall, Englewood Cliffs, New Jersey.

American Fisheries Society Symposium 7:574–580, 1990

Microchemical Analysis of Fish Hard Parts for Reconstructing Habitat Use: Practice and Promise

CHARLES C. COUTANT

Environmental Sciences Division
Oak Ridge National Laboratory[1]
Oak Ridge, Tennessee 37831, USA

Abstract.—This paper describes and evaluates several microscale chemical-analysis techniques that allow the chemical record imbedded in daily and seasonal growth bands of biological skeletal materials to be used as an indicator of habitat history. Electron microprobe techniques offer the ability to scan many elements simultaneously but with low quantitative resolution. Laser-based techniques, developed at Oak Ridge National Laboratory for use in the physical sciences, offer extremely high spatial and quantitative resolution for elements or isotopes, but the elements must be analyzed one at a time. The techniques include several forms of resonance ionization spectroscopy and laser-induced fluorescence. Applications envisioned for fisheries research include differentiating freshwater, saltwater, and estuarine habitats of migratory striped bass *Morone saxatilis*; chemically marking hatchery-raised fish; and determining the pollutant-exposure history of fish.

The ability to determine when an organism was at a particular location by analyzing its hard parts would be a great boon to aquatic ecology and fisheries science. For example, basic research on habitat use and selection would be given a valuable tool. Determination of when and where fish were exposed to a contaminant that now taints their flesh and forces closure of a fishery could allow the contaminant source to be located and corrected. Protection of endangered species would be enhanced if critical habitats occupied by the species could be recognized without destructive sampling of either their populations or their habitats. Anadromous fish restoration efforts would be aided by a knowledge of when in their life cycle these fish move between fresh, brackish, and marine waters. The location and behavior gaps associated with marine mammals, which migrate over great distances in the sea and often are unobserved for months, might be filled. Environmental impact assessments for the U.S. National Environmental Policy Act and other hazard evaluations could be greatly improved if the times when animal populations were in the vicinity of proposed developments could be established without extensive field investigations.

Researchers might also be able to quantify temporal changes in water quality at a single site

where sessile aquatic organisms—such as clams, scallops, and corals—maintain a permanent record of changes in water quality in their shells or skeletal deposits. This information could be used, for example, to assess exposures in toxicology, to track changes in water movements in oceanography, and to read the history of temperature change in climatology.

A biological feature that may be exploited to obtain such information is the tendancy of hard parts such as scales, otoliths, bones, teeth, and shells to incorporate trace materials from the water as they are deposited (Bagenal et al. 1973; Van Coillie and Rousseau 1975; Moreau and Barbeau 1979, 1983; Rosenberg 1980; Rye and Sommer 1980; Lapi and Mulligan 1981; Belanger et al. 1987). The incorporated materials include chemical markers associated with the habitat that was occupied. Among these markers are elements characteristic of local water quality (including contaminants and trace metals) and element and isotopic ratios characteristic of the location and indicative of features such as salinity and temperature. The hard parts tend to be deposited in layers (much like tree rings) that can be correlated with years or even days in the case of otoliths of fish, shells of mollusks, and statoliths of cephalopods (Rhoads and Lutz 1980). Layers deposited in fish sagittae (otoliths) offer particularly good records of growth patterns; gaps and breaks in the regularity of incremental sequences record unusual events (Pannella 1980). There is extended

[1]Operated by Martin Marietta Energy Systems, Inc., for the U.S. Department of Energy under contract DE AC05-84OR21400. Publication 3141, Environmental Sciences Division, Oak Ridge National Laboratory.

discussion of growth markings in this symposium. In some cases, such as fish scales, the samples and the environmental and behavioral history they contain could be obtained without destroying the animal (although scales may be more prone to postdepositional modification than otoliths are; Koenings et al. 1986). Marine mammals deposit layers of dentine and cementum in their teeth, which can be removed without killing the animals and have proved useful for determining age (Scheffer and Myrick 1980; Myrick et al. 1983).

In this paper I introduce some techniques of microchemical analysis of biological hard parts to obtain records of habitat exposure. I also outline a vision for future fisheries uses of these techniques, which are now in the research stage for biological materials even though they have been used in the physical sciences for several years. This paper issues a challenge to fisheries scientists and ecologists to work with physical scientists and commercial analytical firms to adapt the techniques to our problems and materials

The notion that hard parts of aquatic organisms record chemical traces characteristic of their habitat has a long history in the geological sciences, particularly invertebrate paleontology. Variations in the concentrations of Ca, Mn, Sr, Mg, S, B, and other elements in preserved hard parts have been used to infer paleotemperatures, paleosalinities, and even degree of tidal exposure (see the critical review by Rosenberg 1980). Isotopic ratios have been used as indicators of paleotemperatures (Emiliani 1965, 1966). However, Rosenberg (1980) has criticized this past work because integrated analysis of large samples (which incorporate several ages) cannot resolve ontogenetic and seasonal changes in the incorporation of elements at a fine scale. Most early studies used whole-shell chemical analyses and Rosenberg challenged the inferences about habitat occupancy drawn from fossil shells by means of pre-1980 analytical techniques.

Geochemical methods for deciphering animal histories are being used more extensively in contemporary aquatic ecology (Haines and Montague 1979; Peterson et al. 1985), but all studies have been limited by an inability to obtain spatial resolution on body parts comparable to that obtained on structural features. For example, stable carbon isotopes (^{13}C and ^{12}C) are fractionated differently by different types of plant photosynthesis, which allows researchers to differentiate food webs based on marine phytoplankton from those based on estuarine sea grasses (Fry et al. 1977). This characteristic fractionation allowed

Fry (1981) to reconstruct movements of Texas brown shrimp *Penaeus aztecus* between bays and the Gulf of Mexico from analyses of whole-shrimp samples. Aquatic food chains have been followed by Haines and Montague (1979) and by Peterson et al. (1985) through use of isotopic techniques. Oxygen isotope fractionation ($^{18}O/^{16}O$) in the hydrological cycle (McCrea 1950) has been used by paleoecologists to distinguish between animal growth in high and low salinities (Rye and Sommer 1980) and at different temperatures (Urey et al. 1951), but has not been used extensively by contemporary ecologists. Killingley and Berger (1979) used stable isotopes in mollusk shells to detect upwelling events. Radtke (1983) explored the use of stable isotopes of oxygen and carbon in squid statoliths as indicators of average habitat occupancy.

Although the biological foundation for these new capabilities has a firm and long-standing basis, the problem has been that microchemical methods for detailed reading of the chemical records of environmental and biological events in an animals's life history have not been readily available. Layers of tissue that hold temporal information of value for many applications are often extremely thin and are not amenable to dissection and wet chemistry. In situ methods are preferable.

Electron Microprobe Spectroscopy

Electron microprobe spectroscopy, an analytical tool developed from scanning electron microscopy and energy-dispersive X-ray analysis, is one approach to in situ analysis at a fine spatial scale. Bombardment of a tissue sample with a focused beam of electrons in an electron microscope causes the elements in the sample to absorb energy, change electron state, and then radiate X-rays with element-specific energy levels as the elements return to their original stable-energy states. The technique has been used for identification of elements at a fine spatial resolution in fish scales and otoliths (Lapi and Mulligan 1981 and others, below) and in such diverse structures as cnidarian statoliths (Chapman 1985) and polychaete jaws (Valderhaug 1985). The value of fish otoliths for information storage has been recognized because of analyses with both electron microprobe and atomic absorption and mass spectrometry. Dahl examined otoliths of Atlantic cod *Gadus morhua* and explored the use of Sr/Ca ratios (measured with electron microprobe analysis) and $^{18}O/^{16}O$ ratios (measured with whole ground samples and wet chemistry) as habitat markers. Radtke and Targett (1984) used the electron microprobe technique to examine rhythmic structures and Sr/Ca ratios in otoliths of the Antarctic fish *Notothenia larseni* for age determination. Sponge spicules were analyzed by Laghi et al. (1984) with an electron-dispersive spectrometer for several elements, including Sr, Fe, S, K, Mg, and Mn. Mulligan (1987) and Mulligan et al. (1988) have used electron microprobe analyses for identifying the various stocks of white perch *Morone americana* and striped bass *M. saxatilis* in Chesapeake Bay. The elements most important in separating stocks of striped bass were Na, V, Sr, Sm, Ni, and Te. Using this technique, Mulligan (1987) and Mulligan et al. (1988) were able to correctly identify the river of origin in 60–80% of the specimens through discriminant analyses. These papers provide the most up-to-date techniques for preparing otolith specimens and comparing results from different samples.

The principal advantage of the electron microprobe method over more conventional techniques is that is provides data for a spectrum of chemicals from a small location on a sample. The method can analyze simultaneously for all elements with atomic numbers from 11 (Na) through 92 (U) (Mulligan et al., 1988). It analyzes for a wide range of elemental concentrations and is nondestructive. This is in contrast to atomic absorption spectroscopy, for which a sample must be ashed before analysis and only one element can be analyzed at a time. Neutron activation analysis, another analytical alternative, not only is destructive but is expensive and logistically complicated in that it requires exposure to a strong neutron source in a nuclear reactor.

The disadvantage of electron microprobe analysis is that the ability to quantify the concentration of any single chemical is poor because the output is in the form of a continuous energy spectrum (see spectra reproduced in Lapi and Mulligan 1981; Mulligan et al. 1988). A spectrum must be separated by statistical methods into signals representing the concentrations of separate elements (Lapi and Mulligan 1981). The minor elements are dwarfed by calcium, the principal element in the hard parts of fish. Thus, although a full spectrum of all elements present is obtained, only the major elements can be assigned quantitative values. The method's chemical sensitivity may be too low for many applications.

Laser Techniques for Microprobe Analysis

Advanced laser-based techniques developed for highly localized and very sensitive microchemical analyses in the physical sciences offer another set of methods for in situ analysis of biological hard parts (Gustafson and Wright 1979; Parks et al. 1983, 1987). These techniques have the capability of analyzing zones of material on a micrometer scale, and they provide exceptional sensitivity for selected individual elements or isotopes (in the extreme, they have "one-atom detection" ability). The physicists who developed these techniques indicated that the equipment and protocols could be adapted to handle biological materials, although this is not as simple as submitting a different material for routine analysis (Moore et al. 1987).

There are several promising laser-based analytical techniques. Each is a variant of the basic technique, which relies on highly focused beams of energy (ion beams or lasers) to ablate material from a very small point on a sample surface, followed by analysis of the chemical content of the vaporized cloud by means of selective resonant excitation of the chemical species of interest, as discussed in more detail below. The various techniques use different focused beams and different techniques for analyzing the vaporized sample.

In current practice, there are two in situ microprobe sampling techniques (i.e., means of removing material from a small, micrometer-scale point on a sample of biological material) and three detection techniques for the elements or for molecules of more complex constituents. The sampling techniques are ion beam sputtering and laser ablation. The detection techniques are resonance ionization spectroscopy (RIS) without mass analysis, RIS with mass analysis, and laser-induced fluorescence.

For biological samples to be used as habitat or water quality indicators, they must be analytically probed on a scale of tens of micrometers; the accuracy of sampling location and repeatability must be of the same magnitude. In all of the laser-based techniques, this goal can be met through use of a steerable, highly focused laser beam or a microprobe ion beam to remove and vaporize a very small sample at a controlled spot on a ground surface of a fish scale or otolith. A few microseconds later, a pulsed, tunable laser beam (or beams) traverses the small sample vapor plume for resonance ionization or laser-induced fluorescence analysis of the chosen constituent. The process is repeated after the first beam is moved a short distance on the sample.

For spectroscopic chemical analysis of a spatially chosen sample, resonant atomic or molecular excitation of the chemical species of interest is carried out with a tunable laser. Resonance excitation is a critical feature of each of the spectroscopic techniques. If an atom is illuminated with a quantum of light of a wavelength that matches a resonant transition, there is a high probability that the quantum will be absorbed, thereby exciting the atom to the resonant electronic state. Because each chemical species has its own characteristic set of energy states, only photons of light of specified wavelengths may be absorbed. The efficient resonant excitation of a selected element gives laser spectroscopic techniques their high sensitivity and high selectivity.

In RIS, the atoms or molecules that have been selectively excited through the resonant photon absorption phase are subsequently ionized either by photons from the same laser beam that provided the excitation or by a second high-intensity laser pulse. Resonance ionization spectroscopy was developed in 1975 by scientists at Oak Ridge National Laboratory (ORNL) to provide a basis for new ultrasensitive analytical techniques (Hurst et al. 1975). The method was used, for example, to demonstrate that one atom of cesium could be detected in 10^{19} atoms of other elements (Hurst et al. 1977). Further developments of RIS as an analytical method have been made at ORNL and Atom Sciences, Inc. (a company that holds patent license agreements with the U.S. Department of Energy for RIS commercialization). In one of these developments mass spectroscopy is coupled with RIS to permit sensitive isotopic analysis of atomic and molecular species (Payne et al. 1981). Sputter-initiated resonance ionization spectroscopy (SIRIS) is a further development involving ionbeam sputtering of samples (Parks 1986); SIRIS is patented by ORNL and Atom Sciences, Inc. Initial study indicates that RIS can be modified to detect complex organics (preliminary work at ORNL on the pesticide Dinaseb has indicated a sensitivity on the order of 5 femtograms for a single laser pulse).

An alternative to RIS is laser-induced fluorescence spectroscopy. Here, the initial selective resonance excitation mentioned above is not followed by additional laser excitation; rather, the excited species is allowed to fluoresce and emit photons characteristic of a given chemical species. This method provides no isotopic information because emissions from various isotopes of a given atomic species are not easily distinguishable, but it can provide high sensitivity with relatively simple technology. The method is capable of detecting on the order of a thousand atoms of a chemical species of interest within a laser-vaporized sample. Laser-induced fluorescence spectroscopy offers simplicity and high sensitivity for several metallic species and also offers promise for identifying organic contaminants.

The key advantage of the laser-based techniques is their extremely high analytical sensitivity. They have high selectivity and efficiency for all elements in the periodic table except helium and neon (used in the lasers), including most isotopes (Parks 1986). The main disadvantage is that only one element can be analyzed at a time. The techniques are also very experimental and lack the history of demonstrated usefulness of other microprobe techniques (Lapi and Mulligan 1981; Mulligan et al. 1988).

These and other laser-based analytical spectroscopic methods are under development in the Biological and Radiation Physics Section and Chemical Physics Section of the Health and Safety Research Division of ORNL, and at Atom Sciences, Inc., in Oak Ridge, Tennessee. The methods described here are being evaluated for biological materials in consultation with the staff of the Environmental Sciences Division and the Biology Division of ORNL. Further development of equipment and techniques will be required to tailor the application of these analytical technologies to biological materials and to provide standardized analyses at reasonable costs.

Future Applications

In practice and in promise, the microanalytical techniques of electron microprobe spectroscopy and the several forms of laser-based microanalysis have much to offer aquatic ecology and fisheries science. Fisheries scientists and managers often are maligned for relying on only the most direct techniques, which were practiced thousands of years ago (e.g., fish netting), and for repeating population surveys ad nauseam. The discipline often fails to take advantage of indirect ways of obtaining useful information or to make inferences from these indirect measures. In contrast, geologists, paleontologists, and oceanographers, who cannot directly measure most of what interests them, have honed their inferential abilities to a fine degree.

I envision several applications for microanalysis in fisheries in the near future and present three of them below. However, directed research and technique development will be necessary before we can go beyond enthusiastic speculation.

Differentiation of Freshwater, Saltwater, and Estuarine Striped Bass

Analyses of the chemical composition of fish scales have already been used successfully to differentiate striped bass reared in fresh water and salt water. Belanger et al. (1987) digested entire scales and used standard flame atomic absorption spectrophotometry for chemical analyses. Estuarine striped bass, which have intermediate chemical characteristics because of their ontogenetically variable habitat selection, were poorly distinguished. The technique did not differentiate the ages when estuarine fish occupied freshwater and saline habitats. It should be possible to associate chemical characteristics that distinguish water type with datable banding, thereby indicating the ages and times of year at which fish moved from one water type to another. This more detailed information is important for the resolution of questions of estuarine residence and pollution-induced limitation of summer habitat for subadult and adult stages in estuaries such as Chesapeake Bay and Albemarle Sound (Kohlenstein 1981; Coutant 1985; Coutant and Benson 1988).

Marking Fish with Rare Earth Elements

Augmentation of natural populations of fishes by hatchery production has expanded recently, but a definitive means for recognizing hatchery fish among wild stocks has eluded managers. Muncy et al. (1988) experimentally fed small quantities of the rare earth elements europium and terbium to juvenile striped bass and attempted to detect these additives by neutron activation of whole scales. Soon after feeding of the tracers was discontinued, the europium and terbium signals dropped below detectable levels because growth reduced their concentrations per unit of body mass. Microchemical analysis techniques seem appropriate for scanning scales from the origins to the margins to detect bands containing the rare earth elements.

Exposure of Fish to Water Pollutants

When and where organisms were exposed to a toxicant are persistent questions in evaluations of aquatic animals whose contaminated tissues might be eaten by humans. The ultimate desire is to find the source of contamination and correct it. Methods for relating toxicant intake to a fish's age and the habitat it occupied at the time of exposure would help answer the when and where questions. Advances in microanalytical instrumentation may provide means to establish the time and dosage of environmental contamination.

Conclusions

With improved analytical capabilities, the future should give us new keys to an organism's past. The future understanding and management of our aquatic and fishery resources beg that we actively pursue the promises that microanalytical techniques offer for tracing the habitat history of individuals and populations.

But major questions confront us. Which techniques should be selected? How confident can one be of successfully answering important questions for research and management? What is the cost? The use of microanalysis to discern habitat history is in its infancy. We do not have enough experience to judge which of the two families of approaches—electron microprobe and laser—will be more informative. Whether we prefer a broad spectrum of elements present in the largest concentrations or detailed quantitative information on a few selected trace elements or isotopes ultimately will be decided by the application. The relative values of the techniques will be determined by the development effort devoted to refining each technique for specific purposes.

Because application of the techniques is still in the research phase, routine analytical services are not generally available. Cost estimates on a per-sample basis are only now being developed by firms such as Atom Sciences, Inc. Fishery researchers and managers, physical scientists, and commercial firms that would offer analytical services must get together (along with the funding agencies) to jointly develop the technologies that seem to hold immense potential for fisheries management.

Acknowledgments

I especially thank K. B. Jacobson, W. R. Garrett, M. G. Payne, L. R. Shugart, C. H. Chen, E. T. Arakawa, and J. E. Parks (Atom Sciences, Inc.) for the exciting interaction of physical and biological disciplines at ORNL relative to laser-based chemical analysis, and T. J. Mulligan (now at the University of Washington, Seattle), who shared his experiences with electron microprobe techniques. B. G. Blaylock and G. F. Cada critically reviewed the draft manuscript.

References

Bagenal, T. B., F. J. H. Mackereth, and J. Heron. 1973. The distinction between brown trout and sea trout by the strontium content of their scales. Journal of Fish Biology 5:555–557.

Belanger, S. E., D. S. Cherry, J. J. Ney, and D. K. Whitehurst. 1987. Differentiation of freshwater versus saltwater striped bass by elemental scale analysis. Transaction of the American Fisheries Society 116:594–600.

Chapman, D. M. 1985. X-ray microanalysis of selected coelenterate statoliths. Journal of the Marine Biological Association of the United Kingdom 65:617–627.

Coutant, C. C. 1985. Striped bass, temperature, and dissolved oxygen: a speculative hypothesis for environmental risk. Transactions of the American Fisheries Society 114:31–61.

Coutant, C. C., and D. L. Benson. 1988. Linking estuarine water quality and impacts on living resources: shrinking striped bass habitat in Chesapeake Bay and Albemarle Sound. U.S. Environmental Protection Agency, Report 503/3-88-001, Washington, D.C.

Dahl, E. 1984. Codfish otoliths: information storage structures. Pages 273–289 in E. Dahl, D. S. Danielssen, E. Moksness, and P. Solemdal, editors. The propagation of cog Gadus morhua L. Institute of Marine Research, Fløedevigen Biological Station, Flødevigen rapportserie 1, Arendal, Norway.

Emiliani, C. 1965. Isotopic paleotemperatures. Science (Washington, D.C.) 154:851–857.

Emiliani, C. 1966. Paleotemperature analysis of Caribbean cores P6304-8 and P6304-9 and a generalized temperature curve for the past 425,000 years. Journal of Geology 74:109–126.

Fry, B. 1981. Natural stable carbon isotope tag traces Texas shrimp migrations. U.S. National Marine Fisheries Service Fishery Bulletin 79:337–345.

Fry B., R. S. Scalan, and P. L. Parker. 1977. Stable carbon isotope evidence for two sources of organic matter in coastal sediments: seagrasses and plankton. Geochimica et Cosmochimica Acta 41:1875–1877.

Gustafson E. J., and J. C. Wright. 1979. Trace analysis of lanthanides by laser excitation of precipitates. Analytical Chemistry 51:1762–1774.

Haines E. B., and C. L. Montague. 1979. Food sources of estuarine invertebrates analyzed using $^{13}C/^{12}C$ ratios. Ecology 60:48–56.

Hurst, G. S., M. H. Nayfeh, and J. P. Young. 1977. A demonstration of one-atom detection. Applied Physics Letters 30:229–231.

Hurst, G. S., M. G. Payne, M. H. Nayfeh, J. P. Judish, and E. B. Wagner. 1975. Saturated two-photon resonance ionization of He (^{21}As). Physical Review Letters 35:82–85

Killingley, J. S., and W. H. Berger. 1979. Stable isotopes in a mollusk shell: detection of upwelling events. Science (Washington, D.C.) 205:186–188.

Koenings, J. P., P. J. Lipton, and P. McKay. 1986. Quantitative determination of oxytetracycline uptake and release by juvenile sockeye salmon. Transactions of the American Fisheries Society 115:621–629.

Kohlenstein, L. C. 1981. On the proportion of the Chesapeake Bay stock of striped bass that migrates into the coastal fishery. Transactions of the American Fisheries Society 110:168–179.

Laghi G. F., G. Martinelli, and F. Russo. 1984. Localization of minor elements by EDS microanalysis in aragonitic sponges from the St. Cassian beds, Italian dolomites. Lethaia 17:133–138.

Lapi, L. A., and T. J. Mulligan. 1981. Salmon stock identification using a microanalytic technique to measure elements present in the freshwater growth region of scales. Canadian Journal of Fisheries and Aquatic Sciences 38:744–751.

McCrea, J. M. 1950. On the stable isotope chemistry of carbonates and a paleotemperature scale. Journal of Chemical Physics 18:849–857.

Moore, L. J., J. E. Parks, E. H. Taylor, D. W. Beekman, and M. T. Spaar. 1987. Medical and biological applications of resonance ionization spectroscopy. Institute of Physics Conference Series 84:239–244.

Moreau G., and C. Barbeau. 1979. Différenciation de populations anadromes et dulcicoles de Grands Corégones (Coregonus clupeaformis) par la composition minérale de leurs écailles. Journal of the Fisheries Research Board of Canada 36:1439–1444.

Moreau, G., and C. Barbeau. 1983. Zinc, manganese, and strontium in opercula and scales of brook trout (Salvelinus fontinalis) as indicators of lake acidification. Canadian Journal of Fisheries and Aquatic Sciences 40:1685–1691.

Mulligan, T. J. 1987. Identification of white perch (Morone americana) stocks in Chesapeake Bay based on otolith composition and mitochondrial DNA analysis. Doctoral dissertation. University of Maryland, College Park.

Mulligan, T. J., F. D. Martin, R. A. Smucker, and D. A. Wright. 1988. A method of stock identification based on elemental composition of striped bass Morone saxatilis (Walbaum) otoliths. Journal of Experimental Marine Biology and Ecology 114:241–248.

Muncy, R. J., N. C. Parker, and H. A. Poston. 1988. Marking striped bass with rare earth elements. Proceedings of the Annual Conference Southeastern Association of Fish and Wildlife Agencies 41:244–250.

Myrick A. C., Jr., E. W. Shallenberger, I. Kang, and D. B. MacKay. 1983. Calibration of dental layers in seven captive Hawaiian spinner dolphins, Stenella longirostris, based on tetracycline labeling. U.S. National Marine Fisheries Service Fishery Bulletin 82:207–225.

Pannella, G. 1980. Growth patterns in fish sagittae. Pages 519–560 in D. C. Rhodes and R. A. Lutz, editors. Skeletal growth of aquatic organisms. Plenum, New York.

Parks, J. E. 1986. Laser applications to materials and surface analysis. Optics News 12:22–27.

Parks, J. E., H. W. Schmitt, G. S. Hurst, and W. M. Fairbank, Jr. 1983. Sputter-initiated resonance ionization spectroscopy. Thin Solid Films 108:69–78.

Parks, J. E., and seven coauthors. 1987. Progress in analysis by sputter initiated resonance ionization spectroscopy. Institute of Physics Conference Series 84:157–162.

Payne, M. G., C. H. Chen, G. S. Hurst, and G. W. Foltz. 1981. Application of resonance ionization spectroscopy in atomic and molecular physics. Advances in Atomic and Molecular Physics. 17:229–274.

Peterson, B. J., R. W. Howarth, and R. H. Garritt. 1985. Multiple stable isotopes used to trace the flow or organic matter in estuarine food webs. Science (Washington, D.C.) 227:1361–1363.

Radtke, R. L. 1983. Chemical and structural characteristics of statoliths from the short-finned squid *Illex illecebrosus*. Marine Biology 76:47–54.

Radtke R. L., and T. E. Targett. 1984. Rhythmic structural and chemical patterns in otoliths of the Antarctic fish *Notothenia larseni*: their application to age determination. Polar Biology 3:203–210

Rhoads, D. C., and R. A. Lutz, editors. 1980. Skeletal growth of aquatic organisms. Plenum, New York.

Rosenberg, G. D. 1980. An ontogenetic approach to the environmental significance of bivalve shell chemistry. Pages 133–168 *in* D. C. Rhoads and R. A. Lutz, editors. Skeletal growth of aquatic organisms. Plenum, New York.

Rye, D. M., and M. A. Sommer II. 1980. Reconstructing paleotemperature and paleosalinity regimes with oxygen isotopes. Pages 169–202 *in* D. C. Rhoads and R. A. Lutz, editors. Skeletal growth of aquatic organisms. Plenum, New York.

Scheffer V. B., and A. C. Myrick, Jr. 1980. A review of studies to 1970 of growth layers in the teeth of marine mammals. Report of the International Whaling Commission, Special Issue 3:51–63.

Urey H. C., H. A. Lowenstam, S. Epstein, and C. R. McKinney. 1951. Measurement of paleotemperatures and temperatures of the upper Cretaceous of England, Denmark, and the southeastern United States. Geological Society of America Bulletin 62:399–416.

Valderhaug, V. A. 1985. Population structure and production of *Lumbrineris fragilis* (Polychaeta: Lumbrineridae) in the Oslofjord (Norway) with a note on metal content of jaws. Marine Biology 86:203–211

Van Coillie, R., and A. Rousseau. 1975. Distribution minérale dans les écailles des poissons d'eau douce et ses relations avec le milieu aquatique. Internationale Vereinigung für Theoretische und Angewandte Limnologie Verhandlungen 19:2440–2447.

American Fisheries Society Symposium 7:581–587, 1990

DESIGN, PROGRAM CONTRIBUTION, AND ANALYSIS

Design of Survival Experiments with Marked Animals: A Case Study

ERIC REXSTAD

Colorado Cooperative Fish and Wildlife Research Unit, Colorado State University
Ft. Collins, Colorado 80523, USA

KENNETH P. BURNHAM AND DAVID R. ANDERSON

Colorado Cooperative Fish and Wildlife Research Unit
U.S. Fish and Wildlife Service, Colorado State University

Abstract.—The statistical theory for survival experiments with marked animals is new. To elucidate guidelines for the design of such treatment–control experiments, we present a hypothetical case study involving estimation of total fish mortality for a project consisting of a reservoir, a turbine system, and three downstream recovery sites. Animals are allocated to two treatment groups and one control group, and the sampling effort at recovery sites varies. Repeated capture and release, rather than single captures of marked animals, is shown to be superior for parameter estimation and model selection. Alternative methods of variance estimation such as replication and quasi-likelihood are considered. Monte Carlo simulation aids in the design of such complex experimental studies. A comprehensive computer package, RELEASE, is available to design and analyze such experiments.

Planned treatment and control experiments are fundamental to scientific endeavor. In investigations of biological systems, it is often difficult to apply rigorous enough controls to ascertain the effect of treatments on the organisms of interest. Consequently, determination of mortality caused by specific agents often is confounded by numerous unmeasured variables.

However, when marked animals can be exposed to a treatment, while others are not exposed to it, assessment of the risk posed by the treatment can be made. For this paper, we defined the treatment as the passage of marked fish through a reservoir and its associated dam. The theory we describe is less restrictive than our example; Burnham et al. (1987) described several examples in which treatment effects were temporal rather than spatial, and they dealt with a variety of treatments.

The theory for analysis of these release–recapture experiments is analogous to the theory of generalized linear models in statistics. It encompasses studies involving single treatment–control pairs, multiple treatments, unequal numbers of individuals in treatment groups (unbalanced designs), and ordered treatments.

In this paper, we present a comprehensive, realistic (albeit hypothetical) case study of a system-wide experiment to assess the effect of a dam on the survival of migrating salmonids. We describe the design of such a study, address such issues as sizes of treatment and control groups, replicate lots, and allocation of sampling effort at downstream recapture sites, and touch on data analysis. For the last, we used program RELEASE to simulate a set of data under the assumed conditions and constructed design.

Notation and Theory

The notation used throughout this paper emphasizes the parameters of interest and associated data collected in capture–recapture studies.

R_{vi} = the number of fish of treatment group v released at dam i.

m_{vij} = the number of fish of treatment group v recaptured at dam j from the cohort of R_{vi} fish released at dam i.

P_{vi} = the capture probability at site i for fish in treatment group v that reach the site alive.

ϕ_{vi} = the survival probability between release site i and recapture site $i+1$ for fish in treatment group v.

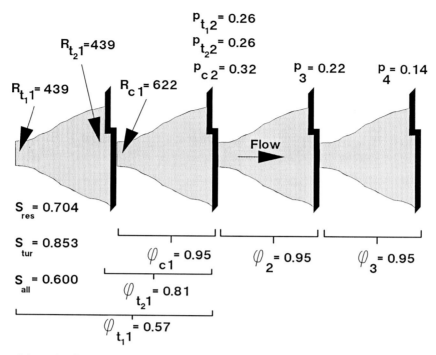

FIGURE 1.—Schematic of a hypothetical survival experiment involving marked animals in a lotic system of dams and reservoirs. There are three release sites in the vicinity of the first dam and three downstream recapture sites. The true parameter values used in the simulation are shown: R_{vi} are numbers of individuals released, p_{vi} are recapture probabilities, ϕ_{vi} are survival probabilities, and S_v are treatment effects. Subscripts c and t denote control and treatment groups. Subscripts on treatment effects (S) are reservoir (res), turbine (tur), and total (all).

S = treatment effect, defined as the ratio of survival rates between treatment and control groups, ϕ_{ti}/ϕ_{ci}.

The theory of Burnham et al. (1987) supports a series of models to describe the treatment effect. The null model (H_0) assumes no treatment effect, i.e., the parameters p_{vi} and ϕ_{vi} do not differ between groups. Model $H_{1\phi}$ assumes an acute treatment effect manifested by a difference between groups in survival rate between the release site and the first recapture site (ϕ_{v1}). Model H_{2p} assumes that differences between groups exist both in ϕ_{v1} and the recapture probabilities at the first recapture site (p_{v2}). Other models ($H_{2\phi}$, H_{3p}) are even more general to allow chronic treatment effects to be modeled. For purposes of our hypothetical study, we assume model H_{2p} to be the true model.

Case Study

In our hypothetical study, we wish to measure total survival at the project and partition the estimated mortality into components attributable to

passage of fish through a reservoir and to passage of fish through a turbine system at the dam creating the reservoir. There are three dams downstream of the dam at which the mortality to be estimated occurs; these serve as sampling sites for fish released during the experiment. A schematic diagram of the system is presented in Figure 1.

Experimental Constraints

We assume that our fish are tagged with passive integrated transponders (PITs). Use of PIT tags influences capture probabilities (p_{vi}), making them higher than could be achieved by physical recapture. We also assume cost constraints which limit the number of PIT-tagged fish to 12,000. The parameter values (p_{vi} and ϕ_{vi}) chosen for this hypothetical study are based on a release–recapture experiment conducted by the National Marine Fisheries Service (A. Giorgi, NMFS, personal communication) on the Snake and Columbia rivers. These values are shown in Figure 1.

All parameters are assumed to be heterogeneous to impose the realism of variability in survival and recapture probabilities within groups due to factors such as body size. Survival and recapture probabilities vary among individual fish; their mean values are presented in Figure 1. These probabilities are modeled as β-distributed random variables. The resulting coefficients of variation are on the order of 25% for the survival probabilities and 60% for the recapture probabilities. The interested reader is referred to the Appendix for the input specifications for the simulation study and to Burnham et al. (1987) for discussion of the mechanics of PROC SIMULATE in program RELEASE.

Design

The first design issue we address is the need for replication. Repeatability is fundamental to scientific inquiry. When sampling variation is substantial, as in most fisheries experiments, there should be many repeated releases of treatment and control fish. If the treatment effect varies with conditions, such as flow regimes or power generation schedules, a sample of those conditions should be observed.

How many replicates are necessary? For this study, we assume that releases take place at weekly intervals, and that a 2-month period encompasses the migration by the salmonids of interest. Therefore, eight replicate releases are required. The true variance of estimated treatment effect ($\hat{S}$) is unaffected by the number of replicates, given a fixed total number of fish released in all replicates. The shape of the t-distribution, which dictates the magnitude of confidence intervals for estimates of treatment effects, changes little beyond replicates greater than six (Burnham et al. 1987). Hence use of eight replicate lots, each of 1,500 released fish, is satisfactory in our hypothetical study.

The next design issue is allocation of the fish in each lot among two treatment groups and one control group. The optimum allocation of fish to treatment groups, based on minimization of the sum of var($\hat{S}_v$), can be determined as follows (Burnham et al. 1987):

$$R_{c1} = \frac{R_{.1}}{1 + \sqrt{V - 1}}$$

and

$$\frac{R_{v1}}{R_{c1}} = \frac{1}{\sqrt{V - 1}};$$

$R_{.1}$ is the total number of fish released in a single lot (1,500). In our hypothetical study, there are three groups ($V = 3$), so the calculations result in an allocation of 622 fish to the control group and 439 fish to each of the treatment groups.

If the number of fish available for the experiment were not fixed, we could pursue an iterative process of choosing sample size (total number of fish), calculating numbers per lot and allocation among groups, and simulating test power and estimator precision until the design achieved the study objectives.

We release one-sixth of the fish in each group at 4-h intervals during the course of a day chosen randomly within each week. This minimizes confounding caused by the diurnal cycle. Each treatment and control group is released at multiple sites to minimize the confounding possible if all releases take place at a single location. To ensure that fish released at dam 1 travel downstream simultaneously with fish released at the head of the first pool, radio transmitters are placed on several fish released at the head of the pool to synchronize releases of the other groups. These fish will be excluded from the estimation of treatment effects because the radios may affect their survival. Installation of PIT-tag sensors at dam 1 is another way to coordinate releases of fish at the dam with the arrival of fish from the pool.

It is critical to randomly assign fish to treatment groups. The implantation of PIT tags should be randomized also. Thus, a particular person should not mark all fish in one group while another person marks all fish in another group. Quality assurance must be a major concern in all steps of the experiment.

The fish are held for a time after they are tagged, and any that die are subtracted from the numbers marked. Further suggestions for the practical conduct of experiments with marked fish are given in Burnham et al. (1987) and citations therein.

Our last design consideration is allocation of recapture effort at downstream sites. If treatment effect is assumed to be acute (as it is under model H_{2p}), most of the effort needs to be focused at the first recapture site (dam 2). However, effort at dams 3 and 4 is needed for model selection tests to have adequate discrimination capability. We used PROC SIMULATE in program RELEASE to investigate various allocations of recapture effort. These investigations showed that greater effort than we had anticipated needed to be allocated at dam 3 to detect treatment effects on recapture probabilities at dam 2 (p_{v2}). Manipulation of recapture effort with PIT tags involves deployment of sensors to read the tags as the fish pass by.

TABLE 1.—Recapture data for a hypothetical case study involving three release (R) groups of fish (pool head, above dam, and in tailrace) for eight replicate lots. Rows within lots (designated with i subscript) represent release at dams 1, 2, and 3; m denotes recaptured fish. These data were analyzed by program RELEASE to estimate mortality due to the reservoir and turbines of dam 1.

| | Release group | | | | | | | | | | |
| | Pool | | | | Above dam | | | | Below dam | | |
Lot	R_{1i}	m_{i1}	m_{i2}	m_{i3}	R_{2i}	m_{i1}	m_{i2}	m_{i3}	R_{3i}	m_{i1}	m_{i2}	m_{i3}
1	439	59	45	18	439	95	55	26	622	189	79	53
	59		13	8	95		21	9	189		40	20
	58			7	76			10	119			20
2	439	55	46	19	439	98	56	26	622	185	87	31
	55		14	8	98		16	7	185		37	17
	60			10	72			12	124			15
3	439	77	47	22	439	102	59	23	622	170	94	40
	77		16	10	102		20	10	170		38	12
	63			9	79			10	132			22
4	439	62	45	15	439	104	54	32	622	184	89	43
	62		16	5	104		30	12	184		40	13
	61			10	84			12	129			19
5	439	64	37	18	439	89	60	27	622	205	90	35
	64		11	5	89		20	12	205		52	18
	48			8	80			6	142			22
6	439	67	37	16	439	87	50	24	622	164	90	45
	67		15	9	87		16	12	164		31	12
	52			7	66			6	121			18
7	439	64	34	20	439	81	52	25	622	188	93	41
	64		19	4	81		15	7	188		40	15
	53			11	67			9	133			18
8	439	70	28	16	439	87	54	25	622	178	77	40
	70		16	4	87		15	7	178		30	24
	44			12	69			13	107			20

Saturation of bypasses, turbine intakes, raceways, and other key points is the best way to increase recapture probabilities at recapture sites. Deflection structures can also enhance detection probabilities as fish pass recapture sites.

Evaluation of Design

If additional funds became available for the hypothetical study at this stage of design, the issue would arise whether to use them to purchase more tags and fish for release or to further enhance recapture probabilities. Burnham et al. (1987) presented an illuminating table that indicated greater gains are achieved by increasing recapture probabilities than by increasing numbers of releases.

In planning an experiment or evaluating a particular design, the use of numerical expected values in program RELEASE represents a powerful tool. Approximate bias in estimators of S, ϕ_{vi}, and p_{vi} can be quickly evaluated by the procedures explained in Burnham et al. (1987). For example, it is possible to evaluate the expected bias in S if the true model is H_{2p} but tests suggest model $H_{1\phi}$ is adequate and S is estimated under the assumption that $H_{1\phi}$ is true. Equally important is estimation of test power. Here, RELEASE allows analysis of numerical expected values to produce a test statistic asymptotically distributed as χ^2. This statistic is the noncentrality parameter of the corresponding noncentral χ^2 power curve of the test. In particular, the power of tests 1.R1 and 1.T2 should be considered in the design of any fish survival experiment. Prior to execution of an experiment, bias and test power can and should be evaluated through the use of numerical expectations computed by program RELEASE. This can prevent the undertaking of an expensive experiment from which no information will be gained.

Data for the three treatment groups of eight replicate lots were generated by program RE-

TABLE 2.—Summary of model selection test χ^2 values produced by RELEASE for the eight lots of fish from the hypothetical case study. Models in order of increasing acuity of treatment effects are $H_{3\phi}$, H_{3p}, $H_{2\phi}$, H_{2p}, $H_{1\phi}$, and H_0. H_0 assumes no treatment effect, so test H_0 versus $H_{3\phi}$ is an overall test of the existence of treatment effect.

Lot or statistic	H_0 versus $H_{3\phi}$	H_0 versus $H_{1\phi}$	$H_{1\phi}$ versus H_{2p}	H_{2p} versus $H_{2\phi}$	$H_{2\phi}$ versus H_{3p}
1	68.2	60.6	4.2	0.5	2.2
2	63.7	49.1	8.4	4.7	0.5
3	28.3	25.8	0.4	0.5	0.9
4	65.0	56.5	2.1	4.0	2.2
5	86.0	71.9	7.0	2.3	1.4
6	51.8	47.7	0.1	2.4	0.3
7	76.1	70.4	2.3	1.4	0.2
8	57.1	50.2	3.3	0.7	1.3
Pooled χ^2	496.1	432.2	27.7	16.6	9.0
df	80	16	16	16	16
P (pooled)	<0.001	<0.001	0.034	0.414	0.915

LEASE under model H_{2p} and are presented in Table 1. The first performance consideration of the RELEASE theory is the model selection procedure. A series of statistical tests was performed by RELEASE to select the model most appropriate for the data being analyzed. TEST 1 is an omnibus test for the existence of a treatment effect. This test is divided into components, each with specific null and alternative models (Burnham et al. 1987). For our hypothetical case study, the test of greatest interest is 1.T2, which has model $H_{1\phi}$ as the null and model H_{2p} as the alternative. The individual tests for each lot can be summed to arrive at a pooled χ^2 test for model selection. Table 2 shows that RELEASE correctly chose model H_{2p} as the best model for these data. Survival rates between the release site and the first recapture site, as well as recapture probabilities at the first recapture site, differ between groups in this model.

The parameters of interest in our case study are treatment effects ($\hat{S}_{res}$ for reservoir survival, $\hat{S}_{tur}$ for turbine survival, and $\hat{S}_{all}$ for total project survival). Table 3 summarizes estimates of these parameters for each lot and the means over lots. When these estimates are compared with actual parameter values given in Figure 1, it is clear that these parameters are well estimated. Using empirical standard errors derived from replicate lots, we constructed confidence intervals for treatment effects. A t-statistic with 7 df for a 0.05 α-level confidence interval is 2.365, resulting in a confidence interval of (0.54, 0.65) for $\hat{S}_{all}$.

If interest lies in interval estimates, the simple procedures suggested by Burnham et al. (1987) are recommended, especially for estimates from individual lots. This procedure takes into account the nonnormal sampling distribution of the estimators associated with small sample sizes.

Discussion

Our hypothetical case study offers several lessons for design and analysis of mark–recapture survival experiments. We highlight several of these to point the reader to pertinent literature.

Tag loss.—Tag loss or failure of PIT tags is a potential problem in tagging studies. In usual estimation of parameters in the Jolly–Seber framework, tag loss is a source of potential bias (Seber 1982:488–489). However, in a treatment–control experiment, it seems reasonable that failure of PIT tags is random with respect to group membership. If this mild assumption can be met, the estimators of treatment survival will be unbi-

TABLE 3.—Estimates and standard errors (SE, in parentheses) of treatment effect ($\hat{S}_v$) between release site and first recapture site for the three treatment groups. $\hat{S}_{res}$ represents reservoir survival, $\hat{S}_{tur}$ represents turbine survival, and $\hat{S}_{all}$ represents total system survival. The estimates were produced by RELEASE under model H_{2p} for the eight lots from the hypothetical case study.

Lot or statistic	$\hat{S}_{res}$	$\hat{S}_{tur}$	$\hat{S}_{all}$
1	0.731(0.083)	0.817(0.071)	0.600(0.062)
2	0.728(0.084)	0.908(0.081)	0.664(0.071)
3	0.813(0.087)	0.860(0.075)	0.702(0.068)
4	0.665(0.074)	0.882(0.073)	0.589(0.060)
5	0.650(0.075)	0.853(0.073)	0.559(0.058)
6	0.725(0.088)	0.767(0.072)	0.560(0.060)
7	0.718(0.087)	0.751(0.069)	0.542(0.058)
8	0.612(0.076)	0.873(0.082)	0.538(0.061)
Mean	0.705	0.839	0.594
SE (mean)	0.022	0.020	0.021

ased (because treatment survival is a ratio of ϕ_{vi} parameters).

Sophisticated models.—In studies involving four to seven downstream dams, it is possible that the dam-to-dam survival rate, ϕ_{vi}, is approximately constant. Thus, instead of estimating, say, ϕ_{v4}, ϕ_{v5}, ϕ_{v6}, and ϕ_{v7}, one might entertain the null hypothesis that these rates are equal. A likelihood ratio test (Burnham et al. 1987) can be used to test this hypothesis. If the null hypothesis is supported, then a model with one parameter replacing these four parameters could be considered. This procedure would allow increased precision and test power. Alternatively, the finite rates ϕ_{v4}, ϕ_{v5}, ϕ_{v6}, and ϕ_{v7} might differ, but the standardized rate (survival per 100 km) might be approximately constant. Again, estimation of one standardized rate would be beneficial. Both of these extensions can be examined by using program RELEASE's PROC SURVIV interface to a numerical optimization program, SURVIV (White 1983).

Heterogeneity.—Heterogeneity in fish size can be dealt with in several ways. First, simulation studies in Burnham et al. (1987) indicated that the effect of heterogeneity on estimators of treatment effect is weak. Second, deletion of the smallest and largest extremes of fish available for marking might substantially reduce the heterogeneity. We routinely recommend deletion of the smallest 10% and largest 10% of the fish entering the experiment.

Alternatively, if a subobjective is to examine treatment survival as a function of fish size, the researcher might consider nine lots—three of small fish, three of medium-sized fish, and three of large fish. This represents a "blocking" on fish size, as in a randomized complete-block analysis of variance. Programs RELEASE and SURVIV (particularly the likelihood ratio tests between models) could be employed for a full analysis of such experiments.

Variance estimation.—Theoretical estimates of variance, derived from the theory of maximum likelihood, often underestimate the true variation present in survival experiments. Other techniques can be employed to better estimate true variation in such studies. Replication is one method already discussed. Data on PIT tags at dams 2, 3, and 4

should include, for each tag number, the time and date of that tag's passage past the sensor. This information allows a temporal comparison of the various treatment–control groups as they pass downstream sampling locations. In addition, if data include the specific detector or sensor identification, the use of quasilikelihood methods to assess variability will be facilitated (Burnham et al. 1987). Such variance estimates could be compared to empirical variance estimates obtained from replicate lots. Further insight into experimental variability could come from partitioning data by the last digit of the PIT tag number. This partitioning could be done for the entire data set or stratified by lots. The partitioned data could be jackknifed or bootstrapped (Efron 1982) to obtain alternative variance estimates.

Conclusions

Statistical theory exists for a general class of experiments involving marked fish (Burnham et al. 1987). Questions regarding reservoir or turbine or bypass mortality can be addressed with replicated experiments involving two or more groups of marked fish. Many sampling protocols are possible under the recently developed theory.

Computer software (e.g., programs RELEASE and SURVIV) have been developed that afford comprehensive analysis capability. Many features are available to aid in the design of experiments, in addition to analysis algorithms for use with empirical or simulated data.

References

Burnham, K. P., D. R. Anderson, G. C. White, C. Brownie, and K. H. Pollock. 1987. Design and analysis methods for fish survival experiments based on release–recapture. American Fisheries Society Monograph 5.

Efron, B. 1982. The jackknife, the bootstrap, and other resampling plans. SIAM (Society for Industrial and Applied Mathematics), SBMS-NSF (National Science Foundation) Regional Conference Series in Applied Mathematics 30, Philadelphia.

Seber, G. A. F. 1982. The estimation of animal abundance and related parameters, 2nd edition. Macmillan, New York.

White, G. C. 1983. Numerical estimation of survival rates from band-recovery and biotelemetry data. Journal of Wildlife Management 47:716–728.

Appendix: Input for the Simulation Study

Input to program RELEASE to generate hypothetical case study discussed in this paper.
PROC TITLE Seattle paper;
PROC SIMULATE OCCASIONS=4 GROUPS=3 NSIM=8 SEED=7654321 UNBIAS
DETAIL SUMMARY;
GLABEL(1)=Pool;
GLABEL(2)=Above dam;
GLABEL(3)=Tailrace;
phi(1)=beta(4.7,3.54561,0,1) beta(4.6,1.07901,0,1) beta(4.55,0.23947,0,1);
phi(2)=beta(4.55,0.23947,0,1) beta(4.55,0.23947,0,1) beta(4.55,0.23947,0,1);
phi(3)=beta(4.55,0.23947,0,1) beta(4.55,0.23947,0,1) beta(4.55,0.23947,0,1);
p(2)=beta(.5,1.42,0,1) beta(.5,1.42,0,1) beta(.5,1.06,0,1);
p(3)=beta(.5,1.77,0,1) beta(.5,1.77,0,1) beta(.5,1.77,0,1);
p(4)=beta(.5,3.04,0,1) beta(.5,3.04,0,1) beta(.5,3.04,0,1);
R(1)= 439 439 622;
R(2)= 0   0   0;
R(3)= 0   0   0;
PROC STOP;

American Fisheries Society Symposium 7:588–603, 1990

Use of Tag-Recovery Information in Migration and Movement Studies

Carl James Schwarz

Department of Statistics, University of Manitoba
Winnipeg, Manitoba R3T 2N2, Canada

A. Neil Arnason

Department of Computer Science, University of Manitoba

Abstract.—The use of tag-recovery information to describe the movement or migration of fish has traditionally been limited to qualitative descriptions. The major problem in using tag-recovery information to quantify migration or movements has been that unequal tag-recovery rates from different areas confounds migration rates with tag-recovery rates. We consider three models for which we make various assumptions about the fidelity of the fish to the tagging and recovery locations over time. For each model, we demonstrate which biological parameters of interest can be estimated from ordinary tag-recovery data and indicate which assumptions are necessary to resolve the problem of confounding. We then consider what modifications to current experimental designs are required to estimate the fundamental migration and movement rates and such derived rates as relative immigration rates (stock composition) and relative harvest-derivation rates (catch composition).

The use of marking to study movement and migration of fish started on a large scale in the early 1900s. Today, thousands of hatchery-raised fish are tagged routinely with coded wires. These studies provide information on homing behavior of one-time and repeat spawners to the parent area, ground, or bed; migration of larval, juvenile, and young fish from spawning areas to nursery areas, within nursery areas, and from nursery areas; migration of older fish among spawning, feeding, and wintering areas; composition of fisheries; and survival rates in subpopulations.

The following problems inherent in using tag-recovery data to quantify migration patterns are well known.

• The number of tags returned depends upon the product of migration rate (or homing rate) and tag-recovery rate (harvest intensity).

• Multiyear studies must account for year-to-year survival, which is likely to differ among migration destinations.

• Migration routes are seldom straight-line movements between release and recovery areas.

• Tag loss and tagging-induced mortality can be severe.

In this paper, we examine the use of data from multiyear tag-recovery studies to investigate seasonal migrations among geographical areas. For example, North Sea plaice *Pleuronectes platessa* have several spawning areas (Flamborough, Southern Bight, and German Bight), move to overlapping feeding areas in the North Sea, and spawn repeatedly, usually in the same spawning area (Harden Jones 1968). If tagged fish were released in the spawning areas and recoveries occurred in the feeding area, then it would be of interest to investigate not only the migration rates from each spawning area to each of the feeding areas, but the contribution of each spawning area to each of the fisheries.

We also review a comprehensive study (Schwarz 1988) of the use of tag-recovery models to estimate survival and migration rates in geographically stratified populations. The purpose of that study was to develop several classes of biologically plausible models and then determine to what extent migration rates, survival rates, and other parameters of interest could be identified from standard multiyear tag-recovery experiments. When estimation results were unsatisfactory because parameter estimates were confounded, had extremely poor precision, or were sensitive to untestable assumptions, models and estimates were sought that could be applied to more informative data. In general, resolution of estimation problems was seen to require the additional information gained by combining some live recapture (and live returns) of animals with tag-recovery programs.

Schwarz (1988) used two classes of models to investigate seasonal migrations. In the first class, the tagging areas are distinct from the recovery areas, as when tags are applied in spawning areas and recoveries occur in one or more fishing areas. In the second class, tagging and recoveries take place in the same set of discrete areas and migrations occur among these areas, as when year-to-year interchange of fish among spawning areas is of interest. We present only the results from the first class of models because they are somewhat simpler and probably have wider applicability to fish populations. Within this class, we investigate three migration mechanisms that differ in their treatment of fidelity to the tagging and recovery areas. In theory, one super model incorporating all three models could be developed; however, such a model would be complex, would require (usually unobtainable) detailed data, and would be difficult to use. Instead, we consider three biologically plausible models that span the range of fidelity assumptions, do not require complex data structures, and are conceptually simple:

(1) the complete-fidelity model, for which fish are assumed to exhibit complete fidelity to both the tagging and recovery areas;

(2) the partial-fidelity model, for which fish are assumed to be faithful to the tagging area but may choose a new recovery area each season;

(3) the nonfidelity model, for which fish are assumed to be faithful neither to the tagging areas nor to the recovery areas from season to season.

Models of both classes have been examined in a different context by Darroch (1961) and Arnason (1972, 1973), who considered using capture–recapture models to estimate migration rates. However, assumptions of capture–recapture models are that captures or recaptures occur at a point in time and that all animals alive at that time have the same probability of capture. This is not true in studies of fish migration because recoveries occur over an entire fishing season, the capture rates may vary among subpopulations, and there may be behavioral or physical differences among the subpopulations.

Tag-recovery experiments are particularly suited to large-scale experiments in highly mobile or unconfined populations. If tagging and recovery are carried out over many years, batch marks identifying the year and location of tagging are adequate, and each fish is seen at most twice—the second time when tags from dead fish are returned. Thus tag returns from a commercial or sport fishery can be used in place of the experimenter-directed sampling needed to obtain live recaptures. Data collection and statistical treatment are greatly simplified over what is required for capture–recapture experiments. Tag-recovery models, unlike capture–recapture models, do not permit estimation of abundance, only of survival. This is also true when these methods are extended to migration models. The models of this paper potentially permit estimation of survival, migration, and various stock-composition parameters, but not of abundance. However, a further advantage of the tag-recovery models is that we do not need to assume that all destinations to which the fish migrate are sampled or exploited to provide tag returns. The capture–recapture models of Darroch (1961) and Arnason (1972, 1973) allow for permanent emigration to areas outside those sampled, in which case emigration becomes a component of the mortality rates but temporary migration to nonsampled areas can produce serious bias.

Despite the advantages of tag-recovery estimation methods, they have not been used by fisheries biologists until lately (Burnham et al. 1987). Perhaps one reason for this is that many tag-recovery data have been generated by migrating populations and methods of analysis (and their limitations) have never been explicitly stated for multiyear data. The limitations have been implicitly recognized, which is why many tag-recovery studies are restricted to recoveries made in the year after tagging. Even then, it is often assumed that mortality, tag-recovery, and tag-reporting rates of the subpopulations are the same (Melvin et al. 1986). Graphical or descriptive methods (Campbell 1986; Green 1987) also involve the implicit assumption that harvest rates are equal in the recovery areas so that migration rates can be extracted from the gross recovery rates. Crissey (1955) outlined the problems of using multiyear tag-recovery data to study migration routes of birds; similar comments apply to fish populations.

Burnham et al. (1987) reviewed the advantages of using maximum-likelihood estimates (MLEs) of parameters for formally parameterized models of the population dynamics and the sampling process. Briefly, the advantages include increased insight into the meaning and identifiability of parameters, ability to impose restrictions on the model to improve identifiability and precision of the estimates, ability to test the goodnesses-of-fit of a restricted against a more general model, and assurance that the estimates from an adequately fitting model have the highest possible precision (which can be quantified by coefficients of variation or confidence intervals). Schwarz (1988) used the maximum-likelihood approach, and all these advantages apply to the models discussed here.

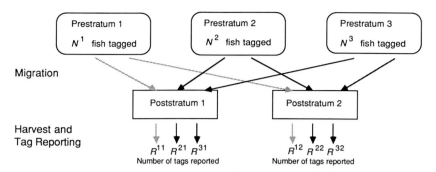

FIGURE 1.—Illustration of the experimental protocol. Fish are tagged in three prestrata and migrate to two poststrata, where they are harvested and some tags are reported.

The dynamics of survival and migration can be incorporated into these models in many ways, and many different sampling strategies can be considered. The models used here are constructed to include the main parameters of interest to fisheries biologists, and the sampling methods are either in common use or involve extended sampling plans that are feasible in some population-monitoring situations. Even if these models are not immediately applicable to fish populations, they are sufficiently general to demonstrate the sorts of problems that are bound to occur in the interpretation of tag-recovery data from migrating populations.

In this paper, we first illustrate the concepts and problems of using tagging experiments to study migration. We then outline the experimental protocol and give the notation used in the remainder of the paper. The complete-, partial-, and nonfidelity models are discussed in turn. For each model, we outline the assumptions, demonstrate what can be estimated with ordinary tag-recovery data, and discuss what assumptions must be made to estimate parameters. Our emphasis is on model concepts and parameter meanings, not on the development of formulae.

Experimental Protocol, Data Organization, and Notation

Protocol and Data Organization

The experimental protocol for all three models is similar and is illustrated in Figure 1. Many fish are tagged in year i. These taggings are stratified into subpopulations on the basis of attributes (e.g., sex or species), geographical location (e.g., spawning area), or any other attribute that will divide the population into groups to be compared (e.g., experimental and control fish). For conve-

nience, we denote these strata determined at the time of release as prestrata. The tag should allow (at a minimum) identification of the year of release and the prestratum of release. (More complex models, involving multiple recaptures of fish, require individually numbered tags to distinguish individual fish.) The number of fish tagged and released in year i in prestratum s is denoted as N_i^s.

After release, the fish are free to migrate. Some of the tagged fish may be harvested and their tags may be returned. Fish that survive the year in the recovery area migrate back to the tagging areas. Restrictions on migration and return destinations are described by the fidelity assumption of the model. For fish that survive, the cycle of migration, recovery or survival, and return is repeated in the next year, combined with another batch of releases in each of the prestrata. The recovery of the tagged fish may be stratified by attributes such as the area of recovery (fishery), type of recovery (commercial or sport fishing), or any other attribute that can only be determined at recovery time. For convenience, we denote strata formed at recovery time as poststrata. The number of fish (out of N_i^s) harvested in poststratum t and whose tag is reported in year j is denoted by R_{ij}^{st}. Usually a tag can be recovered only once because the fish is killed at recovery. In more complex capture–recapture studies, however, the fish may be recaptured several times, possibly in a different poststratum each year. Individually numbered tags allow a record to be kept of the complete capture history for individual fish.

The set of N_i^s and R_{ij}^{st} from a study can be arranged as shown in Table 1. The number of fish whose tag is never recovered is obtained by subtracting the row totals from the number released.

TABLE 1.—Data representation for simple tag-recovery experiments in the complete-, partial-, and nonfidelity models illustrated in the case of $k = 3$ years of releases, $l = 4$ years of recovery, $a = 2$ prestrata, and $b = 2$ poststrata.[a]

Year of release	Number released	Number recovered in poststratum 1				Number recovered in poststratum 2			
		Year 1	Year 2	Year 3	Year 4	Year 1	Year 2	Year 3	Year 4
Releases from prestratum 1									
1	N_1^1	R_{11}^{11}	R_{12}^{11}	R_{13}^{11}	R_{14}^{11}	R_{11}^{12}	R_{12}^{12}	R_{13}^{12}	R_{14}^{12}
2	N_2^1		R_{22}^{11}	R_{23}^{11}	R_{24}^{11}		R_{22}^{12}	R_{23}^{12}	R_{24}^{12}
3	N_3^1			R_{33}^{11}	R_{34}^{11}			R_{33}^{12}	R_{34}^{12}
Releases from prestratum 2									
1	N_1^2	R_{11}^{21}	R_{12}^{21}	R_{13}^{21}	R_{14}^{21}	R_{11}^{22}	R_{12}^{22}	R_{13}^{22}	R_{14}^{22}
2	N_2^2		R_{22}^{21}	R_{23}^{21}	R_{24}^{21}		R_{22}^{22}	R_{23}^{22}	R_{24}^{22}
3	N_3^2			R_{33}^{21}	R_{34}^{21}			R_{33}^{22}	R_{34}^{22}

[a] N_i^s is the number of tagged fish released in year i in prestratum s; R_{ij}^{st} is the number of tagged fish released in prestratum s in year i whose tag is reported from poststratum t in year j.

For convenience, releases in all prestrata occur over the same number of years, and recoveries from all poststrata occur over the same number of years. The modification to allow unequal numbers of years or releases or recovery is straightforward. The time divisions are years but can be any other period.

Notation

Estimates of the parameters are denoted by a circumflex. A dot in a superscript or a subscript implies that the parameter is homogeneous across all values of that index.

The following are fundamental parameters.

k = number of years of releases of tagged fish.
l = number of years of recovery of tagged fish.
a = number of tagging areas (prestrata).
b = number of recovery areas (poststrata).
M_i^{st} = migration rate in year i of prestratum s fish to poststratum t;

$$0 \le M_i^{st} \le 1; \quad 0 \le \sum_{t=1}^b M_i^{st} \le 1.$$

S_i^{st} = poststratum-specific survival rate—the probability that a prestratum s fish alive at the time of tagging in year i will survive in poststratum t until the time of tagging in year $i+1$;

$$0 \le S_i^{st} \le 1.$$

f_i^{st} = poststratum-specific tag-recovery rate— the probability that a prestratum s fish alive at the time of tagging in year i will be harvested and its tag will be reported in poststratum t between the time of tagging in years i and $i+1$;

$$0 \le f_i^{st} \le 1.$$

N_i^{s*} = population size of prestratum s at the time of tagging in year i.
λ_i^{st} = poststratum-specific tag-reporting rate— the conditional probability that a prestratum s tagged fish that is harvested in poststratum t in year i will have its tag spotted, removed, identified, and reported. Hence f_i^{st}/λ_i^{st} is the probability that a fish alive in prestratum s at the time of tagging in year i will be harvested in poststratum t in the next year;

$$0 \le \lambda_i^{st} \le 1.$$

The following are derived parameters.

$S_i^{s\cdot}$ = overall survival rate—the probability that a prestratum s fish alive at the time of tagging in year i will survive to the time of tagging in year $i+1$; summed over all migration routes,

$$S_i^{s\cdot} = \sum M_i^{st} S_i^{st};$$

$$0 \le S_i^{s\cdot} \le 1.$$

I_i^{st} = relative immigration rate of prestratum s fish into poststratum t in year i (discussed and defined later);

TABLE 2.—Parameters used for a numerical example illustrating concepts and problems in the use of tagging experiments to study migration. It is assumed that tagging takes place only in prestrata 1 and 2, and that harvesting occurs only in poststrata 1 and 2; na = not applicable.

Prestratum	Population size	Number tagged	Fraction of fish migrating to poststratum			Harvest rate[a] in poststratum			Tag-reporting rate[b] in poststratum			Survival rate in poststratum		
			1	2	3	1	2	3	1	2	3	1	2	3
1	100,000	1,000	0.40	0.40	0.30	0.10	0.20	0.00	0.50	0.25	na	0.50	0.60	0.80
2	200,000	1,000	0.40	0.30	0.30	0.20	0.30	0.00	0.40	0.40	na	0.50	0.60	0.40
3	50,000	0	0.10	0.60	0.30	0.20	0.10	0.00	na	na	na	0.60	0.60	0.40

[a]The harvest rate is the conditional probability that a fish will be killed and retrieved.
[b]The tag-reporting rate is the conditional probability that a tag from a harvested fish will be spotted and reported to a central reporting site.

$$I_i^{st} = \frac{N_i^{s*} M_i^{st}}{\sum\limits_{s=1}^{a} N_i^{s*} M_i^{st}};$$

$$0 \le I_i^{st} \le 1.$$

D_i^{st} = relative derivation (contribution) of pre-stratum s fish to the harvest in poststratum t in year i (discussed and defined later);

$$D_i^{st} = \frac{N_i^{s*} M_i^{st} f_i^{st}/\lambda_i^{st}}{\sum\limits_{s=1}^{a} N_i^{s*} M_i^{st} f_i^{st}/\lambda_i^{st}};$$

$$0 \le D_i^{st} \le 1.$$

The following are statistics. A dot in a super-script or a subscript implies summation over the corresponding index.

N_i^s = number of fish tagged and released in prestratum s in year i.

R_{ij}^{st} = number of fish tagged and released in year i in prestratum s that are harvested and reported in poststratum t in year j.

$R_{i\cdot}^{st}$ = total number of fish tagged and released in year i in prestratum s that are harvested and reported in poststratum t over all years of the experiment;

$$R_{i\cdot}^{st} = \sum_{j=i}^{l} R_{ij}^{st}.$$

$R_{\cdot j}^{st}$ = total number of fish tagged and released in all years in prestratum s that are harvested and in poststratum t in year j;

$$R_{\cdot j}^{st} = \sum_{i=1}^{\min(j,k)} R_{ij}^{st}.$$

T_j^{st} = total number of fish tagged and released in poststratum s that are harvested and re-ported in poststratum t in year j or later;

$$T_1^{st} = R_{1\cdot}^{st};$$

$$T_j^{st} = T_{j-1}^{st} + R_{j\cdot}^{st} - R_{\cdot j-1}^{st}, \quad j = 2 \ldots k;$$

$$T_j^{st} = T_{j-1}^{st} - R_{\cdot j-1}^{st}, \qquad j = k+1 \ldots l.$$

Z_j^{st} = total number of fish tagged and released in poststratum s that are harvested and for which the tag is reported in poststratum t after year j;

$$Z_j^{st} = T_j^{st} - R_{\cdot j}^{st}, \quad j = 1 \ldots l.$$

Examples of the computation of these summary statistics from recovery data were illustrated by Schwarz et al. (1988).

Concepts, Terminology, and Potential Problems

We use a simple numerical example to illustrate the concepts, terminology, and problems involved in using tag-recovery data for migration studies. Table 2 gives our parameter values for this exam-ple. We assume, for the sake of demonstration, there is no variation about the expected values and we consider only the movements in one year. We can construct tables of the expected number of fish migrating to each poststratum, the ex-pected number of fish harvested in each poststra-tum, the expected number of tags reported, and the expected number of fish surviving to the next year if we assume these rates are representative of the fish in the respective populations (Table 3).

TABLE 3.—Expected number of fish migrating to, harvested in, reported from, or surviving in poststrata.

Prestratum	Expected number of all fish in poststratum			Expected number of tagged fish in poststratum		
	1	2	3	1	2	3
Migration						
1	40,000	40,000	20,000	400	400	200
2	80,000	60,000	60,000	400	300	300
3	5,000	30,000	15,000	0	0	0
Total	125,000	130,000	95,000	800	700	500
Harvest						
1	4,000	8,000	0	40	80	0
2	16,000	18,000	0	80	90	0
3	1,000	3,000	0	0	0	0
Total	21,000	29,000	0	120	170	0
Tags reported						
1	Included at right			20	20	0
2				32	36	0
3				0	0	0
Total				52	56	0
Survival						
1	20,000	24,000	16,000	200	240	160
2	40,000	36,000	24,000	200	180	120
3	3,000	18,000	9,000	0	0	0

Suppose that the population of fish consists of three subpopulations, only two of which have fish that are captured, tagged, and released. The population size of each prestratum is assumed to be known or has been estimated from other studies, and a known number of fish from each prestratum is tagged as shown in Table 2. Suppose also that the fish must migrate to one of the three poststrata in the proportions shown in Table 2. Fish from each prestratum may have a different migration pattern. Note that if the complete set of poststrata is unknown, a new poststratum corresponding to "other" can always be constructed. Because the tagged fish are assumed to have the same migration rates as the rest of the prestratum populations, the distribution of the tagged fish from each prestratum among the poststrata will mirror that of the total populations, as shown in Table 3. For example, the expected numbers of tagged fish from prestratum 2 that migrate to the three poststrata occur in the ratio 400:300:300 (40%:30%:30%), compared to the expected numbers from the total population in prestratum 2 that migrate to the three poststrata in the ratio 80,000:60,000:60,000 (40%:30%:30%). However, because of different tagging intensities, the numbers of tagged fish that migrate into a poststratum from the prestrata will not reflect the relative immigration of fish from the total population. For example, in Table 3, the expected numbers of fish migrating into poststratum 1 from the three pre-strata occur in the ratio 40,000:80,000:5,000 (32%:64%:4%), whereas the expected numbers of tagged fish that migrate into poststratum 1 from the three prestrata occur in the ratio 400:400:0 (50%:50%:0%).

When fish migrate to the poststrata, they may be harvested. Harvesting may not always occur in all poststrata for reasons of regulation or lack of knowledge of migration destinations. Moreover, the harvest rates may vary among poststrata and even among fish from different prestrata within a poststratum because of differences in behavior, in characteristics of fish from different prestrata, or in the timing of migration relative to the harvest. The rates in Table 2 reflect these possibilities.

As presented in Table 3, the expected numbers of fish from the total population in prestratum 2 that are harvested in the two poststrata occur in the ratio 16,000:18,000 (47.1%:52.9%). Since tagged fish are assumed to migrate and get harvested at the same rates as untagged fish, the expected numbers of tagged fish from prestratum 2 that will be harvested in the two poststrata also occur in the same 80:90 ratio (47.1%:52.9%). Note, however, that the ratio is not equal to the relative immigration rates into these poststrata, 0.40:0.30 (57.1%:42.9%), because the harvest rates (20%, 30%) differ among the poststrata. Shifting our point of view to the poststratum, we examine the contribution to the harvest from the prestrata. From Table 3, the expected contribu-

TABLE 4.—Expected number of tagged-fish recoveries in the complete-fidelity model illustrated in the case of $k = 3$ years of releases, $l = 4$ years of recovery, $a = 2$ prestrata, and $b = 2$ poststrata.[a]

Year	Number released	Expected recoveries in poststratum 1			
		Year 1	Year 2	Year 3	Year 4
		Releases from prestratum 1			
1	N_1^1	$N_1^1 M_1^{11} f_1^{11}$	$N_1^1 M_1^{11} S_1^{11} f_2^{11}$	$N_1^1 M_1^{11} S_1^{11} S_2^{11} f_3^{11}$	$N_1^1 M_1^{11} S_1^{11} S_2^{11} S_3^{11} f_4^{11}$
2	N_2^1		$N_2^1 M_2^{11} f_2^{11}$	$N_2^1 M_2^{11} S_2^{11} f_3^{11}$	$N_2^1 M_2^{11} S_2^{11} S_3^{11} f_4^{11}$
3	N_3^1			$N_3^1 M_3^{11} f_3^{11}$	$N_3^1 M_3^{11} S_3^{11} f_4^{11}$
		Releases from prestratum 2			
1	N_1^2	$N_1^2 M_1^{21} f_1^{21}$	$N_1^2 M_1^{21} S_1^{21} f_2^{21}$	$N_1^2 M_1^{21} S_1^{21} S_2^{21} f_3^{21}$	$N_1^2 M_1^{21} S_1^{21} S_2^{21} S_3^{21} f_4^{21}$
2	N_2^2		$N_2^2 M_2^{21} f_2^{21}$	$N_2^2 M_2^{21} S_2^{21} f_3^{21}$	$N_2^2 M_2^{21} S_2^{21} S_3^{21} f_4^{21}$
3	N_3^2			$N_3^2 M_3^{21} f_3^{21}$	$N_3^2 M_3^{21} S_3^{21} f_4^{21}$

[a] N_i^s is the number of tagged fish released in prestratum s in year i; M_i^{st} is the migration rate in year i from prestratum s to poststratum t; S_i^{st} is the survival rate in year i in poststratum t for fish from prestratum s; f_i^{st} is the recovery rate in year i in poststratum t for fish from prestratum s.

tions within poststratum 1 from the three prestrata occur in the ratio of 4,000:16,000:1,000 (19.0%: 76.2%:4.8%), whereas the expected number of tagged fish in the harvest occurs in a different ratio, 40:80:0 (33.3%:66.7%:0.0%), because of different tagging intensities.

Not all tags from the harvest of tagged fish are spotted and reported. The tag-reporting rate probably will differ among poststrata, and it may differ among prestrata within a specific poststratum. For example, rewards may increase the tag-reporting rate overall in a specific poststratum, but fishing quotas may discourage fishermen from reporting tags from certain sizes or species. The rates in Table 1 reflect these possibilities. Table 3 illustrates the expected number of tags reported from the tagged fish in the harvest. No tags can be returned that represent strata where no tagging takes place (prestratum 3). In addition, the tags that are reported reflect neither the migration of the total populations to the poststrata nor the contribution of the harvest within each poststratum. For example, the expected numbers of tags reported in the two poststrata from those tagged in prestratum 2 occur in the ratio 32:36 (47.1%: 52.9%), compared to the 80,000:60,000 (57.1%: 42.9%) ratio of the migrations from prestratum 2. Similarly, the expected number of tags reported in poststratum 1 from the three prestrata occur in the ratio 20:32:0 (38.5%:61.5%:0.0%), compared to the 4,000:16,000:1,000 (19.0%:76.2%:4.8%) ratio from the contribution to the harvest within poststratum 1 from the three prestrata.

Fish that are not harvested or have not died of natural causes leave the poststrata at the end of the season. The poststratum-specific survival rates will likely differ among the poststrata for fish from the same prestratum; conversely, they may differ among fish from different prestrata within a specific poststratum. The expected numbers of fish from the original populations and from the tagged fish surviving within each poststratum is shown in Table 3. For example, 100,000 (40,000 + 36,000 + 24,000) of the fish from prestratum 2 survive the year for an overall survival rate of 50% (100,000/200,000), which is reflected in the proportion of tagged fish that survive (500/1,000 = 50%).

This cycle of migration, harvest, and tag-reporting may be repeated in subsequent years. However, additional complications introduced by fidelity considerations arise. The surviving tagged fish may migrate back to prestrata other than where they were tagged, and may migrate in later years in a different way than in the year of release. The harvest and tag-reporting rates also are likely to differ in the following years. Additional tagged fish may be released in subsequent years, and they may or may not behave as do the surviving tagged fish of previous years. Nevertheless, the consequences of any well-defined behavior can be worked out in much the same way as in the above examples.

Several derived quantities of interest relate to emigration, immigration, and harvest-derivation rates. The total emigration rate from a prestratum to all the harvested poststrata is computed as the

TABLE 4.—Extended.

Year	Number released	Expected recoveries in poststratum 2			
		Year 1	Year 2	Year 3	Year 4
		Releases from prestratum 1			
1	N_1^1	$N_1^1 M_1^{12} f_1^{12}$	$N_1^1 M_1^{12} S_1^{12} f_2^{12}$	$N_1^1 M_1^{12} S_1^{12} S_2^{12} f_3^{12}$	$N_1^1 M_1^{12} S_1^{12} S_2^{12} S_3^{12} f_4^{12}$
2	N_2^1		$N_2^1 M_2^{12} f_2^{12}$	$N_2^1 M_2^{12} S_2^{12} f_3^{12}$	$N_2^1 M_2^{12} S_2^{12} S_3^{12} f_4^{12}$
3	N_3^1			$N_3^1 M_3^{12} f_3^{12}$	$N_3^1 M_3^{12} S_3^{12} f_4^{12}$
		Releases from prestratum 2			
1	N_1^2	$N_1^2 M_1^{22} f_1^{22}$	$N_1^2 M_1^{22} S_1^{22} f_2^{22}$	$N_1^2 M_1^{22} S_1^{22} S_2^{22} f_3^{22}$	$N_1^2 M_1^{22} S_1^{22} S_2^{22} S_3^{22} f_4^{22}$
2	N_2^2		$N_2^2 M_2^{22} f_2^{22}$	$N_2^2 M_2^{22} S_2^{22} f_3^{22}$	$N_2^2 M_2^{22} S_2^{22} S_3^{22} f_4^{22}$
3	N_3^2			$N_3^2 M_3^{22} f_3^{22}$	$N_3^2 M_3^{22} S_3^{22} f_4^{22}$

sum of the emigrations to the poststrata divided by the population size. For example, the total emigration rate from prestratum 2 to the first two poststrata is $(80,000 + 60,000)/200,000 = 70\%$ (Table 3). The relative emigration rates from a prestratum to each of the poststrata where recoveries occurred are the ratios of the number of fish migrating to each poststratum to the total migrating to all the poststrata where recoveries occur. For example, the relative emigration rates from prestratum 2 to the first two poststrata are $80,000/140,000 = 57\%$ and $60,000/140,000 = 43\%$, respectively (Table 3). The tag recoveries reflect neither of these rates due to the confounding effects of the harvest and tag-reporting rates.

The total immigration rate into a poststratum from all of the prestrata where tagging occurred is the ratio of the total immigrants from the prestrata where tagging occurred to the total of the immigrants from all prestrata (including those where no tagging occurred). For example, the total immigration rate into poststratum 1 from the first two prestrata is $(40,000 + 80,000)/125,000 = 96\%$ (Table 3). Note that the number of fish that immigrated from prestrata where no tagging occurred is required to compute the denominator for the total immigration rates. This information normally is not available, so the total immigration rates cannot be computed. To compute the relative immigration rate into a poststratum from a prestratum where tagging occurred, the ratio of the number of immigrants from the prestratum to the total immigrants from all the prestrata where

tagging occurred is used. For example, the relative immigration rates into poststratum 1 from prestrata 1 and 2 are $40,000/120,000 = 33\%$ and $80,000/120,000 = 67\%$, respectively (Table 3). The relative immigration rates only require information from prestrata where tagging occurred. The relative number of tag recoveries from each poststratum do not reflect the relative immigration rates because of the confounding effect of the harvest and tag-reporting rates.

The total harvest-derivation rate within a poststratum from all of the prestrata where tagging occurred is the ratio of the total harvest from the prestrata where tagging occurred to the total harvest from all prestrata, including those where no tagging occurred. For example, the total harvest-derivation rate within poststratum 1 from the first two prestrata is $(4,000 + 16,000)/21,000 = 95\%$ (Table 3). Computation of the total harvest-derivation rates within a poststratum requires information from prestrata where no tagging occurred, so usually it cannot be computed. The relative harvest-derivation rate within a poststratum is the ratio of the harvest from the prestrata to the total harvest from all prestrata where tagging occurred. For example, the relative harvest derivations within poststratum 1 from the first two poststrata are $4,000/20,000 = 20\%$, and $16,000/20,000 = 80\%$, respectively (Table 3). These rates do not require information from prestrata where no tagging occurred. The relative harvest derivation can be computed from the tag recoveries by weighting the recoveries by the prestratum population sizes and the inverse of the tag-reporting rates; that is, (tag recovery rate) × (prestratum population) × (1/tag-reporting rate). For example, the harvest derivations within poststratum 1 from the first two prestrata are,

for prestratum 1,

$$(20/1{,}000) \times 100{,}000 \times (1/0.50) = 4{,}000$$

and, for prestratum 2,

$$(32/1{,}000) \times 200{,}000 \times (1/0.40) = 16{,}000$$

for a total of   20,000.

The relative harvest derivation rates thus are 4,000/20,000 = 20% for prestratum 1 and 16,000/20,000 = 80% for prestratum 2. Harvest derivations may be the most valuable quantities to managers because they indicate the relative importances of prestrata (stocks) to each fishery.

Survival rates cannot be estimated for prestrata in which no fish were tagged. Similarly, poststratum-specific recovery and survival rates cannot be estimated for poststrata from which no tags were reported. The prestratum net survival rate over all poststrata is the fraction of all fish tagged in one year that survive to the time of tagging the next year, regardless of where the fish migrated (even if they migrated to poststrata where no tags were recovered). It is computed by weighting each poststratum-specific survival rate by the respective emigration rate and summing the products. For example, the net survival rate over all poststrata for prestratum 1 is

$$\sum_{\substack{\text{all} \\ \text{poststrata}}} (\text{emigration rates} \times \text{poststratum survival rates}),$$

or

$$0.40(0.50) + 0.40(0.60) + 0.20(0.80) = 0.60.$$

This implies that 60% of the prestratum 1 fish tagged in year 1 will survive to the time of tagging in year 2, as can be seen by the proportion of the total prestratum survivors (60,000/100,000, Table 3). Unless tagged animals are released in more than one year, it is impossible to estimate the survival rates from tag-recovery data (Brownie et al. 1985).

In a typical tagging study, only the number of fish tagged, the number of tags from each prestra-tum that are recovered in each poststratum, and the year of recovery are known. The original prestratum population sizes and the tag-reporting rates must be estimated from other studies. The harvest rates are rarely known. It appears that some confounding among the emigration, harvest, and tag-reporting rates will take place. Consequently, not all of the fundamental or derived parameters of interest can be estimated.

Basic Assumptions

So that inferences drawn from the tagged fish are valid for the population from which they are drawn, we make several basic assumptions that are common to all three of our models.

• Tagging has no effect on subsequent survival. This is a major assumption because it is well known that poor tagging procedures can dramatically increase subsequent mortality. However, proper tagging procedures and post-tagging holding and monitoring of fish to remove obviously weakened individuals can keep tagging-induced mortality to low levels.

• Tagging has no effect on subsequent migration patterns. This is difficult to verify or control. A typical violation might occur if migration routes are determined by undersea currents and tagged fish are released at a depth different from where they were taken. If migration routes depend on olfactory cues, the anesthetic used at tagging might interfere with subsequent detection of those cues.

• There is no tag loss. Modern tagging procedures have reduced this problem to acceptable levels. Any tag loss will bias survival downwards, but if tag loss is independent of migration route, it will have no effect on the estimation of migration routes.

• All tags are correctly recorded with respect to year of release, prestratum of release, year of recovery, and poststratum of recovery.

• All fish act independently of each other. This can be a serious problem if schooling effects are large and fish released together migrate together.

We also make the following assumptions about data from other studies.

• Estimates of the absolute or relative prestratum s population sizes at the time of release in year i ($N_i^{s^*}$) are available along with estimates of their precision. These are needed to estimate the relative immigration into a poststratum and the contribution to the harvest in a poststratum.

• Estimates of the poststratum tag-reporting rates (λ_i^{sr}) are available along with estimates of

TABLE 5.—Numerical illustration of the expected number of tagged-fish recoveries in the complete-fidelity model in the case of $k = 3$ years of releases, $l = 4$ years of recovery, $a = 2$ prestrata, and $b = 2$ poststrata.[a]

Year of release	Number released	Number recovered in poststratum 1				Number recovered in poststratum 2			
		Year 1	Year 2	Year 3	Year 4	Year 1	Year 2	Year 3	Year 4
				Releases from prestratum 1					
1	1,000	20.00	10.00	5.00	2.50	20.00	12.00	7.20	4.32
2	1,000		20.00	10.00	5.00		20.00	12.00	7.20
3	1,000			20.00	10.00			20.00	12.00
				Releases from prestratum 2					
1	1,000	32.00	16.00	8.00	4.00	36.00	21.60	12.96	7.78
2	1,000		32.00	16.00	8.00		36.00	21.60	12.96
3	1,000			32.00	16.00			36.00	21.60

[a] The expected values were computed with the formulae in Table 4 and the parameter values given below. For simplicity, the parameters are assumed to be constant over time. The row-wise similarities are a result of this assumption.

$M_1^{11} = M_2^{11} = M_3^{11} = 0.40 \qquad M_1^{12} = M_2^{12} = M_3^{12} = 0.40 \qquad f_1^{11} = f_2^{11} = f_3^{11} = f_4^{11} = 0.05 \qquad f_1^{12} = f_2^{12} = f_3^{12} = f_4^{12} = 0.05$

$M_1^{21} = M_2^{21} = M_3^{21} = 0.40 \qquad M_1^{22} = M_2^{22} = M_3^{22} = 0.30 \qquad f_1^{21} = f_2^{21} = f_3^{21} = f_4^{21} = 0.08 \qquad f_1^{22} = f_2^{22} = f_3^{22} = f_4^{22} = 0.12$

$S_1^{11} = S_2^{11} = S_3^{11} = 0.50 \qquad S_1^{12} = S_2^{12} = S_3^{12} = 0.60 \qquad N_1^1 = N_2^1 = N_3^1 = 1{,}000$

$S_1^{12} = S_2^{12} = S_3^{12} = 0.50 \qquad S_1^{22} = S_2^{22} = S_3^{22} = 0.60 \qquad N_1^2 = N_2^2 = N_3^2 = 1{,}000$

their precision. In many studies, these can be assumed to equal 1 (i.e., all tags from harvested fish are spotted and reported). In some studies, particularly studies involving returns from commercial fisheries, these must be estimated, often by simple random sampling. They are required for estimates of prestratum contributions to the harvest in a poststratum.

Complete-Fidelity Model

In the complete-fidelity model, we assume that fish are faithful to both the prestratum of tagging and their potential recovery stratum.

Because a fish can only be recovered once, it is sufficient to model only the expected number of recoveries in terms of the migration, survival, and recovery rates, provided the assumptions of the model are valid. According to arguments outlined by Schwarz et al. (1988), the numbers of recoveries in each year after release can be modeled as a multinomial distribution with the expected number of recoveries as shown in Table 4. Recoveries from each release are independent of recoveries from other releases. A numerical example of recovery expectations is shown in Table 5 and discussed below.

The migration rates (M_i^{st}) measure the fraction of the tagged prestratum s fish that migrate to poststratum t in year i. We allow $\Sigma_{t=1}^b M_i^{st} \leq 1$ to accommodate migration to poststrata not part of the study. Because complete fidelity is assumed for recovery areas, it is only necessary to enter the migration rates from the pre- to the poststra-

tum in the year of release. Any fish that survive the year and return to the tagging sites must again, by assumption, migrate back to the same recovery area in all future years. The migration rates are allowed to be prestratum- and calendar year-specific. For example, in Table 5, $M_1^{21} = 0.40$ implies that 40% of the prestratum 1 fish alive at the time of tagging in year 1 will migrate to poststratum 2.

The poststratum survival rate (S_i^{st}) measures the fraction of the prestratum s tagged fish that survive the migration to poststratum t and return to the prestratum of release between the time of tagging in year i and the time of tagging in year $i+1$. Because of the fidelity assumption, any fish recovered in poststratum t in any year after release must have been subject to this poststratum survival rate in all years between release and recovery. We allow the survival rate to be specific for calendar year, prestratum, and poststratum because the survival rate will likely vary among poststrata, among prestrata (if there are different species or tagging locations), and by calendar year (environmental factors and fishing effort). We do assume that the survival rate in a calendar year is the same for all fish of the same prestratum that migrate to the same poststratum, regardless of time of release. This is unlikely to be true if the survival rates are highly age dependent and releases are primarily of one age-group (e.g., only 2-year-old fish are tagged in any release). For example, in Table 5, $S_1^{21} = 0.50$ implies that 50% of the prestratum 2 fish alive at the time of tagging

TABLE 6.—Expected number of recoveries in the partial-fidelity model illustrated in the case of $k = 3$ years of releases, $l = 4$ years of recovery, $a = 2$ prestrata, and $b = 2$ poststrata.[a]

Year	Number released	Expected recoveries in poststratum 1			
		Year 1	Year 2	Year 3	Year 4
Releases from prestratum 1					
1	N_1^1	$N_1^1 M_1^{11} f_1^{11}$	$N_1^1 S_1^1 M_2^{11} f_2^{11}$	$N_1^1 S_1^1 S_2^1 M_3^{11} f_3^{11}$	$N_1^1 S_1^1 S_2^1 S_3^1 M_4^{11} f_4^{11}$
2	N_2^1		$N_2^1 M_2^{11} f_2^{11}$	$N_2^1 S_2^1 M_3^{11} f_3^{11}$	$N_2^1 S_2^1 S_3^1 M_4^{11} f_4^{11}$
3	N_3^1			$N_3^1 M_3^{11} f_3^{11}$	$N_3^1 S_3^1 M_4^{11} f_4^{11}$
Releases from prestratum 2					
1	N_1^2	$N_1^2 M_1^{21} f_1^{21}$	$N_1^2 S_1^2 M_2^{21} f_2^{21}$	$N_1^2 S_1^2 S_2^2 M_3^{21} f_3^{21}$	$N_1^2 S_1^2 S_2^2 S_3^2 M_4^{21} f_4^{21}$
2	N_2^2		$N_2^2 M_2^{21} f_2^{21}$	$N_2^2 S_2^2 M_3^{21} f_3^{21}$	$N_2^2 S_2^2 S_3^2 M_4^{21} f_4^{21}$
3	N_3^2			$N_3^2 M_3^{21} f_3^{21}$	$N_3^2 S_3^2 M_4^{21} f_4^{21}$

[a] N_i^s is the number of tagged fish released in prestratum s in year i; M_i^{st} is the migration rate in year i from prestratum s to poststratum t; S_i^s is the survival rate in year i for fish from prestratum s over all poststrata; f_i^{st} is the recovery rate in year i in poststratum t for fish from prestratum s.

in year 1 that migrate to poststratum 1 will survive and return to prestratum 2 by the time of tagging in year 2.

The tag-recovery rate (f_i^{st}) measures the probability that a fish tagged in prestratum s will be harvested in poststratum t in year i, and that its tag will be found, read, and reported. The recovery rate is the product of two components: (1) A component for the probability that a tagged fish will survive to be recovered in poststratum t in year i, denoted as the harvest component; and (2) a component for the conditional probability that a tagged fish that has been harvested will have its tag spotted, read, and reported, denoted as the tag-reporting component (λ_i^{st}).

The tag-recovery rates are allowed to be specific for calendar year, prestratum, and poststratum. The recovery rate is likely a function of the total effort, which may vary by calendar year within each poststratum. Also, it is possible that the prestrata have different tag-recovery rates because of innate prestratum differences (species, sex), the timing of the migration relative to the harvest period, or other factors such as quotas. Note that this differs from capture–recapture studies, in which recoveries occur at a point in time and all fish alive and present in the poststratum at that point are assumed to have the same probability of capture. We assume that the recovery rate in a calendar year is applicable to all releases of the prestratum. Again, this assumption might be suspect if the recovery rates depend highly on age or size. For example, in Table 5, $f_1^{21} = 0.08$ implies that 8% of the fish tagged in

prestratum 2 that migrate to poststratum 1 will be harvested and their tags will be found, read, and reported between the time of tagging in years 1 and 2.

Given the above assumptions, in Table 5 the expected number of tags returned from fish released in year 1 in prestratum 2 and recovered in poststratum 1 in year 1, $E[R_{11}^{21}]$, is computed as:

$$E[R_{11}^{21}] = N_1^2 M_1^{21} f_1^{21}$$

or

$$E[R_{11}^{21}] = 1,000 \times 0.40 \times 0.08 = 32.00.$$

Because the poststratum-specific survival rate in poststratum 1 for fish released in prestratum 2 is 50% (Table 5), only 200 of the 400 tagged fish that migrate to poststratum 1 ($1,000 \times 0.40$) will survive to return to prestratum 2. Under the complete-fidelity assumption, all 200 fish will migrate in year 2 back to poststratum 1. Hence the expected number of tags recovered in poststratum 1 in year 2 from those released in prestratum 2 in year 1, $E[R_{12}^{21}]$, is

$$E[R_{12}^{21}] = N_1^2 M_1^{21} S_1^{21} f_2^{21}$$

or

$$E[R_{12}^{21}] = 1,000 \times 0.40 \times 0.50 \times 0.08 = 16.00.$$

In this model, there are abk migration parameters, abl tag-recovery parameters, and $a(l - 1)b$ poststratum-specific survival parameters, for a total of $ab(21 + k - 1)$ fundamental parameters. As shown by Schwarz (1988), this simple tagging experiment does not provide sufficient informa-

TABLE 6.—Extended.

Year	Number released	Expected recoveries in poststratum 2			
		Year 1	Year 2	Year 3	Year 4
		Releases from prestratum 1			
1	N_1^1	$N_1^1 M_1^{12} f_1^{12}$	$N_1^1 S_1^1 M_2^{12} f_2^{12}$	$N_1^1 S_1^1 S_2^1 M_3^{12} f_3^{12}$	$N_1^1 S_1^1 S_2^1 S_3^1 M_4^{12} f_4^{12}$
2	N_2^1		$N_2^1 M_2^{12} f_2^{12}$	$N_2^1 S_2^1 M_3^{12} f_3^{12}$	$N_2^1 S_2^1 S_3^1 M_4^{12} f_4^{12}$
3	N_3^1			$N_3^1 M_3^{12} f_3^{12}$	$N_3^1 S_3^1 M_4^{12} f_4^{12}$
		Releases from prestratum 2			
1	N_1^2	$N_1^2 M_1^{22} f_1^{22}$	$N_1^2 S_1^2 M_2^{22} f_2^{22}$	$N_1^2 S_1^2 S_2^2 M_3^{22} f_3^{22}$	$N_1^2 S_1^2 S_2^2 S_3^2 M_4^{22} f_4^{22}$
2	N_2^2		$N_2^2 M_2^{22} f_2^{22}$	$N_2^2 S_2^2 M_3^{22} f_3^{22}$	$N_2^2 S_2^2 S_3^2 M_4^{22} f_4^{22}$
3	N_3^2			$N_3^2 M_3^{22} f_3^{22}$	$N_3^2 S_3^2 M_4^{22} f_4^{22}$

tion, on its own, to estimate the fundamental parameters. The functions of the parameters that can be estimated, and their maximum-likelihood estimators (MLEs), were derived by Schwarz (1988); they are

$$\widehat{M_i^{st} f_i^{st}} = \frac{R_{i\cdot}^{st}}{N_i^s} \frac{R_{\cdot i}^{st}}{T_i^{st}},$$

$$i = 1 \ldots k, \; t = 1 \ldots b, \; s = 1 \ldots a;$$

$$\frac{\widehat{M_i^{st}}}{M_{i+1}^{st}} S_i^{st} = \frac{R_{i\cdot}^{st}}{N_i^s} \frac{Z_i^{st}}{T_i^{st}} \frac{N_{i+1}^s}{R_{i+1\cdot}^{st}},$$

$$i = 1 \ldots k - 1, \; t = 1 \ldots b, \; s = 1 \ldots a.$$

The first identifiable function measures the net probability that a prestratum s fish tagged in year i will have its tag reported in poststratum t in the next year. The individual migration and recovery rates are confounded and cannot be individually estimated (they are said to be nonidentifiable). The second identifiable function has no simple interpretation and is a mixture of survival and migration rates that cannot be individually estimated; consequently, the relative immigration rates cannot be estimated. However, the relative harvest derivations (contributions) to the poststratum t harvest can be estimated as:

$$\widehat{D_i^{st}} = \frac{\widehat{N_i^{s*}} \, \widehat{M_i^{st} f_i^{st}} / \widehat{\lambda_i^{st}}}{\sum_{s=1}^{a} \widehat{N_i^{s*}} \, \widehat{M_i^{st} f_i^{st}} / \widehat{\lambda_i^{st}}};$$

$\widehat{N_i^{s*}}$ and $\widehat{\lambda_i^{st}}$ are estimates of the prestratum population sizes and poststratum tag-reporting rates from other studies. Estimates of the variances and

covariances can be obtained by the straightforward application of the delta method (Schwarz 1988).

In this type of study, the relative contribution of the tagging areas to the harvest in a recovery area can be estimated, but the actual migration rates to the recovery areas cannot be estimated. Schwarz (1988) proposed additional assumptions under which the individual migration, tag-recovery, and poststratum survival rates could be estimated. For example, if the tag-recovery rates in year i in poststratum t are independent of the prestratum of release—that is, if $f_i^{st} = f_i^{\cdot t}$—the relative immigration rates can be estimated because the homogeneous (across prestrata) recovery rates will cancel even though they are nonidentifiable. If the tag-recovery rates in year i for prestratum s fish are homogeneous among poststrata—that is, if $f_i^{st} = f_i^{s\cdot}$—the emigration rates still cannot be estimated, but the relative emigration rates can be estimated as $M_i^{st}/\sum_{t=1}^{b} M_i^{st}$, because the homogeneous (across poststrata) recovery rates will cancel even though they are nonidentifiable. Weaker but similar assumptions are needed to test if the migration, survival, or immigration rates are changing over time or among prestrata. Simple tag-recovery data do not permit testing of the assumptions needed for these tests, nor do they permit testing the homogeneity of recovery rates required for identifiability of emigration and immigration rates.

Migration parameters can also be estimated if an estimate of the tag-recovery rate is available. In capture–recapture studies, this is obtained by comparing the number of animals captured before, in, and after year i to the number of animals captured before, not in, and after year i. The only difference between the two groups is that the first group was captured at time i, so the ratio of their numbers is an estimate of the capture probability. In tag-recovery studies, no such comparison can be made because tag recoveries typically occur from dead fish that are never sighted in the future. Schwarz (1988) considered a modification of the experimental design in the complete-fidelity model involving the resighting of live marked animals in the poststrata by independent observers. (His thesis was concerned primarily with tag-recovery models for birds, for which live resightings are easily obtained.) By comparing the numbers of animals seen but not recovered, seen and recovered, and recovered but not seen in year i, one can estimate the tag-recovery rate and, consequently, the migration and survival parameters of interest. In many studies it is difficult to capture a large number of fish, screen them for tags, record the tag numbers, and then release them without harming the fish. Some tagging methods require dissection of the fish to read the tag (e.g., coded wires).

Some of the above information can be acquired by the use of electronic tags, in addition to regular tags, on a subset of the fish. Then a resighting of the live fish would be comparable to detecting the electronic signature of the fish in the poststratum by telemetry. If 100% of the electronically tagged fish are detectable in all poststrata, other tags become superfluous and the electronically tagged fish alone can be used to estimate the parameters. If the detectability is less than 100%, both types of tags are required.

If we are willing to assume that the tag-recovery rate is directly proportional to recovery effort (with the same constant of proportionality in all the poststrata and for all prestrata), then the effort can serve as a surrogate for the tag-recovery rate. This assumption would be tenable only if comparable fishing equipment were used in all poststrata, all prestrata were equally vulnerable (i.e., stocks were randomly mixed in space and time), and if the tag-reporting rates were comparable. In this case, the migration rates still would not be estimable (because of an unknown constant of proportionality), but the relative migration rates could be estimated.

Partial-Fidelity Model

In the partial-fidelity model, we assume that fish are faithful only to the prestratum of release, and that when they return to their prestratum of release, they choose migration routes independently of previous migration choices. This model is also applicable if the prestrata are determined by unchanging attributes of the fish (e.g., species), in which case the fish are clearly faithful to their prestratum.

As in the complete-fidelity model, the numbers of recoveries in the poststrata can be modeled as a multinomial distribution with the expected number of recoveries, as shown in Table 6 and numerically exemplified in Table 7.

As before, the migration rates (M_i^{st}) measure what fraction of the tagged prestratum s fish migrate to poststratum t in year i. Unlike for the complete-fidelity model, no fidelity to the migration destination is assumed. This implies that a migration parameter must appear in every calendar year. We assume that the migration rates in year i are the same for all fish within a prestratum, regardless of when the fish are released. This assumption would be suspect if migration rates were age-dependent and if releases were of certain age-classes only. For example, in Table 7, $M_i^{21} = 0.40$ implies that 40% of the prestratum 2 fish tagged in year 1 will migrate to poststratum 1. This rate applies to all fish alive in the prestratum, regardless of any prior migration destinations.

The overall survival rate ($S_i^{s\cdot}$) measures the fraction of the prestratum s tagged fish that survive from the time of tagging in year i to the time of tagging in year $i+1$, regardless of migration destination. Because the fish are not assumed to be faithful to the poststrata, a recovery in a poststratum gives no information about the migration route taken in previous years. Indeed, the fish may have migrated to poststrata outside the study area. (In the complete-fidelity model, this still may happen, but because of the fidelity assumption, no recoveries will ever be received from those other poststrata.) Consequently, we cannot enter a poststratum-specific survival rate (S_i^{st}) into the model; a recovery in year i only indicates that the fish survived from year to year in all years between release and recovery. The overall survival rate is a convolution of the migration rate to all poststrata (including those outside the study) and the poststratum-specific survival rates of these poststrata. In some studies, all possible migration destinations are believed to be included in the study; then, we can replace $S_i^{s\cdot}$ by the convolution to all the poststrata. We allow the overall survival rate to be calendar-year and prestratum specific because it is unlikely that migration rates to the poststrata and survival rates in

TABLE 7.—Numerical illustration of the expected number of recoveries in the partial-fidelity model in the case of $k = 3$ years of releases, $l = 4$ years of recovery, $a = 2$ prestrata, and $b = 2$ poststrata.[a]

Year of release	Number released	Number recovered in poststratum 1				Number recovered in poststratum 2			
		Year 1	Year 2	Year 3	Year 4	Year 1	Year 2	Year 3	Year 4
		Releases from prestratum 1							
1	1,000	20.00	12.00	7.20	4.32	20.00	12.00	7.20	4.32
2	1,000		20.00	12.00	7.20		20.00	12.00	7.20
3	1,000			20.00	12.00			20.00	12.00
		Releases from prestratum 2							
1	1,000	32.00	16.00	8.00	4.00	36.00	18.00	9.00	4.50
2	1,000		32.00	16.00	8.00		36.00	18.00	9.00
3	1,000			32.00	16.00			36.00	18.00

[a] The expected values were computed with the formulae in Table 6 and the parameter values given below. For simplicity, the parameters are assumed to be constant over time. The row-wise similarities are a result of this assumption.

$$M_1^{11} = M_2^{11} = M_3^{11} = M_4^{11} = 0.40 \qquad M_1^{12} = M_2^{12} = M_3^{12} = M_4^{12} = 0.40 \qquad f_1^{11} = f_2^{11} = f_3^{11} = f_4^{11} = 0.05$$

$$M_1^{21} = M_2^{21} = M_3^{21} = M_4^{21} = 0.40 \qquad M_1^{22} = M_2^{22} = M_3^{22} = M_4^{22} = 0.30 \qquad f_1^{21} = f_2^{21} = f_3^{21} = f_4^{21} = 0.08$$

$$f_1^{12} = f_2^{12} = f_3^{12} = f_4^{12} = 0.05 \qquad S_1^{1\cdot} = S_2^{1\cdot} = S_3^{1\cdot} = 0.60 \qquad N_1^1 = N_2^1 = N_3^1 = 1,000$$

$$f_1^{22} = f_2^{22} = f_3^{22} = f_4^{22} = 0.12 \qquad S_1^{2\cdot} = S_2^{2\cdot} = S_3^{2\cdot} = 0.50 \qquad N_1^2 = N_2^2 = N_3^2 = 1,000$$

the poststrata will be constant from year to year. Again, additional complications arise if survival is age dependent and releases are a single age-class. For example, in Table 7, $S_1^{2\cdot} = 0.50$ implies that of the prestratum 2 fish alive at the time of tagging in year 1, 50% will survive to return to prestratum 2 by the time of tagging in year 2.

The tag-recovery rate has the same interpretation as in the complete-fidelity model. The expected number of tags returned from fish released in prestratum 2 in year 1 and recovered in poststratum 1 in year 1 is

$$E[R_{11}^{21}] = N_1^2 M_1^{21} f_1^{21}$$

or

$$E[R_{11}^{21}] = 1,000 \times 0.40 \times 0.08 = 32.00.$$

The expected number of tags returned from prestratum 2 fish released in year 1 and recovered in poststratum 1 in year 2 is

$$E[R_{12}^{21}] = N_1^2 S_1^{2\cdot} M_2^{21} f_2^{21}$$

or

$$E[R_{12}^{21}] = 1,000 \times 0.50 \times 0.40 \times 0.08 = 16.00.$$

There are abl migration parameters, abl tag-recovery parameters, and $a(k - 1)$ overall survival parameters for a total of $a(2bl + k - 1)$ parameters. Surprisingly, the partial-fidelity model, with fewer assumptions about migration, allows estimates of the overall survival rate; however, the migration rates still cannot be estimated

and remain confounded with the tag-recovery rates. Schwarz (1988) showed that the function of the parameters that can be estimated and their MLEs are

$$\widehat{M_i^{st} f_i^{st}} = \frac{R_{i\cdot}^{s\cdot} \cdot R_{\cdot\cdot i}^{st}}{N_i^s \, T_i^{s\cdot}},$$

$$i = 1 \ldots k, \, t = 1 \ldots b, \, s = 1 \ldots a;$$

$$\widehat{S_i^{s\cdot}} = \frac{R_{i\cdot}^{s\cdot} \, Z_i^{s\cdot} \, N_{i+1}^s}{N_i^s \, T_i^{s\cdot} \, R_{i+1\cdot}^{s\cdot}},$$

$$i = 1 \ldots k - 1, \, s = 1 \ldots a.$$

As with the complete-fidelity model, the relative immigration rates cannot be estimated because the migration rates are nonidentifiable. The relative harvest-derivation rates can again be estimated as

$$\widehat{D_i^{st}} = \frac{\widehat{N_i^{s*}} \cdot \widehat{M_i^{st} f_i^{st}} / \widehat{\lambda_i^{st}}}{\sum_{s=1}^{a} \widehat{N_i^{s*}} \, \widehat{M_i^{st} f_i^{st}} / \widehat{\lambda_i^{st}}}.$$

Estimates of the variances and covariances were given by Schwarz (1988).

Under the assumptions of the partial-fidelity model, this simple tag-recovery study allows the overall survival rates and the relative contribution of a tagging area to the harvest in a poststratum to be estimated. The actual migration rates still cannot be estimated. Schwarz (1988) also examined additional assumptions under which the individual migration and tag-recovery rates could be estimated and tested for homogeneity in time or among pre- and poststrata. The results were similar to those of the complete-fidelity model.

Again, if estimates of the tag-recovery rates are available from live resightings or electronic tags, the migration rates can be estimated. Refer to Schwarz (1988) for more details.

Nonfidelity Model

In the nonfidelity model, no assumptions are made about the fidelity of the fish to the prestratum of release, the poststratum of recovery, or both. Not only may fish migrate to a different poststratum in each year, but they may select a prestratum different from where they were released. The concept of a prestratum is less well defined in the nonfidelity model because the prestratum of a fish at tagging time is known only in the year of release, but it is still convenient to refer to the strata where tagging and releases occur as prestrata, and to the strata where recoveries occur as poststrata.

As with the previous models, the number of recoveries in each year and poststratum from each release can be modeled as a multinomial distribution. However, because the fish may migrate to new tagging areas each year, the individual cell probabilities for these multinomial distributions are computed as a convolution of the migration rates from pre- to poststrata, the survival rates in poststrata, the migration rates from post- to the prestrata in all years between release and recovery, and a convolution of this product with the tag-recovery rates in the year of recovery. As shown by Schwarz (1988), these expression are complex but can be easily expressed in a matrix notation. For this reason, the interested reader is referred to Schwarz (1988).

Two complications arose in model formulation. First, because no fidelity is assumed to any pre- or poststratum, fish may migrate to poststrata outside of the study and may later enter prestrata outside the study. This indicates that the model must contain pseudo-prestrata and pseudo-poststrata corresponding to destinations outside the study. These pseudo-strata require their own migration and tag-recovery parameters. Second, additional migration parameters representing migration back from the poststrata to the prestrata are required, but such an approach requires many parameters whose additional components are likely to be nonidentifiable. For this reason, Schwarz (1988) used a slightly different set of

parameters. They included the usual tag-recovery and migration parameters, plus new parameters representing the net yearly pre- to prestratum movements for surviving fish. These new parameters incorporated the pre- to poststratum migrations, the poststratum survival rates, and the post- to prestratum migration rates. This resulted in a more parsimonious model.

Schwarz (1988) found that the complex structure of the model made it difficult to explicitly state the set of identifiable parameters. As with the complete- and partial-fidelity models, the tag-recovery and pre- to poststratum migration rates are confounded. However, the presence of pseudostrata implies that the set of identifiable functions depends on the number of prestrata where tagging occurred and the number of poststrata where recoveries occurred. In any case, the MLEs of the identifiable functions must be determined numerically. As before, tag-recovery data are inadequate for inference in the nonfidelity model. The migration rates to the poststrata are confounded with the tag-recovery rates. The yearly pre- to prestratum migration rates may be estimated under certain circumstances, depending upon the choice of the number of prestrata in which tagging occurs and the number of poststrata in which recoveries occur.

The confounding of the pre- to poststratum migration and poststratum recovery rates can be resolved, as in previous models, by obtaining additional sightings of live fish in the poststrata. However, these additional sightings do not resolve the problem of estimating the migration parameters from the pseudo-prestrata. These parameters may be estimated in the general case by obtaining additional live sightings in the prestrata at the time of tagging. These sightings can be used to estimate the migration rates among the prestrata by methods analogous to those of Arnason (1972, 1973) and Seber (1982).

The use of additional sightings in both the pre- and poststrata introduces many new parameters for the sighting rates, which makes the model more complex to formulate and difficult to use. Probably it is more cost-effective to employ electronic tags that have nearly 100% detectability in the pre- and poststrata. Note that it is not necessary to continuously track the fish; all that is required is a high probability that a fish present in the pre- or poststratum will be detected. (This is very similar to employing sightings in the pre- or poststrata, except that electronic tags are easier to "sight.") As a result, more detailed movement histories for each fish are obtained, and fewer fish need to be tagged. Some information will be missing for fish that migrate to poststrata or to prestrata where no sampling effort takes place, and the models for such data will be similar to the capture–recapture formulation of Arnason (1972, 1973). If the monitoring rates are close to 100% with only a small percentage of temporary outmigrations, the model can be approximated by a Markov-chain model, in which case inference is particularly straightforward (Basawa and Rao 1980).

Discussion

Although many multiyear tag-recovery data have been generated for fish populations, there are many limitations on their use. Chiefly, most parameters of interest are confounded unless untestable assumptions are made. The problems can be overcome with the addition of resighting or telemetry data, but even then the experiments must be carefully designed and carried out (Brownie et al. 1985). Some considerations follow.

• If derivations are to be estimated, the prestratum population sizes at tagging time and the tag-reporting rates must also be estimated. Resightings in the prestrata (when new fish are captured to be tagged) may be of assistance in estimating the prestratum population sizes and yearly survival rates, but they are of little use in estimating migration rates. For questions of stock composition of large marine fisheries, other methods (based on morphometric or electrophoretic data) may be more cost-effective.

• The tagging operation should select fish representative of the entire population; size-selective captures or tagging of specific age-classes will likely lead to situations of severe bias (Brownie et al. 1985).

• Resightings should occur in the poststrata and should be nearly simultaneous with, or precede, the recovery period.

Our models help, but they still do not address the problems inherent when non-Markovian behavior and age effects are present. If, however, experiments like those described here are feasible, there is a great deal more that can be done. For example, it is now possible to test for age-dependent effects or non-Markovian behavior, to test for homogeneity of parameters across time or among subpopulations, and to determine appropriate sample sizes needed to estimate the parameters with a given precision or to detect a specified

heterogeneity in the parameters with a given power. Schwarz (1988) presented numerical examples, based on simulated data, that demonstrate the degree of precision and power available from these types of experiments. Computer programs also are available for the analysis of simple and extended tag-recovery data.

References

Arnason, A. N. 1972. Parameter estimation from mark–recapture experiments subject to migration and death. Researches on Population Ecology (Kyoto) 12:97–113.

Arnason, A. N. 1973. The estimation of population size, migration rates, and survival in a stratified population. Researches on Population Ecology (Kyoto) 13:1–8.

Basawa, I. V., and L. S. P. Rao. 1980. Statistical inference for stochastic processes. Academic Press, New York.

Brownie, C., D. R. Anderson, K. P. Burnham, and D. S. Robson. 1985. Statistical inference from band-recovery data—a handbook, 2nd edition. U.S. Fish and Wildlife Service Resource Publication 156.

Burnham, K. P., D. R. Anderson, G. C. White, C. Brownie, and K. H. Pollock. 1987. Design and analysis methods for fish survival experiments based on release–recapture. American Fisheries Society Monograph 5.

Campbell, A. 1986. Migratory movements of ovigerous lobsters, *Homarus americanus,* tagged off Grand Manan, eastern Canada. Canadian Journal of Fisheries and Aquatic Sciences 43:2197–2205.

Crissey, W. F. 1955. The use of banding data in determining waterfowl migration and distribution. Journal of Wildlife Management 19:75–84.

Darroch, J. N. 1961. The two-sample capture–recapture census when tagging and sampling are stratified. Biometrika 48:241–260.

Green, P. E. J. 1987. New graphical techniques for analyzing salmon migration in mark–recapture experiments. Canadian Journal of Fisheries and Aquatic Sciences 44:327–340.

Harden Jones, F. R. 1968. Fish migration. Edward Arnold, London.

Melvin, G. D., M. J. Dadswell, and J. D. Martin. 1986. Fidelity of American shad (*Alosa sapidissima*) to its river of previous spawning. Canadian Journal of Fisheries and Aquatic Sciences 43:640–646.

Schwarz, C. J. 1988. Postrelease stratification and migration models in band-recovery and capture–recapture models. Doctoral dissertation. University of Manitoba, Winnipeg.

Schwarz, C. J., K. P. Burnham, and A. N. Arnason. 1988. Postrelease stratification in band-recovery models. Biometrics 44:765–785.

Seber, G. A. F. 1982. The estimation of animal abundance and related parameters, 2nd edition. Macmillan, New York.

American Fisheries Society Symposium 7:604–612, 1990

Determining Movement Patterns in Marine Organisms: Comparison of Methods Tested on Penaeid Shrimp

PETER F. SHERIDAN

National Marine Fisheries Service
4700 Avenue U, Galveston, Texas 77551-5997, USA

REFUGIO G. CASTRO MELENDEZ

Instituto Nacional de la Pesca
Apartado Postal 197
Tampico, Tamaulipas, Mexico C.P. 89240

Abstract.—Spatial and temporal variations in fishing effort are consistently ignored or overlooked as factors that influence patterns of recapture of tagged organisms. Perceived directional movement or migration of populations of recaptured organisms thus may be incorrect. We compared several analytical methods while trying to interpret recaptures of brown shrimp *Penaeus aztecus* and pink shrimp *P. duorarum* marked and released in 1986. Experiments were conducted off southern Texas, USA, and northern Tamaulipas, Mexico. Most recaptures were recorded north and south of release sites (alongshore), rather than east or west (offshore or inshore). Octant analysis (direction only) indicated strong southward movement of brown shrimp and pink shrimp off Tamaulipas, and equally strong northward and southward movements off Texas for both species. Analysis of mean vector angles (direction plus distance) indicated southward movement of both species off Tamaulipas but easterly movement off Texas. Recaptures per unit fishing effort indicated significant northward movement of pink shrimp only off Tamaulipas, which reflected the nonuniform distribution of fishing effort around release sites. We recommend that studies of movements of tagged organisms account for variations in fishing effort.

Mark–recapture studies of aquatic organisms are used to estimate stock range, growth, mortality, and movements. The latter topic is the subject of this article. Analyses of movements occasionally are used to relate animal distribution to an international border (Sheridan et al. 1987), to an artificial border derived from a management strategy (Booth 1979; Gitschlag 1986), or to a series of fisheries that may or may not be harvesting the same stock (Ruello 1975; Winters and Beckett 1978; Moore and McFarlane 1984). Quite often, though, the objectives of mark–recapture studies are poorly defined variations of the phrase "to investigate movements of species A off locale B." This in itself is a major shortcoming of many of the mark–recapture studies reviewed for this article.

Another serious deficiency of tagging reports is a general lack of an experimental design that can address factors that influence recapture patterns. Perhaps this stems from ill-defined objectives, yet these are mensurative experiments (Hurlbert 1984) and should be treated as such. One factor often not addressed is time; many analyses incorporate all recaptures regardless of how much time has elapsed (weeks, months, years) since the release of tagged animals. Environmental conditions change at least seasonally, and certain physical characteristics such as bottom water temperature or direction of current flow may affect directional movement. A second factor that influences recapture patterns and thus evidence for movement or migration is the effort devoted to recapture, whether it be commercial fishing or fishery-independent sampling. Catch and effort are either ignored or presumed to be uniform in time and space, which certainly is not the case for commercial fisheries. Among 29 articles we surveyed, we identified four categories of mark–recapture studies, based on consideration of recapture effort (Table 1): (1) no mention of catch or effort, (2) recognition of nonuniform catch and effort but no use of catch statistics, (3) presentation of limited catch statistics but little or no direct use of such data in interpreting recaptures, and (4) direct use of such catch statistics as recaptures per unit landings or per unit effort. The researchers used five general methods to document or prove directional movement: (1) maps illustrating release and recapture locations, often connected by straight lines or curves (which we term "connect the dots"), (2) reference to number or percentage of total recaptures in compass octants or

TABLE 1.—Selected mark–recapture studies grouped by their use of catch statistics (1–4), and by how they documented movement or migration (C = "connect the dots"; O = number or percent in octants or sectors; V = vector and circular statistics; A = analysis of variance analogue; R = recaptures per unit effort or per unit landings).

(1) No mention of fishery-dependent or -independent catch per effort
    Kroger and Guthrie (1973)—C
    Moores et al. (1975)—C
    Oesterling (1976)—C
    Ruello (1977)—C
    Uzmann et al. (1977)—C
    Glaister (1978)—C
    Winters and Beckett (1978)—C
    Booth (1979)—C
    Davis and Dodrill (1979)—C
    Annala (1981)—C
    Cody and Fuls (1981)—O
    Lyon and Boudreaux (1983)—C
    Munro and Therriault (1983)—C
    Moore and MacFarlane (1984)—C
    Underwood and Chapman (1985)—A

(2) Nonuniform catch per effort recognized but no use of catch statistics
    Jones (1959)—V
    Saila and Flowers (1968)—V
    Gotshall (1978)—O
    Fogarty et al. (1980)—V
    Bennett and Brown (1983)—V
    Campbell and Stasko (1985)—V

(3) Catch statistics presented but little or no use of them
    Ruello (1975)—C
    Phillips (1983)—C
    Somers and Kirkwood (1984)—O

(4) Catch statistics presented and used
    Bayliff and Rothschild (1974)—V,R
    Bayliff (1979)—V,R
    Wheeler and Winters (1984)—R
    Gitschlag (1986)—R
    Sheridan et al. (1987)—R

other areal divisions, (3) use of vectors and vector angles, (4) analysis of variance (ANOVA) analogues, and (5) recaptures per unit landings or per unit effort.

The objective of our study was to compare tests for directional movement of marked animals across an international border. We used three methods on a single data set. Our mark–recapture experiment, conducted in 1986, dealt with brown shrimp *Penaeus aztecus* and pink shrimp *P. duorarum* native to the adjoining states of Texas (USA) and Tamaulipas (Mexico) in the western Gulf of Mexico. The U.S. National Marine Fisheries Service (NMFS) and Mexico's Instituto Nacional de la Pesca (INP) cooperated in this research.

Methods

Collection and tagging of shrimp.—Shrimp were collected by trawl at night off the Texas and Tamaulipas coasts. All collections were made in 16–20-m water depths within 5 km of release sites. Shrimp were held in flowthrough tanks before and after tagging and until they were released.

Shrimp were marked with colored, numbered polyethylene streamer tags as described by Marullo et al. (1976). Shrimp between 80 and 140 mm, total length, were selected because these sizes represented new recruits to the fishery. Tagged shrimp were released at 18-m depths within 12 h of collection from expendable, delayed-release canisters (Emiliani 1971). Each plastic canister was weighted, filled with 50–75 tagged shrimp, sealed with a salt block, and released overboard. The salt block dissolved after being underwater 10 to 15 min, and the canister sprang open, releasing the shrimp on the sea floor.

Ten releases of tagged shrimp were made at eight sites between 24°44′N 97°31′W and 25°57′N 97°04′W off Tamaulipas. These releases were made during 30 May–8 June 1986 from the INP ship *BIP-IX*. Twelve releases were made at six sites between 26°05′N 97°05′W and 26°55′N 97°17′W off Texas. The Texas releases were made during 21–27 June 1986 and 7–11 July 1986 from the National Oceanic and Atmospheric Administration ships *Chapman* and *Oregon II*. The order of release sites was randomized given the following restrictions: the 21 June release site was fixed because of vessel cruising speed, and each Texas site was visited once before visits to any site were repeated (this was not possible off Tamaulipas). Releases were confined to sites within 150 km of the USA–Mexico border (25°57′N), based on shrimp movement speeds that averaged 2.5 km/d during 1978–1980 (our unpublished data) over a maximum 60-d closure of the fishery. Following the 1978–1980 experiments, 90% of all transborder recaptures resulted from releases within 120 km of the border (Sheridan et al. 1987).

Collection of recaptured shrimp and fishing information.—Port agents employed by NMFS and INP interviewed commercial fishermen and processors in U.S. and Mexican ports to collect recaptured shrimp and information on fishing locations, landings, and effort. All recaptures during the period 30 May–31 August 1986 were checked for accuracy of date and location and were identified to species and measured (total length) when possible. Although recaptures were made after 31 August, only recaptures during the 94-d reference period were analyzed to best reflect summer environments. Recaptures returned with the following inconsistencies were omitted from analyses of movement: (1) recaptures were not identified as brown shrimp or pink shrimp, (2) recapture dates were after 31 August 1986, (3) recapture dates were prior to or the same as release dates, (4) incomplete latitude and longitude coordinates were given, (5) no depth information was available, (6) recapture date was inaccurate, (7) sex was not specified, or (8) shrimp were recaptured in trawl tows over distances exceeding 9 km. These restrictions reduced the number of usable recaptures from 5,639 (as of the date of last recapture, 5 December 1986) to 3,032.

Interviews of fishermen by port agents throughout the U.S. waters of the Gulf of Mexico were used to estimate total brown shrimp and pink shrimp fishing effort off Texas during the period 1 June–31 August 1986. These data were collected by 9-m depth zones within squares of 1° latitude and longitude, which is too coarse a scale for detailed examination of shrimp movements. Logbooks were voluntarily kept by captains of 47 Texas shrimp vessels for the duration of the recapture period; the logs contained precise information on starting and stopping points and times, depths, tow durations, and landings. Logbook data were assumed to reflect fishing activities of all vessels off Texas and were used to estimate the total brown shrimp fishing effort (which includes pink shrimp) within grids measuring 10 minutes of latitude by 10 minutes of longitude along the Texas coast.

Port agents in Tamaulipas interviewed all vessels returning to the primary port of Tampico. An unknown amount, presumed to be relatively small, of catch and effort may have been reported in more southerly ports. The interviews compiled catch and effort data by depth and 10-minute lines of latitude between 26 and 20°N (Tamaulipas and Veracruz). These data were then used to calculate effort within grids, as was done off Texas.

Data analysis.—To evaluate directional movement of brown shrimp and pink shrimp, we employed three methods: octant analysis, vector analysis, and recaptures per unit fishing effort. For octant analysis, uniform fishing effort around each release site (in time and in space) and straight-line movement from release site to recapture site are assumed. The 360° compass was divided into 45° octants with midpoints of 0°, 45°, 90°, 135°, 180°, 225°, 270°, and 315° (N, NE, E, SE, S, SW, W, and NW). The compass heading for each recapture was calculated and assigned to one of these octants. The hypothesis that shrimp moved equally into all octants, which would indi-

cate no directional movement, was tested by χ^2 analysis ($P_\alpha = 0.05$; Batschalet 1965).

Vector analysis also requires assumptions of uniform fishing effort and straight-line movement. The following descriptors of net movement of a population of tagged shrimp were calculated according to Jones (1959).

Mean vector angle (degrees from true north):

$$\bar{\theta} = \arctan \frac{\Sigma r \sin \theta}{\Sigma r \cos \theta} \; ;$$

north-south component (km/d, positive = north):

$$V = \frac{\Sigma r \cos \theta}{\Sigma t} \; ;$$

east–west component (km/d, positive = east):

$$V' = \frac{\Sigma r \sin \theta}{\Sigma t} \; ;$$

Rayleigh test (for uniform circular distribution):

$$Z = R^2/n;$$

r = distance traveled from release site;
θ = direction traveled from release site;
t = days before recapture;
$R = [(\Sigma \sin \theta)^2 + (\Sigma \cos \theta)^2]^{1/2}$;
n = number of recaptures.

The Rayleigh test for uniform circular distributions (i.e., no preferred direction) was used to test these data (Saila and Flowers 1968).

We also tested differences in shrimp movement away from release sites by examining patterns in recaptures per unit fishing effort (R/f), which correct for temporal and spatial variations in fishing effort around each release site and integrate the effects of distance and direction traveled. For each release, recaptures per 10^4 h of effort after each release date were calculated north, within, and south of the release grid. "North" was defined as all grids lying between the northern latitude of the release grid and the northern latitude of the grid containing the northernmost recapture. "South" was defined as all grids lying between the southern latitude of the release grid and the southern latitude of the grid containing the southernmost recapture. "Within" was defined as the release grid and all grids directly east and west of it (recaptures in these grids did not show alongshore movement). Two-factor, mixed-model ANOVA with balanced cell sizes was used to test the hypothesis that there

TABLE 2.—Octant analysis of 1986 brown shrimp mark–recapture experiments. Significant differences from expected uniform distributions were tested by χ^2 analysis (** indicates $P_\alpha < 0.001$). Sites are numbered from south to north.

Release area and date	Site	Number recaptured in compass octant								χ^2
		N	NE	E	SE	S	SW	W	NW	
Tamaulipas										
May 30	6			1		7	5			
May 31	4	17	11		8	28	12			
Jun 1	5	11	7	1		21	6			
Jun 2	8	1	6	10	2			1		
Jun 3	1	5	1	2	1	30	2			
Jun 4	2		10	12	8	8				
Jun 5	3	1	9	9	10	29	6			
Jun 6	3		7	7	15	48	3			
Jun 7	5	1	14			3	14			
Jun 8	7		1			3				
Total		36	66	42	44	187	48	1	0	455.74**
Texas										
Jun 21	6				1	4				
Jun 22	2	11	1		1	9		5		
Jun 23	4	1	1		2	5			1	
Jun 24	5	9	5	2						
Jun 25	1	21	16	16	5	1				
Jun 26	3	26	24	24	31	22			2	
Jun 27	4	2	1		1	2				
Jul 7	3	10	5	8	4	3	1			
Jul 8	5		2	1	2	20				
Jul 9	1	67	13	13	11	5		10	6	
Jul 10	2	16	8	4	13	13	5		9	
Jul 11	6	16	24	22	26	72				
Total		179	100	90	97	136	6	15	38	300.61**

were no detectable differences in shrimp recapture patterns for each species off each state as indicated by R/f values. This was a randomized, complete-blocks design for paired comparisons of R/f values as fixed treatments (north or south) and for releases of tagged shrimp (10 releases off Tamaulipas, 12 off Texas) as randomly chosen blocks (Sokal and Rohlf 1969; Underwood 1981).

Results

Octant Analysis

Octant analysis indicated that brown shrimp exhibited strong southward movement off Tamaulipas (8 of 10 releases), as did pink shrimp (7 of 10 releases) (Tables 2, 3). Both brown shrimp and pink shrimp exhibited nearly equal northward and southward movement off Texas (Tables 2, 3). Chi-square analyses all indicated unequal directional movement for pooled data. Octant analysis was useful in pointing out that the distribution of recaptures may not have been unimodal (recaptures after Texas releases), and that some directions may not have been followed by shrimp before they were recaptured (W and NW off Tamaulipas; SW, W, and NW off Texas). Shrimp preferred alongshore movement and showed no trend of returning to shallower waters.

Brown shrimp and pink shrimp released on the same dates tended to move in the same directions before recapture, more so for Tamaulipas releases (7 of 10 with both species heading southward) than for Texas releases (7 of 12). Releases in the same grid location on different dates indicated that different cohorts of both species did not necessarily move in similar directions. Among the two duplicated releases off Tamaulipas (at sites 3 and 5) and six duplicated releases off Texas (at sites 1–6), brown shrimp exhibited similar directional movements in five of the eight cases, whereas pink shrimp moved similarly in only three comparisons.

Vector Analysis

Mean vector angles after 10 Tamaulipas releases indicated preferred southerly movements for both brown shrimp (in 7 cases) and pink shrimp (in 8 cases). After 12 Texas releases, equal north and south movements were indicated for both species (Table 4). Rayleigh tests indicated that not all releases were followed by significant

TABLE 3.—Octant analysis of 1986 pink shrimp mark–recapture experiments. Significant differences from expected uniform distributions were tested by χ^2 analysis (** indicates $P_\alpha < 0.001$). Sites are numbered from south to north.

Release area and date	Site	Number recaptured in compass octant								
		N	NE	E	SE	S	SW	W	NW	χ^2
Tamaulipas										
May 30	6	1	4	1		29	49			
May 31	4	6	4			6	9		1	
Jun 1	5	2				1				
Jun 2	8	1								
Jun 3	1	1	3			10	1			
Jun 4	2		3	1		1				
Jun 5	3		7	15	9	23	2			
Jun 6	3		3	2	5	15	1			
Jun 7	5	4	14			26	33			
Jun 8	7	1	6	6	1	33	3			
Total		16	44	25	15	144	98		1	435.61**
Texas										
Jun 21	6	10	1	2	13	63	12	6	1	
Jun 22	2	94	1	4	10	96	7	31	18	
Jun 23	4	70	32	10	14	89	8	10	102	
Jun 24	5	48	9		1	1			6	
Jun 25	1	94	142	65	24	85			14	
Jun 26	3	9	4	1	4	3				
Jun 27	4	40	12	14	10	38	2	10	62	
Jul 7	3	4	2	1	1	3				
Jul 8	5		5	2					11	
Jul 9	1	73	2	1	3		4		1	
Jul 10	2	24	19	4	1	12		10	10	
Jul 11	6	1	1	1	4	2				
Total		467	230	105	85	392	33	67	225	885.31**

nonuniform directional movement. Over all releases, shrimp released in Tamaulipas waters exhibited significant southward movement, and shrimp released in Texas waters exhibited net eastward movement. These trends were detected by octant analysis.

Brown shrimp and pink shrimp released on the same dates tended more strongly to move in similar directions off Texas (9 of 12 releases) than off Tamaulipas (6 of 10 releases). This is a reversal of the trends noted by octant analysis. Releases in the same grid on different dates again indicated that cohorts do not necessarily move in similar directions. Only five of eight repeated brown shrimp releases and four of eight repeated pink shrimp releases resulted in mean vector angles within 45° of each other. These results are comparable to those from octant analysis.

R/f Analysis

Analysis of recaptures per unit fishing effort indicated northward movement by pink shrimp (8 of 10 releases) but not by brown shrimp (5 of 10 releases) off Tamaulipas, whereas Texas releases were followed by southward movement after 9 of 12 brown shrimp releases and only 6 of 12 pink

shrimp releases (Table 5). These results differed from both octant and vector analyses. Over all releases, however, ANOVA tests indicated no detectable differences in north versus south R/f values for brown shrimp off either state or for pink shrimp off Texas. Only pink shrimp released off Tamaulipas exhibited significant directional movement (northward).

Brown shrimp and pink shrimp released on the same dates tended to move in similar directions, according to R/f analyses (7 of 10 Tamaulipas releases, 9 of 12 Texas releases). Similar results were noted from vector analyses but not from octant analyses. Releases in the same grid on different dates resulted in different cohorts moving in opposite directions after two of eight brown shrimp releases and four of eight pink shrimp releases. All three tests indicated this result.

Comparison of Methods

Octant and vector analyses yielded similar results (southward movement) for shrimp released in Tamaulipas, but R/f analyses indicated no net movement for brown shrimp and northward movement for pink shrimp (Table 6). In Texas, however, each analysis indicated a different result: for both

TABLE 4.—Vector analysis of 1986 brown shrimp and pink shrimp mark–recapture experiments off Tamaulipas and Texas. $\bar{\theta}$ = mean vector angle (degrees) from magnetic north; V, V' = directed movement (km/d) in north–south and east–west components, respectively (positive values are north and east); Z = Rayleigh test statistic for uniform circular distribution of recaptures (* indicates $P_\alpha < 0.05$). Sites are numbered from south to north.

Release area and date	Site	Brown shrimp				Pink shrimp			
		$\bar{\theta}$	V	V'	Z	$\bar{\theta}$	V	V'	Z
Tamaulipas									
May 30	6	200.5	−1.51	−0.57	9.68*	201.7	−1.16	−0.46	58.73*
May 31	4	90.5	−0.01	0.15	5.32*	182.8	−0.79	−0.04	0.72
Jun 1	5	163.0	−0.32	0.10	2.77	18.1	1.42	0.47	0.33
Jun 2	8	64.8	0.69	1.46	12.81*	349.0	3.14	−0.61	1.00
Jun 3	1	174.9	−0.56	0.05	16.00*	167.5	−0.45	0.10	3.19*
Jun 4	2	100.8	−0.08	0.44	18.62*	153.7	−0.43	0.21	1.67
Jun 5	3	168.1	−0.50	0.11	233.34*	153.0	−0.51	0.26	20.69*
Jun 6	3	168.0	−0.49	0.10	37.37*	170.2	−0.61	0.11	11.33*
Jun 7	5	193.3	−1.58	−0.37	4.43*	196.5	−2.96	−0.88	22.66*
Jun 8	7	191.5	−1.14	−0.23	1.23	195.1	−3.17	−0.86	16.56*
Total		167.1	−0.45	0.10	84.21*	191.9	−1.30	−0.27	100.94*
Texas									
Jun 21	6	165.6	−1.02	0.26	4.90*	166.8	−1.38	0.32	45.26*
Jun 22	2	336.6	0.64	−0.28	1.84	329.6	0.49	−0.29	18.57*
Jun 23	4	154.2	−0.74	0.36	2.52	160.1	−0.42	0.15	10.27*
Jun 24	5	38.4	0.30	0.24	12.85*	23.4	1.06	0.46	50.95*
Jun 25	1	29.2	0.86	0.48	30.84*	57.3	0.24	0.38	109.27*
Jun 26	3	66.5	−0.09	0.20	228.66*	350.7	0.77	−0.13	2.49
Jun 27	4	145.5	−0.33	0.21	0.60	164.8	−0.54	0.15	12.13*
Jul 7	3	70.9	0.04	0.10	7.92*	163.7	0.33	0.09	1.30
Jul 8	5	148.9	−0.04	0.01	13.31*	15.6	0.22	0.06	8.38*
Jul 9	1	21.6	0.32	0.13	43.67*	358.1	0.38	−0.01	58.84*
Jul 10	2	357.1	0.55	−0.03	0.44	352.6	0.27	−0.14	13.42*
Jul 11	6	153.6	−1.22	0.60	54.99*	157.2	−1.29	0.54	3.83*
Total		111.8	−0.10	0.26	64.30*	111.7	−0.05	0.14	55.98*

species, northward movement by octant analysis, offshore movement by vector analysis (essentially no net north or south trend), and no net movement by R/f analysis. Yet recaptures after any given release date indicated that brown shrimp and pink shrimp tended to move in the same direction regardless of analytical method. These results suggest that shrimp in Tamaulipas waters were not subjected to the same factors that influence recapture as shrimp in Texas waters.

Discussion

Comparison of the three methods commonly employed to detect and describe animal movements or migrations revealed serious differences in results due to violations of underlying assumptions. Both octant analysis and vector analysis assumed uniform fishing effort in time and space as well as equal likelihood of movement in any direction. Catch and effort, however, are not uniform, either on a small scale around release sites for short periods of time (Somers and Kirkwood 1984; Gitschlag 1986), or along the length of a fishing ground (Sheridan et al. 1987). An examination of the distribution of fishing effort along

the Texas and Tamaulipas coasts during this study period (our unpublished data) indicated that the Tamaulipas effort was only 13% of the Texas effort; further, the Tamaulipas effort was concentrated around estuary passes or river mouths, whereas the Texas effort was spread in diffuse bands paralleling the coast. Fishing effort thus has a critical influence on movement patterns that are estimated from recaptures by fishing fleets. One way to avoid this problem may be to recapture all tagged organisms. Underwood and Chapman (1985) did so to delineate factors influencing gastropod movements in rocky intertidal habitats. Their method requires the presumption that marked organisms do not move far or fast, and thus it has limited application. However, effort is expended even in collecting intertidal gastropods, and there may be differences in effort necessary to locate tagged organisms in variable rocky intertidal areas.

Octant analysis suggested that marked organisms may not be equally likely to move in any direction (because very few recaptures were made inshore of the release sites), and that directional distributions may not be unimodal. The use of chi-square and Rayleigh tests is thus of dubious

TABLE 5.—Analysis of recaptures of brown shrimp and pink shrimp per 10^4 h of fishing effort (R/f) north and south of 1986 release grids. Significant differences between paired north and south R/f values (north = south?) were tested by analysis of variance (ANOVA). Sites are numbered from south to north.

Release area and date	Site	Brown shrimp R/f		Pink shrimp R/f	
		North	South	North	South
Tamaulipas					
May 30	6	0.0	3.4	23.3	11.0
May 31	4	63.2	11.3	24.8	2.0
Jun 1	5	46.0	11.1	5.4	0.7
Jun 2	8	3.7	0.0	0.1	0.0
Jun 3	1	2.7	23.1	1.7	7.2
Jun 4	2	5.9	7.2	2.2	0.5
Jun 5	3	3.2	15.1	11.7	13.8
Jun 6	3	13.7	13.9	7.8	4.9
Jun 7	5	38.0	9.3	48.9	16.0
Jun 8	7	8.1	1.8	48.4	11.2
ANOVA		$P_\alpha = 0.25$		$P_\alpha = 0.05$	
Texas					
Jun 21	6	0.0	1.2	1.0	6.9
Jun 22	2	1.9	6.2	10.8	83.3
Jun 23	4	0.7	1.4	13.4	17.0
Jun 24	5	0.8	0.0	1.9	0.5
Jun 25	1	2.0	6.4	20.8	515.5
Jun 26	3	2.9	13.7	2.3	0.0
Jun 27	4	1.1	0.6	18.7	10.2
Jul 7	3	0.9	2.0	0.0	1.4
Jul 8	5	10.6	3.7	3.7	0.0
Jul 9	1	9.4	16.7	64.3	0.0
Jul 10	2	1.8	48.0	3.5	6.0
Jul 11	6	3.3	15.4	0.8	0.7
ANOVA		$P_\alpha = 0.14$		$P_\alpha = 0.34$	

value. Directional movement of organisms marked and released on a flat, featureless plain could be tested with either statistic. In reality, organisms are faced with shorelines, depth gradients, substrate variations, and other features that physically and biologically prevent equal distribution in all directions. This study was conducted on newly recruited shrimp, which are not

known to return to estuaries (Sheridan et al. 1987), so the presumption of equal likelihood for movement was negated.

The Rayleigh test statistic can detect unimodal distribution of recaptures away from an expected uniform circular distribution (Batschalet 1965). Octant analysis indicated that the distributions of recaptures in this experiment usually were bimodal, which reflected alongshore movement both north and south of release sites. Bimodality can be corrected but only if the modes are separated by 180° (Batschalet 1965), and such was not always the case along the curving Texas–Tamaulipas coastline. Use of the Rayleigh test in this study was thus inappropriate and could have led to wrong conclusions—e.g., that there was one preferred direction when actually there were two, as noted for Texas brown shrimp and pink shrimp. Similar problems in interpretation could result from the use of the mean vector angle.

To our knowledge, only one other article (Sheridan et al. 1987) has employed statistical testing of the octant analysis method. The authors suggested that better methods were available (recaptures per unit landings) and that octant analysis should be restricted to qualitative investigations of directional movement. Use of octant analysis to make a definitive statement on preferred movements is likely to result in error.

The use of mean vector angles and the associated Rayleigh test has lent some statistical credence to analyses based on them. Each author employing them, however, has added a qualifier to the effect that the results of mark–recapture experiments would be affected by nonuniform distribution of fishing activity around the release sites (Jones 1959; Saila and Flowers 1968; Bayliff and Rothschild 1974; Bayliff 1979; Fogarty et al.

TABLE 6.—Comparison of directional movements of marked brown shrimp and pink shrimp after their release as indicated by octant, vector, and R/f (recaptures per 10^4 h of effort) analyses. For octant and vector analyses, north includes compass headings from 292.5° to 067.4°, east includes 067.5° to 112.4°, south includes from 112.5° to 247.4°, and west includes 247.5° to 292.4°.

Release area	Preferred direction	Brown shrimp releases			Pink shrimp releases		
		Octant	Vector	R/f	Octant	Vector	R/f
Tamaulipas	North	1	0	5	3	2	8
	South	8	7	5	7	8	2
	East–west	1	3	0	0	0	0
	Net	South	South	None	South	South	North
Texas	North	6	6	3	9	7	6
	South	4	5	9	2	5	6
	East–west	2	1	0	1	0	0
	Net	North	East	None	North	East	None

1980; Bennett and Brown 1983; Campbell and Stasko 1985). Thus the need to collect, analyze, and use fishery statistics has been recognized for many years without being regularly incorporated into experimental designs. Fogarty et al. (1980) also presented Rayleigh test statistics indicating that recaptured American lobsters *Homarus americanus* released near barriers (coastlines of rivers, bays, and sounds) invariably exhibited directional movement away from the barriers, but that recaptured lobsters released on the open continental shelf where there were no physiological or physical barriers exhibited no significant directional movement. The assumptions necessary to uphold judgments based on vector analysis (uniform fishing activity and equal likelihood of directional movement) are thus lacking in many coastal fisheries.

We recommend adjusting recapture data by effort or landings; we prefer effort over landings because organisms usually are not uniformly distributed in space (one unit of landings does not equal one unit of effort). Wheeler and Winters (1984) used recaptures and recaptures per unit landings to document the return of Atlantic herring *Clupea harengus harengus* to spawn near bays where they were marked and released. Because effort data were not available, Wheeler and Winters assumed that one unit of catch required the same effort in all bay areas. Their results indicated that recaptures alone did not document the capacity of spawning Atlantic herring to return to their home bays. Only recaptures per unit landings indicated a pattern of declining recapture rates with increasing distance from tagging sites. Homing intensity was nearly 90% for spawning herring. Consequently, it was postulated that management of a spawning group could prevent overfishing to extinction. Because of the high degree of homing, repopulation would only occur by straying from adjacent spawning grounds.

Sheridan et al. (1987) also used recaptures per unit landings, in this case to address possible losses of brown shrimp and pink shrimp from U.S. commercial harvest because of movements across the USA–Mexico border in the Gulf of Mexico. They used landings to adjust tag returns because effort was not available for the entire study period. Their analysis indicated that brown shrimp tended to move south after release in waters of both countries and that pink shrimp had a variable response. Their study primarily addressed long-distance, prolonged returns over a wide range of latitudes

and seasons, whereas our work has been more focused and short-term in nature.

Bayliff and Rothschild (1974) and Bayliff (1979) reported migratory patterns of yellowfin tuna *Thunnus albacares* in terms of recaptures weighted by fishing effort in the eastern Pacific Ocean (Mexico to Ecuador). No preference for offshore movement with growth was noted, but a tendency for southerly movement was noted during spring months. However, the tagging results were not tested for directional movement after individual releases, and interpretation of those results could have been confounded by the long recapture periods (up to 1 year after each release) and the large areas of ocean surface addressed (0–25°N, 80–150°W). Gitschlag (1986) employed recaptures per unit effort to assess movements of pink shrimp near the Tortugas Sanctuary, a management area off southwest Florida in which shrimp fishing is prohibited. Pink shrimp recruited to commercial fishing grounds instead of moving into untrawlable or protected waters where they would be lost to the fishery. Our R/f analyses will be used to investigate potential losses to the U.S. shrimp fishery from shrimp movements during and after the closed fishing season off Texas.

In conclusion, recapture data should be adjusted for fishing effort when they are used to determine movement patterns of species comprising commercial or recreational fisheries. The added expense for data collection would be offset by improvements in the quality of the results.

References

Annala, J. H. 1981. Movement of rock lobsters (*Jasus edwardsii*) tagged near Gisborne, New Zealand. New Zealand Journal of Marine and Freshwater Research 15:437–443.

Batschalet, E. 1965. Statistical methods for the analysis of problems in animal orientation and certain biological rhythms. American Institute of Biological Sciences, Washington, D.C.

Bayliff, W. H. 1979. Migrations of yellowfin tuna in the eastern Pacific Ocean as determined from tagging experiments during 1968–1974. Inter-American Tropical Tuna Commission Bulletin 17:447–506.

Bayliff, W. H., and B. J. Rothschild. 1974. Migrations of yellowfin tuna tagged off the southern coast of Mexico in 1960 and 1969. Inter-American Tropical Tuna Commission Bulletin 16:1–64.

Bennett, D. B., and C. G. Brown. 1983. Crab (*Cancer pagurus*) migrations in the English Channel. Journal of the Marine Biological Association of the United Kingdom 63:371–398.

Booth, J. D. 1979. North Cape—a "nursery area" for the packhorse rock lobster, *Jasus verreauxi* (Decapoda: Palinuridae). New Zealand Journal of Marine and Freshwater Research 13:521–528.

Campbell, A., and A. B. Stasko. 1985. Movement of tagged American lobsters, *Homarus americanus,* off southwestern Nova Scotia. Canadian Journal of Fisheries and Aquatic Sciences 42:229–238.

Cody, T. J., and B. E. Fuls. 1981. Mark–recapture studies of penaeid shrimp in Texas, 1978–1980. Texas Parks and Wildlife Department, Management Data Series 27, Austin.

Davis, G. E., and J. W. Dodrill. 1979. Marine parks and sanctuaries for spiny lobster fisheries management. Proceedings of the Gulf and Caribbean Fisheries Institute 32:194–207.

Emiliani, D. A. 1971. Equipment for holding and releasing penaeid shrimp during marking experiments. U.S. Fish and Wildlife Service Fishery Bulletin 69:247–251.

Fogarty, M. J., D. V. D. Borden, and H. J. Russell. 1980. Movements of tagged American lobster, *Homarus americanus,* off Rhode Island. U.S. National Marine Fisheries Service Fishery Bulletin 78:771–780.

Gitschlag, G. R. 1986. Movement of pink shrimp in relation to the Tortugas Sanctuary. North American Journal of Fisheries Management 6:328–338.

Glaister, J. P. 1978. Movement and growth of tagged school prawns, *Metapenaeus macleayi* (Haswell) (Crustacea: Penaeidae), in the Clarence River region of northern New South Wales. Australian Journal of Marine and Freshwater Research 29:645–657.

Gotshall, D. W. 1978. Northern California Dungeness crab, *Cancer magister,* movements as shown by tagging. California Fish and Game 64:234–254.

Hurlbert, S. H. 1984. Pseudoreplication and the design of ecological field experiments. Ecological Monographs 54:187–211.

Jones, R. 1959. A method of analysis of some tagged haddock returns. Journal du Conseil, Conseil International pour l'Exploration de la Mer 25:58–72.

Kroger, R. L., and J. F. Guthrie. 1973. Migrations of tagged juvenile Atlantic menhaden. Transactions of the American Fisheries Society 102:417–422.

Lyon, J. M., and C. J. Boudreaux. 1983. Movement of tagged white shrimp, *Penaeus setiferus,* in the northwestern Gulf of Mexico. Louisiana Department of Wildlife and Fisheries, Technical Bulletin 39, New Orleans.

Marullo, F., D. A. Emiliani, C. W. Caillouet, and S. H. Clark. 1976. A vinyl streamer tag for shrimp (*Penaeus* spp.). Transactions of the American Fisheries Society 105:658–663.

Moore, R., and J. W. MacFarlane. 1984. Migration of the ornate rock lobster, *Panulirus ornatus* (Fabricius), in Papua New Guinea. Australian Journal of Marine and Freshwater Research 35:197–212.

Moores, J. A., G. H. Winters, and L. S. Parsons. 1975. Migrations and biological characteristics of Atlantic mackerel (*Scomber scombrus*) occurring in Newfoundland waters. Journal of the Fisheries Research Board of Canada 32:1347–1357.

Munro, J., and J.-C. Therriault. 1983. Migrations saissonières du homard (*Homarus americanus*) entre la côte et les lagunes des Îles-de-la-Madeleine. Canadian Journal of Fisheries and Aquatic Sciences 40:905–918.

Oesterling, M. J. 1976. Reproduction, growth, and migration of blue crabs along Florida's gulf coast. University of Florida, Sea Grant Publication SUSF-SG-76-003, Gainesville.

Phillips, B. F. 1983. Migrations of pre-adult western rock lobsters, *Panulirus cygnus,* in Western Australia. Marine Biology 76:311–318.

Ruello, N. V. 1975. Geographical distribution, growth and breeding migration of the eastern Australian king prawn *Penaeus plebejus* Hess. Australian Journal of Marine and Freshwater Research 26:343–354.

Ruello, N. V. 1977. Migration and stock studies of the Australian school prawn *Metapenaeus macleayi.* Marine Biology 41:185–190.

Saila, S. B., and J. M. Flowers. 1968. Movements and behaviour of berried female lobsters displaced from offshore areas to Narragansett Bay, Rhode Island. Journal du Conseil Permanent International pour l'Exploration de la Mer 31:342–351.

Sheridan, P. F., F. J. Patella, Jr., N. Baxter, and D. A. Emiliani. 1987. Movements of brown shrimp, *Penaeus aztecus,* and pink shrimp, *P. duorarum,* relative to the U.S.–Mexico border in the western Gulf of Mexico. U.S. National Marine Fisheries Service Marine Fisheries Review 49(1):14–19.

Sokal, R. R., and F. J. Rohlf. 1969. Biometry. Freeman, San Francisco.

Somers, I. F., and G. P. Kirkwood. 1984. Movements of tagged tiger prawns, *Penaeus* spp., in the western Gulf of Carpentaria. Australian Journal of Marine and Freshwater Research 35:713–723.

Underwood, A. J. 1981. Techniques of analysis of variance in experimental marine biology and ecology. Oceanography and Marine Biology: an Annual Review 19:513–605.

Underwood, A. J., and M. G. Chapman. 1985. Multifactorial analyses of directions of movement of animals. Journal of Experimental Marine Biology and Ecology 91:17–43.

Uzmann, J. R., R. A. Cooper, and K. J. Pecci. 1977. Migration and dispersion of tagged American lobsters, *Homarus americanus,* on the southern New England continental shelf. NOAA (National Oceanic and Atmospheric Administration) Technical Report NMFS (National Marine Fisheries Service) SSRF (Special Scientific Report Fisheries) 705.

Wheeler, J. P., and G. H. Winters. 1984. Homing of Atlantic herring (*Clupea harengus harengus*) in Newfoundland waters as indicated by tagging data. Canadian Journal of Fisheries and Aquatic Sciences 41:108–117.

Winters, G. H., and J. S. Beckett. 1978. Migrations, biomass and stock interrelationships of southwest Newfoundland—southern gulf herring from mark–recapture experiments. International Commission for the Northwest Atlantic Fisheries Research Bulletin 13:67–79.

American Fisheries Society Symposium 7:613–622, 1990

Use of Coded Wire Tag Data to Estimate Aggregate Stock Composition of Salmon Catches in Multiple Mixed-Stock Fisheries

LEON D. SHAUL AND JOHN E. CLARK

Alaska Department of Fish and Game
Post Office Box 20, Douglas, Alaska 99824, USA

Abstract.—A method of estimating the total contribution of stock aggregates to multiple mixed-stock fisheries is introduced and a stochastic model to study the bias and variability of the estimates is developed. Monte Carlo methods are used to evaluate changes in variability and bias caused by changes in tagging rate, catch-sampling rate, catch level, stock abundance in the fisheries, and distribution of stocks across fisheries. The overall variability of the Monte Carlo estimates is surprisingly high and depends principally on variation in tag recovery and stock-specific distribution of probability of harvest across the catch strata. The use and limitations of the model are illustrated by the estimation of the contribution of coho salmon *Oncorhynchus kisutch* stocks to a southeast Alaska gill-net fishery. Although the model is not appropriate for all stock-composition problems, careful experimental design and interpretation could make it useful in problems of coastwide allocation and management of salmon stocks.

Estimation of the stock composition of mixed-stock fisheries is essential when the productivity of individual stocks or groups of stocks is monitored, when existing and alternative management strategies are evaluated, and when data are acquired to address questions of equity and allocation among user groups. Nowhere have mixed-stock management problems received more attention than along the Pacific coast of North America, where stocks of salmon *Oncorhynchus* spp. are harvested as they intermingle and migrate through political and management jurisdictions. The Pacific Salmon Treaty between the USA and Canada specifies coastwide conservation and allocation guidelines that affect the harvest of many fisheries. Implicit in these catch restrictions is an understanding of stock composition of the catches.

One means of estimating stock composition of a catch is to measure differences in natural biological characteristics among fish of known stock origin; the catch is subsequently attributed to these stocks, based on proportions of the same characteristics in catch samples. Age composition (Van Alen et al. 1987; Wood et al. 1987), egg diameter (Craig 1985), and parasites (Margolis 1965; Wood et al. 1987) are biological markers of some value in separating groups of stocks. Techniques that have been employed with success include scale pattern analysis (Henry 1961; Cook and Lord 1978; Marshall et al. 1987) and electrophoresis (Utter et al. 1980; Beacham et al. 1985). However, serious limitations exist when the dis-

tributions of biological characteristics cannot be adequately separated among stocks of interest. Cost can also be a major consideration because extensive catch and escapement sampling and sample processing are expensive.

Mark–recapture studies can also provide information on the stock composition of the catch. By capturing and tagging fish in the immediate vicinity of a fishery and subsequently recovering the tags in terminal fisheries or on spawning grounds, researchers often provide qualitative data on stock composition of adults (Vania et al. 1964; Clark et al. 1986). Marking juvenile salmon with coded wire tags (CWTs) is employed coastwide to measure the contribution of both wild and hatchery stocks. The methodology consists of inserting a small binary-coded tag into the nose cartilage of juvenile salmon, removing the adipose fin as a means of identifying tagged fish as adults, examining catches containing returning tagged adults to recover CWTs, and expanding data from recovered CWTs over untagged fish, unsampled catch, lost heads, and lost tags to estimate total contribution of stocks represented by the code to the fishery. These expansions are routinely employed coastwide. Statistical properties of the estimates have been studied by Webb (1985), de Libero (1986), and Clark and Bernard (1987), among others. Although the use of CWTs can provide detailed information on contribution of stocks represented by CWTs, data on productivity and distribution of untagged stocks in catches are generally not available.

We here introduce a new approach to estimating total abundance of designated stock aggregates for a group of defined fishery strata. We have developed a probability density function to describe the underlying random variability of population estimates. We examined a simple three-stock and three-fishery example by Monte Carlo methods to evaluate changes in variability due to changes in tagging rates, catch sampling levels, catch levels, stock abundances in the fisheries, and distribution of stocks across fisheries. We used the results of this evaluation to provide guidelines for assessing feasibility and limitations of our approach to specific applications. We used this methodology to estimate the contribution of coho salmon *Oncorhynchus kisutch* stock aggregates to the gill-net catch in southeast Alaska, and to evaluate the level of certainty associated with our estimates. The estimation procedures are, of course, not restricted to three-stock, three-fishery problems, but can be used in a wide variety of applications, including coastwide assessment of contribution rates.

Methods

Estimation of production unit abundance.— Given that catches in designated fishery strata consist entirely of fish from defined aggregates of production, also known as stocks (our definition of stock conforms to that of Larkin 1972: a group of one or more populations whose contributions are similarly distributed over a set of catch strata), the total contribution of each stock to these catches can be estimated by dividing the catch into its component stocks for each catch stratum and then summing the stock-specific components across all catch strata. Although the stock composition of the catch is unknown, the proportion of a stock harvested in a given catch stratum can be estimated with CWT recovery data. Each proportion is calculated as the estimated number of coded-wire-tagged fish of a given stock harvested in a given catch stratum, divided by the total estimated number of tagged fish of the stock harvested in all catch strata. For stock i and fishery j, the proportion Ω_{ij} is estimated as

$$\hat{\Omega}_{ij} = \frac{\hat{A}_{ij}}{\sum_j \hat{A}_{ij}};$$

$\hat{A}_{ij}$ is the estimated number of CWTs from stock i caught in fishery j, which is generally calculated as the number of CWTs recovered, expanded over unsampled catch and lost tags. If the total

number of fish in stock i ($S_{i.}$) were known, an estimate of stock i fish in fishery j ($\hat{s}_{ij}$) would be the product of $\hat{\Omega}_{ij}$ and $S_{i.}$. The estimated catch ($\hat{C}_j$) would be the sum of $\hat{s}_{ij}$ over all stocks contributing to that catch stratum:

$$\hat{C}_j = \hat{\Omega}_{1j}S_{1.} + \ldots + \hat{\Omega}_{ij}S_{i.} + \ldots + \hat{\Omega}_{t_s,j}S_{t_s}.$$

for each catch stratum j and for all t_s stocks. Therefore, given known catches C_j and estimates $\hat{\Omega}_{ij}$, and under the condition that the number of fishery strata (t_f) is greater than or equal to t_s, $S_{i.}$ can be estimated by solving t_f simultaneous equations when $t_f = t_s$, or by minimizing the sum of squared deviations, $\sum_j (\hat{C}_{.j} - C_{.j})^2$, when $t_f > t_s$.

Stochastic catch and sampling process.—The uncertainty inherent in estimates of $S_{i.}$ is a function of variability in catch of untagged fish and recovery of CWTs in the catch, which in turn depend upon the distribution of untagged fish in the catch, the distribution and number of tagged fish in the catch, and the number of fish from each catch stratum examined for CWTs. Let $A_{i.}$ be the total number of tagged fish representing stock i. Fish are assumed to be distributed in the catch strata (a_{ij} being the number of stock-i tagged fish in catch stratum j) similar to and independent of the untagged fish. If Ω_{ij} is the probability that a tagged or an untagged fish of stock i will be harvested in catch stratum j, the joint probability of catching s_{ij} untagged fish and a_{ij} tagged fish in fishery j is

$$p(s_{ij};a_{ij}|\Omega_{ij};S_{i.};A_{i.}) = \binom{S_{i.}}{s_{ij}} \Omega_{ij}^{s_{ij}} (1 - \Omega_{ij})^{(S_{i.} - s_{ij})}$$
$$\cdot \binom{A_{i.}}{a_{ij}} \Omega_{ij}^{a_{ij}} (1 - \Omega_{ij})^{(A_{i.} - a_{ij})}; \quad (1)$$

$S_{i.}$ and $A_{i.}$ are the sums of s_{ij} and a_{ij}, respectively, over all catch strata. Note that $S_{i.}$ represents only the total abundance of stock i in catches from the designated catch strata and not catches of stock i outside of these strata. The catch, C_j, by definition is composed exclusively of fish from designated stocks.

The total catch is seldom examined for the presence of CWTs. Sampling programs generally provide estimates of a_{ij} from examinations of part of the catch (n_j being the number of fish examined from catch stratum j), and expansions of the tags recovered from stock i (designated as m_{ij}) over the unexamined catch. Because a large proportion of the catch originates from only some of the strata being sampled, we chose a finite sampling distri-

TABLE 1.—Parameter values used in the Monte Carlo evaluation of the variance and bias of stock-size estimates. Constants included the probabilities that fish of stocks 1–3 would be harvested in fisheries 1–3, the stock abundance of untagged fish, the number of tagged fish, and the sampling rates. Other modeled variables were the sampling rate (25, 50, and 75%) and the number of fish tagged (10, 20, 30, 40, 50, 100, 150, 200, 300, 400, 500, 600, 800, 1,000, and 3,000).

| Fishery | Stock 1 | | Stock 2 | | Stock 3 | | Total expected catch |
	Probability	Expected catch	Probability	Expected catch	Probability	Expected catch	
1	0.60	600	0.30	1,200	0.10	500	2,300
2	0.20	200	0.40	1,600	0.10	500	2,300
3	0.20	200	0.30	1,200	0.80	4,000	5,400
Total		1,000		4,000		5,000	10,000

bution to describe the probability of recovering m_{ij}. We ignored the small effect on the estimation procedure caused by lost heads and lost tags. Therefore, given a_{ij} tags in the catch, the probability of recovering m_{ij} CWTs is

$$p(m_{ij}|a_{ij};C_j;n_j) = \frac{\binom{a_{ij}}{m_{ij}}\binom{C_j - a_{ij}}{n_j - m_{ij}}}{\binom{C_j}{n_j}}. \quad (2)$$

The joint probability density for s_{ij}, a_{ij}, and m_{ij} in catch stratum j is the product of equations (1) and (2). The marginal probability of getting s_{ij} and m_{ij} fish is this product summed over all possible values of a_{ij}:

$$p(m_{ij};s_{ij}|\Omega_{ij};A_{i.};S_{i.};C_j;n_j)$$

$$= \binom{S_{i.}}{s_{ij}}\Omega_{ij}^{s_{ij}}(1 - \Omega_{ij})^{(S_{i.} - s_{ij})}$$

$$\cdot \sum_{a_{ij}=m_{ij}}^{A_{i.}} \binom{A_{i.}}{a_{ij}}\Omega_{ij}^{a_{ij}}(1 - \Omega_{ij})^{(A_{i.} - a_{ij})}$$

$$\cdot \frac{\binom{a_{ij}}{m_{ij}}\binom{C_j - a_{ij}}{n_j - m_{ij}}}{\binom{C_j}{n_j}}. \quad (3)$$

Equation (3) is the probability density function that describes the probability of realizing any given set of estimates. The random variables are defined as m_{ij} and s_{ij}. It is the variance of and covariance between m_{ij} and s_{ij} that determine the variability in total contribution estimates. Given equation (3), maximum-likelihood estimates are obtainable numerically if not explicitly. We chose first to look at the precision and bias of such estimates by Monte Carlo methods to acquire

some general understanding of the behavior of these estimates relative to tagging and sampling levels.

Monte Carlo simulations.—We studied variability in and bias of the estimates of $S_{i.}$ using Monte Carlo techniques with a three-fishery, three-stock example. Parameter values used in simulations are given in Table 1. The probability matrix of Ω_{ij}'s and stock abundances, $S_{i.}$'s, remained the same for all simulations. The number of fish tagged, $A_{i.}$, varied between sets of 1,000 to 2,000 simulations but remained unchanged and equal across all stocks within a set of simulations. Sampling rates also varied between sets of simulations but remained unchanged and equal across all fisheries within a set. Therefore, for a given set of simulations, the number of tagged fish representing each stock was set between 10 and 3,000 fish; the sampling rate was set at 25, 50, or 75%; and for each simulation, each of the $S_{i.}$ unmarked and $A_{i.}$ marked fish were randomly assigned a stratum of catch based on the Ω_{ij}'s associated with each stock. These tagged and untagged fish were randomly selected from the catch until the sampling rate was achieved, and the resulting catch, number of fish sampled, and number of recovered tags (C_j, n_j, and m_{ij}, respectively) were used to obtain estimates of the true stock abundances for all three stocks.

In some simulations, a singular matrix of catches and estimated Ω_{ij}'s was obtained. These cases were simply noted, and estimation of $S_{i.}$ by another method was not attempted. At low sampling rates and low numbers of fish tagged (10–50 fish), a relatively large number of simulations resulted in no solution. This necessitated an increase in number of simulations (over 40% for simulations with a 25% sampling rate and tagging level of 10 tags/stock) in the set to obtain relatively accurate estimates of the average value,

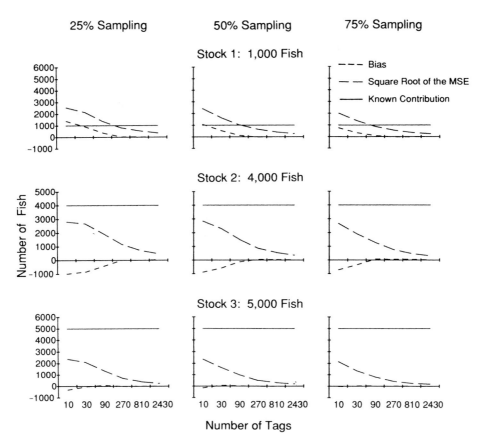

FIGURE 1.—Average bias and square root of the mean squared error (MSE) of estimated contributions by three fish stocks to three fisheries, based on Monte Carlo simulations for stocks of 1,000, 4,000, and 5,000 fish, three sampling rates, and a varying number of individuals tagged.

mean squared error, and bias of $\hat{S}_{i.}$. In these sets, the number of simulations was increased to between 1,500 and 2,000 simulations. Many simulations produced an estimate of $S_{i.}$ that was less than 0. In these instances, the negative $S_{i.}$ was set equal to 0, and the remaining $S_{i.}$'s were estimated by minimization of the sum of the squared deviations of observed catch and catch calculated from the sum of $\hat{\Omega}_{ij}\hat{S}_{i.}$ over all i.

The expected value of each $\hat{S}_{i.}$ was calculated as the average $\hat{S}_{i.}$ over all simulations that were solvable. The bias is the difference between the true value of $S_{i.}$ *(Table 1) and the average value of* $\hat{S}_{i.}$. The mean squared error is the average squared difference between $\hat{S}_{i.}$ and $S_{i.}$. The contribution of variation in m_{ij} and in s_{ij} to the total variability was examined by assuming that catch is constant and letting only the m_{ij}'s vary and by assuming that Ω_{ij} are known parameters and letting the s_{ij}'s vary.

Results of Monte Carlo Study

A relatively high level of variability and significant bias were found to be intrinsic in the estimation procedures, especially at low tagging and sampling levels (Figure 1). Estimates of stock 1 (1,000 individuals) were biased by +1.4 times the true value at the lowest tagging and sampling levels. The variability of these same estimates, as measured by the square root of the mean squared error, was over 2.5 times the population size, resulting in a coefficient of variation of 250%. Although stock 2 was 4 times the size of stock 1, the levels of bias and variability were of approximately the same absolute magnitude (square root of the mean squared error and bias of 2,807 and −1,060 individuals, respectively). Although stock 3 was the largest of the stocks, the bias was much smaller (less than −7% of the stock size), and the

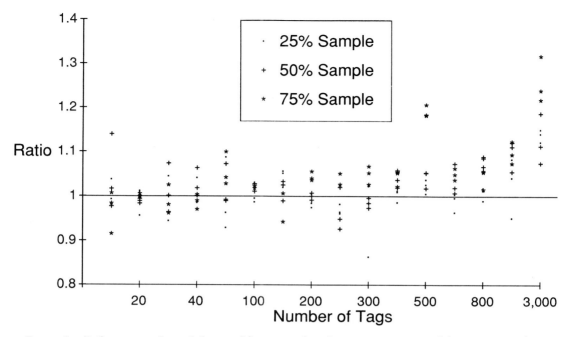

FIGURE 2.—Ratios—uncertain catch (assumed)/known catch—of average square roots of the mean squared error of stock contribution estimates, as functions of the number of tagged fish and sampling intensity. Ratios above 1 correspond to larger errors caused by variability in the catch of untagged fish.

variability was similar in magnitude (2,368 fish) to that of the other two stocks at low tagging and sampling rates.

The bias in the estimates rapidly became insignificant as tagging rates increased. At the 25% sampling rate, the bias approached 0 at 250 tags/ stock for stocks 1 and 2, and at 40 tags/stock for stock 3. Increased sampling rates also reduced the bias but not as markedly as did higher tagging levels. Bias at the 25% sampling level was reduced by an average of 36% and 51% over all stocks at the 10–20-tags/stock level for increases to 50% and 75% sampling levels, respectively.

Variability of the estimates also decreased with increased sampling and tagging rates. The level of tagging required to achieve a square root of the mean squared deviation equal to half the population size decreased from 1,000 to 400 tags/stock for stock 1, from 100 to 30 tags/stock for stock 2, and remained at less than 10 tags/stock for stock 3 as sampling levels increased from 25% to 75%. The variability of estimates at the 25% sampling level decreased by an average of 22% when the sampling level increased to 50%, and by an average of 34% when the sampling increased to 75%.

Variability in estimates also depended upon the underlying distribution of probabilities across fisheries and stocks. Estimates of stock abundance for stocks with a relatively even distribution over several fisheries was more variable than corresponding estimates of stocks harvested predominantly by one fishery. However, because of the constraints placed on population estimates (they must be greater than zero and sum to the total catch), estimates that are highly variable will convey some of this variation to estimates of other stocks. The variability inherent in abundance estimates of large stocks may contribute unacceptable uncertainty to estimates of smaller stocks, even if the smaller stocks are tagged in relatively high numbers.

Simulations in which catch was held constant at its expected value resulted in estimates of variability that were observably smaller only at high tagging and sampling levels, as indicated by consistent increases in the ratio of variable catch to constant catch square roots of the mean squared error for increased tagging levels and sampling levels (Figure 2). In simulations based on a 25% sampling rate, 26 of the 48 comparisons of constant catch to variable catch variability (16 tagging levels and 3 stocks) were characterized by smaller variation in the stock estimates with constant catch, and the average decrease in the variability

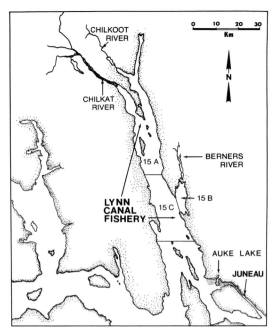

FIGURE 3.—Map of Lynn Canal, southeast Alaska, showing the major spawning rivers and fishing areas (management district 115, the subareas of which are abbreviated here to 15A, 15B, and 15C).

2, compared to 128 fish and 86 fish for populations 1 and 3, respectively). Bias in the estimates was not evident.

Abundance of Lynn Canal Coho Salmon Stocks

We used the drift gill-net catch of coho salmon in Lynn Canal, southeast Alaska to test our method of estimating stock abundances. The fishery operates in those waters of Lynn Canal north of Little Island (Figure 3). This management area (district 115) is divided into three principal subareas (115-A, 115-B, and 115-C), of which part or all may be opened to salmon harvest from 1 to 5 d each week. Although the fleet targets sockeye salmon *Oncorhynchus nerka* and chum salmon *O. keta* for most of the mid-June through late September fishing season, it also harvests many coho salmon from mid-August through late September. District 115 gill-net catches averaged 66,221 coho salmon annually during 1980–1986.

Most of the Lynn Canal coho salmon harvest consists of stocks from the Chilkat, Chilkoot, and Berners rivers; other Lynn Canal and more southern systems (such as Auke Lake) also contributing to the catches (see Shaul et al. 1986, 1987). The distribution of these stocks in southeast Alaska fisheries was first measured by marking juvenile fish with fluorescent pigments in 1972 (Gray et al. 1978). Since 1976, coded wire tagging and recovery programs have been used to study the contribution of these stocks to southeast Alaska commercial and sport fisheries.

In 1981 and 1982, juvenile coho salmon from Chilkat River, Chilkoot River, Berners River, and Auke Lake drainages were coded-wire-tagged (Shaul et al. 1986). The surviving fish returned predominantly as 4-year-olds in 1983 and were harvested in the district-115 gill-net fishery at rates estimated to be from 37% for lower Lynn Canal stocks to 59–74% for upper Canal stocks. Altogether, 163 CWTs from fish tagged in these three areas were recovered, and an additional 7 tags were recovered from wild and hatchery fish

was 0.6%. This compared to 34 of 48 and 39 of 48 comparisons that employed sampling levels of 50% and 75%, respectively, and had more precise estimates and average decreases in the square roots of the mean squared error of 2.5% and 4.4%, respectively. At the higher tagging rates of 500 to 3,000 fish, the increase in precision was greater; decreases in precision ranged from −5% to 24%.

Simulations in which the catch was allowed to vary but the probabilities of fish being harvested in each catch stratum were known produced estimates with small but non-zero variability. The variability was larger for populations more evenly distributed across catch strata (a square root of the mean squared error of 165 fish for population

TABLE 2.—Tag-recovery information used to estimate the numbers of coho salmon from Chilkat, Chilkoot, and southern drainages taken in the 1983 district 115 drift gill-net catch, Alaska.

Time period (fishery)	Number of tags recovered			Estimated number of tags in fishery			Sampling rate
	Chilkat	Chilkoot	Southern	Chilkat	Chilkoot	Southern	
Jun 19–Sep 10	24	4	24	46	6	50	0.480
Sep 11–17	16	8	39	31	15	67	0.544
Sep 18–Oct 10	16	21	11	46	53	29	0.355
Total	56	33	74	123	74	146	

TABLE 3.—Proportional contributions to the 1983 fishery estimated for three Alaskan coho salmon stocks, by time period.

Time period (fishery)	Stock (proportion)			Catch (number of fish)
	Chilkat	Chilkoot	Southern	
June 19–Sep 10	0.3740	0.0810	0.3424	26,379
Sep 11–17	0.2520	0.2027	0.4590	22,826
Sep 18–Oct 10	0.3740	0.7163	0.1986	20,305
Total	1.000	1.000	1.000	69,510

released farther south. Of the total catch of 69,510 coho salmon, 48% were sampled. These data were used to estimate the total abundance in the district-115 catch of stocks represented by the Chilkat and Chilkoot rivers in northern Lynn Canal and the Berners River and Auke Creek near southern Lynn Canal.

Data were stratified temporally into three strata based on changes in management of the fishery and recovery of CWTs. From August 14 through September 4 (no tags were recovered from weeks prior to August 14), during which time 50% of the coho salmon harvest was taken, all of district 115-C and much of district 115-A remained open. The fall chum salmon management program began the week of September 11. During the following week, a portion of district 115-C was closed to protect Berners Bay coho salmon stocks. In the third time stratum (September 18–October 10), relatively large numbers of Chilkat and Chilkoot CWTs were recovered. Data, stratified by time and stock, are presented in Table 2. Estimated stock proportions and total catches by time strata are given in Table 3. The estimation procedures resulted in a negative abundance for Chilkoot River stock. We then set the number of Chilkoot fish in the catch equal to 0 and estimated the Chilkat and southern stock abundances by allocating the total Lynn Canal gill-net catch between these two stocks, thereby minimizing the sum of squared deviations between estimated catch and observed catch. This resulted in a catch composition estimate of 39,615 coho salmon of southern stock origin and 29,895 coho salmon of Chilkat River origin (Table 4).

Sensitivity of the estimates was studied by the Monte Carlo methods previously described. The proportion of each stock that contributed to each of the time strata, the total number of CWTs representing each stock in the fishery, and the estimated stock sizes were set equal to the estimated values of these parameters. Sampling rates were the same as observed sampling rates. Three thousand simulated catch and sampling events were generated, and the resulting distributions of estimated stock-sizes were compared. These distributions are depicted in Figure 4.

Estimates of the contribution of Chilkat and southern stocks to the gill-net fishery were variable and biased. The average estimated value for Chilkat stocks was 35,605 fish with a standard deviation of 24,291 fish. The corresponding value for southern stocks was 33,884 fish with a standard deviation of 24,270 fish. Approximately 10% of the simulations resulted in estimates of Chilkat stocks equal to zero, and 18% of the time the southern stocks were estimated to be zero. The Chilkoot stock was consistently estimated to be zero, only 0.5% of the simulations resulting in estimates greater than zero.

Variability such as that depicted in Figure 4 would preclude any definitive allocation of catch across the three stocks. The Chilkoot stock certainly appears to be the smallest of the three, and its estimate was relatively insensitive to variation in tag recovery. Most of the catch was of fish returning to Chilkat-like drainages or southern areas. However, attribution of this majority to

TABLE 4.—Estimated total harvests (numbers of fish) of three Alaskan stocks of coho salmon in all fisheries, 1983.

Method	Stock			Total
	Chilkat	Chilkoot	Southern	
Solution of equations	106,880	−42,343	4,972	69,510
Minimum sum of squares	29,895	0	39,615	69,510

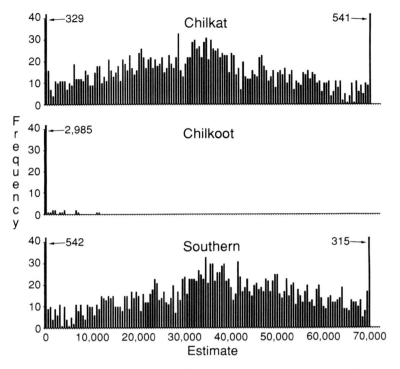

FIGURE 4.—Frequency distributions of catch contribution estimates for three coho salmon stocks in the 1983 Lynn Canal drift gill-net fishery. The distributions were constrained by a lower limit of 0 (no catch) and an upper limit of 69,510 fish, which was the total catch of all stocks.

Chilkat and southern stocks must be qualified because of the large uncertainty associated with these estimates.

Contribution estimates obtained from the model agree by and large with available information on the distribution of coho salmon production. Independent CWT contribution estimates are available only for the Chilkoot River stock, which contributed only an estimated 2,711 fish to the total harvest of 69,510 coho salmon (Shaul et al. 1986). The Chilkat River is the largest river system associated with Lynn Canal and has extensive spawning and rearing habitat. In addition, the drift gill-net fishery effort is relatively high near the Chilkat River, which suggests that Chilkat River stocks are probably harvested at a higher rate than other stocks and undoubtedly contribute a large number of fish to the total Lynn Canal catch. Southern stocks are also major contributors. Berners River, which accounts for a small portion of the southern production, was estimated to contribute 5,819 coho salmon (Shaul et al. 1986).

Discussion

Because of model limitations, our method has relatively limited application. Its high degree of sensitivity to underlying stock distributions indicates that substantial information on stock distributions is essential to evaluate its applicability to any particular problem. In many cases, tagging at the high rates necessary to achieve acceptable resolution may be logistically and economically unfeasible. However, because no immediate alternative solution is available for several stock-composition problems in Pacific salmon fisheries, we will continue to refine and evaluate our method.

Coded-wire-tagging and -recovery programs exist for all five species of Pacific salmon and for steelhead *O. mykiss*. Of these species, the most extensive annual tagging and sampling programs are conducted coastwide for coho salmon and chinook salmon *O. tshawytscha*. Data presently available may be adequate to determine the applicability of the model and, if feasible, develop programs to estimate stock composition. The coho salmon has been one of the most difficult species to assess for stock composition because of its mixed-stock fisheries, widely scattered production, and geographical inconsistencies in biological characteristics used for stock identification (Utter et al. 1980). Even so, the life history of coho salmon is conducive to stock-contribution estimates based on CWT data because of the high survival rates of tagged juveniles and a relatively strong relationship between the harvest distribution of stocks and the geographical location of their natal streams. Recent coded wire tag data are available throughout the range of coho salmon stocks that contribute to ocean fisheries from northern California to southeast Alaska (Johnson 1987). It may be feasible to develop coastwide estimates for the continuum of ocean fisheries, which would complement other techniques that provide better resolution within restricted geographical areas. Our method of estimating stocks may also be applicable to chinook salmon, although tagging of wild stocks has been less successful for that species than for coho salmon, and in some cases chinook salmon from adjacent stream systems have different migratory patterns and harvest distributions in the ocean fisheries.

References

Beacham, T. D., R. E. Withler, and A. P. Gould. 1985. Biochemical genetic stock identification of chum salmon (*Oncorhynchus keta*) in southern British Columbia. Canadian Journal of Fisheries and Aquatic Sciences 42:437–448.

Clark, J. E., and D. R. Bernard. 1987. A compound multivariate binomial-hypergeometric distribution describing coded microwire tag recovery from commercial salmon catches in southeastern Alaska. Alaska Department of Fish and Game, Commercial Fisheries Division, Informational Leaflet 261, Juneau.

Clark, J. E., A. J. McGregor, and F. E. Bergander. 1986. Migratory timing and escapement of Taku River salmon stocks, 1984–1985. *In* Completion Report 85-ABC-00142 to Joint U.S. and Canadian Interception Studies, Juneau, Alaska.

Cook, R. C., and G. E. Lord. 1978. Identification of stocks of Bristol Bay sockeye salmon, *Oncorhynchus nerka*, by evaluating scale patterns with a polynomial discriminant method. U.S. National Marine Fisheries Service Fishery Bulletin 76:415–423.

Craig, P. C. 1985. Identification of sockeye salmon (*Oncorhynchus nerka*) stocks in the Stikine River based on egg size measurements. Canadian Journal of Fisheries and Aquatic Sciences 42:1696–1701.

de Libero, F. E. 1986. A statistical assessment of the use of the coded wire tag for chinook (*Oncorhyn-chus tshawytscha*) and coho (*Oncorhynchus kisutch*) studies. Doctoral dissertation. University of Washington, Seattle.

Gray, P. L., J. F. Florey, J. F. Koerner, and R. A. Marriot. 1978. Coho salmon (*Oncorhynchus kisutch*) fluorescent pigment mark–recovery program for the Taku, Berners, and Chilkat rivers in southeastern Alaska (1972–1974). Alaska Department of Fish and Game, Commercial Fisheries Division, Informational Leaflet 176, Juneau.

Henry, K. A. 1961. Racial identification of Fraser River sockeye salmon by means of scales and its applications to salmon management. International Pacific Salmon Fisheries Commission Bulletin 12.

Johnson, K. E. 1987. Pacific salmonid coded wire tag releases through 1986. Pacific Marine Fisheries Commission, Portland, Oregon.

Larkin, P. A. 1972. The stock concept and management of Pacific salmon. Pages 7–15 *in* R. C. Simon and P. A. Larkin, editors. The stock concept in Pacific salmon. H. R. MacMillan Lectures in Fisheries, University of British Columbia, Vancouver.

Margolis, L. 1965. Parasites as an auxiliary source of information about the biology of Pacific salmons (genus *Oncorhynchus*). Journal of the Fisheries Research Board of Canada 22:1387–1395.

Marshall, S., and nine coauthors. 1987. Application of scale patterns analysis to the management of Alaska's sockeye salmon (*Oncorhynchus nerka*) fisheries. Canadian Special Publication of Fisheries and Aquatic Sciences 96:307–326.

Shaul, L. D., P. L. Gray, and J. F. Koerner. 1986. Coded-wire tagging of wild coho salmon (*Oncorhynchus kisutch*) stocks in southeastern Alaska, 1983–1984. Alaska Department of Fish and Game, Commercial Fisheries Division, Technical Data Report 162, Juneau.

Shaul, L. D., P. L. Gray, and J. F. Koerner. 1987. Coded-wire tagging of wild coho salmon (*Oncorhynchus kisutch*) stocks in southeastern Alaska, 1984–1985. Alaska Department of Fish and Game, Commercial Fisheries Division, Technical Data Report 218, Juneau.

Utter, F. M., D. Campton, S. Grant, G. Milner, J. Seeb, and L. Wishard. 1980. Population structure of indigenous salmonid species of the Pacific Northwest. Pages 285–304 *in* W. J. McNeil and D. C. Himsorth, editors. Salmonid ecosystems of the north Pacific. Oregon State University Press, Corvallis.

Van Alen, B. W., K. A. Pahlke, and M. A. Olsen. 1987. Abundance, age, sex, and size of chinook salmon (*Oncorhynchus tshawytscha* Walbaum) catches and escapements in southeastern Alaska. Alaska Department of Fish and Game, Commercial Fisheries Division, Technical Data Report 215, Juneau.

Vania, J. S., G. J. Paulik, and D. B. Siniff. 1964. Salmon tagging in northern southeastern Alaska. Pages 1–7 *in* Studies to determine optimum escapement of pink and chum salmon in Alaska. Final Summary Report (Contract 14-17-0007-22) to U.S. Fish and Wildlife Service. Juneau, Alaska.

Webb, T. M. 1985. An analysis of some models of variance in the coded wire tagging program. Contract Report to Canadian Department of Fisheries and Oceans, Vancouver, Canada.

Wood, C. C., B. E. Riddell, and D. T. Rutherford. 1987. Alternative juvenile life histories of sockeye salmon (*Oncorhynchus nerka*) and their contribution to production in the Stikine River, northern British Columbia. Canadian Special Publication of Fisheries and Aquatic Sciences 96:12–24.

American Fisheries Society Symposium 7:623–630, 1990
© Copyright by the American Fisheries Society 1990

Use of Abdominal Streamer Tags and Maximum-Likelihood Techniques to Estimate Spotted Seatrout Survival and Growth

ALBERT W. GREEN[1]

Texas Parks and Wildlife Department
4200 Smith School Road, Austin, Texas 78744, USA

LAWRENCE W. MCEACHRON

Texas Parks and Wildlife Department
100 Navigation Circle, Rockport, Texas 78382, USA

GARY C. MATLOCK

Texas Parks and Wildlife Department
4200 Smith School Road, Austin, Texas 78744, USA

ED HEGEN

Texas Parks and Wildlife Department
100 Navigation Circle, Rockport, Texas 78744, USA

Abstract.—Tag release and recapture data permitted estimation of time-specific survival rates and von Bertalanffy growth parameters for spotted seatrout *Cynoscion nebulosus* in Texas bays. Abdominal streamer tags placed in spotted seatrout, which were subsequently recaptured by fishermen, furnished the data used to estimate survival and growth. Spotted seatrout (N = 13,044) were tagged and released during summer (June–September) and winter (January–April) in eight bay systems from June 1981 through April 1984. Fishing effort to obtain fish for tagging was distributed equally among the bay systems so the distribution of tagged fish was proportional to relative fish densities within the bay systems. Recapture data were adjusted to account for incorrect reports and nonreporting of recovered tags and for handling mortality during tagging. We used 880 recoveries in the survival analyses and 622 in the growth analyses. Maximum-likelihood techniques were used to estimate 6-month survival rates, and nonlinear regression was used to estimate values for von Bertalanffy model. Effect of temperature on growth was examined by removing periods (total days) of coldest temperature. Mean 6-month survival rates (±SE) ranged from 38 ± 4 to 55 ± 15%. We obtained best growth estimates with a model that removed the coldest 60 d of the year from total time tagged fish were at large. Von Bertalanffy parameter estimates (±SE) were K = 0.288 ± 0.041 for the growth coefficient and L_∞ = 6.91 ± 41 mm for asymptotic length.

Spotted seatrout *Cynoscion nebulosus* are economically important in the Gulf of Mexico. Recreational fishermen landed an average 3,600,000 kg/year during 1981–1984, and commercial fishermen landed an average 2,800,000 kg/year during 1973–1977 (USNMFS 1973, 1974, 1978, 1980, 1984, 1985a, 1985b). Historically, management of these fisheries has involved gear and time restrictions, area restrictions, and size and bag limits. Most of the specific regulations have been based on limited information about population dynamics for the species (Perret et al. 1980; Williams et al. 1980). Available information has included mortality estimates and von Bertalanffy growth param-

eters (VBGP) for populations in Chesapeake Bay (Mercer 1984), Florida (Iversen and Moffett 1962; Mercer 1984), Alabama (Tatum 1980), and Texas (Baker et al. 1986).

Researchers in previous studies, however, usually ignored or violated underlying assumptions of the models they used to estimate parameters. Tag recoveries were not adjusted for reporting rates and handling mortality, nor were tagged fish distributed throughout the geographical area for which estimates were made. Seasonal variation in growth estimates was not considered, so yield analyses were based on incorrect assumptions of constant growth throughout the year. In this paper, we present practical techniques to satisfy assumptions of mortality and growth models that rely on tagging data. We used data from spotted

---

[1]Present address: Texas Parks and Wildlife Department, Fountain Park Plaza 1, Suite 320, 3000 IH 35, Austin, Texas 78704, USA.

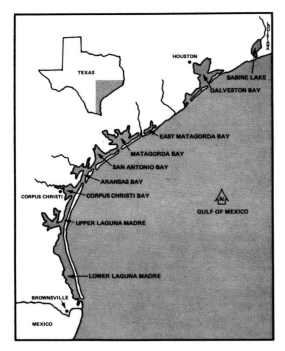

FIGURE 1.—Texas bays considered in this study.

seatrout tagged by the Texas Parks and Wildlife Department (TPWD) to demonstrate the results of these techniques.

Methods

Tagging procedures and data description.—Spotted seatrout were caught by TPWD biologists using rods and reels in eight Texas bay systems (Figure 1). They were tagged and released from June 1981 through April 1984. Plastic internal abdominal tags (25.4 × 6.4 × 0.8 mm) that trailed a 75-mm yellow plastic streamer (2 mm in diameter) were placed in the abdominal cavity of appar-

TABLE 1.—Sample sizes (N) and mean total lengths (TL, mm) for spotted seatrout released and recaptured in Texas bays.

Bay system	Releases		Recaptures	
	N	TL±SE	N	TL±SE
Galveston	1,695	394±26	154	401±70
East Matagorda	1,410	376±72	96	364±58
Matagorda	1,858	346±61	137	346±65
San Antonio	1,610	326±48	66	346±47
Aransas	1,724	307±42	104	320±51
Corpus Christi	1,604	323±59	125	337±51
Upper Laguna Madre	1,532	348±54	87	351±57
Lower Laguna Madre	1,567	346±69	111	338±59
Coastwide	13,000	346±67	622	350±63

ently healthy fish. At least 300 spotted seatrout were tagged in each bay system every summer (mid-June to mid-September) and winter (mid-January to mid-April) over 3 years. Each bay system was divided into as many as eight zones and up to 100 spotted seatrout were tagged in each zone each season. These procedures ensured distribution of tagged fish in proportion to the population within each bay system because spotted seatrout generally do not migrate among bay systems (Moffett 1961; Iverson and Tabb 1962; Green 1986; Marwitz 1986).

From mid-June 1981 through mid-April 1984, 13,044 spotted seatrout (13,000 with measured lengths) were tagged and released (Table 1). There were 880 returns (622 with usable lengths) from recreational and commercial fishermen. All recaptures were pooled for analysis, because pooling results in the least-biased estimates (Doerzbacher et al. 1988).

Data screening.—Only returns with all necessary information (date tagged, date recaptured, length when tagged, length when recaptured) were included in the growth analyses. Standard lengths were converted to total lengths for 77 spotted seatrout according to preestablished formulae (Harrington et al. 1979). All growth data were screened for outliers. Growth per day was computed as the difference between total length at recapture and at tagging, divided by the number of days fish were at large (exclusive of the tagging). Returns ($N = 22$) with either positive or negative growth greater than 2 mm/d were deleted because this was at least three times greater than reported rates for spotted seatrout of comparable sizes. Growth analyses were based on 622 spotted seatrout (Table 1).

Survival estimates.—Six-month spotted seatrout survival was estimated with maximum-likelihood estimators (MLE). A computer program described by Brownie et al. (1978) estimates survival rates from multiple tag releases and recovery data. The program fits each of four theoretical models to the observed data by a goodness-of-fit (χ^2 statistic) test. Each model differs with respect to assumptions about the variability of survival (S) and recovery rates (f). For the most generalized model (fewest assumptions), true value of each parameter is assumed to differ from one recovery period to another, and the first recovery rate for a newly tagged cohort is assumed to differ from recovery rates for previously tagged cohorts in the same recovery period. For the most restrictive model (most assumptions), survival and re-

TABLE 2.—Six-month recovery rate estimates (%±SE) for tagged spotted seatrout in Texas bays, based on the assumption that survival rates and recovery rates are time-specific; ND = not estimated because there were no recoveries.

Bay system group	Time period						Overall mean recovery rate
	1 Nov 1981–30 Apr 1982	1 May–30 Oct 1982	1 Nov 1982–30 Apr 1983	1 May–30 Oct 1983	1 Nov 1983–30 Apr 1984	1 May 1984–30 Oct 1984	
East Matagorda	6.2±1.6	3.1±1.1	3.7±1.5	3.2±1.0	2.6±1.0	1.5±1.6	3.7±0.6
Galveston–lower Laguna Madre	1.8±0.6	9.8±1.3	1.5±0.5	8.1±1.1	1.2±0.4	8.7±2.3	4.5±0.4
Matagorda–Aransas–Corpus Christi	3.3±0.6	6.2±0.8	4.5±0.7	5.4±0.8	2.2±0.5	4.8±1.2	4.3±0.3
San Antonio–upper Laguna Madre	1.5±0.6	3.3±0.7	1.3±0.5	6.0±1.1	1.1±0.5	ND	3.1±0.4

covery rates are assumed to be constant among all recovery periods. All four models require that tagged fish be placed instantaneously into the population at the beginning of each recovery period. Budget and personnel schedules prevented this requirement from being met because tagging took 2–3.5 months. Green et al. (1985) found that when tagging periods took longer than 17% of the recovery period (more than 1 month in a 6-month period), the most generalized model with the least precision for N was selected. To reduce this effect, the beginning date for the recovery period was adjusted to the date immediately following release of the last tagged fish for a specific cohort.

Survival rates estimated with MLE are not affected by the number of tagged fish released during a period, provided enough fish are tagged to assure some recoveries. However, estimated recovery rates and variances are affected. Some tagged fish are caught and returned during the tagging period and some handling mortality occurs. Therefore, we calculated an "effective number released" by subtracting tagged fish returned during tagging and tagged fish that would have died within 7 d of release from the total number released. We estimated deaths during the first 7 d

by multiplying the number of tagged fish released by 17% (Hegen et al. 1984, 1987). We estimated number of tagged fish caught (including those not reported) by dividing the number recovered by the recreational anglers' reporting rate of 0.33 (Green et al. 1983). Both events were assumed to be mutually exclusive and calculated as:

$$E_i = N_i - \left(\frac{R_i}{0.33} + 0.17 \, N_i \right);$$

i = time period;
E = effective number of tagged fish released;
N = total number of tagged fish released;
R = total number of tagged fish recovered.

We used data grouped from bays that had statistically similar release and recovery data to estimate 6-month survival rates. Homogeneity of tag release and recovery data among bay systems was tested (Brownie et al. 1978). Bay system groups, with statistically similar matrices for tag release and recovery (Tables 2, 3) were (1) Galveston–lower Laguna Madre, (2) Matagorda–Aransas–Corpus Christi, (3) San Antonio–upper Laguna Madre, and (4) East Matagorda. Bays were grouped primarily by their patterns of recovery

TABLE 3.—Results of bay-to-bay tests of homogenity (χ^2) among releases and recoveries of tagged spotted seatrout in Texas. An A following a number indicates that tag-recovery data resembled data for the bay system listed in the left column.

Bay system	East Matagorda	Matagorda	San Antonio	Aransas	Corpus Christi	Upper Laguna Madre	Lower Laguna Madre
Galveston	59.6	34.0	38.4	43.2	38.0	22.9	18.7A
East Matagorda		27.4	23.9	25.9	26.4	18.4	31.5
Matagorda			21.4	9.1A	16.3A	10.2A	18.5A
San Antonio				16.3A	36.2	8.4A	17.6A
Aransas					13.2A	12.8A	22.8
Corpus Christi						22.2	22.6
Upper Laguna Madre							2.9A

rates. All bay systems except East Matagorda showed a high recovery rate in summer and a low recovery rate in winter. Galveston–lower Laguna Madre and Matagorda–Aransas–Corpus Christi differed in the proportions of tags returned during winter and summer. In Galveston–lower Laguna Madre, approximately four times more tags were recovered during summer than during winter; that differential was two times in Matagorda–Aransas–Corpus Christi. In San Antonio–upper Laguna Madre two to five times as many tags were recovered during summer than during winter, but—unlike the other bay systems—no tags were returned after a coastwide freeze in December 1983 (McEachron et al. 1984). Because previously published studies did not estimate spotted seatrout survival for a 6-month period, we converted the estimates in our study to annual estimates for comparison purposes. A winter survival rate was multiplied by a subsequent summer rate when survival rates varied among seasons; alternatively, the 6-month estimate was squared when survival rate was constant among seasons.

Recovery periods (6 months) corresponded with seasonal changes in fishing pressure (McEachron et al. 1985), which is high in summer (1 May–30 October) and low in winter (1 November–30 April).

Mortality components.—We computed 6-month instantaneous fishing (F_i) and natural (M_i) mortality rates from estimated survival rates, adjusted tag release numbers, and observed recoveries:

$$F_i = \left\{ \frac{\sum_{j=1}^{i} R_{ij}}{[(E_i + \sum_{j=1}^{i-1} E_j S_j)(1 - S_i)] \, 0.33} \right\} Z_i,$$

and

$$M_i = Z_i - F_i;$$

R = observed tag recoveries;
i = time period;
j = cohort;
E = effective number of tagged fish released;
S = estimated 6-month survival rate;
0.33 = adjustment for unreported tags;
Z = $-\log_e (S_i)$; and
$\sum_{j=1}^{i-1} E_j S_j = 0$ for $i = 1$.

Growth rates.—Parameters of L_∞ and K in the von Bertalanffy growth model were determined

with methods discussed by Doerzbacher et al. (1988), who used Fabens's (1965) equation for tag and recapture data when time t is not known:

$$l_r = l_m + (L_\infty - l_m) [1 - \exp(-Kp)];$$

L_∞ = average asymptotic total length in a population of fish allowed to grow indefinitely in accordance with the model;
K = Brody's growth constant (Ricker 1975);
l_m = total length at marking;
l_r = total length at recapture;
p = time lapse between time at marking (m) and recapture (r).

The parameters K and L_∞ can be estimated by a least-squares procedure. We used the SAS nonlinear procedure (NLIN) with the Marquardt option (SAS 1982) to make estimates and obtain asymptotic confidence intervals. The initial-time parameters t_0 for the usual von Bertalanffy growth model can be obtained only if some known size-at-age data are supplied; therefore, it could not be estimated.

We computed time between tagging and recapture in two ways to examine the effect of temperature on the model (Doerzbacher et al. 1988). First, p was expressed as number of days each fish was free; second, p was expressed as a measure of temperature exposure (degree-days) calculated by summing predicted average temperatures for each day a fish was free. Monthly mean water temperatures (1978–1984), measured coastwide by the TPWD during routine bag-seine sampling (McEachron and Green 1985), were used as the annual water temperature cycle. The model was based on the sine function and was fitted by the NLIN procedure (SAS 1982). The model,

$$T = \langle -8.3 \, \{\sin [0.017 \, (D - 5772)]\} + 1 \rangle + 31.6,$$

predicted an average temperature (T) for each day (D) of the year measured from 1 January 1960. The coldest predicted temperature was 14.9° C on 19 and 20 January. The effect of cold was further examined by constructing intervals of plus and minus 10, 20, 30, . . ., 120 days about the coldest days and computing an adjusted p for each tagged fish for each time interval in which fish spent some portion of its time at large. An adjustment was made by subtracting the number of days or degree-days at large during cold interval from the total days or degree-days at large. The von Bertalanffy growth model was fitted for each measure of p and for the interval-adjusted p's. "Best" fit was determined from the reduction in mean

TABLE 4.—Probability that observed release and recovery data for spotted seatrout in Texas bays fit models based on different assumptions. An A following a number indicates that survival and recovery rate estimates for the bay system group are based on the respective assumptions.

	Model assumption			
Bay system group	Initial recovery, recovery, and survival rates vary in time	Recovery and survival rates vary in time	Recovery rates vary in time; survival rates constant	Recovery and survival rates constant in time
East Matagorda	0.00	0.01A	0.00	0.00
Galveston–lower Laguna Madre	0.56	0.46	0.09A	0.00
Matagorda–Aransas– Corpus Christi	0.17	0.23A	0.00	0.00
San Antonio–upper Laguna Madre	0.00	0.49	0.79A	0.00

square error. Growth parameters K and the corresponding SE were multiplied by either the total number of remaining warm days or the degree-days in a calendar year to adjust them to an annual basis.

Results

Estimates of survival rates for Galveston–lower Laguna Madre and San Antonio–upper Laguna Madre were made under the assumption of time-specific recovery rates and constant survival rates. The Matagorda–Aransas–Corpus Christi model was more complex and required only estimates of time-specific survival rate (Table 4). Estimated 6-month survival rates ($\pm$SE) for Galveston–lower Laguna Madre and San Antonio–upper Laguna Madre were 42 $\pm$ 5% and 48 $\pm$ 10%, respectively (Table 5). Matagorda–Aransas–Corpus Christi 6-month estimates ranged from 8 $\pm$ 3% during 1 November 1983 through 30 April 1984 to 72 $\pm$ 22% during 1 May through 30 October 1983. None of the model assumptions adequately predicted East Matagorda survival rates. Best-fit (time-specific) survival rates for East Matagorda ranged from 7 $\pm$ 5% to 192 $\pm$

78%; the latter estimate indicated problems with the East Matagorda data. Annual survival rates in our study ranged from 17 to 23%.

Six-month fishing mortalities were lower during winter than summer (Table 6). Estimates of F ranged from 0.06 to 0.22, accounting for 6–22% of total mortality during winter. In summer, F ranged from 0.15–0.44, accounting for 19–61% of Z.

We obtained a best-fit growth model (lowest mean square error) for estimating VBGP by using the number of days tagged fish were at large minus the coldest 60 days ($\pm$30 of coldest day) of the year. Parameter estimates $\pm$1 SE were $L_\infty = 691 \pm 41$ mm and adjusted $K = 0.288 \pm 0.041$ per year. Mean daily growth rates computed from the model for fish with total lengths of 250, 400, and 550 mm were 0.42, 0.27, and 0.13 mm/d, respectively. Total lengths for recaptured fish ranged between 253 and 605 mm when released, and between 289 and 705 mm when recaptured. There was no indication that size at release influenced the probability of recapture. Released fish with total lengths of 230–350 mm ($N = 7, 960$), 351–450 mm ($N = 4,018$), and greater than 450 mm ($N =$

TABLE 5.—Six-month survival estimates (% $\pm$ SE) for tagged spotted seatrout in Texas bays, based on the assumption that survival rates and recovery rates are time specific; ND = not estimated because there were no recoveries.

	Time period					Overall mean survival rate
Bay system group	1 Nov 1981– 30 Apr 1982	1 May– 30 Oct 1982	1 Nov 1982– 30 Apr 1983	1 May– 30 Oct 1983	1 Nov 1983– 30 Apr 1984	
East Matagorda	33 $\pm$ 19	14 $\pm$ 7	192 $\pm$ 78	30 $\pm$ 17	7 $\pm$ 5	55 $\pm$ 15
Galveston–lower Laguna Madre	44 $\pm$ 10	51 $\pm$ 15	40 $\pm$ 10	64 $\pm$ 23	13 $\pm$ 5	42 $\pm$ 5
Matagorda–Aransas– Corpus Christi	21 $\pm$ 6	52 $\pm$ 11	36 $\pm$ 8	72 $\pm$ 22	8 $\pm$ 3	38 $\pm$ 4
San Antonio–upper Laguna Madre	46 $\pm$ 16	40 $\pm$ 15	39 $\pm$ 12	66 $\pm$ 40	ND	48 $\pm$ 10

TABLE 6.—Six-month estimates of natural (M) and fishing (F) mortality for spotted seatrout by Texas bay system group and recovery period; ND = not estimated because there were no recoveries. Values in parentheses are fishing mortalities as percentages of total mortality ($F/[M + F]$).

Bay system group	1 Nov 1981– 30 Apr 1982		1 May– 30 Oct 1982		1 Nov 1982– 30 Apr 1983		1 May– 30 Oct 1983		1 Nov 1983– 30 Apr 1984	
	M	F	M	F	M	F	M	F	M	F
Galveston–lower Laguna Madre	0.76	0.08 (10)	0.40	0.44 (52)	0.77	0.07 (8)	0.47	0.37 (44)	0.77	0.07 (8)
Matagorda–Aransas– Corpus Christi	1.36	0.20 (13)	0.40	0.25 (38)	0.80	0.22 (22)	0.13	0.20 (61)	2.36	0.16 (6)
San Antonio–upper Laguna Madre	0.73	0.07 (9)	0.65	0.15 (19)	0.74	0.06 (8)	0.55	0.25 (31)	ND	ND

1,022) were recovered at rates of 6.2, 7.7, and 7.3%, respectively.

Discussion

The tagging methodology and computer programs used in the present study have distinct advantages over other methods for estimating fish survival and growth parameters and their variances. Tagging is generally easy and rapid, and many fish often can be tagged in a short time. Tagging allowed us to estimate Z, F, M, L_∞, and K. No other single methodology, to our knowledge, allows estimation of all these parameters. To determine F and M, we required additional information on tagging mortality and reporting rate by fishermen. The models used in this study did not depend on age estimates for fish. This was an advantage because protracted spawning and differential growth make the age-groups of spotted seatrout difficult to discern.

The computer models we used are relatively simple to run because they accept standard data-entry formats, are relatively simple to interpret, and allow for seasonal and annual estimates of Z, F, and M. Simultaneous evaluation of different assumptions about patterns of time-specific survival and recovery rate allows managers to make the most precise estimate for a given N. Trimming of data recovered prior to a recovery period resulted in selection of models with simpler assumptions about survival and recovery rates. However, trimming of data required more tags for a population than were originally needed to attain a preselected precision level. The methodology and model theory we used offer an accepted procedure (Brownie et al. 1978) for determining survival in terrestrial wildlife studies.

Mortality estimates for spotted seatrout in Texas were higher than those for the species in Alabama (46–58%; Tatum 1980) and Florida (45%; Iverson and Moffett 1962). However, rates were within the range of 67–80% reported for Galveston Bay, Texas (Baker et al. 1986). Estimates of Z, F, and M in the present study also agreed reasonably well with those by Baker et al. (1986).

Temperature can significantly limit the growth of spotted seatrout (Tabb 1958). The model described by Doerzbacher et al. (1988), which we also used, includes temperature effects on growth. Direct comparisons of our growth rates for Texas spotted seatrout with published estimates were difficult because most other studies have not estimated growth rate per day, eliminated cold days, or used fish of the same length range (250–700 mm). However, growth rates in our study and those reported by Colura et al. (1984) and Baker et al. (1986) revealed that, for similar-sized fish, Texas spotted seatrout grow faster than fish in Alabama (Tatum 1980) and Florida (Perret et al. 1980). Estimates of growth rate should take into account days of "cold" water temperatures. Methods developed by Doerzbacher et al. (1988) have general application to growth-rate estimation and do not require fish of known age. Determination of the coldest days during a year may also be used to assess the validity of annulus formation in scales and for other growth-estimation methods.

Texas Parks and Wildlife Department tagging procedures permitted direct measurement of mortality caused by a catastrophic weather event. From late December 1983 to early January 1984, there was an extensive freeze along the Texas coast, and bay waters remained below 4° C for 13 d. This prolonged freeze caused a massive fish kill (McEachron et al. 1984). Because of very low survival during the freeze, three of the six bay systems within the coldest area required estimates based on the model assumption that survival rates varied among periods. East Matagorda Bay and the upper Laguna Madre yielded no tag returns after the freeze, and survival estimates during the freeze could not be made. Estimates for Galveston Bay and lower Laguna Madre were based on the assumption of a constant survival rate among time periods. These two bays were in areas of less severe cold; moreover, Galveston Bay had extensive deep-water areas in which fish could find refuge from extreme cold. However, even these bay systems revealed lower survival-rate estimates (although not statistically lower) during the freeze.

Tag recovery patterns generally reflected fishing pressure patterns as described by Osburn and Ferguson (1985). High fishing pressure during summer resulted in high tag recoveries; the reverse occurred in winter. An exception was East Matagorda Bay, which had more recoveries during winter than summer. This may have been caused by winter movements of many spotted seatrout to the nearby Colorado River and Caney Creek systems, where there is concentrated fishing effort (M. G. Weixelman, TPWD, personal communication). Baker et al. (1986) noted a tendency for spotted seatrout to move to rivers in late fall and early winter. If a large proportion of spotted seatrout go to riverine areas during winter as suggested by Baker et al. (1986), and riverine fisheries vary among bay systems, tag-recovery rates will vary accordingly.

References

Baker. W. B., Jr., G. C. Matlock, L. W. McEachron, A. W. Green, and H. E. Hegen. 1986. Movement, growth and survival of spotted seatrout tagged in Bastrop Bayou, Texas. Contributions in Marine Science 29:91–101.

Brownie, C., D. R. Anderson, K. P. Burnham, and D. S. Robson. 1978. Statistical inference from band recovery data—a handbook. U.S. Fish and Wildlife Service Resource Publication 131.

Colura, R. L., C. W. Porter, and A. F. Maciorowski. 1984. Preliminary evaluation of the scale method for describing age and growth of spotted seatrout (Cynoscion nebulosus) in the Matagorda Bay system, Texas. Texas Parks and Wildlife Department, Coastal Fisheries Branch, Management Data Series 57, Austin.

Doerzbacher, J. F., A. W. Green, H. R. Osburn, and G. C. Matlock. 1988. A temperature compensated von Bertalanffy growth model for tagged red drum and black drum in Texas Bays. Fisheries Research (Amsterdam) 6:135–152.

Fabens, A. J. 1965. Properties and fitting of the von Bertalanffy growth curve. Growth 29:265–289.

Green, A. W., G. C. Matlock, and J. E. Weaver. 1983. A method for directly estimating the tag-reporting rate of anglers. Transactions of the American Fisheries Society 112:412–415.

Green, A. W., H. R. Osburn, G. C. Matlock, and H. E. Hegen. 1985. Estimated survival rates for immature red drum in northwest Gulf of Mexico bays. Fisheries Research (Amsterdam) 3:263–277.

Green, L. M. 1986. Fish tagging on the Texas coast, 1950–75. Texas Parks and Wildlife Department, Coastal Fisheries Branch, Management Data Series 99, Austin.

Harrington, R. A., G. C. Matlock, and J. C. Weaver. 1979. Standard–total length, total length–whole weight, and dressed–whole weight relationships for selected species from Texas bays. Texas Parks and Wildlife Department, Coastal Fisheries Branch, Technical Series 26, Austin.

Hegen, H. E., G. E. Saul, and G. C. Matlock. 1984. Survival of handled and tagged spotted seatrout held in wood and wire cages. Texas Parks and Wildlife Department, Coastal Fisheries Branch, Management Data Series 61, Austin.

Hegen, H. E., G. E. Saul, and G. C. Matlock. 1987. Survival of hook-caught spotted seatrout. Proceedings of the Annual Conference Southeastern Association of Fish and Wildlife Agencies 38:488–494.

Iversen, E. S., and A. W. Moffett. 1962. Estimation of abundance and mortality of a spotted seatrout population. Transactions of the American Fisheries Society 91:395–398.

Iversen, E. S., and D. C. Tabb. 1962. Subpopulations based on growth and tagging studies of spotted seatrout, Cynoscion nebulosus, in Florida. Copeia 1962:545–548.

Marwitz, S. R. 1986. A summary of fish tagging in Texas bays, 1975–1982. Texas Parks and Wildlife Department, Coastal Fisheries Branch, Management Data Series 66, Austin.

McEachron, L. W., and A. W. Green. 1985. Trends in relative abundance and size of selected finfish in Texas bays: November 1975–June 1984. Texas Parks and Wildlife Department, Coastal Fisheries Branch, Management Data Series 79, Austin.

McEachron, L. W., A. W. Green, and G. E. Saul. 1985. Increasing sampling efficiency in creel surveys. Proceedings of the Annual Conference Southeastern Association of Fish and Wildlife Agencies 37:376–384.

McEachron, L. W., G. Saul, J. Cox, C. E. Bryan, and G. C. Matlock. 1984. Fish kill. Texas Parks and Wildlife Magazine 42(4):10–13.

Mercer, L. P. 1984. A biological and fisheries profile of spotted seatrout, Cynoscion nebulosus. North Carolina Department of Natural Resources, Special Scientific Report 40, Raleigh.

Moffett, A. W. 1961. Movement and growth of spotted seatrout, Cynoscion nebulosus (Curvier), in west Florida. Florida Board of Conservation Marine Laboratory Technical Series 36.

Osburn, H. R., and M. O. Ferguson. 1985. Trends in finfish catches by private sport-boat fishermen in Texas marine waters through May 1984. Texas

Parks and Wildlife Department, Coastal Fisheries Branch, Management Data Series 78, Austin.

Perret, W. S., and six coauthors. 1980. Fishery profiles of red drum and spotted seatrout. Gulf States Marine Fisheries Commission, Report 6, Ocean Springs, Mississippi.

Ricker, W. E. 1975. Computation and interpretation of biological statistics of fish populations. Fisheries Research Board of Canada Bulletin 191.

SAS (Statistical Analysis System). 1982. SAS users guide: statistics. SAS Institute. Cary, North Carolina.

Tabb, D. C. 1958. Differences in the estuarine ecology of Florida waters and their effect on populations of spotted weakfish, *Cynoscion nebulosus* (Curvier and Valenciennes). Transactions of the North American Wildlife Conference 23:392–401.

Tatum, W. M. 1980. Spotted seatrout (*Cynoscion nebulosus*) age and growth: data from annual fishing tournaments in coastal Alabama, 1964–1977. Gulf States Marine Fisheries Commission, Special Report 5:89–92. (Ocean Springs, Mississippi.)

USNMFS (U.S. National Marine Fisheries Service). 1973. Fishery statistics of the United States, 1973. U.S. National Marine Fisheries Service Statistical Digest 67.

USNMFS (U.S. National Marine Fisheries Service). 1974. Fishery statistics of the United States, 1974. U.S. National Marine Fisheries Service Statistical Digest 68.

USNMFS (U.S. National Marine Fisheries Service). 1978. Fishery statistics of the United States, 1975. U.S. National Marine Fisheries Service Statistical Digest 69.

USNMFS (U.S. National Marine Fisheries Service). 1980. Fishery statistics of the United States, 1976. U.S. National Marine Fisheries Service Statistical Digest 70.

USNMFS (U.S. National Marine Fisheries Service). 1984. Fishery statistics of the United States, 1977. U.S. National Marine Fisheries Service Statistical Digest 71.

USNMFS (U.S. National Marine Fisheries Service). 1985a. Marine recreational fishery statistics survey, Atlantic and Gulf coasts 1983–1984. U.S. National Marine Fisheries Service Current Fishery Statistics 8324.

USNMFS (U.S. National Marine Fisheries Service). 1985b. Marine recreational fishery statistics survey, Atlantic and Gulf coast, 1983–1984. U.S. National Marine Fisheries Service Current Fishery Statistics 8326.

Williams, R. O., J. E. Weaver, and F. A. Kalber, editors. 1980. Proceedings: colloquium on the biology and management of red drum and spotted seatrout. Gulf States Marine Fisheries Commission, Special Report 5, Ocean Springs, Mississippi.

American Fisheries Society Symposium 7:631–646, 1990

Jeopardized Estimates of the Contribution of Marked Pacific Salmon to the Sport Fishery of the Strait of Georgia, British Columbia, Due to Awareness Factor Variability

R. V. PALERMO

Department of Fisheries and Oceans, Fisheries Branch
Vancouver, British Columbia, V6B 5G3, Canada

Abstract.—The awareness factor—the ratio of marked Pacific salmon *Oncorhynchus* spp. voluntarily reported by anglers to the total number of marked fish in a time–area stratum—was calculated from Strait of Georgia sport creel survey data for nine areas and monthly time periods from July 1980 to July 1984. Estimated variances of awareness factors were very high; coefficients of variation typically exceeded 80%. Restratification and smoothing of the raw data marginally improved these estimates. The awareness factor statistic also exhibited significant differences among times and areas. This analysis indicated that calculated mean awareness factors for the Strait of Georgia sport fishery are statistically unreliable and should be used with caution for estimates of stock contributions to the sport harvest.

In British Columbia, several million chinook salmon *Oncorhynchus tshawytscha* and coho salmon *O. kisutch* fingerlings are individually marked at hatcheries with binary coded wire tags inserted into the nose cartilage. Because the tags are externally invisible, the adipose fin is clipped at tagging to identify fish that carry them. Typically, only a fraction of a production group is tagged.

During a normal life cycle, a production group contributes to fisheries in specific areas at specific times. The commercial fishery catch is sampled for fish without adipose fins, and pertinent catch information is recorded. Tags are then removed surgically and decoded.

Mark–recovery data are commonly used to estimate relative contributions of marked groups to the catch for a specific area and time (Kimura 1976). No attempt is made to correct for mark-induced mortality, therefore, the estimated contribution of a production group is somewhat conservative.

The standard analysis of stock contribution to commercial harvest is based on sampling within various area and time strata. Let i denote the area stratum and j denote the time stratum; for each stratum, then, the following data are obtained:

C_{ij}, total catch in numbers;
S_{ij}, size of sample examined for marks;
m_{ij}, number of marks found in the sample;
K_{ij}, number of unique tag groups in the sample;
t_{kij}, $k = 1$. .K_{ij}, size of tag-group k in numbers.

The estimated proportion of marked fish (or mark incidence) is

$$\hat{I}_{ij} = m_{ij}/S_{ij}, \tag{1}$$

and

$$\hat{M}_{ij} = \hat{I}_{ij} \cdot C_{ij} \tag{2}$$

is the simple ratio estimate of the number of marks in the catch. The relative contribution by tag code is then estimated by weighting the $\hat{M}_{ij}$ by the fraction of tag code k among the total number of unique tag groups. Thus, the estimated contribution of tag k in stratum ij is

$$\hat{T}_{kij} = \hat{M}_{ij} \cdot t_{kij}/\sum t_{kij}. \tag{3}$$

This estimate can now be expanded to estimate the total contribution of the production group represented by tag code k.

The catch from the sport fishery is not actively sampled for marks. Instead, estimates of tag-groups in the catch are made from voluntary returns (VR) of adipose-clipped fish (or their heads) caught by anglers and information about mark incidences information gathered during creel surveys. The estimated total marks caught by anglers in stratum ij is

$$\hat{V}T_{ij} = VR_{ij}/\hat{P}A_{ij}; \tag{4}$$

$\hat{V}R_{ij}$ is the number of marked fish returned voluntarily by anglers, and $\hat{P}A_{ij}$ is the ratio of voluntary mark returns to the total number of marked fish caught by anglers in stratum ij. The $\hat{P}A_{ij}$ is termed the "awareness factor" (Kimura 1976; Argue et al. 1977); it measures the "sampling rate" achieved by voluntary tag returns from anglers.

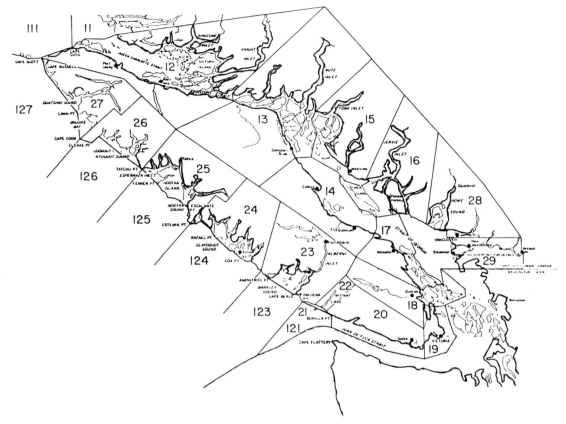

FIGURE 1.—Statistical areas used by the Department of Fisheries and Oceans for the Strait of Georgia.

The relative contribution of tag-group k can then be calculated by weighting the VT_{ij} by the fraction $t_{kij}/\Sigma t_{kij}$ as before.

The parameter $\hat{P}A_{ij}$, which can also be considered as the probability that an angler will return a marked fish, is commonly calculated by

$$\hat{P}A_{ij} = VR_{ij}/m_{ij}/(S_{ij} \cdot \hat{C}_{ij}), \qquad (5)$$

$\hat{C}_{ij}$ being the estimated sport catch in the stratum ij. It is reasonably assumed that anglers do not selectively fish for or retain marked fish.

Adequate sampling of the sport catch often is impossible or too expensive. In such circumstances, the awareness factor for each stratum can be combined with voluntary mark returns to estimate the total marked fish caught in a particular stratum. Then, by rearrangement of equation (5), an estimate or revision of the stratum sport catch can be derived. The results can also be summed across strata to arrive at overall contributions.

Awareness factors have been used in this manner by Argue et al. (1977) for the Strait of Georgia sport fishery. They applied awareness factors from the 1974 sport fishery in Puget Sound (Kimura 1976) to the statistical areas of the Strait of Georgia, based on qualitative fishery comparisons with the Puget Sound areas (within-area variability was not considered). They then reestimated the Strait of Georgia sport catch for the years 1972–1976, based on the voluntary head returns for those years. The awareness factors assigned by Argue et al. (1977) to the Strait of Georgia fishery continue to be used to estimate mark contributions to the harvest and and to revise sport catch estimates. That practice prompted the present report.

Anglers in the Strait of Georgia account for approximately 90% of the sportfishing effort in British Columbia (Pearse 1982), and the percentage has steadily increased since the mid-1970s. This upward trend precipitated an urgent management need for

TABLE 1.—Mean awareness factors (MAF) and associated variances (Var) and coefficients of variation (CV = $100 \cdot$ SD/mean), by statistical area, for raw and smoothed data from the Strait of Georgia creel survey, July 1980–July 1984.

Statistic	Statistical area								
	13	14	15	16	17	18	19–20	28	29
	Raw data: chinook salmon								
MAF	0.185	0.279	0.047	0.632	0.263	0.337	0.216	0.175	0.034
Var	0.056	0.123	0.020	0.588	0.097	0.434	0.049	0.042	0.010
CV	129.0	126.6	304.7	122.1	111.0	195.6	103.0	117.2	289.9
	Raw data: coho salmon								
MAF	0.147	0.316	0.156	0.149	0.405	0.319	0.403	0.268	0.029
Var	0.022	0.078	0.091	0.041	0.389	2.488	0.866	0.297	0.008
CV	100.7	89.2	195.6	136.3	144.7	497.9	323.5	204.9	310.4
	Smoothed data: chinook salmon								
MAF	0.243	0.376	0.076	0.323	0.243	0.301	0.216	0.241	0.055
Var	0.010	0.034	0.018	0.038	0.035	0.025	0.021	0.027	0.008
CV	41.3	49.3	179.8	60.8	77.9	52.9	67.1	68.9	169.6
	Smoothed data: chinook salmon								
MAF	0.229	0.393	0.292	0.261	0.333	0.177	0.293	0.265	0.060
Var	0.009	0.039	0.013	0.023	0.031	0.015	0.033	0.012	0.006
CV	42.4	50.4	38.9	58.5	53.5	68.6	62.4	42.2	133.1

timely, accurate sport-catch and stock-contribution data. Creel surveys of the sport catch in the Strait of Georgia were begun in 1980 (DPA 1982). Using data from the 1980 survey, I have calculated awareness factors and their associated variances by stratum. I anticipated that these estimates could be used to (1) back-calculate historical sport catches for the Strait of Georgia, (2) prove a measure of the mark contribution of tagged groups, and (3) provide a parameter (a mean awareness factor) to expand voluntary mark-return data in the coastwide mark–recovery program's data base.

The findings reported here indicate that the calculated mean awareness factors are statistically unreliable for estimating mark contribution and hence for revising sport-catch estimates.

Methods

The data for awareness factor calculations for chinook and coho salmon were taken from several sources. They covered monthly time periods from July 1980 to July 1984 in Department of Fisheries and Oceans statistical areas 13 through 19/20, 28, and 29 (Figure 1). Data from July 1980 to December 1982 came from the DPA final report (DPA 1982), Shardlow (1983), and LGL and ESSA (1982). Details and scope of the creel survey may be found in these reports. Data through July 1984 came from monthly reports prepared by the DFO staff responsible for the creel-survey program. The data from these sources were for each area-

time stratum and included estimated sport catch ($\hat{C}_{ij}$), numbers of fish sampled for marks (S_{ij}), numbers of marks found in the sample (m_{ij}), calculated mark incidence (i.e., m_{ij}/S_{ij}), and the simple binomial variance of the mark incidence [var (m_{ij}/S_{ij})]. The voluntary returns (VR_{ij}) from the same strata were provided from the coastwide mark–recovery program's data base.

The first step in the analysis was calculation of the awareness factor and its mean, variance, standard deviation, standard error, and coefficient of variation (CV) for each area stratum over time. Jackknife estimates (Mosteller and Tukey 1977; Efron 1982) were then calculated for these parameters. Quantile analysis (Chambers et al. 1983) was applied to assess the distributional qualities of the awareness factors.

The awareness factor parameter is assumed to be invariant over time. To test this assumption, I used spectral analysis (Box and Jenkins 1976) to construct a spectrum plot of each selected species–area combination over time.

I calculated point estimates of the variance and CV of the awareness factors for each stratum. By assuming there is no variance in the voluntary returns (VR_{ij}), and then applying the guidelines for combining variances (Goodman 1960; Frishman 1975) to equation (5), I estimated the point variance of the awareness factor ($\hat{P}A_{ij}$) as follows (subscripts are omitted for clarity):

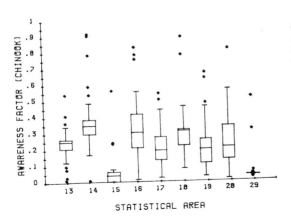

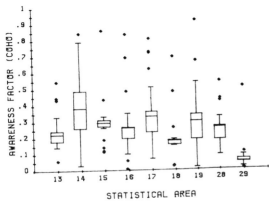

FIGURE 2.—Box plots of smoothed angler awareness factors for marked Strait of Georgia chinook and coho salmon, July 1980–July 1984. Horizontal lines of each box indicate (top to bottom) the 75th, 50th, and 25th data percentiles; extenders indicate 90th and 10th percentiles. Isolated points are data outliers.

$$\mathrm{var}(\hat{P}A) = VR^2/\{[(m/S)\hat{C}] \cdot \mathrm{var}[(m/S)\hat{C}]\}, \quad (6)$$

which expands to

$$\mathrm{var}(\hat{P}A) = VR^2/\{[(m/S)\hat{C}]^4 \\ \cdot [(m/S)^2\mathrm{var}(\hat{C}) + \mathrm{var}(m/S)\hat{C}^2 \\ + \mathrm{var}(m/S)\mathrm{var}(\hat{C})]\}. \quad (7)$$

The variance of the mark incidence [var(m/S)] was calculated by the simple binomial formula (Cochran 1977)

$$\mathrm{var}\ (m/S) = (m/S)[1 - (m/S)]/(S - 1). \quad (8)$$

Variance estimates for the estimated sport catch were not always available, so the following procedure was used to estimate var($\hat{C}_{ij}$). If sport

TABLE 2.—Mean awareness factors (MAF) and associated coefficients of variation (CV = $100 \cdot$ SD/mean) for restratification scenarios for the Strait of Georgia creel survey, July 1980–July 1984.

Scenario and statistical area combination	Chinook salmon		Coho salmon	
	MAF	CV	MAF	CV
Bimonthly–bisplit				
13 + 14 + 15 + 16	0.44	116.1	0.43	116.0
17 + 18 + 19–20 + 28 + 29	0.29	83.2	0.78	171.0
Bimonthly–trisplit				
13 + 14 + 15 + 16	0.51	98.0	0.38	191.0
17 + 18 + 28 + 29	0.26	70.1	0.71	205.0
19–20	0.36	97.0	0.77	153.0
Monthly combined				
All nine	0.32	71.2	0.41	98.3

catch is estimated to be within 10% of the true sport catch, then

$$1.96[\mathrm{var}(\hat{C}_{ij})]^{0.5} = 0.1\ \hat{C}_{ij},$$

which can be rearranged so that

$$\mathrm{var}(\hat{C}_{ij}) = 0.0026\ \hat{C}_{ij}^2. \quad (9)$$

This procedure allowed me to examine the effect on the point variance of different awareness factors that resulted from changing error bounds of the estimated sport catch. The computer programs were designed to offer this option.

When the estimated sport catch ($\hat{C}_{ij}$) and voluntary returns (VR$_{ij}$) are very large, and the mark incidence ($\hat{I}_{ij}$) is very small, the calculated awareness factor can be greater than 1. An awareness factor greater than 1, however, indicates that more marked fish were voluntarily returned than were caught! This situation can arise if either the sport catch is greatly overestimated or the mark incidence is greatly underestimated or both. With creel survey data, the most common cause of this problem was underestimation of mark incidence. Consequently, when the awareness factor was calculated to be greater than 1, I smoothed the data by replacing the sampled mark incidence with the running mean mark incidence up to the period before the time of interest. This procedure was adopted because mark incidence and sport catch are assumed to be independent.

Results and Discussion

Calculated mean awareness factors and associated statistics are presented in Table 1 by species and statistical area over the time period. There are marked differences in mean awareness factor for both species of salmon over all areas, but there

TABLE 3.—Annual mean awareness factors (MAF) and their coefficients of variation (CV = 100·SD/mean), by statistical area, from the Strait of Georgia creel survey, July 1980–July 1984.

Year and statistic	Statistical area								
	13	14	15	16	17	18	19/20	28	29
	Chinook salmon								
1980									
MAF	0.158	0.387	0	0.410	0.119	0.395	0.223	0.154	0.025
CV	101.4	76.3	0	59.8	65.8	75.1	67.0	118.3	113.8
1981									
MAF	0.136	0.109	0.043	0.121	0.135	0.596	0.168	0.164	0.011
CV	111.8	96.38	214.1	141.5	79.9	182.3	97.1	133.5	178.0
1982									
MAF	0.329	0.254	0.029	0.615	0.17	0.195	0.136	0.1	0.004
CV	134.5	149.1	291.7	116.4	123.8	147.4	54.0	147.2	291.7
1983									
MAF	0.173	0.346	0.060	0.038	0.473	0.387	0.326	0.233	0.009
CV	81.3	137.8	324.1	101.8	104.4	225.3	103.3	118.2	230.7
1984									
MAF	0.095	0.308	0.094	1.094	0.309	0.097	0.191	0.209	0.151
CV	114.9	128.5	225.2	111.2	64.5	131.3	130.3	95.8	146.5
	Coho salmon								
1980									
MAF	0.146	0.433	0.124	0.159	0.211	0.139	0.199	0.146	0.028
CV	62.1	71.4	161.6	78.1	98.1	234.8	116.6	90.3	114.5
1981									
MAF	0.204	0.248	0.124	0.114	0.477	1.379	0.287	0.384	0.010
CV	101.9	56.4	114.5	95.1	94.6	268.3	120.8	159.4	246.3
1982									
MAF	0.178	0.284	0.203	0.255	0.322	0.139	0.243	0.121	0.020
CV	111.3	97.6	216.3	136.0	129.9	129.4	91.3	149.9	199.7
1983									
MAF	0.117	0.323	0.112	0.094	0.614	0.002	0.311	0.266	0.063
CV	99.1	115.8	252.2	191.9	161.7	324.1	119.4	178.0	257.9
1984									
MAF	0.089	0.317	0.228	0.134	0.246	0	1.074	0.431	0.004
CV	116.7	94.0	194.7	114.5	64.6	0	215.3	255.2	179.3

does not appear to be any spatial trend. Especially disturbing are the very high CVs associated with each estimated mean awareness factor. For all the analyses, the jackknife estimates of the parameters were very similar to the estimates found by classical methods.

The high CVs indicate that severe problems will arise when the awareness factor is used to estimate mark contributions of marked fish to the catch or to revise catch estimates. The problem is with the uncertainty of each estimate. For example, assume a voluntary return of 100 marked chinook salmon for statistical area 13 in some period. The mean awareness factor for chinook in statistical area 13 is 0.185 and the calculated CV is 129% (Table 1), so the estimated number of marked fish caught would be 541 (rounded). If 95% confidence limits are placed around the mean awareness factor, then the estimated number of marked fish caught is between 312 and 1,124. The problem is much worse if the goal is to back-calculate sport catch. When equation (5) is rear-

ranged to yield a revised sport-catch estimate (given a mark incidence of 3% and the 0.185 awareness factor), the estimated sport catch becomes 18,018 fish with upper and lower bounds of 37,453 and 10,417. These wide bounds are unacceptable for management purposes.

When raw data are smoothed by running means, which constrain awareness factor estimates greater than 1, the CVs are about halved but remain large (Table 1). For area 13, the smoothed data yield a mean awareness factor of 0.243 for chinook salmon; the estimate of 406 marked fish in the sport catch has upper and lower bounds of 763 and 262 fish. The reestimated sport catch becomes 13,717 with upper and lower bounds of 25,335 and 8,726 fish. Again, the confidence interval is impracticably large for management.

Quantile analyses performed on the calculated awareness factors for the smoothed data reveal that, for most areas, the awareness factors are asymmetrically distributed (skewed to higher values). The box plots for these data are presented in

TABLE 4.—Basic data and awareness factor (AF) calculations for the monthly combined scenario, with associated variances (Var) and coefficients of variation (CV = 100·SD/mean), for chinook and coho salmon. The statistical calculations have been rounded.

	Catch				Voluntary returns	Mark incidence		Awareness factor		
Time	Number	Var[a]	Marks	Sample		Frequency	Var	AF	Var	CV
Chinook salmon										
1980										
Jul	68,400	12.164	58	3,474	260	0.017	0.047	0.228	0.001	14.002
Aug	50,800	6.710	74	2,295	204	0.032	0.136	0.186	0.001	12.538
Sep	33,700	2.953	24	652	178	0.037	0.545	0.144	0.001	20.720
Oct	17,900	0.833	12	588	89	0.020	0.341	0.244	0.005	29.096
Nov	16,650	0.721	11	266	51	0.040	1.431	0.078	0.001	30.796
Dec	16,650	0.721	11	266	61	0.040	1.431	0.093	0.001	30.796
1981										
Jan	17,700	0.815	10	451	66	0.022	0.482	0.168	0.008	31.775
Feb	17,700	0.815	10	451	23	0.022	0.482	0.059	0.000	31.775
Mar	10,500	0.287	1	13	25	0.080	59.172	0.001	0.001	102.188
Apr	7,300	0.139	14	102	36	0.137	11.724	0.036	0.000	25.557
May	32,300	2.713	53	1,074	81	0.049	0.437	0.051	0.000	14.356
Jun	34,100	3.023	68	1,216	147	0.056	0.435	0.077	0.000	12.860
Jul	40,122	4.185	124	1,970	510	0.063	0.300	0.202	0.000	10.091
Aug[b]	37,517	3.660	66	1,227	279	0.054	0.415	0.138	0.000	13.036
1982										
May[b]	12,854	0.430	7	197	95	0.036	1.749	0.208	0.006	37.657
Jun	17,183	0.768	17	617	203	0.028	0.435	0.429	0.011	24.514
Jul	29,800	2.309	42	774	637	0.054	0.664	0.394	0.004	15.881
Aug	27,121	1.912	24	769	606	0.031	0.394	0.716	0.022	20.773
Sep	20,300	1.071	29	523	359	0.055	1.003	0.319	0.004	18.802
Oct	4,184	0.060	9	360	131	0.025	0.679	1.089	0.132	33.417
Nov	6,159	0.099	26	574	103	0.045	0.742	0.376	0.006	20.067
Dec	6,159	0.099	26	574	151	0.045	0.742	0.551	0.012	20.067
1983										
Jan	4,029	0.042	19	598	163	0.032	0.516	1.272	0.087	23.199
Feb	4,029	0.042	19	598	113	0.032	0.516	0.882	0.042	23.199
Mar[c]	9,394	0.229	11	443	97	0.025	0.548	0.416	0.016	30.296
May[c]	6,695	0.117	13	237	211	0.055	2.197	0.575	0.025	27.561
Jun	27,659	1.989	26	828	316	0.031	0.368	0.364	0.005	20.005
Jul	37,248	3.607	50	1,185	568	0.042	0.341	0.361	0.003	14.775
Aug	37,438	3.644	27	1,101	542	0.025	0.218	0.590	0.014	19.716
Sep	37,637	3.683	18	743	271	0.024	0.319	0.297	0.005	23.888
Oct	20,474	1.090	13	806	130	0.016	0.197	0.394	0.012	28.040
Nov	4,993	0.065	14	874	47	0.016	0.181	0.588	0.025	27.054
Dec	4,993	0.065	14	874	53	0.016	0.181	0.663	0.032	27.054
1984										
Jan	4,407	0.050	10	651	71	0.015	0.233	1.048	0.112	31.867
Feb	4,407	0.050	10	651	29	0.015	0.233	0.428	0.019	31.867
Mar	5,363	0.075	12	507	65	0.024	0.457	0.512	0.022	29.055
Apr	17,809	0.825	3	59	89	0.051	8.321	0.098	0.003	57.275
May	35,380	3.255	32	1,785	294	0.018	0.099	0.464	0.007	18.275
Jun	96,794	24.360	69	4,280	551	0.016	0.037	0.353	0.002	13.001
Jul	95,904	23.914	83	3,547	622	0.023	0.064	0.277	0.001	12.001
Coho salmon										
1980										
Jul	204,900	110.000	633	11,653	2,441	0.054	0.044	0.219	0.000	6.402
Aug	113,000	33.199	435	6,524	1,776	0.067	0.095	0.236	0.000	6.893
Sep	62,200	10.059	113	1,844	670	0.061	0.312	0.176	0.000	10.458
Oct	9,600	0.240	27	440	156	0.061	1.312	0.265	0.003	19.385
Nov	1,900	0.009	0	13	19	0.000	0.000	0.000	0.000	0.000
Dec	1,900	0.009	0	13	17	0.000	0.000	0.000	0.000	0.000
1981										
Jan	2,550	0.017	2	36	25	0.056	14.991	0.177	0.015	70.456
Feb	2,550	0.017	2	36	33	0.056	14.991	0.233	0.027	70.456
Mar	8,700	0.197	0	6	52	0.000	0.000	0.000	0.000	0.000
Apr	14,200	0.524	27	254	417	0.106	3.755	0.276	0.003	18.971
May	35,400	3.258	187	2,854	1,002	0.066	0.215	0.432	0.001	8.725
Jun	79,900	16.598	421	4,958	1,939	0.085	0.157	0.286	0.000	6.914
Jul	100,870	26.454	457	4,750	3,472	0.096	0.183	0.358	0.001	6.770
Aug[b]	72,921	13.845	209	2,245	892	0.093	0.376	0.131	0.000	8.339

TABLE 4.—Continued.

Time	Catch		Marks	Sample	Voluntary returns	Mark incidence		Awareness factor		
	Number	Var[a]				Frequency	Var	AF	Var	CV
1982										
May[b]	63,008	10.322	28	836	574	0.034	0.388	0.272	0.003	19.306
Jun	71,862	13.427	70	2,629	1,275	0.027	0.099	0.666	0.007	12.865
Jul	134,852	47.281	141	3,505	1,792	0.040	0.110	0.330	0.001	9.710
Aug	87,204	19.772	74	1,929	986	0.038	0.191	0.295	0.001	12.506
Sep	41,730	4.528	16	786	454	0.020	0.254	0.535	0.018	25.319
Oct	8,361	0.182	8	444	113	0.018	0.399	0.750	0.071	35.509
Nov	2,335	0.014	7	304	53	0.023	0.742	0.986	0.139	37.845
Dec	2,335	0.014	7	304	10	0.023	0.742	0.186	0.005	37.845
1983										
Jan	912	0.002	2	214	12	0.009	0.437	1.405	0.990	70.902
Feb	912	0.002	2	214	12	0.009	0.437	1.405	0.990	70.902
Mar[c]	3,228	0.027	4	229	30	0.018	0.753	0.532	0.071	50.050
May[c]	32,399	2.729	65	1,066	640	0.061	0.538	0.324	0.002	13.079
Jun	151,358	59.564	216	5,112	1,847	0.042	0.079	0.289	0.001	8.395
Jul	102,946	27.554	146	3,490	2,154	0.042	0.115	0.500	0.002	9.583
Aug	50,847	6.722	77	1,366	1,031	0.056	0.390	0.360	0.002	12.207
Sep	39,510	4.059	20	682	413	0.029	0.418	0.356	0.007	22.665
Oct	15,627	0.635	11	392	147	0.028	0.698	0.335	0.010	30.254
Nov	224	0.000	1	30	5	0.017	5.846	1.320	3.559	144.144
Dec	224	0.000	1	30	5	0.017	5.846	1.320	3.559	144.144
1984										
Jan	600	0.001	2	37	12	0.041	10.805	0.494	0.161	81.896
Feb	600	0.001	2	37	22	0.041	10.805	0.905	0.542	81.896
Mar	3,700	0.036	15	387	55	0.039	0.965	0.384	0.010	25.904
Apr	30,909	2.484	2	125	219	0.016	1.270	0.443	0.098	70.842
May	58,223	8.814	142	3,572	804	0.040	0.107	0.347	0.001	9.687
Jun	86,436	19.425	157	4,217	1,059	0.037	0.085	0.329	0.001	9.354
Jul	136,076	48.143	168	5,843	1,387	0.029	0.049	0.355	0.001	9.164

[a]In millions. Multiply entries by 10^6 to get actual variances, which have been rounded to the nearest 1,000 for this presentation.
[b]No catches during September 1981–April 1983.
[c]No catch data for April 1983.

Figure 2. The plots for the raw data exhibit even greater asymmetry. When so plotted, the biasing effect of the higher outliers on the estimated mean awareness factors can be appreciated.

In consultation with departmental management biologists, a consensus regarding spatial strata for the Strait of Georgia was reached. We mutually agreed that restratification would reduce the variability in the awareness factors. Of many possible restratification schemes, those that made sense to managers were created and analyzed. In each scheme, however, species strata were not combined. Four restratification scenarios were identified.

(1) *Bimonthly scenario*: each area stratum was kept separate, but the time periods were combined into bimonthly periods.

(2) *Bimonthly–bisplit scenario*: the Strait of Georgia was stratified spatially into a northern area comprising statistical areas 13–16 and a southern area comprising statistical areas 17 through 19–20, 28, and 29; time periods were combined into bimonthly periods.

(3) *Bimonthly–trisplit scenario*: the Strait of Georgia was stratified into three areas, a northern

one comprising statistical areas 13–16, a second comprising statistical areas 17, 18, 28, and 29, and a third area consisting of areas 19–20; time periods were bimonthly.

(4) *Monthly combined scenario*: all of the statistical areas of the Strait of Georgia were combined and time periods were monthly.

Results for the bimonthly scenario with smoothed data were similar to those for the monthly smoothed data. Awareness factors for chinook salmon ranged between 0.05 and 0.96 with CVs in the range of 70–230%. Coho salmon awareness factors ranged between 0.06 and 1.0 with CVs in the range of 88–400%. Results for the bimonthly–bisplit, bimonthly–trisplit, and monthly combined scenarios are summarized in Table 2. None of these restratification schemes significantly improved the variability in the calculated mean awareness factors. Several other scenarios were run with similar results. Two interesting stratification schemes emerged. In one, time was stratified into yearly periods but area strata were kept separate; in the other, all area strata were combined in yearly stratifications. In both cases, lower CV values were observed for each time

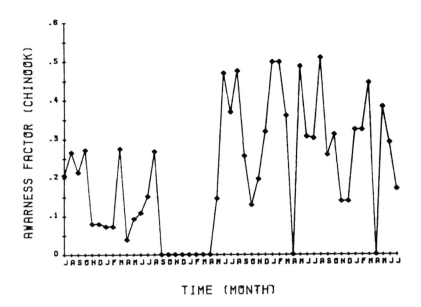

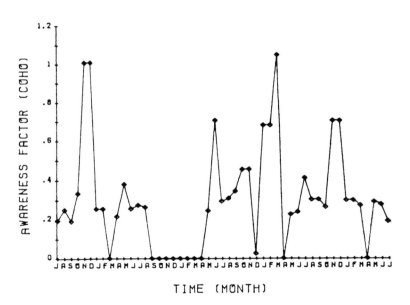

FIGURE 3.—Time-series plots of angler awareness factors (combined scenario) for Strait of Georgia chinook and coho salmon, July 1980–July 1984.

period, but there were significant yearly variations in calculated awareness factors, both within and among years, that made these yearly roll-ups inappropriate (Table 3).

I concluded from the foregoing analyses that variation in the calculated awareness factors is caused by factors other than space and time. Plausible interactive causes of this high variability are:

(1) errors in sampling for marks, (2) real-time dynamics whereby awareness factors change over time, (3) errors in the estimation of sport catch, and (4) errors in assigning recovered marks to an area.

At present, there is no independent method by which errors in assigning voluntarily returned marks to a particular statistical area can be assessed. However, creel survey interviews are

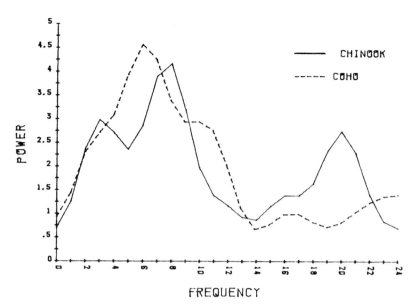

FIGURE 4.—Spectrum plots of smoothed angler awareness factors (combined scenario) for Strait of Georgia chinook and coho salmon.

conducted to determine in what area an angler fished; because most anglers rarely fish in more than one statistical area in a given day, errors in assigning marks probably are rare.

In equation (5), the only known quantity is the number of voluntary returns by stratum. Estimates of mark incidence and sport catch are subject to both sampling and measurement error. Problems in the sampling program can lead to biased calculation of the awareness factor.

Point estimates of awareness factor variance (equation 7) and CV were calculated for each stratum in each scenario; those for the monthly combined scenario are listed in Table 4 along with estimates of the sport catch and catch variance (calculated with 10% error bounds). In any stratum, the estimated sport catch was generally within 10–20% of the real sport catch in the Strait of Georgia (T. Hoyt, Department of Fisheries and Oceans, personal communication). Gaming with the error bounds for the sport-catch estimate within the 10–20% range produced only minor differences in the overall results. This suggests that errors in estimated sport catch are reasonably small and contribute little to the variability in the awareness factor estimates. The qualitative results of the data in Table 4 are typical for any scenario.

Estimated awareness factor variances and CVs were surprisingly low for most time periods, indi-

cating relatively good estimates of awareness factor for most periods. For the monthly combined scenario, the CVs for chinook salmon ranged from 10 to 102% and a majority were below 25% (Table 4). For coho salmon, CVs ranged from 6 to 144% and, again, a majority were less than 25%. The calculated awareness factors themselves, however, varied widely over time (Figure 3).

Spectral analysis (Box and Jenkins 1976) was performed on both the calculated awareness factors and their associated sampling rates to expose any time dynamics (periodicity or trend) in the data. The awareness factor data (Figure 3) were detrended with a third-order polynomial, and a time series analysis was done on the residuals. Spectrum results for the unsmoothed chinook salmon awareness factors show peaks at frequencies 3, 8, and 20 that correspond to the periodicities of 16, 6, and 2.2. months (Figure 4). When Fourier smoothing with aliasing correction (Aubanel and Oldham 1985) was applied before detrending, the only significant peak was at frequency 8, which corresponds to a periodicity of 6 months. The coho salmon data showed a spectral peak at frequency 6, corresponding to a periodicity of 8 months, but exhibited more "noise" than chinook salmon data (Figure 4). When spectral analysis was applied to sampling-rate data (Figure 5), significant peaks resulted at frequency 4 for chinook salmon and frequency 3 for coho salmon,

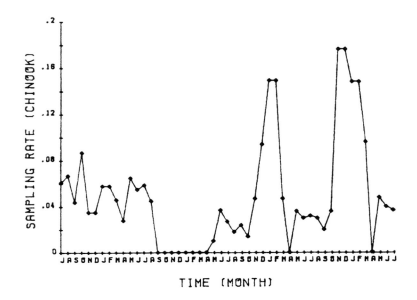

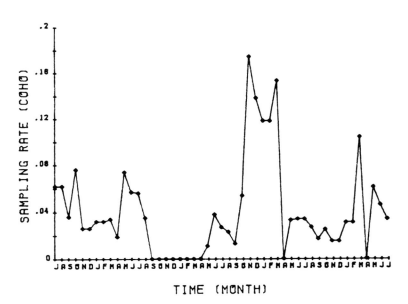

FIGURE 5.—Time-series plots of creel survey sampling rates (combined scenario) for the Strait of Georgia chinook and coho salmon sport fisheries, July 1980–July 1984.

implying periodicities of around 12 months (Figure 6). The same 12-month periodicity was found for mark incidence data. These plots show that the sampling effort for marks was reasonably consistent from year to year in the creel survey, as was the incidence of marks in the samples. Further, the results show periodic behavior of 6–8 months for the calculated awareness factors.

Cross-spectral analysis (Box and Jenkins 1976) between awareness factors and sampling rates did not reveal any significant cross-correlations at any lag within a 12-month time shift. Sampling rates and mark incidence also did not show any significant cross-correlation, even though they had similar spectrums with a 12-month cycle. I conclude that similar periodic components existed in

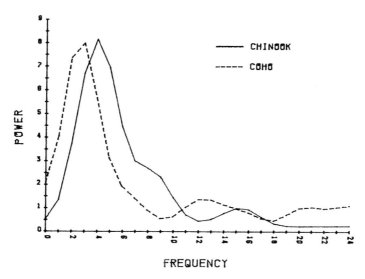

FIGURE 6.—Spectrum plots of creel survey sampling rates (combined scenario) for the Strait of Georgia chinook and coho salmon sport fisheries.

the awareness factor data for both species. Although this analysis does not conclusively identify the cause of the periodicity, the yearly periodicity in sampling rates and mark incidence may have been responsible.

Sampling rates for chinook and coho salmon both averaged mean value of 0.054 (5.4% of the catch was sampled); standard errors were 0.007 and 0.008, respectively. Average mark incidence values were slightly but not significantly different for each species: 0.029 for chinook and 0.036 for coho salmon. Their associated CVs, however, were quite large: 64.6% and 68.2% for chinook and coho salmon respectively.

Mark incidence for each species was independent of voluntary returns (Figure 7), which supports the assumption that anglers do not selectively fish for marks. However, the data suggest that, for each species, mark incidence is inversely proportion to catch (Figure 8). This finding disturbs me because it suggests that fewer marks are encountered at higher sport catches or when sampling effort is highest. Sample size was linearly related to estimated sport catch for each species (Figure 9). Voluntary returns were similarly (if more noisily) related to estimated catch (Figure 10), indicating that anglers encountered marks at a relatively steady rate. These data are not consistent with an inverse relationship between mark incidence and sport catch. It is my belief that such an incon-

sistency could occur if, in the creel survey, mark incidence were underestimated at higher catches.

To test this hypothesis, I fitted a negative binomial model with varying element sizes (Bissel 1972) to data from the monthly combined scenario; the number of marks found in a sample was the measured event and sample size was the varying element. This fit indicated that, at high sample sizes, the observed number of recovered marks was less than the expected number of marks projected by the model (Figure 11).

The effect of underestimating mark incidence at high sport catches, given the increase in voluntary returns with higher catches, is to overestimate awareness factors. From equation (5), even a difference in mark incidence of one percentage point makes a great difference in calculated awareness factor. For example, let 120 marks be returned in a stratum and the estimated catch be 10,000. If the true mark incidence is 3%, the awareness factor is 0.4. If the mark incidence is mistakenly estimated to be 4%, the awareness factor is 0.3. When the new values are expanded back to calculate the estimated mark contribution, a difference of 100 marks results.

Given the high CVs found in mark incidence data, I conclude that the high variability in mean awareness factor is caused by errors in the estimation of mark incidence that bias the calculations of awareness factors. The apparent time

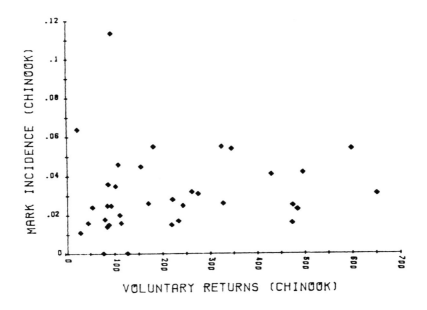

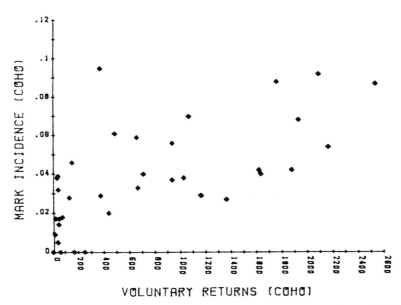

FIGURE 7.—Voluntary returns of chinook and coho salmon marks by anglers in relation to the incidence of marked fish in the Strait of Georgia sport-fish samples, July 1980–July 1984.

dynamics that indicate seasonal changes in awareness factor may be due to this bias.

The cause of the periodicity in estimated awareness factor is difficult to assess. I showed that the sampling effort is reasonably proportional to catch, as are the voluntary returns. However, as the catch increases, mark incidence decreases. This situation could lead to a periodic trend in awareness factors because higher catches occur in summer months. The reasons for decreased mark incidence at higher catches may be due to several problems in the sampling design of the creel survey. Chief among them are limited personnel, flawed sampling, and mark dilution.

Personnel limitations are obvious in the creel survey. As catch increases, fewer marks are ob-

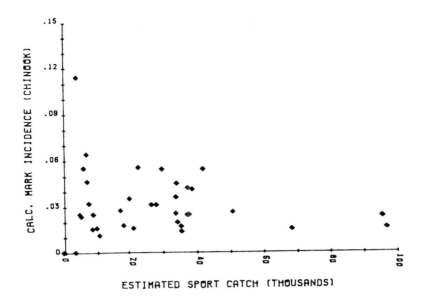

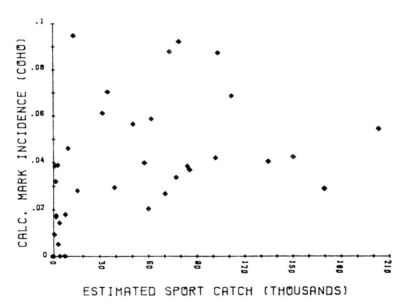

FIGURE 8.—Calculated incidences of marked chinook and coho salmon in the Strait of Georgia sport-fish samples in relation to the number of fish caught by anglers, July 1980–July 1984.

served and counted because of an increased work load for the sampling crews.

A paneling problem (Williams 1978) in the sampling design also affects summer data. Creel survey crews conduct interviews at selected marinas and boat ramps. In the intense fishing months of summer, a great many boats are permanently on the water, and many boats land at private residences and marinas. Few if any of these are sampled. Statistical area 15 is one where sampling is very difficult because of this problem (T. Hoyt, personal communication). If the sampled groups represent a significantly larger proportion of the angler population in the summer, and if these groups

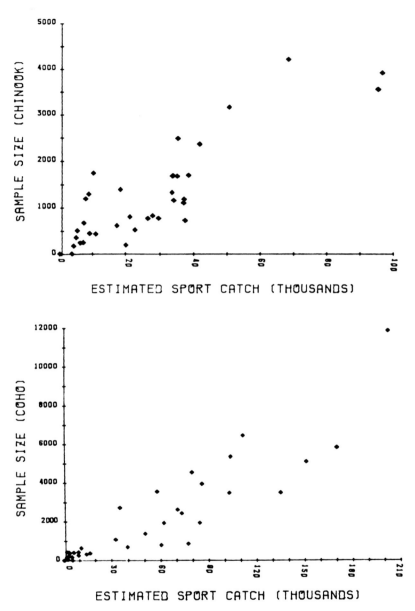

FIGURE 9.—Sample sizes of sport-caught chinook and coho salmon examined for marks in relation to the number of these fish caught by Strait of Georgia anglers, July 1980–July 1984.

have a higher catch per unit effort than the unsampled population, the catch estimate will be biased. Secondly, if each such group encounters and returns marks at the same rate, an apparent underestimate in mark incidence will occur, resulting in a biased, variable calculated awareness factor.

A mark dilution phenomenon also is possible. Recovery of marks is rare at best, and a biased

underestimate of mark incidence can easily result from low sampling effort. The 5% sampling rate is undoubtedly too low. Reliable estimation of mark incidence can only be accomplished by increased sampling effort.

Although a possible solution to errors in mark incidence estimates may be to use calculated mark incidence from the commercial hook-and-

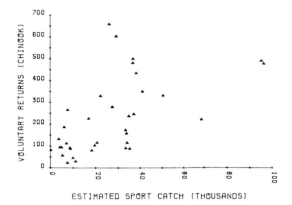

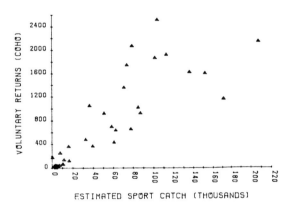

FIGURE 10.—Voluntary returns of chinook and coho salmon marks by anglers in relation to the number of these fish caught in the Strait of Georgia sport fishery, July 1980–July 1984.

line fisheries, this is inappropriate because different age-classes contribute to the sport fishery. Without knowing the age distribution in each fishery, the mark incidence from the commercial hook-and-line fishery is badly biased with respect to the sport fishery.

Conclusion

It is clear from this study that calculated mean awareness factors for the Strait of Georgia sport fishery are highly variable. Consequently, their use in estimates of mark contribution and back-calculations of sport catches is statistically unreliable. The mechanisms causing most of the variability remain unclear, but the evidence points to biased mark incidence estimations as the most likely cause.

Finding a mark is a rare event, and low sampling effort has badly biased the estimation of mark incidence. I conclude that the sampling effort for mark incidence is dangerously low. In the commercial fisheries, 20% of the landed catch is sampled, and this should probably be the goal of the creel survey also.

The survey design for the Strait of Georgia creel survey should be reexamined for paneling problems.

Acknowledgments

I thank M. Birch, P. Starr, D. Bailey, K. Pitre, S. Argue, and J. Mason for their assistance and valuable comments and T. Shardlow and T. Hoyt for supplying data from the creel surveys.

References

Argue, A. W., J. Coursley, and G. D. Harris. 1977. Preliminary revision of Georgia Strait and Juan de Fuca Strait tidal salmon sport catch statistics, 1972 to 1976, based on Georgia Strait head recovery program data. Canadian Department of Fisheries and Oceans, Technical Report Series PAC/T-77-16, Ottawa.

Aubanel, E. E., and K. B. Oldam. 1985. Fourier smoothing without the fast fourier transform. Byte 10(2):207–218.

Bissel, A. F. 1972. A negative binomial model with varying element sizes. Biometrika 59:435–441.

Box, G. E. P., and G. M. Jenkins. 1976. Time series analysis: forcasting and control. Holden-Day, San Francisco.

Chambers, J. M., W. S. Cleveland, B. Kleiner, and P. A. Tukey. 1983. Graphical methods for data analysis. Wadsworth, Belmont, California.

Cochran, W. G. 1977. Sampling techniques, 3rd edition. Wiley, New York.

DPA (DPA Consulting Ltd.) 1982. The Georgia Strait sport fishing creel survey: final report. Report to Canadian Department of Fisheries and Oceans, Vancouver, Canada.

Efron, B. 1982. The jackknife, the bootstrap, and other resampling plans. SIAM (Society for Industrial and Applied Mathematics) Review 38.

Frishman, F. 1975. On the means and variances of products and ratios of random variables. Pages 401–406 in G. P. Patil, editor. Statistical distributions in scientific work, volume 1. Reidel, Dordrecht, The Netherlands.

Goodman, L. A. 1960. On the exact variance of products. Journal of the American Statistical Association 55:708–713.

Kimura, D. K. 1976. Estimating the total number of marked fish present in a catch. Transactions of the American Fisheries Society 105:664–668.

LGL (LGL Environmental Research Associates) and ESSA (ESSA Environmental Systems Analysts Ltd). 1982. Information for management of the Strait of Georgia salmon sport fishery: data analysis, problem analysis, and recommendations. Sidney, Canada.

Mosteller, F., and J. W. Tukey. 1977. Data analysis and regression: a second course in statistics. Addison-Wesley, Reading, Massachusetts.

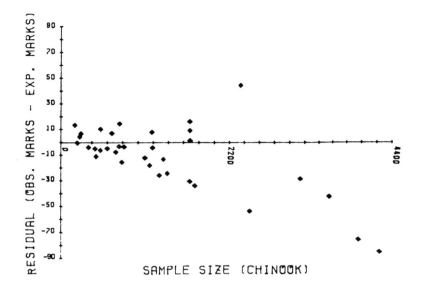

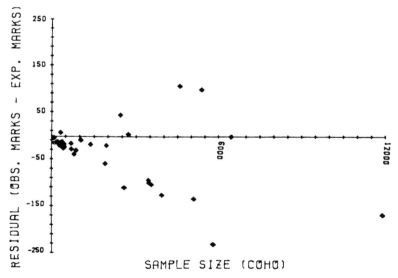

FIGURE 11.—Residuals of the difference between observed and expected chinook and coho salmon marks returned by anglers in relation to sample size for the Strait of Georgia sport fishery. The residuals resulted from fits of observed marks to a negative binomial model in which sample size was the varying element.

Pearse, P. H. 1982. Turning the tide: a new policy for Canada's Pacific fisheries. The Commission on Pacific Fisheries policy: final report. Canada Ministry of Supply and Services, FS23-18/1982, Ottawa.

Shardlow, T. 1983. Georgia Strait creel survey 1981–1982: interim report. Canada Department of Fisheries and Oceans, South Coast Division, Nanaimo.

Williams, B. 1978. A sampler on sampling. Wiley, New York.

American Fisheries Society Symposium 7:647–654, 1990
© Copyright by the American Fisheries Society 1990

Angler Use of Black Crappie and the Effects of a Reward-Tag Program at Jamesport Community Lake, Missouri

STEPHEN EDER

Missouri Department of Conservation, Cameron, Missouri 64429, USA

Abstract.—The effect of angler harvest on black crappie *Pomoxis nigromaculatus* was evaluated from 1979 through 1987 at Jamesport Community Lake, a 12-hectare public impoundment in northwest Missouri. Total annual angler hours spent on black crappies ranged from 64.5 to 166.3 h/hectare (mean, 116.6 h/hectare). Annual harvests of black crappies ranged from 5.5 to 36.0 kg/hectare (mean, 18.7 kg/hectare). Angler exploitation rates for black crappies (≥ 203 mm) were estimated from returns of reward-tagged fish from 1983 through 1987. Reward values ranged from US$5 to $100. Total annual fishing pressure for black crappies increased significantly ($P < 0.08$) after rewards were offered. Total harvest of black crappies averaged 20.5 kg/hectare before rewards and 16.7 kg/hectare during the reward period. The presence of reward tags did not alter the harvest behavior of crappie anglers: 98% of the crappies (≥ 203 mm) caught were kept both before and during the reward program. First-year exploitation of tagged fish ranged from 52% to 89% and averaged 72%. Total exploitation of tagged groups ranged from 75% to 92% and averaged 84%. There was no significant difference ($P > 0.05$) in the rate of return of tagged black crappies by 25-mm size-groups. The rate of noncompliance for anglers who caught tagged fish but did not submit the tags was estimated at 7%, based on actual creel-clerk contacts. Population assessments based on recruitment, growth, relative abundance, and size distribution revealed that a good black crappie fishery persisted despite the heavy exploitation of adult fish.

Populations of crappie *Pomoxis* spp. commonly show high recruitment, slow growth of fish under 200 mm in length during the third or fourth year of life, then high mortality by age 5 (Ellison 1984). Numerous techniques have been tried to improve growth and size structure of high-density crappie populations. Introductions of such forage species as gizzard shad *Dorosoma cepedianum* (Anderson 1983) and threadfin shad *Dorosoma petenense* (Mosher 1984) proved beneficial in some instances. Mechanical reduction of populations with fyke nets (Hanson et al. 1983), chemical culling (Rutledge and Barron 1972), commercial harvest (Schramm et al. 1985), and high densities of largemouth bass *Micropterus salmoides* (Gabelhouse 1984; Boxrucker 1987) have also been employed to improve slow-growing crappie populations. There are few descriptions in the literature of sustained high-quality crappie populations in small impoundments. The objectives of this study were to determine the age composition, growth, and size-structure of a population of black crappie *Pomoxis nigromaculatus* in a small impoundment in northwest Missouri and to measure angler exploitation of adult black crappies in this popular fishery.

mately 129 km northeast of Kansas City. The lake covers 12 surface hectares, has a maximum depth of 5 m, and has a 290-hectare watershed consisting of timber, pasture, and row crops.

The lake was stocked initially with fingerlings of largemouth bass, bluegills *Lepomis macrochirus*, and channel catfish *Ictalurus punctatus* on September 17, 1956, and was opened to public fishing on May 29, 1958. Black crappies were not stocked by the Missouri Department of Conservation but were present in 1968 when annual electrofishing surveys began. Gizzard shad were stocked in 1972 to broaden the forage base, and a population of redear sunfish *Lepomis microlophus* was established with stockings in 1972 and 1974. Supplemental stockings of 203- to 254-mm channel catfish have been made annually since 1962 at the rate of 25 to 100 fish/hectare. The largemouth bass population has been managed from the opening under a variety of regulations, beginning with no length limit and progressing through a 305-mm minimum length limit and a 305- to 380-mm protected length range to the current 381-mm minimum length. Daily limits are 4 fish each for largemouth bass and channel catfish and 30 fish for black crappie.

Study Site

Jamesport Community Lake was built by the Missouri Department of Conservation in 1956. It is 3 km northwest of Jamesport and approxi-

Methods

A randomized creel census by a roving clerk was used to obtain annual daytime angling estimates, April through October, from 1979 through 1987.

From 1979 through 1983, 2-h angler surveys were
made once daily for an average of 15 d/month. From
1984 through 1987, 1-h angler surveys were made
twice daily for an average of 20 d/month. For the
purpose of this study, local anglers were defined as
those who resided within 32 km of the lake.

Annual spring electrofishing collections were
made with a boat equipped with a 220-V DC
system. Four to six sampling trips were made
each spring. A sampling trip consisted of one
complete shoreline circuit. Size-structure evaluations of electrofishing collections were based on
proportional stock density (PSD) and relative
stock density (RSD) indices; minimum stock,
quality, and relative stock sizes of black crappie
were set at 127 mm, 203 mm, and 254 mm total
length, respectively (Anderson 1980; PSD is the
numerical ratio of quality- to stock-size fish and
RSD is the proportion of any size-group in the
population; both ratios are multiplied by 100).
Scale samples were collected prior to or at the
onset of annulus formation each spring, and
growth of black crappies was determined annually
from mean lengths recorded at capture for the
various age-groups.

Beginning in 1983, annual spring trap-netting
collections of black crappies were made. Four to
six nets were set each trip, and total annual effort
ranged from 35 to 37 net-days. The quality of the
black crappie population was assessed with the
method proposed by Colvin and Vasey (1986) for
Missouri reservoirs based on October trap-net collections, except that age-1 fish collected in the
spring at Jamesport Community Lake were substituted in the assessment for age-0 fish in the recommended fall sample. Spring trap-netting was substituted for fall sampling because of personnel
availability and to avoid overwinter loss of tagged
fish. Some bias may have been incorporated in the
assessment system because of this change.

Exploitation of black crappies was estimated
from anglers' tag returns. One hundred black
crappies at least 203 mm long were marked each
spring, 1983–1986, with numbered FD-68B Floy
tags, which were anchored between the posterior
pterygiophores of the dorsal fin. All tags carried
the inscription that the reward value could range
from $5 to $100. Of the 100 tags, 1 carried a
US$100 value, 5 carried a $20 value, and the rest
were worth $5 apiece, for a total of $670. Anglers
were not aware of the exact value of a tag until it
was returned. Signs were posted on bulletin
boards, and newspaper articles were distributed
to local newspapers to inform anglers of the

study. Tag-return cards were distributed to anglers by the county conservation officer, the creel
clerk, and fisheries personnel. Cards provided
space for information on tag number, capture
date, length, and specific location of capture.
Seven capture sites were designated. They consisted of the dam, three cove areas, and brush
piles in each of the three coves.

Annual springtime estimates of the adult black
crappie population ($\geq$203 mm) were based on tag
returns reported in the creel census and on the
Petersen equation

$$N = \frac{M \cdot C}{R};$$

N = estimated number of black crappies $\geq$203
mm;
M = 100 tagged fish;
C = total number of fish $\geq$203 mm recorded by
the creel clerk from April through June;
R = total number of tagged fish recorded by
the creel clerk from April through June.

I made another population estimate, again using
tag returns and creel census data, for which

$$N = \frac{M \cdot E}{T};$$

N = estimated number of black crappies $\geq$203
mm;
M = 100 tagged fish;
E = estimated angler catch of fish $\geq$203 mm
from April through June;
T = number of tags received from April–June
angler catches.

Statistical analyses of the data were performed
with the Statistical Analysis System (SAS) and
Statistix II.

Results

First-year angler exploitation of tagged black
crappies ranged from 52% to 89% and averaged
72% (Table 1). By the end of 1987, total angler
exploitation of tagged groups ranged from 75% to
92% and averaged 84% (Table 1). Total mortality
of tagged fish appeared to be 100% within 3 years
of marking because no tags were returned from
the 1983 and 1984 groups in 1986 and 1987 (Table
1). Exploitation figures were not adjusted for tag
loss, tagging mortality, or nonreporting by anglers. Angler compliance was estimated at 93%
from 1983 through 1987, during which time I

TABLE 1.—Annual and total angler exploitation of tagged adult black crappies ($\geq$203 mm) at Jamesport Community Lake.

Year of tagging	Number of fish tagged	Number of fish caught in					
		1983	1984	1985	1986	1987	Total
1983	100	67	13	1	0	0	81
1984	100		52	22	1	0	75
1985	100			79	5	2	86
1986	100				89	3	92
Total	400	67	65	102	95	5	334

received 39 tags from the 42 tagged fish recorded by the creel clerk.

The size of tagged fish did not appear to bias my estimates of the exploitation of the adult black crappie population. Even though there were noticeable differences in the rate of return by size (Table 2), these differences were not statistically significant ($P > 0.05$).

There was a significant increase ($P < 0.08$) in the fishing hours spent on black crappies after the reward-tag program began (Table 3). Crappie anglers averaged 96 h/hectare on the lake before the use of reward tags and 137 h/hectare afterwards. The increase could not be attributed to the presence of larger fish, because no meaningful correlation was observed between fishing pressure and black crappie PSD ($r = -0.36$) or RSD ($r = -0.10$), which were computed from electrofishing data (Table 3). The increase in black crappie fishing pressure after rewards were offered was not attributable to increased angler success because there was a weak negative relationship ($r = -0.46$) between catch rates and fishing pressure. Therefore, it appears that monetary rewards were primarily responsible for the observed increase in black crappie fishing pressure.

Reward tags did not change the composition of the lake's angling clientele. Local anglers accounted for 78% and 79% of the black crappie fishing prior to and during the reward-tag period, respectively.

Rewards did not alter the seasonal fishing patterns of black crappie anglers (Figure 1). With no tags present, 80% of the crappie fishing occurred in April and May, versus 85% when the reward-tag program was in effect.

Reward tags did not change the harvest ethic of Jamesport anglers. They kept 98% of their black crappie catch ($\geq$203 mm) before and after rewards were offered. Garner et al. (1987) pointed out that exploitation can be grossly overestimated from tagged fish if there is a significant amount of catch-and-release fishing. This was not a factor at James-port Community Lake because 100% of the tagged black crappies recorded by the creel clerk were harvested. Anglers also kept most black crappies (88%) in the 178–202-mm size range, but only 38% of the smaller fish in the catch were harvested.

As expected, most of the annual black crappie harvest occurred in the spring during the period of greatest angling pressure (Figure 2). Eighty-three percent of the black crappie harvest occurred during April and May from 1979 through 1982, and 90% during the same months from 1983 through 1986. A greater proportion (97%) of the tagged-fish harvest took place in April and May from 1983 through 1986. According to the tag-return cards, 84–95% of the harvested tagged fish were caught along the dam from 1983 to 1987. Tag-return cards also revealed that 70% of the fish caught within 4 weeks of tagging were measured or estimated within 12 mm of their actual total length by anglers.

Annual disbursements ranged from $400 to $725 and averaged $554. For the entire study period, 83% of available funds were used.

According to the trap-net assessment method of Colvin and Vasey (1986), the black crappie fishery remained in good condition from 1983 to 1986, despite high angler exploitation (Table 4). The population received a poor rating in 1987, primarily because of slow growth. The 1984 year-class failed to reach at least 203 mm by age 3, which led to a low proportion of fish 229 mm and larger. The

TABLE 2.—Size range and first-year rate of return of the 400 black crappies tagged at Jamesport Community Lake from 1983 through 1986.

Total length (mm)	Number tagged	First-year catch	
		Number	%
203–228	88	53	60.2
229–253	208	152	73.1
254–278	90	71	78.9
279–304	12	6	50.0
330–354	2	1	50.0

TABLE 3.—Characteristics of the black crappie sport fishery and size-structure indices from annual spring electrofishing samples at Jamesport Community Lake.

	Year							
	Without reward tags				With reward tags			
Statistic	1979	1980	1981	1982	1983	1984	1985	1986
Total angling pressure (h/hectare)	625	618	620	748	795	700	638	513
Crappie angling pressure (h/hectare)	92	122	65	106	156	166	132	94
Crappie angler catch rate (number/h)	1.24	1.36	1.46	0.35	1.02	0.51	0.34	0.79
Crappie harvest (kg/hectare)	21.6	36.0	19.0	5.5	30.7	13.8	8.8	13.5
Crappie harvest (number/hectare)	113	187	93	39	179	83	50	76
Mean length harvested (mm)	218	216	231	208	221	218	224	226
Crappie proportional stock density	20	22	77	35	52	27	33	35
Crappie relative stock density	15	0	8	2	11	4	11	5

slower growth appeared to be density dependent because it occurred after the production of the two largest black crappie year-classes, in 1984 and 1985 (Table 4).

According to the Petersen estimates, black crappie ($\geq$203 mm) densities varied from 58 to 408 fish/hectare from 1983 through 1986 (Table 5). The proportion estimator had a smaller range: 52–208 fish/hectare. Both estimators showed strong positive correlations, $r = 0.97$ and $r = 0.94$, respectively, to estimated springtime catches of black crappies of 203 mm and larger.

Trap-net catch rates were more reliable ($r = 0.84$) than electrofishing catch rates ($r = 0.54$) for predicting estimated angler catches (Table 5). Electrofishing catch rates and trap-net catch rates did not provide comparable estimates of relative abundance for adult black crappies ($\geq$203 mm; $r = -0.38$) or young black crappies ($<$127 mm; $r = 0.08$).

Discussion

Angler cooperation is a major concern when tag-return data are used to estimate exploitation.

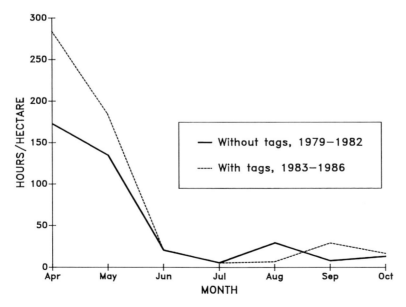

FIGURE 1.—Mean monthly angler-hours spent on black crappies before (1979–1982) and during (1983–1986) the reward-tag program.

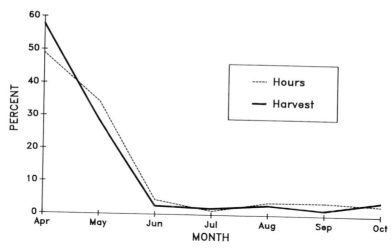

FIGURE 2.—Mean monthly distributions of black crappie angler-hours and harvest from Jamesport Community Lake, 1979–1986.

Rawstron (1971) compared the return rate of reward tags to nonreward tags for bluegills, largemouth bass, and white catfish *I. catus* at Folsom Lake, California, and he estimated nonresponse to be 69%, 46%, and 39%, respectively. In a similar study on largemouth bass at Merle Collins Reservoir, Rawstron (1972) reported a mean nonresponse of 34% over 4 years. Folmar et al (1980), after documenting 66% nonresponse by anglers who captured largemouth bass with nonreward tags, concluded that a variable reward system was a more effective and economical means of stimulating angler response. However, Garner et al. (1987) found that the nonreporting rate for reward tags with $1 to $100 values ranged from 29% to 31% at two North Carolina reservoirs. Weaver and England (1986) observed higher returns for reward tags that bore no specific dollar value, and they suggested that angler curiosity about the value of the tags may have contributed to a better response. The variable reward system used at

Jamesport Community Lake appeared to provide the necessary incentive to encourage good angler cooperation, because estimated noncompliance was only 7%. I did not estimate tag retention or tagging mortality. The rapid and high recapture rates for tagged fish minimized any loss in accuracy of the exploitation estimate.

Black crappie exploitation estimates probably would have been lower if fish less than 203 mm had been used in the study. Jamesport anglers were more inclined to release smaller fish, particularly those measuring less than 178 mm. Adding sexually immature fish less than 203 mm to the marked pool at Jamesport Community Lake could have reduced the spring exploitation rate because of reduced vulnerability to bank anglers. In addition, when fish between 203 and 216 mm were tagged, greater care was needed to insure the proper insertion angle without penetration of the skin on the opposite side of the dorsal fin. I felt that tagging smaller fish would have produced

TABLE 4.—Population variables and assessment of the black crappie fishery as reflected by spring trap-net samples at Jamesport Community Lake.

Year	Net-days	Number of age-1 fish per net-day	Number of age-2 and older fish per net-day	Percent over 229 mm[a]	Mean total length at age 3 (mm)	Percent age 4 and older[a]	Assessment points	Rating
1983	36	2.0	4.5	46.0	257	2.4	26	Good
1984	37	6.1	5.1	33.5	234	3.7	32	Good
1985	35	9.0	3.7	44.5	241	10.3	36	Very good
1986	35	2.2	2.7	35.8	234	1.9	26	Good
1987	36	2.0	3.8	6.6	201	5.1	12	Poor

[a]Excludes age-1 black crappies.

TABLE 5.—Measurements of relative abundance from trap nets, electrofishing, and angler catch rates, of estimated densities, and of estimated catch of black crappies (≥203 mm) at Jamesport Community Lake.

Statistic	Year				
	1983	1984	1985	1986	1987
Number/net-day	2.9	2.4	1.7	1.6	0.7
Number/electrofishing-hour	8.0	3.3	5.6	13.3	2.8
Number/angler-hour	0.84	0.40	0.29	0.75	0.14
Petersen population estimate (number/hectare)	408	114	58	107	
Proportion estimate (number/hectare)	208	110	52	87	
Estimated catch (number/hectare)	139	56	41	77	14

additional trauma that could have increased tagging mortality.

Studies of angler exploitation of crappies are few. Early studies by Wegener and Clugston (1967) and Copeland and Huish (1965) in Florida indicated statewide tag-return rates for black crappies of 6% and 9%, respectively. More recently, Schramm et al. (1985) estimated the annual sport harvest of black crappies to be 10% at Lake Okeechobee, Florida. At Branched Oak Lake in Nebraska, 12–25% of the black crappies and white crappies *Pomoxis annularis* were caught by sport fishermen over a 3-year period (S. C. Shainost, Nebraska Game and Parks Commission, unpublished data). At Truman Lake, Missouri, 28% of reward-tagged crappies (≥203 mm) were captured by anglers in 1986, and 46% of tagged crappies (≥229 mm) were caught in 1987 (R. J. Dent, Missouri Department of Conservation, unpublished data). First-year exploitation of reward-tagged crappies was 48% at Stockton Lake and 42–56% at Pomme de Terre Lake in Missouri (M. A. Colvin, Missouri Department of Conservation, unpublished data). In comparison to these studies, use of the black crappie fishery at Jamesport Community Lake was intensive.

Because of their selection of spawning sites, the majority of adult black crappies at Jamesport Community Lake were subjected to intensive fishing pressure from bank fishermen, who made up 78% of the angling clientele. The dam area was the only location along the shoreline where spawning concentrations were found during our spring electrofishing surveys, and local anglers were well aware of these concentrations. Mitzner (1987) found that two important components of high-quality crappie spawning habitat at Rathbun Lake, Iowa, were a firm substrate and protection from heavy wind and wave action. Jamesport black crappies may be attracted to the dam area in the spring because it offers both a firm bottom substrate off the riprapped face and protection from the prevailing southerly winds.

Even though informational releases were confined to local newspapers, mean annual crappie angler-hours increased by 42% after rewards were offered at Jamesport Community Lake. If fishing pressure had been positively related to exploitation, rewards would have created an overestimate of angler impact on the fishery. However, during the reward period the highest first-year exploitation of tagged black crappies coincided with the lowest amount of fishing pressure in 1986 (Tables 1, 3), and a negative relationship ($r = -0.94$) was observed between the two variables. Consequently, I felt that the reward-tag program provided an accurate representation of normal angler exploitation.

There was a negative relationship ($r = -0.85$) between first-year exploitation and the percentage of black crappies at least 229 mm long available the following spring, but it was not enough to seriously affect the quality of the fishery. With the exception of 1987, the spring black crappie population was rated from good to very good after high exploitation of adults (Table 4). Deterioration in the fishery from 1986 to 1987 appeared to be more a product of excessive recruitment and slow growth than of overharvest because there was a strong correlation ($r = 0.99$) between the availability of large black crappies (≥229 mm) and mean length at age 3.

Crappie populations have benefited from major reductions in standing crop. Rutledge and Barron (1972) found that a 29% reduction in the standing crop of white crappies in a 28-hectare Texas impoundment improved the size structure and condition of the population. Hanson et al. (1983) documented an appreciable improvement in black crappie growth rates when the density of black crappies longer than 120 mm was reduced to 10% (271/hectare) of the number removed by trap netting (2,822/hectare). Schramm et al. (1985) reported that an annual combined exploitation rate of 65% for sport and commercial harvests at Lake Okeechobee produced rapid improvements in the growth of black crappies without signs of overharvest. Even though the 84% average exploitation of black crappies at Jamesport Community Lake limited the proportion of 229-mm and larger fish, high angler use could be beneficial for removing the slow-growing segment of the population.

Adding reward tags might promote angling interest in poor-quality crappie fisheries. After a reward system similar to that of Jamesport Community Lake was established in the spring of 1987 at Ray County Community Lake, a 10-hectare impoundment, spring fishing on a stunted white crappie population increased from less than 2 h/hectare in 1986 to 98 h/hectare in 1987. With the increase in fishing pressure, I documented a corresponding increase in harvest from 0.8 to 22.5 kg/hectare despite deterioration in the size-structure from a PSD of 21 in 1986 to a PSD of 4 in 1987.

Conclusion

The variable tag values of $5 to $100 produced a reliable, unbiased estimate of angler exploitation of adult black crappies at Jamesport Community Lake. High spring exploitation resulted because anglers fished intensively when black crappies gathered to spawn along a small stretch of the dam's shoreline. When high exploitation of adults was combined with moderate recruitment and satisfactory growth rates, a good-quality black crappie fishery was maintained at Jamesport Community Lake. When combined with creel census data, tag returns provided estimates of black crappie densities that are difficult to obtain from recapture efforts dependent on netting and electrofishing. Based on my findings at Jamesport Community Lake and Ray County Community Lake, I conclude that increases in fishing pressure generated by reward tags are more likely to affect annual harvests when angler interest has been low prior to rewards.

Acknowledgments

My special thanks go to Carolyn Fewins and Howard Cox, who served as creel clerks during the study. I appreciated the able assistance provided by J. B. Lowe, Rob Goin, Tom Elliott, and Tom Reed during field sampling. I also thank Pam Haverland for her help with the statistical analyses of the data and Jake Allman and Fred Vasey for their manuscript reviews.

References

Anderson, R. O. 1980. Proportional stock density (PSD) and relative weight (Wr): interpretive indices for fish populations and communities. Pages 27–33 in S. Gloss and B. Shupp, editors. Practical fisheries management: more with less in the 1980's. American Fisheries Society, New York Chapter. (Available from New York Cooperative Fish and Wildlife Research Unit, Ithaca.)

Anderson, W. M. 1983. Effect of stocking gizzard shad on the fishery in Little Dixie Lake, Missouri. Pages 58–76 in D. Bonneau and G. Radonski, editors. Proceedings of small lakes management workshop "pros and cons of shad." Iowa Conservation Commission, Des Moines.

Boxrucker, J. 1987. Largemouth bass influence on size structure of crappie populations in small Oklahoma impoundments. North American Journal of Fisheries Management 7:273–278.

Colvin, M. A., and F. W. Vasey. 1986. A method of qualitatively assessing white crappie populations in Missouri reservoirs. Pages 79–85 in G. E. Hall and M. J. Van Den Avyle, editors. Reservoir fisheries management: strategies for the 80's. American Fisheries Society, Southern Division, Reservoir Committee, Bethesda, Maryland.

Copeland, J. B., and M. T. Huish. 1965. A description and some results of a Florida statewide fish tagging program. Proceedings of the Annual Conference Southeastern Association of Game and Fish Commissioners 16:242–246.

Ellison, D. G. 1984. Trophic dynamics of a Nebraska black crappie and white crappie population. North American Journal of Fisheries Management 4:355–364.

Folmar, H. G., W. D. Davies, and W. L. Shelton. 1980. Factors affecting estimates of fishing mortality of largemouth bass in a southeastern reservoir. Proceedings of the Annual Conference Southeastern Association of Game and Fish Commissioners 33:402–407.

Gabelhouse, D. W., Jr. 1984. An assessment of crappie stocks in small midwestern impoundments. North American Journal of Fisheries Management 4:371–384.

Garner, K. E., F. A. Harris, and S. L. Van Horn. 1987. Catch/release bias in reward tag exploitation studies. Proceedings of the Annual Conference Southeastern Association of Game and Fish Commissioners 38:579–582.

Hanson, D. A., B. J. Belonger, and D. L. Schoenike. 1983. Evaluation of a mechanical population reduction of black crappie and black bullheads in a small Wisconsin lake. North American Journal of Fisheries Management 3:41–47.

Mitzner, L. R. 1987. Classification of crappie spawning habitat in Rathbun Lake, Iowa with reference to temperature, turbidity, substrate and wind. Iowa Department of Natural Resources, Technical Bulletin 1, Des Moines.

Mosher, T. D. 1984. Responses of white crappie and black crappie to threadfin shad introductions in a lake containing gizzard shad. North American Journal of Fisheries Management 4:365–370.

Rawstron, R. R. 1971. Nonreporting of tagged white catfish, largemouth bass, and bluegills by anglers in Folsom Lake. California Fish and Game 57:246–252.

Rawstron, R. R. 1972. Nonreporting of tagged largemouth bass, 1966–1969. California Fish and Game 58:145–147.

Rutledge, W. P., and J. C. Barron. 1972. The effects of the removal of stunted white crappie on the remaining crappie population of Meridian State Park Lake, Bosque, Texas. Texas Parks and Wildlife Department, Technical Series 12, Austin.

Schramm, H. L., Jr., J. V. Shireman, D. E. Hammond, and D. M. Powell. 1985. Effect of commercial harvest of sport fish on the black crappie population in Lake Okeechobee, Florida. North American Journal of Fisheries Management 5:217–226.

Weaver, O. R., and R. H. England. 1986. Return of tags with different rewards in Lake Lanier, Georgia. North American Journal of Fisheries Management 6:132–133.

Wegener, W. L., and J. P. Clugston. 1967. Florida's state-wide tagging program. Proceedings of the Annual Conference Southeastern Association of Game and Fish Commissioners 18:239–247.

American Fisheries Society Symposium 7:655–659, 1990

Effects of Monetary Rewards and Jaw-Tag Placement on Angler Reporting Rates for Walleyes and Smallmouth Bass

ROBERT C. HAAS

Michigan Department of Natural Resources, Lake St. Clair Fisheries Station
Mt. Clemens, Michigan 48045, USA

Abstract.—I conducted two tag studies to examine aspects of angler tag-reporting behavior. I used large numbers of walleyes *Stizostedion vitreum* and smallmouth bass *Micropterus dolomieui*. The fish had been tagged in Lake St. Clair from 1974 through 1985 as part of a population dynamics study, and anglers were asked to voluntarily report recoveries of the monel metal jaw tags used to mark both species. I conducted a 3-year reward-tag study to examine the extent of nonreporting of tag recoveries. A chi-square test showed that significantly more reward than nonreward tags were reported by anglers for both fish species ($P < 0.05$). I found little or no detectable differences in reporting for rewards of US$2, $4, $6, and $8. To estimate the reporting rate for nonreward tags, I combined reward denominations for both species, which allowed adjustment of exploitation estimates generated from 11 years of voluntary tag returns. In a separate but related 2-year study, I examined whether tag placement on the upper or lower jaw influenced angler recovery and reporting rates. The recovery data indicated that anglers reported more walleyes with tagged lower jaws, but no corresponding differences appeared for smallmouth bass.

Tagging was used by the Michigan Department of Natural Resources (MDNR) on Lake St. Clair and Lake Erie to assess the population dynamics of and fisheries for walleye *Stizostedion vitreum* and smallmouth bass *Micropterus dolomieui*. During this research, questions arose concerning two of the six assumptions for tagging studies listed by Ricker (1975): that mortality characteristics were similar for tagged and untagged fish, and that tags on harvested fish were seen and reported by fishermen. Also, I wanted to examine whether voluntary reports by Lake St. Clair and Lake Erie anglers reflected true conditions in the fishery. To accomplish this, I manipulated segments of the tagged populations of walleyes and smallmouth bass so that comparisons of recapture records would provide pertinent information. Manipulations included tagging some fish with reward tags and changing the tag location from the lower to the upper jaw.

Extensive sport fisheries occur on Lake St. Clair and western Lake Erie, where the annual harvest of walleyes exceeded 8,000,000 fish from 1985 through 1987 (Lake Erie Committee, Great Lakes Fishery Commission, unpublished data). Henny and Burnham (1976) found that reporting rates for banded mallards were lower near banding sites because experienced hunters were familiar with the study. This was not considered to be a significant problem with the Lake St. Clair and Lake Erie studies, particularly for walleyes, because the harvest was spread over so many anglers that the probability of one angler catching more than one tagged fish was very small.

The data reported here were gathered during extensive netting and tagging studies. Eleven surveys were carried out on Lake St. Clair from 1975 through 1985 and 10 surveys on Lake Erie from 1978 through 1987. The average Lake St. Clair survey began May 23 and lasted 29 d, whereas the average Lake Erie survey began April 17 and lasted 20 d. The yearly mean surface water temperature at the Lake St. Clair net site varied from 14 to 20°C; the overall weighted mean was 16°C. The Lake Erie temperature varied from 9 to 15°C and averaged 11°C overall.

Methods

The surveys on Lake Erie and Lake St. Clair were very similar in time and duration each year, and the nets were tended on all weekdays if weather permitted. Five trap nets were set at the same location each year during the spring and fished through the entire survey. We minimized potential bias from annual variations in the fish populations and fisheries by tagging fairly large numbers of each species over a long period and allowing several years for tag recovery. Still, there was concern about whether the voluntary, nonreward-tag recovery data represented the real populations.

Walleyes and smallmouth bass were removed from the nets and placed on board the boat in a live tank equipped with circulating lake water. Fish were individually tagged without anesthetia

TABLE 1.—Numbers of Lake St. Clair walleyes with reward and nonreward tags, and the number of recoveries reported by anglers in subsequent years.

Reward value (US$)	Number tagged	Recovery year							Percent reported
		1981	1982	1983	1984	1985	1986	1987	
1981 tagging									
$2	79	5	4	1	0	1	0	0	14
$4	77	6	2	3	0	2	1	0	18
$6	80	9	3	0	0	0	0	0	15
$8	81	4	4	1	3	1	0	0	16
Reward	317	24	13	5	3	4	1	0	16
No reward	650	22	22	6	3	2	1	0	9
1982 tagging									
$2	50		3	1	1	1	1	0	14
$4	48		3	3	2	4	0	0	25
$6	51		9	2	1	0	0	0	24
$8	49		5	2	1	0	0	0	16
Reward	198		20	8	5	5	1	0	20
No reward	259		15	10	9	1	3	0	15
1983 tagging									
$2	95			12	3	1	0	1	18
$4	100			12	3	3	1	0	19
$6	100			9	8	2	1	0	20
$8	99			17	5	3	1	1	27
Reward	394			50	19	9	3	2	21
No reward	1,134			94	40	9	3	5	13

and released at the net site. The total length of each tagged fish was measured. Weight measurements and scale samples were taken yearly from representative portions ranging from 36% to 100% of the net catch. If any scale samples were taken, all fish in that trap net were processed; and some net lifts were sampled throughout each survey period for scales.

Fish were tagged with monel metal bands, which were wrapped firmly around the dentary bone of the lower jaw or the maxillary bone of the upper jaw. The size-10 and size-12 tags were inscribed with the local MDNR address and an individual tag number.

A reward-tag study of walleyes and smallmouth bass was carried out on Lake St. Clair from 1981 through 1983 to provide an estimate of the reporting frequency for traditional nonreward tags. Altogether, 909 walleyes were tagged with reward tags and 2,043 with nonreward tags; similar numbers of walleye were tagged each year. The angler recovery patterns for walleye reward and nonreward tags were consistent through the 7-year recovery period (Table 1). There were 738 smallmouth bass tagged with reward tags and 1,585 tagged with nonreward tags. Smaller numbers of smallmouth bass than walleye were tagged and

higher numbers were caught by anglers and reported to MDNR (Table 2).

Reward tags worth $2, $4, $6, and $8 were employed. Each carried an additional inscription that stated its value, such as "Reward $8.00." Every third fish received a reward tag; reward tags were applied in repeating sequence from $2 through $8 to eliminate bias. Funds to pay the rewards were solicited from local conservation organizations.

Most of the walleyes and smallmouth bass were tagged with nonreward tags, and recovery information was solicited on a voluntary basis. An angling fishery predominated throughout the harvest area and accounted for 92% of all walleye tag recoveries and 100% of smallmouth bass recoveries.

Results

Walleye Reward Study

There were 1.6 walleye reward tags reported for every nonreward tag. To determine recovery frequencies, I combined recovery data for the five seasons without regard to calendar year, because the ratio of one reward to two nonreward tags was maintained throughout the 3-year tagging period. This resulted in five sequential recovery periods for each tagged cohort (5 years was essentially the

TABLE 2.—Numbers of Lake St. Clair smallmouth bass with reward and nonreward tags, and the number of recoveries reported by anglers in subsequent years.

Reward value (US$)	Number tagged	Recovery year							Percent reported
		1981	1982	1983	1984	1985	1986	1987	
				1981 tagging					
$2	88	6	4	2	1	0	0	0	15
$4	93	8	9	2	1	0	0	0	22
$6	88	13	2	4	0	0	0	0	22
$8	85	8	4	2	1	0	1	0	19
Reward	354	35	19	10	3	0	1	0	19
No reward	644	43	38	13	1	2	1	0	15
				1982 tagging					
$2	43		5	2	1	0	0	0	19
$4	44		11	4	2	0	0	0	39
$6	39		3	7	1	0	0	0	28
$8	44		4	3	0	0	0	0	16
Reward	170		23	16	4	0	0	0	25
No reward	214		30	10	4	1	0	0	21
				1983 tagging					
$2	54			12	3	1	0	0	30
$4	52			17	6	1	0	0	46
$6	53			13	2	1	0	0	30
$8	55			17	5	0	0	0	40
Reward	214			59	16	3	0	0	36
No reward	727			102	41	13	4	0	22

maximum survival period for any tagged fish). The observed recovery frequency for nonreward tags was then used to calculate the expected recovery frequency for the reward tags, and frequencies of reward recoveries were higher than nonreward frequencies during all recovery seasons (Figure 1). The reporting frequency of walleye reward tags was significantly higher than the frequency of nonreward tags ($\chi^2 = 20.9$, $P < 0.005$).

The reporting rate for walleye reward tags was thought to be related to the monetary value of tags. The recovery frequency for $2 tags was compared with that for each of the larger denominations. The recoveries of the $4 (20.0%), $6 (19.1%), and $8 (21.0%) tags were slightly higher than the $2 (15.6%) tags but did not vary among themselves. There was no detectable difference in the reporting rates of the four walleye reward-tag denominations ($\chi^2 = 1.3$, $P > 0.75$).

Smallmouth Bass Reward Study

There were 1.3 smallmouth bass reward tags reported for every nonreward tag, which reflected a significantly higher rate of reporting ($\chi^2 = 9.7$, $P < 0.005$). Smallmouth bass tags were recovered from anglers at a higher frequency than walleye tags during the first and second fishing seasons but

declined much more rapidly during the next three seasons (Figure 1). The difference in reporting rates apparently occurred only during the first two seasons, because returns of smallmouth bass reward and nonreward tags were similar during the third and subsequent seasons. Small sample sizes during the third through fifth recovery seasons may have precluded detection of real differences.

Recovery frequencies for the $4 (33.0%), $6 (25.6%), and $8 (24.5%) tags were higher than that for $2 (20.0%) tags, but the differences in reporting rates for tagged smallmouth bass did not differ significantly among the four reward denominations ($\chi^2 = 5.4$, $P > 0.25$).

Tag Placement

For the first 2 years of the Lake St. Clair and Lake Erie studies, tags were applied to lower jaws. During 1984, walleye and smallmouth bass were tagged on either the upper or lower jaw to determine the effects of tag placement on reporting rates. Anglers on both lakes captured walleyes tagged on the lower jaw significantly more often than those tagged on the upper jaw (Table 3; $\chi^2 = 6.2$, $P < 0.025$). There was no statistical evidence that smallmouth bass tags placed on the upper jaw

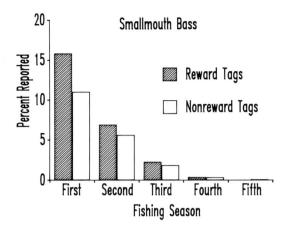

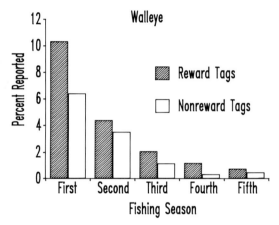

FIGURE 1.—Recoveries by anglers of reward and nonreward jaw tags from smallmouth bass and walleyes during the 5 years after tagged fish were released in Lake St. Clair.

were reported at different frequencies than lower jaw tags ($\chi^2 = 0.6$, $P > 0.50$).

Scale Sampling

Scale samples were collected for analyses from the major portion of the net catch of walleyes and smallmouth bass in Lake St. Clair. Removing scales and gathering weight data took more than twice the time required to tag and release fish, and these processes damaged the fish externally. I compared tag-reporting frequencies for fish that had scales removed versus those that had not been subjected to scale sampling. Tagged walleyes from the two groups were captured and reported by anglers at nearly identical rates (Table 4). Tagged smallmouth bass from the group subjected to scale sampling were reported more frequently. There was no evidence that the additional handling of tagged walleyes and smallmouth bass decreased the probability that they would be recaptured and reported by fishermen because the proportion of these fish in recovery samples was equal to or higher than it was in the tagged population.

Discussion

These studies were carried out to assess the validity of assumptions that tagged fish had essentially the same mortality rates as the unmarked population and that anglers voluntarily reported the nonreward tags they recovered. Different tagged populations were established so that simple comparisons could be made between tag-reporting frequencies. It should be kept in mind that variations in reporting rates may have resulted from additional tagging mortality, tag loss, altered vulnerability to fishing, or different angler response.

The higher reporting rates for reward tags showed that a substantial number of Lake St. Clair walleye

TABLE 3.—Numbers of upper and lower jaw tags applied to walleyes and smallmouth bass, and the numbers subsequently captured and reported by anglers.

Jaw tagged	Number tagged	Recovery year				Total reported	Percent reported
		1984	1985	1986	1987		
Lake St. Clair walleyes							
Upper	686	30	16	5	3	54	7.9
Lower	668	33	24	9	8	74	11.1
Lake Erie walleyes							
Upper	837	17	12	8	5	42	5.0
Lower	834	34	13	8	4	59	7.1
Lake St. Clair smallmouth bass							
Upper	1,042	127	47	15	9	198	19.0
Lower	1,036	109	42	17	14	182	17.6

TABLE 4.—Numbers (*N*) of walleyes and smallmouth bass tagged in Lake St. Clair that were either scale-sampled or not scale-sampled, and the frequency of their reported capture by anglers.

Species	Tagged sample			Reported sample		
	Scale-sampled (*N*)	Not scale-sampled		Scale-sampled (*N*)	Not scale-sampled	
		N	%		*N*	%
Walleye	6,359	3,306	34.2	723	341	32.0
Smallmouth bass	6,737	3,170	32.0	997	332	25.0

and smallmouth bass tags were being recovered by anglers and not reported to MDNR. There was some difference in the walleye anglers reaction to the $2 rewards versus the larger rewards. However, reporting rates for the $4 and larger reward tags were very similar, which indicates that most of these tags were probably being reported. Therefore it seems reasonable to assume that the 1.75 ratio between the recovery rate for $8 walleye tags and the nonreward recovery rate would provide the best factor to expand numbers of nonreward tags reported by Lake St. Clair anglers.

For smallmouth bass, there was less evidence that substantial numbers of tags were not reported. Because the $6 and $8 tags were not reported at a higher rate than the $2 tags, the overall ratio of 1.33 reward to nonreward recaptures should provide the best factor for expanding the reporting rate of smallmouth bass nonreward tags in Lake St. Clair.

Placement of tags on the lower jaws of walleyes was better than upper jaw tagging because walleyes tagged on the upper jaw were reported less frequently. This should be considered for future walleye tagging studies regardless of whether the differences were caused by induced mortality, tag loss, or modified angler response.

Scale-sampled fish were recovered at frequencies essentially identical to those of fish not subjected to scale sampling. This indicates that handling mortality was not a major influence in the tagging studies of Lake St. Clair walleyes and smallmouth bass.

Acknowledgments

I thank W. Bryant, J. Hodge, and L. Shubel (Michigan Department of Natural Resources) for their skilled and diligent tagging effort throughout these surveys. The reward study was sponsored by the Blue Water Sportfishing Association, Lake St. Clair Advisory Committee, Lake St. Clair Walleye Association, and Sterling Heights Bass Anglers.

References

Henny, C. J., and K. P. Burnham. 1976. A reward band study of mallards to estimate band reporting rates. Journal of Wildlife Management 40:1–14.

Ricker, W. E. 1975. Computation and interpretation of biological statistics of fish populations. Fisheries Research Board of Canada Bulletin 191.

American Fisheries Society Symposium 7:660–666, 1990

Comparison of Two Methods for Replicating
Coded Wire Tag Studies

E. A. PERRY

Department of Fisheries and Oceans, 555 West Hastings Street
Vancouver, British Columbia V6B 5G3, Canada

H. L. BLANKENSHIP

Washington Department of Fisheries, 115 General Administration Building
Olympia, Washington, 98504, USA

R.V. PALERMO

Department of Fisheries and Oceans, 555 West Hastings Street
Vancouver, British Columbia V6B 5G3, Canada

Abstract.—Since the late 1970s, fishery agencies on the Pacific coast of the USA and Canada have used replicated releases of hatchery salmon *Oncorhynchus* spp. marked with coded wire tags (CWTs) to estimate variance in tag recovery data caused by random sampling error. Initially, replicates were designed so that each replicate group within a population had a different tag code. This method requires detailed attention to ensure that groups are treated equivalently with respect to tagging, handling, rearing, and release. In 1985, prototype CWTs with replicate codes imbedded within each spool of wire in a repeating format were made available by the manufacturer to the Canadian Department of Fisheries and Oceans and the Washington Department of Fisheries. Studies were designed jointly by the two agencies to evaluate the feasibility of using these new tags and to compare the estimates of variance obtained from them with those obtained from designed replicates involving multiple tag codes. Tagged yearling coho salmon *O. kisutch* were released from the Capilano and Chilliwack hatcheries in British Columbia and from the Nooksack and Skagit hatcheries in Washington during 1986. Recovery data were obtained from the fisheries and hatcheries in 1987. The two methods of replication yielded very similar estimates of CWT sampling variance without consistent bias. Imbedded replicate tags are acceptable markers for salmonid release and recovery studies.

An identification system based on coded wire tags (CWTs), described by Bergman et. al. (1968), has been used extensively in salmonid research and management by fishery agencies on the Pacific coast of North America since the early 1970s. In 1986, for example, over 50 of these agencies collectively tagged and released 42 million salmonids (Johnson 1987). Recovery of the tags is conducted by many fishery agencies that sample a multitude of fisheries and spawning populations.

Errors, both random and nonrandom, are of great concern to those who use the coastwide CWT release and recovery data for resource management. Biases (nonrandom errors) that affect accuracy of the data and random errors that affect precision of the data are both known to occur, but their magnitudes are often unknown.

Several authors have proposed theoretical and often complex models for determining variances for use with CWT statistics (Comstock 1979; Clark 1982; PMFC 1983; Eames 1986). None of

these models has been universally accepted, mainly because of the uncertainties surrounding the assumptions that had to be made. Alternatively, de Libero (1986) proposed that replication be used as a direct measure of the variance due to random error because replication requires minimal assumptions and computations. The International Working Group on Mark–Recovery Statistics for the Pacific Salmon Treaty also favors replication. The group lists four reasons. First, replicate tags are already being introduced in normal tagging operations. Second, any theoretical variance estimate ideally should be checked by comparison with an empirical estimate from replicate tags. Third, an elaborate theoretical model may be difficult to develop and may remain controversial. Finally, and perhaps most important, the use of replicate tags lends itself to a compromise between the two approaches. Replicate codes could be used to develop empirical variance rules that could then be applied as theoretical models when no replicate data are available (T. Mulligan, Canadian Department of Fisheries and Oceans, unpublished).

Designed replication for the purpose of estimating variances began in the late 1970s and has increased in use since. Of the tag codes released by Washington Department of Fisheries (WDF) for the 1980 brood, 22% (14 of 64) were replicated groups. For the 1986 brood, by comparison, designed replicate codes made up 81% (65 of 80) of the release codes (WDF, unpublished data).

At a February 1985 meeting of the Pacific Marine Fisheries Commission (PMFC) on mark coordination, Keith Jefferts of Northwest Marine Technology, Inc. (manufacturers of coded wire tags and tagging equipment), indicated that tag-code replication (i.e., a series of repeating codes) on a single piece of wire appeared to be possible. Such replication would eliminate the need to design replicate codes on-site by manually rotating separate tag codes among tagging personnel and machines. Because it was felt that replicated binary codes would be difficult to read, even though they simplified the replication procedure, the PMFC directed WDF and the Canadian Department of Fisheries and Oceans (CDFO) to design and implement a study to evaluate the feasibility of using imbedded replicate tags, and to compare estimates of variance obtained from imbedded replicates with those obtained from designed replicates. This paper presents the joint study.

Methods

Experimental design.—The original tags described by Bergman et al. (1968) were color coded. Binary-coded tags now in common use are similar in dimension to the original color-coded tags but are etched with a six-digit binary code and a seventh position for an error or parity check (PMFC 1983).

The imbedded, replicate, coded-wire tags manufactured for this study used the three parity bit positions (one each in data 1, agency, and data 2) to represent a new three-bit binary number. This allowed up to eight replicates to be imbedded on a spool of tag wire with a single data 1, agency, data 2 code. Each tag cut sequentially from the wire spool during implantation contained a different replicate number (one through five in this study) in a self-repeating format, and then recycled to replicate number one. Replication was therefore automatic, occurring as tags were cut and implanted. All replicate tags were designated Agency 42, a previously unused agency code, for

positive identification as a replicating code. The prototypes were manufactured in four lots of 50,000 tags with five imbedded replicates per lot.

Coded wire tags for the replicated release groups were standard, commercially available tags (Northwest Marine Technology, Inc.). The wire was purchased in five lots of 10,000 tags per tag code for each hatchery in the study.

Coho salmon *Oncorhynchus kisutch* were selected for the study because of their high survival rate and relatively short life span. The number of fish tagged in each replicated group was determined under the assumption that at least 20 observed recoveries per stratum were required for meaningful statistical analyses. To relate observed recoveries to numbers of fish released, we calculated maximum-likelihood estimations of negative binomial parameters for variant element sizes (Bissell 1972) using 1978–1980 brood-year data for Capilano Hatchery coho salmon. The probability of recovering at least 20 tags for release-group sizes of 1,000–20,000 fish was then calculated for British Columbia. We estimated the required number of tagged fish at 6,000 for 20 recoveries in the sport fishery, 15,500 for 20 recoveries in the commercial fishery, and 4,000 for 20 recoveries in all fisheries combined. Replicate group size was set at 10,000 to ensure a high probability of recovering at least 20 tagged fish from any release ($P = 0.992$) and in all release groups ($P = 0.95$).

The number of replicate groups was set at five as dictated by a ceiling of 50,000 marks per treatment per hatchery, and by the group size of 10,000. This ceiling reflected availability of prototype replicated tags and cost constraints; also, it represented the number of coho salmon normally marked at Washington and British Columbia hatcheries to assess fishery contribution and survival.

Study locations.—We selected the Chilliwack and Capilano hatcheries in southern British Columbia and the Nooksack and Skagit hatcheries in northern Puget Sound for this study. They were chosen to maximize the number of recoveries of marked fish. All four hatcheries were known to contribute heavily to the fisheries, to receive high proportions of the returning runs, and to have good escapement sampling.

Fish tagging and release procedures.—Procedures for replication tagging varied among the four hatcheries. At Chilliwack, fish involved in the study were reared in three sections of one gravel-lined rearing channel and in two sections of another, for a total of five rearing groups. However, each section con-

TABLE 1.—Observed recoveries per 10,000 marked coho salmon in the fishery and in escapement samples. Fish were marked with five designed replicate and five imbedded replicate tag codes and released from four hatcheries. Data with asterisks were not used in the analyses.

Replicate type and tag code	Number released	British Columbia			Washington			Hatchery escapement
		Net	Troll	Sport	Puget Sound net	Other	Total	
Capilano Hatchery								
Designed								
023452	10,172	24.6	36.4	102.2	12.8*	3.9*	16.7	187.8
023453	10,294	20.4	27.2	98.1	20.4*	1.0*	21.4	158.3
023454	10,155	17.7	42.3	91.6	20.7*	3.0*	23.6	170.4
023455	10,150	25.6	35.5	83.7	10.8*	5.9*	16.7	150.7
023456	10,092	17.8	39.6	91.2	10.9*	2.0*	12.9	158.5
Imbedded								
420122r1	10,051	25.9	39.8	83.6	22.9*	4.0*	26.9	158.2
420122r2	10,051	22.9	34.8	74.6	12.9*	4.0*	16.9	169.1
420122r3	10,051	18.9	36.8	78.6	14.9*	9.0*	23.9	162.2
420122r4	10,051	14.9	35.8	93.5	16.9*	5.0*	21.9	179.1
420122r5	10,051	15.9	46.8	88.5	16.9*	3.0*	19.9	163.2
Chilliwack Hatchery								
Designed								
023457	10,027	31.9	104.7	106.7	15.0*	6.0*	20.9	137.6
023458	10,030	30.9	120.6	93.7	18.9*	5.0*	23.9	203.4
023459	9,971	23.1	118.3	67.2	15.0*	13.0*	28.1	173.5
023460	9,960	28.1	121.5	96.4	19.1*	9.0*	28.1	181.7
023461	9,977	31.1	116.3	101.2	21.0*	7.0*	28.1	147.3
Imbedded								
420121r1	9,957	32.1	129.6	89.4	22.1*	5.0*	27.1	171.7
420121r2	9,957	31.1	104.4	86.4	24.1*	11.0*	35.2	152.7
420121r3	9,957	28.1	117.5	93.4	13.1*	7.0*	20.1	177.8
420121r4	9,957	29.1	110.5	111.5	18.1*	8.0*	26.1	149.6
420121r5	9,957	34.1	102.4	103.4	16.1*	6.0*	22.1	151.7
Nooksack Hatchery								
Designed								
633144	10,591	34.9	75.5	68.0	70.8	20.8	91.6*	128.4
633145	10,669	31.9	79.7	62.8	73.1	17.8	90.9*	129.3
633146	10,639	29.1	72.4	78.0	77.1	16.0	93.1*	111.9
633147	10,412	41.3	80.7	74.0	68.2	20.2	88.4*	143.1
633148	10,807	34.2	87.9	66.6	68.5	13.9	82.4*	118.4
Imbedded								
420116r1	10,643	31.9	80.8	67.7	89.3	23.5	112.8*	130.6
420116r2	10,643	37.6	79.9	69.5	65.8	16.9	82.7*	133.4
420116r3	10,643	31.9	67.7	83.6	76.1	21.6	97.7*	126.8
420116r4	10,643	25.4	71.4	66.7	78.9	16.0	94.9*	125.0
420116r5	10,643	27.2	78.0	64.8	89.3	16.9	106.2*	109.9
Skagit Hatchery								
Designed								
633149	10,061	47.7	100.4	17.9	66.6	47.7	114.3*	938.3
633150	10,132	50.3	76.0	13.8	71.1	48.4	119.4*	912.9
633151	10,061	47.7	82.5	14.9	69.6	54.7	124.2*	934.3
633206	10,085	47.6	82.3	12.9	67.4	50.6	118.0*	949.9
633207	10,085	38.7	89.2	17.8	65.4	55.5	121.0*	913.2
Imbedded								
420119r1	10,096	38.6	105.0	13.9	85.2	46.6	131.7*	843.9
420119r2	10,096	46.6	92.1	18.8	68.3	46.6	114.9*	950.9
420119r3	10,096	48.5	103.0	15.8	68.3	58.4	126.8*	866.7
420119r4	10,096	44.6	100.0	21.8	72.3	46.6	118.9*	917.2
420119r5	10,096	40.6	88.2	20.8	75.3	36.6	111.9*	915.2

tained different numbers of fish. To apply 50,000 imbedded replicate tags and 50,000 designed replicate tags proportionately, we calculated the ratio of the inventory in each section to the total inventory in the five sections and used it to allocate the appropriate number of each tag type to each subpopulation. Fish in the first section were tagged with their proportion of the imbedded replicate tags, then with their proportion of the first group of designed replicate tags, then with their proportion of the second group of designed replicate tags. This process was continued until the fifth group of designed replicate tags had been applied. Thus a section containing 20 percent of the total population received 10,000 embedded replicate tags and 2,000 each of the five design replicate tags. Fish in sections 2–5 were tagged in the same manner. All tags were implanted with the same machine.

At Capilano Hatchery, the tagging procedure was identical to that at Chilliwack, including the use of a single machine, except that the fish were held in four concrete ponds.

At Nooksack and Skagit hatcheries, approximately 100,000 fish were seined from a large rearing pond and randomly transferred to two separate concrete raceways. Fish in one raceway were tagged with approximately 50,000 imbedded replicate tags, whereas fish in the other raceway were tagged with five separate codes each inscribed on approximately 10,000 tags (designed replicate). Both groups were tagged in a mobile trailer; five tagging machines were used. Each separate tag code in the designed replicate group was rotated between machines every 2 h so that each machine and operator implanted approximately 2,000 of the 10,000 tags. After a tagging period of approximately 1 week, both groups of fish were returned to their original rearing pond.

Adipose fins were clipped from all tagged fish to facilitate later recognition of experimental animals. Coded wire tagging procedures were those recommended by PMFC (1983). Tagging was done between November 1985 and March 1986, but all tagging at a particular hatchery was done within an 18-d period. At each hatchery, all fish in this study were released at the same time, but release dates ranged from May to June among hatcheries.

Data analysis.—Observed recovery data were used for the analyses because our purpose was to compare replication methods. Observed recoveries were subject to only one level of sampling error and offered the most direct comparison of variation between types of replication. Data expanded by the catch-to-sample ratio may more accurately reflect relative contribution of different tag codes, but they include the added variation associated with estimation of catch and sample size. To use observed recoveries, one must assume (1) that all the tagged fish from a particular stock are homogeneously distributed and exposed to the same fishing pressure (2) that the probability of sampling a tag code is proportional to its abundance, and (3) that the probability of sampling any replicate of a tag group is equal for all the replicates. To compare variation in recoveries for the different methods of replication, we calculated coefficients of variation (CV = 100 × standard deviation/mean) for each replicate type by hatchery and fishery (Sokal and Rolf 1969).

Results and Discussion

Sample data from commercial and sport catches were obtained through the mark–recovery data bases in Nanaimo, British Columbia, and Seattle, Washington. All commercial fisheries and coastal Washington sport fisheries were sampled for mark incidence and recovery at a rate of approximately 20%. Puget Sound and Georgia Strait sport fisheries were sampled at approximately a 5% rate for mark incidence; the program relied on voluntary returns of heads from fish with clipped adipose fins for mark recovery. Escapement sampling at each hatchery was done by CDFO or WDF staff. All observed marks were recorded and all heads were taken for tag recovery. Altogether, 19,944 study tags were recovered from the approximate total of 400,000 released. The recoveries were distributed among Washington fisheries (2,667), British Columbia fisheries (3,684), and escapement to the four hatcheries (13,593).

To check for errors in decoding tags, it is the policy among the agencies that sample the fisheries to read recovered tags twice or to have the codes verified with a computer program that checks for proper species, length, etc. The program check is, however, useful only for the primary code, not for the associated replicate number of the prototype imbedded replicate tags.

We determined the decoding error rate for all study tags recovered in the Washington fisheries and escapements by comparing the first and second readings. The error rates were 0.4% for the standard tags used in designed replication and 3.3% for the prototype imbedded replication. The nearly 10-fold larger reading error rate for the new imbedded replicates was probably a consequence

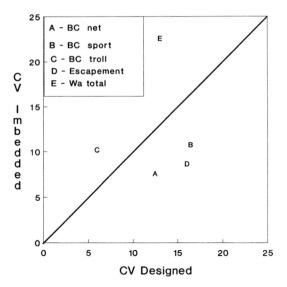

FIGURE 1.—Coefficients of variation (CV) for recoveries of designed and imbedded replicates of coded wire tags in coho salmon released from Chilliwack Hatchery and caught in various fisheries or the escapement. Recovery locations: BC = British Columbia; Wa = Washington.

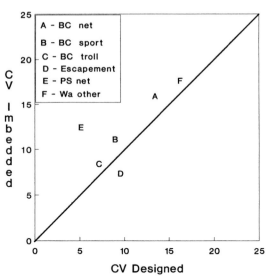

FIGURE 3.—Coefficients of variation (CV) for recoveries of designed and imbedded replicates of coded wire tags in coho salmon released from Nooksack Hatchery and caught in various fisheries or the escapement. Recovery locations: BC = British Columbia; PS = Puget Sound; Wa = Washington.

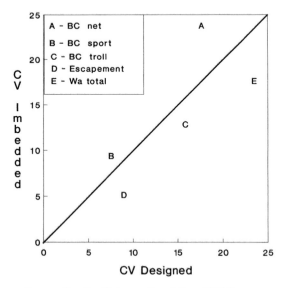

FIGURE 2.—Coefficients of variation (CV) for recoveries of designed and imbedded replicates of coded wire tags in coho salmon released from Capilano Hatchery and caught in various fisheries or the escapement. Recovery locations: BC = British Columbia; Wa = Washington.

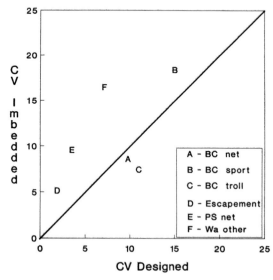

FIGURE 4.—Coefficients of variation (CV) for recoveries of designed and imbedded replicates of coded wire tags in coho salmon released from Skagit Hatchery and caught in various fisheries or the escapement. Recovery locations: BC = British Columbia; PS = Puget Sound; Wa = Washington.

TABLE 2.—Coefficients of variation for designed and imbedded replicate tag codes in various fisheries, expressed for all hatcheries combined (a) and in each hatchery with all fisheries combined (b).

Combination[a]	Replicate type	
	Imbedded	Designed
Fishery (a)		
BC net	30.42	31.06
BC sport	46.65	53.43
BC troll	36.22	36.80
PS net	65.51	61.92
Wa other	89.83	93.57
Hatchery (b)		
Chilliwack	83.22	85.07
Capilano	87.12	96.40
Nooksack	47.85	45.77
Skagit	50.85	45.79

[a]BC=British Columbia; PS=Puget Sound; Wa=Washington.

of the new reading format, which has one more data point to remember before the code and its replicate number are recorded. Corrected data are used in this report. It was felt that the error rate was minimal once personnel had sufficient experience reading the new code (approximately 500 tags per reader) and would not be a problem in the future (P. Crumley, WDF, personal communication).

Detailed release and recovery data were provided by hatchery, tag code, and recovery components (Table 1). We normalized the observed recoveries to recoveries per 10,000 smolts released to compensate for differences in release numbers. Recoveries for each tag code were well in excess of the number (20) specified in the experimental design. Because of this, it was possible to maintain several recovery components for each tag code, which allowed greater analytical flexibility. For example, all British Columbia (B.C.) troll recoveries were combined, but B.C. troll and B.C. net catch were kept as separate strata. Fishery recoveries for the Chilliwack and Capilano stocks were identified as B.C. net, troll, sport, and Washington total. For the Nooksack and Skagit stocks, further subdivision of the Washington catch into Puget Sound net and other Washington catch was possible. Using these catch strata plus the escapement, we identified 220 data cells. In 197 of these cells, 20 or more recoveries were observed, and the lowest number of recoveries observed in any cell was 13 (Table 1).

We compared coefficients of variation for observed recoveries in the fisheries and in escapements for designed and imbedded replicate tag groups by plotting them against each other (Figures 1–4). The variation was relatively low for a biological sampling program (range, 1.75–23.65%). Further, for each hatchery and fishery, variation within the two replicate types was similar. There were no consistent biases between the two methods. For example, within each hatchery there were recovery components in which the CVs for the designed replicates were greater, or less, than the CVs for the imbedded replicates. These results were consistent for all hatcheries, indicating the two replication methods are interchangeable.

These results were reinforced when the recovery data were consolidated for all the hatchery groups and for all the fisheries. Coefficients of variation were generally higher for combined data (range, 30.42–96.40%) because of differences in the recovery distributions for the four hatchery stocks (Table 2a) and because of differences in the magnitude of recoveries in different fisheries (Table 2b). When all hatcheries were combined, there were large differences among the fisheries, but the CVs for the different replication methods were very similar within each fishery. Similarly, when all fisheries were combined, there were large differences among hatcheries, but again the CVs for the two replication methods were very similar within each hatchery. These results demonstrated that designed and imbedded replication studies yield similar estimates of variation in observed recoveries of coded-wire-tagged salmon.

Recommendations

Given the theoretical support for replicated tagging of Pacific salmon and the associated release and recovery programs (de Libero 1986), and given the growing number of replicated releases in the USA and Canada, we recommend the use of imbedded replicate tags in these programs. Based on the results of this study, we conclude that data from designed replication studies are valid, but that imbedded replicates cost less and are easier to use.

Our results indicated that errors in tag decoding will be higher for imbedded replicate tags than for standard tags until personnel are familiar with the new tags. It should be routine practice to provide independent verification of all imbedded replicate tags until error rates are stabilized at less than 1%.

Acknowledgment

We acknowledge the efforts of personnel at the recovery laboratories in Olympia, Washington, and Vancouver, British Columbia. We also acknowledge the contribution of Capilano and Chilliwack hatchery staff. Anne Kling (CDFO) assisted with data analysis. Finally, the generous donation of prototype wire tags by Northwest Marine Technology, Inc., Shaw Island, Washington, is appreciated.

References

Bergman, P. K., K. B. Jefferts, H. F. Fiscus, and R. L. Hager. 1968. A preliminary evaluation of an implanted coded wire fish tag. Washington Department of Fisheries Fisheries Research Paper 3:63–84.

Bissell, A. F. 1972. A negative binominal model with varying element sizes. Biometrika 59:435–441.

Clark, W. G. 1982. Variances of exploitation rates of Klamath River fall chinook salmon estimated from coded wire tag recoveries. University of Washington, School of Fisheries, Seattle.

Comstock, R. M. 1979. Methodology of calculating variances, test of hypothesis, and necessary sample size associated with CWT studies. U.S. Fish and Wildlife Service, Fisheries Assistance Office, Olympia, Washington.

de Libero, F. E. 1986. A statistical assessment of the use of the coded wire tag for chinook (*Oncorhynchus tshawytscha*) and coho (*Oncorhynchus kisutch*) studies. Doctoral dissertation. University of Washington, Seattle.

Eames, M. J. 1986. Variance estimates for estimated salmon coded-wire tag recoveries. Washington Department of Fisheries Technical Report 93.

Johnson, J. K. 1987. Pacific salmonid coded wire tag releases through 1986. Pacific Marine Fisheries Commission, Portland, Oregon.

PMFC (Pacific Marine Fisheries Commission). 1983. Coded wire tagging procedures for Pacific Salmonids. PMFC, Portland, Oregon.

Sokal, R. R. and F. J. Rohlf. 1969. Biometry. Freeman, San Francisco.

American Fisheries Society Symposium 7:667–676, 1990
© Copyright by the American Fisheries Society 1990

Parametric Bootstrap Confidence Intervals for Estimating Contributions to Fisheries from Marked Salmon Populations

HAROLD J. GEIGER

Alaska Department of Fish and Game, Post Office Box 3-2000 Juneau, Alaska 99802, USA

Abstract.—Salmon marked at a hatchery are identified later in samples from fisheries. Then, a reasonable estimate of the hatchery contribution to a particular fishery is the observed number of marks times the reciprocal of the fraction of the original population that was marked times the reciprocal of the proportion of the fishery examined. When similar marking techniques are used on wild salmon, the marking fraction must be estimated by examining adults that return to spawn. To estimate confidence intervals around the contribution estimate, we likened the fishery to an urn containing three colors of marbles: One color corresponded to unmarked hatchery fish, another color to marked hatchery fish, and the third color to fish of other stocks. We likened the hatchery fish in the fishery to a marble-and-urn model wherein a sample was drawn from an urn that contained two colors of marbles, one color representing marked hatchery fish, another color unmarked hatchery fish. We propose a parametric bootstrap technique for confidence interval estimation that involves estimating the parameters of the marble-and-urn model from samples from the fishery and the escapement. Simulations of the marble–urn model generate numerical solutions for confidence intervals. These "Monte Carlo" confidence intervals do not depend on a large-sample approximation to the normal distribution.

The coded wire tag (Jefferts et al. 1963) came into wide use in fisheries research in the early 1970s. Since then, it has become one of the most important tools for the management and allocation of salmon *Oncorhynchus* spp. stocks in the Pacific Northwest. Nearly 13,000 separate tag codes were reported at the end of 1986 to the Pacific Marine Fisheries Commission (Johnson 1987), the agency that coordinates mark–recapture studies of salmonids in western North America. Because most studies involve a single tag code, this number gives some perspective on the enormous investment in this technique. In spite of this tremendous effort, the fisheries and statistical literature is remarkably silent about methods for using this tool beyond variations of the Petersen population estimate.

The Petersen estimate (Seber 1982) is the simplest form of the mark–recapture experiment. A simple random sample of some fish population is captured. The number of fish captured is noted, then these fish are marked and released. At a later time, before there has been any birth, death, immigration or emigration, a second simple random sample is drawn from the population. The total number of fish in the second sample, as well as the marked fish in the sample is noted. The ratio of marked animals to the number of animals examined in the second sample is assumed to be similar to the proportion of marked animals in the entire population. The Petersen estimate of the population total is simply this ratio times the known number of marks in the population.

The Alaska Department of Fish and Game (ADF&G) has routinely used coded wire tags to identify hatchery stocks in mixed-stock fisheries since the mid-1970s. The ADF&G has also frequently tagged wild stocks of chinook salmon *O. tshawytscha* and coho salmon *O. kisutch* and, recently, of pink salmon *O. gorbuscha* and sockeye salmon *O. nerka*. In most of these studies the goal was not to provide a Petersen estimate of the size of some closed population, but to estimate the contribution of hatchery stocks in a fishery.

At first glance the problem of estimating the contribution of a stock to a fishery may seem similar to estimating population size with the Petersen estimate, but the two problems are quite different. In the contribution problem, a simple random sample is drawn from a population of juvenile salmon. These fish are tagged with coded wires, marked with an adipose fin clip, and returned to the larger population. Extremely high mortality occurs at sea as the fish mature and migrate. The fish in the population of interest mix with an unknown (but usually very large) number of fish from other populations. Eventually, a fishery captures some of the marked fish.

---

[1]Contribution PP-002 of the Alaska Department of Fish and Game, Division of Commerical Fisheries, Juneau.

We assume that the fishery captures a simple random sample of the returning adults and that it includes both marked and unmarked fish from the population of interest. The fishery also captures unmarked fish from other stocks that contribute to the fishery. The catch for a particular fishery consists of N fish, of which C fish are from our stock of interest; we also have M marked fish in the fishery.

At the time of tagging, r fish out of a total of H fish from the population of interest were marked. It is sometimes desirable to construct a ratio for the marked fraction. We represent this ratio as P and call it the tagging rate. The notation is listed in Table 1.

In most situations we assume the total catch, N, is known without error. In most commercial fisheries, a government agency attempts to record and publish this value. In sport fisheries, N may be estimated from data collected in creel surveys. (The case with N unknown is a separate problem.)

The catch, C, of all fish from the stock of interest, and the catch of all marked fish, M, are of course unknown. The goal is to estimate C. To do this the fishery is sampled. We assume this is a simple random sample. All fish in the sample are examined for marks. Let s be the number of fish examined and x the number of marks that are observed.

Clark and Bernard (1987) argued that in the Alaska commercial fishery, M should be viewed as following a binomial distribution, in which case x should be viewed as following a hypergeometric distribution. By extending this line of reasoning, we can develop a general way to view the realization of x. Based on our description of the realization of M, we offer the following two probability models:

$$f(M|C,N,P) = \begin{cases} \dfrac{\dbinom{r}{M}\dbinom{H-r}{C-M}}{\dbinom{H}{C}}; & (1.1) \\[2em] \dbinom{C}{M}(r/H)^M[1-(r/H)]^{C-M}. & (1.2) \end{cases}$$

Equation (1.1) is an example of the hypergeometric distribution, which describes sampling without replacement for dichotomous outcomes in a finite population. This sampling situation is idealized by an urn containing two colors of marbles. There are r marbles of one color and $H-r$ of the other color; C marbles are drawn at random, one at a time, without replacement. Given ideal sampling, equation (1.1) gives the exact distribution of M, given perfect knowledge about the number of fish from the population of interest in the fishery and the

number of fish originally in the population at the time of tagging. Equation (1.2) is an example of the binomial distribution. This distribution is idealized by randomly drawing the C marbles from the urn one at a time but replacing each marble before drawing the next.

If $C << H$, the binomial distribution adequately represents the sampling situation. Only a small fraction of juvenile salmon survives to adulthood to be captured in the fishery, so probability that any one fish released from the hatchery will be selected by the fishery is very small. If we allowed sampling with replacement, the probability of any fish being selected twice would be essentially zero. Therefore, we have adopted equation (1.2) because it is easier to use.

Assume we have perfect knowledge about N, the number of fish caught in the fishery; C, the number of fish from the stock of interest; and M, the number of marked fish from the stock of interest. Then the distribution of x, the number of marks in the sample, can be modeled as follows:

$$f(x|M,C,N,P)$$

$$= \begin{cases} \dfrac{\dbinom{M}{x}\dbinom{N-M}{s-x}}{\dbinom{N}{s}}; & (2.1) \\[2em] \dbinom{s}{x}(M/N)^x[1-(M/N)]^{s-x}; & (2.2) \\[2em] \dfrac{\exp[-(sM)/N][(sM)/N]^x}{x!}; & (2.3) \\[2em] \dfrac{\Gamma(x+k)}{\Gamma(k)x!}\left(\dfrac{sM/N}{(sM/N)+k}\right)^x\left(\dfrac{k}{(sM/N)+k}\right)^k. & (2.4) \end{cases}$$

Equation (2.1), again, formally describes a hypergeometric distribution, which is the exact conditional distribution of x. Equation (2.2), again, formally describes the binomial distribution, which could be assumed to be an adequate representation of the conditional distribution of x when $s << N$. In practice, the sampling rate is usually 20% or greater, which leads to a preference for (2.1). Equation (2.3) describes the Poisson distribution, which may be used when s is considered a random variable or if the occurrence of marks is very rare. Equation (2.4) gives the negative binomial distribution, which belongs to a class known as contagious distributions. Mark recoveries conforming to this distribution tend to clump or occur in spurts, like the incidence of a contagious dis-

TABLE 1.—Notation used to describe the marking system in hatchery studies of the contribution of hatchery fish to a single fishery.

From the original population of juveniles

H = the number of fish in the original population of interest
r = the number of fish marked in the population of interest
P = the marking rate (r/H)

From the fishery

N = the number of fish captured in the fishery
C = the number of hatchery fish in the fishery
M = the number of marked and tagged fish in the fishery
s = the number of fish sampled in the fishery
x = the number of recaptured fish with the tag code of interest

From the escapement or hatchery rack

t = the number of fish examined for marks and tags
y = the number of marks and tags discovered

ease. This distribution should be used when a few boats participate in a fishery over a fairly large area, and when fish have some tendency to clump by stock in a fishery. This distribution contains an additional parameter, k, which describes the tendency of marks to clump. Equations (2.1)–(2.4) offer a range of options with which to model the realization of x with increasing variance. Clark and Bernard (1987) recommended the use of equations (1.2) and (2.1), which we will adopt for further discussion.

Because we make inferences about C through the observed outcome of x, we need a probability function for x that does not involve M. From that distribution we can then derive a maximum-likelihood estimator for C. As a first step, consider the joint distribution of x and M:

$$f(x,M|C,P,N)$$
$$= \left[\binom{C}{M} P^M (1-P)^{C-M}\right] \frac{\binom{M}{x}\binom{N-M}{s-x}}{\binom{N}{s}}. \quad (3)$$

To get a distribution for x that does not involve the unobserved quantity M, we need to integrate out M.

$$f(x|C,P,N) = \sum_{M=x}^{C} f(x,M|C,P,N). \quad (4)$$

Finding a closed-form expression for equation (4) is not easy; M occurs in an exponent twice and in complicated ratios of factorials. No obvious solution presents itself. Perhaps we could reparameterize M to work with gamma functions rather than factorials, but this too becomes intractable. In a

sense, we do not actually estimate C with x at all; rather, we estimate the nuisance parameter M with x, then estimate C with the estimate of M.

Most investigators, unhindered by the lack of a maximum-likelihood estimator, estimate C in the obvious and reasonable way. For that reason we call the following the reasonable estimator for C:

$$C' = x (N/s) P^{-1}. \quad (5)$$

This estimator is unbiased, as the following reasoning shows:

$$E(C') = E[x(N/s) P^{-1}]$$
$$= (N/s) P^{-1}E(x)$$

because N, s, and P are constants.

Now,

$$E(x)$$
$$= \sum_{M=0}^{C} \sum_{x=0}^{M} \left[x\binom{C}{M}P^M(1-P)^{C-M}\right] \frac{\binom{M}{x}\binom{N-M}{s-x}}{\binom{N}{s}}$$
$$= \sum_{M}\binom{C}{M} P^M(1-P)^{C-M}\sum_{x} x \frac{\binom{M}{x}\binom{N-M}{s-x}}{\binom{N}{s}}.$$

The inner sum is just the expectation of a hypergeometric random variable, leading to

$$E(x) = \sum \binom{C}{M} P^M (1-P)^{C-M} (sM/N)$$
$$= (s/N) \sum M \binom{C}{M} P^M(1-P)^{C-M}$$
$$= (s/N) E(M)$$
$$= (s/N) P C.$$

So we now have

$$E(C') = (N/s)P^{-1} (s/N) (P C)$$
$$= C.$$

That this estimator is unbiased comes as a surprise to some because the Petersen estimate is biased (Seber 1982). However, the Petersen estimate is biased because the number of recaptures occurs in the denominator, causing the estimate to vanish when zero marks are recovered. In other words, the Petersen estimate is biased because it can not handle the single situation in which there are no recaptures. In equation (5), the number of recaptures appears in the numerator, resulting in an estimate of 0 when there are no recaptures.

Clark and Bernard (1987) suggest an estimator for the variance of C:

$$V(C') = x[(N/s)P^{-1}]^2$$
$$\cdot \{1 - x + [(s - 1)/(N - 1)](xN/s) - P\}.$$

Canceling similar terms and assuming $s - 1 \approx s$ and $N - 1 \approx N$, we have the following approximation:

$$V^*(C') = x[(N/s)P^{-1}]^2$$
$$\cdot \{1 - [(N/s)P^{-1}]^{-1}\}. \tag{6}$$

The Bootstrap Technique

Bootstrap confidence intervals (Effron and Gong 1983) are gaining ever-wider use. Bootstrap techniques were developed to use the information about the variability of an estimate that is contained in the random sample that was used to generate the estimate. Computer simulation repeatedly draws samples, called bootstrap samples, from some sample of actual data, and an estimate is calculated from each sample. A computer can draw thousands of bootstrap samples, and hence we can generate thousands of bootstrap estimates. The bootstrap confidence interval is formed from the percentiles of the simulated, or bootstrap, distribution of the estimate. Under many conditions, bootstrap confidence intervals formed from the percentiles of the bootstrap distribution are exact (Buckland 1985), whereas confidence intervals based on multiples of the standard error of the estimate rarely are so. Parametric bootstrap confidence intervals, also called Monte Carlo confidence intervals (Buckland 1984), are based on a similar idea. In the latter case, specific parametric distributions are proposed to model the sampling situation. The parameters for the Monte Carlo, or bootstrap, distribution are parameter estimates from the data. Each sampling process is simulated by using a computer to draw large samples from the proposed probability distributions with bootstrap parameters.

In our case, we estimate C from the fishery sample; M is fixed for a particular realization of the fishery, although we do not know what it is. Therefore, we first generate a bootstrap realization of M, denoted M_b, using equation (1.2) and substituting C' for C. Next we draw a single bootstrap sample for x, called x_b, using equation (2.1) with M_b in place of M. Thousands of simulated values of x can be drawn in a few minutes on a microcomputer and then expanded to simulated, or bootstrap, values for C'. These bootstrap val-

ues are represented as C'_b. The bootstrap confidence interval is formed from the appropriate percentiles of the distribution of C'_b.

In fisheries research, an analyst often will produce an estimate, then automatically calculate a standard error for that estimate. The analyst will next, unthinkingly, calculate a confidence interval by referring to the Gaussian (normal) distribution for multiples of the estimated standard error. For example, about 95% of the probability is within ± 2 standard deviations of the mean in the standard normal distributions, so 95% confidence intervals are often reported to be ± 2 estimated standard errors—even when the underlying distribution is highly nonnormal. There is ample justification for this approach in large classes of problems. The central limit theorem (Mood et al. 1974) states this approach will work for large sample sizes with sample means. This approach will not work well in every situation; Ricker (1975), for example, discussed why it does not work well for the Petersen estimate.

A Simulated Distribution of C'

Consider a fishery in which 1,000 fish were caught and exactly 100 of these were from the stock of interest. Suppose half the stock was marked by removal of the adipose fin, and injection of a tag. Further suppose that, from the catch of 1,000 fish, a simple random sample of 200 fish is examined for marks. Because we know values of all the parameters in this fishery, we can simulate the values of x, the number of marks found in the sample. Each value of x can then be used to generate a value of C'.

Figure 1 shows a frequency histogram of 1,000 values of C' generated by drawing a value of M from a binomial distribution with parameters $s = 200$ and $P = 0.5$, then drawing a value of x from a hypergeometric distribution with parameters M, $N = 1,000$, and $s = 200$. Superimposed over the histogram is a graph of the probability density function of the normal distribution with the same mean and variance. The histogram is centered over the value of 98.74 with the standard deviation of 30.2.

With 1,000 replications of the simulation, we have a very good numerical approximation of the entire distribution of C'. In our simulated distribution, the 10 percentile falls at 60 and the 90 percentile at 140. Notice that 80% of the estimates cover an interval of 80 fish. We would then hope an 80% confidence interval would similarly cover an interval of 80 fish. See Table 2 for selected percentiles of this distribution.

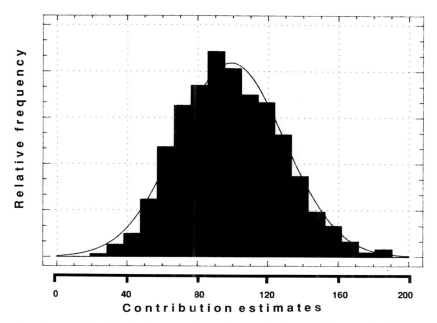

FIGURE 1.—One thousand simulated estimates of a stock's contribution to a fishery. The fishery captured 1,000 fish (N), 200 fish were sampled (s), 100 fish from the stock of interest were actually captured (C), and half of this stock was originally marked (P). The probability density function of the normal distribution is superimposed.

A Simple Bootstrap Example with P Known

Suppose for a moment that the actual value of x that we observe is 8. This is the lower quartile of x under the simulated sampling discussed above, so it seems a good value to choose to represent a typical value of x. To get the classical 80% confidence interval, we first would estimate the standard error of C'. From equation (6),

$$V^* = 8\,[(1{,}000/200)\,(0.5^{-1})]^2$$
$$\cdot\,\{1 - [(1{,}000/200)\,(0.5)^{-1}]^{-1}\}.$$

TABLE 2.—Selected percentiles of a simulated distribution of the contribution estimate. It was assumed that the fishery caught 1,000 fish, a sample of 200 fish was examined for marks, the proportion of the original population marked was 0.5, and the catch of fish from the stock of interest was 100 fish.

Percent of distribution less than or equal to percentile	Percentile
99	170
95	150
90	140
85	130
50	100
15	70
10	60
05	50
01	35

The standard error of C' is calculated to be about 26.8. This leads to a classical 80% confidence interval (i.e., ± 1.282 SE) of (45.6, 114.4). Notice this interval covers 68.8 fish, which is short of our target value of 80 fish.

One could generate a confidence interval by simply estimating C with C' at our observed value of $x = 8$, then resampling new values of C'. When we did this, we generated the distribution depicted in Figure 2a. To avoid confusion we will call this new distribution, which we generated using the observed value of C' as a parameter, the bootstrap distribution. Recall that we already adopted the convention of calling the individual bootstrap estimates of contribution C'_b. Also note that C' and C_b are the same quantity—they differ only if the analyst is considering the statistic from the real world or the parameter from the bootstrap world.

Again, conditioned on observing $x = 8$ and $C' = 80$, we proceed by simulation to get a numerical distribution of C'_b. Drawing 1,000 bootstrap replications from the bootstrap distribution, we find that the bootstrap 80% confidence interval is (40,110). Notice this interval covers 70 units of fish, which is nearly identical to the classical interval.

Alternatively, we could regard the upper quartile of the distribution of C' as a value to be considered as typical under this sampling scheme.

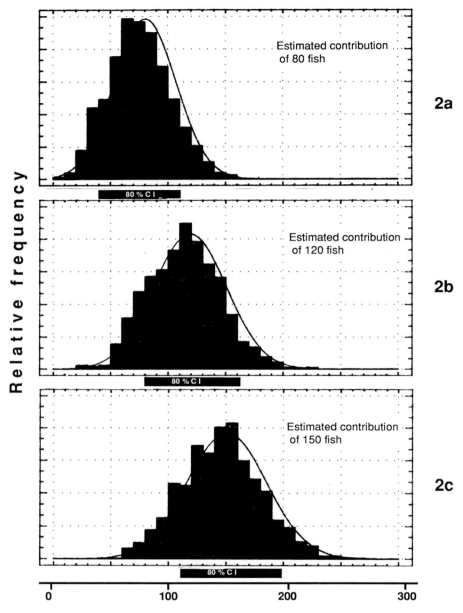

FIGURE 2.—Bootstrap distributions for three fisheries with 1,000 fish caught (N), 200 fish sampled (s), and half of the stock of interest marked (P). The probability density functions of the normal distribution are superimposed. Estimated contributions are of (a) 80 fish, (b) 120 fish, and (c) 150 fish.

In our example, the upper quartile is 120 fish. This value corresponds to $x = 12$ recoveries. Actually, a C' of 120 will lead to an 80% bootstrap confidence interval of width 80. But in general the width of either kind of confidence interval will be less than the target width when the observed number of recoveries is much less than the ex-

pected number of recoveries, and the width will be greater when the observed recoveries are much larger than expected. For example, if C' were 150, the width of the 80% confidence interval would be 150—far greater than our target value of 80. The probability is small that the observed number of recoveries will stray very far from its expected number, but if it does, the confidence interval will also stray from its target width.

The Case in Which P is Estimated

Our single example at least suggests that classical confidence intervals, based on the assumption of normality, are surprisingly adequate. Actually this is our experience after many simulation runs under a variety of assumptions with large x. What, then, is there to recommend the parametric bootstrap in practice when the estimated standard error is computationally much simpler? Sometimes the tagging rate is not known exactly, as when wild stock is tagged. Recall that we assumed we knew N, s, and P without error. Sometimes, however, P is estimated from a sample of fish, usually in the escapement. When the tagging rate is not known, there is no obvious way to estimate the variance of C'.

Notice, though, that although marked fish are likely to stray out of the escapement at some rate, usually only unmarked fish will stray in from neighboring systems. If the escapement is to a very large hatchery, very few fish are likely to stray in relative to the number of fish straying out. If 200,000 fish have returned to the hatchery, and at most 5,000 fish from other stocks have strayed in, it is reasonable to assume that the escapement is a random sample from the original population of interest. Alternatively, if the escapement is in a wild system where 300 spawning fish return and 50 fish of other stocks might stray into the escapement, the inference will not be about the original population. It is not clear what population the sample will represent, but the observed mark rate will be too low, thereby causing a serious upward bias in the estimate.

If we are willing to assume that the escapement is a simple random sample of fish from the population of interest, and if t denotes the size of the sample and y denotes the number of marks in the sample, P can be estimated by y/t. Let y/t be denoted P'. We can still estimate C, but we are now without any guidance as to how to estimate the precision. Let C^* denote the estimate of C when P is estimated:

$$C^* = x \ (N/s) \ P'$$

$$= x \ (N/s) \ (t/y). \tag{7}$$

This distribution of y is easy to find. By making the same assumptions we made when modeling the probability structure of M, we assume y is generated under a binomial law with parameters s and P. Then,

$$f(y|s,P) = \binom{s}{y} P^y \ (1 - P)^{s-y}. \tag{8}$$

Because the sampling of the escapement and the sampling of the fishery are independent, the probability functions of y and x, and of x and M may be factored. To provide confidence intervals for C^* in equation (7) we need to find the variance for the product of two random variables, x and y^{-1}. In the past, investigators overwhelmed by the prospect of the exact variance of the product of random variables have appealed to the delta method (Seber 1982) to give approximate variance estimates for the product of two independent random variables. The delta method only guarantees an approximation, not a fit. Often the fit is very poor (Goodman 1960). The Monte Carlo method can easily accommodate the uncertainty in the estimate of P. The simulation proceeds as before, except at each step a bootstrap value of y, denoted y_b, is drawn from distribution (8) (a binomial distribution with parameters s and P'). This value of y_b is then used to estimate P_b, which is then used in the distribution of M, which is then used in the distribution of x.

Suppose, as before, we have a fishery with $N = 1,000$, $s = 200$, and $C' = 100$. Further suppose we are uncertain about P, and we examine the escapement. First assume we examine 10 fish and discover 5 marked fish. In this situation the bootstrap distribution is very skewed to the right (Figure 3). When we drew 1,000 bootstrap replicates, our 80% confidence interval ran from 57.1 to 183.3 to cover a width of 126.2 fish. Recall that we produced a simulated distribution to benchmark our bootstrap confidence intervals (Figure 1). This benchmark confidence interval ran from 60 to 140 fish (Table 2). Notice that the left point in our recent example interval is only about three fish off our benchmark value of 60, whereas the right end point is over 23 past the benchmark value of 160.

Consider a second example. Assume we examine 50 fish and discover 25 marked fish. Here, the bootstrap distribution we draw gives us an 80% confidence interval that runs from 60.3 to 145.8 to cover a width of 85.5 fish. Notice we again are further off on our right endpoint.

In our final example, assume we examine 100 fish in the escapement and observe 50 marks. When we draw a bootstrap distribution in this case, the 80% interval goes from 61.2 to 140.6 to cover a width of 79.4 fish. Now we begin to see the law of large numbers (Mood, et al. 1974) in action. We have

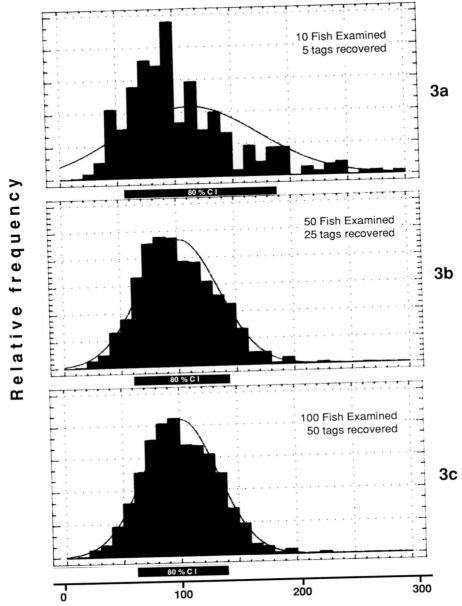

FIGURE 3.—Bootstrap distributions for three fisheries with 1,000 fish captured (N), 200 fish sampled (s), and the marking rate estimated by sampling the escapement. The probability density functions of the normal distribution are superimposed. In (a), 10 fish are examined and 5 tags are recovered; in (b), 50 fish are examined and 25 tags are recovered; in (c) 100 fish are examined and 50 tags are recovered.

essentially the same result as if we assumed the marking rate was known rather than estimated.

Discussion

To generate parametric bootstrap confidence intervals, each stochastic event that leads to the realization of the sample is specified. This involves both specifying that a random variable was generated at each stochastic step and specifying a probability distribution for that random variable. Next, very simple computer programs are written to simulate the process thousands of times. The confidence interval is then formed from the percentiles of the simulated distribution.

In our specific example of the fisheries contribution estimate, one easy way to be misled is to fail to include the sampling event that generates M in the bootstrap sampling regime. One can correctly reason that M is fixed once the fishery has been realized. What is incorrect is to then assume that the imprecision in C' is due only to the random nature of x conditioned on the unknown M. In essence, x and M are drawn simultaneously. As I mentioned above, we estimate M with x, and we estimate C with the estimate of M. Consequently, the event that generates x and the event that generates M both need to be in the bootstrap simulation.

Multiple recovery strata pose no conceptual or programming challenges, but they drastically increase the time required to do the computation. If the estimate is of the contribution to different recovery strata or fisheries, the program must draw a bootstrap contribution for each stratum separately. The contributions then are summed over strata, and the aggregate contribution is stored before the next replicate is drawn.

The parametric bootstrap confidence intervals described here are recommended for several reasons. The underlying theory is strikingly simple. The parametric bootstrap is easy to implement if the numbers of different tag codes and recovery strata remain fairly small, and it is only slightly harder if there are many strata. Computation of these bootstrap intervals is fairly quick on a desktop microcomputer. For example, our simulations took about 5 min or less on a Compaq 386/20 and about 15–20 min on a Compaq Deskpro IV—each with math coprocessors. Most importantly, the parametric bootstrap is appropriate with small samples, unlike methods that require the estimator to be normally distributed. Notice, however, that if the number of recoveries is small in either the fisheries sample or the escapement sample, confidence intervals based on multiples of standard errors should be called into question.

Schenker (1985) reported situations in which nonparametric bootstrap confidence intervals underestimate the true confidence interval width, especially with small sample sizes. His criticisms sometimes cause alarm when bootstrap confidence intervals are discussed. Because of the parametric nature of our bootstrap approach, his criticisms do not apply.

I followed the approach of Clark and Bernard (1987), in which each sampling event leading up to the realization of the number of marks in the fishery is modeled as a marble-and-urn experiment. So, conditional on the total catch in the fishery and on the catch of the stock of interest, each sampling event is likened to drawing several marbles out of an urn; the marbles are either of one color (corresponding to marked fish) or they are of another color (corresponding to unmarked fish). The probability theory for marble-and-urn experiments is quite tractable and found in any probability textbook. Clark and Bernard (1987) furnished a complete explanation, including a discussion of sampling events to account for lost heads and lost tags. More importantly, they provided a multivariate extension that allows simultaneous inference about the contribution of several related tag codes.

Clark and Bernard's method has the definite advantage of computational ease over bootstrap methods. It can be easily implemented on a microcomputer spreadsheet program, even when the number of strata is fairly large. One of the most interesting results of the parametric bootstrap exercise is to show how well the method of Clark and Bernard (1987) works, even when the marking fraction is estimated—provided a large number of fish were used to estimate it. As the number of fish examined in the escapement decreases, the method of Clark and Bernard worsens. It appears from our work that if the number of marks recovered is over 100, the uncertainty in the marking rate can safely be ignored.

De Libero (1986) argued that confidence intervals should be calculated by tagging in replicates or in "interpenetrating subsamples." He felt that only by breaking the tag codes into replicates can the "repeatability" of the result be measured. Although he raised an important issue, replicate tag codes only beg the question. A replicate tag experiment could be carried out by drawing a sample of fish, then randomly assigning each fish in the sample to a group that represents some unrecognized source of error. For example, different tagging machines might cause different rates of tag loss. A sample could be drawn, and a different tag code could be used in each machine. Here, we would say the replication is only over tagging machines. Hundreds of sources of variation could be listed as targets of study with replication; but all—or even a few—sources of error cannot be studied this way in a single experiment. Indeed, only a very few replicates can be used in a single study. Surprisingly, many of the advocates of the interpenetrating subsamples do not advocate replication over any sources of error at all; instead, they advise allocation of the fish within the sample to replicate groups through a random process.

The discussion of whether or not to tag in replicates misses an important point: within the sample of fish to be marked, each marked fish is a replicate in the full sense of the word—each tagged fish will or will not be recovered in a particular sample. To consider how repeatable the experiment was, one needs to judge whether or not *individual* tagged fish were consistently fed into the sampling process (that is, the fishery and then the subsequent sample). The experiment will result in relatively many or relatively few recovered tags. If the sample size was large, and if relatively many tags were recovered, then tagged fish were *repeatedly* entering the sampling process. Here we can confidently infer that there were many tagged fish in the fishery; if we repeated this exact experiment many times, we would repeatedly infer there were many tagged fish in the fishery. Similarly, if few tagged fish were recovered, then the individual tagged fish repeatedly did not occur in the sample, and we can confidently infer that they were absent from the fishery. Alternatively, if the sample size was small in the fishery sample, the experiment cannot be judged very repeatable because individual, replicate tagged fish were not given the opportunity to be consistently present or absent. Confidence intervals based on either the Clark and Bernard (1987) method or the parametric bootstrap estimate formalize the process of judging the repeatability of the experiment by treating it as a function of how many *replicate* tagged fish were recovered, adjusted for sample sizes. In our view, interpenetrating subsamples add an unnecessary level of complexity to the experiment to solve a problem that does not exist.

The tagging study should be viewed as providing an approximation. Estimates of precision provide only one check on the adequacy of the approximation. Other diagnostic checks can and should be performed. For example, individual fishery samples should be compared to see if marks are being recovered at the same rate. We have detected differences on occasion that led to a better understanding of the way the fishery was actually working. The mark rate also needs to be examined at the time of spawning in the hatchery. We have also detected unacceptable assumption failure through this exercise.

In our experience, sampling error, as measured through confidence intervals, ceases to be an issue once examined. Of far greater significance are the basic assumptions of random sampling,

perfect knowledge of the tagging rate, and so forth. Neglect of these can result in catastrophic failure of the study. Worse yet, the investigators may take false comfort from small confidence intervals. A healthy debate is starting to take place on methods to quantify uncertainty in coded wire tag studies. In the end, this will make the coded wire tagging technique an even more valuable tool in fisheries management.

Acknowledgments

I thank John E. Clark for helping me to see the coded wire tagging problem clearly and for many original ideas that I have used here; Steven Buckland gave me invaluable advice on the parametric bootstrap technique; and B. Alan Johnson made several helpful suggestions on the manuscript.

References

Buckland, S. T. 1984. Monte Carlo confidence intervals. Biometrics 40:811–817.

Buckland, S. T. 1985. Calculation of Monte Carlo confidence intervals. Applied Statistics 34:296–301.

Clark, J. E., and D. R. Bernard. 1987. A compound multivariate binomial-hypergeometric distribution describing microwire tag recovery from commercial salmon catches in southeast Alaska. Alaska Department of Fish and Game, Informational Leaflet 261, Juneau.

de Libero, F. E. 1986. A statistical assessment of the use of the coded wire tag for chinook (*Oncorhynchus tshawytscha*) and coho (*Oncorhynchus kisutch*) studies. Doctoral dissertation. University of Washington, Seattle.

Effron, B., and G. Gong. 1983. A leisurely look at the bootstrap, the jackknife, and cross-validation. American Statistician 37:36–48.

Goodman, L. A. 1960. On the exact variance of products. Journal of the American Statistical Association 55:708–713.

Jefferts, K. B., P. K. Bergman, and H. F. Fiscus. 1963. A coded wire identification system for macroorganisms. Nature (London) 198:460–462.

Johnson, K. J. 1987. Pacific salmon coded wire tag releases through 1986. Pacific Marine Fisheries Commission, Portland, Oregon.

Mood, A.M., F. A. Graybill, and D. C. Boes. 1974. Introduction to the theory of statistics, 3rd edition. McGraw-Hill, New York.

Ricker, W. E. 1975. Computation and interpretation of biological statistics of fish populations. Bulletin of the Fisheries Research Board of Canada 191.

Schenker, N. 1985. Qualms about bootstrap confidence intervals. Journal of the American Statistical Association 80:360–361.

Seber, G. A. F. 1982. The estimation of animal abundance and related parameters. Macmillan, New York.

American Fisheries Society Symposium 7:677–683, 1990
© Copyright by the American Fisheries Society 1990

Variance Estimation for Stock-Contribution Estimates Based on Sample Recoveries of Coded-Wire-Tagged Fish

KENNETH B. NEWMAN[1]

Northwest Indian Fisheries Commission
6730 Martin Way East, Olympia, Washington 98506, USA

Abstract.—A variance formula for coded-wire-tag-based estimates of a stock's contribution rate to a fishery is presented. The formula is based on a compound distribution that assumes the distribution of total tag recoveries in a fishery is binomial and the distribution of observed recoveries conditional on total recoveries is hypergeometric. The formula also provides guidelines for tagging rates that permit a particular level of precision in estimates of contribution rate.

Estimates of several stock-specific parameters important to salmon fisheries management are based on recoveries of coded-wire-tagged (CWT) fish. Generally, a sample of a fishery catch is examined for coded-wire-tagged fish, identified by the absence of the adipose fin, which had been removed at the time of tagging. Expansion from the sample to total catch is made to estimate the total number of recoveries. When the sample is a simple random sample of the catch, statistically consistent point estimates for many parameters are often clear. Estimation of variance is generally more difficult, however, because of the influence of random processes other than those induced by sampling the catch.

An estimate of the number of tagged fish in a catch is affected by many random processes; even so, a relatively simple procedure may exist for estimating the variance of that estimate. By randomly dividing the number of fish about to be tagged into two or more groups and using a different tag code for each group, it is possible to calculate two or more estimates of the parameters of interest. If the estimates are independent, the variance of the parameter estimate can be estimated from the sample variance of the mean of the estimates. If the estimates are not completely independent, however, a modification of this approach is necessary. Such an estimator is intuitively appealing, and it sidesteps the problem of determining the influence of the various random processes on the final point estimate. This approach, known as the interpenetrating subsample (IPS), was proposed by Mahalonobis (1946) and described by Cochran (1977), and it has been applied to CWT data by de Libero (1986).

Although IPS-based variance estimates provide a means to calculate variances regardless of the underlying statistical distribution, an exact description of the probability distribution of the number of estimated CWT recoveries is desirable for many reasons. The IPS-based estimates are valuable for variance of parameter estimates after sampling, but their usage prior to tagging to determine the effect of tagging and sampling levels on resulting precision is unclear. A stochastic model that includes terms for tagging level and catch sampling fraction would be a simpler tool for determining the number of fish to tag, and the number to sample from the catch, to achieve a desired level of precision. Furthermore, a stochastic model combined with standard statistical testing procedures, such as likelihood ratio tests, would provide an approach to testing hypotheses about parameter estimates. Currently, for example, there is no established probabilistic method for comparing the contribution rates or harvest rates of two or more tag groups.

This paper presents a statistical model for the probability of recovering tags in a simple random sample of a fishery's catch and for comparing estimates of the variance of contribution rates based on this model with IPS-based estimates. I describe the underlying statistical distribution for CWT recoveries and the associated variance. I then compare IPS-based variance estimates with distribution-based estimates for several groups released from Canadian and Washington state hatcheries.

Compound Distribution for Estimated Recoveries

A salmon smolt released from a hatchery or a wild smolt leaving its stream faces a wide range of

---

[1]Present address: Department of Statistics, GN-22, University of Washington, Seattle, Washington 98195, USA.

possible fates. This range of fates, or the outcome space, can be defined and partitioned in many ways. For example, the outcome space could be defined as the presence or absence of the fish in a particular ocean area during a particular time interval. Another example is death by a particular time. For salmon management and research purposes, the fate of interest is often harvest. This paper will focus on the simple outcome space of a tagged fish that may or may not occur in a particular catch. The parameter of interest is the contribution rate, p, or the number of tagged fish caught in a specific catch divided by the number of tagged fish released.

The following symbols, which refer to a specific catch, will be used.

Y = number of tagged fish in the catch;
X = number of tagged fish in a simple random sample of the catch;
R = number of tagged fish released;
p = contribution rate, Y/R;
q = $1 - p$;
O = number of untagged fish caught;
C = catch of untagged and tagged fish, $Y + O$;
S = percentage of untagged fish caught, O/C;
f = fraction of the catch that is sampled;
n = sample size, $f(O + Y)$;
k = number of interpenetrating subsamples.

For simplicity, both the sample fraction, f, and the catch of other fish, O, are treated as constants. The effect of the latter assumption on the results is uncertain but is not examined in this paper.

The probability that a particular tagged fish will end up in the catch and then in the sample is affected by at least three sources of variation. First, there is the annual variation in ocean conditions and fishery behavior. For any specified fishery, say the Washington state ocean troll catch on July 4 in a particular management zone, the probability of being caught changes from year to year. This between-year-variation is ignored in the proposed formulation of statistical distribution.

Even for a single year or a single set of circumstances, a group of smolts released from a hatchery at a particular time can be viewed as a subsample of a much larger group of possible release groups. The 10,000 salmon smolts reared under treatment A, say, represent a larger, theoretically infinite group of smolts that may have been reared the same way (in the terminology of sampling theory, this larger group is the superpopulation). Variation among these theoretically conceived groups of smolts is analogous to the vari-

ation in the proportion of heads observed between two or more sets of coin tosses with the same coin. This is the type of variation observed in the current situation between IPS groups.

The third source of variation is the ordinary random-sampling variation that arises when a subsample of the catch is used to estimate the number of recoveries in the catch. Additional variation arises in practice because of tag loss and unreadable tags, but these factors are not considered here.

I am proposing a compound distribution model that considers the second and third sources of variation. Estimated recoveries and variance estimates from IPS groups provide a benchmark for comparing such a model because they account for variation caused by brood-year and random-sampling sources. The compound distribution is formulated as follows. A tagged fish is either caught by a particular fishery or it is not. The probability of capture, p, (a realization from the superpopulation of contribution rates) is a value between zero and one, and, assuming independence, the appropriate statistical model for the probability of a single fish being caught is the Bernoulli distribution. The probability of capture of a specified number of fish, Y, follows a binomial distribution. Next, given Y tagged fish in the catch, a simple random sample of size n drawn from the catch will contain X tagged fish and $(n - X)$ untagged fish. In practice, the sample is drawn without replacement and the distribution of X, conditional on the Y tagged fish and O untagged fish in the catch, is hypergeometric. The joint distribution of X and Y can be determined simply by multiplying the binomial distribution for Y times the conditional hypergeometric distribution of X given Y. The probability distribution of X is found by summing the joint distribution of X and Y over all values of Y. More succinctly,

$$P(X) = \sum_Y^R P(X,Y) = \sum_Y^R P(Y)\, P(X|Y).$$

The compound, binomial–hypergeometric distribution for the number of observed recoveries, X, in the catch sample is

$$P(X)$$
$$= \sum_{Y=X}^R \binom{R}{Y} p^Y (1-p)^{R-Y} \binom{Y}{X}\binom{O}{n-X} \Big/ \binom{Y+O}{n};$$
(1)

X ranges from 0 to n.

Total recoveries are estimated by expanding observed recoveries by the inverse of the catch sampling fraction; thus the estimated total recoveries equals X/f. An approximate variance for the estimated recoveries is derived in the Appendix; the final form is

$$\widehat{V(X/f)} = Rp\left[q + \frac{(1-f)O}{f}\left(\frac{(Rp+O)^2 - q}{(Rp+O)^3}\right)\right].$$

(2)

The variance of p, or the estimated contribution rate, is the variance of X/f divided by R^2:

$$\widehat{V(\hat{p})} = p/R\left[q + \frac{(1-f)O}{f}\left(\frac{(Rp+O)^2 - q}{(Rp+O)^3}\right)\right].$$

Substituting C for $Rp + O$:

$$\widehat{V(\hat{p})} = p/R\left[q + \frac{(1-f)O}{f}\left(\frac{C^2 - q}{C^3}\right)\right].$$

(3)

A simplifying approximation to equation (3) that has minimal effect on the estimator results from deleting the q/C^3:

$$\widehat{V(\hat{p})} = p/R\left[q + \frac{(1-f)S}{f}\right];$$

(4)

S is the percentage of untagged fish in the catch.

The number of fish to tag to achieve a desired level of precision can be determined by solving equation (3) for R, the release number, given estimates of contribution rate, sampling fraction, and number of untagged fish in the catch. The resulting formula is the solution to a quadratic equation in R. A much simpler formula that yields slight overestimates of release size results when it is assumed that the percentage of untagged fish in the catch is close to 100% (as is often the case in practice):

$$R = p/V(p)\left[q + \frac{(1-f)}{f}\right].$$

Comparisons with IPS-Based Estimates

The IPS-based variance estimates for contribution rate estimates can provide a benchmark for determining how well the compound distribution formula corresponds to the real world. The IPS-based variance estimator is

$$V(\hat{p}) = \frac{\sum_{i}^{k}(\hat{p}_i - \hat{p}\cdot)^2}{k(k-1)}.$$

(5)

According to computer simulations of IPS samples drawn from the compound distribution, the form of the IPS estimator shown in equation (5) has a positive bias. Additional simulations indicate that the bias is due to the lack of independence in the individual IPS estimates of contribution rate. The simulations do indicate, however, that as the percentage of other fish, O/C, increases, the bias decreases.

TABLE 1.—Comparison of variance estimates based on interpenetrating subsamples (IPS) and the compound binomial–hypergeometric distribution (BH) for estimated stock contributions by a single British Columbia coho salmon release group (Quinsam hatchery, 1978 brood[a]). Variances have been multiplied by 10^4.

Fishery (year, gear code, area code)	Recoveries		Contribution rate (%)	Number of fish		Variance		
	Observed	Expanded		Catch	Sample	IPS	BH	IPS − BH
1981, 33, 057	1	5	0.05	14,459	2,897	0.0022	0.0022	0.0000
1981, 20, 011	2	10	0.09	30,752	6,143	0.0022	0.0044	−0.0022
1981, 15, 011	2	10	0.09	30,752	6,143	0.0088	0.0044	0.0044
1981, 15, 011	1	3	0.03	22,090	6,629	0.0010	0.0010	0.0000
1981, 15, 011	2	19	0.17	54,694	5,887	0.0305	0.0151	0.0153
1981, 15, 011	3	6	0.06	2,232	1,068	0.0012	0.0011	0.0000
1981, 15, 011	2	3	0.03	2,841	1,713	0.0002	0.0005	−0.0002
1981, 32, 004	1	3	0.03	5,396	3,593	0.0008	0.0004	0.0004
1981, 33, 001	2	20	0.19	13,595	4,062	0.0010	0.0010	0.0000
1981, 33, 057	1	4	0.04	55,027	5,370	0.0371	0.0184	0.0186
1981, 15, 011	2	17	0.16	35,456	8,106	0.0017	0.0017	0.000
1981, 32, 004	1	6	0.05	9,897	1,167	0.0063	0.0126	−0.0063
				18,436	3,312	0.0027	0.0027	0.0000

[a]Three tag codes—082020, 082021, and 082022—were applied to this release group; the corresponding numbers of fish released were 3,549, 3,558, and 3,568.

TABLE 2.—Summary of comparisons of variance estimates based on interpenetrating subsamples and the compound binomial–hypergeometric distribution for nine British Columbia release groups. Median variance differences have been multiplied by 10^4.

Tag code	Number of strata	r	Median difference	95% confidence interval for difference
082001, 082007, 082009	26	0.78	−0.00000716	−0.00016652, +0.00001083
082002, 082005, 082006	20	0.81	−0.00000036	−0.00000318, +0.00001001
082003, 082004, 082008	31	0.82	+0.00000311	−0.00000351, +0.00000939
082019, 082024, 082027	14	0.98	−0.0012400	−0.00076662, −0.00002210
082020, 082022, 082021	13	0.91	+0.00000544	−0.00024040, +0.00444658
082021, 082023, 082026	16	0.98	+0.00000544	−0.00000579, +0.00001805
081810, 081843, 081845	46	0.43	−0.00071311	−0.00100501, −0.00028001
081811, 081841, 081844	56	0.73	−0.00000056	−0.00000324, +0.00000542
081812, 081813, 081842	59	0.80	−0.00005316	−0.00031306, +0.00015649

Nineteen sets of release groups, tagged with interpenetrating subsamples, from British Columbia and Washington state hatcheries were used to evaluate estimates of contribution rates based on the compound distribution variance formula. Nine groups of coho salmon *Oncorhynchus kisutch* containing between 10,000 and 12,000 smolts each, were released from a British Columbia hatchery during 1980 and 1981 (Bilton and Coburn 1981; Bilton et al. 1982). Each group was divided into three approximately equal subsamples, and different coded wire tags were applied to each subsample. Equation (4) was used to estimate the variance of the contribution rate because detailed catch and sample information for the recoveries in various catches was available (T. Mulligan and J. Schnute, Canadian Department of Fisheries and Oceans, unpublished data). Based on point estimates of tagged recoveries, the percentage of untagged fish was quite high, in most cases well above 95%, thus minimizing the bias of IPS-based estimates. The IPS-based and compound distribution-based variance estimates were calculated for recoveries in all the commercial fishery catches.

Recoveries in the sport fisheries and in the escapement were not used because these data do not represent simple random samples. Detailed results of the comparisons for a single release group are shown in Table 1, and summarized results for all nine groups are shown in Table 2.

The remaining 10 release groups came from Washington state hatcheries and included both coho salmon and chinook salmon *Oncorhynchus tshawytscha*. These groups, ranging from 50,000 to 200,000 fish, were considerably larger than the British Columbia releases. The Washington releases were split into two approximately equal subsamples and tagged with different CWTs. This data set (de Libero 1986) lacked catch and sample information needed for calculating the compound distribution-based variance estimator with equation (4). Only estimates of the observed and estimated recoveries were available. Furthermore, the data were aggregated across an unknown number of catches—the total observed and estimated recoveries were year-, gear-, and catch area-specific but pooled over several catches. With assumptions that the sampling fraction was

TABLE 3.—Comparison of variance estimates based on interpenetrating subsamples (IPS) and the compound binomial–hypergeometric distribution (BH) for estimated stock contributions by a single Washington State coho salmon release group (Toutle Hatchery, 1978 brood[a]).

Fishery	Recoveries		Contribution rate (%)	Variances		
	Observed	Expanded		IPS	BH	IPS − BH
1	1	1	0.00	0.000002	0.000002	0.000000
2	13	52	0.07	0.000837	0.000341	0.000496
3	2	9	0.01	0.000014	0.000067	−0.000053
4	127	415	0.53	0.019016	0.002219	0.016797
5	38	95	0.12	0.000661	0.000389	0.000272
6	79	174	0.22	0.001473	0.000628	0.000845
7	36	115	0.15	0.001134	0.000602	0.000532
8	29	109	0.14	0.005560	0.000671	0.004889
9	1	10	0.01	0.000160	0.000164	−0.000004

[a]Two tag codes, 631931 and 632058, were applied to this release group; the corresponding numbers of fish released were 38,612 and 39,496.

TABLE 4.—Summary of comparisons of variance estimates based on interpenetrating subsamples and the compound binomial–hypergeometric distribution for 10 Washington State release groups.

Tag codes	Number of strata	r	Median difference	95% confidence interval for difference
631612, 631613	21	0.96	+0.000391	−0.000075, +0.002053
631709, 631710	32	0.44	−0.000007	−0.000032, +0.000007
631717, 631718	35	0.90	−0.000000	−0.000010, +0.000000
150710, 150813	11	0.95	+0.000006	−0.003037, +0.001127
631758, 631913	9	0.67	+0.004963	−0.000041, +0.012031
631931, 632058	9	0.97	+0.000496	−0.000004, +0.004889
632039, 632040	9	0.70	0.000000	−0.000060, +0.000952
632037, 632038	11	0.72	+0.000001	−0.000084, +0.000087
631954, 631955	12	0.83	+0.000001	−0.000666, +0.000238
071954, 071957	7	0.98	+0.000599	−0.001258, +0.047164

relatively constant across the fisheries and that the percentage of other fish remained high, an approximate compound distribution-based variance estimate can be computed. The approximate variance formula is

$$\widehat{V(\hat{p})} = \frac{p}{R}\left[\frac{q + (1 - f)}{f}\right]. \qquad (6)$$

Detailed results for a single Washington release group are shown in Table 3, and summarized results for each of the 10 Washington release-groups are shown in Table 4.

For the British Columbia release groups, the compound distribution variance formula (4) tended to slightly overestimate the variance relative to the IPS-based formula, according to the median value. In seven of the nine groups, however, the 95% confidence interval for the median contained the value zero, which suggested that the two estimators were not significantly different. The degree of correlation between the two estimators was generally quite high, as a close examination of Table 1 indicates, and as the correlation coefficients in Table 2 indicate.

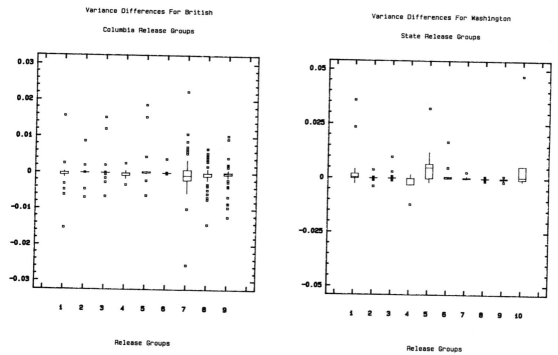

FIGURE 1.—Multiple box-and-whisker plots of paired differences between the interpenetrating subsample and the compound distribution-based variances for contribution estimates of 9 British Columbia and 10 Washington State salmon release groups.

A more detailed examination of the British Columbia data is presented in Figure 1, which shows a slightly modified box-and-whisker plot for each of the nine groups. The vertical lines of the boxed region mark the bounds of the lower and upper quartiles; the horizontal line through the box (not always figured) represents the median value; the vertical lines (whiskers) extend to the largest and smallest observations that are within 1.5 times the interquartile range of their nearest quartiles; outliers are plotted as individual points (Statgraphics 1985). For all nine groups, the differences were concentrated about zero, but a few relatively large negative differences tended to move the median value below zero.

With the Washington State data, despite the assumption of a constant sampling fraction and ignorance of the aggregation of recoveries across an unknown number of catches, the two estimators again matched up very well. In all 10 cases, the 95% confidence interval for the median contained the value zero. Figure 1 shows the box-and-whisker plots for the 10 Washington State groups. With this set, the extreme points tended to be positive.

Discussion

The variance estimator for contribution rate estimates under the compound binomial–hypergeometric distribution was quite similar to the IPS-based estimator when tested on 19 data sets. With each group there were a few catches or fisheries in which the differences were quite large. The compound distribution-based estimator tended to slightly exceed the IPS estimator and pull the median differences (IPS minus compound distribution) downward with the British Columbia groups. The reverse was true with the Washington State data. In 17 of the 19 cases, the confidence interval for the median difference contained zero, which is strong support for the appropriateness of the compound distribution.

The compound distribution for estimated recoveries of tagged fish does more than provide a means to estimate the variances of contribution rate estimates, something IPS-based estimates often can accomplish when multiple tag codes are used. It also provides a structure for determining how many fish to tag to achieve a desired level of precision in contribution rate estimates, and for analyzing the relative effects of changes in tagging level and catch sampling level. Several more extensions based on this distribution can be made as well.

When contribution rates are based on estimated recoveries from more than one catch, there are special problems in calculating the variance of the estimate. This variance is not simply the sum of the individual variances, because covariances occur at the first-stage level. Extending the compound distribution model for this situation involves treating the first-stage variation as multinomial instead of binomial. The compound distribution formula worked well with the Washington State data, which was already pooled to varying degrees, but exactly why it worked remains unknown in the absence of detailed catch and sample information.

A second use of compound distribution is to estimate the variance of a harvest rate estimate, which is roughly the ratio of a group's contribution to a single fishery to its contribution to all fisheries and the escapement combined. Accounting for variances in contribution estimates for sport fisheries and escapement, which are rarely based on simple random samples, will present new problems.

Test statistics for comparing the contribution rates and harvest rates for two or more release groups can be based on the compound distribution. Likelihood ratio test statistics are a possible tool.

More data sets with interpenetrating subsamples will be available in the next few years and more testing can be done. Despite the difficulties described, the compound binomial–hypergeometric distribution for CWT recoveries looks promising, based on analyses thus far.

References

Bickel, P. J., and K. A. Doksum. 1977. Mathematical statistics. Holden-Day, San Francisco.

Bilton, H. T., and A. S. Coburn. 1981. Time and size at release experiment: four releases of three size groups of juvenile coho salmon from Quinsam hatchery, spring of 1980. Canadian Data Report of Fisheries and Aquatic Sciences 252.

Bilton, H. T., A. S. Coburn, and R. B. Morley. 1982. Time and size at release experiment: four releases of three size groups of juvenile coho salmon from Quinsam hatchery, spring of 1981. Canadian Data Report of Fisheries and Aquatic Sciences 329.

Cochran, W. G. 1977. Sampling techniques, 3rd edition. Wiley, New York.

de Libero, F. E. 1986. A statistical assessment of the use of the coded wire for chinook (*Oncorhynchus tshawytscha*) and coho (*Oncorhynchus kisutch*) studies. Doctoral dissertation. University of Washington, Seattle.

Mahalonobis, P. C. 1946. Recent experiments in statistical sampling in the Indian Statistical Institute. Journal of the Royal Statistical Society, 109:325–370.

Statgraphics. 1985. STSC, Rockville, Maryland.

Appendix: Calculation of Variance Based on the Binomial–Hypergeometric Distribution

The estimate of total recoveries under the assumption that a simple random sample was taken from the catch is XC/n or X/f. Rather than working through the algebra of computing the expectation and variance of X/f by using the compound distribution equation directly, I use a conditional variance formula to derive the variance of X/f. The general formula (Bickel and Doksum 1977) is

$$V(X/f) = V_Y E_{X|Y}(X/f) + E_Y V_{X|Y}(X/f).$$

The derivation of the variance of the estimated recoveries, X/f, is broken into two parts. First I solve $V_Y E_{X|Y}(X/f)$, and then $E_Y V_{X|Y}(X/f)$. While the sample size n will be an integer and fC will often be a fraction, fC is treated as though it were rounded to the nearest whole number.

$$V_Y E_{X|Y}(X/f) = V_Y[f Y/f]$$

$$= Rpq.$$

$$E_Y V_{X|Y}(X/f) = E_Y\left[\frac{1}{f^2} \cdot \frac{nY}{C} \cdot \left(1 - \frac{Y}{C}\right) \cdot \frac{(C-n)}{(C-1)}\right]$$

$$\simeq E_Y\left[\frac{1}{f^2} \cdot \frac{fCY(C-Y)}{C^2} \cdot \frac{(C-fC)}{C}\right]$$

$$(n = fC; \text{ let } C - 1 = C)$$

$$= E_Y\left[\frac{YO(1-f)}{(Y+O)f}\right]$$

$$(C = Y + O)$$

$$= \frac{(1-f)O}{f} E_Y\left[\frac{Y}{Y+O}\right]$$

$$\simeq \frac{(1-f)O}{f}\left[\frac{Rp}{Rp+O} \cdot \frac{-Rpq}{(Rp+O)^3}\right]$$

(applying the delta method)

$$= \frac{RpO(1-f)}{f}\left[\frac{(Rp+O)^2 - q}{(Rp+O)^3}\right].$$

Combining the two components yields equation (2):

$$V(X/f) = V_Y E_{X|Y}(X/f) + E_Y V_{X|Y}(X/f)$$

$$\simeq Rpq + \frac{pO(1-f)}{f}\left[\frac{(Rp+O)^2 - q}{(Rp+O)^3}\right]$$

$$= Rp\left[q + \frac{O(1-f)}{f}\left(\frac{(Rp+O)^2 - q}{(Rp+O)^3}\right)\right].$$

American Fisheries Society Symposium 7:684–690, 1990
© Copyright by the American Fisheries Society 1990

Sample-Size Determination for Mark–Recapture Experiments: Hudson River Case Study

Douglas G. Heimbuch

Coastal Environmental Services, Inc.
1099 Winterson Road, Linthicum, Maryland 21090, USA

Dennis J. Dunning

New York Power Authority
123 Main Street, White Plains, New York 10601, USA

Harold Wilson[1]

Versar, Inc.
9200 Rumsey Road, Columbia, Maryland 21045, USA

Quentin E. Ross

New York Power Authority
123 Main Street, White Plains, New York 10601, USA

Abstract.—An approach for determining sample-size requirements and related experimental design elements of mark–recapture experiments is presented. Issues addressed include where and at what times of the year sampling should be conducted, which ages of fish should be sampled, what type of sampling gear should be used, and how many fish must be tagged and examined for tags. We resolve these issues by assessing conditions under which assumptions and data requirements of proposed statistical estimators can be satisfied and by analyzing catch data from preliminary and historical studies to determine likely catch rates for alternative gears. We apply the approach to a mark–recapture study designed to evaluate stocking of hatchery striped bass *Morone saxatilis* in the Hudson River.

The success of failure of a mark–recapture experiment can depend on the number of fish marked and on the number of fish examined for marks. Inadequate sample sizes will cause the precision of estimates and the power of hypothesis tests to be unacceptably low. If this occurs, inferences drawn from the results of the experiment will be questionable at best and cannot be used to provide reliable answers to questions the study intended address. Accordingly, sample-size requirements should be determined before a mark–recapture experiment is begun.

Sample sizes depend on the objectives of the study and on the statistical methods chosen to address those objectives. The feasibility of collecting required sample sizes depends on the field methods used to collect fish for marking and to examine them for marks, and on the abundance of the target fish at the times of sampling and in the locations sampled. Consequently,

determining sample sizes and assessing whether they can be obtained require a broad experimental perspective.

In this paper we describe a systematic approach for determining required sample sizes and the sampling effort necessary to obtain them. The approach provides a means for evaluating various experimental design elements, such as where and when to collect fish and what gear to deploy, and it systematically identifies and avoids study objectives that are not achievable. Application of the approach is illustrated for a mark–recapture study (MMES 1986a) whose main goal was to evaluate stocking of hatchery striped bass *Morone saxatilis* in the Hudson River.

Approach

In our approach, we assume that the overall goal for a study has been identified and that there are several alternative means for reaching that goal. The first step in the approach is to delineate the alternatives. This entails (1) defining the candidate study objectives, including the required

[1]Present address: Coastal Environmental Services, Inc., 1099 Winterson Road, Linthicum, Maryland 21090, USA.

level of precision and power, (2) selecting the statistical methods (estimators or tests) and computing the required sample sizes, and (3) delineating the alternative field methods. After the alternatives are defined, the feasibility of each is evaluated in terms of its ability to satisfy all assumptions of the statistical method and to obtain the required sample sizes.

The first step in delineating alternatives is to generate potential objectives. An objective is an exact statement of the information to be collected, such as the proportion of age-1 fish in the spring or of age-2 fish in the winter that are of hatchery origin. One or both of these objectives might be attainable, and achieving either one could satisfy the needs of a study designed to evaluate a stocking program.

The study objectives should include the level of reliability required. This can be done by specifying the level of precision (for parameter estimates) or the power (for hypothesis tests). Specifying the reliability is as important as specifying what information should be produced because sample-size requirements cannot be determined without it.

The second step in delineating alternatives is to select appropriate statistical methods. The assumptions and data requirements of each possible method should be identified; these will be used as criteria in the subsequent evaluation of alternatives.

The last step in delineating alternatives is to select field sampling methods. Alternatives are selected on the basis of the preliminary data and the type of data required by the statistical methods. Background information on the geographic distribution of the target species and on estimates of the catch rates for the sampling gear are essential to assess the sampling effort required to collect a given number of fish.

Hudson River Case Study

As part of the 1980 Hudson River Cooling Tower Settlement Agreement (Sandler and Schoenbrod 1981), utilities operating power plants on the Hudson River contracted for the construction and operation of a striped bass hatchery at Verplanck, New York. The hatchery was designed to produce 600,000 76-mm-long striped bass per year for stocking in the Hudson River from 1983 through 1990.

The utilities also planned to conduct a recapture program whose goal was to evaluate the contribution of hatchery fish to the Hudson River population. Such a program appears to be straightforward. Nevertheless, several questions had to be

answered to ensure that statistically valid results would be obtained. At what time of year should striped bass be sampled? Where in, or outside of, the Hudson River should striped bass be sampled? How many striped bass should be examined for hatchery tags? What type of gear should be used to capture striped bass, and which ages of striped bass should be examined for tags? We systematically answered these questions by using the approach described above.

Objectives

We considered two alternatives as means of addressing the study goal: estimating the proportion of Hudson River striped bass of hatchery origin, and testing the hypothesis that hatchery-reared fish were present in the Hudson River stock. We defined candidate objectives by identifying a set of points in the life history of a year-class when assessments of the hatchery contribution might be made. We considered ages from age 0, the age at release from the hatchery, to age 9, the maximum for fish caught historically in nontrivial numbers by standard collection gears. Four combinations of time and location were also considered: (1) The mouth of the Hudson River in winter; (2) below the spawning grounds in spring; (3) in the coastal fishery year-round; and (4) in the mainstem of the river in the fall following hatchery release (Figure 1). These combinations were selected because the necessary preliminary information on catch rates and age distributions was available.

From the options reviewed, we selected seven candidate objectives for evaluation, based on their relevance to striped bass management. Six of the objectives were for estimating the proportion of hatchery fish in the population. The age of fish, time of year, and river location of these objectives were (1) Young of year (age 0) during the fall following hatchery release, from river kilometer (RK) 22 to RK 224, (2) yearlings (age 1) in the mouth of the river (RK 0–14) during the winter, (3) yearlings downriver of the spawning grounds (RK 39–63) during the spring, (4) subadults and adults (age 2 and older) in the mouth of the river during the winter, (5) yearlings, subadults, and adults in the coastal fishery year-round; and (6) mature females downriver of the spawning grounds during the spring. The seventh candidate objective was to test for the presence of hatchery striped bass as mature females downriver of the spawning grounds during the spring.

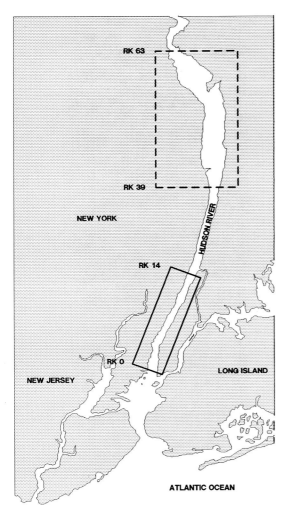

FIGURE 1.—The lower Hudson River estuary. The solid rectangle indicates the area referred to in the text as the mouth of the river. The dashed rectangle indicates the area referred to as downriver of the spawning grounds; RK is river kilometer.

Sample-Size Requirements

Estimation.—For the candidate objectives involving estimation, we chose the criterion recommended by Robson and Regier (1964) for defining adequate sample size. We required that the sample size be just large enough to satisfy the following probability statement

$$\Pr\left(-A < \frac{P - \hat{P}}{P} < +A\right) = 1 - \alpha \;; \quad (1)$$

P is the parameter being estimated, $\hat{P}$ is the parameter estimate, A is the acceptable relative error, and $1 - \alpha$ is the desired confidence level. We chose $A = 0.25$ and $1 - \alpha = 0.95$ as recommended by Robson and Regier (1964).

The proportion of hatchery fish in the population, P, was defined as the ratio of the number of fish of hatchery origin, H, to the total number of wild and hatchery fish in the population, $H + W$. If all fish in the population have the same probability of being captured and no misclassification occurs in differentiating wild from hatchery fish, the number of hatchery fish, h, in a sample of n fish can be viewed as a binomial random variable with parameters n and P. Accordingly, the simplest unbiased estimator of P is h/n.

Substitution of the estimator for P into equation (1) (note that h must be an integer) gives

$$\Pr(B < h < C) = 1 - \alpha; \quad (2)$$

B is the greatest integer less than or equal to $nP(1 - A)$, and C is the greatest integer less than or equal to $nP(1 + A)$. The required sample size was evaluated for fixed values of P by determining the smallest n that validated equation (2). This was done with a computer search algorithm and the binomial cumulative distribution function.

We selected values for H and W, and hence P, by reviewing existing information on the number of hatchery fish stocked and the estimates of age-0 striped bass abundance in the Hudson River. Annually computed estimates of $H + W$ (MMES 1986b) range from a low of 12 million in 1979 to a high of 43 million in 1981, and averaged 30 million from 1976 to 1984. With the intention of ensuring that the actual values of $H + W$ were bracketed in the analyses, we generated tables of required samples sizes for young-of-year abundance which ranged from 1 million to 90 million fish.

Selection of values for H was based on the numbers of fish stocked annually since 1983; we assumed similar mortality for hatchery and wild fish between the times that fish were stocked and estimates were made. Although the production goal for the hatchery is 600,000 fingerlings annually, the number produced has ranged from 61,000 in 1983 to 530,000 in 1986. The selected range of values for number of hatchery fish released was 150,000–600,000.

If a year-class abundance of 30 million age-0 fish and a hatchery release of 300,000 are selected, the required sample size is 6,100 (Table 1). This is

TABLE 1.—Sample sizes required to estimate the proportion of hatchery fish in the population by comparing the number of hatchery fish tagged and stocked with the abundance of young-of-year striped bass in the Hudson River when the hatchery fish were released.[a]

Number stocked (thousands)	Young-of-year abundance (millions)			
	1	30	60	90
150	400	11,700	>20,000	>20,000
300	150	6,100	11,700	18,300
450	80	3,900	7,800	11,700
600	50	3,100	6,100	9,300

[a] Sampling is binomial, the acceptable relative error is 25%, and the confidence level is 95%.

TABLE 2.—Sample sizes required to test for the presence of hatchery fish in the population by comparing the number of hatchery fish tagged and stocked with the abundance of young-of-year striped bass in the Hudson River when the hatchery fish were released.[a]

Number stocked (thousands)	Young-of-year abundance (millions)			
	1	30	60	90
150	10	322	643	965
300	5	161	322	483
450	3	107	214	322
600	2	80	161	241

[a] Sampling is binomial and a power of 0.8 is desired.

the number of fish in a year-class that would have to be examined.

Hypothesis testing.—We required that the power of the test be at least 0.8 for the alternative hypothesis that represented our best (non-zero) guess at the true hatchery proportion. Consequently, even if our guess were correct, there would be one chance in five that we would not reject the null hypothesis (that no hatchery fish were present in the Hudson River stock).

The null hypothesis was to be rejected if any hatchery striped bass were found in samples of striped bass from the Hudson River. Accordingly, the probability of a type I error for this test (i.e., the probability of rejecting the null hypothesis when it is true) was zero, if no misclassification had occurred.

The probability of a type II error (failure to reject the null hypothesis when it is false) equals one minus the power of the test and can be calculated as

$$1 - \text{power} = (1 - P)^n.$$

Therefore, the required sample size is

$$n = \frac{\log_e(1 - \text{power})}{\log_e(1 - P)}. \quad (3)$$

Values for n from equation (3) over the same range of values for the number of stocked and wild age-0 fish as described in the preceding section are listed in Table 2. If year-class abundance is equal to 30 million age-0 fish and the number of hatchery fish stocked is 300,000, the number of fish that must be examined for hatchery tags is 161.

Field Methods

All fingerlings released from the hatchery were tagged with magnetic, coded wire cheek tags prior to stocking. Striped bass collected during the recapture portion of the study were aged and examined with a magnetic detector for hatchery tags. Fish from a subsample were killed and inspected for tags to determine the extent of misclassification.

After the initial examination, all fish suspected of being of hatchery origin were killed to confirm the presence of coded wire tags. Any misclassifications would have produced an underestimate of the hatchery proportion, as would tag losses. For this paper, we assume there were no misclassifications or tag losses.

Because the commercial fishery for striped bass in the Hudson River is closed, all collections have to be made by research vessels and survey gear. Three types of gear were considered as alternatives: 12-m and 9-m high-rise otter trawls and a fly seine. Each was tested and found to be effective for sampling striped bass in the Hudson River.

Average catch rates by size of striped bass, time of sampling, and location of sampling were computed for each of the three gear types. These statistics are referred to in the evaluations that follow.

Feasibility Evaluations

Each of the seven candidate objectives listed previously was evaluated separately. An alternative was judged to be feasible if all underlying assumptions were likely to be satisfied, and if the required sampling could be completed by no more than two boats and crews working over a 3-month period (60 working days).

Age-0 estimates.—For the statistical method we selected, a fundamental requirement for producing valid estimates of proportions is that hatchery and wild striped bass are caught with equal probability. This requirement can be satisfied in two ways. Either the locations of samples must be random or the hatchery and wild fish must have the same distribution patterns at the times samples are taken.

Randomizing the locations of samples does not appear to be feasible in the main stem and mouth of the Hudson River because many portions of the river suitable for age-0 striped bass cannot be fished effectively with gears tested to date. Inaccessible areas in the mouth of the river include shallow water and water under piers; those in the main stem include nearshore and midriver shallows.

Hatchery and wild age-0 striped bass do not appear to have the same distribution patterns at the time samples are taken (unpublished data). This should not be surprising: hatchery fish are raised on feed pellets in relatively small rearing tanks, whereas wild fish feed on live prey in a relatively unbounded environment. Such behavioral differences would tend to produce substantial differences in distribution. The very habitats that resist sampling—shallow waters and pier sites—might induce different behavioral responses in the two groups.

Because wild and hatchery fish of age 0 do not appear to have the same distribution patterns, and because some areas cannot be sampled, the alternative of estimating young-of-year proportions is considered infeasible.

Winter yearling estimates.—By the winter of their second year, hatchery striped bass released as fingerlings would have had more than a full year to acclimate to the Hudson River. Consequently, the assumption that wild and hatchery striped bass have the same distribution patterns is more plausible for yearlings than for young of year.

Catch per tow of yearling-sized striped bass (150–299 mm) averaged 18.9 in 12-m trawls and 5.2 in 9-m trawls during the winter sampling period. If the mix of 12-m and 9-m trawls was even, the average catch per tow was 12. Therefore, approximately 510 tows would be required to catch 6,100 yearling striped bass. At eight tows per day, approximately 64 boat-days would be needed to collect the required number of yearlings. This could be accomplished by one boat and crew over a 3-month period.

Spring yearling estimates.—During the spring, the average catch of yearlings per tow by the 12-m trawl downriver of the spawning grounds was

0.45. Therefore, 13,555 tows would be necessary to collect the required 6,100 yearlings. If eight tows could be completed per day, 1,694 boat-days would be required. Consequently, this alternative was infeasible.

Catch rates for yearlings taken by the fly seine were higher, averaging 1.55. Thus 3,935 hauls would be needed to collect 6,100 fish. With four seine hauls per 8-h day, 983 boat-days would be required. Therefore, this alternative also was infeasible.

Winter subadult and adult estimates.—As with age-0 and age-1 fish, the assumption of equal probability of capture for hatchery and wild fish must be satisfied for subadults and adults. By age 2, hatchery striped bass are likely to have completely adapted to the Hudson River and, therefore, should exhibit distribution patterns similar to those of wild fish. Accordingly, we assumed this assumption was satisfied for subadults and adults.

However, assessing the contribution hatchery releases make to a year-class as that class ages beyond the yearling stage requires consideration of a new factor: immigration. Striped bass spawned in other estuarine systems such as Chesapeake Bay may occur in the mouth of the Hudson River. This problem is negligible for yearling fish, which do not appear to move large distances.

Immigration could be addressed in two ways. Striped bass examined for hatchery tags could also be examined for characteristics used to discriminate between Hudson River and other stocks, in which case a subsample probably would have to be killed for examination. Alternatively, no determination of origin would be required if the management objective was simply to estimate the proportion of striped bass from the Hudson River hatchery in a particular age-group in the mouth of the river, regardless of where other fish originated. We selected the latter approach for this evaluation.

The catch rate of age 2 and older striped bass was considerably lower than for yearlings. Both mortality and migration into the coastal environment reduce inriver densities of larger and older fish. The average catch per tow of age-2 fish (300–399 mm) was 4.4 in 12-m trawls and 1.5 in 9-m trawls. For age-3 fish (400–499 mm), the average catch per tow was 0.7 in 12-m trawls and 0.2 in 9-m trawls. For age-4 and -5 fish (500–599 mm), the average catch per tow was 0.1 in 12-m trawls and 0.02 in 9-m trawls.

Given an equal number of tows with 9-m and 12-m trawls, the average catch per tow for ages 2,

3, and combined 4 and 5 were 3.0, 0.5, and 0.05, respectively. Consequently, 2,034, 12,200, and 122,000 tows would be required to collect 6,100 striped bass of ages 2, 3, and combined 4 and 5, respectively. The level of effort required for ages 3 and combined 4 and 5 seem prohibitively high. Even for age-2 fish, the required level of effort is substantial. At eight tows per day, 255 boat-days would be required. To complete sampling over a 3-month period, over four boats and crews would be required.

Coastal fishery estimates.—The contribution of hatchery releases to the coastal fishery could be assessed if striped bass used as part of the inriver hatchery evaluation program were externally tagged at the time of handling. Tags returned by anglers would provide the data for evaluating the proportion of hatchery fish in the fishery.

Prior to externally tagging striped bass collected during the inriver hatchery evaluation program, each fish would be scanned magnetically for coded wire tags. Each external tag would contain a unique serial number. The serial number of a returned tag would indicate whether the fish was a hatchery or wild fish. Returned external tags would serve the same purpose as striped bass collected in the inriver evaluation program. The estimator for the hatchery proportion is the ratio of the number of tags returned that come from hatchery fish to the total number of tags returned. The sample size in this case is the total number of tags returned. Accordingly, 6,100 external tags would have to be returned.

The Hudson River Foundation (HRF 1985) reported a tag return rate of 7.7% in the first year after tagging for striped bass larger than 300 mm that were tagged in water temperatures less than 14°C. The return rate ranged from 2.4% for 300–399-mm striped bass to 16.1% for striped bass larger than 500 mm.

In this study, we assumed a first-year tag return rate of 10% for subadults and adults, and a 2.5% first-year tag return rate for yearlings. Given these values, 244,000 yearlings would have to be tagged to produce the required 6,100 tag returns. For subadults and adults, 61,000 fish would have to be tagged.

The average catch per tow in the 12-m trawls was 18.9 for yearlings and 4.4 for age-2 striped bass. The corresponding values for the 9-m trawls were 5.2 and 1.5. With the 12-m trawl, 12,910 tows would be required to collect enough yearlings and 13,864 tows would be required to collect enough age-2 striped bass. The required effort

TABLE 3.—Percentages of mature striped bass by age and sex.[a]

Age	Sex	
	Male	Female
2	15	0
3	42	2
4	65	6
5	79	21
6	84	55
7	100	80
8	100	91
9	100	100

[a] Modified from McLaren et al. (1981).

would increase progressively for older age-classes because their abundance decreases with age. If eight tows per boat per day are possible, 1,614 and 1,733 boat-days would be needed for yearlings and age-2 fish, respectively. If deployed over a 3-month sampling period, roughly 30 boats and crews would be required. The high level of effort needed clearly makes such a program infeasible.

Mature female estimates.—Determining the sampling effort necessary to estimate the proportion of hatchery fish in the population of mature females requires information on (1) the catch rate for each age of interest, (2) the proportion of each age-class that is mature, by sex, and (3) the sex ratio. These factors are combined to determine the catch rate for mature female striped bass.

McLaren et al. (1981) presented estimates of the proportion of mature striped bass by age and sex, as well as the sex ratio, for collections made in 1976 and 1977 (Table 3). The sex ratio ranged from 25% in late May 1976 to 77% in mid-April 1977. For planning purposes we assumed a sex ratio of 50%.

Using length-at-age data from McLaren et al. (1981), we determined the approximate age distribution in fly-seine catches. An exponential decay curve with an annual decay rate of 60% was chosen to approximate the age-frequency distribution and to predict the percent of the catch of striped bass larger than 300 mm that would be in each age-class.

The mean catch per fly-seine haul was 16.2 striped bass over 300 mm. However, the predicted catch per haul of mature female striped bass was only 0.29. More than 20,000 hauls would be required to collect 6,100 mature female striped bass of any age. At four seine hauls per 8-h day, the required sampling effort would exceed 5,000 boat-days. This was not a feasible alternative.

For trawl sampling, the average catch per tow for all striped bass greater than 300 mm was only

2.4—less than 15% of the catch rate for the fly seine. Therefore, this alternative also was not feasible.

Test for presence of hatchery fish.—The question of mature female striped bass of non-Hudson River origin and the requirement of equal probability of capture for hatchery and wild fish apply to this alternative just as they did to the alternatives involving estimation of the proportion. Also, the previously discussed catch rates for sampling downriver of the spawning grounds are applicable here. However, the required sample size is an order of magnitude less, which makes hypothesis-testing feasible whereas estimation is not.

The mean catch per fly-seine haul was 16.2 striped bass larger than 300 mm, and the predicted catch per haul of mature female striped bass was 0.29. Therefore, to collect 161 mature females, 555 hauls would be required. At four seine hauls per 8-h day, the required sampling effort would be 139 boat-days. This would require two boats and crews working for about 60 d and is therefore considered a feasible alternative.

With the 12-m trawl, the average catch per tow of all mature female striped bass was approximately 0.05. Therefore, 3,360 tows, or 420 boat-days at eight tows per day, would be needed to collect the required 161 striped bass. This would be an unacceptably high level of effort.

Discussion

The results from this case study illustrate some important points that may be relevant to a wide variety of mark–recapture experiments. The sampling effort required to produce reasonably precise parameter estimates was very high. However, the required precision of such estimates depends on the purpose of the study and the intended use of its results. If a lower level of precision is adequate, the required sampling effort can be reduced. For example, if a relative error of 50%, rather than 25%, had been acceptable in the Hudson River study, we could have satisfied our sampling requirement by inspecting 1,500 rather than 6,100 fish for tags.

Another point worth noting is the trade-off between satisfying the statistical assumptions and being able to collect an adequate number of fish. Obtaining adequate sample sizes would have been much less of a concern if the study had targeted young-of-year fish because they are more abundant and geographically less dispersed than older striped bass. However, newly released hatchery fish appear to exhibit different distribution patterns than their wild brethren, which violates a key assumption of the statistical method. Meeting assumptions only, or satisfying sample size requirements only, is inadequate; both must be considered.

The foregoing case study also illustrates that even with a relatively simple goal, a substantial amount of information may have to be compiled and analyzed to identify attainable objectives. We started with seven alternative objectives, each of which seemed reasonable, and found five to be infeasible. Although the effort required to conduct a thorough planning study of this type appears substantial, it is very small compared to the effort that can be saved by avoiding a sampling program with an unattainable objective.

References

HRF (Hudson River Foundation). 1985. Hudson River striped bass tagging program. HRF, Report 1985-1, New York.

MMES (Martin Marietta Environmental Systems). 1986a. An overall study design for a hatchery evaluation and population assessment for Hudson River striped bass. Report to the New York Power Authority, White Plains, New York.

MMES (Martin Marietta Environmental Systems). 1986b. 1984 year class report for the Hudson River estuary monitoring program. Report to Consolidated Edison Company, New York.

McLaren, J. B., J. C. Cooper, T. B. Hoff, and B. Lander. 1981. Movements of Hudson River striped bass. Transactions of the American Fisheries Society 110:158–167.

Robson, D. S., and H. A. Regier. 1964. Sample size in Peterson mark–recapture experiments. Transactions of the American Fisheries Society 93:215–226.

Sandler, R., and D. Schoenbrod, editors. 1981. Hudson River power plant settlement. New York University School of Law, New York.

American Fisheries Society Symposium 7:691–707, 1990

Random-Sampling Design to Estimate Hatchery Contributions to Fisheries

Robert Ramsay Vreeland

National Marine Fisheries Service, Fishery Management Services Division
7600 Sand Point Way N.E., Seattle, Washington 98115, USA

Abstract.—This report describes sampling methods used to study the effectiveness of hatcheries that rear Pacific salmon *Oncorhynchus* spp. and steelhead *O. mykiss*. Effectiveness is measured as the contribution of hatchery fish to the fishery. To determine fishery contribution, fish from hatcheries must be marked. Fish for marking are obtained by netting all rearing environments or by using mechanical sampling devices. Information recorded for a marking study should include numbers of fish marked, location and dates of marking, species, race, stock, brood, mark type, purpose of marking, mark retention, and numbers of fish released. When fish are marked with a coded wire tag, counters on the tagging machine are used to determine the number of fish marked. Mark retention can be estimated by separately holding marked fish for examination before release; the precision of the estimate will depend on the numbers of fish examined. Release numbers can be determined by electronic counters, by subtracting deaths that occur between the time of marking and release, or by sampling the population at release. Sampling of major fisheries on the Pacific coast of North America is well established and has occurred routinely since 1963, but sampling of hatchery returns is not as well established. All fish returning to a hatchery should be examined for marks. A systematic sample of unmarked fish is recommended to allow an estimate of the age distribution of returning fish. The precision of the estimate will depend on the number of fish sampled. Potential straying of marked fish may be assessed by sampling at other hatcheries and streams adjacent to the expected return site.

Seven species of anadromous salmonids are reared at public and private hatcheries on the Pacific coast of North America: Chinook salmon *Oncorhynchus tshawytscha*, coho salmon *O. kisutch*, sockeye salmon *O. nerka*, pink salmon *O. gorbuscha*, chum salmon *O. keta*, steelhead *O. mykiss*, and cutthroat trout *Salmo clarki* (Clemens and Wilby 1946). The hatcheries are a direct result of fishing pressures greater than natural salmonid runs could sustain, and of destruction or impairment of salmonid habitat through pollution, mining, logging, agricultural practices, and dam construction (Laythe 1948; Craig and Hacker 1950; Hagen 1953; Larkin 1970). To rebuild lost populations, there have been massive expenditures for fish passage and protection facilities as well as for propagation in hatcheries (Laythe 1948; Larkin 1970; Wahle and Smith 1979).

Given the demand for salmonids and the expenditures to meet this demand, it is imperative to measure the success of salmon propagation (Paulik 1963). Recently, the U.S. Congress has mandated evaluation of hatchery projects constructed under the Salmon and Steelhead Conservation and Enhancement Act of 1980 (Public Law 96-561) and the Pacific Northwest Electric Power Planning and Conservation Act of 1980 (Public Law 96-501) (Northwest Power Planning Council 1984; Salmon and Steelhead Advisory Commission 1984). The success of a public hatchery should be measured by its contribution to the fisheries, not by the number of returning adults; returns to a hatchery do not necessarily reflect its contribution to the fisheries (Paulik 1963). This paper reviews sampling design for studies of stock contributions to the Pacific coast fisheries of North America.

Experimental Design

A contribution study consists of (1) specifying objectives, (2) selecting a method to determine fishery contributions, (3) designing statistical analysis, (4) determining numbers of fish to mark, (5) organizing the marking operation, (6) collecting release and recovery data, and (7) conducting the analysis. This report deals specifically with phases (5) and (6).

The experimental methods used in a contribution study depend on the experimental objectives. Some objectives for studies of hatchery contributions to fisheries are to determine when and where a group of fish contributes, to estimate the contribution of a group of fish to a specific fishery, and to estimate total fishery contribution of a group of fish (Pacific Marine Fisheries Commission 1984).

Two methods have been used to determine the fishery contribution of a hatchery's salmonids. In the first method, the estimated catch of marked fish is multiplied by the ratio of total release to marked release to estimate contribution. This method has seven assumptions (Rounsefell and Everhart 1953; Bevan 1959; Worlund et al. 1969; Pacific Marine Fisheries Commission 1984): (1) fish to be marked are representatively sampled and receive the same treatment as unmarked fish before and after marking, (2) marked fish are identifiable throughout their lives, (3) marked and unmarked fish have the same growth and survival rates and maturity schedules, (4) marked and unmarked fish have the same distribution and vulnerability to catch, (5) the probability of a fish being sampled is independent of whether the fish is marked or unmarked, (6) all marked fish in the sample are recognized, correctly identified, and reported, and (7) the mark is not duplicated for any other study.

Assumption (1) is concerned with the hatchery life of the fish and is important because the catch of marked fish is to be extrapolated to total contribution. The remainder of the assumptions are concerned with the life of the fish after they leave the hatchery. Assumptions (3) and (4) are necessary for expansion of marked catch to total fishery contribution. Assumptions (5) through (7) are concerned with fishery sampling. Inherent in assumption (5) is that marked fish do not travel in clumps separate from unmarked fish. Thus recovery of a marked fish does not make a recovery of another mark more likely.

In the second method, the estimated total numbers of marked fish caught in the fisheries are multiplied by the ratio of total return to marked return to estimate the contribution of a hatchery. The seven assumptions previously mentioned also underlie this method. Three additional assumptions are necessary: (8) unmarked fish from sources other than the release facility of interest do not occur at the return facility or their occurrence is estimated and adjustments are made, (9) marked and unmarked fish do not stray differently from the return group of interest, and (10) marks are not lost between harvest and return (Pacific Marine Fisheries Commission 1984). The second method requires information on numbers of marked and unmarked fish that return for the group of interest and on numbers of unmarked fish that stray to the return site (if straying occurs).

Method two is initially attractive because it avoids the hatchery release statistics required for method one, which can be difficult to obtain (see the section on "Marked and Unmarked Releases"). However, required terminal return statistics may be equally or more difficult to obtain (see "Sampling Marked Fish"). Return statistics are affected by release of all or a portion of the production of a hatchery away from the rearing site, straying of returning fish, inability to recover all fish returning to a facility, and by intense fishing or by migration obstacles (such as hydroelectric dams) below a return facility. Consequently, this paper is confined to the assumptions, data needs, and procedures for method one.

Sampling and Marking Prior to Release

To reduce the impact of handling and marking stress on a hatchery population, the time required for marking, and the cost, only a sample of fish from a hatchery population is marked.

Obtaining Fish for Marking

Marked fish should represent the entire population at a facility because catches of marked fish will be expanded to estimate the fishery contribution of that facility. To ensure likeness of marked and unmarked fish, a random sample from all rearing environments at a hatchery should be removed for marking. The sample must be proportional to the numbers of fish in each rearing environment, unless different marks are used for each environment. (For example, remove a 5% sample rather than 5,000 fish from each hatchery pond.) Even if all rearing environments physically appear to be the same, it would not be advisable to randomly select one or more rearing areas (raceways, for instance) to represent the hatchery population. Factors such as time of egg take, hatching, fish ponding, disease history, food conversion, size of fish, water flow patterns, rearing density, and predation likely will differ among rearing environments. Effects of these differences on fish survival and catch are unknown.

A method for randomly sampling fish from each rearing area at a hatchery must be employed. It should give each fish an equal and known chance of being selected for marking. A method commonly used is to dip one or more netfulls of fish from each rearing environment. This is usually achieved with the fish crowded into a small area. This method, however, is likely to be nonrandom, so the relationship between marked and unmarked fish will be unknown (Bevan 1959). Hewitt and Burrows (1948) examined the dip-net method and found that the smaller sockeye salmon in a population were obtained in the first haul of fish; the fish also were stratified in the net, the bigger individuals being on top. Bias caused by this method consistently resulted in overestimates of population size. The sample size had to include at least 38% of all the fish before there was no statistically significant difference between the estimated and actual population sizes.

Work by D. Buchanan (Oregon Department of Fish and Wildlife, personal communication) also has shown the potential for nonrandom sampling with the dip-net technique. However, Buchanan believes that tight crowding of salmon, proper mixing, and several replicate samples can yield a representative sample with a dip net. Repeatability among rearing environments at a hatchery or between hatcheries, particularly if sampling personnel differ between hatcheries, could prove to be a major problem for this more rigorous dip-net technique.

Another problem with the dip-net method is that it does not subject all fish at a hatchery to the same handling. If sampling fish with a dip net affects the survival of the fish netted, a survival bias will be introduced between the marked and unmarked fish.

A more statistically sound method to ensure that each fish has an equal chance to be selected for marking is systematic sampling. This can be accomplished by crowding the fish in the rearing environment, then netting all the fish from the crowded area to another area. If, for example, a 5% sample is desired for marking, every twentieth net of fish can be set aside. If the entire population is sampled in this manner, and the number of times fish are set aside for marking is large, the assurance of obtaining a representative sample is improved over the dip-net method. The ultimate systematic sample would be the removal of every twentieth fish in the sampling process, but this would be too time-consuming and too stressful to the fish.

A systematic dip-net sampling technique may be not applicable in all situations, and it subjects the fish to considerable handling. Because of this, fishery scientists and engineers have attempted to develop mechanical sampling devices for randomly sampling populations. Hewitt and Burrows (1948) described such a device. It consisted of a circular frame divided into four equal sections with a net hung from each section. Three nets were open at the bottom and the fourth was closed. The frame was placed in a tub of water,

fish were added, and when the frame was lifted, the closed pocket retained sample of fish.

This sampler was modified to remove a random 10% sample of fall chinook salmon for a hatchery contribution study on the Columbia River. The "10-part sampler" consisted of a circular metal frame and a cylindrical liner. The frame was divided into 10 pie-shaped sections of equal size. A net pocket with a zipper at the bottom was hung from each section. To obtain a 10% sample, the frame and liner were placed in a water-filled tub and all but one pocket were left open. Fish were then placed inside the liner. When the liner and frame were lifted, the closed pocket retained fish for marking (Wahle and Vreeland 1978).

The above sampling devices were labor intensive and they require more fish handling than do systematic dip-net techniques. Faster and less disruptive devices include one described by Jones (1965), in which fish flowed over an inclined plane, fell onto a rotating cone, and passed through a slot in the cone during each rotation. The device was reported to remove 5.1% of the fish passing down the trap, with good repeatability.

Webb and Noble (1966) described a sampler that removed a constant sample of coho and chinook salmon. The mean percent sample varied from 0.2% to 1.1%. However, the device would not sample fish longer than 16 cm.

A device used to remove a sample of sockeye salmon from spawning channels in Canada was described by Davis and Hiltz (1971). The sampler removed from 4.4% to 5.2% of the fish in tests that had a standard deviation of 0.06 to 0.85.

An incremental sampler developed in 1979 by the Washington Department of Fisheries employs two jets of water controlled by solenoids. The jets are located on either side of the throat of a Y-shaped discharge pipe. All fish from a pond are pumped to a rectangular box where excess water passes through a bar grate and fish pass into the foot of the Y. For a contribution study at a Columbia River hatchery, one jet was set to operate 95% of the time and the other jet 5% of the time. One water jet force fish down the 95% arm for 19 s. The other jet operated for 1 s, forcing fish down the 5% arm. The fish passing through the 5% arm were retained for marking (Foster 1981). The pump reduced handling but the fish still had to be crowded.

A sampler developed by the National Marine Fisheries Service in Portland, Oregon, consisted of an A-shaped inclined plane table. Fish were pumped to the narrow top of the A, passed across a perforated plate that drained excess water, and swept off the wide lower end of the table into a return trough. The foot of the table was divided into 20 equal-width sections. To remove fish for marking, a flume was attached to the foot of the table to pass fish over the return trough. The flume was the width of one or two divisions. Thus a sample representing a desired percentage could be removed from the table for marking by altering the width and number of flumes attached to the table.

Attempts were made to examine the incremental and inclined plane samplers to determine the percentage of fish removed. However, neither sampler has been adequately tested to determine if each fish has an equal chance of being sampled or if the sample fish are representative of the population.

Accuracy of the sampling percentage may not be of great importance unless the percentage is used to determine hatchery releases. "Procedure for Coded Wire Tagging Pacific Salmonids," a manual resulting from a Pacific Marine Fisheries Commission (PMFC) workshop at Silver Falls, Oregon, in September 1981, recommended periodic testing of any sampling device to ensure that each fish has an equal chance of being sampled and that the intended percentage of fish is being removed. In one suggested procedure, a known number of fish are marked, mixed with unmarked fish, and resampled. If there is no significant difference in the proportion of marked fish within each equal portion of the sample, it may be assumed that each fish has an equal chance of being sampled. If there is no significant difference in mean lengths and weights of fish within each portion, it may be assumed that the sampler removes a representative sample of fish. Finally, the sampler may be assumed to remove the intended percentage of fish if there is no significant difference in percentage of fish removed by each section or over a number of trials, and if there is no significant difference among percentages removed by a chosen sampling section (Pacific Marine Fisheries Commission 1984).

Of the previously described samplers, the pie samplers are the most feasible to test. Because the numbers of fish placed in the samplers are small in comparison to the numbers of fish in a pond, it is possible to count, measure, and weigh the marked and unmarked fish retained by each section over several tests. The sampling characteristics of the other samplers can be influenced by the rate of fish delivery and the sample size. Foster (1981)

reported that the incremental sampler operates best with a constant flow of fish and an extended sampling period, which is also true for the inclined plane sampler. Only fall chinook salmon weighing 1.8 to 6.4 g were sampled with the inclined plane sampler; how it would operate with larger or smaller fish is unknown.

The most important factor in selecting a method of obtaining fish for marking is to understand how the results will be used. If the purpose of the contribution study is to show survival and contribution trends over time, the requirement that marked fish represent the entire hatchery population may be less important than it would be for other uses of the data. In a study to compare contributions over time, the same sampling method should be used throughout. Researchers must consider which sampling method their peers or those who will be influenced by the results will accept. Sampling methods vary among agencies on the Pacific coast. Fishery scientists with the U.S. Fish and Wildlife Service and the Department of Fisheries and Oceans in British Columbia use the crowding and dip-net method. Scientists with the Washington Department of Fisheries and the National Marine Fisheries Service use a more rigorous sampling method to ensure that the marked fish represent the entire hatchery population. No studies have been conducted to determine the influence of sampling method on the estimation of hatchery contribution.

More studies of mechanical sampling devices are needed before one of them can be identified as the best method for removing a random sample of fish for marking.

Marking Organization and Timing

The success of a marking experiment depends on equipment and techniques used, timing and organization of marking, and records kept. The coded wire tag (Jefferts et al. 1963) is the marking method most frequently used for salmonid studies on the Pacific coast. Several publications and manuals contain information on equipment needs, set up, and care and maintenance of equipment for coded wire tagging of salmon and steelhead (Moberly et al. 1977; Duke 1980; Jenkinson and Bilton 1981; U.S. Fish and Wildlife Service 1985); this information will not be repeated here.

In choosing appropriate sizes of salmon and steelhead to be marked with a coded wire tag, biologists in Region 1 of the U.S. Fish and Wildlife Service (1980) recommended that fish should be larger than 2.3 g but smaller than 15 g when tagged with a full-length coded wire tag. However, salmon fry 0.9 g and larger are routinely tagged with full-length tags in British Columbia (T. Perry, Canada Department of Fisheries and Oceans, personal communication). Yearling fish should not be tagged close to the time of smolting. The water temperature at the facility during tagging should be lower than 13°C. If the health of the fish is jeopardized, tagging should not take place. The hatchery biologist and manager make the decision about whether tagging should commence or continue (U.S. Fish and Wildlife Service 1985).

Anadromous salmonids normally are marked in spring and fall months, when water temperatures and recovery time are optimal. At some hatcheries it is difficult to match the desired water temperature, fish size, and time needed for marking and recovery. Development of the half-length wire tag (Opdycke and Zajac 1981) has helped to alleviate this situation by allowing successful tagging of small fish. Moberly et al. (1977) reported that chum salmon weighing 1.0 g could be tagged at a rate of 156 to 183 fry/h per person with the half-tag. Fish 2.5 g or larger were tagged at an average rate of 700/h. Rates as high as 1,200/h were achieved as taggers became more experienced. Opdycke and Zajac (1981) reported successful tagging of chum salmon fry that averaged 0.8 g; tag loss was 2% over 41 d of observation and negligible tagging-associated mortality occurred. No tagging rates were given. More recently, unfed chum salmon fry weighing 0.4 g were tagged at a rate of 828 fish/h (K. Crandall, Alaska Department of Fish and Game, personal communication). Thrower and Smoker (1984) tagged pink salmon fry averaging 33 mm in length and 0.25 g in weight. The tagging rate averaged 350 fish/h and approached 600/h by the end of the marking period. After 14 d, tagging-related mortality was 0.15% and tag retention was 95.7%.

Sorting fish by size may improve tag placement and reduce tag loss. Sorting was found to be unnecessary if 98% of the fish fell into one of three length ranges: 50–90, 70–140, or 110–300 mm. If 10% or more of the fish fell in one of the ranges, the benefits of sorting were believed to outweigh the disadvantage of the additional handling (Duke 1980).

The success of any marking program can depend on the records kept and their accuracy. All agencies using coded wire tags have forms for recording pertinent tagging information, examples of which are illustrated in the tagging manuals

previously mentioned. Obvious data to record include numbers of fish marked; location and dates of marking; species, race, stock, and brood of fish; tag code; and purpose of the marking. Other information often is kept on the method and dates of sampling of fish to be marked, holding environment for fish to be marked, size of fish marked, disease history and treatment of fish during the entire rearing period, fish condition, mortalities and water quality during marking, mark loss, and any problems occurring during marking (Duke 1980). When the coded wire tag is used, it is also recommended that a sample of wire from each roll be retained to check the tag code. Some cases of improperly labeled spools of wire have occurred. Some fish should be sacrificed and the placement and tag code should be checked to verify the records (King 1979; Duke 1980; Jenkinson and Bilton 1981).

Notes on naturally missing fins are important if fin removal is part of the mark. Marking personnel should examine and record all cases of missing fins during marking. Fins can be missing for genetic reasons or lost because of aggression by other fish, erosion from disease, or abrasion on concrete pond walls (King 1979). Normally, the adipose fin is clipped on anadromous salmonids that receive coded wire tags. Unusual numbers of fish with naturally missing adipose fins have occurred in some species at several hatcheries and in wild populations in Washington (Blankenship 1981). Unrecorded occurrences of missing fins at hatcheries could result in overestimation of the contribution, so marking supervisors must keep an accurate record of the number of fish found with naturally missing fins.

Treatment of Marked and Unmarked Fish

Handling and marking of fish could introduce bias between marked and unmarked fish. However, handling unmarked fish in the same manner as the fish during marking normally is impractical and unacceptable to hatchery personnel. Consequently, stress associated with marking must be minimized to ensure validity of the assumption of equal survival of marked and unmarked fish. To help guarantee equal treatment, marked fish should be returned to the population of unmarked fish from which they came. This creates difficulties in determining mark retention and numbers of marked fish released, but it also provides an opportunity to estimate the total population in a rearing environment.

Determining Numbers of Fish Marked

In many cases, knowing how many fish were marked is simply a matter of keeping a tally during the marking operation. When marks such as coded wire tags are internal, determining the number of fish tagged is more difficult. The wire tag injector (Northwest Marine Technology model) contains a counter that counts the number of times the injector is cycled (counter A). There are two counters in the quality control device (QCD). One counts the number of magnetized tags passing through the QCD (counter B), and the other counts the number of times the tagging cycle functioned but a tag was not detected (counter C) (Duke 1980; Jenkinson and Bilton 1981). It would seem that the number of times a tag was not detected (counter C) could be subtracted from the count in the wire tag injector (counter A) to give the count of the magnetized tags passing through the QCD (counter B). However, in practice the derived and actual numbers may differ for numerous reasons. These include stuck QCD counters, low water pressure in the QCD, electronic control box malfunction, moisture-caused shorts on the control box connector, large fish that temporarily block the exit from the QCD, fish too large for the water jet to direct to the correct QCD exit channel, a large range of fish sizes resulting in incorrect water jet pressures for the smallest or largest fish, small fish that turn sideways in the pipe entering the QCD so the tag does not become magnetized, fish tagged externally, fish caught in the QCD and washed through with another fish without being separately counted, and tag loss before fish get to the counter (Duke 1980; Jenkinson and Bilton 1981). The extent of these errors is difficult to assess. They can be minimized by proper tag placement, water pressure, QCD slope, and electronic setting (Jenkinson and Bilton 1981). Jenkinson and Bilton (1981) recommended that a separate count be maintained of any fish passed through the QCD a second time to check for the presence of a tag. Duke (1980) recommended that when counts are questionable the adjusted injector count (counter A minus counter C) be used. Despite possible counting mistakes, de Libero (1986) speculated that the incidence of counting errors is less than 0.1%. However, potential counting problems emphasize a need to estimate tag loss percentage and total tagged and untagged populations at the time of release.

Mark Retention

Marks may be lost at the hatchery or in the natural environment after release. Coded wire (and other) tags may be lost because of defective head molds, poor tagging technique (King 1979), and small fish. Bergman and Hager (1969) and Blankenship (1981) found that tag loss increased with a decrease in fish size and that tag loss was essentially complete 1 month after tagging. It is important to know the extent of tag loss so that the ratio of tagged to untagged fish can be corrected; otherwise, errors will ensue in the estimates of a hatchery's contribution to a fishery.

Duke (1980) recommended that 300 to 500 fish be randomly collected from each tag group and examined for tag loss. Of these, five from each group should be killed and their tag position checked. A minimum of 2 weeks should elapse between completion of tagging and the tag-retention check. Each fish should be examined for presence of a tag, quality of the alipose clip, and fish condition.

The U.S. Fish and Wildlife Service (1985) recommended that tag retention rate be determined at least 1 week before release. This allows fish time to recover from effects of the anesthetic used during examination. Bouck and Johnson (1979) found that fish treated with MS222 (a commonly used anesthetic) at a concentration of 100 mg/L suffered 100% mortality when transferred directly to 28‰ sea water, but only 12% mortality if 4 d elapsed before transfer.

If one were to follow the recommendations of Duke (1980) and the U.S. Fish and Wildlife Service (1985), tagging would have to be completed at least 3 weeks prior to release. Because tag loss can occur for up to 1 month, it would be advisable to complete the tagging 1 month prior to release.

The U.S. Fish and Wildlife Service (1985 suggested two ways to obtain fish for tag-retention checks. In the first, several ponds are selected for sampling, the fish are tightly crowded, and netfulls of fish are removed from all crowded areas until the desired number of marked fish is obtained. This sampling method requires the assumption that each netfull of fish is a random sample of the population. As previously mentioned, Hewitt and Burrows (1948) found this may not be true. All fish missing an adipose fin are tested for the presence of a tag. Fish with no tag are passed through the field of a powerful magnet in three different planes, then retested. This is done to ensure that the fish tested negative because they lost their tags, not because the tags lost their magnetism.

An alternative method for collecting marked fish has been used by the Washington Department of Fisheries, the Oregon Department of Fish and Wildlife, and the U.S. Fish and Wildlife Service. A sample of newly tagged fish is periodically collected from all tagging personnel. The fish are checked with a tag detector to verify that each fish contains a tag. This check is done to detect tagging-machine or tag-placement problems. Fish without tags are counted and left in the sample. The sample fish are held separately from other fish in the population, either in hatchery troughs or floating net pens. After an appropriate amount of time, sample fish are examined for tag retention. There are several advantages to this method. First, overall fish handling prior to release is reduced. Second, the tagged sample can easily be retained after release of the other fish; thus, if the tagging was completed less than 2–4 weeks before release, the tagged sample could be held for the recommended time to obtain the most accurate estimate of tag retention. Also, when an emergency or early release is made, the separately held sample remains available for tag-retention examination. Finally, separate holding allows an assessment of the number of fish that receive a tag but do not receive a recognizable adipose clip. This is important if the number of fish tagged minus those that die or lose the tag is used as the number of tagged fish released. If some fish are tagged but do not receive a recognizable adipose clip, the fish cannot be identified as tagged in the fishery or return samples. Thus, the tagged and unclipped fish should be added to the untagged population in calculations of the tagged to untagged ratio at release.

It is also possible that the separately held fish in the tag-retention sample are not representative of the entire tagged population. This could occur if growth and activity of the sample fish differed from those of the other tagged fish in ways that affected tag loss. These possibilities have not been examined.

To determine the appropriate number of fish to examine for tag and mark retention, decisions must be made concerning the maximum tag loss expected and the desired precision for the tag-loss estimate. In recent years, tagging programs at salmonid hatcheries on the Pacific coast of North America generally have had tag losses of between 5 and 10% (Johnson 1987).

If sampling is done without replacement, the hypergeometric distribution best describes the distribution of the estimated proportions of untagged fish obtained from a population whose fish were at one time all tagged (Chapman 1951). Use of the normal approximation to the hypergeometric distribution allows a closed formula to be used for simple calculations with various levels of tag or mark loss and estimates of precision. D. D. Worlund (National Marine Fisheries Service, personal communication) developed an equation that yields the number of fish to examine for tag retention, given a maximum tag-loss rate, the total population tagged, and a desired precision of the tag-loss estimate. If

N = number of fish tagged prior to any loss,
M = actual number of fish without tags in the population,
$P = M/N$ = proportion of tag loss in the population,
n = number of fish sampled from the population,
m = number of fish in the sample without tags, and
k = precision as 1/2 the absolute confidence interval width.

an equation for determining the number of fish to sample to obtain a desired precision can be developed from the expected proportion of the tag loss with the formula

$$E(\hat{P}) = P = M/N .$$

In Cochran (1977), theorem (3.2) states the variance of $\hat{P}$ (the estimated proportion of tag loss) is

$$V(\hat{P}) = \left[\frac{P(1 - P)}{n} \right] \left[\frac{(N - n)}{N - 1} \right].$$

To restrict the precision of the difference between the actual and estimated values of tag loss to some probability, n should be chosen large enough that

$$\text{Prob}\,[-k \le |\hat{P} - P| \le k] \ge (1 - \alpha);$$

k is small (≤ 0.1) and $(1 - \alpha)$ is large (≥ 0.95). Let $Z_{\alpha/2}$ represent the area under a standard normal distribution curve lying outside of $-Z_{\alpha/2}$ and $Z_{\alpha/2}$. Then

$$k^2/V(\hat{P}) = Z^2_{\alpha/2},$$

and

TABLE 1.—Numbers of fish to examine for tag retention to be 95% confident that the true value of tag loss is within ±0.01 or ±0.02 of the estimated value.

Tagged population	Tolerance = ±0.01 for expected tag loss of				Tolerance = ±0.02 for expected tag loss of			
	0.05	0.10	0.15	0.20	0.05	0.10	0.15	0.20
20,000	1,673	2,948	3,935	4,702	447	829	1,154	1,428
40,000	1,746	3,183	4,364	5,328	452	847	1,189	1,480
60,000	1,771	3,270	4,529	5,576	453	853	1,201	1,499
80,000	1,785	3,315	4,616	5,709	454	856	1,207	1,508
100,000	1,793	3,342	4,670	5,791	455	857	1,210	1,514
150,000	1,803	3,380	4,744	5,905	455	860	1,215	1,522
200,000	1,809	3,398	4,781	5,964	456	861	1,218	1,525

$$\frac{k^2 n(N-1)}{P(1-P)(N-n)} = Z^2_{\alpha/2}.$$

Through algebraic manipulation, the above equation becomes

$$n = \{[k^2/Z^2_{\alpha/2}P(1-P)][(N-1)/N] + (1/N)\}^{-1}.$$

An examination of data in Johnson (1987) reveals that tag loss is normally less than 20% and in many cases less than 5%. Using the above equation, I calculated the numbers of fish to sample given four different expected maximum tag-loss levels and the desire to be 95% confident that the true tag loss is within ±0.01 or ±0.02 of the estimated value (Table 1). For example, if the estimated tag loss were 0.05, the 95% confidence interval would be 0.04–0.06 or 0.03–0.07, depending on the desired precision.

The U.S. Fish and Wildlife Service (1985) recommended the following formula for determining the numbers of fish to examine for tag loss:

$$n = [1.96/rq][(1-q)/q];$$

q is the estimated tag-retention rate and r is some percentage of the rate. For this equation, a binomial distribution is assumed for the proportion of tagged fish based on a sample from a population whose fish were at one time all tagged. The precision does not fix the width of the confidence interval. The smaller the tag-retention rate, the smaller the confidence interval. For example the 95% confidence interval for a tag retention of 0.95 is 0.9005–0.9995. For a tag retention of 0.85, the 95% confidence interval is 0.8075–0.8925. Because of this difference in precision, the suggested numbers of fish to sample are less than those listed in Table 1.

As can be seen from the sample sizes in Table 1, if the expected tag loss is 0.05 or less, the true tag loss will be within ±0.02 of the estimated value with a sample size of 500 fish (as recommended by

Idaho Department of Fish and Game). It is also clear, if one samples about 2,000 fish, that the true tag loss will be within 0.02 of the estimated value for normally expected tag losses (5–20%).

Marked and Unmarked Releases

One of the most critical elements of a hatchery contribution study is to determine how many marked and unmarked fish are released. This is critical because to expand the catch of marked fish to the total release, one must know the marked to unmarked ratio at release. It is assumed that the survival and distribution of the marked and unmarked fish is the same after release.

The ideal method for determining marked and unmarked releases is an exact count. Hand counting is too time-consuming and detrimental to fish health, but machine counting of fish carrying coded wire tags is presently being tested by several fishery agencies on the Pacific coast of North America. Fish counters manufactured by Northwest Marine Technology, Inc., and by Smith Root, Inc., have been tested at Washington Department of Fisheries and Washington Department of Game hatcheries to count chinook and coho salmon and steelhead at release. The counter records both coded-wire-tagged and untagged releases. The error rate is less than 5% when the fish are not forced through the counter, but it increases when fish are forced through (Appleby and Schneider 1983). At present, the counter appears to be useful for species that are volitionally released (coho or spring chinook salmon). However, the counter technology is not sufficiently advanced to give 5% accuracy for species normally released en masse (fall chinook, pink, and chum salmon). Thus, with some species, other forms of sampling at or near the time of release may be required.

Because the number of fish marked is recorded at the time of marking, this number minus any mortalities of marked fish prior to release could be used as the release number. There are two problems inherent in this procedure: collecting all dead fish and determining how many of them are marked.

It may be very difficult to collect all fish that die before the release date, particularly if several weeks elapse between marking and release. Dead fish normally collect on pond drain screens, but some may sink directly to the bottom, and some may never appear because of predation. Determining predatory losses can be difficult or impossible; collection and examination of dead fish may be nearly as difficult. The routine at a hatchery may be to collect dead fish daily when several hundred fish are dying per day but only once every second or third day if daily mortalities are 50 or fewer. The longer the time before collection, the greater the chance of losing dead fish to predators or deterioration. If predator problems were nonexistent or minimal, pond bottoms remained clean, and water remained clear enough to see all dead fish, daily collections might yield a reasonably accurate number. However, such ideal conditions are rare.

Dead fish must be examined for tags. Because dead fish deteriorate rapidly, they must be examined soon after they are collected or the fish must be preserved. Fish may be frozen or placed in a preserving solution. Fish must be frozen individually—a block of frozen fingerling salmonids quickly turns into a fish slurry when thawed. Freezing fish individually is time-consuming and takes considerable space. Preserving solutions may be noxious to work with and must be kept away from production facilities.

One might assume that the absence of an adipose fin indicates the dead fish was tagged, but the adipose fin is the first external part of a fingerling salmonid to deteriorate after death. Thus, every fish missing an adipose fin must be passed through a tag detector. Fish that test negative must then be passed through a magnet and rechecked to control for any tags that lost magnetism. When the number of marked fish that died before release has been determined, it can be subtracted from the initial number marked to determine the number of marked fish released.

To expand the catch of marked fish from a hatchery to the total catch of fish, one must know the total release. Records are normally kept of the numbers of fish on station during rearing and at release. These numbers are based on (1) samples taken periodically through the rearing period, (2) subtraction of deaths in ponds from the original counted egg take or the numbers of fish placed in the ponds, (3) application of some standard mortality rate, or (4) a combination of (1)–(3). In the past, these methods have led to overestimates of release numbers (Worlund et al. 1969; de Libero 1986).

A reasonably accurate estimate of total population can be obtained when appropriate sampling procedures are followed. That is, fish in all hatchery rearing environments are weighed, a random sample is removed to estimate fish per kilogram, and the total population is estimated by multiplying the total weight of fish by the estimated fish per kilogram. Subtraction of deaths in ponds is fraught with the previously mentioned problems. The application of a standard mortality rate to estimate the total population is probably the least accurate method to estimate populations because it cannot account for unexpected survival or mortality. Thus a researcher must know the method used to estimate populations before accepting total release figures from hatchery records.

A more accurate method for determining releases requires that all fish be handled very near the time of release. This procedure was used to estimate releases of chinook and coho salmon for contribution studies at Columbia River hatcheries (Worlund et al. 1969; Wahle et al. 1974). The entire population of chinook and coho salmon at each hatchery was sampled with the 10-part sampler. The numbers of marked and unmarked fish retained by the closed pocket were counted. These counts were then divided by the estimated proportion of fish retained in the pocket. The sampler was tested to determine the variance of the proportion of fish retained. This allowed calculation of variances for the number of marked and unmarked fish released. Other sampling devices or procedures (such as those suggested for removing fish for marking) could be used to remove a random sample of fish for examination. However, the device or procedure must be calibrated if an estimate of variance is desired.

Methods developed to estimate populations of animals in the wild could be applied to hatchery fish. To use these methods, either marking or subsequent sampling must be random. If the marked fish are randomly distributed in the population sampled, the subsequent sampling does not have to be random (Ricker 1948; Schaefer 1951). Five assumptions apply to any population estimate based on marking and recapture: (1) the marked fish randomly mix with unmarked fish, (2) the sampling method is not selective for marked or unmarked fish, (3) the marked and unmarked fish suffer equal mortality, (4) the mark is not lost, and (5) all marks are recognizable when fish are recaptured (Ricker 1948; Fredin 1950). These assumptions are difficult to test and are not always reasonable (Chapman 1955).

If a mark–recapture method is used to estimate hatchery populations, the Petersen method is more appropriate for a study designed to determine the contribution of hatchery fish. Multiple marking and recapture methods described by De-Lury (1951) and Ricker (1975) require more handling and tagging of fish, in addition to what is done to estimate the hatchery contribution. Because marking was done randomly to ensure that marked fish represent the total population, subsequent sampling to estimate hatchery populations need not be random, provided one is willing to accept the assumption that marking does not alter the behavior of the fish in a manner that affects the probability of recapture. A dip net could be used to grab-sample an appropriate number of fish for an estimate of any desired precision. This technique has been used routinely since the mid 1970s at salmon hatcheries in British Columbia to obtain estimates of the tagged-to-untagged ratios and the total population sizes (T. Perry, Canada Department of Fisheries and Oceans, personal communication).

All of the previously mentioned five assumptions must be well satisfied if the Petersen technique is to yield an accurate population estimate. If a month or more has elapsed between marking and Petersen sampling, it could be difficult to determine the number of marked fish in the population. Even if resampling occurs within a month of marking, the actual number of marked fish may be difficult to ascertain for the reasons previously mentioned. In addition, fish are not fed during marking, and the stress of handling and marking may reduce their food consumption after marking. This could cause the marked fish to be smaller than the unmarked fish. This difference in size likely would not be made up if the time between marking and release were a month or less. The smaller size of the marked fish could result in a nonrandom mix of marked and unmarked fish because of the selectivity of larger fish for the more favorable habitats in ponds (Senn et al. 1984). Also, size difference could result in selectivity for tagged fish in sampling (Hewitt and Burrows 1948). It is probably also unreasonable to assume that marked and unmarked fish undergo the same rate of mortality, considering the additional handling stress incurred by the marked fish.

If one is unwilling to accept the validity of the assumptions concerning random mixing of marked and unmarked fish, then random sampling at the time of release is required. If the number of marked fish present in the population at the time of sampling cannot be accurately determined, then a Petersen estimate of the total population is not useful.

In summary, although it is desirable to minimize the handling of fish just before release, it is also necessary to obtain an accurate release estimate. Not sampling fish at release increases the chance of inaccurate release estimates. The greater the probability of error in release estimates, the less useful the contribution estimates. In short, to ensure that the funds and time expended for a hatchery contribution study are well spent, it is necessary to obtain the best possible estimate of the number of marked and unmarked fish released. With some species of salmonids, this estimate may be obtained with an electronic counter. With other species, it may be necessary to weigh the entire population at the hatchery and randomly sample to obtain release estimates. In some cases, as when fish are released from large ponds, it may be impossible to obtain an electronic count or to weight the release population, and a Petersen estimate may be the only alternative. It would be best to apply a new mark to a sample of fish a week or so before release, and resample just prior to release. A granule spray dye or a partial clip of the caudal lobe or ventral fin might provide an acceptable mark. However, given all the problems with the Petersen technique, an equally accurate release estimate might be obtained with an electronic counter at the pond outlet, even for mass releases. If electronic counters continue to improve, they may provide the best release estimates for all situations.

Fishery Sampling

Anadromous salmonids on the Pacific coast range from central California to central Alaska (Yonker 1963). They are captured in a variety of commercial and sport fisheries in marine and fresh water, often far from their origin. For example, chinook and coho salmon from hatcheries in the Columbia River Basin are caught in marine fisheries from Alaska to California (Wahle et al. 1974; Wahle and Vreeland 1978). This causes unique problems in sampling the fisheries for marked fish.

The major marine and freshwater sport and commercial salmonid fisheries from Alaska through California have been sampled for fin marks since 1963 (Worlund et al. 1969) and for coded wires since 1974 (Oregon Department of Fish and Wildlife 1976; Heizer and Beukema 1977). The sampling is done by the Department of Fisheries and Oceans in British Columbia and by fishery agencies in Alaska, Washington, Oregon, Idaho, and California.

The Canada Department of Fisheries and Oceans (formerly Canada Fisheries and Marine Service) began examining chinook and coho salmon for coded wire tags in 1973 in the commercial troll fishery in Georgia Strait (Heizer and Argue 1976), and has been sampling salmonids in the sport and commercial fisheries along the coast of British Columbia since 1974. Tag recovery information is available in Heizer and Beukema (1977) for 1974 and at the Pacific States Marine Fisheries Commission's Mark Processing Center for 1975 through the present.

The Oregon Department of Fish and Wildlife housed the Regional Mark Processing Center from 1970 through 1977. The Center assimilated, compiled, and distributed data on recovery of wire-tagged salmonids in U.S. coastal fisheries in Alaska, Washington, Oregon, and California (Oregon Department of Fish and Wildlife 1976, 1977a, 1977b). The Pacific States Marine Fisheries Commission assumed the duties of the center in 1977. Recovery data from 1977 onward can be retrieved on line from the Commission at 2501 S.W. First Avenue, Metro Center Suite 200, Portland, Oregon 97201. Descriptions of fishery sampling may be obtained by writing the Canada Department of Fisheries and Oceans, Mark Recovery Program, 1090 West Pender Street, Vancouver, British Columbia, V6E 2P1, and the Pacific Marine Fisheries Commission. De Libero (1986) discussed fishery sampling errors.

Sampling at the Spawning Site

In contrast to fishery sampling, routine sampling at the return sites does not always occur. Thus, plans to sample hatchery returns and adjacent streams must be developed. The plans must include sampling purposes, location, design, and data requirements.

Purpose of Sampling

Fish are sampled at return to obtain an estimate of the survival of all marked fish. This sampling gives managers a complete picture of the life cycle of marked salmonids. The sampling allows one to determine the harvest-to-return rate (catch to escapement) of marked fish. Return-site sampling also allows one to evaluate the permanence of the mark and the equality of survival of marked and unmarked fish after release. It also gives an indication of the extent to which hatchery salmonids stray.

Sampling Marked Fish

Two types of sampling occur at the spawning site—sampling for marked fish, and sampling to obtain age distributions and average fish lengths. Usually, returns are small enough so the entire population can be sampled for marks. The U.S. Fish and Wildlife Service (1980) recommended examining all returning fish for marks at the time of spawning. The normal spawning procedure is to examine the returning salmonids for maturation one to three times a week during the spawning season. Fish ready to spawn are removed and spawned. Immature and dead fish also are removed from the holding ponds. In some cases immature, and excess male fish are removed from the holding ponds before spawning begins.

Personnel should be present for the specific task of mark sampling. Hatchery personnel are normally too busy with the spawning operation to adequately examine all fish. Each fish with a mark is set aside for later examination and collection of biological data. In coded wire tag sampling, the snout of each marked fish is removed with a cut from the top of the head, behind the eyes, to the back of the mouth. The snout is placed in a plastic bag with a label that notes sampling site, date, species, length, sex, and mark quality (U.S. Fish and Wildlife Service 1980). Each fishery agency on the Pacific coast of North America has a form for recording these data.

It is also recommended that a scale sample be taken from all returning fin-clipped fish. If a fish has lost its tag, ages determined from the scales will allow the fish to be assigned to its mark group. If the fish retains its tag, which will indicate age precisely, the accuracy of scale-reading can be checked and, if necessary, corrected.

If the entire returning population is sampled, no estimates of marked returns are required. In some cases, the entire population cannot be sampled because of inefficient weirs or traps at return sites. These may allow smaller salmon to escape. In other cases, the return exceeds egg-take needs, and the traps may be removed. Then, if a count of the total return population can be obtained, the return to the hatchery is assumed to be a random sample, and the return of marked fish can be applied to the total return to obtain an estimate of the total marked return.

Sampling Unmarked Fish

If fish from one year of marking return to a facility over more than one year, scale samples should be collected to estimate the age of returning unmarked fish. In some cases, as with returning coho salmon, the lengths of returning fish may yield a sufficiently accurate estimate of age at return. In many cases, however, a scale sample from unmarked fish is necessary. Usually, scale removal from all unmarked fish would be impractical because of the expense and time required. Simple random sampling of unmarked fish would be complex and difficult to achieve during a spawning operation that lasts over several days or weeks. However, a systematic sample could easily be drawn and accomplished with less mistakes. Another advantage of systematic sampling is that the sample can be spread more evenly through the population (Cochran 1977). A systematic sample consists of choosing a starting point and then sampling every kth fish from that starting point. The starting point can be chosen from a random number table (Schaeffer et al. 1979). The size of the kth interval will depend on the population size and on the sample size needed.

Suppose 10,000 fish are expected to return to a spawning site. Also suppose a sample of 2,000 fish is needed to estimate the age proportions at return within certain limits. To obtain 2,000 fish, one out of every five fish could be examined starting with a number (1, 2, . . ., 5) given by a random number table. If the starting number were 2, the 2nd, 7th, 12th, 17th, . . . fish spawned would be sampled. If the sampled fish was marked, the sampler could choose the next unmarked fish spawned, then continue sampling every k^{th} unmarked fish.

A systematic sample yields variances that are equal to or less than those yielded by a simple random sample, if the order of the population is random or the measurements are not related to the order of the periods within the population (Williams 1978; Scheaffer et al. 1979). Systematic sampling could lead to bias if there were periodic cycles in the population of spawned fish. For example, if only one person were spawning the fish and the spawning procedure consisted of taking eggs from two female fish and then fertil-

izing the eggs with one male, a systematic sample of every third fish would result in a sample of all male or all female fish.

There is no reason to believe that periodicity occurs in a spawning operation at salmonid hatcheries. The order of spawning depends on fish ripeness. Several hatchery personnel normally spawn fish, so the mix of males and females is not periodic. In most cases, it seems safe to assume that a systematic sample will yield estimates of variance equivalent to those obtained from a simple random sample. If this cannot be assumed, one could repeatedly choose a number from a random number table and count that many fish to choose the one sampled.

It is best not to include marked fish in the age sample of unmarked fish. The age of marked fish is known. Including marked fish reduces the number of unmarked fish examined. This dilution could result in age proportion variances for unmarked fish that are larger than desired; in turn, the larger variances may make it impossible to detect the influence of marking on age of return. To obtain the desired sample of unmarked fish in the previous example, one could sample one in every four fish (rather than one in five) if it was believed that 500 marked fish would return.

The sampling operation is accomplished most efficiently with two or more samplers. One person can examine all fish spawned for marks, while the second person records all data and keeps track of the kth unmarked fish to be sampled for age determination.

The number of fish to sample to obtain an age distribution depends on the expected age of the returning fish and the desired precision of each of the age proportions. The expected age may be based on previous studies of return age at the spawning site, ages of returns to nearby spawning sites, or a reasonable guess. For example, coho salmon return in their second and third years. In the absence of information on age at return, an assumption of 50% 2-year-olds and 50% 3-year-olds could be made. This is probably an unreasonable assumption, given the general knowledge of hatchery personnel. An assumption of 25% or less 2-year-old fish and 75% or more 3-year-old fish might be more appropriate.

After an appropriate age proportion has been assumed, one must decide what precision is desired. The desired precision of a particular age proportion will proscribe the necesary sample size. For example, it would require a much smaller sample size to estimate the age proportion of 3-year-old coho salmon in the previous example to within ±10% of the expected age proportion than would be required to estimate the proportion of 2-year-old fish with the same precision (if 25% are 2-year-old fish and 75% are 3-year-old fish). If a small age proportion is to be estimated very precisely, the entire population may have to be sampled.

Because samples for scale analysis are taken without replacement, the normal approximation to the hypergeometric function best describes age distribution. The numbers of unmarked fish to examine for various possible age proportions and numbers of returns are presented in Table 2 for two different levels of precision (10 and 20% of the expected age proportion). The number to sample comes from an equation developed by Worlund (personal communications). The equation was developed in the same manner as that for the numbers of fish to sample for mark loss, except that the confidence interval around P is not fixed. In Worlund's equation,

N = number of fish returning to a hatchery,
M = number of fish of a specific age returning to the hatchery,
P = M/N proportion of fish of a specific age,
n = number of fish sampled for age, and
k = dP = precision as one-half the absolute confidence interval width.

As was the case for the mark-loss equation,

$$E(\hat{P}) = P = M/N.$$

The variance of $\hat{P}$ is described by Cochran (1977):

$$V(\hat{P}) = \left[\frac{P(1 - P)}{n}\right]\left[\frac{(N - n)}{(N - 1)}\right].$$

Again, n is to be chosen such that

$$\text{Prob } [\,|(\hat{P} - P)| \le dP] \ge (1 - \alpha),$$

dP being small and $(1 - \alpha)$ large. Let $Z_{\alpha/2}$ represent the area under a standard normal distribution curve lying outside of $-Z_{\alpha/2}$ and $Z_{\alpha/2}$. Then,

$$d^2 P^2 / V(\hat{P}) = Z^2_{\alpha/2}.$$

Substituting for $V(\hat{P})$,

$$\left[\frac{d^2 P^2 n(N - 1)}{P(1 - P)(N - n)}\right] = Z^2_{\alpha/2}.$$

Solving for n, the above equation becomes:

$$n = \left[\left(\frac{k}{Z_{\alpha/2}}\right)^2\left(\frac{P}{1 - P}\right)\left(\frac{N - 1}{N}\right) + \frac{1}{N}\right]^{-1}.$$

Because the confidence interval $(-dP \le P \le dP)$ is not fixed, the number of fish to sample decreases as P increases. Also, the less stringent the precision, the smaller the sample size required for a given return number and age proportion.

Because the sample size for age analysis depends on the precision desired, it seems prudent to select a small age proportion, say 0.20, and a reasonable precision level, 0.10. The sample sizes suggested in Table 2 are for readable scales. Some unreadable scales will inevitably occur, so it is wise to set a sampling goal somewhat larger than the tabled values. For example, to estimate the 0.20 age proportion within ±0.1 for an expected return of 750 unmarked fish, a sample size of 504 is necessary. Removing scales from two of every three unmarked fish that are spawned would yield a sample size of 500 fish. Unreadable scales would dilute this sample and make the precision less than desired. Removing scales from three of every four unmarked fish would yield a sample size of 562 fish, which would provide some buffer for unreadable scales and other unforeseeable circumstances.

For each scale collected, data should be taken on spawning return site, sample date, species, record number, sex of fish, and fork length. These data will ensure proper organization of the information, allow application of age proportions to total population, and may aid in reading some scales. It is recommended that the age proportions be applied to total returns by sex, as determined by the spawning crew. If all fish are spawned, total return and total male and female fish are known. If the small, "jack" salmonids are not spawned, a small error in the numbers of male and female fish returning may result. Jack salmonids are almost always males, but small females may be included inadvertently with them. I believe that incorrect sexing of the returning fish causes a smaller error than that introduced by applying age proportions irrespective of sex. Application of age proportions without regard to sex may yield numbers of males and females quite different from those reported by the spawning crew, particularly if the sample size for age at return is small. Each situation should be examined carefully. The decision on how to apply age proportions will ultimately rely on the researchers' knowledge of the percentage of returns han-

TABLE 2.—Numbers of unmarked fish to examine to be 95% confident that the true age proportion is within ±0.10 or ±0.20 of the estimated value.

Hatchery return	Estimated age proportion									
	0.05	0.10	0.15	0.20	0.25	0.30	0.35	0.40	0.45	0.50
Sample size for tolerance = ±0.01 of true age proportion										
100	99	97	96	94	92	90	88	85	83	80
500	468	437	407	377	349	321	294	268	242	217
750	680	616	558	504	455	409	366	326	289	254
1,000	880	776	685	606	536	473	417	366	320	278
1,500	1,244	1,046	888	759	652	561	484	417	358	306
5,000	2,968	2,044	1,517	1,176	937	760	624	517	429	357
7,500	3,699	2,367	1,687	1,275	999	801	652	535	442	365
10,000	4,220	2,569	1,788	1,332	1,033	823	666	545	449	370
20,000	5,348	2,948	1,963	1,427	1,090	858	689	560	459	377
30,000	5,871	3,100	2,030	1,462	1,110	870	697	565	462	379
Sample size for tolerance = ±0.02 of true age proportion										
100	95	90	85	80	74	69	59	59	54	49
500	393	317	261	217	183	155	132	112	95	81
750	532	402	316	254	208	173	144	121	102	85
1,000	646	464	353	278	224	183	151	126	105	88
1,500	824	549	400	306	242	195	160	132	109	90
5,000	1,337	737	491	357	272	215	172	140	115	94
7,500	1,468	775	507	365	277	218	174	141	116	95
10,000	1,543	796	516	370	280	219	175	142	116	95
20,000	1,672	829	530	377	284	222	177	143	117	96
30,000	1,720	840	535	379	285	222	177	143	117	96

dled, numbers of fish spawned (sex known), thoroughness of the spawning crew, size of the jack population, and the size of the sample for age determination.

The variances of the age proportions based on actual returns and sample size may be calculated with formula (3.6) from Cochran (1977):

$$V(\hat{P}) = \left[\frac{(P)(1 - P)}{n}\right]\left[\frac{(N - n)}{(N - 1)}\right];$$

N is the total return of fish and n is the sample of readable scales. A confidence interval (CI) around the age proportion may be calculated by multiplying the square root of the variance by the appropriate Z value:

$$CI - \hat{P} \pm Z_{\alpha/2} [V(\hat{P})]^{1/2}.$$

For a 95% confidence interval, the appropriate $Z_{\alpha/2}$ value is 1.96.

Sampling Adjacent Streams

Sampling of adjacent streams is important, particularly if fish are passed upstream to spawn naturally above a hatchery, or if a hatchery meets the egg-take needs before the run is complete and the ladder is then closed or the weir removed. In such cases, stream surveys are necessary to obtain complete return information. Even when all returning fish are examined, surveys of adjacent streams can be useful because they provide an indication of straying of marked fish.

Stream surveys are fraught with difficulties. It is impossible to observe all the fish in a stream or to sample all the fish found. Deteriorated carcasses complicate sampling by increasing the likelihood that regenerated scales will be collected and by hampering recognition of marked fish. For these reasons, it is important to examine all fish found for marks, to remove the snout from all fish suspected of carrying a coded wire tag, to obtain a scale sample, and to record length and sex of all fish. Sampling other hatcheries near the release sites is also recommended to check for straying of marked fish.

Application of Hatchery Return Data

Hatchery return data are useful for examining two key assumptions: (1) marked and unmarked fish have the same survival rates and maturity schedules, and (2) insignificant loss of marks occurs after release. The age proportions and sizes of fish at the spawning sites can be used to generate not only the survival and growth rates but the maturity schedules of marked and unmarked fish. A comparison of marked to unmarked ratios at return with those at release can be used to document loss of marks and differential mortality between marked and unmarked fish. However, straying of unmarked fish from other sources may influence the comparisons. If marked to unmarked ratios do not differ significantly among the ages of return, or between release and return, the assumption of equal survival and maturity of marked and unmarked fish probably is satisfied. If significant differences occur, further investigation is needed.

Postrelease mark loss can also influence marked–unmarked ratios at return. For studies that involve coded wire tags, it is necessary to carefully examine all fish for missing adipose fins (the external indicator for the presence of a coded wire), and to record which fin-clipped fish did not contain tags. Then, returns of fin-clipped fish with no tag can be applied to the appropriate tag group by age. The assumption of insignificant mark loss after release fails if a significant difference occurs between marked and unmarked ratios at release and return.

Lowering of the marked–unmarked ratio at return relative to release indicates a higher mortality of marked fish. This can be further examined by comparing marked–unmarked ratios by age of return. If the ratio for returning jack salmon is significantly lower than the release ratio, but the jack ratio does not differ significantly from the adult ratios at return, then higher mortality of marked fish likely occurred after release but prior to the first year of return. The possible influence of straying and increased catch caused by the mark must always be considered. Straying among hatcheries on the same river system, as in the Columbia River, can be consequential (Vreeland 1989). Certain types of marks—dangler tags, for example—may create a bias for capture of marked fish because of entanglement in gill nets or other fishing gear. It is recommended that marked–unmarked ratios only be compared when straying and catch bias are believed to be nonexistent or modest.

Returns of marked fish to spawning sites can be combined with fishery recoveries to obtain a total picture of survival. Catch and return data also are useful in developing standard catch to escapement ratios.

Summary

In this analysis, I have outlined the steps necessary to determine the fishery contribution of an individual hatchery with one year of marking, and to compare this contribution with that of other hatcheries and other years. A critical assumption for hatchery contribution studies is that the marked fish are representative of the total release. To ensure that the assumption is correct, methods must be employed to obtain a random sample of fish for marking. Opinions vary as to the appropriate method for obtaining the random sample. Some believe an adequate sample may be obtained by crowding fish in all rearing areas and netting them for marking. Others believe a more rigorous procedure is required, whereby all fish are handled in some fashion and systematic samples are frequently removed. The more rigorous sampling procedure has a better statistical foundation, but a comparison of the procedures has never been made. The variance of the contribution estimates due to fishery sampling procedures and expansion methods may be large enough to mask any possible difference between contribution estimates that result from different procedures for removing fish for marking. Given the stress and potential added mortality placed on hatchery fish by a rigorous sampling procedure, it is appropriate that comparisons be made between contribution estimates from the grab-net sample and from the more rigorous procedures. Until results from this type of comparison are available, it is recommended that a rigorous sampling procedure be employed to obtain fish for marking. This will ensure that the marked fish reflect the total population.

Once the fish have been marked, fishery scientists must obtain the most accurate release statistics possible. To apply the recovery of marked fish to the entire population, one must know the numbers of marked and unmarked fish released. Determining the original number of fish marked can be troublesome, but adherence to meticulous marking procedures should yield reliable numbers of fish marked. However, determining mortality of marked fish between marking and release is fraught with difficulties. Sampling of rearing environments for marked and unmarked fish prior to release also has its difficulties, and it places additional stress on the fish. In studies with coded wire tags, electronic counters collect sound data on releases of tagged and untagged fish, provided the fish are released on their own volition. The data become less reliable when fish are forced through the counters. Given the difficulties of sampling prior to release, electronic counters are recommended to determine the numbers of tagged and untagged fish released.

It is important to determine the extent of mark loss at and after release. The marked to unmarked ratio will be used to apply marked catch to total hatchery contribution. Undetected or unaccounted losses of marks after release will result in an underestimate of hatchery contribution. The recommended numbers of fish to examine for mark loss have ranged from 300 to 2,000; sampling approximately 2,000 fish at release allows the mark loss to be estimated within 1%, provided the mark loss is expected to be equal to or less than 5%.

It is assumed fishery sampling will occur in the fisheries of interest, otherwise a hatchery contribution study should not be undertaken. Random errors occurring during fishery sampling and expansion of the observed catch of marked fish are not addressed here, but they must be assessed if contribution estimates are to be compared among hatcheries and years.

Sampling at the return site allows age structure to be estimated, mortalities to be examined, and maturity schedules for marked and unmarked fish to be charted. All returning fish should be sampled for marks, and information should be collected as described earlier.

A systematic sample of returning unmarked fish is also recommended. The number of fish to sample depends on expected returns, the age proportion to be estimated, and precision of the estimate.

Sampling of hatcheries and streams adjacent to the return facility is also recommended to obtain an indication of straying. The assessment of all returns will allow the best estimate of catch to escapement ratios.

References

Appleby, A., and R. Schneider. 1983. One fish, two fish. Pages 67–72 in G. W. Klontz and E. M. Parrish, editors. Proceedings of the 34th Annual Northwest Fish Cultural Workshop. University of Idaho, College of Forestry, Moscow.

Bergman, P., and R. Hager. 1969. The effects of implanted wire tags and fin excision on the growth and survival of coho salmon (Oncorhynchus kisutch, Walbaum). Washington Department of Fisheries, Olympia.

Bevan, D. E. 1959. Tagging experiments in the Kodiak Island area with reference to the estimation of salmon (Oncorhynchus) populations. Doctoral dissertation. University of Washington, Seattle.

Blankenship, L. 1981. Coded-wire tag loss study. Washington Department of Fisheries Technical Report 65.

Bouck, G. R., and D. A. Johnson. 1979. Medication inhibits tolerance to seawater in coho salmon smolts. Transactions of the American Fisheries Society 108:63–66.

Chapman, D. G. 1951. Some properties of the hypergeometric distribution with applications to zoological sample censuses. University of California Publications in Statistics 1:131–159. (Berkeley.)

Chapman, D. G. 1955. Population estimation based on change of composition caused by a selective removal. Biometrika 42:279–290.

Clemens, W. A., and G. V. Wilby. 1946. Fishes of the Pacific coast of Canada. Fisheries Research Board of Canada Bulletin 68.

Cochran, W. G. 1977. Sampling techniques, 3rd edition. Wiley, New York.

Craig, J. A., and R. L. Hacker. 1950. The history and development of the fisheries of the Columbia River. U.S. Fish and Wildlife Service Fishery Bulletin 49(32):133–216.

Davis, W. E., and H. K. Hiltz. 1971. A constant-fraction sampling device for enumerating juvenile salmonids. Progressive Fish-Culturist 33:180–183.

de Libero, F. E. 1986. A statistical assessment of the use of the coded wire tag for chinook (Oncorhynchus tshawytscha) and coho (Oncorhynchus kisutch) studies. Doctoral dissertation. University of Washington, Seattle.

DeLury, D. B. 1951. On the planning of experiments for the estimation of fish populations. Journal of the Fisheries Research Board of Canada 8:281–307.

Duke, R. C. 1980. Fish tagging mobile unit operation, repair, and service manual. Idaho Department of Fish and Game, Boise.

Foster, R. W. 1981. Incremental fish sampler. Progressive Fish-Culturist 43:99–101.

Fredin, R. A. 1950. Fish population estimates in small ponds using the marking and recovery technique. Iowa State College Journal of Science 24:363–384.

Hagen, W. 1953. Pacific salmon hatchery propagation and its role in fishery management. U.S. Fish and Wildlife Service Circular 24.

Heizer, S. R., and A. W. Argue. 1976. Basic catch sampling and coded wire tag recovery data for Georgia strait chinook and coho fisheries in 1973. Canada Fisheries and Marine Service, Data Record Series PAC/D-76-9, Ottawa.

Heizer, S. R., and J. C. Beukema. 1977. Basic data for the 1974 Canadian chinook and coho catch sampling and mark recovery program, volume 1. Canada Fisheries and Marine Services, Data Record Series PAC/D-77-6, Ottawa.

Hewitt, G. S., and R. E. Burrows. 1948. Enumerating hatchery fish populations. Progressive Fish-Culturist 10:23–27.

Jefferts, K. B., P. K. Bergman, and H. F. Ficus. 1963. A coded wire identification system for macroorganisms. Nature (London) 198:460–462.

Jenkinson, D. W., and H. T. Bilton. 1981. Additional guidelines to marking and coded wire tagging of juvenile salmon. Canadian Technical Report of Fisheries and Aquatic Sciences 1051.

Johnson, J. K. 1987. Pacific salmonid coded wire tag releases through 1986. Pacific Marine Fisheries Commission, Portland, Oregon.

Jones, R. D. 1965. Engineering and construction division. Pages 44–47 *in* D. Reed, editor. Years 1965–66, 75th and 76th annual report. Washington State Department of Fisheries, Olympia.

King, G. 1979. Pacific salmon sampling and tagging a review of current methodology. Pacific Marine Fisheries Commission, Portland, Oregon.

Larkin, P. A. 1970. Management of Pacific salmon of North America. American Fisheries Society Special Publication 7:223–236.

Laythe, L. L. 1948. The fishery development program in the lower Columbia River. Transactions of the American Fisheries Society 78:42–55.

McKervill, H. W. 1967. The salmon people. Gray, Sidney, Canada.

Moberly, S. A., R. Miller, K. Crandall, and S. Bates. 1977. Mark–tag manual for salmon. Alaska Department of Fish and Game, Juneau.

Northwest Power Planning Council. 1984. Columbia River basin fish and wildlife program. NPPC, Portland, Oregon.

Opdycke, J. D., and P. Zajac. 1981. Evaluation of half-length binary coded wire tag application in juvenile chum salmon. Progressive Fish-Culturist 43:48.

Oregon Department of Fish and Wildlife. 1976. 1974 wire tag and fin-mark sampling and recovery report for salmon and steelhead from various Pacific coast fisheries. ODFW, Clackamas, Oregon.

Oregon Department of Fish and Wildlife. 1977a. 1975 wire tag and fin-mark sampling and recovery report for salmon and steelhead from various Pacific coast fisheries. ODFW, Clackamas, Oregon.

Oregon Department of Fish and Wildlife. 1977b. 1976 wire tag and fin-mark sampling and recovery report for salmon and steelhead from various Pacific coast fisheries. ODFW, Clackamas, Oregon.

Pacific Marine Fisheries Commission. 1984. Procedures for coded wire tagging Pacific salmonids. PMFC, Portland, Oregon.

Paulik, G. J. 1963. Are adequate techniques for the evaluation of artificial propagation available? Pages 133–135 *in* R. S. Croker and D. Reed, editors. Report of second governor's conference on Pacific salmon. Washington Department of Fisheries, Olympia.

Ricker, W. E. 1948. Methods of estimating vital statistics of fish populations. Indiana University Publications, Science Series 15:39–52.

Ricker, W. E. 1975. Computation and interpretation of biological statistics of fish populations. Fisheries Research Board of Canada Bulletin 191.

Rounsefell, G. A., and W. H. Everhart. 1953. Fishery science its methods and applications. Wiley, New York.

Salmon and Steelhead Advisory Commission. 1984. A new management structure for anadromous salmon and steelhead resources and fisheries of the Washington and Columbia River conservation areas. National Marine Fisheries Service, Northwest and Alaska Fisheries Center, Seattle, Washington.

Schaefer, M. B. 1951. A study of the spawning populations of sockeye salmon in the Harrison River system, with special reference to the problem of enumeration by means of marked members. International Pacific Salmon Fisheries Commission Bulletin 4.

Scheaffer, R. L., W. Mendenhall, and L. Ott. 1979. Elementary survey sampling, 2nd edition. Duxbury, North Scituate, Massachusetts.

Senn, H., J. Mack, and L. Rothfus. 1984. Compendium of low-cost Pacific salmon and steelhead trout production facilities and practices in the Pacific Northwest. Bonneville Power Administration, Portland, Oregon.

Thrower, F. P., and W. W. Smoker. 1984. First adult return of pink salmon tagged as emergents with binary-coded wires. Transactions of the American Fisheries Society 113:803–804.

U.S. Fish and Wildlife Service. 1980. Anadromous fish tagging procedures U.S. Fish and Wildlife Service region 1. USFWS, Fisheries Assistance Office, Olympia, Washington.

U.S. Fish and Wildlife Service. 1985. Anadromous fish tagging procedures U.S. Fish and Wildlife Service region 1. USFWS, Fisheries Assistance Office, Olympia, Washington.

Vreeland, R. R. 1989. Evaluation of the contribution of chinook salmon reared at Columbia River hatcheries to the Pacific salmon fisheries. Bonneville Power Administration, Portland, Oregon.

Wahle, R. J., and R. Z. Smith. 1979. A historical and descriptive account of Pacific coast anadromous salmonid rearing facilities and a summary of their releases by region, 1960–76. U.S. Department of Commerce National Marine Fisheries Service Special Scientific Report Fisheries 736.

Wahle, R. J., and R. R. Vreeland. 1978. Bioeconomic contribution of Columbia River hatchery fall chinook salmon, 1961 through 1964 broods, to the Pacific salmon fisheries. U.S. National Marine Fisheries Service Fishery Bulletin 76:179–208.

Wahle, R. J., R. R. Vreeland, and R. H. Lander. 1974. Bioeconomic contribution of Columbia River hatchery coho salmon, 1965 and 1966 broods, to the Pacific salmon fisheries. U.S. National Marine Fisheries Service Fishery Bulletin 72:139–169.

Webb, R. D., and R. E. Noble. 1966. A device for randomly sampling juvenile fish populations. Washington Department of Fisheries Fisheries Research Paper 2(4):94–103.

Williams, B. 1978. A sampler on sampling. Wiley, New York.

Worlund, D. D., R. J. Wahle, and P. D. Zimmer. 1969. Contribution of Columbia River hatcheries to harvest of fall chinook salmon (*Oncorhynchus tshawytscha*). U.S. Fish and Wildlife Service Fishery Bulletin 67:361–391.

Yonker, W. V. 1963. The salmon fisheries. Pages 107–119 *in* M. E. Stansby, editor. Industrial fishery technology. Reinhold, New York.

American Fisheries Society Symposium 7:708–713, 1990

Factors That Affect the Recapture of Tagged Sablefish off the West Coast of Canada

M. W. SAUNDERS, G. A. McFARLANE, AND R. J. BEAMISH

Department of Fisheries and Oceans, Pacific Biological Station
Nanaimo, British Columbia V9R 5K6, Canada

Abstract.—Several factors affect recovery rates for sablefish *Anoplopoma fimbria* tagged off the west coast of Canada. Chief among them are tag loss, oxytetracycline injection, variation in recovery effort, and nonreporting of recovered fish. Recovery rates increased with the size of fish at release.

From 1977 to 1982, sablefish *Anoplopoma fimbria* were tagged off the west coast of Canada to determine if the Canadian stock is part of a single North American stock (Beamish and McFarlane 1983, 1988). In addition, sablefish were tagged to examine movement and patterns of recruitment, and to validate a method of age determination. Altogether, 122,715 sablefish were tagged and released. In the analyses of movements and stock identification (Beamish and McFarlane 1983, 1988) and age of validation (Beamish et. al 1983a; McFarlane and Beamish 1987), it was necessary to consider factors that affected recovery percentages.

In this paper we review factors that affected the recovery rates of sablefish, and we present new information on the relationship between fish length at tagging and recovery percentage.

Methods

Fishing and tagging methods.—Descriptions of capture and tagging methods were reported by Beamish et al. (1978, 1979, 1980, 1983b). Adult sablefish were captured with rectangular and Korean-style traps deployed from chartered fishing vessels. Traps were baited with herring and, on some cruises, with squid. The depth of each trap string was approximated by averaging depths recorded at regular intervals while the string was being set. Fishing time was recorded as the time that elapsed between setting of the last trap and retrieval of the first one.

Freshly caught fish were transferred directly to holding tanks equipped with flowing seawater. Fish were taken from tanks and placed on a measuring board, measured for fork length, and tagged with a Floy FD-68 anchor tag. We also applied a suture tag (White and Beamish 1972) to approximately 10% of the fish. Some 15,000 fish were injected with oxytetracycline (OTC) during the program to induce fluorescent marks in calci-

fied body parts; these marks were used to validate ages of the fish.

Records were kept on the general condition of the fish, noticeable injuries, and any problems with the tag. Anaesthetic (tricaine, MS-222) was used only during the first cruise. Only fish in good condition were tagged. Tagged fish were released directly into the ocean.

To test the effect of fish length at tagging on recovery percentage, we used log-linear modeling approach to multidimensional contingency table analysis. This was carried out with a categorical modeling computer program (CATMOD) produced by SAS Institute (SAS 1985).

For analysis, we grouped releases and recaptures by 5-cm length-intervals. The program then constructed the equivalent of a two-dimensional contingency table with one row for each combination of independent variable levels. Columns consisted of the response or dependent variable values, either recovery or nonrecovery. On the basis of significance of effects indicated by the Wald statistic, we tested the hypothesis that recovery proportions are independent of length at tagging.

Results and Discussion

Length at Release

We found, for all cruises examined, that the proportion of sablefish recaptured was not independent of fish length ($P > 0.05$). Cumulative recovery percentages decreased with decreasing length of fish (Figure 1), ranging from 25 to 32% for larger fish and from 6 to 22% for small fish. Recovery of tagged smaller fish often is less than that of larger fish (McCracken 1963; Westrheim and Morgan 1963; and Wise 1963). Size-related recovery rates were attributed to size selectivity of recovery gear (McCracken 1963), size limits on commercial catches (Westrheim and Morgan

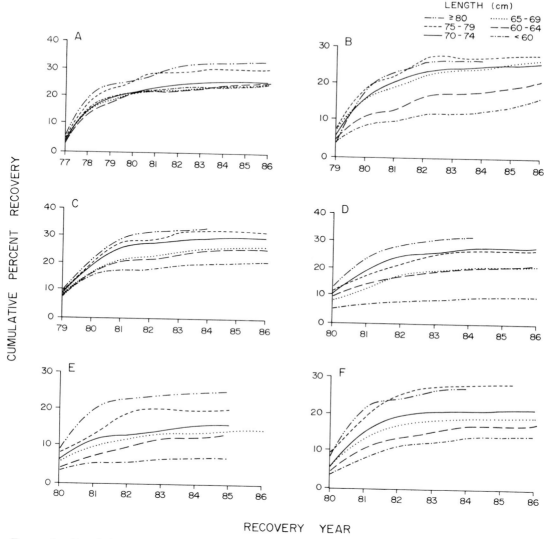

FIGURE 1.—Cumulative recovery percentages for tagged sablefish by cruise and by 5-cm length-intervals at release.

1963), and differential handling or tagging mortality (Wise 1963).

There was no difference in sex ratio for released and recaptured fish in any length category (Table 1). This indicates that differential recovery rates by length were not a result of the different growth rates of male and female sablefish (McFarlane and Beamish 1987).

Beamish and McFarlane (1988) also found no difference in mean size of males or females that moved farther than 200 km from the release area, compared with fish recaptured within 50 km of the release area. This indicates that different recovery

rates by release length were not a result of size-related differences in emigration rate. In addition, the mean age of males or females that moved more than 200 km from the release area was not significantly different from that of fish that moved less than 200 km (Beamish and McFarlane 1988). Similarly, Holmberg and Jones (1954) and Wespestad et al. (1983) found no significant difference in migration distance among tagged sablefish of various lengths at release.

Juvenile sablefish recovered in the nursery area up to 2 years after release had similar recovery rates for all lengths released (our unpublished

TABLE 1.—Percentages of male sablefish released and recovered, 5-cm length-intervals.

Fork length (cm)	Percent males	
	Release samples	Recovery
<60	52.9	60.8
60–64	56.4	60.6
65–69	36.2	36.0
70–74	10.3	9.4
75–79	1.6	2.7
>79	2.7	2.1

data). However, when these juveniles moved off-shore, fish that were longer at time of release showed a higher rate of recapture. The relatively high survival of smaller fish in the nursery area indicates that tag loss or physiological stress were not affecting smaller fish more than larger fish. The relatively higher survival of larger juvenile fish in outside waters suggest that they are less subject to predation. The possibility that possession of a tag contributed to increased predation was examined in the evaluation of the standardization procedure of Beamish and McFarlane (1988). After returns were standardized and corrections were made for tag loss and tagging mortality, the expected number of returns expected given a total mortality rate of 0.2 was only 10% higher than the standardized returns. Thus, if tags do attract predators, the selection for tagged fish appears to be minor.

Tag Loss

As reported by Beamish and McFarlane (1988), tag loss has a great effect on the recovery percent-age for tagged sablefish. Of the 5,076 fish that received two tags, 822 (16.2%) were recovered. There was no significant difference in tag loss between the two tag types (*t*-test, $p \leq 0.01$). This indicates that the loss of either tag can be considered as a loss for an individual fish, provided that the rate of loss of either tag is not a result of interaction between the two tags. Tag loss in the first year after tagging was approximately 10% and approximately 2% per year afterwards (Figure 2).

Tag loss resulted in a significant reduction in marked fish over a 10-year period. Natural (*M*) and fishing (*F*) mortality rates of 0.1 for sablefish (Saunders et al. 1987) cause the most losses of marked fish. Even though many fish were tagged and released, tag loss and annual mortality limited the length of this study to about 10 years. After this period, recoveries were too few to be interpreted.

Standardization

In few studies have attempts been made to standardize recovery percentages. It seems to be generally accepted that standardization is important, but results often are interpreted with the disclaimer that no standardization was undertaken. Standardization can be difficult because catch and effort statistics may not be available for all potential recovery areas throughout the duration of study. For estimates of movement, abundance, and mortality, some form of standardization is necessary.

In this study, we compared rates and direction of movement of adult sablefish by using recaptures standardized for variation in recovery effort

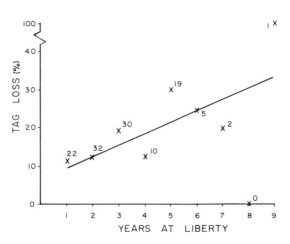

FIGURE 2.—Percentage of sablefish recaptured annually with one tag missing. Numbers of fish recaptured that had lost a tag are indicated for each year at liberty. The line was fitted by eye. (From Beamish and McFarlane 1988.)

TABLE 2.—Actual (Act) and standardized (Stand) recoveries of tagged sablefish released off Vancouver Island (VI), in Queen Charlotte Sound (QCS), off the Queen Charlotte Islands (QCI), and in mainland inlets, by years at liberty.

Years at liberty	Total recaptures							
	VI		QCI		QCS		Inlets	
	Act	Stand	Act	Stand	Act	Stand	Act	Stand
1	973	1,241	1,475	2,556	165	258	472	745
2	1,211	1,452	2,065	3,426	265	395	271	424
3	585	663	1,009	1,557	145	228	236	366
4	410	447	463	711	82	123	143	217
5	175	200	268	406	41	54	169	221
6	103	114	155	253	14	23	30	42
7	120	120	73	117	8	10		
8	14	14	30	40	3	3		
9	6	7	20	32				

and for unreported recaptures (Table 2; Beamish and McFarlane 1988). We tagged and released 72,735 adult sablefish off the west coast of Canada from 1977 to 1982. As of December 31, 1985, 11,121 (15%) of these fish were recaptured. If all fish receiving oxytetracycline (OTC) injections of 100 and 75 mg/kg are excluded, the nominal release was 54,916 adult fish, of which 10,152 (18.5%) were recovered. When we standardized

all recaptures (including OTC-injected fish), 16,539 fish (23%) were considered as recoveries.

The standardization procedure resulted in more recaptures. To evaluate this procedure, Beamish and McFarlane (1988) compared expected and standardized returns for one release each year off Vancouver Island in 1979 and in 1980, and also for one release each year off the Queen Charlotte Islands in 1977 and 1980. The total number of

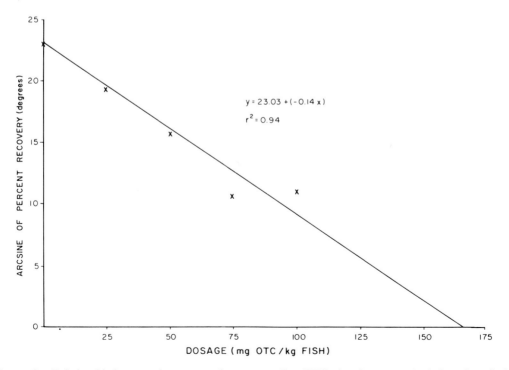

$$y = 23.03 + (-0.14\,x)$$

$$r^2 = 0.94$$

FIGURE 3.—Relationship between dosage rate of oxytetracycline (OTC, x) and recovery (an index of survival, y) for tagged and injected sablefish. (From McFarlane and Beamish 1987.)

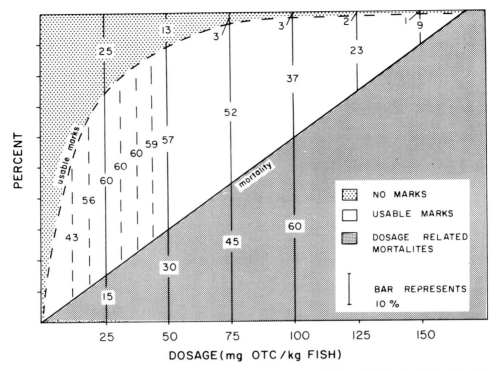

FIGURE 4.—Nomograph of optimal dosage. For example, a dosage of 25 mg OTC/kg fish kills 15% of the fish, imparts usable marks to 60%, and does not mark 25%.

released fish was 23,060 and the expected number of returns was 6,972. The standardized number of returns was 5,576—20% fewer than expected. If only the release group that had been at liberty longest was used from each area, the expected return was 4,510 fish and the standardized number was 4,041 fish—10% fewer than expected. This agreement confirmed the reliability of the standardization procedure.

Chemical Injections

In a previous study, McFarlane and Beamish (1987) reported on the effect of OTC dosage rate on mortality of fish released into the ocean. They tagged 15,183 fish, injected the fish with OTC at varying dosage rates, and released them. The relationship between dosage and survival of injected fish was positive and strongly linear (Figure 3). Even the smallest dosage rates had some level of mortality associated with them, and a dosage rate of 165 mg OTC/kg was the theoretical rate at which 100% mortality occurred (Figure 4).

Although this study specifically examined one chemical, it is clear that there can be a trade-off between recovery percentage and mortality when

a tagging program attempts to carry out several objectives. In the OTC study, even the lowest dosage caused 15% mortality. Although this mortality was acceptable in relation to the importance of developing a method for age determination, it would not be acceptable for studies of mortality rates and abundance.

Other Factors

Factors such as density of fish in the holding tank, depth of capture, fishing time, and tagging procedures were examined and in some cases were found to affect recovery percentages. These factors frequently are associated with other conditions, such as weather, and can be difficult to interpret. Often, investigators are more interested in examining the response of the population than in maximizing the number of recoveries. It is important, however, to appreciate that these other factors can affect recovery percentages and, depending on the objectives of the study, may need to be examined.

Investigators commonly attempt through laboratory study to examine factors that affect mortality. For example, laboratory studies conducted on English sole *Parophrys vetulus* by Manzer (1952) and on Pacific halibut *Hippoglossus stenolepis* by Peltone (1969) showed no differences in mortality of tagged fish by length. We found no effect of OTC in the laboratory but, as previously shown, the response in the natural environment was quite different.

Conclusions

Tag loss, OTC injection, variation in recovery effort, and nonreporting had the greatest effects on recovery rates of tagged sablefish.

There is a strong positive relationship between length at release and recovery rate for tagged sablefish. Investigators attempting to tag specific size components of a stock, to analyze growth over the entire size spectrum, or to validate aging criteria over all ages must take into account the differences in effective releases among sizes tagged.

As indicated by the OTC study, laboratory studies may not accurately mimic the ocean environment. Experimental design should include control releases to evaluate the effects of varying conditions at release.

References

Beamish, R. J., C. Houle, and R. Scarsbrook. 1980. A summary of sablefish tagging and biological studies conducted during 1979 by the Pacific Biological Station. Canadian Manuscript Report of Fisheries and Aquatic Sciences 1588.

Beamish, R. J., C. Houle, C. Wood, and R. Scarsbrook. 1979. A summary of sablefish tagging and exploratory trapping studies conducted during 1978 by the Pacific Biological Station. Canadian Data Report of Fisheries and Aquatic Sciences 162.

Beamish, R. J., and G. A. McFarlane. 1983. Summary of results of the Canadian sablefish tagging program. Pages 147–183 in Proceedings of the second Lowell Wakefield fisheries symposium. University of Alaska, Sea Grant Report 83-3, Anchorage.

Beamish, R. J., and G. A. McFarlane. 1988. Resident and dispersal behavior of adult sablefish (*Anoplopoma fimbria*) in the slope waters off Canada's west coast. Canadian Journal of Fisheries and Aquatic Sciences 45:152–164.

Beamish, R. J., G. A. McFarlane, and D. E. Chilton. 1983a. Use of oxytetracycline and other methods to validate a method of age determination for sablefish. Pages 95–116 in Proceedings of the second

Lowell Wakefield fisheries symposium. University of Alaska, Sea Grant Report 83-8, Anchorage.

Beamish, R. J., and seven coauthors. 1983b. A summary of sablefish tagging and biological studies conducted during 1980, and 1981 by the Pacific Biological Station. Canadian Manuscript Report of Fisheries and Aquatic Sciences 1732.

Beamish, R. J., C. Wood, and C. Houle. 1978. A summary of sablefish tagging studies conducted during 1977 by the Pacific Biological Station. Canadian Fisheries and Marine Service Data Report 77.

Holmberg, E. K., and W. G. Jones. 1954. Results of sablefish tagging experiments in Washington, Oregon and California. Pacific Marine Fisheries Commission Bulletin 3.

Manzer, J. I. 1952. The effects of tagging upon a Pacific coast flounder, *Parophrys vetulus*. Journal of the Fisheries Research Board of Canada 8:(7). (No pages given.)

McCracken, F. D. 1963. Comparison of tags and techniques from recoveries of Subarea 4 cod tags. International Commission for the Northwest Atlantic Fisheries Special Publication 4:101–105.

McFarlane, G. A., and R. J. Beamish. 1987. Selection of dosages of oxytetracycline for age validation studies. Canadian Journal of Fisheries and Aquatic Sciences 44:905–909.

Peltone, G. J. 1969. Viability of tagged Pacific halibut. Report of the International Pacific Halibut Commission 52.

SAS. 1985. SAS users guide: statistics, version 5 edition. SAS Institute, Cary, North Carolina.

Saunders, M. W., G. A. McFarlane, and W. Shaw. 1987. Sablefish. Canadian Manuscript Report of Fisheries and Aquatic Sciences 1930:72–87.

Wespestad, V. G., K. Thorsen, and S. A. Mizroch. 1983. Movement of sablefish, *Anoplopoma fimbria*, in the northeastern Pacific Ocean as determined by tagging experiments (1971–80). U.S. National Marine Fisheries Service Fishery Bulletin 81:415–420.

Westrheim, S. J., and A. R. Morgan. 1963. Results from tagging a spawning stock of Dover sole *Microstomus pacificus*. Pacific Marine Fisheries Commission Bulletin 6.

White, W. J., and R. J. Beamish. 1972. A simple fish tag suitable for long term marking experiments. Journal of the Fisheries Research Board of Canada 29:339–341.

Wise, J. P. 1963. Factors affecting number and quality of returns from tagging cod with different tags and using different methods of capture in ICNAF divisions 4X and 5Y in 1957. International Commission for the Northwest Atlantic Fisheries Special Publication 4:101–105.

American Fisheries Society Symposium 7:714–719, 1990

Improved Data in a Tagging Program through Quality Assurance and Quality Control

PAUL GEOGHEGAN AND MARK T. MATTSON

Normandeau Associates, Inc.
25 Nashua Road, Bedford, New Hampshire 03102, USA

DENNIS J. DUNNING AND QUENTIN E. ROSS

New York Power Authority
123 Main Street, White Plains, New York 10601, USA

Abstract.—Approximately 41,000 striped bass *Morone saxatilis* have been tagged and released since 1985 during Hudson River studies sponsored by electric utilities. Our experience indicates that unless formal documented quality assurance (QA) and quality control (QC) procedures are adopted, the resulting data are of unknown accuracy. We have developed cost-effective QA and QC systems to ensure and document the accuracy of data from these studies. Our quality control calls for compliance of project operations with a procedures manual; compliance is then monitored by our QA system. Mistakes in recording tag numbers in the field was the largest single source of errors in a tagging data base. We eliminated virtually all these errors by changing data acquisition procedures so they included 100% reinspection of tag numbers. Double keypunching and computerized error-checking routines also helped us eliminate transcription errors. The cost of this QA–QC system was approximately 10% of all field labor costs. For 1% of the total field labor costs, a QA–QC program consisting of written procedures, a training program for technicians, QA audits that verify adherence to written field procedures, and double keypunching can reduce recording errors to less than 1.0%. These QA–QC systems will (1) provide a data base of known quality, (2) provide feedback to program management to reduce data-recording errors, and (3) provide documentation for future investigators.

Fisheries science can gain further acceptance by the public and legal system as a quantitative science by adopting formal quality assurance and quality control (QA–QC) procedures that have become standard in engineering and analytic sciences (ANSI and ASME 1983; USEPA 1980). Quality control procedures are specific routine activities that determine and measure the error rate of a work product; an example would be a random audit that compared the data points of a final data base with the original field data. Quality assurance procedures determine the effectiveness of the QC procedures in achieving a desired level of accuracy—for example, an audit of field techniques by an independent observer would ensure that written procedures were followed. Properly applied QA–QC procedures result in a data base of known accuracy and a significant decrease in the error rate.

Approximately 41,000 striped bass *Morone saxatilis* have been tagged and released in the Hudson River since 1985 as part of a program to estimate population abundance and survival. Similar studies conducted from 1976 through 1979 came under rigorous legal and scientific scrutiny (Sandler and Schoenbrod 1981), and there is a high probability that the data collected since 1985 will receive the same review. This paper describes the QA–QC procedures used for the Hudson River striped bass tagging program, and it notes how improvements in data acquisition procedures increased the accuracy of the data base. We identify sources of error in the data base, and we suggest procedures to prevent such errors from occurring.

Methods

Striped bass captured in otter trawls and Scottish seines were tagged and released in the Hudson River between New York Harbor and river kilometer 63 north of the Battery, New York City. The tagging continued from November 1985 through May 1986, from December 1986 through May 1987, and from November 1987 through April 1988. For each fish the location, date, time, and method of capture were recorded as numeric codes, along with the total length, physical condition, tag number (if tagged), and condition of the tag if the fish was recaptured. Despite the similar goals and data collected in each annual program, there were several differences with regard to the QA–QC procedures.

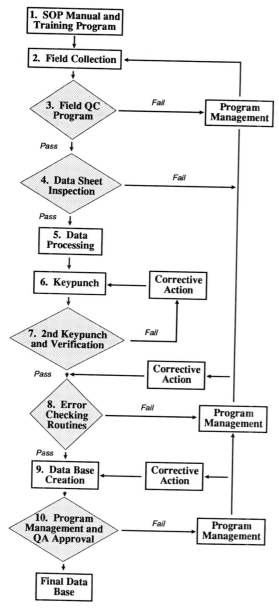

1. SOP Manual and Training Program

2. Field Collection

3. Field QC Program — Fail → Program Management

Pass ↓

4. Data Sheet Inspection — Fail →

Pass ↓

5. Data Processing

6. Keypunch ← Corrective Action

7. 2nd Keypunch and Verification — Fail →

Pass ↓ — Corrective Action

8. Error Checking Routines — Fail → Program Management

Pass ↓

9. Data Base Creation ← Corrective Action ←

10. Program Management and QA Approval — Fail → Program Management

Final Data Base

FIGURE 1.—Quality control flowchart for collecting and processing data for Hudson River striped bass tagging programs. SOP = standard operating procedures; QA = quality assurance; QC = quality control.

which involved (1) a complete reading of the SOP by all project personnel, (2) supervision by project management of field technicians when a new procedure was introduced, and (3) supervision of new technicians by experienced personnel.

During the 1985–1986 program, fish of 200 mm or more in total length (TL) were tagged with internal anchor–external streamer tags (NAI 1986). All data were numeric and were verbally relayed to a second technician who recorded them on a data sheet. Usually, two tagging technicians and one data technician worked simultaneously. Tags were used in random order. Field procedures were subjected to two random QA audits by personnel independent of the project. The purpose of these audits was to document the extent of compliance with the SOP.

The next formal QC procedure during the 1985–1986 program was a visual inspection of the data sheet prior to keypunching (Figure 1). This visual inspection resolved such gross errors as missing sample data for collection sites. The data were double-keypunched, a procedure whereby data are keyed once by a technician and then by a different technician or else by the same technician 24 h later. During the second keying of the data, each record was compared by the computer to the same record the first time it was keyed. Any discrepancy between the first and second keying of the data was presented to the keypunch technician, who resolved it before proceeding to the next record. Double keypunching eliminated virtually all keypunching errors.

After all data were keypunched, they were subjected to computerized error-checking routines (Figure 1). In these routines, each variable was compared to a predetermined range of acceptable values on a univariate, bivariate, and multivariate level. A univariate error check compared an observation against a single specified range of values. A bivariate error check compared the range of a variable conditioned by a second variable. A multivariate error check was similar to a bivariate error check except that the range of the variable being checked was conditioned by three or more other variables. Values outside these specified ranges were rejected as potential errors. Project personnel inspected suspect data singled out by the computer and decided if errors had been made. If project personnel verified the presence of errors, they attempted to determine what the correct values were and to correct the data base and the original data sheets. After all error-checking procedures were completed, the final

1985–1986 program.—The first step in the QA–QC procedures was to develop a manual on standard operating procedures (SOP) that described all field, laboratory, and data-processing procedures. A training program was developed, based on the contents of the SOP (Figure 1),

TABLE 1.—Application of lot sampling plans for an average maximum error of 1 per 100 data in Hudson River striped bass data bases, derived from the hypergeometric distribution for up to 320 data points (Ogden 1968) and from the MIL-STD-105D normal inspection plans for more than 320 data points (ASQC 1981).

Number of data in data base	Number of randomly selected data to be inspected	Number of errors causing rejection of data base
1–6	All	1
7–8	6	1
9–13	8	1
14–20	10	1
21–25	12	1
26–35	15	1
36–45	17	1
46–65	20	1
66–110	25	1
111–320	30	1
321–500	50	2
501–1,200	80	3
1,201–3,200	125	4
3,201–10,000	200	6
10,001–35,000	315	8
35,001–150,000	500	11
150,001–500,000	800	15
>500,000	1,250	22

data base was subjected to a random QC audit that compared the data to the original data sheets. The single sampling plans for normal inspection tables (ASQC 1981) were used to select the sample size for inspection (Table 1). If the data failed the QC audit, they were 100% inspected against the original field data. This sampling plan ensured that no more than one observation in 100 was in error. All the data from the 1985–1986 program were processed as one batch after the field portion of the program was over.

1986–1987 program.—During the first 2 months (December and January) of the 1986–1987 program, data collection and processing procedures were virtually identical to those of the 1985–1986 program and included the development of an SOP and a training program (NAI 1987a). We did make two innovations, however. We processed data in monthly batches, and we used a tag-inventory system. Monthly processing enabled us to identify the types and frequency of recording errors and to reduce those errors by changing procedures. In the new inventory system, we divided tags into unique lots of 100 tags each and recorded the tag lot used for a given sample. When an erroneous tag number was observed, we found the correct tag number by consulting the tag inventory and accounting for the remaining 99 tags in that lot.

To reduce error-checking time and improve the final quality of the data base, we made a major

change in data collection on 17 March 1987. Previously, the tagging technician called out the tag number to the data technician, who wrote it on the data sheet. In the new procedure the data technician recorded the tag number first, then handed the tag to the tagging technician, who verbally confirmed the number.

The last major change in data collection and handling procedures was the implementation on 1 April 1987 of a continuous QC sampling plan for field collection of data (Figure 2). In the field QC procedure, if a datum upon reinspection was not exactly the same as the original datum, the entire record was rejected (except for fish length) and the process was subjected to further inspection. The standard we used for fish lengths allowed them to differ by no more than 3% in duplicated measurements of total length to the nearest millimeter.

1987–1988 program.—The 1987–1988 program was identical to the 1986–1987 program with respect to the QA–QC procedures, except that tags within each inventory lot were used in numeric sequence (NAI 1987b).

Error-rate calculations.—The tag number is probably the most important datum in a mark–recapture program, because it uniquely identifies each fish. This datum was misrecorded more than any other. Therefore, we used the error rate for recording tag numbers as an index of overall data-recording error rate. The raw error rate for recording tag numbers during the 1985–1986 program was calculated by comparing the data files immediately after they were double–keypunched with the final data files. (Double-keypunched files reflected the quality of the data as they left the field—neither double keypunching nor visual inspection of the data sheets resolved errors in tag numbers.) The comparison between the double-keypunched and final data files was accomplished through the merge function of SAS (1985). Observations that were not identical in tag number between the raw keypunch file and the final data base were identified as tag-number recording errors that were remedied by the data QC program. The raw error rates for recording tag numbers during the 1986–1987 and 1987–1988 programs were calculated during the monthly data QC.

Results

1985–1986 Program

The error rate for the 1985–1986 program for recording tag numbers was 9.68/1,000 tag num-

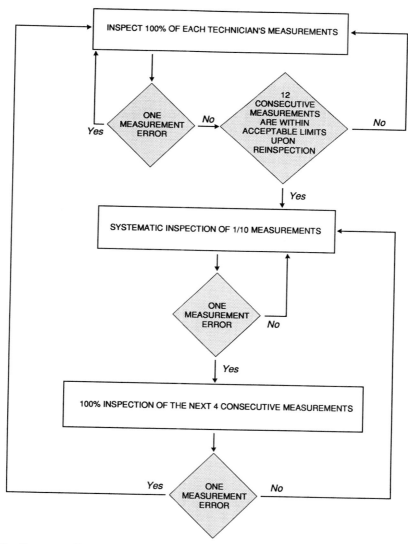

FIGURE 2.—Continuous quality control sampling plan for field data that results in a maximum error of 10 per 100 data for Hudson River striped bass tagging programs.

bers recorded (Table 2). Although this rate was within our 1% tolerance for data bases before any error-checking procedures were applied, we believed it could be decreased. Incorrect tag numbers were identified either as duplicates or as numbers outside the known range of those used. Duplicate tag numbers were only accepted for fish recaptured from an earlier release. An inspection of tags before they were used in the field indicated that duplicated tag numbers did not originate with the tag manufacturer. Because all data were error-checked in one large batch during 1985–1986,

erroneous tag numbers were not discovered until after the data-collection portion of the program was completed. Therefore, it was not possible for program management to introduce any changes to minimize those errors before the program ended.

1986–1987 Programs

Our tag-lot system and feedback to field personnel on the frequency and types of recording errors, reduced the tag-number error rate to 5.3/1,000 during the first 2.5 months of the

TABLE 2.—Cumulative error rates in recording tag numbers in relation to improvements in the field quality assurance (QA) and quality control (QC) procedures for the 1985–1988 Hudson River striped bass tagging programs.

Annual program	Field QA–QC procedures	Month implemented	Tag number errors per 1,000 tags
1985–1986	Random QA audits of field procedures	Nov 1985	9.68
1986–1987	Random QA audits of field procedures	Dec 1986	
	Tag lot inventory system	Dec 1986	5.30
	100% QC of tag numbers in the field	Mar 1987	1.12
	Field QC program	Apr 1987	0.00
1987–1988	Random QA audits of field procedures	Nov 1987	
	Tag lot inventory system	Nov 1987	
	100% QC of tag numbers in the field	Nov 1987	
	Field QC program	Nov 1987	0.40

1986–1987 program (Table 2). The lower error rate and the tag-lot system reduced the time required to correct erroneous tag numbers. However, because the tag-lot system resolved rather than prevented errors, we felt that additional procedural changes were necessary to eliminate causes of recording errors in the field. On 17 March 1987, a major change in tagging procedures was begun that resulted in 100% QC of all tag numbers. The new procedure required little extra time—technicians gathered all field data for each fish in approximately 30 s (mean, 0.98 fish/30 s; $N = 21$ samples). The error rate in recording tag numbers decreased to approximately 1/1,000 (Table 2), and handling mortality rates remained below 1.0% (Dunning et al. 1989).

A field QC procedure was begun on 1 April 1987 for fish-specific data to provide feedback to management on sources of error and to document the effectiveness of the new data-recording procedures (Figure 2). After the field QC procedure began, only five erroneous tag numbers were detected and fish-processing rates did not change. The field QC procedure also allowed the errors in measurement of fish length and in recording of fish-specific tagging data to be calculated. Field QC was also continued in the 1987–1988 program, and the error rate for all fish-specific data as they left the field was less than 0.1%.

Discussion

The QA–QC procedures that evolved in this study resulted because project management reviewed field procedures and made improvements where possible. In the 1985–1986 program, much time was given to resolving errors in the recording of tag numbers. The first step taken to resolve these errors was the new tag inventory. Although the tag inventory decreased the time required to resolve errors, it did not change the data-acquisition procedures that caused tag-number errors in the field. In the 1985–1986 field procedures, the tagging technician verbally called out the tag number to a data technician. This procedure had three possible sources of error: the tagging technician could incorrectly read the tag number, the data technician could misunderstand the numbers called out by the tagging technician, or the data technician could transpose numbers or otherwise incorrectly record the tag number on the data sheet.

Changes in data-acquisition procedures reduced the overall error rate. Specifically, procedural changes minimized the verbal transfer of data and provided 100% inspection of tag-number data. A field QC procedure was implemented so that field technicians were immediately aware of problems with the data. Quality problems were solved by making it easier for biologists and technicians to do their jobs correctly the first time, not through increased inspection or by making it easier to fix mistakes after they occurred.

Feedback provided by the QA–QC system, along with procedural changes, reduced the error rate in recording striped bass tag numbers from 1.0% during the 1985–1986 program to 0.4% in the first 5 months of the 1987–1988 tagging program. It was not possible to measure the raw error rate in recording tag numbers in the absence of all QA–QC procedures because some QA–QC procedures had always been part of this program, including written SOPs, training programs, and visual inspection of data sheets. However, the error rate in recording tag numbers in the absence of QA–QC procedures would probably be much higher than 1.0%. We observed tag returns from similar tagging programs that lacked documented

QA–QC procedures; these programs had error rates in recording tag numbers as high as 23%.

In the Hudson River striped bass mark–recapture program, the number of recaptured fish with valid release data may be a limiting factor for migration, mortality, and abundance estimates. Moreover, anticipated rigorous legal and scientific scrutiny make it desirable to reduce the error rate to the lowest practical level. The cost of a QA–QC system (Figure 1) that reduced errors in tag-number recording from 1.0% to less than 0.1% was approximately 10% of the total labor cost. This cost appears small compared to the cost of collecting the data and the potential cost of fisheries management decisions resulting from errors in the data.

Depending on the objectives of a tagging program, it may not be necessary to achieve such low error rates. The data from the 1985–1986 tagging program suggested that a reduction in the error rate to approximately 1.0% is possible at approximately 1% of the total field labor costs. This can be accomplished through a QA–QC system that includes a written SOP manual, proper training of personnel, QA audits to document that written procedures are followed, and double keypunching.

Conclusions

A QA–QC system such as the one we have described in Figure 1 will (1) provide a data base of known quality, (2) provide feedback to program management during a program to reduce errors in data recording, and (3) provide documentation for future investigators.

Acknowledgments

Funding for this project was provided by Central Hudson Gas and Electric Corporation, Consolidated Edison Company of New York, Inc., the New York Power Authority, Niagara Mohawk Power Corporation, and Orange and Rockland Utilities under the terms of the Hudson River Cooling Tower Settlement Agreement. This project was developed in cooperation with the New York State Department of Environmental Conservation. Fieldwork was conducted by Normandeau Associates Inc., under the direction of Michael Ricci, who was responsible for the training and implementation of the field QC program. Computer programing was done under the direction of John Bonin.

References

ANSI (American National Standards Institute) and ASME (American Society of Mechanical Engineers. 1983. Quality assurance program requirements for nuclear facilities. ANSI and ASME, NQA-1, New York.

ASQC (American Society for Quality Control). 1981. American national standard sampling procedures and tables for inspection by attributes. ASQC, ANSIZ1.4-1981, Milwaukee, Wisconsin.

Dunning D. J., Q. E. Ross, M. T. Mattson, P. Geoghegan, and J. Waldman. 1989 Reducing mortality of striped bass captured in seines and trawls. North American Journal of Fisheries Management 9:171–176.

NAI (Normandeau Associates, Inc.). 1986. 1985–86 Hudson River striped bass program. Fisheries Report R-503-2 to New York Power Authority, White Plains.

NAI (Normandeau Associates, Inc.). 1987a. 1986–87 Hudson River striped bass hatchery evaluation. Fisheries Report R-0012 to New York Power Authority, White Plains.

NAI (Normandeau Associates, Inc.). 1987b. 1987–88 Hudson River striped bass hatchery evaluation/Atlantic tomcod program standard operating procedures. Fisheries Report R-1163 to New York Power Authority, White Plains.

Ogden J. E. 1968. Small-lot sampling plans based upon the hypergeometric probability distribution. American Society for Quality Control, Rockwell International X8-847/201, Milwaukee, Wisconsin.

Sandler, R. and D. Schoenbrod, editors. 1981. The Hudson River power plant settlement. New York University School of Law, New York.

SAS. 1985. SAS user's guide: statistics, version 5 edition. SAS Institute, Cary, North Carolina.

USEPA (U. S. Environmental Protection Agency). 1980. Guidelines and specifications for preparing quality assurance project plans. USEPA, Office of Monitoring Systems and Quality Assurance, QAMS-005/80 Washington, D.C.

American Fisheries Society Symposium 7:720–724, 1990

Data Organization and Coding for a Coastwide Mark–Recovery Data System

Louis Lapi and Marc Hamer

Department of Fisheries and Oceans
Pacific Biological Station, Hammond Bay Road
Nanaimo, British Columbia V9R 5K6, Canada

Bill Johnson

Alaska Department of Fish and Game, Coded Wire Tag Laboratory
Post Office Box 3-2000, Juneau, Alaska 99802, USA

Abstract.—Fisheries agencies from California to Alaska independently developed coded wire tag (CWT) sampling programs and related data-processing facilities in the 1970s. The 1985 Pacific Salmon Treaty between Canada and the USA called for a method by which participating agencies could share CWT-related information. A working group, leaning heavily on an earlier information-sharing system managed by the Pacific Marine Fisheries Commission Regional Mark Processing Center, has devised a detailed methodology for making management information available coastwide through a computer data base. This paper describes how long-established systems used by agencies in British Columbia and Alaska are being integrated with the new data base. It also offers a novel solution to the major problem of encoding unique geographical locations, which could be adopted by other contributing agencies. Issues affecting data consistency and integrity in a data set that is updated by many independent sources are identified, and descriptions of how related problems can be overcome are given.

Fisheries managers in Canada and the USA have been faced with the interception fishery problem since the beginning of this century. When fishermen from one nation intercept salmon migrating through their waters en route to rivers of the other nation, unilateral stock enhancement and stock conservation efforts can be futile. The Pacific Salmon Treaty, signed by Canada and the USA in 1985, was to be the mechanism for dealing with this problem. The Pacific Salmon Commission (PSC), established under the terms of the treaty, is the forum for joint discussion and negotiation.

A key requirement for negotiations of this type is an exchange of fisheries data. The commission formed a Data Sharing Committee and gave it the task of determining what data were needed and how they could be jointly accessed. The principal information to be exchanged consisted of mark recovery data, including releases of coded-wire-tagged (CWT) and unmarked fish; harvest and escapement data, including commercial and sport catches; and associated sampling data. Various agencies within both nations had been collecting these data for many years. A bilateral working group undertook the task of defining formats to make the data usable coastwide. After 2 years of deliberations, the group recommended a PSC coastwide standard for coded wire tag data sets, with duplicate sets to be maintained in Nanaimo, British Columbia, and Portland, Oregon.

In this paper, we explain how two contributing agencies with totally different data bases have been able to provide their data to others and to access all coastwide data themselves. Especially challenging to the agencies were the problems of time and area stratification.

Canadian Data Base

The Canadian mark-recovery system is maintained and accessed via custom-written FORTRAN programs. Custom-written code was chosen over commercial data base software because it gives the system designers complete control over data context and contents. The programs are able to make optimal use of the VAX/VMS operation system and available hardware resources. Users can approach the data from several different perspectives. For example, hatchery and fishery managers can directly access data related to a rearing facility or a management area to see where, when, which, or how many fish were released or were caught.

The system is made up of a series of separate subsystems (e.g., recovery and catch and sample) related to one another through common qualifiers.

For the most part, data files within each subsystem can be updated independently of the other subsystems; within the subsystems, data for one year can be updated independently of data from other years.

Central to the Canadian system is the catch region, which involves more than just a geographical location. It is a location attribute within a specific fishery. Canadian salmon fisheries are managed in part on an area-and-gear basis. Areas open for one gear type are often more tightly defined than areas open for other types; also, a given geographical area may be open to fishing by one gear type, whereas only subsections of that area are open to another gear type. Consequently, two fisheries can be sampled at the same physical location, but the area resolution of the two samples can be very different. In some troll fisheries, a catch region can span over nearly 200 km of coastline while a concurrent net fishery will be restricted to a 20-km section of the coast.

Two types of time stratification are used for Canadian marine fishery data. Commercial fisheries are managed on a statistical week basis. Sport data are recorded by the month, because samples taken over shorter periods often are too small to be of statistical value.

Once fish enter fresh water, the concept of catch region has a different meaning. It now describes only a generic fishery—for example, "freshwater sport." Data are kept for each stream, which is assigned a unique code known to the system as a site code. For a large river system, several such sites may be defined. These same codes are used to qualify release information.

Freshwater time stratification varies according to the generic fishery. For example, escapement may be reported by timing of the run (e.g., fall-run chinook salmon *Oncorhynchus tshawytscha*), whereas sport catch is reported by month.

To use data collected by outside agencies, the Canadian system must first transcribe it into the forms described above.

Alaska Department of Fish and Game Data Base

The Alaska Department of Fish and Game data base is maintained on a dedicated computer that uses the Pick operating system. Pick, which is specialized to support only data base management, has the flexibility to redefine data elements and schemata easily as new needs are identified. There are virtually no restrictions on formats, item sizes, sort-selection criteria, or data types, or on joining records with common attributes. In the area of coded wire tagging, for which methodologies are continually evolving, this flexibility is especially valuable.

Intrinsic to the data base are fourth-generation update and retrieval languages. Almost all ad hoc production reports are accomplished with single pseudo-English statements entered by users or invoked as menu selections or abbreviated commands. The updating process supports a robust level of data control. The common pitfall of this very flexible data-base management system— namely, a high system overhead that results in poor response—was avoided by purchasing high-capacity hardware and dedicating it to the coded wire tag program.

The team charged with Pacific Salmon Treaty research and management, whose needs extend beyond coded wire tag work, separately developed a VAX/VMS computer facility that is accessed through networked personal computers. By installing a Pick coprocessor within the VAX, users are able to select Alaska tag data and feed them into such tools as SAS and Lotus or into FORTRAN applications.

Several area stratifications are simultaneously maintained. The Southeast Alaska Region is composed of four quadrants. Each quadrant is assigned a set of unique districts, each district is subdivided into subdistricts, and flowing into the subdistricts are numbered streams. Data are entered at the lowest known level of area resolution; meanwhile, the system automatically calculates the locations of all data at all levels of the hierarchy, and it generates separate statistics at each level of the hierarchy by pooling appropriate data.

Two levels of time are also used simultaneously—statistical week and period. A period is a group of contiguous weeks assigned to each fishery; for example, the commercial troll fishery in 1985 had a set of periods based on the openings called during that year. This results in each observation having statistics associated with it in up to 10 time–area strata.

The user is free to choose an appropriate stratum for analysis. For example, activity in a cost-recovery fishery, typically open for several days at the mouth of a stream, could be analyzed at the subdistrict–week level. Alternatively, a troller might fish for 2 weeks and cover hundreds of kilometers before being sampled, in which case a quadrant–period would be preferable. The user has the ability to extract statistics at all 10 levels, compare behavior among them all, and then select one for a subsequent purpose.

By coupling multiple stratification with the ability to select data points by any criteria, one can retrieve estimates of, say, the contribution made by Whitman Lake Hatchery to the segments of the chinook salmon fisheries that were delivered to Petersburg during August 1986.

To share data, Alaska needed standards and structure for extracting core information of coastwide interest from its detailed data base.

Problems Relating to Location Coding

Each agency had developed internal structures to fit its own needs. How were agencies to share information? The Pacific Marine Fisheries Commission's Regional Mark Processing Center (RMPC) had developed a system of exchanging information on fin marks in the early 1950s, and was adopted for CWT data in the 1970s.

The RMPC information already contained output from analyses. This made it difficult for people to repeat analyses; moreover the data were not adequate for many of the uses that they were put to. For stock composition estimates, one of the major uses of the CWT data by PSC technical committees, more information had to be exchanged.

The bilateral working group that was assigned the task of exchanging data used the RMPC information formats as the basis for the new data-exchange standard. The group began by addressing such technical details as defining data files, data fields, and valid codes for standard fields within fisheries data. Some of the difficulties identified in the RMPC arrangement included the following. (1) Area codes were allowed to change from year to year; for example, the meaning of area code 80 in 1981 differed from its meaning in 1982. (2) Location coding for some data elements used site names in common, but abbreviated the names many different ways; by 1983, for example, there were over 40 different spellings for Bonneville Dam. (3) Over the years, sampling areas changed size and shape. (4) Different coding schemes had evolved for release and recovery, so that the same physical place had two codes, one for release and one for recovery. (5) It was difficult to relate catch and sample information to recovery information because there was no well-defined link.

Differences in spelling could be cleared up by reference to master lists but this did not assure consistent area definitions, which were a major problem in understanding the data of different agencies. The working group decided to handle all of the above problems at once. The idea was to develop a hierarchical coding strip that would help standardize area location and that would provide links to other information (e.g., linking catch and sample data to recovery information). This coding strip would give each agency a standard way to express its own fisheries data and to interpret data of other agencies.

To accommodate all coding schemes, the coding strip's definition calls for seven subfields and is 19 characters long. This enables each agency to code its individual scheme into the hierarchical scheme. The coding strip is to be used wherever an area definition is needed. The strip is divided as follows.

Level 0   Each state or province has a unique code assigned to it.

Level 1   Currently, this is either fresh water or salt water, but it could be expanded to include estuaries so that sampling programs for juveniles would be covered.

Level 2   In Canada, British Columbia is broken into north or south. Alaska is divided into regions.

Level 3   In Canada, this represents catch regions for saltwater fisheries; it is not currently used in the freshwater category. In Alaska, this is used to represent quadrants within a region.

Level 4   In Canada, this is equivalent to statistical areas for salt water; for fresh waters, it represents production areas and is used to group hatcheries, release sites, and recovery sites within a geographically similar area. In Alaska, this represents districts, which are equivalent to statistical areas.

Level 5   In Canadian sport fisheries, this represents specific sport locations; and for fresh water, it represents hatchery and stream locations. In Alaska, this level represents subareas.

Level 6   In the Canadian commercial fisheries, this represents subareas. In Alaska, this represents the final section in the stream code.

The use of codes rather than names meant that a fourth file was needed to translate codes into common names for hatcheries, stocks, and release sites. However, a similar file would have been necessary had we relied on standard names. Our choice for a coding system has solved all area code problems, and it allows a system to incorporate any agency's recovery data into any other system.

Problems Relating to Data Integrity

The RMPC data-sharing process, which was the first successful attempt to make interagency CWT data widely available, had been in regular use for over 10 years. In defining the PSC coded wire tag data set, special attention was paid to unresolved data control issues identified in the pioneer RMPC coded wire tag arrangement. Although innovative data encoding meets many needs, this in itself does not make the data base useful. The user must be able to trust the validity of values coded into fields.

Data integrity is especially important in the many FORTRAN environments served, where an attempt to perform numerical calculations on a nonnumeric value can cause entire analyses to abort. However, even worse are the insidious errors unknowingly used as a basis for conclusions. Examples include obsolete versions of data, data values whose exact meanings differ among jurisdictions, and data misinterpreted because an obsolete format is referenced. The major problems we addressed are described below, together with the action taken to address each.

(1) Elements of interest to current researchers were not present. For example, the life stage of fish when released and the number of hatchery fish not represented by a tag were unavailable.

The basic RMPC format was defined in the 1970s when CWT methodologies were in an early stage of development. Furthermore, the mandate for rigorous coastwide data analysis codified in the Pacific Salmon Treaty did not exist when the RMPC specifications were devised. The PSC analysis was the first major reassessment of data needs in about 10 years.

(2) Reporting specifications were unsuitably vague in places. Although users reported samples as being random, selective, or voluntary, it was discovered that at least six distinct types of samples occurred among agencies. It is still unclear how each agency decided to translate them into the three.

The PSC format went considerably further in defining the meaning of commonly used values.

Also, it became clear that some data collected by agencies have been assigned values that are not consistent among groups. For example, one agency may release a lot as "fingerlings" while a different agency reports the same release as "fry." These different definitions often are institutionalized within agency procedures, and they extend far into the historical data base. To permit users to interpret the data effectively, an appendix of local definitions is being compiled as part of the final report.

(3) Data specifications were periodically revised on an ad hoc basis. The user community did not formally deliberate on the implications of these changes. This resulted in confusion as to what exact information had to be submitted to the RMPC. A case in point is the method for reporting releases with embedded replicate tags.

The PSC specification is compiled as a set of documents carrying format version numbers. The data standard calls for every record in the system to include the version number corresponding to the definitions used in its creation. The final report will recommend that a standing multilateral committee be maintained to periodically revise specifications and issue new versions.

(4) No rigorous validation was defined for incoming data submissions. Many fields were checked for reasonableness, but this was not enough to uncover all data inconsistencies.

Each field definition in the PSC standard is accompanied by a validation criterion. These criteria consist of detailed rules that determine whether a field contains acceptable data. Before the data base can be updated, each submission should be validated. Failure of any fields to pass this validation is an indication of data control problems on the submitting agency's part and is grounds for rejection of the data set. This protects the data base from corruption, and it ensures a basic integrity of format for all users; for example, users need not check each field for numerics before computation. A complication is introduced by rigorous validation. A submitter could identify a pressing need for a new value. For example, a new agency might begin tagging fish for the first time, yet the fixed validation does not support the new name. In this case a special procedure has been devised: any field that cannot be filled because the format or code structure is incapable of conveying an appropriate value is to be filled with a string of commercial "at" signs (@). Fields of @ are always accepted by the validation process, but individual users must then determine whether or not the affected records in the data set are suitable for specific analysis. By reporting @ records, the validation process creates an agenda of format problems to be addressed at subsequent meetings of the oversight committee.

(5) Corrections were applied to the data by the RMPC, and the agency that provided the data was later notified in a somewhat informal manner. This procedure did not ensure that the users verified the problems and corrected them on their own agency's data base. Besides aggravating the inherent problem of divergence, this allowed future updates to return the problem to the Mark Center.

The responsibility for correcting errors has been firmly placed in the hands of the originator. The validation process catches errors before they become installed in the data base.

(6) There was never a requirement that substantive changes to a specification had to be applied to data previously submitted. For example, in 1984 a field for tag-type code was inserted in the middle of the recovery record. Pre-1984 data typically show a blank for this field, leading the uninitiated user to believe that all recoveries were of standard binary tags.

Now that the recommendations are formally adopted, all historical data will be loaded on the data base in the current version. Although no policy forces resubmission of all data every time specifications are revised, users will always know the exact meaning of data they use because every record carries its explicit version and generation date.

End-User Implications

This coastwide data-sharing facility meets two vital functions. First, it makes available to problem-solvers a comprehensive and consistent data set for the areas outside their geographical juris-diction. Treaty terms, such as the equity principle, require management decisions to consider coastwide implications. For example, before the treaty was in place, the Alaska Department of Fish and Game, when setting a minimum size for troll-caught chinook salmon, would consider the effect on factors such as future Alaskan returns, market response, and induced reallocation among Alaskan user groups. Now, additional factors must be analyzed, such as reallocation among user groups by political jurisdiction, economic impacts on the fishing industry coastwide, and possible effects on specific stocks outside of Alaska that are targeted for rebuilding. Not until the coastwide data set is available can this be done in a highly consistent and reliable manner.

Second, this facility allows all parties access to the same data set. In the past, divergence of basic data regularly caused different individuals to come up with conflicting results, even though they used indentical analyses. Now, although parties may be 2,500 km apart, their use of the standardized coastwide coded wire tag data set can ensure consistency.

Incidental mortality analysis, stock-composition analysis, equity studies, assessment of replicate tagging, and assessment of hatchery distribution and survival can now be attempted on a complete, coastwide data set.

The enumerated shortcomings of earlier data-sharing systems prevented the collected data from standing alone. For many purposes, users had to hunt down individuals within diverse agencies to learn the scope, meaning, and currency of information. Even if the user knew to do this, success in accomplishing it was not guaranteed.

The PSC standard coastwide coded wire tag data base is a concerted attempt to overcome technical problems that chronically afflict information sharing among diverse autonomous agencies. If successful, its scope can be extended to cover escapement and effort data within the PSC context, and its approach may warrant adoption by other agencies.

American Fisheries Society Symposium 7:725–735, 1990

Sizes, Structures, and Movements of Brook Trout and Atlantic Salmon Populations Inferred from Schnabel Mark–Recapture Studies in Two Newfoundland Lakes

Patrick M. Ryan

Department of Fisheries and Oceans, Science Branch
Post Office Box 5667, St. John's, Newfoundland, A1C 5X1, Canada

Abstract.—Schnabel's multiple mark–recapture method was used to census brook trout *Salvelinus fontinalis* and anadromous Atlantic salmon *Salmo salar* each spring and fall over 5 and 6 years in two lakes at the headwaters of the Gander River, Newfoundland. Fish were captured in modified fyke nets, marked with fin clips or fin holes, released, and recaptured for the calculation of population sizes. Stock structures were calculated from the lengths of all fish captured and from fish subsampled for age and weight. Movements of fish to and from the lakes were calculated from differences in stock structures between censuses. The experimental design was tested by the use of bioassay cages and counting fences for migratory fish in streams, and by the examination of relationships between marking frequency and recapture frequency, between catch per unit effort and population size, and between calculated population characteristics and events external to the study. Census requirements were only approximated, but it was concluded that the techniques provided reliable estimates of population sizes, stock characteristics, and movements for these dynamic, riverine stocks.

Ideal censusing of a fish population by use of multiple mark–recapture methods requires that the population is constant during the census with no emigration, immigration, or mortality, that the marked fish are recognizable as such, and that they are in every other way the same as unmarked fish. These requirements are seldom, if ever, completely met in practice, and numerous methods have been developed to prevent or to document and correct for the resulting errors in census data. These methods, primarily applicable to censuses with discrete markings, were reviewed by Ricker (1975), Youngs and Robson (1978), Arnason and Mills (1981, 1987), and Seber (1986). Such correction may not always be possible or necessary, depending on the census method employed or the study objectives. For practical purposes, if the above requirements are approximately met, valuable inferences can be drawn about the status of fish populations (Ricker 1975). However, there is little information available in the published literature about whether or not census requirements are met during multiple mark–recapture estimates with nondiscrete marking.

This paper describes in detail the methods employed during repeated Schnabel population censuses (Ricker 1975) in two Newfoundland lakes inhabited for varying times by migratory riverine populations of brook trout *Salvelinus fontinalis* and anadromous Atlantic salmon *Salmo salar*. Some of the census data have been documented previously in analyses of salmonid population sizes, population structures, and movements (Ryan 1984a, 1986a, 1986b). In this paper, the validity of the censuses is examined for practical application. The relatively simple and rapid methods will likely be useful in studies of other fish populations.

Study Area

Salmonid populations were censused in two small, shallow, brown-water lakes within the Department of Fisheries and Oceans' Experimental Ponds Area at the headwaters of the Gander River, Newfoundland (Table 1). Like most Newfoundland lakes, they are referred to locally as ponds. Headwater Pond drains 3.5 km north into Spruce Pond and then about 155 km northeast to the Atlantic Ocean. In addition to brook trout and Atlantic salmon, both lakes are inhabited by threespine sticklebacks *Gasterosteus aculeatus* (Ryan 1984b) and American eels *Anguilla rostrata*. The Atlantic salmon are part of a sea-run stock, and those individuals censused in the present study were immature and precociously mature forms that had not left fresh water. Young Atlantic salmon enter the lakes from downstream spawning areas, subsequently smoltify, and migrate to sea; the majority of survivors return to the river after 1 year in the ocean (Ash and Tucker

TABLE 1.—Summary of physical and chemical descriptors of the study lakes at the headwaters of the Gander River, insular Newfoundland. Data from water analyses are the means of values obtained. Further description is in Ryan and Wakeham (1984).

Descriptor	Headwater Pond ($N = 87$)	Spruce Pond ($N = 125$)
Latitude	48°16′15″	48°15′40″
Longitude	55°29′39″	55°28′24″
Catchment basin area (km^2)	5.96	20.06
Water area (hectares)	76.12	36.50
Volume (m^3)	871,767	364,282
Maximum depth (m)	3.25	2.13
Mean depth (m)	1.14	1.00
Turbidity (JTU)[a]	1.1	1.3
pH	6.5	6.3
Conductance (μS/cm)	34.7	34.2
Chloride (mg/L)	5.1	6.0
Total EDTA hardness (mg/L)	12.0	11.0
Calcium (mg/L)	3.5	2.3
Total alkalinity (mg/L)	4.4	4.2

[a] Jackson turbidity units.

1984; Ryan 1986a). There has been no indication of anadromy in the brook trout population.

Census Methods

Brook trout and Atlantic salmon were concurrently censused each spring and fall, starting in 1978 in Spruce Pond and 1979 in Headwater Pond, by means of Schnabel's multiple mark–recapture method (Ricker 1975). This paper is based primarily on data collected during 44 censuses conducted up to 1983, but includes small amounts of information obtained in subsequent censuses.

The Schnabel census method relies on nondiscrete marking of fish that are subsequently recaptured. This method was selected over others, which require the use of individually recognizable marks (Youngs and Robson 1978), because of the small size of the fish and the mortality associated with greater handling, particularly in the case of young Atlantic salmon (Bourgeois et al. 1987).

Fish were captured in 13- and 19-mm stretched-mesh fyke nets set from shore at standard locations (Ryan 1984a). Typically, nets were hauled each day. Occasionally, nets were fished for 2–3 d or were collapsed for similar periods. Fish were collected in 20-L plastic pails with frequent water replacement and then concentrated by water removal prior to handling. If fish were obviously distressed by warm weather and could not be processed immediately, they were released and not considered in that day's catch for the computation of population size. All remaining fish were measured for fork length (mm). Unmarked fish

were marked with adipose or pelvic fin clips, dorsal fin holes, or—in most censuses—caudal fin holes. Holes were made with 3-mm diameter, single-hole paper punches; in many cases, only a segment of tissue was removed from the edge of the fin. Weights (g) and scale samples were obtained from a subsample of 100-125 fish selected approximately at random near the beginning of each census; subsequently, selections emphasized less common length intervals. Fish killed accidentally were neither considered as members of the population nor included in that day's catch for the computation of population size. The few fish that were dead when recaptured were assumed to have been captured first the previous day, and a fish of identical length was removed from the previous day's record. Processed fish were released away from their location of capture at various shore sites where measurement and marking took place or by subsequent transport to open water.

Anaesthetics were not necessary for the short period of length measurement and marking (<10 s/fish) or subsampling for age and growth (about 20 s/fish). Fish tended to be lively upon release, and the occasional inactive fish revived quickly after being placed in the lake.

Each census was continued until recaptured fish represented approximately 30–50% of the total catch over several days or until catches remained consistently low after 2–3 weeks. A population estimate with 95% confidence interval was calculated for each day on which recaptures were possible; Schnabel's short formula was used as described by Ricker (1975). The last day's population estimate was employed in subsequent analyses of the population, and data obtained prior to the last day were used to examine the validity of the census methods.

Ages were assigned to the subsampled fish in the manner described by Tesch (1971) with an acceptable degree of reliability (Tesch 1971; Nickerson et al. 1980; Ryan et al. 1981). The age composition of each species in each lake during each population estimate was calculated from the ages and lengths of fish in the subsample, the lengths of captured and released fish, the Schnabel population estimate, and their relative proportions, with the aid of age-length keys (Ricker 1975). The weight of fish in each lake during each census was calculated by summing the products of the mean weight of individuals in each age-

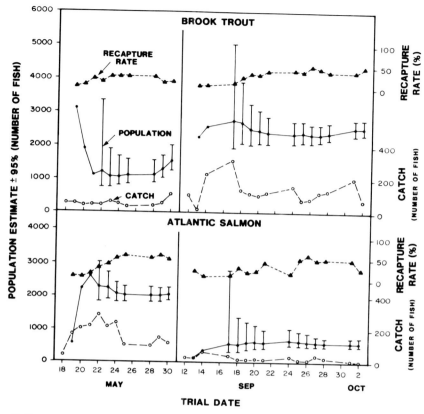

FIGURE 1.—Daily catches, Schnabel population estimates, and recapture rates for brook trout and Atlantic salmon in Headwater Pond in the spring and fall of 1979.

group in the aged subsample and the calculated number in each age-group in the lake.

The net number of fish migrating into or out of the lakes between censuses was calculated as the difference between the number of fish in all age-groups present during a given census and the number present during the subsequent census. Details of this technique have been described by Ryan (1986a). The ancillary information generated contributed to the assessment of census methods.

Census Results

Each census was characterized by sample-to-sample variation in catch, recapture rate, and population estimate, as exemplified by the first four censuses in Headwater Pond (Figure 1). Daily fluctuations in catch rates were influenced by differences in water level, water temperature, and the time the nets were fished between hauls (Hamley and Howley 1985). Recapture rates also fluctuated from day to day, but they exhibited a general pattern of increase over the duration of

each census, the rate of increase tending to slow as a greater proportion of the apparent population was captured. Population estimates tended to stabilize over the course of each census and the corresponding confidence intervals narrowed. Occasional anomalous census results occurred when a stabilizing population was not apparent, as in the spring census of Headwater Pond brook trout in 1979 (Figure 1).

Estimated population sizes for each species in each lake varied considerably over the course of the study (Tables 2, 3). Calculated population structures exhibited a corresponding variation as detailed by Ryan (1986a) and exemplified in Figure 2. Such fluctuations are not atypical of these species. Year-to-year survival is variable (Power 1980; Gibson and Myers 1988). Movements are influenced by such factors as water level, temperature, population pressure, and the degree of smoltification in the Atlantic salmon population (Scott and Crossman 1964; Power 1980; Ruggles 1980).

TABLE 2.—Marking and recapture data, population estimates, and population weights of brook trout and Atlantic salmon in Headwater Pond. Only non-sea-run fish are included.

Dates of population estimate	Captured and released					Recaptured		Population estimate		Population weight (g)
	Number	% of population estimate	Length (mm)			Total number	% of last catch	Number of fish	±95%	
			Mean	Range	SD					
Brook trout										
18–30 May 1979	509	33	157	67–334	57	92	16	1,549	1,257–2,018	103,176
12 Sep–2 Oct 1979	1,413	55	154	56–330	52	563	51	2,556	2,344–2,810	163,373
18 May–6 Jun 1980	754	34	185	76–340	49	142	35	2,252	1,897–2,777	207,293
10–24 Sep 1980	1,630	49	186	71–368	49	521	40	3,316	3,021–3,674	331,255
14–27 May 1981	847	50	180	74–354	52	270	36	1,703	1,502–1,966	133,090
16 Sep–1 Oct 1981	2,205	46	164	70–347	41	613	42	4,761	4,372–5,226	289,587
19 May–2 Jun 1982	1,152	38	180	78–345	47	246	39	3,070	2,692–3,571	258,498
15 Sep–5 Oct 1982	1,410	39	172	80–324	46	315	30	3,645	3,252–4,146	239,657
18–27 May 1983	1,058	33	185	85–323	39	198	29	3,190	2,748–3,801	262,534
14–29 Sep 1983	1,221	42	187	80–309	41	318	32	2,883	2,566–3,289	247,757
Atlantic salmon										
18–30 May 1979	1,254	61	158	70–256	28	589	49	2,065	1,890–2,273	92,980
12 Sep–2 Oct 1979	318	51	172	89–275	30	113	25	622	518–779	38,433
18 May–6 Jun 1980	872	57	187	87–323	38	394	62	1,520	1,367–1,713	106,648
10–24 Sep 1980	234	42	187	92–301	41	59	38	563	436–793	48,494
14–27 May 1981	466	70	149	89–268	34	294	79	668	592–766	26,539
16 Sep–1 Oct 1981	113	30	155	78–230	30	17	14	382	249–819	19,376
19 May–2 Jun 1982	554	57	161	82–243	35	239	56	965	845–1,125	45,718
15 Sep–5 Oct 1982	117	21	167	94–256	42	13	11	551	345–1,361	29,461
18–27 May 1983	157	40	140	92–338	31	34	38	394	284–643	13,682
14–29 Sep 1983	120	23	144	88–351	29	14	11	521	328–1,265	16,101

Methodology

A variety of tests provided evidence as to whether or not census requirements were met. These were the use of bioassay cages and counting fences for migratory fish in streams during censuses, and the examination of relationships between marking frequency and recapture frequency, between catch per unit effort and population size, and between calculated population characteristics and events external to the study.

The mortality of fish in bioassay cages was considered to be an estimate of the mortality caused by handling and marking procedures. During the censuses of 1981 and 1982, 37 brook trout and 27 Atlantic salmon were captured, handled in the usual manner, and placed in the lakes in cages constructed of hardware cloth on wooden frames. No caged fish died during the censuses in spite of nose and tail abrasions from the cages and varying degrees of emaciation. This was interpreted as evidence of negligible differential mortality for marked and unmarked fish.

Fish captured in streams in counting fences adjacent to the lakes provided measures of emi-

TABLE 3.—Marking and recapture data, population estimates, and population weights of brook trout and Atlantic salmon in Spruce Pond. Only non-sea-run fish are included.

Dates of population estimate	Captured and released					Recaptured		Population estimate		Population weight (g)
	Number	% of population estimate	Length (mm)			Total number	% of last catch	Number of fish	±95%	
			Mean	Range	SD					
Brook trout										
18–26 May 1978	481	49	156	72–335	47	134	35	988	824– 1,234	58,500
14 Sep–6 Oct 1978	322	56	169	75–302	50	122	47	574	481 – 712	41,886
18–25 May 1979	517	47	140	61–320	65	151	49	1,094	917 – 1,355	48,845
12–28 Sep 1979	348	47	163	62–305	49	105	35	740	610 – 939	47,209
19 May–6 Jun 1980	428	31	176	79–340	42	72	23	1,378	1,088 – 1,878	137,113
10–24 Sep 1980	619	56	192	78–329	41	230	57	1,100	959 – 1,290	84,234
16–27 May 1981	664	55	156	68–331	50	243	48	1,206	1,053 – 1,411	77,245
16 Sep–1 Oct 1981	1,082	52	171	58–344	45	357	47	2,083	1,863 – 2,362	146,475
18 May–1 Jun 1982	1,043	62	168	62–316	44	504	46	1,680	1,531 – 1,861	125,431
15 Sep–5 Oct 1982	911	67	172	78–310	37	454	47	1,368	1,243 – 1,522	91,585
18–26 May 1983	1,181	50	167	76–328	41	387	42	2,349	2,103 – 2,661	161,932
16–29 Sep 1983	850	51	176	75–310	39	276	34	1,669	1,469 – 1,932	120,555
Atlantic salmon										
18–26 May 1978	1,289	45	132	75–277	35	352	45	2,871	2,557 – 3,273	100,419
14 Sep–6 Oct 1978	501	42	139	77–283	36	127	35	1,198	1,006 – 1,480	44,621
18–25 May 1979	1,572	57	143	74–262	42	620	72	2,755	2,516 – 3,044	123,280
12–28 Sep 1979	478	36	152	74–268	65	101	32	1,328	1,091 – 1,695	62,688
19 May–6 Jun 1980	1,227	63	164	50–294	39	630	68	1,943	1,782 – 2,135	92,922
10–24 Sep 1980	488	31	157	68–270	38	83	27	1,592	1,279 – 2,107	79,179
16–27 May 1981	930	54	140	75–249	38	343	56	1,725	1,537 – 1,965	55,744
16 Sep–1 Oct 1981	661	34	144	78–282	30	126	36	1,949	1,626 – 2,432	66,473
18 May–1 Jun 1982	1,379	65	155	77–246	43	734	60	2,112	1,955 – 2,297	96,220
15 Sep–5 Oct 1982	417	48	142	81–233	33	125	29	874	736 – 1,087	32,233
18–26 May 1983	506	42	128	81–254	26	125	29	1,209	1,002 – 1,523	30,291
16–29 Sep 1983	231	30	135	84–213	26	39	28	759	557 – 1,190	24,681

gration and immigration during the 1978–1982 censuses. Counting fences were constructed of 4.8-mm-square-mesh seine netting mounted on wooden frames and stretched from shore to two-way traps set on the stream bottoms. The fences were designed to catch both upstream and down-stream migrants as described by Craig (1980). Fences frequently were damaged by floating debris and animals, and they were often submerged during the high flows common in spring and fall (Ryan and Wakeham 1984). Consequently, they did not capture all fish entering or leaving the

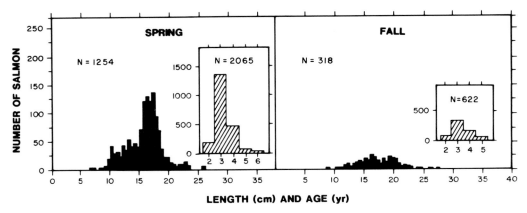

FIGURE 2.—Length–frequency distributions of captured and released Atlantic salmon and calculated age–frequency distributions (inset) of the Atlantic salmon population during censuses in Headwater Pond in 1979.

lakes. The counting fences were discontinued after 1982 because of their inefficiency during high water and because of fish mortalities during floods and high temperatures.

Fish were captured while moving to and from the lakes during 29 of the 36 censuses from 1978 to 1982 (Table 4). During 17 of those 29 censuses, net movement represented an influx of fish to the lakes and a resultant dilution of marked fish. Also, further reductions in the proportion of marked fish in the populations were indicated by captures of emigrant marked fish during four censuses. However, captured migrating fish represented small portions of the corresponding population estimates and were within the range of the 95% confidence interval for individual estimates (Ta-

bles 2, 3). Counting-fence data were regarded as evidence that the lake populations were not constant during most of the censuses, the predominant effect being a reduction in the proportion of marked fish.

The recapture rate in the last day's catch was compared to the marking rate or expected recapture rate in individual and combined censuses. The marking rate was calculated as the number of marked fish in a population on the second-last day of a census, and was expressed as a percentage of the final population estimate. If all census requirements were ideally met, the recapture rate would equal the marking rate.

Large proportions of the estimated populations were sampled during the censuses—47% for

TABLE 4.—Numbers of brook trout and Atlantic salmon that moved into and out of Headwater and Spruce ponds through stream counting fences during population estimates from 1978 to 1982.

Season and year	Headwater Pond				Spruce Pond			
	Brook trout		Atlantic salmon		Brook trout		Atlantic salmon	
	In	Out	In	Out	In	Out	In	Out
Spring 1978							3	
Fall 1978								
Spring 1979	28	11	72	6	76	9	163	27
Fall 1979	17	9	1		15	1	1	6
Spring 1980	6	5	12	62[a]	2	1		68[b]
Fall 1980					4	10[c]	8	1
Spring 1981	20		4		3	4[d]		2
Fall 1981	2	4			1	13		15
Spring 1982	8	14	5	9	26	8	50	1
Fall 1982		1			36	10	10	4

[a] Includes four marked fish.
[b] Includes 29 marked fish.
[c] Includes two marked fish.
[d] Includes two marked fish.

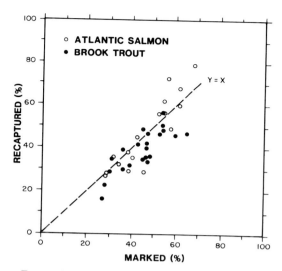

FIGURE 3.—Recapture rates for brook trout and Atlantic salmon in the last day's catch compared with the marking rates during Schnabel population estimates in Headwater and Spruce ponds from 1978 to 1983. Marking rate was calculated as the number of marked fish in a population on the penultimate day, expressed as a percentage of the final population estimate. Five points based on a catch of less than 10 fish on the last day were omitted.

brook trout and 45% for Atlantic salmon, on the average (Tables 2 and 3). The calculated marking rates on the second-last day of the censuses and the recapture rates on the last day tended to approximate a 1:1 ratio, although usually the recapture rate was less than expected (Figure 3). Calculated marking rates averaged 45% for brook trout and 44% for Atlantic salmon, whereas average recapture rates were 39% and 41%, respectively. The differences between average marking and recapture rates were consistent with the small reductions in the proportions of marked fish indicated by the counting fence data. These small differences were regarded as evidence that census requirements were approximated but not ideally met.

The relationship between catch per unit effort and estimated population density was used to judge reliability of all census methods. Such a relationship was documented for the two species by Ryan (1984a). Catch per unit effort was calculated as the average number of fish captured in a single fyke net fished overnight during each census. If census requirements were met, catch per unit effort should be proportional to population number (Ricker 1975; Youngs and Robson 1978).

Catch per unit effort of brook trout was a highly significant ($P < 0.01$) correlate of estimated population density (Figure 4). The catch per unit effort of Atlantic salmon was also a strongly significant ($P < 0.01$) correlate of estimated population density, provided the season of capture was taken into account (Figure 4). Atlantic salmon were easier to capture in the spring, probably because of their enhanced activity during smoltification (Ryan 1984a). In view of the large number of factors that affect fish catchability (Hamley and Howley 1985), the strength of these relationships supported the hypothesis that census requirements were met.

The correspondence between calculated population parameters of Atlantic salmon in the study lakes and observed characteristics of adult salmon returning to the Gander River as grilse after 1 year in the ocean provided an additional measure of the reliability of census methods. This correspondence was examined in detail by Ryan (1986a, 1986b). Relationships were not expected to be precise, because of the strong modifying effect of the commercial fishery (Chadwick 1985). However, reliable census methods should indicate some correspondence between the age composition of the Atlantic salmon calculated to have left the study lakes as smolts and the freshwater age of grilse (Chadwick et al. 1978; Ryan 1986a). Similarly, the methods should indicate proportionality between the calculated abundance of young Atlantic salmon prior to smoltification, the calculated subsequent abundance of smolts, and the resultant abundance of returning sea-run adults (Ruggles 1980; Ryan 1986b). Population estimates from spring censuses were employed as a measure of the abundance of young Atlantic salmon prior to smoltification in the study lakes. The age composition and abundance of smolts were calculated as the decreases in the estimated number of fish in each age group between spring and fall censuses. Comparable data for adults were the age composition and catch per unit effort of grilse angled in the recreational fishery 1 year later.

The calculated modal age of smolts from the study lakes was identical to the varying modal freshwater age of returning adults in each of four comparable years (Figure 5). Although the two distribution sets did not correspond completely, that could not be considered atypical in comparison with other Atlantic salmon rivers, and the differences probably were compounded by an age differential in oceanic survival of smolts (Chad-

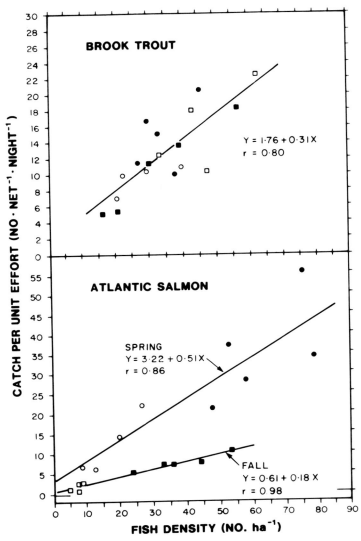

FIGURE 4.—Catch rates for brook trout and Atlantic salmon related to densities estimated by the Schnabel multiple mark–recapture technique in Headwater (open symbols) and Spruce (solid symbols) ponds from 1978 to 1982. Circles and squares indicate spring and fall data, respectively. Redrawn from Ryan (1984a).

wick 1985; Ryan 1986a). In addition, the spring population estimate of Atlantic salmon in the lakes was a strong correlate ($r = 0.99$) of the number calculated to have left as smolts (Ryan 1986b) and of the relative abundance of adults returning the following year (Figure 6). A large part of the additional variation in adult abundance, as estimated from angler success, is likely attributable to an inverse relationship between Atlantic salmon catchability and river discharge (Chadwick 1982; Ryan 1986b). Because of the many variables that affect relationships among the

parameters examined, the degree of correspondence observed in the present study was taken as an indication of the reliability of the census methods for monitoring salmonid populations.

Discussion

Obviously, census requirements were not completely met during the censuses of salmonid populations in Headwater and Spruce ponds. However, the results do suggest that, overall, the requirements for multiple censuses were closely approached and, for practical purposes, the meth-

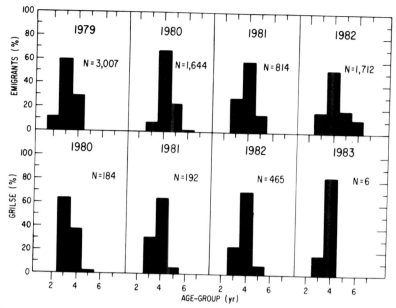

FIGURE 5.—Calculated age composition of Atlantic salmon emigrating from Headwater and Spruce ponds, compared with the composition of river ages (freshwater life) of adults sampled as grilse in the Gander River angling fishery the following year. Emigration from the study lakes was calculated as the decrease in the number of fish in each age-group between spring and fall censuses. Redrawn from Ryan (1986a).

ods provided reliable estimates of salmonid population sizes, structures, and movements.

Actual population sizes probably were underestimated because the smallest fish moved less and

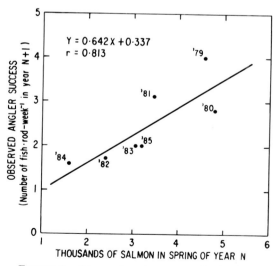

FIGURE 6.—Observed angler success in the Gander River grilse fishery related to the estimated abundance of Atlantic salmon in Headwater and Spruce ponds in the spring of the preceding year. Populations are based on Schnabel estimates from 1978 to 1984. Years of angling are shown. Redrawn from Ryan (1986b).

were caught less often. However, fish measuring 5–10 cm were relatively abundant in catches in some years and rare in others. The smallest of the two fyke-net mesh sizes did not capture smaller fish than the larger mesh. In addition, numerous threespine sticklebacks less than 3 cm long were seined in August of 1978 to 1981 and throughout the summer of 1979 (Ryan 1984b), but those hauls included no salmonids smaller than 5–10 cm. It appears that the smallest salmonids were rarely caught in the lakes because most of them were still occupying stream habitats.

A small degree of overall bias usually was apparent in the less-than-expected recapture rate for the last day's catch of the censuses (average differences of 6% for brook trout and 3% for Atlantic salmon). The survival of all handled and caged fish during the censuses of 1981 and 1982 suggests that a loss of marked fish due to mortality was not the major contributor to this bias. As indicated by the counting-fence data, a major factor accounting for the lower-than-expected recapture rates was probably the migration of fish to and from the lakes.

Attempts to correct for this bias in the present study were not considered practical because malfunctions of the counting fences prevented precise quantification of migration and the internal consistencies of all of the results suggested that the bias was small. In addition, it is likely that factors influencing the accuracy of census results varied in the magnitude of their effect from census to census. For example, the highest number of migrants captured in the counting fences came during the spring of 1979, when 163 Atlantic salmon were intercepted on their way to Spruce Pond (Table 4). This migration would have diluted marks in the population. However, at the end of the census, the higher-than-expected recapture rate (Table 3) suggested that other factors, such as a nonrandom distribution of marked fish, were also affecting the results.

The usually lower-than-expected recapture rates also might have been influenced by differences in catchability between marked fish and fish not captured previously. Marked fish might have required a recovery period before becoming fully mobile. Fin clips or holes might have slowed their movement and thus reduced their susceptibility to capture. Similarly, the transport of marked fish to lake sites away from their location of capture might have reduced their probability of capture the next day.

Thus a variety of factors could have contributed to this bias, but no corrections appear to be necessary even if they were possible. The degree of bias varied from census to census, probably because several factors operated in varying combinations; the ambiguities preclude quantification. Methods developed to correct bias that arises from varying causes (Youngs and Robson 1978; Arnason and Mills 1981, 1987; Seber 1986) require the application of individually recognizable marks. Arnason and Mills (1981) suggested that a correction for bias arising from tag loss is useful if the bias is large and if uncorrected population estimates are highly precise. In the present study, the small differences between the average recapture rates and the expected recapture rates, when compared to the confidence intervals about the population estimates (Tables 2 and 3), indicate that correction for this bias would not be useful.

Recent advances in marking techniques such as the use of coded wire tags or cold branding might allow more precise quantification of the population characteristics of small salmonids in lakes. However, such techniques require more labor than did the present study, which required only three or four people each spring and fall, and they involve a higher rate of tag loss and fish mortality (Pepper and O'Connell 1983; Bourgeois et al. 1987).

In the present study, there was a good correspondence between Atlantic salmon population characteristics derived from the censuses and those derived from sea-run adults returning to spawn. As a result, it was concluded that, for the practical purposes of monitoring the status of salmonid populations, the methods provided reliable estimates and allowed valuable inferences.

Acknowledgments

Many individuals—especially L. J. Cole, D. P. Riche, and D. Wakeham—assisted with the assessment of population structures; H. Mullett and the staff of Peregrine Corporation drafted the figures; R. J. Gibson and M. F. O'Connell constructively reviewed the original manuscript.

References

Arnason, A. N., and K. H. Mills. 1981. Bias and loss of precision due to tag loss in Jolly–Seber estimates for mark–recapture experiments. Canadian Journal of Fisheries and Aquatic Sciences 38:1077–1095.

Arnason, A. N., and K. H. Mills. 1987. Detection of handling mortality and its effects on Jolly–Seber estimates for mark–recapture experiments. Canadian Journal of Fisheries and Aquatic Sciences 44 (Supplement 1):64–73.

Ash, E. G. M., and R. J. Tucker. 1984. Angled catch and effort data for Atlantic salmon in Newfoundland and Labrador, and for other fishes in Labrador, 1982. Canadian Data Report of Fisheries and Aquatic Sciences 465.

Bourgeois, C. E., M. F. O'Connell, and D. C. Scott. 1987. Cold-branding and fin clipping Atlantic salmon smolts on the Exploits River, Newfoundland. North American Journal of Fisheries Management 7:154–156.

Chadwick, E. M. P. 1982. Recreational catch as an index of Atlantic salmon spawning escapement. International Council for the Exploration of the Sea, C.M. 1982/M:43, Copenhagen.

Chadwick, E. M. P. 1985. The influence of spawning stock on production and yield of Atlantic salmon, Salmo salar L., in Canadian rivers. Aquaculture and Fisheries Management 1:111–119.

Chadwick, E. M. P., T. R. Porter, and P. Downton. 1978. Analysis of growth of Atlantic salmon (Salmo salar) in a small Newfoundland river. Journal of the Fisheries Research Board of Canada 35:60–68.

Craig, J. F. 1980. Sampling with traps. EIFAC (European Inland Fisheries Advisory Commission) Technical Paper 33:55–70.

Gibson, R. J., and R. A. Myers. 1988. Influence of seasonal river discharge on survival of juvenile Atlantic salmon, Salmo salar. Canadian Journal of Fisheries and Aquatic Sciences 45:344–348.

Hamley, J. M., and T. P. Howley. 1985. Factors affecting the variability of trapnet catches. Canadian

Journal of Fisheries and Aquatic Sciences 42:1079–1087.

Nickerson, S., P. M. Ryan, G. Somerton, and J. Wright. 1980. A computer program for processing fish age and growth data. Canadian Technical Report of Fisheries and Aquatic Sciences 965.

Pepper, V. A., and M. F. O'Connell. 1983. Coded wire microtag applications in Newfoundland, the first three years. International Council for the Exploration of the Sea, C.M. 1983/M:15, Copenhagen.

Power, G. 1980. The brook charr, *Salvelinus fontinalis*. Pages 141–203 *in* E. K. Balon, editor. Charrs, salmonid fishes of the genus *Salvelinus*. Dr. W. Junk, The Hague, The Netherlands.

Ricker, W. E. 1975. Computation and interpretation of biological statistics of fish populations. Bulletin of the Fisheries Research Board of Canada 191.

Ruggles, C. P. 1980. A review of the downstream migration of Atlantic salmon. Canadian Technical Report of Fisheries and Aquatic Sciences 952.

Ryan, P. M. 1984a. Fyke net catches as indices of the abundance of brook trout, *Salvelinus fontinalis*, and Atlantic salmon, *Salmo salar*. Canadian Journal of Fisheries and Aquatic Sciences 41: 377–380.

Ryan, P. M. 1984b. Age, growth and food of threespine sticklebacks (*Gasterosteus aculeatus*) in three lakes of central Newfoundland, Canada. Canadian Technical Report of Fisheries and Aquatic Sciences 1325.

Ryan, P. M. 1986a. Lake use by wild anadromous Atlantic salmon, *Salmo salar*, as an index of subsequent adult abundance. Canadian Journal of Fisheries and Aquatic Sciences 43:2–11.

Ryan, P. M. 1986b. Prediction of angler success in an Atlantic salmon, *Salmo salar*, fishery two fishing seasons in advance. Canadian Journal of Fisheries and Aquatic Sciences 43:2531–2534.

Ryan, P. M., L. J. Cole, D. P. Riche, and D. Wakeham. 1981. Age and growth of salmonids in the Experimental Ponds Area, central Newfoundland, 1977–80. Canadian Data Report of Fisheries and Aquatic Sciences 304.

Ryan, P. M., and D. Wakeman. 1984. An overview of the physical and chemical limnology of the Experimental Ponds Area, central Newfoundland, 1977–82. Canadian Technical Report of Fisheries and Aquatic Sciences 1320.

Scott, W. B., and E. J. Crossman. 1964. Fishes occurring in the fresh waters of insular Newfoundland. Canada Department of Fisheries and Oceans, Ottawa.

Seber, G. A. F. 1986. A review of estimating animal abundance. Biometrics 42:267–296.

Tesch, F. W. 1971. Age and growth. IBP (International Biological Programme) Handbook 3:98–130 (2nd edition).

Youngs, W. P., D. S. Robson. 1978. Estimation of population number and mortality rates. IBP (International Biological Programme) Handbook 3:137–164. (3rd edition.)

American Fisheries Society Symposium 7:737–745, 1990

LARGE-SCALE MARKING PROGRAMS

Fish-Marking Techniques in New Zealand

TALBOT MURRAY

MAF Fisheries Greta Point
Post Office Box 297, Wellington, New Zealand

Abstract.—Fish-marking techniques used in New Zealand are reviewed. Applications have ranged from studies of population size of brown trout *Salmo trutta* in the 1930s to recent studies of movements and stock discreteness of species on the continental slope at 300–1,200 m, which involved detachable hook tags and parasite faunistic indices. This paper summarizes the chronology of marking techniques and their use with 42 fish species. Of the 107 studies in which marine and freshwater fish have been marked, 42% dealt with movements, 22% with growth, 10% with population size, and 8% with mortality.

Fish-marking experiments have been conducted in New Zealand since 1930 on freshwater fish, since 1954 on marine fish, and since 1973 on marine shellfish. These experiments included studies of movement and growth as well as studies that estimated population parameters and exploitation rates. Types of markers used included external and internal coded tags, clipped fins, vital dyes, and parasite faunas. Early tagging studies were summarized by Allen (1963) for 5 marine and 3 freshwater species and more recently by Crossland (1982b) for 16 marine fish species. The growing importance of fish marking in New Zealand is reflected in the number of species studied since those reviews.

In New Zealand, 35 species of marine sharks and bony fishes, 7 freshwater fish species, and 12 shellfish species have been or are being studied through mark–recapture techniques. Most of these experiments have been conducted on nearshore continental shelf species, but recent applications of marking techniques to study deeper-slope fish (at 300–1,200 m depths) are showing promise.

This paper reviews the New Zealand experience with fish-marking techniques in marine and freshwater environments. Preliminary results of using detachable hook tags to study movements and parasite fauna to study stock heterogeneity of continental slope fish also are presented.

General Features of New Zealand Fish-Marking Experiments

A summary of tagging experiment types, number of species studied, and number of studies of each type is presented in Table 1. The most common studies are of movements and growth, which can be addressed simultaneously and involve the fewest assumptions. Other subjects of interest to researchers differ between freshwater and marine environments.

In New Zealand, most fish marking is conducted in support of fisheries management, and the applications reflect differences in fishery types and potential for dispersal in each environment. In fresh water, recreational fishing interests predominate, but there are small commercial fisheries for eels *Anguilla dieffenbachii* and *A. australis* and ocean ranching for chinook salmon *Oncorhynchus tshawytscha* (called quinnat salmon in New Zealand). In the marine environment, the predominance of commercial fisheries and the greater potential for dispersal increase the importance of stock-related issues. Highly migratory species have been studied effectively only in cooperative international programs. Examples include tagging of skipjack tuna *Euthynnus pelamis* (Argue and Kearney 1983; Kleiber et al. 1987) and studies of albacore *Thunnus alalunga* begun in 1986 (Beardsell 1987).

Freshwater Environments

In studies of brown trout *Salmo trutta* and rainbow trout *Oncorhychus mykiss*, both single and multiple release and recovery designs have been used; the research staff uses fishing or visual sightings during drift diving for recoveries. The largest and longest-running fish-marking experiments in fresh water have been directed at assess-

TABLE 1.—Applications of fish-marking techniques in New Zealand.

Purpose of tagging	Number of species studied		Number of studies
	Freshwater	Marine	
Movement	6	34	46
Growth	7	9	23
Population size	4	2	11
Mortality	1	2	8
Tag shedding	1	3	7
Age validation	0	4	4
Stock heterogeneity	0	3	3
Exploitation rate	0	1	2
Fisheries interactions	0	1	1
Hatchery returns	1	0	1
Trap efficiency	1	0	1

ing returns of sea-run, hatchery-raised chinook salmon. This program is being conducted at several South Island hatcheries. From 1977 to 1983, over 3,000,000 juveniles were adipose fin-clipped and tagged subcutaneously with binary-coded wire tags (Unwin 1985). The purpose of this ongoing program is primarily to determine the return rates of 2-, 3-, and 4-year-old fish as a function of release size. Returns from the recreational sector are encouraged by a reward and by entry into a lottery for those who return the heads of fin-clipped chinook salmon.

Marine Environments

Fish-marking studies of demersal species in the marine environment have largely been short-term, single-release experiments in which recoveries are made opportunistically by recreational and commercial fishermen. A well-publicized nationwide reward and lottery system is the primary inducement to return tags. Differences in underreporting among fisheries generally have not been estimated and few double-tagging experiments have been carried out. The reliance on fishermen to return tags, the lack of estimates of tag shedding, and the nonreporting of tags have limited interpretations of tag returns to studies of growth rate and patterns of movement. Recent tagging experiments, primarily on the New Zealand snapper *Chrysophrys auratus* (family Sparidae) on which spaghetti lock-on tags were used, have focused on the estimation of stock sizes available to recreational and commercial fishing sectors. These experiments include estimation of tag shedding, and initial tagging mortality, and they incorporate tetracycline injections to validate otolith daily growth rings. About 25,000 fish in one experiment and 4,700 fish in another have been

tagged. Recoveries to date have been about 25%, and tag shedding also has been estimated at about 25%. Although New Zealand snapper frequently has been studied by mark–recapture techniques, the present studies are the first to estimate initial mortality and to stratify releases by major commercial fishing methods and recoveries by fishing sector. The high percentage of fish recovered is evidence of the effectiveness of program advertisement.

Pelagic marine tagging experiments have concentrated on coastal species and, more recently, on the highly migratory tuna and game-fish species. Game fish have been tagged opportunistically during a small big-game fishing season since 1975. Initially part of a tagging program run by the U.S. National Marine Fisheries Service's Southwest Fisheries Center, game-fish tagging is now conducted by the New Zealand Ministry of Agriculture and Fisheries. However, relatively few fish have been tagged—3,476 sharks, 1,010 marlins, 343 swordfish, 564 tuna, and 2,584 kingfish between 1975 and 1990—and 101 sharks, 3 marlin, 1 tuna, and 423 kingfish have been recaptured. Recent concern over the by-catch of billfish and potential interactions between the recreational fishery and foreign licensed tuna longline and domestic commercial fisheries led New Zealand to impose mandatory tagging on nonrecreational fishermen. The intention of the new regulations is to increase tagging of billfish.

Tagging of tunas, primarily skipjack tuna and albacore, continues to be part of cooperative international programs. In the case of skipjack tuna, all tagging in New Zealand has been part of the research initiative by the South Pacific Commission's Tuna and Billfish Assessment Programme to study population dynamics and fisheries interactions in the southwest Pacific. Albacore tagging is also part of a south Pacific study by New Zealand, the USA, France, and the South Pacific Commission. The program was begun in 1986 to determine general features of albacore movement in the south Pacific and to validate banding sequences in otoliths and other hard parts. The albacore tagging program will continue for several years before significant numbers of tag returns can be expected. The time lag is due to the age difference between fish captured for tagging by trolling and taken by the major fishery. Young albacore are captured by surface trolling with artificial lures that have single barbless hooks; up to 70% of fish caught in this manner are suitable for tagging. Through 1990, approximately 6,650 albacore had been tagged, and 7 tags recovered.

TABLE 2.—Summary of fish-marking techniques used in New Zealand.

Method of marking	Date used	Species (see Appendix)
External tags and marks		
Group marks		
Fin clipping	1930–present	11, 12, 13, 15
Vital staining	1967–1977	9, 10, 11, 14, 15
Low aspect tags		
Petersen disk	1939–1966	12, 13, 25, 39, 40, 41
Operculum band	1940–1955	25, 29
Loop tags		
Crimped spaghetti loop	1964	40
Tied spaghetti loop	1964–1980	8, 19, 20, 26, 40
Spaghetti lock-on	1974–present	3, 4, 19, 24, 25, 27, 28, 30, 37
Streamer tags		
Tuna dart tags	1972–present	4, 25, 31, 32, 33
Spaghetti streamers	1973–present	7, 11, 12, 13, 21, 25, 29, 37
Small dart tags	1974	25
H-anchor streamers	1975–present	1, 2, 5, 6, 23, 32, 33, 34, 35, 36
Detachable hook tags	1987–present	17, 18, 38
Internal tags and marks		
Subcutaneous coded wire	1956–present	9, 10, 11
Parasite fauna	1980–present	16, 31, 32
Radio transmitters	1985	11

Future directions of this program include the design of experiments to estimate mortality and fishery interactions.

Chronology of Fish-Marking Methods

External Tags and Marks

The chronology of fish-marking applications and the species that have been studied with a particular technique are summarized in Table 2.

The earliest and most widespread fish-marking technique in New Zealand was adipose fin-clipping of trout and salmon. This technique was first used in 1930 on parr during the introduction of chinook salmon to the South Island (Flain 1971). The ease and utility of this technique for identifying a group of fish ensure its continuation. It is extensively used in New Zealand at present to identify hatchery-reared, sea-run chinook salmon that carry coded wire tags. Other fin-clipped species include brown and rainbow trout and the galaxiid *Neochanna apoda*. With trout, removal of the adipose fin has been extensively used to assess tag shedding; with *Neochanna apoda* the top or bottom lobe of the caudal fin has been removed as a batch identification.

Another technique for identifying a group of fish, especially small fish or fry, is the injection of (or immersion in) vital stains or dyes. This technique has been used in fresh water on two species

of eel elvers (Jellyman and Ryan 1983), newly emergent chinook salmon (Unwin 1986b), and juvenile galaxiids ranging in size from 36 to 155 mm total length (Hopkins 1971; Eldon 1978). The purpose of the first two studies was to estimate sampling efficiency and movement, and that of the latter two studies was to estimate growth rate. In the study by Hopkins (1971), patterns of dye were injected to establish individual growth histories of 60 juveniles. The dyes and stains used included India ink, national fast turquoise PT, cadmium sulfide, neutral red, and Bismarck brown Y.

The introduction of Petersen disk tags in 1939 by Allen (1951) in his classic study on brown trout greatly expanded the range of studies that could be carried out through tagging. Petersen disks were in use from 1939 to 1966 with freshwater and marine species (see Table 2) for studies of population size, movement, and growth. Growth studies in particular became more common once these tags allowed individual fish to be recognized after periods at liberty. A related tag type, the opercular band, was used between 1940 and 1955, primarily on snapper and blue cod *Parapercis colias*. This tag was abandoned after independent experiments showed that it had low visibility and retention relative to the Petersen disk tag (Crossland 1982b).

The chief modification in the use of Petersen tags in New Zealand involved methods of attachment. Two disks tied together, one on each side of the fish, to increase the probability of detection during commercial fishing operations. The difficulty of attachment, according to Crossland (1982b), was the reason for discontinuing the Petersen tag.

The introduction of spaghetti loop tags—which are easier to apply, more visible, and less likely to be lost through entanglement in nets or stream debris—also contributed to the replacement of Petersen tags. Spaghetti loop tags are readily attached by passing a coded plastic tube directly through the musculature of the fish and joining the loose ends. This tag type has been in continuous use in New Zealand since 1964. Several methods of attaching the ends have been tested. Crimping was tried in 1964 but abandoned because it was slower than tying (Colman 1978). Tying the two ends was generally employed from 1964 to 1980; a figure-of-eight knot was preferred to an overhand knot because it held better (Crossland 1982b). The ease of attachment improved with the introduction of a preformed locking mechanism on loop tags. These were first employed in New Zealand in 1974 and have been extensively used since 1980 with snapper and other species. Initial locking mechanisms appeared to fail with time and in general had lower retention than tied loops; Crossland (1982b) reported retention up to 2.5 years for lock-on tags, in contrast to 7 years for tied loop tags. Unpublished data from tagging experiments on snapper suggest that tag loss may be as high as 25% with the lock-on loop tag.

The introduction of streamer tags in the early 1970s has been followed by wide application, which continues primarily in studies of movement and growth of albacore. Four types of streamer tags have been used, primarily on marine species. A fifth type is being developed for in situ tagging of fish living at depths of 300–1,200 m. Streamer tag types differ in size and in the method of anchoring the tag, which is inserted into the musculature with a barbed, T-shaped, or H-shaped head. The tag is extremely easy to apply and has good retention, particularly when the anchoring head is lodged between bones or pterygiophores. This tag also has the advantage of being equally suited to bodies that are laterally compressed or rounded in cross-section, whereas loop tags are less suited to the latter.

In fresh water, streamer tags have been used to study brown and rainbow trout, and to identify chinook salmon bearing radio tags. Fourteen species of marine fish have been or are being studied with streamer tags. The majority of these species are being tagged opportunistically in fairly small numbers by recreational game fishermen. The largest program using streamer tags is on albacore and has already been described.

Internal Tags and Marks

Subcutaneous coded tags have been used extensively in studies of movement and hatchery returns. The use of coded internal tags began in 1956 with studies on two species of eels. In these experiments, serially numbered tags were applied successfully and the problem of tag loss was overcome (Burnet 1968, 1969b). These early subcutaneous tags were described by Burnet (1969b) as flat, $10 \times 2 \times 0.6$-mm, stainless steel strips stamped with a series of numbers. To apply them, researchers anesthetized the eel, cut through the skin with a 2-mm-wide knife blade, formed a subcutaneous pocket with a blunt probe, and then inserted the tag. Tags were all inserted in the same relative position and were detected subsequently with a small electronic metal detector. Eels tagged by this method were recovered up to 10 years later.

Since 1977, coded wire tags etched with a binary code have been implanted in chinook salmon juveniles to estimate hatchery returns (Unwin 1985). All salmon bearing a coded wire tag also have the adipose fin removed to facilitate recognition of the tagged fish.

Internal radio transmitters have been only briefly used but provided detailed information for short periods on the movement of migrating chinook salmon adults in response to river flow (Glova and Docherty 1986).

The use of parasite fauna as intrinsic marks successfully demonstrated a recent tropical origin for migrating juvenile skipjack tuna (Lester et al. 1985) and for albacore. These inferences are based on the presence of large numbers of a common group of tropical parasites of scombrids (didymozoid trematodes and others) that are not found in endemic species. The parasites are common in small, migrating skipjack tuna and albacore, but are uncommon or absent in larger tuna caught around New Zealand.

New Developments in Fish Marking in New Zealand

Hook Tagging

As part of a new study of movements of continental slope species, a tagging technique based on detachable hooks has been developed. Initial trials with bluenose *Hyperoglyphe antarctica* and alfonsino *Beryx splendens* showed that even species with partially vented gas bladders will, when caught at depths of 300–600 m and landed in good condition, die before they can be measured and tagged on deck (Horn and Massey 1987). This led to the design of a new streamer tag. The tag consists of a yellow plastic tube 10 cm long, with return information and individual number, attached by a 10-cm length of stainless steel wire to a baited longline circle-hook on a breakaway monofilament trace. The breaking strength of the trace is crucial to the success of this tagging method and should match the strength of the target species. A breaking strain of 5.5 kg is ideal for bluenose (Horn 1988). Because tagging is done in situ at depths below 300 m, the species tagged is unknown until recapture, and initial data consist of location of hook loss and tag number. Control fishing is used to estimate the probable species mix of the tagged population. The technique of hook tagging circumvents two problems inherent in capture of continental slope fish species: trawl damage and gas expansion with subsequent eversion of the viscera. The latter has been a recurrent problem even in continental shelf species able to vent their gas bladders, such as snapper and groper *Polyprion oxygeneios* (Crossland 1976; Johnston 1983).

Horn (1988) described details of hook tagging. He reported that in the first year of tagging, 2,122 hook tags were released and 36 were returned as of 25 February 1988, 34 from the tagging area and 2 that indicated fish movements of 45 and 490 km in 3 and 4.5 months, respectively. Attempts to extend hook tagging to orange roughy *Hoplostethus atlanticus* were unsuccessful because fish in spawning aggregations did not take baited hooks.

Defining Spawning-Stock Discreteness in Deep-Slope Species

Parasite faunistic indices constructed by principal component analysis and other multivariate statistical techniques are being investigated for use on orange roughy. Evidence suggests that orange roughy form large spawning aggregations in discrete localities, although movements and extents of population exchange during the rest of the year are unknown. Studies thus far suggest that fish from widely separated spawning aggregations exhibit definable differences in their parasite fauna. If these preliminary analyses are corroborated, parasite fauna is likely to be useful as a composite biological marker with which to study stock discreteness.

Acknowledgments

I am very much indebted to the scientific staff of the Ministry of Agriculture and Fisheries for helping locate published and unpublished information on fish tagging. Although it is not possible to acknowledge everyone who aided me, I particularly want to thank Larry Paul, Vaughn Wilkinson, Derek Parkinson, and Phillip Kirk for providing information on snapper and other marine species. Details and results of hook tagging were kindly furnished by Peter Horn and information on parasites by Brian Jones. Information on freshwater species was supplied by Don Jellyman, Peter Todd, and Martin Unwin. Information on big game fish tagging was supplied by Peter Saul. I am also grateful for the considerable editorial assistance provided by Bob Kendall, his associates, and two anonymous reviewers.

References[1]

1. Allen, K. R. 1951. The Horokiwi stream, a study of a trout population. New Zealand Marine Department Fisheries Bulletin 10.
2. Allen, K. R. 1963. A review of tagging experiments in New Zealand. International Commission for the Northwest Atlantic Fisheries Special Publication 4:140–141.
3. Annala, J. H. 1987a. Report of the 1986 snapper research workshop held at Fisheries Management Division, Wrightson House, Auckland, 6–7 October 1986. New Zealand Ministry of Agriculture and Fisheries, Fisheries Research Division, Internal Report 62, Wellington.
4. Annala, J. H. 1987b. The biology and fishery of tarakihi, *Nematadactylus macropterus*, in New Zealand waters. New Zealand Ministry of Agriculture and Fisheries, Fisheries Research Division, Occasional Publication 26, Wellington.
5. Anonymous. 1975. James Cook surveys albacore resource. Catch '75 2(3):8–10. (New Zealand Ministry of Agriculture and Fisheries, Wellington.)
6. Argue, A. W., and R. E. Kearney. 1983. An assessment of the skipjack and baitfish resources of New Zealand. South Pacific Commission, Skipjack and Assessment Programme, Final Country Report 6, Noumea, New Caledonia.
7. Bagley, N., and R. Hurst. 1987. Southern shelf species surveyed. Catch 14(7):23, 25–26. (New Zealand Ministry of Agriculture and Fisheries, Wellington.)

---

[1]Numbers are used to facilitate citations given in the appendix.

8. Beardsell, M. 1987. Albacore tagging. Catch 14(2):23–34. (New Zealand Ministry of Agriculture and Fisheries, Wellington.)

9. Boyd, R. 1979. Low tag returns could threaten fishery. Catch '79 6(9):11. (New Zealand Ministry of Agriculture and Fisheries, Wellington.)

10. Burnet, A. M. R. 1968. A study of the relationships between brown trout and eels in a New Zealand stream. New Zealand Marine Department, Fisheries Technical Report 26.

11. Burnet, A. M. R. 1969a. Territorial behaviour in brown trout (*Salmo trutta* L.). New Zealand Journal of Marine and Freshwater Research 3:385–388.

12. Burnet, A. M. R. 1969b. The growth of New Zealand freshwater eels in three Canterbury streams. New Zealand Journal of Marine and Freshwater Research 3:376–384.

13. Burnet, A. M. R. 1970. Seasonal growth in brown trout in two New Zealand streams. New Zealand Journal of Marine and Freshwater Research 4:55–62.

14. Colman, J. A. 1974. Movements of flounders in the Hauraki Gulf, New Zealand. New Zealand Journal of Marine and Freshwater Research 8:79–93.

15. Colman, J. A. 1978. Tagging experiments on the sand flounder, *Rhombosolea plebeia* (Richardson), in Canterbury, New Zealand, 1964 to 1966. New Zealand Ministry of Agriculture and Fisheries, Fisheries Research Bulletin 18.

16. Crossland, J. 1976. Snapper tagging in north-east New Zealand, 1974: analysis of methods, return rates, and movements. New Zealand Journal of Marine and Freshwater Research 10:675–686.

17. Crossland, J. 1980. Population size and exploitation rate of snapper, *Chrysophrys auratus*, in the Hauraki Gulf from tagging experiments, 1975–76. New Zealand Journal of Marine and Freshwater Research 14:255–261.

18. Crossland, J. 1982a. Movements of tagged snapper in the Hauraki Gulf. New Zealand Ministry of Agriculture and Fisheries, Fisheries Research Division, Occasional Publication 35, Wellington.

19. Crossland, J. 1982b. Tagging of marine fishes in New Zealand. New Zealand Ministry of Agriculture and Fisheries, Fisheries Research Division, Occasional Publication 33, Wellington.

20. Cudby, E. J., and R. R. Strickland. 1986. The Manganuioteao River fishery. New Zealand Ministry of Agriculture and Fisheries, Fisheries Environmental Report 14, Wellington.

21. Docherty, C. 1985. "Bugged" salmon movements monitored. Freshwater Catch (Winter 1985):7–8. (New Zealand Ministry of Agriculture and Fisheries, Wellington.)

22. Drummond, K. 1987. Nelson snapper tagging progress. Catch 14(1):14–15. (New Zealand Ministry of Agriculture and Fisheries, Wellington.)

23. Drummond, K., and P. Kirk. 1987. Nelson snapper tagging programme continues. Catch 14(9):27–28. (New Zealand Ministry of Agriculture and Fisheries, Wellington.)

24. Eldon, G. A. 1978. The life history of *Neochanna apoda* Gunther (Pisces: Galaxiidae). New Zealand Ministry of Agriculture and Fisheries, Fisheries Research Bulletin 18.

25. Eldon, G. A. 1986. Nomadic eel. Freshwater Catch 29:36. (New Zealand Ministry of Agriculture and Fisheries, Wellington.)

26. Eldon, G. A., and A. J. Greager. 1983. Fishes of the Rakaia lagoon. New Zealand Ministry of Agriculture and Fisheries, Fisheries Environmental Report 30, Wellington.

27. Field-Dodgson, M. S., and J. R. Galloway. 1985. The Glenariffe Salmon Research Station. New Zealand Ministry of Agriculture and Fisheries, Fisheries Research Division, Fisheries Information Leaflet 13, Wellington.

28. Fish, G. R. 1963. Limnological conditions and growth of trout in three lakes near Rotorua. New Zealand Ecological Society Proceedings 10:1–7.

29. Fish, G. R., R. L. Allen, and H. S. Fairburn. 1968. An examination of the trout population of five lakes near Rotorua, New Zealand. New Zealand Journal of Marine and Freshwater Research 2:333–362.

30. Flain. 1971. Early recoveries of fin-clipped quinnat salmon in New Zealand (note). New Zealand Journal of Marine and Freshwater Research 5:516–518.

31. Francis, M. P. 1988. Movement patterns of rig (*Mustelus lenticulatus*) tagged in southern New Zealand. New Zealand Journal of Marine and Freshwater Research 22:259–272.

32. Gauldie, R. W., and A. Nathan. 1977. Iron content of the otoliths of tarakihi (Teleostei: Cheilodactylidae). New Zealand Journal of Marine and Freshwater Research 11:179–191.

33. Glova, G., and C. Docherty. 1986. Waimakariri-radio tracking of adult salmon. Freshwater Catch 29:4–5. (New Zealand Ministry of Agriculture and Fisheries, Wellington.)

34. Habib, G., I. T. Clement, and K. A. Fisher. 1980a. The 1978–79 purse-seine skipjack fishery in New Zealand waters. New Zealand Ministry of Agriculture and Fisheries, Fisheries Research Division, Occasional Publication 26, Wellington.

35. Habib, G., I. T. Clement, and K. A. Fisher. 1980b. The 1979–80 purse-seine skipjack fishery in New Zealand waters. New Zealand Ministry of Agriculture and Fisheries, Fisheries Research Division, Occasional Publication 29, Wellington.

36. Habib, G., I. T. Clement, and K. A. Fisher. 1981. The 1980–81 purse-seine skipjack fishery in New Zealand waters. New Zealand Ministry of Agriculture and Fisheries, Fisheries Research Division, Occasional Publication 36, Wellington.

37. Hanchet, S. M. 1986. The distribution and abundance, reproduction, growth, and life history characteristics of the spiny dogfish, *Squalus acanthias* (Linnaeus), in New Zealand. Doctoral dissertation. University of Otago, Dunedin, New Zealand.

38. Hardy, C. J. 1968. Freshwater fisheries field techniques: tagging; transportation; mortality and drift sampling. New Zealand Marine Department, Fisheries Technical Report 27.

39. Hopkins, C. L. 1971. Life history of *Galaxias divergens* (Salmonoidea: Galaxiidae). New Zealand Journal of Marine and Freshwater Research 5:41–57.

40. Horn, P. 1988. Tagging bluenose and alfonsino with detachable hooks. Catch 15(1):17–18. (New Zealand Ministry of Agriculture and Fisheries, Wellington.)

41. Horn, P. 1989. An evaluation of the technique of tagging alfonsino and bluenose with detachable hook tags. New Zealand Ministry of Agriculture and Fisheries, Fisheries Technical Report 16.

42. Horn, P., and B. Massey. 1987. Tagging deepwater fish—trials with a new technique. Catch 14(4):21–22. (New Zealand Ministry of Agriculture and Fisheries, Wellington.)

43. Hurst, R. J., and N. W. Bagley. 1989. Movements and possible stock relationships of the New Zealand barracouta, *Thyrsites atun*, from tag returns. New Zealand Journal of Marine and Freshwater Research 23:105–111.

44. Ichikawa, W. 1981. Report of the albacore survey by the RV *Kaio Maru* no. 52 in New Zealand waters, 1981. Japan Marine Fishery Resource Research Center Report 20, Tokyo.

45. Iwasa, K., G. Habib, and G. I. T. Clement. 1982. Report of the albacore survey by the RV *Kaio Maru* no. 52 in New Zealand waters, 1982. Japan Marine Fishery Resource Research Center, Report 18, Tokyo.

46. James, G. D. 1980. Tagging experiments on trawl-caught trevally, *Caranx georgianus*, off north-east New Zealand, 1973–79. New Zealand Journal of Marine and Freshwater Research 14:249–254.

47. Jellyman, D. J. 1977. Freshwater eels—an important resource. Commercial Fishing (April):10–11. (Federation of Commercial Fishermen, Wellington, New Zealand.)

48. Jellyman, D. J., and C. M. Ryan. 1983. Seasonal migration of elvers (*Anguilla* spp.) into Lake Pounui, New Zealand, 1974–1978. New Zealand Journal of Marine and Freshwater Research 17:1–15.

49. Johnston, A. 1983. The southern Cook Strait groper fishery. New Zealand Ministry of Agriculture and Fisheries, Fisheries Technical Report 159.

50. Kearney, R. E. 1983. Assessment of the skipjack and baitfish resources in the central and western tropical Pacific Ocean: a summary of the Skipjack Survey and Assessment Programme. South Pacific Commission Report, Noumea, New Caledonia.

51. Kearney, R. E., and J.-P. Hallier. 1979. Interim report of the activities of the Skipjack Survey and Assessment Programme in the waters of New Zealand (17 February–27 March 1979). South Pacific Commission, Skipjack Survey and Assessment Programme, Preliminary Country Report 16, Noumea, New Caledonia.

52. Kleiber, P., A. W. Argue, and R. E. Kearney. 1983. Assessment of skipjack (*Katsuwonus pelamis*) resources in the central and western Pacific by estimating standing stock and components of population turnover from tagging data. South Pacific

Commission, Tuna and Billfish Assessment Programme, Technical Report 8, Noumea, New Caledonia.

53. Kleiber, P., A. W. Argue, and R. E. Kearney. 1987. Assessment of Pacific skipjack tuna (*Katsuwonus pelamis*) resources by estimating standing stock and components of population turnover from tagging data. Canadian Journal of Fisheries and Aquatic Sciences 44:1122–1134.

54. Lawson, T. A., R. E. Kearney, and J. R. Sibert. 1984. Estimates of length measurement errors for tagged skipjack (*Katsuwonus pelamis*) from the central and western Pacific Ocean. South Pacific Commission, Tuna and Billfish Assessment Programme, Technical Report 11, Noumea, New Caledonia.

55. Lester, R. J. G., A. Barnes, and G. Habib. 1985. Parasites of skipjack tuna, *Katsuwonus pelamis*: fishery implications. U.S. National Marine Fisheries Service Fishery Bulletin 83:343–356.

56. Mace, J., and K. Drummond. 1982. Tagging gives more answers. Catch '82 9(10):27. (New Zealand Ministry of Agriculture and Fisheries, Wellington.)

57. Mace, J., and A. D. Johnston. 1983. Tagging experiments on blue cod (*Parapercis colias*) in the Marlborough Sounds, New Zealand. New Zealand Journal of Marine and Freshwater Research 17:207–211.

58. McKensie, M. K. 1960. Fish of the Hauraki Gulf. New Zealand Ecological Society Proceedings 7:45–49.

59. Palmer, K. L. 1987. Adult trout in the demonstration channels, lower Waitaki River, 1982–85. New Zealand Ministry of Agriculture and Fisheries, Fisheries Environmental Report 81, Wellington.

60. Paul, L. J. 1967. An evaluation of tagging experiments on the New Zealand snapper, *Chrysophrys auratus*, (Forster), during the period 1952 to 1963. New Zealand Journal of Marine and Freshwater Research 1:455–463.

61. Paul, L. J. 1976. A study on age, growth, and population structure of the snapper, *Chrysophrys auratus*, (Forster), in the Hauraki Gulf, New Zealand. New Zealand Ministry of Agriculture and Fisheries, Fisheries Research Bulletin 13.

62. Pepperell, J. G. 1985. Cooperative game fish tagging in the Indo-Pacific region. Pages 241–252 *in* R. H. Stroud, editor. World angling resources and challenges. International Game Fish Association, Fort Lauderdale, Florida.

63. Pollard, J. 1969. Tagging in New Zealand and tagging of elephant fish. Pages 791–793 *in* J. Pollard, editor. Australian and New Zealand fishing. Paul Hamlyn, Sydney, Australia.

64. Rapson, A. M. 1956. Biology of the blue cod (*Parapercis colias* Forster) of New Zealand. Doctoral dissertation. Victoria University, Wellington, New Zealand.

65. Roberts, P. E. 1974. Albacore off the north-west coast of New Zealand, February 1972. New Zealand Journal of Marine and Freshwater Research 8:455–472.

66. Todd, P. 1986. 1985 quinnat salmon hatchery returns. Freshwater Catch 29:10. (New Zealand Ministry of Agriculture and Fisheries, Wellington.)

67. Tong, L. J. 1978. Tagging snapper *Chrysophrys auratus* by scuba divers. New Zealand Journal of Marine and Freshwater Research 12:73–76.

68. Tunbridge, B. R. 1966. Growth and movements of tagged flatfish in Tasman Bay. New Zealand Marine Department, Fisheries Technical Report 13.

69. Unwin, M. J. 1985. Salmon coded-wire tagging results, 1977–84. New Zealand Ministry of Agriculture and Fisheries, Fisheries Research Division, Occasional Publication 47:18–24, Wellington.

70. Unwin, M. J. 1986a. Release programme brings early success. Freshwater Catch 29:9. (New Zealand Ministry of Agriculture and Fisheries, Wellington.)

71. Unwin, M. J. 1986b. Stream residence time, size characteristics, and migration patterns of juvenile chinook salmon (*Oncorhynchus tshawytscha*) from a tributary of the Rakaia River, New Zealand. New Zealand Journal of Marine and Freshwater Research 20:231–252.

72. Unwin, M. J., and F. Lucas. 1984. Salmon tag returns for 1983/84. Freshwater Catch 24:11–14. (New Zealand Ministry of Agriculture and Fisheries, Wellington.)

73. Unwin, M. J., and F. Lucas. 1986. 1985 salmon tag returns. Freshwater Catch 29:7–8. (New Zealand Ministry of Agriculture and Fisheries, Wellington.)

74. Unwin, M. J., D. H. Lucas, and T. Gough. 1987. Coded-wire tagging of juvenile chinook salmon (*Oncorhynchus tshawytscha*) in New Zealand, 1977–86. New Zealand Ministry of Agriculture and Fisheries, Fisheries Technical Report 2.

75. Wood, B. A., M. A. Bradstock, and G. D. James. 1990. Tagging of kahawai *Arripis trutta* in New Zealand, 1981–84. New Zealand Ministry of Agriculture and Fisheries, Fisheries Technical Report 19.

Appendix: New Zealand Mark–Recapture Studies

TABLE A.1.—Phylogenetic list of fish species that have been tagged and marked in New Zealand programs. Species numbers are those used in Table 2. Reference numbers correspond to those in the main bibliography.

Family and species tagged	Reference
Lamnidae	
1. *Isurus oxyrinchus*	18
Alopiidae	
2. *Alopias vulpinus*	Unpublished
Carcharhinidae	
3. *Mustelus lenticulatus*	31
4. *Galeorhinus australis*	7
5. *Prionace glauca*	Unpublished
Sphyrnidae	
6. *Sphyrna zygaena*	Unpublished
Squalidae	
7. *Squalus acanthias*	37
Callorhinchidae	
8. *Callorhinchus milii*	63, unpublished
Anguillidae	
9. *Anguilla dieffenbachii*	3, 10, 11, 25, 26, 47, 48, 63
10. *Anguilla australis*	10, 11, 26, 47, 48, 63
Salmonidae	
11. *Onchorhynchus tshawytscha*	21, 26, 27, 30, 33, 66, 69, 70, 71, 72, 73, 74
12. *Oncorhynchus mykiss*[a]	3, 20, 26, 28, 29, 38, 59
13. *Salmo trutta*	2, 3, 10, 12, 13, 20, 26, 38, 59, 63
Galaxiidae	
14. *Galaxias divergens*	39
15. *Neochanna apoda*	24
Trachichthyidae	
16. *Hoplostethus atlanticus*	Unpublished
Berycidae	
17. *Beryx splendens*	40, 41
Scorpaenidae	
18. *Helicolenus papilosus*	40, 41
Percichthyidae	
19. *Polyprion oxygeneios*	18, 49

Table A.1.—Continued.

Family and species tagged	Reference
Carangidae	
20. *Caranx georgianus*	18, 46
21. *Trachurus* spp.	Unpublished
22. *Seriola lalandei*	18, unpublished
Coryphaenidae	
23. *Coryphaena hippurus*	Unpublished
Arripidae	
24. *Arripis trutta*	75
Sparidae	
25. *Chrysophrys auratus*	3, 5, 9, 16–19, 22, 23, 56, 58, 60, 61, 63, 67
Cheilodactylidae	
26. *Nemadactylus macropterus*	3, 4, 18, 32, 63
Latrididae	
27. *Latridopsis ciliaris*	18
Mugilidae	
28. *Mugil cephalus*	Unpublished
Mugiloididae	
29. *Parapercis colias*	18, 57, 64
Gempylidae	
30. *Thyrsites atun*	43
Scombridae	
31. *Katsuwonus pelamis*[b]	6, 18, 34–36, 44, 45, 50–55
32. *Thunnus alalunga*	6, 8, 18, 44, 45, 65, unpublished
33. *Thunnus albacares*	45, unpublished
Istiophoridae	
34. *Tetrapturus audax*	18, unpublished
35. *Makaira nigricans*	Unpublished
36. *Makaira indica*	18, 62, unpublished
Centrolophidae	
37. *Seriolella brama*	Unpublished
38. *Hyperoglyphe antarctica*	40
Pleuronectidae	
39. *Rhombosolea leporina*	3, 14, 18, 68
40. *Rhombosolea plebeia*	3, 14, 15, 18, 63, 68
41. *Pelotretis flavilatus*	3, 18, 63, 68
42. *Peltorhamphus novaezeelandiae*	18

[a]Formerly *Salmo gairdneri*.
[b]Formerly *Euthynnus pelamis*.

American Fisheries Society Symposium 7:746–764, 1990
© Copyright by the American Fisheries Society 1990

History of the ICCAT Tagging Program, 1971–1986

P. M. Miyake

International Commission for the Conservation of Atlantic Tunas
Príncipe de Vergara 17, Madrid, Spain

Abstract.—The International Commission for the Conservation of Atlantic Tunas (ICCAT) started an International Cooperative Tagging Program in 1971. Most of the tagging cruises have been financed by national funds. However, the commission has funded some cruises, and it provides national laboratories with tagging materials. Sustained international efforts to publicize the program, including a reward system and an annual lottery, have helped to establish a good image for the program among fishermen, longshoremen, and cannery workers. From 1971 through 1986, 140,255 tunas and tuna-like fishes were tagged and released in the Atlantic Ocean, of which 8,957 fish (6.4%) were recaptured. High recovery rates have been recorded often for yellowfin tuna *Thunnus albacares*, bigeye tuna *T. obesus*, and skipjack tuna *Katsuwonus pelamis*, but the times these fish were at liberty usually were relatively short. In contrast, bluefin tuna *T. thynnus* and billfish recoveries often have been made many years after release. The mark–recapture data have provided scientists with valuable information on migration, stock structure, growth, and natural and fishing mortalities.

The International Commission for the Conservation of Atlantic Tunas (ICCAT) was founded in 1969 based on the International Convention for the Conservation of Atlantic Tunas. The commission's objectives, set forth by the convention, are to sustain maximum yields of tunas and tuna-like species in the Atlantic Ocean and its adjacent seas. The commission maintains a limited scientific staff at its secretariat, but most of the research is conducted by scientists at laboratories of the member countries.

This paper reports on the organization, nature, and scope of the commission's International Cooperative Tagging Program, and it introduces recent work with nontraditional tags. It does not present new analyses of tagging results. Scientific names of the species involved are given in Table 1.

The International Cooperative Tagging Program

At the initiation of its scientific work in 1970, the commission decided to start a multipurpose International Cooperative Tagging Program, based on traditional tags, for tunas and tunalike species in the Atlantic Ocean. The data obtained from the program have provided information on stock structure, migration, growth, and natural and fishing mortality.

Even before this program was under way, many tunas and tuna-like fishes had been tagged in the Atlantic Ocean by Norway (Hamre 1963), Canada (Beckett 1970), Spain (Rodríguez-Roda 1969), and the USA (Mather et al. 1973; Mather and Mason 1973). Many tagging cruises had also taken place

in the Pacific Ocean. Of these, the most important were carried out by the Inter-American Tropical Tuna Commission (IATTC) on tropical tunas (yellowfin and skipjack tuna) as early as 1952 (Fink and Bayliff 1970; Bayliff 1979).

Because tunas and tuna-like species are highly migratory, and because they are exploited with various gears at different ages by fishermen of many countries, an international rather than a national tagging program was considered essential. Moreover, to release even one tagged fish is expensive, especially in the case of tunas and billfishes, which are of high commercial value. International funding and internationally shared mark–recapture data would be necessary to obtain meaningful results.

At a meeting of tagging experts held in Lisbon in 1971, the commission agreed on tagging techniques, the types of tags to be used, an accounting system for releases and recaptures, a collaborative planning system, and publicity and incentives (Anonymous 1971). Experience in tagging techniques gained by other organizations, particularly IATTC, served as the basis for developing the ICCAT tagging system. However, there is an important logistical difference between the ICCAT tagging program, which depends mainly on national resources, and IATTC tagging, which is carried out by that commission's own scientific staff with its own funds.

The ICCAT program is set up so that at the meetings of the commission's scientific body, the Standing Committee on Research and Statistics, basic plans are laid out and results are reported. National laboratories are responsible for carrying out tagging activities. The national laboratories that have participated in the program at one time or another are given in the Appendix.

The commission assists with these activities by ordering tagging materials in bulk, storing the materials, and providing them free of charge to member countries upon request. The Secretariat is in charge of publicity and serves as the reporting center for recovered tags. For specific projects, the commission also provides funds to charter fishing boats, to purchase fuel for research boats, to compensate fishermen for fish released in opportunistic tagging (Anonymous 1986a), and to train inexperienced taggers.

Methods

Tags

Tags used.—For tunas and small tunalike fishes, dart tags are used (Figure 1). They consist of nylon dart heads glued to a polyvinyl tube streamer about 10–11 cm long, with an external diameter of 2.3 mm. Because fish caught by long-liners are stored at temperatures of minus 50°C, a special polyvinyl material resistant to temperatures must be used. A serial number (combination of one or two letters and a four- or five-digit number) is printed twice on the streamer. Tags ordered by ICCAT are labeled "ICCAT CP 542 MADRID SPAIN" on one side and "RECOMPENSA REWARD" on the other. Tags ordered by a national laboratory usually bear the laboratory's own name and mailing address. The streamers are generally yellow, but when the fish are injected with tetracycline for growth studies, red tags are used.

For large billfishes, including swordfish, darts made of stainless steel attached to a yellow polyvinyl tube, with the same inscription as that on the nylon dart tags, are used. This tag was formerly used for bluefin tuna as well, until Baglin et al. (1980) found that they shed it slightly more often than the nylon head. Stainless steel darts are also cheaper and easier to apply when fish are brought aboard.

Number of tags applied per fish.—Lenarz et al. (1973) and Baglin et al. (1980) analyzed tag-shedding data for Atlantic bluefin tuna and found that tag loss by shedding was important. Bayliff and Mobrand (1972) also concluded that tag shedding was important for Pacific yellowfin tuna and that the recapture rate could be increased by double

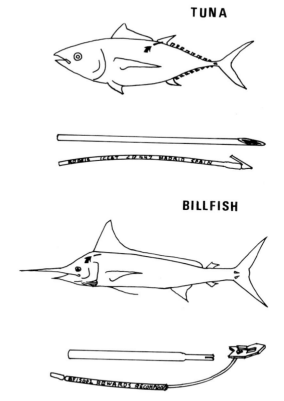

FIGURE 1.—Tags and applicators used by the commission for its International Cooperative Tagging Program, and the location on the fish where tags are applied.

tagging. To minimize tag loss by shedding, the commission recommended applying two tags per fish unless the fish are extremely small (less than 30 cm in length). Double tagging has been practiced widely on bluefin tuna but less frequently on tropical tunas and billfishes. (Yellowfin tuna have routinely been double-tagged by the IATTC in the Pacific Ocean since 1967, however.) Bayliff and Holland (1986) reported that double tagging is useful not only to increase the return rate but to estimate natural and fishing mortality and growth rate.

Applicators.—For the nylon dart tags, researchers use cold-drawn hypoflex stainless steel tubes about 17 cm long and 3.3 mm in inside diameter. For large tuna and billfish, special stainless steel applicators to hold the stainless steel dart heads are used (Figure 1).

Fishing gear used in tagging.—The types of gear used to catch fish for tagging purposes are as follows: For tropical tunas (yellowfin, bigeye, and skipjack), pole and line with live bait and sometimes purse seine; for albacore, troll; for billfishes, rod and reel; for juvenile bluefin tuna, purse seine or pole and line; and for large bluefin tuna, rod and reel or trap.

Tagging techniques.—Tags are applied after fish are brought aboard the vessel, with the exception of large bluefin tuna (over 150 kg) or billfishes. Fish brought aboard are either laid on top of a cradle or held under the arm of a tagger, and then one or two tags are applied. The latter method requires less time but the size of released fish can only be estimated, whereas fish received on a cradle can be more accurately measured. In most large-scale tagging operations, a tape recorder with a remote control switch is used to record tag numbers, species, and sizes. The tags are applied near the posterior end of the base of the first dorsal fin, so that the dart can be well-anchored between pterygiophores (Figure 1). When two tags are applied to one fish, they are placed on both sides.

Large bluefin tuna or billfish caught on a line are kept in the water while being tagged. They are brought close to the side of the vessel and a tag is applied to the base of the anterior part of the first dorsal fin by means of an applicator attached to the end of a long pole, after which the line is cut to release the fish (Figure 1). Fish size (length or weight) is estimated.

Incentives for Reporting Recoveries

Publicity.—One important feature of this International Cooperative Tagging Program is that the tagging publicity is well organized through international collaboration. The ICCAT program has some advantage in obtaining wide publicity because many national offices of major tuna fishing countries are directly involved in tagging activities.

The secretariat prepares posters to announce the commission's tagging programs and the rewards to be paid for reporting recaptures of tagged fish. These posters are printed in English, French, Spanish, Arabic, Chinese, Japanese, Korean, and Portuguese and are distributed widely among fishermen, port authorities, laboratories, agencies at landing and trans-shipment ports, and canneries. An example of one such poster is given in Figure 2.

Rewards.—Two basic rules concerning rewards have been observed by the commission: (1) A uniform reward is paid for all Atlantic tags inasmuch as possible; and (2) tag returns are received by someone at the port who can verify the recovery data and pay the reward on the spot. The

importance of these criteria was discussed by Bayliff and Holland (1986).

When the program started, the commission decided on a reward of US$2 for each tag returned (Anonymous 1971), which was the same as that paid by IATTC. The reward was increased in 1977 to $4 (Anonymous 1978) and has been maintained at that level ever since. If a fish is doubled tagged and both tags are returned, two rewards of $4 are paid. For red tags on fish injected with tetracycline, if returned with the fish, an additional $16 is paid. The increase in the reward to $4 by ICCAT was not discussed with other agencies beforehand and caused some confusion in Puerto Rico, where Pacific tags (which still receive a $2 reward) as well as Atlantic tags are returned. Because one port sampler receives both tags at that port, the problem is not too serious but if two samplers were involved, it would become so. At any rate, the recoverers of Pacific tags are dissatisfied. It would be more desirable if uniform rewards were adopted worldwide because many tuna boats fish in two and often three oceans.

Payment of rewards is, in principle, assumed by the releasing agency. However, for special ICCAT-funded, large-scale tagging programs (e.g., the International Skipjack Year Program and the Yellowfin Year Program), the commission pays all rewards, if so prearranged (Anonymous 1983b).

Arrangements are made so that monetary rewards or T-shirts are given to the recoverers immediately upon receipt of the recovered tags. For this purpose, reward funds (or sometimes T-shirts) are deposited at key ports by the commission, and a person is selected to administer them. At the end of the year, the commission is reimbursed for rewards that should have been assumed by a national laboratory or institute. Collection of recovered tags at ports and payment of rewards on the spot are effective: more than 90% of all tags recovered are collected at ports and only 10% are sent to the address on the tags.

During the International Skipjack Year Program, the standard reward for a tag return was a specially designed T-shirt (Anonymous 1986c). The T-shirts proved more popular than the cash in some areas. Because of accounting difficulties, particularly with the different sizes of shirts that had to be distributed, this reward was abandoned once the Skipjack Year Program was completed in 1983. Some national offices, however, decided to use their own funds to continue the T-shirt rewards. T-shirt rewards are given at certain Afri-

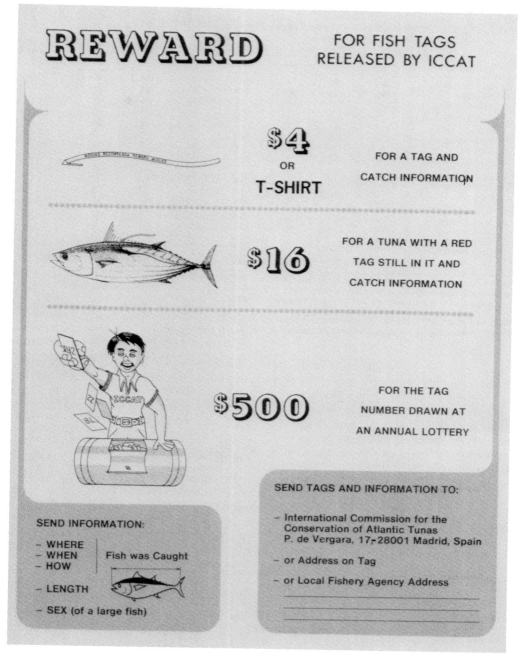

FIGURE 2.—Tag publicity poster prepared for the commission's International Skipjack Year Program. The poster was available in various languages.

can ports, regardless of the origin of tags, and there has not been any problem.

Tag lotteries.—Since the start of the Commission's tagging program, an annual lottery has been held as an additional incentive to return tags. All lottery prizes are paid out of the commission's budget. In 1971 and 1972, a US$300 reward was paid each year to the winning number. Tag recoveries received during one year at the national laboratories or the secretariat, and reported to the

TABLE 1.—Numbers of tunas and tunalike fishes, by species, released in the Atlantic Ocean and Mediterranean Sea under the ICCAT International Cooperative Tagging Program, 1971–1986.

	Species[a]	Year of release								
		1971	1972	1973	1974	1975	1976	1977	1978	1979
(1)	Bluefin tuna	423	280	599	1,633	523	2,434	2,195	1,773	1,124
(2)	Albacore	540	1,618	1,808	240	509	784	595	142	140
(3)	Yellowfin tuna	2,028	3,519	1,601	27	1,120	186	178	236	87
(4)	Bigeye tuna	17	0	450	24	592	3	217	838	10
(5)	Skipjack tuna	821	1,537	175	17	166	34	66	240	108
(6)	Tropical tunas[b]	250	0	0	600	0	0	0	0	0
(7)	Little tunny	36	21	12	4	39	9	12	132	103
(8)	Atlantic bonito	4	7	11	17	15	13	47	408	412
(9)	Blackfin tuna	88	27	71	54	57	44	122	71	20
(10)	Frigate mackerel	0	0	0	0	0	0	0	32	0
(11)	Swordfish	2	0	0	9	4	0	28	110	47
(12)	Blue marlin	121	117	104	95	96	142	172	308	283
(13)	White marlin	1,015	483	280	264	457	310	337	837	739
(14)	Sailfish	1,079	934	934	872	1,024	1,509	1,490	1,594	1,881
(15)	Spearfish	1	0	0	2	0	0	1	21	28
(16)	King mackerels	2	0	2	8	11	16	10	3	1
Total		6,427	8,543	6,047	3,866	4,613	5,484	5,470	6,745	4,983

[a]Bluefin tuna = *Thunnus thynnus*; albacore = *Thunnus alalunga*; yellowfin tuna = *Thunnus albacares*; bigeye tuna = *Thunnus obesus*; skipjack tuna = *Katsuwonus pelamis*; little tunny = *Euthynnus alletteratus*; Atlantic bonito = *Sarda sarda*; blackfin tuna = *Thunnus atlanticus*; frigate mackerel = *Auxis thazard*; (broadbill) swordfish = *Xiphias gladius*; blue marlin = *Makaira nigricans*; white marlin = *Tetrapturus albidus*; (Atlantic) sailfish = *Istiophorus albicans*; spearfish = *Tetrapturus pfluegeri* and *T. belone*; king mackerels = *Scomberomorus* spp.
[b]Yellowfin tuna or bigeye tuna; could not be distinguished at the time of release.

secretariat by a certain deadline date of the following year, are eligible for the lottery. In 1973, the lottery was expanded to two $300 rewards, one for tropical tunas (yellowfin, bigeye, and skipjack) and one for temperate species (the rest of the species, including billfishes). Since 1981, these two lottery prizes have been increased to $500. In addition to the regular annual lottery, the following special lottery prizes ($500 each) have been added: (1) two prizes, one each for east and west Atlantic skipjack tuna tag recoveries, during the International Skipjack Year Program, 1981 to 1983; (2) one prize for yellowfin tuna tag recoveries during the Yellowfin Year Program, 1987 and 1988; and (3) one prize for billfish tag recoveries during the Program of Enhanced Research for Billfish, 1988 and at least until 1989.

After 18 years of operation and publicity, the Atlantic tuna tag recovery lottery has become well known among fishermen and people working at ports. One lottery winner in Africa bought a camel with the reward, and another a house. Often, the ceremony of delivering the winning check is reported in local or national newspapers.

The lottery also serves as an incentive for national laboratories to report all the recoveries they receive to the commission. Were it not for the lottery, the ICCAT tagging data base would include a lesser percentage of the total recoveries.

Feedback to recoverers.—The recoverers are informed of the date and location of releases of the tagged fish they have reported. The commission also reports results of its tagging activities, along with general knowledge obtained from tagging, to the public via a newsletter.

Tag-Accounting System

The tag-accounting system of the commission is rather complicated and has undergone several changes, mostly because tagging cruises have been conducted by national laboratories whereas recaptures come from all the Atlantic ports and canneries. In some cases different procedures have been adopted during a specific project. The most recent accounting system was described by Miyake and Symons (1983) and Anonymous (1983b).

Tag inventory.—Most of the tags used are ordered in bulk by the commission and provided to laboratories upon request. The secretariat keeps a record of all tag numbers distributed to each laboratory. However, there are some laboratories that order tags on their own. Once a year, a survey is made to update the tag inventory, and this is distributed to all people concerned. The inventory assists people who receive tag recovery reports in identifying the agency responsible for releasing the fish.

TABLE 1.—Extended.

				Year of release				
	1980	1981	1982	1983	1984	1985	1986	Total
(1)	3,511	2,092	617	862	141	529	899	19,635
(2)	214	19	65	292	240	21	214	7,441
(3)	2,076	1,083	359	501	537	204	2,581	16,323
(4)	1,416	959	9	138	344	9	1,124	6,150
(5)	8,525	14,019	9,860	293	1,082	270	400	37,613
(6)	0	3,555	0	0	0	0	0	4,405
(7)	68	25	214	5	20	756	238	1,694
(8)	527	3	1	2	6	394	0	1,867
(9)	35	433	326	285	84	71	64	1,852
(10)	0	0	0	0	0	12	0	44
(11)	303	289	95	92	91	202	316	1,588
(12)	490	436	379	417	521	627	784	5,092
(13)	973	779	930	1,023	1,010	898	877	11,212
(14)	2,153	1,862	1,652	1,875	2,239	1,915	2,171	25,184
(15)	18	2	0	2	2	3	5	85
(16)	8	2	2	1	4	0	0	70
Total	20,317	25,558	14,509	5,788	6,321	5,911	9,673	140,255

Tag release information.—Laboratories are asked to report releases of tagged fish after each cruise. In these reports they specify the tag numbers used, the number of fish released by species, the general area of operation, and the time. In addition, at the end of the year, countries are asked to submit a table containing all information on tuna tagging. Reporting at the end of a tagging cruise is irregular, but the annual report is submitted by most member countries. The year-end report should include date, location (in degrees and minutes of latitude and longitude), fishing gears used, species, tag number, whether one or two tags were applied, estimated or measured length or weight of fish, and surface water temperature (Miyake and Hayasi 1978).

Tag recovery information.—Recoveries of tags reported to the responsible person at ports (laboratory researchers or technicians, port authorities, cannery administration, fish-handling agencies, etc.) often can be transmitted directly to the laboratories that released the fish with the help of a tag inventory list. Other recoveries are reported directly or transmitted to the secretariat, whose post office box number is printed on the tag. These tags are then forwarded to the releasing agency by the secretariat.

At the end of the year, national offices report to the commission all tag recoveries received during the year, together with release information, and these are eligible for the lottery. Each recovery report should include tag number, date, location (in degrees and minutes of latitude and longitude), fishing gears employed, species, length or weight of fish, whether one or two tags were found on the same fish, how or where the tag was found (while fishing, in the fish well, cannery, etc.) and the nationality of the recoverer (Miyake and Hayasi 1978).

Tagging Data Base

The secretariat of the commission has been asked to maintain a complete data base of tag releases and recoveries for Atlantic tunas and tuna-like fishes. If ICCAT actually conducted all the tagging cruises—as does IATTC or the South Pacific Commission (SPC)—this could be achieved easily. However, it is not possible in ICCAT because, as mentioned above, not all the laboratories submit complete data to the commission.

Tagging Activities in the Past

Planning of Tagging Cruises

Generally, tagging plans are presented at the scientific meetings of the commission. Once scientists identify the need for tagging, the scope of such activities are evaluated for their effectiveness and costs are estimated.

In organizing tagging, generally four types of tagging cruises are considered. They are cruises with research vessels; charter-boat cruises; opportunistic cruises with commercial fishing boats; and fishermen's voluntary-tagging cruises.

Research vessel cruises (R/V).—Tagging from research vessels requires the least additional ex-

TABLE 2.—Numbers of tunas and tuna-like fishes, by country, released in the Atlantic Ocean and Mediterranean Sea under the ICCAT International Cooperative Tagging Program, 1971–1986.

Country	Year of release								
	1971	1972	1973	1974	1975	1976	1977	1978	1979
Brazil							127	74	18
Canada	33		172	65	185	28		6	2
Cape Verde									
Côte d'Ivoire[a]	2,800	5,000	2,157	600	1,763	116		1,094	
Cuba									
France	524	1,585	1,619	226	545	577	796		137
Japan	17						23		
Korea									
Morocco		60	90		10		121	16	
Portugal									
Senegal	250								
South Africa	35	6	184	36					
Spain		22				211	93	875	584
USA	2,768	1,870	1,825	2,939	2,110	4,552	4,310	4,680	4,242
USSR									
Venezuela									
Total	6,427	8,543	6,047	3,866	4,613	5,484	5,470	6,745	4,983

[a]Some of these tags were released during joint cruises between Côte d'Ivoire and France.
[b]Joint cruises between Côte d'Ivoire and Ghana.

pense because boats are made available by the national government. This type of tagging has the advantage that fish can be tagged at designated times and areas, even outside the fishing area and season. However, most research boats are not designed for tuna fishing, nor do they have experienced fishing crews. Therefore, this type of cruise usually tags fewer fish than a commercial fishing boat.

Charter-boat cruises (charter).—Charter boats are costly, particularly in the case of good tuna-fishing boats. If the plan is to tag fish in the middle of a productive fishing ground at the peak period, an experienced skipper and crew could provide a sufficient number of fish. This type of tagging has the advantage that the boat can be taken to any place at any time. If the plan, as often is the case, is to tag fish outside a fishing area and season, there is a risk of spending a large amount of money only to obtain poor results.

Opportunistic cruises (opportunistic).—Opportunistic tagging is not as costly or risky as charter-boat tagging. Generally, two taggers (one to fish and one to tag) are allowed on a commercial fishing vessel to tag fish while fishermen are engaged in commercial fishing. The limitations are that there is no choice in the tagging location or time; also, fishermen often object to releasing fish (particularly on bait-boats) during a good fishing period, because they believe that fish returned to sea take other fish with them to the depths. Generally, the vessel owner must be compensated for released fish.

Bard et al. (1987) compared opportunistic tagging cruises with research vessel cruises. The advantages and disadvantages of these cruises must be evaluated for each occasion, with various economic elements and tagging objectives taken into consideration.

Voluntary-tagging cruises (voluntary).—Some tuna fishermen and most recreational fishermen are willing to collaborate with scientists to provide a part of their catches for the purpose of tagging. Many recreational fishermen even tag and release the fish. The cost of tagging is thus reduced, but release sites and size estimates must be verified before the data can be used.

Major Tagging Operations

The major tagging schemes that have been or are being carried out by the commission under its cooperative program are as follows:

• juvenile bluefin tuna tagging to determine the exchange rate between eastern and western Atlantic stocks, 1975–1981 (opportunistic, R/V, and voluntary; nationally funded);

• bluefin tuna tagging in the Mediterranean Sea and northeast Atlantic to investigate the relationship between young fish found in the Bay of Biscay and spawning and juvenile fish in the Mediterranean, 1980 to present (R/V; nationally funded);

• yellowfin and bigeye tuna tagging to study stock structure, 1971–1973 (R/V and chartered; nationally funded);

TABLE 2.—Extended.

Country	1980	1981	1982	1983	1984	1985	1986	Total
				Year of release				
Brazil		52	30	10	13			324
Canada								491
Cape Verde		2,685	4,566					7,251
Côte d'Ivoire[a]	1094	4,466		313	1,540	69	2,249[b]	23,261
Cuba		574	673	339				1,586
France			11				1,236	7,256
Japan	7,964	7,519						15,523
Korea	131	399	319					849
Morocco								297
Portugal		11	85			10		106
Senegal	567	2,539	3,002	51		1,132		7,541
South Africa								261
Spain	1,654	1,122	2,240	1,106	480	701	1,396	10,484
USA	8,786	5,315	3,583	3,969	4,288	3,999	4,492	63,728
USSR	121	876						997
Venezuela							300	300
Total	20,317	25.558	14,509	5,788	6,321	5,911	9,673	140,255

- yellowfin tuna tagging to study fishing mortality and interaction between surface and longline gears, 1986–1987 (opportunistic and R/V; partially funded by the commission);
- skipjack tuna tagging during the International Skipjack Year Program to study skipjack stock structure, migration, fishing, and mortality and growth, 1980–1982 (R/V, chartered, and opportunistic; partially funded by the commission);
- billfish (Istiophoridae) tagging by recreational fishermen to identify stocks and their ranges, 1971 to present (voluntary);
- billfish tagging under the Program of Enhanced Research for Billfish, starting in 1988 (opportunistic and voluntary; partially funded by the commission).

Results

Actual Tag Releases

In Table 1 are shown the numbers of fish released, by species, in the Atlantic Ocean and the Mediterranean Sea from 1971 to 1986 under the International Cooperative Tagging Program. This table is based on all the records registered at the commission's headquarters. However, data could be underestimated because of the failure of some national offices to report fish released but not recaptured. Double tags, as far as we know, are counted as one fish. Some reports lack identification of the second tag, and therefore some fish carrying double tags could be counted twice.

In 16 years, over 140,000 fish have been released under this program. The majority of the over 37,000 tagged skipjack tuna were released during the International Skipjack Year Program (1980–1982) (Anonymous 1986c). This program also contributed to major releases of yellowfin and bigeye tunas, because both species are often caught with skipjack tuna.

Temperate species (approximately 20,000 bluefin tuna, 25,000 sailfish, and 11,000 white marlin) have been released more regularly over the 16 years. More of the temperate tunas are recovered after longer periods at liberty than are tropical tunas, so mark–recapture data would give better information on mortality, growth, and movements is available for temperate species.

In Table 2 are shown the total numbers of fish released, by country, for the same period and the same area. The USA released the greatest number of fish (mostly temperate species), followed by the joint French–Côte d'Ivoire tagging program and then Japan (the latter two tagged mostly tropical species). Nearly half of the total releases were made in the 3-year period (1980–1982) when both the International Skipjack Year Program and the juvenile bluefin tuna tagging plan were being carried out.

Recoveries Reported

In Table 3 are shown the total numbers of fish recovered, by species, for each year of release. Recoveries in 1987 are not included in this table because reports were not yet complete.

In Table 4 are shown the recovery rates, by species, for each year of release. The rates for

TABLE 3.—Numbers of tunas and tuna-like fishes, recovered in the Atlantic Ocean and Mediterranean Sea under the ICCAT International Cooperative Tagging Program, 1971–1986.

	Species	Year of release								
		1971	1972	1973	1974	1975	1976	1977	1978	1979
(1)	Bluefin tuna	113	74	108	271	71	333	343	240	69
(2)	Albacore	40	23	11	0	9	8	3	11	2
(3)	Yellowfin tuna	16	161	153	24	14	1	0	19	2
(4)	Bigeye tuna	3	10	113	13	1	1	11	207	0
(5)	Skipjack tuna	5	16	9	14	17	0	0	12	12
(6)	Little tunny	0	0	1	0	0	0	0	1	2
(7)	Atlantic bonito	0	0	0	0	0	0	3	34	11
(8)	Blackfin tuna	1	0	3	2	0	0	0	0	0
(9)	Swordfish	0	0	0	0	0	0	1	10	1
(10)	Blue marlin	0	1	0	1	0	1	2	1	3
(11)	White marlin	14	5	11	4	6	5	3	12	9
(12)	Sailfish	3	6	23	11	19	23	46	27	35
(13)	King mackerels	0	0	0	0	0	0	0	0	0
Total		195	296	432	340	137	372	412	574	146

years when relatively few releases were reported could be considerably overestimated because the failure of a country to report releases could have a serious effect on the calculation of recovery rates. The identification of juvenile yellowfin and bigeye tunas is very difficult, particularly if done hurriedly as is the case at the time of release. Species identification at the time of recovery is more reliable. Consequently, these species were occasionally lumped as tropical tunas or counted as yellowfin tuna in the release data. Therefore, the recovery rate for bigeye tuna, and possibly yellowfin tuna, is most likely considerably overestimated.

In Table 5 are shown the accumulated numbers of fish released and recaptured and the rates of recapture made in the calendar year of release and the next three calendar years. The years do not directly correspond to, but are indicative of, the time at liberty. Although analysis of the results is not the purpose of this paper, it may be noted that the recovery rates for tropical tunas (yellowfin, bigeye, and skipjack tunas) were high in the first year and rapidly decreased with time. Temperate tuna recoveries were distributed over many years after release. The longest period at liberty recorded in the Atlantic was for a white marlin recovered 18 years after release. The longer periods at liberty observed for temperate tunas are most likely attributable to their lower natural mortality or emigration rates, or both. However, many temperate tunas carry metal head tags, which are not used on any of the tropical tunas, and quite possibly these metal tags are retained better. This has to be further investigated.

The high recovery rate for skipjack tuna contradicted the current belief that this species has

TABLE 4.—Percentages of recovery of tunas and tuna-like fishes released in the Atlantic Ocean and Mediterranean Sea under the ICCAT International Cooperative Tagging Program, 1971–1986.

	Species	Year of release								
		1971	1972	1973	1974	1975	1976	1977	1978	1979
(1)	Bluefin tuna	26.7	26.4	18.0	16.6	13.6	13.7	15.6	13.5	6.1
(2)	Albacore	7.4	1.4	0.6	0.0	1.8	1.0	0.5	7.7	1.4
(3)	Yellowfin tuna	0.8	4.6	9.6	88.9	1.3	0.5	0.0	8.1	2.3
(4)	Bigeye tuna	17.6	?	25.1	54.2	0.2	33.3	5.1	24.7	0.0
(5)	Skipjack tuna	0.6	1.0	5.1	82.4	10.2	0.0	0.0	5.0	11.1
(6)	Little tunny	0.0	0.0	8.3	0.0	0.0	0.0	0.0	0.8	1.9
(7)	Atlantic bonito	0.0	0.0	0.0	0.0	0.0	0.0	6.4	8.3	2.7
(8)	Blackfin tuna	1.1	0.0	4.2	3.7	0.0	0.0	0.0	0.0	0.0
(9)	Swordfish	0.0	0.0	0.0	0.0	0.0	0.0	3.6	9.1	2.1
(10)	Blue marlin	0.0	0.9	0.0	1.1	0.0	0.7	1.2	0.3	1.1
(11)	White marlin	1.4	1.0	3.9	1.5	1.3	1.6	0.9	1.4	1.2
(12)	Sailfish	0.3	0.6	2.5	1.3	1.9	1.5	3.1	1.7	1.9
(13)	King mackerels	0.0	0.0	0.0	0.0	0.0	0.0	0.0	0.0	0.0
Total		3.0	3.5	7.1	8.8	3.0	6.8	7.5	8.5	2.9

TABLE 3.—Extended.

	Year of release								
	1980	1981	1982	1983	1984	1985	1986	?	Total
(1)	288	118	15	58	32	13	38	3	2,187
(2)	6	3	0	25	1	0	2	0	144
(3)	160	352	5	41	27	4	89	1	1,069
(4)	112	287	0	3	2	0	64	0	827
(5)	591	1,114	1,971	2	97	27	42	0	3,929
(6)	7	0	2	0	1	2	6	0	22
(7)	35	1	0	0	3	0	0	0	87
(8)	1	6	2	22	11	9	3	0	60
(9)	15	22	4	1	1	1	0	0	56
(10)	1	3	10	4	0	3	1	0	31
(11)	14	15	9	18	15	10	1	1	152
(12)	40	41	30	13	22	31	21	1	392
(13)	0	0	0	0	1	0	0	0	1
Total	1,270	1,962	2,048	187	213	100	276	6	8,957

been underexploited while the other major species have been fully exploited (Anonymous 1986c). Most skipjack tuna recoveries were short-term returns during the International Skipjack Year Program, when the majority of the program's releases took place on important fishing grounds during the height of the season. According to Cayré et al. (1986), 1,502 (80%) of 1,882 recovered skipjack tuna that had been released in the Dakar and Cape Verde areas during the skipjack program were at liberty less than 5 d.

Application of Tag Results to Stock Studies

Stock Structure and Migration

Because mark–recapture data provide direct evidence of migration, stock-structure studies by the commission depend largely on tagging results. Syn-

opses of the commission's latest findings on stock structure of major tuna species are given below.

Billfishes.—Considerable information on migration, stock structure, mortality, and growth are available for billfishes through the recreational fishermen's cooperative tagging program. This program and its results are reported elsewhere in this volume (Scott et al. 1990).

Bluefin tuna.—Atlantic bluefin tuna have two distinct spawning grounds, in the Mediterranean Sea and the Gulf of Mexico. Mather and Mason (1973) summarized data on transatlantic bluefin tuna migrations. Turner (1986) studied stock structure and mixing of eastern and western Atlantic bluefin tuna, based on tag recoveries. The commission's scientific view on the bluefin tuna stock is that "the exchange of fish between the

TABLE 4.—Extended.

	Year of release							
	1980	1981	1982	1983	1984	1985	1986	Total
(1)	8.2	5.6	2.4	6.7	22.7	2.5	4.2	11.1
(2)	2.8	15.8	0.0	8.6	0.4	0.0	0.9	1.9
(3)	7.7	32.5	1.4	8.2	5.0	2.0	3.4	6.5
(4)	7.9	29.9	0.0	2.2	0.6	0.0	5.7	13.4
(5)	6.9	7.9	20.0	0.7	9.0	10.0	10.5	10.4
(6)	10.3	0.0	0.9	0.0	5.0	0.3	2.5	1.3
(7)	6.6	33.3	0.0	0.0	50.0	0.0	0.0	4.7
(8)	2.9	1.4	0.6	7.7	13.1	12.7	4.7	3.2
(9)	5.0	7.6	4.2	1.1	1.1	0.5	0.0	3.5
(10)	0.2	0.7	2.6	1.0	0.0	0.5	0.1	0.6
(11)	1.4	1.9	1.0	1.8	1.5	1.1	0.1	1.4
(12)	1.9	2.2	1.8	0.7	1.0	1.6	1.0	1.6
(13)	0.0	0.0	0.0	0.0	25.0	0.0	0.0	1.4
Total	6.3	7.7	14.1	3.2	3.4	1.7	2.8	6.4

TABLE 5.—Numbers of tunas and tuna-like species released and recaptured, and percentages of recaptures (in parentheses), for 1971–1986 releases.

Species	Total released	Year of recovery				Total recovered
		1st	2nd	3rd	4th	
Bluefin tuna	19,635	913 (4.6)	859 (4.4)	309 (1.6)	106 (0.5)	2,187 (11.1)
Albacore	7,441	74 (1.0)	32 (0.4)	23 (0.3)	15 (0.2)	155 (1.9)
Yellowfin tuna	16,323	845 (5.2)	204 (1.2)	16 (0.1)	4 (.0)	1,069 (6.5)
Bigeye tuna	6,150	609 (9.9)	210 (3.4)	6 (0.1)	2 (.0)	827 (13.4)
Skipjack	37,613	3486 (9.3)	433 (1.2)	10 (.0)	0 0.0	3,929 (10.4)
Little tunny	1,694	18 (1.1)	3 (0.2)	1 (0.1)	0 0.0	22 (1.3)
Atlantic bonito	1,867	63 (3.4)	19 (1.0)	3 (0.2)	2 (0.1)	87 (4.7)
Blackfin tuna	1,852	37 (2.0)	19 (1.0)	2 (0.1)	2 (0.1)	60 (3.2)
Swordfish	1,588	6 (0.4)	15 (0.9)	13 (0.8)	22 (1.4)	56 (3.5)
Blue marlin	5,092	5 (0.1)	18 (0.4)	2 (.0)	6 (0.1)	31 (0.6)
White marlin	11,212	33 (0.3)	66 (0.6)	30 (0.3)	23 (0.2)	152 (1.4)
Sailfish	70	1 (1.4)	0 0.0	0 0.0	0 0.0	1 (1.4)
Tropical tunas[a]	4,405	0	0	0	0	0
Frigate mackerel	44	0	0	0	0	0
Spearfish	85	0	0	0	0	0
Total	140,255	6,252 (4.5)	2,034 (1.5)	467 (0.3)	204 (0.1)	8,957 (6.4)

[a] Yellowfin and bigeye tunas; not distinguished at release.

eastern and western Atlantic has been estimated to be less than 10 percent. Tagging data indicate that such exchanges are variable through time" (Anonymous 1985a, 1985b). Recently, two first-time reports of fish released in the western Atlantic and recaptured in the Mediterranean Sea have been received (Anonymous 1987b). In Figure 3 are presented the major long-distance recaptures throughout bluefin tuna tagging history (1954–1986).

Skipjack tuna.—Cayé et al. (1986) and Miyabe and Bard (1986) studied skipjack tuna stock structure, based on tagging results of the International Skipjack Year Program. There have been no transatlantic tag recoveries. However, the western Atlantic skipjack tuna fisheries have developed only since the late 1970s and are still in the process of development. The commission's Standing Committee on Research and Statistics decided to use a two-stock (east and west) hypothesis. All the findings of the skipjack program were summarized, and the migratory pattern of eastern Atlantic skipjack was presented as shown in Figure 4 (Anonymous 1986c).

Bigeye tuna.—Miyabe (1987) analyzed all tag recoveries for bigeye tuna in the Atlantic. Combining these results with catch distributions, the commission hypothesized a single stock of bigeye tuna in the entire Atlantic Ocean (Anonymous 1987a). In Figure 5 is shown the commission's hypothesis on stock structure.

Yellowfin tuna.—Bard and Cayré (1986) studied all tag recoveries from yellowfin tuna (mostly juveniles) in the Atlantic up to 1983. These recoveries suggested no transatlantic movement of fish, but indicated that fish migrate mostly along the west coast of Africa. In Figure 6 are shown migration patterns of juvenile yellowfin tuna suggested by these tag recoveries. Because the western Atlantic yellowfin tuna fishery started only in the early 1980s, the opportunity to recover fish released in the western Atlantic has been limited. In 1986, however, two yellowfin tuna tagged in the eastern Atlantic were recaptured in the western Atlantic, the first documentation of transatlantic migration (Bard 1987). The commission is currently working under two alternative stock hypotheses, one postulating a single Atlantic yellowfin tuna stock, the other two stocks (one each in the eastern and western Atlantic).

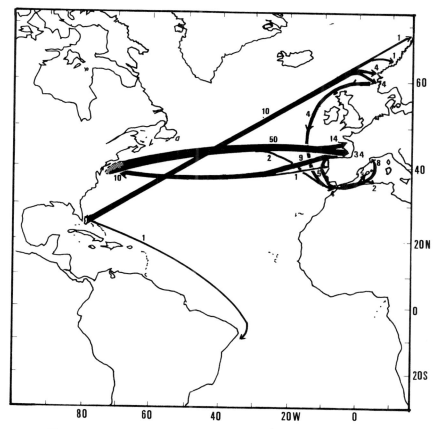

FIGURE 3.—Long-distance recoveries of tagged bluefin tuna, 1954–1986.

Growth

Many growth studies have been based on mark–recapture data. Most of these used estimated length (or weight) of fish at the time of release and at recapture. In addition, there have been over 20 usable recaptures of 566 tetracycline-injected skipjack and yellowfin tunas with special red tags (the tags were returned with adequate data, and the fish were made available for removal of otoliths). Otoliths collected from those fish are being analyzed (Anonymous 1986c).

Bluefin tuna.—Several estimates of bluefin tuna growth have been made from mark–recapture data (e.g., Mather et al. 1973). The growth equations derived by Parrack and Phares (1979) are currently used to age western Atlantic bluefin tuna.

Yellowfin tuna.—It has been known that growth estimated by mark–recapture data for young yellowfin tuna is very slow and does not fit the curve estimated for adult fish. Fonteneau (1980) and Bard (1984) concluded that this slow growth rate

is real, and that two separate curves should be fitted, one to younger fish and one to adults. Miyabe (1984), however, regarded the growth estimates for young fish as biased.

Bigeye tuna.—Cayré (1984) used data on sizes at release and recapture of 243 bigeye tuna to estimate growth curves for northern and southern fish. The curves differed little, and both were close to growth curves estimated from other data sources.

Skipjack tuna.—Bard et al. (1983) presented preliminary estimates of skipjack tuna growth based on mark–recapture data, and these were further developed by Cayré et al. (1986). The estimates generally agreed with those from other analyses, such as of the progression of modal lengths.

Swordfish.—Although data on the sizes at release and recapture for swordfish are limited to 63 fish, including fish from years prior to the start of the ICCAT program, growth rates have been estimated from these data. These show much more rapid growth than do the data from hard

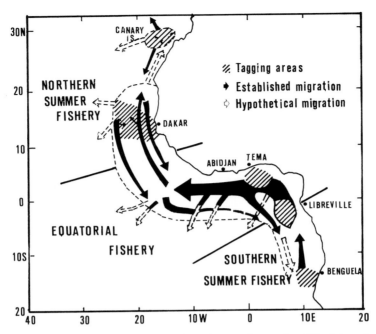

FIGURE 4.—Eastern Atlantic skipjack tuna migration based on tagging results of the International Skipjack Year Program (Anonymous 1986c).

parts. The commission has decided to use the estimates based on mark–recapture data for aging (Anonymous 1988).

Fishing Mortality

Many estimates of fishing mortality have been made from tag recovery data. Because bluefin tuna have been tagged for many years and have produced many recoveries after more than a year at liberty, and because the fishing effort for them was variable during the period under review, they were used for mortality studies (Mather et al. 1973; Farber 1980). Fishing mortality estimates from mark–recapture data have been applied for virtual population analysis of bluefin tuna (Anonymous 1985a, 1987b). Miyabe (1983) used tag recovery data for juvenile bluefin tuna to evaluate the strength of the 1973 cohort, and concluded that it was a dominant year-class.

The objective of the commission's Yellowfin Year Program was to evaluate the effects of a sudden reduction in fishing effort in the eastern Atlantic after 1984, caused by withdrawal of many French purse-seiners to the Indian Ocean. Tagging has been the major method the commission adopted to estimate fishing mortality (Anonymous 1986a). The program was recently terminated, and analyses are now being carried out.

Nontraditional Tags and Marks on Tuna

Parasites

MacKenzie (1983) described a systematic approach to the selection of parasites for use as biological tags to trace bluefin tuna in the Atlantic. He proposed using encysted trypanorhynchan larvae and juvenile acanthocephalans as possible marks.

Microconstituents in Hard Parts

Calaprice (1986) analyzed bluefin tuna collected from fisheries in the western and eastern Atlantic and Mediterranean Sea through ICCAT coordination to estimate the rate at which fish interchange between these areas. He used proton-induced X-ray spectra to analyze inorganic composition in different vertebral formations. Preliminary analysis suggested some interchange between the eastern and western Atlantic and that more fish emigrated from east to west than vice versa. Because these preliminary results were promising, the commission tried to finance further studies, but the high cost of the analyses has delayed this research (Anonymous 1986b).

Acoustic Tagging

Several cruises dedicated to ultrasonic tracking of tuna have taken place in the Atlantic. They

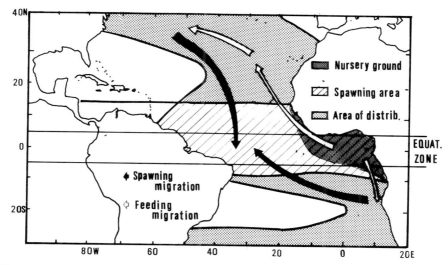

FIGURE 5.—Stock-structure hypotheses adopted for Atlantic bigeye tuna (Anonymous 1987a).

were partially funded by the commission and supported by some national funding. Bard and Pincock (1982) reported that small skipjack tuna were limited to water above the thermocline, and Levenez (1982) reported that large skipjack tuna dive deep across the thermocline.

Others

There are reports that many tuna carrying hooks of Cape Verde fishermen have been caught by Azorean fishermen (Anonymous 1986a). The commission developed posters encouraging fish-

ermen to turn in such hooks, in the hope that recovered hooks could be identified as to origin and used to help explicate fish migration. However, no reports have yet been received from fishermen (Anonymous 1987b).

Various scientists have suggested that other biological marks, such as shark bites, killer whale bites, etc., might be used to obtain better understanding of migration and stock structure. These approaches remain untried.

To estimate the reporting rate for recaptured tagged fish, a tag-seeding experiment was con-

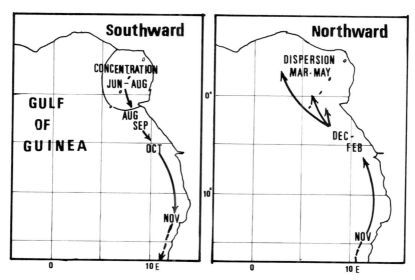

FIGURE 6.—Juvenile yellowfin tuna migrations in the eastern Atlantic, suggested by recaptures of tagged fish (Bard and Cayré 1986).

ducted during the commission's Yellowfin Year Program (Anonymous 1986a). The results were inconclusive because it was difficult to seed tags confidentially (Anonymous 1987b). Also, there was some doubt as to whether tags seeded on dead fish would be retained as well as tags anchored on live fish that remained at sea for a certain period.

Discussion and Conclusions

It would be useful for future planning to compare some aspects of ICCAT tagging with the two most important large-scale tuna-tagging programs carried out in the Pacific Ocean (by IATTC in the eastern Pacific and by SPC in the central southern Pacific). The ICCAT program is unique in the sense that most of its tagging has been conducted by national laboratories with national funds, whereas the other two programs have been carried out by the agencies' own scientific staffs and funds.

The SPC skipjack tuna survey and assessment program covered 4 years (August 1977–September 1981) and used a chartered bait-boat for tagging. The funds available for the entire program (including salaries) amounted to about US$4 million (the charter cost alone was $1.8 million) and came from country contributions. In all, 150,137 tunas (which surpassed the total Atlantic releases over 15 years) were tagged and released, of which 140,433 were skipjack tuna. The tagging results were analyzed in association with results of other activities (e.g., genetic studies, biological sampling, parasite studies, collection of maturity, catch, and effort data), and they produced considerable information on skipjack tuna in the central Pacific (Kearney 1983; Argue et al. 1986.)

The IATTC tagging program started in 1952 and remains under way. Over 300,000 fish have been tagged and released (mostly yellowfin and skipjack tunas). Some cruises have been made on an opportunistic basis, but the major releases have come from chartered cruises. Almost all the tagging is done by IATTC's own staff and funds, but the total cost is not known. This tagging has also provided considerable information on various aspects of the biology and stocks of tunas (Fink 1965; Fink and Bayliff 1970; Bayliff 1973; Bayliff and Rothschild 1974; Bayliff 1979; Bayliff and Holland 1986; Hunter et al. 1986.)

The ICCAT system is the least expensive for an agency because most of the activities are nationally funded. The total ICCAT Skipjack Year Program budget was only US$425,000, of which $85,000 was set aside for tagging. Actually,

$54,699 was spent directly on tagging (Anonymous 1981, 1982, 1983a). This means that most of the 40,000 tropical tunas released during the skipjack program were covered by national laboratories and resources. It is, however, very difficult to evaluate cost-effectiveness without an in-depth analysis. Most likely, the cost per tagged fish is less for a centralized tagging system, such as IATTC or SPC. On the other hand, some national funds would be available for maintaining research boats and scientific staff at the laboratories even if the ICCAT tagging program did not exist. Therefore, these costs cannot not be appraised equally.

The planning of a tagging program with a specific objective is easier in a single organizational set up that has experienced scientific staff. In a multinational scheme, such as ICCAT, it is sometimes difficult to evaluate the overall effectiveness of a proposed tagging program, to plan it at the desired scale, and to carry it out as planned because national interests sometimes conflict. However, such complications often add a broader scope to the tagging program.

A multinational program often requires training for inexperienced scientists; indeed, the commission has sent experts to some areas for the purposes of training. However, participation in the program of individual nations, particularly of developing countries, has stimulated their scientific interests and provided opportunities for inexperienced scientists to receive training. This produces very positive effects in the long run.

As stated earlier, coordination of tagging activities and logistics is somewhat complicated in the ICCAT system, but tagging is more effective through direct involvement of the countries concerned. Another advantage is that some logistical problems (e.g., permission for research vessels to enter jurisdictional waters) can be resolved more easily through participation of all the nations involved.

It is essential that detailed catch and effort data associated with recapture information be available. In the case of ICCAT, the commission has a comprehensive, Atlantic-wide catch–effort data base in 1-degree or 5-degree coordinate rectangles and by month. However, the data by vessels are limited, so some data have to be analyzed at the national level. Such analyses can only be made on part however, if the nation which is analyzing the data cannot obtain all the confidential data of other countries.

The ICCAT tagging program has been somewhat limited in scope, compared to other programs of SPC and IATTC, yet it has provided the only basis for working hypotheses on stock structure and migration of almost all tuna and billfish species. The ICCAT program also has provided a base on which to estimate growth for most tunas and billfishes, and its information has been used to estimate fishing mortality of bluefin tuna. Were it not for the program's tagging data, many analytical stock assessments would have been hampered by the lack of these biological parameters.

Acknowledgments

The author wishes to thank P. Kebe (ICCAT), who assisted in the computer processing of the ICCAT tag files, and P. Seidita (ICCAT) for typing and English editing of the report.

References[1]

Anonymous. 1971. Report of the meeting of the subcommittee on stock identification. Pages 109–128 in ICCAT, report for biennial period, 1970–1971, part 3.

Anonymous. 1978. Report of the standing committee on research and statistics (SCRS). Pages 98–186 in ICCAT, report for biennial period, 1976–1977, part 2.

Anonymous. 1981. Financial report 1980. Pages 14–29 in ICCAT, report for biennial period, 1980–1981, part 1.

Anonymous. 1982. Financial report 1981. Pages 12–29 in ICCAT, report for biennial period, 1980–1981, part 2.

Anonymous. 1983a. Financial report 1982. Pages 11–26 in ICCAT, report for biennial period, 1982–1983, part 1.

Anonymous. 1983b. Report of the ad hoc working group on tagging accounting. Pages 214–219 in ICCAT, report for biennial period, 1982–1983, part 1.

Anonymous. 1985a. Report of the meeting of the bluefin working group. ICCAT, Collective Volume of Scientific Papers 22:1–79.

Anonymous. 1985b. Report of the standing committee on research and statistics (SCRS). Pages 104–197 in ICCAT, report for biennial period, 1984–1985, part 1.

Anonymous. 1986a. Operational plan for the yellowfin year program. ICCAT.

Anonymous. 1986b. Report of the ad hoc working group to review micro constituent analysis and its relation to stock variation. Pages 226–227 in ICCAT, report for biennial period, 1984–1985, part 2.

Anonymous. 1986c. Report of the international skipjack year program conference of ICCAT. Pages 3–32 in Symons et al. (1986).

Anonymous. 1987a. Report of the bigeye day session. Pages 225–231 in ICCAT, report for biennial period, 1986–1987, part 1.

Anonymous. 1987b. Report of the standing committee on research and statistics (SCRS). Pages 133–215 in ICCAT, report for biennial period, 1986–1987, part 1.

Anonymous. 1988. Report of the standing committee on research and statistics (SCRS). Pages 120–206 in ICCAT, report for biennial period, 1986–1987, part 2.

Argue, A. W., P. Kleiber, R. E. Kearney, and J. R. Sibert. 1986. Evaluation of methods used by the South Pacific Commission for identification of skipjack population structure. Pages 242–251 in Symons et al. (1986).

Baglin, R. E., M. I. Farber, W. H. Lenarz, and J. M. Mason. 1980. Estimates of shedding rates of two types of dart tags from northwestern Atlantic bluefin tuna (*Thunnus thynnus*). ICCAT, Collective Volume of Scientific Papers 9:453–462.

Bard, F. X. 1984. Aspect de la croissance de l'albacore est Atlantique (*Thunnus albacares*) à partir des marquages. ICCAT, Collective Volume of Scientific Papers 21:108–114.

Bard, F. X. 1987. Migration transatlantique d'albacore (*Thunnus albacares*). ICCAT, Collective Volume of Scientific Papers 26:27–30.

Bard, F. X., and P. Cayré. 1986. Commentaires sur les migrations de l'albacore (*Thunnus albacares*) en Atlantique est. ICCAT, Collective Volume of Scientific Papers 25:11–29.

Bard, F. X., S. Kume, and L. Antoine. 1983. Données préliminaires sur la croissance, les migrations et la mortalité du listao (*Katsuwonus pelamis*) en Atlantique est obtenues à partir de marquage. ICCAT, Collective Volume of Scientific Papers 18:271–294.

Bard, F. X., M. Mensah, and P. Vendeville. 1987. Etat d'avancement des marquages del'année albacore. ICCAT, Collective Volume of Scientific Papers 26:8–14.

Bard, F. X., and D. Pincock. 1982. Rapport sur une expérience de marquage par microémetteur, de listaos (*Katsuwonus pelamis*) dans le golfe de Guinée, en juillet 1981. ICCAT, Collective Volume of Scientific Papers 17:184–188.

Bayliff, W. H. 1973. Materials and method for tagging purse seine and baitboat-caught tunas. Inter-American Tropical Tuna Commission Bulletin 15:379–436.

Bayliff, W. H. 1979. Migrations of yellowfin tuna in the eastern Pacific Ocean as determined from tagging experiments initiated during 1968–1974. Inter-American Tropical Tuna Commission Bulletin 17:447–506.

Bayliff, W. H., and K. N. Holland, 1986. Material and methods for tagging tunas and billfishes, recovering the tags, and handling the recapture data. FAO

---

[1]All ICCAT publications are available from the International Commission for the Conservation of Atlantic Tunas, Madrid, Spain.

(Food and Agricultural Organization of the United Nations) Fisheries Technical Paper 279.

Bayliff, W. H., and L. M. Mobrand. 1972. Estimates of the rates of shedding of dart tags from yellowfin tuna. Inter-American Tropical Tuna Commission Bulletin 15:441–462.

Bayliff, W. H., and B. J. Rothschild. 1974. Migrations of yellowfin tuna tagged off the southern coast of Mexico in 1960 and 1969. Inter-American Tropical Tuna Commission Bulletin 16:1–64.

Beckett, J. 1970. Swordfish, shark and tuna tagging, 1961–1969. Fisheries Research Board of Canada Technical Report 193.

Calaprice, J. R. 1986. Chemical variability and stock variation in northern Atlantic bluefin tuna. ICCAT, Collective Volume of Scientific Papers 24:222–254.

Cayré, P. 1984. Croissance du thon obèse (*Thunnus obesus*) de l'Atlantique d'après les résultats de marquage. ICCAT, Collective Volume of Scientific Papers 20:180–187.

Cayré, P., T. Diouf, and A. Fonteneau. 1986. Analyse des données de marquages et recaptures de listao (*Katsuwonus pelamis*) réalisés par le Sénégal et la République du Cap-Vert. Pages 309–316 in Symons et al. (1986).

Farber, M. I. 1980. A preliminary analysis of mortality of bluefin tuna (*Thunnus thynnus*) tagged in the northwestern Atlantic Ocean. ICCAT, Collective Volume of Scientific Papers 9:557–562.

Fink, B. D. 1965. A technique, and the equipment used, for tagging tunas caught by the pole and line method. Journal du Conseil, Conseil International pour l'Exploration de la Mer 29:335–339.

Fink, B. D., and W. H. Bayliff. 1970. Migrations of yellowfin and skipjack tuna in the eastern Pacific Ocean as determined by tagging experiments, 1952–1964. Inter-American Tropical Tuna Commission Bulletin 15:1–227.

Fonteneau, A. 1980. Croissance de l'albacore (*Thunnus albacares*) de l'Atlantique est. ICCAT, Collective Volume of Scientific Papers 11:152–168.

Hamre, J. 1963. Tuna tagging experiments in Norwegian waters. FAO (Food and Agriculture Organization of the United Nations) Fisheries Reports 6 (volume 3):1125–1132.

Hunter, J. R., and six coauthors. 1986. The dynamics of tuna movements: an evaluation of past and future research. FAO (Food and Agriculture Organization of the United Nations) Fisheries Technical Paper 277.

Kearney, R. E. 1983. Assessment of the skipjack and baitfish resources in the central and western tropical Pacific Ocean: a summary of the skipjack survey and assessment programme. South Pacific Commission, Noumea, New Caledonia.

Lenarz, W., F. Mather, J. Beckett, A. Jones, and J. Mason. 1973. Estimation of rates of tag shedding of northwest Atlantic bluefin tuna. ICCAT, Collective Volume of Scientific Papers 1:459–471.

Levenez, J. J. 1982. Note préliminaire sur l'opération Sénégalaise de tracking de listao. ICCAT, Collective Volume of Scientific Papers 17:189–194.

MacKenzie, K. 1983. The selection of parasites for use as biological tags in population studies of bluefin tuna. ICCAT, Collective Volume of Scientific Papers 18:834–838.

Mather, F. J., III, and J. M. Mason, Jr. 1973. Summary of recent information on tagging and tag for tunas and billfishes in the Atlantic Ocean. ICCAT, Collective Volume of Scientific Papers 1:501–531.

Mather, F. J., III, B. J. Rothschild, and G. J. Paulik. 1973. Preliminary analysis of bluefin tagging index. ICCAT, Collective Volume of Scientific Papers 1:413–444.

Miyabe, N. 1983. Estimation of recruitment of 1973 cohort of bluefin tuna in the west Atlantic, using tagging results. ICCAT, Collective Volume of Scientific Papers 18:464–467.

Miyabe, N. 1984. On the growth of yellowfin and bigeye tuna estimated from the tagging results. ICCAT, Collective Volume of Scientific Papers 20:117–122.

Miyabe, N. 1987. A note on the movement of bigeye tuna based on tagging experiments. ICCAT, Collective Volume of Scientific Papers 26:105–110.

Miyabe, N., and F. X. Bard. 1986. Movements of skipjack in the eastern Atlantic, from results of tagging by Japan. Pages 342–347 in Symons et al. (1986).

Miyake, M., and S. Hayasi. 1978. Field manual for statistics and sampling of Atlantic tunas and tuna-like fishes, 2nd edition. ICCAT.

Miyake, P. M., and P. E. K. Symons. 1983. ICCAT tag accounting policy. ICCAT, Collective Volume of Scientific Papers 18:704–710.

Parrack, M. L., and P. L. Phares. 1979. Aspects of the growth of Atlantic bluefin tuna determined from mark–recapture data. ICCAT, Collective Volume of Scientific Papers 8:356–366.

Rodríguez-Roda, J. 1969. Resultados de nuestras marcaciones de atunes en el golfo de Cádiz durante los años 1960 a 1969. Publicaciones Técnicas, Junta de Estudios de Pesca 8 (mimeo), Cádiz, Spain.

Scott, E. L., E. D. Prince, and C. D. Goodyear. 1990. History of the cooperative game fish tagging program in the Atlantic Ocean, Gulf of Mexico, and Caribbean Sea, 1954–1987. American Fisheries Society Symposium 7:841–853.

Symons, P. E. K., P. M. Miyake, and G. T. Sakagawa, editors. 1986. Proceedings of the ICCAT Conference on the International Skipjack Year Program. ICCAT. (In English, French, and Spanish.)

Turner, S. C. 1986. An analysis of recaptures of tagged bluefin with respect to the mixing assumption. ICCAT, Collective Volume of Scientific Papers 24:196–202.

Appendix: National Laboratories and Institutes That Are
Past or Present Participants in the ICCAT Cooperative Tagging Program

Centro de Investigação Pesqueira
CX Postal 677
Lobito
Angola

Instituto de Pesca
Avenida Bartolomeu de Gusmão, 192
11.030 - Santos - SP
Brazil

Coordenadoria Regional da SUDEPE
Praça XV de Novembro 2/3° andar
Rio de Janeiro CEP 20010
Brazil

CEPSUL/SUDEPE
Praça Barão do Rio Branco, 3
88300-Itajai - Santa Catarina
Brazil

Marine Fisheries Division
Gulf Region
Department of Fisheries and Oceans
Post Office Box 5030
Moncton, New Brunswick E1C 9B6
Canada

Marine Fisheries Division
Scotia-Fundy Region
Department of Fisheries and Oceans
St. Andrews Biological Station
St. Andrews, New Brunswick E0G 2X0
Canada

Direcção de Biologia Maritima
Casier Postal 30
Praia
Cape Verde

Centre de Recherches Océanographiques
Boite Postale 1286
Pointe-Noire
Congo

Centre de Recherches Océanographiques
Boite Postale V-18
Abidjan
Côte d'Ivoire

Centro de Investigaciones Pesqueras
Avenida Iª y 26
Miramar, La Habana
Cuba

Laboratoire IFREMER
Boite Postale 1049
44037 - Nantes Cédex
France

Laboratoire IFREMER
1, rue Jean-Villar
34200 - Sète
France

Fishery Research Unit
Post Office Box B-62
Tema
Ghana

Far Seas Fisheries Research Laboratory
5-7-1 Orido
Shimizu 424, Shizuoka Prefecture
Japan

National Fisheries Research and Development
  Agency
16, 2-Ga, Namhang-Dong, Yeongdo-Ku
Pusan 606
Korea

I.S.P.M.
11, rue de Tiznit
Boite Postale 21
Casablanca 01
Morocco

Fiskeridirektoratet
Mollendalsvegen, 4
Postboks 185
5001 - Bergen - Nordness
Norway

Instituto Nacional de Investigação das Pescas
Alges - Praia 1400
Lisboa
Portugal

Universidade dos Açores
Departamento de Oceanografía e Pescas
9900 - Horta, Ilha do Faial
Azores
Portugal

Direcção Regional das Pescas
CX Postal 747
9009 - Funchal, Madeira
Portugal

Centre de Recherches Océanographiques de
  Dakar—Thiaroye
Boite Postale 2241
Dakar
Senegal

Sea Fisheries Research Institute
Private Bag X2
Roggebaai 8012
Cape Town
South Africa

Instituto Español de Oceanografía
Centro Costero de Canarias
Apartado 1373
Santa Cruz de Tenerife
Spain

Instituto Español de Oceanografía
Apartado 1552
36080-Santander
Spain

Instituto Español de Oceanografía
Apartado 1552
36080-Vigo
Spain

Instituto Español de Oceanografía
Apartado 285
Fuengirola, Málaga
Spain

Instiuto Español de Oceanografía
Apartado 130
15080-La Coruña
Spain

Instituto Español de Oceanografía
Avenida del Brasil, 31
28020 - Madrid
Spain

Institute of Oceanography
National Taiwan University
Number 1, Section 4, Roosevelt Road
Taipei
Taiwan

Office National des Pêches de Tunisie
Port de la Goulette
Tunis
Tunisia

Industria Lobera y Pesquera del Estado
Rbla. Baltasar Brum y Cnel. Fco. Tajes
Montevideo
Uruguay

Southeast Fisheries Center
National Marine Fisheries Service
75 Virginia Beach Drive
Miami, Florida 33149
USA

Southwest Fisheries Center
National Marine Fisheries Service
Post Office Box 271
La Jolla, California 92038
USA

AtlantNIRO
Dmitrij Donskogo, 5
Kaliningrad
USSR

Centro de Investigaciones Pesqueras
Estación Experimental Sucre - FONAIAP
Avenida Carúpano Caiguire
Apartado 236
Cumaná, Estado Sucre
Venezuela

American Fisheries Society Symposium 7:765–774, 1990

Australian Cooperative Game-Fish Tagging Program, 1973–1987: Status and Evaluation of Tags

Julian G. Pepperell

Fisheries Research Institute
New South Wales Department of Agriculture and Fisheries
Post Office Box 21, Cronulla, New South Wales 2230 Australia

Abstract.—The Australian cooperative game-fish tagging program is coordinated through the Fisheries Research Institute of the New South Wales Department of Agriculture and Fisheries. In its 14-year history, nearly 200,000 tags have been issued, over 66,000 marine game fish have been released by volunteer anglers, and more than 1,300 recaptures have been reported. The main species or species groups tagged are billfishes, yellowtail *Seriola lalandei*, tunas, dolphin *Coryphaena hippurus*, and sharks. A range of tag designs has been used, including several types of nylon barbed darts, T-anchor, and steel anchor tags. Of the nylon darts used on similar species, the tags that showed highest recovery rates and slowest declines in recoveries had robust, stiff-barbed heads, had heads strongly bonded to the streamers, and (in the case of tubular spaghetti tags) were relatively soft and pliable. T-anchor tags produced good initial recovery rates but poor sustained retention, whereas steel anchor tags, although similar or superior to nylon barbed darts in terms of recovery rate, showed better retention with time.

Cooperative game-fish tagging programs, through which volunteer anglers participate in the marking of marine game fish, were pioneered in the USA in the mid-1950s by Frank J. Mather III of the Woods Hole Oceanographic Institution (Mather 1963). Since that time, the concept has been embraced widely, and several hundred thousand game fish have been tagged by anglers in many parts of the world.

Such programs have accumulated information that would otherwise be difficult, if not impossible, to achieve at reasonable cost. For example, seasonal distributions and migrations of billfishes (Squire 1974; Squire and Nielsen 1983; Beardsley 1985) and large sharks (Casey 1985) have been determined with confidence, and theories on longevity of billfishes have been revised following long-term recaptures of recreationally tagged individuals (Mather et al. 1974).

The Australian cooperative game-fish tagging program began in 1973 and was modeled on the successful Woods Hole system. Its history and results to mid-1984 were described by Pepperell (1985). Since then, the numbers of game fish tagged and recaptured by the program have more than doubled.

This program has also been successful in providing new information on game fish in the Australasian region. In particular, data have been gathered on movements of black marlin *Makaira indica* in the western Pacific and on stock structure of yellowtail *Seriola lalandei* along the east-ern Australian coast. The aim of this type of program is to derive information on movements and stock structure of target species. However, unforeseen benefits may emerge during the course of such long-term studies.

In this program, sufficiently large numbers of fish have been tagged over long enough periods of time that the performances of various types of tags can be evaluated. Many variables affect tag performance. This paper concentrates on major differences in recovery rates and on rates of decline of reported recoveries for different tags over time. I hope that this will help other fisheries workers to choose suitable tags and to avoid some of the potential problems inherent in the organization and operation of cooperative tagging programs.

Methods

Tags Used

The program has issued three different types of tags to volunteer anglers. All were plastic streamer dart or anchor tags, shown in Figure 1 and described below. The letter prefixes denoted are those actually used by the program.

Tuna dart tags (type A).—Tuna dart tags have a single barbed nylon head attached to a streamer body that bears the legend. Variations on this theme have been supplied by several manufacturers during the program. Five versions of the type A dart tag have been used. Type A_1 tags have a

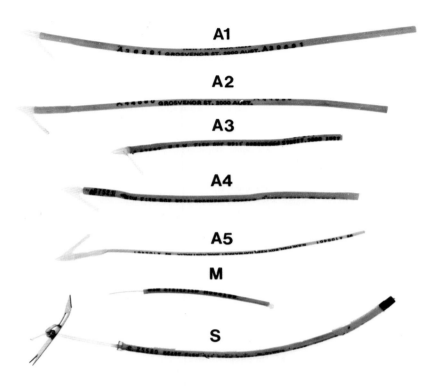

FIGURE 1.—Examples of the seven types of tags used and their designations.

hard polyvinyl tube glued to a rigid barbed nylon head. Manufactured by A. E. King and Co., Sydney, Australia, they have been in use since 1973. Type A_2 tags have a soft polyvinyl tube glued to a rigid barbed nylon head. Also manufactured by A. E. King, they have been in use since 1974. Type A_3 tags have a soft polyvinyl tube glued to a flexible barbed nylon head. These are the standard FT-1 tags manufactured by Floy Tag and Manufacturing, Inc., Seattle, Washington. They have been in use since 1982. Type A_4 tags use the Floy FT-1 tubular bodies glued to A. E. King nylon heads. They have been in use since 1984. Type A_5 tags have a solid thin polyethylene body bonded to an A. E. King nylon head by means of a clear polyethylene outer layer, which also covers the tag body. They have been in use since 1985.

Nylon T-anchor tag (type M).—The T-anchor tag is the standard model FD-68B tag manufactured by Floy. It consists of a nylon T-shaped head running through and glued to a yellow polyvinyl tube 7 cm long and 2 mm in external diameter. A second nylon shaft is glued into the tube from the distal end to prevent slippage. Distribution and use commenced in 1984.

Steel anchor tag (type S).—The standard stainless steel head model FH-69A tag, manufactured by Floy, was issued primarily for use on sharks and billfish. It is the same as the type H tag described by Squire and Nielsen (1983) except that the yellow polyvinyl legend-bearing sleeve is in turn glued inside a clear polyvinyl outer tube. Distribution and use commenced in 1977.

Numbers of each type of tag issued and used are shown in Table 1. All tag bodies were colored yellow except for a red batch of 5,000 A_1 tags.

Tagging Operation

Distribution and methods of application of types A and S tags were described by Pepperell (1981, 1985).

Type M tags were supplied by the manufacturer in magazines of 25, and Monarch model 3030 tagging guns equipped with 2.9-cm needles were

TABLE 1.—Types and numbers of tags issued during the Australian game-fish tagging program, and the numbers used to date.

Tag type[a]	Manufacturer	Year first used	Number issued to anglers	Number used
A₁	A. E. King	1973	45,500	14,226
A₂	A. E. King	1974	14,500	6,115
A₃	Floy	1982	21,000	9,273
A₄	Floy, A. E. King	1984	19,000	7,069
A₅	Hallprint	1985	29,000	8,154
M	Floy	1984	10,000	4,699
S	Floy	1977	32,000	11,858

[a]See Figure 1.

used to apply them. Anglers were requested to anchor types A and M tags behind dorsal pterygiophores (these tags are used mainly on boated fish), but observations of tagging and of recaptured fish indicated that the anglers complied infrequently. Type S tags usually were placed in the anterior dorsal musculature (the "shoulder") of fish as they were held alongside the boat.

Volunteer taggers were trained by means of pamphlets and instruction sheets supplied with tagging equipment, supplemented by field demonstrations and lectures at numerous club meetings and tournaments.

Analysis of Recaptures

Numbers of fish tagged and recaptured (Table 2) were calculated on the basis of tag cards and recovery reports received through 30 June 1987. In comparing recapture variation by tag type, only completed sequences of tag batches were considered. New models of all three tag types (A, S, and M) are now in use, and although some fish tagged and recaptured with these contributed to the program totals of Table 2, insufficient numbers have been used to include them in the comparative analysis.

Results

Numbers Tagged and Recaptured

In every year since the program's inception in late 1973, more game fish have been tagged than in the preceding year (Figure 2). Because all marine species classified as game fish by the Game Fishing Association of Australia are eligible for tagging, more than 50 species have been included in the program to date. However, 15 species or species groups account for over 85% of all fish tagged (Table 2).

Geographic Area

Game fish have been tagged around the Australian coast, off Papua New Guinea, and more recently in the Persian Gulf and off Kenya (Figure 3). Taggings in New South Wales and Queensland dominate the totals. It is also apparent that species composition of tagged fish varies considerably from state to state; for example, billfishes predominate in Queensland (mainly black marlin tagged in the north of the state), and yellowtail and other game fish (mainly small tunas and dolphins) make up the majority of the New South Wales total.

TABLE 2.—The main species tagged and the proportions recaptured from 1973–1974 through 1986–1987 by the Australian game-fish tagging program.

Species	Common name	Number tagged	Recaptured Number	%
Seriola lalandei	Yellowtail	9,594	682	7.1
Makaira indica	Black marlin	9,117	60	0.7
Katsuwonus pelamis[a]	Skipjack tuna	7,091	22	0.3
Coryphaena hippurus	Dolphin	6,545	61	0.9
Euthynnus alletteratus	Little tunny	4,747	18	0.4
Sarda spp.	Bonitos	4,408	68	1.5
Thunnus albacares	Yellowfin tuna	2,748	40	1.5
Istiophorus platypterus	Sailfish	2,726	28	1.0
Carcharhinus spp.	Whaler sharks	2,041	51	2.5
Pomatomus saltatrix	Bluefish	1,918	31	1.8
Thunnus tonggol	Longtail tuna	1,410	28	2.0
Thunnus alalunga	Albacore	1,319	2	0.2
Pseudocaranx dentex	Silver trevally	1,197	20	1.7
Isurus oxyrinchus	Shortfin mako	1,099	17	1.5
Sphyrna spp.	Hammerhead shark	990	9	0.9
All other species		9,947	173	1.7
Total		66,897	1,310	2.0

[a]Formerly *Euthynnus pelamis*.

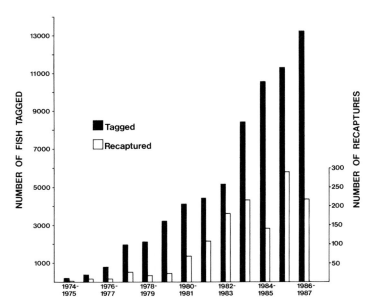

FIGURE 2.—Numbers of game fish tagged and recaptured each year.

Tagging Personnel

Tagging was undertaken by recreational anglers, mostly members of clubs affiliated with the Game Fishing Association of Australia or the Australian National Sportfishing Association. In the case of billfishes tagged from charter boats, as most billfish are tagged in Queensland, tags were struck by experienced deckhands or captains. Over 9,000 anglers have tagged game fish for the program to date.

Evaluation of Tag Types

Tag performances were compared with respect to recovery rate and time of retention.

Recapture rate.—Table 3 summarizes the recovery rates for each of the seven types of tags considered, together with a breakdown by major species groups.

For every tag type, recapture rates for yellowtail were much higher than for any other species or group of species. Consequently, yellowtail provided the best source of data for comparisons of tags.

Recapture rates for yellowtail with different A tags varied considerably. The A_3 tag showed the lowest recovery rate and the A_5 tag the highest—four times that of the A_3 tag. Of the other three A tags, A_1 showed the lowest recovery rate for yellowtail.

The recapture rate for yellowtail carrying the M tag was higher than those of fish carrying all A tags except A_5. Very few yellowtail were tagged with S tags, and numbers recaptured also were small.

Recapture rates for other game fish (a mixed group containing tunas, bonitos, dolphins, mackerels, etc.) showed less variation amongst the seven tag types. The A_3 tag again showed the lowest overall recovery rate, but the S tag in this instance showed the highest. Among A tags, the A_5 model gave the highest recovery rate. The recovery rate for M tags in other game fish was higher than that of A tags.

Recapture rates for sharks and billfishes carrying A tags were difficult to compare because of the relatively low numbers tagged and recaptured. However, when data for A tags were lumped, sharks carrying them showed a lower recapture rate than did sharks tagged with S tags. Insufficient billfishes were tagged with A tags and subsequently recaptured to allow comparisons.

When data for all species were combined, the A_3 model again produced the lowest and the A_5 model the highest recoveries among the A tags recapture rate. The recovery rate for M tags was within the range of A tags. Similar proportions of species were tagged with A and M tags; S tags,

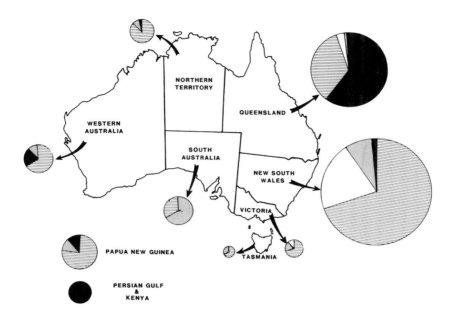

FIGURE 3.—Proportions of species or species groups tagged in each main geographic region. Areas of circles are proportional to numbers tagged. Key: White—yellowtail; black—billfishes; grey—sharks; stripes—other game fish.

however, were used mostly on sharks and billfish, so they were excluded from comparisons of overall recapture rates.

Time of tag retention.—The duration of tag retention is best measured by controlled aquarium or double-tagging experiments, but indirect measures can provide an indication of this important characteristic. Here, I have used proportion of reported recaptures against time after release as a comparative indicator of tag retention.

For all tags except type S, over 60% of all recoveries occurred within 50 d of release (Figure 4). Within 100 d of release, the percentages of total recoveries ranged from 92.8% for type A_3 to 72.4% for type A_5. The striking exception to this high proportion of early recoveries was shown by the S tag (Figure 4g), for which only 40.3% of the recoveries occurred within 100 d and 78.8% within 400 d. This contrasted with 100% and 99% recoveries after 400 d for M and A_3 tags, respectively, and 95% for the other three A types.

The comparison among A tags (Figure 4 a–e) indicated a more rapid decline in likelihood of recovery for the A_3 model than for the other four, although the fewest recaptures were recorded for this tag. Tags, A_1, A_2, and A_5 showed the slowest rates of decline of reported recoveries. All exhibited very similar patterns, after I allowed for the slight differences in species composition.

Of the A tags, A_1 and A_2 were recovered after the longest times at liberty, perhaps not surprisingly because these models have been in use the longest. Tag A_4 exhibited a slightly greater likelihood of recovery with time than did A_3, as was also the case for M tags (Figure 4f). Recaptures of species bearing type A or type M tags were dominated by yellowtail and other game fish, whereas billfishes and sharks accounted for most of the recoveries of S tags.

Times at liberty.—Longest times at liberty for the program to date are 2,199 d for a wobbegong shark *Orectolobus maculatus* tagged with an S tag, 1,884 d for a common whaler shark *Carcharhinus obscurus* tagged with an A_1 tag, 1,790 d for an albacore tagged with an A_2 tag, and 1,481 d for a sailfish tagged with an S tag. Recaptures after more than 1,000 d have also included four black marlin and a leopard shark *Stegostoma fasciatum*, all tagged with S tags, and a yellowtail bearing an A_1 tag.

Damaged tags.—A few tags were illegible after removal from a recaptured fish. The most common causes of illegibility of A tags were shredding or breakage of the tubular body (seven examples) and abrasion (five examples). Another five fish were reported that had only the nylon barbs. For S tags, the legend-bearing tubular section of the tag sometimes slipped off the monofilament

TABLE 3.—Numbers of all species and major species groups tagged and recaptured with each tag type (Figure 1) by the Australian cooperative game-fish tagging program.

Species and statistics	Tag type							Total
	A_1	A_2	A_3	A_4	A_5	M	S	
Yellowtail								
Tagged	3,077	1,500	1,837	1,035	1,145	558	138	9,290
Recaptured	190	121	54	102	141	59	61	673
% recaptured	6.8	8.1	2.9	9.9	12.3	10.6	4.4	7.2
Other game fish								
Tagged	8,823	4,076	6,737	5,500	6,449	4,124	1,483	37,192
Recaptured	80	50	57	53	94	62	28	424
% recaptured	0.9	1.2	0.8	1.0	1.5	1.5	1.9	1.1
Sharks								
Tagged	1,586	263	494	390	344	17	1,506	4,600
Recaptured	29	6	8	2	8	0	43	96
% recaptured	1.8	2.3	1.6	0.5	2.3	0	2.9	2.1
Billfish								
Tagged	740	276	205	144	216		8,731	10,312
Recaptured	4	1	0	0	4		61	70
% recaptured	0.5	0.4	0	0	1.9		0.7	0.7
All species combined								
Tagged	14,226	6,115	9,273	7,069	8,154	4,699	11,858	61,394
Recaptured	303	178	119	157	247	121	138	1,263
% recaptured	2.1	2.9	1.3	2.2	3.0	2.6	1.2	2.1

streamer after loss of the distal brass crimp (seven examples). Two S tags were discarded by anglers as illegible. Figure 5, which shows S tags removed from fish after different periods, demonstrates that the brass crimp may corrode over time.

Discussion

The program has continued to expand rapidly in terms of numbers of fish tagged and recaptured. The recent growth in activity has been so great that, even though the first game fish were tagged in 1973, the last 3 years accounted for more than half of all fish tagged and recaptured. Tags issued at the beginning of the program remain in use, but fewer early tags are used each year.

It is important to be aware of the possible factors that may confound comparisons among the different tags. Recapture rates may be influenced not only by tag design and construction, but by some or all of the following: changes in susceptibility of different species to capture with time; changes in composition of tagged species groups with time; shifts in localities of tagging activities; changes in tagging personnel; and differences in size-groups tagged over time. Because one or more of these factors may have influenced some recapture data presented here, statistical treatment was inappropriate. This evaluation is primarily concerned with large differences between tag types.

Among A tags, the A_3 model showed the lowest recovery rate in nearly all categories. This was so consistent as to suggest a fault in construction; indeed, tag separation may have been a significant factor because loose heads were noted often with this tag early in its use. Consequently, all A_3 tags were checked prior to distribution and loose heads were reglued. Nevertheless, many A_3 tags were returned by anglers because of faulty attachments, and it was for this reason that the decision was made to change to tag A_4 (i.e., the A_3 body glued to the slightly larger and more robust head that had been used for previous A tags).

The modification was successful—the A_4 model gave a higher recovery rate than the A_3 model for yellowtail and for all species combined. Also, rates of decline of recoveries were lower for the A_4 tag than for the A_3 tag, especially for yellowtail (Figure 4). Therefore, it must be concluded that, because the major difference between these tag types was the nylon head, the stiff-barbed larger head was superior when used in this cooperative tagging program.

Another interesting comparison was between the A_1 and A_2 tags—both virtually identical except that A_1 had hard, relatively stiff streamers, whereas A_2 had soft, flexible ones. The softer tag had a higher overall recovery rate, presumably because it offered less resistance in the water and therefore had less effect on the fish and less cause to work loose. Both tags showed very similar rates of decline of recoveries (Figure 4), suggesting that differences in recovery rates were due to relatively constant losses of the hard tag.

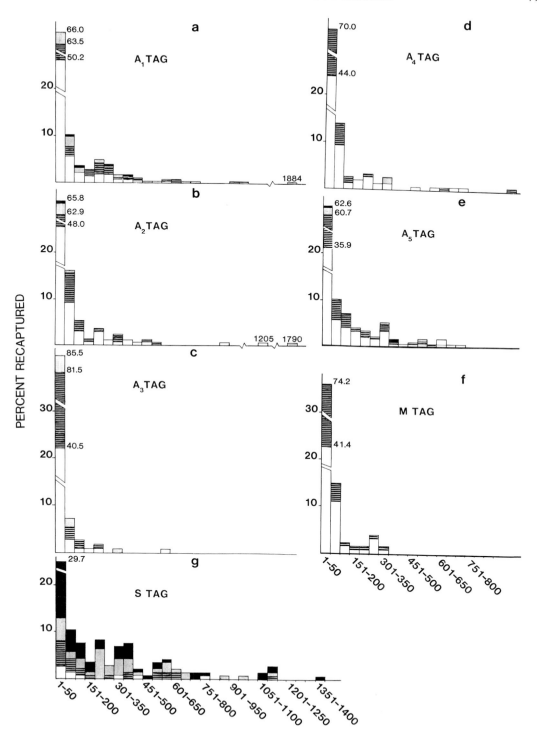

FIGURE 4.—Proportion of completed recaptures against days after release (abscissa scales) for each type of tag. Key: black—billfishes; grey—sharks; stripes—other game fish; white—yellowtail.

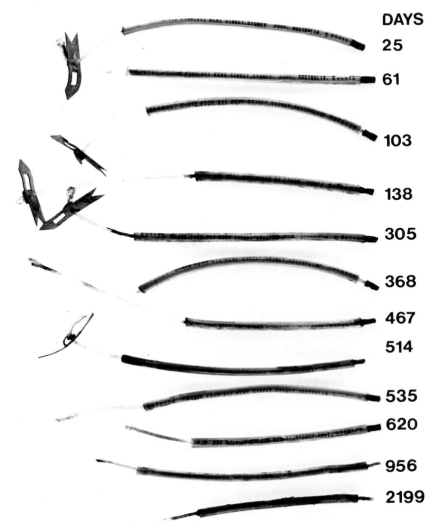

DAYS
25
61
103
138
305
368
467
514
535
620
956
2199

FIGURE 5.—Examples of deterioration of tags with time the fish were at liberty. Numbers of days at liberty are shown.

Among the A tags, the A_5 model represented a radical departure from the traditional polyvinyl spaghetti dart in that the streamer was solid and polyethylene, and no glues were used to attach it to the nylon head. Rather, an outer polyethylene sheath bonded the head to the body. The nylon head used on the A_5 tag was the same as for all other A tags except A_3. The main differences in construction, therefore, were the solid body, its smaller diameter, and the method of attachment to the head.

The use of this new model in 1985 was approached with some trepidation, but results to date have vindicated the decision to introduce it

to the program. In no comparison of recovery rates was the A_5 tag inferior. Rather, it has consistently outperformed most A models. Also, the rate of decline of recoveries of the A_5 model compared very favorably with the other A types (Figure 4), which indicated that the high recovery rates for the A_5 model were not the result of a high recovery rate very soon after release (as seemed to be the case for M tags).

Possible causes of good performance by the A_5 model were a lower tag separation rate and less resistance to water flow, resulting (presumably) in lower tag loss through shearing forces or ulceration, and even in lower tag-induced mortality.

The thin diameter of this tag was initially considered a negative factor in that it would be less visible than thicker tags. This factor remains to be tested, but good results to date tend to alleviate this concern.

The T-anchor M tag also was introduced into the program with some doubts because studies had suggested that its type of head did not anchor as well as the single-barb A type (Davis and Reid 1982). Results presented here showed that, in terms of recovery rates, the M tag performed at least as well as A tags in similar species groups. However, the important difference was that very low proportions of M tags were recovered after 100 d compared with most A tags, and none was recovered after 300 d (Figure 4). This strongly suggested a high rate of tag loss.

Two species groups (other game fish and sharks) allowed comparisons of recovery rates between S and A tags; in both cases, S tags showed the highest rates (Table 3). (The low recovery rate for S tags for all species combined was obviously due to the use of S tags mainly for billfish.)

Higher proportions of fish tagged with S tags were likely to be recaptured after long periods compared with all other tags (Figure 4). This suggests a slower shedding rate for S tags, but the different species compositions tagged with S and A tags warrant caution in this interpretation.

Lenarz et al. (1973) and Baglin et al. (1980) found no significant differences in loss rates between steel-head (S) and nylon-head (A) tags in double-tagging experiments with bluefin tuna *Thunnus thynnus*. In their studies, however, the nylon tag was anchored behind pterygiophores, whereas that did not occur consistently in this study.

Mather (1963), using limited recapture data, suggested that S tags may provide a higher recovery rate than model A_3 tags when used on billfish and bluefin tuna. Mather et al. (1974) noted that steel-head tags, such as the type S used here, were easier to place into sailfish alongside the boat than were nylon-head tags such as type A, but they gave no comparative analysis of recovery rates. Squire (1974) compared recovery rates of S and A tags used on eastern Pacific striped marlin and on eastern Pacific and Atlantic sailfish. He concluded that the nylon head (type A) was at least as satisfactory as the metal head (type S).

The limited evidence available from other studies does not suggest differences in overall recovery rates between type A and type S tags. Evidence presented here also indicates no marked differences between them, although I do recommend S tags for use on billfish and sharks for long-term studies, provided that slippage of the streamer is prevented.

Tag Damage with Time

The few illegible tags returned from recaptured fish gave some insight into deterioration over time; because legends were illegible, however, there was no way of knowing how long it took for the damage to occur. Damaged but decipherable tags provided some clues (Figure 5).

In the case of A tags, abrasion of the tubular body was very evident in tags used on yellowtail as early as 1 week after release. Abrasion increased with time until the tube was worn through and the streamer was shredded and, presumably, broken. This occurred almost exclusively with tags placed in yellowtail. Perhaps this species actually rubs against hard surfaces in attempting to dislodge the tag. Coblentz (1987), after observing a group of yellowtail rubbing against a blue shark, proposed they did so to dislodge ectoparasites. Tags recovered during the program from tunas and other game fish after long periods showed no such rubbing; neither did recoveries of legible A tags from southern bluefin tuna after periods of up to 18 years (Anonymous 1981).

Abrasion has not been observed on S tags in billfish, but the denticles of sharks may scuff the outer plastic sheath. Corrosion and loss of the distal brass crimp appeared to be the main cause of loss of tag information for type S tags. However, legible S tags have been recovered from white marlin *Tetrapterus albidus* after as long as 12 years (E. Prince, National Marine Fisheries Service, personal communication). My limited observations suggest that manufacturers may have altered the composition of the crimp in recent years, because pitting and polishing of the crimp have been observed on tags recovered from billfish after periods as short as 2 weeks. A new S model in which the steel head is attached to the streamer by means of stainless steel wire is now in use. Early results with this modification are promising; tags have shown no sign of separation after up to 440 d in fish. Much longer periods will be required, however, to evaluate its relative performance.

Conclusions

A long-term cooperative tagging program inevitably tries various forms of similar and dissimilar tags. Uncontrollable factors confound evaluations of subtle differences in performance among these tags, but the recovery data can be used to assess clear-cut advantages or disadvantages.

Recapture rates, and rates of decline for recapture of fish tagged with different nylon-headed A tags, varied considerably in the present program. Type A tags with robust stiff-barbed heads, adequate bonding of the head and streamer, and soft pliable tubing (in the case of polyvinyl spaghetti tags) gave the best results. T-anchor tags, which had not been used before in such programs, proved successful as measured by short-term recovery rates. However, a rapid decline in recoveries with time indicated that this tag is unsuitable for long-term studies that rely on tagging by volunteer anglers. Recovery rates for steel heads (type S) were at least as high as for nylon heads, and retention time appeared to be much greater for tags with steel than with nylon heads. Tag separation of the former suggested that their design can still be improved.

Choice of tags for cooperative tagging programs should be made with care. Consideration should be given to maximizing recovery rates and tag-retention time for the target species. Education of anglers is important to the success of such programs, but good tag design and construction are essential.

Acknowledgments

I thank Kerrie Deguara for her untiring attention to the large data base accumulated during this program. I am grateful to Bob Kearney and Johann Bell for their critical reviews of the manuscript. I am also indebted to the thousands of volunteer anglers, the many charter-boat captains and deckhands, and the administrators of sport-fishing in Australia, without whom the program would not be possible.

References

Anonymous. 1981. Southern bluefin tuna sets new tag record. Australian Fisheries 40(10):40.

Baglin, R. E., Jr., M. I. Farber, W. H. Lenarz, and J. M. Mason, Jr. 1980. Shedding rates of plastic and metal dart tags from Atlantic bluefin tuna, *Thunnus thynnus*. U.S. National Marine Fisheries Service Fishery Bulletin 78:179–185.

Beardsley, G. L. 1985. The importance of cooperative game fish tagging programs in marine fisheries research. Pages 233–240 *in* R. H. Stroud, editor. World angling resources and challenges. International Game Fish Association, Fort Lauderdale, Florida.

Casey, J. G. 1985. Transatlantic migrations of the blue shark. Pages 253–268 *in* R. H. Stroud, editor. World angling resources and challenges. International Game Fish Association, Fort Lauderdale, Florida.

Coblentz, B. E. 1987. Yellowtail chafing on a shark: parasite removal? California Fish and Game 73:244.

Davis, T. L. O., and D. D. Reid. 1982. Estimates of tag shedding rates for Floy FT-2 dart and FD-67 anchor tags in barramundi, *Lates calcarifer* (Bloch). Australian Journal of Marine and Freshwater Research 33:1113–1117.

Lenarz, W. H., F. J. Mather III, J. S. Beckett, A. C. Jones, and J. M. Mason, Jr. 1973. Estimation of rates of tag shedding by northwest Atlantic bluefin tuna. U.S. National Marine Fisheries Service Fishery Bulletin 71:1103–1105.

Mather, F. J., III. 1963. Tags and tagging techniques for large pelagic fishes. International Commission for the Northwest Atlantic Fisheries Special Publication 4:288–293.

Mather, F. J., III, D. C. Tabb, J. M. Mason, Jr., and H. L. Clark. 1974. Results of sailfish tagging in the western north Atlantic Ocean. Pages 194–203 *in* R. S. Shomura and F. Williams, editors. Proceedings of the International Billfish Symposium, Kailua-Kona, Hawaii, 9–12 August, 1972, part 2. U.S. Department of Commerce, Seattle, Washington.

Pepperell, J. G. 1981. Use of tagging data to study some aspects of the biology of the black marlin *Makaira indica* on the east coast of Australia. Proceedings of the Annual Conference Western Association of Fish and Wildlife Agencies 61:75–87.

Pepperell, J. G. 1985. Cooperative game fish tagging in the Indo-Pacific region. Pages 241–252 *in* R. H. Stroud, editor. World angling resources and challenges. International Game Fish Association, Fort Lauderdale, Florida.

Squire, J. L., Jr. 1974. Migration patterns of Istiophoridae in the Pacific Ocean as determined by cooperative tagging programs. Pages 226–237 *in* R. S. Shomura and F. Williams, editors. Proceedings of the International Billfish Symposium, Kailua-Kona, Hawaii, 9–12 August, 1972, part 2. U.S. Department of Commerce, Seattle, Washington.

Squire, J. L., Jr., and D. V. Nielsen. 1983. Results of a tagging program to determine migration rates and patterns for black marlin *Makaira indica*, in the southwest Pacific Ocean. NOAA (National Oceanic and Atmospheric Administration) Technical Report NMFS (National Marine Fisheries Service) SSRF (Special Scientific Report Fisheries) 772.

American Fisheries Society Symposium 7:775–781, 1990

Striped Bass Restoration along the Atlantic Coast: A Multistate and Federal Cooperative Hatchery and Tagging Program

Charles M. Wooley

U.S. Fish and Wildlife Service
1825 Virginia Street, Annapolis, Maryland 21401, USA

Nick C. Parker[1]

U.S. Fish and Wildlife Service
Southeastern Fish Cultural Laboratory
Marion, Alabama 36756, USA

Benjamin M. Florence

Maryland Department of Natural Resources
C-2 Tawes Building, Annapolis, Maryland 21401, USA

Roy M. Miller

Delaware Department of Natural Resources and Environmental Control
Post Office Box 1401, Dover, Delaware 19901, USA

Abstract.—Stocks of adult striped bass *Morone saxatilis* were extremely low from 1980 to 1987 along the Atlantic seaboard, especially in Chesapeake Bay. In an effort to rebuild these stocks, the Atlantic States Marine Fisheries Commission developed a coastwide management plan for anadromous striped bass. This plan included a stocking and evaluation program developed by a Technical Advisory Committee composed of representatives from all coastal states from Maine to North Carolina, the U.S. Fish and Wildlife Service (USFWS), and the National Marine Fisheries Service. The committee prepared a report that provided guidance for restoration and tagging programs for striped bass along the Atlantic coast. The USFWS, Maryland Department of Natural Resources, and state of Virginia entered into a cooperative agreement to develop a striped bass stocking and tagging program in Chesapeake Bay in 1985. The USFWS assigned a coordinator to assist the coastal states in implementing this program. By January 1988, 1.35 million striped bass had been tagged with binary-coded wire tags and 23,250 of these fish had also been tagged with internal anchor tags. Tags from this program and others along the Atlantic coast were collected by personnel of the coastal states and returned to the coordinator for processing. The USFWS National Fisheries Research Center at Leetown, West Virginia, helped evaluate tag returns.

Several factors that may have contributed to the decline in abundance of striped bass *Morone saxatilis* in Chesapeake Bay were reviewed in recent studies summarized in the Emergency Striped Bass Research Study Reports of USDI and USDC (1986, 1987). Because of the potentially synergistic and masking effects of interacting causes, no single reason for the decline was identified.

Suggested causes of the decline have included contaminants, predation and competition, availability of acceptable and nutritionally adequate prey for younger fish, nutrient overenrichment,

water-use practices, disease, natural climatic or random environmental events, and overexploitation (Weaver et al. 1986). Recent investigations (USDI and USDC 1986; MDNR 1987) suggested that, of these factors, contaminants, prey availability, nutrient overenrichment, and water-use practices may be important in localized situations, on either a temporary or sustained basis.

However, two primary factors appear to exert significant control over striped bass populations: A large component of random environmental or abiotic events that influence, either positively or negatively, the survival of eggs to the juvenile stage; and overexploitation or excessive fishing mortality, which reduces survival from the juvenile to the spawning adult stage (Weaver et al. 1986).

[1]Present address: U.S. Fish and Wildlife Service, Texas Cooperative Fish and Wildlife Research Unit, Texas Tech University, Lubbock, Texas 79409-2125, USA.

Because of the limited stock of adult striped bass (Goodyear et al. 1985; Van Winkle et al. 1988) and extensive reproductive failure within Chesapeake Bay (Boreman and Austen 1985), the U.S. Fish and Wildlife Service (USFWS) and the Maryland Department of Natural Resources (MDNR) signed a cooperative agreement in 1985. This agreement was to implement an experimental program to tag and evaluate hatchery-reared striped bass in Chesapeake Bay. In 1986, the state of Virginia and the USFWS also signed a cooperative agreement, whose goal was to investigate the feasibility of using artificial propagation to supplement the spawning stocks of striped bass in Maryland and Virginia; it has been considered a pilot program and not a full restoration program based on stocking hatchery-reared fish. Under these cooperative agreements, the USFWS committed six federal hatcheries to the production of striped bass 15–20 cm long, commonly known as phase-II fish, to be reared from fry provided by MDNR (Van Tassel 1986) and Virginia.

The intent of these efforts was to maintain the viability of the resource by artificial means, (i.e., by stocking hatchery-reared fish) until the quality of the habitat improved, the fishery was brought under coordinated control, and natural reproduction and recruitment were restored.

In the present paper, we have three objectives: To explain the organization and development of the cooperative, coastwide, striped bass restoration program; to examine fishery management techniques used to develop and coordinate a large-scale striped bass restoration and tagging program; and to present results on organizational success, coordination, tagging techniques, guidelines, data organization, and preliminary tag-return data.

Methods

During meetings held to develop an Atlantic coast striped bass management plan, biologists and managers voiced concerns about the potential effects of restoration actions on native stocks. The Striped Bass Stocking Subcommittee of the Atlantic States Marine Fisheries Commission (ASMFC) established a Technical Advisory Committee in early 1985 to address these and other concerns, especially in Chesapeake Bay. Members of the committee were selected to represent states bordering the migratory range of striped bass along the Atlantic coast, from Maine to North Carolina. Seven charges were assigned the committee by the ASMFC (Parker and Miller 1986).

The first charge was to develop an inspection system to ensure that no pathogens were present on eggs or larvae shipped into other states and then returned to Maryland to be stocked in Chesapeake Bay. The committee's second charge was to review tagging programs for striped bass and recommend a coordinated tagging system for all stocked fish. The third charge was to develop procedures to evaluate the stocking restoration program to determine its effectiveness and when it should be terminated. In charges four and five, the committee was to assess the threat posed by stocking programs to the genetic integrity of native striped bass along the Atlantic coast, and to make recommendations regarding time, size, and strain of fish to be stocked. The sixth charge was to review stocking programs in each Atlantic state to ensure they did not conflict. The final charge was to develop an evaluation program, based on marked and tagged fish, to determine if hatchery-reared fish stocked along the Atlantic coast would mature, return to the areas where stocked, and spawn.

In this paper, we have treated in detail only the charges that directly affected development of the tagging and evaluation program. Other committee decisions are briefly presented as background information.

Results

After considerable debate among members of the Technical Advisory Committee, advisors from state and federal agencies, and public and private organizations regarding the seven charges, the committee made a series of recommendations. They were approved by the Atlantic States Marine Fisheries Commission and adopted as standard operating practice.

Responses to Charges

Recommendations for pathogen control.—The committee recommended that fish be screened for pathogens, especially the IPN (infectious pancreatic necrosis) virus, and that only disease-free fish be stocked. Through 1987, all adult striped bass used in the restoration program were screened for the IPN virus. Gamete samples were obtained by biologists during manual spawning, and fry samples were collected and analyzed where natural spawning had occurred and gamete samples were not readily available. No positive IPN samples were identified.

Recommendations for a tagging program.—The committee recommended that binary-coded wire tags be used to mark hatchery-reared fish stocked along the Atlantic coast, and that all fish be tagged if fewer than 1 million were stocked in 1 year. If more than 1 million were stocked, then a percentage, based on the number released and the estimated number in the natural stock, should be marked. Each lot of fish should be marked with a unique code to allow recognition among lots.

All striped bass released in 1986–1988 were marked with binary-coded wire tags, a method developed by Jefferts et al. (1963) that allows identification of thousands of different groups. A tagging center was developed to tag fish returning to Maryland from various hatcheries along the Atlantic coast. Large circular holding tanks (2.1 m × 0.9 m) supplied with 10‰ salt water and an extensive recirculating, liquid-oxygen injection system (10–15 mg/L) were used to hold fish for as long as 12 h before tagging. Coded wire tags were placed in the adductor mandibularis muscle (a muscle below the eye) of phase-II striped bass (Klar and Parker 1986).

For the tagging operation, a specially modified trailer (12 m × 2.5 m) was used that held six coded-wire-tagging machines and quality control units, as well as large temporary holding tanks for anesthetizing fish. Large tanks outside of the trailer were used to hold fish for post-tagging recovery from the anesthetic. Tagging equipment (Mark 4 machines, Northwest Marine Technology, Incorporated, Shaw Island, Washington[1]) was used to inject a binary-coded wire tag 1.07 mm long and 0.254 mm in diameter into each fish through a 24-gauge hypodermic needle. Fish were aligned visually to the hypodermic needle, without the aid of a guiding head mold, and impaled manually on the injector needle; the tag was then injected through the hypodermic needle into the adductor muscle. The tagging machine was adjusted so that the needle was extended and stationary at the start of each cycle. Tags were magnetized in the needle before the machine was cycled. Up to 25,000 fish (over 4,000 per machine) were tagged daily with binary-coded wire tags, of which 500 also were marked with internal anchor tags. The typical tagging crew consisted of 12 members—6 to operate the six machines, 3 to handle fish for the machine operators, 2 to insert

---

[1]Reference to manufacturer or trade name does not imply endorsement by the U.S. government nor by the states of Delaware or Maryland.

TABLE 1.—Number and size of fish, and retention of coded wire tags, in phase-II striped bass tagged on 30 November 1987 and checked for presence of tags 10 or 12 d later.

Fish tagged		Days after tagging	Percent tag retention
Number	Size (number/kg)		
275	30	10	95.7
30	30	12	96.6

the internal anchor tags, and 1 to maintain the records. Tagging mortality was less than 1% when water temperature was less than 15°C. Fish were held in highly oxygenated salt water, and 1 mg/L of tricaine methanesulfonate was used as an anesthetic during the tagging. Tags were placed only in the left adductor mandibularis muscle to ensure proper tag placement and to localize the area to be searched for coded wire tags during coastal sampling programs.

All binary-coded wire tags showed agency code, year stocked, hatchery producing the fish, and stocking location. Short-term tag retention of coded wire tags has averaged about 96% (Table 1). More extensive long-term tag retention work is under way. X-rays (USFWS series number 4 taken March 1986; C. Wooley, unpublished data) showed that the coded wire tag was almost immediately encapsulated within the adductor muscle mass, thus the potential for tag loss was low. Apparently the few tags lost during this procedure worked loose within 24 hours of injection.

Subsamples were marked with an internal anchor tag (a 5-mm × 20-mm toggle attached to a 75-mm streamer) inserted just posterior to the left pectoral fin while it was compressed to the body. A scale was removed at the point of tag insertion, and a vertical incision above 5 mm long was made with a curved scalpel blade, through the peritoneum but not deep enough to damage internal organs. The anchor of the tag was inserted through the incision and set in place with a gentle pull on the streamer. All streamers were treated with an algicide. Anchor tags were used to obtain additional information on coded wire tag retention in the wild, exploitation rates, movement, and growth. They also served as indicators of movement of fish marked with binary-coded wire tags outside the sampling areas.

Recommendations for an evaluation program.—The Committee recommended that the stocking and evaluation program continue for 9 years to allow maturation and return of three year-classes of hatchery-reared fish. If stocked fish failed to return, or if they returned, spawned, and their progeny died because of environmental conditions, then the restoration program should be terminated.

Numerous surveys were conducted coastwide to obtain tag returns for program evaluation. Survey techniques included coordinated sampling by state, private, and federal agencies. Sampling in Chesapeake Bay was conducted with beach seines, gill nets, trawls, and by electrofishing. All fish within the possible size-range of stocked tagged fish were checked for the presence of tags. At this point in the cooperative restoration program, fish that tested positive for binary-coded wire tags were not sacrificed to obtain additional data.

Along the Atlantic seaboard, additional surveys and adult striped bass tagging programs have been planned for the eastern end of Long Island, New York; Hudson River, New York; Delaware River and Bay, New Jersey; Point Judith, Rhode Island; and Cape Hatteras, North Carolina. At these sampling locations, all striped bass 4 years old or less are to be surveyed for binary-coded wire tags as part of a coordinated coastal effort to obtain information on movements, migration, and exploitation of hatchery-reared and tagged fish.

Recommendations to maintain genetic integrity.—All hatchery striped bass involved in this large-scale tagging program have been identified through their hatchery phase by lot numbers and parental rivers of origin. All fish are tagged and released only into their natal rivers. Techniques used to ensure this arrangement are careful record keeping, coordination of assigned binary-coded wire tag codes with the tag manufacturer to ensure correct stocking location designation, use of a central tagging location, and coordination to help ensure that the correct tagging codes are used each day.

Recommendations for size, source, and time of stocking.—In the Chesapeake Bay striped bass restoration program, all stocking has occurred with tagged phase II (15–20 cm TL) striped bass except for an experimental stocking in June–July 1987 of phase I (35–50 mm long) fish performed to test marking methods and tag retention. In phase I striped bass, tag placement in the adductor muscle was perpendicular to the body axis. Tagging machines were used with the tag injection needle in the stationary mode. Standard length (1 mm) binary-coded wire tags were used. About 15,800 phase I fish were tagged and released into

the Patuxent River, Maryland. Post-tagging, overnight mortality rates averaged 1% during two test periods. In a control group of 3,000 phase I fish held for long-term tag retention studies by the USFWS, National Fisheries Research Center, Leetown, West Virginia, tag retention was 97.3% after 6 months (J. G. Geiger, U.S. Fish and Wildlife Service, personal communication).

Recommendations to ensure that state programs are nonconflicting.—Each year, coastal states engaged in stocking and tagging of striped bass held a review meeting. This enabled the subcommittee to verify compliance with stocking and tagging guidelines. Problems, questions on techniques, and new data and ideas were presented.

Recommendations for an evaluation program.—All coded wire and internal anchor tag returns have been coordinated for the cooperative tagging program by the USFWS. Most internal anchor tag returns were obtained by collect phone calls to the USFWS, Annapolis, Maryland. Each caller was asked by trained personnel to answer a standard questionnaire over the phone. A central depository and data-base organizational protocol have been developed, with extensive assistance from the USFWS National Fishery Research Center. The computer program PARADOX (2.0) is used to manage the data base, which is then sent to the center for data analysis.

Tagging in 1985

Adult striped bass captured during the spawning period in the Patuxent River, the Upper Bay, and the Nanticoke River produced about 15 million fry 3–5 days old—8 million from the Patuxent, 4 million from the Upper Bay, and 3 million from the Nanticoke River. The fry were transported to private, state, and federal hatcheries to be reared to an age suitable for stocking in the waters from which their parents were taken.

Nearly 400,000 striped bass fingerlings were later stocked back into the bay system. Of these, about 166,000 were stocked as 5-cm fingerlings in early summer—120,000 in the Patuxent, 20,000 in the Upper Bay, and 26,000 in the Nanticoke. The rest were reared to a larger size and released in November and December—126,000 in the Patuxent, 49,000 in the Upper Bay, and 58,000 in the Nanticoke. Nearly 187,000 of these larger fish were tagged with coded wire tags before release (Table 2), and 4,000 were also tagged with internal anchor tags (Table 3).

TABLE 2.—Numbers of hatchery-reared striped bass tagged with binary-coded wire tags, and numbers not tagged, released in Chesapeake Bay in 1985.

Hatchery[a]	Number released	
	Tagged	Not tagged
Manning State Hatchery, MDNR, Maryland	4,723	2,845
Harrison Lake NFH, Virginia	61,840	
Edenton NFH, North Carolina	56,852	12,874
McKinney Lake NFH, North Carolina	7,404	23,000
BG&E, Maryland	40,672	1,697
Horn Point, Maryland	6,405	
Frankfort NFH, Kentucky	5,092	5,930
Orangeburg NFH, South Carolina	3,939	7,061
Total	186,926	53,407

[a]MDNR, Maryland Department of Natural Resources; NFH, National Fish Hatchery; BG&E, Baltimore Gas and Electric Company; Horn Point Environmental Laboratories, University of Maryland.

Tagging in 1986

In 1986 nearly 11.5 million fry 8–10 days old were produced, 8 million from spawners in Patuxent River and 3.5 million from fish in the Upper Bay. Of this total, 8.4 million were distributed to federal hatcheries and the rest to a variety of research stations and state (Maryland DNR) and private hatcheries. In 1986 all phase II fish were tagged with coded wire tags and released between 27 October and 17 December (Table 4). A total of 367,995 fish were tagged with coded wire tags; of these, 9,750 were double-tagged with coded wire and internal anchor tags (Table 3).

Tagging in 1987

In 1987 about 18.5 million fry 6–14 days old were produced by the Maryland Department of Natural Resources. Of this total 14 million were distributed to federal, state, and private utility company hatcheries. In 1987 all phase II striped bass were tagged with binary-coded wire tags and released from 14 October to 11 December (Table

TABLE 3.—Numbers of striped bass marked with binary-coded wire tags, numbers of coded-wire-tagged fish marked with internal anchor tags, and percent of tagged fish marked with both tags in 1985–1987, Chesapeake Bay.

Year	Number of coded wire tags	Number of internal anchor tags	Percent with both tags
1985	187,000	4,000	2.1
1986	368,000	9,750	2.6
1987	801,000	9,500	1.2
Total	1,356,000	23,250	1.7

TABLE 4.—Numbers of striped bass from 12 hatcheries marked with binary-coded wire tags and released in Chesapeake Bay in 1986 and 1987.

Hatchery[a]	Number tagged and released	
	1986	1987
Manning State Hatchery, MDNR, Maryland	21,782	26,897
Bowden NFH, West Virginia	6,497[b]	84,459
Senecaville NFH, Ohio	89,335	105,561
BG&E, Maryland	10,840	116,713
NFC-Leetown, West Virginia	4,874	
PEPCO, Maryland	3,495	237,128
Harrison Lake NFH, Virginia	31,739	96,449
Edenton NFH, North Carolina	117,247	43,359
McKinney Lake NFH, North Carolina	67,601	66,723
Warm Springs NFH, Georgia	3,549	
Millen NFH, Georgia	2,754	20,705
Elkton, Maryland	8,362	1,977
Total	367,995	801,341

[a]MDNR, Maryland Department of Natural Resources; NFH, National Fish Hatchery; BG&E, Baltimore Gas and Electric Company; NFC, National Fisheries Center, U.S. Fish & Wildlife Service, Leetown, West Virginia; PEPCO, Potomac Electric Power Company, Chalk Point, Maryland; Elkton, a private fish hatchery operated by contract from the Maryland Department of Natural Resources.
[b]Phase-I fish were transferred from PEPCO, Maryland, to Bowden NFH where they were reared to phase-II size.

5). A total of 801,341 fish were tagged with binary-coded wire tags, and 9,500 of these were double-tagged (Table 3). By January 1988, more than 1.35 million striped bass had been tagged and stocked in three river systems and the upper Chesapeake Bay (Table 5).

Returns

Between November 1985 and 20 June 1988, 566 internal anchor tags from hatchery-reared striped bass were returned. Tags were recovered primarily from anglers who caught striped bass while fishing for other species, or during biological sampling programs. More than 525 coded-wire-tagged striped bass were captured during limited experimental sampling by survey crews in the bay during 1986, 1987, and early 1988. The program to recover coded wire tags was scheduled to expand in 1988 as the fish approached the size range vulnerable to gill nets, the main sampling gear used in Maryland.

Discussion

Techniques and Problem Identification

IPN.—Because IPN may cause mortality under certain conditions found in a hatchery or a large-scale tagging center, we believe it is extremely

TABLE 5.—The numbers and sizes of striped bass marked with binary-coded wire tags, and the location and time of stocking in Chesapeake Bay.

Year	Period	Size[a]	Location	Number stocked
1985	Nov–Dec	Phase II	Patuxent River	126,000
			Nanticoke River	58,000
			Upper Bay	20,000
1986	Oct–Dec	Phase II	Patuxent River	298,129
			Nanticoke River	8,866
			Upper Bay	59,282
1987	Jun–Jul	Phase I	Patuxent River	15,800
	Oct–Dec	Phase II	Patuxent River	377,242
			Choptank River	324,529
			Upper Bay	31,129
			Nanticoke	68,441

[a]Phase I, 2.5–7 cm long; phase II, 15–20 cm long.

important to have all stocks of striped bass sampled for this pathogen before a tagging program is started. Obviously, we did not want to release infected juvenile striped bass into the natural environment where they could pass the disease to uninfected wild fish. Subsequent undetected releases of large numbers of infected fish would also bias tag returns for that particular cohort. An undetected kill of a large percentage of IPN-positive fish could result in a critical bias if known numbers of tagged striped bass were used in extensive mark–recapture experiments.

Reduced effort effect.—Because of the Maryland moratorium of taking striped bass in Chesapeake Bay, and because of greatly reduced commercial and sport harvests along the Atlantic coast, more help is needed from state and federal agencies to sample the wild stock of coastal migratory striped bass. The temporary loss of samples, formerly supplied by fishermen, has required public agencies to develop a large-scale assessment program to evaluate the status of the striped bass population. Therefore, more time, energy, and money must be spent by state and federal agencies to obtain fishery-independent data on the coastal migratory stock. Excellent interagency cooperation and use of one state's field-sampling program to obtain data to meet another agency's needs must become the normal operational procedure.

Lack of external marks.—The traditional method of marking salmonids that carry binary-coded wire tags by clipping their adipose fins does not work with striped bass, who lack an adipose fin and quickly regenerate other fin clips. Instead, biologists must sample or subsample large numbers of fish to obtain tag returns. Portable detection units must be used by field crews to document the presence or absence of tags in fish.

Because portable tag detectors are capable of detecting minute disturbances in a magnetic field, they are difficult to use in rolling seas, in the presence of large metal booms and winches, and on board vessels whose vibrating diesel engines produce positive readings. Thus, with certain vessels used in fishery research along the Atlantic seaboard, biologist are not able to use existing portable tag detectors.

Because this problem is unique to striped bass being tagged with binary-coded wire tags, biologists have been working closely with the manufacturer of the equipment to document problems encountered in field sampling. The next generation of portable detectors for coded wire tags will contain a shielding mechanism to prevent interference from vibration and movement; also, the new design will enable field crews to quickly sample large numbers of striped bass, whether they are captured commercially or during the biological sampling program.

Application to Other Programs

The pilot restoration program for striped bass has operated smoothly during its first 3 years. Its operational success has been attributed largely to the decision-making process and cooperative efforts of all Atlantic coast states, the federal agencies involved, and the private sector including Baltimore Gas and Electric Company, Potomac Electric Power Company, and Delmarva Ecological Lab. The Technical Advisory Committee and the Striped Bass Board of the Atlantic States Marine Fisheries Commission provided the avenues for the state and federal agencies to jointly develop and modify the program to accommodate special needs or address specific problems. The potential success of the program was greatly increased when the USFWS assigned one employee to work full time as the program coordinator. Other factors in support of the program have included the willingness of the states to collect and share data, to collect and spawn brood fish, and to provide fry to be reared in federal and private hatcheries.

Public support and involvement in the program have been encouraged and maintained through a series of educational activities, including news releases, press conferences, video tapes, and conspicuous involvement of high-level public officials at ceremonial releases of striped bass in the Chesapeake Bay or other coastal waters. The reward system established for return of the external portion of the internal anchor tag also has increased awareness and public support for the program. Establishment of a central processing point for all tags has increased chances for success of the program by reducing the confusion and duplication associated with multiple tag-return sites.

Recommendations for Other Programs

On the basis of the early operational success of this program, we offer the following recommendations for similar ventures.
(1) Involve all affected parties in the decision-making process.
(2) Assign one person to coordinate operations of all involved parties.
(3) Establish a central point for tagging fish or use a mobile crew to tag fish at various locations.
(4) Apply tags in the same location on the fish by standardized techniques.
(5) Tag all hatchery-reared fish to be released.
(6) Establish a reward system to encourage return of tags by the public.
(7) Establish standardized surveys conducted by agency personnel to gather data on internal tags.
(8) Establish a central data-processing point to collect and analyze all data.
(9) Ensure that funding for field surveys, tag rewards, other data collection by agency personnel, and data analyses is at least equal to the funding required for producing, tagging, and releasing fish.

Acknowledgments

We thank all members and advisors to the Technical Advisory Committee for their assistance; E.A. Science and Technology, Middletown, New York, for sharing their knowledge of marking striped bass with coded wire tags; and R. S. Holt, National Marine Fisheries Service, LaJolla, California, and J. G. Loesch, Virginia Institute of Marine Science, Gloucester Point, Virginia, for reviewing the manuscript. The National Fish and Wildlife Foundation provided financial assistance for the tag-reward program.

References

Boreman, J., and H. M. Austin. 1985. Production and harvest of anadromous striped bass stocks along the Atlantic coast. Transactions of the American Fisheries Society 114:3–7.

Goodyear, C. P., J. E. Cohen, and S. W. Christensen. 1985. Maryland striped bass: recruitment declining below replacement. Transactions of the American Fisheries Society 114:146–151.

Jefferts, K. B., P. K., Bergman, and H. F. Fiscus. 1963. A coded wire identification system for macroorganisms. Nature (London) 198:460–462.

Klar, G. T., and N. C. Parker. 1986. Marking fingerling striped bass and blue tilapia with coded wire tags and Microtaggants. North American Journal of Fisheries Management 6:439–444.

MDNR (Maryland Department of Natural Resources). 1987. Second annual report on striped bass, 1986. MDNR, Tidewater Administration, Fisheries Division, Annapolis, Maryland.

Parker, N. C., and R. W. Miller. 1986. Recommendations concerning the striped bass restoration program for the Atlantic coast with emphasis on Chesapeake Bay. Special Report 10 to the Atlantic States Marine Fisheries Commission, Striped Bass Stocking Subcommittee, Washington, D.C.

USDI (U.S. Department of the Interior) and USDC (U.S. Department of Commerce). 1986. Emergency striped bass research study. USDI and USDC Report for 1985, Washington, D.C.

USDI (U.S. Department of the Interior) and USDC (U.S. Department of Commerce). 1987. Emergency striped bass research study. USDI and USDC, Report for 1986, Washington, D.C.

Van Winkle, W., K. D. Kumar, and D. S. Vaughan. 1988. Relative contributions of Hudson River and Chesapeake Bay striped bass stocks to the Atlantic coastal population. American Fisheries Society Monograph 4:255–266.

Van Tassel, J. 1986. Culture of Maryland striped bass. Maryland Department of Natural Resources, Tidewater Administration, Annapolis, Maryland.

Weaver, J. E., R. B. Fairbanks, and C. M. Wooley. 1986. Interstate management of Atlantic coastal migratory striped bass. Marine Recreational Fisheries 11:71–95.

American Fisheries Society Symposium 7:782–816, 1990

Regional Overview of Coded Wire Tagging of Anadromous Salmon and Steelhead in Northwest America

J. Kenneth Johnson

Regional Mark Processing Center, Pacific States Marine Fisheries Commission
2501 S.W. 1st Avenue, Suite 200, Portland, Oregon 97201, USA

Abstract.—Coded wire microtags (CWTs) were introduced in the Pacific Northwest in the late 1960s as an alternative to fin clipping and external tags for identification of anadromous salmonids in the region, particularly those of hatchery origin. Coastwide use of CWTs quickly followed, and fisheries agencies in Alaska, British Columbia, Washington, Oregon, and California established ocean sampling and recovery programs. Now, over 50 federal, provincial, state, Indian, and private entities release over 40 million salmonids with CWTs yearly. Regional coordination of these tagging programs is provided by the Regional Mark Processing Center operated by the Pacific States Marine Fisheries Commission. The center also maintains a centralized data base for coastwide CWT releases and recoveries, as well as for associated catch and sample data. Data are distributed to users via printed reports, magnetic media, and interactive on-line data retrieval. The system works very well despite its piecemeal growth and dependence on cooperative support by all agencies, but it has several problems. These include improperly designed CWT studies for management purposes, lack of standards for tagging levels, unstable long-term funding, inequitable burdens of cost on recovery agencies, sampling of harvest from many catch areas, un- or misreported harvest biases, lack of standardized statistical procedures for estimating variance, and limitations in marking nonhatchery stocks. Progress in solving these problems is reviewed, as are changes introduced by the USA–Canada Salmon Treaty.

The coded wire tag (CWT) is widely used by fisheries agencies on the west coast of North America to gather major information on stocks of salmon *Oncorhynchus* spp. and steelhead *Oncorhynchus mykiss*. Management of the resource is based on hatchery contribution studies, differential treatment studies, fishery contribution studies, and a variety of related studies.

The highly migratory nature of salmonid species necessitated the development of a cooperative coastwide exchange of tag data. This paper presents an overview of the system now in place, of its problems and of its future direction. It also briefly reviews the important role fin marking played as a precursor of the present system.

Fin-Marking Era

Early Coordination Efforts

Fin clipping was the standard marking method for stock identification until the early 1970s, when coded wire tags became popular. To avoid conflicts among the limited number of possible fin clips, agencies voluntarily agreed to abide by mutually established rules for fin-mark studies. These regional agreements were formalized in the early 1950s through efforts of the Committee on Anadromous Fish Marking and Tagging, more commonly known as the Mark Committee.

Work of the Mark Committee was facilitated by the Pacific Marine Fisheries Commission (PMFC; now the Pacific States Marine Fisheries Commission), an interstate fisheries compact created by the U.S. Congress in 1947. Membership on the committee consisted of one representative from each of the major state and federal fisheries agencies engaged in marking. Canadian fisheries agencies also participated on an informal but active basis.

The Mark Committee met annually in January or February to review fin-mark programs of the prior year and to coordinate and approve mark requests for coming years. Regional agreements also were reviewed for possible revision. In addition, committee members served as focal points for the exchange of fin-mark recovery data on an as-needed basis.

Regional Fin-Mark Recovery Efforts

In spite of the early coordination of fin-marking programs, sampling and mark-recovery efforts remained limited to in-state and in-province programs for many years. This changed in 1962 when the first large-scale marking program was initiated to evaluate the contribution of fall chinook salmon *Oncorhynchus tshawytscha* released from 13 production hatcheries in the Columbia River basin

(Worlund et al. 1969; Wahle and Vreeland 1978). Coastwide sampling for the study's sequestered fin marks (adipose, left ventral, right ventral, and maxillary marks in various combinations) began in 1963. Areas sampled included the major ocean commercial and sport fisheries from southeast Alaska to central California, Columbia River fisheries, parent hatcheries, and certain natural spawning grounds.

In 1965, this study was expanded to include coho salmon *O. kisutch*. Twenty hatcheries distributed over much of the Columbia River main stem participated by marking representative 10% samples of their 1965- and 1966-brood coho salmon (Wahle et al. 1974).

All data collected during the recovery phase of the study (1963–1969) were recorded on a standard form and forwarded to the Oregon Fish Commission (now Oregon Department of Fish and Wildlife) Mark Processing Center at Clackamas, Oregon. After appropriate coding, the data were keypunched onto computer cards and then tabulated. The tabulations were forwarded to the Bureau of Commercial Fisheries (now the National Marine Fisheries Service) in Seattle, where annual summary reports were produced (Worlund et al. 1969).

In 1970, Oregon's Mark Processing Center formally became the regional center when it was funded through the Anadromous Fish Act (Public Law 89-304) to establish and maintain a regional data base for mark recoveries. From 1970 through 1976, the center published annual regional summaries for fin-marking and tag releases and subsequent recoveries in the ocean and Columbia River fisheries. This effort promoted better cooperation and coordination among agencies.

Coded-Wire-Tagging Era

Introduction of Coded Wire Tags

The invention of minute coded wire tags (0.25 × 1.00 mm) that could be easily implanted in the tough nasal cartilage of juvenile salmonids (Figure 1) greatly changed marking studies because of this tag's numerous advantages over fin clipping. The first tags were developed in the 1960s (Jefferts et al. 1963; Bergman et al. 1968) and carried up to five longitudinal colored stripes. More than a dozen different colors provided approximately 5,000 different codes, compared to the 15–20 fin-mark codes normally used to identify groups of fish.

Binary-coded tags were later introduced in 1971 by Northwest Marine Technology, Inc. The new

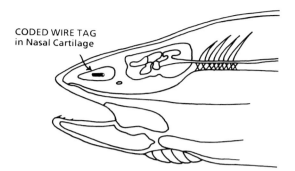

CODED WIRE TAG in Nasal Cartilage

FIGURE 1.—Longitudinal section through the head of a juvenile salmonid showing the correct placement of a coded wire tag in the nasal cartilage. (After Koerner 1977.)

tags quickly replaced color-coded tags because of the greatly improved readability and the enormous number of available codes. Standard-length binary tags (1 mm, 6-bit word) for example, have 63 possible agency codes and 3,969 codes per agency (not using zero), for a total of 250,047 unique codes. This will be adequate for many years at today's tagging rates. However, when necessary, tens of thousands of additional codes can be created with a slight change in formating on the wire.

The large number of available binary codes, low cost per tag, and ease of application opened the way to large-scale experimentation (i.e., multiple experiments on given stocks) by tagging agencies because all experimental groups could be identified accurately regardless of recovery location or time. Another major advantage was that all experimental groups could be treated the same during the tagging process, thus reducing the variability in survival and behavior imparted by clipping different fins.

Restriction of the Adipose Mark

The widespread acceptance and use of the coded wire tag (CWT) made it imperative that a single fin mark be reserved as a flag for tagged salmon and steelhead. Therefore, in February 1977, the Mark Committee recommended that the adipose fin be sequestered for tagged chinook and coho salmon. This recommendation was approved by PMFC's Salmon and Steelhead Committee (composed of fisheries management leaders from each of PMFC's five member states) on March 3, 1977, and implemented in the 1977 tagging season.

TABLE 1. Summary, by Pacific Northwest species, of fin marks that require a coded wire tag (CWT) if the fin marks are used. The adipose-fin-only mark is exclusively reserved as a CWT flag for all species except steelhead. Ad = adipose fin; LV = left ventral fin; max = maxillary.

	Fin mark					
Species	Ad	Ad + max	Ad + LV	Ad + other fin(s)	LV	LV + other fin(s)
Chinook salmon	Yes	Yes	Yes	Yes	No	No
Coho salmon	Yes	Yes	Yes	Yes	No	No
Chum salmon	Yes	Yes	No	No	No	No
Pink salmon	Yes	Yes	No	No	No	No
Sockeye salmon	Yes	Yes	No	No	No	No
Steelhead						
Coastal	No	No	No	No	No	No
Columbia basin	No	No	Yes	No	Yes	Yes

This restriction was expanded later to include chum *Oncorhynchus keta*, sockeye *O. nerka*, and pink salmon *O. gorbuscha* and steelhead, with some exceptions made for geographic areas and use of multiple fin clips. Current restrictions on use of the adipose clip are summarized in Table 1.

Concurrent with the restriction of the adipose clip, the Salmon and Steelhead Committee directed agencies to immediately phase out multiple fin marking of chinook and coho salmon for experiments that required recovery at sea or in major rivers. It was expected that regional sampling would continue for studies already in progress. However, the end result was that 1976 was the last year in which coastwide sampling was conducted for multiple fin marks.

Changes in Regional Coordination

The dramatic upsurge in coded wire tagging in the mid-1970s placed an increasing burden on the data-processing facilities of individual recovery agencies and especially the Regional Mark Processing Center (RMPC) at Clackamas. Therefore, in June 1976, PMFC's Salmon and Steelhead Committee upgraded RMPC operations by establishing a Regional Mark Coordinator position to facilitate interagency coordination and timely exchange of CWT release and recovery data. This was done in May 1977.

The Salmon and Steelhead Committee also recommended that the Mark Center be transferred to PMFC to further facilitate interagency coordination, because PMFC could more effectively serve as intermediary among the various state and federal agencies. This transfer was carried out in July 1977 following unanimous approval of the directors of Alaska Department of Fish and Game (ADFG), Washington Department of Fisheries (WDF), Oregon Department of Fish and Wildlife (ODFW), Idaho Department of Fish and Game (IDFG), and California Department of Fish and Game (CDFG).

Current Procedures for Regional Coordination

Functions and Duties of the RMPC

The reorganization of the Mark Center in 1977 merged the previously separate but closely interrelated functions of regional coordination and data management into a single operation. Duties for each of these functions are summarized below, along with comments on how the tasks are carried out.

Regional coordination tasks.—The Mark Center coordinates tagging programs by (1) establishing regional agreements for fin marking and use of coded wire tags with the assistance of agency coordinators; (2) recommending changes for upgrading the regional CWT data base to meet expanding or changing user requirements; (3) assisting agencies to improve timeliness of reporting, with special emphasis on tag recovery data; and (4) developing recommendations for improving coordination and quality of coded wire tagging studies, with emphasis on experimental design, sampling design, estimation procedures, statistical problems, and documentation.

These tasks are achieved by several methods, including personal contacts by the Regional Mark Coordinator, use of meetings and workshops, and preparation of technical reports. In addition, the Mark Committee plays an invaluable role in facilitating regional coordination efforts of the RMPC.

Data management tasks.—The Mark Center manages data by (1) maintaining and upgrading a regional data base for all CWT releases and recoveries, plus a separate data base for fin marks; (2) ensuring that reported data meet established format standards and pass validation procedures; (3) producing and distributing timely CWT release and recovery data reports; (4) providing magnetic tape copies of data upon request; and (5) implementing recommended changes in the regional data base to meet expanding requirements for new information.

The primary focus of the RMPC's data management activities since 1977 has been to serve as a clearinghouse for CWT release and recovery data, with special emphasis on timely reporting of data, standardization of data formats, and integrity of the data. Analysis of the recovery data, however, has remained the responsibility of the reporting agencies and other interested data users.

Distribution of data is achieved by hard-copy reports, magnetic media, and more recently by interactive on-line data retrieval. Among the regional data reports is the "Pacific Salmonid Coded Wire Tag Release Report," which appears annually in the late spring and documents CWT applications for all Pacific coast salmonid studies that use the adipose fin clip. It includes both the most recent year's releases plus all previous releases back to 1971. It also provides a complete summary of regional agreements on tagging and fin marks.

In September, a midyear CWT Release Report is distributed to tag recovery agencies. The report is limited to tag codes released during the first 6–7 months of the current year. The information is used by tag recovery labs to verify tags recovered from juvenile fish and early returning "jacks."

The RMPC published annual "Recovery Reports" from 1970 through 1982. The reports provided summaries of observed and estimated recoveries and the associated catch and sample statistics, organized by agency–fishery–area strata in 2-week time periods. The data series was ended in 1983 because of continuing problems in getting final data on a timely basis from all recovery agencies for a given year. Consequently, the recovery reports typically lagged 2–3 years or more behind the year of recovery. The delay greatly reduced the value of the data to fishery managers. This problem was resolved in 1983 with the development of interactive on-line data retrieval capability for accessing both final and preliminary CWT recovery data. This afforded users the option of down-loading "brood reports" for any given tag code, with total recoveries reported across all agencies, fisheries, areas, and years. Hard-copy reports are now also available from the RMPC on request.

Each spring the RMPC publishes the "Mark List," which provides a listing of all salmon and steelhead fin marks (other than the adipose and CWT) used for studies not requiring ocean recovery. The report is cumulative and includes fin-mark usage from 1972 onward. Although not directly related to CWT studies, the report plays an important role in coordinating the limited fin marks available to agencies for marking studies.

Role of the Mark Committee

Membership.—All tagging and recovery agencies on the Pacific coast are represented by the 16-member Mark Committee. Membership includes mark coordinators for the five member states of PMFC (Alaska, Washington, Oregon, Idaho, and California), the National Marine Fisheries Service (NMFS), U.S. Fish and Wildlife Service (USFWS), Canada Department of Fisheries and Oceans (CDFO), British Columbia Ministry of Environment and Parks, Fisheries Branch (BCEP), and the Metlakatla Indian Community in southeast Alaska. In addition, the Northwest Indian Fisheries Commission (NIFC) coordinates the tagging and fin-marking activities of 20 treaty tribes in western Washington. Private aquaculture, universities, and other nongovernmental organizations are coordinated through the respective state or provincial coordinator.

Duties.—The Mark Committee provides oversight and guidance to the Regional Mark Coordinator in carrying out the operations of the Mark Center. In addition, the Mark Committee meets each year in February to expedite coastwide coordination of fin-marking and tagging activities. It is during this annual meeting that regional agreements are reviewed and updated if necessary.

Mode of decision making.—Regional agreements and restrictions on fin marking and coded wire tagging are reached by committee consensus after thorough discussion of the issues. A 30-d review period follows publication of the minutes to allow for agency reversal on an issue if necessary. If no objections are raised, the agreement stands as recorded in the minutes.

In those situations where unanimity cannot be achieved, the decision is reached by a 75% or greater affirmative vote. Eleven votes are possible (Table 2); a single vote is assigned to the state level or federal agency level regardless of the respective number of coordinators serving on the committee. Canadian agencies are treated similarly in that the province level (BCEP) and federal level (CDFO) are accorded separate votes.

TABLE 2. Votes (total, 11) assigned to Mark Committee members in the event there is no consensus on an issue involving fin marking or coded wire tagging. Private and other nongovernmental organizations are represented by state or provincial coordinators.

Jurisdiction	Committee representatives (total)	Number of votes
USA		
State agencies		
Alaska	Alaska Department of Fish and Game: Southeast and South Central regions (2)	1
Washington	Washington Departments of Fisheries and of Wildlife (2)	1
Oregon	Oregon Department of Fish and Wildlife (2)	1
California	California Department of Fish and Game (1)	1
Idaho	Idaho Department of Fish and Game (1)	1
Federal agencies		
U.S. Fish and Wildlife Service	Region-wide (1)	1
National Marine Fisheries Service	Alaska and Northwest regions, and Northwest and Alaska Fisheries Center[a] (3)	1
Indian groups		
Southeast Alaska	Metlakatla Indian Community	1
Western Washington	Northwest Indian Fisheries Commission, 20 tribes (1)	1
Canada		
Federal level	Canada Department of Fisheries and Oceans (1)	1
Provincial level	British Columbia Ministry of and Parks, Fisheries Branch	1

[a]Now divided into the Northwest Fisheries Science Center and the Alaska Fisheries Science Center.

Compliance with regional agreements.—The Mark Committee does not have any legal authority to enforce the regional agreements. Therefore, cooperation and compliance are voluntary. This has not proven to be a serious weakness because all agencies benefit from standardized tagging and sampling procedures. In addition, there exists tremendous peer pressure among the agencies to support the system because noncompliance can negatively affect studies of other agencies. The system works wonderfully well without having to resort to extreme measures to resolve problems, such as loss of funding support or embargoes on recovery data.

Current Tagging Program

Scale of Tagging Effort and Cost

Some 54 state, federal, Indian, and private entities in the USA and Canada (Table 3) presently participate in a massive coastwide coded-wire-tagging effort to provide essential data for effective conservation and management of Pacific salmonid stocks. This information forms the basis for monitoring the fisheries, allocating harvest rights among competing domestic users, improvements in productivity of hatchery stocks, establishment of escapement goals, and satisfaction of Indian treaty obligations. These data also play a key role in USA–Canada Salmon Treaty allocations and management of transboundary stocks.

Over 40 million juvenile salmon and steelhead are now tagged annually. Chinook salmon tagging levels are the highest (about 28 million), followed by coho salmon (10 million). Tagging of steelhead and of chum, pink, and sockeye salmon is of minor importance at about 2 million, 1 million, 500,000, and 350,000 fish, respectively.

This massive tagging effort requires approximately 1,600 new tag codes each year. Hundreds of separate studies are involved, many of which include replication groups as part of the basic design. Total cost exceeds US$3.5 million annually. The cost per individual fish ranges between 6 and 10 cents, depending on local labor costs, logistics of tagging, and number of tags purchased for a given code. An additional $7–8 million is expended annually coastwide for tag recovery programs in U.S. and Canadian commercial and recreational fisheries. Tag recoveries from returning adult fish are on the order of 300,000 per year.

Salmon and steelhead feed in the ocean from one to five or more years, depending on the species, before returning to spawn in their natal streams. Consequently, many millions of tagged fish from several brood years are present in the Pacific Ocean at any given time. The multiplicity

TABLE 3. Federal, state, Indian, and private entities in Pacific northwestern North America that use coded wire tags with salmonid fishes.

Acronym	Entity
AAI	Alaska Aquaculture, Inc.
ADFG	Alaska Department of Fish and Game
AKAF	Alaska Aquaculture Foundation, Inc.
AKI	Armstrong Keta, Inc. (Alaska)
ANAD	Anadromous, Inc. (Oregon)
BCEP	British Columbia Environment and Parks
CDFG	California Department of Fish and Game
CDFO	Canada Department of Fisheries and Oceans—Operations
CDFR	Canada Department of Fisheries and Oceans—Research
CEDC	Clatsop Economic Development Committee (Oregon)
CIAA	Cook Inlet Aquaculture Association (Alaska)
COOP	Washington Department of Fisheries—Cooperative
DIPC	Douglas Island Pink and Chum, Inc. (Alaska)
DOMS	Domsea Farms, Inc. (Oregon)
ELWA	Elwha Indian Tribe (Washington)
FWS	U.S. Fish and Wildlife Service
HOH	Hoh Indian Tribe (Washington)
HSU	Humboldt State University (California)
HVT	Hoopa Valley Indian Tribe (California)
IDFG	Idaho Department of Fish and Game
LUMM	Lummi Indian Tribe (Washington)
MAKA	Makah Indian Tribe (Washington)
MIC	Metlakatla Indian Community (Alaska)
MUCK	Muckleshoot Indian Tribe (Washington)
NISQ	Nisqually Indian Tribe (Washington)
NMFS	National Marine Fisheries Service
NSRA	Northern Southeast Regional Aquaculture Association (Alaska)
OAF	Oregon Aquafood, Inc.
ODFW	Oregon Department of Fish and Wildlife
OPSR	Oregon-Pacific Salmon Ranch
OSU	Oregon State University
PNPT	Point No Point Treaty Council (Washington)
PPWR	Puget Power (Washington)
PUYA	Puyallup Indian Tribe (Washington)
PWSA	Prince William Sound Aquaculture Association (Alaska)
QDNR	Quinault Department of Natural Resources (Washington)
QUIL	Quileute Indian Tribe (Washington)
SJ	Sheldon Jackson College (Alaska)
SKOK	Skokomish Indian Tribe (Washington)
SOF	Silverking Oceanic Farms (California)
SQAX	Squaxin Indian Tribe (Washington)
SSC	Skagit System Cooperative (Washington)
SSRA	Southern Southeast Regional Aquaculture Association (Alaska)
STIL	Stillaguamish Indian Tribe (Washington)
SUQ	Suquamish Indian Tribe (Washington)
TULA	Tulalip Indian Tribe (Washington)
UAJ	University of Alaska-Juneau
UI	University of Idaho
USFS	U.S. Forest Service
UW	University of Washington College of Fisheries
VFDA	Valdez Fisheries Development Association (Alaska)
WDF	Washington Department of Fisheries
WDW	Washington Department of Wildlife
YAKI	Yakima Indian Tribe (Washington)

of tagging studies today represents a long-term, multimillion-dollar investment by state, federal, Indian, and private sector entities.

Types of Tagging Studies

Although there are many kinds of tagging studies, they can be divided into three basic types (PMFC 1982a): experimental (e.g., multiple comparison); stock assessment (from the hatchery viewpoint); and stock contribution[1] (from the fishery[2] viewpoint).

Experimental tagging studies are designed to compare the relative survival or contribution of two or more experimental groups to the fisheries. Studies in this category deal with diet comparisons, time or site of release, pond density factors, disease control, and genetics. Most tagging studies to date have been of this type.

Stock assessment studies (hatchery viewpoint) have localized objectives and are designed to measure contributions and distributions of particular stocks among various fisheries, as well as escapement of those stocks. With this information, the success of a hatchery's production or of natural production can be evaluated. The data may also have value to fishery management if adequate numbers of fish are tagged.

Stock contribution studies also are done for stock assessment purposes. However, the focus is from the fishery management perspective. In this case, fishery managers seek information on the contribution rates of key stocks in a given fishery (by time and area strata) in order to better manage harvest rates for conservation of the resource.

The major difference between stock assessment and stock contribution studies is in the number of fish tagged. Stock contribution studies require far more tagged fish to generate meaningful recovery rates on a regional basis.

Principal Tagging Facilities

Tagging programs are carried out at over 330 federal, state, Indian, and private hatcheries and rearing facilities on the west coast. In addition, wild stocks are trapped and tagged at numerous sites. The principal tagging facilities are presented by state and province in Figures 2–8. Unless otherwise noted in the legend, the facilities are operated by the state or province. Sites for tagging naturally produced fish in streams are not plotted because of the large number involved.

---

[1]Contribution is defined as the number of fish of a defined group occurring in a specific fishery.

[2]Fishery, as used here, is defined in a broad sense to include both harvest and escapement (fish that reach their natal streams on spawning runs).

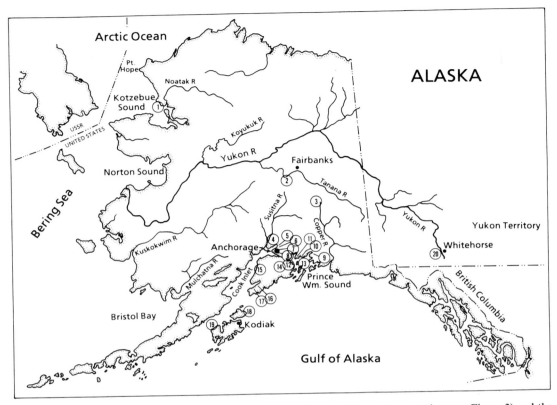

FIGURE 2.—Hatcheries and rearing facilities in Alaska (exclusive of the southeast region; see Figure 3) and the Yukon that release salmonids with coded wire tags.

1 Sikusuilaq	8 Elmendorf	15 Crooked Creek (formerly
2 Clear	9 Solomon Gulch (VFDA)	Kasilof)
3 Gulkana	10 Cannery Creek (ADFG, PWSA)	16 Halibut Cove
4 Big Lake	11 Esther Lake (PWSA)	17 Tutka Bay
5 Eklutna	12 Main Bay	18 Kitoi Bay
6 Fire Lake (closed)	13 Armin F. Koernig (PWSA)	19 Karluk
7 Fort Richardson	14 Trail Lakes	20 Whitehorse (CDFO)

ADFG = Alaska Department of Fish and Game   PWSA = Prince William Sound Aquaculture Association
CDFO = Canada Department of Fisheries and Oceans   VFDA = Valdez Fisheries Development Association

Alaska: south-central region.—Alaskan hatcheries north of the panhandle are concentrated in the Cook Inlet and Prince William Sound areas of the south-central region (Figure 2). Hatchery production in the vast interior is limited to the Clear Hatchery on the Tanana River, a tributary of the extensive Yukon River system. (CDFO also maintains a hatchery far upstream on the Yukon River at Whitehorse where tagged fish are released). Sikusuilaq Hatchery is far to the northwest above the Arctic Circle in Kotzebue Sound.

Recent tagging levels for this region have varied between 700,000 and 1.6 million fish per year. Coho, chum, and pink salmon have received the most emphasis at somewhat equivalent levels (200,000–700,000 range). However, plans for 1988 and 1989 call for tagging 1 million pink salmon but only 100,000–200,000 coho and chum salmon.

Alaska: southeast region.—Twenty-five hatcheries release tagged fish in southeast Alaska (Figure 3). Of these, 11 are operated by regional aquaculture associations (e.g., NSRA, SSRA) and private nonprofit groups (e.g., AKI, DIPC). In addition, the Metlakatla Indian Community operates a large hatchery (Tamgas Creek) on Annette Island in the southernmost part of the state.

Chinook, coho, and chum salmon tagging levels are on the order of 1.4 million, 1 million, and 500,000

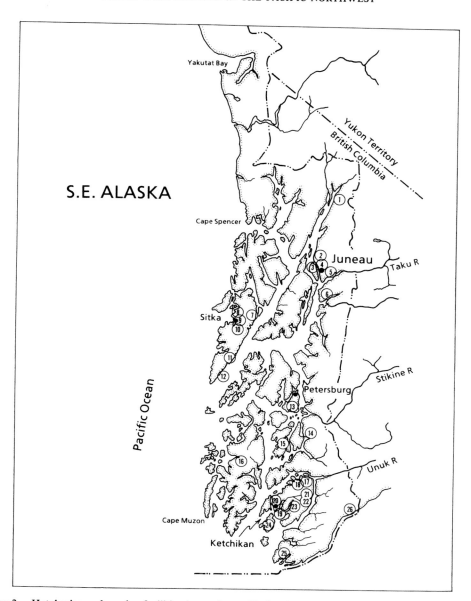

FIGURE 3.—Hatcheries and rearing facilities in southeast Alaska that release salmonids with coded wire tags.

1 Pullen Creek
2 Auke Creek (NMFS, UAJ)
3 Fish Creek
4 Salmon Creek (NSRA, ADFG)
5 Sheep Creek (DIPC)
6 Snettisham
7 Hidden Falls
8 Starrigavan (closed)
9 Sheldon Jackson (SJ)

10 Medvejie CIF (NSRA)
11 Little Port Walter (NMFS)
12 Port Armstrong (AKI)
13 Crystal Lake
14 Earl West Cove (SSRA, ADFG)
15 Burnett Inlet (AKAF)
16 Klawock
17 Shrimp Creek (SSRA)
18 Neets Bay (SSRA)

19 Whitman Lake (SSRA)
20 Deer Mountain
21 Beaver Falls
22 George Inlet (SSRA)
23 Carroll Inlet (SSRA)
24 Tamgas Creek (MIC)
25 Nakat Inlet (SSRA)
26 Marx Creek

AKAF = Alaska Aquaculture Foundation, Inc.
AKI = Armstrong Keta, Inc.
DIPC = Douglas Island Pink and Chum, Inc.
MIC = Metlakatla Indian Community
NMFS = National Marine Fisheries Service

NSRA = Northern Southeast Regional Aquaculture Association
SJ   = Sheldon Jackson College
SSRA = Southern Southeast Regional Aquaculture Association
UAJ = University of Alaska, Juneau

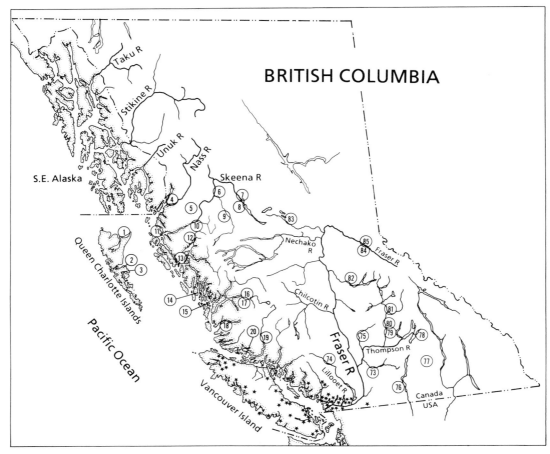

FIGURE 4.—Hatcheries and rearing facilities in British Columbia that release salmonids with coded wire tags. Sites marked with stars in the left panel are numbered in the right panel.

1 Masset (CDP)	16 Bella Coola (PIP)	31 Robertson Creek
2 Sachs Creek (PIP)	17 Snootli Creek	32 Thorton Creek (CDP, PIP)
3 Pallant Creek	18 Owekeeno (CDP)	33 Carnation Creek
4 Kincolith (CDP)	19 Devereux Creek	34 Nitinat River
5 Kitsumkalum River	20 Scott Cove (PIP)	35 Cowichan Lake
6 Kispiox River (CDP)	21 Port Hardy–Stephens (CDP)	36 San Juan River (CDP)
7 Fort Babine (CDP)	22 Port Hardy–Quatse (CDP)	37 Sooke River (PIP)
8 Fulton River	23 O'Connor Lake (BCEP)	38 Goldstream River (PIP)
9 Toboggan Creek (CDP, PIP)	24 Colonial River (PIP)	39 Cowichan River (CDP)
10 Terrace (CDP, PIP)	25 Marble River (PIP)	40 Chemainus River
11 Oldfield Creek (PIP)	26 Nimpkish (CDP)	41 Nanaimo River (CDP)
12 Kitimat River	27 Tahsis (PIP)	42 Pacific Biological Station
13 Hartley Bay (CDP, PIP)	28 Conuma River	43 Millstone River (SPU)
14 Klemtu (CDP)	29 Gold River (PIP)	44 Little Qualicum River
15 Bella Bella (CDP)	30 Clayoquot (CDP)	45 Big Qualicum River

fish, respectively. However, in sharp contrast to the south-central region, pink salmon production is limited and tagging levels are less than 100,000.

British Columbia.—Eighty-four hatcheries and rearing facilities release coded-wire-tagged salmonids in British Columbia (Figure 4). Most of them are on the lower Fraser River system and on the southern half of Vancouver Island. Public participation is strongly encouraged, and 49 of the facilities are operated by CDFO as

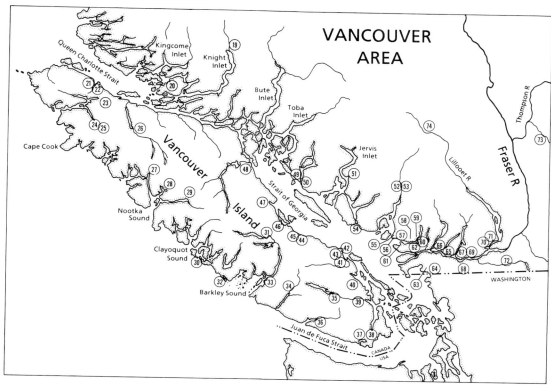

FIGURE 4.—Continued.

46 Rosewall Creek
47 Puntledge River
48 Quinsam River
49 Sliammon River (CDP)
50 Powell River (CDP)
51 Vancouver Bay (SPU)
52 Tenderfoot Creek
53 North Vancouver Outdoor
   School (PIP)
54 Sechelt (CDP)
55 Horseshoe Bay (PIP)
56 West Vancouver Laboratory
57 Capilano River
58 Seymour River (CDP, PIP)

59 Deep Cove Pens (CDP)
60 Noons Creek (PIP)
61 False Creek Pens (CDP)
62 Brunette River (PIP)
63 Surrey Rearing (BCEP)
64 Little Campbell River (PIP)
65 Kanaka Creek (PIP)
66 Alouette River (SPU)
67 Stave Lake (PIP)
68 Fraser Valley (BCEP)
69 Inch Creek
70 Chehalis River (CDP)
71 Chehalis River
72 Chilliwack River

73 Spius Creek
74 Birkenhead River
75 Loon Creek (BCEP)
76 Summerland (BCEP)
77 Shuswap River
78 Eagle River
79 Louis Creek (PIP)
80 Barriere (CDP, PIP)
81 Clearwater River
82 Quesnel River
83 Fort St. James (CDP)
84 Penny Creek
85 Penny Creek (CDP)

BCEP = British Columbia Ministry of Environment and Parks
CDP = Community Economic Development Project

PIP = Public Involvement Project
SPU = Small Projects Unit

either Community Economic Development Programs (CDP), Small Project Units (SPU), or Public Involvement Programs (PIP).

British Columbia facilities release approximately 10 million tagged salmonids yearly. Chinook salmon are the most important species (6.0–6.5 million), followed by coho salmon (about 2.5 million). Steelhead and chum salmon are the only other species of importance, with tagging levels on the order of 1 million and 500,000 fish,

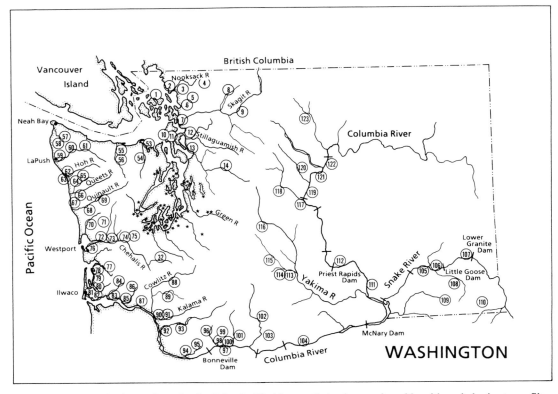

FIGURE 5.—Hatcheries and rearing facilities in Washington that release salmonids with coded wire tags. Sites marked with stars in the left panel are numbered in the right panel.

1 Glenwood Springs	23 Keta Creek (T)	44 Shelton
2 Lummi Sea Pens (T)	24 Totem Marina Pens	45 Enetai (T)
3 Bellingham	25 Puyallup Tribal (T)	46 Hood Canal
4 Nooksack River	26 Puyallup River	47 Gorst Creek
5 Skookum Creek (T)	27 Narrows Pens	48 Manchester (F)
6 Samish River	28 Chambers Creek	49 Grovers Creek (T)
7 Swinomish Raceways (T)	29 Garrison Springs	50 Suquamish Pens (T)
8 Puget Power (U)	30 Kalama Creek (T)	51 Port Gamble Pens (T)
9 Skagit River	31 Deschutes River	52 Quilcene River (NFH)
10 Oak Harbor Pen	32 Skookumchuck River	53 Hurd Creek
11 Whidbey Island Pens	33 Allison Springs	54 Dungeness River
12 Stillaguamish Tribal (T)	34 Squaxin Island Pens (and	55 Lower Elwha (T)
(formerly Armstrong Creek)	South Sound Pens)	56 Elwha Channel
13 Tulalip Creek (T)	35 Filucy Bay Pens	57 Hoko Ponds
14 Skykomish River	36 Shaws Cove Pens	58 Makah (NFH)
15 Montlake (F)	37 Fox Island Pens	59 Lonesome Creek (T)
16 University of Washington (CF)	38 Gig Harbor Pens	60 Soleduck River
17 Seattle Aquarium	39 Minter Creek	61 Bear Springs
18 Elliott Bay Pens	40 Hupp Springs	62 Chalaat Creek (T)
19 Issaquah Creek	41 Coulter Creek	63 Reservation Slough (T)
20 Northwest Steelheaders (S)	42 George Adams	64 Mule Pasture Pond (T)
21 Green River	43 McKernan	65 Shale Creek
22 Crisp Creek		

respectively. Steelhead fall under the jurisdiction of BCEP, whereas CDFO is responsible for managing all salmon species.

Washington.—Over 120 hatcheries and rearing facilities in Washington participate in tagging salmonids, many of which are in the southern half of

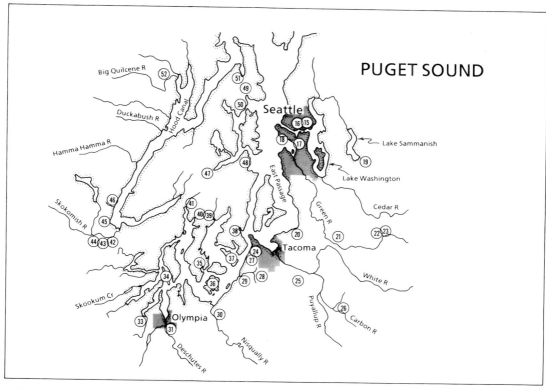

FIGURE 5.—Continued.

66 Salmon River Pond (T)
67 Raft River Pond (T)
68 Quinault River (NFH)
69 Quinault Lake (T)
70 Humptulips River
71 Mayr Brothers Ponds
72 Aberdeen
73 Wynoochee River Pens
74 Satsop Springs
75 Simpson
76 Westport Pens
77 Willapa River
78 Bay Center Springs
79 Nemah River
80 Naselle River
81 Fisher Slough
82 Sea Resources
83 Grays River
84 Weyco Pond
85 Beaver Creek

86 Elokomin River
87 Abernathy Salmon Culture
   Development Center (F)
88 Cowlitz River
89 Toutle River
90 Lower Kalama
91 Kalama Falls
92 Lewis River
93 Speelyai
94 Skamania
95 Washougal River
96 Carson River (NFH)
97 Drano Lake Pens (F)
98 Little White Salmon (NFH)
99 Willard (NFH)
100 Spring Creek (NFH)
101 Big White Salmon (F)
102 Klickitat River
103 Goldendale
104 Rock Creek Ponds (F)

105 Lyons Ferry
106 Little Goose Dam (F)
107 Lower Granite Dam (F)
108 Tucannon River
109 Curl Lake
110 Cottonwood Pond
111 Ringold
112 Priest Rapids
113 Nelson Springs (S)
114 Naches
115 Nile Spring (T)
116 Cle Elum Ponds (T)
117 Rocky Reach
118 Leavenworth (NFH)
119 Turtle Rock
120 Entiat River (NFH)
121 Chelan
122 Wells Channel
123 Winthrop (NFH)

CF   = College of Fisheries, Fisheries Research Institute
NFH = National Fish Hatchery
F   = other federal facilities

S = sportsmen's project
U = public utility

Puget Sound (Figure 5). The majority are state hatcheries operated by WDF (salmon) and WDW (steelhead). However, Indian tribes also play a prominent role and release tagged salmonids at 24 tribal facilities, mostly in coastal Washington and Puget Sound. Another 17 federal facilities (USFWS and NMFS) also release large numbers of tagged salmonids.

As in British Columbia and southeast Alaska, chinook salmon are tagged at the highest rates, followed by coho salmon. The WDF is the primary tagging agency and tags 6 million chinook annually. The Columbia River (3 million) and Puget Sound (2.5 million) receive the most emphasis, followed by the coast (600,000). The WDF releases 2 million tagged coho salmon annually, with somewhat comparable levels of tagging for the Columbia River (525,000), coast (570,000), and Puget Sound (885,000). Much of this tagging is done to meet USA–Canada Salmon Treaty requirements.

The tribes release approximately 2 million tagged chinook and 800,000 tagged coho salmon. Approximately 75% of the tagging is associated with the USA–Canada assessment efforts for indicator stocks.

Federal tagging programs primarily focus on marking chinook salmon in the Columbia Basin. The NMFS is currently tagging 2 million fall chinook and 600,000 spring chinook salmon in the Columbia River basin. The primary objective of the tagging is to evaluate dam passage and transportation programs. The USFWS also tags approximately 2.5 million chinook salmon in Washington, of which 2 million are released from Columbia River hatcheries. Tagging in this case is done primarily for hatchery evaluations.

Tagging of steelhead is relatively limited; WDW tags approximately 500,000 each year. Some steelhead are also tagged by the Indian tribes (125,000) and the USFWS (250,000).

Oregon.—Oregon has 42 hatcheries and rearing facilities that release tagged salmon and steelhead (Figure 6). Most are located on the coast, the lower Columbia River, or Willamette River. In addition, nearly all are state facilities operated by ODFW. However, Oregon Aqua-Foods, a private aquaculture venture, has been an important contributor in terms of both production and tagged fish released. The Warm Springs and Eagle Creek national fish hatcheries are the only federal hatcheries in Oregon.

Chinook salmon tagging levels for ODFW and private agencies are 4.1 million and 500,000, while coho salmon levels are 1.5 million and 350,000, respectively. The ODFW also manages steelhead and releases some 400,000 tagged smolts annually.

Idaho.—In spite of their great distance from the ocean, Idaho's anadromous fish hatcheries (Figure 7) are major contributors in terms of both production and numbers of tagged chinook

salmon (1.7–2.0 million tags). Millions of steelhead also are produced, but their tagging levels are on the order of 300,000–500,000.

Part of the reason for the reduced emphasis on tagging steelhead is that the adipose clip on steelhead was desequestered in the Columbia basin in 1983, so that it could be used for harvest management purposes as a flag signifying a hatchery stock. The left ventral (LV) fin clip was sequestered in its place as the flag for CWT-marked steelhead in the Columbia basin (Table 1). Because there is no coastwide recovery effort for LV-marked steelhead, there is less incentive to make the additional effort to tag the fish.

The Clearwater and Salmon rivers are the only chinook–steelhead-producing waters in Idaho now that Hells Canyon Dam has blocked off the upper Snake River (Figure 7). However, 4 of the 11 hatcheries (McCall, Niagara Springs, Hagerman National Fish Hatchery, and MacKay) are located upstream of Hells Canyon Dam because of superior water quality and other factors. Fish produced by these facilities are trucked to the Salmon River for release.

California.—California has 33 hatcheries and fish rearing facilities in the northern half of the state that release tagged chinook and coho salmon and steelhead (Figure 8). Eleven of these (primarily rearing ponds) are on the Klamath–Trinity rivers and are operated by CDFG, USFWS, and the Hoopa Indian Tribe. Most of the others are on the Eel River (north coast) and the Sacramento–San Joaquin system in the Central Valley. Nearly all are operated by CDFG.

Approximately 3.2 million chinook salmon are tagged annually. The majority of these are Central Valley stocks (1.9 million), followed by Klamath–Trinity rivers (900,000) and Eel River stocks (300,000). Some additional tagging (about 50,000) is done in small coastal streams.

Coho salmon tagging is limited (about 200,000–400,000) and is presently focused on Russian River, Mad River, and Klamath–Trinity river stocks.

Tag Recovery and Estimation Procedures

Regional Sampling Effort

Many agencies release tagged salmonids, but the burden of ocean tag recoveries falls on five agencies: ADFG, CDFO, WDF, ODFW, and CDFG. The primary responsibility for sampling the commercial, tribal, and recreational fisheries in the lower Columbia River is shared by ODFW and WDF. Limited sampling is done by a few of

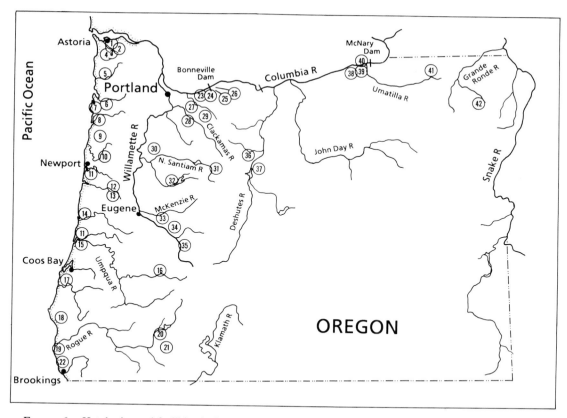

FIGURE 6.—Hatcheries and facilities in Oregon that release salmonids with coded wire tags.

1 Big Creek	14 Domsea Farms (P)	28 Clackamas
2 Klaskanine River	15 Anadromous, Inc. (P)	29 Eagle Creek (NFH)
3 Klaskanine River Ponds, South Fork	16 Rock Creek	30 Stayton-Aumsville Ponds
4 Vanderveldt Ponds	17 Bandon	31 Marion Forks
5 Nehalem River	18 Elk River	32 South Santiam
6 Trask River	19 Indian Creek Pond	33 McKenzie River
7 Whiskey Creek (OSU)	20 Cole Rivers	34 Dexter Ponds
8 Cedar Creek	21 Butte Falls	35 Willamette River
9 Salmon River	22 Oregon Pacific Salmon (P) (formerly Burnt Hill Salmon)	36 Warm Springs (NFH)
10 Siletz River	23 Wahkeena Pond	37 Round Butte
11 Oregon Aqua-Foods (P) (Coos Bay, Yaquina Bay)	24 Bonneville	38 Irrigon
12 Fall Creek	25 Cascade	39 Bonifer Pond
13 Alsea	26 Oxbow	40 Social Security Pens (F)
	27 Sandy River	41 Lookingglass
		42 Wallowa River

NFH = National Fish Hatchery
F = other federal facility
P = private aquaculture
OSU = Oregon State University

the other agencies. The Quinault Indian Nation on the Olympic Peninsula (Washington) samples its fisheries. In Alaska, NMFS and the Metlakatla Indian Community maintain sampling programs for their respective fisheries and escapement. The National Marine Fisheries Service also did extensive sampling of outmigrant juveniles in the lower Columbia River during 1977–1983 (Dawley et al. 1985) and may resume the program if funding is restored. Lastly, the USFWS maintains a sampling program on the Klamath–Trinity River system in northern California, as well as sampling programs at its various hatcheries in Washington, Oregon, Idaho, and California.

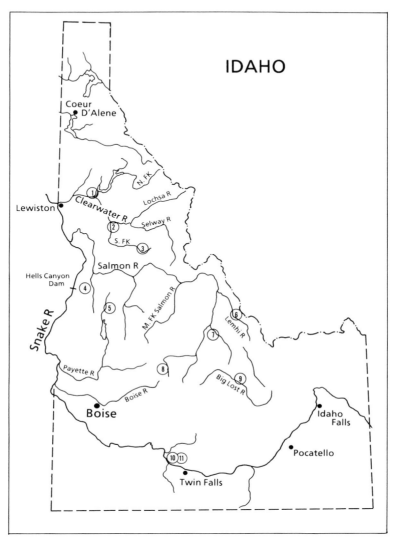

FIGURE 7.—Hatcheries and rearing facilities in Idaho that release salmonids with coded wire tags.

1 Dworshak (NFH)   5 McCall   9 Mackay
2 Kooskia (NFH)   6 Hayden Creek   10 Hagerman (NFH)
3 Red River Pond   7 Pahsimeroi   11 Niagara Springs
4 Rapid River   8 Sawtooth

NFH = National Fish Hatchery

Sampling Design

The sampling programs of the participating agencies are comparable in overall design but differ in many specifics because of constraints imposed by local conditions and differing approaches to mark recovery. There are, however, five common elements of the major recovery programs, discussed below.

Sampling of commercial fisheries.—All recovery programs sample landings of commercial marine and main-stem river fisheries for adipose-clipped chinook and coho salmon. Representative samples are randomly taken at ports throughout the state or province at appropriate intervals to track fluctuations in the species and stocks of interest.

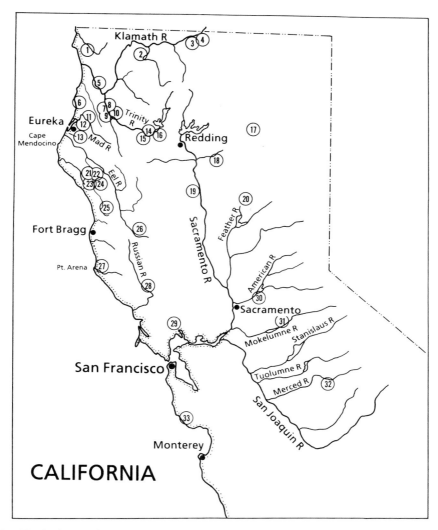

FIGURE 8.—Hatcheries and facilities in California that release salmonids with coded wire tags.

1 Rowdy Creek
2 Kelsey Creek Channel
3 Fall Creek Ponds
4 Iron Gate
5 Chappel (F)
6 Little River Ponds
7 Mill Creek Ponds (T)
8 Supply Creek Ponds (T)
9 Tish Tang Creek Ponds (T)
10 Horse Linto Creek Ponds
11 Mad River

12 Arcata
13 Cochran Ponds
14 Sawmill Pond
15 Ambrose Pond
16 Trinity River
17 Crystal Lake
18 Coleman (NFH)
19 Tehama Colusa Fish Facility (F)
20 Feather River
21 Redwood Creek Ponds
22 Dinner Creek Ponds

23 Marshall Creek Ponds
24 Sprowel Creek
25 Hollow Tree Creek Ponds
26 Van Arsdale
27 Garcia River Ponds
28 Warm Springs
29 Silverado Planting Base
30 Nimbus
31 Mokelumne River
32 Merced River
33 Silverking Oceanic Farms (P)

NFH = National Fish Hatchery
F   = other federal facilities

P = private aquaculture
T = Tribal facility

Sampling of recreational fisheries.—A second major component is the sampling of the recreational fishery. The emphasis typically is focused on sampling day boats and charter boats in marine waters. Creel sampling is also carried out in some inland fisheries.

The chinook and coho salmon fisheries in Washington's Puget Sound and British Columbia's Strait of Georgia pose a special problem because both are geographically widespread and typically open year-round. In addition, there are hundreds of marinas and private and public launch ramps where anglers land their catch, thus rendering representative sampling cost-prohibitive.

To circumvent this problem, CDFO and WDF maintain voluntary return programs in which anglers are encouraged to turn in heads of adipose-clipped salmonids at any of the many "head depots" located throughout the region.

Field samplers also from WDF interview Puget Sound anglers at major marinas and boat ramps. In addition to determining area of catch and number of fish taken, the interviewer assesses the angler's awareness of the CWT program and the monthly drawing and reward system. This awareness factor is then used in the estimation of total sport recoveries (described later). Approximately 40% of the tags recovered in the Puget Sound sport fishery are voluntary returns; field samplers collect 60% (D. O'Connor, WDF, personal communication).

British Columbia does not supplement its voluntary program with a comparable random-sampling program. However, various creel surveys have been in progress since 1977. The creel census design is structured to record regional mark incidence and compute the awareness or compliance of anglers, based on voluntary submissions to the head depots.

Sampling of escapement.—A third common element is the sampling of escapement. This includes both returns to the hatchery and surveys of the spawning grounds. Historically, this has been the weakest component of the sampling coverage by nearly all recovery agencies. However, it has received ever-increasing attention and importance with the implementation of the USA–Canada Salmon Treaty.

Minimum of 20% sampling rate.—All recovery agencies strive to randomly sample at least 20% of marine landings to have a statistically acceptable estimate of total tag recoveries for a given area–time stratum. In many cases, sampling coverage may exceed 50% if port coverage by samplers is high. However, inland sport fisheries may be sampled at less than 20% because of inherent sampling difficulties, coupled with limited staffing. The Columbia and Willamette river sport fisheries, for example, are sampled at the 10–15% rate.

Sampling of chinook and coho salmon emphasized.—Lastly, coastwide sampling coverage is universal only for adipose-clipped chinook and coho salmon. Chum, sockeye, and pink salmon also are sampled extensively but not in all fisheries.

Steelhead are a special case because of the desequestered adipose clip for this species and its replacement by the LV mark for use in the Columbia basin. Basically, the sampling effort for the LV and Ad–LV marks on steelhead (Table 1) is limited to the Columbia River system. However, British Columbia and Alaska do sample steelhead landings for adipose clips and the possible presence of CWTs.

Sampling Procedures

Field samplers typically work on the docks and sample commercial landings at buying stations. Recreational vessels also are sampled as they return to port. The basic sampling unit is the boat-load of fish, not the individual fish. Samplers attempt to randomly sample vessels, whether they are day boats or trip boats. In the latter case, some of the larger vessels must be subsampled because of the size of the catch. Bins of fish then become the sampling units.

Sampled fish missing the adipose fin are set aside for removal of the head or snout. The sampler then records species, sex, and fork length of the fish on a small waterproof label and encloses it with the head in a small plastic bag for later processing. Scale samples and weight information also may be collected.

Information on the sampled unit (boat load or bin) is recorded on a sample form. This typically includes catch location, catch period, gear type, processor, species, total fish sampled, total marks recovered, and sample date.

CWT-Processing Procedures

Heads removed from adipose-clipped salmonids are transported frozen or preserved to the agency's head lab for tag removal and decoding. The tiny tags are recovered by dissection, aided by an electronic metal detector that indicates which portion of the snout the tag is in after each sectioning of the sample. If no tag is found, the sample is passed through a magnetic field to remagnetize the tag (if present). The sample is then passed through a highly sensitive tubular tag detector to confirm the absence of a tag.

Following their removal, the tags are decoded under a low-power microscope. After the initial

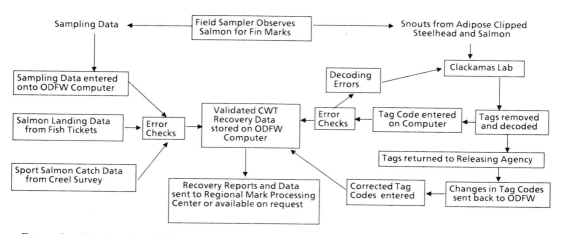

FIGURE 9.—Flowchart for the Oregon Department of Fish and Wildlife (ODFW) coded wire tag (CWT) recovery system. Other recovery agencies follow similar procedures for tag recovery and data management.

reading, a second tag reader makes an independent reading for verification. Several agencies now use a small television camera to project the tag image on a small screen, thereby making it easier to read the code.

Once decoded, the tag code and associated sampling data are entered on the computer for further processing. Several error checks are then run, including verification that the tag code is legitimate (i.e., was previously released) and that the species is correct. Questionable tag codes are re-read by dissection laboratory personnel, and pertinent supplementary data are checked to resolve other errors.

Once validated, the recoveries are designated as "observed recoveries" and made available for use in preliminary reports. This includes expansion of the observed recoveries into "estimated recoveries" for the given area–time stratum once the catch–sample data are available.

A simplified flowchart of this process is shown in Figure 9, which uses ODFW's system as the example.

Catch Data

Total landings are required for a given sampled time–area stratum to estimate total tags recovered. These data for commercial fisheries usually are obtained from fish tickets provided by the buyers. Fish tickets, however, often are not finalized and error-checked until months after the catch was landed and sampled for adipose-clipped fish. Estimates of recreational catch, on the other hand, typically are based on angler surveys and are more timely.

Recovery Estimation Equations

The total number of fish from a particular release group that are caught in a particular area (or landed at a particular port) during a particular time period can be estimated in a two-step process. The first step is to estimate the number of tagged fish in the fishery sample for that area (or port) and time:

$$R_T = aR_O;$$

R_T = the estimated total recoveries of tags bearing the release group's code;

R_O = the observed number of tags of the appropriate code;

a = a sampling expansion factor: (total catch)/(sampled catch).

The second step is to account for the fraction of the release group that was tagged:

$$C = bR_T;$$

C = the total estimated contribution of the release group to the fishery in that area at that time;

b = a marking expansion factor: (total fish released)/(total fish marked).

The contribution estimates then are summed over all relevant area (port) and time strata.

These are the simplest forms of the recovery expansion equations. Typically, the sampling expansion factor is adjusted to account for biases introduced by snouts with no tags, snouts sampled but not taken, lost snouts, and lost tags. In addition, WDF includes an adjustment for angler awareness in the Puget Sound and freshwater sport fisheries because of the large number of voluntary (out-of-sample) recoveries present.

Reporting

Upon completion of this process, the recovery agency forwards the observed and estimated tag recovery data and associated catch and sample data on magnetic tape to the RMPC (Figure 9). The RMPC then checks the data for errors and works with the recovery agency to resolve discrepancies. Once validated, the data (preliminary or final) are combined with those of other recovery agencies to document coastwide recoveries of any given tag code.

Overview of Major Recovery Programs

As noted above, sampling programs of the major recovery agencies have general similarities. However, important differences also exist that must be understood in order to compare approaches. Therefore, I offer the following overview of recovery programs operated by ADFG, CDFO, WDF, ODFW, IDFG, and CDFG. Stratifications used for fisheries, statistical areas, time, and expansion of recovery data also are summarized.

Alaska: South-Central Region

The CWT recovery program in south-central Alaska is unusual in that there are no regional troll fisheries to deal with; instead, terminal net fisheries are the rule. Consequently, the sampling programs are designed to meet localized objectives for the various management units involved in tagging pink and sockeye salmon stocks.

Prince William Sound is the only exception. There, the area's net fisheries harvest both feeders and fish returning to natal streams. Because of the mixed-stock harvest, many of the canneries in the sound are sampled by ADFG (W. Hauser, ADFG, personal communication).

Alaska: Southeast Region

Southeast Alaska's CWT recovery program is designed to sample at least 20% of the chinook, coho, and chum salmon caught in the commercial troll and net fisheries. Additionally, 20% of the sockeye salmon caught in net fisheries in southern southeast Alaska are sampled for the presence of coded wire tags. Tagged fish also are now being recovered from the ocean sport fishery, but the

recovery data have not been expanded because of insufficient catch and sample data.

Samplers are deployed to approximately 12 ports in southeast Alaska each year, based on expected landings. At each port, samplers allocate their time to processor facilities according to the observed level of activity.

Sampling of tenders with catch from more than one area or more than 1 week is avoided when possible. However, depending on the sampling rate, tenders often are the only practical source for sampling. When tender loads or landings are processed faster than samplers can properly observe and handle fish, only a portion of the landing is sampled. In all cases an effort is made to achieve a random subsample of the tender or individual landings.

Area and time stratifications.—The troll-fishery sampling is stratified by 24 statistical areas: 101–116, 150, 152, 154, 157, 181, 183, 186, and 189 (Figure 10). Most of these areas also are used for reporting catch in the gill-net and seine fisheries. Time is stratified by statistical week. Statistical weeks are standard calendar weeks beginning at 12:01 am Sunday and ending at midnight Saturday.

Stratification for expansion-factor calculations.—Samples are pooled into nine geographic strata (Figure 10; Table 4) because of the large number of tags recovered from catches taken in more than one area and because sampling effort is small in some statistical areas. These pooled areas are used to report recovery data to the RMPC. However, conversion to new data formats will likely result in adoption of new catch areas for reporting purposes.

British Columbia

The goal of the commercial CWT recovery program in British Columbia is to randomly sample 20% of the chinook salmon, coho salmon, and steelhead caught in all fisheries throughout the season. Sampling effort also is directed at fisheries anticipated to include coded-wire-tagged chum. A fishery is defined by its operating gear (e.g., freezer troll, ice troll, day troll, gill net, or purse seine) and by a specific catch area.

Samples from sport fisheries originate solely from voluntary submissions of anglers. Voluntary recoveries are solicited by a variety of advertising mechanisms, including radio, fishery publications, and posters displayed at boat ramps and marinas. A network of 234 head depots blankets

SOUTHEASTERN ALASKA

Commercial Catch Areas

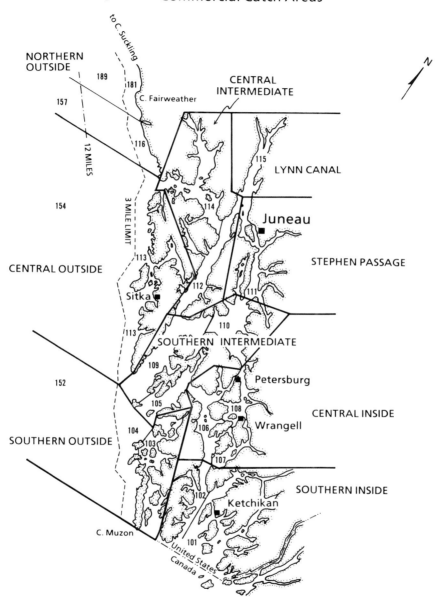

FIGURE 10.—Commercial fishing statistical areas (small numbers) of southeast Alaska and nine pooled area (names) for reporting coded wire tag recoveries to the Regional Mark Processing Center.

areas supporting ocean sport fisheries and terminal freshwater areas.

Ten major coastal ports or port areas are sampled on the basis of their spatial and temporal fisheries. These are listed in Table 5, along with associated regions represented in the catch landings.

Area and time stratifications.—The British Columbia coast is divided into 32 statistical areas (Figures 11, 12). Each statistical area is parti-

TABLE 4.   Southeast Alaska's nine catch regions used for reporting catch and coded wire tag recovery data to the Regional Mark Processing Center (see Figure 10).

Code	Catch region	Commercial statistical areas
SIN	Southern inside	101, 102, 150
SOUT	Southern outside	103, 104, 152
SNTR	Southern intermediate	105, 109, 110
CIN	Central inside	106, 107, 108
COUT	Central outside	113, 154
CNTR	Central intermediate	112, 114
STEP	Stephens Passage	111
LYNN	Lynn Canal	115
NOUT	Northern outside	116, 157, 181, 183[a], 186[a], 189

[a]Areas 183 (Yakutat Bay) and 186 (Icy Bay) are not shown in Figure 10, but are between Cape Fairweather and Cape Suckling.

tioned into a variety of subareas (80 for the coast) that represent localized fishing areas. Subareas are uniquely coded to reflect the type of fishing activity they support inside and outside the surf line; nets are not permitted to operate seaward of the surf line.

Statistical weeks are the time periods used during which catch:sample ratios are computed to obtain the estimated number of marks in the catch. The starting date for each week is Sunday, and the first time period begins the Sunday on or preceding January 1.

Stratification for expansion-factor calculations.—Statistical areas are aggregated to form 14 commercial and 4 sport-catch regions (Table 6). In most cases, the catch region is defined with respect to the operating gear. Hence, some catch regions differ for the troll and net fisheries. Central Troll, for example, includes statistical areas

TABLE 5.   Major coastal sampling locations in British Columbia and associated catch regions represented in the landings.

Port	Catch region
Masset	Queen Charlotte Islands
Prince Rupert	Nass, Skeena River area, central coast, Queen Charlotte Islands
Namu	Central coast
Port Hardy	Northwest Vancouver Island
Winter Harbour	Northwest Vancouver Island
Tofino	Southwest Vancouver Island
Ucluelet	Southwest Vancouver Island
Georgia Strait (minor ports)	Southeast Vancouver Island
Steveston	Georgia Strait mainland, central coast, Johnstone Straits
Vancouver	Georgia Strait mainland, southeast Vancouver Island, Johnstone Straits

6–12, and Central Net is defined as statistical areas 6–11 (Figures 11, 12).

British Columbia is unique in defining its catch regions on the basis of gear type. All other recovery agencies maintain fixed boundaries for their respective catch regions, regardless of the fishery. The CDFO considers its catch data accurate only at the catch-region level and computes expansion factors only at that level.

Washington

Most of Washington's salmon-sampling program is centered on commercial and sport sampling of ocean and Puget Sound fisheries. Major distinctions between coastal and Puget Sound fisheries warrant separate management and sampling teams.

The ocean sampling program focuses on commercial troll and sport landings at the four ports that represent the bulk of the fisheries: Ilwaco, Westport, La Push, and Neah Bay (Figure 13). In recent years, an Indian commercial troll fishery has developed, mostly out of the ports of La Push and Neah Bay. Landings from these vessels now represent a good portion of the total salmon sampled.

The majority of the ocean sport catch is landed at Ilwaco and Westport, where the fleet is divided between kicker (private) and charter vessels. This concentration of effort to the south is most likely due to the proximity to Washington and Oregon population centers. Northern recreational effort is much less intense and consists almost exclusively of kicker vessels.

Sampling in Puget Sound is oriented to management area and gear type (Figure 14). Commercial net fisheries receive first priority in sampling and typically include eight gear types: Indian and non-Indian purse seine, Indian and non-Indian drift gill nets, Indian set gill nets, Indian beach seines, Indian traps, and reef nets. Whenever possible, samples are taken by area and by gear. However, mixed gear types occur and must be sampled as such.

There are over 100 active salmon buyers in Puget Sound, which makes total coverage impractical. Therefore, all major ports and major buyers are sampled, along with as many small dealers as possible.

The sport fishery in Puget Sound is even more complex, with hundreds of public and private sites where landings occur. Therefore, as noted previously, a voluntary return program is used to

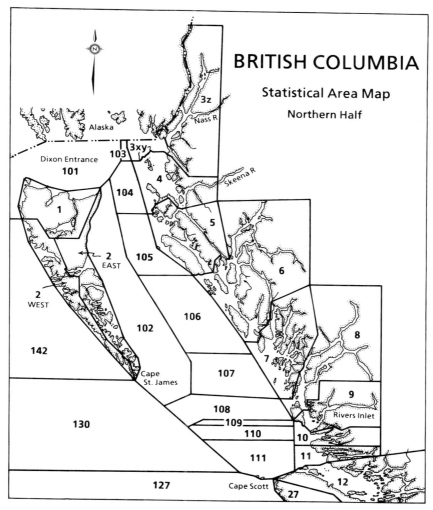

FIGURE 11.—Statistical catch areas for northern British Columbia waters.

supplement WDF's sampling program at the major marinas. Approximately 75 head depots are maintained at marinas, sporting goods stores, and businesses located near major fishing areas throughout Puget Sound.

Area and time stratifications.—The Washington coast is stratified into four catch areas (Figure 13): Columbia River (1), Grays Harbor (2), Quillayute (3), and Cape Flattery (4). Sampling is largely limited to Ilwaco (Chinook), Westport, La Push, and Neah Bay, the four major ports on the coast.

Puget Sound is divided into more than 35 statistical areas because of the complexity of managing local fisheries. Figure 14 shows the management and catch areas currently in use. Several of these are new and have been defined in the past few years for commercial fisheries. They include areas 13C, D, E, F, G, H, I, and K in the southwestern embayments of South Sound, 10F and G in the Lake Washington areas, 8D at Tulalip Bay, and 7E in East Sound on Orcas Island.

Time is stratified by statistical week, as in Alaska and British Columbia. However, Washington differs in that its 7-d statistical week starts on Monday and ends on Sunday rather than running from Sunday to Saturday.

Stratification for expansion-factor calculations.—Sampling rates and stock composition of the ocean sport and troll catches vary somewhat over the season. Therefore, WDF computes expansion factors on the basis of a statistical month.

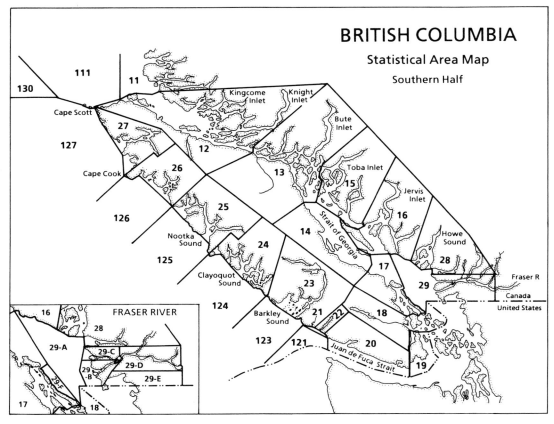

FIGURE 12.—Statistical catch areas for southern British Columbia waters.

Ocean fisheries are further stratified by boat type because day-boat trollers may fish different areas from trip-boat trollers, and charter boats may fish different areas from kicker boats. Both catches and samples for each statistical month are summed by area, species, and boat type and are then used in finding the basic expansion factor for recoveries made in that fishery during that month.

Tag recoveries in coastal gill-net fisheries (e.g., Willapa Bay and Grays Harbor) are estimated separately. In this case, sampling coverage is high and the expansion factor is calculated on the basis of weekly catch and sampling data.

As noted previously, approximately 40% of Puget Sound sport recoveries come from voluntary returns. Therefore, WDF calculates an awareness factor (AF) to estimate the sample size. The AF is the probability that the angler will turn in a head if the fish is missing its adipose fin:

$$ AF = \frac{VR}{\dfrac{SR}{SS} \times (TC - SS)} ; $$

VR = voluntary returns;
SR = sample recoveries by WDF;
SS = sample size taken by WDF;
TC = total catch.

The awareness factor is estimated separately for each area on a seasonal basis. Estimated recoveries are calculated from samples of time strata for statistical months.

Recoveries in Puget Sound net fisheries are expanded by statistical weeks because of more reliable sampling data. However, some pooling across weeks is necessary when samples are inadequate. In addition, some pooling of sampling areas is done to compensate for inadequate sample size or misreported catch areas.

TABLE 6.—British Columbia's catch regions, defined by gear type, and their respective statistical areas (see Figures 11, 12).

Code	Catch region	Gear	Statistical areas
NWTR	Northwest Vancouver Island	Troll	25–27
SWTR	Southwest Vancouver Island	Troll	21, 23, 24
GSTR	Georgia Strait	Troll	13–18, 29A–C
CTR	Central	Troll	6–12
NTR	Northern	Troll	1–5
JFTR	Juan de Fuca Strait	Troll	20
FGN	Fraser	Gill net	29A–E
NN	Northern	Net	1–5
GSN	Georgia Strait	Net	14–18
JSN	Johnstone Straits	Net	12–13
CN	Central	Net	6–11
JFN	Juan de Fuca Strait	Net	20
NWVN	Northwest Vancouver Island	Net	25–27
SWVN	Southwest Vancouver Island	Net	21–24
NSPT	North	Sport	1–5
CSPT	Central	Sport	6–12
GSPT	Georgia Strait	Sport	13–20
WSPT	West coast Vancouver Island	Sport	21–27

Oregon

Oregon's CWT recovery programs are designed to sample 20% or more of the chinook salmon, coho salmon, and steelhead caught in the ocean troll, ocean sport, Columbia River drift gill-net, and Columbia River Indian set-net fisheries. The Columbia and Willamette river sport fisheries are sampled at the 10–15% rate because of the difficulties in sampling these fisheries with the available personnel. Sampling programs for the Columbia River commercial fisheries are reviewed separately (below) because WDF, WDW, NMFS, and USFWS are coparticipants.

Area and time stratifications.—Ocean fisheries sampling is stratified by port of landing and by fishery (Figure 15). Sampling of the commercial troll fishery is conducted at the 10 major ports where most troll salmon are landed. These ports (Table 7) include Astoria, Garibaldi, Pacific City, Depoe Bay, Newport, Florence, Winchester Bay, Charleston, Port Orford, and Brookings. In addition, troll sampling is supplemented on a part-time basis in Gold Beach by recreational samplers.

The sampling period varies by port (Table 7). The major ports of Astoria, Newport, Florence, Winchester Bay, Charleston, Gold Beach, and Brookings are sampled throughout the troll season. Coverage for the remaining ports is limited to the coho season, which accounts for the bulk of Oregon's troll landings. In recent years, efforts have been made to extend sampling coverage on the south coast later in the year because the majority of Oregon's chinook salmon landings are made in this area, particularly after August 1. Sampling is conducted at least 5 d/week or is tailored to the pattern of landings dictated by weather or other factors.

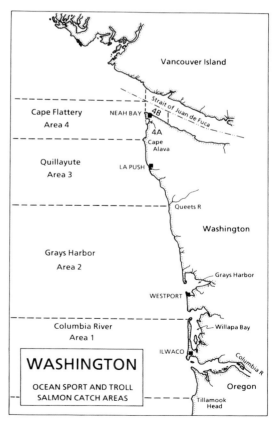

FIGURE 13.—Reporting areas for Washington's coastal sport and commercial troll fisheries for salmon.

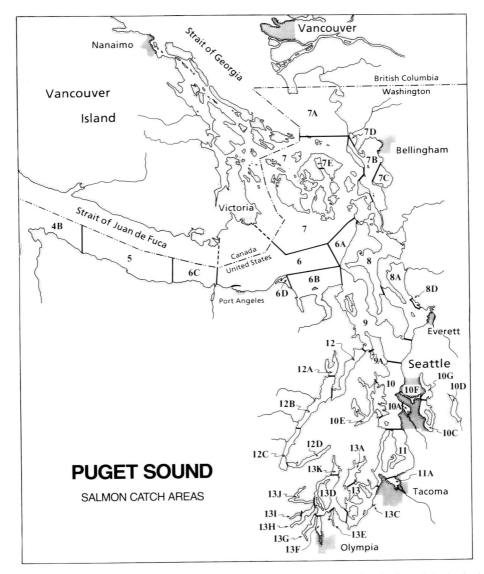

PUGET SOUND

SALMON CATCH AREAS

FIGURE 14.—Management and catch-reporting areas for commercial and recreational salmon fisheries in the Puget Sound area of Washington.

4B Neah Bay	8D Tulalip Bay	12B Dabob Bay
5 Clallam Bay	9 Admiralty Inlet	12C Central Hood Canal
6 Port Angeles	9A Possession Sound	12D South Hood Canal
6A West Beach	10 Seattle	13 Nisqually
6B Discovery Bay	10A Elliott, Shilshole Bay	13A Carr Inlet
6C Crescent Bay	10C South Lake Washington	13C Chambers Bay
6D Dungeness Bay	10D Lake Sammamish	13D South Sound Passages
7 San Juan Islands	10E East Kitsap	13E Henderson Inlet
7A Point Roberts	10F Ship Canal	13F Budd Inlet
7B Nooksack Bay	10G North Lake Washington	13G Eld Inlet
7C Samish Bay	11 East Pass, West Pass	13H Totten Inlet
7D Lummi Bay	11A Commencement Bay	13I Skookum Inlet
7E East Sound	12 North Hood Canal	13J Hammersley Inlet
8 Skagit Bay	12A Quilcene Bay	13K Case Inlet
8A Saratoga Passage		

FIGURE 15.—Oregon's major ports for ocean fisheries. Coded wire tag recovery data currently are expanded by port of landing but soon will be cataloged by catch area.

Sampling of the recreational ocean fishery is carried out at 10 ports at which approximately 98% of the total catch is landed. These ports are Astoria (including Warrenton and Hammond), Garibaldi,

TABLE 7. Oregon's major ports and sampling coverage for troll and sport fisheries (see Figures 15, 16).

Port of landing	Sampling in time periods for	
	Troll	Sport
Astoria	All	All
Garibaldi	Most	All
Pacific City	Most	All
Depoe Bay	Most	All
Newport	All	All
Florence	All	All
Winchester Bay	All	All
Charleston (Coos Bay)	All	All
Port Oxford	Most	None
Gold Beach	All	Most
Brookings	All	All

Pacific City, Depoe Bay, Newport, Florence, Winchester Bay, Charleston, Gold Beach, and Brookings (Figure 15). Sampling is done throughout the season at all ports except Gold Beach (Table 7).

All data are collected and analyzed on the basis of statistical weeks, beginning on Monday and ending on Sunday. The first statistical week of the year ends on the first Sunday of the calendar year, and the weeks are numbered sequentially thereafter. This system is identical to that used by WDF.

Stratification for expansion-factor calculations.—Catch estimates for the commercial troll fishery are based on total weight landed (determined from fish tickets) and average weight information (determined from sampling). This information is obtained by port for each species and grade. The weight landed is divided by average weight to determine number of fish landed. For

FIGURE 16.—Ports and catch management areas for Oregon's ocean salmon fisheries.

ports and time periods not sampled, average weight data are generated from numbers and weights reported on fish tickets or from average weights reported for adjacent ports or time periods.

Catch and effort estimates are reported by both port of landing and catch area. However, CWT data are expanded and reported only by port of landing for sampled ports. This approach, used also by California, was adopted by ODFW's Biometrics Section because of concerns about the accuracy of catch areas reported on fish tickets. In addition, approximately 10% of the fish tickets lacked any catch area information.

It is important to note, however, that ODFW's harvest management staff has always regarded tag recovery data, when expanded to area of catch, as necessary in managing shifting ocean fisheries. As a result, ODFW is now converting to an area-of-catch expansion and reporting system for CWT recoveries. Catch management areas are shown in Figure 16. This new system is expected to be implemented in time to report 1990 ocean recoveries.

Ocean recreational catch is estimated weekly by expanding catch per boat by total effort for each port. Effort counts are made for three vessel categories: salmon charter, bottomfish charter, and pleasure craft. Catch is calculated separately for

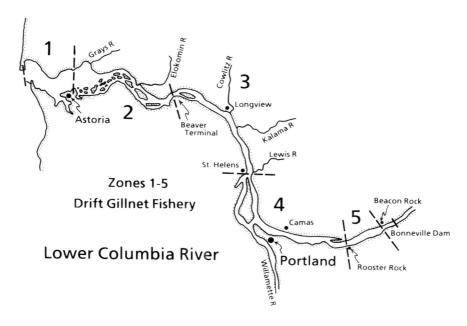

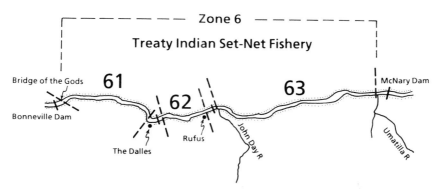

FIGURE 17.—Management and catch-reporting zones for the commercial drift gill-net fishery in the Columbia River below Bonneville Dam (zones 1–5, above) and for the Indian set-net fishery above Bonneville Dam (zone 6, areas 61–63, below).

weekends and weekdays, and the results are summed to derive estimates for each week and port.

Columbia River

The nontreaty drift gill-net fishery below Bonneville Dam and the treaty set-net fishery between Bonneville and McNary dams (Figure 17) are the major commercial fisheries in the lower Columbia River. In the gill-net fishery, most of the fish are taken in zones 1–3. The set-net fishery is conducted by the Yakima, Warm Springs, Nez Perce,

and Umatilla treaty tribes. Most of the effort and catch are in the Bonneville Pool (zone 6, area 61).

Sampling programs are designed to sample at least 20% of the catch of chinook salmon, coho salmon, and steelhead by statistical week and area. Five agencies are involved in the sampling: ODFW, WDF, WDW, NMFS, and USFWS. However, ODFW and WDF are the primary sampling agencies, and they assist each other in all facets of sampling the main-stem fisheries along their respective shores of the lower Columbia

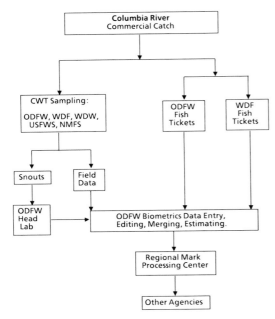

FIGURE 18.—Flowchart for multiagency management of commercial catch, sample, and coded wire tag (CWT) recovery data for the lower Columbia River. ODFW = Oregon Department of Fish and Wildlife; WDF = Washington Department of Fisheries; WDW = Washington Department of Wildlife; USFWS = U.S. Fish and Wildlife Service; NMFS = National Marine Fisheries Service.

River. Tributary terminal fisheries are the responsibility of the state in which they occur.

Sampling data gathered by WDF are forwarded by ODFW, where they are verified and then merged with the Oregon data before tag expansions are done. Following this, the pooled data are reported to RMPC. Figure 18 summarizes the flow of recovery data for Columbia River commercial fisheries.

By a similar process, WDF's and ODFW's sampling data are combined for the recreational fisheries in the lower Columbia River.

Area and time stratifications.—The lower Columbia River below Bonneville Dam is stratified into five management and catch-reporting zones (Figure 17) for the nontreaty drift gill-net fishery. Catch data and tag recoveries in the Indian set-net fishery historically have been reported as a single area (zone 6) that extended from Bonneville Dam to McNary Dam. However, starting in 1988, zone 6 was split into three zones (61, 62, 63) delimited by Bonneville, The Dalles, and McNary dams.

The commercial fisheries are stratified by statistical weeks, which begin on Monday and end on Sunday. The Columbia River sport fishery, however, is stratified by calendar month.

Stratification for expansion-factor calculations.—Tag recoveries in zones 1–5 are expanded after aggregation of catch and recoveries for all five zones. Consequently, recovery data reported to RMPC are defined as being either below Bonneville or above Bonneville. Probably recoveries will be reported by the specific area of catch once the new data-base formats are implemented.

Idaho

Idaho's tag recovery programs historically have received little interest from most other tagging and recovery agencies, because sampling typically encounters only upper Snake River stocks. However, Idaho's contribution of chinook salmon and steelhead to the ocean fisheries has become very important with the passage of the USA–Canada Salmon Treaty. Consequently, all escapement and freshwater sport recoveries are important for evaluating the contribution of Idaho stocks designated as index stocks.

Tag recovery programs rely on sampling of sport fisheries at major fishing sites. This includes a cooperative sampling effort with WDW to sample steelhead harvested on the Snake River where it forms the boundary between Washington and Idaho. Extensive spawning ground surveys also are taken on a regular basis. Hatchery returns are also sampled, along with the use of off-site traps. Sampling and expansions are stratified by calendar month.

Idaho's recovery data do not yet appear in the RMPC data base. However, efforts are under way to report all historical data by 1991.

California

California's CWT sampling programs are designed to sample at least 20% of the chinook and coho salmon landed in the ocean commercial (troll) and recreational (charter boat and skiff) fisheries. Sampling of inland fisheries has been limited to a systematic creel survey on the Klamath–Trinity rivers. Additional inland recoveries are obtained from hatchery returns and spawning-ground surveys. The latter freshwater recovery data have not yet been reported for inclusion in the regional CWT data base.

Area and time stratifications.—Five major ports are sampled for the ocean troll fishery (Figure 19): Crescent City, Eureka, Fort Bragg, San Francisco, and Monterey. In some cases, the

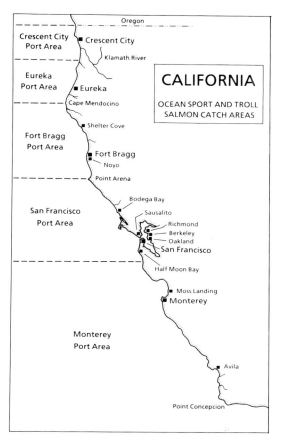

FIGURE 19.—Port management and catch-reporting areas for California's commercial troll and ocean sport fisheries.

major ports may consist of several small adjacent subports. San Francisco is an example of this. Sampling is conducted during the entire season at all five ports.

The same basic five-port design is used to sample the recreational skiff and charter boat fisheries. However, major ports may contain several strata. The major skiff strata in Eureka, for example, include Trinidad Bay (two strata) and Humboldt Bay (two strata). Skiff sample areas consist of the collective landing facilities that can be covered by one or two samplers on any day during the salmon season. Charter boat sample areas include all docks in a port area where landings occur.

Sampling is carried out the entire season at all five ports for the commercial troll fishery and the charter boat and skiff fisheries.

Semimonthly time periods are the time strata used to sample all fisheries. The periods are from

the first to the 15th and from the 16th to the end of the month. California is the only agency to use this sampling stratification.

Stratification for expansion-factor calculations.—Tag recovery expansions for the commercial troll fishery are calculated and reported by port of landing, similar to the Oregon approach. A major reason for this is that fish dealers do not record area-of-catch information on the fish tickets at the time of landing.

Samplers from CDFG have been collecting supplemental area-of-catch information since 1984, but these data are not currently transmitted to the RMPC. However, like Oregon, CDFG is evaluating conversion to an expansion system that would use area of catch instead of the traditional port of landing.

Expansions for the charter boat fisheries are estimated from sampling at least 20% of the landings on randomly selected days within time-strata subsamples. Total vessel trips are estimated from telephone surveys and from log-book data. The log books are required by law, and data from unsampled charter boat fisheries are incorporated into season final estimates.

Recoveries in the skiff fisheries are expanded by means of stratified random sampling scheme wherein the basic sampling unit is a randomly chosen calendar day and subport (launching facility or marina). Sample items are boats, anglers, and salmon. A 100% sample is obtained in the sample unit. Each major port area is treated as a separate population.

Skiff and charter boat recovery data are pooled before being reported to RMPC for regional distribution. However, the data are kept separate on CDFG's computer.

The Achilles' Heel of the Coded Wire Tagging Program

Not unexpectedly, the regional coded wire tagging program has several problems that have reduced its effectiveness. Some problems arose because the regional program is a composite of individual agency programs that have coevolved on a somewhat piecemeal basis over the past two decades. Other problems reflect the complexities associated with marking organisms in an open system where many unknowns exist and can only be approximated.

At the recommendation of the Mark Committee, PMFC sponsored two technical workshops in 1982 to review all aspects of tagging, to identify problem areas, and to make recommendations.

The first workshop dealt with experimental design, and the second focused on tag recovery and estimation procedures (PMFC 1982a, 1982b). Results of the workshops demonstrated that several problem areas were common to all agencies. Most of those areas persist today and are reviewed below.

Improperly Designed CWT Studies for Management Purposes

As noted earlier, the vast majority of tagging studies over the years have been designed for either experimental or hatchery-evaluation purposes. In few of the early studies were enough fish tagged to give reliable estimates of a stock's contribution to a fishery. As a consequence, many fishery management decisions were based on analyses of data from CWT studies inadequately designed to estimate ocean contribution rates.

Considerable progress has been made in this area during the mid-1980s because of the USA–Canada Salmon Treaty. To address interception and harvest allocation, several stocks have been selected as representative of given production areas. These index stocks are being tagged at sufficient levels, and generally on an annual basis, so that the recovery data are meaningful for management applications. Even so, most tagging studies continue to be designed for localized objectives, with Washington as a notable exception.

Lack of Standards for Tagging Levels

Regardless of the type of tagging study, few guidelines exist for determining the minimum number of juvenile fish to tag in a given release to assure scientific validity of recovery results. Similarly, guidelines are needed for determining maximum tagging levels to prevent unnecessary recovery costs. The basic problem encountered is that the number of tagged fish required depends upon too many variables to be accommodated by a few well-defined guidelines. Some of the variables involved include specific objectives of the research study, biology of the stock, expected tag loss, in-river predation rates, and sampling rates expected in the future for specific fisheries.

Several models have been developed, however, and they provide some guidance to the researcher in determining numbers of fish to tag. Most work to date, however, is contained in unpublished draft reports, and it is largely unavailable. Published work includes models by Reisenbichler and Hartmann (1980) and Vreeland (1987). Further work is needed to simplify decisions on correct tagging levels, which are vital to the success of CWT studies.

Unstable Funding for Tag Recovery Programs

Stable, long-term funding is essential to guarantee that tags released in a given year will be recovered at an adequate sampling rate when the fish return two, three, or more years later. Yet funding for these programs continues to be at risk as because most fisheries agencies are under restraint.

Furthermore, most recovery agencies do not closely link their release and recovery programs under a single coordinated budget and administrative system. Consequently, an agency's recovery program typically has little control over within-agency tagging levels. The WDF is an exception in that its budget for tagging and tag recovery passes through one person (L. Blankenship, WDF, personal communication).

Inequitable Cost Burden upon Recovery Agencies

Although many agencies release tagged fish, the cost of tag recovery basically falls upon ADFG, CDFO, WDF, ODFW, and CDFG. This responsibility has been borne willingly in the past by the recovery agencies to avoid bookkeeping and billing demands. However, the number of tags released by nonrecovery agencies has grown greatly in recent years and is expected to continue to expand over the next decade. Thus a more equitable means is needed to distribute recovery costs across all tagging agencies.

Some progress has been made in recent years to ease the burden on recovery programs in Oregon and Washington. The USFWS and the Indian tribes, for example, reimburse ODFW and WDF for recovering their respective tags ($5–7/tag). In addition, the Bonneville Power Administration (BPA) contributes $700,000 annually to support the Oregon and Washington recovery programs in the Columbia basin. These funds are intended to defray recovery costs for tags released by the many BPA-sponsored tagging programs.

The Mark Committee recommended in 1987 that the status quo be maintained rather than switching over to billing all agencies on a cost-recovery basis. However, significant new tagging programs by nonrecovery agencies are to be charged an incremental fee for each tag recovered.

Sampled Harvest from Multiple Catch Areas

Nearly all agencies rely upon estimates of mark contribution by area of catch for management of the ocean fisheries. However, in many cases a vessel will land fish harvested in two or more management areas. This poses a major problem for those engaged in sampling and expanding tag recoveries because individual tags (or fish) cannot be assigned with certainty to a specific catch area. As a result, an expansion cannot be made in most cases for the specific catch area. This is particularly a problem with freezer boats operating in Puget Sound and the Strait of Georgia; although all agencies suffer from the problem to some degree. One solution agencies might try is to aggregate the catch and sample data across two or more adjacent catch areas until the recoveries can be assigned with certainty to a larger catch region. New expansions could then be made. The new CWT data base will facilitate this type of data roll-up.

Unreported and Misreported Harvest Bias

Unreported catches occur in every fishery, and misreporting is a problem in some fisheries. These problems include fish taken home for personal use, some subsistence and ceremonial catches, and incidental catches of one species sold as the target species in a fishery. An example of the latter is the sale of incidental juvenile chinook salmon as pink salmon in purse-seine fisheries of British Columbia and Alaska. These nonreported and misreported catches result in a downward biasing of the estimated total number of tags recovered in a given fishery.

Need for Estimates of Statistical Variability

A sound theoretical framework has not been achieved yet for estimating statistical variability of CWT studies. As a result, only a few studies have had the needed variance estimates and confidence limits. Consequently, it is difficult to determine the statistical significance of the results for a given tagging study. Estimates of total recoveries, for example, may be misleading in strata of limited sampling effort or for releases where only a small proportion was tagged (Clark and Bernard 1987).

Considerable progress has been made in the past 5 years in developing the necessary statistical models and methodologies. Impetus was given to the work in 1984 with the establishment of a CWT

Statistical Committee by PMFC. More recently, this committee was replaced in function by the Pacific Salmon Commission's Working Group on Mark Recovery Statistics.

Several statistical papers (published and unpublished) are now available for CWT applications. These include methods for calculating variances (Reisenbichler and Hartmann 1980; Neeley 1982; Webb 1985; Newman 1990, this volume), use of replication (de Libero 1986; Perry et al. 1990, this volume), determination of sample sizes (Palermo 1984), evaluation of awareness factor variability (Palermo 1990, this volume), contribution and variances (Clark and Bernard 1987; Geiger 1990, this volume), and statistical design of contribution studies (Vreeland 1987 and 1990, this volume). It is expected that this work will continue to accelerate, given the PSC's requirements for statistically sound fisheries information.

Limitation in Marking Nonhatchery Stocks

Although the CWT represents a major technological advance in marking hatchery stocks, it has only limited value for coastwide identification of nonhatchery stocks. This is not a problem so much as an assessment of its limitations as a marking tool. Constraints for marking nonhatchery stocks with CWTs include the following: inaccessibility of many chinook and coho salmon streams; difficulty in collecting statistically significant numbers of representative nonhatchery fish to tag; fragility of wild chinook smolts when handled and marked; a great number of nonhatchery stocks from Alaska to California; and the need for repeated marking on an annual basis.

Other tools are now available or under development that hold promise for overcoming some or most of the above CWT limitations for marking wild spawning stocks. These include the electrophoretic method of genetic stock identification (Milner et al. 1985; Shaklee and Phelps 1990, this volume), the scale pattern analysis method (Conrad 1984; Marshall et al. 1984), and inducement of unique otolith banding patterns (Volk et al. 1990, this volume). These methods, used individually or in concert with CWTs, can provide valuable and timely data on fishery composition for in-season management.

Expanded Role of the RMPC

Upgrading of Operations

The use of coded wire tags for stock identification rose dramatically from 1983 to 1986. On a coastwide basis, CWT releases increased from 24.6 million in 1983 to over 42 million in 1986. During the same period, the RMPC experienced successive budget reductions because of shrinking federal funds. These cutbacks, plus an outdated computer, increasingly limited the RMPC's data-processing capabilities.

Accordingly, when the Mark Committee met during a special meeting in September 1987, it directed PMFC to enhance the Mark Center's operations to improve accessibility and timeliness of the regional CWT data. Approved enhancements included the addition of a second staff member (computer specialist) and various hardware upgrades, notably a DEC MicroVAX minicomputer, a microcomputer, and several high-speed modems (19,200 baud) to expedite on-line data retrieval. Development of menu-driven software also was approved, which would allow end users the flexibility of specifying the type of CWT data needed.

Funding for the upgrade subsequently was provided by nearly all major U.S. tagging and recovery agencies at contribution levels ranging between $6,000 and $10,000. Bonneville Power Administration provided $95,000 for the purchase of the computer.

The new hardware is now installed and operational. Considerable programing, however, is required before a fully functional on-line data retrieval system is available. It is expected this will be completed in 1990.

New CWT Data Formats

A memorandum of understanding to the Pacific Salmon Treaty (January 28, 1985) called for Canada and the USA ". . . to develop a coastwide stock assessment and management data system, including catch, effort, escapement, and coded wire tag data that will yield reliable management information in a timely manner. . . ."

Approximately 1 year later, the Data Sharing Committee of the Pacific Salmon Commission (PSC) met and established a Working Group on Mark Recovery Databases. Assignments included documentation of existing CWT data files and recommendation of a preferred system that could be adopted coastwide.

The first task accomplished by the working group was a description of the existing CWT data sets and the associated limitations of each. Once this was done, the existing RMPC formats for CWT release, recovery, and catch–sample data

were used as the starting point for defining the preferred data base. This second task required several meetings before a consensus was reached on data files, data fields, valid codes, and validation procedures. Fine tuning continues.

Advantages of the new PSC coded wire tag data base over the "old" RMPC data base are discussed at length by Lapi et al. (1990, this volume). Major benefits includes the use of a hierarchical area-coding scheme that standardizes codes and provides links to other information (e.g., catch/sample to recovery information). In addition, special attention has been given to data control issues to ensure the validity of data coded into fields.

The new PSC data formats were presented to the Mark Committee in September 1987 and adopted for use coastwide by both non-PSC and PSC agencies. Conversion to the new formats will be required for reporting 1988 release and recovery data in January–February 1989.

Expanded Tasks of RMPC

Commissioners of the Pacific Salmon Commission agreed in November 1987 that no single USA–Canada coded wire tag data base would be established under the auspices of PSC. Instead, it was agreed that the USA and Canada would each maintain a single data base to expedite exchange of CWT data between the two nations. The U.S. commissioners subsequently considered the sitting issue for the U.S. data base and concluded that PMFC's Mark Center was best suited for the task. This position was supported by the Working Group on Mark Recovery Databases. Advantages of the RMPC cited by the working group included: long-term experience in CWT data administration; coastwide representation of all fisheries agencies; well-established coordination and reporting procedures; no start-up costs to PSC; reduced time for implementation of the new formats; and lack of vested interest in any data interpretation or applications.

The Mark Center's primary role will be to collect, validate, archive, and exchange U.S. data with Canada in the new formats. Additional features of the preferred system recommended by the Working Group will be incorporated as the information systems of the Mark Center and participating PSC agencies continue to evolve.

Oversight of the Mark Center's PSC activities will be provided by selected individuals on PSC's Data Sharing Committee. While details are yet to be worked out, it is not likely that this committee will find itself in disagreement with the Mark Committee. The reason for this is that the role of PSC oversight will focus on data management, whereas the Mark Committee's role is primarily one of regional coordination.

Summary

The coded wire tag is the most important identification tool used on the west coast for salmonid research and management. This paper has attempted to give an overview of its historical development, current regional coordination procedures, agency tagging and recovery programs, major problems, and upcoming improvements. It is important to note, however, that the various agencies' release and recovery programs are considerably more complex than presented here. As such, additional information should be obtained either directly from the agency tag coordinators or from the RMPC.

It is relatively easy to identify problems and shortcomings of any marking tool. In the process, however, one must not lose sight of the tool's positive benefits. Such is the case for coded wire tags. Several problems continue to exist, some of them relatively major, which reduce the CWT's effectiveness as a marking tool. Even so, the CWT has proved invaluable in mass-marking salmonid hatchery stocks and, to a lesser extent, wild stocks. Its widespread and large-scale use on the west coast are ample evidence of this. In addition, coded wire tags are now being used increasingly with other marking techniques (e.g., genetic markers, scale pattern, and otolith banding) to provide a better analysis of salmonid population dynamics. Continual use and expanding research efforts (particularly in the areas of statistical applications) are certain to further strengthen its value as a marking tool.

Acknowledgments

Appreciation is gratefully acknowledged for the help provided by various members of the Mark Committee in supplying timely information and maps when requested. Special thanks are given Erica Jensen, Mary Washkoske, and Theresa Fogg (PMFC) for their many hours of assistance in preparing the manuscript and tables. Of particular help were Karen Crandall (ADFG); Margaret Birch and Louis Lapi (CDFO); Lee Blankenship, Dick O'Connor, and Bill Kinney (WDF); Charlie Corrarino (ODFW); Alan Baracco and Frank Fisher (CDFG); and Tim Cochnauer (IDFG).

References

Bergman, P. K., K. B. Jefferts, H. F. Fiscus, and R. C. Hager. 1968. A preliminary evaluation of an implanted coded wire fish tag. Washington Department of Fisheries, Fisheries Research Paper 3:63–84.

Clark, J. E., and D. R. Bernard. 1987. A compound multivariate binomial-hypergeometric distribution describing coded micro-wire tag recovery from commercial salmon catches in southeastern Alaska. Alaska Department of Fish and Game. Technical Report 202, Juneau.

Conrad, R. H. 1984. Scale pattern analysis as a method for identifying the origins of sockeye salmon (Oncorhynchus nerka) in the waters surrounding the Kodiak Archipelago. Alaska Department of Fish and Game, Information Leaflet 241, Juneau.

Dawley, E. M., R. D. Ledgerwood, and A. L. Jensen. 1985. Beach and purse seine sampling of salmonids in the Columbia River estuary and ocean plume, 1977–1983, volume 1. Procedures, sampling effort, and catch data. NOAA (National Oceanic and Atmospheric Administration) Technical Memorandum NMFS (National Marine Fisheries Service) F/NWC-74, Seattle, Washington.

de Libero, F. E. 1986. A statistical assessment of the use of the coded wire tag for chinook (Oncorhynchus tshawytscha) and coho (Oncorhynchus kisutch) studies. Doctoral dissertation. University of Washington, Seattle.

Geiger, H. J. 1990. Parametric bootstrap confidence intervals for estimating contributions to fisheries from marked salmon populations. American Fisheries Society Symposium 7:667–676.

Jefferts, K. B., P. K. Bergman, and H. F. Fiscus. 1963. A coded wire identification system for micro-organisms. Nature (London) 198:460–462.

Koerner, J. F. 1977. The use of the coded wire tag injector under remote field conditions. Alaska Department of Fish and Game, Information Leaflet 172, Juneau.

Lapi, L., M. Hamer, and B. Johnson. 1990. Data organization and coding for a coastwide mark–recovery data system. American Fisheries Society Symposium 7:720–724.

Marshall, S. L., G. T. Oliver, D. R. Bernard, and S. A. McPherson. 1984. Accuracy of scale pattern analysis in separating major stocks of sockeye salmon (Oncorhynchus nerka) from southern southeastern Alaska and northern British Columbia. Alaska Department of Fish and Game, Information Leaflet 230, Juneau.

Milner, G. B., D. J. Teel, F. M. Utter, and G. A. Winans. 1985. A genetic method of stock identification in mixed populations of Pacific salmon Oncorhynchus spp. U.S. National Marine Fisheries Service Marine Fisheries Review 47(1):1–8.

Neeley, D. 1982. Estimating variances of coded wire tagged salmonid ocean recovery statistics. Pages 176–218 in Workshop on coded wire tag recovery and estimation procedures for Pacific salmon and steelhead. Pacific States Marine Fisheries Commission, Portland, Oregon.

Newman, K. B. 1990. Variance estimation for stock-contribution estimates based on sample recoveries of coded-wire-tagged fish. American Fisheries Society Symposium 7:677–683.

Palermo, R. V. 1984. Determination of sample size requirements to analyze the age structure of the B.C. chinook catch. Canadian Technical Report of Fisheries and Aquatic Sciences 1335.

Palermo, R. V. 1990. Jeopardized estimates of the contribution of marked Pacific salmon to the sport fishery of the Strait of Georgia, British Columbia, due to awareness factor variability. American Fisheries Society Symposium 7:631–646.

Perry, E. A., H. L. Blankenship, and R. V. Palermo. 1990. Comparison of two methods for replicating coded wire tag studies. American Fisheries Society Symposium 7:660–666.

PMFC (Pacific Marine Fisheries Commission). 1982a. Workshop on coded wire tagging experimental design: results and recommendations. Pacific State Marine Fisheries Commission, Portland, Oregon.

PMFC (Pacific Marine Fisheries Commission). 1982b. Workshop on coded wire tag recovery and estimation procedures for Pacific salmon and steelhead. Pacific State Marine Fisheries Commission, Portland, Oregon.

Reisenbichler, R. R., and N. A. Hartmann. 1980. Effect of number of marked fish and years of repetition on precision in studies on contribution to a fishery. Canadian Journal of Fisheries and Aquatic Sciences 37:576–582.

Shaklee, J. B. and S. R. Phelps. 1990. Operation of a large-scale, multiagency program for genetic stock identification. American Fisheries Society Symposium 7:817–830.

Volk, E. C., S. L. Schroder, and K. L. Fresh. 1990. Inducement of unique otolith banding patterns as a practical means to mass-mark juvenile Pacific salmon. American Fisheries Society Symposium 7:203–215.

Vreeland, R. R. 1987. An experimental design for studies of contribution to fisheries by salmonid hatcheries. Master's thesis. University of Washington, Seattle.

Vreeland, R. R. 1990. Random-sampling design to estimate hatchery contributions to fisheries. American Fisheries Society Symposium 7:691–707.

Wahle, R. J., and R. R. Vreeland. 1978. Bioeconomic contribution of Columbia River hatchery fall chinook salmon, 1961 through 1964 broods, to the Pacific salmon fisheries. U.S. National Marine Fisheries Service Fishery Bulletin 76:179–208.

Wahle, R. J., R. R. Vreeland, and R. H. Lander. 1974. Bioeconomic contribution of Columbia River hatchery coho salmon, 1965 and 1966 broods, to the Pacific salmon fisheries. U.S. National Marine Fisheries Service Fishery Bulletin 72:139–169.

Webb, T. M. 1985. An analysis of some models of variance in the coded wire tagging program. Report to Canada Department of Fisheries and Oceans, Vancouver.

Worland, D. D., R. J. Wahle, and P. D. Zimmer. 1969. Contribution of Columbia River hatcheries to harvest of fall chinook salmon (*Oncorhynchus tshawytscha*). U.S. Fish and Wildlife Service Fishery Bulletin 67:361–391.

American Fisheries Society Symposium 7:817–830, 1990

Operation of a Large-Scale, Multiagency Program for Genetic Stock Identification

JAMES B. SHAKLEE AND STEVAN R. PHELPS

Washington Department of Fisheries
115 General Administration Building
Olympia, Washington 98504, USA

Abstract.—Electrophoretic analysis of genetic variation in proteins has been used for over 20 years to characterize natural populations of fishes. The extensive geographic ranges of many species and the multiplicity of agencies involved in their management and conservation make it imperative to establish coordinated programs for genetic stock identification (GSI). However, large-scale, multiagency efforts can provide the information necessary for optimal management only if such programs involve comparable experimental protocols, data sets, and methods of analysis. In this paper, we use studies of chinook salmon *Oncorhynchus tshawytscha* and chum salmon *O. keta* to exemplify some of the major problems encountered in setting up a large-scale GSI program, and we suggest ways to avoid or solve these problems. Important concerns include standardization of experimental techniques, creation of baseline data sets that provide adequate representation of the genetic diversity of stocks contributing to the fisheries, and establishment of criteria and methods for statistical analysis. Research coordination, the development and use of a reference collection of allele-mobility standards, routine exchange and "blind" analysis of replicate samples, and adequate documentation are key elements for the overall success of large-scale GSI efforts.

Successful management of all fisheries depends on biological, economic, and social considerations. Of the many biological factors that directly affect management options and decisions, the continued successful reproduction of all major stocks making up a species is of paramount importance. We define a stock as a randomly interbreeding, self-reproducing subset of a species that is genetically distinct from other such groups. The genetic stocks (spawning populations) that make up a species represent the fundamental units of both reproduction and genetic diversity; thus they determine the ecological and evolutionary potential of the species. Reduction in the number and diversity of a species' reproductive groups can lead to its depletion or extinction. For these reasons, the identification and characterization of fish stocks are extremely important aspects of fisheries biology.

One of the most informative methods for investigating stock structure of fishes is the electrophoretic analysis of protein variation. This approach is based on the principle that populations that exist as discrete reproductive units become genetically different over time. The scientific literature abounds with examples of electrophoretic studies of fish species and populations. Coverage includes freshwater (Avise and Smith 1974; Turner 1983), marine (Winans 1980; Shaklee 1984), anadromous (Okazaki 1982; Gharrett et al. 1987), and catadromous forms (Williams et al. 1984; Shaklee and Salini 1985), species found in the inland or offshore waters of all continents (Christiansen et al. 1976; Vawter et al. 1980; Andersson et al. 1981; Ståhl 1981; McAndrew and Majumdar 1983; Beacham et al. 1985a; Grant 1985; Salini and Shaklee 1988), and fishes from all major oceans and seas (Smith et al. 1978; Grant et al. 1983, 1987; Mork et al. 1985; Milton and Shaklee 1988; Lavery and Shaklee 1989), as well as species supporting major (Grant and Utter 1980; Philipp et al. 1981; Kornfield et al. 1982; Richardson 1983; Ryman et al. 1984), minor (Allendorf et al. 1976; Ihssen et al. 1981; Shaklee and Tamaru 1981), or no commercial or recreational fisheries (Johnson 1975; Cashon et al. 1981).

Although genetic stock identification (GSI) has been used for many years for a variety of species, sustained, large-scale applications of this approach have been rare or generally unsuccessful. The reasons for this general failure stem largely from a diversity of logistical concerns—some attributable to the experimental technique and some to the magnitude of such operations. In this presentation, we outline the most important considerations and components for the successful operation of large-scale, multiagency GSI programs, and provide specific examples based on GSI efforts in North America for chinook salmon *Oncorhynchus tshawytscha* and chum salmon *O. keta*.

Salmon GSI Programs in North America

Utter and his associates at the National Marine Fisheries Service (NMFS) established the first successful, large-scale GSI program during the 1970s and early 1980s (Milner et al. 1985). They used starch-gel electrophoresis to detect and quantify genetic variation in spawning stocks of chinook salmon from California to southern British Columbia, and they developed statistical procedures and computer programs to estimate stock contributions in samples from mixed-stock fisheries (Grant et al. 1980; Milner et al. 1981, 1983). In collaboration with the Washington Department of Fisheries, this large-scale effort was first applied to the management of major coastal chinook salmon fisheries in 1983 (Miller et al. 1983; Utter et al. 1987). Similar programs were developed subsequently in Canada for chum salmon and pink salmon *Oncorhynchus gorbuscha* fisheries (Beacham et al. 1985a, 1985b). Each of these efforts was largely the product of a single group of researchers and a single electrophoresis laboratory. Since 1983, there have been major efforts to expand GSI applications for chinook, chum, pink, and sockeye salmon *O. nerka* across political boundaries and among agencies to encompass the North American ranges of these species. Such expanded efforts provide more comprehensive pictures of population genetics and fishery harvests, and they facilitate implementation of the Pacific Salmon Treaty with regard to interception of fish originating in one country by fisheries in the other. Currently, five government agencies (California Department of Fish and Game in cooperation with University of California at Davis, Canada Department of Fisheries and Oceans, U.S. National Marine Fisheries Service, U.S. Fish and Wildlife Service, and Washington Department of Fisheries) support six laboratories where GSI studies of chinook, chum, pink, and sockeye salmon from California to Alaska are conducted. It is the maturation of GSI from one-laboratory, single-agency, geographically narrow approaches to large-scale, multiagency, coastwide programs that is the focus of this paper.

Development of today's GSI programs has involved innovations in five general areas: standardi-

zation, data management, creation of GSI baselines, mixed-stock fishery analysis, and documentation. Each of these is considered below.

Standardization

Field Sampling

There are three major standardization needs in field sampling for GSI studies. One of these is the need to sample baseline stocks and fisheries in a representative way. In the case of baseline stocks, it is necessary to represent the full range of genetic diversity present in spawning populations. For salmon, this requires representative sampling of all geographic components (e.g., mainstream and tributary spawners), all age-groups, and the temporal spectrum of returning spawners. Distinct spawning runs (e.g., spring, summer, and fall-run chinook salmon populations, and early, normal, and late-run chum salmon stocks) must be included in baseline collections, because important portions of the genetic diversity of the species may be distributed among these major "races." To a lesser degree, it may be necessary to sample the early, peak, and late portions within a spawning run (in proportion to their contribution to the spawning population) if the run shows indications of genetic heterogeneity (Leary et al. 1989). In order to assure representative samples of spawning stocks, we in the Washington Department of Fisheries often intentionally oversample by about 30% during the spawning period and then eliminate excess samples based on the final pattern of return (once it is known) to generate a sample of approximately 100 adult fish per collection that represents all portions of the run.

Life history stages typically sampled in salmon GSI baseline studies include parr, smolt, and adult. Important ontogenetic differences in enzyme expression in fishes have been recognized for some time (Shaklee et al. 1974). With some enzyme systems, including alcohol dehydrogenase in chinook salmon, these ontogenetic differences make it difficult or impossible to collect comparable data from all life history stages. Furthermore, computer simulations indicate that the error associated with estimating the true allele frequencies in a stock is significantly greater with samples of juveniles than it is with samples of adults (Allendorf and Phelps 1981). Waples (1990) also demonstrated that, in most cases, sampling adult salmon resulted in the greatest effective sample. Consequently, we recommend the routine use of adults for all baseline collections.

Adults are also preferable because larger tissue samples can be taken. Samples from adults are generally easier to work with and often make it possible to screen more loci.

Genetic stock identification analysis of fishery catches is usually undertaken either to provide stock-specific estimates of actual harvests by commercial fisheries or to allow predictions of stock-specific harvests based on samples from test fisheries. The methods used to obtain fish for GSI estimates must yield samples proportional to the actual catch in each time–area stratum. This is important, and difficult to achieve, because GSI samples usually represent only 1–15% of the total harvest. As with baseline sampling, it is usually necessary to oversample the fishery and then to eliminate the excess to ensure representative samples.

The number of fishery samples per time–area stratum is also an important consideration and often reflects a compromise between optimal information goals and logistical constraints. The required sample size per stratum should be determined by the level of precision needed for management decisions. Precision, in turn, is a function of the number of stocks or stock groups to be distinguished and the amount of genetic differentiation among them. The greater the number of stocks or stock groups, the larger the fishery sample must be. Conversely, the greater the genetic divergence among stocks or stock groups, the smaller the fishery sample needs to be. Finally, the amount of genetic information available is at least partially determined by the number of variable loci screened.

The particular tissues sampled and the amount of tissue taken also affect electrophoretic analyses. Physiological and biochemical differences among organs and tissues influence the expression of the gene loci used in GSI studies (Shaklee and Keenan 1986). Because it is advantageous to include as many alleles and loci as possible in a GSI study, multiple tissues from each fish are usually sampled. For baseline characterization, we routinely use muscle, heart, eye, and liver tissues to maximize the number of loci we can screen. From troll-caught salmon, which are eviscerated on board, we are only able to sample muscle and eye tissues. Recent investigations have shown that some muscle types exhibit differential locus expression (Shaklee and Keenan 1986). We have found that cheek muscle is the muscle tissue of choice for chinook salmon, primarily because it yields superior banding patterns

for at least three important polymorphic isozymes: mMDH-2 (mitochondrial malate dehydrogenase, IUBNC [1984] number 1.1.1.37), sMEP-1 (supernatant malic enzyme [NADP⁺], 1.1.1.40), and GPI-B2 (glucose-6-phosphate isomerase, 5.3.1.9).[1] For chum salmon, in contrast, body muscle yields clearer patterns for mMEP-2 (an important polymorphic enzyme in this species) than does cheek muscle.

The ability to collect electrophoretic data in the laboratory is directly related to the quality of the tissue samples. Field sampling should be done by trained individuals. We recommend that tissue samples be placed directly into the plastic tubes in which they will be homogenized and centrifuged in preparation for electrophoresis. Dry ice should be used at the time of sampling to guarantee immediate freezing of the samples. The samples should be transported on dry ice and stored in an ultrafreezer at −80°C until they are analyzed. These safeguards minimize postmortem protein denaturation.

Laboratory Methodology and Experimental Techniques

Standardization of all laboratory aspects of electrophoretic analysis will help keep data collection consistent and comparable in large-scale, multilaboratory GSI studies. In particular, methods of tissue extraction and homogenate preparation can significantly affect the quality and quantity of enzyme extracted from a sample. Care must be taken to ensure that tissue disruption is adequate to release sufficient quantities of enzymes for electrophoretic detection and histochemical staining. Enzymes or isozymes restricted to specific subcellular compartments, such as mitochondrial isozymes, often require vigorous extraction techniques. The extraction buffer used can have a major effect on the detection of certain enzymes. For example, the use of an extraction buffer containing pyridoxal-5′ phosphate (Aebersold et al. 1987) makes scoring possible for aspartate aminotransferase isozymes (AAT; 2.6.1.1) in chinook salmon (seven variable loci) and chum salmon (five variable loci). The use of extraction buffers lacking this component

---

[1]Protein-coding genetic loci commonly are named for the proteins they encode, and take the same abbreviations as the proteins. In this paper, italic abbreviations and codes denote loci and alleles (which encode variants of the basic protein). Nonitalic abbreviations refer to the proteins.

makes the routine scoring of many of the *AAT* loci difficult or impossible for these species.

Electrophoretic techniques can affect the ability to detect and score genetic variation in proteins. Electrophoretic media, dimensions of the gels, position of the sample origin, voltage and amperage settings, temperature, and duration of the electrophoretic run all affect final resolution and must be carefully controlled. Some buffer systems are especially sensitive to these variables. Two major factors that affect the final isozyme patterns are the buffers used during electrophoresis and the enzyme-specific histochemical staining procedures. One common buffer component that has a marked effect on the resolution and staining of several isozyme systems is EDTA (ethylenediaminetetraacetic acid). The presence of approximately 0.0005 M EDTA in the gel buffer significantly improves the resolution and staining of several mitochondrial isozymes, such as mitochondrial isocitrate dehydrogenase (mIDHP; 1.1.1.42) and mitochondrial aconitate hydratase (mAH; 4.2.1.2). Although EDTA aids in the study of these isozymes, too high a concentration inhibits staining of other enzymes, such as mannose-6-phosphate isomerase (MPI; 5.3.1.8).

Cooperating laboratories can promote standardization be exchanging buffer and staining recipes and specific screening protocols. These recipes and protocols must be annotated to document important but sometimes subtle or obscure details that affect the results. For example, seemingly minor differences in the buffers used in screening can obscure or bring out distinctions between alleles. In the case of chum salmon malate dehydrogenase (MDH; 1.1.1.37), the two alleles at the *sMDH-B1,2* isoloci (relative mobilities[2] of *72* and *100*) are readily separable at pH 6.0 but indistinguishable at pH 6.8. When the CAM buffer (citric acid–aminopropylmorpholine) is used at both pH 6.0 and pH 6.8 to screen chinook salmon *sIDHP-1,2**, it is possible to distinguish the *127* from the *129* allele. These two alleles exhibit sharply contrasting geographic

distributions, the distinction between them adds to our understanding of population subdivision for this species. Two different buffers must be used to resolve the two most common variant alleles for tripeptide aminopeptidase (PEPB-1; 3.4.11.4) in chinook salmon. Screening with a high-pH buffer allows resolution of the *130* allele from the *100* allele but provides no separation of the *−350* allele. Screening with a low-pH buffer allows resolution of the *−350* allele from the *100* allele but does not allow separation of the *130* from the *100* allele. Thus, in order to collect data for all three alleles, the screening protocol for chinook salmon *PEPB-1** must include both high- and low-pH buffers. Such electrophoretically cryptic alleles are known to be widespread among organisms (Singh et al. 1976; Jamieson and Turner 1978; Coyne 1982). The use of electrophoretic conditions that maximize detection and resolution of cryptic variation extends the power of GSI analysis but increases the need for standardization among laboratories.

To achieve consistency in gel scoring (allele identification), laboratories should exchange allele mobility standards (Shaklee 1983). This exchange promotes consistent detection and correct identification of all allelic isozyme forms used in the analysis of fishery samples. The precise electrophorectic methods and conditions each laboratory uses to attain this end are of secondary importance, a point that cannot be emphasized too strongly. Although standardization of laboratory conditions and methods is one obvious way to accomplish consistent data collection, the best approach is the exchange and use of known allele mobility standards for all variable systems. When each participating laboratory can reliably score all known variation accurately, data comparability and integrity will be assured. Otherwise, multilaboratory and multiagency efforts to use GSI techniques will be of questionable validity.

At the Washington Department of Fisheries (WDF), we use frozen tissues to establish allele mobility standards for both chinook and chum salmon. We take two sets of tissue samples (muscle, heart, eye, and liver) from each fish during baseline sampling. The primary set of samples consists of small pieces of tissue in tubes; the second set consists of large tissue chunks in plastic bags. The biochemical phenotypes of all individuals in the collection are determined by routine electrophoretic processing of the primary samples. Next, the large tissue chunks from individuals with variant phenotypes to be included in

---

[2]The mobility (migration rate) of protein variants through electrophoretic gels is proportional to the variant's net electrical charge. The mobility of the most common variant often (and in this paper) is designated 100; mobilities of other variants are expressed as percentages of the distance migrated by the most common variant (minus signs denote migration in the cathodal direction). When italicized, these numbers refer to the alleles that encode the protein variants.

the standards are subsampled (while still frozen), and the resulting pieces are individually placed in appropriately labelled sample tubes. Each such sample can then be used as an allele mobility standard to assure consistent identification. We have already used these standards in our own laboratory to assure consistent allele identification through time and across starch lots, and have distributed several hundred allele mobility standards to other laboratories involved in GSI studies of chinook salmon.

As a check on their methods of data collection, we recommend that GSI laboratories routinely exchange replicate fishery or baseline samples (or allele mobility standards) and conduct blind analyses of them. Each laboratory would receive and analyze one of the matched sets of tissue samples without knowledge of its geographic origin or of the alleles or phenotypes present. Identical scores among laboratories for all individuals at all loci would indicate complete standardization and comparable analytical skills. Conversely, scoring discrepancies would indicate a need for additional training or standardization. If the discrepancies persisted in spite of training or standardization, the loci or alleles involved should be omitted from the GSI database because of unresolvable analytical inconsistencies.

Nomenclature

Standardization of the biochemical and genetic nomenclature used in GSI studies is also important. Historically, the absence of a recognized international or national genetic nomenclature for fish enzymes, isozymes, loci, and alleles, together with the diversity of scientists involved in biochemical genetic studies of fish, has contributed to the somewhat chaotic state of nomenclature now is use (Table 1). Standardization can improve information exchange, especially among laboratories engaged in large-scale studies, by obviating the tendency of each laboratory to develop its own nomenclature.

In the last few years, most of the major North American laboratories involved in GSI studies of Pacific salmon have begun using a standardized nomenclature for enzymes, loci, and alleles. The common names used for enzymes are those recommended by the Nomenclature Committee of the International Union of Biochemistry (IUBNC 1984). Abbreviations derived from the recommended enzyme name are used to identify both the enzyme and the locus encoding the enzyme.

TABLE 1.—Examples of locus and allele nomenclature used in four genetic stock identification studies of chum salmon.

Enzyme	Locus and allele			
	Washington Department of Fisheries 1990 (unpublished)	Wishard (1981)	Beacham et al. (1985a)	Okazaki (1982)
Glycerol-3- phosphate dehydrogenase	G3PDH-2 100 93	GPD-2 100 85	Agp-2 100 95	α-GDH-2 a b
Isocitrate dehydrogenase	sIDHP-2 100 116 86 65 45 32 31 24	IDH-3 100 90 33 20	Idh-3 100 110 85 40 25	IDH-2 a c b d
Malate dehydrogenase	sMDH-B1,2 100 130 72 50	MDH-3 100 125 75	Mdh-3,4 100 120	MDH-B a c b
Tripeptide aminopeptidase	PEPB-1 −100 −126 −127 −146 −72 −50	LGG-mf 100 100 66 66	Lgg 100 100 75 75	Pep-LGG a a b b

When practical, these abbreviations have standardized endings (e.g., . . . DH for NDA$^+$-dependent dehydrogenases) to identify certain classes of enzymes, and they may include appropriate modifiers to distinguish among multiple loci encoding the same or similar enzymes. Each allele is identified by the relative electrophoretic mobility (under a specific set of electrophoretic conditions) of the homomeric isozyme it encodes.

The creation and use of a uniform system of genetic nomenclature for fish (such as that in use in human genetics) to facilitate standardization and communication has been long overdue. A standardized gene nomenclature for protein-coding loci in fish has been developed and implemented as an editorial standard by the American Fisheries Society (Shacklee et al. 1990). Such a system will benefit large-scale GSI studies and smaller, independent research investigations alike.

Inheritance Testing

By definition, GSI relies on a direct and specific genetic basis for the observed electrophoretic variation. However, inheritance testing of the

electrophoretic variation used in GSI studies of
Pacific salmon has been accomplished only in a
handful of cases (Wishard 1981; Kobayashi et al.
1984; Beacham et al. 1985a, 1985b). Investigators
have relied instead on inferences from the results
of inheritance investigations in other species of
fish (e.g., Clayton et al. 1971, 1973; Whitt et al.
1971; Allendorf and Utter 1973; Utter et al. 1973,
Purdom et al. 1976; Leslie and Vrijenhoek 1977;
Place and Powers 1978; May et al. 1979, 1980,
1982; Morizot and Siciliano 1979, 1983; Stoneking
et al. 1979; Kornfield et al. 1981; Van Beneden et
al. 1981; Pasdar et al. 1984).

Despite the large body of information indicating
that a great deal of the electrophoretic variation
observed in fish has a direct genetic basis, it is
important to conduct inheritance testing and linkage analysis for the specific loci and alleles used in
large-scale, multiagency GSI applications. For
each species, inheritance tests are needed to
establish the genetic basis of observed variations
in the isozyme banding patterns that form the
basis of GSI analyses. The magnitude, expense,
and economic and biological ramifications of
large-scale GSI studies warrant these tests. Moreover, genetic confirmation will encourage coastwide participation in GSI studies and help assure
the reliability of the conclusions.

Inheritance studies are also needed to reveal
any linkage relationships among the variable loci
used in GSI analysis. Discovery of linkage between or among loci would require reassessment
of the present statistical analysis used in GSI
studies, which treats these loci as independent
(i.e., unlinked) characters.

Data Management

Computerization

The amount of data generated by large, multiagency GSI programs requires a computerized
system for data management. The large numbers
of samples processed, loci screened, and alleles
scored make direct data input highly desirable.
The need for in-season GSI fishery estimates, to
be used in management decisions, makes on-line
data input a necessity. Appropriate computer
programs and data-management systems have
been developed and tested by the NMFS laboratory in Seattle (G. A. Winans and associates,
NMFS, personal communication) and by WDF.

Verification and Evaluation of Raw Data

As the volume of data increases, the need to
scrutinize and test it becomes more important.
Verification of the raw data should include confirmation of the identify of all samples by means of
complete labeling and careful handling. A thorough and consistent system for labeling all samples
is especially important in a large-scale, multiagency effort in which the individuals and agencies
collecting the samples may not be the same as
those responsible for the laboratory analyses.

The scoring of electrophoretic patterns also
should be verified for each tissue and fish. We
believe that independent, double-scoring of all
gels and redundant analysis of all highly variable
systems in multiple tissues from the same fish
(where possible) are the two best ways to achieve
accuracy and consistency in isozyme interpretation. For example, when scoring PGK-2 (phosphoglycerate kinase; 2.7.2.3) variability in chinook salmon, we routinely analyze both eye and
muscle samples of fish from troll fisheries and eye,
muscle, and liver samples of baseline fish. We
believe it is especially important to double-score
all duplicate isolocus enzyme systems because the
scoring of doses can be especially difficult. When
scoring discrepancies appear, either they should
be reconciled by reanalysis, or the discordant
scores should be eliminated from the data bases.
The use of allele mobility standards during electrophoretic analysis may also be necessary when
two or more alleles have such similar electrophoretic properties that they cannot be routinely
distinguished from each other in the absence of
such standards.

An additional concern, often overlooked, is the
problem of missing data from unscorable samples
(Shaklee 1983). Except in the case of decomposed
samples, the inability to score particular genotypes is likely to be nonrandom. That is, for
samples having weak enzyme activity, heterozygotes are more likely to be unscorable than homozygotes, because the enzyme activity is spread
among several isozymes instead of being concentrated in a single band. The problem increases
with increasing subunit number: heterozygotes
for monomeric enzymes generally have the total
enzyme activity divided between two isozymes;
heterozygotes for dimeric enzymes have activity
spread among three isozymes; and heterozygotes
for tetrameric enzymes have activity distributed
among five isozymes. We recommend that, for
any baseline collection in which less than 90% of

TABLE 2.—The increase in genetic stock information data for chinook salmon, 1983–1988.

Measure[a]	1983 (Milner et al. 1983)	1985 (NMFS,[b] unpublished)	1988 (Multiagency data, unpublished)
Enzyme systems	16	20	22
Variable loci	27	34	46
Common alleles (p > 0.02)	59	75	111
Rare alleles (p < 0.02)	8	18	48
Laboratories	2	3	4
Electrophoresis buffers	3	3	9
Stocks in baseline	81	96	107

[a] p denotes frequency.
[b] NMFS = National Marine Fisheries Service.

the samples are successfully scored for any given locus, data for that locus be omitted from the data base. The occurrence of low-activity alleles (known for chinook, chum, and pink salmon) can also result in missing data. In such cases, missing scores are likely to be nonrandom with regard to genotype and, thus, lead to incorrect estimates of allele frequencies.

The GSI process requires computer programs to construct genetic profiles from raw electrophoretic data. When new programs are used, the allele and genotypic frequencies they calculate must be checked for accuracy. This debugging is unnecessary with established programs such as BIOSYS-1 of Swofford and Selander (1981) or PHYLIP of Felsenstein (Department of Genetics, University of Washington). In either case, it is imperative that the raw data are correct and apply to the fish in question. The large size and complexity of many GSI data sets necessitate thorough error checking at all steps during summarization and analysis.

Robust GSI analysis of mixed-stock fisheries requires statistically significant genetic differentiation among stocks. Therefore, it is desirable to identify and use as many variable loci and alleles as possible. The number of alleles and loci identified invariably increases as the size of a GSI baseline increases and as the electrophoretic technology matures. In the case of the chinook salmon GSI baseline, the number of variable loci and common alleles almost doubled between 1983 and 1988, and the number of rare alleles increased sixfold (Table 2). Much of this increase was due to greater electrophoretic resolution resulting from the use of diverse buffer systems rather than to an

increase in the number of enzyme systems stained for. The dynamic and expanding nature of long-term GSI studies (increasing numbers of fish, collections, stocks, loci, and alleles) enhances our ability to differentiate individual stocks, but it adds significantly to the burden of data management, analysis, and documentation (see below).

Prior to their use to generate allele frequency data, genotype distributions in each baseline collection should be statistically tested for agreement with Hardy–Weinberg expectations and analyzed for associations with sex and age. Any significant deviation from Hardy–Weinberg expectations or correlations with sex or age should be further examined and, if possible, verified and explained. If deviations from Hardy–Weinberg expectations occur consistently at some loci, these loci should be excluded from any fishery analyses until the genetic basis for the variation is understood. This step is necessary because such deviations may indicate an incorrect genetic model for the locus in question, and because the present method of analysis assumes Hardy–Weinberg proportions to calculate the probabilities of expected multilocus genotypes.

Likewise, genotype and allele frequency distributions for baseline collections should be compared with similar data for previous collections from the same locality and for collections from geographically close localities. Any major differences in frequencies identified by such comparisons, especially among collections expected to be similar, should be further examined and explained or, at the very least, verified.

Creation and Use of GSI Baselines

Use of Allele Frequencies

The stock composition of the baseline used for mixed-stock fisheries analysis can substantially affect the accuracy and precision of stock-contribution estimates generated by maximum-likelihood analysis. For this reason, it is important to use the most appropriate, representative, and accurate baseline possible. We recommend the following approaches.

A GSI baseline is intended to provide an accurate characterization of the type, amount, and pattern of genetic variation among stocks potentially contributing to the fishery under study. Because of the large number of individuals and stocks normally present in a baseline, the genetic data are usually summarized as either genotype or allele frequencies for each locus in each stock.

The baseline is used by the maximum-likelihood program to estimate the contribution of various stocks to a mixed-stock fishery. It is possible to use either genotype or allele frequency data in maximum-likelihood estimates, but the latter are preferred when the number of variable loci and the number of alleles per locus are large.

The use of genotype frequencies might at first seem desirable, because it would eliminate assumptions of Hardy–Weinberg equilibrium. However, this approach has a severe weakness because the number of fish normally sampled for GSI baseline collections is small (100–200). When large numbers of variable loci and alleles are included in the baseline, it is virtually impossible to accomplish sufficient field sampling and laboratory analysis to assure that accurate estimates for all possible (i.e., expected) multilocus genotypes will be obtained. The best solution to this problem is to convert the observed genotype distributions at each locus into observed allele frequencies (Altham 1984). Then it is a direct and relatively simple process to calculate the probabilities of occurrence of all possible multilocus genotypes from the single-locus allele-frequency data, based on Hardy–Weinberg expectations.

To put it another way, when the number of possible genotypes becomes large relative to the number of fish in the baseline collections, it is inappropriate to use the distributions of observed genotypes in the baseline collections for the analysis of fishery samples (Fournier et al. 1984; Wood et al. 1987). Because of this limitation, we strongly recommend the use of allele frequencies in GSI baseline databases.

Sampling

Baseline sampling must include all stocks that are major producers. In addition, it is desirable to sample all genetically distinct populations. We create a baseline by sampling postspawning adults, or in some cases juveniles (but see above), from each location (river, stream, or minor tributary in the case of wild stocks; hatchery or rearing facility in the case of cultured stocks) that has a population large enough to contribute measurably to a mixed-stock fishery. As the number of baseline collections increases, the number of collections that are not genetically distinguishable is apt to increase also. Because GSI should not be used to discriminate among genetically indistinguishable collections or stocks, it is necessary to eliminate all members but one of such pairs or groups

of collections from the baseline before the maximum-likelihood estimates are made.

Occasionally, collections from different geographic areas are electrophoretically indistinguishable. This usually occurs because populations in different areas have recently been part of the same gene pool, but migrated or were transported apart, or because the number of variable loci used for the analysis was too small to differentiate the stocks (chance similarity). In such cases, several options are available. (1) The minor contributors in groups of genetically indistinguishable collections can either be dropped from the maximum-likelihood analysis, or their estimates can be combined, and the proportional contribution of each can be determined by another stock-identification technique or by run reconstruction. (2) The stocks or stock groups can be combined to form larger reporting groups. (3) The number of genetically variable loci analyzed can be increased by more extensive electrophoretic analysis in order to increase the power of stock discrimination.

It is *not* an option to retain the indistinguishable collections as baseline stocks in different reporting groups. The maximum-likelihood program does not average the contribution among indistinguishable baseline stocks; instead, it often assigns most of the contribution to one baseline stock. Situations such as this can sometimes be recognized by large negative correlations among the stock-contribution estimates (Millar 1987).

In many cases, it is neither necessary nor desirable to include data from every collection in the GSI baseline. Rather, only collections for which we can electrophoretically measure a significant genetic difference (i.e., distinct stocks) should be included. Furthermore, stocks unlikely to contribute to a particular mixed-stock fishery, because of geographical remoteness or inappropriate spawning time, should not be included in the baseline used to analyze that fishery. Use of a larger-than-necessary number of baseline stocks is undesirable because of the increased expense, increased potential for bias, and decreased precision. Millar (1987) has pointed out that the number of baseline stocks used to describe the mixed-stock fishery sample should be less than the number of unique multilocus genotype arrays in the fishery sample.

Baseline collections for GSI do not have to be made every year, because allele frequencies tend to be relatively stable from year to year in most fish stocks. However, stocks should be monitored periodically, both to provide more accurate estimates of allele frequencies and to detect any significant changes in their genetic profiles. Changes in allele frequencies are most likely in stocks with small effective population sizes, because of random genetic drift. We also recommend periodic monitoring for strong stocks that are major contributors to fisheries, in order to provide the most reliable fishery estimates possible and to detect stock transfers or hatchery practices that could change the allele frequencies.

Treatment and Testing of Baseline Collections

Baseline collections can be treated in three fundamentally different ways to generate stock-contribution estimates for fisheries. (1) All baseline collections can be used individually. (2) The genetic data from individual collections can be pooled prior to analysis. (3) The stock-contribution estimates of individual collections can be added together after maximum-likelihood estimation.

The decision on whether to use collections individually, to pool collections prior to stock composition estimation, or to sum the individual estimates afterwards should be based on statistical tests of genetic characteristics. There are no obvious or infallible criteria for determining if two or more collections represent multiple stocks or multiple samples from a single stock. Nevertheless, the statistical tests most often used for this purpose are chi-square or G-tests.

Our approach has been to regard collections taken from the same locality in different years as representing separate stocks if they exhibit statistically significant differences at the 0.01 level, based on an overall chi-square or G-test (Sokal and Rohlf 1981) for all variable loci and for at least one individual locus. Another criterion we have used is the genetic distance (Nei 1972) between pairs of collections. Prior to analysis, we combine collections within a reporting group to form a baseline stock if the genetic distance between them is less than 0.0005, a value approximately twice the spurious genetic distance (genetic distance due to sampling error). Wood et al. (1987) recommended a modification of the genetic distance measurement of Nei (1972) so that it would represent the separation potential of the loci assayed, rather than the overall genetic divergence.

Because it seems likely that better statistical methodologies for addressing this question will be developed in the future, the above criteria should be viewed as working approaches. For now, we recommend the following testing.

(1) Test pairs of collections from the same locality taken over time. Pool such collections when they are not significantly different and recalculate allele frequencies (weighted by sample sizes).

(2) Once all temporal collections have been tested and pooled as necessary, test pairs of collections (or aggregates) from geographically adjacent localities (or pairs thought to share common ancestry). Pool these when they are not significantly different and recalculate allele frequencies, weighted by sample sizes.

(3) Test pairs of collections from different geographic areas. Spatially distributed collections should, in general, be combined prior to maximum-likelihood analysis if they are not significantly different genetically (if $P > 0.01$; overall chi-square test). Marginally different collections $(0.05 > P > 0.01)$ from different areas should be put into the same reporting group for stock-contribution estimation if they are not pooled prior to maximum-likelihood analysis. Note that if reporting groups are large enough (i.e., the genetic distances among them are much larger than the genetic distances among stocks within each group), misallocations will tend to occur within rather than among the reporting groups in the stock-contribution estimates.

Because pairs or groups of geographically disparate collections can exhibit genetic similarity, it is necessary to test all possible pairs of baseline collections to detect such cases and to avoid use of such indistinguishable collections for mixed-stock fishery analyses.

Pooling the genetic data from genetically different stocks (e.g., all collections in a management region) prior to maximum-likelihood analysis will create an artificial "stock" characterized by averaged allele frequencies at each locus. When such a pooled "stock" is used to calculate expected genotypes, the result will be distributions of genotypes and probabilities that do not represent any of the stocks originally sampled. Thus, the true contribution of fish from these pooled stocks will be underestimated. In contrast, using "stocks" that do not differ significantly in genetic characteristics can be expected to result in inaccuracy and poor precision. Proper GSI can be achieved only if all stocks or management groups for which estimates are reported have significant genetic differences.

Computer Simulations

Once a baseline is available, simulation analysis should be performed to test how well the baseline provides estimates of stock contribution for fisheries (Pella and Milner 1987). Such computer simulations should use stock proportions expected of the actual fishery and should be aimed at determining the accuracy and precision of the resulting stock estimates. Simulations can also be used to determine how many fish in a fishery must be sampled to yield stock estimates with acceptably small coefficients of variation. The simulation we use creates a randomly sampled mixed-stock fishery from the baseline and produces maximum-likelihood estimates (no resampling), or it uses a random sample from each baseline (resampling) to estimate contributions to the mixed-stock fishery. Resampling the baseline stocks simulates that we only have an estimate of the true allele frequencies of each stock. When the distributions of estimates derived from simulations that use a resampled baseline do not differ substantially from those obtained from a nonresampled baseline, the database is sufficiently robust for fishery analysis. In other words, when the statistical errors associated with finite sampling of baseline stocks are considerably smaller than the genetic differentiation among stocks in the baseline, the resulting stock-composition estimates will be insensitive to random baseline variability.

Simulations can also be used to refine and enhance both field sampling and laboratory analysis by identifying those tissues where the most genetic information can be screened and those loci that provide the greatest discrimination in any given mixed-stock fishery, respectively.

Mixed-Stock Fishery Analysis

When all participating agencies use the same analytical methods for determining stock contributions, their estimates will be comparable, if fishery sampling and laboratory analyses also are equivalent. However, when different agencies use different analytical methods (e.g., different statistical methods or different computer programs) to estimate stock contributions, those methods must be tested for comparability.

In addition to verifying that all agencies are using statistical and computer methods that yield comparable estimates of stock composition, it is desirable to examine the accuracy and precision of the maximum-likelihood estimates. Two approaches are feasible. One is to use computer simulation to examine the properties of the maximum-likelihood technique (see above). The second is to compare the stock-contribution estimates

derived from GSI analysis with other information about the composition of the fishery, such as the information derived from fish marked with coded wire tags. In this case, the tagged fish serve as a "blind" mixture of knowns that can be analyzed by means of GSI. The results can then be compared with data provided by the coded wire tags on the hatchery sources of the fish in the mixture. Such comparisons are presently ongoing in at least two laboratories.

Another problem typical of mixed-stock fishery analysis involves the difficulty of sampling baseline stocks adequately to provide accurate estimates of the occurrence and frequency of rare alleles. The problem increases when some or all baselines stocks are represented by small sample sizes. For example, in order to be 99% certain of observing an allele that has a true frequency of 0.01 in a population, 230 individuals must be analyzed. The sample size required to provide an accurate estimate (e.g., ± 0.005) of the true frequency of that allele is even larger.

Failure to detect any given allele in a baseline stock may indicate the absence of that allele in that stock, or it may result from an inadequate or nonrepresentative sample. If the former possibility is true, then the maximum-likelihood criterion will correctly determine that fish carrying this allele cannot have come from the baseline stock. In most cases, however, inadequate sampling is more likely, and the maximum-likelihood criterion will incorrectly exclude fish carrying this allele from membership in the baseline stock. Several suggestions for circumventing this problem have been proposed. One is to replace each zero allele frequency in baseline stocks with a small but positive value (e.g., 0.0001). A second is to use a correction associated with the actual sample size of each baseline collection. Smouse (1974) recommended a correction value of $1/2N$ (0.005 when $N = 100$).

Documentation

The detailed and dynamic nature of current GSI investigations necessitates thorough documentation of all steps in the process, from field sampling to laboratory analysis to statistical testing. Documentation of the specifics of field sampling provides a permanent record that relates the GSI samples from each fishery to the overall estimated composition of the fishery, and documentation links biological data (such as age and sex) with the genotype of each fish to allow tests of associations between these factors.

Our ability to recognize and resolve genetic variation depends on analytical conditions (sample preparation, specific electrophoresis buffers, enzyme staining recipes, etc.) employed in the screening of the samples. Because these conditions are steadily being improved, the scope of the data sets is constantly expanding, typically in the form of new loci and new alleles at previously studied loci. Such newly recognized variation is continuously being added to the growing GSI database (Table 2).

The dynamic nature of GSI studies requires that investigators include information on how data were collected for baseline. In general, older collections in the baseline have fewer loci and alleles than more recent collections. It is necessary to know if the absence of a particular allele in any given baseline collection or fishery sample means that this allele was tested for and not observed (frequency = 0.0), or that the electrophoretic conditions used in screening the samples were inadequate to resolve this allele (frequency unknown).

The agencies involved in GSI studies of chinook salmon have formalized a process for integrating new electrophoretic methodologies, newly discovered variable loci, and newly resolved alleles into their coastwide data base. In this process, one or more laboratories sponsor recommended changes in a document that describes the appropriate laboratory techniques and resulting isozyme patterns, together with preliminary information concerning their advantages. Whenever possible, this sponsorship is accompanied by allele mobility standards for all newly recognized variants. In the case of a new electrophoresis buffer system, the justification is often the improved resolution of enzyme systems already used in the GSI process, or the resolution of an additional enzyme system or allelic isozyme. Enhancements in tissue collection, homogenate preparation, or staining conditions generally are directed at adding new enzyme systems or at improving the resolution and scoring of problem systems.

Once the sponsorship document has been circulated, each laboratory is expected to evaluate the proposed changes. At the end of a 6-month evaluation period, the laboratories attempt to reach a consensus on whether to adopt, modify, or reject the changes. If the group cannot reach a clear consensus, it may decide to extend consideration for another 6 months.

The details of each fishery analysis must also be thoroughly documented. The statistical tests used in each fishery analysis must be identified. Such variables as the specific stocks included in the baseline, the actual loci and alleles used in the analysis, and the pattern of aggregation used to generate stock-group estimates must be recorded and associated with each fishery summary. Each of these factors can have a large influence on the final estimates.

Summary and Conclusions

Large-scale, multiagency programs for genetic stock identification are likely to become more important as national and international concerns about the harvest and management of fish stocks increase. For GSI programs of this scope to be successful, attention must be given to assuring data comparability (by standardizing laboratory techniques), consistency in baseline development and statistical methods, and coordination among laboratories. Communication among laboratories and agencies, in the form of regular meetings and shared development of work plans and complementary objectives, are essential to the long-term success of such efforts.

Considerable success has already been achieved in establishing coastwide programs of genetic stock identification for several species of Pacific salmon. In the past few years, shared GSI data bases have been developed for chinook and chum salmon and are now being used to provide estimates of stock composition in several mixed-stock fisheries. Similarly, laboratories in Alaska and Washington are currently developing a comprehensive GSI baseline for pink salmon that will encompass Alaskan, Canadian, and Washington stocks and will be used to analyze stock-specific harvests of this species.

Acknowledgments

We thank Craig Busack, Rich Lincoln, Anne Marshall, and Robin Waples for their helpful comments and suggestions on early drafts of the manuscript.

References

Aebersold, P. B., G. A. Winans, D. J. Teel, G. B. Milner, and F. M. Utter. 1987. Manual for starch gel electrophoresis: complete procedures for detection of genetic variation. NOAA (National Oceanic and Atmospheric Administration) Technical Report NMFS (National Marine Fisheries Service) 61.

Allendorf, F. W., and S. R. Phelps. 1981. Use of allele frequencies to describe population structure. Canadian Journal of Fisheries and Aquatic Sciences 38:1507–1514.

Allendorf, F., N. Ryman, A. Stennek, and G. Ståhl. 1976. Genetic variation in Scandinavian brown trout (*Salmo trutta* L.): evidence of distinct sympatric populations. Hereditas 83:73–82.

Allendorf, F. W., and F. M. Utter. 1973. Gene duplication within the family Salmonidae: disomic inheritance of two loci reported to be tetrasomic in rainbow trout. Genetics 74:647–654.

Altham, P. M. E. 1984. Improving the precision of estimation by fitting a model. Journal of the Royal Statistical Society Series B, Methodological 46:118–119.

Andersson, L., N. Ryman, R. Rosenberg, and G. Ståhl. 1981. Genetic variability in Atlantic herring (*Clupea harengus harengus*): description of protein loci and population data. Hereditas 95:69–78.

Avise, J. C., and M. H. Smith. 1974. Biochemical genetics of sunfish. I. Geographic variation and subspecific intergradation in the bluegill, *Lepomis macrochirus*. Evolution 28:42–56.

Beacham, T. D., R. E. Withler, and A. P. Gould. 1985a. Biochemical genetic stock identification of chum salmon (*Oncorhynchus keta*) in southern British Columbia. Canadian Journal of Fisheries and Aquatic Sciences 42:437–448.

Beacham, T. D., R. E. Withler, and A. P. Gould. 1985b. Biochemical genetic stock identification of pink salmon (*Oncorhynchus gorbuscha*) in southern British Columbia and Puget Sound. Canadian Journal of Fisheries and Aquatic Sciences 42:1474–1483.

Cashon, R. E., R. J. Van Beneden, and D. A. Powers. 1981. Biochemical genetics of *Fundulus heteroclitus* (L.). IV. Spatial variation in gene frequencies of Idh-A, Idh-B, 6-Pgdh-A, and Est-S. Biochemical Genetics 19:715–728.

Christiansen, F. B., O. Frydenberg, J. P. Hjorth, and V. Simonsen. 1976. Genetics of *Zoarces* populations. IX. Geographic variation at the three phosphoglucomutase loci. Hereditas 83:245–256.

Clayton, J. W., W. G. Franzin, and D. N. Tretiak. 1973. Genetics of glycerol-3-phosphate dehydrogenase isozymes in white muscle of lake whitefish (*Coregonus clupeaformis*). Journal of the Fisheries Research Board of Canada 30:187–193.

Clayton, J. W., D. N. Tretiak, and A. H. Kooyman. 1971. Genetics of multiple malate dehydrogenase isozymes in skeletal muscle of walleye (*Stizostedion vitreum vitreum*). Journal of the Fisheries Research Board of Canada 28:1005–1008.

Coyne, J. 1982. Gel electrophoresis and cryptic protein variation. Isozymes: Current Topics in Biological and Medical Research 6:1–32.

Fournier, D. A., T. D. Beacham, B. E. Riddle, and C. A. Busack. 1984. Estimating stock composition in mixed stock fisheries using morphometric, meristic, and electrophoretic characteristics. Canadian Journal of Fisheries and Aquatic Sciences 41:400–408.

Gharrett, A. J., S. M. Shirley, and G. R. Tromble. 1987. Genetic relationships among populations of

Alaskan chinook salmon (*Oncorhynchus tshawytscha*). Canadian Journal of Fisheries and Aquatic Sciences 44:765–774.

Grant, W. S. 1985. Biochemical genetic stock structure of the southern African anchovy, *Engraulis capensis* Gilchrist. Journal of Fish Biology 27:23–29.

Grant, W. S., R. Bakkala, F. M. Utter, D. J. Teel, and T. Kobayashi. 1983. Biochemical genetic population structure of yellowfin sole, *Limanda aspera*, of the north Pacific Ocean and Bering Sea. U.S. National Marine Fisheries Service Fishery Bulletin 81:667–677.

Grant, W. S., G. B. Milner, P. Krasnowski, and F. M. Utter. 1980. Use of biochemical genetic variants for identification of sockeye salmon (*Oncorhynchus nerka*) stocks in Cook Inlet, Alaska. Canadian Journal of Fisheries and Aquatic Sciences 37:1236–1247.

Grant, W. S., and F. M. Utter. 1980. Biochemical genetic variation in walleye pollock, *Theragra chalcogramma*: population structure in the southeastern Bering Sea and the Gulf of Alaska. Canadian Journal of Fisheries and Aquatic Sciences 37:1093–1100.

Grant, W. S., C. I. Zhang, T. Kobayashi, and G. Ståhl. 1987. Lack of genetic discretion in Pacific cod (*Gadus macrocephalus*). Canadian Journal of Fisheries and Aquatic Sciences 44:490–498.

Ihssen, P. E., D. O. Evans, W. J. Christie, J. A. Reckahn, and R. L. DesJardine. 1981. Life history, morphology, and electrophoretic characteristics of five allopatric stocks of lake whitefish (*Coregonus clupeaformis*) in the Great Lakes region. Canadian Journal of Fisheries and Aquatic Sciences 38:1790–1807.

IUBNC (International Union of Biochemistry, Nomenclature Committee). 1984. Enzyme nomenclature 1984. Academic Press, Orlando, Florida.

Jamieson, A., and R. J. Turner. 1978. The extended series of Tf alleles in Atlantic cod, *Gadus morhua* L. Pages 699–729 *in* B. Battaglia and J. A. Beardmore, editors. Marine organisms—genetics, ecology and evolution. Plenum, New York.

Johnson, M. S. 1975. Comparative geographic variation in *Menidia*. Evolution 28:607–618.

Kobayashi, T., G. B. Milner, D. J. Teel, and F. M. Utter. 1984. Genetic basis for electrophoretic variation of adenosine deaminase in chinook salmon. Transactions of the American Fisheries Society 113:86–89.

Kornfield, I., P. S. Gagnon, and B. J. Sidell. 1981. Inheritance of allozymes in Atlantic herring (*Clupea harengus harengus*). Canadian Journal of Genetics and Cytology 23:715–720.

Kornfield, I., B. D. Sidell, and P. S. Gagnon. 1982. Stock definition in Atlantic herring (*Clupea harengus harengus*): genetic evidence for discrete fall and spring spawning populations. Canadian Journal of Fisheries and Aquatic Sciences 39:1610–1621.

Lavery, S., and J. B. Shaklee. 1989. Population genetics of two tropical sharks, *Carcharhinus tilstoni* and *C.*

sorrah, in northern Australia. Australian Journal of Marine and Freshwater Research 40:541–557.

Leary, R. F., F. W. Allendorf, and K. L. Knudsen. 1989. Genetic differences among rainbow trout spawned on different days within a single season. Progressive Fish-Culturist 51:10–19.

Leslie, J. F., and R. C. Vrijenhoek. 1977. Genetic analysis of natural populations of *Poeciliopsis monacha*. Allozyme inheritance and pattern of mating. Journal of Heredity 68:301–306.

May, B., M. Stoneking, and J. E. Wright, Jr. 1980. Joint segregation of biochemical loci in Salmonidae. II. Linkage associations from a hybridized *Salvelinus* genome (*S. namaycush* × *S. fontinalis*). Genetics 95:707–726.

May, B., J. E. Wright, Jr., and K. R. Johnson. 1982. Joint segregation of biochemical loci in Salmonidae: III. Linkage associations in Salmonidae including data from rainbow trout (*Salmo gairdneri*). Biochemical Genetics 20:29–40.

May, B., J. E. Wright, and M. Stoneking. 1979. Joint segregation of biochemical loci in Salmonidae: results from experiments with *Salvelinus* and review of the literature on other species. Journal of the Fisheries Research Board of Canada 36:1114–1128.

McAndrew, B. J., and K. C. Majumdar. 1983. Tilapia stock identification using electrophoretic markers. Aquaculture 30:249–261.

Millar, R. B. 1987. Maximum likelihood estimation of mixed stock fishery composition. Canadian Journal of Fisheries and Aquatic Sciences 44:583–590.

Miller, M. P., P. Pattillo, G. B. Milner, and D. J. Teel. 1983. Analysis of chinook stock composition in the May, 1982 troll fishery off the Washington coast: an application of the genetic stock identification method. Washington Department of Fisheries Technical Report 74.

Milner, G. B., D. J. Teel, and F. M. Utter. 1983. Genetic stock identification study. Report (Contract DE-A179-82BP23520) to Bonneville Power Administration, Portland, Oregon.

Milner, G. B., D. J. Teel, F. M. Utter, and C. L. Burley. 1981. Columbia River stock identification study: validation of genetic method. Report (Contract DE-A179-80BP18488) to Bonneville Power Administration, Portland, Oregon.

Milner, G. B., D. J. Teel, F. M. Utter, and G. A. Winans. 1985. A genetic method of stock identification in mixed populations of Pacific salmon, *Oncorhynchus* spp. U.S. National Marine Fisheries Service Marine Fisheries Review 47(1):1–8.

Milton D. A., and J. B. Shaklee. 1988. Biochemical genetics and population structure of blue grenadier *Macruronus novaezelandiae* (Pisces: Merlucciidae), from Australian waters. Australian Journal of Marine and Freshwater Research 38:727–742.

Morizot, D. C., and M. J. Siciliano. 1979. Polymorphisms, linkage and mapping of four enzyme loci in the fish genus *Xiphophorus* (Poeciliidae). Genetics 93:947–960.

Morizot, D. C., and M. J. Siciliano. 1983. Comparative gene mapping in fishes. Isozymes: Current Topics in Biological and Medical Research 10:261–285.

Mork, J., N. Ryman, G. Ståhl, F. Utter, and G. Sundnes. 1985. Genetic variation in Atlantic cod (*Gadus morhua*) throughout its range. Canadian Journal of Fisheries and Aquatic Sciences 42:1580–1587.

Nei, M. 1972. Genetic distance between populations. American Naturalist 106:283–292.

Okazaki, T. 1982. Genetic study on population structure in chum salmon (*Oncorhynchus keta*). Bulletin of the Far Seas Fisheries Research Laboratory (Shimizu) 19:25–116.

Pasdar, M., D. P. Philipp, and G. S. Whitt. 1984. Linkage relationships of nine enzyme loci in sunfishes (*Lepomis*; Centrarchidae). Genetics 107:435–446.

Pella, J. J., and G. B. Milner. 1987. Use of genetic marks in stock composition analysis. Pages 247–276 *in* N. Ryman and F. M. Utter, editors. Population genetics & fishery management. University of Washington Press, Seattle.

Philipp, D. P., W. F. Childers, and G. S. Whitt. 1981. Management implications for different genetic stocks of largemouth bass (*Micropterus salmoides*) in the United States. Canadian Journal of Fisheries and Aquatic Sciences 38:1715–1723.

Place, A. R., and D. A. Powers. 1978. Genetic bases for protein polymorphism in *Fundulus heteroclitus* (L.). I. Lactate dehydrogenase (Ldh-B), malate dehydrogenase (Mdh-A), glucosephosphate isomerase (Gpi-B), and phosphoglucomutase (Pgm-A). Biochemical Genetics 16:576–592.

Purdom, C. E., D. Thompson, and P. R. Dando. 1976. Genetic analysis of enzyme polymorphisms in plaice (*Pleuronectes platessa*). Heredity 37:193–206.

Richardson, B. J. 1983. Distribution of protein variation in skipjack tuna (*Katsuwonus pelamis*) from the central and south-western Pacific. Australian Journal of Marine and Freshwater Research 34:231–251.

Ryman, N., U. Lagercrantz, L. Andersson, R. Chakraborty, and R. Rosenberg. 1984. Lack of correspondence between genetic and morphologic variability patterns in Atlantic herring (*Clupea harengus*). Heredity 53:687–704.

Salini, J., and J. B. Shaklee. 1988. The population genetic structure of barramundi (*Lates calcarifer*) from the Northern Territory, Australia. Australian Journal of Marine and Freshwater Research 39:317–329.

Shaklee, J. B. 1983. The utilization of isozymes as gene markers in fisheries management and conservation. Isozymes: Current Topics in Biological and Medical Research 11:213–247.

Shaklee, J. B. 1984. Genetic variation and population structure in the damselfish, *Stegastes fasciolatus*, throughout the Hawaiian Archipelago. Copeia 1984:629–640.

Shaklee, J. B., F. W. Allendorf, D. C. Morizot, and G. S. Whitt. 1990. Gene nomenclature for protein-coding loci in fish. Transactions of the American Fisheries Society 119:2–15.

Shaklee, J. B., M. J. Champion, and G. S. Whitt. 1974. Developmental genetics of teleosts: a biochemical analysis of lake chubsucker ontogeny. Developmental Biology 38:356–382.

Shaklee, J. B., and C. P. Keenan. 1986. A practical laboratory guide to the techniques and methodology of electrophoresis and its application to fish fillet identification. Australia Commonwealth Scientific and Industrial Research Organization, Marine Laboratories, Report 177, Melbourne.

Shaklee, J. B., and J. P. Salini. 1985. Genetic variation and population subdivision in Australian barramundi, *Lates calcarifer* (Bloch). Australian Journal of Marine and Freshwater Research 36:203–218.

Shaklee, J. B., and C. S. Tamaru. 1981. Biochemical and morphological evolution of Hawaiian bonefishes (*Albula*). Systematic Zoology 30:125–146.

Singh, R. S., R. C. Lewontin, and A. A. Felton. 1976. Genetic heterogeneity within electrophoretic "alleles" of xanthine dehydrogenase in *Drosophila pseudoobscura*. Genetics 84:609–629.

Smith, P. J., R. I. C. C. Francis, and L. J. Paul. 1978. Genetic variation and population structure in the New Zealand snapper. New Zealand Journal of Marine and Freshwater Research 12:343–350.

Smouse, P. E. 1974. Likelihood analysis of recombinational disequilibrium in multiple locus gametic frequencies. Genetics 76:557–565.

Sokal, R. R., and F. J. Rohlf. 1981. Biometry, 2nd edition. Freeman, San Francisco.

Ståhl, G. 1981. Genetic differentiation among natural populations of Atlantic salmon (*Salmo salar*) in northern Sweden. Ecological Bulletin (Stockholm) 34:95–105.

Stoneking, M., B. May, and J. E. Wright. 1979. Genetic variation, inheritance, and quaternary structure of malic enzyme in brook trout (*Salvelinus fontinalis*). Biochemical Genetics 17:599–619.

Swofford, D. L., and R. B. Selander. 1981. BIOSYS-1: a FORTRAN program for the comprehensive analysis of electrophoretic data in population genetics and systematics. Journal of Heredity 72:281–283.

Turner, B. J. 1983. Genic variation and differentiation of remnant natural populations of the desert pupfish, *Cyprinodon macularius*. Evolution 37:690–700.

Utter, F. M., H. O. Hodgins, F. W. Allendorf, A. G. Johnson, and J. L. Mighell. 1973. Biochemical variants in Pacific salmon and rainbow trout: their inheritance and application in population studies. Pages 329–339 *in* J. H. Schroder, editor. Genetics and mutagenesis of fish. Springer-Verlag, Berlin.

Utter, F., D. Teel, G. Milner, and D. McIsaac. 1987. Genetic estimates of stock compositions of 1983 chinook salmon, *Oncorhynchus tshawytscha*, harvests off the Washington coast and the Columbia River. U.S. National Marine Fisheries Service Fishery Bulletin 85:13–23.

Van Beneden, R. J., R. E. Cashon, and D. A. Powers. 1981. Biochemical genetics of *Fundulus heteroclitus* (L.). III. Inheritance of isocitrate dehydrogenase (Idh-A and Idh-B), 6-phosphogluconate dehydrogenase (6-Pgdh-A), and serum esterase (Est-S) polymorphisms. Biochemical Genetics 19:701–714.

Vawter, A. T., R. Rosenblatt, and G. C. Gorman. 1980. Genetic divergence among fishes of the eastern Pacific and the Caribbean: support for the molecular clock. Evolution 34:705–711.

Waples, R. S. 1990. Temporal changes of allele frequency in Pacific salmon: implications for mixed-stock fishery analysis. Canadian Journal of Fisheries and Aquatic Sciences 47:968–976.

Whitt, G. S., W. F. Childers, and T. E. Wheat. 1971. The inheritance of tissue-specific lactate dehydrogenase isozymes in interspecific bass (*Micropterus*) hybrids. Biochemical Genetics 5:257–273.

Williams, G. C., R. K. Koehn, and V. Thorsteinsson. 1984. Icelandic eels: evidence for a single species of *Anguilla* in the north Atlantic. Copeia 1984:221–223.

Winans, G. A. 1980. Geographic variation in the milkfish *Chanos chanos*. I. Biochemical evidence. Evolution 34:558–574.

Wishard, L. 1981. Stock identification of Pacific salmon in western Washington using biochemical genetics. Washington Department of Fisheries Service (Contracts 1176 and 1276), Final Report, Olympia.

Wood, C. C., S. McKinnell, T. J. Mulligan, and D. A. Fournier. 1987. Stock identification with the maximum likelihood mixture model: sensitivity analysis and application to complex problems. Canadian Journal of Fisheries and Aquatic Sciences 44:866–881.

American Fisheries Society Symposium 7:831–840, 1990
© Copyright by the American Fisheries Society 1990

Sixty Years of Tagging Pacific Halibut: A Case Study

ROBERT J. TRUMBLE, IAN R. MCGREGOR, GILBERT ST-PIERRE,
DONALD A. MCCAUGHRAN, STEPHEN H. HOAG

International Pacific Halibut Commission
Post Office Box 95009, Seattle, Washington 98145, USA

Abstract.—The International Pacific Halibut Commission (IPHC) has tagged and released over 350,000 Pacific halibut *Hippoglossus stenolepis* since 1925. Over 35,000 tagged fish have been recovered, and tagging results include estimates of migration, mortality, and growth. Individually numbered metal strap tags were used until 1970, when the program switched to spaghetti tags. Problems associated with the tagging data include tag shedding, tagging mortality, and nonreporting of tags. Differential rates of tag nonreporting, both over time and among areas, appear to be the most serious sources of error when tagging results are used to assess and manage halibut stocks. The IPHC staff is reevaluating the tagging program to determine the most appropriate analytical treatments of existing data and the best way to proceed with future tagging.

The International Pacific Halibut Commission (IPHC) was established in 1923 by a convention between Canada and the USA for the preservation of the Pacific halibut *Hippoglossus stenolepis* fishery of the northern Pacific Ocean and the Bering Sea. The commission's scientific work began in 1925, and soon thereafter two important research areas were identified: migration among various fishing regions, and rates of natural and fishing mortality (Thompson and Herrington 1930). Tagging was chosen as a primary means of addressing these questions. Tagging operations have continued since then, although methods, procedures, and goals have evolved.

The IPHC has tagged and released over 350,000 Pacific halibut, and over 35,000 tagged fish have been recovered. Pacific halibut have been tagged throughout their range from northern California to the Bering Sea (Figure 1).

Tagging results have been an important ingredient in developing management philosophy, regulations, and population biology. Analyses by IPHC staff suggest that tagging data may not be useful for annual assessment of biomass and exploitation rates. Also, differential nonreporting of tags by area and over time confound calculations of migration rates and exploitation rates. Tag shedding and tagging mortality are other factors that limit the use of tagging data. The IPHC staff is reevaluating the tagging program to determine if modifications in the design or additional experiments are needed.

Tagging Procedures

The first Pacific halibut tagging operation (Thompson and Herrington 1930) is significant for the care and forethought that led to many procedures still in use. Tags were placed on the head to avoid damaging the valuable flesh. Although tag visibility would be better on the white (blind) side (because fishermen and processors handle Pacific halibut white side up), tags were placed on the opercular bones of the dark (eyed) side of the fish to prevent them from contacting the bottom which could irritate the fish and cause tag loss. Fish to be tagged were captured with standard commercial longlines deployed from chartered vessels. Only the most viable fish were tagged. The length of each fish was measured and the individual tag numbers were recorded. The IPHC generated publicity to inform fishermen of the tagging program. The aid of fish-processing plants and government agencies was enlisted in redeeming tags, and a tag reward was initiated. All of these procedures are still followed in our tagging operations. Other aspects of halibut tagging, such as tag type, tagging platform, and recovery process, have evolved since the time of Thompson and Herrington.

Tag Type

The first tag extensively used for Pacific halibut was a metal strap (Figure 2) originally used to mark cattle (Thompson and Herrington 1930). It was chosen because of the large amount of opercular bone available, and because the single fastening point made attachment convenient. The individually numbered tags were made of monel and thought to be corrosion resistant. In double-tagging experiments by Thompson and Herrington, button tags that attached to a small area

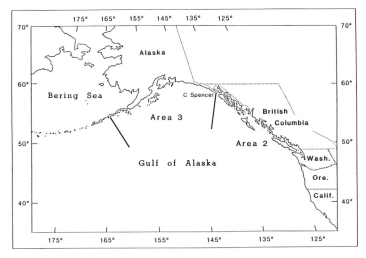

FIGURE 1.—Regional divisions of the Pacific coast that include the range of Pacific halibut.

were shed often compared with strap tags. Subsequently, tatoos, body-cavity tags, and dart tags were tested in double-tagging experiments and rejected (Myhre 1966). Tatoos of India ink faded within 2 years; internal tags, which had been recovered when viscera was used as a source of vitamin A, were often thrown away unnoticed at sea; and dart tags suffered high shedding loss. Spaghetti tags (polyethylene tubes reinforced with nickel-silver wire) were double tagged with metal straps beginning in 1955, and Myhre (1966) found similar retention rates. A double-tagging experiment conducted in 1968 indicated that strap tags suffered higher losses than did spaghetti tags (IPHC, unpublished). Electrolysis may have weakened the bend of the strap tags, but the higher loss was attributed to tag contact with trawl nets. The last strap tag was used in 1970, after which the spaghetti tag (Figure 2) became standard. Spaghetti tags continued to evolve with improved tubing, better legend retention, and more corrosion-resistant wire.

Tagging Platform

For the first 35 years of halibut tagging (through 1960), the IPHC almost exclusively chartered commercial longline vessels. Standard longlines consists of 457- or 549-m-long main lines called skates, each of which has a number of hooks attached to it by shorter lines called gangions. In most cases, vessels longer than 20 m were used to accommodate IPHC staff and to assure ample working space on deck for tagging activities.

Commercial longline gear is selective for large halibut (Myhre 1969); this emphasis on adult fish made commercial techniques and gear appropriate for capture of fish to be tagged.

Beginning in 1947, research trawl gear was used in the tagging and release program (Dunlop et al. 1964). Trawls are selective for small Pacific halibut 3–5 years old. Chartered trawlers usually were typical Pacific seine-type vessels with work space aft, stern-mounted drum, 21–27 m in length, and powered by 200–650 horsepower diesel engines. The standard research gear was a 400-mesh eastern trawl with 21.6-m headrope, 28.6-m groundrope, and 8.9-cm cod-end mesh.

Both longline and trawl vessels continue to be chartered for tagging operations, depending on the location and purpose of the experiment. Trawl operations are restricted from rough bottom (roller gear is not used), but longline gear is not.

Tag Recovery

The IPHC initiated an intensive port sampling program early in its history, and one of the primary objectives was recovery of tags from fishermen. Port samplers inquire about tags when interviewing fishermen, collect tags if present, and pay rewards. Port samplers continually encourage fishermen to look for tags, to mark tagged fish for separate handling during unloading, and to retain tags if port samplers are not available to collect biological data from tagged fish. The primary data desired for each tagged fish are the tag and both otoliths and fish length, the exact loca-

FIGURE 2.—Examples of the strap and spaghetti tags used by the International Pacific Halibut Commission.

tion caught, depth caught, vessel name, and name and address of finder.

Publicity is considered important in recovering tags. Posters (Figure 3) that describe the program and solicit tag returns are displayed at unloading sites and are provided to fishing vessel owners and fishermen's organizations. The program is frequently reviewed at meetings with fishermen. Port samplers carry a notebook with tag data for month, year, and general location of release, so that release information can be relayed immediately to the fisherman who returns a tag. A follow-up letter gives precise release data for date, length, and location.

Tag rewards provide a tangible incentive for fishermen to retain tags. Although rewards seldom are high, they indicate that recovery is important to the releasing agency. Rewards started in 1925 at $0.50 (US or Canadian funds) for tags without associated data and $1 for tags with location, date, and fish length. An additional dollar was added in 1927 for tags still attached to the fish. The $2 reward was standard until 1978, when the reward was increased to $5. To further heighten interest, a lottery with a $100 reward for several preselected tag numbers was established.

The reward program was last modified in 1986 to eliminate the lottery and to offer a "tag reward" hat (Figure 3) as an alternative to the standard $5. The hats have been well received, and in 1986, 83% of the finders chose hats over the $5.

Tagging Objectives

Objectives of tagging programs at IPHC can be summarized in four categories: migration, fishing and natural mortality, growth, and special projects. In many cases, a single tagging operation was designed to provide data for more than one category.

Migration

The earliest tagging programs at IPHC were oriented to migration as an indicator of discrete biological units within the Pacific halibut resource. Thompson and Herrington (1930) initiated tagging in 1925 over the range of fishing activity (Vancouver Island to the Alaska Peninsula), so that both releases and recoveries would be representative of any populations harvested. They concluded that separate populations existed north and south of Cape Spencer, Alaska, and they

TAGGED HALIBUT

The INTERNATIONAL PACIFIC HALIBUT COMMISSION attaches plastic tags to the cheek on the dark side of the halibut. Fishermen should return all tags, even those from halibut below legal size or those caught in trawls.

REWARD

$5.00 will be paid for the return of each tag.

OR

A "Hat" will be paid for the return of each tag.

WHEN YOU CATCH A TAGGED HALIBUT:
1. Record tag numbers, date, location and depth in your log book.
2. Leave the tag on the fish.
3. Mark the fish with a gangion around tail.

WHEN YOU LAND A TAGGED HALIBUT:
1. Report fish to a Commission Representative or Government officer

   or

2. Forward tags to address below and enclose recovery information (see above), your name, address, boat name, gear, length of fish, and, if possible, earstones.

FINDER WILL BE ADVISED OF MIGRATION AND GROWTH OF THE FISH.

International Pacific Halibut Commission
P.O. Box 95009
Seattle, Washington 98145-2009

FIGURE 3.—Poster used by the International Pacific Halibut Commission to solicit returns of Pacific halibut tags.

created separate management areas (IPHC areas 3 and 2, respectively). Tag recoveries were interpreted as showing migrations of several hundred kilometers by mostly older fish in area 3, but limited movements by mostly younger fish in area 2. Thompson and Herrington (1930) discussed several major issues that affected tagging experiments: mortality caused by capture and tagging, tag loss at sea, tags not seen by fishermen or shore workers, and natural mortality. They did not believe these potential problems seriously threatened their conclusions.

Kask (1935) extended the analysis with additional recoveries and provided details of tagging methodologies. He demonstrated that onshore–offshore movements were linked to spawning migrations, and that net migration countered the drift of eggs and larvae from the spawning grounds. Kask concurred with the earlier conclusion that Cape Spencer represented a boundary between Pacific halibut populations, and that potential tagging problems did not jeopardize the reality of this boundary. Although the conclusion that mixing does not occur across the boundary is no longer generally accepted, the Cape Spencer line remains useful for managing Pacific halibut.

The next migration study involved the Bering Sea (Dunlop et al. 1964), where the IPHC was concerned about the effect of increased foreign and domestic fishing pressure on stocks throughout the region. Tagging was performed at selected spots rather than over a broad range. Dunlop et al., after examining tag recoveries, concluded that Pacific halibut migrated freely within the eastern Bering Sea, that virtually none migrated to Asia, and that about 25% emigrated from the Bering Sea to the Gulf of Alaska. The authors acknowledged that concentration of fishing activity on preferred grounds prevented tag returns from representing the true distribution of the fish.

A program to study the basic biology of juvenile Pacific halibut, begun in 1955, included tagging of juveniles in the Bering Sea and the Gulf of Alaska. In summarizing the results of the tagging, Best (1968) said that easterly and southerly movements were evident (Figure 4), but he did not address rates and stock relationship between areas 3 and 2. Skud's (1977) analysis of these data showed that young fish migrated more extensively than did older juveniles, and that interchange occurred from area 3 to area 2. This extended Kask's (1935) claim that eastward and southward migration functioned to counter the drift of eggs and larvae, but it disputed the earlier conclusion that Cape Spencer was a boundary between populations. Skud further concluded that adult halibut moved relatively little from summer to summer while on the feeding grounds, but that extensive migrations noted by previous authors were movements to and from spawning grounds (winter) and feeding grounds (summer).

In 1976, a large-scale, multipurpose resource survey began in which tagged fish were released in a series of approximately 100 × 300-km grids. Migration data were expected from tag recoveries. Deriso and Quinn (1983) and Quinn et al. (1985) estimated migration rates for exploitable fish from grid-survey and earlier tag returns. They found that few adult Pacific halibut migrated from the management area of release, although emigration rates were higher from the Bering Sea than from the Gulf of Alaska.

Recent analyses performed to determine migration rates and exploitation rates pointed out several major problems in the tagging data (R. Deriso, IPHC, unpublished). Foremost was nonreporting of tags by fishermen. Mortality calculated from tag returns was approximately twice that calculated from catch-per-effort and age data, which suggested that only half the expected tags were returned. The return rate was unequal along the coast; apparently, fishermen were more willing to cooperate in some areas than others. Tags were not distributed in proportion to density of fish. Tagging in limited areas for special projects clearly did not represent the distribution of the resource, and possibly the grid surveys were not representative.

A trawl survey designed to index juvenile Pacific halibut abundance in the Bering Sea and Gulf of Alaska began in 1964 and continued through 1986. Tagged fish were released to obtain information on transboundary movements and recruitment to the commercial fishery; peak tagging occurred in 1980 and 1981. More than 67,000 tagged fish were released. Analysis of tags recovered from this experiment has begun and is expected to improve migration-rate estimates.

Mortality

Thompson and Herrington (1930) first calculated mortality for Pacific halibut by following declines in numbers of tag recoveries for a single release year. They estimated an annual fishing mortality and an annual total mortality for area 2. Kask (1935) extended mortality estimates by using more tagging series that also included area 3. Estimated area-2 fishing mortality was almost identical with Thompson and Herrington's estimate; estimated area-3 fishing mortality was much less than in area 2, based on releases over 2 years. As with migration-rate estimates, these authors assumed that potential tagging problems did not jeopardize their analyses.

Calculations from tagging experiments from 1947 to 1953 indicated that natural mortality was similar in areas 2 and 3; fishing mortality, however, was several times larger in area 2 than in area 3 (IPHC 1960). The authors rejected natural mortality rates from tagging results as unreasonably high,

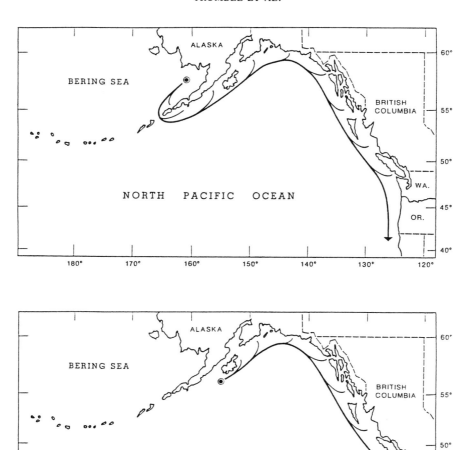

FIGURE 4.—Migratory patterns of juvenile Pacific halibut from different tagging sites (after Best 1968).

because estimates from age-composition analysis suggested much lower rates. The authors also were concerned that tagging results tended to estimate fishing mortality lower than did other methods, and they pointed out that violation of assumptions causes natural mortality to be overestimated and fishing mortality to be underestimated.

Myhre (1963) concluded that tag loss and tagging mortality were not serious problems for analysis of halibut tag data. He further noted that fishing mortality estimated from tagging data correlated well with gear density (number of hooks per unit area of fishing grounds), and that both were good

indices of actual fishing mortality. He could not show that tagging data were bias-free, but he suggested that biases were similar among survey areas. Even so, Myhre believed that tagging data did not provide information on stock condition as good as catch-per-unit effort data unless dependable information on recruitment, growth, and population density was also available.

Myhre (1967) recalculated mortalities using a longer time series (1925–1955). In his analysis, he consolidated all tag losses, other than those to known fishing mortality, and included tag shedding, nonreporting, and losses to trawl bycatch in

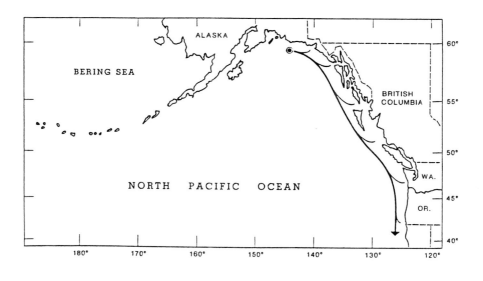

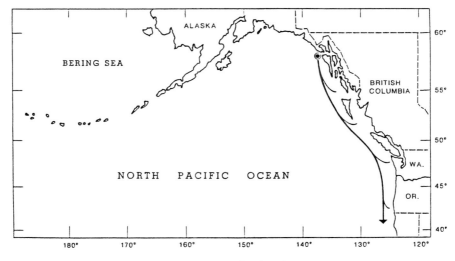

FIGURE 4.—Continued.

his analysis. All the miscellaneous losses were considered small, which made natural mortality the dominant component. He calculated that fishing mortality was higher in area 2 than in area 3. Drawing on previous knowledge that Pacific halibut in areas 2 and 3 were fished at or near maximum sustainable yield, Myhre concluded that either the fishing mortality calculations were not comparable for the two areas, or Pacific halibut in the two areas had divergent productivity.

Dunlop et al. (1964) used the percentage of tags recovered as a measure of exploitation in the eastern Bering Sea. They accounted for annual changes in fishing effort by using tag recoveries per unit of fishing effort as a measure of the number of tags in the population. Tag shedding and tagging mortality were regarded as minor. Foreign fishing was recognized as a potential source of mortality, but only two tags were returned from foreign vessels, which suggested that foreign fishing played a minor role. Dunlop et al. also used tagging data to estimate catchability coefficients, and then calculated instantaneous fishing mortality for the Bering Sea. Fishing mortality values were in the same range as mortalities calculated with age-composition and catch statistics.

As was discussed earlier, recent attempts to calculate mortality or exploitation rates have demonstrated that nonreporting of tags is a major impediment to the use and reliability of tagging data for these purposes.

Growth

Kask (1935) first analyzed tagging data to determine Pacific halibut growth for combined sex samples. He calculated growth rates in centimeters per day by dividing length increment gained while the fish were at large by time at large. Fish from area 2 showed faster growth than did fish from area 3. Kask attributed the difference to separate characteristics of different biological populations. He noted that growth analysis for Pacific halibut was confounded by lack of data on sex composition for commercially caught fish, by comparison of dissimilar size-ranges (because females grow faster than males), and by measurement errors. Kask concluded that faster-growing small fish and faster-growing females were more common in area 2, and he assumed that data problems did not compromise his analysis.

McCaughran (1981) developed statistics for modeling Pacific halibut growth from a subset of the tagging data that included sex of recaptured fish. The natural logarithm of lengths converted the data to a normal distribution. The three-parameter von Bertalanffy equation and the two-parameter power function fit the data equally well when tested with tag return length data for males from the northern British Columbia coast. The power function model, preferred for its simplicity, was then used to estimate growth parameters for females.

McCaughran (1987) extended his previous analysis to males and females throughout the range for year-classes 1935–1965. To avoid gear-selectivity bias, longline-caught fish less than 10 years old when tagged were excluded from the analysis. Growth of either sex varied little since 1935. This result contradicted a conclusion of increased growth since the 1950s, based on otolith back-calculations (Southward 1967) that could not account for different growth rates by sex. McCaughran hypothesized that migratory fish, which may travel from the Bering Sea to the Oregon coast, would use more energy and grow slower than nonmigratory fish. However, he found statistically significant differences in growth parameters for migratory and nonmigratory fish of both sexes, growth being higher for migrating fish.

Special Projects

Several tagging activities that do not fit the above categories have been conducted as special projects. Most of them addressed survival of Pacific halibut caught incidentally and returned by law to the sea. Other special projects have examined survival of tagged Pacific halibut confined to holding pens, and some fish have been tagged to indicate presence of oxytetracycline (OTC) injections.

The IPHC regulations permit Pacific halibut to be harvested only by hook-and-line gear. Hoag (1975) tagged and released Pacific halibut caught by domestic trawlers to estimate survival of discarded fish. By recording physical condition of the fish on a scale of 1 (excellent) to 5 (dead), he was able to relate condition to handling practices on board. Return of tags was used to determine relative survival for fish in each conditions category. Survival was correlated positively with length of fish and negatively with time on deck and weight of catch. Hoag concluded that survival of Pacific halibut released by domestic trawlers was approximately 50%.

Pacific halibut are caught incidentally by salmon troll gear and may be retained during open seasons if of legal size. Most of them are of sublegal size and must be returned to the sea. The IPHC compared survival of sublegal, troll-caught Pacific halibut with halibut released from longline gear. Tags recovered over a 10-year period showed equal return rates, which led to the conclusion that survival was comparable for the two types of gear (IPHC, unpublished).

In 1986, IPHC staff used tags to test the survival of Pacific halibut injured with a device that automatically strips the hook as fish are pulled on board with a longline. The staff tagged and released fish that passed through the automatic hook stripper for comparison with a group of optimally handled fish tagged as a control. Tag returns indicated that optimally released fish had 3.4 times higher survival than fish that had passed through the automatic hook stripper (IPHC, unpublished data).

An experiment in which Pacific halibut were confined to holding pens immediately after tagging was conducted in 1958 and 1960 to evaluate mortality caused by capture and tagging (Peltonen 1969). Surface pens proved unsuccessful because wave action stressed the fish and increased mortality. Underwater pens set on the bottom kept Pacific halibut without mortality for 6 d, and estimated mortality attributed to handling and tagging ranged from 2 to 5% for the 4 weeks of the experiment.

Pacific halibut injected with OTC were tagged and released in 1982 and 1983 as part of an experiment to validate age readings of otoliths. The OTC induces a mark on an otolith, allowing the number of annuli to be compared with time at large. Otoliths from fish with OTC injections can be identified by tag number and processed separately. The limited number of recoveries have demonstrated adequate marking by OTC and have confirmed that time at large is consistent with the number of annuli counted (IPHC 1989).

Current Status

Elements of a Successful Program

A primary requirement for a tagging program that relies on voluntary tag returns by the finders is the use of a conspicuous tag that stays attached. Several models were tried before the spaghetti tags were chosen.

A promotional campaign to create high awareness in the industry is paramount. Both fishermen and shore workers have the opportunity to recover tags, so it is necessary to maintain awareness through personal contacts, posters, and rewards. The IPHC staff has an active waterfront presence to recover and redeem tags. Port samplers inquire about tags at all vessels visited and remind fishermen that all recovery data are confidential. We regularly remind shore workers of the value of tags and the reward for returned tags. Because representation is not possible at all ports, IPHC sends a letter to uninterviewed fishermen requesting data on their fishing activities and return of any tags they possess. Many tags are recovered this way.

When tags are returned by fishermen, the port samplers can immediately respond with location and date of the tag release. Detailed information is sent later by mail. Initially, we included catch location in the letter, but we dropped it to prevent misdelivered letters from revealing confidential fishing locations.

A reward program shows that an agency is serious about tag recovery, and it produces higher tag returns. Each tag reward paid requires a receipt signed by the payee. Small rewards seem to be of little incentive to most fishermen. To keep interest keen, we have tried higher-value ($100) lotteries and desirable merchandise, such as the tag reward hats, but we cannot quantify their effect.

Once tags have been collected, many opportunities exist for degradation of the data set. Maximizing usefulness of the tagging data requires a commitment to consistent processing and editing. At IPHC one staff member is responsible for reviewing all tagging data, for reconciling tag reward payments and receipts, and for maintaining tag-record archives (both computer and physical).

Reevaluation

The 60-year history of tagging by the IPHC has shown mixed results. The determination of general patterns of migration and growth rates has been most successful. Estimating rates for migration and exploitation from tagging has been of limited value, but it is being actively pursued at present. Useful information has been obtained from special-project tagging, whereas large-scale releases with a specific objective have proved less valuable.

Differential nonreporting of tags has been identified tentatively as the main impediment to effective use of tagging data to calculate mortality and migration rates for Pacific halibut. If nonreporting were consistent, it could be corrected for; but the proportion of nonreporting has apparently changed over time and geographic area as the nature of the fishery has changed.

It appears that most tags were returned during the early days of the fishery, but more recently, lower rates of return are suspected. The Pacific halibut fishery has changed from a fleet of several hundred vessels fishing for several months of the year throughout the range to a fleet of several thousand vessels fishing for several days in the major fishing areas. Many new fishermen are not experienced at retaining and reporting tags, except in Canada where limited entry has stabilized the fleet. The short, intense fishery leaves little time for activities other than killing fish.

The potential for bias caused by nonreporting has been recognized since the early tagging studies, but experiments have not been conducted to properly test the amount of nonreporting that occurred over the years. The recent concern for the magnitude of nonreporting resulted from comparison of exploitation rates calculated from catch-at-age analysis[1] and tagging (R. Deriso, IPHC, personal communication); the comparison suggests that 50% of the tags are not reported.

---

[1]Catch-at-age analysis (Quinn et al. 1985) is the standard stock-assessment method at IPHC.

A proper test of nonreporting of Pacific halibut tags is difficult to design and conduct. Tags can be unreported at several stages of processing. The fishermen may not see the tags or may remove them but not turn them in. The same is true for plant workers, both before and after the heads are removed. Further, the expected recovery rate varies among ports and vessels, depending on where fishing occurs and where tags are released. Any results obtained from a test will be applicable only to the year of the test, particularly if incentive or opportunity to return tags changes from year to year.

An agency sampling program designed to recover tags without relying on voluntary returns might eliminate nonreporting problems. Coded wire tags and passive integrated transponder (PIT) tags are being evaluated at IPHC as alternatives to external tags. Coded wire tags have been used in an experiment, the results of which are still being analyzed.

In spite of experiments that indicate tag shedding and tagging mortality are small, these sources of tag loss still have the potential to account for losses now attributed to nonreporting. Any experiment to assess nonreporting should also address these other sources.

A critical review of our tagging program, with an analysis of tags released and recaptured from the grid survey, is now under way at North Carolina State University. A second study to estimate migration rates from the tagged juveniles released by trawlers in 1980 and 1981 is under way at the University of Washington. Until the two contracts are completed, little emphasis will be placed on tagging during survey operations. Fish are still tagged when they are available at little additional cost. Specific tagging studies now in the planning stage, with objectives not compromised by the problems identified above, are anticipated to add to our understanding of Pacific halibut biology and dynamics.

References

Best, E. A. 1968. Studies of young halibut: census of juveniles. Western Fisheries 75(5):38–41, 59–60.
Deriso, R. B. and T. J. Quinn II. 1983. The Pacific halibut resource and fishery in regulatory area 2. II. Estimates of biomass, surplus production, and reproductive value. International Pacific Halibut Commission, Scientific Report 67:55–89.

Dunlop, H. A., F. H. Bell, R. J. Myhre, W. H. Hardman, and G. M. Southward. 1964. Investigation, utilization, and regulation of the halibut in southeastern Bering Sea. Report of the International Pacific Halibut Commission 35.
Hoag, S. H. 1975. Survival of halibut released after capture by trawls. International Pacific Halibut Commission, Scientific Report 57.
Kask, J. L. 1935. Studies in migration, fishing mortality, and growth in length of the Pacific halibut (*Hippoglossus hippoglossus*) from marking experiments. Doctoral dissertation. University of Washington, Seattle.
IPHC (International Pacific Halibut Commission). 1960. Utilization of Pacific halibut stocks: yield per recruitment. Report of the International Pacific Halibut Commission 28.
IPHC (International Pacific Halibut Commission). 1989. Annual report 1988. International Pacific Halibut Commission, Seattle.
Myhre, R. J. 1963. A study of errors inherent in tagging data on Pacific halibut (*Hippoglossus stenolepis*). International Commission for the Northwest Atlantic Fisheries, Special Publication 4.
Myhre, R. J. 1966. Loss of tags from Pacific halibut as determined by double-tag experiments. Report of the International Pacific Halibut Commission 41.
Myhre, R. J. 1967. Mortality estimates from tagging experiments on Pacific halibut. Report of the International Pacific Halibut Commission 42.
Myhre, R. J. 1969. Gear selection and Pacific halibut. Report of the International Pacific Halibut Commission 51.
McCaughran, D. A. 1981. Estimating growth parameters of Pacific halibut from mark–recapture data. Canadian Journal of Fisheries and Aquatic Sciences 38:394–398.
McCaughran, D. A. 1987. Growth in length of Pacific halibut. Pages 507–515 in R. C. Summerfelt and G. E. Hall, editors. Age and growth of fish. Iowa State University Press, Ames.
Peltonen, G. J. 1969. Viability of tagged Pacific halibut. Report of the International Pacific Halibut Commission 52.
Quinn, T. J., II, R. B. Deriso, and S. H. Hoag. 1985. Methods of population assessment of Pacific halibut. International Pacific Halibut Commission, Scientific Report 72.
Skud, B. E. 1977. Drift, migration, and intermingling of Pacific halibut stocks. International Pacific Halibut Commission, Scientific Report 63.
Southward, G. M. 1967. Growth of Pacific halibut. Report of the International Pacific Halibut Commission 43.
Thompson, W. F. and W. C. Herrington. 1930. Life history of the Pacific halibut. (1) Marking experiments. Report, International Fisheries Commission 2.

American Fisheries Society Symposium 7:841–853, 1990

History of the Cooperative Game Fish Tagging Program in the Atlantic Ocean, Gulf of Mexico, and Caribbean Sea, 1954–1987

EDWIN L. SCOTT, ERIC D. PRINCE, AND CAROLE D. GOODYEAR

National Marine Fisheries Service
Southeast Fisheries Center, Miami Laboratory
75 Virginia Beach Drive
Miami, Florida 33149, USA

Abstract.—The history of the Cooperative Game Fish Tagging Program, managed by the Miami Laboratory of the U.S. National Marine Fisheries Service's Southeast Fisheries Center, is reviewed in terms of the program's major objectives and accomplishments from 1954 to 1987. Over 118,000 fish of 78 species have been tagged by recreational fishermen, commercial fishermen, and scientists since the program started in 1954. Tagging of the target species (bluefin tuna *Thunnus thynnus*, blue marlin *Makaira nigricans*, white marlin *Tetrapturus albidus*, sailfish *Istiophorus platypterus*, swordfish *Xiphias gladius*, and greater amberjack *Seriola dumerili*) is evaluated in relation to the types of tags used and the yearly fluctuations in the numbers of fish released and recaptured. Although problems have been encountered with tag construction and retention, recapture information has resulted in substantial increases in our knowledge of movement patterns and age and growth of these species. A new data-management system will allow the program to more efficiently handle the data resulting from increased tagging efforts by the National Marine Fisheries Service, state agencies, and private organizations and from the tagging of new target species (red drum *Sciaenops ocellatus*, king mackerel *Scomberomorus cavalla*, and Spanish mackerel *S. maculatus*).

The Cooperative Game Fish Tagging Program (CGFTP) is a joint research effort by scientists and by recreational and commercial fishermen. It is designed to provide basic information on the movements and biology of game-fish populations in the Atlantic Ocean, Gulf of Mexico, and the Caribbean Sea through the direct participation of the public in scientific research. The program was begun in 1954 by Frank J. Mather III of the Woods Hole Oceanographic Institution. His primary interest was bluefin tuna, but the program soon expanded to include billfishes and jacks (see Table 1 for scientific names of species). In 1973, the program became a combined effort between Woods Hole and the National Marine Fisheries Service (NMFS). After Mather's retirement in 1980, the NMFS Southeast Fisheries Center's (SEFC) Miami Laboratory assumed sole responsibility for funding and operation of the program.

Over 10,000 individual anglers, charter boat captains, and commercial fishermen have participated in the CGFTP since 1954. At present, about 2,500 persons are listed as active cooperators in the program. Participants reside not only in the USA but also in Canada, Mexico, South America, western Africa, Europe, and various Caribbean island countries. Program results and activities are often publicized in popular fishing magazines (Prince 1984; Beardsley and Scott 1987; Dugger 1988).

Through 1987, cooperators tagged and released over 118,000 fish of 78 identified species representing 30 families (Table 1). Current tagging efforts for oceanic pelagic species are directed at billfishes (primarily blue marlin, white marlin, sailfish, and swordfish), tunas (particularly bluefin tuna), and greater amberjack. Sailfish leads the list of tagged oceanic target species with 39,880 releases (Table 1), followed by bluefin tuna (26,795), white marlin (19,689), blue marlin (6,962), greater amberjack (5,643), and swordfish (1,807). In 1986, red drum, king mackerel, and Spanish mackerel were added as target species (Table 1) because of their economic importance and the intense public and political interest in the status of these coastal species. The objective of this paper is to review the history of the CGFTP by examining the program's operational policies, along with the associated problems and successes, and the tagging effort for each of the target species.

Program Materials and Policy

Species Tagged

Tags are distributed by the CGFTP to cooperators for use on the nine target species listed above. However, the species tagged by the cooperators cannot be completely controlled, and in-

TABLE 1.—Numbers of tagged fish released and recaptured in the Cooperative Game Fish Tagging Program, 1954–1987. Species are listed in phyletic order by family and alphabetically by scientific name within family, according to Robins et al. (1980).

Scientific name	Common name	Releases	Recaptures
Orectolobidae	Carpet sharks	5	
Odontaspididae	Sand tigers	39	1
Alopiidae	Thresher sharks	12	
Lamnidae	Mackerel sharks	99	6
Carcharhinidae	Requieum sharks	1,236	25
Sphyrnidae	Hammerhead sharks	97	
Squalidae	Dogfish sharks	39	12
Dasyatidae	Stingrays	1	
Mobulidae	Mantas	2	
Elopidae	Tarpons	1,575	16
Albulidae	Bonefishes	31	
Clupeidae	Herrings	2	
Ariidae	Sea catfishes	1	
Centropomidae	Snooks	6	
Percichthyidae	Temperate basses	101	8
Serranidae	Sea basses	27	2
Pomatomidae	Bluefishes	254	16
Rachycentridae	Cobias	56	4
Echeneidae	Remoras	1	
Carangidae	Jacks		
Seriola dumerili	Greater amberjack	5,643	642
Other carangids		138	3
Coryphaenidae	Dolphins	478	8
Lutjanidae	Snappers	2	
Sciaenidae	Drums		
Sciaenops ocellatus	Red drum	482[a]	10[a]
Other sciaenids		7	
Sphyraenidae	Barracudas	172	2
Trichiuridae	Cutlassfishes	6	
Scombridae	Mackerels		
Scombero-morus cavalla	King mackerel	2,199[a]	58[a]
Scombero-morus maculatus	Spanish mackerel	752[a]	5[a]
Thunnus thynnus	Bluefin tuna	26,795	4,058
Other scombrids		8,448	140
Xiphiidae	Swordfishes		
Xiphias gladius	Swordfish	1,807	84
Istiophoridae	Billfishes		
Istiophorus platypterus	Sailfish	39,880	543
Makaira nigricans	Blue marlin	6,962	29
Tetrapturus albidus	White marlin	19,689	339
Other istiophorids		2,298	6
Bothidae	Lefteye flounders	14	
Balistidae	Leatherjackets	3	
Total		118,607	6,012

[a]The number does not include all 1986 and 1987 releases and recaptures from state and federal agencies.

discriminate tagging, especially in the early days of the program, resulted in data on many other species (Table 1). Only about 1% of the releases in 1987 were for nontarget species. Individuals who are specifically interested in tagging species not targeted by the CGFTP are referred to programs operated by other organizations.

Tag Type

A stainless steel dart tag, developed at the beginning of the program (Mather 1960) and modified in 1981 (see changes in tag design, below), has been used almost exclusively for oceanic pelagic species since the CGFTP began. The tag is composed of a yellow vinyl streamer attached to a stainless steel barb, which is inserted into the flesh of the fish. The streamer is imprinted with the tag identification number, the word "reward," and the return address of the Miami Laboratory. Instances in which other tags were used are discussed under each target species.

The tag distributed to anglers for use on red drum, king mackerel, and Spanish mackerel is a hydrostatic nylon dart tag with a double barb (Fable 1990; Gutherz et al. 1990; both this volume). This tag is discussed further under changes in tag design.

Tag Distribution

Each program cooperator is issued a tagging kit (Figure 1). The kit consists of seven items: (1) a brochure that explains the history of the CGFTP; (2) instructions on how and what to tag and how to fill out the tag release card; (3) a packet of five tags (with consecutively numbered tag release cards); (4) a stainless steel tagging pin; (5) rubber bands to hold the tag in place on the tagging pole; (6) a tagging flag; and (7) a plastic ziplock container for convenient storage of tagging materials. A tagging pole (a broom handle with a hole drilled for the tagging pin and a slot into which the tag sleeve fits) also is provided when requested.

The cooperator is asked to return a tag inventory card included with the kit to verify receipt of the tags and identity of the recipient. Cooperators may request additional tags by marking the appropriate box on the tag report card.

Tags are issued only to individual cooperators who request them directly from the Miami Laboratory. We have found that we increase the probability that a tag will be used if we restrict the distribution of tags to individuals who personally request them. This direct communication also

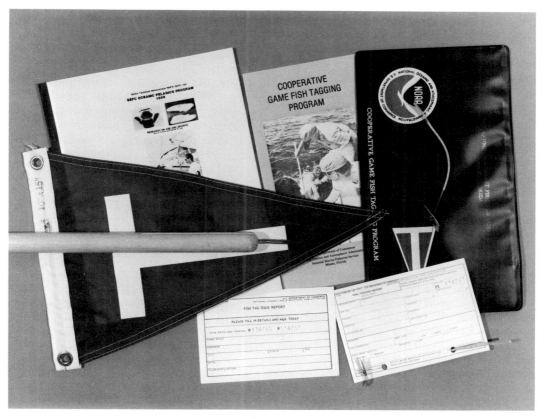

FIGURE 1.—Components of the tagging kit issued to participants in the Cooperative Game Fish Tagging Program.

allows us to maintain an inventory of cooperators, which is useful in responding to questions about releases.

Fishing clubs, tournament sponsors, and other groups often make requests for large numbers of tags. In these cases, we ask that individual club members or tournament participants personally contact us for tags. Alternatively, the club may use our numbering sequence and tag legend and purchase the tags in bulk directly from the manufacturers, along with tagging poles and pins. We supply the numbered tag release and tag inventory cards, incorporate the club's release data into our computerized files, and provide the release and recapture data to the club as requested. This policy was initiated in 1982 with the International Billfish League. Many other organizations have since chosen to purchase their own tags, especially in recent years when the number of tag-and-release billfish tournaments has increased dramatically.

Tagging Procedure

Participants in the CGFTP are instructed to tag a fish while it is in the water alongside the boat. The tag is to be inserted about 5 cm into the anterior dorsal musculature above the lateral line, with the barb and streamer slanted posteriorly. The tagging pole allows cooperators to apply the tag without taking the fish out of the water and with minimal handling. Anglers are not restricted in the size of fish tagged or the location or time of tagging.

A participant is asked to complete and return the tag release card, imprinted with the tag identification number, to the CGFTP soon after a fish is tagged. Information requested from the tagger includes fish species, estimated length, estimated weight, tagging date, tagging location, angler's name and address, captain's name and address, condition of fish, bait type, fishing club affiliation, and additional remarks.

Incentives

The CGFTP has always provided incentives to encourage tagging and return of recovered tags. Cash rewards of $5 are given to persons who return tags and provide recapture information for oceanic species. However, interest in the release data is often sufficient incentive to report recaptured fish. In 1987, the CGFTP began paying a $5 reward for red drum recaptures (Gutherz et al. 1990) and a $10 reward for king mackerel and Spanish mackerel recaptures (Fable 1990). The SEFC also conducts two annual drawings for $1,000 each from all the king mackerel and all the Spanish mackerel tags returned. If the recapture is a billfish or a bluefin tuna, the person who returns the tag is automatically eligible for three $500 yearly lotteries conducted by the International Commission for the Conservation of Atlantic Tunas.

Since 1982, the CGFTP has provided additional incentives to cooperators who tag billfish and tuna. Each year, the CGFTP awards a plaque to the commercial captain who tags and releases the most fish. Recreational fishermen are eligible for annual trophies awarded for tagging the most game fish, sponsored by five major conservation organizations: Sport Fishing Institute for sailfish; The Billfish Foundation for white marlin; National Coalition for Marine Conservation for blue marlin; International Game Fish Association for bluefin tuna; and Florida League of Anglers for king mackerel. In 1987, the CGFTP also began cooperating in a program that is offering over $50,000 in prizes for the tagging and recapture of yellowfin tuna *Thunnus albacares* and bigeye tuna *T. obesus*. The program is sponsored by leading tackle manufacturers and fishing publications, along with east coast fishermen, charter boat captains, fishing clubs, and outdoor writers. Tags with the CGFTP numbering sequence and legend are used, and all data are returned to the CGFTP, which is also responsible for paying the rewards.

Annual Report

Each year, the CGFTP publishes a report to provide program cooperators with a summary of yearly activities. Publication of this report began in 1974, and until 1983 it was circulated as a separate newsletter. Since 1983, it has been integrated with the annual report of the SEFC Oceanic Pelagics Program, which is published as a National Oceanic and Atmospheric Administration Technical Memorandum. The CGFTP report summarizes the numbers of releases and recaptures of the target species, along with other important events of the year. Because we feel that public recognition provides incentive for future cooperation, each participant who tags and releases 10 or more fish is listed by name with the number of each species tagged. The winners of the tagging awards are also announced.

In addition, the annual report occasionally discusses such topics as tagging procedures, the use of lures or live bait versus dead bait, and survival of tagged fish. The report may present innovative ideas to assist taggers in such tasks as estimating the size of fish released.

What Have We Learned?

The tag–recapture information obtained through the CGFTP has greatly improved our understanding of the biology of target species. Data from the program have helped to document seasonal movement patterns, delineate stocks, determine maximum life span, understand growth and survival, and validate aging techniques. Highlights of the tagging efforts and what has been learned about each target species are discussed below and summarized in Tables 1 and 2. If text information is not accompanied by a citation, it is from the unpublished data files of the CGFTP. Because red drum, king mackerel, and Spanish mackerel were added to the program only recently, tag–recapture data on these species are not discussed here, but they are reviewed by Fable (1990) and Gutherz et al. (1990).

Bluefin Tuna

Altogether, 26,795 bluefin tuna, the original target species when the CGFTP was created in 1954, were tagged and released through 1987, and 4,058 have been recaptured. This 15% recapture rate is the highest in the CGFTP and reflects the schooling behavior of this species, as well as the intense commercial fishery for all size categories.

Figure 2 shows the number of bluefin tuna tagged each year by the CGFTP. Most of the tagging has been conducted by NMFS scientific staff on commercial vessels and by recreational fishermen using the stainless steel dart tag. Information obtained from these recaptured fish contributed substantially to the case made by the International Commission for the Conservation of Atlantic Tunas for imposing conservation measures in 1976. Additional harvest quotas, put into effect by the commission in 1983, not only re-

TABLE 2.—Summary data for oceanic pelagic target species of Cooperative Game Fish Tagging Program, 1954–1987. Tagging cooperators, types of tags used, maximum time at large, maximum distance traveled, percent recaptures, and evidence for equatorial and transatlantic crossings are given for each species.

Species	Tagging cooperators[a]	Types of tags used	Maximum time at large	Maximum distance traveled	Percent recaptures	Equatorial Atlantic crossing	Trans-atlantic crossing
Bluefin tuna	Recreational and commercial fishermen, scientific staff	Stainless steel dart, nylon dart, steel harpoon	18.0 years	10,000 km	15.1	Yes	Yes
Blue marlin	Recreational and commercial fishermen, NMFS observers	Stainless steel dart	7.9 years	7,000 km	0.4	No	Yes
White marlin	Recreational and commercial fishermen, NMFS observers	Stainless steel dart	11.8 years	5,000 km	1.7	No	No
Sailfish	Recreational and commercial fishermen, NMFS observers	Stainless steel dart	10.9 years	3,400 km	1.4	No	No
Swordfish	Recreational and commercial fishermen, NMFS observers	Stainless steel dart	15.1 years	2,400 km	4.6	No	No
Greater amberjack	Recreational fishermen	Stainless steel dart	10.1 years	2,400 km	11.4	No	No

[a]NMFS = National Marine Fisheries Service.

duced the harvest but also greatly reduced the tagging effort.

Several types of tags other than the stainless steel dart have been used on bluefin tuna during research projects at the Miami Laboratory. These have included three sizes of hydrostatic nylon dart tags (18, 10, and 6.5 mm in diameter) used on all sizes of fish and a steel harpoon tag tried briefly

for giant bluefin tuna (over 136 kg). The release and recapture data from these trials have been entered into CGFTP files, but the tags have not replaced the standard stainless steel dart, which is provided to program cooperators.

The longest distance traveled by bluefin tuna was over 10,000 km: nine fish tagged and released off Bimini-Cat Cay, Bahamas, were recaptured off

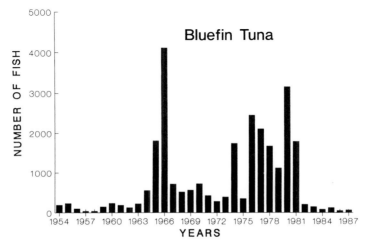

FIGURE 2.—Numbers of bluefin tuna tagged annually in the Cooperative Game Fish Tagging Program, 1954–1987.

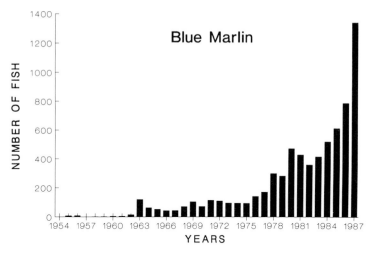

FIGURE 3.—Numbers of blue marlin tagged annually in the Cooperative Game Fish Tagging Program, 1954–1987.

the coast of Norway. Transatlantic movement was first recorded in 1959 when a fish tagged off Massachusetts was recaptured in the Bay of Biscay (Mather 1960). Additional recaptures during the next few years demonstrated transatlantic movements by both large and small fish, often with times at large of only a few months (Mather 1962, 1969; Mather et al. 1967). Subsequent records have been of sufficient detail to document movement patterns for most size categories of bluefin tuna in the North Atlantic Ocean (Rivas 1978). Transequatorial movement was recorded in 1965 when a fish released off Cat Cay was recaptured south of the equator off Brazil.

The longest time at large for a bluefin tuna was 18 years for a fish tagged off Montauk Point, New York, and recaptured off Nantucket Island, Massachusetts. However, a bluefin tuna at large for 16 years provided more valuable information because skeletal structures from this specimen were used to validate the vertebral method of age determination for the species (Lee et al. 1983; Prince et al. 1985). The high quality of age and growth information obtained from this fish also led to a concerted effort to save skeletal structures from fish tagged and recaptured through the CGFTP. This "Save It For Science Program" is discussed in a later section on age and growth data.

Blue Marlin

Although 6,962 blue marlins have been tagged since 1954, only 29 recaptures (0.4%) have been reported. This unusually low recapture rate, the

lowest in the CGFTP, may be related to the lack of a directed commercial fishery for this species, its "rare event" status, large size, great mobility, and extremely large range. In addition, the majority of blue marlin landings in the Atlantic Ocean are by foreign and domestic long-liners (ICCAT 1986); nonreporting of tag recaptures from these commercial landings probably contributes to the low recapture rate. Tagging of blue marlin initially increased in 1963 (Figure 3), and activity has increased steadily since then. Tagging is conducted almost exclusively by recreational anglers using the stainless steel dart tags.

Because of the small number of recaptures, movement patterns of blue marlin are not well documented. However, recaptures do indicate seasonal movement between the U.S. east coast and the Caribbean Sea, and between the east coast and the Gulf of Mexico. Transatlantic movement was also documented from the Caribbean Sea to the coast of western Africa. The blue marlin is the only billfish to demonstrate transatlantic movement. It has traveled as far as 7,000 km from St. Thomas, Virgin Islands, to the Ivory Coast of West Africa. Transequatorial movements have not been documented for this species. The longest time at large for blue marlin is almost 8 years for a fish tagged and recaptured off St. Thomas.

White Marlin

Through 1987, 19,689 white marlins have been tagged and 339 (1.7%) have been recaptured. Large numbers of white marlin were tagged dur-

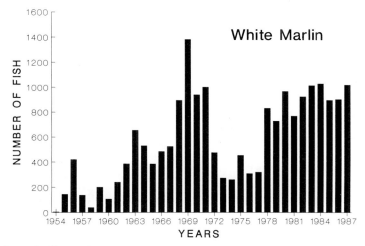

FIGURE 4.—Numbers of white marlin tagged annually in the Cooperative Game Fish Tagging Program, 1954–1987.

ing the late 1960s and early 1970s, and tagging activity has remained steady except for a decline during 1972–1977 (Figure 4). White marlin tagging is conducted mostly by recreational fishermen using the stainless steel dart tag.

Although recapture rates for white marlin were less than 2% (consistent with the low recapture rates for other istiophorids), results indicate seasonal movement patterns from the mid-Atlantic coast to the central Caribbean Sea, Gulf of Mexico, and various locations in the tropical Atlantic Ocean (Mather 1967, 1969; Mather et al. 1972, 1974). The longest distance traveled by a white marlin was about 5,000 km from the Bahamas to waters off the Amazon River. Although no records of transatlantic or transequatorial movements have been documented for this species, the record for the longest time at large for a billfish is held by a white marlin tagged off Maryland in 1970 and recaptured almost 12 years later off New York. Skeletal structures from this specimen were saved for age and growth analyses.

Sailfish

More sailfish have been tagged (39,880) than any other fish in the CGFTP. Tagging effort for sailfish has remained relatively stable throughout the program (Figure 5) and is conducted primarily by recreational anglers using the stainless steel dart tag. The very low recapture rate found for other istiophorids was also documented for sailfish. Only 1.4% (543 fish) of the sailfish tagged have been recaptured, even though sailfish are generally more abundant and are more accessible to fishermen than the marlins.

The longest distance traveled by a sailfish was about 3,400 km from North Carolina to Guyana. The time-at-large record is almost 11 years for a fish tagged off the Florida Keys and recaptured off Boynton Beach, Florida. This specimen was particularly valuable because results of aging analysis (Prince et al. 1986) contributed new information to the life history of the species. No transatlantic or transequatorial movements for sailfish have been documented by the CGFTP.

Swordfish

Program cooperators have tagged and released 1,807 swordfish and recaptured 84 since 1954. The directed commercial fishery for swordfish and the greater abundance of this species probably contributed to its higher recapture rate (4.6%) compared to istiophorids. Most swordfish were tagged with stainless steel darts by commercial fishermen or NMFS observers on domestic and foreign longline vessels. A few have been tagged with a stainless steel dart whose terminal end is in the form of a capsule instead of a vinyl sleeve. Capsule dart tags are issued to domestic long-liners through the NMFS shark-tagging program at the Narragansett Laboratory (Casey et al. 1983). Tagging effort for swordfish has generally increased in several recent years (Figure 6), when more fish weighing less than 10 kg have appeared in the commercial catches. These small fish are not

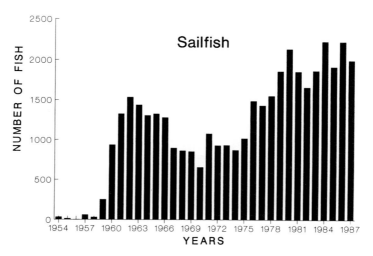

FIGURE 5.—Numbers of sailfish tagged annually in the Cooperative Game Fish Tagging Program, 1954–1987.

commercially valuable and are generally released by the fishermen.

The longest distance traveled by a tagged swordfish was about 2,400 km. This fish was tagged in the southeastern Gulf of Mexico and moved to Georges Bank, east of Cape Cod, Massachusetts, in only 130 d. Transatlantic and transequatorial movements of swordfish have not been documented by the CGFTP. The at-large record for a swordfish is just over 15 years for a fish tagged off Cape Race, Newfoundland, and recaptured off Nova Scotia. Tag–recapture data for swordfish have been used to estimate growth and establish the age distribution of the catches for use in a preliminary stock assessment (SEFC 1986); however estimation of swordfish age and growth from skeletal structures has not been attempted through the CGFTP.

Greater Amberjack

Since 1954, 5,643 greater amberjacks have been tagged and 642 recaptured. The high tag recovery rate of 11.4% is probably due to the abundance of amberjacks in inshore waters and the active recreational fishery, which is responsible for most of the tagging effort. Stainless steel dart tags have been used exclusively on greater amberjacks. The dramatic decline in numbers tagged in the early 1980s (Figure 7) reflects the discontinuance of amberjacks as a CGFTP target species in 1979.

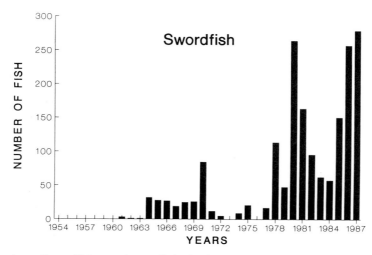

FIGURE 6.—Numbers of swordfish tagged annually in the Cooperative Game Fish Tagging Program, 1954–1987.

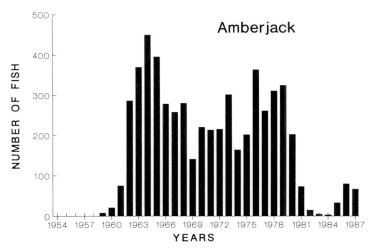

FIGURE 7.—Numbers of greater amberjack tagged annually in the Cooperative Game Fish Tagging Program, 1954–1987.

Tagging efforts for this species were de-emphasized because tagging was not providing new information on movement patterns, despite a high recapture rate, and the increased cost of tagging equipment forced us to become conservative in issuing equipment. Greater amberjack was reinstated as a target species in 1987 because of its increasing recreational and commercial importance.

The longest distance traveled by a greater amberjack was about 2,400 km from central west Florida to Venezuela. The record time at large for this species is just over 10 years for a fish tagged and recaptured off Jacksonville, Florida. No transatlantic or transequatorial movements by greater amberjacks have been documented by the CGFTP.

Age and Growth

Tagging plays an important role in research on age and growth of billfishes and tunas by supplying scientists with skeletal structures from fish of relative known age (i.e., age closely approximated from tagging records of specimens tagged at a small size and at large for extended periods). This approach to age and growth is often the only way to validate the accuracy of the aging techniques used on oceanic pelagic species (Lee et al. 1983; Prince and Pulos 1983; Prince et al. 1985, 1986). However, retrieving otoliths, vertebrae, and spines from recaptured fish is the key to productive results. This is facilitated by the Save It For Science Program started by the Miami Laboratory in 1982 (Prince 1984). Through this program,

fishermen participating in the CGFTP are encouraged to retain carcasses of recaptured fish, as well as those of unusually small or large specimens, and to notify the Miami Laboratory so scientific staff can retrieve the skeletal structures. Alternatively, we may provide the angler with instructions for preserving or sampling the carcass and then assume the costs of obtaining the structures and shipping them back to the Miami Laboratory. The Save It For Science Program has also been recently adopted by the International Commission for the Conservation of Atlantic Tunas (Miyake 1990, this volume) because many of the fish tagged through the CGFTP are recaptured by foreign long-liners.

Table 3 lists the recaptured specimens from which we have been able to recover structures for age and growth analyses. The first bluefin tuna listed was tagged by Canadian scientists and recaptured by an angler. The two albacore *Thunnus alalunga* were tagged by French and Spanish scientists in Europe and recaptured by Japanese long-liners off the U.S. northeast coast. United States observers on these long-liners retrieved these fish for the Save It For Science Program. The recovery of the specimens emphasizes the international cooperation involved in the CGFTP.

Changes in the Program

New Target Species

Because of heavy fishing pressure on red drum, king mackerel, and Spanish mackerel, and the

TABLE 3.—Tagged oceanic pelagic fish whose skeletal structures were recovered through the Cooperative Game Fish Tagging Program for age and growth studies, 1981–1987.

Release data			Recapture data			Time at large (years)	Skeletal structures recovered
Date	Location	Size (kg)[a]	Date	Location	Size (kg)[b]		
White marlin							
Sep 26, 1970	Maryland	15.9	Jul 10, 1982	New York	29.5	11.8	Spines, vertebrae
May 6, 1980	Mexico	11.3	Jun 27, 1981	Louisiana	21.3	1.2	Spines, vertebrae, otoliths
Oct 31, 1981	Florida	22.7	Sep 19, 1982	Florida	23.4	0.9	Spines, vertebrae, otoliths
Jun 17, 1982	Louisiana	24.9	Sep 17, 1982	Florida	27.4	0.3	Spines, vertebrae, otoliths
May 18, 1982	Florida	27.2	Nov 12, 1985	Florida	20.4	3.5	Spines, otoliths
Sep 13, 1984	Florida	18.1	Aug 27, 1985	Alabama	18.6	1.0	Spines, otoliths
Oct 4, 1985	Florida	27.2	Sep 17, 1986	Louisiana	22.2	1.9	Spines, otoliths
Aug 12, 1986	Florida	19.1	Jul 3, 1987	Florida	19.1	0.9	Spines, otoliths
Bluefin tuna							
Aug 5, 1965	New Jersey[c]	11.3	May 28, 1981	Bahamas	223.6	15.7	Caudal vertebrae
Jun 24, 1980	Virginia	11.3	Feb 11, 1984	New Jersey	72.1	3.7	Caudal vertebrae
Jun 21, 1981	Virginia	18.1	Aug 20, 1987	New York	163.3	6.2	Vertebrae, otoliths
Jun 23, 1980	Virginia	11.3	Sep 2, 1987	Massachusetts	247.2	7.3	Vertebrae, otoliths
Jul 9, 1977	Virginia	11.3	Sep 10, 1987	Massachusetts	292.6	10.2	Vertebrae, otoliths
Jul 4, 1977	Virginia	11.3	Sep 10, 1987	Massachusetts	335.7	10.2	Vertebrae, otoliths
Jul 9, 1977	Virginia	13.6	Sep 30, 1987	Massachusetts	200.9	10.2	Otoliths
Sailfish							
Mar 5, 1973	Florida	18.1	Jan 14, 1984	Florida	24.5	10.9	Spines, vertebrae, otoliths
Jan 1984	Florida	22.7	Jan 19, 1986	Florida	20.4	2.1	Spines, vertebrae, otoliths
Jan 25, 1986	Florida	22.7	Apr 27, 1986	Florida	15.6	0.3	Spines, otoliths
Apr 15, 1987	Florida	4.5	Nov 8, 1987	Florida	10.9	0.6	Spines, otoliths
Jan 1, 1987	Florida	24.9	Dec 7, 1985	Florida	28.6	0.9	Spines, otoliths
Albacore							
Aug 17, 1978	Spain[d]	5.0	Dec 30, 1984	New Jersey	23.1	6.3	Spines, vertebrae, otoliths
Jun 23, 1980	France[e]	5.0	Dec 31, 1984	New Jersey	19.1	4.5	Spines, vertebrae

[a] Sizes at release are usually estimated.
[b] Sizes at recapture are measured.
[c] Tagged by Canadian scientists.
[d] Tagged by Spanish scientists.
[e] Tagged by French scientists.

need for biological data on these species, the CGFTP added them to the list of target species in 1986. These three coastal species are the subject of intense public and political interest because of their importance to federal and state economies. The need for more detailed biological data to assess the stocks has resulted in the recent initiation of numerous scientific investigations, many of which include tagging studies in the Gulf of Mexico and the western North Atlantic Ocean. These studies are being conducted cooperatively by the coastal states, the NMFS, Mexico, universities, private laboratories, and recreational groups. All tagging records for these species are not presently in the CGFTP data base, but they are being added as part of a new data-management system (discussed below).

The greater amberjack is a resurrected target species. It was dropped from the program in 1979 and reinstated in 1987 (see discussion of greater amberjack above).

Tag Design

The design of the stainless steel dart tag used by the CGFTP has been modified once since the Miami Laboratory took over the program. Prior to 1981, a brass sleeve was crimped on the end of the monofilament shaft to hold the vinyl tubing in place. Many tags, especially from bluefin tuna, were returned with the tubing and its imprinted legend missing—only the monofilament shaft was projecting from the dorsal musculature. It was found that the brass sleeve often corroded away,

FIGURE 8.—An example of barnacle growth on a Cooperative Game Fish Tagging Program tag. The barnacles abraded the side of this sailfish.

allowing the tubing to slip off and be lost. To prevent this, we designed a plastic sleeve that, when slipped over the doubled end of the monofilament and then heat-shrunk, retained the identification tubing. This new tag replaced the old version in 1981.

The accumulation of marine growth on the tubing and monofilament of the tag also has been a problem (Figure 8). In an attempt to control the growth, we requested that the manufacturer apply an antifouling chemical (copper sulfate) to the tags. This was partially successful, but it has not completely eliminated growth on the tags.

The hydroscopic nylon dart tags were first developed for bluefin tuna (discussed earlier) and later miniaturized for king mackerel and red drum (Fable 1990; Gutherz et al. 1990). These tags were designed as "intermuscular" anchor tags, and the heads were injection-molded with hydroscopic (porous) nylon. The intent was to develop a tag whose surface area would allow tissue to adhere to the nylon anchor, thereby reducing long-term tag shedding. The success of these tags is still being evaluated, but preliminary results show that fish tissue does encapsulate and adhere to the nylon anchor head (Gutherz et al. 1990). The refinement of this approach could be the first step in developing a tag that is both biologically acceptable to the fish and useful for fisheries scientists.

Data Collection

The major items of data collected on tagged fish, and the methods of collecting the data, have remained the same throughout the program. A few new items have been added to the data we request from a fisherman releasing a tagged fish. These include type of bait, whether the fish was tagged in the boat or in the water, if the hook was removed, and if a net was used. This information is used primarily to assess mortality factors that contribute to the recapture rate.

Other modifications to the tag report card were made to lessen the probability that it will be lost in the mail. We now use a bigger card and heavier paper.

Data Management

Because of the proliferation of cooperative tagging studies in the southeastern USA, researchers from many state agencies, as well as the NMFS, are facing the responsibility of collecting, maintaining, retrieving, and analyzing increasing amounts of data. It is apparent that compilation and dissemination of these diverse data sets are necessary to ensure their availability to all research participants for accurate and complete analysis. This is especially important for data analysis on species that migrate throughout inshore waters under various state jurisdictions and

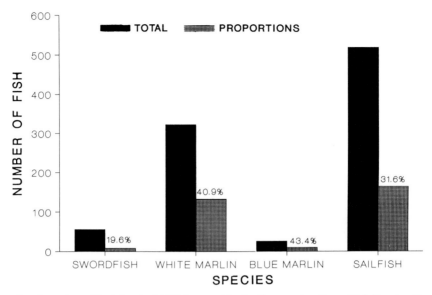

FIGURE 9.—Total numbers (black bars) of billfish tagged in the Cooperative Game Fish Tagging Program during 1954–1987 and subsequently recovered, and numbers (shaded bars) and percentages of those fish for which the estimated weight at tagging was greater than the measured weight at recapture (minimum error).

between inshore waters and offshore waters under federal jurisdiction.

The Cooperative Tagging System (CTS), now being developed as a comprehensive data base to serve as the central archive for data from the participating tagging projects, will efficiently manage these data. The system has been designed as an integrated network of microcomputers linked to a mainframe. The data may be retrieved from the data base by researchers for specific analyses of spatial and temporal distribution, stock identification, age and growth, and other subjects. The CTS eventually will integrate tagging data from state and federal research programs on southeastern coastal species with the federal, recreational, and commercial data on billfish and other oceanic species presently in CGFTP.

Major Drawback of Program

A primary aspect of the operation of the CGFTP is its dependence on volunteers in the recreational and commercial fisheries to tag fish and to return tags from recaptures. This cooperation from the public is critical to the program's success. At the same time, it is the source of one of the program's major problems—inaccurate estimation of the size of a fish when it is tagged and released. This problem does not occur as frequently when scientists tag fish because, in most

cases, they make actual measurements. The tagging procedure for volunteers, while minimizing handling and stress on the fish and also the chance of injury to the tagger, tends to maximize the chance of error when estimating the size at release. The fish is tagged while it is in the water, and its size must be estimated rather than actually measured. This is particularly important with the larger billfish, which are dangerous to bring aboard a boat. Under these conditions, the size of a fish tends to be overestimated because of refraction of light by water. The level of excitement of the angler also contributes to inaccurate size estimates; however, errors in underestimating size at release likely occur as well. The type and magnitude of these errors often depend on the individual.

Figure 9 illustrates the magnitude of the estimation problem for the four billfishes that are target species of the CGFTP. The only tag–recapture records considered are those for which both release and recapture weights are known. For a large proportion of the records (averaging about 30% for the four species), the estimated weight at release was greater than the measured weight at recapture—the release weight in these cases was probably overestimated. The proportions range from 43.4% for blue marlin to 19.6% for swordfish. These figures should be interpreted as minimum errors because more errors of less severity may occur but are not accounted for in this analysis. For example, if a release weight, although overestimated, is still less than the measured weight at recapture, this will not be detected as an error. This means that the tag-recapture records must be evaluated carefully in analyses of growth information.

References

Beardsley, G. L., and E. L. Scott. 1987. Playing tag with ocean gamefish. Florida Sportsman 19(12): 20–25.

Casey, J. G., H. L. Pratt, Jr., and C. E. Stilwell. 1983. Age and growth of the sandbar shark, *Carcharhinus plumbeus*, from the western North Atlantic. NOAA (National Oceanic and Atmospheric Administration) Technical Report NMFS (National Marine Fisheries Service) 8:189–191.

Dugger, A. 1988. The mystery of blue marlin migrations. Marlin 6(6):19–22. (Marlin Magazine, Pensacola, Florida.)

Fable, W. A. 1990. Summary of king mackerel tagging in the southeastern USA: mark–recapture techniques and factors influencing tag returns. American Fisheries Society Symposium 7:161–167.

Gutherz, E. J., B. A. Rohr, and R. V. Minton. 1990. Use of hydrostatic molded nylon dart and internal anchor tags on red drum. American Fisheries Society Symposium 7:152–160.

ICCAT (International Commission for the Conservation of Atlantic Tunas). 1986. Report for biennial period, 1984–85 (part 2, 1985). ICCAT, Madrid, Spain.

Lee, D. W., E. D. Prince, and M. E. Crow. 1983. Interpretation of growth bands on vertebrae and otoliths of Atlantic bluefin tuna, *Thunnus thynnus*. NOAA (National Oceanic and Atmospheric Administration) Technical Report NMFS (National Marine Fisheries Service) 8:61–69.

Mather, F. J., III. 1960. Recaptures of tuna, marlin and sailfish tagged in the western North Atlantic. Copeia 1960:149–151.

Mather, F. J., III. 1962. Transatlantic migration of two large bluefin tuna. Journal du Conseil, Conseil International pour L'Exploration de la Mer 27: 325–327.

Mather, F. J., III. 1967. The trail of the tail-walker. Oceanus 13:10–16.

Mather, F. J., III. 1969. Long distance migrations of tunas and marlins. Underwater Naturalist 6(1): 6–14, 46.

Mather, F. J., III, M. R. Bartlett, and J. S. Beckett. 1967. Transatlantic migrations of young bluefin tuna. Journal of the Fisheries Research Board of Canada 24:1991–1997.

Mather, F. J., III, A. C. Jones, and G. L. Beardsley, Jr. 1972. Migration and distribution of white marlin and blue marlin in the Atlantic Ocean. U.S. National Marine Fisheries Service Fishery Bulletin 70: 283–298.

Mather, F. J., III, J. M. Mason, Jr., and H. L. Clark. 1974. Migrations of white marlin and blue marlin in the western north Atlantic Ocean—tagging results since May 1970. NOAA (National Oceanic and Atmospheric Administration) Technical Report NMFS (National Marine Fisheries Service) SSRF (Special Scientific Report Fisheries) 675.

Miyake, P. M. 1990. History of the ICCAT tagging program, 1971–1986. American Fisheries Society Symposium 7:746–764.

Prince, E. D. 1984. Save it for science, don't throw back recaptured tagged billfish or tuna. Marlin 3(2):50–54. (Marlin Magazine, Pensacola, Florida.)

Prince, E. D., D. W. Lee, and J. C. Javech. 1985. Internal zonations in sections of vertebrae from Atlantic bluefin tuna, *Thunnus thynnus*, and their potential use in age determination. Canadian Journal of Fisheries and Aquatic Sciences 42:938–946.

Prince, E. D., D. W. Lee, C. A. Wilson, and J. M. Dean. 1986. Longevity and age validation of a tag-recaptured Atlantic sailfish, *Istiophorus platypterus*, using dorsal spines and otoliths. U.S. National Marine Fisheries Service Fishery Bulletin 84:493–502.

Prince, E. D., and L. M. Pulos, editors. 1983. Proceedings of the international workshop on age determination of oceanic pelagic fishes: tunas, billfishes, and sharks. NOAA (National Oceanic Atmospheric Administration) Technical Report NMFS (National Marine Fisheries Service) 8.

Rivas, L. R. 1978. Preliminary models of annual life history cycles of the North Atlantic bluefin tuna. Pages 369–393 *in* G. D. Sharp and A. E. Dizon, editors. The physiological ecology of tunas. Academic Press, New York.

Robins, C. R., and six coauthors. 1980. A list of common and scientific names of fishes from the United States and Canada. American Fishereis Society Special Publication 12.

SEFC (Southeast Fisheries Center). 1986. Report of the swordfish assessment workshop, Miami, Florida, April 16–26, 1986. National Marine Fisheries Service, SEFC, Miami Laboratory, Miami.

American Fisheries Society Symposium 7:854–862, 1990

Marine Fish Tagging in South Africa

R. P. van der Elst

Oceanographic Research Institute
Post Office Box 10712, Marine Parade 4056, South Africa

Abstract.—The seas off southern Africa sustain important fisheries and are of zoogeographic interest because they include Atlantic and Indo-Pacific faunas. As a consequence, much research has been undertaken here, including the tagging of fish. This tagging began in 1934 and has included a variety of species. Shark tagging has received much attention, and local research and development has contributed to improved tags and techniques. Tank trials have been used to evaluate tag retention, and limited underwater tagging of sharks and teleosts has been tried. Many tagging projects ended before data were published, hence a nationwide cooperative tagging program was launched in 1984 to ensure continued availability of data. In this program, nearly 2,000 people have tagged 25,000 fish of 254 species. Data management has fallen into two categories: regular newsletters and data printouts sent to anglers, and comprehensive reports distributed to the scientific community.

The southern tip of Africa occupies a strategic zoogeographic position. The more than 2,200 marine fish species recorded here represent 83% of the world's known fish families, second only to Japan in familial diversity (Smith and Heemstra 1986). Although the origins of this ichthyofauna are diverse, they are clearly dominated by Indo-Pacific, Atlantic, and Southern Ocean components. The Cape of Good Hope is an interface between these major faunas, and it has been suggested that this meeting of two great oceans presents a barrier impenetrable to many species and serves to maintain the relative isolation of these major faunas (Penrith and Cram 1974). In sharp contrast is the physicochemical evidence for periodic massive leakage of water from the Indian to the Atlantic Ocean via the Agulhas Current (Lutjeharms and Stockton 1987), which could facilitate interchange between the two fish faunas. Indeed, several species do occur on both sides of the subcontinent, such as the bluefish *Pomatomus saltatrix* and the lesser guitarfish *Rhinobatos annulatus*. Many other features make the region unique, not least of which is its exceptionally high proportion of endemic species, amounting to at least 13% of the total species assemblage (Smith and Heemstra 1986). Fish in the families Sparidae, Coracinidae, Clinidae, Scyliorhinidae, and Gobiidae are almost entirely confined to these waters.

In addition to these ichthyological characteristics, the region also sustains some of the richest commercial fisheries in the southern hemisphere with annual landings in excess of 2 million tonnes (Crawford et al. 1987). Furthermore, there exist a flourishing recreational fishery of 750,000 anglers and many subsistence fishermen, who provide an essential supply of daily protein to numerous coastal communities (van der Elst 1984).

As in many other regions of the world, there have been large-scale fluctuations in the finfish harvest, many of them attributable to overexploitation. The Benguela Current along the west coast, for example, has seen major changes in species composition, with the dominant Cape pilchard *Sardinops ocellatus* replaced by the anchovy *Engraulis japonicus* (Crawford et al. 1987). Similarly, the Agulhas Current along the east coast has seen major changes, with nearly a 90% reduction of endemic sparid species and a corresponding increase in wider-ranging but often less-desirable species (van der Elst, in press).

Because of the importance of this ichthyofauna, the value of the commercial stocks, and (above all) the considerable variation in species composition of the harvest, fisheries research has been undertaken by at least six institutes. Increasingly, these programs have adopted tag and recapture as a technique, though not always with much success. This paper reviews the progress of tagging research in South Africa.

History of Tagging

The first species to fall victim to a tagger's needle was the Cape snoek *Thyrsites atun*. In 1934, South African Division of Sea Fisheries staff tagged 3,755 of these prized food and game fish with Petersen disks and released them off the Cape of Good Hope. Though only 0.5% were recaptured, they did provide evidence of seasonal

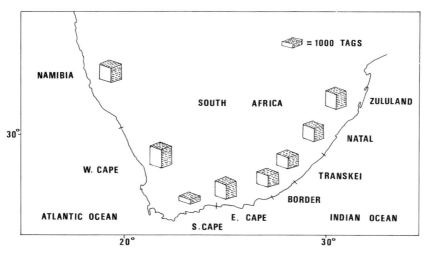

FIGURE 1.—Map of southern Africa with schematic distribution of tagging effort.

migrations between Namibia and South Africa (Figure 1), thereby emphasizing the need for joint management of a shared resource (de Jager 1955). It took researchers another 23 years before tagging studies were resumed, this time on the Cape pilchard. In this project, 141,000 Cape pilchards were tagged with small internal metal tags for later magnetic recovery in processing factories. Over an 8-year period, some 10.7% were recovered, which established migration routes and confirmed intermixing of spawning stocks from different fisheries (Newman 1970). Further Cape snoek tagging was undertaken in 1973–1974 by Nepgen (1979), who recovered 0.6% of 3,139 tagged fish and concluded that their movements were random and local. Another early fish-tagging venture, and one that included tag evaluations, was undertaken by Davies and Joubert (1966) at the Oceanographic Research Institute, Durban. This project—the first to tag elasmobranchs in southern Africa, as well as the first along the Indian Ocean seaboard—was initiated because of shark attacks in this area. The researchers evaluated the performance of different tag types on a variety of species observed in an experimental tank 17 m × 10 m, and then proposed improved techniques. Subsequent results were striking: of 1,001 sharks tagged and released, 38.7% were recaptured in the first year. Shark tagging was taken further by Bass et al. (1973), who tagged several thousand dusky sharks *Carcharhinus obscurus* and documented their migration routes and sexually segregated shoaling behavior.

Progressively more tagging ventures were launched, usually to test hypothesis for species already the subjects of other studies. Not all these projects yielded relevant results, hence many of the tag data remain unpublished. This information clearly is of value, especially to detect long-term trends in recapture rates. A partial remedy to the problem was to establish a South African marine fish-tagging register. This annual inventory of all tagging activities allows researchers to track down unpublished but potentially useful data.

Research and Development of Tagging Techniques

Although most tags used off southern Africa have been imported, there have been several noteworthy local developments. The first involved shark tagging. Davies and Joubert (1966) tested Woods Hole M-dart tags, along with a modified version of the Petersen disk applied by Olsen (1953). Initially, the latter were commercially manufactured sheep-ear tags comprising two elongate polyethylene disks, one with a shaft and the other with a hole, that are pressed together through a perforation in the dorsal fin. These were modified to round disks 26 mm in diameter and locally manufactured (Figure 2). This "ORI" tag was a considerable improvement over the Woods Hole M-dart and sheep-ear tags, especially when applied to smaller sharks. Tank trials on dusky sharks, spinner sharks *C. brevipinna*, and hound sharks *Mustelus mosis* indicated that rates of tag loss were variable but

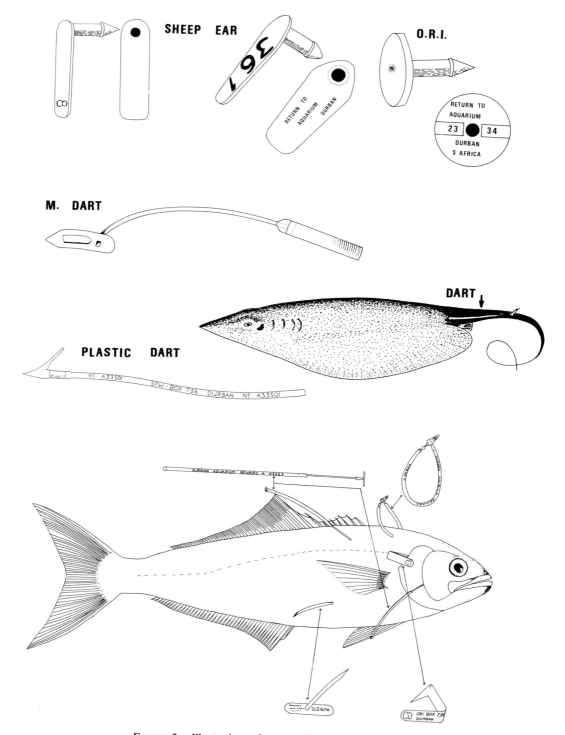

FIGURE 2.—Illustrations of tags used on South African fishes.

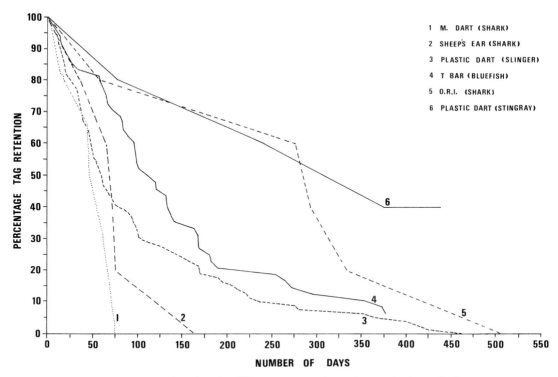

FIGURE 3.—Rates of tag loss for different tags tested on teleosts, sharks, and stingrays.

much improved for the ORI tags—average retention times were 51 d for dart tags, 83 d for sheep-ear tags, and 234 d for ORI tags (Figure 3). The main reason for better performance of the round ORI tag was its reduced vertical movement in the water, an action that caused considerable irritation by and subsequent shedding of the other types tested.

A problem did arise with ORI tags in practice, however, because the tags caught up on netting. With much of the Natal coast (Figure 1) permanently protected by shark gill nets, it became important not to release tagged sharks in proximity to these nets. Although this problem remains to be resolved, the ORI tag has proved suitable for tagging sharks less than 1.5 m long. To date, more than 12,000 of the tags have been used with considerable success.

Since 1970, there have been many teleost tagging ventures in southern Africa, mostly with T-bar or plastic dart tags. Extremely low recapture rates gave rise to fears that the tags were being shed. Consequently, several tag-retention studies were conducted with disturbing results. Tank trials of bluefish tagged by five techniques

(Figure 2) confirmed that retention over 540 d was poor even under optimal conditions. By far the worst problem was necrosis of tissue surrounding the tag, which resulted in encapsulation of the tag's anchor and subsequent rejection. To avoid this effect, T-bar spaghetti tags were inserted between opercular plates, where there is very little fleshy tissue. Although this did improve retention, 1% of the tags were still lost every 4.8 d for an annual loss of 76.5% (Figure 3). Similar results were obtained with captive slingers *Chrysoblephus puniceus*, an endemic sparid of major local importance. With this species, the loss of intramuscular plastic darts was 78% annually (Figure 2; P. A. Garratt, Oceanographic Research Institute, unpublished report). These results imposed major limitations on fish-tagging research.

Although techniques for tagging sharks and teleosts have improved, tagging of skates and rays has had little attention. Despite positive results in tagging thorny skates *Raja radiata* off Newfoundland, Canada (Templeman 1984), the tagging of 700 stingrays off South African yielded no returns. This may have reflected very low fishing mortality, but tank studies on the three species tested

(*Dasyatis marmorata, D. jenkinsi,* and *Gymnura natalensis*) revealed they suffered high mortality when tagged with the body disk. On dissection, these dead fish had inflamed tissue surrounding the tag, especially where there was damage to the pectoral rays. Further trials in which a plastic barb tag was inserted into and parallel with the tail base have shown far better retention in tank trials (Figures 2, 3). This technique is now being introduced in the field.

Another local development involved improvement of tagging under water. Many species are not readily caught by net or line, though they are important to spearfishermen. Underwater tagging was pioneered by Ebert (1964) and Cousteau and Cousteau (1970). Their methods were improved with local modifications by M. M. Griffiths (University of Natal, unpublished report) and M. Roxburgh and R. P. van der Elst (Oceanographic Research Institute, unpublished report), who used a metal-tipped dart tag slotted into the end of a 6-mm-diameter spear with a collar that limited depth of insertion to 60 mm (Figure 2). The spear was fitted to a conventional rubber-powered speargun with strengths of 20 and 53 kg for teleosts and sharks, respectively. To date, this technique has proved particularly effective in tagging pregnant nurse sharks *Carcharias taurus* when they congregate on certain reefs. Altogether, 154 nurse sharks have now been tagged and 6 recaptured (Cliff 1988). Similarly, several knifejaws *Oplegnathus robinsoni* and lampfish *Dinoperca petersii* were tagged in situ in their home territories, which provided proof of their local residency; tagged fish have been continually sighted for up to 473 d.

Underwater tagging holds promise, but it is a difficult technique and its application should be confined to slow-swimming or cave-dwelling species so that optimal insertion of the tags can be assured. In field trials conducted here, the tagged fish were all observed at least once after a lapse of 24 h, to visually confirm tagging success.

Cooperative Tagging Program

Tagging of marine fish has been used widely by biologists in southern Africa, but there have been few published results. There remains a need for data derived from tagging studies, especially for information on stock migrations and rates of fishing mortality of recreational species. The stop–start nature of projects has resulted in confusion and uncertainty amongst fishermen who, after all, are a key component in most tagging studies.

Tagging has become a socially acceptable addition to recreational angling, and there are many who undertake angling purely for the purpose of catch and release.

In response to these needs and problems, a cooperative marine fish tagging program was launched in 1984. This has allowed anglers to tag their own fish, and it has served as a mechanism whereby ad hoc scientific tagging studies can be coordinated and given greater exposure and continuity.

This program appears to be justified. More than 25,000 fish of 254 species have been tagged by nearly 2,000 taggers along a 4,500-km coastline. The program is now a major feature of recreational angling in South Africa.

Modus Operandi

The entire program is sponsored by Stellenbosch Farmers' Wineries, a major liquor company popular with fishermen. This backing provides a computer-based infrastructure and a full-time tagging officer.

Participation is by membership only and applicants are screened for their genuine interest and competence before being admitted. Tagging kits are then issued to individuals in special pouches containing promotional matter, instruction book, tape measures, and one or more of four tagging systems. The choice of tagging system depends on the angler's preference and capabilities. The following systems are available.

System A is used on teleosts in the 1–25-kg range, as well as on batoid fish except for Rhinobatidae. The system consists of plastic, single-barb dart tags 90 mm long. Each is attached to a precoded, self-addressed, prepaid return card. A stainless steel hollow-tube applicator is included, and tags are inserted into musculature below the second dorsal fin; the barb becomes anchored in the interneural spines. The tag is inserted in the tail of batoids, as described earlier, where it is adequately anchored by muscle and skin tissue.

System B is used on large sharks and teleosts in excess of 25 kg, especially billfishes and tunas. It consists of 140-mm, metal-tipped dart tags, together with a stainless steel applicator either mounted for hand-held use or located at the end of a specially manufactured Purglass® tagging pole. These tags are inserted in the main dorsal musculature of fish, the metal tip being anchored by muscle tissue.

System C is used on small sharks under 25 kg and on guitarfishes (Rhinobatidae). It consists of the

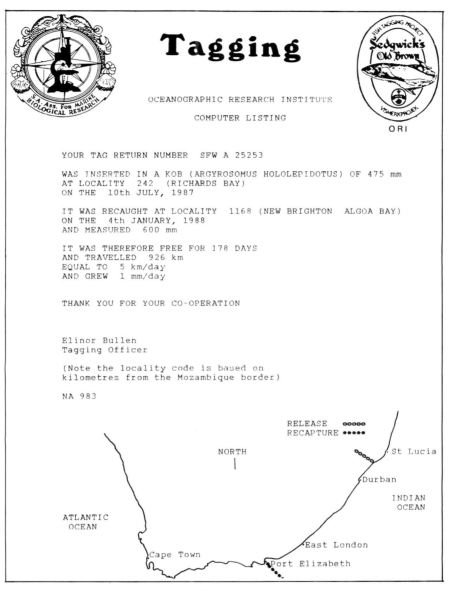

FIGURE 4.—Sample of feedback to taggers on recoveries.

ORI tag, whose two plastic disks are pressed together after being inserted through the dorsal fin of the shark. No applicator is needed, but a leather punch is issued with each kit to punch the pilot hole. The tag is loosely fitted to allow for growth.

System D is used on teleosts weighing less than 3 kg. It is issued only to experienced personnel, such as nature conservation officers, fellow scientists, and other persons capable of handling small fish with care. It is similar to system A, but

consists of 80-mm spaghetti tags with correspondingly smaller applicators.

Once admitted, the participant angler is entitled to a continuing supply of tags at a rate equal to the application rate. Some people tag nearly 1,000 fish per annum, but most manage no more than 50.

Researchers also participate in the tagging program. They are issued tags in bulk, usually at cost, and administration and analyses of their tagging are undertaken centrally.

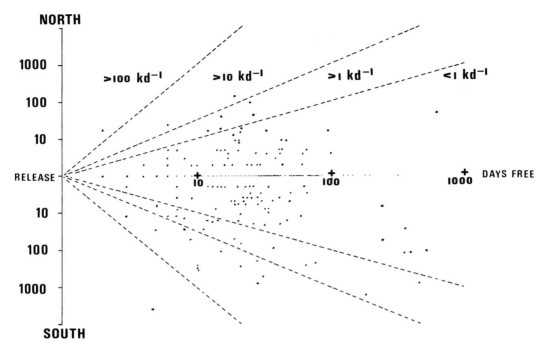

FIGURE 5.—Distance traveled (km/d) versus maximum time at large for 410 bluefish tagged at Durban.

Feedback to Taggers

Much of the program's success stems from its continual feedback to members. Recaptures are immediately reported to the original tagger, and both tagger and recapture angler receive a detailed analysis of the fish's migratory and growth statistics. Where possible, a computer-generated map with a schematic presentation of the fish's movements also is provided (Figure 4). Quarterly newsflashes are mailed to all participants in addition to a comprehensive annual newsletter. Participating members are entitled to draw on information generated by the program, and taggers may request a personal analysis of their own performance or recapture rates for a certain species in a certain area. Other promotional materials include personalized certificates for those who attain tagging targets and videos that demonstrate tagging techniques.

Scientific Analysis

The ultimate objective of this program is to generate data valuable to fisheries management. To achieve this, and to broadcast the available information, tagging data reports are provided on request to researchers throughout the subcontinent. These reports are made available as a service to the community at large and to marine science in particular. They consist of semiprocessed data subjected to various analyses, among them graphs of time elapsed versus distance traveled for recaptured fish (Figure 5). Further analyses provide mean direction of dislocation as described by Jones (1966). Analysis on a time-interval basis (e.g., monthly) quantifies the extent of fish movement and identifies the onset of migrations.

The data reports also compare size-frequency distributions, geographic dispersal, and seasonality of tagged and recaptured specimens. Recapture rates are presented by variable geographic and time frequency, so fishing mortalities can be calculated by region and period. These data have limitations, but an effort is made to minimize them or to quantify their bias. Hence tag-shedding rates, nonreporting, and the relative efficacy of individual taggers are all evaluated.

Progress to Date

Good progress can be reported. Increasing numbers of fish are tagged each year (Figure 6), and they embrace a wide variety of species in 25 families. The 10 most numerous species tagged represent 55% of the total fish tagged and the species composition of tagged fish resembles that of fish caught by

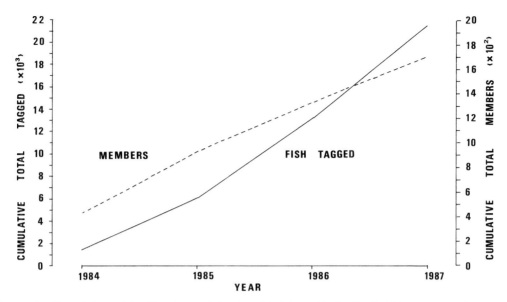

FIGURE 6.—Cumulative totals of tagging participants and fish tagged in the South African cooperative program, 1984–1987.

anglers. Angling effort is distributed optimally, and tagging is evenly distributed from southern Zululand to northern Namibia (Figure 1).

The overall tag recapture rate is 2.8% with a range from 0 to 8%. Highest recaptures are for sharks and resident reef species such as the groupers (Serranidae). The average time to recapture is 132 d, but it ranges between 1 d and 2.8 years.

This tagging program is valuable not only for studying individual species, but also for providing information on the ichthyofauna as a whole. It has, for example, confirmed migrations between Atlantic and Indo-Pacific regions, mostly by endemic species such as the lesser guitarfish and the galjoen. There has been no indication yet of such migrations by species known to have invaded the Indo-Pacific from their Atlantic origins. Hence, tagging of bluefish, garrick *Lichia amia*, and yellowbelly grouper *Epinephelus guaza* on both sides of the Cape has failed to confirm their common ancestry.

Conclusion

Marine fish tagging off southern Africa has a long history marked by uncoordinated tagging and unpublished results. The Oceanographic Research Institute's cooperative tagging program, launched in 1984, has improved this situation considerably, especially in addressing some of the problems that plagued earlier tagging studies,

such as poor tag performance caused by low-quality tag manufacture. Testing tags of different origin on shoals of captive fish proved useful in identifying the best manufacturer and substantially improving tag performance.

The high shedding rate of tags has been a major problem. Nevertheless, continual testing of different tagging techniques has markedly reduced tag loss, especially for the batoid fishes used in this study. Variability due to differences in tagger performance also were identified as potential sources of error, most often in programs in which anglers do the tagging. This problem was addressed by conducting parallel tagging in which the work of volunteers and well-trained staff was compared. Results from these tests indicated higher tag returns for trained staff, but the inherent variance of the data has thus far precluded firm conclusions.

Finally, there has been the problem of nonreporting of tags. This aspect was investigated by conducting face-to-face surveys of anglers, which revealed that on average only 3% of Natal recreational anglers did not return tags. Their recalcitrance was due more to general disinterest than any other factor.

Two important observations can be made from this program. First, wherever possible the tagging study should involve tag evaluations of captive fish. Many of the shortcomings in the South

African studies went undetected until such evaluations were made. Secondly, it is considered most important not to terminate a tagging program prematurely simply because an apparently insurmountable problem has arisen. In many cases, it is better to direct energies to quantifying rather than overcoming the problem. In the case of the ORI tagging program, the problems of nonreporting, tag shedding, and individual tagger performance remain, but knowing their influences has permitted continued progress and generation of data.

The information being generated by the nationwide tagging program adequately justifies its continuation—not only to contribute to fisheries management, but also to investigate wider aspects of marine fish ecology.

References

Bass, A. J., J. D. D'Aubrey, and N. Kistnasamy. 1973. Sharks of the east coast of southern Africa. I. The genus *Carcharhinus* (Carcharhinidae). Oceanographic Research Institute (Durban) Investigational Report 33:1–168.

Cliff, G. 1988. Natal Sharks Board tagging news. Oceanographic Research Institute (Durban) Tagging News 4 (Summer 87/88):6. (Durban, South Africa.)

Cousteau, J.-Y., and P. Cousteau. 1970. The shark: splendid savage of the sea. Cassell, London.

Crawford, R. J. M. 1987. Food and population variability in five regions supporting large stocks of anchovy, sardine and horse mackerel. South African Journal of Marine Science 5:735–757.

Crawford, R. J. M., L. V. Shannon, and D. E. Pollock. 1987. The Benguela ecosystem, part 4. Pages 353–505 *in* M. Barnes, editor. Oceanography and marine biology annual review, volume 2. Aberdeen University Press, Aberdeen, Scotland.

Davies, D. H., and L. Joubert. 1966. Tag evaluation and shark tagging in South African waters. Oceanographic Research Institute (Durban) Investigational Report 12:1–36.

de Jager, B. v. D. 1955. The South African pilchard (*Sardinops ocellata*). The development of the snoek (*Thyrsites atun*) a fish predator of the pilchard. South Africa Division of Sea Fisheries Investigational Report 19:1–11.

Ebert, E. E. 1964. Underwater tagging gun. California Fish and Game 50:28–32.

Jones, R. 1966. Manual of methods for fish stock assessment. Part 4—marketing. FAO (Food and Agricultural Organization of the United Nations) Fisheries Technical Paper 51 (Supplement 1):1–90.

Lutjeharms, J. R. E., and P. L. Stockton. 1987. Kinematics of the upwelling front off southern Africa. South African Journal of Marine Science 5:35–49.

Nepgen, C. S. de V. 1979. Trends in the line fishery for snoek *Thyrsites atun* off the south-western cape, and in size composition, length-weight relationship and condition. Fisheries Bulletin South Africa 12:35–43.

Newman, G. G. 1970. Migration of the pilchard *Sardinops ocellata* in southern Africa. South Africa Division of Sea Fisheries Investigational Report 86:1–16.

Olsen, A. M. 1953. Tagging of school shark, *Galeorhinus australis* (Macleay) (Carcharhinidae) in south eastern Australian waters. Australian Journal of Marine and Freshwater Research 4:95–104.

Penrith, M. J., and D. L. Cram. 1974. The Cape of Good Hope: a hidden barrier to billfishes. NOAA (National Oceanic and Atmospheric Administration) Technical Report NMFS (National Marine Fisheries Service) SSRF (Special Scientific Report Fisheries) 675.

Smith, M. M., and P. C. Heemstra, editors. 1986. Smith's sea fishes. Macmillan, Johannesburg, South Africa.

Templeman, W. 1984. Migrations of thorny skate, *Raja radiata*, tagged in the Newfoundland area. Journal of Northwest Atlantic Fishery Science 5:55–64.

van der Elst, R. P. 1984. Marine sport fishing in South Africa: an up-to-date evaluation. South African Journal for Research into Sport and Recreation 7:59–69. (Pretoria).

American Fisheries Society Symposium 7:863–879, 1990

The Fisheries Research Institute's High-Seas Salmonid Tagging Program and Methodology for Scale Pattern Analysis

NANCY D. DAVIS, KATHERINE W. MYERS,
ROBERT V. WALKER, AND COLIN K. HARRIS

Fisheries Research Institute, School of Fisheries, WH-10
University of Washington, Seattle, Washington 98195, USA

Abstract.—The International North Pacific Fisheries Commission (INPFC), comprising Japan, Canada, and the USA, mandates research to determine continental origins of Pacific salmon *Oncorhynchus* spp. in the area of the Japanese high-seas gill-net fisheries. From 1978 to 1986, 43,339 salmonids—20,627 chum salmon *O. keta*, 10,584 pink salmon *O. gorbuscha*, 7,830 sockeye salmon *O. nerka*, 3,470 coho salmon *O. kisutch*, 518 chinook salmon *O. tshawytscha*, and 310 steelhead (*O. mykiss*), were disk-tagged in the north Pacific Ocean and Bering Sea. The total number of tag recoveries in North America during this period was 134 fish, mostly sockeye salmon (102). The total number of tag recoveries in Asia was 284, mostly chum salmon (197). The Fisheries Research Institute (FRI), University of Washington, has conducted INPFC-related research since 1955, and it serves as a processing center for North American recoveries of high-seas salmonid tags. Although tagging studies have provided invaluable information about oceanic distributions of salmonids, tag recovery data by themselves do not provide the information required to estimate the extent of intermingling of Asian and North American stocks in the high-seas fisheries nor the detail needed for international negotiations. Such detailed information can be provided by means of scale pattern analysis to identify stock origins of salmon in high-seas fisheries catches. Scientists conducting INPFC-related research have developed and agreed upon a standardized methodology for the collection and analysis of scales. These procedures include (1) collection of scales from a preferred area on the body of the fish, (2) exchange of acetate impressions of scales, (3) collection of scale data by a variety of methods, (4) use of either maximum likelihood or classification (with error correction) as a statistical procedure for estimating stock composition of fishery samples, and (5) calculation of stock composition estimates based on unknown-origin samples of 100 or more scales for any established stratum. Scientists at FRI and in Japan are analyzing scales to determine the continental origins of chum, sockeye, coho, and chinook salmon in the area of the Japanese high-seas fisheries.

The International Convention for the High Seas Fisheries of the North Pacific Ocean (North Pacific Treaty, signed in 1952 by the governments of Japan, Canada, and the USA) requires research on Pacific salmon *Oncorhynchus* spp. to determine continental origins of fish in the area of the Japanese high-seas fisheries (Figure 1). The North Pacific Treaty established the International North Pacific Fisheries Commission (INPFC), in part to recommend and review research required for member nations. The research commitment of the U.S. National Section of the INPFC has supported the high-seas salmon research program at the Fisheries Research Institute (FRI), University of Washington, continuously since 1955. During that time, the high-seas program has engaged primarily in tagging studies and scale pattern analyses to determine high-seas distributions of Pacific salmon. The high-seas program at FRI is funded by the U.S. National Marine Fisheries Service (NMFS).

In the early years, FRI's research focused on broad tagging experiments at sea to determine overall oceanic migration routes of major Asian and North American salmonid stocks (Hartt 1962, 1966). Intensive sampling and tagging operations were conducted in an area south of Adak Island (Aleutian Islands) because purse-seine catches in this area provided an index of the run timing and abundance of immature Bristol Bay sockeye salmon *O. nerka* that could be used to forecast the inshore run (Figure 1; Hartt 1962, 1966; Rothschild et al. 1971; French et al. 1976). In addition, tagging experiments in the Gulf of Alaska and eastern Aleutian region examined the migration routes of juvenile salmonids during their early ocean residence (Hartt and Dell 1986).

Results of early INPFC-related tagging studies (INPFC 1959) showed that North American and Asian salmon intermingle in the area of the Japanese high-seas gill-net fisheries (Figure 1). The USA has had a long-standing concern over capture

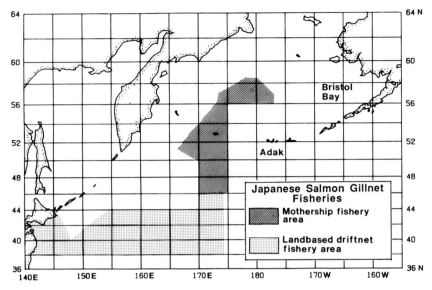

FIGURE 1.—Areas in the north Pacific Ocean and Bering Sea fished by the Japanese high-seas salmon gill-net fisheries in 1987.

of North American salmonids in these fisheries. Japanese high-seas salmon fishing is conducted by two very different fisheries, a mothership fishery and a land-based fishery. The mothership fishery consists of large processing vessels, each with its own discrete fleet of catcher boats. Between 1959 and 1976, the size of the mothership fishery gradually decreased from a maximum of 16 fleets (approximately 28 catcher boats per mother ship) to 10 fleets. From 1978 to 1986, the size of the mothership fishery stabilized at four fleets; and in 1987, the fishery was reduced to three fleets, each with a complement of 43 catcher boats. The land-based fishery consists of vessels that fish independently and deliver their catches to ports in Japan. The number of vessels in the land-based fishery decreased from a maximum of 374 in 1972 to 157 in 1987 (Harris 1988; M. Dahlberg, NMFS, personal communication). Harris (1987, 1988) has summarized the history of these fisheries and their governing treaties.

Prior to 1978, the eastern boundary of the Japanese high-seas fisheries was 175°W longitude. However, in 1978 the North Pacific Treaty was renegotiated to bring it into conformity with the Magnuson Fishery Conservation and Management Act of 1976. The resulting revision of the Annex governing the fisheries prohibited mothership and land-based fishing east of 175°E in waters south of 56°N, and limited fishing effort in the central Bering Sea north of 56°N latitude. In 1986,

further revisions included gradual elimination of the mothership fishery in the central Bering Sea and movement of the eastern boundary of the land-based fishery 1° westward to 174°E longitude (Figure 1; Harris 1988).

The 1978 and 1986 agreements included a mandate for cooperative research to determine origins of salmonids in the land-based area south of 46°N latitude. In response to this mandate, the high-seas program at FRI has used two techniques to investigate origins of salmonids in the salmon fishery area: tagging studies and scale pattern analyses (Cook et al. 1981; Harris et al. 1983; Walker and Davis 1983; Myers et al. 1987).

Knowledge gained from tag recoveries and from scale pattern analyses together provide information required for international negotiations. Tag recoveries proving continent of origin provide invaluable information that increases knowledge about the distributions and migrations of regional stock-groups of salmonids. Over the past 30 years, tagging by the USA, Japan, and Canada has provided a considerable amount of unequivocal information on stock origins of Pacific salmon in the area of the Japanese high-seas fisheries. However, tag recovery data by themselves will not permit estimation of the extent of intermingling of Asian and North American stocks because of the variability in exploitation rate, recovery effort, and reporting of recovered tags and the low number of recoveries for some species (Margolis et al. 1966; Harris 1987, 1988). In addition, high-seas tagging is very expensive. Funds have not been allocated by the USA to enable FRI to conduct its own tagging program since 1982, although scientists at FRI have participated in cooperative tagging cruises with the USSR since 1983. Because large numbers of scales can be obtained and measured, scale pattern analysis can be used to estimate proportions of Asian and North American salmon caught in the high-seas fisheries. The estimates are statistical ranges, however, and their quality and precision depend on factors such as representative sampling, adequate comparison material from known stocks, and adequate sample sizes. They do not prove the presence of particular stocks in the fisheries areas as do tag returns, and there have been disagreements on methodology.

Late in 1977, FRI researchers began to apply scale pattern analysis to determine origins of salmon caught in and near the Japanese high-seas salmon fishery areas. Cook et al. (1981) analyzed the origins of sockeye caught on the high seas between 1972 and 1976, and concluded that Bristol Bay sockeye salmon occur west of 170°E and the Kamchatkan sockeye salmon occur east of 175°E in significant percentages. The results from scale pattern analyses of coho salmon *Oncorhynchus kisutch*, however, were less conclusive; Walker and Davis (1983) thought poor scale sample quality may have introduced bias into the analyses and caused conflicting results for the years sampled. The most recent results from FRI's scale pattern analyses have come from scale samples of chinook salmon *O. tshawytscha* collected on the high seas from 1975 to 1981 (Myers et al. 1984, 1987). They showed that western Alaskan chinook salmon predominated in the Bering Sea and were an important secondary stock in the north Pacific Ocean. Kamchatkan chinook salmon formed an important secondary stock in all the fishery areas. An unexpected result of this analysis was the predominance of central Alaskan chinook salmon in the north Pacific Ocean.

In recent scale pattern studies of chinook salmon distributed in the high-seas fishery area, Japanese and FRI researches used different methodologies and sometimes obtained very different results (Myers et al. 1987). This led to considerable discussion among members and advisers of the INPFC Sub-Committee on Salmon about the best procedures to use for scale pattern analysis. It was decided that methodologies should be jointly agreed to before more analyses were performed. This has led to the recent development of an INPFC standardized methodology for future scale pattern analyses.

In the first half of this paper, we describe high-seas tagging techniques and the tag recovery program at FRI, and we review the INPFC tagging program. In the second half, we summarize FRI's methodology for scale pattern analysis, review the research on which this standardized methodology was based, and discuss some of its advantages and disadvantages. Although our particular application is to determine stock origins of Pacific salmon in the area of the Japanese high-seas salmon fisheries, many of the procedures we describe also apply to any study involving large-scale tagging operations or stock separation analyses based on scale pattern analysis.

High-Seas Tagging

Tagging Techniques

Capture and release of live salmonids on the high seas is accomplished with two types of fishing gear, purse seines and floating longlines. Japan has used longlines exclusively to capture salmon for tagging since 1958. Japanese longlining operations since 1978 have occurred in the western and central north Pacific Ocean, the Bering Sea, and the Gulf of Alaska. Canada also used longlines for high-seas tagging operations from 1961 until 1967. After a 20-year hiatus, Canada resumed high-seas tagging operations with longlines in 1987. Canadian tagging operations occurred in the Gulf of Alaska and the eastern north Pacific, i.e., east of 170°W (Giovando 1969; LeBrasseur et al. 1987). Purse seines were the primary gear used in FRI's tagging experiments from 1955 through 1982 (except in 1979 and 1981, when tagging experiments were not conducted). However, longlines also were used from 1963 through 1970 and in 1980 and 1982. From 1983 to 1986, and in 1988, FRI participated in cooperative high-seas purse seining and tagging cruises with the USSR in the central north Pacific Ocean and the Bering Sea.

Longlining operations were reviewed in detail by Hartt (1962, 1963), Kondo et al. (1965), and Light and LeBrasseur (1986). A floating longline consists of a main line suspended at the water's surface by small floats and 1-m-long branch lines (gangions) spaced about 1.5 m apart. A hook at the end of each gangion is baited with a small salted anchovy. In recent Japanese operations,

2.8–6.9 km of longline are set at dawn or dusk; the soak time is approximately 30 min and gear retrieval takes about 2 h.

Purse-seine operations were described in detail by Hartt (1962, 1963) and Harris (1983, 1984, 1985). In general, seines used over the years varied from 731 m to 970 m long and 37 m to 127 m deep. Small mesh size (51 mm) prevents fish from becoming gilled in the netting. Knotless netting in the bunt avoids chafing and scaling of the fish during brailing. An outrigger and bridle suspended out over the side of the vessel helps to hold the bunt away from the side of the vessel, further minimizing entanglement of fish in the netting. The time needed to set and retrieve the seine is approximately 2–3 h.

Longline and purse-seine gear each have their own inherent design aspects that make them useful for capturing salmon live at sea. Because longlines are lightweight, they can be deployed safely in moderately rough weather, and they can be fished efficiently where salmonid concentrations are low. However, salmon suffer from exhaustion and injury more often when captured on longlines than with a purse seine (Kondo et al. 1965). Use of purse seines is severely limited by the weather conditions frequently encountered in the north Pacific Ocean. When weather permits, the purse seine is capable of catching salmon efficiently in areas of open ocean where fish are known to concentrate. The seine is less size-selective than longlines and, in general, captures fish in good condition (Hartt 1963; Kondo et al. 1965).

Fish judged to be suitable for tagging are tagged and sampled for data as soon as possible after capture. They are brought aboard and put into a live tank. The procedure used by FRI includes anesthetizing the fish. The fish are transferred two or three at a time from a live tank, which is supplied with circulating seawater, to a tub containing the anesthetic, usually tricaine (MS-222). Use of an anesthetic may decrease tagging mortalities, especially among small fish (Hartt 1963). A fish is laid on a cradle or board for length measurement and tagging. A scale sample is removed, and a Peterson disk tag is placed at the base of the dorsal fin. The tag is anchored with either a plastic cinch strap or a metal pin. If anesthetized, the tagged fish is allowed to recover from the anesthetic before release.

Disk tags are used to mark salmonids on the high seas because they are retained better than spaghetti tags (Hartt 1963). Long retention of disk tags is critical to the success of high-seas tagging experiments, because fish may be at liberty for 2 or more years before recovery.

Tag Releases and Recoveries

Several aspects of a high-seas tagging operation influence the number of tags recovered from inshore areas. First, it is difficult to tag large numbers of salmon at sea because of weather and general working conditions and because the fish are sparsely distributed. Second, the tagged fish may be caught by a high-seas fishery before they start their inshore migration. Although it is valuable to learn where such fish moved between tagging and recapture, their continental origin cannot be inferred. Also, some tagged fish may escape inshore fisheries and return to remote spawning areas, decreasing the likelihood that they will be found.

Figure 2 shows the number of tagged salmonids released from 1978 to 1986 by the USA, Japan, and the USSR in cooperation with the USA. The total number of salmonids tagged over this period was 43,339 which included 20,627 chum salmon *Oncorhynchus keta* (48%), 10,584 pink salmon *O. gorbuscha* (24%), 7,830 sockeye salmon (18%), 3,470 coho salmon (8%), 518 chinook salmon (1%), and 310 steelhead *O. mykiss* (less than 1%).

The numbers of tag releases shown in Figure 2 in the north Pacific Ocean, Bering Sea, and Gulf of Alaska generally reflect research priorities of the INPFC. From 1978 to 1986, for example, the region of the north Pacific Ocean south of 46°N latitude had the most releases (21,344, or 49%), and this concentration of tagging was a consequence of the research mandate to investigate origins of salmonids found in the land-based fishery area (south of 46°N latitude). Releases in the north Pacific Ocean north of 46°N latitude (14,401, or 33%) were due to INPFC interest in salmonid tag recovery information from the area between 175°E and 175°W longitude, an area that was eliminated from the fishery in 1978. Releases in the Bering Sea (7,230, or 17%) illustrated a specific interest in chum and chinook salmon distributions in that region. There were few releases in the Gulf of Alaska during the 1978–1986 period because the gulf was not part of the historical or current fishery area, and therefore the data were not required for treaty negotiations.

The tagged fish released by the USA (and the USA in cooperation with the USSR; Figure 2) by species in the north Pacific Ocean (north and

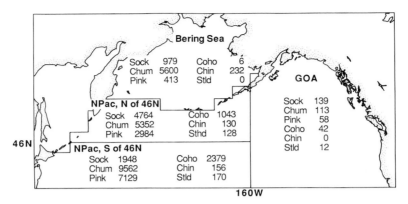

FIGURE 2.—Numbers of high-seas tagged salmonids released by the USA and Japan from 1978 to 1986. NPac: north Pacific; GOA: Gulf of Alaska, Sock: sockeye salmon; Chin: chinook salmon; Stld: steelhead.

south of 46°N latitude) consisted of 3,767 sockeye, 3,331 chum, 1,653 pink, 691 coho, and 134 chinook salmon and 25 steelhead. The remaining 26,144 fish tagged in the north Pacific Ocean were tagged by Japan. In the Bering Sea, the USA (and the USA in cooperation with the USSR) released 152 sockeye, 1,208 chum, 83 pink, and 115 chinook salmon. The remaining 5,672 tagged salmonids were released by Japan. The USA released no tagged coho salmon in the Bering Sea over this period, and no steelhead were released by either country in the high seas; steelhead distributions have not been found to extend into the Bering Sea. The few releases in the Gulf of Alaska shown in Figure 2 were made by Japan.

The coastal recoveries of these high-seas tags are shown in Figure 3. The recoveries from North America total 134 fish, an average of 15 tags per year. The 102 sockeye salmon (76%) were the most numerous North American recoveries. Other salmonids were recovered in fewer numbers: 13 pink salmon (10%), 8 coho salmon (6%), 8 steelhead (6%), and 1 chinook salmon (less than 1%). The total number of salmonid tags recovered in Asia was 284. Chum salmon were the most numerous Asian recovery (197, or 69%), followed by 50 pink salmon (18%), 30 coho salmon (11%), 6 sockeye salmon (2%), and 1 chinook salmon (less than 1%).

Although recoveries of fish tagged on the high seas since 1978 are not numerous, they have significantly increased the known limits of distribution of many Asian and North American salmonid stocks around the Pacific Rim, from Japan and the Primore district of the USSR to Washington and Oregon. Particularly important are recent

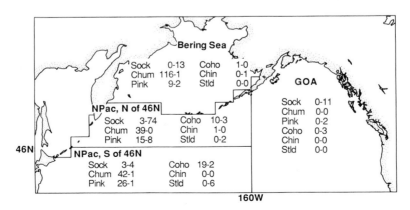

FIGURE 3.—Numbers of high-seas tagged salmonids that were released from 1978 to 1986 and recovered in coastal areas of North America and Asia. NPac: north Pacific; GOA: Gulf of Alaska; Sock: sockeye salmon; Chin: chinook salmon; Stld: steelhead. The left number of each pair is the number of Asian coastal recoveries; the right one is the number of North American coastal recoveries.

recoveries of fish tagged in and near the Japanese high-seas fisheries areas. These have provided the first unquestioned evidence of the presence of North American sockeye, coho, chum, and pink salmon in the area of the north Pacific Ocean closed to the Japanese land-based salmon fleet in 1978, and evidence of USSR sockeye, chum, and coho salmon in the same region. On the basis of recent tag recoveries, we know that North American steelhead are found far to the west, inside of the current land-based fishery area. Chinook salmon have been recovered from Asia and from the Canadian portion of the Yukon River since 1978.

Tag Recovery Program

The functions of FRI's tag recovery program include (1) enlisting the cooperation of fishermen and others in the salmon industry to return tags by informing them about high-seas tagging experiments and about why data from recovered tags are important to them, (2) sending recovered tags and accompanying data back to the agency that released tagged fish, and (3) reporting the recoveries of fish tagged during FRI tagging studies.

The process of informing fishermen and others in the fishing industry about the existence of high-seas tags is accomplished by an informational campaign and by submission of articles to periodicals of interest to fishermen and others in the salmon industry. The informational campaign focuses on salmon and steelhead fishermen and processors in Alaska, British Columbia, Washington, Oregon, portions of Idaho along the Snake River, and northern California. The campaign consists in part of a direct mailing, and it runs on a budget of less than $600 per year with no paid advertising. The informational packet contains a cover letter that explains how recoveries of high-seas tags can benefit the North American salmon industry. In addition, the packet contains a poster and business reply envelopes. The poster is designed to remind people of the appearance of high-seas tags, the reward for tag recoveries, and where to send recovered tags. The postage-paid reply envelopes are printed with requests for information on location, date, and method of recovery and the finder's name and address. The packet is sent to numerous canneries, fish buyers, and processors from Alaska to Washington, as well as to management and research agencies, vessel owners' associations, fishermen's unions, and sport-fishing lodges. Another component of

FRI's advertising campaign is a public service announcement aired during the fishing season by radio stations throughout Alaska.

When FRI receives a recovered high-seas salmonid tag, the individual who found the tag is sent a reward, a letter with information about when and where the fish was tagged, and a map showing the location where the tagged fish was released. These letters communicate to the finder the importance of every recovery to the success of the program.

The reward structure for return of high-seas tags has changed over the years. In the 1950s and 1960s, rewards of $1 for a tag return and $25 for a whole fish with an intact tag were paid. The program for return of whole fish was dropped in the late 1960s, and the reward for return of tags was increased to $3 in the 1970s. Rewards were increased to $15 in 1984 and to $20 in 1987.

If the recovered high-seas tag was not released by FRI tagging operations, the tag and accompanying recovery data are sent to the country, Japan or Canada, that conducted the tagging experiment. All release and recovery data are summarized in reports to the INPFC by the nation that released the tagged fish. Information on tag recoveries from USA–USSR cooperative tagging experiments is provided to the USSR's Pacific Scientific Research Institute of Fisheries and Oceanography (TINRO).

Tag recoveries proving continent of origin provide invaluable information about oceanic distributions and migrations of salmon. In research planning meetings and other arenas, FRI urges the INPFC member nations and also the USSR to continue and, if possible, to expand their high-seas tagging efforts. Coastal recoveries of Pacific salmon tagged in the area of the high-seas salmon fisheries document the presence of specific stocks in the fishery area, and the evidence from tagging data will continue to be useful in international negotiations. Tag recovery data constitute a standard by which other methods of obtaining information on stock origins of Pacific salmon are evaluated. But to provide adequate quantitative estimates on intermingling of North American and Asian fish in the areas of high-seas gill-net fisheries, additional information can be obtained from scale pattern analysis. A standardized methodology for this field has been developed and agreed to by INPFC member nations for future estimates of high-seas fisheries stock composition.

Methodology for Scale Pattern Analysis

The scales of Pacific salmon have been used to identify stock or location of origin since the early 1900s, and scale pattern analysis is now routinely used by management agencies to estimate stock composition of catch in mixed-stock fisheries (Marshall et al. 1987). Scales are relatively inexpensive and easy to collect, store, measure, and analyze, and large samples can be obtained without killing or mutilating the fish. Major et al. (1972) outlined the basic principles and procedures of stock identification of Pacific salmon by analysis of scale characters, and they reviewed research conducted for INPFC in the 1950s and 1960s. In the 1970s, the USA conducted most of the INPFC-related scale pattern research. The NMFS conducted studies on chinook salmon (Major et al. 1975, 1977a, 1977b) and, after 1975, contracted with the high-seas program at FRI to conduct studies on sockeye, coho, and chinook salmon. Scale pattern studies conducted by FRI were reviewed by Hartt et al. (1977, 1979), Harris et al. (1979, 1980a, 1980b, 1981, 1982, 1983, 1984), and the U.S. National Marine Fisheries Service (1985, 1986, 1987). Partially in response to the FRI studies, Canada conducted research in the 1980s on consistency of age determinations and statistical procedures for estimating stock composition (Department of Fisheries and Oceans 1985, 1986, 1987), and Japan conducted scale pattern research on chum, coho, and chinook salmon (Fisheries Agency of Japan 1982, 1985, 1986, 1987). As mentioned previously, in recent scale pattern studies of chinook salmon from the high-seas fishery area, Japanese and FRI researchers used different methodologies and sometimes obtained widely different results (Myers et al. 1987). This led to considerable discussion by the INPFC Sub-Committee on Salmon about the best procedures to use for scale pattern analysis.

In April 1986, the INPFC member nations adopted a new draft Annex to the Protocol and two new Memoranda of Understanding (MoUs; one pertains to research and the other to patrol and surveillance of the fisheries). The 1986 MoU on research mandates, in part, (1) new scale pattern analyses based as much as possible on methodologies jointly agreed upon beforehand by the three countries, and (2) improved collection of adequate (in number and quality) Asian and North American standard (or reference) scale samples.

A scale pattern analysis working group was formed by the Sub-Committee on Salmon to consider the requirements of the 1986 research MoU and to make recommendations for standardizing procedures for the collection and analysis of scales. These procedures will be used in future studies to determine the stock origins of salmon in the area of the Japanese high-seas salmon fisheries (INPFC 1987a, 1987b).

At its 1987 annual meeting, the INPFC recommended (1) that scale data collection not be restricted to any particular method, (2) that either maximum likelihood or classification (with error correction) should be used as the statistical procedure for estimating the stock composition of fishery samples, (3) that stock composition estimates for any established stratum should be based on mixed-fishery samples of 100 or more scales and that this should be a target sample size, (4) that the level of time and area stratification should be month by 5°-longitude subarea (Figure 4), although finer stratification is preferable when sample sizes permit, (5) that all national sections should continue efforts to procure reference scale samples from the USSR, and (6) that the national section should proceed with studies on the continental origins of catches in high-seas salmon fisheries based on these recommendations.

Scale Structure and Age Determination

The cycloid scales of Pacific salmon consist of two layers, an outer calcified layer and an inner fibrous layer (Simkiss 1974). The outer layer of the scale is sculptured by a series of concentric ridges called circuli (Figure 5). The focus or central platelet of the scale forms when the fish is approximately 30–50 mm long. Variation in growth of the fish is reflected in differences in spacing, number, and height of circuli on the scale. Freshwater growth, if any, is recorded near the focus of the scale and is characterized by finer and more closely spaced circuli than those formed during estuarine or oceanic growth. Annuli (bands of closely spaced, narrow, broken, or resorbed circuli) form once a year on the scale and are used to determine the age of the fish. For INPFC-related research, age is designated by the European formula (Koo 1962): the number preceding the decimal point is the number of freshwater annuli, and the number following the decimal point is the number of ocean annuli; for example, an age-1.4 scale has one freshwater annulus and four ocean annuli (Figure 5). Canadian, Japanese, and U.S. scientists have conducted considerable research on problems and methods of age deter-

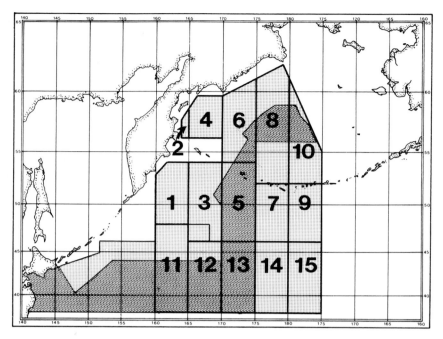

FIGURE 4.—The area of the Japanese high-seas salmon fishery before 1977 (stippled) and during 1977–1986 (crosshatched) and the International North Pacific Fisheries Commission statistical subareas. Statistical areas 1 through 10 are the mothership fishery area, and areas 11 through 15 are the land-based fishery area.

mination (e.g., Machidori 1981; Bilton 1985; Davis and Light 1985; and many others). In 1984, the INPFC sponsored a workshop on fish age determinations. Although the goal of the workshop was to resolve problems in determining accurate and reproducible age data, the participants agreed that much research was still required before this can be achieved. Standardized procedures for age determination in INPFC-related research have not yet been established among the three countries.

Preferred Body Area for Scale Collection

All of the scales used in an analysis should be collected from the same body area of the fish. This is because time of first scale formation, scale size and shape, and patterns of circuli vary with location on the body, so that differences in body area of scale collection can introduce variability or bias or both into scale pattern analyses (Knudsen 1985).

In the 1950s, U.S. and Canadian scientists recommended use of the "preferred" scale, located in the second scale row above the lateral line on the diagonal scale column that extends downward from the posterior insertion of the dorsal fin (Koo 1955; Clutter and Whitesel 1956;

Figure 6). The preferred scale is in the area where scales first form on the body of the fish, thus it has a more complete record of growth than scales from other body areas.

The method of scale collection in the USSR is not standardized (Knudsen 1985; Knudsen and Davis 1985), and we have observed that body area of scale collection can vary from sample to sample. In addition, samples from the Japanese high-seas catches and samples collected by fisheries agencies in Canada and the USA sometimes include nonpreferred scales. Because most agencies do not routinely record the body area of scale collection, researchers have developed criteria for the identification of nonpreferred scales by examination of the appearance of scales taken from various known areas of the body of the fish. Knudsen and Davis (1985) found that visual criteria were effective in identifying nonpreferred scales from area C, but that scales from areas A and B could not be distinguished (Figure 6). Knudsen (1985) recommended that agencies record the body area of scale collection so that nonpreferred scales can be easily identified and eliminated from scale pattern analyses.

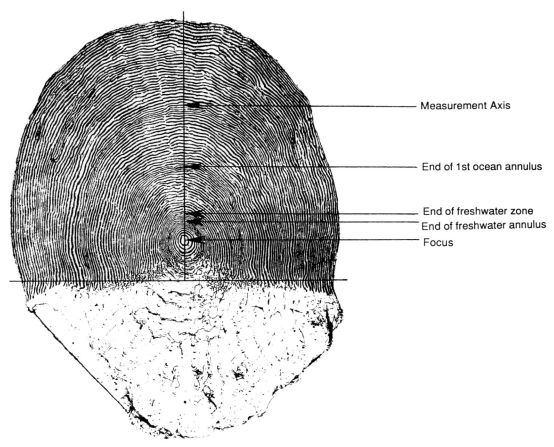

Measurement Axis

End of 1st ocean annulus

End of freshwater zone
End of freshwater annulus
Focus

FIGURE 5.—Age-1.4 chinook salmon scale from the Kamchatka River, USSR (16 June 1980), showing scale features, measurement axis, and life history zones used in scale pattern analyses.

Acetate Scale Impressions

Canadian, Japanese, and U.S. researchers use acetate impressions of scales for INPFC-related studies. Acetate impressions are easy to prepare, store, and measure, and the ability to produce multiple copies of the same sample facilitates exchanges. Scales are individually selected from the preferred body area of the fish with forceps, cleaned, moistened, stuck on gummed cards with the sculptured surface up, and impressed in transparent acetate (0.5 mm thick) with a heated hydraulic press (Koo 1962). The Fisheries Research Institute presently uses a temperature of 100°C and a pressure of 34,474 kN/m² (5,000 lbs/m²) for 3 min, but optimum settings can vary with acetate type. The acetate shrinks slightly (1–2%) as it cools; thus an acetate impression is not an exact duplicate of the scale, and researchers should

avoid pooling data from acetate impressions and actual scales (Rankis 1987).

Sample Exchanges

Arrangements for exchanges of scale samples and associated data among Canada, Japan, and the USA are made through INPFC during the meetings of the Sub-Committee on Salmon or its Ad Hoc Salmon Research Coordinating Group. Reference samples, which will be used to form standards that characterize the scale patterns of Asian and North American stocks, are obtained from fisheries agencies in Canada, Japan, the USA, and the USSR, and are collected from maturing adult salmon in commercial, subsistence, and sport catches and from agency samples on or near the spawning grounds. Samples from the high-seas fishery area (unknowns) are ob-

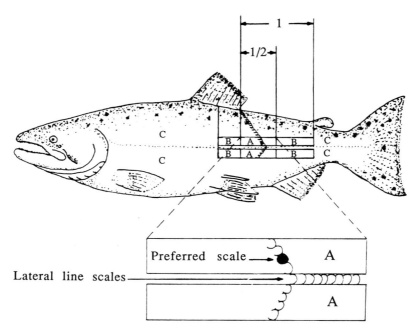

FIGURE 6.—The area of a salmonid's body designated as preferred for scale collection (area A) by the International North Pacific Fisheries Commission (from Knudsen 1985).

tained by Japanese research vessels and from Japanese commercial vessels operating in the Bering Sea and north Pacific in May through July. Each country prepares acetate impressions from samples collected within their respective areas and, upon request through INPFC, forwards copies to scientists in the other two countries. If there are budgetary or time constraints on preparation of duplicate acetate impressions, original acetates or gummed cards may be loaned on a temporary basis. The three countries also exchange impressions of any USSR scale samples that they have been able to obtain with permission from the USSR.

Collection of Scale Data

Scale data are collected for INPFC-related studies by use of microprojection devices, digitizing tablets, or video image analysis systems. Research by Ito (1987) and Walker (1987) showed that scale pattern data collected by use of any of these devices are similar and produce similar results in scale pattern analyses. Therefore, the INPFC Sub-Committee on Salmon agreed that there is no need to standardize the instrument used to collect scale data. However, the collection of scale data with microprojectors (such as the one described by Dahlberg and Phinney 1968)

is much slower than it is with computer-operated digitizing tablets or video image-processing systems, primarily because data must be coded and entered into the computer by hand. We have found that the speed of data collection with a commercially produced video system is similar to that obtained with a digitizing tablet system developed at FRI in 1979 (approximately 3–5 min per scale when incremental distances are measured between all circuli in the freshwater and first ocean zone). The high costs of software development have made video-imaging systems more expensive than digitizing tablet systems at present. The primary advantages of video-digitizing systems over digitizing tablets are that images can be enhanced and measurements automated. Both the Fisheries Agency of Japan and FRI are presently using a video image analysis system (Optical Pattern Recognition System,[1] model OPR-512, manufactured by BioSonics, Inc., Seattle) developed specifically for scale data collection. We hope that the use of the same measurement system by both countries will encourage standardization of measurement procedures and the exchange of scale pattern data.

---

[1] Use of brand-name products does not imply endorsement by FRI or contracting agencies.

Measurement Procedures

Selection of a measurement axis and criteria for inclusion or exclusion of circuli for measurement are not standardized among researchers conducting studies for INPFC. However, within a particular study it is important that the same procedures be used for every scale. Scale data are usually collected along a single radius in the anterior (sculptured) field of the scale. The two most commonly used radii are the longest axis of the scale (Anas and Murai 1969; Tanaka et al. 1969) and the axis that is perpendicular to the boundary of the sculptured and unsculptured fields of the scale (Figure 5). Both of these measurement axes bisect the focus. Researchers should avoid selecting a measurement axis that is close to the dorsoventral axis of the scale, where circuli are often resorbed.

The circuli are usually assigned to life history zones (e.g., freshwater zone, first ocean zone) by eye, and counts and measurements of predetermined characters (e.g., the number of circuli in the freshwater zone, the size of the first ocean zone), or the incremental distances between circuli in each of these zones along the selected measurement axis, are recorded. Some researchers use the criteria of Tanaka et al. (1969) for decisions about whether to include or exclude the measurement of broken or forked circuli. Other researchers measure only those circuli that cross or touch the selected radius (Anas and Murai 1969). Consistent application of one set of criteria is more important than the set of criteria used.

Scale Characters and Variable Selection

The results of scale pattern studies can vary when different scale characters are used (Ishida et al. 1985; Myers 1986). Davis (1987) reviewed scale character types and variable selection methods, which are not presently standardized in INPFC-related scale pattern analyses. The most commonly used scale characters are circulus counts and measurements in the freshwater and first ocean zone (Figure 5). Circulus patterns in this portion of the scale are formed prior to broad mixing of Asian and North American stocks on the high seas; usually they provide the characters that give greatest separability between stocks. The particular characters used in an analysis may be determined beforehand and measured directly from the scales or calculated from data on incremental circulus spacing after the scales have been measured. In some studies, all of the scale characters, whether measured directly or calculated, are included in an analysis. In other studies, a variable selection process, such as the Kruskal–Wallis *H*-statistic (Cook and Lord 1978) or Wilks' lambda (Davis 1987), is used to select a subset of variables for analysis, so that only those characters that best distinguish the stocks are used.

Davis (1987) used scale data collected by Myers et al. (1987) to examine the performance of variable subsets in scale pattern analysis. Directly measured characters based on life history information (e.g., the number of circuli in the freshwater zone, the size of the first ocean zone) produced higher classification accuracies of known-origin fish than did rate characters (e.g., average circulus spacing in the first ocean zone) or ratio characters (e.g., the size of the freshwater zone divided by the total size of the scale) or characters not based on life history information (e.g., the distance between the 8th through the 10th circulus on the scale). A reduction in the number of scale characters from the 48 used by Myers et al. (1987) to 11 directly measured characters decreased overall accuracies of classification by only 1–2%. Two criteria (Wilks' lambda and Mahalanobis distance) selected different variables from the larger data set and the same variables from the smaller data set. Based on these results, Davis recommended that researchers use a small number of directly measured characters that adequately describe scale patterns. Researchers at FRI plan to use this approach in future studies.

Nonrandom Variation in Scale Characters

There are many sources of nonrandom variation in scale data. Variability associated with interpretation of scale patterns and selection of a measurement axis can be significant, but usually it is minimized or eliminated by standardization of procedures among scale readers prior to collection of scale data or by having one individual collect all of the data for a particular analysis in a consistent manner. Variability associated with date of collection, fishing gear, and sexual dimorphism usually is not significant in scale characters that describe growth in fresh water or the first ocean year, but it may be significant in data collected farther out on the scale (Anas and Murai 1969; Tanaka et al. 1969). Year-class, age-group, and the use of scales collected from different areas on the body can introduce significant sources of variation in scale data (e.g., Myers et al. 1984; Knudsen 1985; Knudsen and Davis 1985). Errors can be minimized if all scales are collected from the same body area and if each age-group within a year-class is analyzed separately.

Statistical Techniques

Over the years, several classification procedures have been used to assign Pacific salmon in the Japanese high-seas catches to their continent or region of origin; these include linear and quadratic discriminant analysis (Anas and Murai 1969), polynomial discriminant function analysis (Marshall et al. 1978), and a direct-density, leaving-one-out approach (Cook 1982a, 1982b). Researchers sometimes incorrectly use the percentages of fish assigned to each category (raw classification proportions) as their estimates of final stock composition, but raw classification proportions must be corrected for errors in the classification scheme (Millar 1988). Classificatory errors usually are estimated by a leaving-one-out technique (Lachenbruch 1967) and corrected by the classification-matrix correction procedure (Cook and Lord 1978). Because this estimator is nonlinear, stock composition estimates of less than zero or more than one may be obtained (Cook 1983). In this case, categories with negative estimates are dropped and the analysis is repeated (the collapsed approach; Myers et al. 1987), or procedures that constrain the estimates between zero and one are used (constrained approaches; Cook 1983; Millar 1987). The variance estimator of Pella and Robertson (1979) is used most frequently to calculate confidence intervals for the corrected estimates. Researchers often use 90% rather than 95% confidence intervals because confidence intervals obtained with Pella and Robertson's variance formula are conservative (Cook 1983).

The maximum-likelihood procedure (Milner et al. 1981) has not yet been used for INPFC-related scale pattern analyses because until recently it was thought that this method could not be used with continuous data. Fournier et al. (1984) used both discrete and continuous data in a conditional maximum-likelihood procedure to estimate stock composition in a mixed-stock fishery. Millar (1987) showed that a slight change in notation extends the maximum-likelihood procedure to allow continuous data, and that classification with error correction is a special case of the maximum-likelihood method. Millar (1987) used the infinitesimal jackknife to estimate the mixed-fishery sample's contribution to variability, but at present there is no computationally simple method to calculate the variance due to the standards. Simulations have been used to assess the variability of maximum-likelihood estimates, but the practical applications of this technique are limited (Millar 1987).

Millar (1988) discussed the advantages and disadvantages of the two basic methods of composition estimation—classification and maximum likelihood—and used simulations with real scale data to assess the relative performance of raw classification proportions, Cook and Lord's (1978) corrected classification estimates, Cook's (1983) geometrically constrained classification estimates, Millar's (1987) maximum-likelihood constrained classification estimates, and maximum-likelihood estimates (Millar 1987). Millar (1988) regarded the collapsed approach to classification (Myers et al. 1987) as too difficult to automate and did not include it in his study. The bias of raw classification proportions can be considerable, and Millar dismissed them from consideration. If negative estimates are calculated, only the two constrained classification estimators (Cook 1983; Millar 1987) and the maximum-likelihood estimator (Millar 1987) are legitimate, and Millar found that the performances of these three estimators were similar.

As a result of Millar's work and discussions among Canadian, Japanese and U.S. scientists at meetings of the INPFC scale pattern working group (INPFC 1987a, 1987b), the 1987 Sub-Committee on Salmon recommended the use of either maximum likelihood or classification (with error correction) as statistical procedures for estimating the stock composition of high-seas catches.

Composite Standards

There are numerous salmon stocks along the Pacific Rim, and because scale samples are not collected from all of these stocks, the reference samples used in INPFC-related studies have to be grouped into composite standards that are representative of all Asian and North American stocks likely to occur in the fishery area. With classification procedures, the reference samples must be combined into a few standards (generally six or fewer in INPFC-related studies) because classificatory accuracies decrease substantially as the number of groups is increased (Lachenbruch 1973). With the maximum-likelihood procedure, the number of groups that can be used is apparently limited only by the number of variables used in an analysis or by computer speed, memory, and costs; accuracies can be improved and the complexity of problems reduced, however, by grouping stocks with similar attributes (Wood et al. 1987).

There are presently no standardized procedures or criteria for grouping stocks into composite standards for INPFC-related studies, and this is an important area for new research. Traditionally, scientists have grouped the reference samples by major geographical or political region of origin (e.g., Asia, western Alaska, central Alaska, and southeast Alaska and British Columbia) because the INPFC requires information on the continental origins of fish in the high-seas catches for treaty negotiations. To eliminate potential sources of bias, the standards are usually composed of fish of the same freshwater age and brood year as the fish to be analyzed in the high-seas samples, and the proportions of each stock and age-group included in a standard are weighted in accordance to their relative abundance in inshore runs (Myers et al. 1987). However, Canadian and U.S. studies indicate that biases can result if composite standards include stocks with dissimilar scale patterns and if the proportions of these stocks are different from their respective proportions in the fishery samples (Ishida et al. 1984; Myers 1985, 1986; INPFC 1987b; Myers et al. 1987; Wood et al. 1987). Therefore, both similarity of scale patterns and geographic area of origin should be considered when stocks are grouped into composite standards. Researchers at FRI are presently attempting to develop composite standards that are more representative of Asian and North American stocks of sockeye, coho, and chinook salmon than were the regional groups used in previous analyses.

Sample Sizes and Time–Area Stratification

Sample sizes required to provide satisfactory stock composition estimates vary with the estimation procedure used. For classification procedures, researchers at FRI have used 25 fish as a minimum sample size for unknowns and 200 fish as a maximum sample size for standards (e.g., Myers et al. 1987). However, the variance of stock proportion estimates based on classification procedures is high when standard and unknown samples are less than 100 scales (INPFC 1987b). For the maximum-likelihood estimation procedure, Wood et al. (1987) suggested a minimum standard sample size of 50 fish per stock, and sample sizes for the unknowns must be based on the expected number of contributing stocks (i.e.,

the greater the number of contributing stocks, the larger the sample size). If the amount of separation between stocks is low and there are few discriminating traits, the maximum-likelihood procedure is very sensitive to the effect of sampling error, even with unknown sample sizes of more than 50 fish per stock (Wood et al. 1987). Researchers have found that the cost of collecting more than 100 scales for a particular time–area stratum is often not worth the increase in precision at larger sample sizes (INPFC 1987b). Therefore, the INPFC has adopted 100 scales as a target sample size for any established stratum. To achieve this sample size, high-seas samples probably will have to be stratified by month and 5°-longitude subarea (Figure 4), although finer stratification is preferable when sample sizes permit.

Current and Future Research Plans for FRI's High-Seas Program

Researchers at FRI are aiming to improve the information about migrations of North American salmonids in the high seas. Placing tags on fish caught live on the high seas is an expensive operation that would be of greater value if more tags were returned. The incentive for fishermen or cannery workers to return tags involves two components: their desire to cooperate and a monetary reward. Cooperation is important and is derived from the understanding that the information gained from tag recoveries benefits the fishery in which the fishermen participate. If the level of interest and cooperation are very high among the fishermen, the monetary incentive need not be large. But in a long-term tagging study such as ours, it is hard to sustain interest. Consider, for example, a highly compensated commercial fishermen who is catching and handling many salmon during a short fishing season. It might take a sizable reward to motivate such a person to stop working long enough to collect the tag, record the relevant recovery data, and return the tag and data to FRI. For this reason, we are concerned that fishermen may not be returning tags that they see in their catches. To address this concern, FRI has instituted a drawing for large cash prizes with the aim of increasing the motivation to look for and return tags. The prizes in the drawing consist of one $1,000 prize and two $500 and five $100 prizes. High-seas tags recovered in previous years, but not yet returned to FRI, also are eligible in the drawing. Tags returned to FRI many years after recovery may still show from which continent, Asia or North America, the fish originated. The drawing is an option that increases the financial incentive to return tags while maintaining the reward budget at a predetermined level. Plans include a follow-up analysis of how the drawing affects the number of recoveries received.

Coordinated scientific studies are presently being conducted by INPFC member nations to determine the continent of origin of Pacific salmon and steelhead, primarily in the land-based driftnet fishery area. The information from these studies will be used in negotiations, which must occur prior to the 1991 fishing season, over the location of the eastern limit of that fishery (presently 174°E longitude). Research is continuing on the continent of origin of salmonids in the operating area of the mothership fishery, with emphasis on the identification of areas of abundance of North American chinook, coho, and chum salmon and steelhead. As a part of this research, scientists in FRI's high-seas program are conducting scale pattern studies of coho, chinook, and sockeye salmon and coordinating data exchanges with Japanese scientists who are also analyzing coho and chinook salmon scales in addition to chum salmon scales (INPFC 1987b). Prior to the 1986 field season, FRI provided the USSR's TINRO with information on the preferred body area for scale collection (Figure 6) and requested scale samples from major USSR runs of sockeye, chum, coho, and chinook salmon. Because we anticipate that the 1986 USSR samples may be of improved quality, we are beginning our studies with standard samples collected in 1986 and high-seas unknowns from 1985–1986. Initially, Japan will conduct studies to evaluate the quality of recent inshore samples of chinook salmon scales provided to them by the USSR (INPFC 1987b). Procurement of adequate samples representing USSR stocks continues to be a central problem for INPFC-related studies, and Canada, Japan, and the USA are continuing their efforts to obtain scale samples from the USSR. Results of these coordinated scale pattern analyses, based on adequate high-quality scale samples, will help to increase knowledge about migrations of salmonids in the high seas.

References

Anas, R. E., and S. Murai. 1969. Use of scale characters and a discriminant function for classifying sockeye salmon (*Oncorhynchus nerka*) by continent of origin. International North Pacific Fisheries Commission Bulletin 26:157–192.

Bilton, H. T. 1985. A test on the accuracy of ageing of chinook salmon (*Oncorhynchus tshawytscha*) of known age from their scales. International North Pacific Fisheries Commission Document, Vancouver, Canada. (Available from Department of Fisheries and Oceans, Pacific Biological Station, Nanaimo, British Columbia.)

Clutter, R. I., and L. E. Whitesel. 1956. Collection and interpretation of sockeye salmon scales. International Pacific Salmon Fisheries Commission Bulletin 9.

Cook, R. C. 1982a. Estimating the mixing proportion of salmonids with scale pattern recognition applied to sockeye salmon (*Oncorhynchus nerka*) in and around the Japanese landbased driftnet fishery area. Doctoral dissertation. University of Washington, Seattle.

Cook, R. C. 1982b. Stock identification of sockeye salmon (*Oncorhynchus nerka*) with scale pattern recognition. Canadian Journal of Fisheries and Aquatic Sciences 39:611–617.

Cook, R. C. 1983. Simulation and application of stock composition estimators. Canadian Journal of Fisheries and Aquatic Sciences 40:2113–2118.

Cook, R. C., and G. E. Lord. 1978. Identification of stocks of Bristol Bay sockeye salmon, *Oncorhynchus nerka*, by evaluating scale patterns with a polynomial discriminant method. U.S. National Marine Fisheries Service Fishery Bulletin 76:415–423.

Cook, R. C., K. W. Myers, R. V. Walker, and C. K. Harris. 1981. The mixing proportion of Asian and Alaskan sockeye salmon in and around the landbased driftnet fishery area, 1972–1976. International North Pacific Fisheries Commission Document, Vancouver, Canada. (Available from Fisheries Research Institute, University of Washington, Seattle.)

Dahlberg, M. L., and D. E. Phinney. 1968. A microprojector for use in scale studies. Progressive Fish-Culturist 30:118–119.

Davis, N. D. 1987. Variable selection and performance of variable subsets in scale pattern analysis. International North Pacific Fisheries Commission Document, Vancouver, Canada. (Available from Fisheries Research Institute, FRI-UW-8713, University of Washington, Seattle.)

Davis, N. D., and J. T. Light. 1985. Steelhead age determination techniques. International North Pacific Fisheries Commission Document, Vancouver, Canada. (Available from Fisheries Research Institute, FRI-UW 8506, University of Washington, Seattle.)

Department of Fisheries and Oceans. 1985. Report on research by Canada for the International North Pacific Fisheries Commission in 1984. International North Pacific Fisheries Commission Annual Report 1984:32–37.

Department of Fisheries and Oceans. 1986. Report on research by Canada for the International North Pacific Fisheries Commission in 1985. International North Pacific Fisheries Commission Annual Report 1985:33–40.

Department of Fisheries and Oceans. 1987. Report on research by Canada for the International North Pacific Fisheries Commission in 1986. International North Pacific Fisheries Commission Annual Report 1986:51–56.

Fisheries Agency of Japan. 1982. Report on research by Japan for the International North Pacific Fisheries Commission in 1981. International North Pacific Fisheries Commission Annual Report 1981:37–77.

Fisheries Agency of Japan. 1985. Report on research by Japan for the International North Pacific Fisheries Commission in 1984. International North Pacific Fisheries Commission Annual Report 1984:38–46.

Fisheries Agency of Japan. 1986. Report on research by Japan for the International North Pacific Fisheries Commission in 1985. International North Pacific Fisheries Commission Annual Report 1985:41–49.

Fisheries Agency of Japan. 1987. Report on research by Japan for the International North Pacific Fisheries Commission in 1986. International North Pacific Fisheries Commission Annual Report 1986:57–65.

Fournier, D. A., T. D. Beacham, B. E. Riddell, and C. A. Busack. 1984. Estimating stock composition in mixed stock fisheries using morphometric, meristic, and electrophoretic characteristics. Canadian Journal of Fisheries and Aquatic Sciences 41:400–408.

French, R., H. Bilton, M. Osako, and A. Hartt. 1976. Distribution and origin of sockeye salmon (*Oncorhynchus nerka*) in offshore waters of the north Pacific Ocean. International North Pacific Fisheries Commission Bulletin 34.

Giovando, D. P. 1969. Recoveries of salmon tagged offshore in the eastern north Pacific by Canada, 1960 to 1967. Manuscript Report of the Fisheries Research Board of Canada 1038.

Harris, C. K. 1983. Summary of U.S.–U.S.S.R. cooperative high seas salmonid tagging operations in 1983. International North Pacific Fisheries Commission Document, Vancouver, Canada. (Available from Fisheries Research Institute, University of Washington, Seattle.)

Harris, C. K. 1984. Summary of U.S.S.R.–U.S. cooperative high seas salmonid tagging operations in 1984. International North Pacific Fisheries Commission Document, Vancouver, Canada. (Available from Fisheries Research Institute, University of Washington, Seattle.)

Harris, C. K. 1985. Summary of U.S.S.R.–U.S. cooperative high seas salmonid tagging operations in 1985. International North Pacific Fisheries Commission Document, Vancouver, Canada. (Available from Fisheries Research Institute, University of Washington, Seattle.)

Harris, C. K. 1987. Catches of North American sockeye salmon (*Oncorhynchus nerka*) by the Japanese high seas salmon fisheries, 1972–84. Canadian Special Publication of Fisheries and Aquatic Sciences 96:458–479.

Harris, C. K. 1988. Recent changes in the pattern of catch of North American salmonids by the Japanese high seas salmon fisheries. Pages 41–65 *in* W. McNeil, editor. Salmon production, management, and allocation: biological, economic and policy issues. Oregon State University Press, Corvallis.

Harris, C. K., R. H. Conrad, R. C. Cook, K. W. Myers, R. W. Tyler, and R. L. Burgner. 1981. Monitoring migrations and abundance of salmon at sea—1980. International North Pacific Fisheries Commission Annual Report 1980:71–90.

Harris, C. K., R. C. Cook, R. L. Burgner, and D. E. Rogers. 1979. Monitoring migrations and abundance of salmon at sea—1977. International North Pacific Fisheries Commission Annual Report 1977:59–66.

Harris, C. K., R. C. Cook, R. L. Burgner, and D. E. Rogers. 1980a. Monitoring migrations and abundance of salmon at sea—1978. International North Pacific Fisheries Commission Annual Report 1978:56–64.

Harris, C. K., R. C. Cook, K. W. Myers, R. V. Walker, and R. L. Burgner. 1982. Monitoring migrations and abundance of salmon at sea—1981. International North Pacific Fisheries Commission Annual Report 1981:78–93.

Harris, C. K., S. L. Marshall, R. C. Cook, R. H. Conrad, J. P. Graybill, and R. L. Burgner. 1980b. Monitoring migrations and abundance of salmon at sea—1979. International North Pacific Fisheries Commission Annual Report 1979:55–67.

Harris, C. K., K. W. Myers, C. M. Knudsen, R. V. Walker, and R. L. Burgner. 1983. Monitoring migrations and abundance of salmon at sea—1982. International North Pacific Fisheries Commission Annual Report 1982:83–100.

Harris, C. K., and six coauthors. 1984. Monitoring migrations and abundance of salmon at sea—1983. International North Pacific Fisheries Commission Annual Report 1983:88–108.

Hartt, A. C. 1962. Movement of salmon in the north Pacific Ocean and Bering Sea as determined by tagging, 1956–1958. International North Pacific Fisheries Commission Bulletin 6.

Hartt, A. C. 1963. Problems in tagging salmon at sea. International Commission for the Northwest Atlantic Fisheries Special Publication 4:144–155.

Hartt, A. C. 1966. Migrations of salmon in the north Pacific Ocean and Bering Sea as determined by seining and tagging, 1959–1960. International North Pacific Fisheries Commission Bulletin 19.

Hartt, A. C., R. C. Cook, and R. H. Conrad. 1979. Migration and abundance of salmon at sea, 1976. International North Pacific Fisheries Commission Annual Report 1976:79–88.

Hartt, A., and M. B. Dell. 1986. Early oceanic migrations and growth of juvenile Pacific salmon and steelhead trout. International North Pacific Fisheries Commission Bulletin 46.

Hartt, A. C., G. E. Lord, and D. E. Rogers. 1977. Migration and abundance of salmon at sea, 1975. International North Pacific Fisheries Commission Annual Report 1975:64–68.

INPFC (International North Pacific Fisheries Commission). 1959. International North Pacific Fisheries Commission Annual Report 1958:74–119.

INPFC (International North Pacific Fisheries Commission). 1987a. Report of scale pattern analysis workshop. International North Pacific Fisheries Commission Document, Vancouver, Canada.

INPFC (International North Pacific Fisheries Commission). 1987b. Report of the scale pattern working group. International North Pacific Fisheries Commission Document, Vancouver, Canada.

Ishida, Y., S. Ito, and K. Takagi. 1984. Further analysis of scale patterns of Japanese hatchery-reared chum salmon in the north Pacific Ocean. International North Pacific Fisheries Commission Document, Vancouver, Canada. (Available from Fisheries Agency of Japan, Tokyo.)

Ishida, Y., S. Ito, and K. Takagi. 1985. Stock identification of chum salmon based on scale patterns by discriminant function. International North Pacific Fisheries Commission Document, Vancouver, Canada. (Available from Fisheries Agency of Japan, Tokyo.)

Ito, J. 1987. Measurement of the scale characters of chum and pink salmon by optical pattern recognition system. International North Pacific Fisheries Commission Document, Vancouver, Canada. (Available from Fisheries Agency of Japan, Tokyo.)

Knudsen, C. M. 1985. Chinook salmon scale character variability due to body area sampled and possible effects on stock separation studies. Master's thesis. University of Washington, Seattle.

Knudsen, C. M., and N. D. Davis. 1985. Variation in salmon scale characters due to body area sampled. International North Pacific Fisheries Commission Document, Vancouver, Canada. (Available from Fisheries Research Institute, FRI-UW-8504, University of Washington, Seattle.)

Kondo, H., Y. Hirano, N. Nakayama, and M. Miyake. 1965. Offshore distribution and migration of Pacific salmon (genus Oncorhynchus) based on tagging studies (1958–1961). International North Pacific Fisheries Commission Bulletin 17.

Koo, T. S. Y. 1955. Biology of the red salmon, Oncorhynchus nerka (Walbaum), of Bristol Bay, Alaska, as revealed by a study of their scales. Doctoral dissertation. University of Washington, Seattle.

Koo, T. S. Y. 1962. Age and growth studies of red salmon scales by graphical means. Pages 53–121 in T. S. Y. Koo, editor. Studies of Alaska red salmon. University of Washington Press, Seattle.

Lachenbruch, P. A. 1967. An almost unbiased method of obtaining confidence intervals for the probability of misclassification in discriminant analysis. Biometrics 23:639–645.

Lachenbruch, P. A. 1973. Some results on the multiple group discriminant problem. Pages 193–211 in T. Cacoullos, editor. Discriminant analysis and applications. Academic Press, New York.

LeBrasseur, R., B. Riddell, and T. Gjernes. 1987. Ocean salmon studies in the Pacific Subarctic Boundary Area. International North Pacific Fisheries Commission Document, Vancouver, Canada. (Available from Pacific Biological Station, Nanaimo, British Columbia.)

Light, J. T., and R. LeBrasseur. 1986. Japanese ocean salmon research, synopsis of observations aboard Shin Riasu maru, 11 June to 20 July 1986. International North Pacific Fisheries Commission Document, Vancouver, Canada. (Available from Fisheries Research Institute, University of Washington, Seattle.)

Machidori, S. 1981. Photographic atlas of the scales and otoliths of coho salmon caught in the north Pacific Ocean. International North Pacific Fisheries Commission Document, Vancouver, Canada. (Available from Fisheries Agency of Japan, Tokyo.)

Major, R. L., K. H. Mosher, and J. E. Mason. 1972. Pages 209–231 in R. C. Simon and P. A. Larkin, editors. The stock concept in Pacific salmon. H. R. MacMillan Lectures in Fisheries, University of British Columbia, Vancouver, Canada.

Major, R. L., S. Murai, and J. Lyons. 1975. Scale studies to identify Asian and western Alaskan chinook salmon. International North Pacific Fisheries Commission Annual Report 1973:80–97.

Major, R. L., S. Murai, and J. Lyons. 1977a. Scale studies to identify Asian and western Alaskan chinook salmon. International North Pacific Fisheries Commission Annual Report 1975:68–71.

Major, R. L., S. Murai, and J. Lyons. 1977b. Scale studies to identify Asian and western Alaskan chinook salmon: the 1969 and 1970 Japanese mothership samples. International North Pacific Fisheries Commission Annual Report 1974:78–81.

Margolis, L., F. C. Cleaver, Y. Fukuda, and H. Godfrey. 1966. Salmon of the north Pacific Ocean—Part 6. Sockeye salmon in offshore waters. International North Pacific Fisheries Commission Bulletin 20.

Marshall, S. L., C. K. Harris, D. E. Rogers, and R. C. Cook. 1978. Investigations on the continent of origin of sockeye and coho salmon in the area of the Japanese landbased driftnet fishery. Fisheries Research Institute, FRI-UW-7816, University of Washington, Seattle.

Marshall, S., and nine coauthors. 1987. Application of scale patterns analysis to the management of Alaska's sockeye salmon (Oncorhynchus nerka) fisheries. Canadian Special Publication of Fisheries and Aquatic Sciences 96:307–326.

Millar, R. B. 1987. Maximum likelihood estimation of mixed stock fishery composition. Canadian Journal of Fisheries and Aquatic Sciences 44:583–590.

Millar, R. B. 1988. Statistical methodology for estimating composition of high seas salmonid mixtures using scale analysis. International North Pacific Fisheries Commission Document, Vancouver, Canada. (Available from Fisheries Research Institute, FRI-UW-8806, University of Washington, Seattle.)

Milner, G. B., D. J. Teel, F. M. Utter, and C. L. Burley. 1981. Columbia River stock identification study: validation of genetic method. U.S. National Marine Fisheries Service, Northwest and Alaska Fisheries Center, Annual report of research (FY80), Seattle, Washington.

Myers, K. 1985. Racial trends in chinook salmon (Oncorhynchus tshawytscha) scale patterns. International North Pacific Fisheries Commission Document, Vancouver, Canada. (Available from Fisheries Research Institute, FRI-UW-8503, University of Washington, Seattle.)

Myers, K. 1986. The effect of altering proportions of Asian chinook stocks on regional scale pattern analysis. International North Pacific Fisheries Commission Document, Vancouver, Canada. (Available from Fisheries Research Institute, FRI-UW-8605, University of Washington, Seattle.)

Myers, K. W., C. K. Harris, C. M. Knudsen, R. V. Walker, N. D. Davis, and D. E. Rogers. 1987. Stock origins of chinook salmon in the area of the Japanese mothership salmon fishery. North American Journal of Fisheries Management 7:459–474.

Myers, K. W., D. E. Rogers, C. K. Harris, C. M. Knudsen, R. V. Walker, and N. D. Davis. 1984. Origins of chinook salmon in the area of the Japanese mothership and landbased driftnet salmon fisheries in 1975–1981. International North Pacific Fisheries Commission Document, Vancouver, Canada. (Available from Fisheries Research Institute, University of Washington, Seattle.)

Pella, J. J., and T. L. Robertson. 1979. Assessment of composition of stock mixtures. U.S. National Marine Fisheries Service Fishery Bulletin 77:387–398.

Rankis, A. E. 1987. Factors biasing scale analysis estimates of size selective mortality in Columbia River coho salmon rearing density experiments. Master's thesis. University of Washington, Seattle.

Rothschild, B. J., A. C. Hartt, D. E. Rogers, and M. B. Dell. 1971. Tagging and sampling. International North Pacific Fisheries Commission Annual Report 1969:67–89.

Simkiss, K. 1974. Calcium metabolism of fish in relation to ageing. Pages 1–12 in T. B. Bagenal, editor. Ageing of fish. Gresham Press, Old Woking, England.

Tanaka, S., M. P. Shepard, and H. T. Bilton. 1969. Origin of chum salmon (Oncorhynchus keta) in offshore waters of the north Pacific in 1956–1958 as determined from scale studies. International North Pacific Fisheries Commission Bulletin 26:57–155.

U.S. National Marine Fisheries Service. 1985. Investigations by the United States for the International North Pacific Fisheries Commission in 1984. International North Pacific Fisheries Commission Annual Report 1984:47–59.

U.S. National Marine Fisheries Service. 1986. Investigations by the United States for the International North Pacific Fisheries Commission in 1985. International North Pacific Fisheries Commission Annual Report 1985:50–67.

U.S. National Marine Fisheries Service. 1987. Investigations by the United States for the International North Pacific Fisheries Commission in 1986. International North Pacific Fisheries Commission Annual Report 1986:66–75.

Walker, R. V. 1987. Examination of an image analysis system for collection of scale pattern data. International North Pacific Fisheries Commission Document, Vancouver, Canada. (Available from Fisheries Research Institute, FRI-UW-8711, University of Washington, Seattle.)

Walker, R. V., and N. D. Davis. 1983. The continent of origin of coho salmon in the Japanese landbased driftnet fishery area in 1981. International North Pacific Fisheries Commission Document, Vancouver, Canada. (Available from Fisheries Research Institute, University of Washington, Seattle.)

Wood, C. C., S. McKinnell, T. J. Mulligan, and D. A. Fournier. 1987. Stock identification with the maximum-likelihood mixture model: sensitivity analysis and application to complex problems. Canadian Journal of Fisheries and Aquatic Sciences 44:866–881.